SPE MONOGRAPHS

Injection Molding

Irvin I. Rubin

Nylon Plastics

Melvin I. Kohan

Introduction to Polymer Science and Technology:
An SPE Textbook

Edited by Herman S. Kaufman and Joseph J. Falcetta

Principles of Polymer Processing

Zehev Tadmor and Costas G. Gogos

Coloring of Plastics

Edited by Thomas G. Webber

The Technology of Plasticizers

J. Kern Sears and Joseph R. Darby

Fundamental Principles of Polymeric Materials

Stephen L. Rosen

Plastics Polymer Science and Technology
An SPE Textbook

Edited by Mahendra D. Baijal

PLASTICS POLYMER
SCIENCE AND TECHNOLOGY

Plastics Polymer Science and Technology

Edited By

MAHENDRA D. BAIJAL

Technology and Management Consulting
Painesville, Ohio

A Wiley-Interscience Publication

JOHN WILEY & SONS, New York Chichester Brisbane Toronto Singapore

Library of Congress Cataloging in Publication Data:
Main entry under title:

Plastics polymer science and technology.

  (SPE monographs ISSN 0195-4288)
  "A Wiley-Interscience publication."
  Includes index.
  1. Plastics   I. Baijal, Mahendra D.   II. Series.

TP1120.P544   668.4   81-13066
ISBN 0-471-04044-4   AACR2

Printed in the United States of America

10 9 8 7 6 5 4 3 2 1

CONTRIBUTORS

E. O. Allen
Eastman Kodak Company
Plastics and Metals Products Division
Rochester, New York 14650

M. D. Baijal
Technology and Management Consulting
Plainesville, Ohio 44077

B. Baum
Springborn Laboratories
Enfield, Connecticut 06082

J. J. Bikerman
Department of Chemical Engineering
Case Western Reserve University
Cleveland, Ohio 44102

C. A. Daniels
B. F. Goodrich Chemical Company
Avon Lake Technical Center
Avon Lake, Ohio 44012

C. C. Davis
Department of Mechanical Engineering
General Motors Institute
Flint, Michigan 48502

R. Deanin
Lowell Technological Institute
Lowell, Massachusetts

P. Dubin
Memorex Corporation
Santa Clara, California 95052

A. A. Duswalt
Hercules, Incorporated
Wilmington, Delaware 19899

K. Eise
Werner & Pfleiderer Corporation
Ramsey, New Jersey 07446

R. E. Evans
American Cyanamid Company
Stamford, Connecticut 06904

J. Harris
Werner & Pfleiderer Corporation
Ramsey, New Jersey 07446

A. M. Hassan
Armstrong Cork Company
Lancaster, Pennsylvania 17604

S. Y. Hobbs
General Electric Company
Corporate Research and Development
Polymer Properties and Reactions Branch
Schenectady, New York 12301

I. Klein
Scientific Process and Research, Inc.
Highland Park, New Jersey 08904

J. V. Koleske
Union Carbide Corporation
Chemicals and Plastics
Research and Development Department
South Charleston, West Virginia 25303

M. Kronstadt
Dynapol
Palo Alto, California 94304

S. A. Liebman
Armstrong Cork Company
Lancaster, Pennsylvania 17604

E. B. Nauman
Xerox Corporation
Rochester, New York 14644

D. M. Pai
Xerox Corporation
Webster, New York 14580

C. Parker
Springborn Laboratories
Enfield, Connecticut 06082

G. H. Pearson
Eastman Kodak Company

Research Laboratories
Rochester, New York 14650

J. M. Pochan
Xerox Corporation
Webster, New York 14580

H. Schonhorn
Bell Laboratories
Murray Hill, New Jersey 07974

C. G. Seefried
Union Carbide Corporation
Chemicals and Plastics
Research and Development Department
South Charleston, West Virginia 25303

I. M. Spier
Beacon Plastics and Metal Products
New York, New York 10016

SERIES PREFACE

The Society of Plastics Engineers is dedicated to the promotion of scientific and engineering knowledge of plastics and to the initiation and continuation of educational programs for the plastics industry. Publications, both books and periodicals, are major means of promoting this technical knowledge and of providing educational materials.

New books, such as this volume, have been sponsored by the SPE for many years. These books are commissioned by the Society's Technical Volumes Committee, and, most importantly, the final manuscripts are reviewed by the Committee to ensure accuracy of technical content. Members of this Committee are selected for outstanding technical competence and include prominent engineers, scientists, and educators.

In addition, the Society publishes *Plastics Engineering*, *Polymer Engineering and Science* (*PE & S*), *Journal of Vinyl Technology*, *Polymer Composites*, proceedings of its Annual, National, and Regional Technical Conferences (ANTEC, NATEC, RETEC), and other selected publications. Additional information can be obtained from the Society of Plastics Engineers, 14 Fairfield Drive, Brookfield Center, Connecticut 06805.

ROBERT D. FORGER

Executive Director
Society of Plastics Engineers

PREFACE

Plastics Polymer Science and Technology is structured to deal with material science (Chapters 2 through 9) and product technology (Chapters 10 through 17) aspects. Contributions on material science and product technology are specialized treatments of certain key topics. Chapter 1 provides an overview account of plastics polymer to facilitate the study of technical treatments that follow. Each chapter is a self-contained treatment, and as such may be regarded as a short monograph. The orientation in these contributions is on basics, timely research information, pragmatism, unresolved and debatable issues, and some thoughts on possible future developments. Different chapters have been written with varying focus in keeping with the interests of the contributors. Therefore an attempt has been made here to provide a logical and systematic progression of relevant information. The discussions of various topics take into account (as much as possible) interacting chemical, physical, and technological considerations. This theme has been followed by the editor over the last several years in the practice of plastics technology in academic and industrial environments.

This book is written with a focus on the physical apsects of the product line. Increasing attention is being paid both by academics and industry on material physical details and product performance for the reason that (in many cases) it is easier and more economical to alter product performance by physical design modifications. Construction of a new molecule to meet a given service demand is a time-consuming and costly proposition. Possibilities exist where given structures upon processing (alone or in combination with other materials) may be made to yield a spectrum of performance (polyoelfins, alloys, and others are such candidates). To address the issues involved, material science chapters approach product performance from structure-property considerations, whereas chapters on product technology attempt to relate product performance to production-related aspects. The purpose of this book on plastics polymer is to present an alternative treatment, to provide an understanding of the usual and unusual service response, to provoke thought and generate curiosity for further efforts in reaserch, development, and production for new generation of materials and products, and to add to information base.

This contribution is offered for academic training and industrial practice. As an academic tool, this book can be used as a basis for a graduate or senior undergraduate course (two quarters or one semester) on plastics polymers. Material science could be covered in one quarter and product technology could be addressed in another quarter. Alternately, the whole book could be covered in one semester. It is suggested that the course curriculum include outside readings, problem sessions, laboratories, and industrial visits, as possible. Such a program should ensure the meshing of operating principles with industrial practice. Further, the course could be hosted by the polymer department, chemistry department, chemical engineering department, or materials and metallurgy department. In industry this book should be of value to material scientists, material engineers, product designers, product technology managers, and product managers since it is a ready reference resource that attempts to join theory with practice.

This project has been in progress since 1975.

The editor acknowledges the various authors and their respective organizations for making this book possible. Management of 17 contributions from 26 authors affiliated with 23 organizations has been an overwhelming task. Lots of perspiration and inspiration has been involved in the last few years. Needless to say, this has required patience, understanding, and hard work on the part of all concerned, including Technology and management Consulting, the Society of Plastics Engineers and Wiley-Interscience. Although every precaution has been taken to avoid mistakes, it is requested that the reader forward errors, omissions, and suggestions for improvement to the editor for future editions. Finally, this book is dedicated to my parents (Narayan and Lila), whose quest for knowledge and love unto all has been a constant source of inspiration for me.

M.D. Baijal

Painesville, Ohio
February, 1982

CONTENTS

1   General Considerations   1
    M. D. Baijal

Part One   Material Science

2   Theories of Solutions   27
    C. A. Daniells

3   Molecular Weight Distribution   77
    P. L. Dubin and M. Kronstadt

4   Microstructure and Analysis   133
    S. A. Liebman and A. M. Hassan

5   Thermal Characterization   201
    A. A. Duswalt

6   Morphology   239
    S. Y. Hobbs

7   Mechanical Behavior   297
    C. G. Seefried and J. V. Koleske

8   Dielectric and Photoconductive
    Properties   341
    J. M. Pochan and D. M. Pai

9   Surface and Interfacial
    Characteristics   395
    J. J. Bikerman, C. C. Davis, and
    H. Schonhorn

Part Two   Product Technology

10 Synthesis and Reactor Technology   471
    E. B. Nauman

11 Rheology   519
    G. H. Pearson

12 Processing   571
    E. O. Allen

13 Additives   613
    K. Eise and J. Harris

14 Product Design   653
    I. M. Spier

15 Computer Applications to
    Processing   699
    I. Klein

16 Physical Testing for End-Use
    Applications   799
    R. E. Evans

17 Environmental Resistance   883
    B. Baum, C. Parker, and
    R. Deanin

Index   939

xi

PLASTICS POLYMER
SCIENCE AND TECHNOLOGY

## CHAPTER 1

General Considerations

M. D. BAIJAL

Technology and Management Consulting, Painesville, Ohio

1.1	Introduction	1
1.2	Materials Science	3
1.3	Product Technology	3
1.4	Industry	7
1.5	Summary	12
1.6	References	25

1.1 INTRODUCTION

The subject matter of synthetic organic polymer materials (chain structures of high molecular weight) is rather complex because such materials can be utilized as plastics (polyethylene, polyvinyl chloride, polystyrene), elastomers (styrene-butadiene rubber, polybutadiene, ethylene-propylene copolymers and terpolymers), and fibers (polyethylene terphthalate, nylon, polyacrylonitrile). Generally chain structures range from flexible as found in polyethylene to rigid as in nylon to cross-linked as in epoxy. The molecular weight is normally in the 10^4–10^7 range. The assignment of a specific polymer to the three major categories is a function of bulk physical response. Plastics can be semi-crystalline (or crystalline), and therefore usually rather hard and tough (polycarbonate), or amorphous and therefore brittle and glassy (polystyrene).

Elastomers are linear polymers (amorphous or semicrystalline) that contain cross-links, entanglements, or microcrystalline regions to prevent flow. They exhibit long-range reversible extensibility under small applied stress. Fibers are semicrystalline (or crystalline) polymers with high melting transition and high tensile strength, and are capable of being oriented and spun into filaments. Many crystalline polymers serve as both plastics and fibers (polyethylene terphthalate, nylon, polypropylene). Certain polymers behave as plastics and elastomers (polyvinyl chloride, polyurethanes, polysiloxanes). Polyethylene, polytetrafluoroethylene, polystyrene, polymethyl methacrylate, phenol-formaldehyde, urea-formaldelyde, malamineformaldehyde, and others are solely plastic materials. Some elstomers are styrene-butadiene, polybutadiene, ethylene-propylene (EPDM), butyl, noeprene, nitrile, and isoprene. Thermal transitions (glass transition temperature T_g, room temperature T_r, melting point T_m) and mechanical properties (elastic modulus, tensile strength, elongation) are conveniently utilized to distinguish between plastics, elastomers, and fibers (See Table 1.1).

This book is limited to the discussion of synthetic organic polymers that respond as plastics (plastics polymer). Some relevance to other materials is noted for illustrative purposes. Also, throughout the text the terms polymer and plastic have been used inter-

Table 1.1 Thermal Transitions and Mechanical Properties of Polymers[a]

| Parameter | Polymer | | |
	Plastic	Elastomer	Fiber
	(Low-Pressure polyethylene)	(Styrene-butadiene rubber)	(Nylon 66)
Thermal Transitions[a]	$T_g \geqslant T_r + 75°C$ (glassy) $T_g < T_r, T_m > T_r$ (crystalline)	$T_g + 75°C \leqslant T_r$	$T_m > T_r + 150°C$
Mechanical Properties			
Modulus of elasticity (N/cm^2)[b]	10^3–10^4	10–10^2	10^5–10^6
Elongation $(\%)$[c]	100–200	1000	10–30
Tensile strength $(10^4 \ K \ m^2)$[d]	200–400	280	4600–8700 (depending on the draw ratio)
Elongation at break $(\%)$	15–100	580	19–32

Source.   Adapted from References 2 and 5.

[a]Glass transition (T_g) is the center of temperature range in which an amorphous material changes from glass-brittle to being viscous.

[b]Ratio of stress to strain in elastic deformation. Also called elastic modulus and coefficient of elasticity.

[c]The fractional increase in length of a material stressed in tension. Measure of ductility.

[d]Maximum stress developed in a material in tension. Measure of ultimate strength.

changeably. Since most of the key technical issues are dealt with in material science and product technology sections, this chapter is merely an overview account of plastics polymer. What follows is a commentary on information located elsewhere in this book in varying forms, as modified by industrial implications of the product line.

Broadly speaking, all types of polymers can be classed as inorganic or organic with both classes further distinguished in terms of natural and synthetic origin. Clays and sands are examples of inorganic natural polymers, whereas fiber is an illustration of an inorganic synthetic polymer. Well known organic natural polymers are nucleic acids, proteins, polysaccharides, and polyisoprene. Organic synthetic polymers (hereinafter referred to as "polymers") are available (as previously noted) as plastics, elastomers, and fibers.

Polymers are also known as high polymers, resins, and macromolecules. The terms "polymer" and "resin" have commercial significance while "macromolecule" is commonly used in the scientific arena. The word polymer has Greek origin ("poly" and "meros" meaning

many parts). Resin was coined (in the early days) to describe the (functional and substitute nature of) naturally occuring polymers, such as amber. Macromolecule (giant or large molecule) is of recent origin derived from structural understanding and constituting the most significant definition of polymers.

Plastics polymer can be classified by basis generic type, chemically, by structure property, and commercially. Generic classification is based upon the sources (starting material or monomer) used in the makeup of the polymer. Polyolefins (polyethylenes, polypropylene); vinyls (polyvinylchloride, polystyrene*); polyesters (polycarbonate, polyethylene terphthalate, unsaturated polyesters); aromatic polyethers (phenoxy, epoxy, polysulfones); amino resins (ureaformaldehyde, melamineformaldehyde); phenol resins (phenolics) are examples of some generic classes. Chemical differentiation is based upon the type of polymerization (polyaddition and polycondensation) used in the synthesis of plastics polymer. Polyaddition is mechanistically chaingrowth mech-

*Commonly classed generically as styrene.

anism leading to such addition polymers as polyolefins and vinyls. Polycondensation is governed by stepgrowth mechanism and leads to condensation polymers such as epoxies and phenolics. Structure property considerations divide plastics polymer into thermoplastics (materials workable in the final state, such as polyolefins), and thermosets (materials infusible and insoluble in the final state, such as phenolics). Commercially plastics polymer are available as commodity resins (general purpose and tonnage materials such as polyolefins) and specialty plastics (engineering plastics such as polycarbonates and advanced materials such as blends and alloys). Plastic additivies (lubricants, stabilizers, fillers, and others) are also part of plastics polymer commercial regime (1–10).

1.2 MATERIALS SCIENCE

Plastics polymer form a part of the engineering materials regime. Engineering materials are distinguished as metals and nonmetals. Metals include basic materials (aluminum, magnesium, iron) and alloys (steel, brass, bronze). Nonmetals are polymers, ceramics, glasses, and other materials. Polymer materials include thermoplastics, thermosets, and elastomers. Examples of ceramics would be aluminas, beryllias, and carbides. Silica and paper are examples of glass and other materials respectively.

Material science deals with the understanding of structure-property-performance relationships in solid, melt, and solution physical states and the application of such knowledge to the development of useful materials of commerce. Unlike metals, ceramics are insulators whereas plastics are insulators and semiconducting materials. In metals and ceramics the atom, the ion, and the unit structure control properties and performance attributes. On the other hand, plastics properties are controlled by the type, size, and distribution of polymer chains. Polymer chains are predominantly linear and covalently bonded (as mentioned elsewhere). Plastics are light, flexible, durable, technologically versatile, and commercially cheap

materials that in many instances duplicate or outperform such traditional commercial materials as metals, wood, glass, or stone.

Plastics structure includes configuration, conformation, and morphology. Configuration is spatial arrangement of structure developed during polymerization. It includes chain structure (linear, branched, lightly cross-linked and close network); degree of polymerization (molecular weight average, molecular weight distribution, averaging mechanism due to random events); structured features (head-to-tail and head-to-head, cis and trans and stereoregularity-tacticity). Conformation relates to structure development as a result of rotation about single bonds. Planar zigzag is a local conformation. Other conformations identified are helical, chain folding long-range and random coil long-range. Morphology includes glass transition, amorphous and crystalline order. Intermolecular bonding has a significant effect on plastics structure. Solution thermodynamics, transport, and molecular weight distribution are examples of plastics solution properties. Solid-state properties relate to thermal, mechanical, electrical, interfacial, diffusional, and optical properties. Final-product properties (performance, maintenance, and appearance) are governed by the sum of intrinsic properties (chemical and physical structure and properties) and processing properties (restructurization).

There are 83 generic types of plastics polymer available with 54 known properties (cost, processing, mechanical, thermal, optical, electrical, environmental, barrier, flammability, density, and others) for product designing. In material selection a property assessment audit with respect to usable materials is of significant value (11–19). For an understanding of materials science the reader should read Chapters 2 through 9.

1.3 PRODUCT TECHNOLOGY

Plastic polymer are derived from petrochemicals (ethylene, propylene, benzene, butadiene, and p-xylene) which use natural-gas liquids and naphtha as main energy raw

Table 1.2   Ethylene Product Sheet

Distinguishing Feature	Product Information
Structure	$CH_2{=}CH_2$
Production (billions 1b)	Catalytic or thermal cracking of hydrocarbons ranging from ethane to gas oil (28)
Major products share (%)	Polyethylene (45)
	Ethylene oxide (20)
	Vinyl chloride (15)
	Styrene (10)
	Other derivatives (10)
Major applications share (%)	Fabricated plastics (65)
	Antifreeze (10)
	Fibers (5)
	Solvents (5)
	Other end uses (10)
Trade	Both exports and imports negligible
Price ($/lb)	0.26
Commercial value (billion $)	7.00

Source. Adapted from *Chemical and Engineering News,* November 17, 1980 and November 30, 1981.

materials. Approximately 1.5% of the total hydrocarbon energy resource available is utilized in the production of plastics. In turn, plastics provide a value-added effect of (approximately) 15 times.

Chemical processing (refinery operations, chemical conversions, and polymerizations) is necessary to produce plastics from hydro-carbon feedstocks. Plastics feedstocks (ethylene, propylene, benzene, and derived precursors such as acrylonitrile, butadiene, formaldehyde, phenol, styrene, vinyl chloride, and others) are used as monomers for the production of major plastics (Tables 1.2 to 1.9). Homopolymers (polyethylene) result by using the same monomer (ethylene). Copoly-

Table 1.3   Propylene Product Sheet

Distinguishing Feature	Product Information
Structure	$CH_2{=}CH{-}CH_3$
Production (billions lb)	Coproduct recovered from multiple product steam crackers of olefins plants and fludized-bed catalytic crackers of oil refineries (12.8)
Major products share (%)	Polypropolylene (25)
	Acrylonitrile (15)
	Isopropyl alcohol (10)
	Propylene oxide (10)
	Other derivatives (40)
Major applications share (%)	Fabricated plastics (50)
	Fibers (15)
	Other end uses (35)
Trade	Negligible exports and imports less than 500 million lb (small).
Price ($/lb)	0.22–0.24
Commercial value (billions $)	3.25

Source. Adapted from *Chemical and Engineering News,* November 17, 1980, November 30, 1981.

Table 1.4 Benzene Product Sheet

Distinguishing Feature	Product Information
Structure	
Production (billions gal)	Recovered and extracted from oil refinery catalytic reformate, pyrolysis gasoline of olefin steam-crackers, coking oil light oils, and tolune dealkylation (1.5)
Major products share (%)	Ethylbenzene (50) Cumen (15) Cyclohexane (15) Analine (5) Other derivative (15)
Major applications share (%)	Polystyrene (25) Nylon (20) Other sytrenes (10) Rubber (5) Other end uses (40)
Trade	Both exports and imports negligible
Price ($/gal)	1.55
Commercial value (Billons $)	2.4

Source. Adapted from *Chemical and Engineering News,* November 17, 1980 and November 30, 1981.

Table 1.5 Styrene Product Sheet

Distinguishing Feature	Product Information
Structure	$CH=CH_2$
Production (billions lb)	Ethylebenzene dehydrogenation, recovery from refinery reformate, or propylene oxide coproduction (6.7)
Major products share (%)	Polymers (80) Styrene-butadiene rubber (10) Other derivatives (10)
Major applications share (%)	Fabricated plastics (90) Other end uses (10)
Trade	Exports approximately 750 million lb and negligible imports
Price ($/lb)	0.38
Commercial value (billions $)	2.75

Source. Adapted from *Chemical and Engineering News,* July 7, 1980 and August 3, 1981.

Table 1.6 Vinyl Chloride Product Sheet

Distinguishing Feature	Product Information
Structure	$CH_2=CHCl$
Production (billions lb)	Ethylene dichloride dehydrochloriation (7.1)
Major products share (%)	Vinyls (100)
Major applications share (%)	Fabricated vinyls (100)
Trade	Less than 1 billion lb in exports and negligible imports
Price ($/lb)	0.22
Commercial value (billions $)	1.4

Source. Adapted from *Chemical and Engineering News,* July 7, 1980 and August 3, 1981.

Table 1.7 Butadiene Product Sheet

Distinguishing Feature	Product Information
Structure	$CH_2{=}CH{-}CH{=}CH_2$
Production (billions lb)	Extracted and recovered from olefin steam-cracker coproduct streams; butylenes or butane dehydrogenation hydrocarbons from refinery and natural gas (3.1)
Major Products share (%)	Styrene-butadiene rubber (50) Polybutadiene (20) Hexamethylenediamine (10) Other derivatives (20)
Major applicatioins share (%)	Rubber products (80) Fibers (10) Other end uses (10)
Trade	100 million lb of exports and 500 million lb of imports
Price	0.42
Commercial value (billions $)	1.1

Source. Adapted from *Chemical and Engineering News,* November 17, 1980 and November 30, 1981.

Table 1.8 *P*-Xylene Product Sheet

Distinguishing Feature	Product Information
Structure	 CH_3 (benzene ring) CH_3
Production (billions lb)	Separation from mixed xylenes of oil refineries or from olefin steam-cracker pyrolysis gasoline (3.6)
Major products share (%)	Dimethyl terphthalate (50) Terphthalic acid (50)
Major applications share (%)	Polyester fiber, film, and fabricated items (mostly bottles) (100)
Trade	Exports exceeding 800 million lb and negligible imports
Price ($/lb)	0.31
Commercial value (billions $)	1.1

Source. Adapted from *Chemical and Engineering News,* November 17, 1980 and November 30, 1981.

Table 1.9 Proplylene Oxide Product Sheet

Distinguishing Feature	Product Information
Structure	$CH_2{-}CHCH_3$ with $\backslash O /$ (epoxide)
Production (billions lb)	Made from propylene chlorohydration and subsequent saponification, propylene peroxidation (1.9)
Major products share (%)	Polyols (60) Propylene glycol (25) Other derivatives (15)
Major applications share (%)	Polyurethanes (60) Unsaturated thermoset polyesters (10) Other end uses (30)
Trade	160 million lb exports and imports steady at 40 million lb
Price ($/lb)	0.35
Commercial value (billions $)	0.65

Source. Adapted from *Chemical and Engineering News*, July 7, 1980.

mers (acrylonitrile-butadiene styrene) are formed by using different monomers or comonomers (acrylonitrile, butadiene styrene). Homopolymers are homogenous with respect to repeat units while copolymers (random, block, and graft) are heterogenous in repeat units (equivalent to the monomer in structure). During polymerization (polyaddition or polycondensation) monomers add repeatedly to form long chain molecules of high molecular weight (10^4 – 10^7), which are known as polymers. Polymers are predominantly linear and covalently bonded structures. Polymers are composed of carbon-carbon chains (which is the most fundamental and predominant backbone as found in polyolefins), side chain substituted carbon-carbon chains (as found in vinyl polymers), main chain substitution of carbon-carbon chains by oxygen (carbon-oxygen as in polysolfones), main chain substitution of carbon-carbon chains by nitrogen (carbon-nitrogen as in amino resins), and main chain substitution of carbon-carbon chains by silica (carbon-silica-oxygen as in silicones). Because of chain substitution, copolymerization, chain geometry alteration, molecular order, molecular weight distribution, and reinforcements plastics polymer can be produced in various structure-property profiles.

Physical processing involves processing and shaping of the polymer (alone or in combination with additives) into consumer and industrial end products. Material handling, extrusion, molding and casting, calendering and coating, compounding, and post shaping are examples of physical processing techniques. Technical quality of the end product is largely governed by chemical (composition and structure), physical (property and phase transitions), and engineering (heat, mass, and momentum transfer) response of the raw material feed in the production cycle. The effects of chain substitution, incorporation of additives, and orientation relate to the production of versatile materials. Chain substitution is a way to alter molecular structure during polymerization. Additives and orientation relate to the geometrical arrangement or rearrangement of the particles and molecules

during processing. Multiphase systems such as alloys, composites, polyblends, and foams are examples of materials where geometrical disposition of phases control properties and service response. The effect of molecular orientation in amorphorous plastics, morphology, and molecular orientation in crystallic plastics has been well documented. Polystyrene, rigid polyvinyl chloride, polypropylene, low-density polyethylene, high-density polyethylene, nylon, polyester, and polyoxymethylene are examples of thermoplastics where orientation has been put to use in commercial applications (31). Largely, market determinants are subjected to quality and value engineering (i.e., worth to the customer, and manufacturing cost) and demand considerations for profitable operations (20–22). For details on product technology the reader is referred to Chapters 10 through 17.

1.4 INDUSTRY

The plastics industry is approximately 150 years old and traces its origin to rubber industry. Also, a phenomonal growth has occured in plastics products in the last 50 years. In an historical accounting of the industry one can identify three periods of significant growth. Prior to 1850 the emphasis was on the usage of natural polymers and *composition elucidation*. Hancock's rubber masticator (1820) and Goodyear's vulcanization process (1839) were outstanding events of this period.

From 1850 to 1930 the emphasis was on *structure* determination of natural and synthetic materials. Also, considerable advances were made by the industry in *new products* and *processes*. Outstanding contributions of this period included Ebonite (Hyatt, 1851); the ram extruder (1851); the screw extruder (early 1860s), which is the single most important processing development; celluluse nitrate (Hyatt, 1868), which was the first plastics invention—Hyatt was a pioneer in materials and machines development; injection molding (1870); phenol-formaldehyde (Bakeland, 1909); Bakelite, the first synthetic thermoset plastic developed; Banbury mixer, multiple, and inter-meshing screws (1916). The first thermo-plastics were casein (1919); alkid (1926); cel-lulose acetate (1927); polyvinyl chloride (1927), which was the first synthetic thermo-plastic introduced to the market; and urea-formaldehyde (1929).

From 1930 to 1980 has been truly a *Materials Age*. In this period not only were the majority of plastics introduced but also sig-nificant advances were made in methods and machines. The commercial significance of plastics has been exponential in this age largely owing to the science of polymerization, technology of production, structure-property correlations, and designing for applications.

Outstanding contributions of this period include the discovery of nylon (Carother, 1930), which is the most important event in the growth of plastics—later introduced in 1939 as a fiber and as the engineering thermo-plastic; ethyl cellulose (1935); acrylic and polyvinyl acetate (1936); cellulose acetate bu-tyrate and polystyrene (1938); polyvinylidene chloride (1939); melamine-formaldehydes (1939) were melamines that have utility from aircraft to dinnerware; unsaturated polyester (1942) and low-density polyethylene, which reached a one billion pound production rate per year in 1942; fluorocarbons (1943), which led to Teflon as an accidental discovery; silicones (1943); cellulose propionate (1945); epoxy (1947); acrylonitrile-butadiene styrene (ABS 1948); allylic (1949); polyurethane (1954), plastics that came from Europe to United States; acetal (1956), Delrin, plastic aimed at replacing metals; high-density poly-ethylene, polypropylene, and polycarbonate (1957); chlorinated polyether (1959); polyal-lomer (1962); phenoxy (1962); ionomer, poly-phenylene oxide, polyimide, and ethylene-vinyl acetate (1964); parylene and poly-sulfone (1965); thermoplastic polyester (1970); and nitrile barrier resins (1975). Present under-standing of the product line and growth of the industry can be largely attributed to Carother's discovery of nylon (1930); Staudinger's estab-lisment of macromolecular concept (Nobel Laureate, 1953); rubber research program's discovery of synthetic polybutadiene (1940s); Ziegler-Natta's discovery of stereoregular poly-merization (Nobel Laureate, 1963); Flory's masterpiece treatment of physical chemistry of polymers related to thermodynamics, kinetics, and statistical mechanics (Nobel Laureate 1974); and countless contributions from chemists, physicists, metallurgists, tech-nologists, and engineers from academic and industrial environments. These contributions have resulted in the establishment of major polymer industries that include engineering plastics; fibers and textiles; adhesives, paints, and coatings; composites; rubbers and elas-tomers; films and packaging. The chain of events in plastics and traditional materials has been responsible in the establishment of col-loid science, polymer science and engineering, and materials science and engineering dis-ciplines (32).

Currently the plastics industry is approx-imately a 50 billion dollar business with the basic materials sector contributing one-third, and processed or fabricated products account-ing for the other two-thirds. Plastics pro-duction exceeds steel and iron production in the United States. Also, use follows price—the cheaper the plastic the greater the tonnage. The total production of plastics in 1980 was nearly 35 billion pounds, a drop of 11.8% from 1979. The second drop of the decade was 23% in 1975. There are nearly 11,000 processing plants (U.S. Department of Commerce, Stan-dard Industrial Classification—SIC 3079), which consume 100–15 million pounds of the product per year. There are a total of 16,000 plants that make plastics their business. These plants include 200 plastics resins manufac-turers, 6090 injection molding plants, 3500 extrusion plants, 600 to 800 film and sheet processing plants, 2570 compression molding plants, and 3000 reinforced plastics fabricat-ing plants. The total industry employment is 350,000. Plastics account for nearly 15% of chemical industry's shipments (9). Commod-ity resins and specialty plastics account for (approximately) 90% and 10% share respec-tively. There are almost 7000 extrusion and molding grade resins available commercially. Of these low-density polyethylene, polyvinyl chloride, and polystrene account for nearly 80% of the market share. Major plastics business in the United States is accounted for by five themoplastics including low-density, polyethylene, polyvinyl chloride, high-density polyethylene, polypropylene, polystyrene, and three thermosets including epoxies, polyesters, and phenolics. Worldwide production of plastics is dominated by North America, the European community, and Japan, which account for 75% of total production. The market shares for thermoplastics and thermosets are 75% and 25% respectively. Polyethylene, polyvinyl chloride, polystrene, and polypropylene account for 92% of the market share in thermoplastics business. Amino resins, phenolics, polyurethanes, and unsaturated polyesters constitute 90% of the market share of thermosets.

Relating the energy crisis to the plastics industry, one can generate the following interesting (but approximate) numbers in terms of petrochemical product choice. For instance 100 l of naphtha can provide gasoline for a 600 mile trip. Alternately it can be used to produce ethylene, propylene, butane/butylene, aromatics, and fuel for the production of these monomers feedstocks. Ethylene can be converted to ethylene glycol (21 white shirts), dimethyl terphthalate (DMT), and polyethylene (six garbage cans) or vinyl chloride (520 feet of home water pipe). Propylene can yield acrylonitrile (21 sweaters) or polypropylene (four beer cases). Butane/butylene can be used to produce styrene-butadiene rubber (1 auto tire or 13 bicycle tires) and butyl (17 bicycle tubes). Aromatics lead to caprolactam (2000 panty hose) and DMT (24).

Plastics waste will account for nearly 5–7% of the total collected solid waste in 1990 as recorded by Predicasts Incorporated, and therefore use and production will not be constrained by disposal problems. Further, it takes less energy (fuel and feedstock) to make plastics parts versus parts made from competing materials (steel, stainless steel, aluminum, magnesium, and zinc die cast, see Tables 1.10 to 1.12). The use of plastics seems to go hand in hand with mass production,

Table 1.10 Comparative Materials Energy Requirements

Material	Approximate Energy Needs BTU/in^{3a}
Rigid polyurethane/RIM	1,000
Low-density poiyethylene	1,200 (87 + 13)
Polystyrene	2,000 (94 + 6)
Poly vinyl chloride[b]	2,100 (69 + 31)
Acrylonitrile-butadiene styrene	2,200 (92 + 8)
Polypropylene	2,400 (93 + 7)
Acrylic	2,800
Polycarbonate	3,000
Polyester	3,200
Nylon 66	3,400
Acetal	5,100
Steel	5,200
Zinc die cast	6,100
Aluminum die cast	7,600
Magnesium	10,000

Source. Adapted from a Mobay Chart and paper presented by Ford Motor Company, Materials Conference, 1980 and Reference 30.

[a]Feedstock and fuel: (Proportionate sources of energy: crude oil/natural gas % + electricity/coal %).

[b]Less dependent on ethylene and therefore has an edge on other commodity resins.

Table 1.11 Comparative Materials Total Energy Requirements for Automotive Parts (Headlight and Grille Housings, Fender Molding and Soft Bumper Systems)

		Plastic, Heat Units		
Metal	Heat Unit	FRP (Fiber Reinforced Plastic), SMC (Sheet Molding Compound), Polyester	RIM[a] (Reaction Injection Molding) Polyurethane	EPDM[a] (Ethylene Propylene Diene Monomer) Rubber
Steel	100	60	81	91
Stainless steel	100	48	67	73
Aluminum[a,b]	100	35	48	54
Magnesium[a,b]	100	30	41	46
Zinc die cast	100	67	89	102

Source. Adapted from Goodyear Chart and Reference 27.

[a] Part twice as light with respect to an equivalent steel part.

[b] Energy input twice as much as required by steel.

innovation, and cost sensitivity. Some new developments in plastic materials include high-temperature-high-strength plastics (acetylene-terminated tetraimide or pyrolyzed organometallic polymers); conducting and semiconductive plastics (electroactive resins—pyroelectric or piezoelectric-doped polypehenylene sulfide or polyvinylidene floride); and polymer support resins (polystyrene cross-linked with divinylebenzene and bearing chloromethyl or diphenylphosphine groups or amino acid esters and steroids linked to the same polyphosphazene backbone). These materials have applications in space and high technology, batteries, immobilization, and controlled-release areas. Plastics are frequently used to gain a competitive edge against lumber, paper, iron, aluminum, and other materials to protect environment and to gain energy and cost savings in film and container packaging, pipe and building insulation, automobile and truck components, consumer durables, and others. Major markets of plastics include packaging; building and construction; transportation; furniture and furnishings; consumer and institutional; industry and machinery; adhesives, ink, and coatings; and resellers, exports, and compounders (9, 10, 20, 23) (see Tables 1.13 to 1.23).

Table 1.12 Energy Savings with the Usage of Plastics Instead of Comparative Materials in Various Markets and Applications

Market-Application (Year of Statistics)	Energy Savings in Production, Service, or Potential savings, trillions BTU's. (million barrels of crude oil)
Construction and Building— pipe and insulation (1977)	Production and service 337.1 (58.3)
Transportation—Automobile (1979) and truck air deflectors (1976)	Service and potential 186.5 (32.0)
Packaging-closures (1977); chemicals containers (1977); beverage containers (1978); and lettuce wrap (1977)	Production and potential 38.8 (6.7)
Appliances—household fans (1977) and refrigerators and freezers (1978)	Production 14.3 (2.4)

Source. Adapted from Reference 10.

Table 1.13 Plastics versus Comparative Materials (Steel, Aluminum, Glass, Paper, and Lumber)

Distinguishing Feature	Advantages and Limitations
Structure	Plastics are synthetic organic materials. They are derived from petrochemical feedstocks using insignificant amounts of such energy resources compared to transporation and heating needs. Plastics are built of covalently bonded molecular chains. These chains are flexible, tough, or rigid providing versatility in structure. Plastics are flexible as rubber and stiff as concrete or wood. The structure of plastics combines glassy and amorphous and nonmetallic crystalline phases.
	Metals are natural inorganic materials. The structure of most metals is composed of relatively simple close-packed assemblies of atoms yielding specific properties in a number of ways. Electronic migration and ability to slip are the two most important aspects of metals structure. They are crystalline materials.
	Ceramics are natural inorganic materials with crystal structure. Ceramics contain metallic and nonmetallic phases. Covalent and ionic bonding predominates in ceramics.
	Broadly speaking materials properties and behavior is governed in part due to the freedom of the outermost electrons of the atom.
Properties	
Density	Plastics are lighter compared to metals or ceramics. Therefore they yield more parts per pound of material.
Mechanical characteristics	Plastics have lower modulii (elastic, tensile, or rigidity) than do steel, aluminum, glass, carbon, and wood. But on a volume basis plastics compare well with metals or glass. Chain substitution, reinforcements, alloying, and part design considerably improve plastics strength. Reinforced thermosets compare well with steel. High-density polyethylene fibers compare well with steel, aluminum, carbon, or glass filaments. Several other high-strength plastics are available.
	Plastics are superior to metals or glass in impact resistance.
	Plastics are more brittle than metals or wood but less than glass.
	The ductility of plastics is poorer compared to metals.
	Plastics are poorer in abrasion resistance than are comparable materials.
Thermal behavior	Plastics maximum service temperature is lower than that of metals or ceramics. Both thermal stability and degradation resistance can be improved by structural changes and by the incorporation of additives. High temperature-high strength plastics are available for exotic applications.
	In general plastics have higher thermal expansion than do metals or glass. Thermal expansion can be reduced by close packing, cross-linking, or filling.
	Plastics have lower thermal conductivity compared to metals. This is an advantage in processing and insulation applications.
Electrical resistance	Plastics are poorer conductors of electricity than are metals. But plastics are available as insulators, semiconducting, and conducting materials. Thus static charge buildup is minimized.

Table 1.13 Plastics versus Comparative Materials (*Continued*)

Distinguishing Feature	Advantages and Limitations
Other properties	Plastics have excellent corrosion resistance but poorer stress cracking and permeability than comparable materials. Both stress cracking and permeability is improved by structural and phase changes.
	Plastics incorporate color at low cost. They are more transparent than metals (which are opaque) and less than glass.
	Most plastics are not degradable but can be used in landfills and safely burned with proper incineration facilities. The amount of collected solid waste is small but it is growing with usage. Programmed life cycle, recovery and reuse will minimize this problem.
Production Related Aspects	Plastics are excellent materials for designing by synthesis and processing. The primary structure developed as a result of synthesis can be restructurized by processing (blending, alloying, and orientation). Steels acquire excellent properties as a result of heat treatment and cold forging.
	Alloys, composites, and polyblends are available in plastics, metals, and ceramics. These mixtures could be homogeneous (solid solution) or heterogeneous (multiple phase). Due to the thermodynamic incompatibility of polymers, plastics polyblends are generally multiphase systems. This is explained in terms of chain restrictions and as a result a small gain in entropy upon mixing. Cross-linked plastics polyblends are known as interpenetrating networks. Any alloy or blend (such as thermoplastic and elastomer blends) provides a unique combination of physical properties, processing, and price characteristics.[a]
	Plastics require little secondary finishing with respect to comparable materials.
	Energy efficiency of plastics is generally higher than comparable materials.
Market Considerations	Plastics markets are diversified in surface, subsurface, space, and medical applications. Plastics are poorer in load-bearing applications compared to metals.
	The Wholesale Price Index (% annual growth 1963/65–1975/77) for plastics, aluminum, and paper was the same (5.5) compared with glass (6.2), steel (7.2), and lumber (8.0).
	The competitive edge between plastics and competing materials may depend more on processing cost rather than on specific properties of a given material.

[a]See Reference 33.

1.5 SUMMARY

An elementary discussion of raw materials to end products is presented in this chapter. The decade of the 1980s will bring challenges and marketing opportunities to the plastics industry in respect to the economy-energy-environment scenarios. Innovation and growth will only be materialized through stronger links between the raw materials producers, the processors, and the end users. Third-country sourcing operations in materials will continue to increase. This would mean furthering of exports of polymers produced in the United

Table 1.14 Thermoplastics versus Thermosets *

Distinguishing Feature	Thermoplastic	Thermoset
Structure	Linear plastics that soften under heat and can be remelted and remolded. They are capable of undergoing indefinite elastic deformations below the chemical decomposition temperature. Allow the use of scrap.	Nonlinear-cross-linked and rigid materials that are insoluble and infusible in the final state. They cure, set, or harden during cross-linking chemical reaction. The reaction can be brought about by heat, pressure, catalyst, or other chemical means.
Properties (only a guideline, vary over wide ranges for various grades)		
Specific gravity	0.91–2.2 Generally lower than thermosets	0.11–2.6
Compressive strength (10^3 psi)	0.7–80.0	0.9–50.0
Tensile strength (10^3 psi)	1.5–14.5	4.0–35.0
Impact strength notched IZOD (ft-lb/notch)	No break–0.2–20.0 Generally greater impact resistance than thermosets.	0.2–20
Elongation at break (%)	3–300	<1–5 650 for polyurethane elastomer
Creep resistance		Better than thermoplastics
Heat resistance (°F)	115–550	160–450 Generally greater than thermoplastics
Flammability	Generally slow burning to self-extinguishing except for fluorocarbons, which do not burn.	Generally slow to self-extingushing or nonburning
Electrical resistance	Excellent to good	Good; epoxies and silicones are excellent
Outdoor resistance	Little discoloration	Discoloration generally
Water and chemical resistance	None to low water absorption. Generally little or some solubility in certain solvents. Attacked by strong acids and alkalies.	Very low to low water absorption. Generally little to some attack by solvents. Attacked by strong acids and alkalies.
Production	Bulk or mass, suspension, emulsion, and solution polymerization	Bulk or mass polymerization
	Extrusion, injection molding, blow molding, rotational molding, calendering, and other processing techniques	Liquid, compression molding, transfer molding, injection molding, and other processing techniques
Secondary finishing		Needs more than thermoplastics
Dimensional control		Yes
Permit thinner walls	Yes	
Free of sink marks		Yes
Better surface gloss	Yes	

Table 1.14   Thermoplastics versus Thermosets (*Continued*)

Distinguishing Feature	Thermoplastic	Thermoset
Major Products	Acetals, acrylics, cellulosics, fluorocarbons, polyamides, polyolefins, styrenes, thermoplastic polyester, vinyls, and others	Aminos, casein, epoxies, phenoloics, polyesters, silicones, polyurethanes, and others
Major Markets	Packaging, construction, household and consumer, transportation, textile, and others	Hosewares and consumers, transportation, construction, marine, industrial, appliances, and others
Approximate market share % (total estimated 1980 production in millions of pounds)	88.3 (30735)	11.7 (4085)
Cost Considerations	Generally molding, secondary finishing, color cost, scrap use, and lower specific gravity make thermoplastics cheaper than thermosets	If molding cost can be reduced, thermosets become more economical than reinforced or flame retardants thermoplastics
Price ($/lb)	0.31–0.45 for general purpose major resins	0.38–1.00 for general purpose major resins
Competitive materials (steel, nonferous metals, paper, glass, and lumber) Index of Physical Production	1972/74 = 100	1975/77 = 100
Annual growth (%)	9.6, higher than all competitive materials	10.4, higher than all competitive materials

*For a better discussion of Properties see Chapter 14, *Product Design*, Table 14.1 and 14.2.

States, Western Europe, and Japan to developing nations such as Taiwan, Philippines, Mexico, and others for parts molding and product assembly. The end products will be reexported to United States, Western Europe, Japan, and other markets. International cooperation between industrialized and developing countries by way of multinational corporations will be a key to the well-being of plastics industry. This will require fast technology transfer and end product trade in exchange for raw materials.

Materials science focus will be on the understanding of molecular engineering and tailoring of materials. Structure-property relationships in multiphase systems and in processing will see further research efforts. Conduction in plastics, permeation and diffusion, interfacial morphology, adhesion in polymer/polymer and polymer metal systems, and thermodynamics of polymer fluids (dilute and concentrated solutions and bulk polymers) also appear to be areas that may see further attention.

Product technology will depend on the use of heavier feedstocks; synthesis and substitution including emphasis on reaction engineering; rheology; heat transfer in molten polymers; and processing and manufacturing (further applications of reaction injection molding, disk extrusion, mass application of minicomputers to control the process and entire factory to save energy, to reduce scrap, and to recycle trimmed or excess resin-smart factories.)

Future **industry developments** will be in the areas of increased substitution of plastics in high-technology applications, in structural use, and in medical implants. Automobile usage of plastics will increase from 6% at

Table 1.15 Low-Density Polyethylene (LDPE)[a] Product Sheet

Distinguishing Feature	Product Information
Structure	

$$-(CH_2CH_2)-(CH_2CH)-(CH_2CH)-(CH_2CH)-$$
$$\qquad\qquad\quad\ \ C_2H_5 \qquad C_4H_9 \qquad C_nH_{2n+1}$$

Distinguishing Feature	Product Information
Advantages and limitations	*Advantages*: Excellent chemical resistance. High flexibility. Good fatigue resistance. Toughness. Very good dielectric properties. Tasteless. Odorless. Most grades meet FDA requirements. *Limitations*: Low tensile stress. Very resistant to bonding and printing. High thermal expansion. Susceptible to stress cracking. Poor weathering resistance.
Production	Initiator-aided high- and low-pressure polymerization of ethylene. Processed by all thermoplastic techniques. Estimated 1980, 7200 million lb.
Major applications (fabricated forms, %)	Trash bags, packaging lids, wire and cable coatings, toys, printing paper, communications cable, bowls, garment bags, scratch wrap, agricultural film. (Packaging film, 65; injection molding, 10; coatings, 10; extrusion, 5; others, 10.)
Market share	20.7%
Trade	Export nearly 1 billion lb and negligible imports
Price ($/lb)	0.41 for common grade
Commercial value (billions $)	2.8

Source. Adapted from *Chemical and Engineering News*, October 6, 1980 and February 9, 1981 and References 10, 19, and 21.

[a] Polyolefin (generic name). Dowlex and Dow (trade name and manufacturer). Tonnage (commodity) plastic. Thermoplastic.

Table 1.16 High-Density Polyethylene (HDPE)[a] Product Sheet

Distinguishing Feature	Product Information
Structure	$-(CH_2CH_2)-$
Advantages and limitations	*Advantages*: Excellent chemical resistance. Very good dielectric properties. Most rigid of ethylene plastics. Toughness. *Limitations*: Resistance to bonding and printing. Self-extinguishing grades have low properties.
Production	Metal salts and aklyls catalyzed polymerization of ethylene under moderate pressures. Processed by all thermoplastic techniques. Estimated 1980, 4400 million lb.
Major applications (fabricated forms, %)	Milk bottles, seating, waste baskets, outdoor furniture, luggage, disposable syringes, pallets, shipping pails, containers, piping. (Blow molding containers, 45; injection molding, 20; extrusion, 10; film, 5; other, 20.)
Market share	12.6%
Trade	Export approximately 600 million lb and negligible imports
Price ($/lb)	0.45 for common grade
Commercial value (billions $)	1.75

Source. Adapted from *Chemical and Engineering News*, October 6, 1980 and February 9, 1981 and References 10, 19, and 21.

[a] Polyolefin (generic name). Marlex and Philipps Petroleum (trade name and manufacturer). Tonnage (commodity) plastic. Thermoplastic.

Table 1.17   Polypropylene (PP)[a] Product Sheet

Distinguishing Feature	Product Information
Structure	$-(CH_2CH)-$ $\quad\quad\mid$ $\quad\quad CH_3$
Advantages and limitations	*Advantages*: Light weight. Good impact strength. Rigidity. Toughness. Good resistance to moisture, chemicals, and elevated ...peratures. Excellent dielectric properties. Mold-in hinges possible. Food grades available. Low cost. *Limitations*: Poor weather resistance. Flammable but retarded grades available. Embrittlement at low temperature.
Production	Metal salts and alkyls catalyzed polymerization of propylene. Processed by all thermoplastic techniques. Estimated 1980, 3700 million lb.
Major applications (fabricated forms, %)	Auto fender skirts, battery cases, carpet backing, cordage, pump housings, caps and closures, dishwater, tub and door liners, radio/television/phonograph housings, food and textile wrap, decorative ribbon. (Injection molding mostly for transportation, 40; fibers, 30; film, 10; others, 20.)
Market share	10.6%
Trade	Export nearly 700 million lb and negligible imports
Price ($/lb)	0.36 for standard grade
Commercial value (billions $)	1.25

Source.   Adapted from *Chemical and Engineering News*, October 6, 1980 and February 9, 1981 and References 10, 19, and 21.

[a]Polyolefin (generic name). Rexene PP and Rexene Polymers (trade name and manufacturer). Tonnage (commodity) plastic. Thermoplastic.

Table 1.18   Polyvinyl Chloride (PVC)[a] Product Sheet

Distinguishing Feature	Product Information
Structure	$-(CH_2CH)-$ $\quad\quad\mid$ $\quad\quad Cl$
Advantages and limitations	*Advantages*: Plasticizers provide wide range of flexibility. Nonflammable. Dimensional stability. Good weathering resistance. Low cost. *Limitations*: Higher density than other plastics. Limited thermal stability and decomposition leading to HCl. Susceptible to solvents. Stained by sulfer.
Production	Suspension, emulsion, or mass polymerization of vinyl chloride with initiator such as peroxide. Processed by thermoplastic and plastisol methods. Estimated 1980, 5500 million lb.
Major applications (fabricated forms, %)	Phonograph records, bottles, piping, siding, food wrap, wall covering, flooring, upholstery, table cloths, footwear, toys. (Extrusion, mainly pipe, 60; calendered sheet and film, 10; coating, 10; molding, 10; others, 10.)
Market share	15.8%
Trade	Export nearly 450 million lb and negligible imports
Price ($/lb)	0.31 for general purpose
Commercial value (billions $)	1.75

Source.   Adapted from *Chemical and Engineering News*, October 6, 1980 and February 9, 1981 and References 10, 19, and 21.

[a]Vinyl (generic name). Geon and B.F. Goodrich (trade name and manufacturer). Tonnage (commodity) plastic. Thermoplastic.

Table 1.19 Polystyrene (PS)[a] Product Sheet

Distinguishing Feature	Product Information
Structure	—(CH$_2$CH)— (with phenyl ring)
Advantages and limitations	*Advantages*: Transparency. Gloss. Hardness and rigidity. Dimensional stability. Low cost. *Limitations*: Brittle. Low heat, chemical, and environmental resistance. Flammable.
Production	Styrene polymerization aided by peroxide initiator. Processable by all thermoplastic methods. Estimated 1980, 3500 million
Major applications (fabricated forms, %)	Flower pots, foam and nonfoam cups, interior doors, margarine tubs, egg cartons, lighting covers, air conditioner housings, shutters, caps and closures, disposable cutlery. (Injection molding, 50; extrusion, 30; foamed beds, 10; others 10.)
Market share	10.1%
Trade	Export nearly 200 million lb and negligible imports
Price ($/lb)	0.43 for general purpose
Commercial value (Billions $)	1.5

Source. Adapted from *Chemical and Engineering News*, October 6, 1980 and February 9, 1981 and References 10, 19, and 21.

[a]Styrene (generic name). TMDA and Gulf (trade name and manufacturer). Tonnage (commodity) plastic. Thermoplastic.

Table 1.20 Phenolic[a] Product Sheet

Distinguishing Feature	Product Information
Structure	OH (aromatic ring with CH$_2$ groups)
Advantages and limitations	*Advantages*: Good heat resistance. Dimensional stability over temperature range. High modulus. Good compressive strength. High resistivity. Good solvent resistance. Self-extinguishing. Low cost. *Limitations*: Requires fillers for molding. Low-impact strength. Color limitations. Poor resistance to bases and oxidizers.
Production	Phenol and formaldehyde condensation. One-stage curing by heat and two-stage by heat and hexamethylenetetramine. Processing ease. Estimated 1980, 1500 million lb.
Major applications (fabricated forms, %)	Circuit breakers, distributer caps, fuse blocks, wiring devices, brake linings, plywood, decorative laminates, switch gears. (Plywood, 34; glass fiber insulation, 19; molding compound, 8; foundry resins, 8; fibrous and granular wood, 8; others 23.)
Market share	4.3%
Trade	Export approximately 30 million lb and negligible imports
Price ($/lb)	0.38
Commercial value (billions $)	0.5

Source. Adapted from *Chemical and Engineering News*, September 29, 1980 and February 9, 1981 and References 10, 19, and 21.

[a]Phenolic (generic name). RX and Rogers (trade name and manufacturer). Tonnage plastic. Thermoset.

Table 1.21 Polyester[a] Product Sheet

Distinguishing Feature	Product Information
Structure (cured)	
Advantages and limitations	*Advantages*: Readily cured. High-filler loading. Low-cost tooling. Nonflammable. Halogenated grades. *Limitations*: Low maximum service temperature. Poor solvent resistance.
Production	Polyesterification of maleic and phthalic anhydrides with propylene glycol; cross-linking with styrene. Estimated in 1980, 950 million lb.
Major applications (fabricated forms, %)	Corrugated panels, boat hulls, imitation marble, sanitary ware, hand tools, bowling balls, shirt buttons, shower stalls, cooling towers, motor housings. (Construction, 29; marine, 20; castings, 14; transportation, 9; others, 28.)
Market share	2.7%
Trade	Export nearly 9.5 million lb and negligible imports
Price ($/lb)	0.54 for general purpose
Commercial value (billions $)	0.46

Source. Adapted from *Chemical and Engineering News*, September 29, 1980 and February 9, 1981 and References 10, 19 and 21.

[a] Polyester (generic name). Gafite and GAF (trade name and manufacturer). Tonnage plastic. Thermoset.

Table 1.22 Epoxy[a] Product Sheet

Distinguishing Feature	Product Information
Structure	
Advantages and limitations	*Advantages*: Convenience in cure. Excellent adhesion. *Limitations*: Poor oxidative stability and moisture resistance. Specialty grades expensive.
Production	Bisphenol A and Epichlorohydrin reaction followed by cross-linking. Processed by all thermosetting methods. Estimated in 1980, 310 million lb.
Major applications (fabricated forms, %)	Flooring, primer coatings, vessels and pipes, can linings, anti corrosive paints, adhesives, electrical insulators, circuit boards, concrete repair, laminates. (Coatings, 44; laminates and composites, 18; molding, 9; flooring, 6; adhesives, 5; others, 18.)
Market share	0.9%
Trade	Export nearly 35 million lb and negligible imports
Price	1.0
Commercial value (billions $)	0.295

Source. Adapted from *Chemical and Engineering News*, September 29, 1980 and February 9, 1981 and References 10, 19, and 21.

[a] Aromatic polyether (generic name). Isochem and Isochem (trade name and manufacturer). Specialty plastic. Thermoset.

Table 1.23 Major Commercial Plastics[a]

Plastic-Generic Name—Trade Name—Manufacturer	Major Limitations	Typical Uses
Alkyd—Alkyd Alkydal—Mobay (S, TS)	Poor chemical and solvent resistance. Unsaturated types have high skrinkage during cure.	Enamels, paints, and lacquers; automobile ignition systems, fuses, and light switches; television tuning devices and tube supports
Acrylonitrile butadiene styrene (ABS) —Styrene-Cycolac— Borg Warner (T, TP, EP)	Poor solvent resistance. Low service temperature. Poor dielectric strength. Low elongation.	Piping, luggage, auto dashboards, telephones, sporting goods, camper tops, safety equipment, calculator housings, refrigerator housings, margarine tubs
Cellulose acetate, cellulose acetate butyrate, cellulose nitrate, cellulose propionate, ethyl cellulose— All cellulosics- Ethocel—Dow (S, TP)	Poor solvent and alkali resistance. High moisture pick-up. High permeability. Flammable. High compressive strength.	Extruded tape, tool handles, industrial items, electrical appliance components and housing, automotive parts, tubular packages, blister packaging, fire extinguisher components, signal lenses, pen and pencil barrels, tooth-brush handles, toilet articles, personal accessories, flashlight cases, premium toys
Diallyl phthalate (DAP)— Polyester-Dapex—Acme (S, TS)	Poor resistance to phenols and oxidizing agents. Shrinkage during cure. More expensive than alkyds.	Electrical insulation housings, circuit boards, junction boxes, potentiometers, encapsulating shells, multipronged connectors, decorative wood laminates in furniture and panelling, vacuum impregnation of nonferrous metal castings
Epoxy*—Aromatic Polyether—Epotuf— Reichhold (S, TS)	Poor oxidative resistance. Moisture pickup. Thermal stability 350°–450° F. Specialty grades are expensive.	Flooring, primer coatings, can linings, anticorrosive paints, adhesives, laminates, vessels and pipes, concrete repair, electrical insulators, circuit boards
Chlorotrifluoroethylene (CTFE)— Alcon—Allied	Poor dieletric properties compared to other fluorocarbons. High cost.	Consumer—cookware coating, coffee pot components, urn components, toaster components; medical— catheters and tubing; Electrical—shrinkable tubing, coaxial cable connectors, slip-on insulation, capacitors, insulator wire; Chemical—pipe fittings, expansion joints, valves, seals, gaskets, seatings, laboratory ware, lining for
Polyvinyl Fluoride (PVF)— Gentron—Allied	Toxic thermal decomposition. Lower thermal stability than other fluorocarbons. Acid Attack. High dipole.	
Polyvinylidene fluoride (PVF$_2$)— Kynar—Pennwalt	Very high cost.	

Table 1.23   Major Commercial Plastics (*Continued*)

Plastic-Generic Name—Trade Name—Manufacturer	Major Limitations	Typical Uses
Fluorinated ethylene propylene (FEP)— Fluorlene—Liquid Nitrogen Processing	Very high cost.	valves and pumps, thread sealants; Mechanical—piston rings, valve seats, energy-absorbing devices, cruise control parts, bearing pads for construction, aircraft (typical applications for fluorocabons)
Ethylene Tetrafluoro ethylene (ETFE)— Tefzel—Dupont	Very high cost.	
Ethylene chlorotrifluoroethylene (ECTFE)—Halar—Allied	Very high cost.	
Perfluoroalkoxy copolymerized with tetrafluoroethylene (PFA)— Hostaflon— American Hoechst All Fluorocarbons (S, TP, EP)	Toxic thermal decomposition. Creep. Low compressive strength. Low tensile strength. Low stiffness. High density. High cost.	
High-density polyethylene (HDPE)*—Polyolefin Eitex—Soltex (T, TP)	Poor weathering. Stress cracking. Poor thermal stability. High thermal expansion. Flammable.	Milk bottles, containers, shipping pails, pallets, seating, wastebaskets, disposable syringes, luggage, outdoor furniture, piping
Ethylene—sodium methacrylate copolymer— Ionomer Surlyn—Dupont (S, TP)	Poor weather and acid resistance. Not self-extinguishing.	Blow-molded bottles; packaging—skin, blister, foil; adhesives for laminating metals
Low-density polyethylene (LDPE)*—Polyolefin Elrey—Dart (T, TP)	Stress cracking. Resistance to bonding and printing. Low tensile strength.	Packaging lids, trash bags, garment bags, stretch wrap, agricultural film, bowls, toys, printing paper, wire and cable coatings, communications cable
Melamine-amino— Cascomel—Borden (S, TS)	Fair dimensional stability. Low impact strength.	Dinnerware, mixing bowls, utensils handles, table tops, ash trays, buttons, lavatory bowls, distributor heads
Nitrile and barrier resins— Nitrile-Barex—Vistron (S, TP)	High cost.	Packaging of food and beverages, oil, insecticides, varnishes, waxes
Nylon 6, 66, 610, 11, 12 (Fiber)*— Polyamide-Vydne— Monsanto (S, TP, EP)	Poor U.V. stabilization. Dimensional instability. High shrinkage. Moisture pickup. Electrical and mechanical properties are moisture sensitive.	Textiles, carpeting, tirecord, drapery slides, combs, cigarette lighters, sporting goods, fuel filter bowls, pipe fittings, gears and bearings, brushes, appliance housings, electronic connectors

Table 1.23   Major Commercial Plastics (*Continued*)

Plastic-Generic Name—Trade Name—Manufacturer	Major Limitations	Typical Uses
Phenolic*—Phenolic Phenolls—General Electric (T, TS)	Poor resistance to bases and oxidizers. Color limitation. Low impact strength. Requires fillers for molding.	Circuit breakers, switch gear, distributor caps, fuse blocks, wiring devices, brake linings, plywood decorative laminates
Polyacrylonitrile (PAN) (Fiber)—Acrylic-PAN— Solvayscie (S, TP, EP)	Poor resistance to solvents. Stress cracking.	Sweaters, blankets, carpeting
Polybutadiene (Elastomer)—Diene- Trans 4—Phillips Chemical (T, TS)	Thermal stability dependent on cure. High shrinkage.	Tires, rubbery goods, castings, laminates, encapsulation and potting of electrical components
Polybutylene—Polyolefin- Film Grade—Shell (S, TP)	Poor weather resistance. Not self-extinguishing.	Food packaging film, protective coatings, piping, electrical insulation
Polycarbonate (PCO)*— Polyester-Lexan— General Electric (S, TP, EP)	Poor solvent and scratch resistance.	Bottles, safety glass, auto lenses, battery cases, power tools housings, helmets
Polyethylene terphthalate (PET) (Fiber)— Polyester-Petpac— Celanese (S, TP)	Poor solvent resistance, acids and alkali attack. Difficult to mold.	Fiber blends with cotton or wool, food packaging film, graphic arts, typewritter ribbon, electrical insulation in capacitors, cable-wrapping, printed circuits, motors, transformers, magnetic recording tape, weather balloons and satellites.
Polyimide (PI)— High Temperature Heterocyclic-Azamide— AZ Products (S, TP, EP)	Alkali attack. Dark color. Difficult to fabricate. High cost.	Randomes, printed circuit boards, turbine blades
Polyamide-Imide (PAI)— High Temperature Heterocyclic-Turlon— Amoco (S, TP, EP)	Difficult to process.	Valves, gears, pumps, magnet wire enamel
Polymethyl methacrylate (PMMA)—Acrylic- Lucite—Dupont (T, TP)	Alkali and solvent sensitivity. Stress cracking. Poor scratch resistance. Combustible.	Nameplates for home appliances, outdoor signs, camera lenses, skylights, television shields, airplane canopies and windows
Polymethyl pentene— Polyolefin— (S, TP)	Poor weather resistance. Not self-extinguishing. Low tensile strength.	Food packaging and catering, laboratory and medical ware, lighting

21

Table 1.23   Major Commercial Plastics (*Continued*)

Plastic-Generic Name—Trade Name—Manufacturer	Major Limitations	Typical Uses
Polyoxymethylene (POM)—Acetal-Celcon—Celanese (S, TP, EP)	Poor acid, base, and UV resistance. Not self-extinguishing. Difficult to bond.	Shaver cartridges, zippers, ballcocks, aerosol bottles, sinks and faucets, lawn sprinklers, meat hooks, telephone push buttons, electrical switch gears
Polyphenylene oxide modified (PPO)—Aromatic Polyether-Noryl—General Electric (S, TP, EP)	Solvent sensitivity.	Shower heads, protective shields, appliance housing, pumps, auto dashboards, auto grilles and trim
Polyphenylene sulfide (PPS)—Aromatic polyether-Ryton—Phillips Petroleum (S, TP, EP)	Chlorinated hydrocarbons sensitivity. Low inpact strength. Difficult to process. High cost.	Electrical connectors, coil forms, lamp housings
Polypropylene (PP)*—Polyolefin—Hercocel—Hercules (T, TP)	Poor weather, oxidative and chemical resistances. Embrittlement at low temperatures. Flammable.	Cordage, carpet backing, food and textile wrap, decorative ribbon, dishpan tub, door liners, caps and closures, battery cases, pump housings, radio/television/phonograph housings, auto fender skirts
Polystyrene (PS)*—Polyolefin-E-Z Flow—Polysar (T, TP)	Low heat and solvent resistance. Brittle. Flammable.	Disposable cutlery, foam and nonfoam cups, margarine tubs, egg cartons, flower pots, shutters, interior doors, caps and closures, lighting covers, air conditioner housing
Polysulfone (PSO)—Aromatic Polyether-Udel—Union Carbide (S, TP, EP)	Not resistant to polar solvents. Poor weatherability. Stress cracking. High processing temperature. High cost.	Coffee makers, camera bodies, battery cases, light fixture sockets, meter housings, electrical connectors, auto switch and relay bases, fuel cell components
Polyvinyl Chloride (PVC)*—Rucodur—Hooker	Solvent sensitivity. Thermal decomposition with HCl. Stained by sulfur. Higher density than many plastics.	Food wrap, upholstery, table cloths, footwear, wall covering, siding, flooring, bottles, piping, phonograph records
Polyvinyl acetate (PVA$_c$)-Daratak—W.R. Grace	Low thermal, solvent and chemical resistance.	Adhesives, paints, textile finishes, paper coatings
Polyvinyl alcohol (PVA)—Elvanol—Dupont	Production from PVA$_c$ adds to cost.	Cosmetic manufacture, paper coatings, adhesives, textile wrap sizing, suspending aid for polymerization, polyvinyl butyral manufacture, soil erosion

Table 1.23 Major Commercial Plastics[a] **(Continued)**

Plastic-Generic Name—Trade Name—Manufacturer	Major Limitations	Typical Uses
Polyvinyl butyral—Butacite—Dupont		Adhesive film in the lamination of auto safety glass
Polyvinyl formal—Formvar—Monsanto		Magnetic wire coatings
Polyvinylidene chloride (PVCl$_2$)—Daran—W. R. Grace	Strength lower than PVC. Subject to creep.	Film, coating for food packaging
All vinyls (T, TP)		
Polyurethane foam (PU)—Polyurethane-Stycast—Emerson and Cuming (T, TS)	Poor thermal and chemical resistance. Weather sensitivity. Flammable. Utilizes toxic Isocynates.	Flexible and semiflexible foam—bedding, furniture, packaging, transportation; rigid foam—appliances, construction, marine, transportation, packaging, furniture
Silicon—Silicon-Pyrotex—Raybestos Friction (S, TS)	Halogenated solvents sensitivity. Low strength. Very high cost.	Coupling agents, mold release agents, water proofing agents, paper coatings, sealents, coil forms, switch parts, Induction heating apparatus, motor and generator coil insulation, power cables, medical implants
Styrene acrylonitrile (SAN)—Styrene-Rovel—Uniroyal (T, TP, EP)	Poor scratch resistance. UV light causes yellowing. Brittle.	Dentures, ice buckets, chair parts, bath accessories, lenses, lamps, syringes, closures, hobby kits, appliance knobs, auto parts
Styrene butadiene latex—Styrene-Darafiber—W. R. Grace (T, TP)		Adhesives in packaging, construction and wood products, coatings in water-based paints, paper binder, textile fabric finishes, lamination and fiber bonding
Thermoplastic elastomer—Alloy-TPR—Uniroyal (S, TP, A)		Sporting goods, sealants, gaskets, hose, belting, tubing, textile machinery parts
Thermoplastic polyester (TP)—Polyester-Celanx—Cellanese (S, TP, EP)		Electrical bobbins, television tuners, fuse cases, bearings, cams, auto ignition caps, speedometer frames, pump impellers, housings
Unsaturated polyester*—Polyester-Alpolit—American Hoechst	Poor solvent resistance. Low maximum service temperature.	Corrugated panels, boat hulls, imitation marble, sanitary ware, hand tools, bowling balls,

23

Table 1.23   Major Commercial Plastics[a] (*Continued*)

Plastic-Generic Name—Trade Name—Manufacturer	Major Limitations	Typical Uses
(T, TS)		shirt buttons, shower stalls, cooling towers, motor housings
Urea—Amion-Styplast— FMC (S, TS)	Poor heat resistance. Brittle.	Wall switches, lamp reflectors, handles, stove knobs, circuit breakers, buttons, closures, toilet seats

Source.   Adapted from References 9, 10, and 19.

[a] Key to the table: Tonnage plastic or high-volume resin (T). Speciality plastic or low-volume resin (S). Thermoplastic (TP). Thermoset (TS). Engineering plastic (EP)—These resins have superior thermal, mechanical, electrical, and chemical properties; such properties can be further improved by reinforcements. Reinforced plastics (RP) are filled polymers with glass, asbestos, carbon and other inorganic fibers. Plastics marked with an asterisk * can be RP. Alloy, composite, and blend (A).

present to 12% to attain weight reduction. Aromatic polyesters that exhibit thermotropic behavior, low melt viscosity-processing ease and exceptional mechanical strength in the solid state, will be pressed into service to lower vechicle energy consumption and to increase operating range. Graphite/epoxy composites will be substituted by graphite/ thermoplastic composites to lower fabrication and tooling cost for applications in military and commercial flight vehicles. According to past experience, about 12 polymers may be introduced each year to meet technological,

economic, and regulatcry demands (20). The U.S. plastics industry is in early stages of maturation (relationship between consumption and general economic activity). It is expected to grow at an 8.7% annual rate, which is twice the forecast for real gross national product (GNP). Polyvinyl chloride and most thermosets have matured. Low-density polyethylene, polystyrene, and polyurethane appear to be in early stages of maturation. Engineering plastics, alloys, and composites are targeted for an annual growth rate of 10%. According to Predicasts Incorpor-

Table 1.24   Plastics Production Worldwide—Global Shift

Region	Production, billions lb (Production Rank)		Annual Growth, %, (Rank)
	1975–1977 Average	1990	
European Economic Community (EEC)	31.4 (1)	97.0 (1)	8.4 (7)
North America	27.2 (2)	88.4(2)	8.8 (6)
Asia	14.1 (3)	49.8 (4)	9.4 (5)
Eastern Europe	11.8 (4)	50.6 (3)	11.0 (3)
Non-EEC Western Europe	4.8 (5)	20.8 (5)	11.0 (3)
Latin America	2.4 (6)	15.2 (6)	13.9 (2)
Africa-Mideast	0.8 (7)	7.4 (7)	17.3 (1)
Oceania	0.8 (7)	3.4 (8)	10.1 (4)
Total World	93.4	332.4	9.5

Source.   Adapted from *World Plastics to 1990*, Predicast Incorporated; *Plastics World*, February 1980, p. 12; *Plastics Engineering*, July 1980, p. 12.

Table 1.25   U.S. Plastics Production and Per Capita Projection

Plastic and Pounds per Capita	(Population in Million) Annual Production, (billions lb)		
	1978	1980	1990
	(218.5)	(227.0)	(241.0)
Thermoplastic	29.360	30.735	74.385
Pounds per capita	134.0	135.0	309.0
Thermoset	3.140	4.085	8.705
Pounds per capita	14.6	18.0	36.0
Total plastics	32.5	34.820	83.09
Total pounds per capita	148.6	153.0	345.0

Source.   Adapted from References 30 and Predicast Incorporated.

ated, worldwide production of plastics is to grow at a 9.5% annual rate. Latin America, Africa, and the Middle East will increase production from 14 to 17%. North America, the European economic community, and Japan's share of the world market will drop from 75 to 68%. Thermoplastics production will increase from 75 to 82%. Polyethylene will account for 30%, polyvinyl chloride will have a 10% share, polypropylene and polystyrene will share 13 and 10% respectively. Thermoset production is expected to drop from 25 to 18%. Polyester, polypropylene, epoxy, phenolics, and amino resins are expected to gain slightly (Tables 1.24 and 1.25).

Finally (to respond to changing scenarios), according to Wachtman and Steiner, our economic system, which is based on land, labor, and capital, may find a new basis in materials, energy, and knowledge (9, 20, 25–33).

1.6   REFERENCES

1. M. D. Baijal, *Plastics Technology Seminars*, New York University, New York; Cleveland State University, Cleveland; Center for Professional Advancement, New Brunswick 1973, 1974, 1977, and 1978.

2. Z. Tadmor and C. G. Gogos, *Principles of Polymer Processing*, Wiley-Interscience, New York, 1979, p. 36.

3. W. A. Holmes-Walker, *Polymer Conversion*, Applied Science Publishers, London, 1975, p. 21.

4. A. V. Tobolsky and H. F. Mark, *Polymer Science and Materials*, Wiley-Interscience, New York, 1971, p. 10.

5. H. G. Elias, *Macromolecule*, Vol. 1, Plenum Press, New York, p. 30.

6. K. J. Saunders, *Organic Chemistry of Polymers*, Chapman and Hall, London, 1973, p. 5.

7. H. L. Williams, *Polymer Engineering*, American Elsevier, Amsterdam, 1975, p. 1.

8. H. I. Bolker, *Natural and Synthetic Polymers*, Marcel Dekker, New York, 1974, p. 8.

9. J. A. Rauch, "Plastics within '80s will face new challenges," *Plastics Engineering*, May 1979, pp. 78–79.

10. *Facts and Figures of the Plastics Industry*, The Society of Plastics Industry, New York, 1979.

11. A. M. Houston, "Thermosets and Thermoplastics vie on performance, moldability, cost," *Materials Engineering*, February 1976, p. 42.

12. *Chemical and Engineering News*, December 15, 1980, pp. 30–31.

13. C. A. Harper, *Handbook of Plastics and Elastomers*, McGraw-Hall, New York, 1975, pp. 1–4.

14. W. A. Holmes-Walker, *Polymer Conversion*, Applied Science Publishers, London, 1975, p. 10.

15. H. G. Elias, *Macromolecules*, Vol. 1, Plenum Press. New York, 1979, pp. xiii–xxiii.

16. O. H. Wyatt and D. D. Hughes, *Metals, Ceramics and Polymers*. Cambridge University Press, 1974, p. 273.

17. Z. Tadmor and C. G. Cogos, *Principles of Polymer Processing*, Wiley-Interscience, New York, 1979, pp. 26 and 34.

18. D. W. Van Krevelen, *Properties of Polymers*, Elsevier, Amsterdam, 1976, p. 5.

19. *Plastics 1980*, Desk-Top Data Bank, Book B, Cordura Publications, LaJolla, California, 1980.

20. H. G. Elias, *Plastics Compounding*, September/October 1979, p. 77.

21. R. D. Deanin, *Polymer Structure, Properties and Application*, Cahners, Boston, 1972, p. 415.

22. A. Van den Brekel, "Quality engineering and value engineering—The two need not be at odds," *Plastics Engineering*, November 1974, p. 298.

23. C. W. Smith and G. W. Pearsall, J. H. DuBois, E. Baer, *Materials Engineering*, July 1979, pp. 47, 50, 66, 67.

24. W. T. Cruse, "Is There still a Feedstock Shortage?" *Plastics World*, September 10, 1974, pp. 48–50.

25. C. D. Han, "Polymer Engineering on Campus," *Chemical Engineering Progress*, January 1977, pp. 62, 63.

26. H. McQuiston, "Trends in Processing Machinery for the 80's," *Plastics Eigineering*, December 1979, p. 17.

27. *Plastics World*, "400 Major U.S. Plastics Processing Plants," January 1980, pp. 10, 50–57.

28. *Materials Engineering*, July 1980, p. 80.

29. J. B. Wachtman and B. W. Steiner, R. L. Kusterg, K. J. Wynne, Development of tailored materials and processes; As confidence increases, look for growth in use of graphite/epoxy composites; ultra high-strength polymers will be sought to replace metals; *Materials Engineering*, May 1980, pp. 49, 61, 80, 91, 92.

30. F. E. Hall, Resins 1980: How are we doing?, *Plastics Engineering*, December 1980, pp. 18–28.

31. T. Alfrey, K. J. Cleereman, E. S. Clark, *Applied Polymer Symposia*, 1974, No. 24, pp. 3–8, 31–35, 45–53.

32. E. Baer, *Materials Engineering*, July 1979, pp. 66–67.

33. D. Klempner, K. C. Frisch, and L. W. Kleiner, *Polymer Alloys*, Plenum Press, New York, 1977, pp. v and vi; Private communications, 1980.

PART 1

Material Science

## CHAPTER 2

Theories of Solution

C. A. DANIELS

B. F. Goodrich Chemical Company
Avon Lake Technical Center
Avon Lake, Ohio

2.1   Introduction   29
   2.1.1   Practical Aspects   30
   2.1.2   Some Early Approaches   30
   2.1.3   The Approach in this Chapter   31

2.2   The Thermodynamics of Solutions   31
   2.2.1   The Ideal Solution   31
   2.2.2   Conditions for Phase Equilibrium   33
   2.2.3   Regular Solutions   33

2.3   Scatchard-Hildebrand Solubility
   Concept—from the van Laar Theory   34

2.4   Three-Dimensional Solubility Parameter
   Theory   35

2.5   Free Volume Theories   36

2.6   Calculation of the Entropy from
   Probability-Lattice Model   36

2.7   Flory-Huggins Theory   37

2.8   Dilute Polymer Solutions—The Flory-
   Krigbaum Treatment   40

2.9   Maron's Theory of Polymer Solutions   41

2.10 Prigogine Theory and Its Modifications
   (Flory-Orwell-Vrij)   43

2.11 Recent Methods   45

2.12 Application to Various Polymer-Solvent
   Systems   46
   2.12.1   Natural Rubber-Benzene   46
   2.12.2   Polyisobutylene in Various
      Solvents   49

   2.13.3   Polystyrene-Solvent Systems   54
   2.12.4   Poly (α-Olefin) Solutions   56

2.13 Thermal Properties of Solutions   59

2.14 Polymer Solutions-Frictional Properties   62

2.15 The Effect of Branching on Intrinsic
   Viscosities   66

2.16 Molecular Dimensions   67

2.17 Critical Solution Behavior   69

2.18 Summary   71

2.19 Glossary of Symbols   72

2.20 References   73

2.1   INTRODUCTION

There have been numerous approaches to the development of a theory of solutions applicable where polymer molecules are the solute. Some have been derived strictly to treat long chain molecules, while others are based on a more general approach to binary solutions of any size molecules. Still to be developed is a satisfactory theory to explore the complex behavior of polyelectrolytes in solution.

The basic criterion by which the success of any given theory is judged is its ability to predict accurately experimental observations.

The author wishes to thank the B. F. Goodrich Chemical Company for granting permission to publish this chapter.

At the same time, it would be most helpful that these theories give fundamental information about the size and shape of the solute in the presence of the solvent, and perhaps, in some "ideal" or isolated state.

However, even if a theory proves accurate, many times it may not seem extremely useful from the practical point of view. For example, a scientist in industry looking for the proper solvent for a polymer to measure its molecular weight is more likely to make a choice based on a simple solubility parameter method rather than on a theory derived from statistical considerations. Still, the need for basic information regarding, for instance, the unperturbed dimensions of the chain molecule in solution must be appreciated. With all these considerations numerous theories of empirical origin, and others based on actual solution models, have emerged, the popularity of which differ. No one single theory of solutions has achieved universal acceptance. Still, work using numerous approaches goes on, because of complaints of complexity, inaccuracy, and empirical origin, and depends upon which school of thought the worker prefers.

The scope of this chapter is to highlight some of the more important theories of solution developed to explain the complex behavior of macromolecules in solution. By restricting this chapter to these ground rules, it is not then possible to cover all theories or all the aspects of polymer solution thermodynamics in great detail. To obtain more explicit information regarding polymer behavior in solution, the reader is referred to the several books that explore the wide spectrum of macromolecular solution properties (1–5).

2.1.1 Practical Aspects

There are practical considerations for looking to polymer solution thermodynamics besides those yielding information about the molecular weight or molecular dimensions. For example, solvents for adhesives and coatings must be chosen so that they provide characteristics sufficient to allow not only the dissolving of the polymer, but also after the solution is applied, that solvent should evaporate at a desired rate leaving the polymeric coating intact. For routine laboratory measurements, such as intrinsic viscosity, knowledge of the thermodynamic "goodness" of a solvent for the polymer is essential before such measurements can be made. For multigrade motor oils, polymeric additives must be chosen so that they improve the viscosity of the base oil at high temperatures while at the same time not producing a gelatinous structure at sub-zero temperature.

In addition, certain polymerizations can be carried out in solution, or at least partially so, such as in bulk or mass polymerization of polystyrene or polyvinyl chloride. Removal of residual monomer is not only dictated by the kinetic (diffusion, transport) considerations, but also by the equilibrium thermodynamics of the system itself. The last consideration has achieved considerable attention of late.

Polymer fractionation, one of the earliest methods used to obtain polymer molecules of uniform length, depends strongly on the choice of both a good solvent for the polymer and a poor one to cause fractional precipitation. The reader is referred to Cantow's book on the subject of polymer fractionation (6). The point is that many practical ramifications of polymer solution thermodynamics manifest themselves daily in industry as well as academia, from both the practical and theoretical points of view.

2.1.2 Some Early Approaches

Many treatments of polymer solutions have tended to treat very dilute solutions separately from those more concentrated. The reason for this is often left vague, as if there were no connection between the dilute and concentrated solution of the same polymer-solvent system. From early work in solution theory, this lack of connection or discontinuity of behavior was strongly suggested to be due to the local chemistry of molecules in dilute solution, followed by entanglement as concentration increased. Only recently have workers in the field been approaching the entire concentration range behavior of solutions of polymers from a unified approach, a much needed improvement. Still, as this chapter demonstrates, anomalies and inaccuracies persist, due both to the behavior of the polymer and weaknesses in theory to account for them. This presentation attempts to treat the entire concentration range of solutions as a continuum without divorcing the dilute from the concentrated solution. There are, however, sections devoted to theories that are applicable only over limited concentration ranges. Because of their importance these theories must be explored, since they were, at their time of introduction, the best available methods and have become the basis of further improvements.

2.1.3 The Approach in this Chapter

The method of presentation in this chapter of the various theories of polymer solutions is to develop the relationships for the thermodynamic properties up to working equations, such as for chemical potentials (osmotic pressures or vapor pressure lowering and perhaps heats of solution). The important underlying concepts for each theoretical approach are briefly reviewed so that the reader understands where one theory differs from the others. This is important because in many instances questions are raised about the relationships between what seem to be rather vague parameters in some theories and the real molecular properties in which one may be interested.

Once the various theories are reviewed, the text continues with examinations of various systems of polymers in solvents. Where comparisons between theory and experiment exist, these are explored. Where comparisons of theories on a given polymer-solvent system exist, these too are examined so that the reader can rate success versus failure for the variety of theories employed. An extensive test of systems examined by way of each approach is given where appropriate, with the resulting parameters and literature references.

Also, where theories have been extended to the point where they can be interrelated to, for example, frictional properties of solutions, these interconnections are also demonstrated.

Only a small portion of the text is devoted to detailing the experimental methods, but the equations relating the data to the desired thermodynamic properties are emphasized. The experimental methods are explored in depth only when it becomes imperative to interrelate the information obtained to other sources of thermodynamic data.

Tables of values (for example, polymer-solvent interaction parameters and Mark-Houwink constants) are presented. These tables are not intended to be exhaustive reviews of all the published values, since this would require a text of several hundred pages. Only "typical" values are listed. No attempt has been made to evaluate the accuracy of these values, since in many instances, large variations from worker to worker can and will be found in the literature. The reader is therefore cautioned to review that literature and refer to the original reference before taking these values to be absolute. They do, however, represent trends and are intended to fulfill just that function.

2.2 THE THERMODYNAMICS OF SOLUTIONS

2.2.1 The Ideal Solution

The total Gibbs free energy of a solution, denoted by G, is given by the relationship shown in Equation (2.1)

$$G = n_1\overline{G}_1 + N_2\overline{G}_2 + \cdots n_i\overline{G}_i \quad (2.1)$$

where the $\overline{G}_i$ are the partial molal free energies of components 1, 2 $\cdots$ i, and n_i are the corresponding numbers of moles of these components.

The respective partial molal entropies $\overline{S}_i$ and partial molal enthalpies $\overline{H}_i$ are related to $\overline{G}_i$ by Equation (2.2)

$$\overline{S}_i = \frac{(\overline{H}_i - \overline{G}_i)}{T} \quad (2.2)$$

and the total entropy and enthalpy (heat) of solution by Equations (2.3) and (2.4).

$$S = n_1 \bar{S}_1 + n_2 \bar{S}_2 + \cdots n_i \bar{S}_i \quad (2.3)$$

$$H = n_1 \bar{H}_1 + n_2 \bar{H}_2 + \cdots n_i \bar{H}_i \quad (2.4)$$

The partial molal free energy of any component i can be written as Equation (2.5):

$$\bar{G}_i = \bar{G}_i^0 + RT \ln a_i \quad (2.5)$$

where $\bar{G}_i^0$ is the partial molal free energy component i in some standard state, and a_i is its activity, and of an ideal vapor, its fugacity.

If mole fractions are employed as units of concentration, one can write Equations (2.6) and 2.7)

$$a_i = N_i \gamma_i \quad (2.6)$$

where

$$N_i = \frac{n_i}{\displaystyle\sum_{i=1}^{n} n_i} \quad (2.7)$$

is the mole fraction, and γ_i is the activity coefficient. For an ideal solution $a_i = N_i$, $\gamma_i = 1$.

For an ideal binary solution Equations (2.8) and (2.9) should be considered.

$$\gamma_1 = \frac{a_1}{N_1} = 1 \quad \text{as} \quad N_1 \longrightarrow 0 \quad (2.8)$$

$$\gamma_2 = \frac{a_2}{N_2} = 1 \quad \text{as} \quad N_2 \longrightarrow 0 \quad (2.9)$$

For ideal solutions, Raoult's law is obeyed, as given by Equation (2.10):

$$a_i = N_i = \frac{P_i}{P_i^0} \quad (2.10)$$

where a_i is the activity of constituent i in solution, P_i is the partial pressure of component i over the solution, and P_i^0 is the vapor pressure of pure constituent i.

Then N_i, the mole fraction of constituent i, is equal to the activity a_i. Inherent in this equality is the assumption that the vapor behaves ideally and that the fugacities f_i are equal to the partial pressures P_i.

And so all the partial thermodynamic functions can be written as Equations (2.11)–(2.13):

$$\Delta \bar{S}_i = -R \ln N_i \quad (2.11)$$

$$\Delta \bar{G}_i = RT \ln N_i \quad (2.12)$$

$$\Delta \bar{H}_i = 0 \quad (2.13)$$

and the remaining functions as Equations (2.14)–(2.16). Here subscript m relates to mixing.

$$\Delta S_m = -R \sum_{i=1}^{n_1} n_i \ln N_i \quad (2.14)$$

To demonstrate for ideal solutions, choosing component 1 as an example, one can write Equations (2.15)–(2.19) in the following forms:

$$\Delta \bar{S}_i = -R \ln N_i \quad (2.15)$$

$$\Delta G_m = -T \Delta S_m = RT \sum_{i=1}^{n_1} n_i \ln N_i \quad (2.16)$$

$$\left[\frac{\partial (\Delta G_m / T)}{\partial T} \right]_P = \frac{-\Delta H_m}{T^2} = 0 \quad (2.17)$$

$$\Delta H_m = 0 \quad (2.18)$$

$$\Delta \bar{H}_i = 0 \quad (2.19)$$

Similar equations can be written for all i components. Thus it has been demonstrated that the "ideal solution" should be characterized by a number of specific criteria, not the least of which is that $\Delta H_m = 0$, that is, the solutions are "athermal." This condition "by itself" is insufficient to classify a solution as behaving ideally, but unfortunately, it is many times mistakenly considered as such.

In dilute solutions van't Hoff's law for osmotic pressures is also maintained; that is Equation (2.20).

$$\frac{\pi}{C} = \frac{RT}{M_n} \quad (2.20)$$

where π is the pressure necessary to prevent the migration of solvent through a semi-permeable membrane to dilute a polymer solution containing that solvent, C is concentration, R is gas constant, T is absolute temperature, and M_n is number-average molecular weight. The driving force is the free energy difference ΔG between the pure solvent G_1^0 and that in the corresponding solution G_1. For polymers Equation (2.20) is the limiting case, since the reduced osmotic pressure π/C is related to the number-average molecular weight in a virial expansion of the type shown in Equation (2.21), where A is virial coefficient.

$$\frac{\pi}{C} = \frac{RT}{M_n} + A_2 C + A_3 C^2 + \cdots \quad (2.21)$$

Common methods for examining thermodynamic properties of solutions include calorimetry, vapor pressure lowering, boiling point elevation, and for polymers, osmometry and light scattering. These techniques provide useful information such as molecular weights, radii of gyration, and polymer-solvent interaction parameters. For detail of the experimental methods involved, the reader is referred to Chapter 3 and to the literature regarding specific methods for polymer solutions, such as those on light scattering (3) and vapor sorption (6).

2.2.2 Conditions for Phase Equilibrium

The condition dictating equilibrium between phases of a substance (at constant temperature and pressure) is that its molar free energy be equal in all phases. This condition applies to vapor-liquid, solid-liquid, solid-solid (etc.) equilibria for pure materials. For a substance in solution, however, a similar condition must be met: the partial molal free energy of each substance must be equal in all the liquid phases. That is Equation (2.22)

$$\bar{G}_i^A = \bar{G}_i^B = \bar{G}_i^C = \text{and so on} \quad (2.22)$$

where $\bar{G}_i$ represents the partial molal free energy of component i in phase A, phase B, phase C, and so on. Thus for a solution of say liquid (l) 1 in liquid 2, at equilibrium with their vapors (v), the condition exists such that Equations (2.23) and (2.24) are satisfied.

$$\bar{G}_1^l = \bar{G}_1^v \quad (2.23)$$

$$\bar{G}_2^l = \bar{G}_2^v \quad (2.24)$$

Similar relationships can be written for polymer-rich and polymer-poor phases in equilibrium, at the conditions of partial and critical miscibility. This aspect is expanded in a later section.

2.2.3 Regular Solutions

Early in the development of polymer solution theories, Hildebrand (7) proposed the concept of the regular solution where the partial and total entropies of mixing are those of an ideal solution, as in Equations (2.1)–(2.19). Also, a further assumption was made such that "athermal" solutions were considered ideal. However, these assumptions are not strictly valid, even though instances of "athermal" polymer solutions can be found in the literature. By size considerations, alone, it must be obvious that the entropy changes on dissolving a macromolecule in solvent should result in an entropy change much unlike that for an ideal solution or one nearly ideal, such as one of molecules of similar size.

Thus the concepts presented in the earlier section on ideal solutions can just be set aside, for no polymer solution ever behaves truly ideally. Some propose that under given special conditions, polymer solutions approach ideality, that is, the interactions between polymer and solvent vanish. However, it is difficult to envision any solution process involving a long chain molecule that would have anything approaching the entropy change on mixing of an ideal solution. It is just on that size consideration alone that the magnitude of the problem of explaining polymer behavior can be anticipated.

2.3 SCATCHARD-HILDEBRAND SOLUBILITY CONCEPT—FROM THE VAN LAAR THEORY

Based on the mixing of two liquids of unspecified dimensions, van Laar (8) developed an expression for the energy of mixing ΔE_m. Two assumptions for simplicity were made: the volume change on mixing $\Delta V_m = 0$; and that the entropy of mixing was that of an ideal solution $\Delta S_m = 0$ (a regular solution). Thus, by use of Equations (2.25) and 2.26),

$$\Delta G_m = \Delta E_m + P\,\Delta V_m - T\,\Delta S_m \quad (2.25)$$

$$\Delta G_m = \Delta E_m \quad (2.26)$$

Van Laar showed that the energy of interaction (per mole) w_{ij} molecular species i and j in the binary solution can be given by Equation (2.27).

$$\Delta E_m = Z\left(\frac{w_{ii}}{2} + \frac{w_{jj}}{2} - w_{ij}\right) = Z\,\Delta w \quad (2.27)$$

where Z is the number of nearest neighbor molecules in solution and w_{ii}, w_{jj}, and w_{ij} are the potential energies of the pairs of i and j molecules. At this point, Scatchard and Hildebrand (4) described their assumptions that for molecules whose attraction primarily consist of London dispersion forces,

$$w_{ij} = (w_{ii}w_{jj})^{1/2}$$

which is the geometric mean of the attractive forces.

Equation (2.27) can now be transformed into Equation (2.28)

$$\Delta E_m = Z\left[\frac{w_{ii}}{2} + \frac{w_{jj}}{2} - (w_{ii}w_{jj})^{1/2}\right] \quad (2.28)$$

Now, let the molar heat of vaporization (ΔE_v) be given by Equation (2.29) and (2.30):

$$\Delta E_v^i = \frac{Zw_{ii}}{2} \quad (2.29)$$

for species i, and for species j

$$\Delta E_v^j = \frac{Zw_{jj}}{2} \quad (2.30)$$

by rearrangement, conversion to volume fractions, and generalizing the energy of mixing for n_i moles of i and n_j moles of j it can be shown that Equation (2.28) can be written as Equation (2.31)

$$\Delta E_m = V\phi_i\phi_j\left[\left(\frac{\Delta E_v^i}{v_i}\right)^{1/2} - \left(\frac{\Delta E_v^i}{v_j}\right)^{1/2}\right]^2$$

$$(2.31)$$

where ϕ_i and ϕ_j are the volume fractions of species i and j, respectively, and v_i and v_j are their molar volumes. V is the total volume of the mixture. Equation (3.31) can now be written as Equation (2.32)

$$\Delta E_m = V\phi_i\phi_j(\delta_i - \delta_j)^2 \quad (2.32)$$

where δ is the solubility parameter for the designated species, $(\Delta E_v^i/v_i)$ is called the cohesive energy density C.E.D., and is given by Equation (2.33):

$$(\text{C.E.D.})_i^{1/2} = (\delta_i) \quad (2.33)$$

for species i (or j).

Hildebrand and Scott (4) have used the assumption that $\Delta E_v \approx \Delta H_v$, and so solubility parameters can be approximated by Equation (2.34)

$$\delta_i = \left(\frac{\Delta H_v^i}{v_i}\right)^{1/2} \quad (2.34)$$

where ΔH_v is the heat of vaporization of species i. The assumption made by Hildebrand and Scott concerning the substitution of ΔH_v for ΔE_v leads to some errors, but this substitution allows one the opportunity to calculate solubility parameters from heat of vaporization for liquids. Also, then Equation (2.32) becomes Equation (2.35)

$$\Delta H_m = V\phi_i\phi_j \, (\delta_i - \delta_j)^2 \qquad (2.35)$$

and therefore one should be able to calculate the total heat of mixing of a binary solution, based on solubility parameter data.

Small (9) has proposed a technique for calculating the solubility parameter from the structural formula of the material and its density. Hoy (10) has developed a method for these calculations from vapor pressure measurements. The values from the Hoy approach differ considerably from those of Small. Burell (11) recommended that solubility parameters be obained from either thermal coefficients, pressure-temperature (P-T) data on the solvent, van der Waals constants, critical pressure data, surface tension measurements, or Kauri-Butanol values.

2.4   THREE-DIMENSIONAL SOLUBILITY PARAMETER THEORY

In 1967 Hansen published a series of articles (12) proposing the use of three-dimensional solubility parameters, demonstrating the utility of the method in paint and coating applications. Hansen suggested that the solubility parameter be subdivided into parameters reflecting the effects of dispersion, polar, and hydrogen bonding forces; see Equation (2.36) and (2.37).

$$\delta^2 = \delta_d^2 + \delta_p^2 + \delta_h^2 \qquad (2.36)$$

where

$$\delta^2 = \frac{\Delta E_v}{V} \quad \text{(as before)} \qquad (2.37)$$

and δ_d^2, δ_p^2 and δ_h^2 are the dispersion, polar, and hydrogen bonding components respectively.

By employing a trial and error approach, Hansen was able to evaluate δ_d, δ_p, and δ_h for a number of solvents and polymers.

Bagley and co-workers (13, 14) related the various δ's to more fundamental properties, particularly the internal pressure $P_i = (\partial E/$

$\partial T)_T = T \, (\partial P/\partial T)_V - P$. Thus by measuring the thermal pressure coefficient $(\partial P/\partial T)_V$ Bagley writes Equation (2.38) for the non-combinational free energy (13).

$$\Delta G_V^N = V_m\phi_1\phi_2 \left[(P_i^{1/2})_1 - (P_i)_2^{1/2} \right]^2$$

$$(2.38)$$

Bagley's treatment holds promise for reevaluating the solubility parameter approach in strongly interacting systems where anomalies were seen previously (12), through fundamental thermodynamic investigations. Table 2.1 lists typical solubility parameter data for a few polymers; Table 2.2 contains similar data on common solvents and monomers; Table 2.3 lists three-dimensional solubility parameters for some low molecular weight liquids.

Patterson and co-workers (15) have tied together the solubility parameter theory and that of corresponding states. This work has shown that very similar predictions can be obtained from both corresponding states and solubility parameter methods. This is attributed to the importance of free volume considerations in the Flory-Prigogine approach and its similarly important role in solubility parameters.

Table 2.1   Measured Solubility Parameters of Selected Polymers

Polymer	$\delta \, (\text{Cal}/\text{cm}^3)^{1/2}$
Polybutadiene	8.5
Polybutadiene co-styrene	
BUNA N (18/20)	9.5
BUNA S (85/15)	8.5
BUNA S (75/25)	8.55
BUNA S (60/40)	8.67
Polychloroprene	8.18
1-4-*Cis*-polyiosprene	7.9
Polyethylene	7.9
Polyisobutylene	7.8
Poly-*n*-butyl methacrylate	8.75
Polyethyl methacrylate	8.95
Polymethyl methacrylate	9.5
Polystyrene	9.10
Polyvinylchloride	9.7

Table 2.2   Solubility Parameters of Selected Solvents and Monomers

Solvent	$\delta\,(Cal/cm^3)^{1/2}$
Acetone	9.9
Acrylic acid	12.0
Acrylonitrile	10.5
Benzene	9.2
1–3-Butadiene	7.1
Butylmethacrylate	8.2
Carbon tetrachloride	8.6
Chlorobenzene	9.5
Chloroform	9.3
m-Cresol	10.2
Cyclohexane	8.2
Cyclohexanone	9.9
N,N,dimethylformamide	10.6
Diisopropylketone	8.0
Diethyl ether	7.4
Dimethyl sulfoxide	12.0
Epichlorohydrin	11.0
Ethylene glycol	14.6
Ethylene oxide	8.3
Heptane	7.4
Hexane	7.3
Methanol	14.5
Methyl acrylate	8.9
Methyl ethyl ketone	9.3
Methylmethacrylate	8.8
α-Methyl styrene	8.5
Naphthalene	9.9
Nitrobenzene	10.0
n-Octane	7.6
Styrene	9.3
Tetrahydrofuran	9.1
Toluene	8.9
Vinyl chloride	7.8
Water	23.4
Xylene	8.8

2.5   FREE VOLUME THEORIES

Dayantis (16, 17) has adopted the free volume approach to polymer solutions, theorizing that the entropy of mixing arises from the different free volumes available to solvent molecules and chain segments in the pure state and in solution. By subdividing the Flory interaction parameter χ into its heat and entropy contribution, as shown in Equation (2.39),

$$\chi = \chi_H + \chi_S \qquad (2.39)$$

Dayantis proposes a method of a priori calculating χ_S utilizing Equation (2.40)

$$\chi_S = \frac{1}{\phi_2^2}\left\{ \ln\left[1 - \phi_2\,(1-\rho)\right] + \frac{1-\rho}{1-\rho_2\,(1-\rho)}\phi_1\phi_2 + \frac{C'\,(1-\rho)}{1-\phi_2\,(1-\rho)}\phi_2^2 \right\} \qquad (2.40)$$

where C is an adjustable parameter defined as in Equation (2.41)

$$C' = W\frac{C}{3}\left(\frac{V_1}{V_2}\right) \qquad (2.41)$$

where W = ratio of the translational to rotational degrees of freedom
ρ = ratio of the free volume fractions of the polymer and solvent

Thus values of χ_S, χ_H and their sum have been made calculable and have been checked against experimental values by adjusting C' and ρ.

Yamakawa (18) has also taken a strictly free volume approach to polymer solution thermodynamics. This method has yet to be extensively tested.

2.6   CALCULATION OF THE ENTROPY FROM PROBABILITY-LATTICE MODEL

Consider the situation where a solution of two components can be represented by a three-dimensional lattice where all the cubicles have the same volume (see Figure 2.1). Now, allow the cubicles to hold exactly one molecule of solvent or one of solute, and allow no interaction between solvent and solute. Then, each solvent molecule, being of the same size as every other molecule, on mixing with solute, would cause the entropy change of ideal solutions, and the entropy of mixing would be give by Equation (2.42).

$$\Delta S_m = -\left[n_1 R \ln N_1 + n_2 R \ln N_2\right] \qquad (2.42)$$

Table 2.3   Three-Dimensional Solubility Parameters of Selected Solvents and Monomers

Solvent	δ	δd	δp	δh	$(Cal/cm^3)^{1/2}$
Acetone	9.77	7.58	5.1	3.4	
Benzene	9.15	8.95	0.5	1.0	
Carbon tetrachloride	8.65	8.65	0	0	
Chloroform	9.21	8.65	1.5	2.8	
Cyclohexane	8.18	8.18	0	0	
Cyclohexanone	9.88	8.65	4.1	2.5	
Diethyl ether	7.62	7.05	1.4	2.5	
N,N,Dimethyl formamide	12.14	8.52	6.7	5.5	
Dimethyl sulfoxide	12.93	9.00	8.0	5.0	
Dioxane	10.0	9.30	0.9	3.6	
Ethanol	12.92	7.73	4.3	9.5	
Ethylbenzene	8.80	8.70	0.3	0.7	
Ethylene glycol	16.30	8.25	5.4	12.7	
Methanol	14.28	7.42	6.0	10.9	
Methyl ethyl ketone	9.27	7.77	4.4	2.5	
Nitrobenzene	10.62	8.60	6.0	2.0	
Styrene	9.30	9.07	0.5	2.0	
Tetrahydrofuran	9.52	8.22	2.8	3.9	
Toluene	8.91	8.82	0.07	1.0	
Water	23.5	6.0	15.3	16.7	
Xylene	8.80	8.63	0.5	1.5	

The total number of ways of arranging the n_1 molecules of solvent and the n_2 moles of solute whose sum is n_0 is expressed as Equation (2.43).

$$\Omega = \frac{n_0!}{(n_1!/n_2!)} \qquad (2.43)$$

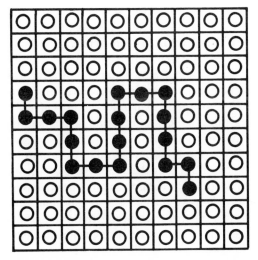

Figure 2.1 Representation of the position of polymer chain segments and solvent molecules in a liquid lattice model of solutions.

Using Stirling's approximation, Equation (2.44)

$$\ln n! \approx n \ln n - n \qquad (2.44)$$

and the Boltzman principle, one can write Equations (2.45) and (2.46).

$$\Delta S_m = k \ln \Omega \qquad (2.45)$$

$$\Delta S_m = k \ [(n_1 + n_2) \ln (n_1 + n_2) - n_1 \ln n_1 - n_2 \ln n_2] \qquad (2.46)$$

For polymer solutions, (quickly) it becomes apparent that the size of the macromolecule dictates that the entropy change on mixing with a solvent must be much different from the ideal solution. To use the lattice approach, then, for polymer solutions, some assumptions must be made regarding the behavior of the system.

2.7 FLORY-HUGGINS THEORY

Flory and Huggins (19, 20) independently derived from lattice considerations an expression for the entropy of mixing based on the following:

1. The lattice theory holds.

2. Each lattice site is of equal volume dictated by the volume of the solvent molecule.

3. The polymer molecule can be subdivided into x segments, defined by its volume v_2 and that of the solvent molecule (on the lattice site) v_1 such that $x = v_2/v_1$.

4. Each segment has z number of sites available, and z is the coordination number.

5. Molecular interactions do not contribute to entropy.

6. Molecular interactions do contribute to the heat of mixing.

From probability, the number of configurations available for the sites in the lattice is given by Equation (2.47)

$$\Omega = \left[\frac{n_0!}{(n_0 + xn_2)! n_2!} \right] \left[\frac{(z-1)}{n_0} \right] n_2 (x-1)$$

$$(2.47)$$

Since, in this case, as shown in Equation (2.48)

$$n_0 = n_1 + xn_2 = n_1 + \left(\frac{V_2}{V_1} \right) n_2 \quad (2.48)$$

Equation (2.45) can now be written as Equation (2.49)

$$\Delta S_m = -R (n_1 \ln \phi_1 + n_2 \ln \phi_2) \quad (2.49)$$

where one defines the volume fractions ϕ_1 of solvent and ϕ_2 of solute by Equations (2.50) and (2.51):

$$\phi_1 = \frac{n_1}{(n_1 + xn_2)} \quad (2.50)$$

$$\phi_2 = xn_2 (n_1 + xn_2) \quad (2.51)$$

knowing that $\phi_1 + \phi_2 = 1$.

Flory (5) pointed out that the entropy defined in Equation (2.49) represented the configurational entropy only and that entropy

effects caused by neighbor interactions were not included.

Flory then chose to define the heat of mixing in terms of the van Laar expression, Equation (2.52):

$$\Delta H_m = Z \Delta w \, n_1 \, \phi_2 \, \chi_1 \quad (2.52)$$

and defined the Flory (and Huggins) interaction parameter of Equation (2.53).

$$\chi_1 = \frac{Z \Delta w \, \chi_1}{kT} \quad (2.53)$$

This gave the heat of mixing expression the form expressed in Equation (2.54):

$$\Delta H_m = RT \chi_1 \, n_1 \, \phi_2 \quad (2.54)$$

and the free energy of mixing as Equation (2.55).

$$\Delta \overline{G}_m = kT [n_1 \ln \phi_1 + n_2 \ln \phi_2 + \chi_1 n_1 \phi_2]$$

$$(2.55)$$

The foregoing equations served as a framework for considerable research in the area of polymer solution thermodynamics. It should be remembered that, according to Equation (2.54), χ_1, the interaction parameter, originally was defined as a heat of mixing parameter, which implies that nearest neighbor interactions would result in changes in ΔH_m, but not in ΔS_m. Flory was indeed aware of the ramifications of such an assumption, which was later to be proven in error. However, the utility of these expressions was indeed a milestone in the treatment and understanding of polymer solutions.

When applied to very dilute solutions, it was apparent that the Flory-Huggins theory did not (and should not, as Flory pointed out) apply. Despite this, even today, a significant number of workers continue to try to use this theory in very dilute solutions, where it is least applicable.

To relate the expressions derived by Flory and Huggins to measureable quantities, differentiation of the free energy of mixing expression is necessary and yields Equation (2.56)

$$\Delta G_1 = RT \left[\ln \phi_1 + \left(1 - \frac{1}{x} \right) \phi_2 + \chi_1 \phi_2^2 \right]$$

$$(2.56)$$

and from the preceding for ideal solutions

$$\Delta G_1 = - \pi V_1^0$$

for very dilute solutions. On rearrangement and substitution the reduced osmotic pressure π / c can be expressed as Equation (2.57).

$$\frac{\pi}{c} = RT \left[\frac{1}{M_n} + \left(\frac{\overline{V}_2^2}{V_1^0} \right) \left(\frac{1}{2} - \chi_1 \right) c + \right.$$

$$\left. \left(\frac{\overline{V}_2^3}{3 V_1^0} \right) c^2 + \cdots \right]$$

$$(2.57)$$

For activity of solvent a_1, when its vapor behaves ideally, one can substitute equation (2.12) into equation (2.56) to yield equations for vapor pressure lowering behavior in the form shown as Equations (2.58) and (2.59):

$$\ln \frac{P_1}{P_1^0} = \ln \phi_1 + \left(1 - \frac{1}{x} \right) \phi_2 + \chi_1 \phi_2^2 \quad (2.58)$$

$$a_1 = \frac{P_1}{P_1^0} \tag{2.59}$$

Alternately, as Scott and Hildebrand (4) showed that χ can be related to solubility parameters δ by Equation (2.60)

$$\chi_1 = \frac{V_1^0}{RT} (\delta_1 - \delta_2)^2 + \beta \tag{2.60}$$

where β is a correction factor necessary when considering high polymer solutions.

For the partial molal heat of mixing and entropy of mixing, Flory writes Equations (2.61) and (2.62).

$$\Delta \overline{H_1} = RT\kappa_1 \phi_2^2 \tag{2.61}$$

$$\Delta \overline{S_1} = R \psi_1 \phi_2^2 \tag{2.62}$$

Thus, knowing that ψ is not a "heat" parameter, but a "free energy parameter," one can write Equation (2.63).

$$\kappa - \psi_1 = \chi_1 - \tfrac{1}{2} \tag{2.63}$$

Then the theta temperature θ, the temperature at which $\chi = \tfrac{1}{2}$ can be defined as in Equations (2.64) and (2.65).

$$\theta = \frac{\kappa_1 T}{\psi_1} \tag{2.64}$$

$$\psi_1 - \kappa_1 = \psi_1 \left(1 - \frac{\theta}{T} \right) \tag{2.65}$$

This term becomes important in dealing with the frictional properties of dilute solution, where it can be shown that the amount the polymer molecule expands in solution α due to interactions is given by Equation (2.66) and associated Equations (2.67) and (2.68)

$$\alpha^5 - \alpha^3 = 2C_m \chi_1 \left(1 - \frac{\theta}{T} \right) M^{1/2} \quad (2.66)$$

where

$$C_m = \left(\frac{3^3}{2^{5/2} \pi^{3/2}} \right) \left(\frac{1}{\gamma_2} \right) \left(\frac{1}{V_1 N_*} \right) \left(\frac{M}{\overline{r}_0^2} \right)^{3/2} \quad (2.67)$$

and

$$(\overline{r}_0^2)^{1/2} = \frac{(\overline{r}^2)^{1/2}}{\alpha} \tag{2.68}$$

and $(\overline{r}^2)^{1/2}$ is the root mean square end-to-end distance of the random coil polymer chain, $(\overline{r}_0^2)^{1/2}$ then represents this end-to-end distance in the unperturbed state (θ conditions), and N_* is the number of molecules per unit volume. Detailed treatment of the frictional properties from methods of experimentally determined chain dimensions in solution is presented in a later section.

Early studies testing in Flory-Huggins theory of solution included natural rubber in benzene [e.g., Gee and Treloar (21), Gee and

Orr (22)], polyisobutylene-benzene [Krigbaum and Flory (23), Jessup (24), Bawn and Patel (25)], polyisobutylene-cyclohexane [Krigbaum and Flory (23)], and numerous others. An extensive list of systems studied and the pertinent references have been summarized by B. A. Wolf in the second edition of the *Polymer Handbook* (26).

2.8 DILUTE POLYMER SOLUTIONS— THE FLORY-KRIGBAUM TREATMENT

Since dilute solution thermodynamic data were inadequately explained by the Flory-Huggins theory, it became necessary not only to understand the reasons for the inadequacy but also to do so in terms of defining chain behavior when their solutions were extremely dilute. Flory postulated that at very high dilutions polymer chains cluster and so sweep out a volume μ, excluding all other molecules. This excluded volume μ must be accounted for in the partition function describing the number of chain arrangements available, such that the total volume available V is reduced by the excluded volume in such a manner that the free energy of mixing be described by Equation (2.69):

$$\Delta G_m = -kT\left[n_2 \ln V + \sum_{i=0}^{i=n_2-1} \ln\left(1 - \frac{i\mu}{V}\right)\right] + c \quad (2.69)$$

where c is a constant.

By simplification and partial differentiation one obtains Equation (2.70):

$$\frac{-\Delta \overline{G}_1}{V_1^0} = \pi = RT\left[\frac{1}{M_n} + \left(\frac{N\mu}{2M_n^2}\right)c\right] \quad (2.70)$$

where N is Avogadro's number, and c (the concetration) is very dilute.

From the virial expansion of π/c in c, one obtains Equation (2.71)

$$\frac{\pi}{c} = \left(\frac{\pi}{c}\right)_0 [1 + \Gamma_2 c + g\Gamma_2^2 c^2 + \cdots] \quad (2.71)$$

which is referred to as the Flory-Krigbaum relation (27). Comparison of Equations (2.70) and (2.71) gives the concentration dependence of the reduced osmotic pressure, and Equations (2.72) and (2.73), the definition of Γ_2, and the limitng reduced osmotic pressure.

$$\Gamma_2 = \frac{N\mu}{2M} \quad (2.72)$$

$$\left(\frac{\pi}{c}\right)_0 = \frac{RT}{M_n} \quad (2.73)$$

Under the proper conditions where $\Gamma_2 = 0$, Equation (2.71) simplifies to Equation (2.74)

$$\frac{\pi}{c} = \left(\frac{\pi}{c}\right)_0 = \frac{RT}{M_n} \quad (2.74)$$

which is independent of concentration. These conditions are indicative of the θ temperature, which, it has been proposed, is the temperature where contributions to $\Delta \overline{G}_1$ from segment-solvent interactions disappear, and the solution behaves "ideally."

From theory and experiment, the parameter g in Equation (2.71) was found to be approximately $5/8$. However, in practice, it was observed that plots of π/c versus c were nonlinear, that is $g\Gamma_2$ became important in dilute solutions. To minimize the nonlinearity, which manifested itself as preventing accurate extrapolation to $c = 0$ for M_n determinations, an assumption regarding g was made, that is, $g \approx 1/4$. Then one can write Equations (2.75) and (2.76)

$$\frac{\pi}{c} = \frac{RT}{M_n}\left[1 + \frac{\Gamma_2}{2c}\right]^2 \quad (2.75)$$

$$\left(\frac{\pi}{c}\right)^{1/2} = \left(\frac{RT}{M_n}\right)^{1/2}\left[1 + \frac{\Gamma_2}{2c}\right] \quad (2.76)$$

Thus plotting $(\pi/c)^{1/2}$ versus c (often) makes the reduced osmotic pressure curves linear, allowing for more reliable M_n determinations, even though, strictly speaking, this is not theoretically justified.

As a final note, some expressions utilizing the Flory-Krigbaum treatment talk of the second virial coefficient A_2, and the relationship of this to Γ_2 defined in the foregoing paragraphs, is given by Equation (2.77)

$$A_2 = \frac{\Gamma_2 M_n}{RT} \qquad (2.77)$$

It is also possible to interrelate the Flory-Huggins χ value to the second virial coefficient A_2 if the solutions are sufficiently dilute, by use of the relationship as given in Equation (2.78).

$$\chi_1 = 0.5 - A_2 \gamma_2^2 \overline{V}_1^0 \qquad (2.78)$$

It has been shown by Berry (28) that A_2 depends on the molecular weight of the polymer, specifically in the case of linear polystyrene in decalin. This work was significant especially in that it defined semiempirically the θ temperature and A_2 relationship. Berry and Casassa (29) explored the effect of molecular weight distribution on A_2 in both good and poor solvents. Table 2.4 lists second virial coefficients for a variety of polymer-solvent pairs.

2.9 MARON'S THEORY OF POLYMER SOLUTIONS

In 1959 Maron (30, 31) introduced a theory of polymer solutions that had as its origin early work on dilute solution rheology of polymer solutions. The theory is based on a fundamental concept regarding the shape of the polymer molecule in solution, namely, that it can be treated as nearly spherical. Maron based his model on the Ree-Eyring theory (32) of dilute solution and concentrated solution viscosities, which said that the viscosity of a solution can be expressed as Equation (2.79).

$$\eta = \Sigma b_i \frac{\sinh^{-1} \beta_i G}{\beta_i G} = \Sigma b_i \theta_i \qquad (2.79)$$

Table 2.4 Second Virial Coefficients of Selected Polymer-Solvent

Polymer	Solvent	$T°C$	Mol. Wt. $\times 10^{-4}$ and Type	$\times 10^4 A_2 (\text{mol. cm}^3/\text{g}^2)$
Cis-1,4-poly (butadiene)	Benzene	28.6	6.0–29.3 (M_n)	~15.3
Poly(butene-1)	Toluene	45	2.63–55.8 (M_n)	(10.8–4.10)
Polyisobutylene	Benzene	24–30	126 (M_w)	78.1
	Cyclohexane	25	55 (M_w)	2.6
Polypropylene	0-chloro-naphthalene	147	~48.4 (M_w)	~4.3
Poly(n-butyl-methacrylate)	Methylethyl ketone	23	25–258 (M_w)	2.08–1.69
Poly(ethyl acrylate)	Acetone	30	32–800 (M_n)	10.52
Poly(ethyl-methacrylate	Methylethyl ketone	23	20–263 (M_w)	4.9–1.83
Polymethyl-methacrylate	Benzene	30	43.3 (M_n)	2.63
	Methylethyl ketone	25	0.5–5.28 (M_w)	7.38–2.18
Polyvinylchloride	Cyclohex-anone	25	11.8 (M_w)	11
	Tetrahydro-furan	25	2.48 (M_n)	6.85
Polystyrene	Benzene	25	10.9–500 (M_w)	~3.83
	Cyclohexane	35–55	320 (M_w)	0–0.54
	Methylethyl ketone	25	0.32–98 (M_w)	3.5–1.05

The parameters b_i and β_i are related to the size of the ith unit in the chain, and depend on concentration and temperature. Maron, Nakajima, and Krieger (33) generalized the expression for known spherical particles given in Equation (2.80) and associated Equations (2.81) through (2.83):

$$\eta = a + b\theta \qquad (2.80)$$

where

$$a = \frac{\eta_0}{(1 + \epsilon_a \phi_2)^2} \qquad (2.81)$$

$$b = \frac{K_b(\epsilon_b \phi_2)^2}{(1 - \epsilon_b \phi_2)^3} \qquad (2.82)$$

and so

$$\beta = \frac{K\beta}{(1 - \epsilon_\beta \phi_2)} \qquad (2.83)$$

where η_0 is the solvent viscosity, and ϵ_a, ϵ_β, and ϵ_b are the effective volume factors for the polymer, and ϕ_2 is its volume fraction. This work was able to show that for a single polystyrene polymer in five different solvents, the rheological behavior in both the Newtonian and non-Newtonian regions could be represented by the Ree-Eyring theory and one could obtain from it the effective hydrodynamic volume occupied by the polymer $\epsilon\phi_2$. For any given solvent, ϵ varied from $[\eta]_{\phi_2}/2$ at $\phi_2 = 0$ ($[\eta]_{\phi_2}$ being the intrinsic viscosity where concentration units used are volume fractions) to $\epsilon \approx 4$. At the point when the viscosity became infinitely high, $\epsilon\phi_2 = 1$, and $\phi_2 \approx \phi_{2_\infty}$. In later work on polymer solutions, Maron and Chiu (34) found that ϵ depended on concentration as expressed in Equation (2.84):

$$\frac{1}{\epsilon} = \frac{1}{\epsilon_0} + \left(\frac{\epsilon_0 - \epsilon_\infty}{\epsilon_0}\right)\phi_2 \qquad (2.84)$$

where

$$\epsilon_0 = [\eta]_{\phi_2}/2, \ \epsilon_\infty \approx 4.$$

The implications of this work are significant. Maron felt that the geometric considerations

of the chain in solution could best be described by volume consideration of the entire chain, rather than those of its individual segments. Moreover, it implies that chain segments do not overlap neighboring chain segments, and the molecules coil up on themselves as their concentration in solution increases.

Maron coupled this concept to one introduced by Hildebrand (35) where the latter described the entropy change on mixing two pure liquids by Equation (2.85):

$$\Delta S_d^m = -R\left[n_1 \ln\left(\frac{n_1 V_1^0 f_1^0}{f_1 V}\right) + n_2 \ln\left(\frac{n_2 V_2^0 f_2^0}{f_2 V}\right)\right] \qquad (2.85)$$

where the final volume of the solution is V, the molar volume of component 1 is V_1^0 and the corresponding number of moles is n_1, and molar volume of component 2 is V_2^0 present as n_2 moles. The parameters f_1^0 and f_2^0 are the packing factors of the two pure components, and f_1 and f_2 are those factors "in solution." Maron showed that, if $\epsilon = 1/f_2$ and $\epsilon^0 = 1/f_2^0$, the entropy change ΔS_d can be written as Equation (2.86).

$$\Delta S_d^m = -R\left[n_1 \ln \phi_1 + n_2 \ln \phi_2 - (n_1 + n_2)\right.$$
$$\left. \ln\left(\frac{V^0}{V}\right) - \frac{n_2 \ln \epsilon}{\epsilon_\infty} + n_2 \ln\left(\frac{\epsilon_\infty}{\epsilon^0}\right)\right] \qquad (2.86)$$

Maron proposed that the energy change on interaction should be taken as a free energy change rather than one involving enthalpy, so that Equation (2.87) applies.

$$\Delta G_{12} = RT \mu_{12} n_1 \phi_2 \qquad (2.87)$$

Here, μ_{12} is the interaction parameter for polymer-solvent pairs.

By combination, the total free energy, enthalpy, and entropy changes on mixing were derived and take the forms shown as Equation (2.88) through (2.91):

$$\Delta G_m = RT[X^0 + n_1 \phi_2] \qquad (2.88)$$

$$\Delta H_m = -RT_2 \, \lambda \, n_2 \, \phi_2 \qquad (2.89)$$

$$\Delta S_m = -R \left[X^0 + (\mu - T\lambda) \, n_1 \, \phi_2 \right] \quad (2.90)$$

where $\lambda = (\partial \mu / \partial T)_{\phi_2, P}$ and X^0 is the term giving the changes in volume on mixing

$$X^0 = n_1 \ln \phi_1 + n_2 \ln \phi_2 + (n_1 + n_2) \ln \left(\frac{V^0}{V} \right)$$

$$+ n_2 \ln \left(\frac{\epsilon}{\epsilon^0} \right) \quad (2.91)$$

For the activities, through partial differentiation of Equation (2.88), and written in condensed form, Maron derived Equations (2.92) and (2.93), where σ is Maron's concentration coefficient.

$$\ln a_1 = X_1 + (\mu - \sigma \phi_1) \, \phi_2^2 \qquad (2.92)$$

$$\ln a_2 = X_2 + (\mu + \sigma \phi_2) \, X_1 \, \phi_2^2 \quad (2.93)$$

Detailed expressions of the volume factors of solvent X_1 and solute X_2 are given in Equations (2.94) and (2.95).

$$X_1 = \ln \left(\frac{V^0}{V} \right) + \left(\frac{V_1^0}{N_1 V_1^0 + N_2 V_2^0} \right)$$

$$- \left(\frac{\bar{V}_1}{N_1 \bar{V}_1 + N_2 \bar{V}_2} \right) + \ln \phi_1 + \left[\left(1 - \frac{\epsilon}{\epsilon_0 x} \right) \right] \phi_2$$

$$(2.94)$$

$$x = \frac{V_2^0}{V_1^0} \qquad (2.95)$$

$$X_2 = \ln \left(\frac{V^0}{V} \right) + \left(\frac{V_2^0}{N_1 V_1^0 + N_2 V_2^0} \right)$$

$$- \left(\frac{\bar{V}_2}{N_1 \bar{V}_1 + N_2 \bar{V}_2} \right) + \ln \phi_2 + \left[\left(\frac{\epsilon}{\epsilon_0} - x \right) \right] \phi_1$$

$$+ \ln \left(\frac{\epsilon}{\epsilon^0} \right) \quad (2.96)$$

While the treatment at this point looks complex, it should be pointed out that in terms of measurables, such as osmotic pressures and light scattering (turbidities), rather concise expressions for obtaining solvent-polymer interaction parameters exist.

For osmotic pressure, Maron's theory yields Equation (2.97) and associated Equations (2.98) through (2.101)

$$\frac{\pi}{c} = \left(\frac{RT}{M_n} \right) \left(\frac{\epsilon}{\epsilon_0} \right) + B_1 \left[\tfrac{1}{2} - (\mu^0 - \sigma^0) \right] C$$

$$+ B_2 C^2 + \cdots B_n C^n \quad (2.97)$$

$$B_1 = \frac{RT}{V_1^0 \, \rho_2^2} \qquad (2.98)$$

$$B_2 = \frac{RT}{3 V_1 \, \rho_2^3} \qquad (2.99)$$

$$B_n = \frac{RT}{(2n - 1) \, V_1^0 \, \rho_2^{n+1}} \qquad (2.100)$$

where ρ_2 is the density of the polymer and $(\mu^0 - \sigma^0)$ is the intercept of a plot of $(\mu - \sigma \phi_1)$ versus ϕ_2 such that

$$\mu = \mu^0 + \sigma^0 \phi_2 \qquad (2.101)$$

So a new function ψ_1 can be defined, which, when plotted against a concentration function y, yields the molecular weight and interaction parameters as in Equations (2.102) and (2.103)

$$\psi_1 = \left(\frac{RT}{M_n} \right) + B_1 \left[\tfrac{1}{2} - (\mu^0 - \sigma^0) \right] y_1 \quad (2.102)$$

$$y_1 = \left(\frac{\epsilon_0}{\epsilon} \right) C \qquad (2.103)$$

2.10 PRIGOGINE THEORY AND ITS MODIFICATIONS (FLORY-ORWELL-VRIJ)

Since χ or μ is a very insensitive interaction parameter, Prigogine offered (2) that terms contributing to χ resulting from pure geometrical factors such as configurational properties, without energetic factors, be called combinatorial. Noncombinatorial effects then include those related to polymer or solute intermolecular forces, chemical nature, and the like. Combinatorial factors are likened to those treated by the athermal solution and should perhaps be presented explicitly as such. In this manner such an approach allows the separation of terms contributing to the thermodynamics and allows the development of equation of state methods. And, furthermore, this approach should be useful for solutions of small as well as large molecules.

In order to examine the expressions developed from the equation of state theories, another term, the "excess" thermodynamic function, must be defined. The total thermodynamic function (ΔH_m, ΔS_m or ΔG_m) less the corresponding function for an "ideal solution," is called the "excess" thermodynamic function (ΔH_m^E, ΔS_m^E, ΔG_m^E).

Flory's (36) treatment of the Prigogine approach has been most successful, and its development is outlined here. A similar treatment has recently been published by Huggins (37). The concept is based on treating the interactions by considering the core of the molecule as surrounded by a continuum of other molecules. Thus we have a departure from the cell model of liquid mixtures in that the interaction potential is described differently from a standard Lennard-Jones model. By defining the partition functions and the equations of state for the two components, Flory derived the total excess heat of mixing as Equation (2.104)

$$\Delta H_m^E = \bar{r} N v^* \left[\frac{\phi_1 P_1^*}{\tilde{v}_1} + \frac{\phi_2 P_2^*}{\tilde{v}_2} - \frac{P^*}{\tilde{v}} \right]$$

$$(2.104)$$

the excess free energy by Equation (2.105)

$$\Delta G_m^E = 3 \bar{r} N v^* \left\{ \phi_1 P_1^* \tilde{T}_1 \ln \left[\frac{(\tilde{v}_1^{1/3} - 1)}{(\tilde{v}_1 - 1)} \right] \right.$$

$$\left. + \phi_2 P_2^* \tilde{T}_2 \ln \left[\frac{(\tilde{v}_2^{1/3} - 1)}{(\tilde{v}^{1/3} - 1)} \right] \right\} + \Delta H_m$$

$$(2.105)$$

and the excess entropy by Equation (2.106)

$$\Delta S_m^E = -3 \left(\frac{N_1 P_1^* V_1^* \tilde{T}_1}{T} \right) \ln \left[\frac{(\tilde{v}_1^{1/3} - 1)}{(\tilde{v}_1^{1/3} - 1)} \right]$$

$$- 3 \left(\frac{N_2 p_2^* v_2^* \tilde{T}_2}{T} \right) \ln \left[\frac{(\tilde{v}_2^{1/3} - 1)}{(\tilde{v}^{1/3} - 1)} \right] \quad (2.106)$$

The partial molal free energy is then given by Equation (2.107)

$$\Delta \bar{G}_1^E = p_1^* V_1^* \left\{ 3 \tilde{T}_1 \ln \left[\frac{(\tilde{v}_1^{1/3} - 1)}{(\tilde{v}_1^{1/3} - 1)} \right] \right.$$

$$\left. + (\tilde{v}_1^{-1} - \tilde{v}^{-1}) + \left(\frac{V_1^* X_{12}}{\tilde{v}} \right) \theta_2^2 \right\} \quad (2.107)$$

Now the reduced variables and other parameters used require defining; see Equations (2.108)–(2.115).

$$P^* = \gamma T \tilde{v}^2 \quad (2.108)$$

where γ is the thermal pressure coefficient,

$$\left(\frac{\partial P}{\partial T} \right)_v \quad \text{at } P = 0 \quad (2.109)$$

and

$$\tilde{v} = \left\{ \left[\frac{(\alpha T/3)}{(1 + \alpha T)} \right] + 1 \right\}^3 \quad (2.110)$$

Here α is the coefficient of thermal expansion at $p = 0$, and $\tilde{v}$ is then the reduced volume. $V^* = V/\tilde{v}$ where V is the molar volume, and so V^* is called the hard-core volume per mole. The reduced temperature $\tilde{T}$ is given by Equation (2.111) and allows the definition of characteristic temperature T^* in Equation (2.112)

$$\tilde{T} = (\tilde{v}^{1/3} - 1)(\tilde{v})^{4/3} \quad (2.111)$$

$$T^* = \frac{T}{\tilde{T}} \quad (2.112)$$

and X_{12} is the intermolecular interaction term given by Equation (2.113):

$$X_{12} = p_1^* \left[1 - \left(\frac{s_1}{s_2} \right)^{1/2} \left(\frac{p_2^*}{p_1^*} \right)^{1/2} \right]^2 \quad (2.113)$$

where s_1 and s_2 are the number of surface

contact sites per segment, and the molecular elements of the two species are r_1 and r_2, given by Equations (2.114) and (2.115).

$$r_1 = \frac{V_1^*}{v_1^*} \qquad (2.114)$$

$$r_2 = \frac{V_2^*}{v_2^*} \qquad (2.115)$$

In essence what one now has is a theory with the volume units described in terms of core volumes and (segment) site volumes as opposed to volumes of segments equal to solvent molar volumes in the lattice theory. In place of mole fractions, one now can define site fractions θ, defined by Equations (2.116) and (2.117):

$$\theta_i = \frac{s_i r_i N_i}{s \bar{r} N} \qquad (2.116)$$

where

$$\bar{r} = \frac{(r_1 N_1 + r_2 N_2 + \cdots)}{N} \qquad (2.117)$$

These expressions, more complex than those seen previously, allow for the calculation of thermodynamic functions from fundamental data and adjustable parameters s, r, θ, and the like.

2.11   RECENT METHODS

Much of the early work on polymer solution thermodynamics involved the use of free energy measurements such as vapor pressure lowering of the solvent solution, or their osmotic pressures. Several workers recently have developed a more rapid technique involving the use of gas chromatography. This advance allows the determination of the thermodynamics of polymer solutions over a wide concentration range. Conder and Purnell (38) developed a theoretical relationship between the retention volume of pure solvent and its solutions on a column using polymer as the stationary phase.

Using the Flory-Huggins theory, Brockmeier et al. (39) have shown the χ can be evaluated from the following Equations (2.118) and (2.199):

$$\chi = \left[\ln \left[\frac{K}{\phi_1/r + \phi_2} \right] - \phi_1 \right] \phi_1^{-2} \qquad (2.118)$$

where

$$K = \frac{P_2}{P_2^0 w_2} \qquad (2.119)$$

P_2 = partial pressure of the solvent above the solution
P_2^0 = vapor pressure of pure solvent at column temperature
w_2 = weight fraction of polymer

$$r = \frac{\rho_2}{\rho_1}$$

Newman and Prausnitz (40) showed that K can be determined using Equation (2.120)

$$K = Vg \rho_2 \qquad (2.120)$$

where Vg is the retention volume of the solvent per gram of polymer. Using the Flory-Huggins theory based on segment fractions, they showed that χ^* can be evaluated using Equation (2.121)

$$\chi^* = \ln \left[\frac{273.2 R}{P_1^s Vg^0 v_1^* \rho_2^*} \right] - \left(1 - \frac{v_1^* \rho_2^*}{M_2} \right)$$
$$- \frac{P_1^s}{RT} \left(B_{11} - v_1 \right) \qquad (2.121)$$

where v^* = the core specific volume
ρ^* = the core density
Vg^0 = the retention volume given by (corrected to be at 0° C) Equation (2.122)

$$Vg^0 = Q(t_g - t_r) \frac{273.2}{T} \frac{1}{w_2}$$
$$\left\{ \frac{3}{2} \left[\frac{(P_i/P_0)^2 - 1}{(P_i/P_0)^3 - 1} \right] \right\} \qquad (2.122)$$

where t_g = retention time from injection to peak maximum for pure solvent

t_r = retention time for air

w_2 = polymer weight on the column

Q = flow rate at column outlet

P_i = inlet pressure

P_0 = outlet pressure

B_{11} = second virial coefficient of pure solvent at column temperature

Thus, with relatively simple apparatus, values of χ or χ^* (or other thermodynamic interaction parameters) can be obtained as a function of the important variables. Results of these investigations are discussed and compared to other methods in a later section.

2.12 APPLICATIONS TO VARIOUS POLYMER-SOLVENT SYSTEMS

2.12.1 Natural Rubber-Benzene

One of the earliest and most extensively studied polymer-solvent systems was that composed of natural rubber dissolved in benzene. Comprehensive experiments on vapor pressure lowering of benzene solutions of natural rubber by Gee and Orr (22) and Gee and Treloar (21) showed that the Flory-Huggins theory was completely applicable to these free energy measurements. The interaction parameter χ_1 was found to be a constant, in fact, over a very wide concentration range, as theory would have predicted. The value of $\chi_1 = 0.40$ reproduced the free energy of mixing data very well and was the first instance of a confirmation of Flory's original hypothesis of the lattice model. However accurate for the free energies, the Flory-Huggins model was inaccurate in its prediction of ΔH_m and ΔS_m, the data for which Gee and Treloar (21) had provided over the temperature range of 25–100° C. Maron and Nakajima (41) reexamined the data of Gee and Trelora using Maron's theory. They were able to obtain μ values as a function of concentration and temperature, and found that μ did in fact vary with ϕ_2 and T in a rather complex manner, as shown in Equation (2.123).

$$\mu = (0.14 - 0.15\phi_2 + 0.04\phi_2^2 - 0.24\phi_2^3)$$
$$+ \frac{74.54 + 44.72\phi_2 - 11.93\phi_2^3 + 71.56\phi_2^3}{T}$$

(2.123)

In a more general form, then, μ was shown to have the form shown in Equation (2.124).

$$\mu = \alpha + \beta/T \qquad (2.124)$$

Equations of this type had been known before to represent the variation of polymer-solvent interaction with temperature.

Utilizing the relationship for μ, Maron and Nakajima were able to represent the free energy, heat, and entropy of mixing data of Gee and Treloar at 25° C and 100° C, and they felt these relationships would hold to as low as 10° C. Curves showing the accuracy of their method as compared to the early work of Flory and Huggins are shown in Figures 2.2–2.4 for partial heats, entropies, and free energies of the rubber-benzene system.

Eichinger and Flory (42) undertook both an experimental and theoretical reexamination of the rubber-benzene system. Their examinations included high-pressure osmometry and vapor sorption measurements. Their data cover 10–40° C and essentially duplicate the earlier experiments of Gee and Treloar. The results of the experiments, when checked against the theoretical predictions of the new Flory theory (based on equation of state data), showed significant disagreement. As shown in Figure 2.5, the χ values predicted from the theory, even with an additional correction term, do not represent the experimental values. Differences occur at low concentrations as well as at high. Eichinger and Flory attribute the discrepancies to either inaccurate estimations of the geometry of the polymer chain or some problem existing in their theoretical approach. The situation then for the system rubber-benzene is summarized in Table 2.5 for the Flory-Huggins theory, the Maron theory, and that of the Flory modification of Prigogine's theory. Other examinations of this system by Bonner and Prausnitz (43) and Tewari and Schreiber (44) are essentially in

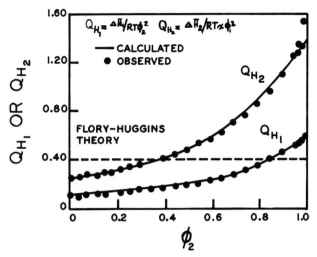

Figure 2.2 Comparison of the partial heats of mixing as predicted by the Maron and Flory-Huggins theories with the experimental data of Gee and Treloar for the solution of natural rubber in benzene (41).

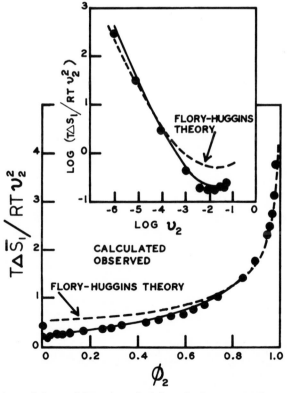

Figure 2.3 Comparison of the partial entropy of mixing of solvent as predicted by the Maron and Flory-Huggins theories with the experimental data of Gee and Treloar for the solutions of natural rubber in benzene (41).

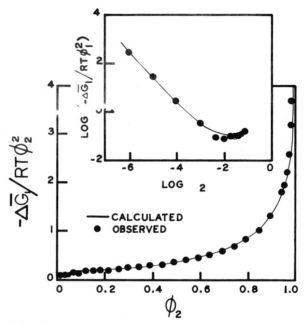

Figure 2.4 Comparison of the partial free energies of mixing of solvent as predicted by the Maron and Flory-Huggins theories with the experimental data of Gee and Treloar for the solutions of natural rubber in benzene (41).

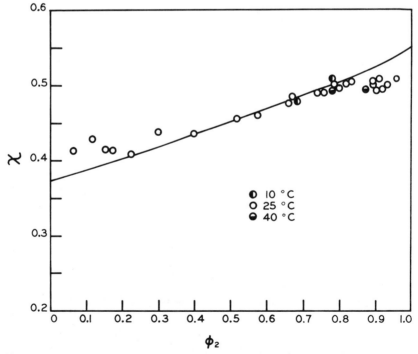

Figure 2.5 Values of the χ parameter as predicted by the new Flory theory compared with those found experimentally for natural rubber in benzene (42).

Table 2.5 Comparison of Results from Three Theoretical Approaches to Selected Polymer-Solvent Systems

System	Flory-Huggins	Maron	Flory-Progogine
Natural rubber–benzene	χ Independent of v_2, T. Therefore predicts free energies fairly well, can not predict heat of mixing.	μ Independent of v_2, dependent on T. Therefore predicts both free energy and heat of mixing.	Predicts free energies in moderately concentrated solution. Fails at low concentration, and at very high. χ now depends on v_2. Heat of mixing prediction inaccurate.
Polyisobutylene-benzene	χ Depends on v_2, T. Does not predict ΔH_m nor vapor pressure curves at high concentration of polymer.	μ Depends on v_2, T. Does not predict ΔH_m^*. Predicts vapor pressures at temp. $> T_m$ polymer. *Suggests some order in the polymer (amorphous).	χ^* Depends on v_2, T. Does not predict ΔH_m, or high-concentration vapor pressure data.
Poly α-olefins	χ Constant—does not predict solvent activities from vapor pressure lowering.	μ Constant—predicts activities for amorphous polymers—predicts order for crystalline polymers.	—

agreement with those early examinations of Gee and co-workers.

2.12.2 Polyisobutylene in Various Solvents

Polyisobutylene-solvent systems have been among the most examined, in particular, polyisobutylene solutions in benzene. In fact, significant amounts of data both in the extreme dilution range and in the very concentrated solution range are available. Among the experimentalists on this polymer are Bawn and Patel (25), Jessup (24), Eichinger and Flory (45), Bonner and Prausnitz (43), Leung and Eichinger (46), and many others.

Jessup and Bawn et al. were among the first to examine both the osmotic pressure and particularly the vapor sorption behavior of polyisobutylene solutions. Their analyses, especially when applying the Flory-Huggins theory, showed the χ parameter to be strongly dependent on concentration. In fact, Jessup found that the relationship necessary to express the dependence of the activity coef-

ficients $\Lambda_1 = a_1/\phi_1$ could be represented by Equation (2.125).

$$\ln \Lambda_1 = \frac{\ln a_1}{\phi_1}$$
$$= \left(1 - \frac{1}{x}\right)\phi_2 + 0.5\phi_2^2 + 0.388\phi_2^3$$
$$+ 0.184\phi_2^{12} \quad (2.125)$$

These data were measured at 26.9°C. Similar high upturns in a_1/ϕ_1 values for benzene in polyisobutylene-benzene solutions were observed by Bawn and Patel in their measurements at 25, 40, but not at 65°C. These deviations from expected behavior occurred at volume fraction of polymer normally approximately 0.80.

Maron and Daniels (47) chose a different analysis approach to these polyisobutylene-benzene solutions. Utilizing the osmotic pressure data of Krigbaum and Flory (23) and Flory and Daoust (48) and analysis using Maron's equations for the osmotic pressures, they found that the interaction parameter μ depended both on temperature and concentration as shown in Equation (2.126).

$$\mu = (1.074 - 1.304 \times 10^{-3}\ T) + 0.188\phi_2$$

$$(2.126)$$

From these data they were able to predict the values of a_1/σ_1 under the conditions measured by Jessup and Bawn and Patel at temperatures ranging from 25–65° C. However, Maron and Daniels (49) also found that their predicted activity coefficients checked extremely well with experiments up to concentrations of approximately 0.80; from there on they were smaller than the measured values (see Figures 2.6 and 2.7). The only exception ws the 65° C data of Bawn and Patel, where exact agreement throughout the concentration range was found, as shown in Figure 2.7.

Instead of attributing these discrepancies to errors in experimental measurement, the authors ascribed the difference between theory and experiment to the fact that most solution theories treated the observed thermodynamic behavior as if both polymer and solvent were liquids. That is, no theory applied to date had considered the effect of ordered regions "in the very concentrated polymer solution," especially those regions that were not truly crystalline in order, but contained some degree of order above the pure amorphous state. Needless to say, this was not a popular concept, for at the time, order in amorphous polymers was an unthinkable concept. Today, however, the situation has changed, especially owing to the work of Geil (50), Yeh (51), Aharoni (52), and others.

Utilizing the differences then between the polymer liquidlike behavior predicted from theory where the concentration of polymer is ϕ_2 and the actual experimental measurements where the actual amount of polymer in solution is ϕ_2, Maron and Daniels calculated that at 25° C, Bawn and Patel's polyisobutylene contained about 10% order called Z_2, about the same amount for Jessup's polyisobutylene polymer, but that at 65° C, for the Bawn and

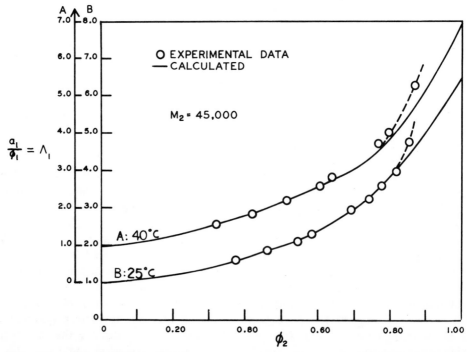

Figure 2.6 Activity coefficients of benzene over solutions of polyisobutylene in benzene as predicted by Maron and Daniels, compared with experimental data at 25 and 40° C. Reprinted from Reference 49, p. 463, by courtesy of Marcel Dekker, Inc.

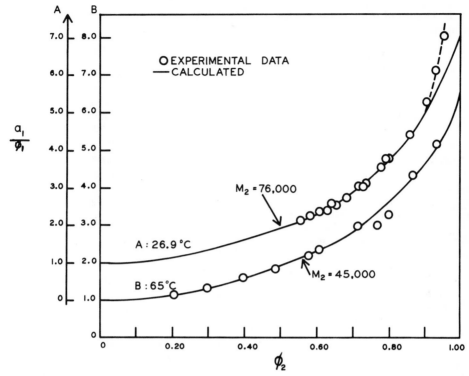

Figure 2.7 Activity coefficients of benzene over solutions of polyisobutylene in benzene as predicted by Maron and Daniels, compared with experimental data at 26.9 and 65° C. Reprinted from Reference 49, p. 463, by courtesy of Marcel Dekker, Inc.

Patel data, no order could be found. Whether fortuitous or not, the 65°C data are at a temperature above the melting point of crystalline polyisobutylene.

To show that these dscrepancies in the vapor pressure lowering were not experimental artifacts, a similar examination by Maron's method was made on the Patel data on polyisobutylene (presumably the same polymer), this time in cyclohexane. Again, from the deviations from predicted vapor pressure lowering data, Maron and Daniels calculated that the polymer contained approximately 14% order at 25°C and about 8% order at 40°C. The tabulated orders for these solutions are given in Tables 2.6 and 2.7.

Returning to the polyisobutylene-benzene system, Daniels and Marion (53) were able to show that the temperature dependence of the interaction parameter $\lambda = (d\mu/dT)\phi_2$, P found to be -1.30×10^{-3} from osmometry, checked well with the λ value found from

examining the heat of dilution data of Kabayama and Daoust (14), which gave $\lambda = -1.22 \times 10^{-3}$ $(\pm 0.18 \times 10^{-3})$. Using the value of $\lambda = -1.304 \times 10^{-3}$, Maron and Daniels attempted to calculate the intregal heats of solution of polyisobutylene in benzene from the Maron Equation (2.127):

$$\Delta H_m' = \frac{-RT^2 \lambda w_1 \phi_2}{M_1 \phi_1} \qquad (2.127)$$

where w_1 is the weight of solvent used, and $\Delta H_m'$ is the heat of solution per gram of polymer, and M_1 is the solvent molecular weight.

These predictions were used to compare with the actual experimental data of Watters et al. (54), who measured the heats of mixing of one sample of polyisobutylene in benzene and chlorobenzene. Again, by assuming that the polymer contained no order, the λ parameter predicted a discrepancy, attributed again

Table 2.6 Degrees of Order Calculated from Observed Activities for Polyisobutylene in Cyclohexane[a]

	25° C			40° C	
ϕ_2	ϕ_2'	$z_2 \times 100$	ϕ_2	ϕ_2'	$z_2 \times 100$
0.500	0.500	0	0.500	0.500	0
0.600	0.600	0	0.600	0.600	0
0.765	0.750	8.0	0.707	0.700	3.3
0.814	0.800	8.8	0.762	0.750	6.4
0.867	0.850	13.3	0.807	0.800	6.9
0.913	0.900	14.4	0.859	0.850	7.1
0.957	0.950	14.7	0.883	0.875	7.8
0.978	0.975	14.4	0.907	0.900	8.1
0.983	0.980	14.1	0.916	0.910	7.9

[a]Method of Maron and Daniels from Reference 49, p. 463, by courtesy of Marcel Dekker, Inc.

to order in the amorphous polymer, accounting for about 4.8% order for the polymer analyzed in benzene, and 5.0% for the same polymer analyzed in chlorobenzene (for which, by osmometry, $\lambda = -0.333 \times 10^{-3}$). For another polymer of higher molecular weight, the authors found about 5.0% order, this time measured only in chlorobenzene.

By using these concepts, Maron and Daniels argued that polymer solution thermodynamic measurements made in dilute and concentrated solutions should only check with predictions from liquid state theory if the polymer were completely in solution, namely free from clustering or ordering. When such order is present, theory can be used with experiment to ascertain the degree of order

contributing to the discrepancy, and should serve as a method of determining small degrees of order in polymers.

Maron and Daniels also examined the vapor pressure lowering data of Prager (55) and co-workers on the system polyisobutylene-n-pentane. These data, measured up to a concentration of polymer of $\phi_2 = 0.87$, covered the temperature range of 15–55°C. Remarkably, the Maron interaction parameter $\mu = 0.625$, independent of concentration and temperature, predicted the experimental values extremely well throughout the concentration range.

Eichinger and Flory (45) carried out an experimental-theoretical program on polyisobutylene-benzene solutions similar to that

Table 2.7 Degrees of Order Calculated from Observed Activities for Polyisobutylene Samples in Benzene[a]

	25° C			26.9° C			40° C	
ϕ_2	ϕ_2'	$z_2 \times 100$	ϕ_2	ϕ_2'	$z_2 \times 100$	ϕ_2	ϕ_2'	$z_2 \times 100$
0.750	0.750	0	0.875	0.875	0	0.750	0.750	0
0.775	0.775	0	0.912	0.910	2.1	0.800	0.792	4.8
0.806	0.800	3.7	0.923	0.920	4.0	0.811	0.800	6.0
0.828	0.820	4.7	0.944	0.940	7.1	0.836	0.820	11.8
0.849	0.840	6.6	0.954	0.950	8.4	0.860	0.840	14.0
0.860	0.850	7.8	0.964	0.960	10.4	0.885	0.860	20.3
0.880	0.870	8.8	0.973	0.970	10.3	0.900	0.875	22.0
0.891	0.880	10.3				0.909	0.886	22.0
0.900	0.800	10.1				0.920	0.900	21.7

[a]Method of Maron and Daniels from Reference 49, p. 463, by courtesy of Marcel Dekker, Inc.

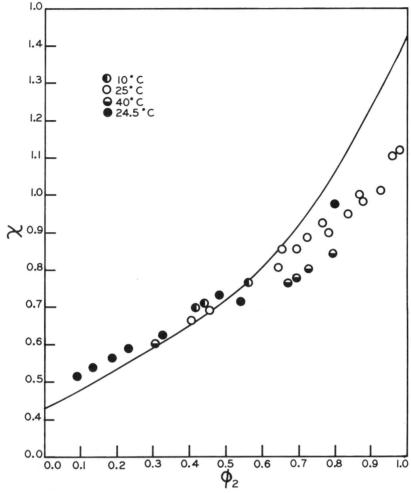

Figure 2.8   Calculated and experimental values of the "new" Flory χ parameter for polyisobutylene solution in benzene (45).

described for natural rubber-benzene. Again, calculations of the χ parameter were made utilizing the relationships from the new Flory theory. The agreement between the calculated values and those from the Eichinger-Flory and Bawn and Patel data is illustrated in Figure 2.8. Again, Eichinger and Flory attribute the poor agreement between their experimental and calculated χ values to problems in estimating the site geometry of the polyisobutylene molecule. Accordingly, attempts to predict the Watters et al. heat of mixing data can not successfully be modeled by this newer approach to solution behavior. Errors could not be ascribed as experimental, since the Eichinger-Flory and Bawn-Patel

data agree remarkably well. Again, the current situation with regard to the comparison of theory with experiment is summarized in Table 2.1.

To further validate that the problems associated with understanding the behavior of polyisobutylene solutions lies with the theoretical interpretation rather than the experiments, Newman and Prausnitz carried out an investigation of polyisobutylene solutions in benzene, n-pentane, and cyclohexane by gas-liquid partition chromatography. Utilizing an entirely different experimental approach, they obtained χ values employing both the Flory-Huggins and Flory-Orwell—Vrij approaches to their data. Virtually identical χ values to those obtained by Eichinger and Flory, Bawn and Patel, and the other experimentalists on this system, were obtained by these gas-liquid partition examinations, and so the experimental data still showed deviations from the corresponding theoretical predictions at high polymer concentrations (56), independent of experimental methodology.

2.12.3 Polystyrene-Solvent Systems

It would be impossible to cover in detail all of the experimental and theoretical work published on polystyrene solutions since this polymer and its solutions have been extensively examined by a wide variety of techniques over the last 25 years. In fact, when B. H. Zimm published his early developments on the use of light scattering measurements (57) for molecular weight determinations, it was polystyrene in butanone and toluene solutions that served as systems for his study. In addition, some of the classic papers on the statistical mechanics of chain molecules in dilute solutions by Flory and Krigbaum were concerned with the θ temperature behavior of polystyrene-cyclohexane solutions (23, 58, 59). Concentrated solution osmometry measurements on polystyrene solutions in four solvents were reported in 1950 by Schick et al. (60) and covered a very wide range in volume fractions of polymer, $\phi \approx 0.002 \rightarrow 0.10$. This, for static osmometry measurements, represented quite an achievement. Two other publications on polystyrene solutions worth detailed examination by the reader are those by Frank and Mark, who summarized the molecular weight, viscosity, and thermodynamic data obtained by several laboratories on four polystyrene fractions. This work, under the sponsorship of the Commission of Macromolecules of the International Union of Chemistry (61, 62), gives the reader an appreciation for the surprising variations found in these solution properties from laboratory to laboratory.

With all the thermodynamic studies available on polystyrene solutions, it was natural that many theories of solution have been applied to investigate their behavior. It was found very early that most polystyrene solution properties could not be explained by the Flory-Huggins theory, especially when χ was assumed to be a constant.

The Flory-Krigbaum relationship was found to be sufficient for many systems; but for a wide range of concentrations, Maron's theory seemed to apply quite well.

Maron and Daniels found that for many polystyrene solutions, the interaction parameter μ varied with temperature, concentration, and molecular weight (63). For example, for polystyrene solutions in benzene and toluene Equations (2.128) and (2.129) apply respectively.

$$\mu = \frac{-8.0 \times 10^{-4} M_2 \times (0.365 \times 10^{-5} \times 10^{-5} M_2) T}{1 - 1.78 \times 10^{-3} M_2 + (0.810 + 10^{-5} M_2) T}$$
(2.128)

$$\mu = \frac{8.55 \times 10^{-5} M_2}{1 + 2.44 \times 10^{-5} M_2} - 0.083 \, \phi_2 \quad (2.129)$$

where M_2 and T are the polymer molecular weight and the temperature respectively. Despite the complexities of these expressions, Maron and Daniels were able to show that Maron's theory applied across the concentration range for polystyrene solutions in benzene, toluene, cyclohexane, methyl ethyl ketone, chloroform, acetone, propyl acetate, ethyl benzene, and chlorobenzene. As evidence see References 63 and 64, which demonstrate the use of the theory to take interaction parameters determined in dilute solutions to predict vapor pressure lowering data on more concentrated polystyrene solutions.

From the interaction parameters derived from dilute solution measurements, Maron and Daniels were also able to evaluate the contributions of the glassy state to heat of mixing measurements on polystyrene solutions (64). The basis for this argument is that the total heat of solution of a glassy polymer, with no crystallinity, is made up of two contributions. The first is the heat of mixing of the liquid-liquid system, and the second contribution is the energy difference between the polymer in its glassy state and what it would be at T_g. In other words, energy "frozen into" the polymer when it cooled through its glass transition temperature will be released and contribute to a calorimetric heat of solution.

To demonstrate this, Maron and Daniels showed that thermal measurements made on a given polymer in the different solvents (65) all had the same residual glassy heat, even though the interaction parameters for the three solvents (toluene, ethyl benzene, and chlorobenzene) are quite different. This energy, which they called ΔHg^0, amounted to -7.73 cal/g when measured in chlorobenzene, -8.14 cal/g in toluene, and -8.18 cal/g in benzene. These were measured on a polymer of molecular weight of 150,000. Hence by showing that the measured heat of solution reflected the physical state of the polymer, Maron and Daniels advocated the use of microcalorimetric measurements on polymer solutions to study the solid-state properties of the polymer.

Similar thermal experiments by Bianchi et al. (66) on polystyrenes showed a ΔHg^0 of -7.07 cal/g, which these authors termed "conformational energy." They also showed that the amount of energy released depended on the thermal history of the polymer, again suggesting this route to determining microstructure in glassy amorphous polymers.

Recently, Flory and co-workers (67–70) have reexamined the behavior of polystyrene solutions in methylethyl ketone and ethylbenzene with the new Flory treatment. The approach used was to obtain the equation of state measurable to calculate χ from the Flory theory to compare with those obtained from their own high-pressure osmometry and the vapor pressure measurements of Bawn et al. At low concentrations, the Flory theory predicted χ values extremely well; but the χ_s at higher ϕ (>0.6) were "higher" than those obtained by experiments. At low concentrations, the χ_H values calculated versus those obtained from the temperature dependence of the osmotic pressure were in very good agreement (see Figure 2.9). The corresponding χ_s values (representing the entropy), as expected from the above, showed excellent agreement at low concentrations, poor agreement as $\phi_2 \longrightarrow 1.0$.

Similar comparisons were made for the system polystyrene-ethylbenzene. Although limited data for comparison with theory were

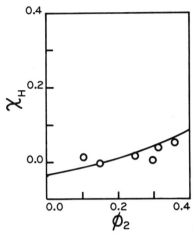

Figure 2.9 χ_H values for polystyrene solutions in methylethyl ketone as a function of segment fraction at 25° C (45).

available (ϕ_2 from 0.1 $\longrightarrow$ 0.3), agreement was superb, shown in Figure 2.10. In both systems χ depended heavily on concentration.

Finally, the thermodynamics of polystyrene and cyclohexane solutions were reanalyzed by Flory and co-workers and compared to the experimental osmometry of Schick et al. (60), Krigbaum and Geymer (59), and Rehage and Palmen (71), the vapor pressure data of

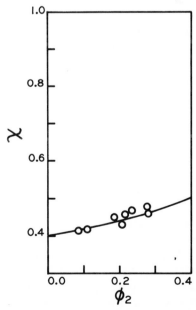

Figure 2.10 χ values for polystyrene-ethyl benzene solutions as a function of segment fraction at 10° and 35° C (45).

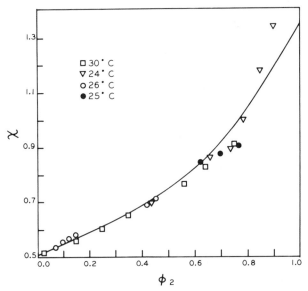

Figure 2.11   Free energy χ values for polystyrene-cyclohexane solutions, the points experimental, the line calculated from the new Flory theory (45).

Krigbaum and Geymer (59) and Schmoll and Jenckel (72), the ultracentrifugation values of Scholte (73, 74), and the miscibility data of Koningsveld et al. (75). Except for the last experiments, the calculations from theory agreed exceptionally well with those from experiments, with only slight deviations at high concentrations of polymer, seen before by Maron and Daniels. The values of χ_H were in good agreement with experiments up to $\phi_2 = 0.6$, beyond that, too low. In all, this experiment-theory comparison was by far the most successful of the new Flory treatment, although surprisingly, for the poorest solvent for polystyrene of those examined. Figure 2.11 shows the comparison of free energy χ values from experiment and theory for polystyrene in cyclohexane.

2.12.4   Poly (α-Olefin) Solutions

Tait and Livesey (76) studied the solution behavior (osmotic pressure, vapor pressure, and intrinsic viscosities) of a series of poly α-olefin in toluene, namely polyheptene-1, polydecene-1, polydodecene-1 and polyoctadecene-1. The Flory-Huggins theory was shown to apply well to the polyheptene-1-toluene solutions, where a constant χ value

reproduced the vapor pressure data extremely well. Maron's theory was quite successful with polyheptene-1 and polydecene-1 solutions. Neither was applicable to polydodecene-1 and polyoctadecene-1 solutions. Application of the theories of Rogers et al. (77) and of Sakurada (78) et al. also failed to explain the behavior of these two polymers.

Tait and Livesey strongly suspected that the failure of all of these theories to explain the solution behavior of polydodecene-1 and polyoctadecene-1 was due to the crystallinity of the polymers. Maron and Daniels (79) had shown previously that such behavior could be expected, and so Daniels, et al. reexamined (80) the polymers and their solution behavior.

From the dilute solution osmometry, Daniels et al. deduced that the Maron interaction parameter for this system followed the relationships at 30°C shown in Equations (2.130)–(2.133). For polyheptene-1

$$\mu = 0.428 \qquad (2.130)$$

For polydecene-1

$$\mu = 0.445 \qquad (2.131)$$

For polydodecene-1

$$\mu = 0.472 + 0.027\,\phi_2 \qquad (2.132)$$

For polyoctadecene-1

$$\mu = 0.460 \qquad (2.133)$$

Daniels confirmed that the μ values from osmometry predicted the vapor pressure values of the toluene solutions of polyheptene-1 and polydecene-1 extremely well (see Figures 2.12 and 2.13). For polydodecene-1 solutions, the Maron equations reproduced Tait and Livesey's vapor pressure data from $\phi_2 = 0.150$ to $\phi_2 = 0.70$; from there on, the calculated activities were less than those observed. Again using the argument that any crystalline portions of the polymer that were insoluble and did not contribute to the vapor pressure lowering, Daniels et al. calculated that the polydodecene-1 polymer was about 41% crystalline. Subsequent differential thermal analysis (DTA) measurements made on the polymer confirmed a high degree of crystalline order. In the polyoctadecene-1 solutions in toluene the agreement between calculated and observed activity values of toluene agreed only up to $\phi_2 = 0.25$. Using the same arguments as before, Daniels and co-workers calculated that two crystalline forms must account for the deviations seen, one representing 29.0% crystallinity and the other an additional 36.2%. Again, subsequent DTA data showed the presence of both backbone and side chain crystalline morphologies, as suggested by Aubrey and Barnatt (81) and Turner-Jones (82). Another polyoctadecene-1 polymer showed the same behavior, and the calculated crystallinities were 42.5 and 18.2% for the two forms, the presence of which were again confirmed by thermal measurements, as shown in Figures 2.14 and 2.15.

Livesey (76) fractionated the first polyoctadecene-1 polymer, and it showed vapor sorption behavior different from the whole

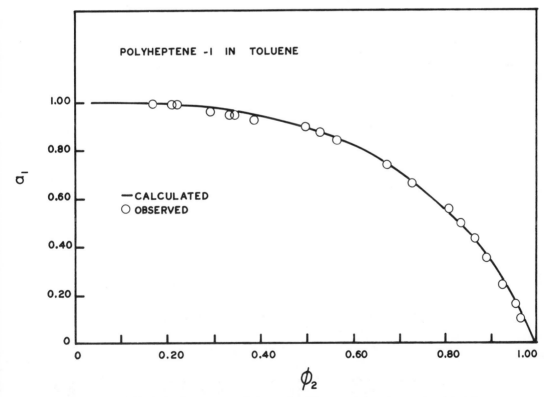

Figure 2.12 Activities of toluene over solutions of polyheptene-1 in toluene; experimental versus theoretical, theory of Maron. Reprinted from Reference 80, p. 47, by courtesy of Marcel Dekker, Inc.

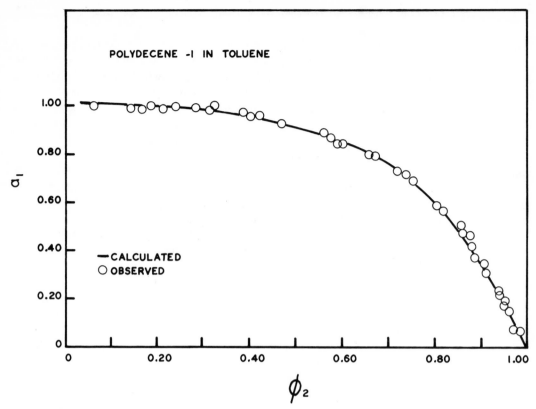

Figure 2.13   Activities of toluene over solutions of polydecene-1 in toluene; experimental versus theoretical theory of Maron. Reprinted from Reference 80, p. 47, by courtesy of Marcel Dekker, Inc.

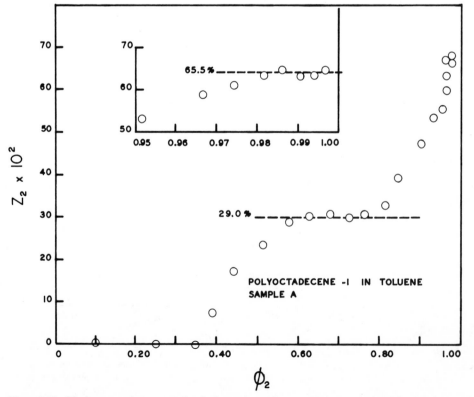

Figure 2.14   The presence of two types of order in a polyoctadecene sample, as calculated from their vapor sorption behavior by Daniels, et al. Reprinted from Reference 80, p. 47, by courtesy of Marcel Dekker, Inc.

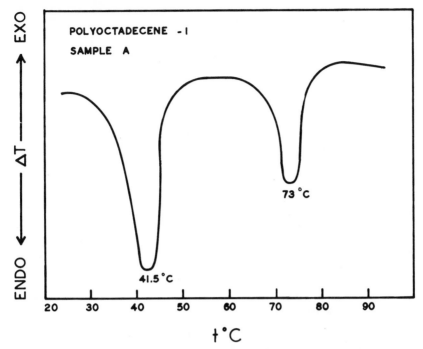

Figure 2.15   DTA thermogram showing the presence of two distinct endotherms signifying two crystalline forms of polyoctadecene-1. Reprinted from Reference 80, p. 47, by courtesy of Marcel Dekker, Inc.

polymer from which it originated. Daniels et al. (80) examined both the polymer and its sorption behavior as before, but this polymer showed only one crystalline-vapor pressure effect, Figure 2.16, instead of two, seen in the parent polymer. The presence of only one crystalline morphology was confirmed by a DTA scan on the polymer, Figure 2.17, further demonstrating the usefulness of this technique for measuring degrees of order in polymer samples.

2.13   THERMAL PROPERTIES OF SOLUTIONS

Since Maron and Daniels and others have demonstrated the applicability of using thermal measurements on polymer solutions to explore polymer properties, this section summarizes the concepts behind such an approach. Also, it should be remembered that significant experimental difficulties exist in making such thermal measurements, and these are well documented (83). In review, the solution pro-

cess for polymers is a lengthy procedure and requires stirring of the system to ensure complete dissolution. Moreover, the thermal changes on mixing a polymer with a solvent are, in general, not large, so that sensitive stable calorimeters capable of measuring microcalorie changes over hours of dissolution time are normally required.

Maron and Filisko have developed a modification of the Tian-Calvet Microcalorimeter that can sufficiently carry out exceedingly long solution process thermal measurements (83).

For a semicrystalline polymer such as polyvinyl chloride, heat of solution measurements made below T_g of the polymer will be a complex mixture of thermal events. The enthalpy change on mixing, per gram of polymer, Δh_m^1, will be comprised of contributions from three separate energy changes. The first of these is the heat of dissolution of the liquidlike polymer, ΔH_m.   The second contribution will be the energy necessary to take the crystalline portions from their ordered to their disordered state ΔH_f.   The third

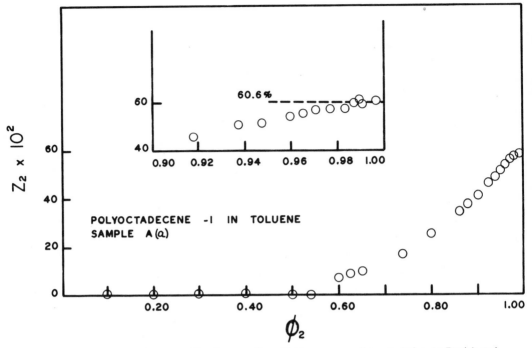

Figure 2.16 Presence of one crystalline form as reflected in the vapor sorption calculations by Daniels et al. on a fractionated polyoctadecene-1 polymer. Reprinted from Reference 80, p. 47, by courtesy of Marcel Dekker, Inc.

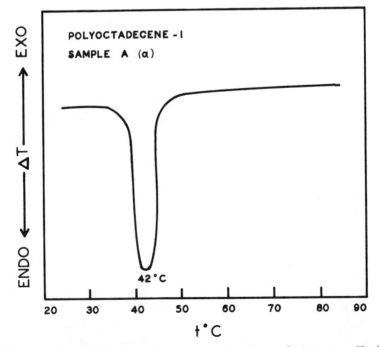

Figure 2.17 Confirming DTA thermogram demonstrating the existence of only one crystalline form in the polyoctadecene-1 fraction examined by Daniels. Reprinted from Reference 80, p. 47, by courtesy of Marcel Dekker, Inc.

contribution is that of the glassy state energy ΔH_g, which has been called by some a "conformational energy." In any case, it represents the energy necessary to change the amorphous polymer from its nonequilibrium glassy state to a truly random liquidlike structure or random coil.

Thus the measured heat of solution Δh_m^1 per gram of semicrystalline polymer is given by Equation (2.134).

$$\Delta h_m^1 = \Delta H_m + \Delta H_f + \Delta H_g \quad (2.134)$$

Maron and Filisko applied this approach to the investigation of the heats of solution of polyethylene oxide (PEO) in chloroform, methylene chloride, and water (84, 85). From heat of dilution measurements, they established that the true heat of solution of "random-coil" PEO in chloroform and methylene chloride depended on concentration, given by the Equations (2.135)–(2.137).

$$\Delta H_m = -\left(\frac{RT^2}{\rho_2 V_1^0}\right)\lambda\phi_1 \quad (2.135)$$

where for chloroform

$$\lambda = 2.36 \times 10^{-2} + 1.62 \times 10^{-2} \phi_2 \quad (2.136)$$

and for methylene chloride solutions

$$\lambda = 1.63 \times 10^{-2} + 1.28 \times 10^{-2} \phi_2 \quad (2.137)$$

Using the above equations and knowing that for a very crystalline polymer $\Delta H_g \approx 0$, Maron and Filisko showed that ΔH_f for PEO of molecular weight 6000 was 56.7 cal/g in chloroform solution measurements and 57.5 from methylene chloride solution measurement. Since ΔH_f is a property only of the polymer, the independence of solvent is expected, and value agreed with published values of ΔH_f for PEO. In PEO-water solutions, these workers observed something unusual, a ΔH_f lower than for the chlorinated solvents. They attributed this difference to a contribution of the conformational difference of PEO in water (helical) versus a random-coil conformation with chlorinated solvents. This

difference, −39.3 cal/g, is on the order of the energy necessary for the transition of isotactic polystyrene (−39.3 cal/g) and polypropylene (−57.1 cal/g) from a helix to a random coil (86) in solution.

For polyvinyl chloride solutions, these same Maron and Filisko showed that the values of $(\Delta H_f + \Delta H_g)$ for various PVCs varied with polymer molecular weight. The values for the sum of the glassy contributions and the ordered (heat of fusion) contribution to the measured heat of solution in tetrahydrofuran are given in Table 2.8. Virtually identical values were obtained in cyclohexanone solutions (see Table 2.9). And, as the authors varied the sample treatment, annealing versus quenching from foregoing T_g, they showed that the exothermic ΔH_g contribution increased on quenching and the endothermic ΔH_f contribution increased on annealing (see Table 2.10). From this they estimated the crystallinity of the PVCs to be 18.4% for the lowest molecular weight, decreasing to 11.8% for the highest.

Other systems have been studied calorimetrically, but have not been examined in as great a detail as those just described. These studies include heat of solution measurements of poly-(1-3 butadiene) in benzene (87, 88), polychloroprene in benzene (87, 89) and heptane (87), polyethylene in α-chloronaphthalene (88), polyisobutylene in benzene (89, 90), carbon tetrachloride (91), and chlorobenzene (90), polymethylmethacrylate in benzene (92), and polystyrene in numerous solvents (93). For a complete listing, the reader is referred to the survey by Booth on

Table 2.8 Thermal properties of PVC Measured by Microcalorimetry in Tetrahydrofuran

M_n	$\Delta H_f + \Delta H_g$ (cal/g)
23,000	−4.86
38,700	−5.21
53,500	−5.24
66,700	−5.47
136,000	−5.77
155,400	−5.86

Source. Reprinted from Reference 85, p. 413, by courtesy of Marcel Dekker, Inc.

Table 2.9   Comparison of $(\Delta H_f + \Delta H_g)$ Values for PVC Obtained from Solutions in Two Solvents at 30° C

M_2	$\Delta H_f + \Delta H_g$ (cal/g)		
	Cyclohexanone	Tetrahydrofuran	Average
23,200	−4.55	−4.86	−4.71
38,700	−5.28	−5.21	−5.25
53,500	−5.33	−5.24	−5.29
66,700	−5.45	−5.47	−5.46
136,000	−6.08	−5.77	−5.92
155,400	−6.03	−5.86	−5.95

Source.   Data of Filisko and Maron. Reprinted from Reference 85, p. 413, by courtesy of Marcel Dekker, Inc.

thermal measurements of polymer solutions (95).

Ichihara and co-workers (94) have investigated the effects of cooling polystyrene glasses at various rates and under elevated pressures. They showed that the heat of mixing in benzene Δh_m^1 becomes more exothermic as the polymer T_g changed. In order to change the values of T_g, cooling rates were varied from 1°C/3 min to 1°C/day. These values and the trends in the data are in essential agreement with the concepts proposed by Maron and co-workers.

Ichihara and co-workers measured Δh_m^1 for poly α-methyl styrene in benzene (96) and found that as the pressure applied to the polymer increased, the measured heat of solution decreased, indicating a densification of the glassy polystyrene. In both the polystyrene and poly α-methyl styrene work, no attempt was made to correct the measurements for true solution heats (amorphous polystyrene in the solvent).

Godovskii and co-workers (97) carried out similar measurements, this time not using a solvent but rather rather simply applying a mechanical deformation to a polystyrene film in a microcalorimeter, and measuring the thermal effects on elongation and relaxation.

2.14   POLYMER SOLUTIONS-FRICTIONAL PROPERTIES

One of the most common dilute solution measurements made to characterize a polymer is its solution viscosity, which is related to the time necessary for selected concentrations of the polymer in a solvent to pass through a capillary of defined length and diameter. The most common mathematical relationships used to interpret the concentration dependence of the viscosity of dilute polymer solutions are those of Huggins and Kraemer. Equations (2.138) and (2.139) express Huggins' equation:

$$\frac{\eta sp}{c} = [\eta] + K_1[\eta]^2 C + \cdots \quad (2.138)$$

where

$$\eta sp = \frac{(\eta - \eta_0)}{\eta_0} \quad (2.139)$$

and η is the viscosity of the polymer solution and η_0 that of the solvent.

The relationship derived by Kraemer is given by Equation (2.140):

Table 2.10   Effect of Sample Treatment on $(\Delta H_f + \Delta H_g)$ at 30° C (cal/g)

Sample	$(\Delta H_f + \Delta H_g)$ (polymer prepared at)		
	−20°C	5°C	55°C
As prepared (powder)	−5.36	−4.44	−4.61
Annealed film	−4.09	−4.64	−4.89
Quenched film	−8.14	−7.90	−7.22

Source.   Data of Filisko and Maron. Reprinted from Reference 85, p. 413, by courtesy of Marcel Dekker, Inc.

$$\frac{\ln \eta_r}{c} = [\eta] - k_2 [\eta]^2 C + \cdots \quad (2.140)$$

where η_r is simply η/η_0, and C is in units of grams of polymer per deciliter of solution.

It can be shown that k_1 from the Huggins relationship and k_2 from the Kraemer relation are related simply by $k_1 + k_2 = \frac{1}{2}$. Thus normal procedure would involve the plotting of both relationships for a given set of viscosity data, yielding the intrinsic viscosity $[\eta]$ at the intercept where $c = 0$, k_1 and k_2 from the slopes. $[\eta]$ This is a measure of the hydrodynamic volume of the polymer molecule.

From the flow properties of the dilute solutions, information about the molecular size and the polymer-solvent interaction can be obtained.

A somewhat different approach to dilute solution viscosities has been suggested by Utracki and Simha (98). Based on the principle of corresponding states, these authors suggest the plot of the reduced viscosity η given by Equation (2.141):

$$\tilde{\eta} = \frac{(\eta/\eta_0 - 1)}{c[\eta]} \quad (2.141)$$

as a function of reduced concentration $\tilde{c} = c/\gamma$. The parameter γ depends on the polymer molecular weight and the temperature. Utracki and Simha have applied this approach to solutions containing a wide variety of polymers such as polyvinyl alcohol, polyvinyl chloride, polystyrene, polyisobutylene, and others. The approach appears to explain the behavior of a number of polymer-solvent temperature conditions, but is somewhat in error when the polymer's molecular weight is very low.

Finally, numerous authors, Maron (99), Quakenbos (100), Schroff (101), Berlin (102), Rudin (103), and others have derived single-point intrinsic viscosity methods based on the Huggins and Kraemer equations. These have met with limited success for some systems.

The foregoing relationships all were derived for and are applicable to the very dilute solution viscosity behavior of polymers. For more concentrated solutions numerous other theories have been proposed, the most widely rec-

ognized being that of Martin (104), who proposed Equation (2.142)

$$\log \frac{\eta sp}{c} = \log [\eta] + K''[\eta]C \quad (2.142)$$

Ott and co-workers (105) have reviewed this and other methods, commenting on the range of applicability. As polymer solutions become very concentrated, Fox and co-workers showed that solution viscosity η_s can be approximated by a rather simple relationship; see Equation (2.143).

$$\eta_s = \eta_p \phi_2 \quad (2.143)$$

Here η_p is the viscosity of the polymer, which of course varies with shear rate, so that the best value to be used in the equation would be the low-shear-limiting viscosity. The actual viscosity behavior of concentrated solutions, as Fox pointed out, depends quite heavily on chain length, branching, molecular weight distribution, and other factors.

Since intrinsic viscosity measurements are relatively easy to measure when compared to absolute measurements (for example, light scattering M_w), they have been used when "calibrated" to determine the molecular weight of polymers. The most common technique is to obtain sharp fractions of polymer of known configuration and measure both an absolute molecular weight and an intrinsic viscosity on each fraction. The preferred molecular weight is light scattering weight average. If the polymer is perfectly monodisperse, it does not matter which "average" one measures, since all should in theory be the same. However, since the narrowest of fractions still contain some polydispersity, it is best to use an average close to that of a viscosity average.

Having measured then $[\eta]$ and M_2 (some workers have chosen osmometry M_n—let us leave the type of molecular weight undefined for the moment), the most common relationship utilized to represent the relationship between these two properties is that of Mark and Houwink, Equation (2.144):

$$[\eta] = KM^a \quad (2.144)$$

where K is a constant for the system and a is the slope. This relationship can also be expressed as in Equation (2.145):

$$\log [\eta] = \log K + a \log M \quad (2.145)$$

Thus a measure of $[\eta]$ for a given polymer-solvent system can be converted to a molecular weight.

Under θ conditions K becomes K_θ and $a = 0.5$. More is discussed regarding these special conditions later. A partial listing of some K and a values, including θ solvents, is given for selected polymers in Table 2.11. No attempt here was made to search the literature exhaustively for K and a values, but rather to contrast polymer-solvent constants in θ solvents and, in most cases, a good solvent. The most extensive review of K and a values is that of Kurata in the *Polymer Handbook* (106), where copolymers, solvent mixtures, and other less common conditions are considered.

Generally a increases with the thermodynamic goodness of the solvents and has a value of 0.5 under θ conditions. The values of K and a are only valid over a limited range of molecular weights, and, as polydispersity of the molecular weight distribution increases, K and a may vary from those found for narrow fractions.

The significance of the Huggins constant k_1 has been the subject of some debate in the literature. Frisch and Simha (107) have reviewed some of the proposals as have Moore, Berger, and Sakai (108).

Peterson and Fixman have argued that k_1 should increase (109), starting at $\frac{1}{2}$ at the theta point and increasing as the thermodynamic "goodness" of the solvent increases.

Sakai (110) summarized the findings of k_1 as follows: In a poor solvent k_1 has a higher value than in a good solvent, but is 0.5 in a θ solvent. The value of k_1 is influenced by the molecular weight distribution and branching in the molecule. The value of k_1 is also affected by the molecular shape in solution, be it of the independent chain or due to the presence of molecular aggregation in dilute solution. Table 2.12 lists k_1 values for some polymer solvent pairs.

Sakai has shown that the k_1 value can be used to get an estimation of the shape of the molecular solution from Equation (2.146):

$$k_1 = \frac{[1 + (1/f)]}{2} \quad (2.146)$$

where for spheres $f = 2.5$, and for ellipsoidal molecules Kuhn and Kuhn predict (111) the relationship expressed in Equation (2.147):

$$f = 2.5 + 0.4075 \, (p - 1)^{1.508} \quad (2.147)$$

where p can assume values from 1 to 15. Extension of the Peterson-Fixman relationship has been published by Sakai for determining the distribution of chain shapes in solution.

From the theory of Riseman and Ullman (112) k_1 should be 0.60 for a coil and 0.733 for

Table 2.11   Values of K and a in the Mark-Houwink Relationship $= KM^d$ for Selected Polymer-Solvent Pairs

Polymer	Solvent	$T°C$	Mol. Wt. Type	$K \times 10^3$ (ml/g)	a
Polybutadiene (98% cis, 2% 1.2)	Benzene	30	M_n	33.7	0.72
Natural rubber	Benzene	25	M_w	18.5	0.74
Polyethylene	Biphenyl	127.5	M_v	323	0.50
Polyisobutylene	Benzene	24	M_v	107	0.50
Polymethyl-methacrylate	Acetone	25	M_v	5.5	0.77
Polyvinylchloride	Cyclohexanone	20–60	M_n	178–159	0.80
Polystyrene	Benzene	25	M_w	9.18	0.74

Table 2.12   Huggins Coefficients k_1 for Selected Polymer-Solvent Systems

Polymer	Solvent	$T°C$	$[\eta]$ (ml/g)	k_1
Natural rubber	Benzene	30	354	0.32
Polychloroprene	Benzene	25	150	0.43
Polyethylene	p-Xylene	100	38	0.71
Polyisobutylene	Benzens	24	128	0.50
	Benzene	40	30–320	0.42
	Cyclohexane	25	19	0.38
	Toluene	25	171	0.36
Polypropylene	Toluene	30	30–120	0.36
Poly(ethylmeth- acrylate)	Benzene	35	145	0.29
	Methyl ethyl ketone	25	158	0.34
Poly(methyl- acrylate)	Benzene	35	50–335	0.38
	Methyl ethyl ketone	25	85	0.56
Poly(methyl- methacrylate)	Benzene	25	4	1.65
	Methyl ethyl ketone	25	56	0.40
Polystyrene	Toluene	25	44–139	0.34
	Benzene	25	105	0.37
	Cyclohexane	34	16	0.50
	Methyl ethyl ketone	~40	44–151	0.54
Polyvinylchloride	Cyclohexanone	25	90–130	0.34
	Tetrahydrofuran	20	171	0.31
Poly(ethylene- oxide)	Benzene	20	48	0.4
	Water	20	3	1.1

rigid rodlike molecules. For spheres, k_1 values equal to and larger than 1.0 have been determined. As far as the dependence of k_1 on polymer molecular weight is concerned, there is considerable debate in the literature. Estimations have been made of the dependence of k_1 on M by Sotobayashi (113) and Utracki and Simha (114), among others.

As was pointed out by Maron and Reznik (115), frequently $k_1 + k_2$ do not equal $\frac{1}{3}$. And, many times, plots of the Huggins and Kraemer equations for a given set of data do not meet at the same intercept, at $c = 0$. Hence the determination of $[\eta]$ becomes somewhat clouded. Numerous workers have pointed out problems utilizing the Huggins and Kraemer relationship. Ibrahim and Elias (116) showed that a plot of the two $[\eta]$ intercepts (one from the Kraemer plot and one from the Huggins plot for the "same data" versus k_1) gave a straight line; and that $[\eta]$ from the Huggins plot equals $[\eta]$ from the Kraemer plot only when $k_1 = \frac{1}{3}$.

Streeter and Boyer (117) pointed out similar difficulties.

In an attempt to allow more accurately the determination of both $[\eta]$ and the k values, Maron and Resnick expanded both the Huggins and Kraemer relationships. Here, we adopt their nomenclature for the k's such that the Huggins relationship is given by Equation (2.148):

$$\frac{\eta sp}{c} = [\eta] + k_1[\eta]^2 C + k_2[\eta]^3 C^2 \quad (2.148)$$

and the Kraemer equation by Equation (2.149).

$$\frac{\ln \eta_r}{c} = [\eta] - k_1'[\eta]^2 C - k_2'[\eta]^3 C^2 \quad (2.149)$$

By expanding the logarithims and taking differences Maron and Resnick define a new relationship: see Equations (2.150 and 2.151).

$$\frac{\Delta}{c^2} = \left(\frac{[\eta]^2}{2}\right) + (k_1 - {}^1/_3)\,[\eta]^3 C \qquad (2.150)$$

where

$$\Delta = \eta sp - \ln \eta_r = (k_1 = +k_1')[\eta]^2 C^2 \\ + (k_2 + k_2')[\eta]^3 C^3 \qquad (2.151)$$

Thus a plot of Δ/c^2 versus C will take a positive slope if $k_1 > \frac{1}{3}$, negative if $k_1 < \frac{1}{3}$, and zero when $k_1 = \frac{1}{3}$. The intercept will be $[\eta]^2/2$ and the slope a measure of the Huggins constant k_1. Maron and Resnick showed that for a series of polymethyl methacrylates and polystyrene in toluene and benzene, these relationships worked extremely well, yielding linear plots from which k_1, k_1', and $[\eta]$ could easily be obtained. In every case $k_1 + k_1'$ was found to equal $\frac{1}{2}$ within experimental error.

Interesting also was the ability of Maron and Resnick to show that, for these polymers, mixtures of the two polymers in one solvent obeyed the relationship derived by Phillippoff (118) expressed as Equation (2.152):

$$[\eta]_{\text{mix}} = x_A\,[\eta]_A + x_B\,[\eta]^B \qquad (2.152)$$

where x_A and x_B are the weight fractions of polymers A and B.

In addition, the Huggins constant k_1 was shown for the blends to follow a similar relationship expressed as Equation (2.153).

$$k_{1(\text{mix})} = x_A\,k_{1(A)} + x_B k_{1(B)} \qquad (2.153)$$

2.15 THE EFFECT OF BRANCHING ON INTRINSIC VISCOSITIES

The Fox-Flory (5) theory of intrinsic viscosity predicted for branched polymers that the intrinsic viscosity $[\eta]_b$, when compared to that of a linear polymer of the same molecular weight $[\eta]_l$, should give the ratio of the mean square radii of gyration of the two as shown in Equation (2.154):

$$\frac{[\eta]_b}{[\eta]_l} = g^{3/2} \qquad (2.154)$$

Experiments by Thurmond and Zimm (119) showed that the preceding equation exaggerated the branching effect. Zimm and Kilb (120) have demonstrated that the following relationship, Equation (2.155), holds.

$$\frac{[\eta]_b}{[\eta]_l} = g^{1/2} \qquad (2.155)$$

This relationship was studied for polystyrene star by Wenger and Yen (121) and by Morton and co-workers (122). The latter work suggested that for a large number of branch points m that Equation (2.156) holds:

$$g^{1/2} = \left[1\left(1 + \frac{m}{7}\right)^{1/2} + \frac{4m}{9\pi}\right]^{-1/4} \qquad (2.156)$$

for trifunctional (three-point) branch junctions.

Zimm and Stockmayer (123) took Equation (2.156) and made the refinement that $\lambda = (m)$ number of branchpoints per (M) unit molecular weight; see Equation (2.157):

$$m = \lambda M \qquad (2.157)$$

and obtained Equation (2.158).

$$g = \left[\left(\frac{1 + \lambda M}{7}\right)^{1/2} + \frac{4\lambda m}{9\pi}\right]^{-1/2} \qquad (2.158)$$

Studies by Zimm and Kilb (120) and Kurata and Fukatsu (124) have shown that Equation (2.159) is satisfied:

$$[\eta]_{\theta_b} = G\,[\eta]_{\theta_l} \qquad (2.159)$$

and that G is defined as in Equations (2.160) and (3.161):

$$G = g^{1/2} \quad \text{(star-type branching)} \qquad (2.160)$$

$$G = g^{3/2} \quad \text{(comblike branching)} \qquad (2.161)$$

For non theta solvents, Equations (2.162) may be written.

$$[\eta]_b = g^{0.6}\,[\eta]_l \qquad (2.162)$$

The case of tetrafunctional branch points can be expressed by Equation (2.163).

$$g^{1/2} = \left[\left(1 + \frac{m}{6}\right)^{1/2} + \frac{4m}{3\pi}\right]^{-1/4} \qquad (2.163)$$

Ordinarily, many workers simply rely on the ratio of $[\eta]_b/[\eta]_l$ to characterize the system for the presence of branching. This, of course, requires the preparation of narrow molecular weight standard materials of known linear structure, aided, say, by fractionation, in order to make such comparisons. However difficult, the fact that the branched molecule of a given molecular weight occupies a smaller volume in solution than a corresponding linear chain of identical molecular weight makes the determination of the relationship between intrinsic viscosity and molecular dimensions even more significant.

As a typical example, Krozer (125) has carried out these kinds of calculations for the effect of branching on low-density polyethylene in several solvents and for polyvinyl acetate as well.

2.16   MOLECULAR DIMENSIONS

Aside from being used strictly as a measure of molecular weight, the intrinsic viscosity, particularly at θ conditions, has been shown to reflect the "unperturbed dimensions" of the polymer chain. To understand this one must first look at some models of polymer chains and show how their dimensions can be experimentally obtained.

M. V. Volkenstein (126) published a monograph describing the rigorous mathematical treatment necessary to describe the properties of chain molecules, averaging them over all the possible rotational angles available. Flory (127) has recently published a detailed book describing the approaches to understanding the statistical mechanics of chain molecules.

It has been shown from random flight statistics that for a freely jointed chain consisting of n bonds of fixed length l, that the root mean square end-to-end distance $\langle r^2 \rangle$ is given by Equation (2.164):

$$\langle r^2 \rangle_0 = nl^2 \qquad (2.164)$$

where the subscript 0 represents the end-to-end distance unperturbed by molecular interactions.

Flory (5) defines a characteristic ratio (C), Equation (2.165),

$$C_n = \frac{\langle r^2 \rangle_0}{nl^2} \qquad (2.165)$$

which for an infinitely long chain of bonds numbering n_i, l_i long, can be defined as Equation (2.166).

$$C = \lim_{n \to \infty} \frac{\langle r^2 \rangle_0}{\Sigma n_i l_i^2} \qquad (2.166)$$

And, Debye has shown that for $n \longrightarrow \infty$, the radius of gyration s is related to the end-to-end distance by Equation (2.167).

$$\langle s^2 \rangle_0 = \frac{\langle r^2 \rangle_0}{6} \qquad (2.167)$$

For freely rotating chains, other relationships have been derived. Extension to real chains were fixed bond angles and independent rotational potentials are encountered has been accomplished, and is detailed in the works of Flory (127) or Volkenstein (126). The important relationships are summarized below.

For freely rotating chain of fixed bond length l, one can write Equation (2.168):

$$\langle r^2 \rangle_f = \frac{nl^2 \, [1 + \cos \theta]}{[1 - \cos \theta]} \qquad (2.168)$$

where θ is the supplement of the valence bond angle.

Kuhn (128), Oka (129), and Taylor, (130) showed that for restricted flexibility, the solution for the mean square end-to-end distance takes the form of Equation (2.169):

$$\langle r^2 \rangle_r = nl^2 \left[\frac{1 - \cos \theta}{1 + \cos \theta}\right] \left[\frac{1 + \cos \Phi}{1 - \cos \Phi}\right]$$

$$(2.169)$$

where Φ is the internal angle of rotation.

Table 2.13   Unperturbed Dimensions of Selected Polymers

Polymer	Solvent	$T°C$	$r_0/M^{1/2} \times 10^4$ (nm)	$s = r_0/r_f$
cis-Polybutadiene	Dioxane	20.2	920	1.68
cis-Polyiosprene	Diisopropyl ether	22	847	1.74
trans-Polyisoprene	Dioxane	47.7	910	1.30
Polyethylene	Decalin	140	1070	1.84
Polyiosbutylene	Benzene	24	740	1.80
Polymethyl-methacrylate	Benzene	21	653	2.12
Polyvinylchloride	Benzyl alcohol	155.4	820	1.08
Polystyrene	Cyclohexane	34	690	2.28

In much of the literature, chain lengths are reported as $r_0/M^{1/2}$, correcting for molecular weight. If σ is defined as the ratio of the mean square end-to-end distance in the unperturbed state $\langle r \rangle_0$ to that of the freely rotating chain $\langle r \rangle_f$, a useful measure of the steric hindrance effect on chain dimensions can be obtained. Table 2.13 lists some typical values of $\langle r_0 \rangle / M^{1/2}$ and σ for some common polymers.

To account for long-range polymer interactions Flory (5) has defined the ratio α as expressed in Equation (2.170).

$$\alpha^2 = \frac{\langle r^2 \rangle}{\langle r^2 \rangle_0} \qquad (2.170)$$

At theta conditions $\alpha = 1$, and the larger the value of α, the "better" thermodynamically is the solvent for the polymer. The two principal methods for obtaining these dimensions experimentally are light scattering and the intrinsic viscosity.

Utilizing viscosity data, Fox and Flory pointed out that $[\eta]$ and $\langle r^2 \rangle$ are related by Equation (2.171) (5).

$$[\eta] = \frac{\Phi \langle r^2 \rangle^{3/2}}{M} \qquad (2.171)$$

For measurement under θ conditions, one can write Equation (2.172).

$$[\eta]_\theta = KM^{1/2} \qquad (2.172)$$

Thus a method from measurements at θ conditions exists for the determination of the polymer molecules dimensions.

Stockmayer and Fixman chose another approach. They showed for many systems (131) one can write Equations (2.173) and (2.174).

$$\frac{[\eta]}{M_2^{1/2}} = K_0 + 0.036\Phi BM_2^{1/2} \qquad (2.173)$$

Table 2.14   Solvents for Some Selected Polymers

Polymer	Solvent	θ Temperature (°C)
Poly(1-4)butadiene	n-Heptane	−1
Poly(1-4—isoprene	Butanone	25
Polybutene-1	Toluene	−46
Polyisobutylene	Benzene	24
Polyisobutylene	Ethylbenzene	−24
Polyethylene	n-Hexane	133
Polypropylene(atactic)	i-Amylacetate	34
PMMA (atatic)	Acetonitrile	30
Polystyrene	Cyclohexane	34

where

$$b = \left(\frac{V_2}{M_2}\right)^2 \left(\frac{1-2\chi}{NV_1}\right) \qquad (2.174)$$

Thus by plotting $[\eta]/M_2^{1/2}$ versus $M_2^{1/2}$, one can obtain K_θ, the value of $[\eta]/M_2^{1/2}$ in θ solvents and, in addition, allow the Flory-Huggins interaction parameter χ to be evaluated. For a variety of polymer systems, θ temperatures have been evaluated. A partial listing is given in Table 2.14, and the reader is referred to the *Polymer Handbook* (106) for a more extensive listing.

2.17   CRITICAL SOLUTION BEHAVIOR

To fulfill the conditions for phase equilibrium stated in a previous section, it is required that there exist two concentrations of polymer ϕ_2 and ϕ_2^1 when its chemical potential G_2 has the same value.

A number of the aforementioned solution theories have been applied to interpret phase equilibria in binary and multicomponent solutions. For presentation, specific examples of the use of some of these theories have been selected to illustrate their use. In all of these theoretical treatments reference is made to binodals and spinodals, terms with which some readers may not be familiar. To represent conditions in a binary solution below its critical temperature T_c, a concentration-composition profile similar to that illustrated in Figure 2.18 is constructed. The line connecting points V_i to V_j corresponds to the molar volumes of i and j in one phase—and V'_i and V'_j the molar volumes of i and j in the second phase. The lines ab and cd are called the binodals. Figure 2.19 illustrates the free energy-volume fraction curve of a binary solution, showing critical miscibility. The curve XBY is the spinodal, compositions between which are unstable. The area between XBY and ABC should be considered metastable compositions.

It was Flory who first developed the equations for the critical behavior of polymer solutions using the Flory-Huggins relationships for the chemical potentials as expressed in Equation (2.175):

$$\phi_{2_c} = \frac{1}{(1 + x^{1/2})} \qquad (2.175)$$

where x is the ratio of the molar volumes of polymer to solvent, V_2^0/V_0^1. In this equation ϕ_{2_c} represents the concentration of polymer at which phase separation first occurs. The value of the interaction parameter χ under these conditions takes the form shown in Equation (2.176)

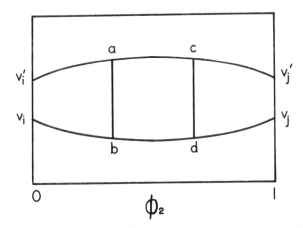

Figure 2.18   A concentration-composition diagram for a binary solution below its critical temperature T_c.

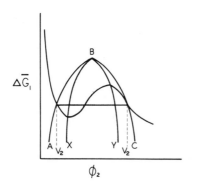

Figure 2.19 A typical free energy-volume fraction profile for a binary solution having critical composition and temperature.

$$\chi_c = \frac{(1 + x^{1/2})^2}{2x} \tag{2.176}$$

and the usual approximation is expressed in Equation (2.177).

$$\chi_c = \frac{1}{2} + \frac{1}{x^{1/2}} \tag{2.177}$$

This states that for infinite molecular weight of polymer, $x \longrightarrow \infty$, and $\chi_c \longrightarrow 0.5$, at the θ temperature.

If one has a situation where two phases are in equilibrium and denote the concentration of polymer in the polymer-rich phase by ϕ_2^1 and in the solvent-rich phase by ϕ_2, then one can write Equation (2.178):

$$\chi = \frac{(\phi_2^1/\phi_2 - 1)(1 - 1/x) + \ln(\phi_2^1/\phi_2)/\phi_2}{2(\phi_2^1/\phi_2 - 1) - \phi_2[(\phi_2^1/\phi_2)^2 - 1]]} \tag{2.178}$$

Maron and Nakajima (133) showed that a similar relationship can be derived from Maron's theory of polymer solutions, since Tompa (132) found experimentally that the Flory-Huggins relation did not exactly reproduce the polystyrene-cyclohexane critical data unless χ was allowed to vary in a power series in concentration ϕ_2.

Maron's equation takes the form (133) shown in Equation (2.179):

$$\mu_c = \frac{4 - 5\phi_{2c}}{6\phi_{1c}^2} + \frac{(9Z_c - 4)\phi_{2c} - (3Z_c - 2)}{6\chi_c\phi_{2c}^2 Z_c^3} \tag{2.179}$$

where

$$Z = \frac{\epsilon_0}{\epsilon} = 1 + (\epsilon_0 - \epsilon_\infty)\phi_2 \approx 1 + \left(\frac{[\eta]\phi_2}{2} - 4\right)\phi_2$$

and the subscripts c represent these quantities (effective volume factors) at the point of critical miscibility.

The best test of these relationships was made by Maron and Nakajima, who used values of μ from dilute solution behavior to compare Maron's relationships with Flory's for the partial and critical miscibility data of Schultz and Flory on polystyrene-cyclohexane (see Figure 2.20). Maron's method was shown to reproduce these polystyrene-cyclohexane data to a much greater degree of accuracy.

The most complete review of the behavior of polymer solutions is given in the article by Koningsveld. In this survey (134) the various systems examined are reviewed, along with a theoretical development based on the Flory-Huggins treatment. Both theory and experimental techniques are examined, and the reader is referred to the exhaustive review for more in-depth detail on the subject. The data on phase equilibria have also been studied by Heil and Prausnitz for polystyrene in eight different solvents, polypropylene glycol in methanol, polyethylene oxide-chloroform, polyethylene glycol-water, cellulose acetate in four solvents, cellulose acetate in acetone, rubber in three different solvents, and polypropylene in two different solvents. The Heil-Prausnitz (135) approach is similar to the Flory-Huggins method but involves a semiempirical theory based on two adjustable parameters, the local volume fraction and the interaction energies. Heil and Prausnitz (135) have demonstrated that their approach applies in systems where much hydrogen bonding is expected. They have, in addition, extended the treatment to polymer-solvent-nonsolvent mixtures.

Patterson and co-workers have studied in depth the use of the Flory-modified Prigogine approach to phase equilibria. A particularly

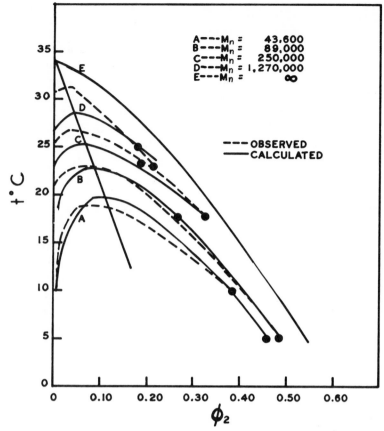

Figure 2.20 Critical miscibility diagram for polystyrene-cyclohexane and predicted values for Maron and Flory-Huggins theories (133).

interesting advancement has been their study of pressure effects on phase equilibria at the upper and lower critical solution temperatures. The important conclusion reached in their studies on polyisobutylene and PDMS, verifying Tompa and Maron, is that a concentration-dependent χ is needed to predict accurately the critical behavior of polymer solutions (136).

Similar studies have been carried out by Cowie et al. (137) on cellulose acetate-acetone using the Prigogine approach and have demonstrated that the Patterson-Delmas method can be very useful (137). The working equation in this theory is Equation (2.180)

$$\chi_c = \frac{C_1 V^2}{1 - \tilde{V}_1^{-1/3}} + \frac{C_1 \tau^2}{2(4/3/\tilde{V}_1^{-1/3} - 1)}$$

$$= \frac{1}{2}(1 + r^{-1/2})^2 \qquad (2.180)$$

where C_1 is the number of degrees of external degrees of freedom of the solvent; r is the value of $= V_2^0/V_1^0$; and $\tau = 1 - T_1^*/T_2^*$, the reduced temperature parameter. Rowlinson and co-workers (138) have shown that lower control solution temperature for polymer solvent mixtures can occur well above the boiling point of the solvent.

In order to predict such behavior, rather complex behavior of the interaction parameter is to be expected.

2.18 SUMMARY

It becomes obvious that from the preceeding arguments, no one solution theory can be considered universal in acceptance or applicability. In some of the theories the models on which they have been based are fundamentally incorrect. In others, they simply have not been extended enough to allow their use throughout all the areas of polymer solution thermodynamics.

Certainly, one would prefer to use a method that allows the prediction of the interaction parameter from fundamental pure component versus equation-of-state data. The Flory-Orwell-Vrij approach is closest to achieving this goal. It, however, has not accomplished the ultimate goal, namely accurate prediction of free energy, entropy, and enthalpy behavior of polymer solution systems. Of those theories that rely on free energy measurements in dilute solutions to predict concentrated solution behavior, Maron's theory has proven very effective. Unfortunately, it has not been extended to as great a number of polymer-solvent systems as would be desirable, and, as a result, has been underrated and, more importantly, misinterpreted. To point out the shortcomings of the Flory-Huggins or Flory-Krigbaum theories would be only to verify what has been documented in the literature. Their usefulness is not to be denied, however, and at their time of introduction guided much of the work on polymer solutions.

Newer theories are no doubt in the future and should be tested as have all those before them. The unsolved problems still remain: multicomponent polymer systems, multiphase equilibria, and the very concentrated polymer solution behavior and its relationship to polymer structure. In this very last point it is well to remember that only when polymer molecules behave as individual random coils can one approach the problem from a uniform point of view. Aggregation in solution complexes the system so that measurements of fundamental properties of the chain become nearly impossible. The presence of microcrystalline regions in solution in polymers such as polyvinylchloride present many interpretive problems. The application of fundamental polymer solution theories to industrial problems must be considered, for every adhesive or plasticized compound is a polymer solution. The problems of polymer-polymer compatibility in the solid state must be addressed from a unified approach.

In all, the area of polymer solutions has shown a great deal of advancement during the last 50 years. Many now famous scientists in the polymer field have, at one time or another, worked with polymers in solution and have attempted to solve the problems presented by these complex systems. This work has attempted to pay tribute and to encourage these scientists (and those to come) in order that we fully understand the behavior of the polymer molecule in solution.

2.19   GLOSSARY OF SYMBOLS

a_i	Activity of component i in solution
A_i	Virial coefficients
B_i	Coefficients in the Ree-Eyring theory
B_{11}	Second virial coefficient (inverse phase gas chromatography method)
Cm	Empirical constant in the expansion expression for molecular size in solution
E	Energy of interaction per mole
E_v	Molar heat of vaporization
f_i	Fugacity of component i
f	Shape factor
g	Coefficient in the Flory-Krigbaum theory, approximately equal to $\frac{1}{4}$
G	Total free energy
$\overline{G}_i$	Partial molar free energy of component i
Δh_m^1	Heat of mixing of binary components, per gram of solute
ΔH_f	Heat of fusion, per gram of constituent
H	Enthalpy (heat)
$\overline{H}_i$	Partial molar enthalpy (heat)
k	Boltzmann constant
k_i	Huggins, Kraemer constants
l	Length of bond between units or segment in the polymer chain
m	Numbers of branch points
M_i	Molecular weight of component i
M_n	Number average molecular weight
n_i	Number of moles of component i
N_i	Mole fraction of component i
P	Total pressure on the system
P_i	Partial pressure of component i
Q	Volumetric flow rate
$\langle r^2 \rangle$	Root mean square end-to-end distance
r_i	Molecular elements in the Flory-Orwell-Vrij theory

R	The gas constant
$\langle s^2 \rangle$	Radius of gyration
s_i	Number of surface contact sites of component i
S	Entropy of the system
S_i	Partial molal entropy of the system
T	Temperature, degrees absolute, unless otherwise indicated
t_g	Retention time of the probe molecule in inverse phase gas chromatography
t_r	Retention time for the reference molecule in inverse phase gas chromatography
V	Total volume of the system
v_i	Volumes of solute and solvent in the lattice theory
w_{ii}	Potential energy of the pair of molecules ii (subscripted jj and ij for the other binary pairs)
x_i	Weight fraction of component i
x	Ratio of the molar volumes of solute to solvent
X_i	Volume change on mixing terms in the Maron theory
Z	Number of nearest neighbor contact sites, number of available lattice sites
α	Polymer chain expansion coefficient
γ	Activity coefficients in concentration units of grams per milliliter
Γ	Second virial coefficient
δ_i	Solubility parameter
ϵ	Effective volume factor
η	Apparent viscosity of polymer solution
$[\eta]$	Intrinsic viscosity
θ	Flory "ideal" temperature
κ	Flory heat of interaction parameter
λ	Temperature coefficient in Maron theory
Λ	Activity coefficient in volume fractions
μ	Maron interaction parameter
π	Osmotic pressure
ρ_i	Density of component i
σ	Maron concentration coefficient
ϕ_i	Volume fraction of component i
$\underline{\phi}$	Viscosity coefficient in the Stockmayer-Fixman relation
χ	Flory interaction parameter
ψ	Flory entropy of interaction parameter
Ω	Probability function

2.20   REFERENCES

1. H. Morawetz, *Macromolecules in Solution,* 2nd ed., Wiley, New York, 1975.

2. I. Prigogine, *The Molecular Theory of Solutions,* Interscience, New York, 1975.

3. M. B. Huglin, *Light Scattering from Polymer Solutions,* Academic, New York, 1972.

4. J. M. Prausnitz, *Molecular Thermodynamics of Fluid-Phase Equilibria,* Prentice-Hall, Englewood Cliffs, New Jersey, 1969.

5. P. J. Flory, *Principles of Polymer Chemistry,* Cornell University Press, New York, 1953.

6. D. C. Bonner, *J. Macromol. Sci.,* Rev. Macromol. Chem., **C13**, 263 (1975).

7. J. H. Hildebrand, *J. Am. Chem. Soc.,* **51**, 66 (1929).

8. J. J. Van Laar, *Z. Physik. Chem.,* **72**, 723 (1910).

9. P. A. Small, *J. Appl. Chem.,* **3**, 71 (1953).

10. K. L. Hoy, *J. Paint Technol.,* **42**, 76 (1970).

11. H. Burrell, "Solubility Parameter Values," in J. Brandrup and E. H. Immergut, Eds., *Polymer Handbook,* 2nd ed., Wiley, New York, 1975, pp. IV—337 to IV—339.

12. C. M. Hansen, *Three Dimensional Solubility Parameters and Solvent Diffusion Coefficients,* Danish Technology Press, Copenhagen, 1967.

13. E. B. Bagley, T. P. Nelson, J. W. Barlow, and S. A. Chen, *Ind. Eng. Chem., Fundam.,* **9**, 93 (1970).

14. E. B. Bagley, T. P. Nelson, J. M. Scigliano, *J. Paint Technol.,* **43(555)**, 35 (1971).

15. J. L. Biros, L. Zeman, and D. Patterson, *Macromolecules,* **4**, 30 (1971).

16. J. Dyantis, *J. Phys. Chem.,* **76(3)**, 400 (1972).

17. J. Dyantis, *J. Phys. Chem.,* **77(25)**, 2977 (1973).

18. H. Yamakawa, *Modern Theory of Polymer Solutions,* Harper & Row, New York, 1971.

19. M. L. Huggins, *J. Chem. Phys.,* **9**, 440 (1941); *J. Phys. Chem.,* **46**, 151 (1942); *Ann. N.Y. Acad. Sci.,* **41(1)**, (1942).

20. P. J. Flory, *J. Chem. Phys.,* **9**, 660 (1941); *J. Chem. Phys.,* **10**, 51 (1942).

21. G. Gee and L. R. G. Treloar, *Trans. Faraday Soc.,* **38**, 147 (1942).

22. G. Gee and W. J. C. Orr, *Trans. Faraday Soc.,* **42**, 507 (1946).

23. W. R. Krigbaum and P. J. Flory, *J. Am. Chem. Soc.,* **75**, 1775, 5254 (1953).

24. R. S. Jessup, *J. Res. Natl. Bur. Stand.,* **A60**, 47 (1958).

25. C. E. H. Bawn and R. D. Patel, *Trans. Faraday Soc.,* **52**, 1664 (1956).

26. B. A. Wolf, "Polymer-Solvent Interaction Parameters," in J. Brandrup and E. H. Immergut, Eds., *Polymer Handbook,* 2nd ed., Wiley-Interscience, New York, 1975, Chapter **4**.

27. P. J. Flory and W. R. Krigbaum, *J. Chem. Phys.,* **18**, 1086 (1950).

28. G. C. Berry, *J. Chem. Phys.,* **44**, 4550 (1966).

29. G. C. Berry and E. F. Casassa, *J. Polym. Sci., Rev.,* **4**, 1 (1970).

30. S. H. Maron, *J. Polym. Sci.,* **38**, 329 (1959).

31. S. H. Maron and N. Nakajima, *J. Polym. Sci.,* **42**, 327 (1960).

32. T. Ree and H. Eyring, *J. Appl. Phys.,* **26**, 793, 800 (1955).

33. S. H. Maron, N. Nakajima, and J. M. Krieger, *J. Polym. Sci.,* **37**, 1 (1959).

34. S. H. Maron and T. T. Chiu, *J. Polym. Sci.,* **A(1)**, 2651 (1963).

35. J. H. Hildebrand, *J. Chem. Phys.,* **15**, 335 (1947).

36. P. J. Flory, R. A. Orwell, and A. Vrij, *J. Am. Chem. Soc.,* **86**, 3507, 3515 (1964).

37. M. L. Huggins, *Polym. J., Jpn,* **4**, 502 (1973).

38. J. R. Corder and J. H. Purnel, *Trans. Faraday Soc.,* **64**, 1505 (1968).

39. N. F. Brockmeier, R. W. McCoy, and J. A. Meyer, *Macromolecules,* **5(2)**, 130 (1972).

40. R. D. Newman and J. M. Prausnitz, *J. Phys. Chem.,* **76**, 1492 (1972).

41. S. H. Maron and N. Nakajima, *J. Polym. Sci.,* **40**, 59 (1959).

42. B. E. Eichinger and P. J. Flory, *Trans. Faraday Soc.,* **64**, 2035 (1968).

43. D. C. Bonner and J. M. Prausnitz, *J. Polym. Sci., Phys.,* **12**, 51 (1974).

44. Y. B. Terari and H. P. Schreiber, *Macromolecules,* **5**, 329 (1972).

45. B. E. Eichinger and P. J. Flory, *Trans. Faraday Soc.,* **64**, 2053 (1968).

46. Y. K. Leung and B. E. Eichinger, *J. Phys. Chem.,* **78**, 60 (1974).

47. S. H. Maron and C. A. Daniels, *J. Macromol Sci., Phys.,* **B2(3)**, 449 (1968).

48. P. J. Flory and H. Daoust, *J. Polym. Sci.,* **25**, 429 (1957).

49. S. H. Maron and C. A. Daniels, *J. Macromol. Sci., Phys.,* **B2(3)**, 463, Marcel Dekker, Inc., N.Y., (1968)

50. G. S. Y. Yeh and P. H. Geil, *J. Macromol. Sci., Phys.,* **B(1)**, 235 (1967).

51. G. S. Y. Yeh, *J. Macromol Sci., Phys.,* **B(6)**, 462 (1972).

52. S. M. Aharoni, *J. Appl. Polym. Sci.,* **19**, 1103 (1975).

53. S. H. Maron and C. A. Daniels, *J. Macromol Sci., Phys.,* **B2(4)**, 591 (1968).

54. C. Watters, H. Daoust, and M. Rinfret, *Can. J. Chem.,* **38**, 1087 (1960).

55. S. Prager, E. Bagley, and F. A. Long, *J. Am. Chem. Soc.,* **75**, 2742 (1953).

56. R. D. Newman and J. M. Prausnitz, *J. Phys. Chem.,* **76**, 1492 (1972).

57. B. H. Zimm, *J. Chem. Phys.,* **16**, 1099 (1948).

58. W. R. Krigbaum, *J. Am. Chem. Soc.,* **76**, 3758 (1954).

59. W. R. Krigbaum and D. O. Geymer, *J. Am. Chem. Soc.,* **81**, 1859 (1959).

60. M. J. Schick, P. Doty, and B. H. Zimm, *J. Am. Chem. Soc.,* **72**, 530 (1950).

61. H. P. Frank and H. F. Mark, *J. Polym. Sci.,* **72**, 530 (1950).

62. H. P. Frank and H. F. Mark, *J. Polym. Sci.,* **17**, 1 (1955).

63. S. H. Maron and C. A. Daniels, *J. Macromol. Sci., Phys.,* **B2(4)**, 743 (1968).

64. S. H. Maron and C. A. Daniels, *J. Macromol. Sci., Phys.,* **B2(4)**, 769 (1968).

65. E. Jenckel and K. Gorke, *Z. Elektrochem.,* **60**, 579 (1956).

66. U. Bianchi, C. Cuniberti, E. Pedemonte, and C. Rossi, *J. Polym. Sci.,* **A-2(5)**, 743 (1967).

67. H. Hocker, G. J. Blake, and P. J. Flory, *Trans. Faraday Soc.,* **67**, 2251 (1971).

68. P. J. Flory and H. Hocker, *Trans. Faraday Soc.,* **67**, 2258 (1971).

69. H. Hocker and P. J. Flory, *Trans. Faraday Soc.,* **67**, 2270 (1971).

70. H. Hocker, H. Shih, and P. J. Flory, *Trans. Faraday Soc.,* **67**, 2275 (1971).

71. G. Rehage, H. J. Palmen, D. Moller, and W. Wefen, papers presented at the IUPAC Symposium on Macromolecules, Toronto, 1968.

72. K. Schmoll and E. Jenckel, *Z. Elektrochem.,* **60**, 756 (1956).

73. Th. G. Scholte, *J. Polym. Sci.,* **A-2(8)**, 841 (1970).

74. Th. G. Scholte, *Eur. Polym. J.,* **6**, 1063 (1970).

75. R. Koningsveld, L. A. Kleintzeus, and A. R. Schultz, *J. Polym. Sci.,* **A-2(8)**, 1261 (1970).

76. P. I. J. Tait and P. J. Livesey, *Polymer,* **11**, 359 (1970).

77. C. E. Rogers, V. Stannett, and W. Szwarc, *J. Phys. Chem.,* **63**, 1406 (1959).

78. I. Sakurada, A. Nakajima, and H. Fujiwara, *J. Polym. Sci.,* **35**, 497 (1959).

79. S. H. Maron and C. A. Daniels, *J. Macromol. Sci., Phys.,* **B2(3)**, 463 (1968).

80. C. A. Daniels, S. H. Maron, and P. J. Livesey, *J. Macromol. Sci., Phys.,* **B4(11)**, 47, Marcel Dekker, Inc., N.Y. (1970).

81. D. W. Aubrey and A. Barnett, *J. Polymer. Sci.,* **A-2(6)**, 241 (1968).

82. A Turner-Jones, *Macromol. Chem.,* **71**, (1964).

83. S. H. Maron and F. E. Filisko, *Analytical Calorimetry,* Plenum, New York, 1968.

84. S. H. Maron and F. E. Filisko, *J. Macromol. Sci., Phys.,* **B6(1)**, 79, Marcel Dekker, Inc., N.Y. (1972).

85. S. H. Maron and F. E. Filisko, *J. Macromol Sci., Phys.,* **B6(2)**, 413, Marcel Dekker, Inc., N.Y. (1972).

86. M. Kobayashi, K. Tsumura, and H. Tadokoro, *J. Polym. Sci.,* **A-2(6)**, 1493 (1968).

87. A. A. Tager and V. Sanatina, *Rubber Chem. Technol.,* **24**, 773 (1951).

88. A. A. Tager, M. V. Tsilipotkina, and V. K. Doronina, *Zh. Fiz. Khim.,* **33**, 335 (1959).

89. G. Gee and W. J. C. Orr, *Trans. Faraday Soc.,* **42**, 507 (1946).

90. H. P. Schreiber and M. H. Waldman, *J. Polym. Sci.,* **A-2(5)**, 555 (1967).

91. M. A. Kabayama and H. Daoust, *J. Phys. Chem.,* **62**, 1127 (1958).

92. M. Senez and H. Daoust, *Can. J. Chem.,* **40**, 734 (1962).

93. A. A. Tager and M. V. Tsilipotkina, *Vysokomol. Soedin.,* **A5**, 87 (1963).

94. S. Ichihara, A. Komatsu, and T. Hata, *Polym. J. Jpn.,* **2**, 640 (1971).

95. C. Booth, "Heat, Entropy and Volume Changes for Polymer-Liquid Mixtures", in J. Brandrup and E. H. Immergut, Eds., *Polymer Handbook,* 2nd ed., Wiley-Interscience, New York, 1975, pp. IV-323 to IV-333.

96. S. Ichihara, A. Komatsu, and T. Hata, *Polym. J. Jpn.,* **2**, 650 (1971).

97. Yu. Godovskii, G. L. Slonimskii, and V. F. Aleksujev, *Vysokomol. Soedin.,* **A11(5)**, 1181 (1969).

98. L. Utracki and R. Simha, *J. Polym. Sci.,* **A-1**, 1089 (1963).

99. S. H. Maron, *J. Appl. Polym. Sci.,* **5**, 282 (1961)

100. H. M. Quakenbos, *J. Appl. Polym. Sci.,* **13**, 341 (1964).

101. R. N. Schroff, *J. Appl. Polym. Sci.,* **12**, 2741 (1968).

102. A. A. Berlin, *Vysokomol. Soedin.,* **8**, 1336 (1966).

103. A. Rudin and R. A. Wagner, *J. Appl. Polym. Sci.,* **19**, 3361 (1975).

104. A. F. Martin, *Tappi,* **34**, 363 (1951).

105. E. Ott, H. Spulin, and M. W. Grafflin, Eds., *Cellulose and Cellulose Derivatives, Part III,* Interscience, New York, 1955, p. 126.

106. M. Kurata, Tsunashima, M. Iwama, and K. Kamada, in J. Brandrup and E. H. Immergut, *Polymer Handbook,* 2nd ed., Wiley-Interscience, New York, 1975.

107. H. L. Frisch and R. Simha, "The Viscosity of Colloidal Suspension and Macromolecular Solutions", in F. R. Eirich, Ed., *Rheology, Theory and Applications,* Vol. I, Academic, New York, 1956, p. 525.

108. T. Sakai, *J. Polym. Sci.,* **A-2(6)**, 1535 (1968).

109. J. M. Peterson and M. Fixman, *J. Chem. Phys.,* **39**, 2516 (1963).

110. T. Sakai, *J. Polym. Sci.,* **A-2(6)**, 1659 (1968).

111. W. Kuhu, H. Kuhu, and P. Buckner, *Ergeb. Exakt. Naturwiss.,* **25**, 1 (1951).

112. J. Riseman and R. Ullman, *J. Chem. Phys.,* **19**, 578 (1951).

113. H. Sotobayashi, *Makromol. Chem.,* **73**, 235 (1964).

114. L. Utracki and R. Simha, *J. Phys. Chem.,* **67**, 1052 (1963).

115. S. H. Maron and R. B. Reznik, *J. Polym. Sci.,* **A-2**, 309 (1969).

116. F. Ibrahim and H. G. Elias, *Makromol. Chem.,* **76**, 1 (1964).

117. D. J. Streeter and R. F. Boyer, *J. Polym. Sci.,* **14**, 5 (1954).

118. W. Phillippoff, *Bericht,* **70**, 827 (1937).

119. C. D. Thurmond and B. H. Zimm, *J. Polym. Sci,* **8**, 477 (1952).

120. B. H. Zimm and R. W. Kilb, *J. Polym. Sci.,* **37**, 19 (1959).

121. F. Wenger and S-P. S. Yen, *Makromol. Chem.,* **43**, 1 (1961).

122. M. Morton, T. E. Helminiak, S. D. Gadkary, and F. Bueche, *J. Polym. Sci.,* **57**, 471 (1962).

123. B. H. Zimm and W. H. Stockmayer, *J. Chem. Phys.,* **17**, 1301 (1949).

124. M. Kurata and M. Fukatsu, *J. Chem. Phys.,* **41**, 2934 (1964).

125. S. Krozer, *Makromol. Chem.,* **175**, 1905 (1974).

126. M. V. Volkenstein, *Configurational Statistics of Polymeric Chains,* Interscience, New York, 1969.

127. P. J. Flory, *Statistical Mechanics of Chain Molecules,* Interscience, New York, 1969.

128. W. Kuhn, *Kolloid Z.,* **68**, 2 (1934).

129. S. Oka, *Proc. Phys.—Math. Soc. Jap.,* **24**, 657 (1942).

130. W. J. Taylor, *J. Chem. Phys.,* **15**, 412 (1947).

131. W. H. Stockmayer and M. Fixman, *J. Polym. Sci.,* **C1**, 137 (1963).

132. H. Tompa, *Polymer Solutions,* Academic, (New York, 1956), p 177.

133. S. H. Maron and N. Nakajima, *J. Polym. Sci.,* **64**, 587 (1961).

134. R. Koningsveld, *Adv. Colloid Interface Sci.,* **2**, 151 (1968).

135. J. F. Heil and J. M. Prausnitz, *A. I. Chem. Eng. J.,* **12(4)**, 678 (1966).

136. L. Zeman, J. Biros, G. Delmas, and D. Patterson, *J. Phys. Chem.,* **76**, 1206 (1972).

137. J. M. G. Cowie, A. Maconnachie, and R. J. Ranson, *Macromolecules,* **4**, 57 (1971).

138. P. I. Freeman and J. S. Rowlinson, *Polymer,* **1**, 20, (1960).

CHAPTER 3

Molecular Weight Distribution

P. L. DUBIN

Memorex Corporation
Santa Clara, California

M. KRONSTADT

Dynapol
Palo Alto, California

3.1	Introduction	78
3.2	Predictions from Theory	78
	3.2.1 Distribution Moments	78
	3.2.2 Polymerization and Molecular Weight Distribution	81
	3.2.2.1 Linear Condensation Polymers	81
	3.2.2.2 Linear Addition Polymers	82
	3.2.2.3 Nonlinear Polymers	82
3.3	Absolute Methods	83
	3.3.1 Molecular Weight Ranges	84
	3.3.2 End-Group Analysis	85
	3.3.3 Thermodynamic Colligative Methods	85
	3.3.3.1 Theory	85
	3.3.3.2 Membrane Osmometry	87
	3.3.3.3 Vapor Phase Osmometry	89
	3.3.3.4 Cryoscopy and Ebulliometry	92
	3.3.4 Light Scattering	93
	3.3.4.1 Principles	93
	3.3.4.2 Technology	96
	3.3.5 Ultracentrifugation	97
	3.3.5.1 Principles	97
	3.3.5.2 Technology	98
3.4	Viscosity	98
	3.4.1 Principles	100
	3.4.1.1 Viscosity and Molecular Dimensions	100
	3.4.1.2 Empirical Relations	100
	3.4.2 Technology	103

	3.4.2.1 Capillary Viscometers	103
	3.4.2.2 Extrapolation to Zero Shear	103
3.5	Fractionation	104
3.6	Gel Permeation Chromatography	104
	3.6.1 Principles	105
	3.6.1.1 Universal Calibration	106
	3.6.1.2 Theoretical Models of the Separation Process	107
	3.6.1.3 Relation of Chromatogram Shape to MWD	109
	3.6.2 Technology	110
	3.6.2.1 Materials	110
	3.6.2.2 Methodology	114
	3.6.2.3 Hardware	115
	3.6.3 Interpretation of Data	116
	3.6.3.1 Measured Parameters	116
	3.6.3.2 Calibration	117
	3.6.3.3 Resolution	117
	3.6.4 Adsorption and Partition	119
	3.6.4.1 Semirigid Cross-Linked Gels	119
	3.6.4.2 Inorganic Substrates	121
	3.6.4.3 Polysaccharide Gels	122
3.7	Determination of Branching	122
	3.7.1 Molecular Weight Distribution in Branched Polymers	123
	3.7.2 Indices of Branching	124
	3.7.3 Measurement	125
	3.7.3.1 Fractionation	125

P. L. Dubin is currently at Indiana-Purdue University, Indianapolis, Indiana.
M. Kronstadt is currently affiliated with Oximetrix, Mountain View, California.

3.7.3.2	GPC/Viscometry	125
3.7.3.3	Spectroscopy	127
3.8	Summary	127
3.9	Glossary	128
3.10	References	129

3.1 INTRODUCTION

Second only to chemical composition, molecular weight distribution (MWD) is a major factor determining the properties of plastics. Consequently the measurement of molecular weight is often the critical link in correlating synthesis and processing conditions with the ultimate desirable properties of polymeric products. Suppliers and consumers of commercial resins once relied heavily on concentrated solution viscosity measurements for molecular weight control. The development and routine application of more refined techniques—most notably gel permeation chromatography (GPC)—has gone hand-in-hand with a greater appreciation of the multiformity of molecular weight distribution and consequent effects on polymer behavior.

Thorough discussions of the physicochemical principles of light scattering, osmometry, ultracentrifugation, and viscosity can be found in the classic reference texts of Flory (1) and Tanford (2); the former also provides a comprehensive treatment of polymer fractionation. Several more recent books on polymer chemistry (3, 4) deal also with vapor phase osmometry and GPC. The second method is developing rapidly, and current reviews and bibliographies may provide a more contemporary perspective. In this chapter we do not attempt the depth of theory available in certain books or the latitude of a general review. The intention is to provide the reader with a critical comparison of available molecular weight methods, focusing on the particular advantages and difficulties of each, and to describe recent developments that enhance their usefulness as routine tools. Fundamental principles are presented with the objective of clarifying the limitations of the methods as well as the processes of data analysis.

Because of speed and versatility, GPC is rapidly becoming the polymer molecular weight method of choice in both industrial and academic laboratories, and this technique has been emphasized here. The number of published articles dealing with specific applications, processing of chromatographic data, and new column materials is legion. In order to assist in the selection of GPC columns, we present here an enumeration of commercial packings. An attempt has been made to emphasize, however, the limitations of what is, after all, a relative and not an absolute method, with special attention to the issues of resolution and adsorption effects.

The final section deals with the analysis of branched macromolecules. This subject presents the greatest gap between theoretical models and experiments, reflecting in part the enormous difficulty of describing the distribution of mass over simultaneous variables of molecular weight and branching density. Presented here are the applications of some new experimental methods that may accelerate understanding of the structure and properties of branched polymers.

3.2 PREDICTIONS FROM THEORY

3.2.1 Distribution Moments

A hypothetically complete description of the molecular weight distribution would involve listing the amount of each individual species present, a formidable task. In practice, one treats the degree of polymerization (DP) or the corresponding molecular weight as a continuous (rather than integer-valued) function, allowing its distribution to be described by mathematical expressions. This distribution is normally expressed in one of two forms. The frequency or number distribution describes the number of molecules or moles n_i of species with degree of polymerization i; the corresponding number fraction x_i is shown as Equation (3.1).

$$x_i = \frac{n_i}{\Sigma_i n_i} \qquad (3.1)$$

Expressed as a continuous function, the frequency distribution $f(x)$ represents the number fraction of species with DP between x and $x + dx$. One normalizes $f(x)$ such that $\int f(x)\,dx = 1$. One may also consider the weight or weight fraction W_i of species of DP_i. One then obtains a corresponding weight distribution $W(x)$ representing the weight fraction of species between x and $x + dx$. This last point is emphasized because of considerable confusion that arises in the interpretations of gel permeation chromatography data. These chromatograms are of the form $W(\ln x)$, corresponding to the weight fraction of species with $\ln$ (DP) between $\ln x$ and $\ln x + d\,(\ln x)$. When viewed from a linear scale, this function considers much broader intervals of x at high DP than at low. The effect of this, as Berger and Shultz show (5) from consideration of the two normalization conditions, Equations (3.2)–(3.4):

$$\int_0^\infty W(x)\,dx = 1 \qquad (3.2)$$

$$\int_{-\infty}^\infty W(\ln x)\,d\,(\ln x) = 1$$

$$= \int_0^\infty W\,(\ln x)\,\frac{dx}{x} \qquad (3.3)$$

$$W(\ln x) = x W(x) \qquad (3.4)$$

is to make the GPC chromatogram, $W(\ln x)$, appear to be skewed towards high MW relative to a conventional plot of the weight distribution $W(x)$. Thus, since molecular weight distributions are commonly plotted on either linear or logarathmic scales, it is important to specify the axis of such representations.

Theoretical treatments of polymerization lead to equations describing the distribution, and some techniques such as fractionation or GPC also yield a complete MWD. Most experimental methods, however, provide only a single number—an average molecular weight. Since there are numerous ways of calculating an average value from a distribution function, one many consider it fortunate that different experimental methods furnish different molecular weight averages. While one such value alone is a poor description of an unknown distribution, combining two or more averages provides some insight into the shape of the distribution. On the other hand, questions about a particular property of a polymer—whether tensile strength, viscosity under processing conditions, or solubility at a specific temperature—can often be answered by reference to a single appropriate molecular weight average.

Consider first an experiment that counts the actual number of molecules (or moles) in a disperse polymer. Then, an average molecular weight is simply obtained by dividing the actual weight of polymer by the number of moles present. Because the total weight of polymer is simply the sum of the weights of each species of degree of polymerization i, one may write this number average molecular weight,

$$\overline{M}_n = \frac{\text{Total weight}}{\text{Total number of moles}} \qquad (3.5)$$

$$= \frac{\Sigma N_i M_i}{\Sigma N_i}$$

where N_i is the number of moles of species i and M_i the corresponding molecular weight. Alternatively, if W_i is the weight of each species present $W_i = N_i M_i$, Equation (3.5) may be transformed into Equation (3.6).

$$\overline{M}_n = \frac{\Sigma W_i}{\Sigma W_i / M_i} \qquad (3.6)$$

Similar expressions can be obtained for the number average DP as follows:

$$\overline{DP}_n = \frac{\Sigma N_i \cdot i}{\Sigma N_i}, \text{ and so on}$$

One can write these and subsequent expressions in integral form, thusly obtaining Equation (3.7).

$$\overline{DP}_n = \frac{\displaystyle\int_0^\infty f(x) \cdot x\,dx}{\displaystyle\int_0^\infty f(x)\,dx} \qquad (3.7)$$

While colligative properties depend only upon the number of particles present and not their mass, consider the measurement of a property such as light scattering to which every molecule contributes according to the square of its mass. (It is interesting to note (6) that properties such as heat capacity that depend on the first power of molecular mass do not permit the determination of molecular weight at all.) For a monodisperse system, the actual molecular weight could be obtained by dividing this product $N_i M_i^2$ (obtained by measurement of the property involved) by the weight of polymer $N_i M_i$. For a polydisperse system such a procedure yields the weight average molecular weight shown as Equations (3.8)–(3.10):

$$\overline{M}_w = \frac{\Sigma N_i M_i^2}{\Sigma N_i M_i} \tag{3.8}$$

$$= \Sigma w_i M_i \tag{3.9}$$

$$= \frac{\Sigma c_i M_i}{c} \tag{3.10}$$

where w_i is the weight fraction and c_i the concentration of species i, with c the total polymer concentration. Similar equations exist for weight average degree of polymerization $\overline{DP}_w$. Comparing these equations with those for $\overline{M}_n$, one notes that $\overline{M}_w > \overline{M}_n$ for disperse polymers because higher MW species are given greater statistical weight by $\overline{M}_w$.

By analogy, one may define higher molecular weight averages, such as the $\overline{M}_z$ and $\overline{M}_{z+1}$ averages given by Equations (3.11) and (3.12).

$$\overline{M}_z = \frac{\Sigma N_i M_i^3}{\Sigma N_i M_i^2} \tag{3.11}$$

$$\overline{M}_{z+1} = \frac{\Sigma N_i M_i^4}{\Sigma N_i M_i^3} \tag{3.12}$$

Of these higher averages, only $\overline{M}_z$ is of interest in normal molecular weight methods.

These averages can also be defined in terms of molecular weight moments. The nth moment of a distribution $F(r)$ is given by Equation (3.13)

$$\mu_n = \int_{-\infty}^{\infty} r^n F(r) \, dr \tag{3.13}$$

For molecular weights r is confined to positive (integral) values and one may write either Equation (3.14):

$$\mu_n = \Sigma x^n F(x) \tag{3.14}$$

or Equation (3.15):

$$\mu_n = \int_0^{\infty} x^n F(x) \, dx \tag{3.15}$$

where $F(x)$ is the frequency distribution discussed. Thus $\overline{M}_n = \mu_1/\mu_0$ and $\overline{M}_w = \mu_2/\mu_1$.

One should also briefly note a molecular weight average that does not correspond to a ratio of moments of the distribution. As is discussed later, measurement of the viscosity of a polymer, all of whose components obey the Mark-Houwink relationship, expressed as Equation (3.16),

$$[\eta]_i = K M_i^a \tag{3.16}$$

yields a viscosity average molecular weight defined by Equation (3.17):

$$\overline{M}_v = \left[\frac{\Sigma N_i M_i^{1+a}}{\Sigma N_i M_i} \right]^{1/a} \tag{3.17}$$

Note that when $a = 1$, $\overline{M}_v = \overline{M}_w$. For linear polymers, however, a is generally in the range 0.5–0.8.

The weight and number average molecular weights are the most important averages, and their relationship deserves further comment. Consider a mixture of an equal number of moles of two polymers with DP = 100 and 1000 respectively. Then one can compute the following $\overline{DP}$s:

$$\overline{DP}_n = \frac{n \cdot 100 + n \cdot 1000}{2n} = 550$$

$$\overline{DP}_w = \frac{n \cdot 100^2 + n \cdot 1000^2}{n \cdot 100 + n \cdot 1000} = 918$$

On the other hand, equal weights of the two

polymers lead to the following $\overline{DP}$s:

$$\overline{DP}_n = \frac{2W}{W/100 + W/1000} = 182$$

$$\overline{DP}_w = \frac{W \cdot 100 + W \cdot 1000}{2W} = 550$$

In the first case both polymers contribute an equal number of moles, but most of the weight of the mixture corresponds to DP = 1000. In the second case, the two polymers contribute the same weight, but the DP = 100 polymer provides 10 times as many moles.

An analogy may demonstrate the relationship of these averages to different molecular weight methods. Consider a box of pearl necklaces, half with 100 pearls and half with 1000 (i.e., the first case). Each necklace also has a clasp at one end only. If one selected a clasp from the box at random, each necklace would have an equal chance of selection. For a large number of samplings, the average number of pearls selected would be the mean of the two sizes, 550. If one got to keep the necklace, however, the correct course would be to select a pearl at random and withdraw the necklace attached to it. Since 10 times as many pearls belong to long necklaces, the expectation value for the size of the necklace increases to 918. Similarly, some methods, such as osmometry, give equal weight to each molecule and furnish $\overline{M}_n$. Other experiments, such as light scattering, give greater weight to larger molecules and yield higher molecular weight averages.

3.2.2 Polymerization and Molecular Weight Distribution

We review here, without derivation, the MWD and molecular weight averages expected for various types of polymerizations. A far more extensive listing of MWD equations and associated averages has been compiled in a lucid and thorough manner by Peebles (7).

3.2.2.1 *Linear Condensation Polymers*
The earliest and most complete theory for the MWD in condensation polymers, as devel-

oped by Flory (8), Shultz (9), and Stockmayer (10), is based on the "principle of equal reactivity," which states that the reactivity of a functional group is independent of the size of the chain to which it is attached. With no other assumptions, one can show that all condensations of the type A—A + B—B → A—AB—B, and so on, or A—B + A—B → A—BA—B, and so on, give rise to MWDs of one type, known as the "most probable distribution" because it describes so many different condensation systems.

For systems with equal stoichiometry of A and B type groups, and with the added assumption that no rings are formed in the reaction, the frequency and weight distribution can be expressed in terms of a single parameter p by Equations (3.18) and (3.19).

$$f(x) = p^{x-1}(1-p) \tag{3.18}$$

$$w(x) = xp^{x-1}(1-p)^2 \tag{3.19}$$

Here p, the extent of reaction, is the fraction of A- or B-type functional groups reacted. This distribution can be shown to apply also to the random degradation of high molecular weight condensation polymers or any equilibration of a previously formed condensation polymer.

From these equations, one may then determine various measurable parameters of the system. Specifically, the number and weight average molecular weights are as shown in Equations (3.20) and (3.21):

$$\overline{M}_n = \frac{M_o}{(1-p)} \tag{3.20}$$

$$\overline{M}_w = \frac{M_o(1+p)}{(1-p)} \tag{3.21}$$

where M_o is the (average) molecular weight of a repeat unit. Moreover, the polydispersity or $\overline{M}_w/\overline{M}_n$ equals $(1+p)$, or approximately 2 for high conversion (hence high molecular weight) polymers.

Some other features of this distribution are also worth noting. The frequency distribution $f(x)$ decreases exponentially with x (since $p <$

1), and hence there are always a greater number of monomers than dimers, dimers than trimers, and so on. On the other hand, the weight fraction is very small for small x, and reaches a maximum at approximately $1/(1 - p)$, that is, the number average DP. In other words, even through more molecules are monomer than any other single species, most of the mass is centered around the average molecular weight. Finally, as has been noted, GPC chromatograms are plots of $W(\ln x) = xW(x)$. Thus, for the most probable distribution, the GPC peak will correspond to $\overline{M}_w$.

This treatment can be extended to account for inexact stoichiometry of functional groups, the presence of monofunctional monomers that cap the chain and permit no further growth (10), the formation of ring structure (11), and, with considerably greater difficulty, deviation from the principle of equal reactivity (12–15).

3.2.2.2 Linear Addition Polymers

For addition polymers, a treatment similar to that of the preceding section can be applied if the rate constants of growth and termination are independent of molecular weight. This will be the case where initiation rate and monomer concentration are constant, where there is no radical transfer to monomer, and where termination is by disproportionation only. In a typical polymerization, however, monomer concentration changes with time and so accordingly will the ratio of growth to termination. In that case the resulting MWDs may be seen as the sum of distributions produced in succeeding instants of time. Each such distribution will be a most probable distribution, but the sum will be considerably broader.

In that case, rather than predicting a MWD equation for the system, one generally fits the distribution data (e.g., from GPC) to a useful equation. The most common of these, compiled in the previously mentioned reference (7) are the log-normal distribution, the Schultz-Zimm two-parameter distribution (a generalization of the most-probable distribution), and the Weibull-Tung distribution. The latter two may be considered (16) as members of a continuum of generalized exponential distribution that can be fit to most linear MWDs. It is best to regard equations as empirical descriptions of the distribution rather than sources of insight into the nature of the polymerization itself.

If a fixed number of active growth sites are introduced simultaneously (as by very rapid initiation) and growth then proceeds without termination (until monomer is exhausted), then every chain will grow for the same length of time. This yields an extremely narrow MWD (known as the Poisson distribution) in which the frequency and weight distributions are virtually identical, as are $\overline{M}_n$ and $\overline{M}_w$. In practice such polymers (called "living polymers" because the active sites live until quenched) can be achieved with anionic initiators and monomers, such as ethylene oxide or styrene, which are not susceptible to transfer reactions as anions. Polystyrene produced by such a process has become familiar as a calibration standard for GPC.

3.2.2.3 Nonlinear Polymers

A general description of the processes of branching and gelation in condensation and addition polymers is well beyond the scope of this chaper. Such a description has, in any case, been presented with commendable clarity by Flory (17). Instead, we consider a few of the most important cases.

Star Polymers. Consider a polycondensation of A-B monomers with a small amount of A-A or R-A$_f$ added. Each molecule of the latter type will serve as a center for a star of f branches with A-type end groups pointing outward. Thus if A-groups only react with B-groups, there is no way for two such stars to unite: molecules can grow large but never form an infinite network. Each arm of the star will be a chain with a most-probable distribution, but as the number of arms increases, it becomes unlikely that any star will have only very long or very short arms—instead one very long arm is likely to be balanced out to some extent by more probable average-length arms. The result is a sharpening of the molecular weight distribution, with $\overline{M}_w/\overline{M}_n = 1 + 1/f$. A similar sharpening (corresponding to

$f = 2$) occurs for the most probable distribution for addition polymers discussed earlier if termination is by combination of two growing chains rather than by disproportionation.

Polyfunctional Condensation Polymers. The general effect of branching, on the other hand, is to broaden the MWD. The equations for these systems are provided by Stockmayer (10). In general, one defines a branching probability, which can then be related to the extent of reaction p and the nature and stoichiometry of polyfunctional monomers. One finds a critical branching probability below which no infinite network exists and above which the amount of gel increases rapidly. For all such systems $\overline{M}_w/\overline{M}_n$ increases rapidly as the gel point is approached, becoming infinite at that point. Most of the molecules are of low molecular mass; the few of very high mass cause $\overline{M}_w$ to increase rapidly with conversion but have little effect on $\overline{M}_n$. Beyond the gel point, high MW molecules with many end groups will be far more likely to react with the existing infinite network, hence while gel mass increases, there are essentially no soluble molecules of extremely high mass. The treatment of addition polymerization containing divinyl monomers is completely analogous.

Cross-Linking of Linear Molecules. The addition of cross-links to previously formed linear ("primary") molecules can be shown (18) to be equivalent to polymerization with tetrafunctional monomer units. Thus the considerations of the preceding section and the equations developed in previously cited references apply here, whether the primary molecules are addition or condensation polymers.

Long Chain Branching. Branching may also occur in addition polymers by means of chain transfer to polymer, in which one polymer chain ceases to grow while a new active site is formed in the middle of a previously formed chain. Because there is no way to link large chains into infinite networks by this mechanism, gelation will not occur. However, a broadening of the distribution will occur because large, branched molecules present more sites for chain transfer and are more likely to be the site of new branches. This "rich get richer" situation may be contrasted to star polymers where the MWD is sharpened with increased branching.

3.3   ABSOLUTE METHODS

A number of measurable properties of polymer solutions vary sensitively and monotonically with molecular weight. Some of these are furthermore insensitive to the specific chemical nature of polymer or solvent. Molecular weight methods based on these properties are termed "absolute" and thus distinguished from the measurement of quantities that are related to molecular weight in a way that depends specifically on the compounds involved. This section deals with the following techniques: membrane and vapor phase osmometry (VPO), cryoscopy and ebulliometry, light scattering, and ultracentrifugation. End-group analysis must also be included in a list of absolute molecular weight methods. In contrast, measurements such as the intrinsic viscosity and the GPC retention volume are primarily reflections of molecular volume and so depend on factors that influence chain configuration. These "secondary" methods, convenient and versatile as they are, yield data that are related to the molecular weight in a manner that must be determined independently for each case.

Characterization of the molecular weight of a polydisperse sample by any absolute method generally leads to a single value, either the weight or number average, according to how the polymeric components contribute to the measured property. The polydispersity $\overline{M}_w/\overline{M}_n$ may be as low as 1.01 for the products of ionic addition polymerizations, or may be greater than 10 for randomly branched polymers. The choice of a molecular weight method should then be influenced by the relevance of the measured average to the properties of interest. For example, polymer properties related to chain entanglement, such as bulk viscosity and elasticity, correlate best with $\overline{M}_w$. On the other hand, the number average is the key molecular weight parameter for features sensitive to the diluent effects of chain ends—brittleness, melting and glass transition temperatures, and impact strength. Of course, factors other than molecular weight, such as tacticity and crystallinity, are also major determinants of the foregoing properties.

3.3.1 Molecular Weight Ranges

The various methods are limited with regard to molecular weight range by restrictions of sensitivity and instrumental design. End-group determination by chemical methods clearly becomes innaccurate with increasing DP and is generally only considered useful for molecular weights below 20,000. (Detection of end groups by spectroscopic techniques is usually less sensitive, whereas radiolabeled chain termini can be determined much more accurately.) Since all polymer molecules contribute identically to the concentration of end groups, the molecular weight is determined as W_2/N_2 where W_2 is the mass of polymer in the sample and N_2 the total number of solute molecules. The molecular weight so obtained is a number average, that is, $W_2/N_2 = \overline{M}_n$.

Thermodynamic colligative methods also yield a number average. They do so through measuring the total number of solute molecules by means of the depression of solvent activity. Consequently the sensitivity of these methods decreases with increasing molecular weight. One often finds in the literature comparisons of the magnitudes of the relevent physical changes; for example, a 1% solution of a 50,000 MW polymer in benzene exhibits a boiling point elevation of 0.0006°C, a freezing point depression of 0.0012°C, a vapor pressure decrease of 0.0018 mm Hg, and an osmotic pressure of 5.0 g/cm². These comparisons are useful only when considered

Table 3.1   Molecular Weight Methods Reported for Some Commercialized Polymers [a, b, c]

	EG	VPO	MO	LS	IV[d]	GPC
Polyethylene			X	X	X	X
Polypropylene			X	X	X	X
Polyisoprene			X	X	X	X
Polystyrene		X	X	X	X	X
Polybutadiene			X	X	X	X
Polyvinyl acetate			X	X	X	X
Polyvinyl alcohol					X	X
Polyvinyl chloride				X	X	X
Polyacrylonitrile				X	X	X
Polyacrylamide			X	X	X	X
Polyacrylates			X	X	X	X
Polyethers	X	X	X	X	X	X
Epoxide resins						X
Phenolic resins		X		X		X
Polyesters	X	X			X	X
Polysulfones			X			
Polyamides	X	X			X	X
Polycarbonates				X	X	X
Polyurethane		X				X
Silicones				X		X

[a]   This table summarizes a cursory review of the total literature and informal discussions with instrument manufacturers: it is not intended to be rigorously complete.

[b]   Abbreviations: EG, end group analysis; VPO, vapor phase osmometry; MO, membrane osmometry; LS, light scattering; IV, intrinsic viscosity

[c]   Although ultracentrifugation is described in this chapter, its use for the characterization of commercial synthetic polymers is rare.

[d]   Polymers for which intrinsic viscosity-molecular weight relations are reported (reference 41).

in the light of the limitations of instrumental design and technique, so the actual upper molecular weight limits of the methods are elusive quantities. Cryoscopic and ebulliometric methods are in general innaccurate for $\overline{M}_n$ over 5000 although nonroutine applications for molecular weight as much as five times greater are reported. Recent improvements in the design of commercial VPO devices has extended the molecular weight range of that method up to 5×10^4. The most sensitive colligative method, membrane osmometry, is useful for molecular weights up to about 2.5×10^5. On the other hand, a lower limit is imposed on this technique by the permeability of the membranes: 5,000–10,000 for typical commerical products. This restricts the method to samples with $\overline{M}_n$ of 10,000 or more.

In contrast to the preceding techniques, light scattering and ultracentrifugation increase in sensitivity with the molecular weight. Losses of accuracy for lower molecular weight samples are related to the specific chemical nature of polymer and the medium inasmuch as they affect the differential refractive index.

The choice of an absolute molecular weight method thus depends on its range and facility as well as the relevance of the measured average to the property of interest. In practice, VPO is the method of choice for samples in the 100–5000 range. Samples with $\overline{M}_n$ of 20,000–200,000 are suitable for membrane osmometry. Light scattering and ultracentrifugation are usually applied to polymers whose molecular weights exceed 50,000. Measurements in the range 5000–20,000 have until recently been difficult, requiring either membrane osmometry with judicious membrane selection, or VPO with modified electronics and careful calibration methods. Modern VPO instruments now offer good performance in this molecular weight region. In Table 3.1 some of the polymers of greatest commercial significance are listed along with techniques reported for their molecular weight analysis.

3.3.2 End-Group Analysis

Because of its simplicity, end-group analysis was the earliest and most compelling molec-

ular weight method. Condensation polymers are especially suited to this technique since their molecular weights are rather low and their unreacted termini lend themselves to the classical titrations for functional groups such as carboxyl, amine, and hydroxyl. The principal difficulty has been the limited solubility of certain condensation polymers, and recent advances in the use of heterogeneous titration and multistep techniques (as well as new reagents and automation) are described by Garmon (19). In contrast, the application of end-group methods to vinyl addition polymers requires the introduction of heteroatomic or radiolabeled termini, and as such are not very useful for conventional materials.

3.3.3 Thermodynamic Colligative Methods

The techniques of membrane osmometry, vapor phase osmometry (VPO), freezing point depression or cryoscopy, and boiling point elevation or ebulliometry differ greatly in methodology, but all rest on a single principle: that the thermodynamic activity of the solvent in a two-component system at constant temperature and pressure depends on the mole fraction of solute. Consequently, the molecular weight obtained is the number-average. Because of the fundamental interconnection of these methods, it is appropriate to consider them in a common section.

3.3.3.1 *Theory*

As a consequence of the Gibbs-Duhem relation, dilute two-component solutions that obey Henry's law—Equation (3.22)—for the solute, also follow Raoult's law—Equation (3.23)—for the solvent. Thus at very low concentrations:

$$a_2 = k_2 x_2 \tag{3.22}$$

$$a_1 = x_1 = 1 - x_2 \tag{3.23}$$

where a_1, x_1 and a_2, x_2 are activities and mole fractions for solvent and solute, respectively. If W and M are defined as the mass and molecular weight of either solvent or solute, then one can write Equation (3.24):

$$x_2 = \frac{W_2/M_2}{W_1/M_1 + W_2/M_2} \quad (3.24)$$

For dilute solutions where $W_2/M_2 \ll W_1/M_1$ and $x_2 \ll 1$, Equation (3.24) can be reduced to Equations (3.25) and (3.26):

$$x_2 = \frac{(W_2/M_2)}{(W_1/M_1)} \quad (3.25)$$

$$\ln(1 - x_2) \simeq -x_2 \quad (3.26)$$

Combining Equations (3.23), (3.25), and (3.26) yields

$$\ln a_1 = \frac{-(W_2/M_2)}{(W_1/M_1)} = \frac{-W_2M_1}{W_1M_2} \quad (3.27)$$

Thus the value of a_1 in the limit of infinite dilution of solute where the foregoing assumptions are valid yields M_2, the molecular weight of the solute. The large molecular volumes of polymers lead to intermolecular effects at relatively modest concentrations; therefore data are required at concentrations under 1% w/w in order to make valid extrapolations toward infinite dilution.

In any real synthetic material, with N_i moles per liter of each molecular weight M_i, the mole fraction of polymer is given by Equation (3.28):

$$x_2 = \frac{N_2}{N_2 + N_1} = \frac{\sum_i N_i}{\sum_i N_i + N_1} \quad (3.28)$$

where N_1 is the molar concentration of solvent. At high dilution N_1 is nearly equal to the reciprocal of the molar volume of the solvent, so one can reduce Equation (3.28) to Equation (3.29):

$$x_2 = \frac{\sum_i N_i}{\sum_i N_i + 1/v_1} = v_1 \sum_i N_i \quad (3.29)$$

The total weight concentration of solute species $c = \sum_i N_i M_i$, and therefore one obtains Equation (3.30):

$$\frac{x_2}{c} = v_1 \frac{\sum_i N_i}{\sum_i N_i M_i} \quad (3.30)$$

where the term $\sum_i N_i M_i / \sum_i N_i$ is recognizable as the number-average molecular weight. Colligative (i.e., "counting") methods evaluate the mole fraction of solute per weight concentration (in the limit of infinite dilution) and yield the molecular weight as Equation (3.31):

$$\left(\frac{x_2}{c}\right)_{c=0} = \frac{v_1}{M_n} \quad (3.31)$$

Colligative methods determine the mole fraction of solute from the activity of the solvent according to Equations (3.23), (3.26), and (3.32):

$$-\ln a_1 = -\ln(1 - x_2) \simeq x_2 \quad (3.32)$$

The depression of solvent activity upon solute addition is measured at sufficient dilutions such that Henry's law is followed, and a_2 is proportional to c. One obvious measure of solvent activity is by way of vapor pressure, that is, $a_1 = p_1/p_1^0$. Other properties thermodynamically related to a_1 prove to be more useful however. Freezing point depression and boiling point elevation, for example, are related to solvent activity by the Equations (3.33) and (3.34):

$$-\ln a_1 = \left(\frac{\Delta H_f}{RT^2}\right) \Delta T_f \quad (3.33)$$

$$-\ln a_1 = \left(\frac{\Delta H_v}{RT^2}\right) \Delta T_b \quad (3.34)$$

where ΔH_f and ΔH_v are the solvent's heats of fusion and vaporization, respectively. The osmotic pressure π, the pressure that must be exerted on the solution to raise the activity of the solvent component to that of the pure solvent, can also be related to a_1 as shown in Equation (3.35):

$$\ln a_1 = \int \left(\frac{\partial \ln a_1}{\partial P}\right)_{T,N_i} dP = \frac{\pi v_1}{RT} \quad (3.35)$$

Since c, the polymer concentration in grams per milliliter, is given by W_2M_1/W_1V_1 at the high dilutions commonly employed, Equations (3.27) and (3.35) may be combined to yield Equation (3.36):

$$\left(\frac{\pi}{c}\right)_{c=0} = \frac{RT}{\overline{M}_2} \qquad (3.36)$$

As noted in Equation (3.31), the value of M_2 obtained for a polydisperse solute will be $\overline{M}_n$, the number-average molecular weight.

Analogous relationships for the cryoscopic and ebulliometric methods are obtained from Equations (3.33) and (3.34) as Equations (3.37) and (3.38):

$$\left(\frac{\Delta T_f}{c}\right)_{c=0} = \frac{(RT^2 v_1 / \Delta H_f)}{M_2} \qquad (3.37)$$

$$\left(\frac{\Delta T_b}{c}\right)_{c=0} = \frac{(RT^2 v_1 / \Delta H_v)}{M_2} \qquad (3.38)$$

3.3.3.2   Membrane Osmometry

Membrane osmometry is the single most used absolute molecular weight technique. Relative to other colligative methods, the osmotic effect is large for polymers of moderate molecular weight so that an osmotic pressure of 10–15 mm of solvent head can be observed with polymer solutions for which ΔT_f and ΔT_b would be less than 0.005° C. While conventional light scattering provides similar accuracy in the molecular weight range 2×10^4–2×10^5 (and is in fact more precise at higher molecular weights), its relative disadvantages include the requirements of very careful sample preparation, more elaborate measurements, more complex data analysis, and greater equipment costs. Similar observations apply to analytical untracentrifugation, especially with regard to the last two factors. The efficiency of membrane osmometry is limited in principle by the properties of the membranes for low molecular weight samples and by the inherent insensitivity of the instrument at high molecular weight. Regenerated cellulose and cellulose acetate membranes are manufactured by Schleicher and Schuell with stated lower molecular weight limits ranging from 5000–20,000. These values are of course functions of polymer dimensions, sample polydispersity, and degree of membrane swelling in the chosen solvent. Limitations for high molecular weight samples arise from the reduction of the osmotic effect and the need for lengthy extrapolations for the higher concentrations required. Reproducibility of $\pm$ 10% is feasible up to $\overline{M}_n = 200,000$, but precision falls off rapidly with increasing molecular weight.

Instrumental Design.   Two cells separated by a semipermeable membrane, each connected to a vertical capillary, would constitute a simple working osmometer. If the two cell compartments are filled to the same levels with solvent and polymer solution, the reduced activity of solvent in the latter leads to a net

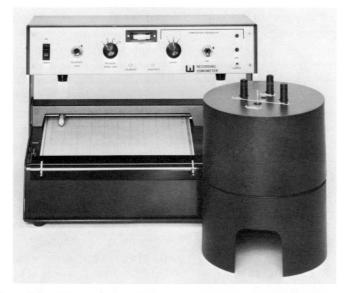

Figure 3.1   Wescan Model 231 Recording Membrane Osmometer. (Courtesy Wescan Instrument Company.)

flow from the solvent cell. The osmotic pressure may then be determined from the equilibrium excess height on the polymer side. Modern membrane osmometers (see Figure 3.1) have been designed to permit sensitive electronic detection of osmotic pressure, provide for efficient filling and flushing of sample and reference cells, minimize solvent equilibration time, and allow for precise temperature control. Since hours or days could be required for solvent equilibration, modern membrane osmometers are of the "dynamic" design type. The Mechrolab and Hallikainen osmometers both utilize servomechanisms to oppose hydraulically the flow of solvent and thus "hunt" until equilibrium is attained. Specific details of these two instruments, are provided in References 20 and 21. However, neither is currently manufactured. The only domestically produced membrane osmometer (Wescan Instruments, Santa Clara) is fortunately of an advanced design. Rather than utilizing a null balance servo, the osmotic flow is measured directly by a strain gauge coupled by way of a diaphragm to the solvent cell. Calibration is accomplished by adjusting the recorder response while a precisely known height of solvent is connected to the solvent cell.

Perhaps most serious operational difficulties in osmometry arise from assymetry, deformation or leakage of the membrane. The choice of membrane and solvent and the technique of membrane preparation and placement are often key to reliable data. Membrane problems may be manifested by finite pressures observed at zero concentration, very long equilibration times, or drifting signals. It is probably fair to say that membrane technology lags behind instrument design. Selection of membranes is dictated by considerations of solvent compatibility and equilibration rate versus permeation of solute components. Desirable properties of solvents for membrane osmometry are low volatility, viscosity, and surface tension. It is presumably the last factor that makes measurements in aqueous solutions more demanding.

Analysis of Data.   Results are obtained as excess solvent heights h (in centimeters) versus polymer concentration (grams per cubic centimeter). Solvent heights are converted to units of mass/area by multiplying by solvent density ρ (grams per cubic centimeter), and then to units of pressure by multiplying by the gravitational constant d. The units of $\pi = h\rho$d are thus dynes per square centimeter. The observed values of π are divided by polymer concentration, and the units of π/c are thus dyne per centimeter per gram. The molecular weight is obtained using Equation (3.39), as illustrated in Figure 3.2.

$$\overline{M}_n = \frac{RT}{(\pi/c)_{c=0}} \qquad (3.39)$$

Hence the appropriate units for R are erg per mole per degree, that is dyne-centimeter per mole per degree.

The effect of polymer concentration in osmometry has been thoroughly examined from thermodynamic considerations. The limiting expression for the osmotic pressure, $\pi/c = RT/M$, is completely analogous to the ideal gas law $P/n = RT/V$ and may similarly be expanded by a virial function (with A's as virial coefficients) to account for nonideality, as shown in Equation (3.40):

$$\frac{\pi}{c} = RT \left(\frac{1}{M} + A_2 c + A_3 c^3 + \cdots \right) \qquad (3.40)$$

The theory of dilute polymer solutions developed by Flory yields expressions for these virial coefficients in terms of the thermodynamics of polymer-solvent interactions and the statistical mechanics of intramolecular segmental forces. While a description of the theory is beyond the scope of this chapter, it may be helpful to observe that the physical exclusion of polymer chains from each other's spatial domains favors the dilute solution over the concentrated one. This effect corresponds to a thermodynamic driving force for the addition of pure solvent to polymer solution, which thus increases the osmotic pressure. To the extent that polymer segments attract each other, and so reduce or eliminate the expansive effects due to excluded volume and solvation, these positive nonideal contributions to the osmotic pressure are similarly reduced or elim-

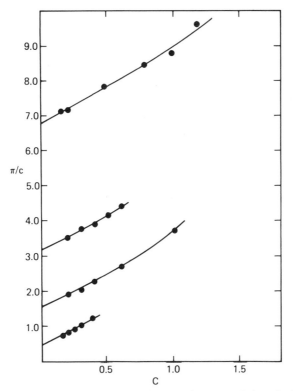

Figure 3.2 Typical plots of membrane osmometry data showing extrapolation of π/c to $c = 0$. Data for polyisobutylene fractions in cyclohexane (top to bottom: increasing molecular weight.)

inated. In addition to explaining the physical meaning of the empirical terms in Equation (3.40), the theory guides the experimentalist by predicting conditions that can minimize the difference between the measured and limiting values of π/c and by providing expressions that facilitate the extrapolation to infinite dilution. Thus ideal solution behavior, where the virial coefficients of Equation (3.40) vanish and π/c becomes independent of c, can usually be obtained by appropriate choice of solvent and temperatures. Such theta conditions have been established for all commercially significant homopolymers.

Under nonideal conditions, Equation (3.40) is closely approximated (22) at low concentrations by Equation (3.41):

$$\frac{\pi}{c} = \left(\frac{\pi}{c}\right)_{c=0} (1 + \Gamma_2 c + g\Gamma_2^2 c_2^2) \quad (3.41)$$

Where g is a constant usually having a value close to 0.25. The coefficient Γ_2 depends in a complex manner on a number of configura-

tional and thermodynamic interaction parameters. Two aspects of Equation (3.41) are relevant to the present discussion. First, reduction of solvent affinity, which lowers the slope Γ_2 of π/c versus c plots, is seen to eliminate rapidly the positive curvature explicit in the third term. Second, since the value of g is typically close to 0.25, Equation (3.41) may be expressed as Equation (3.42):

$$\left(\frac{\pi}{c}\right)^{1/2} = \left(\frac{\pi}{c}\right)^{1/2}_{c=0} \left[1 + \frac{\Gamma_2}{2} c\right] \quad (3.42)$$

Data obtained at moderate concentrations in nonideal solutions can usually be treated according to Equation (3.42) as long as the measured values of π/c do not exceed the limiting value by a factor of 3 or more. Thus plots of $(\pi/c)^{1/2}$ versus c often facilitate extrapolation for measurements made far from theta conditions.

3.3.3.3 Vapor Phase Osmometry

Formerly limited to the molecular weight range below $\overline{M}_n = 5000$, commercial VPO instruments with improved design and electronic circuitry can now be viewed as tools of polymer characterization, particularly for condensation polymers. Some of the processes ocurring in VPO are not readily amenable to theoretical treatment and considerable uncertainty has accompanied the application of VPO for $\overline{M}_n > 5000$. However, it is now possible to compare data from VPO and membrane osmometry (in the region 10,000 < $\overline{M}_n$ < 100,000), and these comparisons greatly enhance confidence in both techniques (23).

Principles. Hill (24) is generally credited with the first description of the principles of VPO some 50 years ago; commercial instruments have been available for about 20 years. The technique is based on the isothermal distillation of solvent from a saturated vapor phase to a drop of solution. The driving force for this process is the depression of the solvent mole fraction, and hence activity, by the solute in the drop. Accompanying heats of condensation and dilution cause the temperature of the drop to rise until heat losses by conductive, evaporative, and convective routes lead to a thermal pseudoequilibrium. The steady-state temperature difference between a suspended drop of pure solvent and a drop of polymer solution thus provides a measure of the solvent activity in the sample. To determine this very small temperature difference, drops of solvent and polymer solution are placed on respective thermistor beads that comprise two arms of a wheatstone bridge. A resistance difference or bridge imbalance Δr is then measured that is linearly proportional to the temperature difference ΔT.

The evaporation and condensation processes that take place on the solvent and sample drops are complex. Consequently, a useful theoretical model for VPO demands a knowledge of the relative heat losses from thermistors and thermistor wires, the drop geometries and surface areas, and the coefficients of surface heat transfer and mass transfer (25, 26), as well as the more accessible values of the solvent's heat of vaporization and molar volume. Kamide et al. (26) for example, obtain the following expression shown as Equation (3.43)—completely analogous to Equation (3.38) for ebulliometry—for the VPO process:

$$\left(\frac{\Delta T}{c}\right)_{c=0} = \frac{RT^2 v_1/\Delta H_v}{M_2(K_e/K_s)} \qquad (3.43)$$

where the left-hand term is the limiting value of the temperature difference per unit solute weight concentration; v_1 and ΔH_v are the molar volume and heat of vaporization of solvent, respectively; and M_2 the solute molecular weight.

K_e/K_s is a function of the kinetic and thermodynamic parameters that determine the efficiency of the heat transfer processes. In practice it is assumed that all of the quantities contained in the function $RT^2 v_1$ $[\Delta H_v (K_e/K_s)]^{-1}$ can be kept constant by operating the instrument using a fixed temperature, solvent, and thermistors. Then the preceding function can be combined with the thermistor constant $(\Delta T/\Delta r)$ to yield an in-

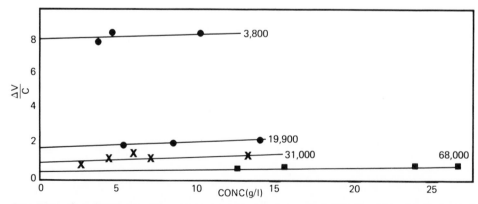

Figure 3.3 Plot of VPO data: reduced bridge resistance versus concentration for plystyrene standards of indicated molecular weight in toluene. (Courtesy Wescan Instrument Company.)

strument calibration constant as given in Equation (3.44):

$$K = M_2 \left(\frac{\Delta r}{c} \right)_{c=0} \qquad (3.44)$$

According to Equation (3.44) the value of the instrument constant can be obtained by measuring the bridge imbalance for solutions of a solute of known molecular weight over a range of concentrations (see Figure 3.3). Similar measurements for the sample lead in a straightforward way to an evaluation of its molecular weight. As noted in Equation (3.31), the use of a technique sensitive to solvent activity to determine M_2 provides a number-average molecular weight.

Reliability and Limitations. There has been considerable disagreement on the reliability of VPO (25–28), largely because confirmatory molecular weight methods, such as membrane osmometry or conventional light scattering, were not applicable in the critical molecular weight range 200–20,000. In the absence of convenient molecular weight standards, experimenters as a rule could only compare their results to those obtained by VPO in other laboratories or examine the consistency of measured molecular weights when operating variables were changed. As improved instrumental design has provided the sensitivity required for molecular weight measurements in the 10^4–10^5 range (23), more information has become available on the accuracy of the method. These results have prompted workers in the field to focus in particular on the behavior of the calibration constant of Equation (3.44).

Meeks and Goldfarb (28) found that the solvent dependence of k could not be accounted for by v_1 and ΔH_v alone. According to Kamide and co-workers, the function K_e/K_s of Equation (3.43) may be described as shown in Equation (3.45).

$$K_e/K_s = 1 + \left(\frac{k_1 A_1 + k_2 A_2}{k_3 A_1} \right) \left(\frac{RT^2}{\Delta H_v^2 P_1(T)} \right)$$

$$(3.45)$$

Where $P_1(T)$ is the solvent vapor pressure, and the first parenthetical term describes the geometry and heat transfer properties of the solution drop. Thus $k = M_2 (\Delta T/c)_{c=0} = RT^2 v_1 \Delta H_v (K_e/K_s)^{-1}$ is strongly dependent on solvent, temperature, and instrument design. Dependence of the calibration constant k on solvent or temperature does not pose a problem in the measurement of M_2 by way of Equation (3.43) since these variables are typically fixed during a series of measurements. Bersted (25), however, has claimed that a significant solute dependence of k arises from the influence of the solute's diffusion constant on the rate of solvent condensation. This effect vitiates the normal calibration procedure that employs a low molecular weight standard such as benzil, and Bersted describes as a consequence of this problem some measured differences between $\overline{M}_n$ by VPO and by membrane osmometry—ranging from −20% at $\overline{M}_n = 40,000$ to −50% at $\overline{M}_n = 80,000$. On the other hand, Morris (27) found only a modest and erratic dependence of k on M_2 that did not correlate with the more substantial variations of $\overline{M}_n$ with solvent and temperature. Using benzil to evaluate the calibration constant, Kamide et al. (26) obtained $\overline{M}_n$ within 10% of membrane osmometry data up to molecular weights of 4×10^5, and observed that k varied by less than 10% over the molecular weight range 10^3–10^5. Bersted and Morris both used the Mechrolab/Hewlett-Packard instrument, while Kamide et al. employed a unit of the Hitachi design, and some of the differences in results may arise from differences in instrument design. Only the latter type of osmometer is currently marketed in the United States (as the "Corona/Wescan"—see Figure 3.4), but the most common laboratory instrument is still the Mechrolab type.

Two other operational variables that have received considerable attention are the drop size and the time required to reach the thermal psuedoequilibrium. With regard to the former, Δr is found to be altered by only 10% when drop size is doubled (27, 28), a variance far in excess of typical drop size fluctuations. The approach to the pseudoequilibrium value of Δr was qualitatively examined by Meeks and Goldfarb (28) who recommended extrapola-

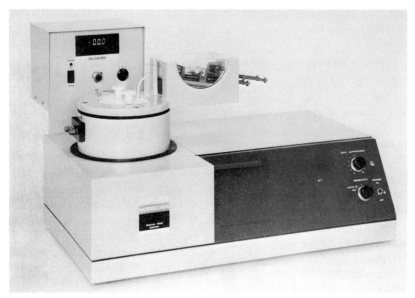

Figure 3.4 Corona-Wescan Model 232A Vapor Phase Osmometer. (Courtesy Wescan Instrument Company.)

tion to zero time from data obtained over 0–10 min. Morris (27) found the dependence of r on time to be influenced by solute concentration and drop size; no clear pattern could be identified, but 4 min. was generally sufficient for a stable reading. Kamide et al. analyzed in detail the behavior of t_s, the time required to obtain a steady state. While their calculated values of t_s ranged from 6–40 sec., the experimentally determined times were 300–600 sec. The approach to thermal steady state is evidently not well understood, and the use of an output recorder to monitor equilibration is generally recommended.

Concluding Remarks. The field of vapor phase osmometry is undergoing progress along the lines of instrumental design and theory. Two VPO instruments are now commercially available from Knauer and Corona/Wescan. Improvements in thermistor design and temperature control in the Corona/Wescan have resulted in a tenfold increase in sensitivity, and the manufacturer quotes an upper molecular weight limit of 100,000 for toluene or chloroform, or 25,000 for less volatile solvents such as water or dimethyl formamide (DMF). This instrument employs a platinum mesh to hold the liquid on the thermistor which obviates questions of drop size control (24).

The theory of VPO is complicated by the dynamic nature of the process. Nonetheless, considerable progress has been made toward elucidating the nature of the instrument constant. An exact treatment of this calibration factor would strengthen the theoretical basis of VPO. Strict resolution of thermodynamic and kinetic factors may be some way off, as is the interpretation of VPO data along thermodynamic lines, for example, in regard to the relationship between concentration dependence and second virial coefficients.

3.3.3.4 Cryoscopy and Ebulliometry
Freezing point depression and boiling point evaluation are basic techniques for the molecular weight determination of organic solutes. These methods have been applied to polymers but have been routinely successful only in the low molecular weight range, that is, below 5000. Very accurate and sensitive temperature detection is required for solutes of greater molecular weight, and all reports of application to such substances involve individually constructed devices. In addition to the low values of ΔT in Equations (3.37) and (3.38) at moderate molecular weight, certain technical difficulties arise in utilizing these techniques for true macromolecules. Superheating and foaming are special problems in ebulliometry

with polymers, whereas solubility considerations frequently limit the choice of solvents in
cryoscopy.

It would appear that the applications of
ebulliometry and cryoscopy as polymer molecular weight methods are mainly limited to
those laboratories with specially designed
instruments. A discussion of the design of
such devices and a thorough review of these
techniques in general is found in the literature
(29).

3.3.4 Light Scattering

Light scattering by dust particles in the air—
the Tyndall effect—is a well known phenomenon. Although not visible to the eye, light
scattering also occurs from solutions of polymer molecules—"particles" that are extremely
small in macroscopic terms, but quite large
compared to the molecules of the solvent in
which they are immersed.

Measurement of light scattering from polymer solutions is a powerful tool, yielding
information on not only molecular weight,
but also molecular size in solution, as well as
the thermodynamically important second virial
coefficient. Until recently, however, such measurement was restricted to relatively few labs:
equipment was costly and often home-built,
experimental techniques were extremely demanding and tedious, and data analysis was
complex. Recent technology, taking advantage
of laser light sources and modern data processing units, has brought light scattering measurements into the realm of the "almost routine," and the next 10 years should see a great
increase in the use of light scattering as an
analytical tool.

3.3.4.1 Principles

The theory behind the analysis of light scattering is certainly the most formidable of any
presented in this chapter. Nevertheless, some
discussion of this theory is warranted in order
to understand how the technique provides
information, not simply on molecular weight,
but on molecular dimensions and on thermodynamic quantities such as the second virial
coefficient. Accordingly, the simplest case for
polymer light scattering will be considered in

detail and results presented for the rest. An
excellent and complete presentation may be
found in work done by Flory (30).

One can start with two assumptions.

1. The scattering particles are spherical and
much smaller than the wavelength of light (not
true for all polymer molecules).

2. The solution is very dilute, with particles
randomly distributed and no interparticle
interaction.

For a beam of light vertically polarized (z
axis), its instantaneous amplitude E is defined
by a sine wave with wavelength λ, Equation
(3.46):

$$E = E_{0,z} \cos \left[\frac{2\pi}{\lambda} (\tilde{c}t - \tilde{n}_0 x) \right] \quad (3.46)$$

where $\tilde{c}$ is the speed of light, $\tilde{n}_0$ the refractive
index of the medium, t the time, and x the
distance from the (arbitrary) origin. This light
wave will induce a dipole moment p in the
scattering particle proportional to its amplitude and the particle's polarizability. Because
this experiment compares scattering from
solution to that from pure solvent, we consider
only the *excess* polarizability α of the particle
compared to the solvent. Then

$$p = \alpha E = \alpha E_{0,z} \cos \frac{2\pi}{\lambda} \tilde{c}t \quad (3.47)$$

By the first assumption, the entire particle has
negligible size and so may be placed at the
origin, $x = 0$ as in Figure 3.5. This induced
dipole oscillates with the same frequency as
the incident light and hence emits light of the
same wavelength—but in all directions. Thus
we may observe a scattered beam at some
angle with respect to the original direction of
light. This scattered light will still be polarized,
and the intensity observed will then depend on
the angle between the direction of observation
and the original polarization axis, as well as
the angle to the incident light beam. If the
incident light is not polarized, we may treat it
as the sum of vertically and horizontally

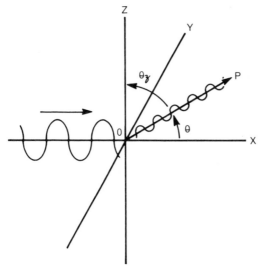

Figure 3.5 Schematic representation of light scattering of a beam of vertically polarized light.

polarized beams, but the result will no longer depend on any angle except θ, the angle between the direction of the scattered beam and the emerging light beam. The measurable intensity (the square of the amplitude) at a distance r from the particle can then be shown to be given by Equation (3.48):

$$i = \frac{8\pi^4 I_0 \alpha^2}{r^2 \lambda^4} (1 + \cos^2 \theta) \qquad (3.48)$$

where I_0 is the incident intensity. If all particles contribute independently, the total scattering intensity by N/V particles per unit volume is given by Equation (3.49):

$$i_\theta^\circ = \frac{I_0 N}{V} \frac{8\pi^4 \alpha^2}{r^2 \lambda^4} (1 + \cos^2 \theta) \qquad (3.49)$$

where the superscript $\circ$ indicates the simplifying assumptions.

The excess polarizability, moreover, can be related to the refractive index of the solvent n_0 and its change upon addition of solute dn/dc, shown as Equation (3.50):

$$\alpha = \frac{\tilde{n}_0 c V}{2\pi N} \left(\frac{d\tilde{n}}{dc}\right)_{c=0} \qquad (3.50)$$

where c is the weight concentration (grams per cubic centimeter). We may substitute for

c in terms of the molecular weight M and Avogadro's number N_A and find for the ratio of scattered to incident intensity Equation (3.51).

$$\frac{i_\theta^\circ}{I_0} = \frac{2\pi}{N_A \lambda^4 r^2} \tilde{n}_0^2 \left(\frac{d\tilde{n}}{dc}\right)_{c=0}^2 (1 + \cos^2 \theta) Mc$$

$$(3.51)$$

Finally, we may define an experimentally observable quantity, the Rayleigh factor according to Equation (3.52):

$$R_\theta^\circ = \frac{r^2 i_\theta^\circ}{I_0} \qquad (3.52)$$

One then obtains Equations (3.53) and (3.54) where K^* depends on experimentally measurable quantities,

$$R_\theta^\circ = K^* (1 + \cos^2 \theta) Mc \qquad (3.53)$$

$$K^* = \frac{2\pi^2 \tilde{n}_0 (dn/dc)_0}{N_A \lambda^4} \qquad (3.54)$$

The quantity $(d\tilde{n}/dc)_0$ must be separately determined by differential refractometry. Having measured K^*, one may now determine the molecular weight of small particles by measuring the light scattered by dilute solutions. Note that values of R_θ° are symmetric about $\theta = 90°$.

If the particles vary in molecular weight, one may assume each species contributes independently and that $d\tilde{n}/dc$ is insensitive to molecular weight, Equation (3.55).

$$R_\theta^\circ = K^* (1 + \cos^2 \theta) \sum_i M_i c_i \qquad (3.55)$$

From the relationship $\overline{M}_w = \sum M_i c_i / c$, one obtains Equation (3.56).

$$R_\theta^\circ = K^* (1 + \cos^2 \theta) \overline{M}_w c \qquad (3.56)$$

Thus, light scattering on polydisperse samples gives the weight-average molecular weight.

We now remove the restrictions imposed by the earlier assumptions, presenting only the results. In doing so, we actually increase the utility of light scattering experiments: consideration of scattering by large particles will provide information on molecular dimensions, while the concentration dependence of scattering will provide a theoretical link to osmometry through the second virial coefficient.

Polymer molecules in solution occupy a large volume. A polystyrene molecule of $M = 10^6$ has a root mean square end-to-end distance of 735 Å in a θ-solvent, and is larger still in a good solvent. For such molecules, when particle dimensions exceed about 5% of the wavelength of light, the induced dipoles at different locations on the particle will be out of phase. The destructive interference in the total observed scattered intensity increases with increasing θ and leads to a decrease in observed intensity. This effect is observable as a loss of symmetry in R_θ with respect to the $\theta = 90°$ axis and can be expressed by a dissymmetry coefficient that is a function of the angle β between the observation and the $90°$ axis, see Equation (3.57).

$$z_\beta = \frac{i_{90° - \beta}}{i_{90° + \beta}} \qquad (3.57)$$

Although z_β can be calculated for various polymer configurations (random coil, rod, etc.), one generally simply extrapolates data to $0°$. Before the advent of laser light sources, this extrapolation was often a problem, as measurements could not be obtained below $30°$ owing to interference from the imperfectly collimated incident beam. Laser light sources permit observation as close as $2°$ to the forward direction, in most cases obviating the need for any extrapolation at all.

If one does measure the dissymmetry coefficient, however, one can directly determine the radius of gyration $(\overline{s^2})^{1/2}$. Assuming some polymer configuration, the theoretical calculations mentioned then furnish the end-to-end distance $(\overline{r^2})^{1/2}$. Unfortunately, determination of $(\overline{r^2})^{1/2}$ in this manner is limited to polymers of a size for which destructive interferences occur, namely $\lambda/10 < (\overline{r^2})^{1/2} < \lambda$. The upper bound is rarely a problem; the lower bound corresponds to molecular weights on the order of 10^5–10^6.

The theory of concentration dependence in light scattering is based on thermodynamic treatments of concentration fluctuations de-

veloped by Smoluchowski and Einstein. For small particles one can write Equation (3.58):

$$R° = \frac{K^* R T v_1 c (1 + \cos^2 \theta)}{(-\partial \mu_1 / \partial c)} \qquad (3.58)$$

where μ_1 and v_1 are the chemical potential and molar volume of the solvent, respectively. Because the osmotic pressure $\pi = (\mu_1 - \mu_1°)/v_1$, one can relate $R_\theta°$ to π, for which one knows the concentration dependence as given in Equation (3.59):

$$\frac{\pi}{c} = \left(\frac{\pi}{c}\right)_0 (1 + \Gamma_2 c + g\Gamma_2^2 c^2) \qquad (3.59)$$

where Γ_2 is the second virial coefficient. Thus, can obtain Equation (3.60).

$$\frac{c}{R_\theta°} = \left(\frac{c}{R_\theta°}\right)_{c=0} (1 + 2\Gamma_2 c + 3g\Gamma_2^2 c^2)$$

$$(3.60)$$

(Note that the coefficient of c is $2\Gamma_2$, not Γ_2 as for osmotic pressure.)

If a low-angle laser system is used and no extrapolation of θ is required, $c/R_\theta°$ is simply measured at several concentrations and extrapolated to zero concentration, with the slope giving the second virial coefficient. Alternatively, if the second virial coefficient is known, measurement may be made at one concentration and the above equation used for determination of $\overline{M}_w$.

The simultaneous extrapolation to both zero concentration and zero angle can be conveniently achieved by means of a Zimm plot (31) as shown in Figure 3.6. One plots K/R_θ against $\sin^2 (\theta/2) + kc$, where k is simply a scaling factor to make $\sin^2 (\theta/2)$ and c of the same order of magnitude. For each angle θ, a set of concentrations is extrapolated to $c = 0$, while for each concentration, extrapolation is to zero angle. The common intercepts equal $1/\overline{M}_w$. The slope of the zero angle line gives the concentration dependence (and Γ_2), while the slope of the zero concentration line gives the angular dependence, and hence $(\overline{s^2})^{1/2}$

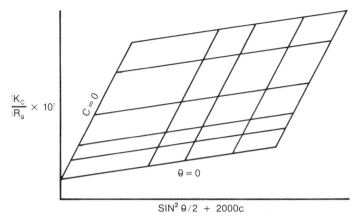

Figure 3.6   Zimm plot of light-scattering data, showing extrapolation to zero angle and zero concentration.

In this context, we should mention a relatively new technique, the use of inelastic ("Doppler Shifted" or Brillouin) light scattering for the investigation of polymer solutions. Brillouin spectroscopy measures light scattered at very slightly shifted frequencies above and below the incident (Rayleigh) frequency. This scattering is analogous to Raman scattering in which frequency differences measure internal molecular vibrations. Here, the frequency shifts are much smaller and correspond to vibrations in the long-range-order of the solute molecules, which can be considered as compression waves moving through the solution. Nordhaus and Kinsinger (32) first looked at this technique for measuring poly-

mer molecular weights, but it currently offers no advantages over conventional light scattering for such an application.

Similarly, the use of laser optics permits the study of the line broadening of the unscattered peak—commonly referred to as quasi-elastic light scattering. Although providing additional information on molecular dynamics, such as a measure of diffusion coefficients, it cannot be considered a practical molecular weight method.

3.3.4.2   Technology

A true revolution in the use of light scattering as a molecular weight method has been the use of laser optics, recently commercialized in

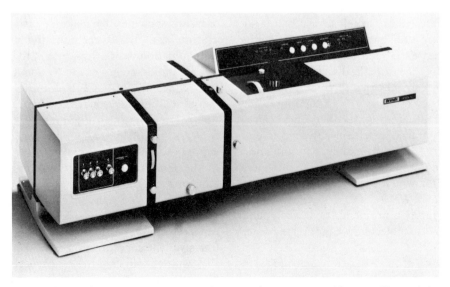

Figure 3.7   Chromatix KMX-6 low-angle laser light scattering photometer. (Courtesy Chromatix.)

the Chromatix KMX-6 low-angle laser light-scattering photometer (undoubtedly to be soon joined by others). This instrument is shown in Figure 3.7. The major advantage is the ability to collect data down to 2° from the forward direction, making extrapolation to zero angle generally unnecessary. Additional advantages are high sensitivity and speed, even to the extent of allowing its use as a detector for GPC. An advantage of great importance to the experimentalist is the use of a flow cell to separate out transient contributions from dust particles, which in a stationary cell must be scrupulously eliminated. All in all, this and similar machines should finally make light scattering a routine analytical tool.

3.3.5 Ultracentrifugation

Although the ultracentrifuge has been in use for the investigation of macromolecules for well over 50 years, it continues to be primarily applied to biological macromolecules. Such molecules, unlike synthetic polymers, are frequently monodisperse and hence give sharp centrifugation boundaries. Moreover, compact molecules such as globular proteins are much less subject to intermolecular interactions, which are evident as concentration-dependent "nonideal" behavior in the ultracentrifuge. Nevertheless, various theories (33, 34) provide mechanisms for dealing with such complications, and recent work (35) on slightly branched dextrans demonstrates the ability of this method to generate MWDs (as well as second virial coefficients) on polydisperse random coil polymers. The ultracentrifuge is unlikely to become a routine molecular weight technique (inspection of Reference 33, 34, 35 will indicate the complexity both of experimental technique and of requisite calculations). However, it can provide—as can light scattering—MWDs for very high molecular weight polymers beyond the exclusion limits of current GPC substrates, as well as for other polymers for which GPC is inappropriate.

3.3.5.1 Principles

The two primary modes of ultracentrifuge analysis are sedimentation velocity, in which

the rate of sedimentation at very high speeds is measured, and sedimentation equilibrium, in which sedimentation at lower speeds is balanced by molecular diffusion and a steady state is obtained. In sedimentation equilibrium the sample is centrifuged at speeds of, typically, 4000–40,000 rpm for periods of one day to a week or more. Concentration is measured (Equation 3.61) as a function of the distance from the bottom of the cell. When equilibrium is attained, the centrifugal force on the molecule is just balanced by its tendency to diffuse in solution. For an ideal solution, one can write Equation (3.61):

$$M_w = \frac{2RT \ln (c_2/c_1)}{(1 - \bar{v}\rho)\omega^2 (r_2^2 - r_1^2)} \quad (3.61)$$

where c_1 and c_2 are the polymer concentrations at distances r_2 and r_1 from the axis of rotation, ω is the angular velocity, v the partial specific volume of the polymer, and ρ the solution density. Moreover, higher molecular weight averages, such as M_z and M_{z+1}, can be calculated by making detailed measurements of concentration as a function of r.

In practice, it is difficult to obtain solutions dilute enough to behave ideally yet concentrated enough to be detected by the usual refractive index measuring detectors. Thus measurements must be made at several concentrations and speeds to permit extrapolation back to ideal conditions (35).

Operation at θ conditions may eliminate many of these problems, as may the use of modern spectrophotometric scanners that permit use of dilute solutions of suitably absorbing polymers.

A similar method is the "approach to equilibrium" method first developed by Archibald (36), which eliminates the long equilibration time by examining concentration at the meniscus and bottom of the cell. A mathematical treatment of the boundary conditions at these two locations permits evaluation of $\bar{M}_w$ and the second virial coefficient at any time.

Sedimentation velocity measures the ratio of sedimentation with a sedimentation constant s defined in Equation (3.62)

$$s = \frac{1}{\omega^2 r} \frac{dr}{dt} = \frac{m(1 - \bar{v}\rho)}{f} \quad (3.62)$$

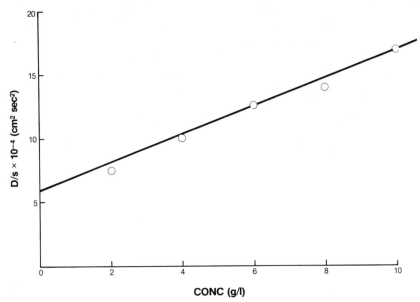

Figure 3.8 Determination of molecular weight from ultracentrifugation data; extrapolation of D/s to zero concentration.

where f is the frictional coefficient for a polymer molecule moving through the solution. This method is difficult to apply to synthetic polydisperse polymers, which do not give a sharp moving boundary. Moreover, for random coil polymers the frictional coefficient is highly concentration dependent. At infinite dilution, however, f can be related to the diffusion coefficient D, measured independently,

$$D = \frac{RT}{f} \qquad (3.63)$$

thus making it possible to extrapolate

$$\frac{D}{s} = \frac{RT}{M(1 - \bar{v}\rho)} \qquad (3.64)$$

to zero concentration (Figure 3.8); see Equations (3.63) and (3.64). The value of D may also be determined by ultracentrifugation at very low speeds where diffusion dominates centrifugation, but the method still appears to be less useful than the equilibrium approach.

3.3.5.2 Technology
Several commercial ultracentrifuges are available (see for example Figure 3.9). The key

component of the machine is a well-thermostatted rotor spun in a vacuum at high speed. The rotor holds one or more solution cells with windows permitting detection by one of several optical systems. Two common detectors measure concentration from the refractive index difference between polymer solution and solvent: Rayleigh optics generate a pattern of interference fringes whose spacing is proportional to refractive index (RI) differences, while Schlieren optics yield a curve that is the derivative of RI with respect to position in the cell. Data from both systems are obtained as photographs which can be scanned and digitized automatically by appropriate equipment. For polymers with suitable ultraviolet (UV) absorption, a spectrophotometric scanner can provide direct concentration data.

3.4 VISCOSITY

Viscosity is one molecular weight method for which practice has consistently outdistanced theory. Since Staudinger (37) first pointed out the dependence of viscosity upon molecular weight, polymer scientists have made increasing use of empirical rules relating the two. While theoreticians have not neglected the problem of dilute solution viscosities, they

Figure 3.9 Beckman Model E Analytical Untracentrifuge. (Courtesy Beckman Instruments.)

have frequently been in the position of rationalizing already determined relationships. Nevertheless, an understanding of the basic principles of the viscosity behavior of dilute polymer solutions can aid immeasurably in the proper design and interpretation of a viscosity-molecular weight experiment. Although similarly rigorous theoretical interpretations of the viscosities of concentrated polymer solutions or molten polymers do not exist, empirical relationships enable such experiments also to be used as tools for molecular weight determination.

Viscosity can be defined (38) as "a measure

of the frictional resistance that a fluid in motion offers to an applied shearing force." (Some commonly used terms for viscosity are listed in Table 3.2). For a liquid flowing through a capillary, this frictional drag is greatest at the walls and is passed on, in decreasing amount, by each concentric layer of fluid to the faster flowing molecules closer to the center. The net effect of this, for a polymer solution flowing through a capillary viscometer, is an increased flow time. For a sufficiently dilute solution, η_{rel} is simply the ratio of polymer-to-solvent flow time.

The key quantity for molecular weight

Table 3.2 Definitions of Viscosity Terms^a

	Common Name	IUPAC Name
η	Viscosity	Viscosity coefficient
$\eta_{rel} = \eta / \eta_0 \approx t / t_0$	Relative viscosity	Viscosity ratio
$\eta_{sp} = \eta_{rel} - 1$	Specific viscosity	None
η_{sp} / c	Reduced viscosity	Viscosity number
$[\eta] = \lim\limits_{c \to 0} \eta_{sp} / c$	Intrinsic viscosity	Limiting viscosity number
$\quad = \lim\limits_{c \to 0} (\ln \eta_{rel} / c)$		

$^a \eta_0$ = viscosity of solvent; c = polymer concentration (usually grams per deciliter); t = flow time through a capillary viscometer; t_0 = solvent flow time.

determination is the intrinsic viscosity $[\eta]$, which one may think of as the fractional increase in viscosity caused by the addition of a unit weight of polymer under conditions in which polymer molecules do not interact. This quantity is obtained by extrapolation of measured viscosities to zero concentration (and, if necessary, zero shear).

3.4.1 Principles

3.4.1.1 Viscosity and Molecular Dimensions

The dependence of inherent viscosity on molecular size has been discussed in Chapter (2). The important results for random coil polymers are reviewed here.

Kirkwood and Riseman (39) and Flory (40) found that the viscosity behavior of polymers could best be explained by assuming that a random coil polymer molecule carried along "trapped" solvent molecules as it moved through the bulk solvent. Thus each molecule behaves as an "equivalent hydrodynamic sphere" with a radius proportional to the root mean square end-to-end distance of the coil. By making some severely simplifying assumptions, which nevertheless appear to agree with experiments, instrinsic viscosity can be shown to be proportional to the volume of this sphere divided by the molecular weight; see Equation (3.65).

$$[\eta] = \Phi \frac{(\overline{r^2})^{3/2}}{M} \quad (3.65)$$

Theory provides a value for Φ of 3.6×10^{21} (with $[\eta]$ in deciliters per gram) but experiments indicate a value of approximately 2.5×10^{21}. Flory state that Φ was constant within experimental error for all polymers of sufficiently high molecular weight. One may write Equation (3.65) in terms of unperturbed dimensions $(\overline{r_0^2})^{1/2}$, and the expansion factor α, where $(\overline{r^2})^{1/2} = \alpha(\overline{r_0^2})^{1/2}$ and $(\overline{r_0^2})^{1/2} \propto M$, thereby obtaining Equation (3.66).

$$[\eta] = \Phi\alpha^3 (\overline{r_0^2})^{3/2} M^{1/2} = K\alpha^3 M^{1/2} \quad (3.66)$$

At θ conditions the molecule has its unper-

turbed dimensions $\alpha = 1$, and Equation (3.66) reduces to Equation (3.67).

$$[\eta]_\theta = KM^{1/2} \quad (3.67)$$

This result, at least, is unquestionably confirmed by the data.

In good solvents, Flory's treatment gives Equation (3.68).

$$\alpha^5 - \alpha^3 \propto M^{1/2} \quad (3.68)$$

As an upper limit, where $\alpha^5 \gg \alpha^3$, one would expect $\alpha^5 \propto M^{1/2}$ or $\alpha \propto M^{0.1}$. Substituting, one finds, for a highly expanded linear polymer, Equation (3.69).

$$[\eta] = K(M^{0.1})^3 M^{0.5} \quad (3.69)$$

$$= KM^{0.8}$$

Indeed, 0.8 is the largest value commonly observed experimentally as the exponent of M for normal random coil polymers. Unfortunately, the theory does less well for predicting intermediate values of a and is virtually useless for predicting K, even at θ conditions.

3.4.1.2 Empirical Relations

In spite of the difficulty of establishing a priori the dependence of dilute solution viscosity on molecular weight (or molecular size), practical use of viscosity depends on a simple empirical relationship. If the intrinsic viscosities of narrow molecular weight fractions of a linear homopolymer are plotted against their molecular weights in a log-log plot, a straight line generally results above a certain minimum molecular weight. This line is represented by Equation (3.70) known as the Mark-Houwink-Sakurada equation.

$$[\eta] = KM^a \quad (3.70)$$

Extensive compilations of K and a for various solvent-polymer pairs are available in the literature (41). For linear polymers, a reaches a lower limit of 0.50 in θ solvents and rarely exceeds 0.80, except for rodlike polymers or for polyelectrolytes at low ionic strength.

For polymers with published values of K and a, calculation of average molecular weights from instrinsic viscosities measured in appropriate solvents is a simple matter. In general, the application of literature K and a values is best confined to the molecular weight range over which they were determined. Such values are usually measured over a molecular weight range of at least two orders of magnitude, thus making them useful for a large number of samples.

In θ solvents where $a = 0.5$, theory and experiments both justify using measured K values at higher and lower molecular weights. In other solvents, however, one should not in general apply published K and a values at molecular weights much lower than those in the original experiment, as curvature of the $M - [\eta]$ relationship usually occurs at sufficiently low M.

Frequently, independent determinations of K and a for a given polymer-solvent combination will seem to differ greatly. This is a consequence of the two adjustable parameters in the Mark-Houwink-Sakurada equation, and normal experimental error, especially when data are obtained in a narrow molecular weight range. One often finds in such cases that the resulting $[\eta]$-molecular weight relationships give similar results over their common molecular weight range. Thus an apparent conflict between K and a values from different sources may prove to be inconsequential.

For a heterogeneous polymer sample, one can only measure a single intrinsic viscosity and hence only an average molecular weight. Consider a polymer made up of species of molecular weight M_i, at concentration c_i, and with specific viscosities $(\eta_{sp})_i$. In the limit of noninteracting polymer molecules, each molecule contributes individually to the viscosity, and one can write Equation (3.71).

$$\eta_{sp} = \sum_i (\eta_{sp})_i \qquad (3.71)$$

If all species are within the molecular weight range corresponding to common K and a values, one can write Equations (3.72)–(3.74):

$$(\eta_{sp})_i = [\eta]_i c_i = K M_i^a c_i \qquad (\lim c_i \to 0)$$

$$(3.72)$$

$$\eta_{sp} = K \Sigma M_i^a c_i \qquad (\lim c_i \to 0) \qquad (3.73)$$

$$[\eta] = \frac{\eta_{sp}}{c} = K \Sigma M_i^a c_i / c \qquad (3.74)$$

where c_i/c is the weight fraction w_i of species i. If one then wishes to identify the viscosity average molecular weight M_v with the molecular weight calculated from the Mark-Houwink-Sakurada equation, that is,

$$[\eta] = K \overline{M}_v^a$$

One must define $\overline{M}_v$ as Equation (3.75).

$$\overline{M}_v = [\Sigma w_i M_i^a]^{1/a} \qquad (3.75)$$

$$= \left[\frac{\Sigma N_i M_i^{1+a}}{\Sigma N_i M_i} \right]^{1/a}$$

M_v differs from other molecular weight averages in several important respects. It depends on the choice of an empirical value a (except in θ solvents where a equals 0.5) and its value is thus not fixed absolutely. For linear polymers with a typical range of $a = 0.5$ to 0.8, however, $\overline{M}_v$ is rather insensitive to a. Thus, for the "most probable" distribution, $\overline{M}_v/\overline{M}_n$ can vary only from 1.77 to 1.91 (42).

Nevertheless, the inability to determine $\overline{M}_v$ except by way of viscosity measurements makes it difficult to establish a viscosity molecular weight relationship for a new polymer. Unless one has narrow polymer fractions (say $\overline{M}_w/\overline{M}_n < 1.1$) for which $\overline{M}_n \approx \overline{M}_v \approx \overline{M}_w$, one must plot $[\eta]$ versus some experimentally determined molecular weight average instead of the necessary $\overline{M}_v$. Although $\overline{M}_w$ is in general a better approximation of $\overline{M}_v$, $\overline{M}_n$ may be more easily determined. If the fractions used have approximately equal ratios of $\overline{M}_v$ to $\overline{M}_n$ (or $\overline{M}_w$), then a will be determined correctly though K will be in error by the factor $\overline{M}_n/\overline{M}_v$. For fractions of irregular breadth, K and a may both be considerably in error, or indeed a linear $[\eta]$-molecular weight relationship may not be obtained. One solution is to employ θ conditions, for which a is known. $\overline{M}_v$ can then be calculated from appropriate GPC data and K then obtained from the measured intrinsic viscosities.

Viscosity as an Independent Method. Establishment of an empirical $[\eta]$-molecular weight relationship for a new polymer requires a carefully fractionated polymer and a means of determining the true molecular weight of those fractions. Thus viscosity measurements on a polymer in the absence of other information reveal little about its molecular weight. If one measures intrinsic viscosity at θ conditions, then $a = 0.5$ and relative molecular weights ($\overline{M}_v$) can be established for different samples. Theory suggests that K depends only on repeat unit molecular weight and unperturbed dimensions $(\overline{r^2})$ which, as a first approximation, ought to be calculable from bond lengths and angles; see Equation (3.76).

$$K = \Phi \frac{(\overline{r_0^2})^{3/2}}{M} \qquad (3.76)$$

However, Flory presents calculations of $(r_0/M)^{1/2}$ from known values of K that are in error by a factor of 1.5–2.5. Back calculation shows that values of K calculated from true bond lengths would be in error by an order of magnitude. Clearly, a priori calculation of K, even at θ conditions, is not satisfactory for establishing viscosity as an independent molecular weight method, except in a qualitative way.

Concentration Dependence. The instrinsic viscosity is obtained by extrapolating either

η_{sp}/c or $\ln \eta_r/c$ to zero concentration as shown in Figure 3.10. Because the plot of $\ln \eta_r/c$ versus c is generally flatter, this extrapolation is sometimes recommended. However, since it can be shown mathematically that the two slopes in Figure 3.10 must differ by 0.5, the best procedure is to extrapolate both sets of points to a common intercept, with slopes differing by 0.5.

The plot of η_{sp}/c versus c, which is usually linear for $\eta_{sp} < 1$, can be described by the Huggins equation; see Equation (3.77):

$$\frac{\eta_{sp}}{c} = [\eta] + k'[\eta]^2 c \qquad (3.77)$$

where k', the Huggins constant, is usually between 0.3 and 0.4. Substantially larger values of k' for nonelectrolytes are generally a sign of aggregation. The corresponding relationship for η_r is given by Equation (3.78):

$$\frac{\ln \eta_r}{c} = [\eta] + k''[\eta]^2 c \qquad (3.78)$$

where the Kraemer constant k'' equals $k' - 0.5$ as mentioned.

Single-Point Methods. It is tempting to try to avoid measuring the viscosity of a polymer at several concentrations by assuming linearity of the concentration behavior. The simplest method is to assume a value for the Huggins constant based on values for a similar polymer and solve for $[\eta]$. This is not a bad procedure, for different fractions of the same polymer, since k' is reported to vary little with molec-

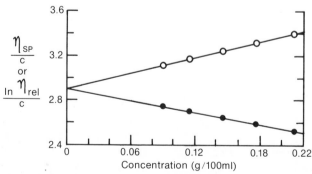

Figure 3.10 Extrapolation of reduced viscosity to infinite dilution. The ordinate intercept is $[\eta]$.

ular weight. However, the method ignores the possibility of curvature in the concentration dependence. A more elegant approach (43) is to eliminate k' and k'' in Equation (3.77) and (3.78) to obtain Equation (3.79):

$$[\eta] = \frac{\sqrt{2}\,(\eta_{sp} - \ln \eta_r)^{1/2}}{c} \qquad (3.79)$$

but it suffers similarly if the concentration dependence is not linear.

Since Ubbelohde viscometers, (discussed in the next section), make serial dilution extremely simple, such single-point methods cannot in general be recommended.

3.4.2 Technology

3.4.2.1 Capillary Viscometers

The basic principle of capillary viscometers, several of which are illustrated in Figure 3.11 (44), is the relationship of viscosity to flow time shown as Equation (3.80):

$$\eta = \alpha'\rho\left(t - \left(\frac{\beta}{\alpha}\right)t\right) \qquad (3.80)$$

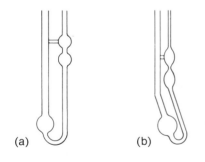

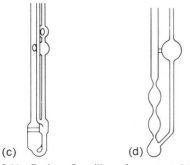

Figure 3.11 Design of capillary viscometers: (a) Ostwald; (b) Cannon-Fenske; (c) Ubbelohde; (d) variable shear rate viscometer.

ρ is the density and α' and β are constants of the instrument, which may be evaluated by measuring flow times of solvents of known viscosity. However, for dilute solutions with flow time exceeding 100 sec., the second term is negligible, and one can utilize Equation (3.81).

$$\eta_r \equiv \frac{\eta}{\eta_0} \equiv \frac{t}{t_0} \qquad (3.81)$$

As the values of η must be extrapolated to zero concentration, viscometric measurements should be made at three or four concentrations. The standard Ostwald viscometer requires constant volumes of solution and must be refilled after each dilution. This is avoided in the Ubbelohde viscometer, which allows serial dilution in its reservoir and yet maintains a constant pressure head. Automatic viscometers make use of such construction to automatically dilute the sample and record flow times.

To ensure that the extrapolation to zero concentration is as nearly linear as possible, the highest value of η_r should be less than 1.5, preferably 1.3, and the lowest about 1.1. If one has an estimate of the expected $[\eta]$, the appropriate initial concentration can be solved from the Huggins equation, assuming a typical Huggins constant of 0.4; see Equation (3.82).

$$c = \frac{0.27}{[\eta]} \qquad (3.82)$$

To ensure sufficient accuracy at such low values of η_r, the temperature must be controlled to $\pm 0.01°$ C (at most $0.02°$ C) by immersion in a constant temperature bath.

3.4.2.2 Extrapolation to Zero Shear

For high molecular weight polymers, or for very extended polymers, viscosity depends on the rate of shear and must be extrapolated to zero shear as well as to zero concentration. Although random coil polymers do not generally exhibit such behavior at intrinsic viscosities less than 3 dl/g, it is appropriate to check for it whenever $[\eta] > 2$ dl/g. There is no agreement on the proper means of extrapolation, but several references all provide reasonable approaches (45–48). The first step, of course, is to determine whether $[\eta]$ indeed varies with shear, and the modified Ubbelohde viscometer provides a simple means for measuring the same sample at different shear rates by varying the pressure head.

Another solution in the case of very high molecular weight or molecularly rigid polymers is to employ a rotating viscometer. Although not normally used to measure dilute solution viscosities, such an instrument provides much greater control over shear rate.

3.5 FRACTIONATION

Both the methods (49, 50) and the principles (51) of fractionation are well established and documented. Prior to the application of GPC, fractionation was the only method for obtaining complete MWDs. Tedious at best, and superseded by GPC for all routine measurements, it still has some important applications in the general field of molecular weight determination.

Basically, polymer fractionation depends on the very slight decrease in solubility with increasing molecular weight for series of high polymers (52) in poor solvents. The most common method is precipitation from very dilute solution in a mixture of solvent and nonsolvent. By gradually increasing the fraction of nonsolvent, a series of fractions ranging from high to low molecular weight can be generated. Similar results can be obtained by maintaining a constant solvent system and decreasing the temperature. Column fractionation (53) essentially reverses this procedure by precipitating the polymer onto a solid substrate and eluting with progressively better solvent to yield fractions in ascending molecular weight order. Once the fractions are obtained, their molecular weight can be determined by any convenent method. The narrowness of their distribution ensures that all molecular weight averages will be approximately equal and thus makes the choice of molecular weight method inconsequential. From the actual masses of the fractions, a cumulative weight distribution can be constructed that can be converted to the weight distribution W_x by graphical or numerical differentiation. The amount of work involved in the procedure precludes its use as a routine tool, especially as a substitute for GPC. However, in cases where data from GPC are suspect, for example, where adsorption may occur, fractionation may be the only certain method of establishing the molecular weight distribution for linear polymers. Recently column fractionation, using a temperature rather than a solvent gradient, has been applied to the determination of a short chaining branching in polyethylene (54). As discussed in Section 3.7, GPC of branched polymers may be subject to considerable ambiguity.

Fractionation also remains an important method of generating standards for GPC or viscometry. When a particular polymer composition is the subject of detailed or longrange study, the labor involved in preparing and characterizing fractions may be justified by the consequent improvement in the accuracy of GPC and viscosity molecular weights.

3.6 GEL PERMEATION CHROMATOGRAPHY

It has been seen previously that the experimental determination of the entire molecular weight distribution requires the separation of the sample into fractions that may then be weighed and characterized. Despite major improvements in the design of fractionation equipment, this approach is too laborious for routine application. On the other hand, a device that simultaneously performs the separation and characterization steps would be an invaluable analytical tool. This capability is the goal of GPC technology. When it is achieved on a routine basis, GPC will have superseded all other molecular weight methods.

Flodin (55) is commonly credited with the discovery of liquid exclusion chromatography, which he called gel filtration. He found that the elution of proteins through cross-linked polysaccharide gels was governed by their molecular weights. Chromatography with such

Sephadex® or Agarose® gels, or with synthetic polyacrylamide gels, has remained largely a tool of protein chemistry. The requirements of synthetic polymer chemistry are better met by rigid or semirigid substrates that are organic-compatible and that lend themselves to operation in modern high-pressure-liquid chromatography systems. The two techniques of gel filtration and gel permeation chromatography (GPC) have thus developed different applications and literatures, sometimes obscuring their fundamental identity. The nomenclature of liquid exclusion chromatography has been further confused by the frequent introduction of superfluous acronyms. In our opinion, common usage justifies defining by "GPC" any liquid column chromatography that operates by way of molecular size exclusion, regardless of the nature of substrate or mobile phase.

A review of the burgeoning literature dealing with GPC is beyond our scope and purpose. While we attempt to describe recent technological advances, the principle theme is a discussion of the factors that affect the interpretation of GPC data in terms of molecular weight distributions. The relationship between GPC chromatograms and MWDs is far from trivial and obvious and depends on a number of assumptions. The major ones are: that the nature of the separation mechanism is consistent and nonspecific, that the correlation between molecular weight and retention volume may be unambiguously established, and that one may either neglect or make corrections for chromatographic band-spreading. Given the validity of these assumptions, we must still question the congruence of the "raw" chromatogram and our concepts of a graphical depiction of the MWD.

3.6.1 Principles

Separation in GPC takes place between two well-defined eluant volumes as shown by the calibration curve of Figure 3.12. The exclusion volume or void volume, V_0, can be determined from the elution volumes V_e of very large polymers such as the highest molecular weight commercial polystyrene standards, "blue dextran," or viruses. Such polymers cannot enter the pores of the GPC packing and are confined to the solvent outside of the beads. Hence, V_0 may also be considered as the interstitial volume or mobile phase volume. Conversely, species with molecular dimensions similar to the solvent's have, in principle, free access to all pores and elute at the total volume V_t. The total pore volume is then given by $V_p = V_t - V_o$. This volume is also sometimes denoted as V_i (internal volume) or—by analogy with liquid chromatography—the "mobile phase volume." Molecular species of intermediate size have limited access to the internal volume of the pores. Discussion of the mechanism of their exclusion can be postponed, but

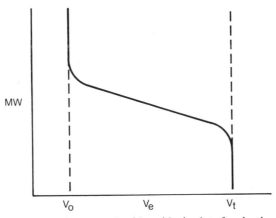

Figure 3.12 Typical GPC calibration curve: Semi-logarithmic plot of molecular weight versus elution volume. V_o and V_t corresponds to complete exclusion and complete permeation, respectively.

one can now state that the average concentration of these species within the pores during chromatography will be less than in the interstitial volume. This fractional occupancy of the pore volume can be expressed in the form of a partition coefficient as given in Equation (3.83):

$$K_{\text{GPC}} = \frac{V_e - V_0}{V_t - V_0} \qquad (3.83)$$

or

$$V_e = V_0 + K_{\text{GPC}} V_p$$

Since V_e falls between V_0 and V_t, the values of K_{GPC} lie between 0 and 1.

All models for GPC are consistent with Equation (3.83), although they differ considerably in the description of the exclusion process. A satisfactory theoretical treatment would be expected to allow prediction of K_{GPC} from the measurable properties of the polymer solution combined with a knowledge of substrate pore geometry and distribution. Furthermore, a fully adequate model would have to explain the influence on K_{GPC} of operating variables such as flow rate and polymer concentration. That no single, accepted treatment exists is partly a reflection of experimental difficulties. For example, the measurement of pore size and its distribution for siliceous packings is complicated by uncertainties regarding pore geometry; while for organic supports no direct quantitative porosimetric techniques exist. However, the effect of pore geometry can be eliminated by comparing the behavior of various polymers on a single set of columns, and a wealth of data of this sort is available. The central observation unifying such data is the universal calibration concept, and consistency with this principle is an important criterion for theoretical treatments.

3.6.1.1 Universal Calibration

In practice, molecular weights obtained by GPC are usually reported as apparent values relative to some standard, typically polystyrene. However, it has long been recognized that the dependence of V_e on M must be sensitive to polymer composition and structure. Initial attempts to take polymer structure into account focused on the relationship between contour length and molecular mass (i.e. "Q factor"), and failed to recognize the role of solvent affinity in determining molecular size.

A good example of sensitivity of molecular size to structure and composition should be branched polymers. Benoit and co-workers (57, 58) studied a wide variety of narrow-distribution linear and nonlinear vinyl polymers. The molecular weights of the comb, star, and graft polymers were two to three times greater than those of linear polymers eluting at the same V_e, but the values of $[\eta]\,\overline{M}_w$ were always coincident (Figure 3.13). Thus a plot of $\ln\,[\eta]\,\overline{M}_w$ versus V_e appeared to be "universal" for a given set of columns, regardless of polymer composition and structure.

This foregoing observation was rationalized partly on the grounds of the Flory-Fox equation, which states that, for linear polymers, the product $[\eta]\,\overline{M}_w$ is proportional to the cube of the mean square end-to-end distance $(\overline{r^2})^{1/2}$ (or the cube of the radius of gyration R_G) as given in Equation (3.84):

$$[\eta]\,\overline{M}_w = \phi(\overline{r^2})^{3/2} \qquad (3.84)$$

Combination of the Flory-Fox equation with the relationship of Zimm and Kilb (109), $(R_G)_{\text{branched}}/(R_G)_{\text{linear}} = g^{1/2}$, shows that $[\eta]\,M$ for branched polymers also increases linearly with the cube of the dimensions. The product $[\eta]\,M$ carries a measure of intuitive appeal for many because it appears in the Einstein equation for the viscosity of impenetrable spheres, Equation (3.85):

$$[\eta]\,M = \frac{5}{2}\,N_{AV}\nu \qquad (3.85)$$

where ν is the volume in cubic centimeters of a single sphere. Equating ν with $(4/3)\,\pi\,R_{\text{eff}}^3$ thus defines an equivalent hydrodynamic radius for a hypothetical sphere that would contribute the same viscosity to the solution as the polymer in question.

The universal calibration concept has had a major impact because it appears to provide a

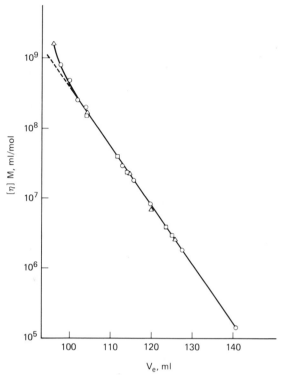

Figure 3.13 Universal calibration curve. Linear polystyrene (O); comb-branched polystyrene (Δ); star-branched polystyrene (□). (57).

route to calibrating GPC for polymers of interest without preparing characterized fractions (59). For a narrow-distribution unknown polymer, a GPC analysis and a single viscosity determination would suffice: if the dependence of $J \equiv [\eta] M$ on V_e had been determined from polystyrene data, the molecular weight of the sample could be obtained as $J/[\eta]$. For broad distribution samples, a number of computer programs have been written to fit GPC data to measured bulk sample viscosities using known Mark-Houwink values of K and a. When these parameters are unknown an iterative procedure can be employed, but the resulting MWDs must be considered less definitive. It would be difficult to assess the degree to which calculated MWDs are made more accurate by the application of universal calibration when K and a are uncertain or viscosity data are limited. Despite any such practical objections, the concept of universal calibration seems to be empirically correct for narrow-distribution polymers and is taken into account in all current GPC theories.

3.6.1.2 Theoretical Models of the Separation Process

The mechanism of GPC separation has been of considerable theoretical interest. Treatments of this phenomenon rest on models developed for gas chromatography, as well as on the statistical theory of polymer chain dimensions. Without discussing the mathematical details of the various GPC theories, we can compare the respective physical models and the extent to which the theoretical results are confirmed by experiment.

Steric Exclusion. According to the most simplified concept of the exclusion process, each polymer molecule sees two kinds of pores: those too small for it to enter and those large enough to accommodate it. In this model—ascribed to Flodin (55) and Porath (60)—the concentration of solute within each pore is either zero or that of the mobile phase. K_{GPC} is then simply determined by the fraction of pores larger than the solute. This model neglects the statistical nature of polmer dimensions and also implies, incorrectly, that the calibration curve is a direct reflection of the pore size distribution. Haller (61), in fact, found that all of the pores of a narrow pore size distribution glass packing were accessible to a virus molecule that eluted at $K_{GPC} = 0.5$.

Equilibrium Theories. The perception of GPC separation as an equilibrium partitioning of solute between pore and mobile phase is supported by the frequently observed agreement between K_{GPC}, the chromotographic "partition constant," and the solute partition constant determined by static measurements. Also, the sensitivity of K_{GPC} to flow rate is usually small, especially for porous glass, which seems to rule out a separation based on flow. Certainly, an equilibrium model must apply in the limit of low flow rate (62) where the effect of transport processes is limited to band-spreading. Last, it may be argued that any gross nonequilibrium between "trapped" and "interstitial" solute would be incompatible with the high resolution typically observed.

Theoretical treatments of the equilibrium GPC partition constant K were put forth simultaneously by Giddings (63) and Casassa (64, 65). Giddings developed a statistical mechanical treatment for the equilibrium distribution of spherical and rodlike macromolecules between pores and bulk liquids. The pore network was characterized simply by the surface area per unit of pore volume, since the fraction of polymer chain configurations that are forbidden turns out to depend solely on this ratio. Pursuing this point of view—that of exclusion as a geometrical surface effect—a system of random pores was modeled as an array of randomly intersecting planar surfaces. Calculated values of K were then obtained as a function of several solute dimension parameters. The dimensional parameter that correlated with K in a manner least sensitive to molecular geometry was L, the mean projection length of the polymer chain on the network surface.

Casassa, meanwhile, focused his efforts on the calculation of K for flexible chains, using as a model a stationary phase with pores of defined geometry. Thus, whereas Giddings'

depiction might be prompted by the chromatography of proteins and nucleic acids on Sephadex® Casassa's model appears more nearly relevant to synthetic polymers on porous glass. Both theories explicitly recognize the decrease in effective solute concentration in the pores relative to the bulk solution, arising from the requirement that the centers of mass maintain a distance from the pore surface on the order of one molecular radius or more. In Casassa's treatment, the fraction of solute-occupied space in a pore is calculated from the probability of each point within the pore being able to serve as the origin of random flights that do not intersect the wall. Attempts to calculate K_{GPC} from the ratio of polymer dimensions to pore size on the basis of this theory have, however, ultimately proven unsuccessful (66). This perhaps reflects the difficulty of interpreting porosimetry data in terms of an idealized pore model.

A more encouraging consequence of this theory was the finding of a single dimensionless variable, directly proportional to $[\eta] M$, that governed the value of K for both branched and linear chains (67). One source of experimental difficulty in relating porosimetry data to effective pore radii is presumably the "ink bottle" effect, which leads to a depression of porosimetry values for cavities with constricted inlets. It is interesting to note that Van Kreveld (68) calculated "effective pore radii" based on measured total surface areas and pore volumes by using a model of randomly sintered microspheres for the bead structure. Values of K were calculated from these pore dimensions and the polymers' radii of gyration, and were found to be in good agreement with measured K_{GPC}.

Flow and Diffusion Theories. Capillary tubes exhibit velocity profiles, with flow most rapid at the center. On the grounds of the surface exclusion effects discussed before, large particles travel preferentially away from the walls and thus may exhibit more rapid flow, as in fact was observed for proteins on columns packed with solid glass spheres (69). Guttman and Dimarzio (70) developed a model based on mass flow through channels or tubes of large and small dimensions, the former corresponding to the interstitial volume. While this treatment leads to an expression for V_e of the correct form, the model suffers from two major defects: the effective channel dimensions are not readily compared to porosimetry data, and the successful GPC separation at very low flow is not accounted for.

Given the probably tortuosity of the pore structure in organic gels, it seems apparent that the dependence of diffusion coefficients on molecular weight should contribute to the separation process, at least for Styragel® type columns at normal flow rates. On the other hand, any treatment based on diffusion-controlled permeation must approach an equilibrium model at low flow rate. The treatment of Yau and Malone (71), in which separation was viewed as a lateral diffusion from mobile phase into gel, failed to do this, in fact, K_{GPC}—being limited only by the time dependence of diffusion and flow—would become equal to one in the limit of zero flow. In order to bridge this gap, Yau subsequently proposed (72) that K_{GPC} could be formally resolved into steric exclusion and restricted diffusion terms. Data recently obtained on porous glass by Giddings (73) reveal very little effect of diffusion coefficients per se, in contradiction of restricted diffusion models.

As Casassa has pointed out (67), the influence of nonequilibrium processes on V_e is probably small under normal conditions, but peak breadth and shape should be quite sensitive to diffusion and flow effects. In this regard Ouano (74) investigated the effects on band-spreading of bead and pore size, flow rate, and diffusion coefficients, and obtained computer-generated chromatograms in good agreement with experimental results.

Concluding Remarks.   Considerable progress has been made in GPC theory, but many questions are unanswered. Equilibrium treatments of GPC have been successful in approximating the dependence of the K_{GPC} on the proportionality of molecular dimensions to pore size, and in rationalizing the universal calibration principle. A more detailed analysis is obstructed by uncertainties in the character-

ization of pore structures (66). The relative contributions of equilibrium and flow effects to K_{GPC} appear to be related to pore geometry (75), but this dependence is not clearly understood. Also lacking is a rigorous treatment of the relationship between the pore size distribution and the calibration curve, although some empirical correlations have been established in the case of controlled pore glass (76). Last, a satisfactory explanation of the seemingly irregular dependence of peak dispersion on K_{GPC} has not been produced.

3.6.1.3   Relation of Chromatogram Shape to MWD

Most computational programs for GPC analysis convert the chromatogram to a cumulative or integral MWD (fraction of mass W above M_x versus M_x) and then generate the differential MWD as $dW/d \log M$ versus $\log M$. Such calculations are correct within the limits of band-spreading, as are determinations of $\bar{M}_w$ and $\bar{M}_n$ in which the chromatogram is divided into "fractions" of known mass and molecular weight. On the other hand, the shape of the chromatogram may differ drastically from the shape of the differential MWD, depending on the form of the calibration curve. This is because the mass contained within each volume ΔV element corresponds to the molecular weight range d ($\Delta \log M$), where d, as in Equation (3.86), is the reciprocal slope of the calibration curve.

$$\log M = A - \left(\frac{1}{d}\right) V_e \qquad (3.86)$$

$$V'_e - V_e = - d \left(\log M' - \log M\right)$$

Shultz (5) pointed out this distinction between the chromatogram and usual descriptions of the MWD and concluded that the GPC curve could be interpreted as a plot of $M_i \cdot W_i$ versus $\log M_i$ if the slope of the calibration curve were constant. Typically, however, the calibration curve is not perfectly linear, and this curvature can profoundly affect the appearance of the chromatogram, as shown by Yau and Fleming (77). As noted, the concentration of the sample eluting in a given volume increment ΔV increases with the slope of the calibration curve, corresponding to an increase in the range of $\Delta \log M$. Thus conversion of the chromatogram height to $dW/d(\log M)$—the usual ordinate of a distribution plot—is accomplished by multiplying the former by the reciprocal slope of the calibration curve as follows:

$$\frac{dW}{d(\log M)} = hd = \left(\frac{dW}{dV}\right)\left[\frac{dV}{d(\log M)}\right]$$

(3.87)

Yau and Fleming provide a simple graphical method for this conversion. This process is not simply equivalent to converting the abscissa of the chromatogram from V_e to $\log M$ since the ordinate is altered as well. Consequently, the relationship between the true distribution—$dW/d(\log M)$ versus $\log M$—and the chromatogram shape is extremely sensitive to seemingly modest changes in the slope of the calibration curve, particularly for bimodal distributions.

3.6.2 Technology

GPC technology appears to be advancing rapidly. Much of the intense current commercial activity is, however, a reflection of the high pressure liquid chromatography (HPLC) "explosion": a substantial portion of the dozen-or-so major HPLC suppliers find it advantageous to offer—through rather minor alterations in column selection and data-processing peripherals—GPC instruments as well. These developments are in fact, for the most part, tangential to the interests of GPC practitioners for whom good column performance is paramount. Thus in an extensive "materials" section, focus is on the heart of GPC devices, the column substrates, with brief attention to the choice of solvents and molecular weight calibration standards. While most GPC users do not prepare their own columns, a description of column packing procedures is offered under "methodology" inasmuch as it may also prove

useful in the selection of commercial columns. In view of the wealth of manufacturers' literature, the description of hardware is brief.

3.6.2.1 Materials

Substrates. GPC stationary phases are porous solids prepared from a wide variety of materials through a number of different processes. Several criteria determine the usefulness of these substrates. (1) To provide separation in both high and low molecular weight regions, a given type of packing must be available over a range of pore sizes. A primary requirement then is the precise control of pore size distribution in manufacture. (2) Stability of separation demands some degree of mechanical rigidity so the substrate does not settle or deform under flow. (3) Solvent "compatibility" broadly defines other boundaries of successful operation. For pore penetration to occur, the solvent must swell or at least wet the packing, but excessive solvent power can sometimes result in substrate degradation. On the other hand, marginal substrate-solvent affinity can promote the adsorption or partition of solutes. (4) Ideally, the substrate is inert with respect to the solute and strong specific interactions are absent. With highly polar inorganic stationary phases, attempts are often made to "deactivate" the surface absorption sites.

Many commercial products have been introduced that meet these criteria to varying degrees and find certain specific applications. None is universally applicable, manufacturers' claims notwithstanding. Relatively few substrates, in fact, have encountered broad commercial success. Let us discuss these materials according to type, with reference to areas of most frequent application.

Sephadex®, a chemically cross-linked bacterial polysaccharide, gained wide acceptance 10–20 years ago for size separation of biological macromolecules. The extensive accrued literature has been capably documented by the manufacturer, Pharmacia, from whom detailed bibliographies may be obtained. The aqueous compatibility and relatively modest surface activity of Sephadex® make it especially suitable for proteins and polypeptides. In addition, a hydroxypropylated derivative, LH-Sephadex®, is commercially available for GPC separations in organic solvents. Disadvantages of Sephadex® are the compressibility of the higher pore size gels, precluding operation under pressure, and the large particle size, which interferes with efficient packing. The use of Sephadex® columns is thus typically characterized by low flow rates and poor chromatographic resolution. These deficits are particularly significant in the industrial laboratory; long assay times are incompatible with the demands on most analytical support groups. Poor resolution interferes with the determination of the molecular weight distributions of synthetic polymers, a measurement not relevant for monodisperse biological macromolecules.

On the other hand, the ease of preparation and operation of Sephadex® columns, and the wealth of information on their physicochemical properties render them very useful for certain analytical or preparative tasks. In one of the author's laboratories, for example, separations of ionic aromatic polymers from low molecular weight analogs are based on the ability to control adsorption of these solutes on the gel by adjusting solvent composition.

In terms of ubiquity, the counterpart of Sephadex® in the industrial labratory is cross-linked polystyrene resin beads, for example, Styragel® or μ-Styragel® (Waters Associates), Bio-Beads (Bio-Rad), and TSK-gel (Toyo Soda, Ltd.). The first substrate, marketed as packed columns along with complete liquid chromatography units, currently dominates the field of GPC for synthetic organic polymers. The mechanical rigidity of the gels enables them to withstand pressure drops of approximately 500 psi/ft. The diversity of available pore sizes provides for good size separations, with 300–500 theoretical plates per foot, for high polymers or for low molecular weight solutes. Columns packed with micron-sized beads, introduced 5 years ago as μ-Styragel provide a tenfold increase in resolving power and substantial reductions in assay time. Toyo Soda (TSK Gel—Type H) columns, currently sold in the United States

by Varian Associates, offer similar or higher efficiency and, in addition, appear to be very stable in long-term use.

Styragel®-type columns are usually operated in Tetrahydrofuran (THF), $CHCL_3$, or toluene at ambient temperatures, although crystalline polymers require high-temperature solvents such as orthodichlorobenzene. In these solvents the substrate usually behaves as if inert to the polymer, neither retaining it nor excluding it beyond size effects. In the authors' experience, only amines appear to contribute to special polymer-gel affinity, but no study of this effect has been published. Nonspecific adsorption or partition effects have been reported by several authors and are discussed below. This nonexclusion separation mechanism is probably significant only when the polarity of the solvent differs from that of the Styragel®-type matrix, as in the cases of dimethyl formamide (DMF) or cyclohexane.

Other cross-linked organic polymers have been utilized as GPC substrates; only a few have become commerical products. The heterogeneous copolymerization of polyethylene glycol dimethacrylate and ethylene glycol dimethacrylate forms a porous polyethylene glycol dimethacrylate (PGM) gel. OR-PGM 2000, manufactured by Merck in Germany and marketed in the United States by EM Labs, is a low-molecular-weight resolving substrate with solvent compatibility ranging from benzene to water, although the manufacturer recommends primarily THF and water. Stable at pressures of up to 500 psi, it has an upper molecular weight limit of about 2000. The same company supplies a series of cross-linked polyvinylacetates with upper molecular weight limits ranging from 500 to 1×10^6. These OR-PVA gels are compatible with most organic solvents and withstand back pressures of 500 psi. In particular, they are useful in polar organic solvents that would promote partition separation with Styragel®. Some other gels manufactured in Europe but not marketed in the United States are described in the literature. Spheron, a copolymer of 2-hydroxyethyl methacrylate and ethylenedimethacrylate is produced by LaChema, Brno, Czechoslovakia. It is reported to provide resolution of dextrans and proteins in water over a molecular weight range 10^4–10^6. Unlike Sephadex®, the gel withstands moderate pressures and can be efficiently packed. Enzacryl gels are manufactured by Koch-Light, Colnbrook, Great Britain by cross-linking polymerization of acryloylmorpholine. They can be used in both water and organic solvents and provide resolution from 10^2 to 10^4 mol. wt. A new semirigid gel specifically designed for high-speed, high-resolution aqueous GPC has recently been introduced by Toyo Soda. Designated as "TSK, Type PW," it is principally comprised of a cross-linked hydroxylated polyether and provides very efficient resolution for polyethylene glycol oligomers (78), as well as separating proteins without adsorption (79).

Rigid inorganic substrates offer two apparent advantages over compressible or semirigid gels. First, the absence of deformation should provide, in principle, for greater long-term column stability. Second, their tolerance for pressures in excess of 1000 psi facilitates the use of very small narrow-size-distribution particles which are required for high resolution (see below) but which also provide considerable resistance to flow. Alumina has been described as a GPC substrate but is not much used. The commercially significant inorganic packings are porous glass, marketed in the United States by Electro-Nucleonics as CPG; and porous silica, sold in the United States as Lichrospher (EM-Merck), Porasil (manufactured by Pecheny-St.-Gobain, sold by Waters Associates), and as prepacked SEC or PSM columns (DuPont).

Although overall chemical compositions of porous glass and silica may be similar, the modes of preparation differ greatly. Borosilicate glasses of certain compositions (i.e, $SiO_2 : B_2O_3 : Na_2O$) undergo, during heat treatment, separation into silica-rich and boric-oxide-rich intermingled phases (80). The continuous boric acid phase may be leached out with acid to leave a porous structure with about 30% pore volume. The removal of siliceous material by a subsequent mild caustic treatment can be employed to enlarge the pores. The resultant product is 96% silica and 3–5% B_2O_3 (80). The surface properties of the glass reflect the ratio of silanol hydroxyls to siloxane groups, and the presence of boron Lewis acid sites. These compositional variables depend in turn on the thermal history of the glass.

Porous silica has been long employed for adsorption chromatography and thus might seem inapplicable to GPC. Under certain conditions, however, such specific interactions are negligible, and separations are size controlled. Silica beads for GPC can be prepared by precipitation of an aqueous silicate solution under controlled pH and cosolvent conditions followed by heat fusion. The hydrolytic polycondensation of ethoxysiloxane (81) is another route to silica supports. In DuPont's process for PSM packings, an organic polymer is used to precipitate silicagel; subsequent firing burns off the polymer and fuses the silica into a porous microsphere. These various methods can produce particles of very uniform demensions and pore size distributions. Being smaller and more porous than CPG, silica GPC packings in general provide considerably more resolution but operate at much higher pressures and are more surface active.

Silica, like glass, has a measurable solubility in aqueous media that increases strongly as the pH exceeds ≈ 9. The rate of solution is probably greater for silica because of its enhanced surface area. Thus silica GPC particles may gradually erode in aqueous buffer. At time there seem to be no definite reports on the stability of siliceous columns used for aqueous GPC.

The silanol groups on silica and glass provide reactive sites for coupling to various chemical groups. The resultant derivatized products possess the rigidity and porosity of the untreated substrates, along with the surface properties of the reagent. By this means, siliceous supports may be rendered compatible with virtually any solute-solvent system. Silanization is the standard means of surface modification and involves the condensation of an alkoxy- or alkylchloro-silane as follows:

Numerous silanizing procedures have been developed for the preparation of HPLC phases and some are applied in commercial preparation of GPC substrates. Derivation of CPG with glycerolpropylsilane yields a hydrophilic bonded surface considerably less anionic than untreated glass. The product, marketed by Electronucleonics as Glyceryl-CPG, and by Pierce as Glycophase-G, is useful for characterization of cationic or ampholytic biopolymers in aqueous media. The same silanizing reagent confers a "diol" surface to porous silica and permits chromatography of proteins. Lichrospher (EM-Merck) treated in this manner is being marketed by both Synchropak and Merck. Toyo Soda sells a similar substrate (TSK gel, Type SW) in the form of high-speed, high-resolution columns pariculary offered for aqueous GPC of proteins (82). Treatment of porous glass with alkylsilanes provides an apolar surface more suitable for GPC with organic solvents. While such silanized glasses are not commercially available, the derivation process is readily carried out in the laboratory. Given the versatility of silane chemistry, it is likely that new treated siliceous GPC packings will appear on the market.

Eluants. Considerations of stability and solvent affinity primarily dictate the choice of solvents for GPC. The eluant must dissolve the polymer and penetrate the pores of the gel and have no degrading effects on either. Low viscosity is important especially for densely packed columns at high flow rates. The breadth of spectral transparency is significant in terms of the detector used. Within these constraints many solvents prove useful. Tetrahydrofuran (THF) meets these criteria admirably for apolar polymers, and the proximity of its solubility parameter to that of Styragel® minimizes adsorption (see below). Chloroform and toluene are used somewhat interchangeably with THF. Crystalline polymers soluble only at high temperatures are commonly analyzed in hot *m*-cresol and orthodichlorobenzene. DMF and DMAC have wide ranges of affinities and are expecially useful for polar compounds such as polyacrylonitrile. Other polar organic solvents described in the literature include hexamethylphosphotriamide (HMPT), dimethyl sulfoxide (DMSO), and hexafluoroisopropanol. For water-soluble nonionic polymers such as dextrans, polyethylene glycol, and polyvinylpyrrolidone pure water is a suitable eluant, but ionic polymers require a supporting electrolyte, most commonly phosphate buffer or sodium sulfate (alkalihalide salts corrode stainless steel).

Calibrators. Molecular weight standards for GPC must be of very narrow distribution so that the molecular weight obtained by absolute methods can be identified with the species eluting at the chromatographic peak. The polystyrene samples sold by Pressure Chemical Company are the basis for calibration of most instruments operating in organic solvents. Other polymers available as narrow fractions are polyisoprene and polytetrahydrofan (Polymer Laboratories, Ltd., Stow, Ohio); polyvinylacetate and polypropylene (National Physical Laboratory, Teddington, United Kingdom).

Rather narrow polyethyleneoxides can be

brought from Union Carbide and Hoechst, but their absolute molecular weights are not well documented. Characterized dextrans of somewhat broader distribution have been supplied by Pharmacia for some time as aqueous GPC standards. Recently, sulfonated polystyrene standards for aqueous GPC have become available from Pressure Chemical Company. Broad distribution samples of a wide variety of polymers, usually characterized by either light scattering or osmometry (occasionally both) may be obtained from Scientific Polymer Products (formerly Cellomer).

3.6.2.2 Methodology

Column Packing. Major new developments in GPC are directed toward improving resolution and assay time, and many of the recent advances in column technology made in high pressure liquid chromatography (HPLC) are now being applied to GPC. The least sophisticated packing methods are dry-packing with column vibration, which is applicable to rigid inorganic substrates, and slurry-packing (with the future eluant) under pump pressure for semirigid organic gels. For large ($> 100 \mu$m) or irregular particles where eddy diffusion and wall effects would lead anyway to considerable solute dispersion, such techniques are acceptable and can provide columns with up to 500 theoretical plates per foot (tpf). Basic chromatographic principles suggest improved efficiency with decreasing substrate particle size—as low as approximately 10 μm, but modifications in packing techniques are required to obtain the full benefit of small particle size substrates. Also, the resultant columns operate under greater back pressure when run at typical flow rates, requiring more sophisticated hardware.

Methods for packing inorganic GPC substrates are based on technology developed for HPLC silica columns. The principal objectives for improved resolution with inorganic substrates are high porosity and maximum density, both of which increase the effective portion of the total column volume. Clearly, diminished particle size enhances substrate density, and particle diameters as low as 5–10 μm are currently employed. Slurry-packing methods are required for such small particles. Although broad particle size distributions could provide denser columns, resistance to flow during packing prevents uniform deposition, and narrow distribution substrates appear to pack more evenly. Ideal column-packing procedures would thus involve the very rapid deposition of a perfectly homogeneous suspension to form a highly uniform array. Uniform suspensions of porous glass can be stabilized by means of a high-density solvent such as tetrabromoethane, a high-viscosity medium, or an ammoniacal solvent in which the substrate is dispersed partly by electrostatic effects. All three methods have disadvantages, that is, the noxious qualities of the halogenated solvents, the low linear packing velocity and concommitant loose deposition in viscous media, and the partial solution of silica or glass in highly basic solvents. Such procedures, however, can provide columns with efficiencies on the order of 3000 tpf.

Techniques for producing high-resolution columns with semirigid gels are more demanding since the particles and the gel bed are more easily deformed. As in the case of glass or silica, efficiency varies inversely with particle size, and the manufacturers of high-resolution cross-linked polystyrene gels probably employ particles with diameters on the order of 5 μm. For organic gels, porosity seems to be less the limiting factor in resolution than is irregular microscopic flow patterns arising from uneven packing. Balanced density slurries are easier to prepare than for porous glass, and somewhat slower rates of packing are thus probably acceptable.

Both organic and inorganic substrates may be packed by methods described in the open literature to yield efficiencies of 200–500 tpf. Higher resolving power, that is, 1000–10,000 tpf, can be achieved only with very small particles of narrow size distribution. The techniques for packing these substrates are to some extent proprietary. The art involved is sufficiently demanding that few individual chromatographers find it worthwhile to attempt to duplicate the resolving power of commercial columns. On the other hand, high-resolution GPC columns are relatively new in the marketplace, and the principal domestic suppliers, for example, Waters, DuPont, and Varian (i.e. Toyo Soda columns), are sometimes unable to supply complete information on constraints in column use or expected stability.

Detection. Many of the numerous reviews on HPLC discuss the relative merits of various chromatography detectors. Refractive index (RI) is the "universal detector" for GPC. Sensitive to samples as small as 100 μg and easy to use and maintain, its only defect is its response to delicate changes in solvent composition, even those arising from preferential solvation effects. More selective are spectrophotometric detectors of either the fixed or variable ("tunable") wavelength type. The former are frequently offered by the manufacturer (e.g. Altex, Pharmacia, LKB) with filters that narrow the mercury lamp source to a specified wavelength band in the range 254–500 nm. The Schoeffel (now Kratos) UV-visible monitor is the variable wavelength spectrophotometric detector of choice, providing high sensitivity and minimal band-spreading from 200 to 800 nm. Maintenance of source and optics is, however, not negligible. Another type of spectral monitor with special potential for copolymers is the variable wavelength infrared (ir) detector available, for example, from Wilkes-Miran. Since some of the most common GPC solvents—THF, DMF, *m*-cresol—are largely ir-opaque, the utility of these devices is limited mainly to halogenated solvents. Several companies offer fixed or variable wavelength fluorescence detectors in addition to flow cells that adapt standard bench-top spectrofluorometers. Enormously sensitive to certain low molecular weight species, their application to polymers might be greatest in the analysis of some specific impurities in polymer samples.

Two types of "detectors" unique to GPC can provide data leading to absolute MWD determinations. Hence they are especially useful for the analysis of branched polymers, where molecular weights based on polystyrene may be particularly inaccurate. The utility of monitoring by viscosity is based on the principal of universal calibration (as discussed in Section 3.6.1.1), which states that the dependence of elution volume V_e on $[\eta] M$ for a given GPC system is insensitive to polymer type. This relationship may then be established with some convenient standards, for example, polystyrene fractions. A knowledge of $[\eta]$ as a function of V_e leads in principle to the absolute MWD. Various in-line viscometers have been designed and used successfully in laboratories at CNRS (Strasbourg), AKZO (Arnhem, Holland), Exxon (Linden), and Northwestern (Evanston). Conventional viscometers employed as GPC monitors demand large efflux volumes and long run times. Ouano (83) built a viscometric flow monitor based on a pressure transducer, but attempts to improve sensitivity posed a design problem. None of these concepts has been successfully commercialized.

A second approach towards absolute molecular weight determination, also developed by Ouano (84) has met with greater commercial success. The low-angle light-scattering (LALLS) detector now available from Chromatix, can provide—with appropriate data processing—the molecular weight of the solute within the detector cell. In contrast to all other GPC techniques, LALLS detection obviates the need for calibration, a major advance in that the considerations of conformation, adsorption, and partition usually relevant to GPC interpretation become moot.

Finally, the analysis of eluant fractions by scintillation counting deserves mention, even though no practical on-line radioisotopic detector has been designed. Such techniques are very familiar to biophysical chemists but have not been much utilized for synthetic polymers. Analysis of the distribution of radio-labeled moieties as a function of molecular weight or molecular size can, however, provide some unique insights into polymer reactivity or structure (85).

3.6.2.3   Hardware
In view of the wealth of information available in commercial brochures from chromatographic suppliers, little need be said on the technical specifications of instrumental components. Figure 3.14 shows the best known

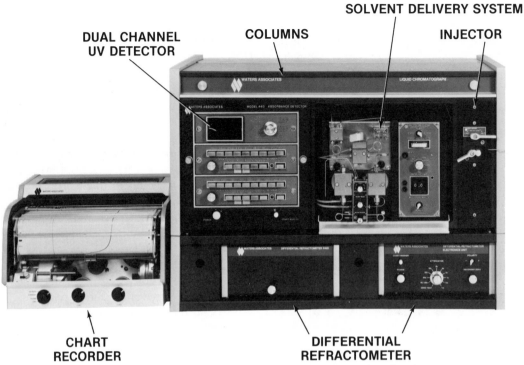

SOLVENT DELIVERY SYSTEM

DUAL CHANNEL   COLUMNS   INJECTOR
UV DETECTOR

CHART   DIFFERENTIAL
RECORDER   REFRACTOMETER

Figure 3.14   Waters ALC/GPC system. (Courtesy Waters Associates.)

GPC system from Waters Associates. In general, the acquisition of a GPC system resembles the purchase of a 10-speed bicycle in two significant regards: first, all instruments should be viewed as a collection of perhaps a half-dozen major components; second, an educated choice depends on a good appraisal of the user's demands and intensity. The criteria for pumps and injectors rests on the need for reproducible flow rates at a pressure that depends on column selection. The columns, themselves the heart of the system, typically represent only some 10–20% of overall price but deserve the greatest critical concern. Detectors have been reviewed. Certain systems offer special features such as high-temperature column operation. Automatic sample injection and solvent recycling are valuable assets for heavy analytical sample loads.

3.6.3   Interpretation of Data

3.6.3.1   Measured Parameters

The simplest GPC result is the peak molecular weight (M_p) a measurement that can be useful when the polydispersity is not very large or the differences among samples are gross. M_p,

however, may fail to differentiate adequately among preparations of large and variable polydispersity. Furthermore, the value of M_p depends to some extent on the shape of the calibration curve, that is, it is a chromatographic average rather than a true moment of the distribution.

Weight and number-average are frequently calcualted from GPC data in an attempt to provide values more readily correlated with relevant properties. Manual calculations are somewhat tedious—about 1/2 h per sample—and integrated data processing units are available (from Waters, or Spectra-Physics, for example) to compute these moments. In contrast to M_p, $\overline{M}_n$ and $\overline{M}_w$ are insensitive to the shape of the calibration curve but may be strongly influenced by band-broadening. The availability of these values in the form of direct printout has been known to distract some operators from the need for critical examination of both the effect of band-broadening and the relevance of the calibration curve to the material studied.

The use of calculated $\overline{M}_n$ and $\overline{M}_w$ values may be partly historical since only these moments could be evaluated prior to GPC. Apart from comparisons of GPC data with classical measurements, industrial applications of GPC results may be enhanced by using other parameters, such as the fraction of the chromatogram above or below some specified apparent molecular weight. The choice of an optimal process or material control parameter depends on the probable MWD variations in production and on the way in which the key properties depend on the MWD. A knowledge of these two relations is prerequisite to selecting GPC parameters that provide the clearest link between the ultimate desirable properties and controllable synthesis and processing variables.

3.6.3.2   Calibration

Since no more than three or four homopolymers are available as satisfactory calibration standards, GPC systems are usually calibrated with polymers different from the samples to be analyzed. Few laboratories undertake the preparation of characterized fractions of their polymers of interest. As described in Section 3.6.1.1., the universal calibration principle states that $([\eta]M_v)_{sample} = ([\eta]M_v)_{standard}$. This enables one to predict the difference between true and relative molecular weights from the difference in viscosity behavior of sample and standard. Such viscosity differences typically arise from differences in solvent affinity, ionic character, chain stiffness, mass per unit contour length, or branching. If samples and standards are reasonably similar in all these categories, it is unlikely that relative and true molecular weights would differ by more than a factor of 2. On the other hand, inaccuracy of an order of magnitude is not unlikely if the sample is, for example, highly branched or charged and the standards are not.

Techniques for calibrating with unfractionated polymers have been discussed by several authors. These typically correspond to iterative processes in which the calibration curve eventually selected provides, in conjunction with the chromatograms, values of $\overline{M}_n$ or $\overline{M}_w$ in agreement with those measured directly. Alternatively, comparisons may be made with measured intrinsic viscosities, given a knowledge of K and a, the Mark-Houwink coeffi-

cients. It is commonly assumed — at least for the first iteration — that the correct calibration curve is linear. These methods are, of course, only as reliable as the accuracy and the range of molecular weight of the primary characterization data. A calibration curve based on light-scattering and osmometry data for three or four well separated fractions could be viewed with some confidence, whereas one based on viscosity data for one or two broad fractions would be tenuous.

3.6.3.3   Resolution

A precise definition of resolution is harder to arrive at for GPC than for conventional gas or liquid chromatography. Most workers would view poor resolution in GPC as synonymous with distortions in the calculated MWD arising from band-spreading. GPC resolution depends on two factors: the retention volume difference corresponding to a given change in molecular weight, that is, the slope of the calibration curve; and the spreading of monodisperse solute components passing through the columns. Volume separation can always be improved by increased column lengths, within the limits of operating pressure. Bandspreading depends in a more complex way on system design and column-packing techniques. To optimize the last two factors, a quantitative and operationally meaningful measurement of overall resolution is necessary.

A useful resolution parameter should describe the ability of the instrument to provide accurate MWD data. This variable should depend on factors that can be readily measured and altered, such as particle size or flow rate. Ultimately, an expression for the resolution parameter should allow the chromatographer to determine the number of columns and degree of column efficiency required for his purposes and enable him to predict the consequences of limited resolution on the accuracy of his results.

Balke and Hamielec (86) and Yau et al. (87) have obtained expressions for resolution that may be combined as in Equation (3.88).

$$\frac{p^*}{p} = \exp\left[\frac{1}{4}(\ln 10)^2 \left(\frac{\sigma}{d}\right)^2\right] \quad (3.88)$$

Here p is the true polydispersity $\overline{M}_w/\overline{M}_n$, and p^* the value obtained by GPC; σ is the standard deviation of the eluted peak, commonly obtained as one-fourth the Gaussian base width; and the column selectivity variable d is the reciprocal slope of the calibration curve, e.g. milliliters per decade, over the molecular weight range of interest. While resolution may also be expressed in terms of peak overlap for monodisperse samples of different molecular weight, the preceding expression is more relevant to typical GPC usage. Inspection of Equation (3.88) reveals that the influence of peak-spreading on the measured polydispersity can be compensated for by increased column length L, that is, by way of increased d. This is because d increases linearly with L, while σ increases only as $L^{1/2}$. For example, even with a peak dispersion as large as $\sigma = 11$ ml, p^*/p is less than 1.3 if the value for d is as big as 35 ml/decade (89). σ/L thus provides a measure of the accuracy in p achievable with a given column length—or a given analysis time—and may thus be regarded as a column efficiency parameter.

According to Equation (3.88) the value of σ/d should correlate well with the accuracy of GPC results. This has been shown for samples of varying molecular weight eluted from a given column set in regions of different calibration curve slope (89). The significance of σ/L as a column efficiency parameter was also confirmed for the same sample on different column sets.

Other measurements of solute dispersion are also functions of the peak standard deviation. Tung and Runyon (88) define a dispersion parameter $h = 1/\sigma\sqrt{2}$. The number of theoretical plates is given by $N = V_e^2/\sigma^2$ where V_e, the peak retention volume, and σ are both usually measured with a low molecular weight solute.

As noted, peak-spreading increases with the square root of retention time. For polymeric solutes, however, σ also depends on peak position relative to the calibration curve. The maximum values of σ are found for molecular weight fractions that elute around the middle of the calibration curve (89); solutes that are either completely excluded or that fully penetrate the pores are spread less than resolved species. This may explain the poor correlation between polymer dispersion (i.e., p^*/p) and the band-spreading of low molecular weight solutes (i.e. N) sometimes observed when different column packings are compared. The value of p^*/p, in conjunction with Equation (3.88), indicates best what column length is required for a given MWD accuracy. Unfortunately, p^*/p is rarely reported, because N (obtained with a low molecular weight solute) is traditionally accepted and also more easily measured. The number of theoretical plates N, while not a suitable guide for comparing column systems of different pore structure or calibration curve shapes, is nonetheless useful for roughly gauging efficiency or for monitoring column stability.

Since about 1970, advances in column technology, initiated by HPLC development, have led to a new generation of high-resolution GPC systems with improved hardware and column packings. For a given retention time, these columns exhibit one-third to one-fifth the peak dispersion of early Styragel® or Corning Glass packings. According to Equation (3.88), this corresponds to an order of magnitude improvement in assay time (i.e. decrease in required magnitude of $1/d$ at fixed p^*/p). These improvements are based on the use of very small particle-size ($5-10\mu m$) GPC packings with narrow particle-size distribution. Such materials facilitate microscopically uniform packing which minimizes "eddy" or multipath diffusion arising from irregular flow patterns. N is thought to increase continuously with decreasing particle size, but in practice the use of yet smaller-size packings would pose special handling problems. Another benefit of smaller particles may be increased GPC capacity by way of enhanced porosity. Porosity can be defined as V_p/V_0, the ratio of pore volume to interstitial volume. Resolution, as expressed in the reciprocal slope of the calibration curve, increases with porosity, but mechanical strength is lost. Unger and Kern (90) working with several preparations of porous silica found for the most efficient packing $V_p/V_0 = 1.3$. The design the current GPC hardware has also been improved to realize the benefits of modern packings: zero dead-volume column end fittings and injectors, capillary tubing in all connections, minimal cell volume in detectors, and reduced column diameters in proportion to reduced particle size.

To recover the expense of the advanced technology of particle synthesis, most manufacturers of "new-generation" packings prefer to market prepacked columns. In fact, few users are prepared to pack columns with state-of-the-art methods. So the factors discussed, for example particle size and porosity, are not within the direct control of the average chromatographer. Actual operational variables, then, are flow rate, temperature, and sample size and concentration.

In general, N varies inversely with flow rate; much chromatographic literature focuses on the dependence of plate height, $H = L/N$, on flow velocity since this relationship determines the trade-off between resolution and assay time. For porous glass, high flow rate (1–2 ml/min) and increased column length L are preferable to the converse situation since increased solute dispersion is more than compensated for by the ability to increase separation. The feasibility of increasing L is determined by the pressure limitations of the system, which are generally high for rigid supports like porous glass. Loss in column efficiency with flow rate is thought to be less for Styragel® because spreading due to varying pore residence time ("mass transfer" spreading) is less significant. This factor would seem to suggest the use of higher flow rates, but pressure constraints for semirigid gels are more severe.

In general, then, operating with a larger number of columns at flow rates near the system's pressure limit should yield optimum resolution per unit of assay time. Of course, any given set of columns will provide highest resolution at low flow rates, say on the order of 0.1 ml/min.

Apparent molecular weights from GPC tend to decrease with increasing sample concentration. This effect, which results from intermolecular steric repulsion, increases with molecular size, hence with molecular weight

and the goodness of the solvent. Resolution is also adversely influenced by high concentrations because pore capacity can be exceeded. Since these effects do not vanish at low concentrations, it is generally advisable to work with the lowest conveniently detectable sample masses. This admonition goes hand-in-hand with the need for small injection volumes to match modern high-speed systems: about 0.3% of the sample retention volume.

Increased temperature enhances column efficiency by increasing the diffusion coefficient of the solute and by lowering the viscosity of the solvent. Both effects increase mass transfer into and out of the pores, and the second also facilitates the use of high flow rates by reducing the pressure drop. The convenience of working at ambient temperatures, however, is such that high-temperature operation is generally reserved for polymers with limited solubility behavior.

3.6.4   Adsorption and Partition

Attractive interactions between polymers and solid surfaces are well known, and "pure" exclusion chromatography—governed by steric (entropic) factors alone—may be the exception and not the rule. It is useful to consider these effects as of two types: adsorption, in which specific site interactions (e.g. hydrogen bonding) occur between functional groups on polymer and substrate; and partition, in which weak reversible interactions such as dispersion and/or "solvophobic" forces result in favorable association of solute and gel. The former type of binding (adsorption) is cooperative because an interaction at one polymer residue reduces the entropy loss involved in the adsorption of a neighboring group on the polymer chain. Therefore adsorption can increase strongly with polymer molecular weight. The second type of interaction (partition) is too short-lived and non-specific to be cooperative and may actually decrease with molecular weight because the accessible substrate surface is concomitantly reduced.

3.6.4.1   Semirigid Cross-Linked Gels

Styrene-divinylbenzene gels are particularly susceptible to the second type of interaction, and the partitioning of polymers and low molecular weight solutes onto Styragel®-type substrates has now been thoroughly invest-igated. According to the simplified scheme presented in Equation (3.89),

polymer $\cdots$ solvent + gel $\cdots$ solvent $\longrightarrow$
polymer $\cdots$ gel + solvent $\cdots$ solvent   (3.89)

retention arises from the replacement of weak polymer-solvent and gel-solvent interactions with stronger polymer-gel and/or solvent-solvent interactions. These interactions may be formally taken into account by introducing as partition term into the chromatographic distribution coefficient K_d (91); as follows in Equation (3.90):

$$V_e = V_0 + K_d V_p$$
$$K_d = K_d^s K_d^p \qquad (3.90)$$

where superscripts s and p denote the steric and partitioning contributions to the observed retention. K_d can be regarded as an equilibrium constant and analyzed in thermodynamic terms. Specifically, solubility parameters seem to provide an accurate assessment of the interactions in Equation (3.90), and K_d^p is found to vary in a linear manner with $Y = (\delta_{\text{polymer}} - \delta_{\text{solvent}})^2 - (\delta_{\text{polymer}} - \delta_{\text{gel}})^2$ (91). The preceding expression represents the preference of polymer for the gel phase over the solvent phase; rewritten as $Y = \delta_s^2 - \delta_g^2 + 2\delta_p(\delta_d - \delta_s)$, it suggests that partition effects vanish when solubility parameters of gel and solvent become identical, as was first predicted by Altgelt and Moore (93). In contrast, however, polycyclic aromatic hydrocarbons exhibit adsorption effects on Styragel® in benzene, elution volumes increasing with molecular weight (94). This retention apparently cor-responds to relatively intense dispersion forces.

The most striking examples of partition are observed for rather apolar gels, like cross-linked polystyrene or polyvinylacetate, in combination with highly polar solvents such as DMF or hexafluoroisopropanol (HFIP). For polystyrene/Styragel®/DMF $Y = 9$; for the same polymer/gel in cyclohexane $Y = 1$.

As a general rule, polystyrene exhibits parti-tioning onto Styragel® in solvents in which its Mark-Houwink exponent a is below about 0.66 (95). Since polystyrene fractions are routinely used to calibrate Styragel® columns, the consequences of its partitioning can be severe even though the polymers being char-acterized are not strongly retained.

The utility of solubility parameters in de-scribing partition on semirigid gels has sug-gested to some authors that the polymer-gel interactions are analogous to the mixing phenomena that determine the compatibility of soluble polymers. Regardless of whether or not chains interpenetrate, it is probably the case that the thermodynamic impetus for mixing—as in compatibility—must be a fa-vorable enthalpy of replacing weak polymer-solvent and gel-solvent interactions with strong polymer-gel associations. Under ideal condi-tions, the transfer of polymer from solvent to gel is governed only by entropic changes. Thus partitioning will usually be exenthalpic and will be reduced at higher temperatures. For polystyrene/Styragel®/decalin, for example, partitioning is strong at ambient temperature but negligible at 100° C (91). However, temper-ature elevation does not usually assure com-plete elimination of partitioning.

Semirigid organic gels bearing functional groups can exhibit retention in organic sol-vents owing to weak interactions—van der Waals or dipole-dipole—as well as hydrogen bonding. The retention of small molecules on hydroxyethylmethacrylate (Spheron) gels in THF increases in a linear fashion with the solubility parameter of the solute (96). How-ever, strongly hydrogen-bond-accepting sol-utes such as Pyridine, DMF, or DMSO exhibit especially high retention, presumably by effectively competing with THF for the gel's hydroxyl groups. In a more polar solvent such as methanol, apolar partitioning effects are enhanced while solute gel hydrogen bond-ing is reduced or eliminated. Similar behavior may be expected for polymeric solutes with hydrogen-bonding functionality.

Partitioning can be ruled out if universal calibration is found to be valid for polymers differing widely in polarity. A simpler diagnostic tool is the comparison of the retention behavior for low molecular weight solutes of varying polarity. In the absence of partition or adsorption, these should elute according to their molar volumes, taking solvation into account as required. Last, partition enhances peak-spreading, and increased apparent polydispersity of narrow MWD fractions can provide circumstantial evidence for nonideal separation.

The affinity of solute and gel can be especially strong in aqueous media, where attractive forces among solvent molecules provide an additional nonspecific force for retention. This force is utilized in "hydrophobic chromatography," where Agarose or Sepharose gels substituted with apolar groups are to separate proteins of varying hydrophobic affinity. Since these gels also resolve by exclusion, sophisticated and complex separations of large and small molecules can be effected.

3.6.4.2 Inorganic Substrates

Specific site interactions are characteristic of nonideal retention effects for rigid GPC supports. The surface silanol groups of glass and silica are highly active because of hydrogen bonding and electrostatic interactions, hence such substrates can irreversibly or reversibly bind polymers with basic or H-bond accepting functionality even in a strongly competing solvent like water. While partitioning in semirigid gels has been the focus of some systematic studies, the adsorption behavior of porous glass GPC supports has been dealt with only on an *ad hoc* basis. The mechanism of adsorption is, in general, unresolved with regard to contributions from H-bonding, electrostatic interactions, or Lewis acid-base pairing. Furthermore, the composition or history of the glass can affect the surface concentration of the various active sites, namely isolated silanol (partly ionized in aqueous medium), neighboring H-bonded silanols, and boron Lewis acid sites. Thus the irreversible adsorption of proteins on untreated porous glass could arise from a multiplicity of effects.

H-bonding by surface silanol groups underlies irreversible adsorption (97) of polyethyleneoxide onto porous glass in water. (The strength of this interaction led to the use of "Carbowax treatment" as a means to deactivate glass packing prior to chromatography of proteins—a somewhat uncertain practice owing to the instability of the resultant surface.) In weaker H-bonding media the surface activity should be even greater; for example, polyamides usually exhibit adsorption behavior on porous glass in polar organic solvents. The magnitude of these effects can be quite sensitive to the structure and solvent affinity of the polymer: amylose and poly-N-vinylacetamide are adsorbed from water onto glass while dextran and polyacrylamide are not. In the case of polyamides, ir spectra show the predominant interaction to involve amide carbonyl and surface silanol; the amide hydrogen is essentially unperturbed.

Electrostatic effects presumably play a significant role in adsorption; glass or silica bears a negative surface charge that increases strongly with pH. It may, however, be difficult to distinguish between electrostatic adsorption and H-bonding. The influence of salt concentration on retention certainly reflects to some extent the effect of ionic strength on the magnitude of the electrical double layer residing on the pore surfaces. Since ionic strength also controls the molecular volume of polyelectrolytes, the effect of salt on polyion retention cannot be ascribed uniquely to "ion exclusion" by porous glass. At ionic strengths above one, the universal calibration behavior of ionic and nonionic polymers is insensitive to salt concentration (98), so, in this case, the glass surface charge seems unimportant. For low-molecular-weight ionic solute at low ionic strength, increases in retention with an added supporting electrolyte do appear suggestive of compression of the substrate's electrical double layer.

Adsorption onto siliceous packings may arise from effects other than strong H-bonding or electrostatic interactions. Polystyrene eluted from Spherosil in a number of solvents does not exhibit congruent universal calibration curves. Rather, elution volumes are seen to decrease in those solvents that adsorb most strongly to the substrate, for example, MeOH-CHCl₃ (99). The explanation advanced is that strong adsorption of solvent on the packing competes with polymer-substrate interactions thus leading to more ideal chromatography and earlier elution. While the mechanism of polymer adsorption is not confirmed, weak H-bonds may be formed between surface silanols (donors) and the aromatic π clouds (acceptors).

Control of adsorption with siliceous GPC packings has been approached in three ways. Temporary surface treatment with polyethyleneoxide, previously mentioned, leads to unstable and rather irreproducible surface conditions. Cosolutes such as sodium dodecyl sulfate sometimes weaken the adsorption of proteins, possibly through the formation of a polymer-small-molecule complex. The most promising approach is a permanent surface treatment to yield an "inert" surface. Glyceryl-CPG (Electro-Nucleonics) and Glycophase-G/CPG (Pierce) are modified—presumably by way of glycerol-propylsilane—porous glasses that are reported to elute proteins without adsorption (97, 100), as well as basic polysaccharides (101). However, the adsorption of a cationic protein to Glyceryl-CPG at low ionic strength has been presented as evidence for the residual siliceous ionic sites (97). Silane chemistry, in general, facilitates preparation of a wide variety of surface-modified GPC substrate. Porous silica has been surface treated with a number of hydrophilic organosilicon compounds, some of which, in fact, exhibit pure exclusion chromatography corresponding to full shielding of silanol groups and minimal reversed-phase character (102).

3.6.4.3 Polysaccharide Gels
Polysaccharide gels such as cross-linked dextran (i.e. Sephadex®) are employed for GPC in water or polar organic solvents such as DMF and DMSO. Polymers most commonly resolved on these gels include proteins, nucleic acids, polypeptides, and polysaccharides, as well as polar synthetic polymers such as polyethyleneoxide (PEO), polyvinylalcohol (PVA) and polyvinylpyrrolidone (PVP). In the typical applications of polysaccharide gels, the solute, solvent, and gel, therefore, all possess strong H-bonding potential. Thus, according to Chitumbo and Brown (103), partitioning on polysaccharide gels reflects the thermodynamic change that accompanies the desolvation of polymer and gel, as well as the direct interaction of the last two components. Consequently, nonideal retention may be either endenthalpic or exenthalpic. Several workers have alluded to hydrophobic solute-substrate interactions on Sephadex®, identifying the epichlorohydrin cross-links as apolar regions in the gel. The partitioning of solute of low polarity is apparently diminished by the addition of alcohols, urea, or guanidine hydrochloride. Since these cosolutes are known to reduce hydrophobic bonding in other systems, their effect on partitioning has been considered evidence for hydrophobic partitioning in Sephadex®, but Chitumbo and Brown (103) have argued that alterations in polymer or gel solvation could account for these observations. On the other hand, the pronounced retention of aromatic solutes on Sephadex® is not explained by the H-bond interactions that predominate in Chitumbo and Brown's description. The ability of carbohydrates to modify the local structure of water might underlie the preference of aromatic solutes for the gel phase in an aqueous medium.

Last, the weakly anionic character of Sephadex® needs to be mentioned with regard to the enhancement or limitation of adsorption of ionic solutes. Sephadex® has a very low but characteristic concentration of carboxyl groups, which, at neutral pH, attract cations and repel anions. The magnitude of this effect is inversely proportional to the ionic strength. Above pH 10 dissociation of saccharide hydroxyls can greatly increase the anionic character of the gel.

3.7 DETERMINATION OF BRANCHING

The presence of branches in a polymer chain can have profound effects on the properties of the material. Thus high-density (linear) polyethylene and low-density (branched) polyethylene differ in such varied properties as compressibility, stiffness, chemical resistance, permeability, tensile strength, and softening temperature. If branching influences such important commercial properties and, if one can establish how synthesis parameters determine the nature and amount of branching, the measurement of branching can provide a key link in the manufacture of certain materials. Quantitation of branching is often difficult and typically requires comparison with the linear polymer homolog (which may not be available). Moreover, in some systems it is not even clear what measurable branching parameters will furnish a useful description. Nevertheless, recent developments in technology and theoretical analysis permit the measurement of branching and its relationship to the MWD for a wide variety of polymers.

Polymer branching is generally classified as long chain branching or short chain branching. Long chain branching—in which branches are typically as long as the "primary" chain—arises from a variety of reactions: polycondensations or vinyl polymerizations with polyfunctional monomer units, addition polymerization in which radical transfer to polymer occurs, and chemical or radiation-induced cross-linking of either condensation or addition polymers. Short chain branching, on the other hand, produces very brief side chains along a clearly defined main chain and typically results from radical transfer back along the growing chain of an addition polymer through what is assumed to be a cyclic transition state.

3.7.1 Molecular Weight Distribution in Branched Polymers

The theory for MWD in branched polycondensations has been thoroughly described by Stockmayer (104), Flory (105), and others. Flory has shown that the MWD described for these systems applies equally to any cross-linked polymer system simply by considering the cross-links as tetrafunctional branch points. The most notable feature of this description is the fact that, above a rather low concentration of multifunctional monomer or cross-linking agent, an infinite network of insoluble polymer (gel) is formed. While this infinite network corresponds to such important polymer systems as cured thermosetting resins, it does not have a measurable molecular weight. Thus one needs to emphasize only the description of nongelled systems.

A thorough description of the MWD in cross-linked polymers is beyond the scope of this chapter but can be found in a variety of sources, such as Peebles (7). However, certain details deserve mention. The critical concentration of cross-linking agent at which gelation first occurs is one cross-link for every two primary molecules (before reaction) for a monodisperse system, or one cross-link for every four primary molecules for a most probable distribution of uncross-linked polymer chains. As the concentration of cross-links (or of multifunctional monomer in a branched condensation) increases past this value, the fraction of polymer included in the infinite network rises extremely rapidly. Even so, the remaining polymer chains are essentially unaffected. There are virtually no chains of very high molecular weight–only starting material (linked perhaps in twos or threes) and infinite network.

On the other hand, long chain branching in addition polymers that occurs by way of chain transfer to polymer cannot lead to an infinite network. This is because chain transfer provides no mechanism by which existing chains can be linked together into such a network. Similarly, certain special polycondensations, most notably those yielding star polymers, can lead to highly branched chains but never an infinite network. Such systems, nevertheless, show MWDs similar to the early stages of the randomly branched polycondensations previously mentioned, subject only to the limitation that they can never reach the critical gelation point.

One other key feature of the MWD in systems with long chain branching is the breadth of the distribution. Where the polydispersity ($\overline{M}_w / \overline{M}_n$) for a linear polycondensation is approximately 2 at all degrees of conversion, and may be considerably less than 2 for some addition polymers, it increases considerably as branching or cross-linking appears, approaching infinity at the gelation point. Thus a branched polyethylene may have $\overline{M}_w/\overline{M}_n$ equal to 50 or more. In fact, a very broad distribution is frequently the first indication that a novel polymer system is branched.

Short chain branching, on the other hand, has no such dramatic effect on MWD or on the solution properties of the polymer. Its primary importance lies in the irregularity it induces along the main chain and the resulting loss of crystallinity of the solid polymer. Long chain branches also disrupt crystallinity in proportion to their number. Since short chain branches, when present, usually far outnumber long chain branches, the contribution from the latter can be ignored when both are present. No theory treats the distribution of short chain branches or their effect on measurable properties; short chain branches must, in general, be quantitated directly, for example, by nuclear magnetic resonance detection of branch points.

3.7.2   Indices of Branching

The adequate description of a branched polymer system is by no means easy. For a heterogeneous polymer, one measures average values. Unlike the case for molecular weight, however, it is not always clear what sort of average (weight-, number-, etc.) is given by any such measurement. The most common index of branching is the number of branch points per molecule. If this is strictly proportional to molecular weight, we can define a branching frequency or branching coefficient as follows:

$$\lambda = \frac{b}{M} \qquad (3.91)$$

where b is the number of branch points per molecule. While a large number of systems appear to possess such a constant λ, others apparently do not (106). Unfortunately, most methods for measuring branching assume a constant λ and thus may sometimes be susceptible to considerable error.

Methods for measurement of branching generally yield an indirect index of branching based on the ratio of some property of a branched polymer to that of a linear polymer of the same molecular weight and composition. The most obvious difference between such polymers is that the branched polymer will be more compact in solution, even though each individual chain behaves as a random coil. Thus solution properties that depend on the average molecular size should vary in a predictable way with the degree of branching.

A commonly used indirect branching index g may be defined in terms of the radius of gyration as follows:

$$g = \frac{\overline{(s^2)}_b}{\overline{(s^2)}_l} \qquad (3.92)$$

where subscripts b and l refer to branched and linear chains of equal molecular weight respectively. Theoretical treatments frequently assume θ-conditions and indicate this parameter as g_0. Although there is no basis for assuming the branched and linear polymers will have identical θ-temperatures in any solvent, Bohdanecky (107) notes that such is the case for all measured systems. The advantage of this parameter is that it is easily treated from a theoretical standpoint (108, 109) and can be related to light-scattering measurements.

For heterogeneous polymers with trifunctional or tetrafunctional branchpoints, one can write Equations (3.93) and (3.94) as follows:

$$\langle g_3 \rangle_w = \frac{6}{n_w}\left[\frac{1}{2}\left(\frac{2+n_w}{n_w}\right)^{1/2} \cdot \right.$$

$$\left. \ln\left(\frac{(2+n_w)^{1/2} + n_w^{1/2}}{(2+n_w)^{1/2} - n_w^{1/2}}\right) - 1\right] \qquad (3.93)$$

$$\langle g_4 \rangle = \frac{1}{n_2} \ln(1+n_w) \qquad (3.94)$$

where the symbol on the left is the weight-average value of g for trifunctional or tetrafunctional branchpoints respectively, and n_w is the weight-average number of branchpoints per molecule. For narrow molecular weight fractions, with a constant number n of branchpoints per molecule one may obtain Equations (3.95) and (3.96):

$$g_3(n) = \left[1 + \left(\frac{n}{7} \right)^{1/2} + \frac{4n}{9\pi} \right]^{-1/2} \quad (3.95)$$

$$g_4(n) = \left[1 + \left(\frac{n}{6} \right)^{1/2} + \frac{4n}{3\pi} \right]^{-1/2} \quad (3.96)$$

The other major indirect measure of branching depends on the intrinsic viscosity, which also varies predictably with polymer volume in solution. The nomenclature for the viscosity branching index varies, but it is most frequently defined as follows in Equation (3.97):

$$G = \frac{[\eta]_b}{[\eta]_l} \quad (3.97)$$

or for θ-condition as in Equation (3.98).

$$g' = \frac{[\eta]_{\theta,b}}{[\eta]_{\theta,l}} \quad (3.98)$$

If a linear polymer of equal (average) molecular weight is available, G may be measured directly, or it may be deduced from GPC measurements as discussed below. Relating G to a direct measure of branching, however, generally reduces to relating G to g (based on radius of gyration) because the latter parameter is more amenable to theoretical analysis. The usual form given is shown in Equation (3.99):

$$G = g^\nu \quad (3.99)$$

Zimm and Stockmayer first suggested (108) on the basis of the Flory-Fox equation that $\nu = \frac{3}{2}$, and this value does seem to apply to comblike polymers with only short chain branches. Zimm and Kilb later analyzed the theory of starlike polymers (109), in which all branches radiate from a central point, and found $\nu = \frac{3}{2}$. It was suggested that this value of ν should also apply to randomly branched polymers. The value of ν in any given instance is not conclusive, and various values can be found in the current literature for the same class of polymers. More recently, Park and Graessley (110) used polymer fractions of previously measured branching density to

determine an empirical relationship between G and g. Bohdanecky (107) also proposes an empirical approach and finds a value of $\nu \approx 0.8$ from a number of previously published studies.

Because the analysis of branching by GPC assumes some value for ν, this problem is of some importance. The most commonly used exponent appears to be 0.5, but a recent tabulation of published studies by Constantin (111) shows that values calculated for branching frequency are no more dependent on the choice of ν in the range 0.5–1.5 than on any of the other assumptions necessary to that method (such as contant λ).

3.7.3 Measurement

3.7.3.1 Fractionation
Classical fractionation followed by measurement of, say, intrinsic viscosity has essentially been superseded as a branching method by GPC in conjunction with on-line or off-line measurement of a molecular-size-dependent property. Both methods suffer from a similar problem: neither separates solely on the basis of molecular weight. Hence for a heterogeneous sample with both linear molecules and species of various degrees of branching, one has in each fraction an unknown mixture of molecular weights. Classical fractionation is even less susceptible than GPC to a theoretical analysis of this mixture, but since it separates by a different mechanism, the errors generated in the molecular weight and branching distributions by the two methods should be independent. Thus when the two agree, the results can be viewed with considerable confidence. Williamson and Cervenka (112) compared, for several polyethylenes, the results from fractionation with those from a Drott and Mendelson GPC analysis (see Section 3.7.3.2). They found good agreement in molecular weight regardless of whether or not θ-conditions were employed, even though the degree of branching λ varied with molecular weight.

3.7.3.2 GPC/Viscometry
Drott and Mendelson (113) first noted that the principle of universal calibration (57) provided

a GPC method for analyzing both the MWD and branching frequency. Universal calibration states that all polymer molecules eluting at a given volume are characterized by a constant hydrodynamic volume $J = [\eta]M$. Hence a branched polymer, eluting at a volume corresponding to some linear polymer standard, is related to it by Equation (3.100).

$$[\eta]_b M_b = [\eta]_1 M_1 \qquad (3.100)$$

This is what one would qualitatively expect, with a more compact branched molecule eluting at the same volume as a lower molecular weight but more expanded linear molecule.

Consider first the GPC of a narrow fraction of a branched polymer with measured intrinsic viscosity $[\eta]_b$. Calibration of the GPC with linear standards yields a universal calibration curve. The elution volume of the branched fraction will then yield M_b directly. Beyond this, one must know the Mark-Houwink viscosity molecular-weight relationship for the linear analog of the branched polymer; see Equation (3.101):

$$\eta_1 = KM_1^a \qquad (3.101)$$

One then calculates η_1 for a linear polymer of the same molecular weight as the branched polymer, and thus $g' = \eta_b/\eta_1$. If one then assumes an appropriate model for branching (tri- or tetrafunctional, mono- or polydisperse), and the appropriate exponent in the equation $g' = g^v$, the equations of Zimm and Stockmayer (108) permit the calculation of the branching frequency λ.

If short chain branching is known to exist (see below), one sets Equation (3.102):

$$g' = g'_{LCB} g'_{SCB} \qquad (3.102)$$

and can use an approximate formula devised by Stockmayer and cited by Billmeyer (114) to factor out the contribution of the short chain branches as given in Equation (3.103):

$$g_{scb} = \frac{1}{s+1} \left\{ 1 + s\,(1 - 2f + 2f^2 - 2f^3) + \right.$$
$$\left. s^2\,(-f + 4f^2 - f^3) \right\} \quad (3.103)$$

Here s is the number of branches and f the fractional length of the branch. For the case of a polydisperse branched polymer. Drott and Mendelson assume a constant branching frequency λ. The intrinsic viscosity of the whole polymer is then set equal to the sum over narrow volume intervals of the GPC chromatogram as follows in Equation (3.104):

$$[\eta]_{\text{measured}} = \sum_i w_i[\eta]_i = \sum \frac{h_i}{\Sigma h_i}[\eta]_i$$

$$(3.104)$$

where h_i is the height of the GPC chromatogram over interval i. One then has Equation (3.105):

$$[\eta]_{\text{measured}} = \frac{K}{\Sigma h_i} \sum_i h_i M_i^a g'_{SCL} g'_{LCB} \quad (3.105)$$

An iterative computer program then varies λ until a fit to the measured viscosity is achieved.

Wild et al. (115) have recently compared Drott and Mendelson's method with another method (116) for treating the same experimental data on LDPE with two variable parameters and no assumption of constant λ. They found no essential difference in the molecular weight distribution calculated. (The latter method gives no description of branching distribution).

Wild et al. also fractionated the same polymer and then used GPC and viscosity measurements to characterize its distribution more accurately. They found the true MWD to be reasonably close to that found by both iterative methods, even though λ was found not to be constant for the polymer in question.

The use of on-line viscometry obviates the need for such an iterative procedure or an assumption of constant λ and has been used with success by a number of workers including Park and Graessley (110) and Constantin (111). As mentioned earlier, such methods are not available for routine use.

A variation of this method that may become more generally available was reported by Hamielec et al. (117) who used a commercially available low-angle light-scattering (LALLS) detector to monitor the GPC of branched polyvinylacetate.

One problem still remains for both on-line methods: any elution volume contains a heterogeneous mixture of polymers of varying molecular weight and degree of branching. Hamielec (117) was the first to point out that the appropriate average molecular weight to use in the universal calibration was $\bar{M}_n$. Since light scattering yields $\bar{M}_w$, they were forced to use an iterative procedure to transform their data by way of an extended Mark-Houwink relationship. However, the method seems to be potentially extremely promising for routine analysis.

3.7.3.3 *Spectroscopy*
Short chain branching, which is generally treated separately only for low-density polyethylene, may be measured (118–121) by ^{13}C-nmr detection of both the branch carbons and the short chain carbons themselves. This technique also distinguishes between two-carbon and four-carbon branches (the major species involved).

3.8 SUMMARY

Though not subject to a "technological revolution," the theoretical treatment of MWDs is an area of current advances. These efforts focus on distribution for branched systems (122–124) and for polymerizations involving more than one set of reactive groups (125–127). Progress is also being made in the theoretical basis of the newest applications of GPC such as branched polymers.

The extension of GPC and its concurrent technical improvements is reducing the role of membrane and vapor phase osmometry in many cases to supporting absolute molecular weight methods. Modern osmometers offer design improvements and operating simplifications, but both techniques still require considerable skill, and their applications are made more laborious by the need for extrapolations to zero concentration. Osmometric techniques are most valuable in those cases where accurate molecular weights values are required (as

opposed to trends or changes) and also when exlusion chromatography results are obscured by adsorption or partition effects.

Light scattering is a fast growing technique largely because of dual advances in laser optics and data handling systems (128). Some areas of notable progress are (1) reduction of labor because of accurate single-concentration measurements at high dilution, (2) extension down into the molecular weight range 10^3–10^4, and (3) coupling with GPC. Instrumentation is still expensive, and further cost reductions would seem to be required for widespread use.

The ultracentrifuge is well suited for biological macromolecules. It can also provide useful information on random coil polymers with molecular weights too high for GPC or osmometry. Cost is a major factor in limiting the role of ultracentrifugation.

Viscometry does not provide absolute molecular weight values; its use is based on empirical relations rather than theoretical predictions. Viscometry continues to be an important tool (especially in quality control applications) because of its extreme simplicity, low cost, and ability to measure small differences in molecular weight for a polymer of given chemical composition and polydispersity. Automation is a boon for this wearisome technique, and a variety of devices have been produced by FICA, Schott, and Wescan.

GPC is still the most rapidly growing molecular weight method, with progress being made in column technology and in peripheral equipment and data processing. New columns — both polymer gels and treated glasses — offer a wider range of applications based on more diversified solvent affinity and adsorption behavior, coupled with higher resolution. Several new detectors are available based on fluorescence and ir adsorption or scintillation counting. The practitioner continues to benefit from the overall advances in HPLC pumps, injectors, and automated collection and injection devices. Last, progress continues on the principles of separation. While ideal GPC — corresponding to the universal calibration approach — is reasonably well reconciled with classical polymer solution theory (66), foundations need to be better developed for partitioning behavior (129), for ionic effects (130), and for the chromatography of associating systems (131).

The measurement of branching (132, 133) suffers from difficulties in defining quantitative parameters, but useful measurements can be made. New methods, notably GPC/light scattering (134), would seem to be limited more by lack of theoretical understanding than by instrumental factors. In particular, the implications of the heterogeneity of species eluting at any given instant have not been fully explored (117).

3.9 GLOSSARY

Distribution Theory

c	Polymer concentration (g/cm^3)
f	Number of branches in a star polymer
n_i	Number of molecules with degree of polymerization i
N_i	Number of moles with degree of polymerization i
$\bar{M}_n$	Number-average molecular weight
$\bar{M}_w$	Weight-average molecular weight
$\bar{M}_z$	z Average molecular weight
p	Extent of reaction of condensation polymer
W_i	Weight fraction of molecules with degree of polymerization i
α	Branching probability
μ_n	nth Moment of distribution
$[\eta]$	Intrinsic viscosity

Osmometry

a	Thermodynamic activity
a	Gravitational constant
A_n	Osmotic virial coefficient
A_1, A_2	Drop geometry parameters for VPO
g	Empirical osmotic virial coefficient factor
h	Excess osmotic height (cm)
ΔH_f	Heat of fusion of solvent
ΔH_v	Heat of vaporization of solvent
k_2	Henry's law constant
k_c, k_s	Heat transfer efficiency parameters for VPO

k	VPO instrument calibration parameter
p_1^0	Solvent vapor pressure
P	Pressure
Δr	Wheatstone bridge imbalance
R	Gas constant
ΔT_f	Freezing point depression
ΔT_b	Boiling point elevation
x	Mole fraction
v_1	Solvent molar volume
Γ	Virial osmotic coefficient
π	Osmotic pressure
ρ	Solvent density

Light Scattering

$\tilde{c}$	Speed of light
E	Amplitude of incident light beam
g	As defined for osmometry
i_θ^0	Intensity of scattered light (superscript 0 indicates particle much smaller than wavelength of light)
I_0	Intensity of incident light
K^*	Constant in Rayleigh factor equation
n	Solvent refractive index
N_A	Avogadro's number
N/V	Solute concentration in particles per unit volume
p	Dipole moment
r	Distance between point of measurement and scattering particle
$(\overline{r^2})^{1/2}$	Root mean square end-to-end distance of polymer chain
R_θ^0	Rayleigh factor (superscript 0 as for i_θ^0)
α	Excess polarazibility
z	dissymmetry coefficient
Γ	As defined for osmometry
λ	Wavelength
μ	Chemical potential
π	Osmotic pressure
θ	Angle between line of observation and path of incident beam

Ultracentrifugation

D	Diffusion coefficient
f	Frictional coefficient
r	Distance from axis of rotation
s	Sedimentation constant

$\bar{v}$ Partial specific volume of polymer

ρ Solution density

ω Angular velocity

Viscosity

a Mark-Houwink exponent

k' Huggins constant

k'' Kraemer constant

K Mark-Houwink coefficient

α Flory expansion coefficient

α',β Viscometry instrument constants

Φ Flory universal constant

η Viscosity (see Table 3.1)

ρ Density

GPC

A Calibration curve intercept constant

d Reciprocal slope of calibration curve

H Plate height L/N

J Universal calibration factor $[\eta]M$

K_{GPC} GPC partition constant

K_d Adsorption/partition term in K_{GPC}

L Column Length

N Number of theoretical plates

p True polydispersity $\bar{M}_w/\bar{M}_n$

p^* Value of p measured by GPC

V_e Elution volume

V_0 Exclusion volume

V_p Total pore volume

V_t Total column volume

Y Solubility parameter partition term

δ Solubility parameter

σ Standard deviation of peak

Branching

b Number of branches per molecule

g Branching index, $(\bar{s})^2_b/(\bar{s})^2_l$

g_0 Value of g at θ-conditions

g' Value of G at θ-conditions

G Branching index $[\eta]_b/[\eta]_l$

h_i Height of GPC chromatogram at $V_e = i$

n_w Weight-average value of b

(s^2) Mean square radius of gyration

λ Branching coefficient b/M

v Branching index constant, $\ln G/\ln g$

3.10 REFERENCES

1. P. J. Flory, *Principles of Polymer Chemistry*, Cornell University Press, Ithaca, New York, 1953.

2. C. Tanford, *Physical Chemistry of Macromolecules*, John Wiley, New York, 1961.

3. B. Vollmert, *Polymer Chemistry*, Springer-Verlag, New York, 1973.

4. H. G. Elias, *Macromolecules Vol. 1 Structure and Properties*, Plenum Press, New York, 1977.

5. H. L. Berger and A. R. Shulz, *J. Polym. Sci. A*, **2**, 3643 (1965).

6. Reference 1, p. 292.

7. L. H. Peebles, Jr., *Molecular Weight Distribution in Polymers*, Interscience Publishers, New York, 1971.

8. P. J. Flory, *J. Am. Chem. Soc.*, **58**, 1977 (1936).

9. G. V. Shultz, *Phys. Chem.*, **B30**, 379 (1935).

10. W. H. Stockmayer, *J. Polym. Sci.*, **9**, 69 (1952).

11. J. H. Jacobson and W. H. Stockmayer, *J. Chem. Phys.*, **18**, 1600 (1950).

12. G. Challa, *Makromol. Chem.*, **38**, 105 (1960).

13. A. Amemiya, *J. Phys. Soc. Japan*, **17**, 1245 (1962).

14. L. H. Peebles, Jr., *Macromolecules*, **7**, 872 (1974).

15. R. Goel, S. K. Gupta, and A. Kumar, *Polymer*, **18**, 851 (1977).

16. W. E. Gloor, *J. Appl. Polym. Sci.*, **19**, 273 (1955).

17. Reference 1, Chapter 9.

18. Reference 1, p. 359.

19. R. G. Garmon, "End Group Determinations," in P. E. Slade, Jr., Ed., *Polymer Molecular Weight Methods, Part 1*, Marcel Dekker, New York, 1975.

20. J. L. Armstrong, "Critical Evaluation of Commercially Available High-Speed Osmometers," in *Characterization of Macromolecular Structure*, National Academy of Sciences, 1968.

21. R. D. Ulrich, "Membrane Osmometry," in P. E. Slade, Jr., Ed., *Polymer Molecular Weights, Part 1*, Marcel Dekker, New York, 1975.

22. Reference 1, p. 280

23. D. Burge, *J. Appl. Polym. Sci.*, **24**, 293 (1979).

24. A. V. Hill, *Proc. R. Soc. (London)*, **A-127**, 9 (1930).

25. B. H. Bersted, *J. Appl. Polym. Sci.*, **17**, 1415 (1973).

26. K. Kamide, T. Terakawa, and H. Uchiki, *Makromol. Chem.*, **177**, 1447 (1976).

27. C. E. M. Morris, *J. Appl. Polym. Sci.*, **21**, 435 (1977).

28. A. C. Meeks and I. J. Goldfarb, *Anal. Chem.*, **39**, 908 (1967).

29. C. A. Glover, "Absolute Colligative Methods," in P. E. Slade, Jr., Ed., *Polymer Molecular Weights, Part 1,* Marcell Dekker, New York, 1975.

30. Reference 1, Chapter 7.

31. Application note LS-1, Chromatix Corp., Sunnyvale, California.

32. D. E. Nordhous and J. B. Kinsinger, *J. Polym. Sci., Symp. No. 43,* 251 (1973).

33. H. Fujita, *J. Phys. Chem.*, **73**, 1759 (1969).

34. H. W. Osterhoudt and J. W. Williams, *J. Phys. Chem.,* **69**, 1050 (1965).

35. P. J. Wan and E. T. Adams, Jr., *Biophys. Chem.*, **5**, 207 (1976).

36. W. J. Archibald, *J. Phys. Coll. Chem.*, **51**, 1204 (1947).

37. H. Staudinger and W. Heuer, *Chem. Ber.*, **63**, 222 (1930).

38. W. J. Moore, *Physical Chemistry,* Prentice-Hall, Englewood Cliffs, New Jersey, 1962, p. 222.

39. J. G. Kirkwood and J. Riseman, *J. Chem. Phys.,* **16**, 565 (1948).

40. Reference 1, Chapter 14.

41. J. Brandrup and G. H. Immergut, Eds., *Polymer Handbook,* 2nd ed., John Wiley, New York, 1975, Chapter 4.

42. Reference 1, p. 313.

43. O. F. Solomon and I. Z. Cuita, *Bul. Inst. Polit. "Georghe Gheorgiudy"*, **30**, 7 (1968).

44. D. K. Carpenter and L. Westerman in P. E. Slade, Jr., Ed., *Polymer Molecular Weights, Part 2,* Marcel Dekker, New York, 1975, p. 449.

45. U. Lohmander, *Makromol. Chem.*, **72**, 159 (1964).

46. J. T. Yang, *Adv. Protein Chem.*, **16**, 323 (1961).

47. C. Mussa and V. Tablino, *Polymer*, **1**, 266 (1960).

48. J. Schurz, *Rheol. Acta*, **3**, 43 (1963).

49. L. H. Tung, Ed., *Fractionation of Synthetic Polymers,* Marcel Dekker, New York, 1977.

50. M. J. D. Cantow, Ed., *Polymer Fractionation,* Academic Press, New York, 1967.

51. Reference 1, pp. 339–345, 559–563.

52. J. H. Hildebrand and R. L. Scott, *The Solubility of Nonelectrolytes,* 3rd ed. Dover, New York, 1964, p. 370.

53. E. M. Barrall II, J. F. Johnson, and S. R. Cooper, in L. H. Tung, Ed., *Fractionation of Synthetic Polymers,* Marcel Dekker, New York, 1977.

54. L. Wild and T. Ryle, *Polym. Prepr.*, **18(2)**, 182 (1977).

55. P. Flodin, *Dextran Gels and Their Applications in Gel Filtration,* unpublished doctoral dissertation, Uppsala (1962).

56. T. C. Laurent and J. Killander, *J. Chromatogr.*, **14**, 317 (1964).

57. H. Benoit, Z. Grubisic, P. Rempp, D. Decker, and J. G. Zilliox, *J. Chim. Phys.*, **63**, 1507 (1966).

58. Z. Grubisic, P. Rempp, and H. Benoit, *J. Polym. Sci. B*, **5**, 753 (1967).

59. A. R. Weiss and E. Cohn-Ginsburg, *J. Polym. Sci. A-2,* **8**, 89 (1970).

60. J. Porath, *Pure Appl. Chem.*, **6**, 233 (1963).

61. W. Haller, *J. Chromatogr.*, **32**, 676 (1968).

62. E. F. Casassa, *J. Phys. Chem.*, **75**, 3929 (1971).

63. J. C. Giddings, E. Kucera, C. P. Russell, and M. N. Meyers, *J. Phys. Chem.*, **72**, 4397 (1968).

64. E. F. Casassa, *J. Polym. Sci. A-2,* **5**, 773 (1967).

65. E. F. Cassassa and Y. Tagami, *Macromolecules*, **2**, 14 (1969).

66. E. F. Casassa, *Macromolecules*, **9**, 182 (1976).

67. E. F. Casassa, *J. Phys. Chem.*, **75**, 3929 (1971).

68. M. E. van Kreveld and N. van den Hold, *J. Chromatogr.*, **83**, 111 (1973).

69. K. O. Pedersen, *Arch. Biochem. Biophys. Suppl. No. 1,* **157** (1962).

70. C. M. Guttman and E. A. DiMarzio, *Macromolecules*, **3**, 681 (1970).

71. W. W. Yau and C. P. Malone, *J. Polym. Sci. B-2,* **5**, 663 (1967).

72. W. W. Yau, *J. Polym. Sci. A-2,* **7**, 483 (1969).

73. J. C. Giddings, L. M. Bowman, and M. N. Meyers, *Macromolecules*, **10**, 443, (1977).

74. A. C. Ouano and J. A. Barker, *Sep. Sci.*, **8**, 673 (1973).

75. M. Kubin, *J. Chromatogr.*, **108**, 1 (1975).

76. W. Haller, A. M. Basedow, and B. Konig, *J. Chromatogr.*, **132**, 387 (1977).

77. W. W. Yau and S. W. Fleming, *J. Appl. Polym. Sci.*, **12**, 2111 (1968).

78. Y. Kato, H. Sasaki, M. Aiura, and T. Hashimoto, *J. Chromatogr.*, **153**, 546 (1978).

79. T. Hashimoto, H. Sasaki, and Y. Kato, *J. Chromatogr.*, **160**, 301 (1978).

80. H. M. Weetall and A. M. Filbert, "Porous Glass for Affinity Chromatography Applications," in W. B. Jakoby and M. Wilcheck, Eds., *Methods in Enzymology,* Volume 34, Academic Press, New York, 1974.

81. K. Unger, J. Schick-Kalb, and K.-F. Krebs, *J. Chromatogr.*, **83**, 5 (1973).

82. S. Rokushika, T. Ohkawa, and H. Hatano, paper presented at U.S.-Japan Seminar on Advanced Techniques of Liquid Chromatography, University of Colorado, Boulder, June, 1978.

83. A. C. Ouano, *J. Polym. Sci. A-1,* **10**, 2169 (1972).

84. A. C. Ouano and W. Kaye, *J. Polym. Sci, A-1,* **12**, 1151 (1974).

85. S. deKeczer, P. Dubin, R. Hale, M. Kronstadt, and A. R. Read, *Am. Chem. Soc., Polym. Prepr.*, 18(2), 221 (1977).

86.   S. T. Balke and A. E. Hamielec, *J. Appl. Polym. Sci.,* **13**, 1381 (1969).

87.   W. W. Yau, J. J. Kirkland, D. D. Bly, and H. J. Stoklosa, *J. Chromatogr.,* **125**, 219 (1975).

88.   L. H. Tung and J. R. Runyon, *J. Appl. Polym. Sci.,* **13**, 2397 (1969).

89.   R. Kotoka, *Makromol. Chem.,* **56**, 77 (1976).

90.   K. Unger and R. Kern, *J. Chromatogr.,* **122**, 345 (1976).

91.   J. V. Dawkins and M. Hemming, *Makromol. Chem.,* **176**, 1815 (1975).

92.   D. H. Freeman and D. Killian, *J. Polym. Sci. Polym. Phys. Ed.,* **15**, 2047 (1977).

93.   K. H. Altgelt and J. C. Moore, in M. J. R. Cantow, Ed., *Polymer Fractionation,* Academic Press, New York, 1967.

94.   M. Pope, J. Fahnrich, and M. Stejskal, *J. Chromatogr. Sci.,* **14**, 537, (1976).

95.   J. V. Dawkins and M. Hemming, *Polymer,* **16**, 554 (1975).

96.   J. Borak and M. Smrzv, *J. Chromatogr.,* **144**, 57 (1977).

97.   A. L. Spatorico and G. L. Beyer, *J. Appl. Polym. Sci.,* **19**, 2933 (1975).

98.   A. Campos and J. E. Figueruelo, *Makromol. Chem.,* **178**, 3249 (1977).

99.   H. D. Crone and R. M. Dawson, *J. Chromatogr.,* **129**, 91 (1976).

100.   P. J. Blagrove and M. J. Frenkel, *J. Chromatogr.,* **132**, 399 (1977).

101.   A. C. M. Wu, W. A. Bough, E. C. Conrad, and K. E. Alden, Jr., *J. Chromatogr.,* **128**, 87 (1976).

102.   H. Englehardt and D. Mathes, *J. Chromatogr.,* **142**, 311 (1977).

103.   K. Chitumbo and W. Brown, *J. Chromatogr.,* **8**, 17 (1973).

104.   W. H. Stockmayer, *J. Polym. Sci.,* **9**, 69 (1952).

105.   Reference 1, pp. 347–398.

106.   F. M. Mirabella, Jr. and J. F. Johnson, *J. Macromol. Sci. — Rev. Macromol. Chem.,* **12**, 81 (1975).

107.   M. Bohdanecky, *Macromolecules,* **10**, 971, (1977).

108.   B. H. Zimm and W. H. Stockmayer, *J. Chem. Phys.,* **17**, 1301 (1949).

109.   B. H. Zimm and R. W. Kilb, *J. Polym. Sci.,* **37**, 19 (1959).

110.   W. S. Park and W. W. Graessley, *J. Polym. Sci. Polym. Phys. Ed.,* **15**, 85 (1977).

111.   D. Constantin, *Eur. Polym. J.,* **13**, 907, (1977).

112.   G. R. Williamson and A. Cervenka, *Eur. Polym. J.,* **10**, 295 (1974).

113.   E. E. Drott and R. A. Mendelson, *J. Polym. Sci., Part 2,* **8**, 1361 (1970).

114.   F. W. Billmeyer, *J. Am. Chem. Soc.,* **75**, 6118 (1953).

115.   L. Wild, R. Ranganath, and A. Barlow, J. Appl. Polym. Sci., **21**, 3331, (1977).

116.   A. Ram and J. Miltz, *J. Appl. Polym. Sci.,* **15**, 2639 (1971).

117.   A. E. Hamielec, A. C. Ouano, and L. L. Nebenzahl, *J. Liq. Chromatogr.,* **1**, 527 (1978).

118.   E. P. Otocka, R. J. Roe, M. Y. Hellman, and P. M. Muglia, *Macromolecules,* **4**, 507 (1971).

119.   D. E. Dorman, E. P. Otocka, and F. A. Bovey, *Macromolecules,* **5**, 574 (1972).

120.   J. C. Randall, *J. Polym. Sci., Polym. Phys. Ed.,* **11**, 275 (1973).

121.   F. A. Bovey, F. C. Schilling, F. L. McCracken, and H. L. Wagner, *Macromolecules,* **9**, 76 (1976).

122.   E. M. Valles and C. W. Macosko, *Macromolecules,* **12**, 521 (1979).

123.   P. Luby and L. Kuniak, *Makromol. Chem.,* **180**, 2207 (1979).

124.   R. S. Whitney and W. Burchard, *Makromol. Chem.,* **181**, 869 (1980).

125.   D. R. Miller and C. W. Macosko, *Macromolecules,* **11**, 656 (1978).

126.   S. K. Gupta, A. Kumar, and A. Bhargava, *Polymer,* **20**, 305 (1979).

127.   E. Ozizmir and G. Odian, *J. Polym. Sci., Polym. Chem. Ed.,* **18**, 2281 (1980).

128.   H. Suzuki, *Br. Polym. J.,* **11**, 35 (1979).

129.   J. E. Figueruelo and V. Soria, *J. Liq. Chromatogr.,* **3**, 367 (1980).

130.   S. N. E. Omorodion, A. E. Hamielec, and J. L. Brash, in R. Provder, Ed., *Size Exclusion Chromatography,* ACS Symposium Series 138, Washington, D.C. 1980.

131.   P. L. Dubin, *J. Liq. Chromatog.,* **3**, 623 (1980); P. L. Dubin, in T. Provder, Ed., *Size Exclusion Chromatography,* ACS Symposium Series 138, Washington D.C., 1980.

132.   W. Burchard and M. Schmidt, *Am. Chem. Soc. Polym. Prepr.,* **20** (2), 164 (1979).

133.   G. C. Berry, *Am. Chem. Soc. Polym. Prepr.,* **20** (2), 168 (1979).

134.   J. E. Girard, D. S. Cretek, P. E. Gundlach, and S. R. Weissman, *Am. Chem. Soc. Polym. Prepr.,* **20** (2), 134 (1979).

CHAPTER 4 ———————————————

Microstructure and Analysis

S. A. LIEBMAN and A. M. HASSAN

Armstrong Cork Company
Lancaster, Pennsylvania

4.1	Introduction	134
	4.1.1 Calculation of Microstructure	135
	4.1.1.1 Copolymer Compositions	135
	4.1.1.2 Conformational and Configurational Analysis	136
	4.1.2 Molecular Orbital Theory Applied to Polymers	137
4.2	High-Resolution Nuclear Magnetic Resonance (NMR) Spectroscopy	138
	4.2.1 ^{13}Carbon Spectroscopy	138
	4.2.2 Chain Statistics of Vinyl Polymers	138
	4.2.3 Chain-Sequence Probabilities	141
	4.2.3.1 Chain Conformation	141
	4.2.4 Nmr Spectra of Poly (methyl methacrylate), PMMA	143
	4.2.4.1 β-Methylene Spectrum	143
	4.2.4.2 α-Methyl Spectrum	146
	4.2.4.3 Observation of Longer-Sequence Configurations	147
	4.2.4.4 ^{13}C Nmr	150
	4.2.5 Nmr Spectra of Poly (vinyl chloride), PVC	151
	4.2.5.1 β-Methylene Spectrum	152
	4.2.5.2 Methine Proton Spectrum	154
	4.2.5.3 Calculation of Nmr Spectra	155
	4.2.5.4 ^{13}C Nmr	155
	4.2.6 Chlorinated PVC, Chlorinated Polyethylene (PE), and Hydrogenated Polybutadiene (BD) Spectra	158
	4.2.7 Polypropylene (PP) Proton Spectrum	158
	4.2.8 Theoretical Calculation of Nmr Spectra of PVC and PP.	159
	4.2.9 Nmr of Binary Copolymers	159
	4.2.10 Branch Content of PE and PVC	159
4.3	Infrared (ir) and Raman Spectroscopy	162
	4.3.1 Factor-Group Analysis	163
	4.3.2 Crystallinity	163
	4.3.3 Polarization Phenomena	164
	4.3.4 Conformational and Configurational Analysis	165
	4.3.5 Vibrational Coupling and Dispersion Curves	167
	4.3.6 Polymer-Defect Analysis	171
	4.3.7 Branch Defects	175
4.4	Pyrolysis Gas Chromatography (pgc)	176
	4.4.1 Pyrolysis Instrumentation and Identification of GC Effluents	176
	4.4.2 Applications and Interpretation	179
	4.4.3 PGC of Polymers and Copolymers	180
	4.4.3.1 General Polymer Systems	180
	4.4.3.2 Vinyl-Type Copolymers	181
	4.4.3.3 Polyolefin Systems	182
	4.4.3.4 Pyrolysis of PVC	185
	4.4.4 PGC of PE and Reduced PVC for Branch Content	186
	4.4.4.1 PE Pyrolysis	189
	4.4.4.2 PGC of LAH-Reduced PVC in Helium	190
	4.4.4.3 Pyrolysis Hydrogenation GC (PHGC) of LAH-Reduced PVC	190
	4.4.5 PGC of PVC for Branch Content	191
4.5	Summary	192
4.6	Glossary of Symbols	193
4.7	References	193

S. A. Liebman is currently at the Chemical Data System, Inc., Oxford, Pennsylvania.
A. M. Hassan is currently at the Kuwait Institute for Scientific Research, State of Kuwait.

4.1 INTRODUCTION

We approach the topic of polymer micro-structure from the standpoint of organic analysis, using the theory and practices common to organic and physical chemists for proof-of-structure of materials. What are the current and potential means available to polymer chemists to define in some detail the chemical architecture of complex macro-molecular systems?

The tailoring of chemical structure in high polymers has shown significant advances within the last few years. Predicting and defining chemical units and their distributions from theoretical polymerization kinetics have been augmented by analytical tools in spectroscopy, thermal analysis, and chro-matography. Synthesis of high-performance polymers for thermal- or mechanical-demand areas now reinforces the need for adequate structural analysis to guide future design criteria. The processing and performance behavior of certain synthetic polymers has been shown to be a function of seemingly minor and heretofore undefined chemical details in microstructure; that is, the presence of chemical defects in the form of branching, end-group anomalies, activated or oxidative sites, and so on.

The means to define chemical structure have undergone significant advances using the theoretical concepts of atomic bonding, interaction potentials, symmetry, and quan-tum mechanical considerations. The limits to which these predictive tools can be applied have now widened owing to the computer technology revolution and the fact that calculations of three-dimensional polymer structures can be realistically achieved. The detailed irregularities, however, which may be in the form of branches, cross-links, or defect linkages, remain to be clearly delineated. Also, analysis of the presence and distribution of small amounts of A monomer in a copolymer $(AB)_x$ challenges all current experimentalists. Hence, the need to confirm major or minor predicted structures, detect and quantify defects, and correlate analytical results with specific structural features of a polymer has been established in the field of instrumental chemical analysis.

General instrumentation and methodology have made such advances with regard to treating a diverse array of polymers, that former restrictions of sample size, solubility, or bulk form are no longer insurmountable problems to the analyst. Means of estab-lishing polymer purity (polymerization ad-ditives, solvent occlusions, etc.) have now allowed significant information on the more defined details of the actual polymer micro-structure. For obvious reasons, it was neces-sary to narrow the analytical tools that are included in this discussion to the major areas in spectroscopy and chromatography that have been successfully utilized by many work-ers: high-resolution nuclear magnetic reso-nance (nmr), infrared (ir), Raman, and pyrol-ysis-gas chromatography (pgc).

The capability of both ^{1}H and ^{13}C$_{nmr}$ for polymer microstructural analysis from basic theoretical considerations is coupled with spe-cific, detailed experimental findings on a few chosen systems. The ^{13}C Fourier-transform instrumentation developments allow unprec-edented studies into the complex architecture of high polymers in solution and, as seen recently, as solids.

Similarly in the ir and Raman fields, spec-tral theory and interpretation of new experi-mental results are shown to be active fields. The current development of computer-as-sisted research efforts, such as Fourier-trans-form ir, provides a major contribution to polymer analysis and relationship of structure to properties.

In addition, the uses of pgc in elucidation of polymer microstructure have been diverse and increasingly impressive. The innovative ana-lytical instrumentation currently used by many laboratories throughout the world en-ables researchers to perform detailed, ac-curate experiments that can be designed for obtaining highly specific information. The determination of tacticity, sequence length, copolymer content, defects, and branch con-tent in many systems illustrates the inherent capability of pgc and its potential contri-butions to the polymer characterization field.

Hence, the concern in this chapter is with both the theoretical, ab initio approaches to bonding, configurational and conformational predictions, and the practical means of experimental verification. The role of modern organic analysis in this interdisciplinary study is evident as theoretical design and synthesis of high-performance polymer systems become reality.

4.1.1 Calculation of Microstructure

It is significant that monomeric sequences in copolymer systems and stereochemical arrangements in homopolymer chains have been studied extensively by theoreticians. The utilization of such information is continually being developed by active polymer researchers in a variety of applications to establish structure-property relationships.

4.1.1.1 *Copolymer Compositions*

The ability to predict structural compositions of copolymers has been developed since the 1940s with varying degrees of success. The classical copolymer equation (1) allowed calculation of copolymer composition at any instant during the reaction by use of reactivity ratios. The Alfrey-Price Q and e scheme (2) has been utilized and improved most recently by Greenley (3) by a linear least squares evaluation of statistically significant reactivity ratios from the literature. This compilation of parameters represents the improved characterization of monomer copolymerization behavior as the available experimental data allow. The work by Kennedy et al. (4) gives a graphic method of determining the reactivity ratios of monomers to develop a relationship to better serve as a diagnostic tool for copolymerizations that follow cationic or free-radical mechanisms. With such advancements, free-radical copolymerization of vinyls will be more exact, and the copolymerization equation will continue to be a useful mathematical approximation, where applicable, to calculate structure from monomer feeds and their respective radical reactivities.

Work by Harwood et al. (5, 6) has allowed calculation of uninterrupted monomer sequences (runs), expressed as run-number per 100 monomer units. For olefins, the run-number R may be calculated from reactivity ratio values and M, the monomer feed in moles. The appropriate Fortran computer programs are available to perform these sequence-distribution calculations (7). A treatise edited by Lowry in 1970 (8) has detailed the theory and application of Markov chain statistics and Monte Carlo calculations in polymer science for a more fundamental understanding of sequence-distribution treatments.

Noah et al. (9) have studied the question as to what length and number of model chains are required to describe the compositional heterogeneity in sufficient detail and, thus, relate to the actual reaction mechanism. The multiplet method (10) was used to follow compositional heterogeneity (11) by the comparison of several solutions generated by varying the Markovian approximations (first, second, third order) that were applied to the model. Helfand in 1972 reported (12) the use of a modified Markov transition matrix to provide a generation function in the calculation of mole fractions of M_1 monomer blocks with certain lengths in component M_2. The results agreed with the original work by Frensdorff (13) who, following Monte Carlo statistics for predictive composition of the block copolymer, noted monomer sequences get into "blocks" long enough to separate and provide domains within which distinct physical properties are exhibited; for example, ethylene interrupts in polypropylene chains (14). Tsuchida et al. (15) in 1974 developed a simulation model of the copolymerization reaction for n-dimensional copolymers using the Frensdorff method of Markov probabilities with a termination reaction and an added recombination reaction in order to calculate the compositional distribution in a terpolymer system. Thus, for applications of "well-behaved" polymerizations, the ability to calculate microstructural details of monomer placements and numbers has been shown to be successful for many years.

4.1.1.2 Conformational and Configurational Analysis

In a more demanding situation, that of conformational and configurational details, the compositional calculations have been extended only recently to polymer systems from the simple molecules traditionally handled by organic chemists.

Initially, conformational analysis as applied to polymers by Markov and Volkenstein (16) assumed the chain to be a set of repeating units with fixed lengths that have a defined number of rotational states. The rotational isomeric state (ris) approach is a significant means of handling polymer conformational calculations in conjunction with matrix-multiplication methods (17).

An extended approach is presented in the all skeletal rotations scheme (asr) that takes into account other conformational possibilities even though they may not correspond to true energy minima, as in the ris method. Using a reasonable van der Waals radius of the methyl group, good agreement was obtained with experimental parameters for isotactic polypropylene (PP) with adjustment of parameters between nonbonded atoms, as reported by Allegra et al. (18) with the asr method. Boyd and Breitling (19) had used the ris scheme with several geometric parameters included to also successfully treat isotactic PP. Additionally, Suter and Flory (20) reported the calculated conformational energies of PP diads (accounting for full interactions between all pairs of atoms in the ris scheme) to obtain an accurate intramolecular energy for the PP chain. Ten energy minima were delineated for the meso diad and ten for the racemic diad. Statistical weight matrices were developed that gave excellent agreement with the experimental values obtained from equilibration of the conformers in appropriate oligomers.

Correlation of ris theory with experimental results for the configuration-dependent properties was reported by Mark and Ko (21) for several polymers—poly(dimethylsilmethylene), poly(dimethylsiloxane), polyisobutylene, and polyethylene (PE). Improved correlation was possible by noting subtle molecular parameters, such as ascribing larger partial charges for skeletal methylene groups than for the pendant methyl groups. Tadokoro et al (22) analyzed the most stable conformation of typical helical isotactic polymers—PP, poly(4-methyl-l-pentene), poly(3-methyl-l-pentene), polyacetaldehyde, and poly(methyl methacrylate) (PMMA)—with the inclusion of the intramolecular potential energy from internal rotational barriers, van der Waals, and electrostatic interactions. The molecular conformations of all systems, except PMMA, were governed primarily by steric interactions of the side chains and were modified also by the intermolecular interactions in the crystal. In 1975 Tonelli (23) reported the application of the ris model of polymer chains to evaluate the conformational entropy by utilization of semiempirical potential-energy functions.

The monumental work by Flory et al. (24–26) brought such computational capabilities to vinyl-polymer microstructural studies using the ris method. A disubstituted vinyl chain (PMMA) was treated with the use of Monte Carlo methods to generate a 200-unit molecular chain for random sequencing of two structural conformers according to Bernoullian statistics. Agreement with experimental data suggested that the theoretical methods were qualitatively useful in predicting conformation of the ester group in the PMMA system and the resulting energy surface over various rotational angles.

Hence, linear-chain configurational studies have used the ris approximations successfully, and attempts to compute configurational-partition functions from a priori probabilities and conditional probabilities for a branched polymer were reported by Tonelli (27) and Mattice (28). It was shown that calculations can be made for butyl branches along a chain to estimate the number of bonds between branch points before the n-butyl groups are sterically independent. The work by Mattice (28) on PE calculations indicated that bonds in the main chain that are one bond away from the branch site have enhanced a priori probability of being in the trans conformation, whereas the conformation of the butyl group itself is essentially independent of position. Whether Tonelli's (27) approach (which used approximate functions for individual linear chains within the branched PE) or Mattice's (with exact configuration-partition function for the whole polymer) is applied to PE, the indicated bond in the main chain at a branch point has a reduced a priori probability of being in a trans state, and the next bond away from the branch point has enhanced probability of being in the trans form. Extensions of these approaches can be made to vinyl-polymer diads for the calculation of preferred conformations.

4.1.2 Molecular Orbital Theory Applied to Polymers

To establish more accuracy in the calculated conformational energies and resultant physical properties, the application of quantum chemistry to polymer systems has been made successfully. Several methods in quantum theory have emerged that are categorically divided into the ab initio (nonempirical) methods and semiempirical methods (29–33). The former treatments rely on calculated molecular orbitals (MO) obtained from a linear combination of atomic orbitals (LCAO). Slater-type orbitals (STO) involve a different radial contribution to the wave function, whereas Gaussian-type orbitals (GTO) utilize still another function for the orbital radial term. General linear combinations of other functions are used, all of which are referred to as "basis sets" in the calculations. The self-consistent field approach (SCF) imposes mathematical treatments in which an electron is in an average potential field of all other electrons and iterations are used to develop the total potential field. Ab initio calculations imply that all the integrals involved in the orbitals designated are calculated, but often these methods do not provide results superior to the semiempirical methods, which can be parameterized to more closely calculate the desired properties.

Semiempirical methods may include: Huckel and extended-Huckel (EHM), and Pariser-Parr-Pople (PPP) methods, both of which neglect electron-repulsion terms and are, therefore, not SCF methods, CNDO (complete-neglect-of differential overlap); IN-DO (intermediate neglect of differential overlap); MINDO (modified INDO); MINDO/2, MINDO/3 modifications of MINDO (34, 35) are actively used approaches with an SCF treatment. The modified CNDO/2 method (36–38) is used extensively, and deals only with the valence orbitals and a minimum Slater basis set, where many integrals are neglected or evaluated semiempirically. The INDO and CNDO/2 SCF methods have given reliable prediction of total energy, geometries, and electron distributions for small molecules and represent realistic compromises for polymer calculations. Generally, it has been found in the calculations for total electron densities, dipole moments, and MO energies that if parameters are changed to improve certain spectral or optical calculations, the predicted geometries are not as successful. These computer programs are available from the Quantum Chemistry Program Exchange, Indiana University, Bloomington, Indiana.

Imamura in 1970 used EHM with a tight-binding approximation to get conformational potential-energy surfaces, minimum energy conformations, and charge distributions in a polymer system (39). The electronic structure of PE was calculated, and a search for the most stable conformation was made using semiempirical potential functions for barriers of internal rotations, nonbonded interactions, and adjustable parameters that influenced the conformations with the minimum energy. The CNDO/2 method by Pople and Segal was applied by Morokuma (40) for PE, and agreement between experimental data and the molecular calculations gave the conclusion that the fully extended, all-trans conformation is the most stable one and confirmed the significance of this approach.

Andre et al. (41) developed a computer program POLYMOL that utilizes a contracted basis set of Gaussian orbitals in the calculation of polymer structure. Application of MO theory to infinite polymer chains by O'Shea and Santry (42) involved solving 50 equations in the CNDO/2 method to satisfy convergence problems peculiar to SCF calculations for the $(CH)_n$ polymer. Translational symmetry was used to simplify the SCF matrix equations. The idealized geometry for the $(CH)_n$ model was planar and regular for calculating energy minima in a symmetric system.

4.2 HIGH-RESOLUTION NUCLEAR MAGNETIC RESONANCE (NMR) SPECTROSCOPY

The major successful application of high-resolution nmr to polymers came in 1960 when it was demonstrated by Bovey and Tiers that nmr is a powerful method for studying the stereoregularity in polymers (43). Since then, with the impressive improvement in instrumentations, especially ^{13}carbon nmr, the application of high-resolution nmr to study polymer structure has been expanding at a rapid pace and has warranted many reviews and books (44–56). In this section two polymer systems, poly(methyl methacrylate) (PMMA) and poly(vinyl chloride) (PVC), are treated in detail for illustration of the nmr analysis in microstructural studies, with emphasis on polymer configuration, conformation, and branching.

The nmr experiment provides detail in molecular structure by means of probing the relative relationships of nuclei by virtue of their chemical shift δ and coupling-constant J parameters. These terms are based upon the local magnetic fields that are modified from the imposed instrumental magnetic field H_0 by interaction of other magnetic nuclei (^{1}H, ^{13}C, ^{19}F, etc.) and their local magnetic or electronic environment. The coupling-constant term arises from electron-coupled nuclear spin-spin interactions. This effect causes the respective nucleus to resonate at slightly altered frequencies, which results in a multiplicity of absorption lines in the spectrum. The basic theory and nomenclature of nmr are treated elsewhere (45, 46, 49, 50) in detail.

4.2.1 ^{13}C Spectroscopy

The instrumental advances in Fourier-transform computer methods for ^{13}C nuclei have made it an important area of nmr for structural determination (53–56). In a H_0 field of 10,000 gauss, an isolated ^{13}C nucleus resonates at $\nu = 10.7$ MHz, but the intensity relative to a ^{1}H in the same field is lowered by a factor of about 5700, since the ^{13}C nucleus has a lower magnetic moment and natural abundance. The chemical shift and coupling constant are used similarly as in proton magnetic resonance (pmr), but it is the former parameter that is most significant for structural determinations by ^{13}C. It is important that spin decoupling (or removal of the spin-spin interactions) of the ^{13}C with nearby ^{1}H is utilized to give singlet spectra, although this method does alter the intensity of the absorption band by virtue of the nuclear Overhauser enhancement. As with ^{1}H spectra, the ^{13}C band intensities are dependent on their relaxation times T_1, and comparisons of band areas must take this factor into account.

4.2.2 Chain Statistics of Vinyl Polymers

Information of chain configuration is derived from the chemical-shift parameter, while information from spin-coupling measurements permits determination of polymer conformation. The terms "nuclear symmetry equivalence" and "nuclear magnetic equivalence" are important in this regard. All nuclei with symmetry equivalence (same elements of symmetry or exchange positions when symmetry operations are performed on them) also have the same chemical shift. Magnetic equivalence, however, is established only when all nuclei with symmetry equivalence couple equally to all other magnetically equivalent nuclei within the same molecule.

Consider a polymer chain with the repeat structural unit $(CH_2CR_1R_2)_n$ in which all units are linked head-to-tail. The chain configuration is determined by the arrangement of R groups about the chain, as seen in Figure 4.1. Since the α-carbons constitute an asymmetric center (actually pseudoasymmetric), the chain configuration can be described in terms of the stereochemical configuration of α-groups as d or l. A further simplification may be made by adopting the stereochemical configuration of meso (m) for a physically identical pair, ll or dd; and racemic (r) for a mirror-image pair, dl or ld. Through this notation, chain configuration may be described in terms

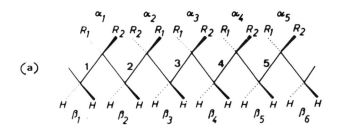

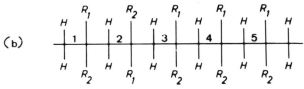

Figure 4.1 Stereochemical representations. (*a*) A segment of the extended chain for a homopolymer with the monomer unit $-CH_2-CH^1R^2-$. The α and β-carbon atoms are numbered according to their respective monomer unit. The two equivalent configurations of monomer units 1 and 2 are termed a racemic diad of monomer units. (*b*) A segment of the same chain in Fischer projection (45). Reprinted with permission from Verlag Chemie GMBH.

of the m and r diad placements (Figure 4.1). Completely tactic chains will have configuration . . . rrrr . . . for syndiotactic structures and . . . mmmm . . . for completely isotactic ones. Completely atactic (heterotactic) chains have an equal number of m and r configurations that are randomly distributed throughout the chain. In partly tactic chains the number of m configurations does not necessarily equal the r configurations. The arrangement of the two sets of units in configurationally different sequences can be treated as a statistical one, and its solution may be verified by comparison to the observed nmr results (44, 45, 47).

If only the R substituents on the nearest α-carbons exert deshielding influence on the β-CH_2 protons in the structure shown in Figure 4.1, then the nmr measurement can discriminate only between the two (meso and racemic) configurations. The nmr spectrum of the chain will be fully described by these two diad placements. Symmetry consideration leads to the conclusion that all the CH_2 groups in racemic configuration are equivalent. Furthermore, the two protons within each group are also equivalent; consequently, all the CH_2 protons in racemic-diad configuration have the same chemical shift. In the absence of vicinal coupling (coupling between protons

on two adjacent carbons), the spectrum of $-CH_2-$ in racemic configuration consists of one singlet. By contrast, although all the CH_2 groups in meso configurations are equivalent, the two protons within each CH_2 group are magnetically nonequivalent. One of the protons is flanked by two R_1 groups, the other is flanked by two R_2 groups (Figure 4.1). In meso configuration, therefore, the two methylene protons are expected to have two different chemical-shift values, and in the absence of vicinal coupling, should give rise to an AB-type or AX-type spectrum with a geminal coupling constant $J_{AB} \simeq 15$ Hz. Hence, from the nmr measurement it should be possible to distinguish equivocally between meso (isotactic) and racemic (syndiotactic) diad configurations. The ratio of the two configurations in the chain can be determined from the relative area of their nmr absorption signal. Thus, in addition to providing for an absolute means of detecting tactic structures in the chain, the nmr measurement also allows for a quantitative determination of the degree of tacticity in polymers.

The preceding argument considers only the deshielding influence exerted by the nearest two α-substituents. Assume that the proton magnetic environment is influenced by the deshielding effect exerted on it by substituents on further removed α-carbons. To treat this problem, symmetry considerations require that the proton must be located at the center of a sequence made of an even number of monomer units (e.g., diad, tetrad, etc.) for the β-CH$_2$ case and of an odd number (e.g., triad, pentad, etc.) for the α-protons case. In tetrad sequences the methylene could be at the center of one of the following tetrad configurations: mmm, mmr=rmm, rmr, rrm—mrr, mrm, and rrr (Table 4.1). The β-CH$_2$ environment in the configurational sequence mmr and rmm are equivalent and hence are indistinguishable by nmr. This is also true of the β-CH$_2$ in the tetrads rrm and mrr. As shown in Table 4.1, which gives a Fisher projection of the configuration of the various tetrads, each of the two central β-CH$_2$ protons in each of the tetrads rrr and mrm are equivalent, and only one chemical-shift value is expected from each. For the remaining tetrad configurations, however, the two protons within each central methylene group are nonequivalent and give rise to two different chemical-shift values. For the tetrad model a total of 10 different chemical shift lines should be expected (Table 4.1). The nmr measure-

ment is expected to discriminate between many of these tetrad configurations.

The chain-sequence configurations can be determined also from the α-methine proton nmr or from that of protons present on the α-substituent. In order to treat this problem, symmetry considerations again require that the sequence consists of an odd number of monomer units, for example, triads and pentads. For the triad case only three different configurations that may be distinguished by nmr are possible: mm, mr=rm, and rr, which are frequently referred to, respectively, as isotactic i, heterotactic h, and syndiotactic s. For the pentad case, 10 configurations with an expected 10-line spectrum are possible (Table 4.1). Fortunately, as can be established from the nmr measurement itself, it was found that the substituent R in the structure shown in Figure 4.1 exerts a strong deshielding influence on the resonating nuclei only when it is located within distances of about 2 to 3 monomer units away. While this limits the nmr ability to distinguish configurationally different sequences to the pentad size or less, it has the advantage of dealing only with sequences of the form shown in Table 4.1.

Table 4.1   Generation of Configurational Sequences (44).

	α Substituent				β-CM$_2$		
	Designation	Projection	Bernoullian probability		Designation	Projection	Bernoullian probability
Triad	Isotactic, mm (i)		P_m^2	Dyad	meso, m		P_m
	Heterotactic, mr (h)		$2P_m(1-P_m)$		racemic, r		$(1-P_m)$
	Syndiotactic, rr (s)		$(1-P_m)^2$				
				Tetrad	mmm		P_m^3
Pentad	mmmm (isotactic)		P_m^4		mmr		$2P_m^2(1-P_m)$
	mmmr		$2P_m^3(1-P_m)$		rmr		$P_m(1-P_m)^2$
	rmmr		$P_m^2(1-P_m)^2$		mrm		$P_m^2(1-P_m)$
	mmrm		$2P_m^3(1-P_m)$		rrm		$2P_m(1-P_m)^2$
	mmrr		$2P_m^2(1-P_m)^2$		rrr		$(1-P_m)^3$
	rmrm (heterotactic)		$2P_m^2(1-P_m)^2$				
	rmrr		$2P_m(1-P_m)^3$				
	mrrm		$P_m^2(1-P_m)^2$				
	rrrm		$2P_m(1-P_m)^3$				
	rrrr (syndiotactic)		$(1-P_m)^4$				

Source.   Reprinted with permission of Academic Press.

Detection and discrimination of the sequences of larger size should become easier by the use of instruments with higher magnetic field strength and by the use of ^{13}C nmr.

4.2.3   Chain-Sequence Probabilities

From the previous discussion, configurationally different sequences in a vinyl polymer can be identified from their nmr chemical shift. Their probabilities in the chain P_{obs} can be determined from their nmr area ratios. The type of propagation mechanism followed in generating the chain may be established by chain-sequence statistical approaches.

The degree of stereoregularity that is normally measured by the nmr is determined by the way in which a monomer unit adds a growing chain to form m or r placements. The addition is normally either dependent or completely independent of the stereoregularity of the growing chain or that of the group at its end. The former case requires the use of complex forms of chain statistics (Markovian) to describe the mechanism of chain propagation and the distribution of configurationally different sequences in the chain. For example, to describe fully the chain propagation for the first-order Markov chain, four conditional probabilities are required (44). Discussions of this type of chain statistics and other forms of higher statistics are described elsewhere (44, 45, 47) as they are applied to polymer microstructure.

By contrast, vinyl chains, in which the addition of monomer unit to form m or r placements is independent of the stereochemistry of the growing chain or the end unit, propagate through a Bernoullian-trial process. Only one conditional probability, that is, the probability of meso placement in the chain P_m, is required to fully describe the chain configuration. This automatically fixes the probability of forming r placement to $P_r = 1 - P_m$. In this type of propagation, the stereochemistry of the final monomer group on a growing chain is established only after the addition of a new monomer unit to that chain. Practically all vinyl polymers initiated by free radicals are known to form through the Bernoullian-trial process.

The Bernoullian probabilities P_{calc} for various length sequences of interest to this discussion, up to pentads, and their configurations are shown in Table 4.1. The effect of varying P_m on triad and tetrad probabilities are shown by the solid lines in Figure 4.2a,b. For any given polymer formed through the Bernoullian-trial propagation process, the observed probabilities (as obtained from nmr measurement) for the triad sequences mm, mr, and rr should fall on those solid lines. Futhermore, they should lie on a single vertical line initiating at the P_m-axis at the corresponding value of P_m for the process.

Normally, in polymer preparation, varying the polymerization conditions affects the stereochemical configuration of the polymer chain; for example, formation of the syndiotactic configuration in PVC is favored by lowering the polymerization temperature (57). A typical Arrhenius plot of the stereoregularity $(P_m/1-P_m)$ for PVC polymerized at different temperatures, is plotted versus $1/RT$ in Figure 4.3 (58). All values, excluding those for D series prepared in butyraldehyde solution, fall on a straight line. The difference in ΔHp, the activation enthalpy, of 510 cal/mol indicates that in PVC, formation of syndiotactic placements r is preferred over isotactic placements m by that much energy.

4.2.3.1   Chain Conformation

It is well established that in both saturated and unsaturated molecules vicinal coupling is very much dependent on molecular geometry. For example, in saturated structures with the staggered conformation (Newman projection), trans coupling J_t is much larger than gauche coupling J_g. Karplus (59) has shown that for saturated systems vicinal coupling J_{vic} varies with the dihedral angle. A determination of vicinal coupling in polymers could, therefore, allow for the determination of polymer conformation. The reliability of the determination, however, is very much dependent on the accuracy of determining the nuclear-coupling constant.

If each of the monomer units in a vinyl chain has the conformation observed for the 2,4-disubstituted pentane model compounds (TG′ or GT conformation for meso and GG or

(a)

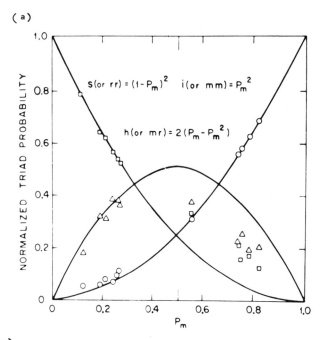

(b)

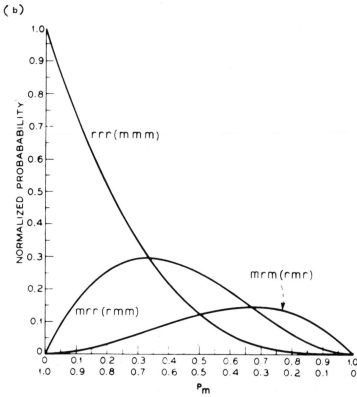

Figure 4.2 Triad and tetrad sequence probabilities as a function of P_m. (a) Probabilities of isotactic (mm), heterotactic (mr), and syndiotactic (rr) triads as a function of P_m, the probability of m placement. Experimental points to the left side are for PMMAs prepared with free-radical initiators and those on the right side with anionic initiators. The solid lines are calculated from theory. (b) Tetrad sequence probabilities as a function of P_m. For rrr, mrr (rrm), and mrm the upper P_m scale is used; for mmm, rmm (mmr), and rmr the lower P_m scale is used (47). Reprinted with permission from Academic Press.

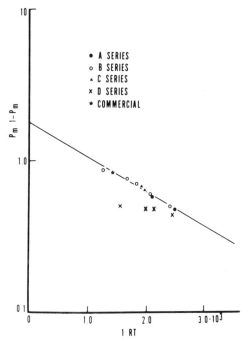

Figure 4.3   Arrhenius plot of the stereoregularity in PVC as a function of the reciprocal of the polymerization temperature (58).

TT for racemic) (Figure 4.4), then isotactic vinyl polymers will be made of a succession of TG' or GT conformers. The chain will have a 3_1 helix in isotactic conformation. By contrast, the syndiotactic chain will consist of a succession of placements, all having either GG or TT conformation, and the chain will have a syndiotactic planar-zigzag conformation. The nmr measurement should distinguish both chain conformations (in the isotactic and syndiotactic chain) by the variation in their vicinal-coupling constants. For polymers, however, the chain mobility about the skeletal bond is much more restricted than in their model compounds. The spectral line width is normally larger than that for small molecules, which masks the difference in coupling constant ΔJ, if any, in polymers. In certain favorable cases, by the use of good solvent, high temperature, and simulated spectra, it is possible to obtain reliable information on the vicinal-coupling constant and, therefore, chain conformation can be obtained from the nmr measurement (60).

For illustration, in PVC chains the vicinal coupling for meso configuration $J_{AX} = J_{BX}$ is 6.5–7.0 Hz (Table 4.2). This value is consistent

with the presence of TG' and GT conformers (Figure 4.4). For the racemic sequences (r configuration) the two vicinal couplings (depending on the solvent employed) vary considerably and have the values of 10–13 Hz in one case and 2–3 Hz for the other. The results can be explained, as was done for model compounds, only by assuming that the r diad configuration has TT or GG conformation. From Table 4.2, and considering $J_t = 12$ and $J_g = 22$ in chlorobenzene, nearly 90% of the r configurations are made of TT conformers.

4.2.4   Nmr Spectra of Poly (methyl methacrylate), PMMA

PMMA is an α,α'-disubstituted polymer (CH_2—CRR'), where R=CH_3 and R'= COOCH$_3$. No vicinal coupling is expected to occur between the β-CH$_2$ protons and those of R and R', which simplifies the spectrum of this system considerably.

The 60 MHz spectra of PMMA measured in chlorobenzene at 150°C are shown in Figure 4.5. The spectrum in Figure 4.5a is that of a predominantly syndiotactic polymer (PMMA-I) which was prepared by a free-radical initiator. The spectrum shown in Figure 4.5b is that of a predominantly syndiotactic polymer (PMMA-II), which was prepared by anionic polymerization in toluene at −78°C in the presence of the initiator phenyl magnesium bromide. The tactic structures in these two polymers were predetermined from other measurements.

4.2.4.1   β-Methylene Spectrum
From the simple diad treatment, a singlet and AB-type spectra are expected from PMMA in completely syndiotactic and completely isotactic configurations, respectively. The expected singlet and AB-spectrum should also constitute the dominating features of the —CH$_2$ spectrum in the PMMA-I (predominantly syndiotactic) and PMMA-II (predominantly isotactic configuration), respectively. The observed results are in agreement with these expectations (Figure 4.5). The β-CH$_2$ spectrum of PMMA-I consists mainly of a singlet, although complicated by the nmr ab-

Conformers and vicinal couplings of m and r diads.

m

configuration						
diad conformers	(H_A H_B / H_X R / H_X R)	(H_A H_B / R / H_X R H_X)	(H_A H_B / H_X R / R)	(H_A H_B / H_X R R / H_X R) ⇌ (H_A H_B / H_X R / R H_X)	(H_A H_B / R H_X / H_X R) ⇌ (H_A H_B / H_X R / R H_X)	(H_A H_B R / H_X / R H_X) ⇌ (H_A H_B H_X / H_X R / R)
chain conformation	TT	GG′	G′G	TG.G′T	TG′.GT	G′G′.GG
vicinal couplings	$J_{AX} = J_g$ $J_{BX} = J_t$	$J_{AX} = J_t$ $J_{BX} = J_g$	$J_{AX} = J_{BX} = J_g$	$J_{AX} = J_g$ $J_{BX} = (J_t + J_g)/2$	$J_{AX} = J_{BX}$ $= (J_t + J_g)/2$	$J_{AX} = (J_t + J_g)/2$ $J_{BX} = J_g$

resulting coupling constants

$$J_{AX} = p_{TT} \cdot J_g + p_{GG'} \cdot J_t + p_{G'G} \cdot J_g + (p_{TG} + p_{G'T}) \cdot J_g + (p_{TG'} + p_{GT}) \cdot (J_t + J_g)/2 + (p_{G'G'} + p_{GG}) \cdot (J_t + J_g)/2$$

$$J_{BX} = p_{TT} \cdot J_t + p_{GG'} \cdot J_g + p_{G'G} \cdot J_g + (p_{TG} + p_{G'T}) \cdot (J_t + J_g)/2 + (p_{TG'} + p_{GT}) \cdot (J_t + J_g)/2 + (p_{G'G'} + p_{GG}) \cdot J_g$$

configuration m | | | r | |

diad conformers* 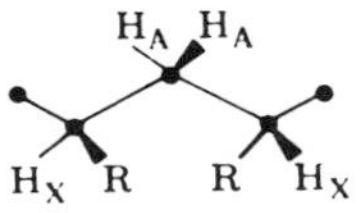 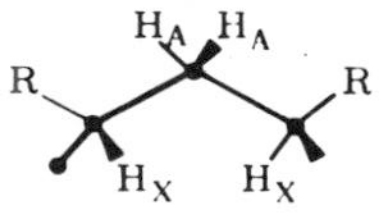  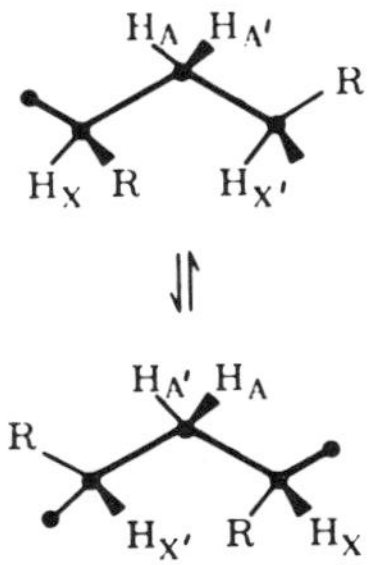 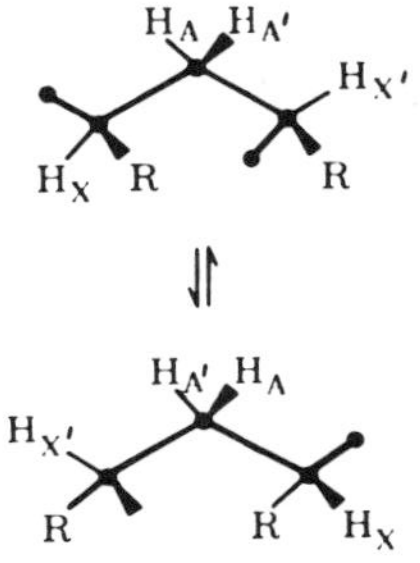 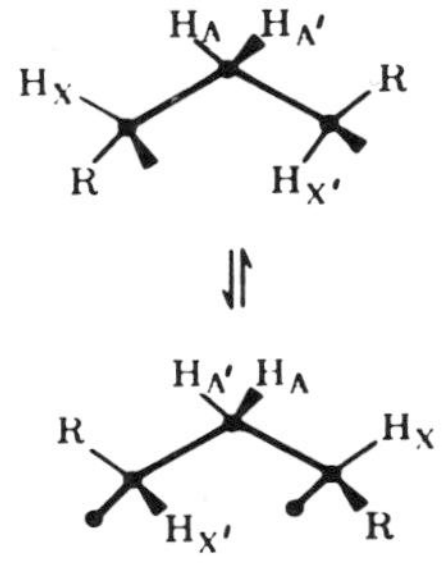					
chain conformation	TT	GG	G'G'	TG, GT	TG', GT'	G'G, GG'
vicinal couplings	$J = J_t$ $J' = J_g$	$J = J_g$ $J' = J_t$	$J = J' = J_g$	$J = J' = (J_t + J_g)/2$	$J = (J_t + J_g)/2$ $J' = J_g$	$J = J_g$ $J' = (J_t + J_g)/2$

resulting coupling constants

$$J = J_{AA'} = J_{A'X} = p_{TT} \cdot J_t + p_{GG} \cdot J_g + p_{G'G'} \cdot J_g + (p_{TG} + p_{GT}) \cdot (J_t + J_g)/2 + (p_{TG'} + p_{G'T}) \cdot (J_t + J_g)/2 + (p_{G'G} + p_{GG'}) \cdot J_g$$

$$J' = J_{A'X} = J_{AX} = p_{TT} \cdot J_g + p_{GG} \cdot J_t + p_{G'G'} \cdot J_g + (p_{TG} + p_{GT}) \cdot (J_t + J_g)/2 + (p_{TG'} + p_{G'T}) \cdot J_g + (p_{GG'} + p_{G'G}) \cdot (J_t + J_g)/2$$

* A series of 9 additional conformers can be constructed which are mirror images of the series shown.

Figure 4.4 Conformations and configurations with their corresponding vicinal coupling constants of m and r diads (45). Reprinted with permission from Verlag Chemie GMBH.

145

Table 4.2   Chemical Shifts and Coupling Constants for PVC at 220 MHz[a]

		Solvent		
	Chlorobenzene	o-Dichloro-benzene	Penta-chloroethane	Relative intensity
m centered tetrads				
mmm ν_A	−44.5	−40.5	−48.5	0.080
ν_B	−46.0	−32.5	−42.5	
mrm ν_A	−46.0	−42.5	−48.5	0.208
ν_B	−28.5	−28.5	−32.5	
rmr ν_A	−55.0	−49	−59.5	0.141
ν_B	−10.5	−10.5	−16.5	
J_{AB}	−15	−15	−15	
J_{AX}	7.5	7.5	7.5	
J_{BX}	6.5	6.5	6.5	
Line width	5.0	5.0	5.0	
r centered tetrads				
mrm	0	0	0	0.105
mrm, τ	7.96	7.90	7.87	
mrr ν_A	−4.0	−3.5	−4.5	0.281
ν_B	17.5	17.0	9.5	
rrr	12.5	12.5	1.0	0.185
$J_{AA'}$	−15	−15	−15	
$J_{AX'}$	11.0	11.0	10.0	
$J_{AX'}$	2.0	2.0	3.0	
Line width	7.0	6.0	5.0	
Pentads				
mrrm	−543		−527	0.060
mrrr	−541		−525	0.160
rrrr	−537		−526	0.105
mmrm	−517		−513.5	0.091
mmrr	−517		−513.5	0.119
rmrmr	−513		−507.5	0.119
rmrr	−513		−505.5	0.160
mmmm	−485		−492	0.035
rmmr	−477		−479	0.060
Line width	5.0		5.0	

[a] Chemical shifts of tetrad and pentad protons are expressed in hertz with respect to mrm as zero in each solvent. Vicinal *J*'s are given in hertz (44).

Source.   Reprinted with permission of Academic Press.

sorption of β-CH$_2$ in nonsyndiotactic structures, while that of PMMA-II consists mainly of the expected AB-spectrum at 60 MHz, which consists of an AB-quartet with $J_{AB} =$ 15 Hz.

4.2.4.2   α-Methyl Spectrum
Because no vicinal coupling is expected to occur between the β-CH$_2$ and α-methyl groups, each of the α-methyl (C—CH$_3$) and the ester methyl (—COOCH$_3$) groups is ex-

pected to give rise to a singlet band. This is only true if the magnetic environment of the methyl group is not influenced by chain configuration. In chlorobenzene, the ester methyl group in both polymers gives rise only to a singlet at $\delta = 3.4$. In other solvents, however, such as benzene, it was found (61) that the ester methyl group nmr absorption consists of three singlets corresponding to ester methyl in the three triad configurations (mm, mr, and rr). In chlorobenzene the α =

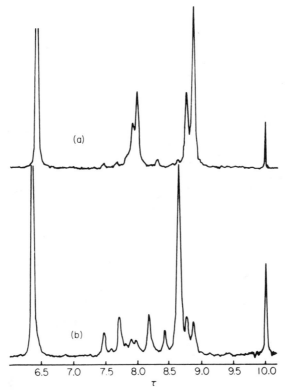

Figure 4.5 The 60 MHz spectra of PMMA, observed as 15% (w/v) solutions in chlorobenzene and prepared with (a) a free-radical initiator and (b) anionic initiator (PhMgBR). The ester methyl resonance is near 6.5τ, the β-methylene protons near 8.0τ, and the α-methyl protons give three peaks between 8.5 and 9.0 τ (44). Reprinted with permission from Academic Press.

CH_3 resonance consists of three singlets between 8.5 and 9.0τ (Figure 4.5). This means that the α-CH_3 magnetic environment is sensitive to chain configuration, and the presence of the three singlets could be explained in terms of having three different triad configurations.

It is only by correlating the α-CH_3 spectrum to that of the β-CH_2, which unequivocally indicates configuration, that the latter can be established. By comparison to the β-CH_2 spectrum (Figure 4.5), the singlet band at 8.9τ ($\delta = 1.10$) is most intense in PMMA-I spectrum and thus was assigned to the resonance of α-CH_3 at the center of rr (syndiotactic) triads. By the same argument, the singlet at 8.7τ ($\delta = 1.33$), which is most intense in predominantly isotactic PMMA-II spectrum, was assigned to α-CH_3 in mm (isotactic) triad configuration. The remaining singlet at 8.8τ ($\delta = 1.21$) was assigned to those in heterotactic triads with rm or mr configurations. The observed probabilities of the three triads in PMMA-I, which were determined from nmr measurement, are consistent with Bernoullian-chain statistics with $P_m = 0.24 \pm 0.01$, for example, $P_{rr} = 0.56$, and so on. All the observed nmr probabilities for the triads in this polymer fall on a single vertical line, Figure 4.2a. By contrast, the observed triad probabilities for PMMA-II do not fall on a straight line (Figure 4.2a), and therefore the triad distributions in this polymer are inconsistent with Bernoullian statistics. This is usually the situation since polymers prepared with free-radical initiators have Bernoullian-sequence statistics, whereas those prepared with anionic initiators do not.

4.2.4.3 Observation of Longer-Sequence Configurations

The preceding diad treatment allowed for general explanation of the β-CH_2 spectrum, but it did not provide an assignment for

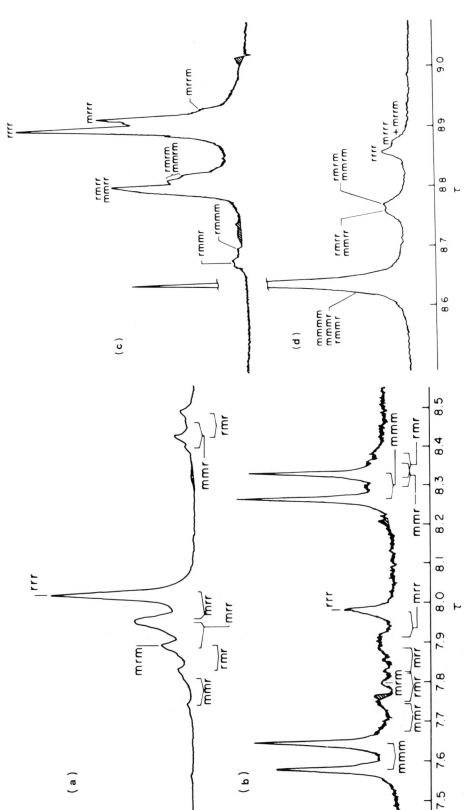

Figure 4.6 The 220 MHz spectra of PMMA (I) and (II) observed as 10% (w/v) solutions at 135° C in chlorobenzene. (a) β-methylene proton spectrum of predominately syndiotactic PMMA (I). (b) β-methylene proton spectrum of predominately isotactic PMMA (II). (c) α-methyl proton spectrum of predominately syndiotactic PMMA (I). (d) α-methyl proton spectrum of predominately isotactic PMMA (II) (44). Reprinted with permission from Academic Press.

some of the fine structures in the spectrum, for example, the broadening of the singlet in the β-CH$_2$ PMMA-I spectrum. Instead of limiting the substituent deshielding influence to diad configuration, consider that the longer tetrad-sequence configurations exert deshielding influences on the central β-CH$_2$ proton. This model predicts, in agreement with observed spectrum, Figure 4.6a and b, a 10 line spectrum singlet: one from each rrr and mrm; and two-line spectrum, AB-type, from each of the remaining mmm, mmr=rmm, rrm—mrr, and rmr tetrads.

In Figure 4.6a and b are shown the 220 MHz spectra of two PMMA polymers measured in chlorobenzene solution at 135°C. These are the same polymers whose spectra are shown in Figure 4.5a and b. As expected, peak separation is vastly improved by using the 220 MHz instrument instead of the 60

MHz. The AB spectrum due to the absorption of the mmm tetrad, which appeared as a quartet when using 60 MHz, consists of two widely separated doublets in the 220 MHz spectra (Figure 4.6). The assignment of the various lines to the configurationally different tetrads is as shown in the spectra. In making these assignments it was assumed that PMMA-I followed Beroullian-sequence statistics, as was established from the diad treatment. The area under each line should be as given by the tetrad Bernoullian probability (Table 4.3).

In similar fashion one could argue that the c and d, a 10-line spectrum should be expected from this model (Table 4.3).

The observed splitting, which is not due to coupling, within each of these three bands will be due to the deshielding influence exerted on the β-CH$_2$ proton by the two monomer units at the end of the pentad. For example, a three-

Table 4.3   Comparison of Observed Sequence Probabilities of PMMA and Probabilities Calculated According to Bernoullian Statistics (44).

sequence probabilities	observed by ^{13}C resonance				calculated by Bernoullian statistics
	β-CH$_2$	α-C	α-CH$_3$	C=O	
P(m)	0.20[a]				0.21
P(r)	0.80				0.79
P(mm)		0.05	0.05	0.04	0.04
P($\overline{mr}$)		0.33	0.34	0.33	0.33
P(rr)		0.62	0.61	0.63	0.63
P(mmm)	⎱ 0.06				0.01
P($\overleftrightarrow{mmr}$)	⎰				0.07
P(rmr)	0.14				0.13
P(mrm)	0.03				0.04
P($\overline{mrr}$)	0.24				0.26
P(rrr)	0.53				0.49
P(mmmm)				⎱ 0.04	0.002
P($\overline{mmmr}$)					0.015
P(rmmr)				⎰	0.028
P($\overline{mmrm}$)				⎱ 0.08	0.015
P($\overline{mrmr}$)				⎰	0.055
P($\overline{mmrr}$)				0.05	0.055
P($\overline{rmrr}$)				0.20	0.207
P(mrrm)				0.03	0.028
P($\overline{mrrr}$)				0.18	0.207
P(rrrr)				0.41	0.390

[a] derived from triad probabilities

Source.   Reprinted with permission of Academic Press.

line band is expected from mm-centered pen-α-CH$_3$ is sensitive to pentad configuration. In agreement with observed spectra (Figures 4.6 tads (mmmm, mmmr=rmmm, and rmmr) as compared to a three-line band from rr-centered and four-line band from mr-centered pentads (Figure 4.6c and d). Only when the α-CH$_3$ spectrum was treated as a pentad case and β-CH$_2$ spectrum as a tetrad case could the PMMA nmr spectrum be fully explained.

4.2.4.4 ^{13}C Nmr

The ^{13}C nmr (cmr) spectra of PMMA prepared by anionic polymerization are shown in Figure 4.7 (62). As can be deduced from their proton nmr (pmr) spectra, which are shown in the Figure 4.7a, polymer-I is predominantly syndiotactic (mainly one band for β-CH$_2$), and polymer-II is predominantly isotactic (β-CH$_2$ spectrum consists mainly of an AB-quartet).

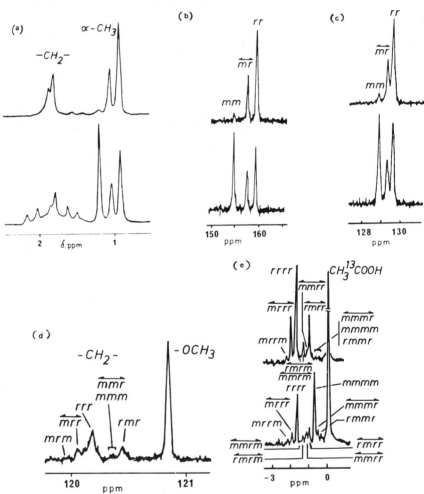

Figure 4.7   Proton and ^{13}C spectra of PMMA. (*a*) Proton resonance spectra of β-CH$_2$ and α-CH$_3$ groups of two PMMA samples prepared by anionic polymerization; 10% (*w/v*) in nitromethane solution at 110° C and 100 MHz. Top spectrum is that of predominately syndiotactic PMMA; lower spectrum, that of blocklike PMMA. (*b*) ^{13}C spectra of α-CH$_3$ groups in PMMA, 40% (*w/v*) in *o*-dichlorobenzene solution at 140° C and 25.14 MHz, 256 scans. Top spectrum is that of predominately syndiotactic PMMA; lower spectrum, that of blocklike PMMA. (*c*) ^{13}C spectra of skeletal α-carbons in PMMA, conditions as in (*a*), 512 scans. Top spectrum, that of predominately syndiotactic PMMA; the lower spectrum, that of blocklike PMMA. (*d*) ^{13}C spectrum of −CH$_2$− and −OCH$_3$ groups in predominately syndiotactic PMMA, conditions in (*a*), 256 scans. (*e*) ^{13}C spectra of the carbonyl group in PMMA, conditions as in (*a*), 512 scans. Top spectrum that of predominately syndiotactic PMMA, the lower spectrum blocklike PMMA (45 and 62). Reprinted with permission from Verlag Chemie GMBH.

The repeat unit in PMMA has five different and magnetically nonequivalent carbons. As shown below, the cmr for each of these different carbons, with the exception of that of the OCH$_3$ group, is sensitive to sequence configuration. For each carbon, however, the Overhauser effect is the same, allowing for area comparison to be used in estimating line intensity from the absorption of that particular carbon.

For the α-CH$_3$ carbon, the cmr consists of a three-band spectrum that is very much similar in appearance and intensity to that of α-CH$_3$ pmr. It is noted that the α-CH$_3$ carbon magnetic environment is sensitive to variations in sequence configuration in a manner similar to that of the proton linked to this same group. This allows for the assignment of α-CH$_3$ cmr spectrum to the various triad configurations as indicated in Figure 4.7b.

The skeletal α-carbon cmr spectrum also consists of three lines (Figure 4.7c). Although the three lines, which appear between 128 to 130 ppm, are less separated than for the α-CH$_3$ pmr case, nevertheless, they resemble the pmr for an α-CH$_3$ group, (Figure 4.7a), and thus they were similarly assigned in Figure 4.7c. The α-ester methyl carbon (OCH$_3$) gives rise to a singlet (Figure 4.7d) in o-dichlorobenzene solution; this carbon is insensitive to chain-sequence configuration. This is similar to its pmr resonance, where it gives rise to a singlet band in all solvents, except in some aromatic solvents or when a paramagnetic shift reagent is added. In general, the further removed the group from the polymer backbone the less likely its spectrum will respond to chain-sequence configuration.

The β-CH$_2$ cmr is resolved into tetrads. The assignment in Figure 4.7d was made on the basis of the peak area and the fact that the sequence distribution in the chain is consistent with Bernoullian statistics (62). Unlike the pmr spectrum, which allows for the assignment of the spectral lines in tetrad configuration, the cmr by itself cannot distinguish tetrad configurations. Only by comparison to the pmr spectra was the assignment of cmr lines possible.

The most striking aspect of PMMA cmr is the —COO group absorption, which is clearly

resolved into a many-line spectrum (Figure 4.7e). The spectrum is fully explained in terms of configurationally different pentad sequences. Since polymer I has Bernoullian-sequence statistics, the line assignments in the spectrum of polymer I were established by comparing their relative cmr areas to their expected Bernoullian probabilities, Table 4.3. Polymer II does not conform to Bernoullian statistics, and line assignment for this polymer was obtained by comparison to cmr spectrum of polymer I.

4.2.5   Nmr Spectra of Poly (vinyl chloride), PVC

In the preceding section, a discussion of the nmr spectra of α,α'-disubstituted polymers, —(CH$_2$CR$_1$R$_2$)$_n$—, was presented. The spectra were relatively simple to interpret since no vicinal coupling between the chain protons occurs. By contrast, vicinal coupling occurs in the α-monosubstituted vinyl polymer, (CH$_2$CHR)$_n$. Although presence of vicinal coupling leads to more complexity in the spectrum, it allows for establishment of chain conformation from the nmr measurement. The nmr spectrum of PVC, α-monosubstituted vinyl in which α-protons do not couple to the substituent, is treated first, followed by that of PP, where coupling between the α-methine protons and the methyl substituent occurs.

In terms of diad configuration, that is, where only the nearest two Cl substituents exert deshielding influence on the β-CH$_2$, the β-CH$_2$ could be either at the center of a meso or a racemic placement. The spectrum consists of AA'XX' for placements with racemic configuration and AB$_2$ for those with meso configurations. For the tetrad case, that is, when the nearest for Cl atoms are assumed to exert deshielding influences on the β-CH$_2$ at the center of the tetrad sequence, the β-CH$_2$ could be at the center of any of the tetrad sequences shown in Table 4.1. There are three configurationally different r-centered tetrads (rrm=mrr, mrm, rrr) and a similar number for m-centered tetrads (mmm, mmr=rmm, rmm). Each of the β-CH$_2$ r-centered tetrads has an AA'XX' spectrum, while each of the m-centered tetrads has an ABX$_2$ spectrum. From

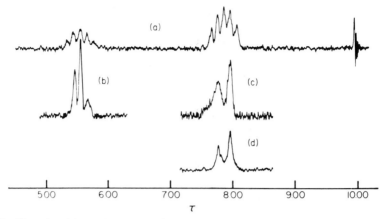

Figure 4.8   Normal and decoupled spectra of PVC, observed in 15% (w/v) chlorobenzene solution. (*a*) Normal spectrum at 170° C, (*b*) α-protons upon irradiation of β-protons at 150° C, (*c*) β-protons upon irradiation of α-protons at 150° C, (*d*) poly-α-d_1-vinyl chloride at 156° C (44). Reprinted with permission from Academic Press.

this it can be argued that the PVC β-CH$_2$ spectrum should be quite complex. This is indeed the case (Figure 4.8) except for the yet unsynthesized cases of completely isotactic or completely syndiotactic PVC, where the spectrum should consist of AB$_2$ for the former and AA'XX' for the latter.

In a similar manner, triad treatment for the α-CH proton spectrum leads to an expected three-line spectrum (mm, mr=rm, and rr). Each of these three lines will be further split by coupling to the protons of the two neighboring β-CH$_2$ groups. Pentad treatment leads to a 10-line spectrum (Table 4.1), and again each is split by the proton on the two nearest β-CH$_2$ groups.

The 60 MHz nmr spectrum of PVC measured in chlorobenzene is shown in Figure 4.8. From chemical shift and area consideration the lower multiplet between 5.0–6.0τ ($\delta \simeq$ 4.5 ppm) is assigned to the α-CH protons, while the broad multiplet at 7.8–8.0τ ($\delta \simeq$ 2.2 ppm) is assigned to the β-CH$_2$ resonance. As is noticed, the spectrum is quite complex and does not lend itself to simple interpretation.

4.2.5.1 β-Methylene Spectrum
Both spin-decoupling and replacement of the α-hydrogen techniques eliminate the vicinal coupling, thereby allowing for extension of the treatment described earlier for PMMA to the present case. Using the simple diad treat-

ment a three-line spectrum is expected. The observed spectrum consisting of an apparent two-line absorption (Figure 4.8c and 4.8d) may fit into this model; if, for example, at 60 MHz the difference in chemical shift between HA and HB in meso configuration is not sufficiently large and therefore HA and HB lines overlap. This overlapping should be removed by increasing the magnetic field strength to 220 MHz (Figure 4.9). The spectrum for poly α-d_1-vinyl chloride is much too complex to be explained from this simple diad model. Furthermore, the diad model that predicts a three-line spectrum does not explain the eight-line spectrum for α-*cis*-β-d_2-PVC (Figure 4.10).

Recalling that the spectrum of PMMA was fully explained only when the β-CH$_2$ resonance was treated as a tetrad-sequence problem, we applied the same treatment to PVC. Yoshino and Komiyama (63) were able to explain fully the spectrum of α-*cis*-β-d_2-PVC in terms of tetrad treatment. Synthesis and interpretation of nmr spectra of this compound constitute a significant step in the understanding of PVC configuration. For this polymer there are 10 possible tetrad configurations that can be distinguished by nmr; two for each of the mmm, mmr, rmr, and rrm, but only one for each of the rrr and mrm. For the case of the mmm pair the central proton is flanked by Cl groups in one configuration and by deuterium groups in the second configura-

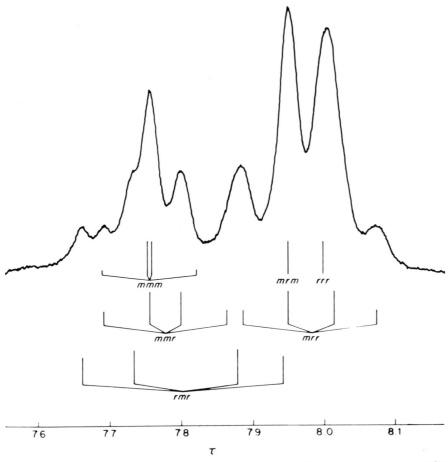

Figure 4.9 220 MHz proton spectrum of poly-α-d_1-vinyl chloride in 15% (w/v) chlorobenzene solution at 140° C (44, 64). Reprinted with permission from Academic Press.

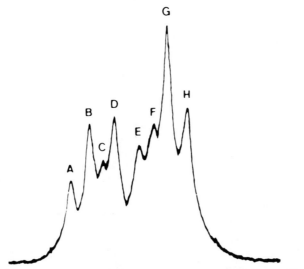

Figure 4.10 100 MHz spectrum of α-cis-β-d^2-PVC dissolved in chloroform (45, 63). Reprinted with permission from Verlag Chemie GMBH.

tion. The assignments of the various peaks were obtained from their relative areas and the assumption that their distribution in the chain follows Bernoullian statistics. Thus, the eight lines in Figure 4.10 were assigned: A to rmr; B to mmr and mmm; C to mmm; D to mrr; E to rmr; F to mrm; G to rrr and mrr; and H to mrr. The tetrad probabilities in this PVC chain were consistent with a Bernoullian distribution for which $P_m = 0.44 \pm 0.02$.

From the analysis of the α, cis-β-d_2-PVC spectrum it is possible to conclude that the β-CH_2 PVC spectrum can be similarly treated as a tetrad problem. For the simpler case of α-d_1-PVC the β-CH_2 spectrum should be similar in type to that of PMMA and consist of a singlet for each of rrr and mrm tetrads and of an AB quartet for each of the remaining tetrads mmm, mmr, rrm, and rmr. In fact, by using chemical-shift values that were consistent

with those of Yoshino and Komimomo, geminal coupling constants of $J_{AB} = -15$ Hz, and a Bernoullian tetrad distribution having $P_m = 0.43$, Heatley and Bovey (64) were able to calculate the nmr spectrum of α-d_1-PVC (65) that was in good agreement with the observed spectrum (Figure 4.9).

4.2.5.2 Methine Proton Spectrum

Figures 4.11a and 4.11b show a comparison of the methine proton spectra measured under the conditions indicated for β,β-d_2-PVC. The methine protons are seen as a three-band nmr spectrum. These bands could be assigned to the absorption of central α-CH in the triads mm, mr=rm, and rr; or they likely correspond to that of —mm, —mr—, and —rr— centered pentads.

Although triad treatment explains the general features of the methine spectra, it does not

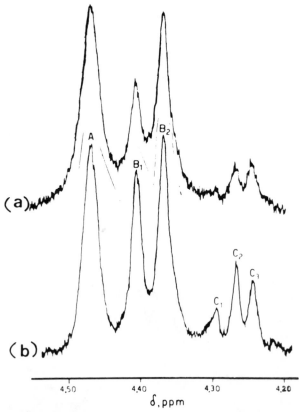

δ, ppm

Figure 4.11   100 MHz spectra with H—D decoupling of β,β-d_2-PVC dissolved in pentachloroethane (15% w/v) at 120° C. (a) Sample prepared at −40° C. (b) Sample prepared at +40° C (45, 66). Reprinted with premission from Verlag Chemie GMBH.

account for the fine splitting within each of the three general bands, especially the upperfield "C" band in the expanded spectra of β,β-d_2-PVC (Figure 4.11*b*). From relative area measurements Cavalli et al. (66) were able to explain these line structures in terms of pentad-sequence configurations. It is obvious that the intensity of the "C" bands decreases in relation to the rest of the spectrum as the polymerization temperature is lowered (Figure 4.11*a*). In both cases of low-temperature ($-40°C$) and high-temperature ($+40°C$) polymerized PVC, the pentad distribution in the chain is consistent with Bernoullian-sequence statistics with $P_m = 0.43 \pm 0.01$ for the latter sample as compared to $P_m = 0.32 \pm 0.02$ for the former one (66).

4.2.5.3   Calculation of Nmr Spectra

The various nmr investigation carried out on PVC, partially deuterated PVC, and PVC-model compounds allowed for the determination of both the chemical shift and coupling-constant values for the α-CH and β-CH$_2$ protons in PVC. From these measurements it was concluded that PVC has Bernoullian sequence statistics that can be characterized by a single P_m value. For partially deuterated PVC the spectrum was fully explained in terms of tetrad and pentad configurations for the β-CH$_2$ and α-CH protons, respectively. From all this information with some minor modification, Heatley and Bovey (64) were able to utilize the nmr parameters given in Table 4.2 to calculate the 220 MHz PVC spectrum that agreed well with the observed results (Figure 4.12). Similar calculations were made by Hassan (67) for the 100 MHz nmr spectra of PVC in pentachloroethane (Figure 4.13) using the nmr parameters shown also in Table 4.2. The geminal coupling constant J_{AB} is -15 Hz. The vicinal-coupling constant has only one value (of the order of 6.5–7.5 Hz) for m-centered tetrads; however, it has two values $J_{AX} = J_{A'X'} = 10.5 \pm 0.5$ Hz and $J_{A'X} = J_{AX'} = 2.5$–3.0 Hz, for r-centered tetrads (Table 4.2). Determination of these vicinal-coupling constants are of significance to the chain conformation.

4.2.5.4 ^{13}C Nmr

The expanded cmr spectrum of commercial PVC (Diamond 450) is shown in Figure 4.14 and is discussed in other work by Carman (68) and Ando et al (69). The spectrum was measured at 25.2 MHz using a Varian XL-100-15 in the cmr mode. Unlike the PVC pmr spectrum, which shows extensive overlapping (Figures 4.8 and 4.12), the cmr spectrum is quite well resolved. The concentration of the various configurationally different sequences in the chain were determined directly from the relative areas of their cmr absorption lines. Basically, the α-carbon spectrum displays three main absorption regions centered at $\delta = 57.10$, $\delta = 56.20$, and $\delta = 55.3$ ppm. Within experimental error, their observed areas have the proportions: 0.21–0.48–0.31, which are consistent with Bernoullian triad distribution for which $P_m = 0.45 \pm 0.01$. The three regions can be assigned to the mm, mr, and rr triad configuration. Since each of the three regions, however, consists of several lines (Figure 4.14), pentad configuration must be effective in modifying the α-carbon cmr absorption.

Using arguments similar to those used earlier in assigning the PVC pmr spectrum, the three regions are better assigned to mm-, mr-, and rr-centered pentads. The final assignment of the lines within each region to their respective pentads is as shown in Figure 4.14 and was established by comparing P_{obs}, obtained from cmr measurement, to P_{calc}, computed according to a Bernoullian pentad-sequence distribution for which $P = 0.44$.

The methylene carbon cmr absorption (Figure 4.14) consists of a six-line spectrum (at $\delta = 45.40$, 45.59, 46.33, 47.01, 47.35, and 47.74 ppm) which is due to the absorption of the methylene carbons at the center of the six, configurationally different, tetrad sequences. The assignment of the lines to the tetrads (mmm, mmr, and rmr; and rrm, mrm, and rrr) is shown in the figure and differs somewhat from those given in References 68 and 69 for the reasons detailed by Ando (69). From the relative intensity of the peaks, the tetrad distribution in the chain follows a Bernoullian statistic with $P_m = 0.45$. For a large molecular

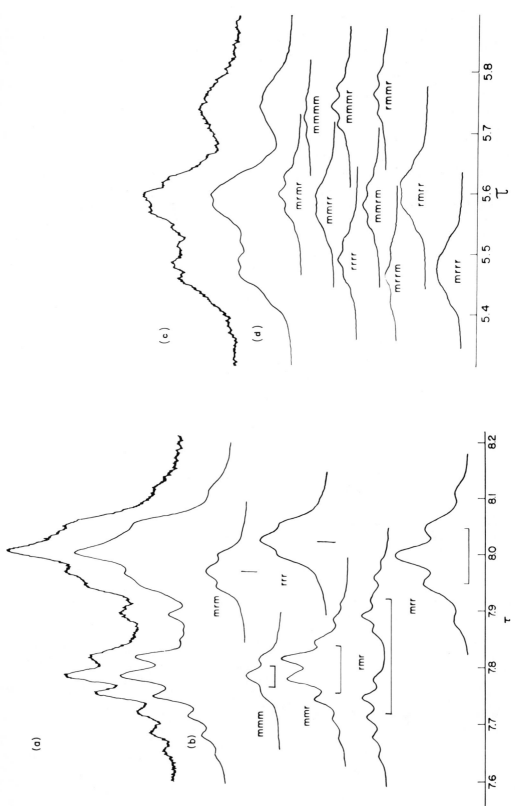

Figure 4.12 (a) 220 MHz β-proton spectrum of PVC, observed in 5% chlorobenzene solution at 140° C. (b) Calculated spectrum with the component tetrad subspectra. (c) 220 MHz α-proton spectrum of PVC, observed as in (a). (d) Calculated spectrum with the component pentad subspectra (44, 64). Reprinted with permission from Academic Press.

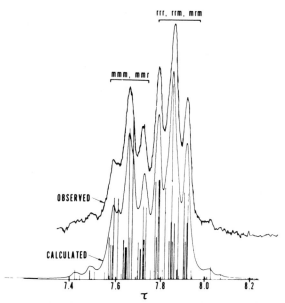

Figure 4.13 Observed and calculated 100 MHz α-proton spectra of PVC, observed as 5% (w/v) pentachloroethane solution at 150° C (67).

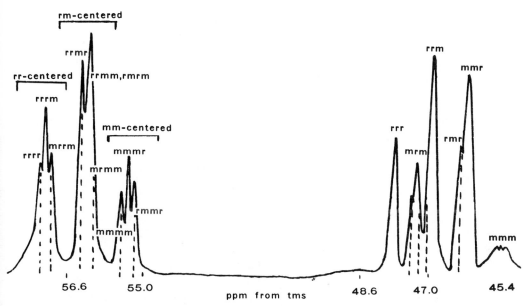

Figure 4.14 The 25.2 MHz pulsed Fourier transform ^{13}C nmr spectrum of the —CH—and—CH$_2$—groups in PVC. Chemical shifts are in ppm from tetramethylsilane. (A. Hassan, unpublished work).

weight PVC, the mmr and mrm lines were inseparable.

4.2.6 Chlorinated PVC, Chlorinated Polyethylene (PE), and Hydrogenated Polybutadiene (BD) Spectra

The spectra of PVC and its chlorinated derivatives were determined by Sobajima, et al. at 100 MHz in o-dichlorobenzene solution (70). Tacticity information on the PVC parent system has been well established so that chlorination sites are identifiable by observation of decreased peak intensities of the CH_2 groups in the isotactic portion, with concurrent increase of peak intensities in the lower field region. The CH_2 protons with neighboring C—Cl bonds (3.0 ppm) had increased absorption as did the number of CHCl protons (4.95 ppm). The increased chlorination of PVC may thus be followed directly by nmr in the CH_2 and CHCl regions. The 69.9% Cl-PVC gave the same essential peak positions and intensities in the Cl-PVC series obtained by Liebman et al. (71). Using precursor PVCs of differing tacticity it was demonstrated that the structural influences of the chain do, in fact, alter the chlorination pattern, although the total Cl content may be similar in the series. The report of Svegliado and Grandi (72) determined from a study of PVC chlorination that among the unchlorinated units more racemic units were present than expected for a completely random chlorination. It was concluded that either the stereoregularity or the crystallinity of the system influences the chlorination process.

High-resolution nmr of chlorinated PEs were likewise examined by Saito et al. (73) in 1970 and, by associating the resonances experimentally determined with the distribution of Cl atoms along the polymer chain, they were able to resolve them into triad or pentad units for their quantitative determination. The sequence length of CH_2 units was also given as calculated using the method of Frensdorff and Ekiner (74) for general substituted polymers.

In 1975 Randall reported the ^{13}C nmr quantitative determination of monomer-sequence distribution in hydrogenated BDs (75). Hydrogenation of the BD chain made possible the study of the sequence distributions in terms of simple 1,2 or 1,4 monomer units, rather than the 1,4-cis, trans, and 1.2-vinyl terpolymer analysis. Since the ^{13}C chemical shifts reflect the number and types of carbons up to four or five bond lengths, the maximum information is associated with five monomer additions. The chemical-shift sensitivity to sequence length depends on the nature of additions and is calculated with the parameters from Grant and Paul (76) for any sequence combination.

4.2.7 Polypropylene (PP) Proton Spectrum

The nmr spectrum of PP is more complex than that of PMMA or PVC because in PP vicinal coupling occurs not only between the methylene α-methine protons, but also between the latter and those of the α-methyl protons. Additionally, the chemical-shift separation between the methyl, methylene, and methine protons is considerably smaller than that observed for PVC or PMMA. One major difference between PP and PVC is that the latter can be prepared as a completely isotactic or as a completely syndiotactic structure. The same nmr procedures described for PVC were also followed in the analysis of nmr spectra of PP. Thus, the nmr spectra of partially deuterated PP, PP-model compounds, and PP with complete tactic structures (isotactic or syndiotactic) were fully examined (77). Use of high-strength magnetic field and PP with regular tactic configuration enabled a more complete understanding of the spectrum. More recent work in pmr at 100, 220, and 300 MHz by Stehling and Knox (78) has led to further detailed PP microstructural information. Tetrad methylene protons were assigned in agreement with previous work (79–84) for five of the possible six tetrads (mmm, mmr, rmr, rrm, and rrr). It was shown that the concentration of r diads and their distribution among the large number of m units in PP may be determined from nmr data. Further assessment of pentad stereochemical information was given by Zambelli et al. (85) using ^{13}C nmr examination of 93% enriched ^{13}C hydrocarbon models.

4.2.8 Theoretical Calculation of Nmr Spectra of PVC and PP

The theoretical calculation of nmr spectra from a molecular orbital semiempirical approach was reported for PVC in 1975 by Ando et al. (86). The ^{1}H chemical shifts of diads, triads, and tetrads were computed as a function of the racemic unit fraction in the chain. The calculations, which were based on Pople's approximation (46), accounted for diamagnetic contributions, magnetic anisotropy, and polar effects. The electron densities on H atoms were calculated by the CNDO/2 method, and the statistical weights of the preferred conformations were obtained by generating a set of 100 Monte Carlo chains using Flory's matrix method.

Likewise, the proton chemical shifts of diad, triad, and tetrad units in an isolated PP chain were calculated (87) taking into account the configuration and conformational structure of the units. Additionally, ^{13}C nmr analysis by these workers reported detection of chemical inversions (head-to-head units) in PP.

Generally, it is shown from these studies on computational nmr results for PVC and PP that the field is a promising one for detailed microstructural information. The combination of a conformational matrix method with the molecular orbital approach to chemical-shift calculations indeed represents a significant contribution to the theory and application of nmr to polymer characterization.

4.2.9 Nmr of Binary Copolymers

In addition to quantitative and qualitative determinations of copolymer composition by nmr methods, it is possible to detect linkage formation, for example head-to-head, tail-to-head, and so on. Sequence composition, length, and sequence distribution, as well as sequence configuration can be established

also from nmr measurements. The subject is reviewed in detail elsewhere (44, 45) and only a brief treatment, using the simple case of vinylidene chloride-isobutylene copolymer (VDC-IB) as an example, is given here.

The copolymer has the two structural units: vinylidene chloride, to be defined the A monomer, and isobutylene, the B monomer (88,89). The nmr spectrum for this copolymer is simple to treat since the PVDC homopolymer exhibits a single absorption at $\delta = 3.88$ ppm and is well separated from that of the isobutylene homopolymer, which gives rise to the two singlets at $\delta = 1.46$ and $\delta = 1.08$. An additional simplification is that the copolymer has no asymmetric center; that is, the α,α'-disubstituents are the same in each of A and B.

Experimentally, the spectrum is resolved into tetrad bands. In a low-field band, three lines at $\delta = 3.86, 3.66$, and 3.47 were assigned, respectively, to the AA-centered tetrads $AAAA$, $AAAB$, and $BAAB$. The four lines in a middle band at $\delta = 2.89, 2.68, 2.54$, and 2.37 were assigned, respectively, to the AB-centered tetrads $AABA$, $BABA$, $AABB$, and $BABB$. An upperfield band, which is overlapped by the α-CH$_3$ absorption, was similarly assigned to BB-centered tetrads. For the α-CH$_3$ group, which should exhibit less response to the substituent deshielding effect, the spectrum consists of a three-line nmr band assigned to the triad configurations: ABA, $BBA = ABB$, BBB.

In the study of the spectrum of α-monosubstituted copolymers or copolymers with an asymmetric center, the monomer stereochemical configuration must be taken into account. The chain configuration can be treated as for the case of the homopolymer in terms of meso and racemic diad configuration. The statistical treatment, however, will be more involved.

4.2.10 Branch Content in PE and PVC

The study of branch content using ^{13}C nmr has been developed by Dorman et al. (90) and Randall (91). In Figure 4.15 the possible branch types in a PE chain are shown and listed with the calculated ^{13}C chemical shifts. It has been shown that a typical low-density

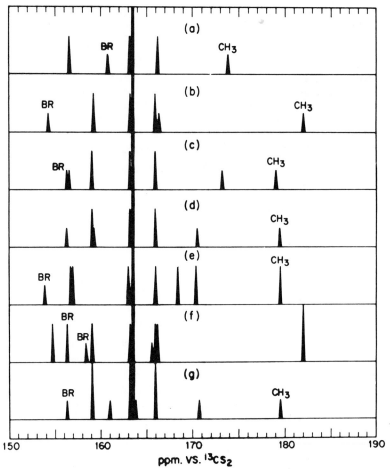

Figure 4.15 Calculated ^{13}C spectra (CS$_2$ as zero reference resonance) for branch structures in PE. The heavy line at 163.5 ppm represents the resonance of all —CH$_2$— groups more than three carbons removed from a branch point or chain end. It is the dominant peak in all experimental spectra. (*a*) Isolated methyl branch. (*b*) Isolated ethyl branch. (*c*) Isolated *n*-propyl branch. (*d*) Isolated trifunctional butyl branch. (*e*) Isolated tetra-functional butyl branch. (*f*) A block of three ethyl branches. (*g*) An isolated "long" branch (90). Reprinted with permission from *J. Polym. Sci.* Copyright by American Chemical Society.

PE (LDPE) chain contains predominantly *n*-butyl branches (5–6/1000 CH$_2$), ethyl (1/1000 CH$_2$), and amyl (2/1000 CH$_2$) branches according to Dorman et al. (90). Likewise, Hama et al. (92) have shown by ^{13}C nmr that various LDPEs could be analyzed using the empirical relations derived by Grant and Paul (76) from model ethyl-α-olefin copolymers. They found mostly *n*-butyl branches (as well as a small amount of methyl branches in commercial samples) in agreement with Dorman et al. This also supported the "back-biting" mechanism of Roedel (93). A com-

parison of LDPE and HDPE by ^{13}C analysis is shown in Figure 4.16 (94) to show the high branch content of the former material, as indicated by the labeled peaks along with the CH$_2$ units at 29.90 ppm.

More recently Cudby and Bunn determined chain branching in LDPE by both ^{13}C and ir examination (95). It was clearly demonstrated that the presence of ethyl, butyl, and longer-chain branches can be detected by pulsed FT-^{13}C analysis, when careful instrumental conditions were observed for the study. This is necessary to obtain the relative quantitative

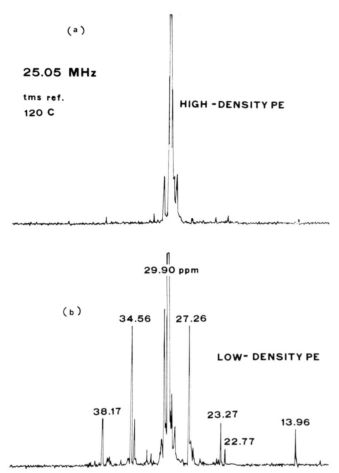

Figure 4.16 ^{13}C nmr spectra of high- and low-density PE, observed at 25.05 MHz in *o*-dichlorobenzene solution, tetramethylsilane reference, 120° C, 4000 pulses (10 μ sec, 90°) heteronoise decoupled. (*a*) High-density PE. (*b*) Low-density PE (94). (S. Liebman, unpublished work.)

values of the branches without distortion from nuclear Overhauser effects that result from differing relaxation times of ^{13}C nuclei in different environments. Comparison was made of ir analysis for total methyl content for the various commercial PEs examined, with the results clearly demonstrating that there is, contrary to previous comments (70), a significant proportion of ethyl branches present in certain LDPEs. Work discussed in the chromatographic section using pgc and ir has indicated that comparison of HDPE and LDPE showed that ethyl branches were a significant component in the later type samples. It is necessary to point out, however, that distinct limitations and careful assessments must be made in the three characterization methods (nmr, ir, and pgc).

Extension of this background information from nmr for reduced PVC systems has likewise been accomplished by Bovey, et al. (Figure 4.17) (96). From the nmr results of PVC reduced to PE by lithium aluminum hydride and with lithium aluminum deuteride, they determined there were approximately 2.8 methyl branches per 1000 CH_2 units in the examined sample. Additionally, the characteristic splittings and isotope shifts in the ^{13}C spectrum of the reduced PVC showed that the branch structure in the original PVC was $-CH_2C(Cl)H-C(CH_2Cl)H-CH_2-C(Cl)H)-$.

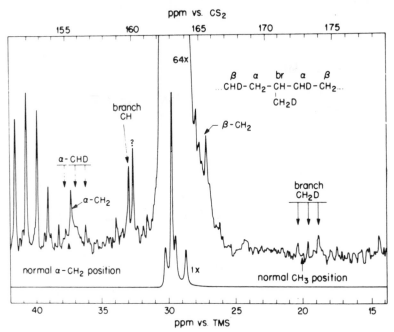

Figure 4.17 ^{13}C nmr spectrum of reduced PVC (96). Reprinted with permission from *Macromolecules.* Copyright by American Chemical Society.

4.3 INFRARED (IR) AND RAMAN SPECTROSCOPY

Vibrational spectroscopy can provide a wide range of detailed information to the polymer chemist about chemical structure and bonding, configuration, conformation, crystallization, and branch or defect content of materials. The theory of ir and Raman has been treated in major contributions throughout the years for both simple and complex molecules at varying levels of sophistication (97–126). The significance of symmetry in polymers is much greater than when dealing with the interpretative problems of small molecules. Symmetry, or lack of it, in the arrangement of repeat units in a polymeric material also has a decided influence on many physical and chemical properties.

The classical dynamics of harmonic oscillators are used to describe the atomic motions within a polymer system and combine with the appropriate symmetry requirements to form the basis of "selection rules" that guide spectral interpretation. The selection rules allow decisions to be made as to whether absorption of ir (or Raman scattering) is "allowed" or

"forbidden" for the system according to the physical and steric factors that are associated with the specific chemical structures involved.

The Wilson GF matrix method (98) is used with powerful computers (111) and takes as input the internal symmetry coordinates of the polymer chain—such as those derived from bond stretching, angle bending, and internal rotational motions—to develop the "normal coordinate analysis" approach. An efficient computer program using mass-weighted Cartesian coordinates was contributed by Gwinn (114) and is available from Quantum Chemistry Program Exchange (113). The rule of mutual exclusion states that for a molecule possessing a center of symmetry a vibrational transition will not be allowed (or "active") in both the Raman and ir. This fact is useful in comparing the ir/Raman spectral data and ascertaining the number of band coincidences in order to decide the presence (or absence) of that molecular-symmetry element. In the normal coordinate analysis of *cis*-1,4-butadiene, as reported by Coleman, et al. (115), the vibrational frequencies were assigned with the aid of laser Raman and Fourier-transform ir analysis. The rule of mutual exclusion was utilized in the interpretative study of this system.

Without restrictions stated in the selection rules one would expect an almost impossible number of potential absorption bands from a molecule such as PVC. Even a low-molecular-weight species would be calculated to have many thousand vibrational modes. However, many of the atomic groupings have equivalent force constants and undergo sterically similar motions, so that a simplicity in the spectrum results, with less than 50 bands being measurable. The semiempirical correlations of characteristic absorption frequencies with specific groups of atoms are more generally utilized by experimentalists to provide interpretation of the common ir/Raman spectra (107, 116–125). Convenient compilations of reference spectra are available for many classes of materials. However, some in-depth understanding of the band origins and their relationship to molecular structure permits the true value of vibrational spectroscopy to be recognized. It must be acknowledged that the complementary nature of the ir/Raman data is basic to band assignments and correlation with molecular structure.

4.3.1 Factor-Group Analysis

Both ir and Raman patterns depend on the structure of the chemical repeat unit (mers) in polymers. Stereoregular polymers have a continuing, regular unit that offers a further dimension and, therefore, additional symmetry elements not found in the point groups for simple molecules. The elements, such as translation and screw-rotation axes and centers within a repeat unit, form a line group, wherein all axes, centers, and planes pass through a line that is the chain axis. Line-group characteristics manifest in ir studies are maintained only to about five to seven repeat units. The symmetry species of the repeat unit are predicted from the symmetry elements of its constituents and these elements form the subgroup of the line group. This subgroup is termed the "unit cell" or "factor group." Factor-group vibrations have corresponding atoms in all the repeat units moving in phase, and only these modes are ir/Raman active.

The factor group is independent of the translational group, so that infinite-space groups have the same factor group. A character table (compilation of the appropriate symmetry elements and their behavior when symmetry operations are performed on them) of a point group is the same as that of a factor group. The line group of an infinite polymer chain is shown to correlate to the polymer "site group" and then to a factor group. The work of Fateley et al. (126) describes in detail the correlation of site groups to factor groups, and the use of factor analysis by the correlation method to characterize the symmetry properties of normal modes. Thus, calculations of the number of normal modes belonging to each symmetry species allows deduction of the expected absorptions using the selection rules for the system.

4.3.2 Crystallinity

Factor-group analysis may also be applied to three-dimensional studies in such a way that information relating to intermolecular interactions is obtained. For illustration, if there are two molecular chains in a unit cell (as is true in PE), a given vibrational mode of one molecule splits into two different crystal modes according to the phase relation (in or out-of-phase) between the two molecular chains. It is estimated that 8–10 associated chains confer "crystallinity" characteristics by the ir vibrational "sensor." True crystallinity bands in the ir are thought to originate in these intermolecular vibrational relationships (127). Therefore, if a unit cell has more than one chain segment, splittings of certain bands may be due to a difference in the phases between like vibrations in the adjacent polymer chains, as well as possible symmetry-induced changes when two chains occupy distinct lattice sites.

The study of crystallinity by low-frequency ir/Raman analysis has recently been discussed by Willis and Cudby (128, 129). Also, much work by Chantry (130), Piseri (131), and Zerbi (132) on polytetrafluoroethylene has been reported in the low-frequency region. The latter two groups of workers attributed activation of internal chain modes at low temperature to conformational disorder in the system rather than to a change in crystal form (from one molecule/unit cell to two) in a temperature-dependent transition, a view held by the former workers.

Low-frequency bands observed in solid *n*-paraffins have fundamental frequencies that are inversely proportional to chain length, according to the 1967 contribution by Schaufel and Shimanouchi (133). They assigned the band to longitudinal acoustic (accordianlike) vibrations of the chain. Similar low-energy absorptions in crystals of PE and polyethylene oxide systems (134) from the ir/Raman data show that lower-molecular-weight members in the series crystallize in chain-extended form, whereas higher members crystallize in chain-folded form.

Lebedev et al. (135) reported on the separation of stereoregularity and crystallinity for PVC systems from different polymerization sources using ir and x-ray data. They found a minimal three to four units of planar syndiotactic conformation was needed for crystallizable segments to occur, spaced at random throughout the noncrystallizable isotactic portions. For the PVC system, Tasumi and Shimanouchi (136) calculated the normal vibrational frequencies of extended syndiotactic PVC and its deuterated analogs using a Urey-Bradley force field. Their assignment of crystalline bands was made from the comparison to experimental frequencies, the dichroic properties, and potential-energy distribution calculations.

The 1975 work of Moore and Krimm (137) has included a normal vibrational analysis for crystalline syndiotactic PVC using a general force field derived from model secondary chlorides. It was calculated that intermolecular ir vibration results from chain rotation in opposite directions to the two chains within the unit cell. Band-splitting occurred and low-frequency ir/Raman bands resulted from the normal modes of intermolecular interaction. Additional data are given in the later work with low-frequency Raman data by Robinson et al. (138).

A unique method for the study of vibrational spectra of semicrystalline polymers was reported by Coleman et al. (139). Spectral manipulation using a Fourier-transform ir

spectrometer enabled a "separation" of the preferred conformation of *trans*-poly (chloroprene) from the semicrystalline polymer. Spectral subtraction of the less-crystalline sample from a more-crystalline sample gave a resultant difference spectrum of the crystalline form, as the absorptions from the amorphous portion had been spectrally removed from the ir pattern. The use of these experimentally determined frequencies of the "pure" crystalline system may then be made in a full normal coordinate approach.

4.3.3 Polarization Phenomena

It is known that the absorption of incident ir radiation occurs only when there is a change in the dipole moment during a vibration. It must also be added that there is a directional limit to this statement; for oriented material there has to be a component acting in the same direction as the dipole vector (ir) or polarizability (Raman) tensor. A convenient listing of dichroic (ir) or depolarization (Raman) behavior with the pertinent symmetry involved is given in Table 4.4 for several polymers when polarized ir radiation is utilized in the examination.

The polarized ir data are shown in Figure 4.18 for PP (140). The 867 cm^{-1} band has been used as a measure of the syndiotactic structure that appears only in helical PP and shows a strong parallel dichroism. The assignment is made to the CH$_2$-group rocking mode between two adjacent gauch skeletal bonds coupled with the rocking mode of CH$_2$ groups between two trans skeletal bonds. The corresponding CH$_2$ rocking bands of planar zigzag and of isotactic PP are assigned to the 831 and 841 cm^{-1} peaks, respectively. The 974 cm^{-1} absorption in isotactic PP as well as those at 977 and 964 cm^{-1} are assigned to the CH$_3$ rocking mode. The parallel 977 and 964 cm^{-1} bands appear characteristic of the helical and planar forms, respectively.

Raman scattering and polarization studies by Vasko and Koenig (141) were also reported for isotactic PP microstructural data.

Ir dichroism is a useful indicator of the orientation factor of polymers (142) with results reported for PP films (143). Other

Table 4.4   Infrared and Raman Activity, Dichroism (IR) and Depolarization Behavior (Raman) of Various Polymer Geometries (109)

Structure	Classes of vibration and symmetry of isolated chain		I.R. and Raman activity and polarization sensitivity		Examples
Syndiotactic vinyl	A_1		R(p)	IR σ	
Planar	A_2	C_{2v}	R(dp)	—	polyvinyl
	B_1		R(dp)	IR σ	chloride
	B_2		R(dp)	IR π	
Helical	A		R(p)	—	Syndiotactic
	A_1	D_2	R(dp)	IR π	
	B_2		R(p)	IR σ	
2 fold 2_1	B_3		R(dp)	IR σ	polypropylene
Helical	A_1		R(p)	—	
3_1	A_2	D_3	—	IR π	—
3 fold	F		R(dp)	IR σ	
Helical	A_1		R(p)	—	
n_1	A_2		—	IR π	PTFE and
	F_1	D_n	R(dp)	IR σ	polyethylene
Multifold	F_2		R(dp)	—	oxide
Isotactic vinyl	A_1	C_s	R(p)	IR σ	
Planar	A^{11}		R(dp)	IR π	—
Helical	A	C_3	R(p)	IR π	
3 fold	F		R(dp)	IR σ	—
Isotactic vinyl	A		R(p)	IR π	
Helical	$B(F_2)$	C_n	R(dp)	—	polybutene
Multifold	F		R(dp)	Ir σ	
Atactic vinyl	Various		R(p)	IR π	Numerous
			R(p)	IR σ	e.g.
			R(dp)	IR π	polystyrene
			R(dp)	IR σ	polypropylene
Centrosymmetric	Various; mutual exclu-		R(p)	—	Polyethylene
systems	sion always applicable		R(dp)	—	
planar D_{2h} or			—	IR π	Polyethylene
helical C_{2h}			—	IR σ	sulphide

Source.   Reprinted with permission of Verlag Chemie GMBH.

workers showed agreement with x-ray orientation data was possible by using the 998 and 842 cm^{-1} peaks in the PP spectrum by correction for imperfect polarization and proper baseline determinations (144). The two bands are crystallinity sensitive because of their parallel vibration in the crystalline region.

4.3.4   Conformational and Configurational Analysis

The description and study of rotational bond positions that are transformed into one another by overcoming rather low energy bar-

riers is the field of "conformational analysis" (145). It is possible to apply this detailed analysis to polymers with significant results from the theoretical and experimental viewpoints (146).

The chain of *n* bonds has been treated as if it were a freely-jointed chain of independent vibrating units that are a length of a "statistical segment." For some real polymers the mathematical analogs of molecules of 100 bonds in length may serve as useful models, in conjunction with the rotational isomeric states (ris) model noted earlier (see Section 4.1.1.2) in this discussion. One-dimensional

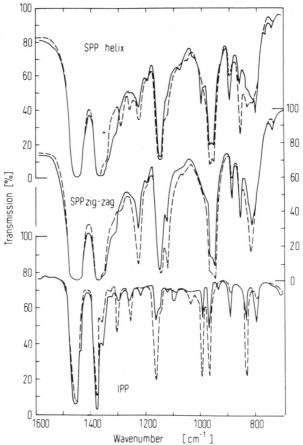

Figure 4.18 Polarized infrared spectra of PP. The solid-line spectrum is that with the electric vector perpendicular to the crystal axis; the dotted-line spectrum, parallel (109). Reprinted with permission from Verlag Chemie GMBH.

crystallinity may be thought of as conformational regularity repeated along the chain or a sequence of conformations. These forms are dependent on the physical state of the polymer chain and are generally noted to be sensitive to solvents, temperatures, or changes in state. Transitions between conformers may thus be studied by ir/Raman spectroscopy by following the loss of so-called "regularity bands," which may be characteristic of the transition. Conformational bands originate from such defined short- or long-range conformational order, and it is this microstructural regularity that is responsible for many of the ultimate macromolecular chain properties. Correlation of ir/Raman bands with conformational details of the polymer chain is certainly a major achievement of the theoretists and experimentalists.

Polymer configurations are part of the

chemical arrangement of the system, and one transforms them into one another only by breaking and remaking chemical bonds; that is, cis-to-trans conversions about a double bond are an illustration of geometrical isomerization concerned with the configurational status of the unit. Isotactic, atactic (heterotactic), and syndiotactic forms for a vinyl-polymer chain have all been noted in the previous discussion. The application of semiempirical molecular-orbital schemes (CNDO/2) and/or the ris model have been used in comprehensive calculations of total energies for the specific configurations and preferred conformations of polymers; that is, Asakura et al. for PP (147); Flory et al. for PVC (148); Stokr et al. for PVC (149); Tosi and Ciampelli for PBDs (150); and Schneider, et al. for PMMAs (151).

Using information from nmr and ir on the

Table 4.5   Carbon-Chlorine Stretching Vibrations in PVC (109, 154)

cm^{-1}	intensity		taxis		chain	conformation
	s	a				
603	vst	w	s	long chains	S_{HH}	TTTT
613	vw	vst	s	short chains	S_{HH}	TT
624	vw	w	i		S_{HH}	TGTTG'T
633	vw	w	i		S_{HH}	
639	vst	vw	s	long chains	S_{HH}	TTTT
647	w	w	s			
677	sh	m	s		S_{HC}	TGGT
695	sh	m	i		S_{HC}	TGTG

s: syndiotactic   i: isotactic (diads)   a: atactic   T: *trans*   G: *gauche*

Source.   Reprinted with permission of Verlag Chemie GMBH.

model stereoisomers of 2,4-dichloropentane and 2,4,6-trichloroheptane, Schneider et al. (152) have interpreted the ir of PVC in the frequency region of the C—Cl stretching vibration. The symbols S_{HH} and S_{CH}, corresponding to a secondary Cl conformation with the H or C atoms trans to the Cl atom, were used in the assignment of bands. The expected shape of an S_{HH} band of PVC in solution for different tacticities was calculated, and comparison of the theoretical spectrum from the C—Cl conformational models with the experimental spectrum allowed the configurational content of PVC to be estimated. It was shown that the interpretation of PVC spectra can be made on the basis of such sterically favored conformations studied in reference trimer molecules and can be experimentally verified. Also, Krimm and Enomoto (153) had established basic conformational information on PVC in 1964.

Work by Pohl and Hummel (154) on the complex 750–550 cm^{-1} region involved measurement of the quantitative temperature dependence of the C—Cl bands, and assignment of eight stretching vibrations from previous work (Table 4.5). A minimum length of approximately five monomer units in a syndiotactic sequence is needed to form the preferred planar-zigzag conformation. The experimental band separation of overlapping peaks ("band deconvolution") was performed using assigned shapes (Lorentzian, Gaussian, or skewed) and is limited to the certain assumptions that must be made in that approach (Figure 4.19). Similarly, application to isotactic PS and its deuterated analogs was recorded for sequence length studies using ir

by Kobayashi et al. (155). Generally, analysis of ir band shapes for sequence distributions has been well documented for a variety of polymers, as discussed by Kumpaneko and Kazanskii (156).

4.3.5   Vibrational Coupling and Dispersion Curves

Vibrations of regular chains are described with the semiempirical concept of independent vibrational units along the chain. However, coupling within such units may be strong and give rise to additional bands, particularly if the atoms are of equivalent masses. It has been well established that light-atom vibrations (such as those of H) individually do not couple strongly with heavy-atom vibrations; that is, coupling is much less between atoms of very different masses or whose bonds have widely varied force constants. Also, steric relationships are important. The CH_2 wag motion, for instance, is a good illustration of strong coupling between CH_2 units if they are in a trans relation, but which give only weak coupling in a gauche relation.

Regular chains undergo vibrations that are studied by dispersion curves. These curves relate the phase difference δ between neighboring chemical groups and the frequency of the vibration ν. The manner in which the neighboring coordinates are vibrationally coupled determines the shape of the dispersion curve. Each of the normal modes is a standing wave that extends over all the oscillators along the chain, and when all the particles move in phase it corresponds to a rigid translation of the whole chain. Certain vibrational modes are sensitive to the planar

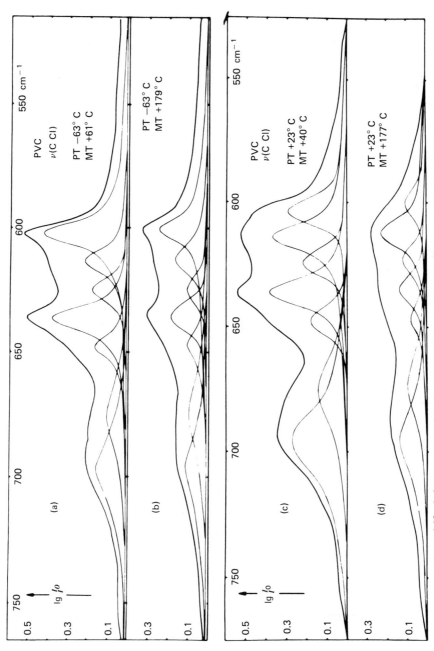

Figure 4.19 Infrared spectra of the C—Cl stretching region in PVC samples of differing tacticity prepared at different polymerization temperatures *PT* and measured at different temperatures *MT*. (*a*) *PT* = −63°C, *MT* = 41°C; (*b*) *PT* = −63°C, *MT* = 179°C; (*c*) *PT* = 23°C, *MT* = 40°C; (*d*) *PT* = 23°C, *MT* = 177°C (109). Reprinted with permission from Verlag Chemie GMBH.

168

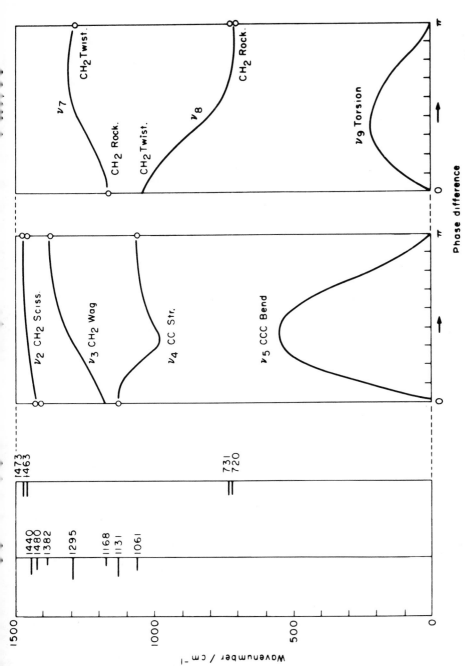

Figure 4.20 Dispersion curve for a PE chain in the extended zig-zag conformation. The frequencies are given as functions of the phase difference between the neighboring CH₂ groups. The observed Raman and ir bands on the left side correspond to the circles given in the dispersion curves. The CH₂ symmetric ν_1 and antisymmetric ν_6 vibrations are in the 3000 cm⁻¹ region and are omitted. The CH₂ wagging vibration ν_3 with zero phase difference is infrared active, but too weak to be observed. The C—C stretching and CCC bending vibrations are coupled with each other and appear as ν_4 and ν_5. The CH₂ rocking and twisting vibrations are strongly coupled in a certain region of the phase difference. The CH₂ wagging vibrations with zero and π phase differences and the CH₂ rocking vibration with π phase differences are split into doublets in the crystalline state (133). Reprinted with permission of John Wiley.

169

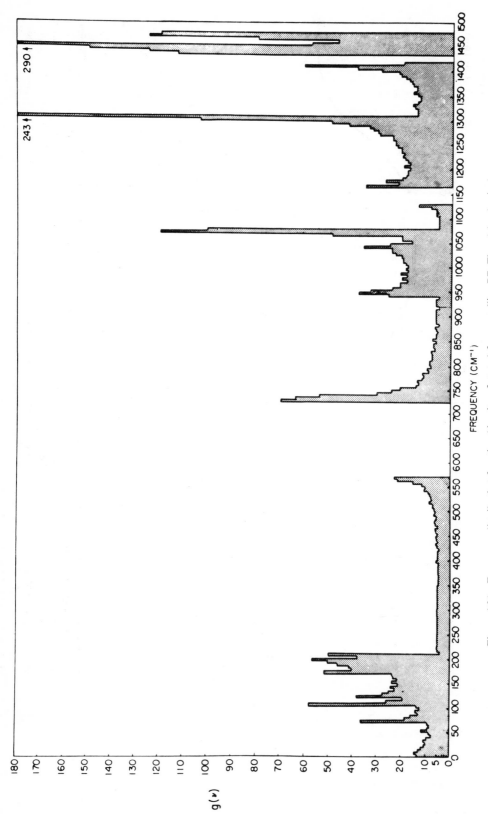

Figure 4.21 Frequency distribution function (density-of-states) for crystalline PE. The $g(v)$ value is the relative number of frequencies in a 5 cm^{-1} interval (158). Reprinted with permission from *J. Polym. Sci.* Copyright by the American Chemical Society.

or helical forms of the chain and hence to the conformation of the system (157). Other modes may be insensitive to their neighboring environment and result in rather flat dispersion curves. The frequency-phase relation of PE is shown in Figure 4.20 with the calculated dispersion curve arrived at from a Urey-Bradley force field. The $\delta = 0$, $\delta = \pi$ are ir active fundamentals; the ν_2 to ν_5 are due to in-plane vibrational modes; the ν_7 to ν_9, to out-of-plane modes.

The relation $g(\nu)$ is the frequency-distribution function and when plotted versus (ν), the result is referred to as the "density-of-states." The density-of-states for PE is given in the work by Opaskar and Krimm (158) and shows how the vibrational modes are distributed throughout a frequency region at 5 cm^{-1} intervals (Figure 4.21). Broad bands are commonly seen from strongly coupled systems in the frequency ranges covered by dispersion curves for extended chains (as are C—C bands in PE). These nonlocalized C—C vibrations are not particularly useful for analytical, semiempirical characterization, or diagnostic work, but are fundamental to a more complete understanding of spectra-structure correlations in polymers.

It may be seen from the preceding discussion that the utilization of ir/Raman spectral data necessitates some acknowledgment of the previous in-depth theoretical developments. Because of the breakdown of the selection rules and the complexity of real polymer ir/Raman analysis, it is necessary to look more closely at the nature of the disturbing features that bring the theoretical, infinitely perfect, regular chains into reality.

4.3.6   Polymer-Defect Analysis

The model of a polymer made up of configurational and conformational regular segments (159) may be interrupted by defect structures or junctions along the chain. Homopolymers may have head/head or tail/tail arrangements in a normally head/tail sequence; asymmetric atoms or copolymer segments may provide joints between sections of mers of a different nature. Conformational defects such as rotational isomers or degree of helical coiling are interruptions controlled by the physical factors of temperature, solvents, and so on noted previously. This disorder gives rise to a widening and shifting of bands and the appearance or disappearance of individual peaks throughout the pattern. There is a wide range of possible types of defects, and each form needs a structure-force constant approximation from which to develop a full interpretative approach. There is insufficient knowledge about the geometry, microstructure, and defects in real polymer systems to apply a proper model in the approximations, so generally semiempirical treatments must be followed in practice (160). With all this diversity there is, however, an approach to the theory of disordered chain vibrations (161) that allows some classification of ir spectra of real polymers.

When one goes from an analysis of an infinite to a finite regular system, a new type of defect arises from "local" vibrations. The selection rules fail, and bands may appear corresponding to all phase angles. Even though local and chain vibrations are mixed in real polymers, bands can be interpreted as arising from local-type defects; conformationally-related defects (trans, gauche) from models; configurationally-related types (isotactic, syndiotactic, head/head, which are independent of the state of the sample); and extraneous defects, as in copolymer segments or blocks. If there is a large distance between them, each is a source of local vibrations, and the frequency of the corresponding vibrations of the two defects is equal. As the distance between the defects decreases, the band splits into two components that are symmetric with respect to the initial frequency. Thus, a local vibration displays splitting when a sample is disordered. In contrast to local vibrations, the chain vibrations increase in peak width and split as the length of the regular segments of the chain increases. The split components get narrower and are sensitive to phase shifts. These vibrations have a high degree of interaction between adjacent units. However, in sequences of only 2 mers, the intensity of the forbidden bands is an order of magnitude less than those allowed absorptions.

Experimental analysis of irregular polymers has been treated by Koenig and Roggen (160) using a band asymmetry method in application to the study of PP. The degree of asymmetry of the band contour with reference to the corresponding vibrational frequency of a highly regular sample is related to the magnitude of the irregularity in the unknown system. Band asymmetry is one parameter that is useful for average regularity characteristics, such as isotacticity, block, or random segments. Several papers (160–165) have described microstructural determinations from the intensity or shape of absorption bands that have been previously identified with local structures and defined sequences in model spectral comparisons. Alternatively, within the semiempirical approach, a band-splitting method may be followed, in which one studies the correlation between the length of the regular sequence and the magnitude of the band-splitting; that is, type II bands of Onishi and Krimm (166). These bands correspond to chain vibrations with large interactions between adjacent mers, as determined in the experimental ir spectra of PE, PP, and PVC. The average block concentration is determined in this manner.

In addition to a band asymmetry or splitting method, there is also one of local vibrations that is used to follow trends in irregular polymers by ir. In a copolymer AB a band corresponding to a vibration of an AB group is determined with its absorption coefficient, and the calculation of the average length of the monomer sequences in average-sized blocks is possible. Rather than determining the complete length of a block segment, a method using the absorption coefficient or threshold sensitivity may be used to find the average length of only the configurationally ordered segment. A defect in this sense is due to an interruption in tacticity, which disturbs the normal tactic sequences. One would use oligomers to get the threshold sensitivity from models at low temperatures (to decrease the conformational, higher energy defects).

It may be seen that any of these methods (band-splitting, shape, local vibrations, or threshold sensitivity) will aid in converting experimental spectra to usable interpretative structures. The number of ethyl groups in PP

has been determined by identification of peaks from vibrations of certain lengths from models; branched paraffins are generally used for the appropriate CH_2 segment. The critical length of sequences for a given band appearance depends on the vibrational type. One may calculate ΔH, ΔS for conformational transitions of a diad (two mers) from an ordered to a disordered state by the use of band contours (iso-PP, chlorinated-PS) and then extend the calculation for triads, tetrads, and so on, transitions using temperature-dependency data (148).

The work by Opaskar and Krimm (158) showed that defect-induced activation of lattice modes can explain certain spectral features of PVC. The normal vibrations calculated with a valence force field for the planar-zigzag form for secondary Cl atoms predict peaks at 607 cm^{-1} and 634 cm^{-1}. Prediction is made of a C—Cl stretching mode at 693 cm^{-1} for the S_{CH} and/or $S_{CH'}$ conformation, so that it is likely a localized defect mode. The observed intensity changes near 639 cm^{-1} in pressed PVC is also likely from defect-induced absorption.

Rubcic and Zerbi (167) have also treated the structurally disordered PVC system in terms of its vibrational specturm. The dynamics of a PVC chain from an infinite perfect model were studied to determine some of the theoretical quantities to be used in the defective or disordered chain. The knowledge of the dispersion curve from the infinite model is needed to assign the peaks in the density-of-states for both the infinite and finite models. It was shown that introduction of defects in a perfect lattice give rise to out-of-band gap and resonance modes. Ir spectral studies can approach both conformational and configurational defects, unlike nmr, which was seen to successfully determine configurational disorder—but only with difficulty handle the complexities of conformational defects.

Commercially available PVC certainly contains major amounts of conformational and configurational defects. The concentration of these defects is so great that the homopolymer is treated essentially as a disordered material. The density-of-states from a model chain of 200 chemical units with perfect conformational and configurational structure was found to be identical with that calculated for the infinite model. The approach used by Zerbi is that of a numerical method of "island analysis," which consists of associating density-of-state peaks with particular segments or "islands of defects" in an otherwise regular host lattice. This method developed the dynamic analysis of a single-chain model with no interchain forces and standing waves propagating only along the chain axis.

The calculations by Zerbi emphasized the 600–700 cm^{-1} region in the C—Cl stretching frequency area and treated the three most likely PVC structures: extended zigzag, syndiotactic (TTTTT), which has a line group isomorphous with the C_{2v} point group; isotactic threefold helical structure (TGTGTG); and folded syndiotactic form (TTGGTTGG) analogous to the D_2-point group. The introduction of defect forms in the calculations (Figures 4.22 and 4.23) will generate gap and resonance modes, and the attempt to determine the corresponding shape of the normal modes was undertaken. The gap frequencies that are generated will shift and split by a different amount with different force fields used in the calculation. The most elaborate valence force field was developed by Opaskar

Figure 4.23 Configurational defects intoduced in an otherwise syndiotactic PVC chain. The conformational changes required by the introduction of a configurational defect are also indicated (167). Reprinted with permission from *Macromolecules*. Copyright by American Chemical Society.

and Krimm (168) and was that used in Zerbi's work.

When the defects are introduced into the linear chain, gap frequencies are generated, and the corresponding shape of the density-of-states in the band is modified (Figures 4.24 and 4.25). This fact shows that resonance modes are thus being generated. When the mutual distance between the defects is shortened, a coupling between gap modes results that gives rise to frequency splittings. The coupling occurs only when the defects are adjacent or nearly so. In the C—Cl frequency range the gap frequencies from the conformational defects occur often at the values close to those from configurational defects and result in very complex spectrum.

The modes corresponding to gap frequencies are clearly localized in space since the vibrational amplitudes decay rapidly along the chain, with the maximum intensity at the defect. The number of chemical units involved in the vibrational gap modes extend throughout 7 to 12 chemical units.

It was noted that the resonance frequencies are grouped together and give rise to sharp peaks in the density-of-states within the frequency band. It is important, therefore, to consider local and cooperative motions as well as their cooperative interactions when long-chain vibrations are studied. The force field used in these calculations gave qualitatively the same general results as those

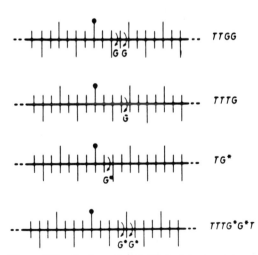

Figure 4.22 Conformational defects introduced in an otherwise all trans syndiotactic PVC chain. The dots refer to the starting point of the conformational sequence considered as defect (167). Reprinted with permission from *Macromolecules*. Copyright by American Chemical Society.

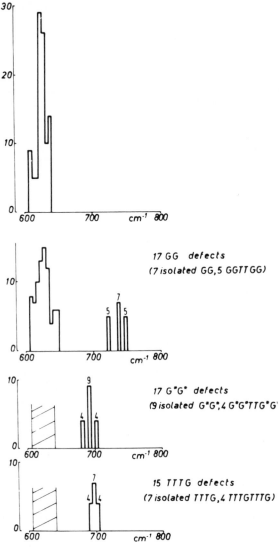

Figure 4.24 Comparison between the density-of-states in the C—Cl stretching region of a conformationally and configurationally regular syndiotactic PVC chain with that of conformationally impure chains (167). Reprinted with permission from *Macromolecules*. Copyright by American Chemical Society.

obtained from other fields that were considered. Zerbi also compared a dipole unweighted density-of-states calculation with the experimental ir spectrum to emphasize frequency fitting and most-probable structures in the chain (Figure 4.25).

In PVC the highest local symmetry C_{2v} is achieved with a GG conformational defect whose ir modes are all inactive. It was likely that all modes were activated and a complex dynamical system resulted. Shown in Figure 4.26 the density-of-states of configurationally

very disordered PVC is calculated and compared to the experimental spectrum in the C—Cl stretching region. In agreement with Krimm, the introduction of conformational disorder gave rise to 2 bands at 693 and 635 cm^{-1}. The former band is associated with the gap modes arising from the TTTG and/or G$^+$G$^+$ defects according to Zerbi. Generally, most workers agree that configurational defects in syndiotactic PVC give rise to a broad ir absorption at approximately 690 cm^{-1} and Raman scattering bands. These calculations

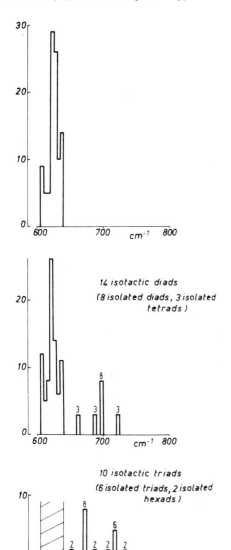

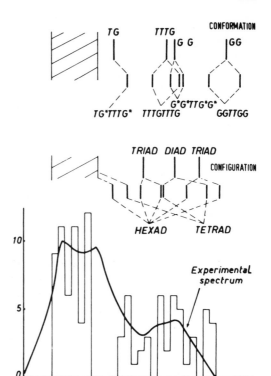

Figure 4.26 Density of vibrational states for a realistic model of configurationally disordered PVC (Bernoullian parameter $P_m = 0.47$) compared to the experimental spectrum. Calculated gap frequencies are schematically indicated at the top of the figure (167). Reprinted with permission of *Macromolecules*. Copyright by American Chemical Society.

Figure 4.25 Comparison between the density-of-states in the C—Cl stretching region of a conformationally and configurationally regular syndiotactic PVC chain with that of configurationally impure chains (167). Reprinted with permission of *Macromolecules*. Copyright by American Chemical Society.

by Zerbi show the 690 cm^{-1} peak is associated specifically with isolated isotactic diads.

It may be concluded from the preceding discussion that vibrational interactions between defects can be predicted by calculations that are based on geometries of classical, unstrained molecules and associate density-of-states peaks with "islands" of defects in an otherwise regular host lattice. The actual

geometric complexity of the PVC chain is clearly seen by an overlapping, complex band pattern in both the calculated and experimental spectra. It would appear that this theoretical study is one that describes rather closely what physically real polymers contain in terms of detailed microstructural features. The development of such an approach will be a major contribution to the interpretive value of ir spectra to polymer characterization.

4.3.7 Branch Defects

Determination of chain branching in low-density PE (LDPE) has been reported by several workers over the years. Generally, the branch content may range between 20 and 40 methyls per 1000 carbon atoms, whereas for high-density PE (HDPE) less than a tenth of that may be present. The ratio of ethyl/butyl branches as reported by Willbourn in 1959 (169) was from data limited to the ir capabilities at that time with respect to resolution and accuracy of extinction coefficients. Work recorded by Baker and Maddams (170) in 1976 gave an improved determination of the methyl content in PE compared to the conventional ir method applied to very low levels of branch content. As is generally done, standards for the absorption coefficients for the branch-content calculations were determined from synthetic branched-alkane reference compounds and allowed the reported detection of 0.1 methyl units per 1000 carbons. This result is an order of magnitude improvement over the previous experimental reports. Other work by Nerheim (171) provides a density-corrected circular polymethylene (PM) wedge for accurate methyl content in PE by ir spectroscopy. Also, results for PE using ir methodology have been reported by Kranbuehl et al. (172).

Not only the total amount of branch content is significant, but also the nature of the alkyl branches present is important detail in characterization of the system. The tenfold greater branch level in LDPE compared to HDPE attests to this fact. It is known that eight methyl branches per 1000 carbons can cause a decrease in crystallinity in PM to approximately 74%, and the presence of eight ethyl branches per 1000 carbons decreases it to approximately 42% of its original highly crystalline level. Recently, long- and short-branch content in LDPE was studied by Domareva et al. (173), and further extension of ir studies by McRae and Maddams attempted to determine specific types of alkyl branches in PE and copolymers (174). Determination of the amount and nature of these branches using ^{13}C nmr was noted earlier in the discussion, and data from pgc further delineates this area in the following discussion. The ir results by Domareva et al. and McRae and Maddams indicated that no specific ethyl band could be found, but propyl and longer branches did have a peak at approximately 890 cm^{-1}, which was assigned to a rocking mode. The methyl branches in the polymers studied gave a 935 cm^{-1} peak, also assigned to a methyl-rocking mode, and whose intensity was proportional to the methyl-branch content in nearly linear PE.

4.4 PYROLYSIS GAS CHROMATOGRAPHY (PGC)

It has only been within the last ten years that pgc has become a relatively common means for polymer analysis and microstructural information. Many books and reviews (175–193) have discussed the development of analytical gc from the most basic systems and theory to the highly sophisticated complex instrumentation that is applied to demanding problems in organic analysis. The wide diversity of applications has on one hand made this innovative, sensitive method most readily adapted to problem-solving and trace analyses, but on the other hand has hindered standardization of the methodology.

The general use of pgc in polymer studies (181) may be grouped into analyses of microstructural details (sequence length, chemical defects, end groups) or simple "fingerprinting" (chromatographic patterns of a known system compared to the unknown pattern). Both approaches rely on some interpretive analysis of the molecular fragments that result on pyrolysis of the macromolecular system in question. Additionally, the mechanisms of decomposition under inert or reactive, catalytic exposures may be an important adjunct to the structural effects and the inferences made from them regarding the thermal degradation.

The use of controlled pyrolysis or reaction gc in the identification of chromatographic effluents is a further application area. This may include providing data on the nature of the functional group(s) present (194–202), the elemental analysis (201–208), or the location of a radiolabeled fragment in a tracer experiment performed on the macromolecular system (209–212).

4.4.1 Pyrolysis Instrumentation and Identification of GC Effluents

Figure 4.27 schematically depicts the varied arrangements of analytical pyrolysis or reaction gc that have been utilized for quali-

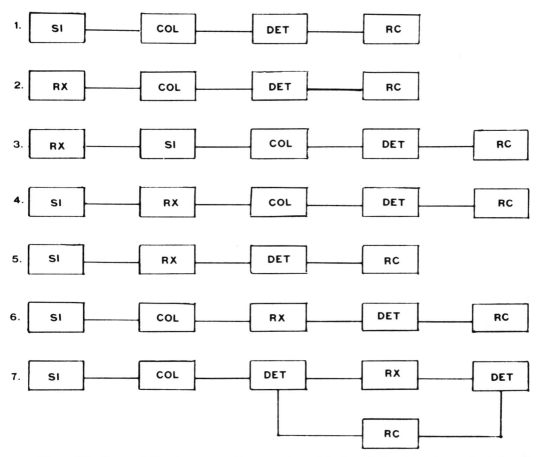

Figure 4.27 Representative chromatographic systems in analytical pyrolysis and reaction gc. Symbol representations: SI = sample inlet; COL = chromatographic column; RX = reactor or pyrolysis unit; DET = detector; RC = recorder or computer (179).

tative and quantitative analysis of materials (179) (see Table 4.6). Most early pyrolysis techniques involved long exposure times in microfurnaces or specialty pyrolyzers at very high temperatures. These conditions were needed to generate enough volatile products for detection or further analysis to provide information on the initial structure that produced them. However, this often resulted in complex, uninformative product mixtures. The advent of rapid, flash pyrolyzers and the ability to analyze trace levels (10^{-6} moles or less) of volatile fragments that contain the structural information of significance has occurred within the past decade.

The Curie-point pyrolyzers represent one of these advances (213–215). These units consist of a ferromagnetic wire that is inductively heated to a specific temperature when ex-

posed to a radio-frequency field. The chosen filament alloy determines the Curie temperature and ranges between approximately 300 and 1000° C at set intervals in commercially available units (216). Other pyrolytic devices (217) utilize a coil (with a quartz sample insert) or a ribbon filament (upon which the sample may be deposited) that are resistively heated to provide flash pyrolysis to any chosen temperature from 100 to 1200°C in milliseconds or, at slower rates of 5, 10, or 20°/ min., to permit thermal degradation studies with programmed pyrolysis.

Initially, limitations to widespread applications of pgc to detailed polymer microstructure studies were the lack of instrumental reproducibility and poor interlaboratory correlations because of these varied and innovative arrangements. The sensitivity of the

Table 4.6   Identification Methods of GC Effluents

```
RETENTION TIME
REACTION GC
        Selective Subtraction or Derivatization
        Controlled Pyrolysis
                Noncatalytic - Carbon Skeleton, Functional Group
                Catalytic    - (De)Hydrogenation, Elemental Analysis
SPECIALIZED GC DETECTORS
        Coulometric
        Sensitized Flame Ionization
        Flame Photometric
        Electron Capture
        Flame Ionization
        Photoionization
        Gas Density
SPECTROSCOPIC ANCILLARY INSTRUMENTATION
        Infrared
        Raman
        Ultraviolet and Visible
        Mass
PEAK TRAPPING, CONCENTRATION, ISOLATION, TRADITIONAL CHEMICAL ANALYSIS
```

pyrograms to thermal risetimes and precise final temperatures was noted early by workers in the field and discussed by Levy et al. in 1972 (218). Most recently, Gough and Jones (219) reported on the assessment of pgc precision from the results of a correlation trial by the Pyrolysis Subgroup of the Chromatography Discussion Group (London). It was determined that with strict observance to operational parameters, the interlaboratory reproducibility is of high order. Other workers (220) have utilized a method, earlier reported by Gross (221) in 1971, to generate a series of reference compounds in situ during the pyrolysis (or "copyrolysis") from which gc retention indices are developed. The copyrolysis of an *n*-hydrocarbon (C_{12}, C_{24}, or C_{44}) or a polyethylene (PE) with the unknown

polymer under study produces a reference series of peaks in the pyrogram that is useful for introducing hydrocarbon standard retention indices during the solid-state pyrolysis.

In a typical indentification process, the information from the mass-spectral (ms) base peak, molecular weight, and the retention index on the analytical system used provides a number of possible species that would be structural candidates for the unknown peak. When these data and other confirmatory information, such as the ir spectrum or ms fragmentation patterns, are searched in compiled computer collections [Sadtler files (222), Aldermaston, NIH-mass spectral files, or others], the identity of the peak is completed. Other means of polymer identification using the pyrolysis method coupled with on-line molecular-weight chromatography and vapor-phase ir spectroscopy was reported by Kiran and Gillham (223). An analytical systems approach to polymer characterization from pyrograms was earlier developed that utilized still another data-base combination consisting of the gc retention time of the pyrolytic peak(s), their functional group or small-molecule fragmentation patterns ("controlled thermolytic dissociation") (194), and/or elemental (C,H,N,O, halogens, S) content (202).

4.4.2 Applications and Interpretation

A wide range of problem-solving capabilities with gc equipment and thermolytic methods is given throughout the literature that is of concern to the polymer analyst, such as determination of residual volatiles and monomers (224–228), or plasticizers and additives (229–233). In conjunction with detection and identification of volatiles, additives, or pyrolytic degradation products, a means to concentrate ultra-trace levels of these species is provided by component concentrator units (234–238).

Reaction gc is an ancillary field in which the sample or products of pyrolysis (or thermolysis) are made to undergo further reaction by catalysis, derivatization, or other treatment. These techniques have been used extensively for identification of functional groups by subtractive or removal operations, or conversions to more easily chromatographed fragments by derivatization and other chemical methods (179, 182, 183, 189, and 191). Reaction gc is the means by which on-line elemental analysis of gc effluents has been achieved (202, 239). Extension to on-line radiochemical tracing for gc effluents has been demonstrated with a ^{14}C-labeled species being converted by reaction gc to $^{14}CO_2$ and counted with a simple Geiger tube (212). "Chromatographic stop-flow," accomplished by the use of appropriate column/valving combinations (239), can be combined with reaction-gc capability to permit the conduct of these specialty analyses directly on the effluent as it emerges from the chromato-graphic separation column. In later discussion, it is shown that a polymer may be synthesized with positions radioisotopically labeled for structural or mechanistic studies and pyrolyzed to provide diagnostic gc peaks. [14]C in the species constituting these peaks can be converted to $^{14}CO_2$, analyzed as above, and related to specific sites of bond scission, fragmentation patterns, or sequence distribution in the original polymer (209). Additionally, in many applications the hydrogenation of pyrolytic products, either directly after the pyrolyzer unit or after they emerge from the chromatographic column, has been an important means of aid to researchers in the interpretation of pyrograms (240–242).

One of the difficult aspects of analyzing any complex chromatogram or pyrogram is the evaluation of closely similar patterns with many (20–200 or more) individual peaks at the level of fingerprint comparisons. The quantification of such a mixture is even more demanding, but with the current data systems available, it is possible to adjust peak-detection logic and various chromatographic parameters to obtain highly accurate and reproducible results (within 1% on minor peaks). Efforts have been made to evaluate just such difficult pyrograms toward interpretatively utilizing the data in macromolecular microstructure studies. Computer programs have been devised to test patterns statistically between chromatograms that have meaningful information related to the biochemistry of normal and abnormal metabolism (243). Automatic comparison of complex pyrograms was reported in 1972 from a study of microorganisms (244), and in 1973 a statistical gc profile analysis was also reported (245). The use of a multivariate approach to pattern analysis in the identification of oils was given by Mattson et al. (246). In that instance, the ir spectral patterns were obtained and 21 absorption bands of interest were further studied with the aid of a normalization procedure and a weighted χ test. This statistical approach was applied to patterns of spectral data in a library comparison series.

A treatment of gc data similarly used such statistical evaluation, but extended and specified the differentiation process in the chromatographic comparisons (247). A means was developed (248) to evaluate experimental pyrograms using a computer program to calculate F and T^2 statistics to determine: if two pyrograms in question were, in fact, identical within the limits of the total system reproducibility; and, if not identical, which peak(s) in the pattern was (were) the differentiating one(s). Each pattern was compared on the basis of up to 28 chosen peaks (limited in number only by the size of the matrix calculation handled by the computer) and their quantitative levels. The total pyrogram was compared by the global response calculations using Hotelling's T^2 statistics. The comparisons of nearly identical pyrograms may be thus conducted in order to detect minor, but often important, differences that result from microstructural details in the original macromolecule.

4.4.3 PGC of Polymers and Copolymers

The use of pgc to elucidate microstructural details has been extended from simple diagnostic fingerprinting of fragmentation patterns from homopolymers to detection and identification of minor structures in complex copolymer systems. Both structure and degradation mechanisms are possible to study within this field of analytical instrumentation and methodology.

4.4.3.1 General Polymer Systems
The thermal degradation of polymers has been studied for many years from various mechanistic and theoretical approaches on a diversity of structures (249–270). For polymers undergoing thermal degradation, generally a radical mechanism is followed, and the products of pyrolysis may be monomer and oligomers or complex side-chain splitting and macroradical rearrangement products. The presence of weak links (defects, branches, etc.) in the chain or reactive atoms, such as halogens or tertiary hydrogen, significantly influence the decomposition pathways and resultant products. Pgc data on a number of systems were reported in 1968 and 1969 (255, 256) and illustrate this point. Statistical chain-breaking and depolymerization take place in

the pyrolysis of polyisobutene, PP, poly(α-methyl styrene), polystyrene (PS), 1,4-*cis*-BD, and poly(2,6-dimethylphenylene oxide). These mechanisms are discussed in detail from the pyrolysis field-ion mass spectroscopy results by Hummel et al. (258).

In the work by Zaitsev and Poddubnyi (259), the thermal degradation of PS was investigated over a temperature range 300–570°C using pgc in the pulsed mode and separate stages in the decomposition were deduced from the quantitative analysis of degradation products. The small (milligram) specimen size and strict temperature control allowed determination of the activation energy for PS thermal decomposition and, thus, direct estimation of the bond energy of carbon atoms in the polymer chain. In a comprehensive study of the pyrolysis and combustion of Nylon 6, it was shown that a first-order decomposition is followed to yield ϵ-caprolactam as the primary product with an activation energy of approximately 46 kcal/mole, indicating a homolytic cleavage (260).

Similarly, many workers over the years have utilized the varied experimental pyrolytic treatments noted and established pgc information on microstructure for polyisoprenes (261, 262); cyclopolyisoprene (263); cross-linking in ion-exchange resins (264, 265); differences between elastomeric materials (266), hydrogenated PVC (267), diene polymers (268), coumaroneindene and cyclopentadiene polymers (269), end groups in siloxanes (270), urethanes (271–273), chlorinated vinyls (274, 275); PP (276); and sequence distribution/microstructure in various copolymer and terpolymer systems (277–288).

The differentiation of pgc patterns from block and random copolymers is an important contribution to the characterization methodology as reported by several workers. Random and block systems of ethyl acrylate and methyl methacrylate were studied in 1974 by Wallish (289) who noted that pyrograms depended on pyrolytic temperature, ratio of copolymerized monomers, degree of conversion, and polymerization method. Larger amounts of ethyl methacrylate and methyl acrylate were formed on pyrolysis of the random copolymers than of block copolymers; the presence of mixed dimers indicated random copolymerization. Additionally, it was noted that random copolymers produced less ethanol than ethyl acrylate on pyrolysis, while homopolymers and block copolymers produced more ethanol and less ethyl acrylate, as noted by earlier workers (290). The apparent availability of the tertiary H atom in the abstraction step of the degradation mechanism by the alkoxy oxygen of a neighboring acrylate unit governed the pyrolysis pattern. Hence, the sequence distribution of the acrylate/methacrylate units is paramount in the interpretative understanding of the relationship of thermal degradation with microstructural details from pgc of the copolymer system.

Likewise, Seymour et al. (291) have reported block systems of acrylonitrile with styrene, methyl methacrylate, ethyl methylacrylate, and vinyl acetate, and the utilization of pgc in their characterization of varied polymerization techniques used in the copolymer preparations. Hurduc et al. (292) studied the thermal degradation of styrene-methylmethacrylate random, alternating, and block copolymers and compared their pgc patterns with those of the homopolymers with the appropriate compositions. It was possible to distinguish random and alternating systems from block copolymers, as well as block systems from homopolymer blends. In contrast to random and alternating copolymers, which had specific thermal behaviors, the block copolymers and polymer blends were essentially superpositions of the degradation processes of the corresponding homopolymers.

4.4.3.2 Vinyl-Type Copolymers

Pioneering efforts by Tsuge and coworkers demonstrated the valuable role pgc has played in detailed microstructural polymer analysis. Vinyl-type copolymers of acrylonitrile/m-chlorostyrene (AN-m-CSt) and acrylonitrile/p-chlorostyrene (AN-p-CSt) systems were studied (293) from a theoretical and experimental standpoint in characterizing diad-sequence distribution. The theory developed by Wall and Flynn (294) relating the monomer yield from vinyl-type copolymers to the arrangement of monomer units in the system introduced a concept of the "boundary effect,"

caused by the influence of neighboring groups on the stability of depropagating macroradicals. The extension of the theory to a two-sided boundary effect was developed and applied by Okumoto, Tsuge et al. (293) from the experimental pyrograms seen in Figure 4.28. The primary pyrolysis products (at 530° C, about 0.5 mg under N_2) from polyacrylonitrile (PAN), P-m-CSt, and P-p-CSt are monomers and dimers, while the copolymers yield characteristic hybrid dimers also. The observed and theoretical concentrations (from application of the copolymerization theory) of diads in the copolymers were determined and a dimer "formation probability constant" was obtained from the experimental pyrolysis results. Excellent agreement was reported of observed dimer yields and the calculated diad concentrations plotted against the relative monomer feeds for copolymerization, as well as the theoretical diad concentrations from copolymerization theory. In addition, this approach allowed evaluation of the boundary effect for the formation of dimers from vinyl-type copolymers in terms of the formation probability constant, which could be extended for formation of monomers from copolymer chains.

The sequence distribution in methyl acrylate-styrene copolymers were studied over a range of compositions and conversions using pgc by these same researchers (295). Development of their microstructural pgc applications made possible the analysis of diad and triad distributions in these copolymers of high conversions. Dimers and hybrid dimers were used for diad characterization, and trimers and hybrid trimers were used for triad characterization. The diad concentrations were in relatively good agreement with the theoretical values over the full range of copolymer composition. However, the observed triad distributions showed significant deviation from the theoretical for copolymers with high styrene content and were likely due to the low yields of SSS trimers from long sequences of styrene units according to this report.

Earlier work by Tsuge et al. employed pgc for the first reported results on the triad distributions in vinylidene chloride-vinyl chloride copolymers (296), wherein specific degra-

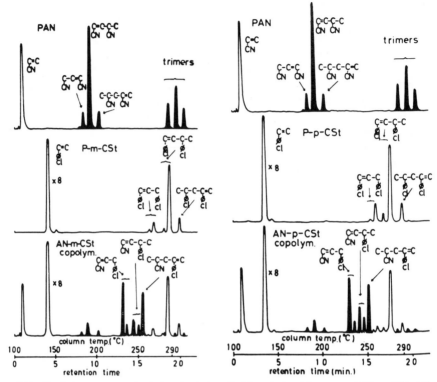

Figure 4.28 Typical pyrograms of poly(acrylonitrile), (PAN); poly (*m*-chlorostyrene) (P-*m*-CSt); AN-*m*-CSt copolymer; poly (*p*-chlorostyrene) (P-*p*-CSt); and AN-*p*-CSt copolymer. Pyrolysis at 530°C (293). Reprinted with permission from *Macromolecules*. Copyright from American Chemical Society.

dation products (benzene and chlorinated benzenes) that are formed from dehydrochlorination and cyclization reactions were analyzed in these studies. The microstructures of chlorinated PEs (297) and chlorinated PVCs (298) similarly followed from application of the pgc methodology and the preceding interpretive approach.

4.4.3.3 Polyolefin Systems

Flash pgc was applied by Verdier and Guyot (299) in the comprehensive examination of propylene-1-butene (PB) copolymers prepared with varying catalysts. They utilized the well established background of polyolefin degradation mechanisms (300, 301) in their evaluation of the resultant pyrograms. In general, the mechanism is a random chain-breaking step that initiates a radical chain reaction with intramolecular transfer on the fifth C atom (251) followed by chain-breaking between the 6–7 carbons to produce a trimer molecule.

For a copolymer, the trimer distribution is related to the sequence distribution of the monomer units along the chain. Pyrograms of the homopolymers and copolymers in this series by Verdier and Guyot showed that new peaks (C, D, E, F) appeared in the pyrolysis products that were generated from two propylene units (C and D peaks, C_{10} olefins) or two butene units (E and F peaks, C_{11} olefins). Hydrogenation of the products simplified the pyrogram Figure 4.29*a* to give corresponding paraffins, wherein from the analysis the alternate sequences (PBP, BPB) were distinguishable from the block (PPB, BBP) unit (Figure 4.29*b*). The latter units are differentiated only by the position of the substitution along the chain and are unlikely to be separated easily in gc analysis. The utilization of conditional probabilities calculated according to Pyun (302) showed from comparison with the experimental trimer concentrations that the distribution was non-Markovian; that is, first-order

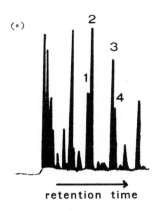

(a)

retention time

(b) Pyrolysis Fragments Related to Trimers in the Copolymer Sequences (299).

Sequence[a]	Olefins	Paraffins	Fragment Peak Number

PPB Block	C=C–C–C–C–C
	    | | |
	    C C C
	            |
	            C
		C–C–C–C–C–C–C–C
		    | |
		    C C
			1

	C=C–C–C–C–C
	    | | |
	    C C C
	    |
	    C
		C–C–C–C–C–C–C–C
		        | |
		        C C
			1

PBP Alternate	C=C–C–C–C–C
	    | | |
	    C C C
	        |
	        C
		C–C–C–C–C–C–C
		    | |
		    C C
		        |
		        C
			2

BPB Alternate	C=C–C–C–C–C
	    | | |
	    C C C
	    | |
	    C C
		C–C–C–C–C–C–C–C
		        | |
		        C C
			3

BBP Block	C=C–C–C–C–C
	    | | |
	    C C C
	    | |
	    C C
		C–C–C–C–C–C–C–C
		    | |
		    C C
		        |
		        C
			4

	C=C–C–C–C–C
	    | | |
	    C C C
	        | |
	        C C
		C–C–C–C–C–C–C–C
		    | |
		    C C
		        |
		        C
			4

[a] P = propene unit, B = butene unit.

Figure 4.29 Pyrogram from propylene-1-butene (PB) copolymer (pyrolysis and hydrogenation under H_2). (a) C_8–C_{13} hydrocarbon products. (b) Predicted trimers from the copolymer sequences and their corresponding pyrogram peaks in (a) (299).

Markovian process calculations did not adequately describe the sequence distribution in these copolymer systems. Complexities in the reaction were also demonstrated by scatter in the Fineman-Ross plots (303) and discrepancies between reported reactivity ratios in the Ziegler-Natta catalysis of olefin copolymerization.

Definitive work by Seno et al. (304) established significant pgc background for PP and ethylene-propylene copolymers (EP) using a prechlorination step to direct the degradation-product distribution to be interpretatively useful. From the defined dehydrochlorination and cyclization steps in chlorinated-chain decompositions, the predicted combinations of triads and associated aromatic compounds produced on pyrolysis were presented. Formation of m-xylene and toluene from the triad PPE is illustrated in Figure 4.30a. A summarized relationship of triad combinations and the most likely aromatic pyrolysis products is given in Figure 4.30b. For a limited range of "degree of chlorination" (DC) which is the number of chlorine atoms contained per four carbon atoms along the chain, both pgc and nmr data were in agreement in the copolymer

compositional calculation. The relationships were determined between the observed triad concentrations and the calculated composition of the EP copolymers. Utilizing the pgc data, the monomer reactivity ratios ($r_E = 19 \pm 4$ and $r_P = 0.22 \pm 0.03$) are in reasonable agreement with those reported for EP copolymers prepared by Ziegler-Natta catalyst ($r_E = 15.7$ and $r_P = 0.11$, respectively).

Extension of this approach made possible the estimation of the quantitative amount of chemical inversions in monomer placement in isotactic and atactic PP, wherein head-to-head defects occur in an otherwise head-to-tail system (304). Using the pgc method on the chlorinated polymer, the molar proportion of mesitylene/pseudocumene was determined, in addition to trimethylbenzene isomer products, for triad content in the original PP. The expected pseudocumene pyrolysis product is produced from a triad sequence containing one inverted unit as shown in Figure 4.31. It was thus demonstrated that mesitylene was associated with three regular placements and pseudocumene with one inverted and two regular ones. From these experimentally measured products, the amounts of regular

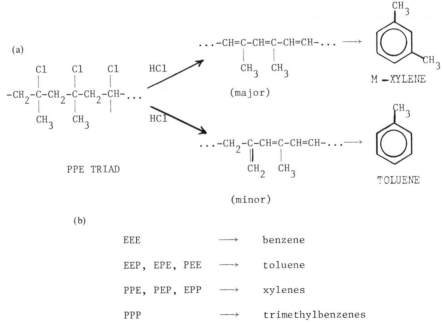

Figure 4.30 Degradation products from differing triads in chlorinated ethylene-propylene copolymer (304).

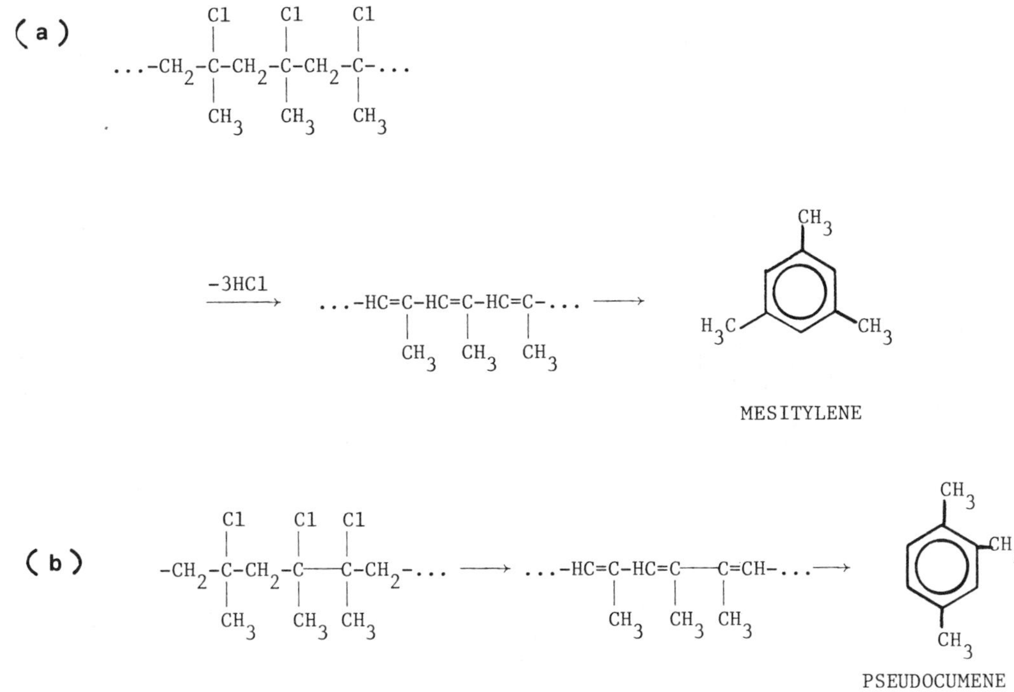

(a)

$$Cl \quad Cl \quad Cl$$
$$... -CH_2-C-CH_2-C-CH_2-C-...$$
$$CH_3 \quad CH_3 \quad CH_3$$

$$\xrightarrow{-3HCl} \quad ...-HC=C-HC=C-HC=C-... \longrightarrow$$
$$CH_3 \quad CH_3 \quad CH_3$$

MESITYLENE

(b)

$$Cl \quad Cl \quad Cl$$
$$-CH_2-C-CH_2-C\quad C-CH_2-... \longrightarrow ...-HC=C-HC=C\quad C=CH-...\longrightarrow$$
$$CH_3 \quad CH_3 \quad CH_3 \qquad\qquad CH_3 \quad CH_3 \quad CH_3$$

PSEUDOCUMENE

Figure 4.31 Degradation products from head-to-head linkages and from inverted monomer linkages in chlorinated PP. (*a*) Normal head-to-head sequences give rise to mesitylene product. (*b*) Inverted monomer sequences give rise to pseudocumene product (304).

and inverted monomer placement in isotactic and atactic PP were determined. It was reported that iso-PP and atactic-PP produced from anionic polymerizations contain the inverted monomer unit in approximately 2.5 and 9.5% levels, respectively. This may relate to the noncrystallinity of the latter system since its stereoirregularity and relatively high content of chemical inversions likely influence this parameter. Research by Audisio and Bajo on stereoisomers of PP has further defined the microstructure of these systems (305).

4.4.3.4 Pyrolysis of PVC

A significant effort has been made by many workers (231, 306–313) to analyze the pyrolysis products from PVC and its composites because of the importance of this polymer in commercial applications and the concern with its smoke, flammability, and toxicity behavior. The relationship of polymerization conditions to the tactic structure of the polymer was noted in our discussion on nmr studies and it has been supplemented by pgc investigations. The pyrolysis of poly(vinyl halogens) is notably

different from those decompositions followed by other vinyl-type polymers by virtue of the side-chain stripping and subsequent chain scissions and intramolecular cyclizations. Figure 4.32 gives the experimental verification by pgc examination of PVC under the denoted conditions.

The specific monitoring of benzene, the major aromatic pyrolysis product, has been accomplished by appropriate gc valve and pyrolyzer arrangements (314). The temperatures of initial detection of benzene and lighter hydrocarbon fragments were monitored during thermal degradation using an automatic, repetitive sampling system in an application of "time-resolved" pgc. Other products could likewise be chromatographically analyzed using optimized column and instrumental parameters. The on-line dynamic analysis of pyrolysis/ combustion products was additionally reported using Fourier-transform ir for identifications and relative quantification (314). A multitude of workers have recorded the aromatic, paraffinic, and olefinic complex mixtures that result on PVC pyrolysis as seen in Figure 4.32

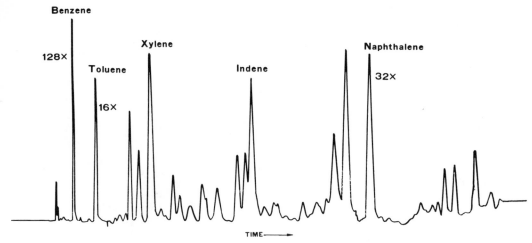

Figure 4.32 Pyrogram of PVC, pyrolysis temperature of 650° C, 10 sec pulse, helium carrier gas, flame ionization detector, 50 ft × 0.020 in i.d. support-coated open tubular column (SCOT) (322).

and which may be summarized in Figure 4.33 from the work of Hummel et al. (315) using pyrolysis field-ionization gc/ms instrumentation. It is the deviation of this "generalized" pyrogram from commercially available homopolymers that demands the utmost in pgc capability to carefully define microstructural details. For example, a vinyl chloride-propylene copolymer of 92:8 molar-percent content gave predominatly benzene, toluene, and naphthalene as expected, with increased levels of C_2H_4, C_3H_6, C_4H_8, and C_5H_8 being measured (315). The pentadiene (C_5H_8) fragment (mass 68) appeared to be characteristic of hetero diads (Figure 4.34).

In 1970, Ito et al. reported (316) the pgc results on the structure of chlorinated 1,4-PBDs for the purpose of using the system as a model for so-called head-to-head PVC. Pyrograms were obtained from PBDs with varying degrees of chlorination (DC) (with DC = 2 corresponding to nearly complete head-to-head PVC). Since it has been established that head-to-tail PVC (CH₂—CH) normally yields
$$\overset{|}{\text{Cl}}$$
benzene as a major product, the Cl—PB head-to-head structures (CH—CH), subject to
$$\overset{|}{\text{Cl}}\quad\overset{|}{\text{Cl}}$$
greater chain scission, gave measurably lower benzene yield on pyrolysis. The presence of chlorobenzene in the pyrolyzate from the latter systems may be explained from the higher-chlorinated units as in Figure 4.35. The

other chlorinated o-, m-, and pCl benzenes may be similarly explained from study of the detailed chlorinated units along the chain. Other workers have concluded differently from pgc studies on Cl-vinyl systems (317).

The minor chlorination (1–2%) of a parent PVC system has been shown (275) to increase remarkably the chlorobenzene pyrolysis product and would suggest this to be a very sensitive monitor of a chlorine atom "defect" structure in the PVC backbone. Higher chlorination of PVC was shown to produce di-, tri-, and tetra-substituted benzenes. These results were of interest since the volatile decomposition products of other various chlorinated polymers (PE, PP) correlated with their tendency to produce smoke and altered their thermal behavior (275). It appears that the relative position and amount of chlorine atoms along the polymer-backbone chain are important aspects in determining their mechanisms of thermal decomposition and the resultant experimental pyrograms.

4.4.4 Pgc of PE and Reduced PVC for Branch Content

The determination of branch content in polyolefin systems has challenged the most sophisticated instrumental polymer analysts. Pgc has shown significant potential in delineating the details of the type and amount of branch content.

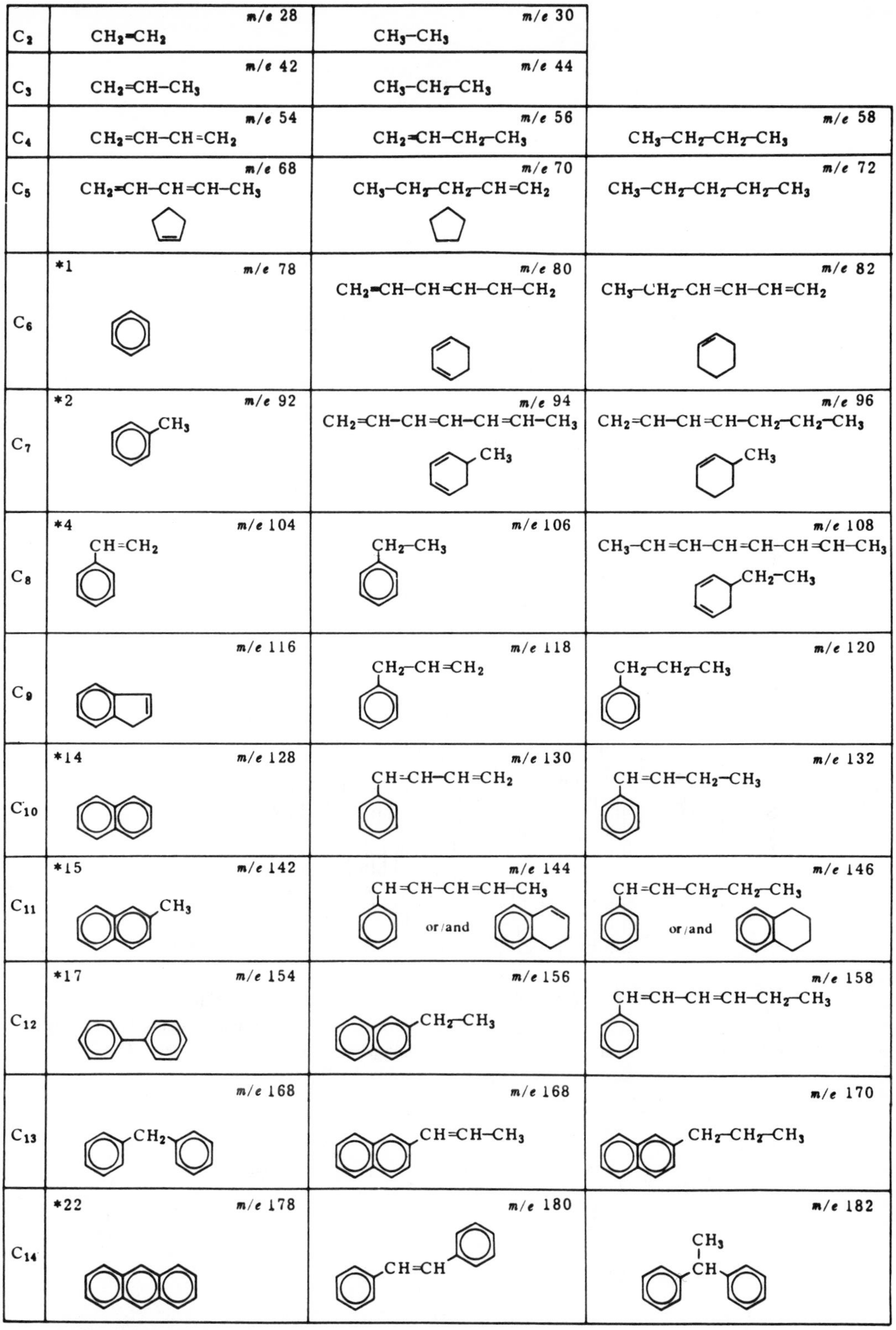
C2
CH2=CH2 m/e 28
CH3-CH3 m/e 30
C3
CH2=CH-CH3 m/e 42
CH3-CH2-CH3 m/e 44
C4
CH2=CH-CH=CH2 m/e 54
CH2=CH-CH2-CH3 m/e 56
CH3-CH2-CH2-CH3 m/e 58
C5
CH2=CH-CH=CH-CH3 m/e 68
CH3-CH2-CH2-CH=CH2 m/e 70
CH3-CH2-CH2-CH2-CH3 m/e 72
C6
*1 m/e 78
CH2=CH-CH=CH-CH-CH2 m/e 80
CH3-CH2-CH=CH-CH=CH2 m/e 82
C7
*2 m/e 92
CH3
CH2=CH-CH=CH-CH=CH-CH3 m/e 94
CH3
CH2=CH-CH=CH-CH2-CH2-CH3 m/e 96
CH3
C8
*4 m/e 104
CH=CH2
CH2-CH3 m/e 106
CH3-CH=CH-CH=CH-CH=CH-CH3 m/e 108
CH2-CH3
C9
m/e 116
CH2-CH=CH2 m/e 118
CH2-CH2-CH3 m/e 120
C10
*14 m/e 128
CH=CH-CH=CH2 m/e 130
CH=CH-CH2-CH3 m/e 132
C11
*15 m/e 142
CH3
CH=CH-CH=CH-CH3 m/e 144
or/and
CH=CH-CH2-CH2-CH3 m/e 146
or/and
C12
*17 m/e 154
CH2-CH3 m/e 156
CH=CH-CH=CH-CH2-CH3 m/e 158
C13
CH2 m/e 168
CH=CH-CH3 m/e 168
CH2-CH2-CH3 m/e 170
C14
*22 m/e 178
CH=CH m/e 180
CH3
CH m/e 182

* identified by gas chromatography

Figure 4.33 Pyrolysis products of PVC as determined by pyrolysis field ion ms (315). Reprinted with permission from Verlag Chemie GMBH.

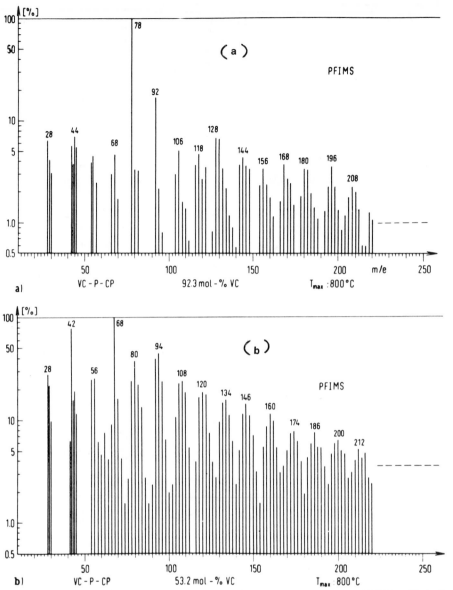

Figure 4.34 Pyrolysis field ion ms patterns from vinyl chloride-propylene copolymers. (*a*) 92.3 mol% vinyl chloride. (*b*) 53.2 mol% vinyl chloride (315). Reprinted with permission from Verlag Chemie GMBH.

Dehydrochlorination and cyclization

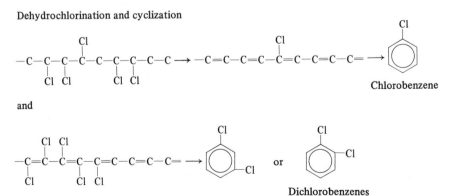

Figure 4.35 Pyrolysis products of chlorinated PVC (316).

Table 4.7 Linear and Branched Fragments with a Higher Probability in PE Pyrolysis (318, 319)

PE BRANCH TYPE	BACKBONE SCISSION cyclic or random (r)			SIDE CHAIN α-SCISSION
	α	β	σ	
LINEAR	--	--	$n\text{-}C_6$	--
METHYL	$n\text{-}C_7$	$2\text{-M }C_7$	$3\text{-M }C_7$	C_1
ETHYL	$n\text{-}C_8$	$3\text{-M }C_8$	$3\text{-E }C_8$	C_2

```
      β α α
  σ ↓ ↓ ↓ β σ
  -C-C-C-C-C-
        |  ←α
        C
        |
        C
```

Source. Reprinted with permission from *J. Polym. Sci.* Copyright by American Chemical Society.

4.4.4.1 PE Pyrolysis

Structural investigations of PE using pyrolysis-hydrogenation gc (phgc) (240–242) have shown that methyl, ethyl, and butyl branches can be distinguished and that quantitative differences in the amounts of these branches can be determined for LDPE (Ziegler) and Phillips HDPE (318, 319; Tables 4.7 and 4.8). Evidence that the products formed from PE pyrolysis and EP copolymers agreed with theoretically predicted products was reported by Seeger et al. (320, 321). This work provides a firm background on the relationship of pyrolysis products to chain branching in PE. Figure 4.36a gives the characteristic pyrogram of PE, with the major products being the *n*-alkanes, α-olefins, and α, ω-diolefins that correspond to the labeled peaks 1, 2, 3 of the evenly spaced sets of triplets. The intermediate, smaller peaks are due to the branch content.

The interpretation by Ahlstrom et al. (322) of the pyrograms obtained for PE and reduced

Table 4.8 Linear and Branched Fragments with a Higher Probability in PE Pyrolysis (318, 319)

PE BRANCH TYPE	BACKBONE SCISSION cyclic or random (r)			SIDE CHAIN SCISSION	
	α	β	σ	α	β
PROPYL	$n\text{-}C_9$	$4\text{-M }C_9$ (r) 4-M alkanes	$4\text{-E }C_9$	C_3	C_2 and 2-M alkanes
BUTYL	$n\text{-}C_{10}$	$5\text{-M }C_{10}$ (r) 5-M alkanes	$4\text{-E }C_{10}$	$n\text{-}C_4$	C_3 and 2-M alkanes

```
  σ β α α β σ
  -C-C-C-C-C-C
        |  ← α
        C
        |  ← β
        C
        |
        C
```

Source. Reprinted with permission from *J. Polym. Sci.* Copyright by American Chemical Society.

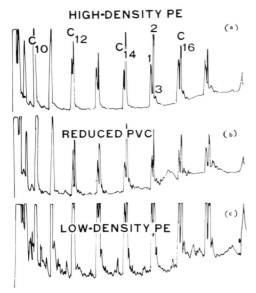

Figure 4.36 Pyrolysis products of PEs and reduced PVC. (*a*) HDPE. (*b*) Reduced PVC. (*c*) LDPE; pyrolysis temperature 650° C, 10 sec pulse, SCOT column, temperature programmed to 175° C; fragments C_9-C_{18} hydrocarbon.

PVC (Figure 4.36) are based on the preceding discussion. It was shown from these data that HDPE contains mainly methyl branches. Additionally, in this study (322) the number of ethyl branches in the examined LDPE were found to be significant relative to the number of butyl branches. These results are in good agreement with the pyrolysis data of van Schooten and Evanhuis (242) and the gamma-radiolysis results of Kamath and Barlow (323) for LDPE. They do not agree closely with the recent work by Bovey et al. (324) using ^{13}C nmr, but are substantiated by the ^{13}C results of Cudby and Bunn (95).

In a detailed study of pyrolysis mechanisms, and the products formed from the pyrolysis of HDPE and polymethylene, Tsuchiya and Sumi (325) proposed that the major hydrogen-abstraction reaction is due to an intramolecular cyclization and preferred macroradical generation. Cyclic intramolecular-hydrocarbon abstraction should proceed further along the chain for less-branched HDPE than for the more highly branched LDPE, since it is acknowledged that HDPE contains longer uninterrupted methylene sequences than does LDPE and that termination of the above

cyclic reaction likely occurs at a branch site. From this it was concluded (322) that the cyclic intramolecular-hydrogen abstraction is more favorable at longer-chain lengths for HDPE, and certain observed differences in peak maxima in the pyrolysis of LDPE and HDPE were accounted for accordingly.

4.4.4.2   Pgc of LAH-Reduced PVC in Helium

The fact that PE is a good model for the study of reduced PVC can be seen from a comparison of the pyrolysis patterns of a reduced PVC with HDPE and LDPE (Figure 4.36). In this same study (322) the pyrograms of a series of reduced PVCs were determined to be similar in the C_9-C_{18} hydrocarbon range. However, quantitative differences in the total branch content were detected.

More information about the specific type of short-chain branch in PVC was found from an examination of the C_1-C_{11} hydrocarbons, where the qualitative differences between the reduced PVCs became more apparent. The most obvious variation occurred in the amounts of iso-C_7 and iso-C_8 products formed, which indicates differences in the total amounts of methyl and ethyl side chains. As the amount of short-chain branching in the reduced PVCs increased, there was a decrease in the amount of *n*-alkenes formed, with a corresponding increase in the amount of isoalkenes formed.

4.4.4.3   Pyrolysis Hydrogenation Gc (Phgc) of LAH-Reduced PVC

As with PE, on-line hydrogenation of the pyrolytic products greatly simplifies the pyrogram, and, therefore, its interpretation. An examination of the overall pyrolytic pattern in the C_1-C_{11} region, coupled with a determination of the 3-methyl/2-methyl alkane ratios for these polymers, indicated that the short-chain branches in PVC are mainly methyl with some ethyl groups. However, the possibility of the presence of minor amounts of propyl, butyl, or longer-chain branches cannot be completely discounted from these data, since some differences were also observed in the iso-C_8, iso-C_9, and iso-C_{10} products. The reduced PVCs with the least branching in this series as determined by pgc has slightly greater

3-methyl/2-methyl alkane ratios than did HDPE (0.95 g/cc). This ratio generally increased for the more highly branched polymers. Recent ^{13}C nmr results showed most of the short-chain branches are pendant chloromethyl groups on the original PVC chain (96).

Resolution of the 3-methyl/2-methyl isomers is shown in pyrograms (Figure 4.37) comparing Marlex 50, Nordforsk E-54, and LDPE. Marlex 50, a Phillips PE, is thought to be highly linear and to contain minimal ethyl branches (242, 326). Thus the 3-methylalkanes formed are due to statistical chain cleavage and the presence of a few methyl branches along the polymer backbone. By comparison to Marlex 50, the increased amounts of 3-MC$_6$ and 3-MC$_7$ from the reduced PVCs is more than can be attributed directly to statistical chain cleavage; therefore, PVC must contain some ethyl branches. In summary, these pgc data (322) indicate the presence of mainly methyl with some ethyl branches in LAH-reduced PVC. An inspection of the C$_{10}$ region of the pyrogram revealed minimal C$_4$-branch content in these samples. This is in general agreement with ^{13}C nmr data (96).

4.4.5   Pgc of PVC for Branch Content

Unlike PE, the effect of branch structure of the polymer on the hydrocarbon pyrolytic products from the parent PVC chain is not as well defined.

The data in Table 4.9 show the amounts of C$_1$–C$_5$ hydrocarbons, benzene, and naphthalene formed from these PVC polymers (322) with the calculated benzene/toluene ratio. The order of branching, determined from pyrolysis of the corresponding LAH-reduced PVCs and the number of methyl groups per 1000 carbon atoms determined from the ir measurements, is included in the table for comparison. As the amount of branching along the polymer backbone increases, fragmentation increases, and results in the formation of increased amounts of C$_1$–C$_5$ hydrocarbons. A further consequence of reduced sequence length is to decrease the amounts of benzene and naphthalene formed. The specific effect of branch type on the pyrolysis products has not yet been determined; however, if it is assumed that the formation of toluene is at least in part an indication of methyl branching,

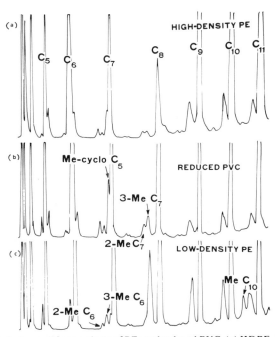

Figure 4.37   Pyrolysis hydrogenation products of PEs and reduced PVC. (*a*) HDPE. (*b*) Reduced PVC. (*c*) LDPE; pyrolysis temperature, 650° C, 10 sec pulse, *n*-octane/Porasil C column, temperature programmed to 170° C; fragments C$_1$–C$_{11}$ hydrocarbons.

Table 4.9   PVC Pyrolysis Products (322)

Sample	C_1-C_5 hydro-carbons wt-%	Benzene/ toluene ratio	Benzene, wt-%	Naph-thalene, wt-%	$CH_3/$ 1000 CH_2[b]	Relative branch order from reduced PVC[a]
Nordforsk E-80	14.1	11.2	59.9	5.7	4.1	2
Nordforsk S-80	14.0	11.1	59.3	5.6	4.1	3
Pevikon R-341	14.8	10.7	57.9	5.6	5.4	1
Nordforsk S-54	14.9	10.5	57.7	6.0	7.8	4
Nordforsk E-54	16.9	9.7	56.2	4.7	8.0	5
Shin-Etsu TK1000	16.0	9.2	55.1	3.9	5.6	6

[a] Relative to 1 = least branched by PGC.
[b] Infrared.
Source. Reprinted with permission from *J. Polym. Sci.* Copyright by American Chemical Society.

then the benzene/toluene ratio should serve as an indicator of the amount of methyl branching in PVC. If these criteria are used as an indication of the degree of branching, then the two pyrolytic methods and the infrared method agree within the experimental limits (322). Although these data do not necessarily represent optimized experimental results for the determination of branching, nevertheless, the potential of pgc to gauge the relative branch content of PVC is indicated. Also, a time-resolved pgc study of PVC homopolymers and reference copolymers indicated a means to detect low levels of methyl branches along an otherwise vinyl chloride chain (314b).

4.5   SUMMARY

The difficult combination of discussing the role of theoretical calculations of polymer microstructure with the available means of experimental, analytical verification has been attempted in this discussion. Computer-assisted computational studies present an almost unbelieveable array of hardware/software tools in the hands of the theorists for information on polymer bonding, conformation, and configuration. The experimentalists have demonstrated that applications of commercially available analytical instrumentation can, in fact, now delineate in a rather fine manner what the theorists predict—or vice versa, interpret analytical results from a developing

background of mechanistic or theoretical background.

In the near future, low glass-transition (Tg) systems will be examined in a routine manner directly as solids, as illustrated by Gray (327) with high-resolution nmr instrumentation. Extension of such applications will advance one of the most limiting features of the nmr method, that of sample form, solubility, and sensitivity. Today, milligram and microgram amounts may generally be studied as solution spectra, but it appears that high-resolution ^{1}H and/or ^{13}C informative patterns from solids with "magic-angle" experiments (328) constitute one of the more important aspects to the development of nmr in wide-ranging applications for polymer characterization.

It is not unlikely, in the near future, that chemical or polymeric systems could be analyzed in a computer-assisted routine manner for general identification or for structurally significant features of some detail. A form of such concerted analysis is in the process of development by coupling statistical evaluation of pyrograms with data from the on-line organic analysis system using gc retention time, functional-group/small-molecule fragmentation, and elemental (C, H, N, O, and S) content (329, 330). The latter elemental data are used for input with the appropriate experimental-error limit into the computer program adapted from the literature (331) to provide possible molecular formulas of the gc peak with a "realistic" search routine. Further identification from nmr, gc-ms, and gc-ir complete the analytical array of instrumentation aids (332–336). It may be seen that the development and utilization of statistical comparison of data profiles can be accomplished for chromatographic, spectral, and microchemical results. With such combined key data bases developed in accompanying computer programs, these three structurally informative analyses may form the framework for potential automatic organic or polymer identifications.

That these combined problem-solving approaches will lead to new and improved materials is almost without question. Better understanding of the conditions or variations needed to synthesize polymer systems of specific structures will certainly aid our ability to design and architecturally tailor polymers of tomorrow. To impose performance stresses and relate them to their effects on microstructural changes should guide the improvement of the materials in a more rational and predictive manner. The polymer analysts have already shown that the tools of modern organic analysis may be profitably applied to realistic problem-solving situations and have interfaced their efforts with the predictive and interpretive abilities of the theorists. Future efforts in this direction will most certainly be sustained.

4.6   GLOSSARY OF SYMBOLS

R	Run number, uninterrupted monomer sequences per 100 monomer units
r_E	Reactivity ratio of monomer E
H_0	Instrumental magnetic field strength, in gauss (nmr)
MHz	Megahertz (nmr oscillator frequency at given H_0 strength)
ν	Frequency, in Hertz (nmr)
J	Nuclear spin-spin coupling constant, in Hz (nmr)
J_t	Trans coupling constant (nmr)
J_g	Gauche coupling constant (nmr)
J_{AB}	Geminal coupling constant for magnetically nonidentical protons A, B, on same atom (nmr)
J_{vic}	Vicinal coupling constant for magnetically

	nonidentical protons on adjacent atoms (nmr)
s	Syndiotactic configuration
i	Isotactic configuration
h	Heterotactic configuration
m	Meso configuration of physically identical pair (dd, ll)
r	Racemic configuration of mirror-image pair (dl, ld)
δ	Chemical shift, in ppm; $H_{sample} - H_{reference} / H_{reference} \times 10^6$ (nmr)
τ	Tau value, defined as the chemical shift expressed relative to tetramethylsilane resonance line at 10 ppm ($\tau = 10.000 + \delta_{ppm}$)
P_m	Sequence probability of meso placements according to Bernoullian statistics
δ	Phase difference between neighboring chemical groups (ir)
ν	Frequency, in cm^{-1} (wavenumbers, ir)
$g(\nu)$	Frequency distribution function (ir)
S_{HH}	Secondary Cl conformation with H atom trans to the Cl atom
C_{2v}	Symmetry group with the element of a C_2 rotation axis and *no* planes containing the C_2 axis
D_2	Symmetry group with one C_n axis and $2C_2$ axes perpendicular to the C_n axis (a molecule with a single C_n axis symmetry element belongs to the C_n group)

4.7   REFERENCES

1. F. R. Mayo and F. M. Lewis, *J. Am. Chem. Soc.*, **66**, 1594 (1944).

2. T. Alfrey and C. C. Price, *J. Polym. Sci. A*, **2**, 101 (1947).

3. R. Z. Greenley, *J. Macrmol. Sci.-Chem.*, **A9**, 505 (1975).

4. J. P. Kennedy, T. Kelen, and F. Tudos, *J. Polym. Sci., Polym. Chem. Ed.*, **13**, 2277 (1975).

5. H. J. Harwood and W. M. Ritchey, *J. Polym. Sci., Polym. Chem. Ed.* **B**, **2**, 601 (1964).

6. H. J. Harwood, *Angew. Chem. Intern. Ed.*, **4**, 394 (1965).

7. H. J. Harwood, N. W. Johnston, and H. Piotrowski, *J. Polym. Sci., Part C, 23*, 37 (1968).

8. G. G. Lowry, Ed., *Markov Chains and Monte Carlo Calculations in Polymer Science,* Marcel Dekker, New York, 1970.

9. O. V. Noah, A. L. Toom, N. B. Vasilyev, A. D.

Litmanovich, and N. A. Plate, Vysokomol. Soyed., **A16**, 412 (1974).

10. E. R. Cohan and H. Reiss, *J. Chem. Phys.*, **38**, 680 (1963).

11. N. A. Plate, A. D. Litmanovich, O. V. Noah, A. L. Toom, and N. B. Vasilyev, *J. Polym. Sci., Polym. Chem. Ed.*, **12**, 2165 (1974).

12. E. Helfand, *Macromolecules*, **5**, 301 (1972).

13. H. K. Frensdorff, *Macromolecules*, **4**, 369 (1971).

14. S. Davison and G. L. Taylor, *Br. Polym. J.*, **4**, 65 (1972).

15. E. Tsuchida, J. Aoyagi, K. Shinzo, and I. Shinohara, *Polymer*, **15**, 479 (1974).

16. M. V. Volkenstein, *Conformational Statistics of Polymer Chains*, transl. S. N. Timasheff and M. J. Timasheff, Wiley-Interscience, New York, 1963.

17. P. J. Flory, (a) Statistical Mechanics of Chain Molecules, Wiley-Interscience, New York, 1969; (b) *Macromolecules*, **7**, 381 (1974).

18. G. Allegra, M. Calligaris, L. Randaccio, and G. Moraglio, *Macromolecules*, **6**, 397 (1973).

19. R. H. Boyd and S. M. Breitling, *Macromolecules*, **5**, 279 (1972).

20. U. W. Suter and P. J. Flory, *Macromolecules*, **8**, 765 (1975).

21. J. E. Mark and J. H. Ko, *Macromolecules*, **8**, 874 (1975).

22. H. Tadokoro, K. Tai, M. Yokoyama, and M. Kobayashi, *J. Polym. Sci., Poly. Phys. Ed.*, **11**, 825 (1973).

23. A. E. Tonelli, *Macromolecules*, **8**, 544 (1975).

24. P. J. Flory, P. R. Sundararajan, and L. C. DeBolt, *J. Am. Chem. Soc.*, **96**, 5015 (1974).

25. P. R. Sundararajan and P. J. Flory, *J. Am. Chem. Soc.*, **96**, 5025 (1974).

26. D. Y. Yoon, U. W. Suter, P. R. Sundararajan, and P. J. Flory, *Macromolecules*, **8**, 785 (1975).

27. A. E. Tonelli, *J. Am. Chem. Soc.*, **94**, 2972 (1972).

28. W. L. Mattice, *Macromolecules*, **8**, 645 (1975).

29. V. G. Dashevskii, *Russ. Chem. Rev.*, **42**, 970 (1973).

30. J. A. Pople, and D. L. Beveridge, *Approximate Molecular Orbital Theory, McGraw-Hill*, New York, 1970.

31. H. F. Hameka, *Quantum Theory of the Chemical Bond*, Hafner Press-MacMillan, New York, 1975.

32. I. N. Levine, *Molecular Spectroscopy*, Wiley-Interscience, New York, 1975.

33. L. Salem, *Molecular Orbital Theory of Conjugated Systems*, Benjamin Press, New York, 1966.

34. R. C. Bingham, M. J. S. Dewar, and D. H. Lo, *J. Am. Chem. Soc.*, **97**, 1285 (1975).

35. M. J. S. Dewar, *Chem. Br.*, **11**, 97 (1975).

36. J. A. Pople, G. P. Santry, and G. A. Segal, *J. Chem. Phys.*, **43**, 5129 (1965).

37. J. A. Pople, D. L. Beveridge, and P. A. Dobosh, *J. Chem. Phys.*, **47**, 2026 (1967).

38. K. B. Wiberg, *J. Am. Chem. Soc.*, **90**, 59 (1968); *Tetrahedron*, **24**, 1083 (1968).

39. A. Imamura, *J. Chem. Phys.*, **52**, 3168 (1970).

40. K. Morokuma, *J. Chem Phys.*, **54**, 962 (1971).

41. J. M. Andre, *J. Chem. Phys.*, **50**, 1537 (1969).

42. S. O'Shea and D. P. Santry, *J. Chem. Phys.*, **54**, 2667 (1971).

43. F. A. Bovey and G. V. O. Tiers, *J. Polym. Sci.*, **44**, 173 (1960).

44. F. A. Bovey, *High Resolution NMR of Macromolecules*, Academic Press, New York, 1972.

45. E. Klesper and G. Sielaff, in D. O. Hummel, Ed., *Polymer Spectroscopy*, Verlag Chemie GmbH, Weinheim/Bergstr, 1974, Chapter 3.

46. J. A. Pople, W. G. Schneider, H. J. Bernstein, *High Resolution NMR*, McGraw-Hill, New York, 1959.

47. F. A. Bovey, *Polymer Conformation and Configuration*, Academic Press, New York, 1972.

48. I. Ya. Slonim and A. N. Lyubimov, *The NMR of Polymers*, Plenum Press, New York, 1970.

49. J. W. Emsley, J. Feeney, and L. H. Sutcliff, *High Resolution NMR Spectroscopy*, Pergamon Press, New York, 1965.

50. F. Kasler, *Quantitative Analysis by NMR Spectroscopy*, Academic Press, New York 1973.

51. A. M. Hassan, *CRC Crit. Rev. Macromol.*, **399**, 1 (1972).

52. L. D. Hall, *Chem. Soc. Rev.*, **S, 401**, 4 (1975).

53. G. C. Levy and G. L. Nelson, *Carbon-13 NMR for Organic Chemists*, Wiley-Interscience, New York, 1972.

54. J. B. Stothers, *Carbon-13 NMR Spectroscopy*, Academic Press, New York, 1972.

55. E. Breitmaier, K. H. Spohn, and S. Berger, *Angew. Chem. Int. Ed.*, **14**, 144, (1975).

56. A. R. Katritzsky and D. E. Weiss, *Chem. Br.*, **12**, 45 (1976).

57. F. A. Bovey, F. P. Hood, E. P. Anderson, and L. C. Snyder, *J. Chem. Phys.*, **42**, 3900 (1965).

58. A. M. Hassan, *J. Polym. Sci.*, **12**, 655 (1974).

59. M. Karplus, *J. Chem. Phys.*, **30**, 11 (1959); *J. Am. Chem. Soc.*, **85**, 2870 (1963).

60. B. Schneider, J. Stoks, D. Doskocilova, S. Sykora, J. Jakes, and M. Kolinsky, *J. Polym. Sci., Part C*, **22**, 1073 (1969).

61. M. Nagai and A. Nishioka, *J. Polym. Sci., Al, 6*, 1655 (1968).

62. Y. Inoue, A. Nishioko, R. Chujo, *Polym. J.*, **2**, 535 (1971).

63. T. Yoshino and J. Komiyamo, *J. Polym. Sci. B*, **3**, 311 (1965).

64. F. Heatley and F. A. Bovey, *Macromolecules*, **2**, 241 (1969).

65. T. Shimanouchi, M. Tasumi, and Y. Abe, *Makromol. Chem.,* **80,** 43 (1965).

66. L. Cavalli, G. C. Borsini, G. Corrard, and G. Confalonier, *J. Polym. Sci.* **A-1, 8,** 801 (1970).

67. A. M. Hassan, *J. Polym. Sci.* **12,** 655 (1974).

68. C. A. Carman, *Macromolecules,* **6,** 727 (1973).

69. I. Ando, Y. Kato, and A. Nishioka, *Makromol. Chem.,* **177,** 2759 (1976).

70. S. Sobajima, N. Takagi, and H. Watase, *J. Polym. Sci., Part A-2,* **6,** 223 (1968).

71. S. A. Liebman, D. H. Ahlstrom, E. J. Quinn, A. G. Geigley, and J. T. Meluskey, *J. Polym. Sci., Part A-1,* **9,** 1921 (1971).

72. G. Svegliado and F. Z. Grandi, *J. Appl. Polym. Sci.,* **13,** 1113 (1969).

73. T. Saito, Y. Matsumura, and S. Hayashi, *Polym. J.,* **1,** 639 (1970).

74. H. K. Frensdorff and O. Ekiner, *J. Polym. Sci., Part A-2,* **5,** 1157 (1967).

75. J. C. Randall, *Polym. Phys. Ed.,* **13,** 1975 (1975).

76. D. M. Grant and E. G. Paul, *J. Am. Chem. Soc.,* **86,** 2984 (1964).

77. R. C. Ferguson, *Macromolecules,* **2,** 237 (1969); *Macromolecules* **4,** 324 (1971).

78. F. C. Stehling and J. R. Knox, *Macromolecules,* **8,** 595 (1975).

79. P. J. Flory and Y. Fujiwara, *Macromolecules,* **2,** 327 (1969); *Macromolecules,* **3,** 613 (1970).

80. F. Heatley and A. Zambelli, *Macromolecules,* **2,** 618 (1969).

81. L. F. Johnson, F. Heatley, and F. A. Bovey, *Macromolecules* **3,** 175 (1970).

82. F. C. Stehling, *J. Polym. Sci., Part A,* **2,** 1815 (1964).

83. A. Zambelli, D. E. Dorman, A. I. R. Brewster, and F. A. Bovey, *Macromolecules,* **6,** 925 (1973).

84. A. Zambelli, C. Wolfogruber, G. Zannoni, and F. A. Bovey, *Macromolecules,* **7,** 750 (1974).

85. A. Zambelli, P. Locatelli, G. Bajo, and F. A. Bovey, *Macromolecules,* **8,** 687 (1975).

86. I. Ando, A. Nishioka, and T. Asakura, *Makromol. Chem.,* **176,** 411 (1975).

87. T. Asakura, I. Ando, and A. Nishioka, *Makromol. Chem.,* **177,** 523 (1976); *Makromol. Chem.,* **178,** 791 (1977).

88. K. H. Hellwege, W. Johnsen, and K. Kolbe, *Kolloid-Z.,* **214,** 45 (1966).

89. J. B. Kinsinger, T. Fischer, and C. W. Wilson, III, *J. Polym. Sci., Part B,* **4,** 379 (1966); *J. Polym. Sci., S-285 (1967).*

90. D. E. Dorman, E. P. Otocka, and F. A. Bovey, *Macromolecules,* **5,** 574 (1972).

91. J. C. Randall, *J. Polym. Sci.,* **11,** 275 (1973).

92. T. Hama, T. Suzuki, and K. Kosaka, *Kobunshi Ronb., Eng. Ed.,* **4,** 110 (1975).

93. M. J. Roedel, *J. Am. Chem. Soc.,* **75,** 6110 (1953).

94. S. A. Liebman and Application Laboratory, JE01 Analytical Instruments, Inc., Cranford, New Jersey, on JE01 FX-100 FT/NMR Spectrometer, 25.05 MHz, ^{2}D Lock, 120° C, *o*-dichlorobenzene solvent, trimethylsilane reference.

95. M. E. Cudby and A. Bunn, *Polymers,* **17,** 345 (1976).

96. F. A. Bovey, K. B. Abbas, F. C. Schilling, and W. H. Starnes, *Macromolecules,* **8,** 437 (1975).

97. G. M. Barrow, *Introduction to Molecular Spectroscopy,* McGraw-Hill, New York, 1962.

98. E. B. Wilson, Jr., J. C. Decius, and P. C. Cross, *Molecular Vibrations,* McGraw-Hill, New York, 1955, Chapter 4.

99. R. Zbindin, *Infrared Spectroscopy of High Polymers,* Academic Press, New York, 1964.

100. G. Zerbi, in E. Brame, Jr., Ed., *Applied Spectroscopy Reviews,* Vol 2, Marcel Dekker, New York, 1968, pp. 193–251.

101. J. L. Koenig, in E. Brame, Jr., Ed., *Applied Spectroscopy Reviews,* Vol. 4, 1971, Marcel Dekker, New York, pp. 233–305.

102. A. Elliot, *Infrared Spectra and Structure of Organic Long-Chain Polymers,* St. Martins Press, New York, 1969.

103. D. Steele, *Theory of Vibrational Spectroscopy,* W. B. Saunders, Philadelphia, Pennsylvania, 1971.

104. F. J. Boerio and J. L. Koenig, *Vibrational Spectroscopy of Polymers, J. Macromol. Sci. Rev., Macromol. Chem.,* **C7,** 209–249 (1972).

105. V. Fawcett and D. A. Long, "Vibrational Spectroscopy of Macromolecules," in *Molecular Spectroscopy,* Vol. 1, Specialist Periodical Reports, Chemical Society, Burlington House, London, 1973.

106. I. N. Levine, *Molecular Spectroscopy,* John Wiley, New York, 1975.

107. N. B. Colthup, L. H. Daly, and S. E. Wiberley, *Introduction to Infrared and Raman Spectroscopy,* 2nd ed., Academic Press, New York, 1975.

108. H. A. Szymanski, *Raman Spectroscopy, Theory and Practice,* Vols. 1 and 2, Plenum Press, New York, 1970.

109. D. C. Hummel, Ed., *Polymer Spectroscopy,* Verlag Chemie GmbH, Weinheim/Bergstr, Germany, 1974, Chapter 2.

110. T. R. Gilson, P. J. Hendra, *Laser Raman Spectroscopy,* Wiley-Interscience, New York, 1970.

111. V. Z. Kompaniets, E. F. Oleinik, and N. S. Yenikolopyeu, *Vysokomol. Soyed,* **A14,** 669–679 (1972).

112. F. A. Cotton, *Chemical Applications of Group Theory,* 2nd ed., John Wiley, New York, 1970.

113. Quantum Chemistry Program Exchange, University of Indiana, Bloomington, Indiana, 47401.

114. W. D. Gwinn, *J. Chem. Phys.,* **55,** 477 (1971).

115. M. M. Coleman, D. L. Tabb, B. L. Farmer, J. L. Koenig, *J. Polym. Sci., Polym. Phys. Ed.,* **12**, 445–454 (1974).

116. S. K. Freeman, *Applications of Laser Raman Spectroscopy,* Wiley-Interscience, New York, 1974.

117. L. J. Bellamy, *Advances in Infrared Group Frequencies,* Vol. 2, Meuthen, London, 1968.

118. L. J. Bellamy, *The Infrared Spectra of Complex Molecules,* Vol. 1, Chapman & Hall, Halsted Press, New York, 1975.

119. A. J. Barnes and W. J. Orville-Thomas, Eds., *Vibrational Spectroscopy—Modern Trends,* Elsvier, New York, 1977.

120. J. R. Dyer, *Applications of Absorption Spectroscopy of Organic Compounds,* Prentice-Hall, Englewood Cliffs, New Jersey, 1965.

121. Sadtler Research Laboratories, Inc., 3316 Spring Garden St., Philadelphia, Pennsylvania, 19104.

122. C. J. Pouchert, *Aldrich Library of Infrared Spectra,* 2nd Ed., 1975, Aldrich Chemical Company, Milwaukee, Wisconsin.

123. D. O. Hummel, *Infrared Analysis of Polymers, Resins & Additives, An Atlas,* Vol. I, Part 2 (1969), John Wiley, New York.

124. R. S. McDonald, "Infrared Spectroscopy," *Fundam. Rev., Anal. Chem.,* **48**, 196R (1976).

125. K. Holland-Moritz and H. W. Siesler in E. G. Brame, Jr., Ed., *Applied Spectroscopy Review,* Vol. 11, Marcel Dekker, 1976.

126. W. G. Fateley, R. F. Dollish, N. T. McDevitt, and F. F. Bentley, *Infrared and Raman Selection Rules for Molecular and Lattice Vibrations: The Correlation Method,* Wiley-Interscience, New York, 1972.

127. G. Zerbi, F. Ciampelli, and V. Zamboni, *J. Polym. Sci., Polym. Symp., Part C,* **7**, 141–151 (1964).

128. H. A. Willis, and M. E. A. Cudby, in K. J. Ivin, Ed., *Structural Studies of Macromolecules by Spectroscopic Methods,* John Wiley, New York, 1976, Chapter 7.

129. M. Goldstein, M. E. Seeley, H. A. Willis, and V. J. I. Zichy, *Polymer,* **14**, 530 (1973).

130. G. W. Chantry, J. W. Fleming, E. A. Nicol, H. A. Willis, M. E. A. Cudby, and F. J. Boerio, *Polymer,* **15**, 69 (1974).

131. L. Piseri, B. M. Powell, and G. Dolling, *J. Chem. Phy.,* **58**, 158 (1973).

132. G. Zerbi and M. Sacchi, *Macromolecules,* **6**, 692 (1973).

133. R. F. Schaufele and T. Shimanouchi, (a) *J. Chem. Phy.,* **47**, 3605 (1967); (b) in K. J. Ivin, Ed., *Structural Studies of Macromolecules by Spectroscopic Methods,* John Wiley, New York, 1976, Chapter 5.

134. J. F. Rabolt, K. W. Johnson, and R. N. Zitter, *J. Chem. Phys.* **61**, 504–506 (1974).

135. V. P. Lebedev, D. Ya. Tsvankin, and Yu. V. Glazkovskii, *Vysokomol. Soyed.,* **A14**, 1010–1016 (1972).

136. M. Tasumi and T. Shimanouchi, *Polym. J.,* **2**, 62–73 (1971).

137. W. H. Moore and S. Krimm, *Makromol. Chem. Suppl.,* **1**, 491–506 (1975).

138. M. E. R. Robinson, D. I. Bower, and W. F. Maddams, *Polymer,* **17**, 355–357 (1976).

139. M. M. Coleman, P. C. Painter, D. L. Tabb, and J. L. Koenig, *Polym. Lett. Ed.,* **12**, 577–581 (1974).

140. H. Tadokoro, M. Kobayashi, S. Kobayashi, K. Yasufuku, and K. More, *Rep. Prog. Polym. Phys. Jap.,* **9**, 181 (1966); in D. O. Hummel, Ed., *Polymer Spectroscopy,* Verlag Chemie GmbH, Weinheim/ Berstr, Germany, pp. 90–91.

141. P. D. Vasko and J. L. Koenig, *Macromolecules,* **3**, 597–601 (1970).

142. A. Cunningham, G. R. Davies, and I. M. Ward, *Polymer,* **15**, 743–756 (1974).

143. Yu. V. Kissin, I. A. Lekaye, Ye. A. Chernova, N. A. Davydova, and N. M. Chirkov, *Vysokomol. soyed.,* **A16**, 677–683 (1974).

144. Y. Kobayashi, S. Okajima, and A. Narita, *J. Appl. Polym. Sci.,* **11**, 2515–2523 (1967).

145. E. L. Eliel, N. L. Allinger, S. J. Angyal, and G. A. Morrison, *Conformational Analysis,* John Wiley, New York, 1965.

146. P. J. Flory, *Statistical Mechanics of Chain Molecules,* Wiley-Interscience, New York, 1969.

147. T. Asakura, I. Ando, and A. Nishioka, *Makromol. Chem.,* **176**, 1151–1161 (1975).

148. P. J. Flory and C. J. Pickles, *J. Chem. Soc., Faraday Trans.,* **2**, (Part 5) 632–642 (1973).

149. J. Stokr, B. Schneider, M. Kolinsky, M. Ryska, and D. Lim, *J. Polym. Sci. Part A-1,* **5**, 2013–2021 (1967).

150. C. Tosi and F. Ciampelli, *Eur. Polym. J.* **5**, 759–766 (1969).

151. B. Schneider, J. Stokr, S. Dirlikov, and M. Mihailov, Macromolecules, **4**, 715 (1971).

152. B. Schneider, J. Stokr, D. Doskocilova, M. Kolinsky, S. Sykora, and D. Lim, *J. Polym. Sci. Part C,* 3891–3900 (1968).

153. S. Krimm and S. Enomoto, *J. Polym. Sci., Part A2,* 669–678 (1964).

154. H. U. Pohl and D. O. Hummel, *Makromol. Chem.,* **113**, 190 (1968); D. O. Hummel, Ed., *Polymer Spectroscopy,* Verlag Chemie GmbH, Weinheim/ Bergstr, Germany, p. 121.

155. M. Kobayashi, K. Akita, and H. Tadokoro, *Makromol. Chem.,* **118**, 324–342 (1968).

156. I. V. Kumpaneko and K. S. Kazanskii, *J. Polym. Sci., Sympos. No.* **42**, 973–980 (1973).

157. J. L. Koenig, Polymer Characterization Conference,

Cleveland State University, Cleveland, Ohio, May 1974, 73–93.

158. C. G. Opaskar and S. Krimm, *J. Polym. Sci., Part A-2*, **7**, 57–75 (1969).

159. G. Natta, P. Corradini, and P. Ganis, *J. Polym. Sci.*, **58**, 1191 (1962).

160. J. L. Koenig and A. V. Roggen, *J. Appl. Polym. Sci.*, **9**, 359 (1965).

161. I. V. Kumpanenko, K. S. Kazanskii in Z. A. Rogovin, Ed., *Advances in Polymer Science*, Halsted Press, New York, 58–90 (1974).

162. F. Ciampelli and C. Tosi, *Spectrochem. Acta*, **A24**, 2157 (1968).

163. H. V. Drushel, J. S. Ellerbe, R. C. Cox, and L. H. Lane, *Anal. Chem.*, **40**, 370 (1968).

164. P. Simak and G. Fahrach, *Angew. Makromol. Chem.*, **16–17**, 309 (1971).

165. J. Millan and J. G. Perote, *Eur. Polym. J.*, **12**, 299–304 (1976).

166. T. Onishi and S. Krimm, *J. Appl. Phys.*, **32**, 2320 (1961).

167. A. Rubcic and G. Zerbi, *Macromolecules*, **7**, 754–759; 759–767 (1974).

168. C. G. Opaskar and S. Krimm, Spectrochem. Acta, Part A, **23**, 2261 (1967).

169. A. H. Willbourn, *J. Polym. Sci.* **34**, 569 (1959).

170. C. Baker and W. F. Maddams, *Makromol. Chem.*, **177**, 437–448 (1976).

171. A. G. Nerheim, *Anal. Chem.*, **47**, 1129 (1975).

172. D. E. Kranbuehl, T. V. Harris, A. K. Howe, D. N. Thompson, *J. Chem. Educ.*, **52**, 261–264 (1975).

173. N. M. Domareva, A. L. Gol'denberg, N. Ya. Tumarkin, and L. A. Chepurko, *Int. Polym. Sci. Technol.*, **2**, T/19-T/20 (1975).

174. M. A. McRae and W. F. Maddams, *Makromol. Chem.*, **177**, 449–459 (1976).

175. R. Kaiser, *Gas Phase Chromatography*, Vols. 1–3, Butterworth, London, 1963.

176. S. Dal Nogare and R. S. Juvet, Jr., *Gas-Liquid Chromatography*, Wiley-Interscience, New York, 1962.

177. L. S. Ettre and Z. Zlatkis, *The Practice of Gas Chromatography*, Wiley-Interscience, New York, 1967.

178. T. H. Gouw, Ed., *Guide to Modern Methods of Instrumental Analysis*, Wiley-Interscience, New York, 1972, Chapters 1 and 9.

179. V. G. Berezkin, *Analytical Reaction Gas Chromatography*, Plenum Press, New York, 1968.

180. L. S. Ettre and W. H. McFadden, *Ancillary Techniques of Gas Chromatography*, Wiley-Interscience, New York, 1969.

181. M. P. Stevens, *Characterization and Analysis of Polymers by Gas Chromatography*, Vol. 3, Marcel Dekker, New York, 1969.

182. D. A. Leathard and B. C. Shurlock, *Identification Techniques in Gas Chromatography*, Wiley-Interscience, New York, 1970.

183. S. Siggia, Ed., *Instrumental Methods of Organic Functional Group Analysis*, Wiley-Interscience, New York, 1972, Chapter 2.

184. T. S. Ma and A. S. Ladas, *Organic Functional Group Analysis by Gas Chromatography*, Academic Press, New York, 1976.

185. V. G. Berezkin and V. S. Tarinskii, *Gas Chromatographic Analysis of Trace Impurities*, Consultants Bureau, Plenum Press, New York, 1973.

186. H. Hachenberg, *Industrial Gas Chromatographic Trace Analysis*, Heyden & Son, Ltd., New York, 1973.

187. R. C. Crippen, *Identification of Organic Compounds with the Aid of Gas Chromatography*, McGraw-Hill, New York, 1973.

188. G. Guiochon and C. Pommier, *Gas Chromatography of Inorganics and Organometallics*, Ann Arbor Science Publishers, Ann Arbor, Michigan, 1973.

189. V. G. Berekin, K. I. Sakodynskii, I. Klesment, and J. Janak, Eds., First Soviet Symposium on Reaction Gas Chromatography, Tallin (Estonian, SSR) July 1971, *J. Chromatogr.*, **69**, No. 1, Special Issue, 1972.

190. E. E. Kugucheva and A. V. Alekseeva, *Russ. Chem. Rev.*, **42**, 1048 (1973).

191. H. Purnell, Ed., *New Developments in Gas Chromatography*, Vol. 2, Wiley-Interscience, New York, 1973.

192. M. Beroza, *Acc. Chem. Res.*, **1**, 33 (1970).

193. R. L. Levy, *J. Gas Chromatogr.*, **5**, 107 (1967).

194. E. J. Levy and D. G. Paul, *J. Gas Chromatogr.*, **5**, 58 (1968).

195. N. Iglauer and F. F. Bentley, *J. Chromatogr. Sci.*, **12**, 23 (1974).

196. S. F. Sarner, G. D. Pruder, and E. J. Levy, *Am. Lab.*, **1971**, 57.

197. C. Merritt, Jr. and C. Dipietro, *Anal. Chem.*, **44**, 57 (1972).

198. W. R. Feairheller, Jr. and F. F. Bentley, Pittsburgh Conference on Analytical Chemistry and Applied Spectroscopy, Cleveland, Ohio, March 1972, Paper 21.

199. J. K. Haken, D. K. M. Ho, and M. K. Withers, *J. Chromatogr. Sci.*, **10**, 566 (1972).

200. T. N. Higgins and W. E. Harris, *J. Chromatogr. Sci.*, **11**, 588 (1973).

201. A. S. Ladas and T. S. Ma, *Mikrochim. Acta*, **1973**, 853.

202. S. A. Liebman, D. H. Ahlstrom, C. D. Nauman, R. Averitt, J. L. Walker, and E. J. Levy, *Anal., Chem.* **45**, 1360 (1973) and references therein.

203. O. Mlejnek, *Chem. Zvesti,* **27**, 421 (1973).

204. E. Haeberli, *Mikrochim. Acta,* **4**, 597 (1973).

205. V. Rezl and J. Janak, *J. Chromatogr.,* **81**, 233 (1973).

206. W. R. McLean, D. L. Stanton, and G. E. Penketh, *Analyst* (*London*), **98**, 432 (1973).

207. J. Franc, J. Senkyrova, and K. Placek, *J. Chromatogr.,* **65**, 197 (1972).

208. E. Pella and B. Colombo, *Mikrochim. Acta,* **4**, 697 (1973).

209. J. Exner, M. Seegar, and H. J. Cantow, *Angew. Chem., Int. Ed.,* **10**, 346 (1971).

210. L. Schutte and E. B. Koenders, *J. Chromatogr.,* **76**, 13 (1973).

211. R. Blomstrand, J. Gurtler, et al., *Anal. Chem.,* **44**, 277 (1972).

212. S. A. Liebman and C. I. Sanders, Armstrong Cork Company, unpublished work.

213. H. Giacobbo and W. Simon, *Pharm. Acta Helv.,* **39**, 162 (1964).

214. Ch. Buhler and W. Simon, *J. Chromatogr. Sci.,* **8**, 323 (1970).

215. H. L. Meuzelaar and R. A. in't Veld, *J. Chromatogr. Sci.,* **10**, 213 (1972).

216. Pye Unicam Ltd., York Street, Cambridge, England.

217. Chemical Data Systems, Inc., Oxford, Pennsylvania, 19363.

218. R. L. Levy, D. L. Fanter, and C. J. Wolf, *Anal. Chem.,* **44**, 38 (1972).

219. T. A. Gough and C. E. R. Jones, *Chromatographia,* **8**, 696 (1975).

220. K. J. Voorhees, F. D. Hileman, I. N. Einhorn, *Anal. Chem.,* **47**, 2385 (1975).

221. D. Gross, *Z. Anal. Chem.,* **256**, 40 (1971).

222. Sadtler Research Laboratories, 3316 Spring Garden Street, Philadelphia, Pennsylvania, 19104.

223. E. Kiran and J. K. Gillham, *J. Appl. Polym. Sci.,* **20**, 931 (1976); **20** 2045 (1976).

224. O. Mlejnek, *J. Chromatogr.,* **65**, 271 (1972).

225. K. Kurosaki and J. Murano, *Kobunsh Kag.,* **29**, 411 (1972).

226. B. V. Shiryeav and K. Kozhukno, (Rus) *Zavod. Lab.,* **38**, 1303 (1972).

227. B. V. Shiryeav, *Sov. Plast.,* **9**, 78 (1972).

228. M. S. Kleshcheva, V. I. Nesterova, and T. V. Smirnova, (*USSR*) *Plast. Massy,* **11**, 67 (1972).

229. A. R. Berens, L. B. Crider, C. J. Tomanek, and J. M. Whitney, *J. Appl. Polym. Sci.,* **19**, 3169 (1975).

230. A. E. Gavany, Jr. and H. Senman, *Am. Lab.,* **1976**, 49.

231. M. M. O'Mara, *J. Polym. Sci., Part A-1,* **8**, 1887 (1970); **9**, 1387 (1971).

232. M. Suzuki, S. Tsuge, and T. Takeuchi, *J. Polym. Sci.,* **10**, 1051 (1972).

233. L. S. Frankel, P. R. Madsen, R. R. Siebert, and K. L. Wallisch, *Anal. Chem.* **44**, 2401, (1972).

234. J. P. Mieure and M. W. Dietrich, *J. Chromatogr. Sci.,* **11**, 599 (1973).

235. 320 Concentrator, Chemical Data Systems, Inc. Oxford, Pennsylvania, 19363.

236. D. H. Ahlstrom, R. J. Kilgour, and S. A. Liebman, *Anal. Chem.,* **47**, 1411 (1975).

237. Vapor Trace Sampler (VTS), Model 104 M, MDA Scientific, Inc., Park Ridge, Illinois.

238. Component Concentrator, Model 216, SKC, Inc., P.O. Box 8538, Pittsburgh, Pennsylvania, 15220.

239. S. A. Liebman, D. H. Ahlstrom, C. D. Nauman, G. D. Pruder, R. Averitt, and E. J. Levy, *Res./Dev.,* **23**, 24 (1972).

240. L. Michajlov, P. Zugenmaier, and H. J. Cantow, *Polymer,* **9**, 325 (1968).

241. B. Kolb and H. Kaiser, *J. Gas Chromatogr.,* **2**, 233 (1964).

242. J. Van Schooten and J. K. Evanhuis, *Polymer,* **6**, 343 (1965); **6**, 561 (1965).

243. A. B. Robinson, D. Partridge, R. Teranishi, and L. Pauling, *J. Chromatogr.,* **85**, 19 (1973).

244. F. M. Menger, G. A. Epstein, D. A. Goldberg, and E. Reiner, *Anal. Chem.,* **44**, 423 (1972).

245. A. Dravnieks, H. G. Reilich, and J. Whitfield, *J. Food Sci.,* **38**, 34 (1973).

246. J. S. Mattson, M. J. Spencer, and S. A. Starks, Pittsburgh Conference on Analytical Chemistry and Applied Spectroscopy, Cleveland, Ohio, March 1976, Paper 326.

247. S. A. Liebman, ACS 6th Northeast Regional, August 1974, University of Vermont, Burlington, Vermont, Paper 93, Symposium on Computers in Chemistry.

248. S. A. Liebman, D. H. Ahlstrom, and A. T. Hoke, *Chromatographia,* **11**, 427 (1978).

249. L. A. Wall, "Mechanisms of Pyrolysis, Oxidation, and Burning of Organic Materials," Proc. 4th Materials Research Symposium, October 1970, NBS Spec. Publ. 357; *J. Elastoplastics,* **5**, 36, (1973).

250. R. T. Conley, Ed., *Thermal Stability of Polymers,* Vols. 1 and 2, Marcel Dekker, New York, 1970.

251. L. Reich, S. S. Stivala, *Elements of Polymer Degradation,* McGraw-Hill, New York, 1971.

252. V. V. Korshak, *Chemical Structure and Thermal Characteristics of Polymers,* Halsted Press, New York, 1971.

253. W. L. Hawkins, Ed., *Polymer Stabilization,* Wiley-Interscience, New York, 1971.

254. G. Geuskens, Ed., *Degradation and Stabilization of Polymers,* Halsted Press, New York, 1975.

255. D. Noffz, W. Benz, and W. Pfab, *Z. Anal. Chem.,* **235**, 121 (1968).

256. S. Toshiyuki, *Sci. Instr. News,* **2**, (1), 1 (1969).

257. K. V. Alexeeva, L. P. Khramova, and L. S. Solomatina, *J. Chromatogr.*, **77**, 61 (1973).

258. D. O. Hummel, H. D. Schuddemage, and K. Rubenacker, in D. O. Hummel, Ed., *Polymer Spectroscopy*, Verlag Chemie GmbH, Weinheim/Berstr., 1974, Chapter 5.

259. N. B. Zaitsev and I. Ya. Poddubnyi, *Vysokomol. soyed*, **A17**, 1130 (1975).

260. T. J. Reardon and R. H. Barker, *J. Appl. Polym. Sci.*, **18**, 1903 (1974).

261. M. Galin, *J. Makromol. Sci.-Chem.*, **A7**, 873 (1973).

262. M. J. Hackathorn and M. J. Brock, *Rubber Chem. Tech.*, **1972**, 1295.

263. P. G. M. Van Stratum and J. Dvorak, *J. Chromatogr.*, **71**, 9 (1971).

264. E. Blasius, H. Lohde, and H. Haeusler, *Z. Anal. Chem.*, **264**, 278 (1973).

265. J. R. Parrish, *Anal. Chem.*, **47**, 1999 (1975).

266. G. G. Esposito and R. G. Jamison, AD 748805, CCL Report No. 309, U.S. Army, Aberdeen Proving Ground, Maryland.

267. T. Okumoto and T. Takeuchi, *Jap. Anal.*, **22**, 931 (1973).

268. V. D. Braun and E. Canji, *Angew. Makromol.*, **33**, 143 (1973) (Ger.); **29/30**, 491 (1973) (Ger.).

269. B. G. Luke, *J. Chromatogr. Sci.*, **11**, 435 (1973).

270. E. R. Bissell and D. B. Fields, *J. Chromatogr. Sci.*, **10**, 164 (1972).

271. R. Takeuchi, S. Tsuge, and T. Okumoto, *J. Gas Chromatogr.*, **6**, (1968).

272. F. D. Hileman, K. J. Voorhees, L. H. Wojcik, M. M. Birkey, P. W. Ryan, and I. N. Einhorn, *J. Polym. Sci., Polym. Chem. Ed.*, **13**, 571 (1975).

273. H. H. Jellinek and K. Takada, *J. Polym. Sci., Polym. Chem. Ed.*, 2709.

274. T. Okumoto, H. Ito, S. Tsuge, and R. Takeuchi, *Makromol. Chem.*, **151**, 285 (1972).

275. E. J. Quinn, D. H. Ahlstrom, S. A. Liebman, *Polym. Prepr.*, **14**(2), 1022 (1973).

276. D. Deur-Siftar and V. Svob, *J. Chromatogr.*, **51**, 59 (1970).

277. Y. Yamamoto, S. Tsuge, and T. Takeuchi, *Macromolecules*, **5**, 325 (1972).

278. J. K. Haken and T. R. McKay, *J. Chromatogr.*, **80**, 75 (1973).

279. J. K. Haken and T. R. McKay, *Anal. Chem.*, **45**, 1251 (1973).

280. R. R. Alishoev, V. B. Berezkin, et al., *Polym. Sci., USSR*, **13**, 3123 (1973).

281. A. S. Tryukina, V. V. Chebotarev, et al., *Sov. Plast.*, **2**, 89 (1973).

282. R. W. May, E. F. Pearson, J. Porter, and M. D. Scothern, *Analyst (London)*, **98** (1166), 364 (1973).

283. A. Krishen, *Anal. Chem.*, **46**, 29 (1974).

284. S. Karayenev, G. Kostow, R. Milina, M. Mikhailov, *Vysokomol. Soyed*, **A16**, 2162 (1974).

285. D. Ya. Vyakhirev, L. E. Reshetnikova, L. I. Slyusareva, N. V. Stankova, and T. N. Shuvalova, *Int. Polym. Sci. Tech.*, **2**, T/71 (1975).

286. K. V. Alekseeva, *Int. Polym. Sci. Tech.*, T/95.

287. M. Blazso and T. Szekely, *Eur. Polym. J.*, **10**, 733 (1974).

288. N. Sellier, C. E. R. Jones, and G. Guiochon, *J. Chromatogr. Sci.*, **13**, 383 (1975).

289. K. L. Wallisch, *J. Appl. Polym. Sci.*, **18**, 203 (1974).

290. K. J. Bombaugh and C. E. Cook, *Anal. Chem.*, **35**, 1834 (1963).

291. R. B. Seymour, D. R. Owen, G. A. Stahl, H. Wood, and W. N. Tinnerman, *Appl. Polym., Symposium*, **25**, 69 (1974).

292. N. Hurduc, C. N. Cascaual, I. A. Schneider, and G. Riess, *Eur. Polym. J.*, **11**, 429 (1975).

293. T. Okumoto, S. Tsuge, Y. Yamamoto, and T. Takeuchi, *Macromolecules*, **7**, 377 (1974).

294. L. A. Wall and J. H. Flynn, *Rubber Chem. Technol.*, **35**, 1157 (1962).

295. S. Tsuge, S. Hiramitsu, T. Horibe, M. Yamaoka, and T. Takeuchi, *Macrtomolecules*, **8**, 255 (1975) and references therein.

296. S. Tsuge, T. Okumoto, and T. Takeuchi, *Makromol. Chem.*, **123**, 123 (1969).

297. S. Tsuge, T. Okumoto, and T. Takeuchi, *Macromolecules*, **2**, 277 (1969),

298. T. Okumoto, H. Ito, S. Tsuge, and T. Takeuchi, *Makromol Chem.*, **151**, 285 (1972).

299. J. C. Verdier and A. Guyot, *Makromol. Chem.*, **175**, 1543 (1974).

300. J. van Schooten and J. K. Evenhuis, *Polymer*, **6**, 343 (1965).

301. Y. Tsuchiya and K. Sumi, *J. Polym. Sci., Part A-1*, **7**, 1599 (1969).

302. C. W. Pyun, *J. Polym. Sci., Part A-2*, **8**, 1111 (1970).

303. R. Laputte and A. Guyot, *Makromol. Chem.*, **129**, 234 (1969).

304. H. Seno, S. Tsuge, and T. Takeuchi, *Makromol Chem.*, **161**, 185 (1972); **161**, 195 (1972).

305. G. Audisio and G. Bajo, *Makromol. Chem.*, **176**, 991 (1975).

306. Y. Tsuchiya and K. Sumi, *J. Appl. Chem.*, **17**, 364 (1967).

307. C. N. Cascaval, I. A. Schneider, and I. C. Poinescu, *J. Polym. Sci., Polym. Chem. Ed.*, **13**, 2259 (1975).

308. W. D. Woolley, *Br. Polym. J.*, **3**, 186 (1971).

309. K. Goto and T. Iida, *Plast. Ind. News*, **1973**, 181.

310. E. A. Boettner, G. Ball, and B. Weiss, *J. Appl. Polym. Chem.*, **13**, 377 (1969).

311. S. A. Liebman, D. H. Ahlstrom, E. J. Quinn, A. G. Geigley, and J. T. Meluskey, *J. Polym. Sci., Part A-1*, **9**, 1921 (1971).

312. E. P. Chang and R. Salovey, *J. Polym. Sci., Poly. Chem. Ed.,* **12**, 2927 (1974).

313. M. Suzuki, S. Tsuge, and T. Takeuchi, *J. Polym. Sci., Part A-1,* **10**, 1051 (1972).

314. S. A. Liebman, D. H. Ahlstrom, and P. R. Griffiths, *J. Appl. Spectr.* **30**, 355 (1976); (b) S. A. Liebman, D. H. Ahlstrom, and C. R. Foltz, *J. Polym. Sci., Polym. Chem. Ed.,* **16**, 3139 (1978).

315. D. O. Hummel, Ed., *Polymer Spectroscopy,* Verlag Chemie GmbH, Weinheim/Bergstr., 1974, pp. 374–385.

316. H. Ito, S. Tsuge, T. Okumoto, and T. Takeuchi, *Makromol. Chem.,* **138**, 111 (1970).

317. T. Iida, M. Nakanishi, and K. Goto, *J. Polym. Sci., Polym. Chem. Ed.,* **12**, 737 (1974); **13**, 1381 (1975).

318. M. Seeger and E. M. Barrall, II, *J. Polym. Sci., Polym. Chem. Ed.,* **13**, 1515 (1975); **13**, 1541 (1975).

319. M. Seeger, R. J. Gritter, J. M. Tibbitt, M. Shen, and A. T. Bell, 172nd ACS National Meeting, San Francisco, California, August 1976; OCPL Preprints, p. 27.

320. M. Seeger, J. Exner, and H. J. Cantow, XXIIIrd IUPAC, Boston, 1971, Macromolecules, Preprint II, 739.

321. J. Exner, M. Seeger, and H. J. Cantow, *Angew. Chem. Int. Ed.,* **10**, 346 (1971).

322. D. H. Ahlstrom, S. A. Liebman, and K. A. Abbas, *J. Polym Sci., Polym. Chem. Ed.,* **14**, 2479 (1976).

323. P. Kamath and A. Barlow, *J. Polym. Sci., Part A-1,* **5**, 2023 (1967).

324. F. A. Bovey, F. C. Schilling, F. L. McCrackin, and H. L. Wagner, *Macromolecules,* **9**, 77 (1976).

325. Y. Tsuchiya and K. Sumi, *J. Polym. Sci., Part B,* **6**, 356 (1968); *Part A-1,* 415 (1968).

326. A. H. Wilbourn, *J. Polym. Sci.,* **34**, 569 (1959).

327. G. A. Gray, *Anal. Chem.,* **47**, 546A (1975).

328. J. Schaefer, E. O. Stejskal, and R. Buchdahl, 172nd ACS National Meeting, San Francisco, California, August 1976; Polymer Chemistry Paper 4; *Macromolecules,* **10**, 384 (1977).

329. S. A. Liebman, D. H. Ahlstrom, T. C. Creighton, G. D. Pruder, and E. J. Levy, *Thermochim. Acta,* **5**, 403 (1973); and unpublished work.

330. S. A. Liebman, D. H. Ahlstrom, and E. J. Levy, GC Symposium, 27th Pittsburgh Conference Analytical Chemistry and Applied Spectroscopy, Cleveland, Ohio, March 1976, Paper 101.

331. D. A. Usher, J. G. Gouhoutas, and R. B. Woodward, *Anal. Chem.,* **37**, 330 (1965).

332. JEOL News, **12a** (1), 8 (1975).

333. W. McFadden, *Techniques of Combined GC/MS,* Wiley-Interscience, New York, 1973.

334. P. R. Griffiths, *Chemical Infrared Fourier Transform Spectroscopy,* Wiley-Interscience, New York, 1975.

335. G. Oehme, H. Bandisch, and H. Mix, *Makromol. Chem.,* **177**, 2657 (1976).

336. P. C. Uden, D. E. Henderson, and R. J. Lloyd, *J. Chromatogr.,* **126**, 225 (1976).

CHAPTER 5

Thermal Characterization

A. A. DUSWALT

Hercules Incorporated
Wilmington, Delaware

5.1	Introduction	201
5.2	Amorphous Transitions	202
	5.2.1 The Glass Transition T_g	202
	5.2.1.1 Theory	202
	5.2.1.2 Relation to Structure and Composition	204
	5.2.2 Secondary Glass Transitions	206
	5.2.3 Measurement of Glass Transitions	206
5.3	Crystalline Transitions	207
	5.3.1 Degree of Crystallinity	208
	5.3.2 Multiple Crystalline Forms	210
	5.3.3 Fusion	211
	5.3.4 Nucleation and Crystalline Growth	213
5.4	Physical Properties	216
	5.4.1 Dimensional Changes	216
	5.4.2 Viscoelastic Bahavior	217
	5.4.3 Thermal and Electrical Properties	219
5.5	Polymer Composition	220
	5.5.1 Qualitative Analysis	220
	5.5.2 Quantitative Analysis	220
5.6	Polymer Reactions	222
	5.6.1 Polymerization and Curing	224
	5.6.2 Polymer Degradation and Oxidation	225
	5.6.3 Thermal Runaway Reactions and Hazards	227
5.7	Thermal Techniques and Instruments	229
	5.7.1 Differential Thermal Analysis and Differential Scanning Calorimetry	229
	5.7.2 Thermogravimetry	230
	5.7.3 Thermomechanical Analysis	231
	5.7.4 Thermal Evolution Analysis	231
	5.7.5 Thermal Optical Analysis	232
	5.7.6 Torsional Braid Analysis	232
	5.7.7 Dynamic Mechanical Analysis	233
5.8	Summary	234
5.9	Glossary of Symbols	234
5.10	References	235

5.1 INTRODUCTION

Whether or not a polymer can be used in a commercial product depends, to a great degree, on its thermal properties. Flexibility, dimensional stability, durability, and other physical and chemical characteristics are temperature dependent. Milk bottles, for example, should be tough and flexible at below ambient temperature. Therefore semi-crystalline plastics such as polyethylene and polypropylene, which have relatively low glass transition temperatures, are suitable. Polystyrene, on the other hand, is a brittle glass under milk storage conditions, and is therefore less suitable in this respect. None of these polymers is usable for high-temperature applications because of their relatively low melting and softening temperatures. The purpose of this chapter is to discuss these important thermal characteristics of polymers and the methods that measure them. Since a great many polymer properties discussed elsewhere in this book are functions of heat and temperature, the material presented is a natural extension of these discussions. Visco-elastic, structural, molecular weight, compositional, crystallinity, and chemical reactivity effects are described and related to thermal measurements.

Greatly increased activity in the thermal analysis of polymers can be traced to improvements in quantitative instrumentation such as that in differential thermal analysis (DTA) in the mid 1960s. Progress in instrumentation and methods for DTA, thermogravimetry (TG), thermal mechanical analysis (TMA), thermal evolution analysis (TEA), torsional braid analysis (TBA), and dynamic mechanical analysis (DMA), have greatly aided the characterization of polymers. These methods differ in operating principle and in the properties measured. They have in common the measurement of polymer behavior as a function of temperature or temperature change.

The chapter is organized in terms of polymer properties rather than thermal methods. A separate section briefly describes thermal techniques and instrumentation.

5.2 AMORPHOUS TRANSITIONS

All polymers on cooling from the melt will reach a temperature where large-scale molecular motion ceases in the noncrystalline component. This component then becomes a glass, and the temperature is called the glass transition temperature T_g. Some limited motion can exist for short segments and side chains below the T_g. The temperature at which these motions cease are called secondary glass transition temperatures, usually designated as T_β, T_γ, and T_δ.

5.2.1 The Glass Transition T_g

The importance of the T_g to the mechanical and thermal behavior of a polymer are seen in the changes that occur during the transition. The elastic modulus can increase many fold; density, dielectric constant, refractive index, mechanical damping, heat capacity, and thermal coefficient of expansion are different above the transition from below. Polymers

designed for automotive, refrigeration, clothing, or construction purposes, for example, would have to have T_g's compatible with good impact resistance, dimensional stability, and/or brittleness properties.

5.2.1.1 Theory

Experimentally determined values of the T_g are rate dependent. As a glass-forming liquid is cooled more and more slowly, the observed T_g appears at lower and lower temperatures. This behavior is illustrated by the idealized dilatometric curves in Figure 5.1. Such curves show a discontinuity in the rate of thermal expansion as the liquid becomes glass. The extrapolated intersection of the linear section of the curve is taken as the T_g. A rapid cooling rate would result in a high $T_g(f)$ measurement and a slower rate, the lower value at $T_g(s)$. If the liquid could be cooled at an infinitely slow rate, a limiting values $T_g(\infty)$ would be approached. As a rule of thumb, Nielson (1) has observed that a tenfold change in the measuring time scale will result roughly a $7°C$ shift in the T_g for typical polymers. Materials cooled through their glass transition temperatures at different rates form glasses with different specific volumes.

The time-dependent behavior of the glass transition adds to the difficulty in determining its temperature and influence on other properties. Numerous attempts have been made to derive relationships that describe thermal and mechanical properties of polymers near this T_g. Because the transition involves no change in heat content it has been described as a thermodynamic second-order transition. Ehrnfest (2) and others (3, 4) have described a second-order relationship that results in Equation (5.1) as follows:

$$\frac{dT_g}{dP} = \frac{\Delta K}{\Delta \alpha} = \frac{TV \Delta \alpha}{\Delta C_p} \quad (5.1)$$

where T, V, and P are the absolute temperature, specific volume, and pressure, respectively. Here K is the bulk modulus (compressibility), α the coefficient of thermal expansion, and C_p the heat capacity at constant pressure. However, for most polymers it

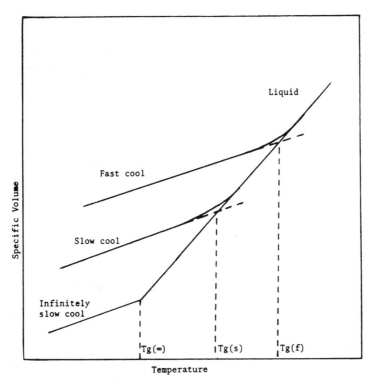

Figure 5.1 Cooling effect on the glass transition temperature.

has been found that Equation (5.2) is the case

$$\frac{\Delta K}{\Delta \alpha} \neq \frac{dT_g}{dP} \qquad (5.2)$$

and that the characterization of the glass transition behavior by Equation (5.1) is not satisfactory (5). The failure of the second-order transition model lies in the assumption of a reversible equilibrium process. The time-dependent nature of a glass transition is in fact irreversible and so does not fit this model.

A useful description of the glass transition can be derived in terms of the free volume theory (6). The free volume in this context refers to the extra room required for large-scale chain vibrations of the polymer molecule above that occupied by the volume of the molecule itself. The free volume expansion is related to measurements of the polymer relaxation processes in the general expression, Equation (5.3)

$$\ln a_T = \frac{-C_1 (T - T_0)}{C_2 + T - T_0} \qquad (5.3)$$

where a_T is a ratio of mechanical to electrical relaxation times. For viscosities $a_T = \eta(T)/\eta(T_0)$. C_1 and C_2 are constants related to the fractional free volume and are generally taken to be 20.4 and 101.6, respectively. Equation (5.3) is referred to as the WLF equation (7). If T_0 becomes T_g the constants C_1 and C_2 become 40 and 50.1, respectively, and the Equation (5.3) becomes Equation (5.4) as follows:

$$\ln a_T = \frac{-4.0(T - T_0)}{51.6 + T - T_g} \qquad (5.4)$$

According to Arridge (4) a WLF equation, although not universally true, fits a broad variety of systems.

In view of the ambiguity of experimentally determined glass transitions, it would be highly desirable to have a thermodynamically sound criterion for a value of $T_g(\infty)$. Christensen (3) has recently derived such a criterion. He found that when the ratios of a polymer's linear creep functions are constant, it is possible to have a zero Gibbs free energy change for a process even though mechanical properties are time dependent. The temperature at which this condition is satisfied can be taken as the glass transition temperature. The derived expression is simply Equation (5.5)

$$\frac{\alpha(t)}{J_2(t)} = \frac{C(t)}{\alpha(t)} = \text{constant} \qquad (5.5)$$

where $\alpha(t)$, $J_2(t)$, and $C(t)$ are the creep function generalizations of the coefficient of thermal expansion, volumetric compliance, and specific heat, respectively. The value of the T_g is found by observing the effects of pressure $J_2(t)$ and temperature $\alpha(t)$ on volumetric response at different polymer base temperatures. Shifts in the $J_2(t)$ function with temperature are greater than those of $\alpha(t)$, as illustrated in idealized curves in Figure 5.2. The value for the polymer T_g would be that above which thermal relaxations are faster—and below which mechanical relaxations are. The validity of such a definition of the glass transition temperature lies in its derivation

from irreversible thermodynamic considerations.

A great many papers on the relation of the T_g to polymer properties have been published. The T_g data in these studies are above $T_g(\infty)$ and, as such, are a function of the rate of measurement. If the investigators use the same measuring rate for all T_g determinations, the relationships found will have an internal consistency, but the T_g's will be valid only for that rate. Different thermal techniques have different optimum rates for observing the glass transition. To compare data from these techniques it is necessary to have the conditions of the measurements stated.

5.2.1.2 Relation to Structure and Composition

The relationship between the T_g and polymer composition and structure is of great practical importance. The relationship can be used to alter T_g to fit specific applications, to study internal structure, and to quantitatively evaluate the effects of polymer alteration, that is, cross-linking and plasticizing effects.

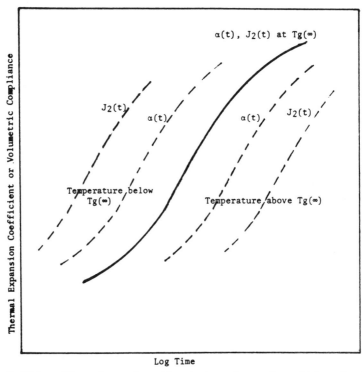

Figure 5.2   Coefficient of thermal expansion α (t) and volumetric compliance $J_2(t)$ as a function of base tempertaure (3).

The effect of molecular weight on the glass transition has been observed in a number of studies (8). The increase of T_g with molecular weight is most evident below about 20,000 and tends to level off at higher values. This effect has been ascribed to the greater proportion of the more flexible chain ends in low molecular weight polymers. The relationship is described as follows by Equation (5.6):

$$\frac{1}{T_g} = \frac{1}{T_{W\infty}} + \frac{K}{M_n} \qquad (5.6)$$

where $T_{W\infty}$ is the T_g at infinite molecular weight, M_n the number average molecular weight, and K a constant.

The T_g can be changed markedly by the addition of certain branches or side chains to the polymer backbone. Thus flexible alphatic chains can decrease T_g while more rigid aromatic side groups will increase it. A series of methacrylic polymers with pendant groups ranging from the methyl ester to the n-dodecyl ester show the T_g decreasing monatomically from + 105°C for the former to − 65°C for the latter (1). Generally the size of the side group is not as important in lowering T_g as its flexibility. Large groups can in fact increase T_g if stearic hindrance or entanglement results. The cohesive energy density of a polymer molecule or side groups plays a large roll in determining the T_g. Increasing the polarity, hydrogen bonding, or ionic nature of the polymer will raise the T_g (1).

The T_g's of copolymers provide insight into their compatibility. Compatible polymers have a single T_g that varies with fractional composition. As the homogeneous nature of the polymer decreases, the damping peak associated with T_g broadens (9). A completely heterogeneous copolymer will exhibit two T_g's that are the same as those observed for the individual homopolymers. In the latter case the size of the T_g signals will be proportional to the amount of each component present. The Fox Equation (5.7) (10) is used to estimate the T_g of polymer mixtures and copolymers as follows:

$$\frac{1}{T_g} = \frac{W_A}{T_{gA}} + \frac{W_B}{T_{gB}} \qquad (5.7)$$

where W_A, W_B, and T_{gA}, T_{gB}, are the weight fractions and glass transition temperatures of components A and B, respectively. A relationship that offers a better fit for most polymers is the Gordon-Taylor-Wood expression, Equation (5.8) (11, 12), which can be written as follows:

$$K \frac{W_A}{W_B} = \frac{T_g - T_{gA}}{T_g - T_{gB}} \qquad (5.8)$$

where K is a constant specific for each copolymer and related to the free volume of each component. Ballesteros, et al. (13) have concluded that Wood's expression adequately describes the T_g copolymer composition relationships for many systems were $T_A > T_g > T_B$. However, Johnson (14) points out that some AB sequence interactions can give rise to data scatter and less accurate values of K. He proposes an expanded Fox Equation (5.9), which considers the contribution of the individual AA, AB, and BB dyads for an AB copolymer,

$$\frac{1}{T_g} = \frac{W_A P_{AA}}{T_{gAA}} + \frac{W_A P_{AB} + W_B + W_B P_{BA}}{T_{gAB}} + \frac{W_B P_{BB}}{T_{gBB}} \qquad (5.9)$$

where P_{AA}, P_{AB}, P_{BA}, P_{BB} are the probabilities and T_{gAA}, T_{gAB}, and T_{gBB} are the glass transition temperatures of the respective dyads. Techniques for calculating dyad probabilities and T_{gAB} have been described (15, 16). The proposed relation can be expanded for terpolymers, and good agreement has been found between calculated and experimental results for acrylonitrile-methyl methacrylate-methylstyrene systems. Other expressions have been recommended for less frequent cases where the T_g versus composition graph shows a maximum or minimum (13, 17).

Plasticizers generally are low molecular weight liquids that are added to polymers to improve low-temperature flexibility. They work by lowering the glass transition temperature. The Fox Equation (5.7) is a commonly used relation between the weight fraction and glass transition temperature of the plasticized polymer.

Cross-linking and curing are similar processes that change the T_g by introducing restrictions to the motion of the polymer chain. The curing process of epoxy resins has been followed by observing the consequent change in the T_g (18, 19). Nielsen (20) has described an empirical relationship for cross-linked polymer as Equation (5.10):

$$T_g - T_{g0} = \frac{3.9 \times 10^4}{M_c} \qquad (5.10)$$

where T_{g0} is the initial T_g of the uncross-linked resin and M_c the number average molecular weight between cross-linked points. Other theoretical relationships have been described relating these parameters (20, 21, 22).

5.2.2 Secondary Glass Transitions

Secondary glass transitions, often designated T_β, T_γ, and T_δ, occur below the normal T_g. They are associated with low-energy motions of short-backbone segments and side-group rotation. They have also been related to phase separation of impurities or diluents, head-to-head polymerization, and polymer crystallinity (23). Multiple secondary transitions have been characteristic of many semicrystalline polymers (9). Because molecular motion can dissipate impact energy, the presence of secondary transitions has been related to low-temperature impact resistance (9). The specific source of motion giving rise to a secondary transition signal is often unclear. A few have been well defined. For example, the presence of more than three CH_2 groups in the polymer backbone or side groups will give rise to a secondary transition between $-150°C$ and $-120°C$. The cyclohexyl transition is well known at about $-80°C$ when measured at a rate of 1 Hz (24). Other examples are given by Nielsen (25).

Two other polymer transitions are related to gelation and liquid-liquid relaxation phenomenon. A mechanical damping signal has been related to the gel point of curing epoxy resins systems (26). It is the stage where the resin develops an intermolecular network. The storage temperature of active thermosets

should be such that the vitrification occurs before the onset of gelation. The existence of a liquid-liquid transition T_{11} for amorphous polymers is controversial. Stadnicki and Gillham have summarized the evidence for such transitions and presented their findings on the T_{11} of amorphous polystyrene (27). The transition has been described as a fixed liquid state and related to entanglement of long-chain segments.

5.2.3 Measurement of Glass Transitions

Measurements of the temperature and magnitude of glass transitions are based on the changes that occur in mechanical damping, heat capacity, refractive index, viscosity, thermal expansion, dielectric, and nmr properties. Dynamic mechanical techniques measure energy dissipation in terms of damping (such as tan δ); these techniques are generally considered to be the most sensitive for studies of glass and secondary glass transitions. Free vibration dynamic mechanical studies have been carried out by torsional pendulum, which measures the damping characteristics of a freely oscillating samply in the shape of a bar or rod. The use of a torsional braid is a variation of the pendulum technique that allows samples to be impregnated on a glass braid from solution (28). The braid technique was an improvement in sample preparation and extended the range of study in the rubber and liquid state. A later improvement in the torsional pendulum technique (29) encased the finely divided test polymer in a supporting bar or rod of cured phenolic resin.

Dynamic mechanical studies are also carried out through forced vibration by applying an oscillating drive signal to the sample. Loss factors due to chain or side-group motion are measured as phase differences between stress and strain. A commercial instrument, the Rheovibron (9), has become well known for determining modulus as well as loss factors. Figure 5.3 shows the effect of side chains on the mechanical damping curves of polyethylene. Nielsen (30) has reported many references concerning the techniques and instruments for determining dynamic mechanical properties of polymers.

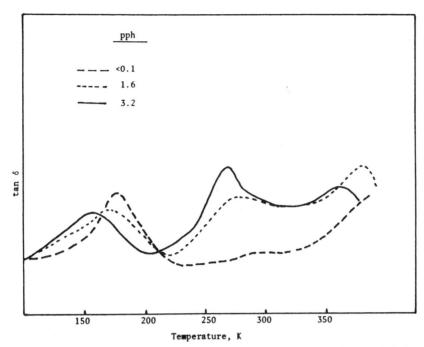

Figure 5.3   Branching effects on mechanical damping of polyethlene pph of CH_3 to CH_2 (30).

Changes in the thermal optical properties have been the basis for determining glass transitions (31). Shultz (32) has observed that birefringence induced by scratching a polymer will disappear near the T_g. He recommended this phenomenon as a simple, sensitive way for detecting the transition.

Amorphous polymers can be rapidly scanned for the T_g by differential thermal analysis. This technique observes the increase in heat capacity with temperature at the T_g. However, the presence of appreciable amounts of crystallinity will tend to mask the effect. The change in viscosity at the T_g can be observed by thermal mechanical analysis (33). Penetration into the polymer by a loaded, sharply pointed probe signals the T_g. This technique is also less sensitive for semicrystalline polymers. A flat probe with zero load enables detection of T_g from the change in expansion rate with temperature. However, measurement of T_g from expansion data is usually done by dilatometry (34). Chiu (35) has listed a number of references for observing changes in the resistivity for dielectric behavior as a function of temperature. A gas chromatographic technique has been described (36) for the determination of T_g. The polymer is the solid support, and

the T_g is signaled as a break in the curve of retention time versus column temperature.

It should be borne in mind that the preceding techniques meassure different properties at different rates and will give varying results for the same transition. Until a unifying theory appears that can relate thermal and mechanical time-temperature properties of a polymer's T_g, caution must be used in relating test T_g's to long-term, real life behavior.

5.3   CRYSTALLINE TRANSITIONS

Crystallinity in polymers varies from zero in rubbery or glassy material to over 97% for fractionated polyethylene (37). Chain entanglements and restricted motion under normal crystallizing times and temperatures produce defects that reduce the perfection and degree of crystallinity. Spherulitic structures are usually formed that contain crystallites in folded-chain lamellae separated by an amorphous region. A diffuse interfacial layer on the crystalline surface has been postulated (38). The form, degree, and size of crystalline material have great influence on thermal and mechanical properties. In practical terms the crystalline nature of a polymer will affect extrusion, molding, sealing, and orientation behavior. With increasing crystallinity, stiffness, strength, hardness, inertness, and brittle failure can increase but solubility and compatibility of dyes, plasticizers, stabilizers, and other additives decrease. Crystallinity also affects permeability, thermal conductivity, expansion, dielectric behavior, and optical properties.

5.3.1 Degree of Crystallinity

Thermal studies of the degree of crystallinity are normally carried out by Differential Scanning calorimetry (DSC). Advantages of this approach are speed, high sensitivity, and precision. In addition, other information such as temperature melt range is obtained. In the usual procedure, approximately 5–10 mg of polymer are linearly temperature programmed from a point well below the melt region to a

temperature where there is no additional evidence of melting. The energy absorbed during the melting process is recorded as an endothermic peak as shown for a polypropylene sample in Figure 5.4. A good estimate of the polymer heat of fusion ΔH_f is obtained from the melt peak area. If the heat of fusion from prefectly crystalline polymer ΔH_μ is known, the degree of crystallinity is simply calculated as follows from Equation (5.11):

$$\%\chi = \frac{\Delta H_f}{\Delta H_\mu}(100) \qquad (5.11)$$

If ΔH_μ is not known, it may be estimated if the degree of crystallinity χ can be determined by another technique. Then one can use Equation (5.12).

$$\Delta H_\mu = \frac{\Delta H_f}{x} \qquad (5.12)$$

A value for ΔH_μ may be obtained from a DTA study of the effect of a diluent on the polymer

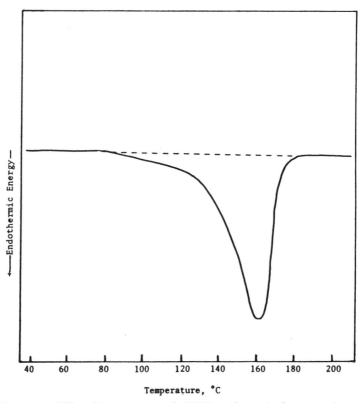

Figure 5.4   Differential thermal analysis (DTA) melting peak of polypropylene.

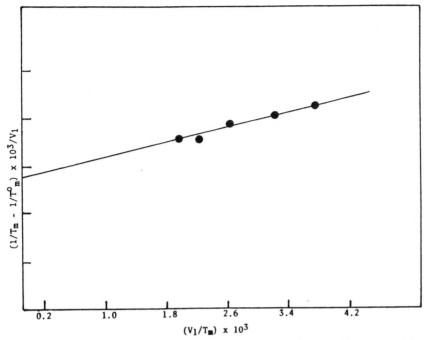

Figure 5.5 Plot for determining ΔH_μ of a polymer, hydroxypropyl cellulose in p-menthane (45).

melt temperature (39). The relationship can be expressed as follows in Equation (5.13):

$$\frac{(1/T_m) - (1/T_m^0)}{v_1} = \left(\frac{RV_\mu}{\Delta H_\mu V_1}\right)\left(1 - \frac{BV_1 v_1}{RT_m}\right)$$

(5.13)

where T_m and T_m^0 are the melt points of the polymer with and without diluent; V_μ and V_1 the molar volumes of polymer and diluent respectively; v_1 the diluent volume fraction; B the polymer diluent interaction energy density, and R the gas constant. The relation assumes crystallization has taken place under conditions approaching equilibrium. In practice, T_m^0 is found by extrapolating a plot of $1/T_m$ versus volume fraction of diluent to $v_1 = 0$. A plot of $(1/T_m - 1/T_m^0)/v_1$ versus V_1/T_m can then be made and a straight line extrapolated to $v_1/T_m = 0$. The intercept at this point is equal to $RV_\mu/\Delta H_\mu v_1$ and is solved for ΔH_μ in kilocalories per mole. Figure 5.5 shows results for hydroxypropyl cellulose (45). Table 5.1 shows the results obtained by the DTA diluent method for a number of common polymers.

Although the spread of ΔH_μ values for polypropylene is broad, the more recent deter-

Table 5.1   Values of ΔH_μ, by Flory's Method (70)

Polymer	ΔH_μ, Kcal/mol	Reference
Polyethylene	0.960	40
Natural Rubber	1.050	40
Polyethylene Oxide	1.980	40
Polypropylene Oxide	0.900	41
Poly(ethylene terephthalate)	6.60	42
Polypropylene	1.84-2.60	37, 43, 44
Hydroxypropyl Cellulose	2.54	45

minations center around fusion heats of 1.85–2.0 kcal/mol (43). These are controversial results. Fatou (46) has plotted fusion heat versus specific volume of polypropylene fractions and extrapolated a value for ΔH_μ of 1.39 kcal/mol. He indicates good agreement between x-ray, calorimetric, and density results for the degree of crystallinity in various polypropylene samples. The author finds a slightly higher value of 1.5 kcal/mol enables good correlation of density, x-ray, and infrared results with calorimetric measurements. The scatter of Fatou's data is sufficient to incorporate the 1.5 value. Difficulties in applying the melt point depression technique to polypropylene may stem from the large discrepancy between the T_m^0, estimated to be as high as 220°C (47, 48) and the much lower values of 165–176°C extrapolated from the plots of T_m versus diluent fraction.

Roberts (49) showed that for polyethylene terephthalate (PET), ΔH_μ values derived from density data (27 ± 2 cal/g) agreed well with melt depression results (29.0 cal/g). The degree of crystallinity was calculated from density data by variously anealed PET samples by the relation shown in Equation (5.14)

$$\rho = X\rho_C + (1 - X)\rho_A \qquad (5.14)$$

where ρ, ρ_C and ρ_A are the measured, crystalline, and amorphous densities, respectively. Equation (5.12) is then used to calculate ΔH_μ. The density of the crystallinity fraction can be estimated from x-ray determinations of the unit cell dimensions (50) and its atomic makeup. Densities of the amorphous fraction are usually obtained from measurements on quenched amorphous polymer or extrapolated from measurements of the molten material.

Common methods for calculating the degree of crystallinity are discussed by Meares (51) and Billmeyer (52). The slope of the plot of $(1/T_m - 1/T_m^0)v_1$ versus v_1/T_m enables calculation of the diluent polymer interaction parameter B. If T_m^0 is known, the melt entropy per chain unit is $\Delta S_\mu = \Delta H_\mu/T_m^0$.

5.3.2 Multiple Crystalline Forms

Some polymers will show one or more crystalline transitions at temperatures between the T_g and the final fusion. These transitions can be due to a true crystal-crystal transformation or to partial melting and reorganization of metastable crystallites. The importance of characterizing such behavior lies in possible effects on thermal, optical, dimensional, and modulus properties. For example, the presence of the hexagonal β-form in normal monoclinic crystalline polypropylene can result in poor optical clarity and mechanical behavior. The slow conversion of poly-1-butene from a metastable to a more stable form causes shrinkage over several days at ambient temperatures (53).

Examples of polymers that undergo reorganization from a metastable state to a more stable form are poly-1-butene, polyoxymethylene (54), and polypropylene (55). Polypropylene (56, 57, 58) may in fact exist in several forms, a metastable and stable hexagonal (β) form, the normal monoclinic (α) form, a triclinic (γ) modification associated with high-pressure crystallization, and paracrystalline smectic form. If polypropylene is crystallized isothermally from the melt at temperatures higher than about 135°C, double melt peaks are observed, suggesting some differentiation in the monoclinic form (47, 48). Other polymers showing polymorphic behavior are tetrafluoropolyethylene and trans-1-4-polybutadiene, 66 nylon (59), poly(p-xylene) (60), and poly(ethylene terephthalate) (61).

The identification of polymer crystalline forms is normally made from x-ray diffraction patterns. Thermal studies have been concerned with the formation, annealing, nucleation, transformation, and fusion of crystalline polymers. Thermal-optical and DTA techniques have been used. The optical methods can observe changes in birefringence pattern of the intensity of depolarized light transmitted through spherulitic crystals as a function of time, temperature, or previous thermal history (62). Optical procedures using a programmable hot stage and differential scanning calorimetry (DSC) studies are convenient, allowing imposition of specific thermal histories under controlled environments and rapid observation of results.

5.3.3 Fusion

It is common practice to report the melt behavior of a polymer in terms of a single temperature. This temperature is almost always that of the DTA peak maximum or the disappearance of birefringence. These temperatures have physical significance and offer a convenient way to simply characterize the melting behavior. However, unlike low molecular weight materials, a pure crystalline polymer does not have a single melt temperature. As seen in Figure 5.4 a typical polypropylene will be observed to start melting at about 80° C, and finally end at about 170° C. Moreover, a polymer's melt temperature may change depending upon how the polymer was crystallized, its aging or annealing history, and the rate of temperature increase upon melting. This behavior reflects the presence of a broad spectrum of crystallites that differ in size and degree of perfection and consequently, melt temperature. A relationship between the crystallite size (thickness of folded lamella) and the melt point is described by Hoffman and Lauritzen's Equation (5.15) (63).

$$T_m = T_m^\circ \left(\frac{1 - 2\,\sigma_e}{\Delta H_f L} \right) \qquad (5.15)$$

where σ_e is the surface free energy, T_m° the equilibrium melt temperature, ΔH_f the heat of fusion, and L the lamella thickness. Thus the lowering of the crystallite melt temperature is inversely proportional to the lamella thickness. Wunderlich (64) has described equations for the effect of defects on melt behavior but indicates (65) that the effect is small for well crystallized polymers.

The melting range of polymers is usually determined by observing changes in enthalpy, birefringence, or density with increasing temperature. However, the rate of temperature increase can influence the observed melting temperature. Low-melting crystallites may readily anneal during a slow heating process. At an appropriate temperature crystalline lamellae will have sufficent internal mobility to thicken, forming more temperature-stable crystallites. This annealing, or thickening, process is time-temperature dependent. As temperature increases, the time to form more ordered crystals becomes greater. Figure 5.6

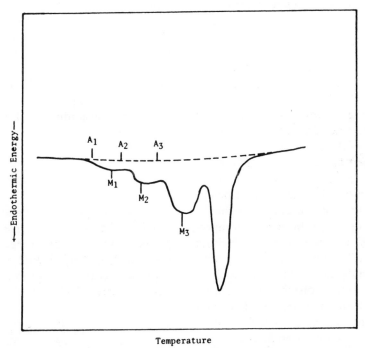

Figure 5.6 Melting peaks of polyethylene M_1, M_2, and M_3, produced by successive annealings at temperatures A_1, A_2, and A_3.

shows the melting curve of a low-density polyethylene sample annealed isothermally at successively lower temperatures. The affected crystallites at each annealing temperature have increased their melting temperature several degrees and appear as distinct peaks. A second process that can occur involves the complete fusion of low-melting, meta stable crystallites and subsequent recrystallization in a more stable form. This process is also temperature-time dependent. At a given heating rate, reorganization will proceed until a temperature is reached where the time for the most easily recrystallizable melt fraction is inadequate. As the heating rate is decreased, and more time is available for reorganization, this limitation occurs at higher temperatures. Since the observed melting temperature of a polymer is a direct function of its crystallization temperature, as discussed later, a relationship between heating rate and melt temperature should be expected. Duswalt (47) has shown that between 157°C and 172°C the melting peak temperatures of polypropylene decrease linearly with the increasing log of the heating rate. The observed change appears to follow the relationship shown as follows in Equation (5.16):

$$\Delta T_m = 4.0 \ln \beta \qquad (5.16)$$

where ΔT_m is the change in melt peak temperature and β the linearly programmed heating rate in degrees Celsius per minute.

The preceding processes show that melt temperatures can increase with heating rate as a result of annealing and recrystallization. Wunderlich (66) has observed that for extended-chain crystals or slow-cooled, well-ordered spherulitic crystals, experimental heating rates can supply heat faster than the polymer's rate of fusion. In such cases super-heating occurs, and the polymer's melting temperatures will appear to increase with heating rate. In the case of highly crystalline, high molecular weight polyethylene the super-heating effect can produce apparent final melt temperatures of just below 150°C; several degrees above its theoretical equilibrium melting temperature.

The fact is well established that the melt temperatures of a crystalline polymer are directly related to crystallization temperature. Mandelkern (67) and Hoffman (68) have described an extrapolation method to estimate the equilibrium melting temperature. Figure 5.7 shows the use of this technique to estimate an equilibrium melt temperature of 220°C for polypropylene (47, 48). Bair (69) has used the Hoffman and Lauritzen Equation (5.15) to estimate T_m° of polyethylene. A plot of observed melt points (corrected peak temperatures) versus the inverse lamella thickness of irradiated solution-grown single crystals was extrapolated in $1/L = 0$ for T_m°. His results ($T_m^\circ = 145.8 \pm 1°C$), agreed well with that of Flory and Vrij ($T_m^\circ = 145 \pm 1°C$) who extrapolated melting points of normal paraffins to large chain lengths (70).

The melting behavior of a polymer also depends upon its structure and composition. The greater the polar attractive forces (e.g., hydrogen bonding) holding the crystal together the higher the melting temperature. Polymer chains incorporating monomer units with even carbon numbers tend to be more stable than those with odd numbers (71). As the number of pendant groups on the main polymer chain increases, the melting point decreases. This behavior has been shown for the addition of methyl or ethyl groups on polymethylene (72). However, long side chains may favor side-chain crystallization, diminishing the melt depression effect. Random copolymerization will decrease the melting point of polymer according to Flory's Equation (5.17) (73) as follows:

$$\frac{1}{T_m} - \frac{1}{T_m^\circ} = \frac{-R \ln X_c}{\Delta H_\mu} \qquad (5.17)$$

where R is the gas constant and X_c is the mole fraction of crystalline polymer X in the copolymer. The equilibrium heat of fusion for polymer X is ΔH_μ. Kamel et al. (74) have expressed X_c in terms of the heat of fusion data, Equation (5.18)

$$\frac{1}{X_c} = 1 + \left(\frac{M_X}{M_Y}\right)\left(\frac{\Delta H_\mu}{\Delta H_X}\right) - 1 \qquad (5.18)$$

where M_X and M_Y are the molecular weight of the repeat unit for crystalline polymer X and

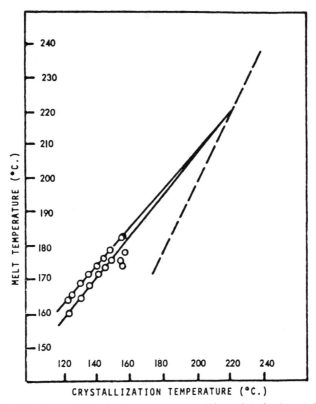

Figure 5.7   Effect of crystallization temperature on melting point of polypropylene (47).

the comonomer Y, respectively. ΔH_X is the measured heat of fusion of the copolymer. If plasticizers are added to the copolymer, the melting point will be depressed according to Equation (5.13).

The melting temperature of a crystalline polymer is related to its molecular weight by Equation (5.19) as follows:

$$\frac{1}{T_m} - \frac{1}{T_m^\circ} = \frac{2\,RM_0}{\Delta H_\mu\,M_n} \qquad (5.19)$$

where M_n is the number average molecular weight and M_0 is the molecular weight of the polymer repeat unit (75). As M_m approaches that of many commercial polymers, the affect of molecular weight on melting behavior becomes very small.

5.3.4   Nucleation and Crystalline Growth

For crystals to grow from a polymer melt or solution, nuclei of sufficient size must be present. The formation of such nuclei is a function of temperature. Above the melt temperature thermal energy disperses small nuclei that form and prevents their growth. As the temperature decreases from the melting point, the kinetic effects become smaller and the chances for larger nuclei greater. At each temperature below the melt point, there is a nuclei size above which the energy gain for adding new molecules becomes greater than that for dispersing them. This critical size is large just below the melt point and decreases with temperature. Thus, as the temperature is lowered the average nuclei size increases, and the critical size for crystal growth decreases. At some point in a rapidly cooling polymer melt, many small nuclei suddenly reach critical size, and crystallization proceeds rapidly with many small crystallites forming simultaneously. If the polymer is held at higher temperatures, fewer nuclei reach critical size and take longer to do so. Crystallization is slower but produces larger crystallites. Figure 5.8 shows how the increase of temperature slows the crystallization rate.

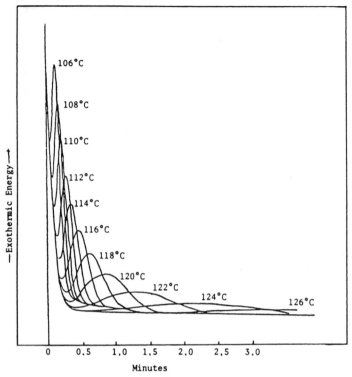

Figure 5.8   Temperature effect on polypropylene crystallization time.

The rate at which crystallization proceeds at constant temperature has been described by the Avrami Equation (5.20) (76) as follows:

$$\ln (1 - \theta) = Kt^n \qquad (5.20)$$

where K is a constant, θ the degree of crystallinity, t a crystalline time interval, and n an integer whose value depends on the form of crystal growth and thus, nucleation mechanism. The crystallization mechanisms for various values of n are given in Table 5.2.

Table 5.2   Relation of Avrami Exponent n to Crystallization Process (77)

n	Nucleation Type	Growth Form
1	Predetermined	Fibrilar
2	Sporatic	Fibrilar
3	Predetermined	Discoid
4	Sporatic	Discoid
5	Predetermined	Spherulitic
6	Sporatic	Spherulitic
7	Predetermined	Sheath-like
8	Sporatic	Sheath-like

Sporadic nucleation mentioned in the table refers to nuclei formed homogeneously from the melt. Predetermined nucleation results from the number of heterogeneous nuclei present in the polymer. In practice the Avrami equation has not been very successful in distinguishing the different crystallization mechanisms. Often a nonintegral value for n is obtained. In some cases nucleation can be partly homogeneous and partly heterogeneous. For spherulitic crystallization, values of n between 3 and 4 may be obtained and remain almost constant for most of the crystallization process (77).

In addition to nucleation effects, temperature can influence the crystallization rate by its effect on melt viscosity. As the melt viscosity increases, the mobility of the polymer chain decreases and crystal growth slows. There is an optimum temperature for crystallization for many polymers. Above this temperature crystallization is limited by slow nucleation, and below by high viscosity. The temperature effect on crystallization rate has been described (78) as expressed in Equation (5.21):

$$K = B - \frac{4E_D}{KT_c} \frac{DT_m^2}{T_c (T_m - T_c)^2} \quad (5.21)$$

where K is the crystallization rate constant, B and D are constants for a given polymer, E_D is the activation energy for segmental diffusion. The melt and crystallization temperatures are T_m and T_c, respectively. Equation (5.21) predicts a maximum in K as the crystallization temperature is decreased from T_m.

If, upon cooling, the melt viscosity inhibits molecular motion before the nuclei reach critical size, the polymer will solidify as a glass. Polyethylene terephthalate is a crystalline polymer that can be transformed into an amorphous solid by rapidly cooling its melt. Nucleation adequate for crystal growth may occur, but the growth is inhibited by high-melt viscosity. As the amorphous polymer is heated, crystal growth begins as soon as the molecules have sufficient kinetic energy to initiate movement. This phenomenon is called cold crystallization and can be observed as an exothermic peak on a DSC thermogram. (See Section 5.7.1, Figure 5.18.)

The rate of crystallization of a polymer can also be affected by its previous melt history. Nylon 6 (79) and polypropylene (47) have been observed to crystallize more rapidly when cooled from temperatures just above their melt than when cooled from higher temperatures. The evidence suggests that elements of molecular order persist after the bulk of the polymer has melted and that these elements aid in the nucleation process upon subsequent cooling. Figure 5.9 shows how the DSC isothermal crystallization rate of polypropylene changes as the previous melt temperature is increased in steps.

Crystallization rates of polymers are influenced by a number of factors other than thermal history. Molecular weight affects crystallizability at both extremes. Low molecular weight chains have higher mobility at a given temperature and therefore enhance crystal growth. Conversely, very high molecular weight chains have restricted mobility, and crystals grow more slowly. This effect has been used to correlate the size of the crystallization peak with the molecular weight

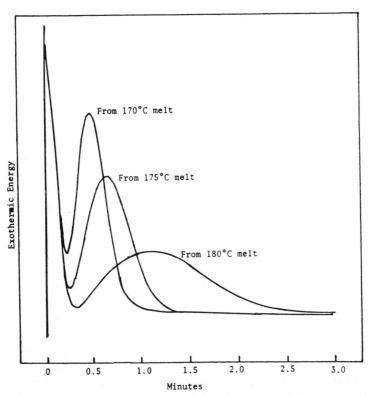

Figure 5.9 Polypropylene crystallized at 120°C after holding 1 min at 170°C, 175°C, and 180°C.

polytetrafluoroethylene (80). However, as discussed, crystalline high molecular weight chains are potential nucleating elements that enhance crystallization. Plasticizers aid crystallization by increasing mobility. Solid fillers, if they restrict motion, may have the opposite effect. Heterogeneous nucleators (81) and dyes, pigments (82), and other fillers that can act as nucleators increase crystallizability. Mechanical stress, by aligning the polymer molecules, promotes crystallization (83). Shear stress has been observed to increase the nucleation rate (84). Since regularity, flexibility, and mobility enhance crystallization, branches, particularly of bulky, rigid side groups or grafts, may be expected to retard or prohibit crystal growth.

The measurement of the isothermal crystallization rate of polymer melts is usually carried out by observing changes in dimension, optical density, or heat. The dilatometric technique measures the decrease in volume with time (85). About a gram of sample is required, and running time is in the order of hours. Families of curves may be generated to show the effects of temperature. Polarizing microscopes with hot stage mounted polymer samples have been used to follow the crystallization process (86, 87). Spherulitic growth rates are followed by observing the increasing birefringent light. The light is measured photographically or recorded as a millivolt signal from a photovoltaic cell. The hot stage arrangement allows observation of very fast as well as very slow crystallizations. Observations can be made under isothermal or temperature programmed conditions. A limitation of the optical technique is that it is confined to the early phase of crystallization. Isothermal crystallizations carried out in a DTA instrument give rise to exothermic peaks as in Figures 5.8 and 5.9. The simplest measure of such peaks is the time to peak maximum T_p which is a sensitive indicator of the crystallization rate (82). A plot of the log T_p values of Figure 5.8 versus temperature is essentially linear and indicates that a 2.6° change in crystallization temperature will cause a sixfold change in the rate of polypropylene crystallization (82). DTA crystallization rates have also been obtained by integrating peak areas and plotting percent

conversion versus time (88). The DTA isothermal technique may not measure slow crystallization directly if the rate of heat evolution is low.

5.4   PHYSICAL PROPERTIES

A number of important polymer properties can be studied through the measurement of vertical displacement of a probe in contact with the polymer surface. The technique is called thermal mechanical analysis (TMA) (89, 90). By varying the type of probe to measure compression, extension, or penetration much information may be obtained.

5.4.1   Dimensional Changes

A flat-ended nonpenetrating probe enables measurement of the linear expansion of a solid polymer as its temperature is changed. The linear coefficient of expansion α may be calculated from a plot of probe displacement versus temperature as expressed in Equation (5.22):

$$\alpha = \frac{dY}{dT}\left(\frac{1}{L}\right) \qquad (5.22)$$

where dY/dT is the change in unit length per degree of temperature change and L is the sample thickness. TMA results have agreed with those obtained by the American Society for Testing and Materials ASTM D-696 procedure (91). Expansion measurements on different planes of a polymer specimen have been used to show nonisotropic behavior. A study of cross-link polyethylene cable showed normal expansion in a transverse direction but contraction in the radial direction due to built-in compressive forces (92). Structural orientation in polyvinyl chloride was shown by differences in the expansion characteristics of the X, Y, and Z axes (91). Volume coefficients of expansion have been obtained using a special dilatometric probe. Studies of the swelling and dissolution rates of various polymer films in organic solvents were carried out using a flat-ended probe (93). For certain systems straight-line relationships were demonstrated when functions of reduced volume swell were plotted against appropriate functions of time. Values for diffusion coefficients could be estimated. The investigators concluded that the simplicity of the TMA procedure in some cases renders it a suitable alternate to the more generally used optical measurements. Swelling and heats of absorption measurements were carried out on cellulose with dimethyl formamide (DMF) and methanol solvents (94). The study was able to relate the relative influence of absorption energy and solvent molecular size to swelling behavior. Initial swelling rates were most influenced by the filling of high-energy sites and thereafter by the affect of molecular size on penetration of the cellulose fibers.

5.4.2 Viscoelastic Behavior

The flow rate of polymer melt under compression is related to its viscosity. This relation was applied by Dienes and Kenn (95) for a polymer compressed between two parallel plates. The relationship is expressed as Equation (5.23):

$$\eta = \frac{4F/3\pi R^4}{(dh^{-2})/dt} \quad (5.23)$$

where η is a viscosity in poise, F the compressive load in dynes, R the plate radius in centimeters, h the distance between plates in centimeters, and t is time in seconds. The relationship assumes the spaces between the plates are completely filled with the polymer and that R/h is greater than 10. The wall shear rate γ has been calculated by Cessna (96) using Equation (5.24).

$$\gamma = \frac{3Rh}{2} \cdot \frac{(dh^{-2})}{dt} \quad (5.24)$$

A small parallel plate accessory has been used with TMA instruments (89, 91, 96) to determine viscosities of polymer materials in the 10^4–10^9 poise range. The microparallel plate device is shown in Section 5.7. Noel (91) has shown that observing creep relaxation when a compressive load is removed can define the temperature range where deformation changes from viscoelastic to viscous behavior.

Using a spherically tipped probe Machin and Rogers (93) were able to estimate creep behavior from penetration results on homogeneous isotropic polymers. Hwo and Johnson (97) used a spherically tipped TMA probe to measure Young's modulus for elastomers. A modulus was calculated from the depth of indentation after a time period using Finkin's Equation (5.25) (98) as follows:

$$E = \frac{3PR(1 - \gamma^2)}{4H^3}[x^{1/2} + 0.252x + 0.159x^{3/2}$$
$$+ 0.225x^2 + 0.307x^{5/2} + 0.298x)]^{-3}$$
$$x = \frac{Rd}{H} \quad (5.25)$$

where E is Young's modulus, P is the load on the sample, R is the radius of the spherical probe, H is the sample thickness, d is measured penetration depth, and γ the Poisson's ratio of the elastic layer. Results from the TMA measurements were in good agreement with the literature values for acrylic and silicon elastomers.

Young's modulus has also been determined by using TMA penetration probes (99, 100). The samples are linearly increased in temperature and the Young's modulus determined as a function of temperature from Equation (5.26) as follows:

$$E = \frac{3F(1 - 0.75 \, R/D)}{8 \, Rd} \quad D \gg R \quad (5.26)$$

where F is the load on the sample, R the probe radius, d the penetration depth, and D the sample thickness. Wood and Vervloet (101, 102) have demonstrated the relationship of Young's modulus with temperature for cross-linked elastomers as Equation (5.27).

$$E = \frac{3\rho RT}{M_c}\left(1 - \frac{2 \, M_c}{M_n}\right) \quad (5.27)$$

where ρ is the elastomer density, M_c the molecular weight between cross-linking, and M_n the number-average molecular weight. Thus it is possible to calculate a degree of cross-linking from the determined Young's modulus. The TMA modulus measurements may be correlated with two widely accepted standard tests of the American Society for Testing and Materials (ASTM) (99, 100).

The ASTM test for the deflection temperature under load (DTUL) (103) defines the temperature at which a sample beam of specified dimensions and load has a given deflection. It is therefore the temperature at which the sample has a specific Young's modulus value. Given the specified ASTM loading and modulus value, the penetration depth of the TMA probe can be calculated from Equation (5.26). The temperature at which the probe penetrates to this depth in the heat programmed sample corresponds to the DTUL. The ASTM's VICAT softening temperature is that at which a probe of 1 mm diameter penetrates 1 mm into a 12.7 cm thick sample under a 1 kg load (104). From Equation (5.26) this penetration corresponds to Young's modulus of 66 kg/cm^2. As with the DTUL measurement, substitution of the Young's modulus and TMA parameters back into Equation (5.26) enables calculation of penetration depth corresponding to the VICAT softening temperature. This temperature is then determined from a heat programmed run. Yanai et al. (99) have indicated good agreement between TMA and ASTM results on various PVC samples. The advantages of the TMA technique for modulus measurements lie in the much smaller sample requirements and shorter test times.

TMA penetration measurements were used by Prest and Porter (105) to observe viscoelastic properties of polypropylene oxide (PPO)-polystyrene (PS) blends. The thermal mechanical response of the polymers and a blend are shown in Figure 5.10. The polymers exhibit only expansion until the T_g where penetration begins. The plateau after the initial drop is a function of chain entanglements and the molecular weight. The degree of penetration to the plateau level is related to an entanglement modulus and is inversely proportional to the molecular weight between entanglements. The temperature range of the plateau is related to the number of entanglements and is a function of the molecular weight. In Figure 5.10 the dashed line shows that the higher molecular weight polystyrene (41,000 versus 97,200 mol. wt.) lengthens the plateau. The smaller degree of penetration to the plateau level of the blend and PPO sample shows that PPO is a highly entangled polymer and that this effect is dominant in PPO-PS blends. Commercial accessories are available for TMA (89) that provide data on thermal deformation of fibers and stress relaxation of fibers and films of polymers. The fiber tension unit measures the force generated by the fiber at constant elongation with changing temperature. For some fibers and films a modified fiber tension accessory may be used to observe stress decay of a polymer sample as a function

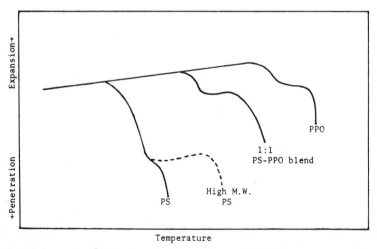

Figure 5.10 Thermomechanical response of polypropylene oxide (PPO), polystyrene (PS), and a blend (105).

of time. Elastic moduli might be calculated for various times at a given temperature from Equation (5.28)

$$E = \frac{(W - KS_t)gL}{AS_t} \qquad (5.28)$$

where the sample parameters are weight W in grams, L length in mils, and A the cross-sectional area in square centimeters. S_t is the displacement at time t in mils, K is the force gradient of the load cell in grams per mil, and g is the gravitational constant 980 cm/sec (106). A series of similar studies was carried out by Buchanan and Hardegree (107) on the stress relaxation of spun, drawn, and textured yarns of polyester, polypropylene, and nylon 66 (107). The results show that the stress temperature fingerprint of the yarn can be significantly affected by orientation, aging, drawing and texturing temperature, and dyeing conditions.

5.4.3   Thermal and Electrical Properties

Other physical polymer properties that may be studied by thermal methods include specific heat, electrical resistance, and dielectric behavior. Specific heat measurements are made on a DSC and involve the difference in thermal lag between sample and blank programmed runs. Figure 5.11 shows blank and sample curves and the measurements necessary for determining specific heat. The thermal lag of the sample, corrected for the lag of the container, is proportional to the sample's specific heat at the temperature measured, see Equation (5.29).

$$C_P = \frac{(\text{sample lag, cal/sec})}{(\text{weight, g}) (\text{heating rate, } ^\circ C/\text{sec})}$$

$$= \frac{(\text{cal})}{(\text{g}) (^\circ C)} \qquad (5.29)$$

Electric thermal techniques are those that measure resistivity or dielectric properties as a function of temperature. Although in some cases these measurements have been shown to correlate with degradation, plasticizer properties, transitions, and structure of polymers, they have not been widely used. A review of

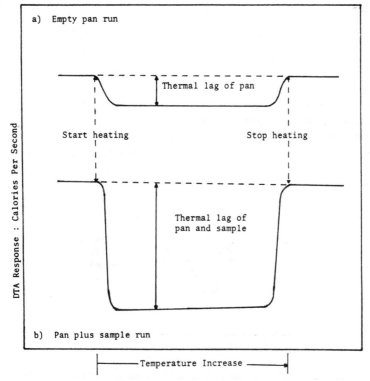

Figure 5.11   DTA Curves for Determining Specific Heat.

the work in this area and descriptions of the apparatus are given by Chiu (108).

5.5 POLYMER COMPOSITION

Thermal techniques for identifying and determining the composition of polymers are convenient. Analyses can often be carried out rapidly by more than one thermal method and on small sample sizes.

5.5.1 Qualitative Analysis

A number of thermal properties can be used to identify a polymer or its constituents. Programmed DSC runs for example, may show glass transitions, melting peaks, or crystallization peaks. The presence and the temperature of one or more of these events may identify a polymer, or at least eliminate some possibilities. Data have been published on the glass transition temperature (109, 110) and melt temperatures (110, 111) for various polymers. The values in published tables must be considered approximations since transition temperatures vary with the rate of heating, thermal history, and polymer properties such as molecular weight and composition. Figure 5.12 shows how a physical mixture of polymers may be separated and tentatively identified on the basis of melting peaks (112). It is also possible to analyze a polymer by passing its thermal degradation products to a gas chromatograph (113). Polystyrene, for example, is a particularly easy polymer to identify by this

technique. Its breakdown products consist of normal butane, benzene, toluene, and styrene—all easily separated and characterized by retention time. Most degradation patterns are more complex but can serve as a fingerprint for comparative purposes. Identification of homopolymers or copolymers may also be accomplished by examination of thermal decomposition products by a mass spectrometer. Although not specifically designed for polymer identification, a DSC-mass spectrograph (ms) technique described by Dugan et al. (114) should work. A small sample is hermetically sealed in a metal capsule thermally treated in a DSC and then transferred to an ms sampling device. The volatile fragments from the polymer sample are released into the ms inlet by puncturing the metal capsule with a needle. Chiu et al. (115) have described a TEA-ms technique. Evolved gases from a program-heated polymer are sensed by the TEA, condensed in traps representing different temperature ranges, and separately analyzed by the ms. A thermal pyrolysis, molecular weight chromatograph, vapor-phase infrared spectrometer combination has been used to examine the decomposition gas of polysulfones (116). This and other combination systems should be highly useful in qualitatively analyzing polymer systems.

5.5.2 Quantitative Analysis

The quantitative analysis of a polymer system is essential for all phases of polymer develop-

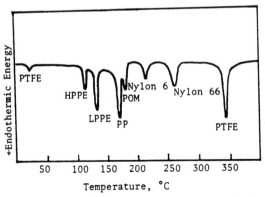

Figure 5.12 DTA melt curve of polymer mixture: High-pressure polyethylene (HPPE), low-pressure polyethylene (LPPE), polypropylene (PP), polyoxymethylene, (POM), Nylon 6, Nylon 66, and polytetrafluoroethylene (PTFE) (112).

ment from the research stage to production quality control. Analyses are carried out to determine the level of a polymer in formulations as well as the amounts of stabilizer, diluent, fillers, pigment, and other additives. There are a number of techniques that may be used to estimate the level of a particular polymer in a multicomponent system. If the fusion or crystallization peaks of a polymer are resolved and its heat of transition is known, this parameter is related to the amount of polymer present. Errors can occur in measuring transition heat if other constituents change the normal degree of crystallinity. For example, the presence of viscous polyterpene in polypropylene may inhibit crystallization of the latter. On the other hand, a viscosity-lowering plasticizer or nucleating agent may actually enhance the degree of crystallinity above a slowly crystallizing polymer's normal level.

The change in a T_g of a noncrystalline polymer is related to its concentration in a multicomponent system, as shown previously, in Equations (5.17) and (5.18). If the volatility or thermal or oxidative stability of a polymer differs from other components it may be determined by an appropriate DSC, TG, or TEA procedure. For example, a few hundredths of 1% polypropylene extracted by a diluent may be determined by volatilizing the diluent and degrading the polymer in the sensitive TEA instrument. Alternately, the diluent may be volatilized in a DSC under nitrogen and the polymer subsequently heated in oxygen. The high combustion heat gives this approach high sensitivity.

Similar analyses can be made to determine the level of nonpolymer materials in a polymer formulation. Many materials including nucleators, pigments, and fillers can significantly influence the crystallization rate according to their concentration in the polymer (117). Additives that plasticize a polymer generally lower the polymer's crystalline melting point or, for noncrystalline polymers, the T_g depending upon the level of plasticizer. Inorganic fillers and impurities may be easily analyzed for by thermogravimetry. Figure 5.13 illustrates how a carbon-filled polyolefin might be

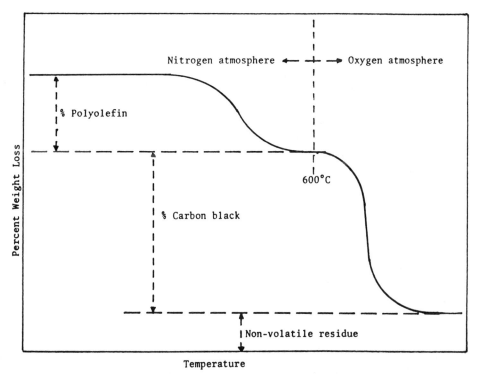

Figure 5.13 Thermogravimetric curve of carbon-filled polyolefin.

analyzed for polymer content, percent carbon black, and residual ash in one run. Antioxidant level may be determined by the length of time a polymer is protected in an oxygen environment at some high-test temperature. When the stabilizer is expended, the heat of oxidation is easily detected by a DSC. The concentration of volatile additives as well as the rate of loss from the polymer may be determined by TEA (118). The sensitive flame ionization detector measures the amount of organic material volatilizing from the sample at a chosen ambient temperature. Using standards with known additive levels, we can directly relate the signal response to the additive concentration in the polymer.

5.6 POLYMER REACTIONS

Reactions of polymers including polymerization, curing, and decomposition can be very complex. Simplified procedures for estimating relative polymer stability, such as determining the temperature for a given weight loss, are not usually satisfactory. Since polymers can and do exhibit multiple reaction steps with varying heats or weight losses, straightforward stability comparisons are sometimes not possible. Attempts to characterize polymer reactions generally center on the obtaining of the Arrhenius kinetic constants that predict the rates of reaction as a function of temperature. A considerable quantity of literature exists on the theory and practice of obtaining kinetic constants for polymeric and nonpolymer reactions by thermal techniques. The diversity and complexity of the many approaches are considerable and overwhelming for the uninitiated. A thorough discussion and evaluation of the many methods, good and bad, proposed in the literature is beyond the scope of this work. An excellent discussion of the methods and principles in this area has been written by J. H. Flynn (119, 120). In general, isothermal techniques are well accepted because they are simple in concept and interpretation. They are also considerably more tedious and time-consuming than temperature programmed methods. Both isothermal and programmed

techniques normally assume that two relationships are valid; see Equations (5.30) and (5.31).

$$k = Z \exp\left(-E/RT\right) \qquad (5.30)$$

$$k = \frac{(dC/dT)}{(1 - C)^n} \qquad (5.31)$$

where k is the specific reaction rate, E Arrhenius activation energy, Z the preexponential factor, C the fraction of conversion, t the reaction time, and n is called the reaction order. Equation (5.30), the Arrhenius equation, is usually a valid measure of the temperature dependence of the reaction rate. The value for E may in fact represent the average effect of several simultaneous rate steps in a complex reaction and is sometimes referred to as a global activation energy. In Equation (5.31) the validity of the term $(1 - C)^n$ as the proper function of the degree of conversion $f(C)$ is a riskier assumption. This is especially so for complex polymer reactions. In isothermal methods the data obtained enables a test of the $f(C)$ relationship as discussed below.

For isothermal thermogravimetry, Equation (5.31) may be written in terms of weight loss as follows in Equation (5.32):

$$k = \frac{dC/dt}{(1 - C)^n} = \frac{(dW/dt)/(W_o - W_f)}{[(W - W_f)/(W_o - W_f)]^n}$$
$$(5.32)$$

where W is the sample weight W_0 and W_f are the initial and final weights, respectively, for the reaction being studied. If Equation (5.32) is valid, a straight-line plot $(1 - C)^n$ versus dC/dT will be obtained for the correct value of n, and k may be calculated from the slope. Rate constants determined for more than one temperature then allow calculations of the Arrhenius activation energy from Equation (5.33) as follows:

$$\frac{d \ln k}{d(1/T)} = E/R \qquad (5.33)$$

where R is the gas constant. The preexponential factor is then determined from Equation (5.30). If a straight-line plot cannot be obtained, then the $F(C) = (1 - C)^n$ relation is not a sufficient description of the process being measured. Rate constants may be approximated in such cases by measuring the weight change close to the initial weight loss or extrapolating the change to zero loss. At this point the conversion function is minimized and Equation (5.34) can be used.

$$k \cong \frac{dC}{dt} \cong \frac{dW/dt}{W_o - W_f} \qquad (5.34)$$

Isothermal kinetic data by DSC are usually obtained by a stepwise procedure. Samples are aged for different times at constant temperature. The extent of reaction as a function of aging time is then determined by the amount of reaction heat observed in a subsequent program DSC run. In terms of thermal measurements Equation (5.31) becomes Equation (5.35) as follows:

$$k = \frac{dC/dt}{(1 - C)^n} = \frac{d[(\Delta H_0 - \Delta H)/\Delta H_0]/dt}{(\Delta H/\Delta H_0)^n}$$

$$(5.35)$$

where ΔH and ΔH_0 are the measured heats per unit weight for the aged and unaged polymer sample. As in the isothermal TG method, linear plots of dC/dT versus $(1 - C)^n$ will confirm the $f(C)$ relationship and enable calculation of the rate constants. The Arrhenius constants are obtained from Equations (5.33) and (5.30).

In recent years temperature program techniques have been frequently used because of the greater speed of such methods and the need of fewer samples. Currently, one of the most popular techniques is that described by Freeman and Carroll (121). The derived expression is expressed as follows in Equation (5.36):

$$\Delta \ln \left(\frac{dC}{dt} \right) = n \ln (1 - C) - \frac{E}{R} \left(\Delta \right) \frac{1}{T}$$

$$(5.36)$$

Plotting $\Delta \ln (dC/dt)/\Delta(1/T)$ versus $\Delta \ln (1 - C)/\Delta(1/T)$ yields n as the slope and E/R as the intercept. The technique has value in that a single run is required and that changes may be observed in reaction order and activation energy with degree of conversion. However, as pointed out by Flynn (120), results from this and similar techniques depend upon the validity of the $f(C) = (1 - C)^n$ relation. Polymers that, for example, undergo degradation by a random scission process will not yield correct results by the Freeman and Carroll technique (122). This method should be applied to complex polymer reactions and degradation processes with caution and, if possible, with an isothermal check of the kinetic parameters.

Attractive alternates to the preceding nonisothermal method are those based upon the integral expression, Equations (5.37) as follows:

$$f(C) \cong \int_0^c \frac{dC}{(1 - C)} - \frac{Z}{\beta} \int_0^T \exp \left(\frac{-E}{RT} \right) dT$$

$$(5.37)$$

where β is a constant heating rate. Equation (5.37) may be approximated in its simplest form by Equation (5.38) (123)

$$F(C) \cong \log \left(\frac{ZE}{R} \right) - \log \beta$$

$$- 2.315 - 0.457 \frac{E}{RT} \qquad (5.38)$$

and in its differential form by Equation (5.39)

$$\frac{d \log \beta}{d (1/T)} \cong 0.457 \frac{E}{R} \qquad (5.39)$$

Thus a plot of the heating rate versus the heating temperature for a selected constant degree of conversion yields the value for the activation energy. An advantage of the method is that it does not depend upon simplifying assumptions concerning the decomposition mechanism. The problem is avoided simply by making measurements at constant conversion. The method can be applied to thermogravimetric curves (119, 124, 125) as well as those of DSC (126, 127). In practice, samples are programmed at different heating rates and the temperature at which a constant degree of conversion occurs recorded. For the intregal TG method, these can be read directly off the weight-loss curve. For DSC, the reaction peak maximum is generally considered to occur at a constant degree of conversion. Figure 5.14 shows typical TG curves representing decomposition runs at different heating rates. The "constant" of 0.457 in Equation (5.39) is actually slightly dependent upon the value of E. A reiteration calculation to refine the value of the activation energy may be carried out using published tables of the constant values (119, 128). Alternate calculation methods have been described (127, 128, 119).

A preexponential value can be estimated from the preceding heating rate measurements using Equation (5.40).

$$Z = \frac{\beta E \exp (E/RT)}{RT^2} \qquad (5.40)$$

This is a first-order form of a more general equation derived by Murray and White (129). If used with reactions other than first order the error is small compared to that resulting from small deviations in the calculated activation energy. For DSC the temperatures measured are those of the peak maximum. For TG the temperature is measured at the inflection point of the weight-loss curve. The complexity of polymerization and degradation reactions requires that quantitative calculations based on any one model of behavior be accepted with caution. Alternate methods for verifying conclusions based on the data should be investigated. At a minimum, kinetic data based upon nonisothermal methods should be checked by an isothermal test (127).

5.6.1 Polymerization and Curing

The kinetic constants of three reaction stages of a thermosetting resin were successfully determined by DSC by measuring peak maximum temperature at different heating

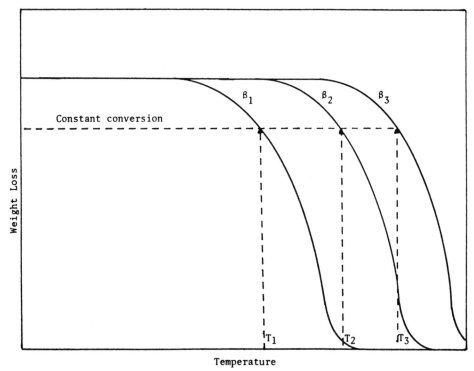

Figure 5.14 Effect of heating rate on the temperature for a constant degree of conversion where $\beta_1 < \beta_2 < \beta_3$ °C/min.

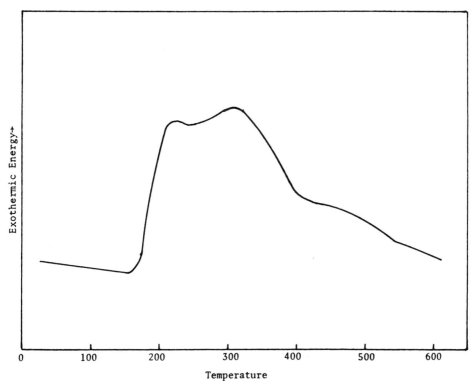

Figure 5.15 Complex DTA reaction curve for thermosetting resin.

rates (127). The agreement with subsequent thermal tests was good. The complexity of the thermosetting reaction and the severe overlapping of the DSC reaction peaks are shown in Figure 5.15. It is doubtful that other program techniques would have been successful.

Fava (130) has used both isothermal and programmed methods to characterize the curing rate of an amine catalyzed, epoxy-anhydride mixture. Sourour et al. (131) obtained kinetic data on the curing of Bisphenol A with m-phenylenediamine. An Arrhenius plot was made of the maximum rate of heat generated in an isothermal reaction versus the inverse cure temperature. A review by Era and Mattila (132) describes recent work on the resinification and curing of thermoset resins. According to Era, most activation energies of epoxy hardening reactions determined by DSC are consistently within the range of 58–71 kJ/mol obtained from gel point and electrical resistance measurements.

Heat measurements become difficult in the important last stages of epoxy polymeriza-

tion. However, cross-linking at this point has a relatively large effect on the T_g. A number of papers have described the relationship of the T_g to density, degree, timing, and temperature of cross-linking (130, 133, 134, 135, 136).

Other thermal techniques have been used to characterize the polymerization reaction. Studies have been made relating dilatometric readings to molecular weight and degree of cross-linking (137). Considerable recent interest has been generated in the study of photo-polymerization reactions using the DSC to monitor reaction rates, heats, and induction times (138, 139, 140, 141). The apparatus involves a rather easy modification of the DSC cell to emit ultraviolet light to the sample as shown in Figure 5.16. Studies to date have been concerned with the cross-linking of various acrylate formulations.

5.6.2 Polymer Degradation and Oxidation

Polymer degradation studies by thermal methods have been discussed in the literature (119, 120, 123, 124, 142, 143). TG methods

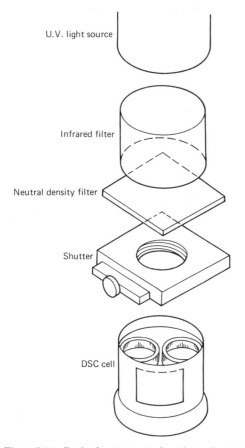

U.V. light source

Infrared filter

Neutral density filter

Shutter

DSC cell

Figure 5.16 Device for photo calorimetric studies (141).

using multiple runs at different heating rates appear to be preferred (120, 124). DSC methods sometimes show considerably more events taking place, which although informative, complicates the determination of kinetic data. Ozawa (124) has attempted to describe more realistically the degree of conversion function based upon a random depolymerization process. However, the degradation of most polymers involve more than this simple mechanism, and the utility of this approach is not clear. It seems hard to argue with Flynn's position that several different techniques, which measure various properties and test the influence of several reaction variables, may be necessary to characterize properly a given degradation process (120).

Thermal kinetic methods can be applied to the decomposition rates of polymerization initiators (144). If the reaction is first order and straightforward, such as that of benzoyl-

peroxide, most thermal techniques are applicable, including the DSC modification of the Borchardt and Daniels original technique (145).

A stabilization of polymers against oxidation is achieved through the addition of antioxidants. It is important to be able to compare the effectiveness of antioxidants, measure their longevity, and determine the influence of concentration on both. Historically, the effectiveness of antioxidants in a polymer formulation has been measured by oven aging at temperatures below the melting point. The development of color, brittleness, or rapid increase in the rate of oxygen uptake are indicators for the depletion of stabilizing effectiveness. By aging at various temperatures, data obtained were extrapolated to ambient conditions in order to estimate a polymer's useful lifetime. The oven test is lengthy and lacking in good precision. It was hoped that a high-temperature DSC test might shorten the test period and produce reliable data. The results have been mixed. The DTA tests carried out at high temperature on molten polymer do not duplicate actual use conditions, and results sometimes do not agree with oven aging predictions or rankings. However, the test has proved useful in quality control applications (146, 147). It is used to insure that the proper level of stabilizer has been added to the raw product and that extrusion, molding, or other heat processing has not severely depleted the stabilizer's effectiveness. It can also be used to screen the effect of other additives and metals on a stabilizer's performance, but care must be taken in extrapolating conclusions to lower temperatures. Variations of the DTA oxidative stable stability test have been described in the literature (147, 148, 149). The test involves (*a*) exposing the polymer sample at some high temperature to oxygen until the stabilizer is depleted and an oxidation exotherm is detected (the induction time necessary for stabilizer depletion is a measure of its protective ability; (*b*) temperature programming the polymer in air or oxygen under ambient or high pressure until an exotherm is initiated. The temperature at which the polymer starts to oxidize is a measure of its

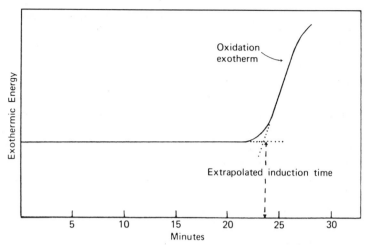

Figure 5.17 DTA curve for isothermal oxidative stability test.

stability. Figure 5.17 shows a thermogram of a typical polymer test by method (*a*). A test based on method (*a*) has been worked out by Western Electric (150) and is one of the most widely used tests for polymer stability.

An interesting apparatus for testing polymer stability has been described by Stapfer (151). Strips of polymer are moved from a heated, flat oven at a slow, constant speed. Gradual color changes of the polymer strips are observed as it emerges from the oven. The degree and rate of change correlates with polymer stability or the effectiveness of a given stabilizer. Spectroscopic measurements have also been made on strips to observe the formation of carbonyl groups. The technique has been applied to polyvinyl chloride, polyolefins, and chlorinated rubbers. Stapfer points out that unlike the lengthy and unreliable oven tests, his results are rigorously reproducible and are typically in the 30 min to 1 hr range. This technique appears to be an attractive compromise between the long oven aging test and the accelerated DTA treatment.

DTA has been used to determine the degree of substitution of cellulose derivatives (152). Exothermic or endothermic peaks were observed in the degradation of cellulose derivatives associated with the cleaving of the substituent groups. Linear relations between the peak height and area and degree of substitution were found for cyanoethyl and trityl cellulose. DTA results agreed well with those of wet chemical methods.

5.6.3 Thermal Runaway Reactions and Hazards

In common with nonpolymers, if the rate of heat generated by reaction (polymerization, decomposition) exceeds that escaping to the environment, temperature and reaction rate will accelerate in an uncontrollable manner. Depending upon the availablity of energy, equipment damage, charred polymer, fire, and explosion are possible. For example, nitrocellulose-based propellant, polyacrylonitrile, and many thermosets may undergo exothermic reactions that can become runaways if improperly carried out. An unstirred exothermic reaction taking place in an nonadiabatic container will always produce a temperature gradient. The temperature will be closest to ambient at the container wall and increase toward the center to a degree depending upon the material's thermal conductivity and the heat and rate of reaction. If ambient temperature rises, the average reaction rate will increase exponentially according to the Arrhenius relationship. At some ambient temperature the heat generated at the center will be greater than that dissipated through the sample bulk, and the reaction will accelerate, releasing all the remaining energy. To prevent a runaway condition, the critical perimeters of the reaction must be known. An equation relating these perimeters has been described by Frank-Kamenetskii (153) and discussed by other (154, 155). See Equation (5.41).

$$\delta = \frac{QZEr^2C_0}{\lambda RT^2} \cdot \exp\left(\frac{-E}{RT}\right) \quad (5.41)$$

where δ is a dimensionless parameter (depending upon geometry), Q the reaction heat, Z and E the Arrhenius kinetic constants, r the radius or half thickness, C_0 the concentration or density of reactant, and λ the thermal conductivity of the reacting material. If the planned parameters of sample bulk, ambient temperature, and concentration result in a calculated δ higher than its critical value, the equation predicts a runaway reaction. The critical values for δ for the slab, cylinder, or spherical configurations are 0.88, 2.00, and 3.32, respectively.

Thermally accelerating reactions are hazardous if the reaction has sufficient energy to cause the ignition of gases or condensed phase matter. Although each case will be different, some estimates of potential hazard can be made by estimating the temperature rise. Thus, if reaction heat is measured at 30 cal/g for a system with the heat capacity of about 0.3 cal/deg, the total potential rise in temperature is 90°C—enough to volatilize some solvents but fairly safe in terms of ignition possibilities. A heat of reaction of 100 cal/g in the same system could result in a maximum instant temperature rise approaching 300°C, which might cause a pressure rise, trigger off a secondary reaction, and so on. The hazardous consequences increase with increasing available reaction heat. Estimates of available heat, heat capacity, and the Arrhenius constants may be made by DSC data, as discussed previously. Table 5.3 lists critical temperatures for the runaway reaction of a sample of polyacrylonitrile fibers in air.

Miller et al. (156) have described a test for the autoignition of polymers in a hot air environment. Ignition takes place when combustible degradation products reach a critical concentration in the air environment around the sample. Samples in fabric form are wrapped around a wire screen. Bulk samples are impaled on a ceramic prong. A test sample is lowered into a hot air furnace and its temperature is measured by a thermocouple.

Table 5.3   Frank-Kamenetskii Calculations for Poly(acrylonitrile)

Parameters:

E = 27 kcal/mole
Z = 9.5 log (1/min)
Q = 135 cal/g
C = 1.18 g/cc
λ = 0.0003 cal/(sec-cm-deg)

Critical Radius, cm	Critical Temperature, °C		
	(Slab)	(Cylinder)	(Sphere)
1	207	223	233
3	170	183	192
5	155	167	175
7	145	157	165
9	139	150	157
11	133	144	151
13	129	140	147
15	125	136	143
17	122	133	139
19	120	130	136
21	117	127	134
23	115	125	131
25	113	123	129
27	111	121	127
29	110	119	126
31	108	118	124
33	107	116	122
35	105	115	121

The "stability time" from the initial heatup to the sudden temperature rise at the point of ignition, is recorded as a measure of polymer stability. Stability times are given in terms of an extrapolated zero mass. The preceding test can yield useful information in ranking the ignition potential of the gaseous decomposition products of some plastics. However, the results may be misleading. Polymers that decompose exothermically in the absence of air, or combustible polymers with high surface areas in air (small particles or fine denier) can self-heat. In sufficient bulk such materials may ignite at temperatures significantly below those predicted by the preceding test.

5.7 THERMAL TECHNIQUES AND INSTRUMENTS

The operational principle of thermal techniques is simple. A controlled amount of heat is applied to a sample, and its effect is recorded as a function of time (isothermal operation) or temperature (programmed temperature operation). Most commercial instruments allow analyses to be made above or below ambient pressure and in various gas environments. Wendlandt (157) and Daniels (158) have written good review books on the principles and applications of thermal analysis methods. The descriptions that follow are thumbnail sketches of the major techniques discussed in this chapter.

5.7.1 Differential Thermal Analysis and Differential Scanning Calorimetry

Differential scanning calorimetry (DSC) and Differential thermal analysis (DTA) are very similar techniques, and the terms are used interchangeably in this chapter. Both measure changes in heat capacity and enthalpy of a sample. The techniques have been used to measure the temperatures and heats of fusion, recrystallization, crystal-crystal transition, volatilization, and reaction. Studies have been made of reaction and crystallization rates, T_g's, and specific heat. Figure 5.18 is an idealized thermogram of quenched, poly(ethylene terephthalate) showing some of the events observable by DTA and DSC.

A DSC operates by compensating electrically for a change in sample heat. Power is distributed to the resistance heaters of the sample and reference holders so as to keep their temperatures the same. The amount of electrical energy neded to compensate for sample events is recorded as a measure of the energy change of the event. The Y-axis units are expressed as calories per second.

DTA instruments operate by simply measuring a change in sample temperature with respect to an inert reference. As used in this chapter, DTA refers to quantitative or "calorimetric" DTA. Early DTA instruments were designed with thermocouple sensors placed within the sample. In this arrangement, the heat path to the environment varied according to packing density, particle shape, and other factors, rendering the technique, at best, semiquantitative. The more modern design, pioneered by Boersma (159), is based upon an externally attached thermocouple with a reproducible heat path. This arrangement improves calorimetric precision considerably, and results are comparable with those of DSC. The temperature measured is that of the

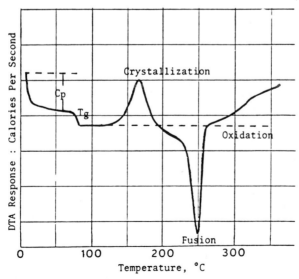

Figure 5.18 DTA curve of quenched poly (ethylene terephalate).

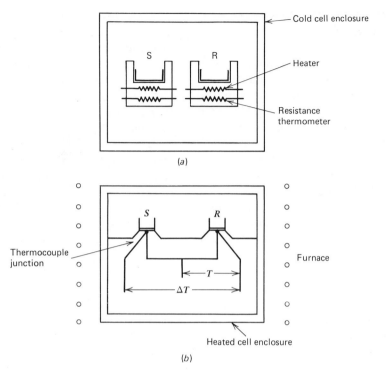

Figure 5.19   Design of DSC and calorimetric DTA cells. (*a*) DSC cell; (*b*) calorimetric DTA cell.

sample holders, and the effects of the heat capacity of the sample are insignificant. The instrument may be calibrated in terms of the heat release of known standards in calories per second, as with the DSC arrangement. Figure 5.19 shows a schematic version of current DSC and DTA commercial designs.

5.7.2   Thermogravimetry

Thermogravimetry (TG) measures the change in mass of a sample due to volatilization, reaction, or absorption from the gas phase. Studies on the polymers have dealt with the amounts and loss of moisture or diluent, and the rates and temperatures of reactions. The instruments consist of a thermally isolated electrobalance and a suspension system that positions the sample in a programmable oven. Commercial instruments generally can electrically suppress sample weight so that an initial, small weight change produces a large recorder signal. Sample size varies with instrument design but is usually less than a gram. The arrangement of a current thermogravimetric analyzer is shown in Figure 5.20.

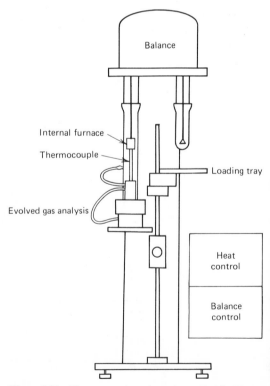

Figure 5.20   Thermogravimetric analyzer, Perkin-Elmer TGS-2.

5.7.3 Thermomechanical Analysis

A thermomechanical analyzer (TMA) operates by a very simple principle. A probe in contact with a sample is moved in a vertical direction as the sample experiences a transitional, dimensional, or spacial change. The vertical movement of the probe is sensed by a linear variable differential transformer (LVDT) and translated as a millivolt signal to the Y-axis of a recorder. Figure 5.21 shows a schematic diagram of a thermomechanical analyzer. A variety of studies can be made on a polymer depending on the configuration of the sample and probe. The types of studies can be divided into four categories: compression, penetration, extension, and flexure. The probe and sample arrangements for TMA studies are shown in Figure 5.22.

In compression mode flat probes or plates in contact with the sample are used to observe expansion behavior or measure viscosity depending on the temperature range. The fine-tipped or pointed probes measure penetration

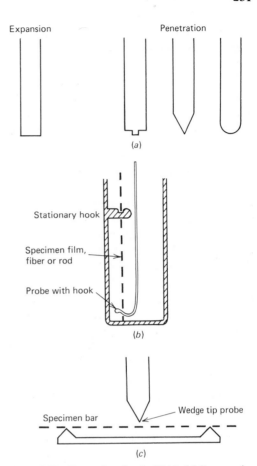

Figure 5.22 Types of probes for TMA. (*a*) Compression mode. (*b*) Extension mode. (*c*) Flexure mode.

under load as a polymer goes through a glass transition or softens and flows at higher temperatures. In the extension mode fibers, films, or rods of polymer can be held in tension under constant stress. Shrinkage or strain can then be measured with changing temperature. An accessory is commercially available to measure sample stress at constant elongation (160). In the flexure mode a sample strip is supported on its ends by two knife edges. The strip is loaded in the middle by the probe. The deflection of the strip under load is related to the elastic modulus.

5.7.4 Thermal Evolution Analysis

A thermal evolution analyzer (TEA) consists of a sample holder in a programmable oven and a flame ionization detector. The detector can sense small amounts of organic carbon

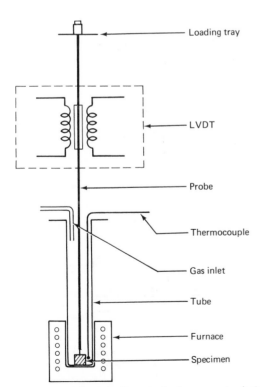

Figure 5.21 Schematic diagram of a thermomechanical analyzer, Du Pont 942.

liberated from the sample. The technique can study volatilization rate of stabilizers from polymers, determine vapor pressure of additivies, and estimate low amounts of extracted polymer in various solvents. The detector is insensitive to water vapor or inorganic gases, which can be an advantage for identification purposes.

5.7.5 Thermal Optical Analysis

Crystalline polymers will convert plane polarized light into elliptically polarized light because of refractive index differences along crystal axes. The intensity of this depolarized light transmitted through a sample is a function of the level of crystallinity. Melting and recrystallization phenomena are observed as intensity changes in the light reaching a photoconductive cell (161). The apparatus requires a polarizer, sample holder, analyzer, and detector as shown in Figure 5.23. The technique is reported to be as accurate and rapid as DTA and, at times, more sensitive to premelt

phenomena (161). It does not appear to be sensitive to glass transitions (162). Thermal optical analysis (TOA) is also referred to as thermal depolarization analysis (TDA) and the depolarized light intensity method (DLI).

5.7.6 Torsional Braid Analysis

Torsional braid analysis (TBA), pioneered and developed by Gillham (28), is an improved variant of the older torsion pendulum methods. A fiberglass braid is soaked in a solution of the polymer, the solvent removed, and the resulting composite subjected to free torsional oscillation. A schematic diagram of the apparatus is shown in Figure 5.24. The natural frequency and decay of the oscillations

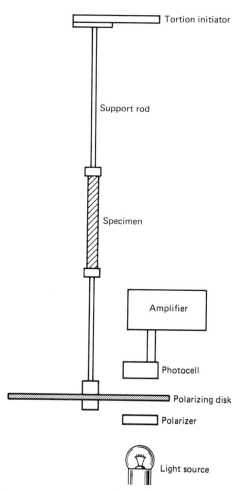

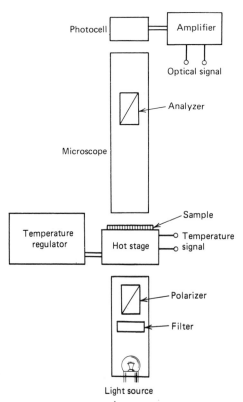

Figure 5.23 Apparatus for thermal optical analysis.

Figure 5.24 Arrangement for tortional braid or tortional pendulum apparatus.

are measured and interpreted in terms of polymer behavior. The shear modulus G of the composite is related to the oscillating period p by Equation (5.42) as follows:

$$G = \frac{8\pi LI}{r^4 p^2} \qquad (5.42)$$

where L is the length, r the radius, and I the moment of inertia of the specimen. The damping factor Δ is related to the oscillation peak amplitude by Equation (5.43)

$$\Delta = \frac{1}{n} \ln \left(\frac{A_0}{A_n} \right) \qquad (5.43)$$

where A_0 is some reference peak amplitude and A_n is the peak amplitude n cycles later. These values for modulus and damping are those of the composite and are generally used as relative estimates of the true polymer value. Plots of loss modulus versus temperature show peaks associated with the onset of motion of internal structural elements, for example, side groups and extended segments of the polymer chain. Glassy state relaxations, glass transitions, crystal-crystal transitions,

and melt transitions may be observed. Vitrification and gelation studies have been made on thermosets (163). Since experiments may require several days and generate many hundreds of damped sine waves, reduction of data by computer is almost essential.

5.7.7 Dynamic Mechanical Analysis

The dynamic mechanical technique stresses a thin film or fiber in forced oscillation. Current instruments take two approaches to obtain dynamic information. The Rheovibron (164) measures the extent that stress is out of phase with strain at constant frequency. The phase lag $\tan \delta$ is directly measured by the instrument. The damping term, $\tan \delta$, is plotted as a function of temperature. The stress-strain data can be used to calculate tensile modulus. The Du Pont DMA (165) measures the resonant frequency and mechanical damping as a function of temperature. The resonant frequency is related to sample modulus. DMA instruments have proved useful in studies involving low-temperature glass transitions, modulus, crystallinity, and effects of additives. Figure 5.25 shows schematically the two instruments discussed.

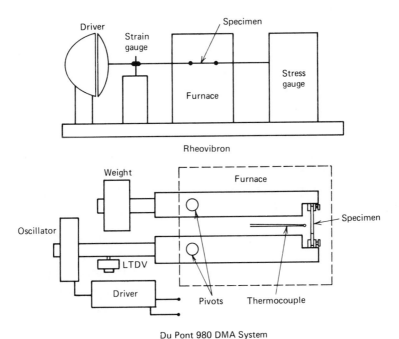

Du Pont 980 DMA System

Figure 5.25 Dynamic mechanical analyzers.

5.8 SUMMARY

Temperature and heat are important functions in all phases of polymer technology. They influence polymerization and processing steps and affect the physical, structural, and chemical properties of polymers. Accordingly, thermal methods of analysis have broad application in characterizing these materials. Various thermal techniques have been used to study crystalline and noncrystalline fractions, physical properties such as dimensional change and visco-elasticity, and chemical behavior such as curing and degradation. Many other characteristics have been examined.

The advantages of thermal techniques are generally speed, sensitivity, and good precision of results. Small sample size requirements can be an advantage if little material is available or if nonuniform distribution is being studied. It can be a disadvange in obtaining single results representative of a large lot of nonuniform material. Future instrumental progress lies in the area of automated data processing, guidance, and control of experimental parameters. Some plant use of thermal methods for quality control is evident. An example is the use of oxidative stability tests of polymers by the cable and automotive industries. The trend would increase with instruments specially designed for reliable, automatic operation.

5.9 GLOSSARY OF SYMBOLS

a_T	Ratio of electrical to mechanical relaxation time
A	Area, amplitude
B	Interaction energy density, a constant
C	Conversion, a constant
C_p	Heat capacity at constant pressure
$C(t)$	Creep function at specific heat
D	Depth, a constant
E	Young's modulus, activation energy
F	Compressive force
g	Gravitational constant
G	Shear modulus
h, H	Thickness
ΔH_f	Observed heat of fusion
ΔH_μ	Heat of fusion of crystalline fraction
I	Moment of inertia
$J_2(t)$	Creep function of volumetric compliance
k	Specific reaction rate constant
K	Bulk modulus, a constant
L	Thickness
M	Molecular weight
n	An integer
p	Period of time
P	Pressure, load, probability
Q	Reaction heat
r	Radius, half-thickness
R	Gas constant, radius
S	Displacement
ΔS_μ	Entropy per chain unit
t	Time
T	Temperature
T_g	Glass transition temperature
T_μ	Liquid-liquid transition temperature
T_m, T_m°	Melt temperature, theoretical or extrapolated melt temperature
$T_{w\infty}$	T_g at infinite molecular weight
$T_{\beta,\gamma,\delta}$	Secondary glass transition temperatures
v_1	Volume fraction of diluent
V	Volume
V_1, u	Molar volume of polymer, diluent
W	Weight fraction, weight
x	Degree of crystallinity
X_c	Mole fraction of crystalline polymer
Y	Unit of length
Z	Arrhenius preexponential factor
α	Coefficient of expansion
$\alpha(t)$	Creep function of α
β	Heating rate, degrees per minute
γ	Shear rate
δ	Stress-strain phase difference, heat balance parameter
Δ	Damping factor
η	Viscosity
θ	Degree of crystallinity
λ	Thermal conductivity
ρ	Density
σ_e	Surface free energy

5.10 REFERENCES

1. L. E. Nielson, *Mechanical Properties of Polymers and Composites*, Vol. 1, Marcel Dekker, New York, 1974, pp. 19–21.

2. P. Ehrenfest, *Collected Scientific Papers*, North-Holland Publishing Company, Amsterdam, 1959; R. M. Christenson, *Tran. Soc. Rheol.*, **21**, 164 (1977).

3. R. M. Christensen, *Trans. Soc. Rheol.* **21**, 163 (1977).

4. R. G. C. Arridge, *Mechanics of Polymers*, Clarenden Press, Oxford, 1975, pp. 28, 36.

5. G. Gee, *Contemp. Phys.*, **11**, 313 (1970).

6. J. D. Ferry, *Viscoelastic Properties of Polymers*, 2nd. ed., John Wiley, New York, 1970.

7. M. L. Williams, R. F. Landel, and J. D. Ferry, *J. Am. Chem. Soc.*, **77**, 3701 (1955).

8. R. D. Deanin, *Polymers Structure, Properties and Applications*, Maple Press, York, Pennsylvania, 1972, pp. 88–93.

9. G. P. Koo, *Plast. Eng.* **30**, 33 (1974).

10. T. G. Fox, *Bull. Am. Phys. Soc.*, **1**, 123 (1956).

11. M. Gordon and J. S. Taylor, *J. Appl. Chem.*, **2**, 493 (1952).

12. L. A. Wood, *J. Polym. Sci.*, **28**, 319 (1958).

13. J. Ballesteros, G. J. Howard, and L. Teasdale, *J. Macromol. Sci.-Chem.*, **A11**, 39 (1977).

14. N. W. Johnson, *Appl. Polym. Symp.*, **25**, 19 (1974).

15. H. J. Harwood, N. W. Johnson, and H. Pitowski, *J. Polym. Sci.-C*, **23**, 25 (1968).

16. N. W. Johnson, *Polym. Prep., Am. Chem. Soc., Div. Polym. Chem.*, **14**, 46 (1973).

17. K. Marcincin and R. Romanov, *Polymer*, **16**, 173 (1975).

18. K. Horie, H. Hiura,, M. Sawada, I. Mitka, and H. Kambe, *J. Polym. Sci.*, **A-1**, 1357 (1970).

19. A. P. Gray, *Thermal Applications Studies*, **2**, Perkin-Elmer Corp., Norwalk, Connecticut, 1972.

20. L. E. Nielsen, *J. Macromol. Sci.—Rev. Macromol. Chem.*, **C3**, 69 (1969).

21. E. A. D. Marzio, *J. Res. Natl. Bur. Stand.*, **68A**, 611 (1964).

22. Y. Diamant, S. Welmer, and D. Katz, *Polymer*, **11**, 498 (1970).

23. L. E. Nielsen, Mechanical Properties of Polymers and Composites, Vol. 1, Marcel Dekker, New York, 1974, p. 219.

24. D. S. Verlag, *Kolloid Z.*, **148**, 36 (1956) through L. E. Nielsen, *Mechanical Properties of Polymers and Composites*, Vol. 1, Marcel Dekker, New York, 1974, p. 223.

25. L. E. Nielsen, *Mechanical Properties of Polymers and Composites*, Vol. 1, Marcel Dekker, New York, 1974 p. 222.

26. J. K. Gillham, *Polym. Eng. Sci.*, **16**, 353 (1976).

27. S. J. Stadnicki and J. K. Gillham, *J. Appl. Polym. Sci.*, **20**, 1245 (1976).

28. J. K. Gillham, *AIChE J.*, **20**, 1066 (1974).

29. R. A. Fava and C. E. Chaney, *J. Appl. Polym. Sci.*, **21**, 791 (1977).

30. L. E. Nielsen, *Mechanical Properties of Polymers and Composites*, Vol. 1, Marcel Dekker, New York, 1974, p. 140.

31. R. Kaneko, *Kobunshi Kagaku*, **26**, 288, 253 (1969).

32. A. R. Shultz and B. M. Beach, *J. Appl. Polym. Sci.*, **21**, 2035 (1977).

33. K. F. Baker, Applications Brief No. TA50, Du Pont Company, Instrument Products Division, Wilmington, Delaware.

34. N. Bekkerdahl, *J. Res. Natl. Bur. Stand.*, **43**, 145 (1949).

35. J. Chiu, *J. Macromol, Sci.-Chem.*, **A8**, 3 (1974).

36. P. L. Hsuing and D. M. Cates, *J. Appl. Polym. Sci.*, **19**, 3051 (1975).

37. J. G. Fatou, *Eur. Polym. J.* **1**, 1057 (1971).

38. L. Mandelkern, *J. Polym. Sci.*, Symposium No. 43, 1 (1973).

39. P. J. Flory, *Principles of Polymer Chemistry*, Cornell University Press, New York, 1953, p. 568.

40. L. Mandelkern, *Rubber Chem. Technol.*, **32**, 1932 (1959).

41. F. E. Karasz, J. M. O'Reilly, H. E. Bair, and R. A. Kloge, in R. S. Porter and J. F. Johnson, Eds., *Analytical Calorimetry* Vol. 1, Plenum Press, New York, 1968, p. 59.

42. P. E. Slade and T. A. Orofine, in *Analytical Calorimetry*, Vol. 1, Plenum Press, New York, 1968, p. 63.

43. J. A. Currie, in R. S. Porter and J. F. Johnson, Eds., *Analytical Calorimetry*, Vol. 3, Plenum Press, New York, 1974, p. 569.

44. J. R. Knox, *Analytical Calorimetry*, Vol. 3, Plenum Press, New York, 1974.

45. R. J. Samuels, *J. Polym. Sci.* **A-2**, 1197 (1969).

46. J. G. Fatou, *Eup. Polym. Sci.*, **7**, 1057 (1971).

47. W. W. Cox and A. A. Duswalt, *Polym. Eng. and Sci.*, **7**, 1 (1967).

48. R. J. Samuels, *J. Polym. Sci., Polym. Phys. ed.*, **13**, 1417 (1975).

49. B. W. Roberts, *Polymer*, **10**, 113 (1969).

50. C. W. Bunn and J. C. Alcock, *Trans. Faraday Soc.*, **41**, 317 (1945).

51. P. Meares, *Polymers Structure and Bulk Properties*, D. Van Nostrand Co. Ltd., New York, 1965, pp. 125–131.

52. F. W. Billmeyer, *Textbook of Polymer Sciences*, Wiley-Interscience, New York, 1962, pp. 162–165.

53. R. D. Deanin, *Polymers Structure, Properties and*

Applications, Maple Press, York, Pennsylvania, 1972, p. 218.

54. B. Wunderlick, *J. Therm. Anal.*, **5**, 131 (1973).

55. A. A. Duswalt, *Polymer Characterization: Interdisciplinary Approaches*, Plenum Press, New York, 1971, p. 147.

56. F. J. Padden and H. D. Keith, *J. Appl. Phys.*, **30** 1479 (1959).

57. G. Natta, *Soc. Plast. Eng. J.*, **15**, 368 (1959).

58. A. Turner-Jones, J. M. Aizelwood, and D. R. Beckett, *Macromol Chem.*, **75**, 134 (1964).

59. B. Ke, *Newer Methods of Polymer Characterization*, Interscience, New York, 1964, pp. 397–398.

60. D. J. David, *Techniques and Methods of Polymer Evaluation*, Vol. 1, Marcel Dekker, New York, 1966, pp. 81–82.

61. E. L. Lawton and D. M. Cates, in R. S. Porter and J. F. Johnson, Eds., *Analytical Calorimetry*, Vol. 1, Plenum Press, New York, 1968, pp. 89–97.

62. G. W. Miller, *Analytical Calorimetry*, Vol. 2, Plenum Press, New York, 1970, pp. 397–415.

63. W. Thompson, *Phil. Mag.*, **42**, 448 (1971).

64. B. Wunderlich, *J. Polym. Sci.*, Symposium No. 43, 32 (1973).

65. B. Wunderlich, *Polymer*, **5**, 125, 611 (1964).

66. B. Wunderlich, *Thermochim. Acta*, **4**, 25 (1972).

67. L. Mandelkern, *J. Polym. Sci.*, 494 (1960).

68. J. D. Hoffman and J. J. Weeks, *J. Res. Natl. Bur. Stand.*, **66A**, 13 (1962).

69. H. E. Bair, T. W. Huseby, and R. Salovey, in R. S. Porter and J. F. Johnson, Eds., *Analytical Calorimetry*, Vol. 1, Plenum Press, New York, 1968, pp. 31–40.

70. P. J. Flory and A. Vrij, *J. Am. Chem. Soc.*, **85**, 3548 (1963).

71. J. Brandrup and E. H. Immergut, *Polymer Handbook*, Interscience, New York, 1966, pp. 347–358.

72. A. S. Kenyon and I. O. Salyer, *J. Polym. Sci.*, **43**, 127 (1960).

73. P. J. Flory, *Trans. Faraday Soc.*, **51** 848 (1955).

74. I. Kamel, R. P. Kusy, and R. D. Cornliussen, *Macromolecules*, **6**, 53 (1973).

75. L. E. Nielsen, *Mechanical Properties of Polymers*, Reinhold, New York, 1962, pp. 56–57.

76. M. Avrami, *J. Chem. Phys.*, **9**, 177 (1941).

77. L. B. Morgan, *J. Appl. Chem.*, **4**, 160 (1964).

78. P. Meares, *Polymers Structure and Bulk Properties*, Van Nostrand, New York, 1965, p. 154.

79. J. H. Magill, *Polymer*, **3**, 43 (1962).

80. D. R. Beckett, *Plast. Rubber: Mater. Appl.*, **1**, 168 (1976).

81. J. A. Brydson, *Plastics Materials*, D. Van Nostrand, Princeton, N.J., 1966, p. 55.

82. A. A. Duswalt, *Soc. Plast. Eng.: Tech. Pap.*, **17**, 223 (1971).

83. R. D. Deanin, *Polymers Structure, Properties, and Applications*, Maple Press, York, Pennsylvania, 1972, pp. 218–219.

84. R. D. Ulrich, *J. Appl. Polym. Sci.*, **20**, 1095 (1976).

85. J. H. Griffith and B. G. Ranby, *J. Polym. Sci.*, **38**, 107 (1959).

86. J. R. Collier and E. Baer, *J. Appl. Polym. Sci.*, **10**, 1409 (1966).

87. J. H. Magill, *Polymer*, **2**, 221 (1961).

88. J. R. Knox in R. S. Porter and J. F. Johnson, Eds., *Analytical Calorimetry*, Vol. 1, Plenum Press, New York, 1968, pp. 45–50.

89. 943 Thermomechanical Analyzer (TMA), Product Bulletin, Du Pont Instrument Products Division, Wilmington, Delaware.

90. Instrument News, Vol. 19, No. 4, Vol. 20, No. 1, Perkin-Elmer Corp., Instrument Division, Norwalk, Connecticut.

91. F. Noel, CIC Symposium, Sarnia, Ontario, October 18–20, 1971.

92. J. L. Haberfeld, J. F. Johnson, and R. C. Gaskill, *Thermochem. Acta*, **18**, 171 (1977).

93. D. Machin and C. E. Rogers, *Polym. Eng. Sci.*, **10**, 300 (1970).

94. Colombo and E. H. Immergut, *J. Polym. Sci.-C*, **31**, 137 (1970).

95. G. J. Dienes, and H. F. Klenn, *J. Appl. Phys.*, **17**, 458 (1946).

96. L. Cessna, Hercules, Inc., communication.

97. C. H. Hwo and J. F. Johnson, *J. Appl. Sci.*, **18**, 1433 (1974).

98. E. F. Finkin, *Wear*, **19**, 227 (1972).

99. H. S. Yanai, W. J. Freund, and O. L. Carter, *Thermochim. Acta*, **4**, 199 (1972).

100. J. Creedon, Application Brief TA 37, Du Pont, Instrument Products Division, Wilmington, Delaware.

101. L. A. Wood, *J. Res. Natl. Bur. Stand.*, **77A**, 171 (1973).

102. C. Vervoloet, *Rev. Gen. Caoutch. Plast.*, **46**, 469 (1969).

103. ASTM Standards, Part 27, June 1969, D-648, ASTM, Philadelphia, Pennsylvania, p. 214.

104. ASTM Standards, Part 27, June 1969, D-1525, ASTM, Philadelphia, Pennsylvania, p. 527.

105. W. M. Prest and R. S. Porter, *J. Polym. Sci.*, **A-2**, 1639 (1972).

106. R. L. Hassel, Applications Brief TA57, Du Pont Instrument Products Division, Wilmington, Delaware.

107. D. R. Buchanan and G. L. Hardegree, *Text. Res. J.*, **47**, 732 (1977).

108. J. Chiu, *J. Macromol. Sci.*, **A8**, (1974).

109. W. J. Roff and J. R. Scott, Eds., *Fibers, Films, Plastics and Rubbers, Handbook of Common Polymers*, Butterworth, London, 1971.

110. J. Brandrup and E. H. Immergut, *Polymer Handbook*, 2nd Ed., Wiley-Interscience, New York, 1975.

111. L. E. Nielsen, *Mechanical Properties of Polymers*, Reinhold, New York, 1962, pp. 527–531.

112. J. Chiu, *Macromol. Sci.-Chem.*, **A8**, 3 (1974).

113. H. H. Willard, L. L. Meritt, and J. A. Dean, *Instrumental Methods of Analysis*, Van Nostrand, New York, 1965, p. 520.

114. G. Dugan, J. D. McCarty, and R. J. Friant, *Analytical Calorimetry*, Vol. 2, Plenum Press, New York, 1970, pp. 417–427.

115. J. Chiu and A. B. Beattie, *Thermochim. Acta*, **21**, 263 (1977).

116. E. Kiran and J. K. Gillham, *J. Appl. Polym. Sci.*, **21**, 1159 (1977).

117. A. A. Duswalt, *Soc. Plast. Eng.: Tech. Pap.*, **17**, 223 (1971).

118. P. S. Gill, Applications Brief TA 43, Du Pont Instrument Products Division, Wilmington, Delaware.

119. J. H. Flynn and L. A. Wall, *Polym. Lett.* **4**, 323 (1966).

120. J. H. Flynn, in H. H. G. Jellinek, Ed., *Aspects of Degradation and Stabilization of Polymers*, Elsevier, Amsterdam, 1978.

121. E. S. Freeman and Carroll, *J. Phys. Chem.*, **62**, 394 (1958).

122. G. C. Cameron and G. P. Kerr, *Vac. Microbal, Tech.*, **7**, 27 (1970).

123. C. D. Doyle, *Nature*, **207**, 290 (1965).

124. T. Ozawa, *Bull. Chem. Soc. Jap.*, **38**, 1881 (1965).

125. R. Auderbert and C. Aubineau, *Eur. Polym. J.*, **6**, 965 (1970).

126. T. Ozawa, *J. Therm. Anal.*, **2**. 301 (1970).

127. A. A. Duswalt, *Thermochim. Acta*, **8**, 57 (1974).

128. K. H. Baker, Application Brief TA 73, Du Pont Instrument Products Division, Wilmington, Delaware.

129. P. Murray and J. White, *Trans. Br. Ceram. Soc.*, **54**, 204 (1955).

130. R. A. Fava, *Polymer*, **9**, 137 (1968).

131. S. Sourour and M. R. Kamal, *Thermochim. Acta* **14**, 41 (1976).

132. V. A. Era and A. Matila, *J. Therm. Anal.*, **10**, 461 (1976).

133. K. Horie, H. Hivra, M. Sawada, L. Mitra, and H. Kambe, *J. Polym. Sci.*, **A-1**, 1357 (1970).

134. L. J. Gough and J. T. Smith, *J. Appl. Polym. Sci.*, **3**, 362 (1960).

135. B. Miller, *J. Appl. Polym. Sci.*, **10**, 217 (1966).

136. A. P. Gray, *Thermal Application Studies*, **2**, Perkin-Elmer Corp. Norwalk, Connecticut.

137. A. Adicoff and R. Yee, *J. Appl. Polym. Sci.*, **20**, 2473 (1976).

138. I. Mitsura, Y. Teramodo, and M. Yasutake, *J. Polym. Sci.: Polym. Chem. Ed.*, **16**, 1175 (1978).

139. F. R. Wright, *J. Polym. Sci.: Polym. Lett. Ed.*, **16**, 121 (1978).

140. J. E. Moore, S. H. Schroefer, A. R. Schultz, and L. D. Stang, "Ultraviolet Light Induced Reaction in Polymers," A.C.S. Symposium Series 25, ACS, Washington, D.C., 1976, p. 90.

141. F. R. Wright, 35th SPE ANTEC Meeting, Montreal, Quebec, April 25–28, 1977.

142. D. J. Toop, *IEEE Trans. Electr. Insul.*, **6**, 2 (1971).

143. D. J. Toop, *IEEE Trans. Electr. Insul.*, **7**, 25, 32 (1972).

144. K. E. J. Barrett, *J. Appl. Polym. Sci.*, **11**, 1617 (1967).

145. M. Uricheck, *Perkin-Elmer Instrum. News*, No. 2, 17 (1966).

146. D. I. Marshall, E. J. George, and J. M. Turnipseed, *Polym. Eng. Sci.*, **13** 415 (1973).

147. F. R. Wright, *Polym. Eng. Sci.*, **16**, 652 (1976).

148. J. B. Howard, *Polym. Eng. Sci.*, **13**, 429 (1973).

149. ASTM D20.17 Committee's "Proposed Standard Specification for Polyethylene Plastic Pipe and Fittings Materials," 1973.

150. Western Electric Manufacturers Standard, 17000, Section 1230, August 1971.

151. C. H. Stapfer, *Plast. Eng.*, **33**, 35 (1977).

152. P. K. Chattergee and R. F. Schwenker, Jr., *TAPPI*, **55**, 111 (1972).

153. D. A. Frank-Kamenetskii, *Diffusion and Heat Transfer in Chemical Kinetics*, Plenum Press, New York, 1969, pp. 344–347.

154. P. H. Thomas, *Trans. Faraday Soc.*, **54**, 60 (1958).

155. T. Kimbara and K. Akita, *Combust. Flame*, **4**, 173 (1960).

156. B. Miller, J. R. Martin, and C. H. Meiser, Jr., *J. Appl. Polym. Sci.*, **17**, 629 (1973).

157. W. W. Wendlandt, *Thermal Methods of Analysis*, John Wiley, New York, 1974.

158. T. Daniels, *Thermal Analysis*, Anchor Press, London, 1973.

159. S. L. Boersma, *J. Am. Ceram. Soc.*, **38**, 281 (1955).

160. 943 Thermomechanical Analyzer, Product Bulletin, Du Pont Instrument Products Division, Wilmington, Delaware.

161. E. M. Barrall, J. F. Johnson, and R. S. Porter, *Appl. Polym. Sym.*, **8**, 191 (1969).

162. G. W. Miller and R. S. Porter *Analytical Calorimetry*, Vol. 2, Plenum Press, New York, 1970, p. 407.

163. J. K. Gillham, *AIChE J.*, **20**, 1066 (1974).

164. G. P. Koo, *Plast. Eng.*, **30**, 33 (1974).

165. Du Pont Instruments Thermal Analysis Review: Dynamic Mechanical Analysis, Du Pont Instrument Products Division, Wilmington, Delaware.

## CHAPTER 6

Morphology

S. Y. HOBBS

General Electric Company
Corporate Research and Development
Schenectady, New York

6.1	Introduction	239
6.2	The Morphology of Crystalline Homopolymers	240
	6.2.1 Crystals from Dilute Solution	242
	6.2.2 Crystallization from the Melt	243
	6.2.3 Chain Folding	247
	6.2.4 Chain Extension under Pressure	248
	6.2.5 Annealing	250
	6.2.6 Epitaxial Crystallization and Crystallization under Shear	252
	6.2.7 Deformation and Drawing	254
6.3	The Morphology of Glassy Polymers	256
	6.3.1 Ordered Structures	258
6.4	The Morphology of Copolymers	261
	6.4.1 Amorphous Random Copolymers	262
	6.4.2 Crystalline Random Copolymers	262
	6.4.3 Amorphous Block Copolymers	264
	6.4.4 Crystalline Block Copolymers	269
	6.4.5 Graft-Block Copolymers	271
6.5	The Morphology of Polymer Blends	274
6.6	The Morphology of Processed Polymers	279
6.7	Polymer Microscopy	282
	6.7.1 Staining	283
	6.7.2 Decoration	284
	6.7.3 Etching and Extraction	284
6.8	Summary	287
6.9	Glossary of Symbols	289
6.10	References	289

6.1 INTRODUCTION

The discussion of polymer morphology presented in this chapter deals broadly with the microscopic and macroscopic structural features of high polymers. For the most part, these features originate directly from some form of molecular ordering although, in some cases, they may develop as a result of incompatibility and phase separation. In many instances the prevailing morphology depends strongly on variations in molecular architecture, stereoregularity, and preferred conformational states. More extensive treatments of this topic and related subjects that are not explicitly addressed herein may be found in references cited in the text.

The first detailed studies of polymer morphology were focused almost exclusively on crystalline or semicrystalline materials. Even before the long-chain nature of high polymers had gained universal acceptance or the concepts of lamellar crystallization had been developed, the ability of many polymers to crystallize in spherulitic habitats not too much different from those observed for low molecular weight compounds was recognized. The discovery of polymer single crystals in the late 1950s opened the door for a flood of experiments that, during the succeeding ten years, were to elucidate the structure of crystalline polymers in immense detail.

These research activities continue although the morphologies of glassy polymers, block copolymers, and polymer blends are receiving increased (and long overdue) attention. This shift has come about partly because of the development of more sophisticated techniques for probing the structure of the glassy state such as high-resolution solid-state nuclear magnetic resonance spectroscopy and neutron diffraction techniques, as well as from a realization that the behavior of semicrystalline polymers is in large measure a reflection of the behavior of their amorphous components. More subtle pressure, at least in the industrial environment, stems from a desire to make more effective use of morphological observations and the data base that has been developed on polymer structure to understand and improve the mechanical performance of currently available materials. The growing body of research on the morphology of polymer blends and processing/morphology/property interrelationships is a reflection of this interest.

This chapter is divided into six major sections:

6.2   The morphology of crystalline homopolymers

6.3   The morphology of glassy polymers

6.4   The morphology of block copolymers

6.5   The morphology of polymer blends

6.6   The morphology of processed polymers

6.7   Polymer microscopy

The morphology of crystalline polymers, which has been discussed at length in a number of review articles, is summarized here in a somewhat abbreviated form to give the reader some perspective on the immense literature on this subject. The sections on glassy polymers, copolymers, and blends provide a much needed reveiw of morphological studies in these important areas where the complete picture is still under development. In Section 6.6 the morphology of polymers processed by injection molding is discussed with reference to the effects of structure on material performance. The last section is included since it appears that microscopy, both optical and electron, will continue to play a critical role in the elucidation of polymer structure. Of particular interest here is the necessity of developing improved sample preparation and staining techniques in order to make maximum use of curent instrument capabilities. In the summary an attempt is made to identify some of the research areas where major emphasis and advances may be expected in the next several years.

6.2   THE MORPHOLOGY OF CRYSTALLINE HOMOPOLYMERS

The ability of synthetic homopolymers to undergo crystallization has been recognized since the original work of Staudinger in 1927 (1). Early x-ray studies revealed the presence of unit cells similar to those found in low molecular weight materials. The longest unit cell dimensions were found to be considerably shorter than the length of a typical chain, however, and it was soon appreciated that a single molecule must traverse many unit cells. This conclusion, together with the observation of substantial line broadening in wide-angle x-ray patterns and the presence of a relatively large fraction of amorphous material in most crystalline samples, led to the development of the fringed micelle model of polymer crystallization (2). In this model, which is shown schematically in Figure 6.1, crystallographic ordering of chain segments is proposed to take place over regions several hundreds of angstroms in length with any given chain participating in several crystal "micelles." No allowance is made for chain folding. The micelles themselves were thought to be randomly dispersed in the surrounding amorphous glass.

Although the fringed micelle model satisfactorily accounted for many of the microstructural features of bulk crystalline polymers, it did not adequately explain the development of the spherulitic superstructure that is a general feature of both polymeric and

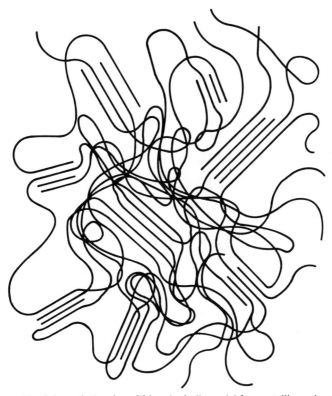

Figure 6.1   Schematic drawing of fringed micelle model for crystalline polymers.

nonpolymeric crystalline solids. In spherulites the chain axes lie perpendicular or nearly perpendicular to the spherulite radius and crystallization proceeds by rapid addition of chains in the lateral rather than the longitudinal direction. The fringed micelle model, in contrast, suggests that crystal growth most likely proceeds by an extension of order along the chain axis.

An alternative model, in which the high amorphous content of typical crystalline polymers was accounted for in a defect crystal or *paracrystalline* structure, was proposed by Hosemann (3). In this model the line broadening, characteristic of wide-angle x-ray diffraction patterns of semicrystalline polymers, was assumed to result from a defect lattice structure rather than from crystal micelles of very small size. The development of a more definitive description of the crystal morphology of high polymers was delayed until it became generally recognized that chain folding was a fundamental feature of the crystallization process.

As early as 1938 Storks suggested that thin films of gutta-percha crystallized in a chain-folded configuration (4). Evidence for the universality of this growth habit began to develop, however, only with the extensive investigations of solution-grown single crystals by several research groups in the 1950s (5–7). The first of these studies showed that it was possible to grow very thin (100 Å) plate-like lamellae by slowly cooling highly dilute solutions of polyethylene. Electron diffraction patterns obtained from these platelets revealed that the chain axes were oriented perpendicular to the lamellar basal planes and implied that some sort of chain folding must occur. Similar observations were soon made on other polymer systems (8–10). About the same time Claver et al. and others reported the presence of comparable lamellar structures in studies of the surface texture of bulk crystallized polyethylene and nylon specimens (10, 12). During the next 10 years the fine structure of crystalline homopolymers was elucidated in great detail through parallel studies of crystals grown from dilute solution and from the melt. Some of the most important of these studies are considered in the following sections. More extensive summaries have been presented in several excellent reviews (13–16).

6.2.1 Crystals from Dilute Solution

As noted, folded-chain lamellae are the basic morphological entities obtained when a wide variety of polymers are crystallized from dilute solution. At high dilutions ($< 0.1\%$) single isolated lamellae are formed. These crystals, which appear flat in the electron microscope, frequently grow as hollow pyramids in solution (17, 18) and collapse as the solvent is removed. Large pleats are often visible on the surfaces of such lamellae as a result of this collapse (see Figure 6.2). The pyramidal structures are postulated to arise from a regular displacement of the folds as successive molecular layers are added to the growing lamellae (19).

Many of the defects characteristic of small molecule crystals are also found in polymer lamellae. These increase in number with increasing concentration and supercooling. Spiral growths resulting from tears or impinging growth faces (20) from neighboring crystals as well as stacks of lamellae are commonly observed. The latter may originate from crystallization on imperfect chain folds or chain ends that exit from the fold surface. When viewed in solution these multilayer crystals often splay apart as shown in Figure 6.3 (21). Twinned crystals, which grow in a specific crystallographic relationship to one another, have been described by several authors and represent a major route to the more complicated crystalline forms that develop in concentrated solutions and melts (22, 23). Blundell and Keller (24) and Wittman and Kovacs (25) have noted that a large variety of twins can be prepared by a self-seeding technique in which the dissolution temperature and time are kept sufficiently short that many

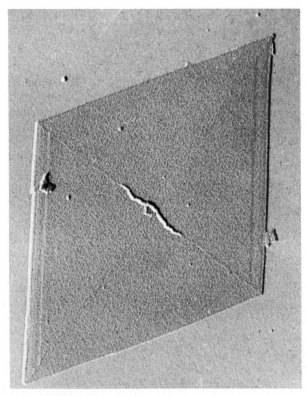

Figure 6.2 Collapsed polyethylene single crystal grown in xylene solution showing central pleat and separate growth quadrants (54).

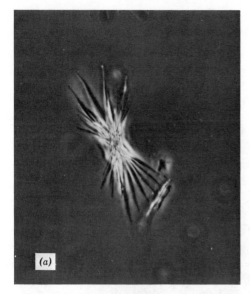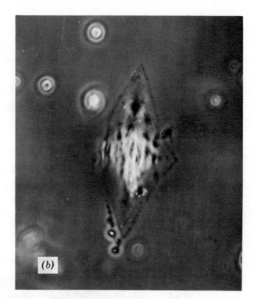

Figure 6.3 Multilayer polyethylene crystal viewed (*a*) edge on in solution, and (*b*) perpendicular to the basal plane showing lamellar displaying (21).

submicroscopic crystal nuclei can persist and initiate crystallization during subsequent cooling.

Nonpolyhedral crystal structures called dendrites begin to appear at higher crystallization rates. This structural transition can be a rather sharp function of crystallization temperature and concentration, with one range of conditions giving almost exclusively dendritic growth and another lamellar growth (26). While dendritic crystals may be initiated by many of the defects decribed in the preceding paragraph, their growth is controlled primarily by the concentration and impurity gradients that build up at the crystal-melt interface during crystallization. The development of dendritic crystals has been discussed in depth by Saratovkin (27) and recently reviewed by Wunderlich (16). It appears that when crystallization is sufficiently rapid the concentration of polymer molecules in solution near the growth face is depressed while that of noncrystallizing impurities is enhanced. Since these gradients are shallower near a corner or apex, growth in these areas is favored over that on adjacent planar faces, and needlelike dendritic structures are formed. Geil and Reneker have noted that spiral growths may develop preferentially on dendrites where reentrant growth faces occur (28).

Keith has proposed that the transformation in habit occurs when single crystals reach the size δ given by

$$\delta = \frac{D}{G_r} \qquad (6.1)$$

where D is the diffusion coefficient in the crystallizing medium and G_r is the rate of crystal growth (29).

Depending on the relationship between these variables, crystalline aggregates showing a wide range of complexity may be obtained (29, 30, 31), as shown in Figure 6.4. In highly concentrated solutions volume-filling structures with spherulitic symmetry have been observed (32, 33), although true spherulitic crystallization is generally restricted to polymer melts. This limitation appears to arise from the fact that the impurity buildups and high viscosities that enhance noncrystallographic branching and spherulite development (34) are not readily obtainable in solution.

6.2.2 Crystallization from the Melt

With relatively few exceptions, polymers crystallized from their equilibrium melts exhibit spherulitic morphologies. These struc-

Figure 6.4   Interference micrograph of polyethylene dendrite growth from toluene solution (31).

tures are themselves microcrystalline and composed of ribbonlike lamellae that grow radially from the spherulite nuclei. The problem of filling a spherical volume with a bundle of small platelike crystals is overcome to some extent by extensive low-angle branching of the lamellae. Presumably many of the same processes that give rise to the multilayer and dendritic crystals obtained from solution contribute to lamellar branching during melt crystallization. A detailed description of the development of both the central body and radiating superstructure of polymer spherulites was given by Keller and Waring in 1955 (35). To the authors' credit the discussion is still highly relevant despite the fact that chain-folded lamellar crystallization was not well understood at the time of the study.

The size of the spherulites in a given sample is uniquely determined by the nucleation density and radial growth rate, but under most conditions diameters ranging from a micron to perhaps a hundred microns are obtained. These structures are thus readily visible in the optical microscope, and a great deal of structural information can be obtained by examination between crossed polarizers. In polarized light spherulites typically appear as circularly birefringement regions exhibiting a characteristic Maltese cross extinction pattern running parallel and perpendicular to the polarization direction. The origins of this pattern depend on the orientation of the optical indicatrix of the birefringent crystals within the spherulite (36). A coarser structure is often superimposed on the extinction cross resulting from slight variations in orientation of the radiating fibrils as shown in Figure 6.5. In some materials a characteristic concentric ring pattern may also be present. Electron micrographs of such spherulites show a regular cooperative twisting of the lamellae in the radial direction (11).

The observation of negative birefringence in polyethylene spherulites by Bunn and Alcock (37) was taken as early evidence for the tangential orientation of the polymer chain since the largest refractive index was expected to be along the chain axis. These results were confirmed in microbeam x-ray experiments (38). Later on, however, it was found that both positive and negative spherulites commonly exhibit tangential orientation of the chain backbones (39) with the optical sign of the

Figure 6.5   Optical micrograph, crossed polarizers, of polypropylene spherulites crystallized from the melt at 127° C.

spherulite being determined by the relative polarizabilities of functional groups along the backbone.

The intrinsic lamellar charcter of spherulites of linear polyethylene was confirmed in fracture surface studies by Anderson in 1963 (40) and studies of partially digested crystal fragments by Palmer and Cobbold (41) and by Keller and Sawada in 1964 (42). Anderson was able to identify both folded-chain and extended-chain lamellae in melt crystallized samples, depending on their molecular weight. Whereas the extended-chain material fractured in a brittle manner with little microscopic deformation, extensive fibrillation was observed for the folded-chain material. Fibril formation was attributed to unfolding of the polymer chains.

By selectively degrading the amorphous regions of polyethylene spherulites with fuming nitric acid, Palmer and Cobbold (41) and Keller and Sawada (42) were able to make detailed investigations of the structures of their crystalline subunits (Figure 6.6). Single lamellae as well as more complicated multilayer crystals and twisted lamellar bundles

could be identified. Electron diffraction patterns of the lamellae showed that the chains were oriented perpendicular to the basal planes as in the case of solution-grown lamellae. Elongation of lamellae along the crystallographic b axis which lies in the radial direction in polyethylene was observed in many cases.

Following these early studies a number of polymers including polyamides, polyolefins, and polyesters were investigated using a variety of etching techniques. A paper by Miyagi and Wunderlich in 1972 describes the hydrolytic etching of crystalline poly(ethylene terephthalate) in which the reaction was carried far enough to isolate individual chain segments from the lamellae (43). At relatively long hydrolysis times the chain lengths as measured by viscosity average molecular weights were found to be comparable to the crystal thickness calculated from low-angle x-ray diffraction measurements (see Figure 6.7).

An adequate explanation for the observed strength and ductility of high polymers in light of increasingly convincing evidence for lamellar crystallization from the melt demanded

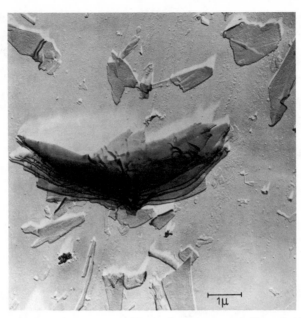

Figure 6.6   Lamellar fragments from melt crystallized polyethylene sample digested with fuming nitric acid (42).

that some molecules traverse the structure to provide a degree of mechanical integrity. Such tie molecules were postulated by several authors (44–46). In 1966 Keith, Padden, and Vadimsky conducted an elegant series of experiments that conclusively demonstrated the presence of intercrystalline links (47). In their studies melt mixtures of fractionated polyethylene and a C_{32} linear paraffin were isothermally crystallized. The paraffin was subsequently removed by washing with xylene at room temperature, and the films were observed by transmission electron microscopy.

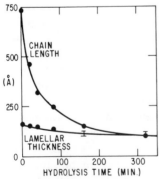

Figure 6.7   Change in viscosity average molecular weight and crystal thickness with hydrolysis time for poly-(ethylene terephthalate) (43).

At high magnification numerous microfibrils 50–150 Å in diameter could be seen extending between neighboring spherulites and among the radiating arms of individual spherulites (Figure 6.8). The number of links was found to increase with increasing molecular weight of the fractions, and the maximum length of the links was found to vary with the square root of the molecular weight. The authors proposed that the links were formed by nucleation of additional chains on molecules that were initially extended as the result of simultaneous incorporation in two growing crystallites.

Evidence for intercrystalline links composed of single molecules has recently been reported by Phillips and Edwards in studies of crystallization in thin films of polyisoprene (48). These authors found that secondary lamellae that were nucleated near the surfaces of existing lamellae were bent toward their neighbors as crystallization took place. The bending was attributed to the "reeling in" of tie molecules incorporated in adjacent lamellar surfaces. The thicker fibrillar intercrystalline links reported by Keith and Padden were absent and attributed to phase separation and secondary nucleation processes characteristic of the blended systems.

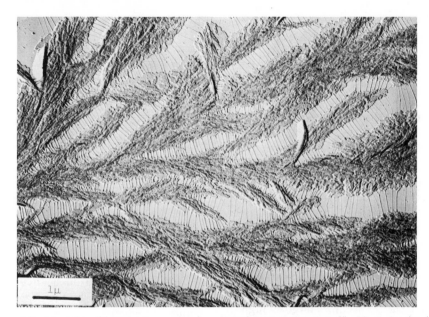

Figure 6.8   Interspherulitic boundary region in extracted polyethylene/paraffin blend showing intercrystalline links (47).

At this time many of the essential elements of bulk crystallization have been elucidated. Individual chain-folded lamellae are accepted as the basic structural feature from which more complicated crystal structures evolve. The role of crystal defects, concentration gradients, and impurity rejection in diversifying the crystalline microstructure are appreciated. Spherulites are known to be made up to a large extent of noncrystallographically branched lamellar ribbons tied together by molecules or groups of molecules that have been simultaneously incorporated in more than one crystallite. The delineation of these features provides an effective framework for the characterization of more complicated morphological forms that develop under shear or deformation and for understanding the mechanical properties of crystalline homopolymers.

6.2.3   Chain Folding

Both thermodynamic and kinetic arguments have been put forth to explain the tendency for macromolecules to undergo chain-folded crystallization. In the former, advanced by Peterlin and his co-workers, it is assumed that folded chains are in a minimum free energy state (49, 50). In the latter, which has gained more general acceptance, it is argued that the fold period is established by the energy barrier that must be traversed to achieve an appreciable rate of crystallization (51–54).

Since chain folding is such an integral aspect of the polymer crystallization process, a brief description of the basic kinetic model is presented here. The model has been expanded and modified to treat heterogeneous nucleation, secondary crystallization processes, and variations in the chain configuration and nucleus shape. The reader is referred to reviews by Price (55), Binsbergen (56–69), and Wunderlich (60) for more comprehensive treatments of this subject.

Fischer and Turnbull have shown that the rate of nucleation or crystal growth can be expressed in the form

$$I = I_0 \exp\left(\frac{-\Delta G}{kT}\right) \exp\left(\frac{-\Delta E}{kT}\right) \quad (6.2)$$

where I_0 is a temperature independent constant, ΔG is the free energy for formation of a nucleus of critical size, and ΔE is the activation energy for transport across the liquid-crystal interface (61). ΔE is essentially a viscosity term and may be represented with fair accuracy by several empirical expressions (62). The free energy change for nucleus formation may be expressed as

$$\Delta G = V G_V + A(\sigma_A) \qquad (6.3)$$

where V and A are the nucleus volume and surface area and ΔG_v and σ_A are the bulk free energy of fusion and interfacial free energy for a nucleus of lateral dimension a and b and height l (assumed to be the fold period). Equation (6.3) can be rewritten as

$$\Delta G = -abl(\Delta G_v) + 2al\sigma(010) \\ + 2bl\sigma(100) + 2ab\sigma(001) \qquad (6.4)$$

The free energy for nucleus formation thus passes through a maximum as the decrease in volume free energy becomes large enough to compensate for the energy increase associated with the fomulation of additional surface. The dimensions of the critical nucleus can be readily obtained by differentiation of Equation (6.4) as follows:

$$\frac{d\,\Delta G}{da} = 0 = -bl\,\Delta G_v + 2b\sigma(010) \\ + 2b\sigma(001) \qquad (6.5a)$$

$$\frac{d\,\Delta G}{da} = 0 = -al\,\Delta G_v + 2l\sigma(100) \\ + 2a\sigma(001) \qquad (6.5b)$$

$$\frac{d\,\Delta G}{da} = 0 = -ab\,\Delta G_v + 2b\sigma(100) \\ + 2a\sigma(010) \qquad (6.5c)$$

giving upon rearrangement and substitution

$$l^* = \frac{4\sigma(001)}{\Delta G_v} \qquad (6.6a)$$

$$a^* = \frac{4\sigma(100)}{\Delta G_v} \qquad (6.6b)$$

$$b^* = \frac{4\sigma(010)}{\Delta G_v} \qquad (6.6c)$$

At the melting point T_m°

$$\Delta S_f = \frac{\Delta H_f}{T_m^\circ} \qquad (6.7)$$

where H_f and S_f are the latent heat and entropy of fusion. The free energy change during crystallization may be rewritten as

$$\Delta G_v = \Delta H_f(1 - T/T_m^\circ) \qquad (6.8)$$

and the fold period is calculated to be

$$l^* = \frac{4T_m^\circ \sigma(001)}{\Delta H_f \,\Delta T} \qquad (6.9)$$

Thus the fold period is predicted to be uniquely determined by the degree of super-cooling at which crystallization takes place, with longer folds occurring at higher temperatures. An increase in lamellar thickness with crystallization temperature in polymer samples crystallized from dilute solution and from the bulk has been noted by several authors (Fig. 6.9; 63–66). Similar increases in thickness have been reported by Makarewicz and Wilkes during the solvent-induced crystallization of poly(ethylene terephthalate) at elevated temperatures (67).

6.2.4   Chain Extension under Pressure

Since the rate of crystallization decreases sharply with increasing crystallization tem-

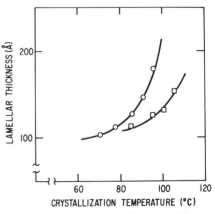

Figure 6.9 Variation in lamellar thickness with crystallization temperature for polyethylene crystallized from xylene (O) and hexadecane (□) (63).

perature, the maximum lamellar thickness that may be obtained in a realistic time period is normally limited to a few hundred angstroms. As a result chain-folded morphologies are normally observed for solution-crystallized material even at elevated pressures. In contrast, high-pressure crystallization of many polymer melts gives rise to extremely thick ($> 10,000$ Å) lamellae which are composed largely of extended chains (68–70). Although the fracture surfaces of such specimens are characteristically different from those obtained from chain-folded lamellae, many of the features of spherulitic morphology, including their unique extinction pattern in polarized light, are retained as shown in Figure 6.10. The lamellae, viewed edge on, appear to have a striated structure in which the striations run parallel with the chain axes. The chains, themselves, are tangential to the spherulite radii. Relatively large variations in lamellar thickness are observed for isother-

mally crystallized samples, and some molecular weight segregation may be indicated (71).

The actual process of extended-chain crystallization is still not fully understood. Mellilo and Wunderlich have cited the rounded ends of individual lamellae as well as the increase in lamellar thickness with distance from the center of the spherulites as evidence for a two-stage crystallization process in polytetrafluoroethylene samples crystallized at elevated pressure (72). These authors proposed that the molecules first crystallize in a folded configuration and then extend in a second, solid-state process. A second hypothesis has been advanced by Hatakeyama et al. who suggested that larger (thicker) lamellae were formed at an early stage of the crystallization process and that the smaller lamellae subsequently nucleated and grew on their surfaces (73). There is also evidence that alternate crystalline forms of some polymers may develop only at elevated pressures (74–76). The extended-

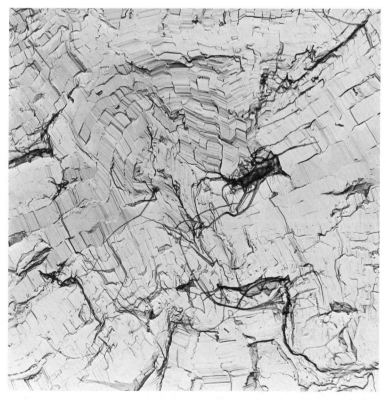

Figure 6.10 Fracture surface of polyethylene spherulite crystallized at elevated pressure showing extended chain crystals (98).

chain crystal morphologies formed during
high-pressure extrusion (77) are quite different in character from bulk, high-pressure
crystallized materials and are considered in a
later section.

6.2.5 Annealing

Distinct morphological changes are observed
on annealing both solution and bulk crystallized samples below their equilibrium melting
points. With few expections (78) considerable
thickening of the individual lamellae, as
measured by the change in x-ray long period,
is observed (79–81). Analysis of polyethylene
single-crystal mats has shown that the increase
in lamellar thickness is logarithmic in time at
constant temperature, with faster annealing
occuring at higher temperatures (81). The rate
of annealing can be markedly affected by the
initial rate of heat transfer to the sample (79),
suggesting that several types of reorganization
processes may be involved. Unless considerable melting occurs, the density of annealed
samples increases appreciably with time and
temperature. The consequences of lamellar
thickening are perhaps most visually evident
in the case of polymer single crystals. Upon
annealing, large voids are seen to develop
within the perimeter of the crystals as material
is removed from the interior and incorporated
in the larger fold period (79). In extreme cases
only the outline of the original platelet may
remain in the form of a "picture frame" (82).
When mats of agglomerated lamellae are
prepared by filtration from dilute solution
and subsequently annealed, considerable interpenetration of molecules from neighboring
lamellae is observed (83, 84). Apparently the
vacancies formed in one lamella by the increase in fold period are rapidly occupied by
polymer chains that emerge from the surface
of an adjacent crystal. A possible concept of
how this might occur has been offered by
Statton and is illustrated schematically in
Figure 6.11 (84). Blackadder and Lewell have
found that the mechanical properties of single-
crystal mats as well as their resistance to
swelling are considerably improved by annealing (83). Similar processes are believed to

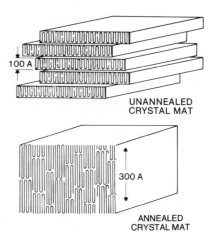

Figure 6.11 Possible mode of chain reorganization
during annealing of single crystal mats (84).

be involved in the annealing of solid samples,
with amorphous material residing between
lamellae being incorporated into the crystalline regions. As the same time an increase in
crystalline perfection is achieved by elimination of defects such as kinks and irregular
folds from the crystal structure (see Figure
6.12).

Several schemes have been proposed to
account for the high level of molecular mobility exhibited during annealing. Reneker suggests that point defects may be propagated
through the crystal with a relatively small
expenditure of energy until they terminate at a
fold or chain end (85). The appearance of each
defect at the fold surface increases the lamellar
thickness by one unit. A similar argument, in
which collective jumps of chain segments transport new material into the interior of the
crystal, has been advanced by Peterlin (86).
Etching experiments carried out by Keller and
Priest have indicated that thickening in single
crystals of polyethylene occurs through a
mutual rearrangement of chain folds while the
chain ends remain outside of the crystal lattice
(87). Sanchez et al. have developed theoretical
expressions relating the driving force for lamellar thickening to the initial crystal thickness l_0,
the equilibrium thickness l^*, and a relaxation
time τ characteristic of the degree of super-
cooling, which adequately describe the thickening kinetics of fluorocarbon polymer crystals (88).

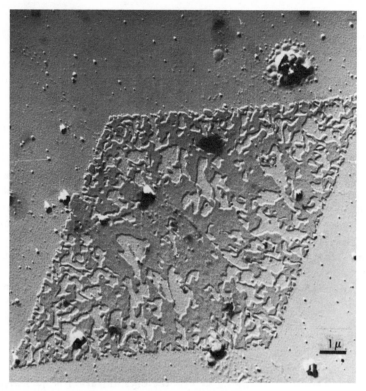

Figure 6.12   Annealed polyethylene single crystal showing internal voiding that accompanies lamellar thickening (14).

There is considerable evidence that, in addition to solid-state reorganization processes, melting and recrystallization play an important role in the changes that occur during annealing (89–91). Fischer and Hinrichsen found that polyethylene single crystals that were annealed above 130°C exhibited a low-temperature melting peak in addition to the primary melting endotherm (91). This peak, which was attributed to that part of the sample that melted during crystallization and recrystallized on quenching, decreased in size with annealing time as further recrystallization took place at elevated temperatures. The development and disappearance of multiple melting peaks during annealing has been observed in other polymers as the result of partial melting and recrystallization (92–94). Baer et al. were able to differentiate the different processes contributing to lamellar thickening through studies of irradiated polyethylene single crystals precipitated from xylene at 80°C (89). DSC scans showed that

between 118°C and 126°C solid-state chain extension was the dominant reorganization process while above 127°C the lamellae increased in thickness through melting and recrystallization.

High-pressure annealing of folded-chain crystals prepared from solution or from the bulk can produce extended chain crystals that retain the gross texture and molecular orientation of the original material (95). In polyethylene this process is accompanied by a phase transition from an orthorhombic to a hexagonal unit cell. Basset and his co-workers suggest that melting and recrystallization are responsible for this phase transition and are a critical factor in achieving the very large increases in lamellar thickness (96). It should be noted that the opposite process of annealing extended-chain crystals at atmospheric pressure produces overgrowths of folded-chain lamellae that subsequently anneal with modest increases in thickness (97, 98).

It is clear from the preceding discussion that

the annealing process is highly complicated and still only partially understood. Both solid-state rearrangements and true melting and recrystallization inevitably occur during any annealing experiment. The relative contribution of each of these factors is, to a large extent, a function of the degree of perfection inherent in the original crystalline sample. This parameter is difficult to quantify although density determinations, x-ray diffraction studies, and "zero-entropy" melting (99) experiments provide important information. The application of new spectroscopic techniques such as high-resolution solid-state nmr may offer new insight into the types of molecular motion that play a critical role in the reorganization process. The effects of residual stress amounts of retained solvent on annealing behavior warrant more extensive investigation.

6.2.6 Epitaxial Crystallization and Crystallization under Shear

Epitaxial growth of polymer crystals on various foreign substrates as well as on the surfaces of previously crystallized samples of the same polymer have been reported (100–108). In many cases an exact matching of lattice parameters is not necessary. Polyethylene single crystals can be grown from solution on a variety of alkali halide substrates with quite different unit cell dimensions (101, 102). The orientation of these crystals appears to be governed primarily by the arrangement of positive or negative ions in the substrate. During the early stages of crystallization the lamellae appear to grow with the basal plane normal to the substrate, suggesting that crystallization is initiated by the adsorption of a folded polymer chain on the substrate. The observed fold thickness (in excess of 300 Å) is substantially greater than that normally found in solution crystallized samples. The lamellae, when allowed to grow to a sufficient height, are often observed to topple over with the chain axis normal to the substrate when the solvent is removed.

Epitaxial growth of folded-chain polyethylene lamellae on substrates of extended-chain crystals has been reported by Wunderlich et al. After annealing the extended-chain material at atmospheric pressure, small ripples are observed to develop on the substrate (108). The ripples have been shown to be individual lamellae lying perpendicular to the original surface. The original orientation of the chain axis is maintained. Similar growths have been observed on extended-chain samples of polytetrafluoroethylene (109).

Epitaxial crystallization from the bulk melt appears to proceed by a similar mechanism, although isolated lamellae cannot be readily observed. Studies of polypropylene crystallization on graphite fiber substrates have shown that dense transcrystalline growths develop on the fiber surfaces and continue to grow outward until they impinge on neighboring spherulites (see Figure 6.13). The nucleation density is controlled by the size of the graphite basal planes (106). In all essential respects transcrystalline growth is identical to normal spherulitic crystallization, except that the crystallization is nucleated on a surface rather than at a point, and the nucleation density is sufficiently high that growth is restricted to one dimension. Transcrystallinity that is not epitaxial in origin may also develop on a variety of nucleating surfaces or from surfaces that are rapidly cooled to initiate crystallization (110, 111).

Very unusual morphologies which also involve epitaxial crystallization may be produced from polymer solutions or melts under conditions of high shear. Perhaps the most well known of these are the "shish-kabob" structures reported by Pennings and his coworkers in studies of stirred polyethylene solutions (107, 112) as shown in Figure 6.14. These structures are believed to develop from fibrous nuclei that are formed by partial extension of polymer chains in the direction of flow. An epitaxial overgrowth of folded-chain lamellae is then formed on the fibrous core giving rise to the characteristic "meat on a skewer" appearance when viewed in the electron microscope. An alternate explanation for the origin of the shish-kabob morphology has recently been proposed by Nagasawa and Shinomura (113). These authors suggest that

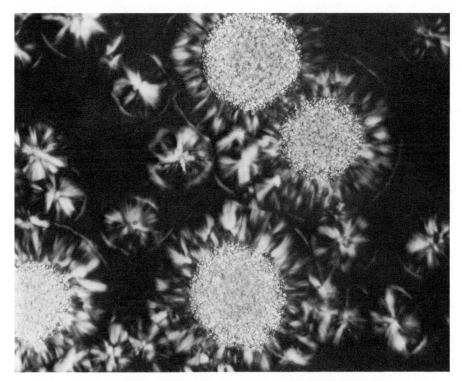

Figure 6.13 Optical micrograph, crossed polarizers, of polypropylene crystallites nucleated and grown on dispersed poly(butylene terephthalate) particles.

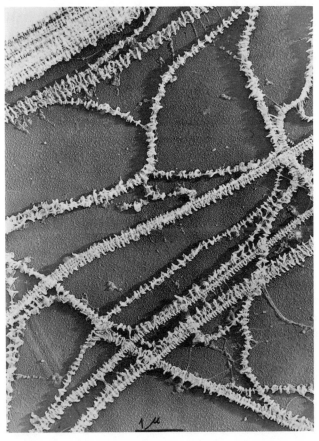

Figure 6.14 Shish-kabobs of polyethylene crystallized from stirred xylene solution (107).

such crystals grow from screw dislocations in lamellae that are deformed under shear. Macroscopic models indicate that the observed longitudinal c axis orientation of the overgrowths may be retained in such growths without the requirement of epitaxial crystallization.

Crystallization in flowing polymer melts has been investigated by a number of authors and appears to occur in much the same way as crystallization from stirred solutions (114–119). The development of rodlike nuclei in the oriented melt is followed by transcrystalline growth of folded-chain lamellae perpendicular to the flow direction. Both shear and extensional flow fields seem to be capable of initiating this type of crystallization (116). Ulrich and Price have observed that the large increases in the observed crystallization rate at relatively low shear rates is indicative of orientation due to melt elasticity rather than flow (118, 119). At higher shear rates molecular orientation appears to result primarily from the longitudinal velocity gradients established in the flowing system (120). In all cases, chain entanglements contribute significantly to the elongation process, and higher molecular weight molecules have been observed to markedly increase both the melt shear stress (at a given shear rate) and the nucleation density (115). The break-up of crystalline aggregates in the flow field may decrease the rate of crystallization if the particles are smaller than the size of the critical nucleus or increase the rate if they are large enough to grow spontaneously (115).

6.2.7 Deformation and Drawing

Many complex morphological changes are observed when semicrystalline polymers are deformed. These changes can be broken down conveniently into those processes that occur within individual lamellae and those that occur between neighboring lamellae. The extent to which each of these processes contributes to the deformation of a bulk sample is determined by the overall crystallinity, the molecular weight of the sample, the time-temperature schedule under which the sample is crystallized, the deformation conditions, and the overall extent of deformation.

The primary deformation mechanisms available to polymer single crystals are analogous to those observed in crystals of low molecular weight compounds and include slip, twinning, and martensitic transformations (121). Stress-induced twinning does not appear to be a general phenomenon in semicrystalline polymers but has been reported for both polyethylene and nylon 6,6 at relatively small deformations. Twinning is characterized by a rapid rotation of the crystal axes and is limited to those crystals having less than hexagonal symmetry. Only small molecular translations are necessary to cause twinning as shown in the schematic representations in Figure 6.15 of 110 and 310 twins formed from an orthorhombic crystal lattice. Both types of twin formation have been observed in polyethylene crystals (122, 123). Twinning of the 110 type has been reported in nylon 6,6 (124).

Transformations from an orthorhombic to

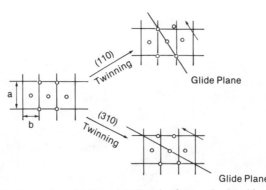

Figure 6.15   Formation of (110) and (310) twins from orthorhombic crystal lattice.

a monoclinic and to a hexagonal unit have both been reported for polyethylene under high pressure (74, 76, 125). Formation of the monoclinic unit cell appears to be accompanied by repeated twinning. Kiho et al. have found that the monoclinic form disappears spontaneously at atmospheric pressure and at elevated temperatures and is thus a metastable structure (126). Other cases of martensitic transformations in high polymers are lacking at present.

At larger deformations, intermolecular slip is observed. This process appears to be a much more general mode of deformation for crystalline polymers than twinning or phase changes, although the critical shear stress for slip is higher than that for twinning or martensitic deformation (127). The available glide planes in folded-chain lamellae are established by the orientation of the molecular axes. Stresses imposed parallel to the lamellar basal plane result in both thinning and elongation of the crystal as well as a rotation of molecules or molecular bundles in the direction of draw. Hansen and Rusneck have proposed that the natural draw ratio of many polymers is determined in part by number of slips required to align the chains in the draw direction (128).

Interchain slip becomes energetically feasible only if a sufficient number of dislocations are available to permit easy movement through the crystal matrix. A rough calculation of the number of such defects has been made by Reneker (85). The Peierls stress, or the stress required to propagate a dislocation through the crystal, has been estimated for polyethylene (129). The results indicate that slip on planes parallel with the crystal b axis is greatly favored over corresponding movement along the a axis.

Electron microscopy studies of the tensile deformation of polymer single crystals have shown that the lamellae begin to break up at elongations of around 20% with thin fibrils extending across the voids (130). The fibers themselves often have a "string-of-pearls" appearance resulting from blocks of material that have been torn loose from the lamellae (131). Variations in the size of these blocks are thought to arise from fracture along randomly distributed imperfections in the original crystal, and the blocks themselves have been postulated to be inherent structural features in the original lamellae (132). While there is general agreement that the molecules in the fibrils are aligned with the fibril axes, evidence for and against alignment of the individual blocks has been presented (130–133). An increase in the observed fold period in the fibrillar core has been attributed to local melting and recrystallization arising from adiabatic heating of the fibrils during extension (133). The simultaneous appearance of micronecks and large undeformed lamellar regions indicates that there are large variations in the local stress field.

In more complicated crystalline structures such as spherulites, the breakdown of lamellar ribbons is preceded by or occurs in conjunction with *inter*lamellar reorganization processes. These rearrangements are dictated to a large extent by the character of the amorphous interlayers between neighboring crystallites and in particular by the density of intercrystalline links. In early studies of the deformation of polyethylene spherulites, Keith and Padden noted that samples crystallized slowly at high temperatures failed in a brittle manner at very low strains, with the fracture plane passing both through and around spherulite boundaries (134). At lower crystallization temperatures, interspherulitic fracture occurred at higher strains but with negligible yielding. Samples crystallized at the lowest temperatures exhibited considerable plastic flow prior to fracture. Later experiments have shown that these changes were a direct reflection of the increase in the number of intercrystalline links, with increasing crystallization rate and the preferential formation of such links first within individual spherulites and then between neighboring spherulites (135).

When the molecular weight of the sample is high enough and crystallization is sufficiently incomplete that brittle failure does not occur immediately on the application of stress, spherulitic polymers pass through a series of deformation steps. At very low elongations ($< 1\%$) the deformation is elastic and recoverable when the stress is removed. Plastic flow occurs at larger strains, but affine deformation may persist to elongations of several hundred percent (136). During this elongation a transformation from spherulitic to fibrillar morphology takes place. The type of reorganization that occurs within individual spherulites depends on the orientation of the lamellar ribbons with respect to the applied stress and the allowable modes of crystal reorganization. In general, in the zones of highest shear stress (45° to the stress axis) interlamellar and intralamellar slip predominate, the lamellae thin, and the chains tilt toward the stress axis. In polyoxymethylene compressive stresses that develop along the spherulite equator result in buckling of the lamellar ribbons at elongations greater than 20% (137). In polypropylene the spacing between the lamellae has been observed to increase (138). In polyethylene, in contrast, equatorial lamellae first twin and then undergo a phase transformation to a monoclinic unit cell before catastrophic disruption occurs (139). A similar phase change occurs in ribbons lying along the stress axis in polyethylene at relatively low strains. At higher strains the polar lamellar ribbons in polyethylene as well as in polypropylene and polyoxymethylene and other crystalline polymers break up into fibrils with the chains aligned in the direction of stress. Ultimately the spherulites lose their identity and are replaced with a fibrillar morphology.

Crystal orientation functions of around 0.95 are normally obtained for drawn polymers in which the change to fibrillar morphology is complete. Under such conditions the modulus is increased from approximately 10^9 to 10^{10} Pa (140). Under very special conditions, however, involving high-pressure solid-state extrusion very close to the melting point, fibers showing moduli approaching 10^{11} Pa and crystal orientation functions in excess of 0.99 can be produced. The preparation and morphology of these ultra-high-modulus polymers has been extensively investigated by Southern and Porter and thier co-workers (77, 141, 142). In the case of polyethylene, the fibers appear to be composed primarily of folded-chain lamellae 200 and 300 Å in thickness. Approximately 10% of the core material consists of extended chains, and the overall level of crystallinity is measured to be 80–90%. The extremely high modulus is thought to arise from the exceptional degree of chain alignment rather than from chain extension. The fibers are transparent or nearly so and exhibit good lateral integrity owing to the inclusion of folded-chain material (77). The potential utility of such fibers either by themselves or in composite materials continues to spur interest in improved methods of fabrication.

6.3 THE MORPHOLOGY OF GLASSY POLYMERS

The subject of order in glassy polymers has been one of increasing interest and controversy as new theoretical and experimental data have become available. Historically, polymeric glasses were assumed to be completely amorphous in structure with neither long- nor short-range order. Early calculations and experiments on the mean end-to-end distance of polymer chians and the dependence of these parameters on temperature led Flory and his co-workers to postulate that the conformations of individual molecules in glassy solids were equivalent to those observed in corresponding θ solutions (143). Flory's conclusions have received considerable support from other investigations. In 1965 Krigbaum and Goodwin carried out low-angle x-ray scattering studies on monodisperse polystyrene samples in which each chain end was terminated with a disilver carboxylate salt (144). The unperturbed displacement length $(r_0)^{1/2}$ of this sample, calculated from osmotic pressure measurements of molecular weight, was 217 Å. In comparison, values of 220 Å and 269 Å were obtained from the x-ray scattering data before and after desmearing. Thus the mean end-to-end distance for these chains in the bulk appeared to be comparable to, although slightly greater than, that observed in a θ solvent.

In a more indirect study, Alberino and

Graessely measured the relative degrees of inter- and intramolecular cross-linking in irradiated samples of bulk polystyrene and polystyrene dissolved in ethylbenzene (a θ solvent) (145). No observable differences were obtained. The authors cited this result as an indication that little if any contraction of the random coil occurred in the glassy solid. A similar conclusion was reached by Semlyen and Wright in a series of ring cyclization studies of polydimethylsiloxanes (146). The dimensions of the dimethylsiloxane chains were computed from measured values of the molar cyclization equilibrium constants using a theoretical treatment developed by Jacobson and Stockmayer (147). For molecules having a contour length > 50 Å the values agreed with those predicted to occur in a θ solvent. Significant deviations were observed for shorter polydimethylsiloxane (PDMS) chains.

Neutron diffraction studies, which are increasing in number and sophistication with the availability of more numerous research facilities, generally confirm the statistical chain nature of glassy solids although they do not rule out the possibility of very localized segmental ordering. In most cases these experiments have been carried out on partially or wholly deuterated molecules since the coherent scattering length for a CD_2 moiety is $2.00/0.74$ times that of a corresponding CH_2 group. Ballard et al. found that in a mixture of 5% protonated polystyrene in 95% deuterated

polystyrene the radius of gyration was 90 ± 5 Å whereas that in a low molecular weight solvent was 84 Å (148). These results were supported by Benoit and his co-workers who showed that narrow molecular weight fractions of deuterated polystyrene dissolved in protonated styrene matrices displayed radii of gyration equivalent to those observed in cyclohexane θ solutions (149). Neutron diffraction experiments on molten polyethylene samples of 81,000 mol. wt. have indicated a radius of gyration of 126 Å which is very close to that of 123 Å calculated for the unperturbed chain (150). Subsequent studies on these and other polymers such as poly(methyl methacrylate) and poly(dimethyl siloxane) have also indicated (see Figure 6.16) that the bulk radii of gyration approximate those observed in θ solvents (151–157). A detailed review of the present status of neutron diffraction studies of high polymers has recently been published by Maconnachie and Richards (158).

It should be emphasized that the preceding observations do not imply that much shorter chain segments follow Gaussian statistics or that significant pairwise correlations dominate local configurational states. Some information of this kind can be obtained from scattering intensities at wider angles, although considerable ambiguity exists in such analyses. Flory has attempted to show that rotational isomeric state calculations for short-chain sequences predict the proper scattering func-

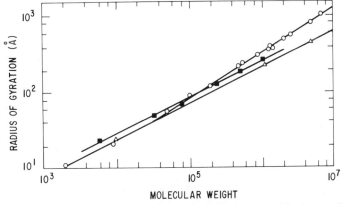

Figure 6.16 Variation in radius of gyration as a function of molecular weight for: ■, vitrious PMMA (neutron scattering); O, acetone (light and x-ray scattering); and △, butyl chloride-θ solvent (light and x-ray scattering (151).

tions (153), but this correlation is not necessarily a unique one. Bendler has made some recent calculations that indicate that significant statistical deviations in local chain conformation are possible at a relatively small cost in free energy and with very little perturbation of the root mean square radii of gyration (159). Ongoing research in this area will undoubtedly help clarify some of these points.

The results of light-scattering experiments on glassy polymers agree well with those obtained by neutron diffraction. By using traditional light-scattering methods to measure the size of styrene domains in styrene/ butadiene block copolymers, Hoffman was able to show they followed the $[M_w]^{0.58}$ molecular weight dependence predicted from a random chain model (160). More detailed information has been obtained from depolarized Raleight scattering studies. Using this technique a molecular anisotropy factor γ^2 can be determined from an analysis of the molecular optical polarizability tensor components α_{ij} using the relationship

$$\gamma^2 = \frac{(\alpha_{11} - \alpha_{22})^2 + (\alpha_{22} - \alpha_{33})^2 + (\alpha_{33} - a_{11})^2}{2}$$

(6.10)

This parameter provides a direct measure of the deviations in chain shape from those predicted from random statistics.

The polymers that have been studied by this technique have been found to have very low anisotropy factors. In poly(methyl methacrylate) (PMMA) an upper limit of 28 Å has been placed on the size of the anisotropic structures present with a note from the authors that these entities might be considerably smaller (161). Studies on polycarbonate and polystyrene also reported by Dettenmaier and Fischer reveal a similar lack of anisotropic structures (162). In the case of polystyrene, values of γ^2 obtained on small molecule model compounds such as 1,3 diphenylbutane and low molecular weight oligomers are almost identical with those of the high polymer (163). Patterson and Flory have shown that experimentally measured anisotropy factors for low molecular weight n-alkanes agree well with those cal-

culated from the rotational isomeric state model using a value of 500 ca/mol for the trans-gauche energy difference (164).

6.3.1 Ordered Structures

In spite of general agreement over the absence of long-range order or significant perturbations of random chain dimensions in glassy polymers, there is still considerable controversy over the existence of short-range correlations or organized subcrystalline microstructures. The most direct evidence for some form of supermolecular organization in the glassy state has come from observations of small, ball-like structures in the electron microscope. These were first reported by Schoon et al. (165, 166) in studies of surface replicas and stained thin films of solvent cast polystyrene and PMMA. Similar structures were later found in both glassy poly(ethylene terephthalate) (167–169; see Figure 6.17) polycarbonate (170, 171), and polystyrene (172). In each case the ball-like structures averaged approximately 75 Å in diameter and appeared to be composed of clusters of macromolecules. The existence of some degree of order in these domains in poly(ethylene terephthalate) is suggested by dark field microscopy experiments carried out by Yeh and Geil (167) as shown in Figure 6.18. The diffuse character of the diffraction maxima associated with these entities is cited as evidence for the presence of highly disordered crystals. Some movement of the nodules is seen on annealing at temperatures below T_g. Higher annealing temperatures bring about two-dimensional alignment of the balls into fibrillar ribbons having spherulitic organization. The authors propose that this phenomenon is a general mechanism for crystallization from the glassy state. Coincident decreases in the x-ray long period with annealing time at constant temperature are attributed to improved molecular packing within the balls. Stress-induced crystallization appears to take place through a similar rotation, alignment, and partial merging of the ball-like structures (168). Kashmiri and Sheldon report the presence of a very small endotherm in the DSC traces of quenched

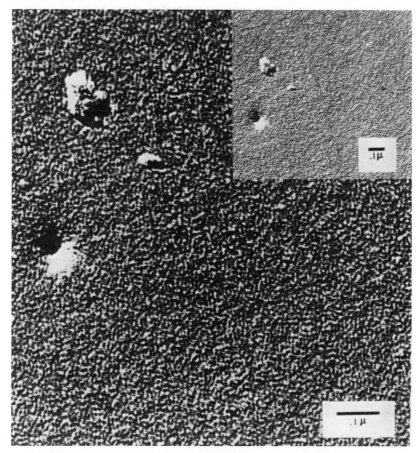

Figure 6.17   Replica surface of amorphous bulk poly(ethylene terephthalate) showing nodular structures (167).

poly(ethylene terephthalate) which they believe to be associated with nodular development (169). The level of "crystallinity" in these samples is estimated to be 1% or less.

The resolution of similar ball-like structures in BPA polycarbonate after ion milling for 10 min with oxygen ions has been noted by Frank et al. (170). These grains could be observed to increase in size and agglomerate on annealing at temperatures as low as 110°C (T_g of polycarbonate = 140°C) and to disappear on reheating to 160°C. Selective etching of these areas by the ion beam is believed to reflect changes in local order and density. No structures were observed in unannealed free surfaces. In contrast, Carr et al. report the presence of nodules whose size depends on both the casting solvent and subsequent annealing conditions on unetched BPA poly-

carbonate surfaces (173). As in the case of poly(ethylene terephthalate) samples, spherulites appear to be formed by association and migration of the nodular structures.

Significant questions still exist over the proper interpretation of these observations. In a recent paper Zingsheim and Bachmann show that in defocused micrographs of negative stained polystyrene/dextran films 30–70 Å domains appear to exist, while in properly focused pictures these structures were absent (174). It is not clear from this work that the structures observed in surface replicas such as those used in the poly(ethylene terephthalate) and polycarbonate studies can be associated with improper focusing procedures, but it seems unlikely. On the positive side a number of researchers have shown that it is possible to prepare microglobular samples of

Figure 6.18   Incipient spherulites formed by reorganization of ball-like structures in annealed poly(ethylene terephthalate) (167).

various polymers by precipitation from dilute solution (175–177). Individual molecules as well as aggregates having various degrees of interpenetration have been observed, depending on the preparation technique. Reasonable molecular weight determinations can be made from the micrographs. While these experiments have no clear implication on bulk structure they do indicate that various forms of molecular bundling are possible in amorphous systems.

Both positive and negative evidence for small-scale molecular ordering has been obtained from electron diffraction studies of glassy polymers. Ovchimikov and Markova (178) have presented data indicating the presence of domains of parallel chains approximately 50 Å in diameter in molten polyethylene. In a more careful analysis Voight-Martin and Mijlhoff argued that the sharp intermolecular peak in the 4–6 Å region,

which has been quoted as proof for the existence of molecular bundles, was an experimental artifact and that no molecular ordering was observable in polyethylene melts down to a level of 10 Å (179). Similar differences in interpretation have arisen in x-ray scattering experiments (180, 181). In a recent review Fischer points out that all these studies suffer from the fact that the radial distribution functions are insensitive to the long-range correlations that differentiate the coil and bundle models (182).

An interesting commentary on the degree of randomness in a glassy polymer has been provided by Robertson in some theoretical calculations of chain-packing characteristics (183). According to Robertson's calculations, molecules retaining the dimensions of an unperturbed chain could pack to no greater than 65% of the observed crystalline density even when effective coil interpenetration is achieved. Since bulk densities reaching 85–95% of that of an efficiently packed crystal are commonly observed, some degree of cooperative ordering must be invoked. The specific configurational restrictions that might be imposed are conjectural, although one might expect straight segments to be favored over right-angle bends in the bulk because they require a lower degree of interchain cooperation. Robertson points out that local alignment is compatible with retention of the end-to-end distance predicted for the random coil and suggests that a one-dimensional ordering exists similar to that proposed by Hosemann for a paracrystalline substance (3).

A more precise description of ordering in glassy polymers will continue to evolve as more sophisticated probing techniques such as neutron diffraction see wider use and as improved models for treating scattering phenomena are developed. Electron microscopy studies, packing calculations, and some depolarized light-scattering experiments plus an inherent "feel" for the way in which polymer chains might be expected to cooperatively interact make some short-range models highly attractive. It seems clear, however, that the size of these ordered domains is in the range of approximately 20 Å and not several hundreds of angstroms and that whatever ordering is present is accomplished with little, if any, perturbation of statistical chain dimensions. Similar conclusions have been drawn by Robertson, whose highly detailed review amplifies many of the points discussed in this summary (184).

6.4   THE MORPHOLOGY OF COPOLYMERS

The morphology of copolymers is dictated by a number of factors including their overall composition, the sequence-length distribution of their components, the ability of one or both of the components to undergo crystallization, and the conditions under which polymerization is carried out. Several of the most important classes of copolymers are illustrted schematically in Figure 6.19.

The majority of copolymers of academic and commercial interest fall into one of these groups, and they provide a logical outline for the following discussion. More detailed treat-

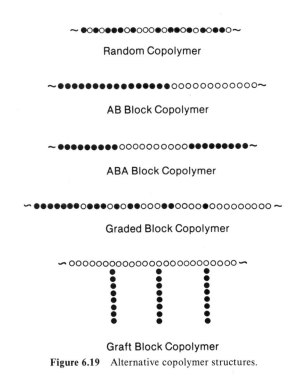

Random Copolymer

AB Block Copolymer

ABA Block Copolymer

Graded Block Copolymer

Graft Block Copolymer

Figure 6.19   Alternative copolymer structures.

ments of materials of special interest such as multiblock copolymers, ionomers, and interpenetrating networks have been presented in books by Aggarwal (185) and Manson and Sperling (186).

6.4.1 Amorphous Random Copolymers

In general, no phase separation is observed in rubbery or glassy random copolymers. Theoretical calculations have shown that the probability of having an extended repeat sequence of one monomer unit or a large fluctuation in molecular composition due to statistical fluctuations is very small (187, 188). As a result, domain formation does not occur, and each monomer appears to retain its characteristic homopolymer free volume (189). An expression for the T_g of a random copolymer can be developed from this assumption and has been shown to successfully describe the behavior of a number of systems (190, 191). The relevant equation is

$$\frac{1}{T_g} = \frac{1}{(W_1 + BW_2)} \frac{W_1}{T_{g_1}} + \frac{BW_2}{T_{g_2}} \quad (6.11)$$

where W is the weight fraction, the subscripts 1 and 2 refer to the different comonomers, and B is a numerical constant close to unity. When B is equal to one, Equation (6.11) reduces to the convenient form

$$\frac{1}{T_g} = \frac{W_1}{T_{g_1}} + \frac{W_2}{T_{g_2}} \quad (6.12)$$

In the rare cases when appreciable compositional disparity occurs in copolymers synthesized from the same two monomer units, phase separation may be observed. This effect has been reported by Molau in styrene-acrylonitrile copolymers (192). In contrast to block copolymers in which each phase contains only one type of monomer (see Section 6.4.4), the domains in these materials are composed of groups of copolymer molecules that are rich in one monomer or the other. Only small differences in composition (less than 5%) are required for such segregation to occur. The two phases can be easily dis-

tinguished by phase-contrast optical microscopy as a result of differences in refractive index.

6.4.2   Crystalline Random Copolymers

The structural constraints on compatibility are much more severe when one or both of the comonomer units are capable of crystallizing. If these moities are comparable in size and polarity, isomorphic replacement of the minor component in the crystal lattice of the major component may occur with some distortion of the lattice parameters. A continuous change in lattice dimensions with composition may be observed if the two unit cells are similar. When different crystal structures are involved, that of the major component is usually found at the ends of the composition range, with both crystals appearing side by side at intermediate concentrations. Comonomer units that are not capable of crystallizing or that are too large to fit in the prevailing crystal lattice may be accommodated as defect structures at low concentrations and be rejected from the crystals at higher concentrations. In some cases very low levels of crystallinity are observed over the composition midrange while appreciable crystallization of the individual components occurs when their respective concentrations are high. A number of systems that exhibit these features to a greater or lesser degree have been studied.

Wunderlich and Poland examined the effect of inclusion of random carbonyl and methyl comonomer units on the crystallization of polymethylene (193) as shown in Figures 6.20 and 6.21. No variation in crystallinity or melting temperature was observed in samples containing up to 13.2% carbonyl groups, while a steady decrease in crystallinity and melting temperature were observed with methyl group concentration. In each case a slow linear increase in the length of the crystal a axis occurs with no change in crystal morphology. The authors conclude that the carbonyl groups are accommodated in the lattice positions occupied by methylene units while the larger $CHCH_3$ units exist as point defects in the crystal structure. Similar results

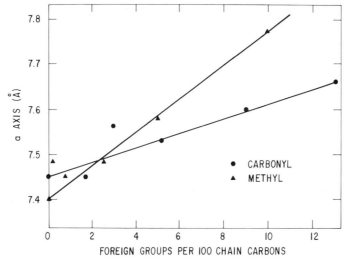

Figure 6.20 *a* axis dimensions of polyethylene unit cells containing randomly distributed CO and CH$_3$ side groups (193).

were obtained by Baker and Mandelkern in studies of polymethylene samples having random methyl or *n*-propyl side groups (194).

Several other studies are worthy of note. Natta et al. examined the effects of ring substitution on samples of isotactic polystyrene (195). Copolymers of styrene and *o*-fluorostyrene were found to be crystalline and to exhibit a monotonic increase in melting point from 235 to 270° C over the entire range of compositions. No change in crystal symmetry was observed. This behavior contrasts with that of *p*-fluorostyrene/styrene co-

polymers in which those compositions enriched in styrene crystallize in a threefold helix while those enriched in *p*-fluorostyrene crystallize in a fourfold helix. Both crystalline phases appear to be present at intermediate compositions. A large distortion in the styrene lattice is noted when even small amounts of *p*-fluorostyrene are present, showing that the *para* position is much more sensitive than the *ortho* to intermolecular contacts. Significant decreases in crystallinity are observed when larger moieties such as CH$_3$ or Cl are substituted for hydrogen.

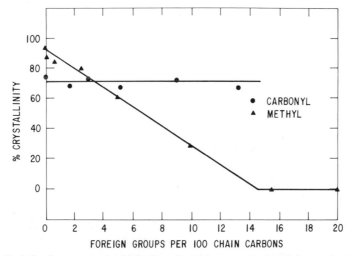

Figure 6.21 Variation in percent crystallinity for polyethylene samples containing randomly distributed CO and CH$_3$ side groups (193).

Holdsworth and Keller have investigated the structure of single-crystal mats of ethylene-propylene and ethylene-butene copolymers by selective etching with fuming nitric acid (196, 197). In these experiments each of the co-monomer units contained one ^{14}C atom, and a comparison of sample radioactivity before and after etching permitted analysis of the branch content. Comparison of the etching rates of the copolymers with that of poly-ethylene homopolymer showed that the co-polymers initially lost weight must faster although the changes in x-ray long period were comparable. This difference was attrib-uted to selective degradation of branches *within* the lamellae aided by expansion of the crystal lattice around the defect regions. Some branches were intact after lengthy exposure to the nitric acid indicating that encapsulation of isolated defects in regions of high crystallinity prevented attack by the etchant. Somewhat greater etching rates were observed for the samples containing the more bulky butene comonomer units. When the comonomer unit is rejected from the growing crystal a rapid decrease in the crystallinity, crystallization rate, and melting point of the copolymer and an appreciable broadening of the melting range are observed. The thermodynamic theory describing these phenomena has been exten-sively developed by Flory and his co-workers (198). In general the equilibrium melting point of a copolymer in which only one unit is capable of crystallizing is given in terms of the sequence propagation probability p as

$$\frac{1}{T_m} - \frac{1}{T_m^\circ} = \frac{R}{\Delta H_f} \ln p \qquad (6.13)$$

where ΔH_f is the molar heat of fusion. Inspec-tion of Equation (6.13) shows that the most severe melting point depression is expected for randomly dispersed comonomers. This rela-tionship can be reexpressed in terms of the activity of the crystallized units X_A as follows (199):

$$\Delta H_f \left[\frac{1}{T_m} - \frac{1}{T_m^\circ} \right] = R \ln X_A \qquad (6.14)$$

Experiments by Bodily and Wunderlich on polyethylene samples containing various co-polymerized side groups have shown that large side groups lower the activity of crystal-lizable units in the melt to an extent greater than the mole fraction of such groups (200). As a result the melting point is depressed to a larger extent than predicted by theory. The reduced melting point appears to be a con-sequence of amorphous defects within the crystals rather than large-surface free energies associated with very thin lamellae. Support for this conclusion comes from data presented by Holdsworth and Keller on branched poly-ethylene (201). Crystals of these materials were grown by slow cooling from 0.1% xylene solution. All were lamellar and showed single-crystal diffraction patterns. Both the lamellar thickness and the change in lamellar thickness with temperature for these materials were comparable to those commonly observed for corresponding homopolymers.

6.4.3 Amorphous Block Copolymers

Block copolymers, in which neither component is crystalline, form characteristic microdomain structures in both cast films and moldings. In many cases these domains are so small that they do not scatter visible light, and the materials appear to be transparent or highly translucent rather than opaque. When the dispersed phase is glassy, as in styrene/buta-diene/styrene (SBS) block copolymers, the materials form tough, thermoplastic rubbers in which the high T_g phase provides effective cross-links that become mobile at elevated temperatures. The morphology of these block copolymers has been extensively investigated beginning with the classic studies of Matsuo et al. in 1968 (202). A great deal of information has been obtained from transmission electron microscopy (TEM) investigations of thin sec-tions in which contrast is developed by selective staining of the butadiene phase with osmium tetroxide (203; see Figures 6.22, 6.23, and 6.24). In compression molded sheets of SBS copolymers having a styrene/butadiene mole ratio of 60–40, Matsuo et al. found spherical butadiene domains 300–350 Å in diameter

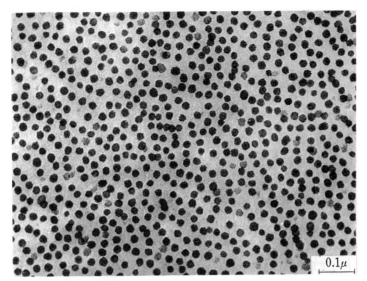

Figure 6.22 Ultrathin section of SBS block copolymer film cast from toluene (styrene/butadine = 80–20). (204).

that were partially agglomerated. Reduction in butadiene concentration at constant overall molecular weight was observed to reduce the domain size. SB and BSB copolymers of similar composition and molecular weight were observed to contain irregular rod-shaped structures. These results suggested that both the morphology and physical properties of these polymers were dependent on the arrangement of blocks within the chain. In a sub-

sequent paper in which the morphologies of solvent cast films of SBS and BSB block copolymers were investigated in more detail, Matsuo concluded that the final microstructure was dominated by the overall polymer composition rather than the block sequence. This conclusion was supported by independent studies on styrene/isoprene block copolymers carried out by Inoue et al. (205). These investigators also noted that the casting pro-

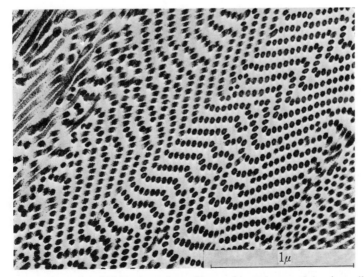

Figure 6.23 Ultrathin section of SB block copolymer film cast from toluene and showing rodlike structures viewed end on and laterally (top left) (styrene/butadiene = 60–40) (204).

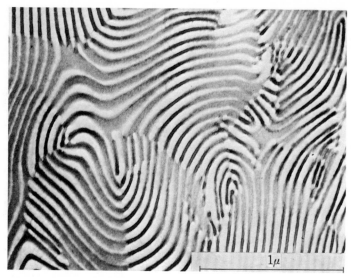

Figure 6.24 Ultrathin section of SBS block copolymer film cast from toluene and showing sheetlike domains (styrene/butadiene = 40–60) (204).

cess itself has a major influence on the observed structure with elongated structures consistently lying in the plane of the cast film.

Uchida and his co-workers identified a whole range of micromorphologies that could be developed in solvent cast films of SBS block copolymers (206) by varying the fractional compositions of the two components in a given solvent. These included:

1. Spherical domains of A in the matrix of B
2. Rodlike domains of A in a matrix of B
3. Alternating lamellae of the two components
4. Rodlike domains of A in a matrix of B
5. Spherical domains of A in a matrix of B

When a better solvent for the minor phase was chosen as the casting solvent, an expansion of the spheres and a shift from spherical to rodlike morphology was noted. Somewhat larger domains were observed for AB block copolymers compared to ABA block copolymers of comparable molecular weight and composition. Very similar morphologies are exhibited by graded block copolymers although the interface as observed by transmission electron microscopy is more diffuse in character. This effect is clearly shown in Figure 6.25 (207).

The existence of some short-range order in block copolymers, particularly those showing rodlike or lamellar morphologies, is evident from the micrographs presented in Figures 6.23 and 6.24. Low-angle diffraction studies by a number of researchers have demonstrated that considerable long-range order is often present as well. McIntyre and Campos-Lopez assigned an orthorhombic macrolattice having unit cell dimensions of 676, 676, and 566 Å to a triblock SBS copolymer cast from a 90–10 tetrahydrofuran/methyl ethyl ketone (THF/MEK) solvent mixture (208). The styrene domains that made up 38.5% by weight of the material were found to be spherical in shape with a mean diameter of 356 Å. Significantly the interdomain distance in these polymers was considerably shorter than that expected for an unperturbed butadiene chain comparable in length to the butadiene block. A direct implication of this observation is that a significant number of molecules run between domains that are not nearest neighbors.

Keller at al. found that the styrene domains in extrudates of Kraton®*102, an SBS block copolymer containing 25% by weight styrene, took the form of very long hexagonally packed cylinders (209). The cylinders were found to be

*Kraton® is a registered trademark of Shell Company.

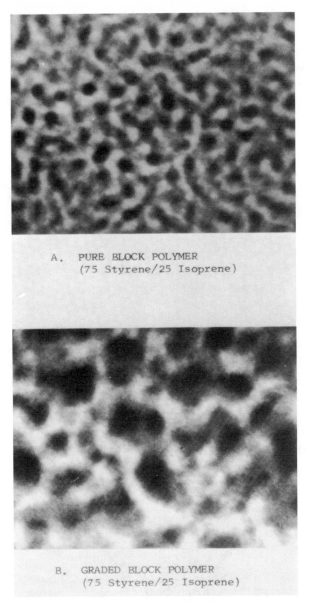

A.   PURE BLOCK POLYMER
(75 Styrene/25 Isoprene)

B.   GRADED BLOCK POLYMER
(75 Styrene/25 Isoprene)

Figure 6.25   Osmium stained thin sections of "pure" (*a*), and "graded" (*b*) styrene/isoprene/styrene block copolymers (207).

150 Å in diameter and to have a lateral lattice spacing of 300 Å. The authors proposed that the extraordinary regular structure resulted from the butadiene midblocks acting as "entropy springs." In this model long-range order is thought to be established through compensation of compressive forces arising from butadiene segments running between nearest neighbors by extensional forces from connections to nonnearest neighbors. Some doubt was cast on this hypothesis by subsequent birefringence studies that showed that only form contributions were important in these samples and that both phases consisted of randomly dispersed chains (210). In a comparable study of Kraton® 1107 containing significantly lower fraction of styrene (10%), Pedemonte et al. report that extrusion has no effect on morphology (211). Moreover, a body-centered cubic lattice of polystyrene spheres was observed in both the as-received polymer and extruded plugs after annealing. Clearly a

more careful analysis of extrusion conditions, sample rheology, and sample composition will have to be made before a definitive association betwen molecular architecture, sample fabrication, and micromorphology can be made.

The unusual star block copolymers synthesized by Fetters (212) also exhibit a high degree of long-range order as shown in Figure 6.26. As in the case of linear block copolymers, both cylindrical (213, 214) and spherical (212) morphologies have been reported for polymers of this type. In a recent paper Bi and Fetters have noted that the overlap of polystyrene spheres in different layers of the films can produce striated structures having the appearance of rods (212). From an extensive analysis of over 250 micrographs, the authors conclude that equilibrium films of both linear and star

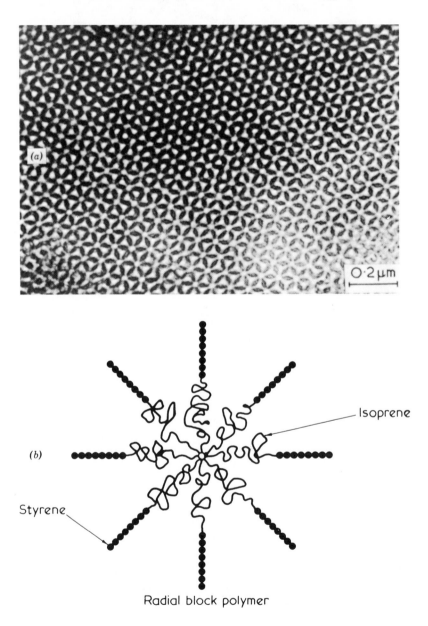

Radial block polymer

Figure 6.26 (*a*) Osmium stained thin section showing morphology of styrene/isoprene star block copolymer; and (*b*) schematic diagram of star block structure (207).

block SIS copolymers exhibit spherical, body-centered cubic morphologies. Lattice dimension data are presented.

6.4.4   Crystalline Block Copolymers

The morphology of block copolymers having a crystallizable component has been the subject of numerous investigations beginning with the studies of Skoulios, Kovacs, and Lotz in the 1960s on the polystyrene-poly(ethylene oxide) (PS-PEO) system (215–218). Surprisingly these copolymers exhibit many of the morphological features characteristic of crystalline homopolymers even when the noncrystallizing component is present in high concentration.

Lamellar single crystals are formed on slow cooling of dilute colutions of PS-PEO copolymers when the dissolution temperature is above the PEO melting point (217). The unit cell dimensions in these crystals are identical to those observed in PEO homopolymer single crystals, and the chain axes are oriented perpendicular to the lamellar basal plane. Since the lamellar thickness (~100 Å) is substantially smaller than the PEO block length, chain folding occurs, and the rejected PS blocks are thought to form an amorphous overlayer approximately 25 Å thick on the crystal surfaces. An analysis of shadow lengths in the electron micrographs showed that both single- and double-layered crystals could be obtained (215), while more complicated

dendritic growths were evident at higher concentrations and lower temperatures. Crystal et al. (219) reported the presence of both single- and double-layer lamellae in their studies of crystalline thin films of PEO-PS copolymers. X-ray diffraction measurements indicated thicknesses of 95 and 180 Å respectively. These authors proposed the alternate sandwich structures shown in Figure 6.27 to account for these observations.

The melting points of these materials show a small but regular decrease with increasing styrene content. This drop has been attributed to imperfections associated with styrene segments in the PEO crystals (216). In contrast, two crystallization temperatures are observed for the PEO component. Lotz and Kovacs suggest that this behavior is associated with the presence or absence of heterogeneous nuclei. A similar effect has been reported by Price in studies of crystallization and polyethylene homopolymers from silicone oil aerosols (220, 221). The volume contraction that accompanies crystallization of the PEO suggests somewhat greater mobility in the PS regions than is observed in the amorphous homopolymer.

Similar morphologies were observed by Kojima and MaGill in crystals of multiblock poly(tetramethyl-*p*-silphenylene-dimethyl-siloxane) TMPS-DMS copolymers grown from benzene and ethyl acetate/methyl alcohol solutions (222). Higher levels of DMS were

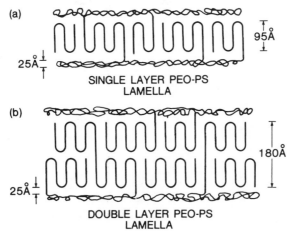

Figure 6.27   (*a*) Single, and (*b*) double layer sandwich structures proposed for PEO-PS block copolymer lamellae.

found to inhibit the formation of dislocations and growth spirals and to depress the crystallization rate. The authors believe that the DMS blocks are preferentially excluded from the crystal lattice. See Figure 6.28.

LeGrand has reported the growth of single crystals of polydimethylsiloxane-polycarbonate copolymers from dilute methylene chloride solutions (223). Electron diffraction studies showed that, in contrast to PEO-PS single crystals, these crystals exhibited a unit cell different from that of either of the components. An unusually high melting temperature was also observed. The authors suggested that these observations may indicate that siloxane blocks are buried within the lamellae and do not form a separate amorphous overlayer. A photomicrograph is presented in Figure 6.29.

Spherulitic superstructures are commonly observed when thin films of block copolymers are crystallized from the melt or cast from solution. As in the case of noncrystallizing block copolymers, considerable variation in morphology occurs when different casting solvents are employed. In general, lower levels of crystalline perfection are obtained when the solvent system is preferential for the amorphous block. Skoulios et al. have noted that considerable molecular organization may already occur in the casting solution before evaporation (218). These structures have been collectively referred to as mesomorphic gels.

Kovacs (217) has found that well formed spherulites can be grown from PEO-PS copolymer melts when the styrene content is as high as 50%. Similar results were obtained by Li and Magill on TMPS-DMS block copolymers, although rougher textures were observed at high DMS levels (224). It is important to remember that Magill's materials were short-sequence multiblock copolymers in contrast to the longer AB blocks studies by Kovacs. In both cases the overall crystallinity decreased rapidly as the fraction of noncrystallizing comonomer increased. For the TMPS-DMS block copolymers the maximum overall crystallization rate was displaced downward in temperature and reduced in magnitude as the concentration of DMS was increased. A similar decrease was observed in the equilibrium melting point obtained from Hoffman-Weeks plots. These drops were attributed to higher crystal surface free energies associated with a more diffuse interface. Although similar kinetic

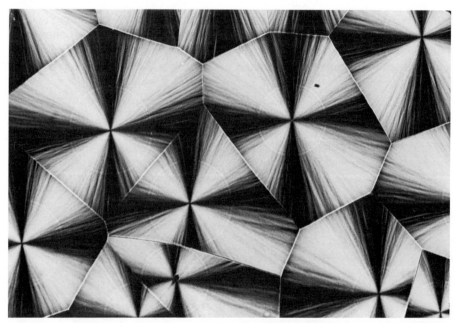

Figure 6.28   Spherulitic thin film of poly(tetramethyl-*p*-silphenylene-dimethylsiloxane) (TMPS-DMS) block copolymer cast from methylene chloride. Photograph courtesy of R. W. LaRochelle, General Electric Corporate Research and Development.

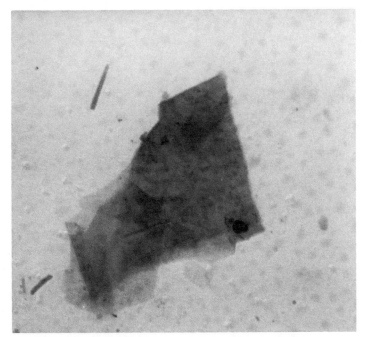

Figure 6.29 Single crystals of polydimethylsiloxane-polycarbonate multiblock copolymers crystallized from methylene chloride (223).

data is not available for bulk crystallized PEO-PS copolymers, studies of dilute solution crystallization show that the rates are not too much different from those observed for homopolymer PEO (225). The effects of block length, multiple blocking, and composition on the bulk crystallization kinetics of block copolymers merit further investigation.

6.4.5 Graft-Block Copolymers

Block copolymers in which the second component is grafted onto the chain backbone rather than polymerized linearly from its ends, form the basis of a class of engineering thermoplastics of major economic importance. HIPS (high-impact polystyrene) and ABS (acrylonitrile-butadiene-styrene) resins are noteworthy examples. These materials are usually prepared by bulk polymerization of styrene monomer containing dissolved rubber followed by phase inversion or by various types of emulsion polymerization (226–228). An extensive patent literature exists describing various modifications in the preparation procedure and the reader is referred to the reviews by Fetters and Maclay (229) and by Amos (230) for further details.

The morphology of graft-block copolymers of this type has been the subject of numerous investigations (228, 231–236). The primary tools have been optical and electron microscopy although light scattering, coulter counter measurements, and gel separation techniques have also been employed. Many of these studies have been summarized by Bucknall (237). In general these materials are distinguished from the linear block copolymers described by the larger domain size of the rubber inclusions (generally 0.5–$5\,\mu$) and from highly dispersed blends by the presence of numerous glassy occlusions within the rubbery domain. These features are readily apparent in Figure 6.30 (238).

Rubber particle formation during the polymerization of HIPS has been studied in detail by Keskkula and his co-workers (233, 234) and by Freeguard (235). These authors found that the intensity of mixing during the polymerization has a dominant effect on the final morphology of the copolymer. If little agitation is available, phase inversion may not take place, and the rubber phase ends up as a continuous network containing embedded polystyrene particles. These structures prevent the material from behaving as a thermoplastic

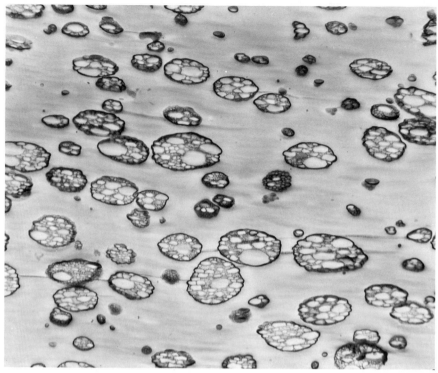

Figure 6.30 Osmium stained thin section of HIPS showing rubbery domains and occluded styrene particles.

polymer and can only be broken up by extensive mastication. Keskkula and Traylor note that the fine structure of rubber particles produced from a well mixed styrene solution is not substantially altered if the amount or type of rubber is changed or if the toughened polymer is blended with homopolyer styrene (234).

The overall size of the rubber inclusions in bulk polymerized HIPS can be markedly affected, however, by a number of factors including the relative viscosity of the components at the time of inversion, the intensity of mixing, and the addition of prepolymerized polystyrene to the reaction mixture (238–240). An example of the changes that occur in the size distribution of rubber particles under different mixing conditions is shown in Figure 6.31 (238). Freeguard and Karmarkar have reported the agglomeration of rubbery domains in HIPS during testing in a cone and plate viscometer at shear rates of 130/sec suggesting that some changes in dispersion may be expected during processing (239). The

agglomeration of occluded styrene domains appear to be rare in all cases.

The size of the dispersed particles in block-graft copolymers prepared by emulsion polymerization, such as ABS, is governed primarily by the surfactant concentration during the initial butadiene polymerization. The subsequent grafting of styrene and acrylonitrile monomers, however, produces a unique core-shell structure in which polybutadiene spheres approximately 0.5 μ in diameter are encap-

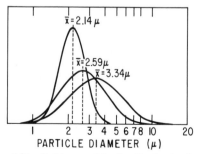

Figure 6.31 Rubber particle size distribution functions in high-impact polystyrene samples prepared under different stirring conditions (238).

sulated in 0.1 μ thick styrene-acrylonitrile shells (241–242). As in the case of bulk polymerized HIPS, small glassy subinclusions are present in the rubber. See Figure 6.32.

As noted at the beginning of this section, graft-block copolymers represent the major commercial route to toughened engineering thermoplastics. This situation is in part economic since the synthesis of these materials is simpler and less expensive than that of linear block copolymers of similar composition. More important, however, the unique microstructure of graft block copolymers such as HIPS plays a critical role in the impact toughening processes. Since the interplay of morphology, mechanical behavior, and engineering utility of these materials is so intricate, a short discussion of their deformation and failure follows. For a more comprehensive treatment the reader is encouraged to consult the excellent review recently given by Bucknall (237).

Polymers typically fail by one of two mechanisms: crazing or shear yielding. The first process, which is favored under conditions of high-dilational stress, is characterized by the formation of bands of highly voided material running perpendicular to the applied stress. The density of crazes is normally 40–60% of

that of the bulk polymer, and their structure consists of a series of elongated strands of resin punctuated by microvoids (243). Shear yielding, in contrast, is volume conservative and results from plastic flow of the resin along the lines of maximum shear stress, which lie at an angle of 45° to the load. Both crazing and shear yielding are capable of soaking up large amounts of stored energy and are enhanced by the presence of a discontinuous phase (particles or voids) which acts to localize and concentrate the stress field. When crazing is the preferred mode of deformation, as it is in the case of a brittle polymer like polystyrene, it is important to maximize the size of the craze field while effectively terminating craze growth in order to dissipate the largest possible amount of energy before catastrophic cracking occurs. Several criteria must be satisfied to optimize the toughness of such a two-phase system.

The dispersed rubber particles must be well bonded to the matrix in order to permit effective stress transfer between the two phases and to provide an effective craze termination mechanism. If the particle is pulled away from the matrix, the craze is regenerated on the opposite side of the void and continues to grow in size until a crack develops and the

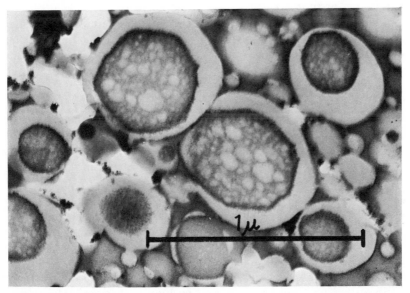

Figure 6.32   Osmium stained thin section of ABS particles prepared by emulsion polymerization showing core-shell structure and occluded styrene domains (242).

material fails. The requirement for good adhesion is an important reason for the superior impact resistance of graft-block co-polymers over blends (244, 245).

When crazing is the dominant mode of deformation there is also an optimum rubber particle size for effective toughening. Although the number of crazes increases roughly in proportion to the number of particles present, craze termination drops off rapidly as the particle diameter becomes comparable to the craze thickness and the rubbery domains become encapsulated in the growing crazes. Thus, for a given matrix and a fixed concentration of dispersed rubber, the size of the craze field is established by the tradeoff between craze initiation and termination. There is considerable variation in the most effective rubber particle size depending on the matrix resin and its tendency to craze or shear yield. A diameter of $1.0\ \mu$ or larger is required in HIPS whereas particles only $0.3\ \mu$ in diameter are sufficient in ABS. These size ranges, which are considerably larger than those found in linear block copolymers, are easily obtainable with current graft-block polymerization techniques. Moreover, cross-linking of the rubbery phase, which occurs in the last stages of polymerization in materials such as HIPS, permits retention of domain size during processing. Degradation of rubber particle size during melt blending remains a major concern in homopolymer and linear block copolymer blends (244) as noted in the following section.

The internal rubber particle morphology may have a considerable effect on its toughening potential. By raising the modulus of the rubbery phase, glassy occlusions improve stress transfer between the matrix and dispersed phase, and a greater fraction of the load can be born by the rubber. A higher modulus material is obtained, but crazing efficiency is reduced (246). In some cases this tendency may be offset by an increase in the volume of a given rubber particle due to inclusions to constant overall rubber concentration that reduces the distance between rubbery domains and inhibits craze growth (247). Comparison of HIPS specimens having composite rubber particles with the best block copolymer blends

in which no glassy occlusions occur in the rubber phase has shown that pure rubber particles give the largest improvements in impact strength (244).

The strong correlation between the micromorphology and physical properties of these materials and the continued interest in developing tougher thermoplastics with delicately tailored properties assures that active research on the structure and performance of graft-block copolymers will continue.

6.5 THE MORPHOLOGY OF POLYMER BLENDS

Interest in polymer blends has expanded greatly in recent years as plastics producers have become aware that blending can improve the physical properties, facilitate the processing, and lower the cost of existing materials. For the most part polymer pairs are incompatible, and blends of commercial thermoplastics are typically two-phase composite materials (248). The physical properties of these materials are highly dependent on both the morphology of the dispersion and the interfacial adhesion. These features are in turn dictated by the rheological characteristics of the components and by the way in which they are mixed rather than by thermodynamic or structural considerations. The latter factors come into play, however, in blends exhibiting a limited degree of compatibility or the ability to crystallize. Both types of materials are considered in this review.

The general inability of long-chain molecules of different chemical structure to form equilibrium solutions is thermodynamic in origin and results from the requirement that the free energy of mixing must be zero or negative. In thermodynamic terms

$$\Delta G_m = \Delta H_m - T\Delta S_m \leq 0 \qquad (6.15)$$

where ΔG_m is the Gibbs molar free energy and ΔH_m and ΔS_m are the molar enthalpy and entropy of mixing. In most cases ΔH_m is positive, and mixing takes place only if the entropy increase due to the disordering of the two phases is large enough to offset the heat effect. This requirement is often met in mixtures of small molecules. In high polymers, however, a large part of the entropy increase is configurational in nature, and this quantity does not increase in proportion to the molecular volume as do the enthalpy and interaction entropy. As a result, the free energy of mixing is dominated by the enthalpic contribution, and most polymer pairs are incompatible. The condition for polymer compatibility is often expressed in terms of the Flory-Huggins interaction parameter χ_{12}, using the relationship

$$\Delta G_m = RT(n_1 \, ln\phi_1 + n_2 ln\phi_2 + \chi_{12} n_1 \phi_2)$$

$$(6.16)$$

where n_1 and n_2 and ϕ_1 and ϕ_2 are the mole and volume fractions of the two components, respectively. For mixing to occur χ_{12} must be very small or negative.

A great deal of the work on the morphology of two-phase blends has been phenomenological in nature. A notable exception is the work of Van Oene (250). The theoretical relationships developed in this study provide a framework for predicting whether the composite will have a dispersed or a stratified morphology. Van Oene shows that the interfacial tension $\gamma_{\alpha\beta}$ between two phases in flow can be expressed as:

$$\gamma_{\alpha\beta} = \gamma_{\alpha\beta}^\circ + \frac{1}{6} a_\alpha \left[(\sigma_2)_\alpha - (\sigma_2)_\beta \right] \quad (6.17)$$

$$\gamma_{\alpha\beta} = \gamma_{\alpha\beta}^\circ + \frac{1}{6} a_\beta \left[(\sigma_2)_\alpha - (\sigma_2)_\beta \right] \quad (6.18)$$

where $\gamma_{\alpha\beta}^\circ$ is the static interfacial tension, a_i is the droplet size, and $(\sigma_2)_i$ is the second normal stress difference. Several conclusions may be drawn from this analysis:

1. When $(\sigma_2)_\alpha > (\sigma_2)_\beta$ phase α will always form dispersed droplets in phase β. The droplets may or may not be elongated into fibers.

2. Phase β will form droplets in phase α only if the initial droplet radius is less than or equal to $6\gamma_{\alpha\beta}^\circ / [(\sigma_2)_\alpha - (\sigma_2)_\beta]$. For polymer melts

this quantity is the range $0.1-1 \, \mu$. If this condition is not met, stratification occurs.

3. Droplet formation in phase β occurs when α_β is less than a micron and leads to composite droplets containing dispersed droplets of the α phase.

4. The homogeneity of the dispersion, but not the mode of dispersion, is affected by differences in viscosity, shear rate, temperature, and residence time.

Experimental data collected by Van Oene on blends of polystyrene, poly(methylmethacrylate), and a polyethylene based ionomer, support his hypothesis and indicate that this theory may have general applicability to a variety of polymer-polymer and polymer-plasticizer blends. Similar but less explicit conclusions were reached by Starita from empirical studies of dispersion in polystyrene-polyethylene blends of different molecular weight and viscosity (251). Starita found that coarse dispersions were formed when the viscosity and elasticity of the minor component exceeded that of the major component, and fine dispersions were formed in the reverse situation. Fine dispersions were also formed when the two materials exhibited similar viscoelastic characteristics regardless of which was present in excess. Subsequent studies of other melt-blended polymer pairs have led to similar conclusions (252, 253).

When the discontinuous phase can agglomerate and stratify, laminated structures may be formed. When lamination occurs in molded parts both the physical properties and aesthetic appearance of the article may be harmed. For reasons as yet unclear, blends containing a saturated rubber copolymer seem more susceptible to lamination than those containing a corresponding unsaturated rubber. Koiwa has found that delamination in blends of styrene-acrylonitrile and ethylene-vinylacetate copolymers can be eliminated by grafting and cross-linking the particles prior to extrusion (265). Studies (see Figure 6.33) of the dispersion of Kraton G®* (saturated SBS block copolymer)

---

*Kraton G® is a registered trademark of Shell Company.

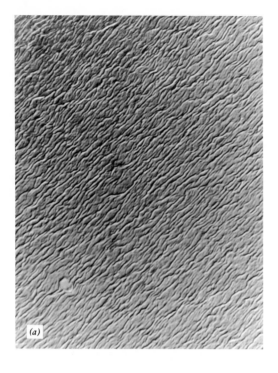

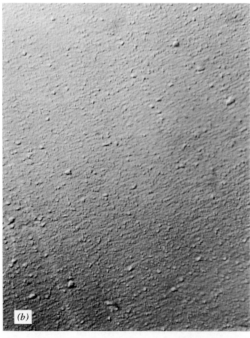

Figure 6.33 Nomarski interference micrographs of (a) 90–10 polystyrene/Kraton® G (low molecular weight) blend showing elongated particles, and (b) 90–10 polystyrene/Kraton® G (high molecular weight) blend showing spherical particles.

samples of different molecular weight and block length in homopolymer polystyrene indicate that a transition from a stratified to a particulate morphology occurs as the molecular weight of the copolymer is increased (266). These studies, although limited, imply that chemical or physical stabilization of the dispersed phase may be required to insure that lamination is not encountered when processing blends containing a saturated rubbery block copolymer.

Heikens et al. noted that the degree of dispersion in polystyrene-polyethylene blends may be diminished if the concentration of the low viscosity dispersed phase is sufficiently high that the overall viscosity of the system is reduced (253–255). Increased coalescence is also observed under these conditions. Likewise, the dispersion of a high viscosity discontinuous phase improves with concentration as the overall viscosity of the system increases, although the improvement is ultimately limited by increasing agglomeration of the dispersed particles. In all cases much finer dispersions are obtained if a small amount of copolymer surfactant is added to the binary blend (253, 254; see Figure 6.34). The yield stress, modulus, and impact strength of these blends is also improved as a result of increased adhesion between the two phases (255).

The control of the morphology, and particularly the size of the dispersed phase in melt-mixed polymer blends, is of primary importance in effective impact modification. This requirement was emphasized in the preceding discussion of graft copolymer morphology. Early attempts to toughen polystyrene by blending with rubber were of limited success because of poor bonding between the two components and because of inadequate control over the level of dispersion (256). Blending with block copolymers gave improved adhesion and impact strength (257). Likewise, the addition of small amounts of peroxide during mixing resulted in enough cross-linking of the rubber to limit the level of dispersion (258) and increase the impact strength. More recently Durst et al. (244) found that SBS block copolymers of sufficiently high molecular weight to resist breakup during mixing were

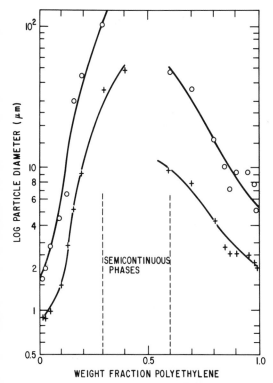

Figure 6.34 Average particle dimensions in polystyrene/low-density polyethylene blends after Brabender mixing (253).

even more effective in toughening the homopolymer. For example with $\overline{M}_n = 65,000$ the average particle diameter was $0.2\,\mu$m and the notched Izod impact strength only 1.0 ft-lb/in. When $\overline{M}_n = 250,000$ the particle diameter increased to 4 μm and the Izod impact strength to 7.5 ft-lb/in. Bucknall has noted that these results indicate that pure rubber particles are more effective in impact modification than are the composite particles characteristic of graft copolymers (237).

The morphologies of blends of polystyrene and styrene-isoprene block copolymers precipitated from solution rather than melt blended have been investigated by Kawai and his coworkers (259). In cases where the homopolymer molecular weight was comparable to the styrene block length, the homopolymer was incorporated directly into the existing styrene domains with a corresponding increase in size. When higher molecular weight styrene was used, the copolymer phase separated without solubilization of the homopolymer in the

occluded styrene domains. There was evidence, however, of penetration of external styrene end blocks into the higher molecular weight matrix. Similar results have been obtained by Niinomi et al. (260). Addition of polybutadiene to the copolymer apparently has the same effect as that of styrene, with low molecular weight polymer being taken up in the existing rubber domains and higher molecular weight polymer phase separating (261). Similar results are expected, but as yet unconfirmed, in melt-mixed systems.

In the unusual situation where one of the components of the copolymer is miscible with a chemically dissimilar homopolymer the properties of the blend may be markedly superior to those of either material (262, 263). An excellent illustration is provided by the PPO®* resin-HIPS blends manufactured and sold under the tradename Noryl®† resin by the General Electric Company (264). The morphology of these blends is very similar to that of HIPS although the matrix is in this case a solid solution of polystyrene and polyphenylene oxide. The high-heat-distortion temperature and ductility of PPO is largely retained while the rubbery domains offer substantial improvement in impact strength. As in comparable blends with homopolymer polystyrene, excellent adhesion between the two phases results from penetration of the styrene blocks into the surrounding PPO matrix. Moreover, a broad range of physical properties can be obtained by adjusting the composition of the blended resin. Cross-linking in the rubber phase assures retention of the size and shape of these domains during processing.

A number of rubber-modified acrylic polymer blends, which appear to be one-phase materials because the refractive indices of the two components are matched, form two-phase systems similar to those described previously. Bauer and Pallai have shown that both the particle size and interparticle separation distance in these materials can be effec-

---

*PPO® is a registered trademark of the General Electric Company.
† Noryl® is a registered trademark of the General Electric Company.

tively characterized using laser light-scattering techniques (267, 268). For samples of a saturated acrylic plastic containing from 25 to 50% of an acrylic graft rubber, these authors found that the average domain radius remained constant at approximately 0.25 μ while the distance between neighbors decreased from 1.05 to 0.78 μ. These results were found to agree well with those calculated on the basis of pure geometrical considerations for blends of these compositions.

The morphology of incompatible polymer blends in which one or both components is capable of crystallizing is not altered significantly by the crystallization process since the materials remain phase-separated in the melt and subsequently crystallize independently in their own domains (252–254). Both the crystal structure and crystallization kinetics are identical to those observed for the homopolymers. Exceptions to this behavior are noted only if one polymer serves as a nucleating substrate for the other (104, 105, 269) or if partial mixing of the two materials occurs (270–273). Wahrmund et al. found that for blends of BPA polycarbonate and poly(butylene terephthalate) amorphous phases for the pure components as well as for solutions of the two materials could be detected depending on the blend composition (270). Some dependence of the composition of the amorphous phases on prior thermal history was indicated and the possibility of chemical reaction was noted by the authors. Crystallization rates were not reported, but a small depression in the PBT melting point in blends rich in polycarbonate was ascribed to the diluent effect of the amorphous component. Solvent etching of injection-molded polycarbonate-polybutylene terephthalate (PBT) blends carried out by Hobbs has shown that the polycarbonate phase separates from the PBT matrix at concentrations up to 40% by weight, forming domains 1–10 μ in diameter (274).

A much higher degree of miscibility has been observed in blends of poly(vinylidinefluoride) (PVF$_2$) and PMMA (275). Both dynamic mechanical measurements and glass transition temperature data indicate that seg-

mental mixing occurs in these blends (276). At concentrations of PVF$_2$ greater than 50% some crystallization of this component occurs. This transformation appears to be a true liquid-solid phase transformation that is not preceded by demixing of the two components. A more careful analysis of the crystallization of PMMA-PVF$_2$ films cast from dimethyl formamide (DMF) has been carried out by Nishi and Wang (277). Plots of T_m and T_g for these films showed a steady decrease in the PVF$_2$ melting point with increasing PMMA concentration as a result of the diluent effect. The magnitude of this depression was satisfactorily accounted for using expressions developed by Scott from the Flory-Huggins treatment of crystallization in polymer mixtures (278). Surprisingly, the T_g's measured for these samples remained more or less constant, indicating that the uncrystallized material contained very little residual PMMA. An analysis of the crystallization kinetics showed that complete suppression of crystallization occurred at cooling rates in excess of $10°$ C/min for samples containing up to 50% PVF$_2$. No detailed morphological descriptions of the PVF$_2$ crystallites were given.

The morphology and crystallization kinetics of isotactic polystyrene crystallized from blends of isotactic and atactic polystyrene (APS) have been investigated in some detail (279–281). Yeh has noted that the spherulitic structure becomes more coarse with increasing percentages of APS in the blends and with increasing APS molecular weight (29). X-ray line-broadening studies indicate that the noncrystallizing component resides between the spherulitic fibrils rather than within the crystallite domains. The spherulite growth rate is observed to decrease with the molecular weight of the APS in the ranges between 4800 and 19,800 and between 51,000 and 1,800,000. An anomolous increase in growth rate, which is observed in the presence of APS of intermediate molecular weight, is ascribed to entrapment of the noncrystallizing diluent. The authors do not postulate a more detailed mechanism for this process. A linear increase in growth rate with isotactic polystyrene content is observed in blends containing APS of a given molecular weight. In contrast, the crystallization rate of poly-ε-caprolactone (PCL) from PCL-poly(vinyl chloride) (PVC) blends, increases nonlinearly with PCL content because of the variation of T_g with composition (282). The spherulite growth kinetics are adequately represented by substitution of the Williams, Landell, and Ferry (31) equation into a modified version of the expression for spherulite growth rate developed by Hoffman (284).

A few unusual cases have been noted in which two crystallizable polymers are not only thermodynamically miscible in the melt but also exhibit isomorphic replacement on crystallization. PVF and poly(vinylidene fluoride) are one such pair. Natta et at. found that mixed crystals of these polymers are formed over the entire range of compositions where the average level of crystallinity remains constant at about 70% (285). The type II PVF_2 crystal structure is observed at all concentrations.

Segmental mixing of normally incompatible polymers can be achieved by dissolving the pair in a common solvent and rapidly quenching the solution below its glass transition temperature. Intimately blended samples of PMMA and poly(vinylacetate) and PS and PMMA have been obtained from benzene and napthalene solutions using this technique (286, 287). In both cases the solvent was subsequently removed by freeze-drying. Shultz has reported that PMMA-PS samples prepared by this method show one glass transition temperature that is intermediate between those of the two components. Phase separation appears to occur spontaneously on passing through T_g, and the normal glass temperatures for the isolated components are observed on reheating in the differential scanning calorimeter (DSC) (288). The changes in local molecular order and morphology that accompany this transition have not yet been elucidated. It is hoped that future work in this area will provide information on the kinetics of demixing as well as improved values for thermodynamic parameters.

6.6 THE MORPHOLOGY OF PROCESSED POLYMERS

Variations in temperature, pressure, and shear and elongational flow, which are intrinsic features of almost all polymer processing operations, have a marked effect on the morphology of the fabricated parts. These structural changes in turn may strongly influence the performance and mechanical properties of the final product. As thermoplastics find increasing use as metal substitutes in engineering and load-bearing applications, it has become increasingly important to characterize these morphological changes, to understand how and why they develop, and to define their relationship to the behavior of the material. The following discussion focuses on injection molding since, with the exception of solid-state deformation processes such as drawing or biaxial orientation, this process exposes the polymer to almost all of the effects commonly encountered in processing (289).

The morphologies that develop during injection molding depend on flow, thermal history, hydrostatic pressure, and shear deformation. Under elevated pressure, for example, both the melting point and glass transition temperature are displaced upward causing solidification to occur more quickly than under ambient conditions. This effect is often more evident near the gate region where higher pressures are experienced. Polymers that vitrify under pressure at temperatures higher than T_g have somewhat lower free volumes at room temperature than those prepared under ambient pressure and may thus exhibit markedly different mechanical properties and thermal aging characteristics (290). If the polymer is crystalline, the nucleation density increases and the spherulite size decreases with increasing pressure. Fine control of morphology may thus be obtained by systematic variation of the injection pressure. Unfortunately neither of these effects has been investigated in sufficient detail, although significant increases in density and dynamic modulus have been observed in samples injection molded at elevated pressures (289). In a more recent study Reinshagen and Dunlap have made use of the Clausius-Claperpon equation to quantify both temperature and pressure effects on the crystallization process in terms of the degree of undercooling (291). Experiments on isotactic polypropylene under conditions approximating those encountered during injection molding show that this parameter can be used to predict nucleation density, spherulite diameter and growth rate, and lamellar thickness with considerable success.

Injection molding is a dynamic process, and temperature equilibration of the polymer melt is almost never achieved in the mold cavity. As a result injection-molded parts invariably display a characteristic skin/core morphology. In amorphous polymers this may amount to no more than a thin band of oriented material near the cool mold surface. In multiphase and crystalline polymers, however, distinct mor-phological zones of varying degrees of complexity may develop (292–297, 237), as shown in Figure 6.35. Three primary zones are often identified: an oriented but noncrystalline skin; a subsurface zone showing a high amount of shear orientation or, as sometimes the case in crystalline polymers, considerable row nucleation and transcrystalline growth; and a core zone in which the morphology approximates that of the quiescent material. Generally, the thickness of the skin and shear zones increases relative to the core zone with decreasing melt and mold temperature. Significant variations in the detailed morphology are observed, however, depending on the polymer and processing conditions.

In polypropylene injection moldings a change in crystal structure can be induced in the high-shear region. Kantz et al. note that row-nucleated, metastable type III spherulites

Figure 6.35   Typical skin-core morphology of injection molded poly(butylene terephthalate) (297).

having a hexagonal crystal structure often predominate just below the surface skin (296, 298). Under other conditions this transcrystalline morphology may be less well defined, with a steady decrease and then increase in spherulite size being observed on passing from the skin to the core region (295). This unexpected behavior has been explained by assuming that the temperature at which the maximum nucleation rate occurs is established below the surface of the molding as a result of the thermal gradient set up by the cold mold wall. Lower nucleation densities and larger spherulite sizes are observed on both the high- and low-temperature sides of this isotherm. Variations in local crystal orientation, which are not obvious in optical micrographs of thin cross sections, have been detected in wide-angle x-ray diffraction patterns. (295) In contrast, moldings of PBT show a gradual transition from a nonspherulitic to a volume-filling spherulitic morphology on passing from the skin to the core region without evidence of shear crystallization (297). Both the viscosity and elasticity of the PBT melt are significantly lower than those of polypropylene, and the morphology suggests that stress relaxation may be relatively complete before crystallization takes place.

Studies of the morphology of injection-molded polyacetal carried out by Clark have shown that a considerable variation in skin thickness can be observed along the length of the part (292). The increase in skin thickness near the gate region is attributed to a decrease in spherulite size and faster rate of crystallization brought about by increased molecular alignment and an upward displacement of the melting point as a result of flow constriction and pressure buildup at the gate. The effect is particularly severe in large molds that are improperly gated. Regardless of the skin thickness, the cross-sectional morphology of injection-molded polyacetal consists of an outer layer of fibrillar lamellae oriented perpendicular to the surface and parallel to the injection direction, a transcrystalline region of twisted lamellae, and a spherulitic core. It is significant that the transcrystalline growth is nucleated on oriented crystals near the wall

but develops in a region where the residual axial stress is low, as evidenced by the random orientation of crystals with respect to the flow direction in this region. A comparison of the subsurface morphologies of polypropylene, PBT and polyacetal indicates that intrinsic differences in stress relaxation rates as well as differences in crystallization rate can influence the structure of injection-molded parts.

Heterogeneous nucleating agents are often added to injection-moldable thermoplastics to expand the range of processing conditions under which satisfactory parts may be obtained (299). These agents bring about an increase in the crystallinity and overall crystallization rate and a corresponding decrease in the average spherulite size, but do not significantly alter the spherulite growth characteristics. Recent studies of the crystalline texture of nylon 6 moldings, however, indicate that heterogeneous nucleation may also result in an increase in crystal anisotropy (308). The origins of this anisotropy are unclear but appear to arise from a prior alignment of the nucleating particles in the flow field rather than from preferential crystal growth along the flow axis. More careful microscopic analysis will be required before a less ambiguous model can be developed.

In multicomponent polymers the dispersed phase acts as a microscopic tracer, and the morphology to a large degree reflects the characteristics of the flow field. Highly elongated particles are commonly observed near the mold surface. In the subsurface shear region less elongation is observed since a greater time is available for relaxation, and the major axis of these eliptical particles is oriented roughly 45° to the flow direction. Largely undeformed particles are observed near the core. Since the system is two-phase, particle migration away from the mold walls can occur (301). The rate of migration increases with the shear gradient, particle size, and distance from the center of flow (302). The lateral movement of dispersed particles often results in partial agglomeration and "necklace" formation, creating stratified structures parallel to the mold surfaces (303). Some physical properties such as tensile and impact strength may be enhanced by such structures because in the usual test configurations the direction of stress runs parallel to the orientation. Much poorer properties are generally observed in the transverse direction, however. As a result, careful correlation between morphology and test procedures must be made to insure that the data reflect real material properties rather than local structural variations.

Most large injection moldings are fabricated using molds with multiple gates, and weld lines invariably form where the separate flows come together. Surface defects as well as discontinuities in internal morphology reduce the strength of these regions compared to that in the rest of the material. Several discussions of knit-line morphology have appeared in the literature (304–306); see Figure 6.36. Extensional flow along the parabolic melt front results in molecular orientation parallel to the knit line when two melt fronts meet (307). In crystalline polymers transcrystalline regions are often nucleated along this line of deformed molecules and grow perpendicular to the knit line. More complicated crystal growths may develop at futher distances from the knit line as a result of local fluctuations in temperature and deformation (305).

In most cases fracture occurs at the knit line itself because of limited intermolecular diffusion across the boundary as well as stress concentrations that build up at the surface notch. Microscopic studies have shown that the failure zone in crystalline polymers can be displaced away from the knit line, however,

when there is an inherent weakness between the transcrystalline zone and the surrounding material (305). A similar phenomenon may occur in polymer blends when there is agglomeration and buildup of the dispersed phase in a region adjacent to the knit line (306). More extensive studies of the kinetics and ultimate extent of molecular diffusion across boundaries of this type, together with morphological and mechanical property data, are needed to improve current understanding of the factors effecting knit-line development and behavior.

6.7 POLYMER MICROSCOPY

A great deal of our current knowledge of polymer morphology has been obtained through the use of optical and transmission electron microscopy (TEM) and, more recently, through the use of scanning electron microscopy (SEM). In the course of such investigations a number of specialized sample preparation and observation techniques have been developed that are unique to the study of high polymers. Some of these methods are extensions of those procedures used for the study of metallographic and biological systems, whereas others have been developed exclusively for various classes of synthetic thermoplastics. Surprisingly, there has been relatively little effort to summarize these methods for those actively working in this area, although the microscopy of rubbers has

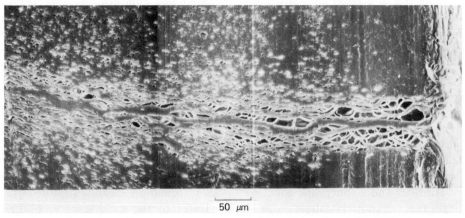

50 μm

Figure 6.36 SEM photograph of hexane extracted knit line in polypropylene-EPDM (85–15) blend (306).

been discussed by Krause (308) and that of multiphase polymers by Thomas (309). A limited review focusing primarily on staining, decoration, and etching procedures is presented in this section.

6.7.1 Staining

One of the most popular methods of developing morphological contrast has been through the use of heavy metal staining. Selective absorption of such stains in amorphous regions, or regions of unsaturation, increases the electron density of these areas of the sample, making them distinguishable in the electron and, in some cases, the optical microscope. Some of the first successful staining experiments were reported by von Hess in studies of cellulose and poly(vinyl alcohol) and later of nylon 6 (310, 311). In the first case von Hess was able to identify periodic crystal structures of the order of a few hundred angstroms after preferential staining of the noncrystalline regions with a mixture of iodine and sulfur (310). A similar level of contrast was obtained in melt-crystallized nylon 6 by first soaking the samples of hydrochloric acid and then treating with phosphotungstic acid, $H_3[P(W_3O_{10})_4]$ (311). Photomicrographs showed that lamellae approximately 70 Å thick were resolvable in nylon 6 thin sections stained in this manner.

Kanig has more recently used a chlorosulfonation technique to differentiate the crystalline and amorphous regions of polyolefins (312). Sulfonic acid groups are introduced into the amorphous regions of the polymer by treating with chlorosulfonic acid for 16 hr at 60°C. Uranyl sulfonate is subsequently formed by the addition of aqueous uranyl acetate. This methods has also been used with considerable success by Hodge and Bassett for investigating the fine structure of pressure-crystallized extended-chain polyethylene (313).

Peterlin et al. were able to preferentially stain the amorphous domains in fibrils obtained from deformed polyethylene single crystals be exposure to iodine vapor (314). Diffusion of the stain was found to be slow at room temperature, and the best results were obtained after 24 hr at 60°C. By staining the crystals prior to deformation or annealing, it was possible to observe changes in crystal thickness resulting from such treatment.

The osmium tetroxide staining technique popularized by Kato has been particularly successful in elucidating the morphology of two-phase polymer systems in which one phase contains residual unsaturation (315, 316). The staining is accomplished by exposing the sample to OsO_4 vapor (317) or to an aqueous or organic solution (318). The use of an organic solvent is especially efficacious when the matrix material is heavily cross-linked as in thermosetting resins. In the stained material the osmium appears to be present as a complex of the form

Significant advantages are obtained by staining the samples prior to microtoming since the osmium hardens or "fixes" the rubbery phase, reducing compression and deformation and making the sample easier to cut. Osmium staining has been successfully used in HIPS (233) and ABS, rubber-modified PVC (319), propylene-EPDM blends (320), and epoxy systems (321), as well as in a variety of block copolymer systems (see Section 6.7.3).

In general, saturated rubber systems are not ammenable to this staining technique (322) although in certain cases it appears possible to develop contrast as a result of preferential absorption of OsO_4 rather than through actual chemical reaction. This effect has been observed in polycarbonate blends and block copolymers (323) and in semicrystalline poly-(ethylene terephthalate) (324). Niimoni et al. have also found that sufficient phase contrast for TEM observation can be achieved in copolymer systems in which the components have different levels of unsaturation or contain functional groups such as —OH, —O—, or —NH₂ that may display different levels of reactivity with the osmium vapor (322).

The "ebonite" staining technique described by Smith and Andries likewise depends on slight differences in the levels of unsaturation and reactivity for contrast (325). In this method, a sample of mixed rubber is immersed in a solution of molten sulfur, zinc stearate, and an accelerator at 120°C for 8 hr. After this treatment the compound is sufficiently hardened to permit microtoming at room temperature. The higher sulfur and zinc content of the harder phase renders it darker in the electron microscope. Contrast is further improved by the fact that the harder phase shows less tendency to thin out as a result of heating in the electron beam.

Another technique for differentiating the two phases in rubber-SBR blends has been developed by Hess and his co-workers (326). In this method, the vulcanizates are mixed with blends of acrylic monomers that are subsequently polymerized. A higher content of the cured acrylate is obtained in the natural rubber phase as a result of the difference in monomer solubility in the two rubbers. Thin sections are cut directly from the embedded samples for observation by TEM. When the acrylate depolymerizes in the electron beam, the phase containing the excess embedding medium thins relative to the other phase, and the resulting difference in thickness allows the two phases to be distinguished in much the same way as if the samples were stained.

6.7.2 Decoration

TEM relies primarily on local fluctuations in electron density for contrast and as such is not sensitive to small differences in surface topography. Surface relief can be accentuated by standard low-angle shadowing and replication procedures (327), however, and these methods have been used extensively in studies of the surface morphology of crystalline polymers the limit of vertical resolution provided by shadowing is about 25 Å.

A much more sensitive technique for investigating variations in surface roughness has been developed by Bassett and his co-workers (328, 329). These investigators found that very thin layers (~3 Å) of gold and several other

metals that are deposited by vacuum evaporation nucleate in islands on the surfaces of lamellar single crystals without forming continuous films. These islands tend to align along any vertical steps or imperfections on the surface. The nucleation density, and hence the level of resolution obtained, is found to depend on the type of metal used as well as the surface roughness. The gold decoration pattern may be observed directly if the crystals are very thin or removed by the subsequent evaporation and stripping of carbon films. A number of features including growth spirals, collapsed pleat patterns, crystal growth sectors, and areas high in amorphous content have been visualized using this technique. Gold decoration has also been found to be useful in characterizing the morphology of solution cast nylon films (330) as well as shish-kabob structures produced from stirred polyethylene solutions (331). In the latter studies distinct Au periodicities characteristic of folded-chain domains were observed in the central "shish."

A novel self-decoration technique has been employed by Kovacs and Gonthier to study the morphology of low molecular weight PEO crystals grown in thin films (332), as shown in Figure 6.37. The crystals, which reach lateral dimension in excess of 200 μ and lie well within the range of optical microscopy, are normally invisible because of their extreme thinness (100–500 Å). If, however, during the course of crystallization the temperature is suddenly dropped, bands of much smaller crystals are nucleated at the perimeters of the original lamellae. The optical axes of these overgrowths are no longer perpendicular to the viewing direction, and they appear highly birefringent under crossed polarizers. The outline of the parent crystals as well as various imperfections such as spiral growths can be visualized using this self-decoration process.

6.7.3 Etching and Extraction

Variations in the morphology of both crystalline homopolymers and multiphase polymers can often be highlighted by preferential etching with strong acids or bases, solvents, or

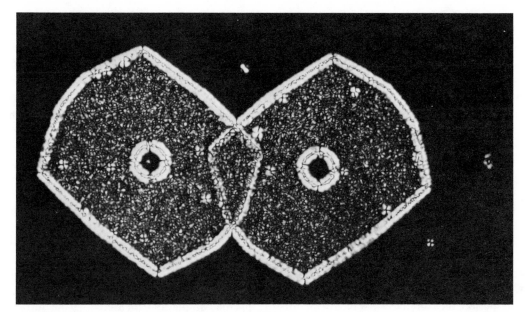

Figure 6.37 Self-decorated single crystals of PEO. Photograph courtesy of A. J. Kovacs, Centre de Recherches sur les Macromolecules.

various oxidizing agents. In the investigation of surface structures careful control of the etching process is required to avoid artifacts associated with overly vigorous attack by the etchant. More severe conditions can be employed for the elucidation of the internal morphology of bulk polymers, although only remnants of the original crystals may be left for microscopic investigation. The degradation of polyethylene by exposure to fuming HNO_3 is such an example (41–42).

Reding and Walter were able to follow changes in the size and structure of spherulites in molded polyethylene specimens by selective etching with hot carbon tetrachloride, toluene, or benzene (333). Satisfactory levels of etching were achieved by monitoring the loss of surface gloss on polished samples. Mackie and Rudin found that samples treated under similar conditions often developed non-representative surface deposits as a result of reprecipitation of dissolved polymer from the etching solvent (334). Similar effects have been observed in Mylar®* polyester films etched with xylene and trichloroethylene

*Mylar® is a registered trademark of E. I. duPont deNemours Company.

(335). Other studies indicate that changes in the crystal morphology of the polyethylene surface itself may occur in conjunction with solvent etching processes (336). Successful solvent etching of crystalline polymers without the introduction of morphological artifacts appears to be possible if the solvent is sufficiently poor and etching is carried out at temperatures well below those where significant reorganization is possible. Comparison of the morphologies of nylon 6, nylon 6, 6, polychlorotrifluoroethylene, and polypropylene, mildly etched with aromatic and chlorinated hydrocarbons, with those revelaed by microscopy of microtomed thin sections has shown them to be equivalent (337). Preferred solvents and etching times and temperatures for each polymer have been noted (337).

Davidson and Wunderlich have used hot nitric acid to remove fibrils that adhere to the fracture surfaces of pressure crystallized polyethylene (338). This treatment appears to leave the extended-chain substructure intact. In materials of lower crystallinity such as polypropylene, however, etching with oxidizing acids may be accompanied by extensive oxidative stress cracking. Chromic and sulfuric acid mixes appear to be particularly severe in this respect (295). More selective etching of polypropylene appears to be possible through the use of chromic acid alone or by sequential exposure to concentrated nitric and chromic acid (339). Etching with an even milder agent such as ozone may result in further improvement (340), although microscopic investigations of ozone etched crystalline polymers appear to be lacking.

Etching of both crystalline polymers and blends by ion and electron bombardment has been reported (341–343). Both nylon and poly(ethylene terephthalate) fibers were observed to develop rodlike structures perpendicular to the drawing direction after ion etching in an argon atmosphere at a pressure of approximately 20 mm Hg (341). The spacing of the rods could be correlated with the fold period measured by x-ray diffraction. Excessive degradation and carbonation of the fiber surfaces occurred when the etching was carried out in an air atmosphere. Cross-linking was observed after lengthy exposure to the ion beam. Phase contrast in microtomed

thin sections of PMMA-SAN copolymers can be developed by irradiation in the electron beam (343). Under such conditions the PMMA regions are selectively thinned by chain scission reactions whereas the SAN regions are stabilized by cross-linking. The differential thickness of the two regions provides enough contrast to identify domains of each material in the electron microscope.

Many of the ambiguities associated with the examination of etched homopolymers are avoided when similar techniques are applied to polymer blends since the structural details are considerably coarser, and the samples may be more conveniently examined by SEM rather than TEM. Williams and Hudson found that the size and shape of the rubber particles in HIPS could be characterized by microtoming the sample and then etching the surface with isopropanol to leave the rubber protruding above the surface (344). Metal replicas of these surfaces were viewed in transmission. Radiation hardening of the samples prior to microtoming reduced the amount of deforma-

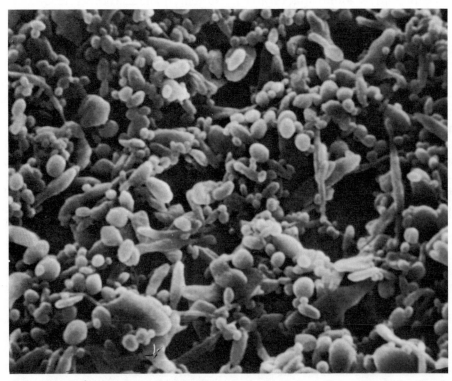

Figure 6.38   Polyolefin particles extracted from injection-molded bar of impact-modified nylon 66.

tion. Resolution of the occluded styrene domain in the rubber particles was poor.

Improved results were obtained by Bucknall et al. by exposing the samples to a bath containing 400 ml H_2SO, 130 ml H_3PO_4, 125 ml H_2O, 20 g CrO_3 for three minutes at 70°C in lieu of the isopropanol wash used by Williams (345). The samples were viewed in the SEM. Preferential etching of the butadiene phase resulted in outstanding resolution of both the rubber particles and their styrene occlusions. Successful etching of ABS with a similar acid solution has been reported by Freeguard (346). More recently Hobbs has found that a similar degree of surface relief can be obtained by low-temperature microtoming without subsequent etching (347). This discovery is particularly advantageous in characterizing saturated rubber systems, which are not easily etched.

When the components of a polymer blend show appreciable differences in solubility it may be possible to extract the dispersed phase for independent examination. Keskkula and Traylor have found that the rubber particles in impact polystyrene and ABS can be removed by dissolving the matrix in a cyclohexane (for IPS) solution of OsO_4 (35). The osmium stabilizes the rubber particles and allows them to be removed by mild centrifugation. The separated particles are then readily viewed in the SEM. Similar procedures have been used to characterize impact-modified nylons and polyesters (36) as shown in Figure 6.38. The results agree well with those obtained by complementary x-ray and microscopic techniques. Improved methods for developing textural differences in blends whose components exhibit similar solubilities and chemical reactivities continue to be sought.

6.8 SUMMARY

Large advances in the understanding of polymer morphology have been made in the last 15 years. Chain folding in crystalline polymers has become a generally accepted phenomenon. The complex polycrystalline structure of bulk crystallized polymers has been largely unravelled as a result of the intensive studies of lamellar single crystals carried out in the 1960s. Many aspects of deformation behavior have been adequately explained in terms of behavior of individual lamellae and the tie molecules running between them. Impurity and molecular weight segregation and their effects on crystallization kinetics, morphology, and physical properties have been well documented.

In spite of this progress a number of unanswered questions remain. Considerable controversy still exists over the exact nature of the fold surface. Both adjacent and nonadjacent (switchboard) models have been proposed (see Figure 6.39). While there is increasing evidence that regular folding may occur in solution-grown lamellae, it seems likely that much greater disruption of the fold surface takes place in samples crystallized from the melt. The complex reorganization processes that occur during annealing are still only partially understood. Further advances in this area depend on improved methods of characterizing samples before and after annealing. New spectroscopic and diffraction techniques may provide a dynamic way of looking at specific types of molecular rearrangements that take place on annealing. With the possible exception of true epitaxial crystallization, our understanding of heterogeneous nucleation is still largely phenomenological. The specific surface effects that control the nucleation process require better definition. Improved analytical methods, such as FTIR and ESCA, for looking at adsorption on surfaces will provide valuable information on this subject.

The level of molecular organization in glassy polymers has been much more difficult to characterize than that in crystalline or multiphase polymers where direct microscopic examination is possible. There appears to be little doubt that the volume occupied by an individual chain in a polymeric glass is comparable to that in a θ solvent. Measurements of radii of gyration and root mean square end-to-end distances confirm this fact. The energy costs for nonrandom trans bond sequences appear to be relatively small, how-

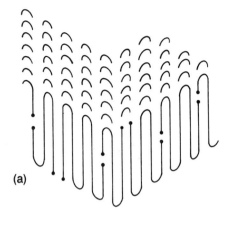

(a)

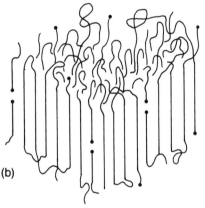

(b)

Figure 6.39 (*a*) Adjacent and (*b*) nonadjacent "switchboard" models for lamellar crystals.

The micromorphology of multiphase polymers has been extensively characterized by a number of academic and industrial research groups. In block copolymers, spheres, rods, and sheetlike structures commonly occur depending on the composition, block length, and precipitating solvent. Mechanical models have been developed to describe the physical behavior of these materials based on their composite structure. When one component is capable of crystallizing, crystallization proceeds in much the same way as for homopolymers except that the resulting structures are somewhat coarser and lower in crystallinity because of the forced inclusion of the amorphous component. Much of the current interest in copolymers and homopolymer blends is focused on toughened thermoplastics having one rubbery phase. Both the morphology and physical properties of these composites are strongly influenced by the way in which they are processed. Although some of the rheological factors affecting dispersion have been investigated, adequate control of particle size and delamination in two-phase systems is a continuing practical problem. Further complementary studies of the morphology, rheology, and processing behavior of such systems are needed to solve some of these problems.

Microscopy will continue to play an important role in morphological characterization although improved sample preparation techniques, especially for multiphase systems, are needed. Scanning electron microscopy and scanning-transmission electron microscopy have not yet reached their full potential as tools for the determination of polymer structure, and further advances may be expected in these areas. New techniques such as neutron diffraction, ESCA, FTIR, and "magic angle" NMR of solids should provide additional insight into surface and interfacial morphology, low level molecular-ordering phenomena, and reorganization processes. Many of these activities will have an increasingly practical focus as the need for high-performance polymers in engineering applications continues to increase during the next decade.

ever, and local segmental ordering may be accomplished with relatively little perturbation of the random coil dimensions. It thus seems likely that there must be some cooperative interaction over distances of 20 Å or so. A more careful study of energy transfer and relaxation processes in glassy polymers will help to verify or reject this hypothesis. Techniques such as high-resolution solid-state NMR and neutron diffraction, which are capable of examining the behavior of specific functional and atomic groups, are beginning to provide such information. To a large degree their success depends on the development of appropriate mathematical expressions to relate the observed spectroscopic relaxation behavior to the actual molecular relaxation processes. Several avenues are currently being pursued.

6.9   GLOSSARY OF SYMBOLS

δ	Crystallite size
D	Coefficient of diffusion
G_r	Crystal growth rate
I	Nucleation rate
I_0	Temperature independent constant in Fischer-Turnbull Equation
ΔG	Free energy for nucleus formation
k	Boltzmann constant
T	Temperature
ΔE	Activation energy for transport
V	Volume
G_v	Bulk free energy of fusion
A	Surface area
σ	Interfacial free energy
l	Lamellar thickness
l^*	Fold period
ΔS_f	Latent entropy of fusion
ΔH_f	Latent heat of fusion
T_m°	Equilibrium melting point
ΔT	Supercooling
τ	Relaxation time
$\overline{M}_n$	Number average molecular weight
$\overline{M}_w$	Weight average molecular weight
γ	Molecular anisotropy factor
α	Optical polarizability
T_g	Glass transition temperature
B	Numerical constant in Gordon-Taylor equation
W	Weight fraction
P	Sequence probability
G_m	Free energy of mixing
H_m	Enthalpy of mixing
S_m	Entropy of mixing
χ	Flory-Huggins interaction parameter
n_i	Mole fraction of component i
ϕ_i	Volume fraction of component i
R	Gas constant
$\gamma_{\alpha\beta}$	Dynamic interfacial tension between components α and β
$\gamma_{\alpha\beta}^\circ$	Static interfacial tension between components α and β
σ_2	Second normal stress difference
a	Droplet size

6.10   REFERENCES

1. H. Staudinger, H. Johner, R. Singer, G. Mie, and J. Hengstenberg, *Z. Phys. Chem.*, **126**, 425 (1927).
2. W. M. D. Bryant, *J. Polym. Sci.*, **2**(6), 547 (1947).
3. R. Hosemann, R. Bonart and G. Schoknecht, *Z. Phys.*, **146**, 588, 1956.
4. K. H. Storks, *J. Am. Chem. Soc.*, **16**, 1753 (1938).
5. W. Schlesinger and H. M. Leper, *J. Polym. Sci.*, **11**, 203 (1953).
6. A. Keller, *Phil. Mag.*, **2**, 1171 (1957).
7. E. W. Fischer, *Z. Naturforsch.*, **12a**, 753 (1957).
8. P. H. Geil, N. K. J. Symons and R. G. Scott, *J. Appl. Phys.*, **30**, 1516 (1959).
9. P. H. Geil, *J. Polym. Sci.*, **44**, 449 (1960).
10. B. G. Ranby, F. F. Morehead, and N. M. Walter, *J. Polym. Sci.*, **44**, 349 (1960).
11. R. Eppe, E. W. Fischer, and H. A. Stuart, *J. Polym. Sci.*, **34**, 721 (1959).
12. G. C. Claver, Jr., R. Buchdall, and R. L. Miller, *J. Polym. Sci.*, **20**, 202 (1956).
13. D. A. Blackadder, *J. Macromol. Sci. Rev. Macromol. Chem.*, **C1**(2), 297 (1967).
14. P. H. Geil, *Polymer Single Crystals*, Wiley-Interscience, New York, 1963.
15. P. Ingram and A. Peterlin, "Morphology" in H. F. Mark and N. G. Gaylord, Eds., *Encyclopedia of Polymer Science and Technology* Wiley-Interscience, New York, 1968, pp. 204–274.
16. B. Wunderlich, *Macromolecular Physics*, Vol. 1, Academic Press, New York, 1973.
17. D. Bassett, F. C. Frank, and A. Keller, *Phil. Mag.*, **8**, 1739 (1963).
18. W. D. Niegisch, *J. Polym. Sci.*, **40**, 263 (1959).
19. D. H. Reneker and P. H. Geil, *J. Appl. Phys.*, **31**, 1916 (1960).
20. N. Hirai, *J. Polym. Sci.*, **59**, 321 (1962).
21. A. Keller and S. Mitsuhashi, *Polymer*, **2**, 109 (1961).
22. F. Khuory and F. J. Padden, Jr., *J. Polym. Sci.*, **47**, 455 (1960).
23. R. D. Burbank, *Bell Syst. Tech. J.*, **39**, 1627 (1960).
24. D. J. Blundell and A. Keller, *J. Macromol. Sci.*, **B2**, 337 (1968).
25. J. C. Wittman and A. J. Kovacs, *Ber Bernsenges*, **74**, 901 (1970).
26. B. Wunderlich, E. A. James, and S. W. Shu, *J. Polym. Sci.*, **A2**, 2759 (1964).
27. D. D. Saratovkin, *Dendritic Crystallization*, (translated by J. E. S. Bradley Consultant Bureau, New York, 1959.

28. P. H. Geil and D. H. Reneker, *J. Polym. Sci.*, **51**, 569 (1961).

29. H. D. Keith, *J. Polym. Sci.*, **A2**, 4339 (1964).

30. H. D. Keith, *J. Appl. Phys.*, **35**(11), 3115 (1964).

31. B. Wunderlich and P. Sullivan, *J. Polym. Sci.*, **61**, 195 (1962).

32. M. J. Schick, unpublished, quoted by P. H. Geil in *Polymer Single Crystals*, Interscience, New York, 1963.

33. N. K. J. Symons, *J. Polym. Sci.*, **51**, 521 (1961).

34. H. D. Keith and F. J. Padden, Jr., *J. Appl. Phys.*, **33**(8), 2409 (1963).

35. A. Keller and J. R. S. Waring, *J. Polym. Sci.*, **27**, 447 (1955).

36. E. A. Wood, *Crystals & Light*, D. Van Nostrand, Princeton, New Jersey, 1964.

37. C. W. Bunn and T. C. Alcock, *Trans. Faraday Soc.*, **41**, 317 (1945).

38. A. Keller, *J. Polym. Sci.*, **17**, 351 (1955).

39. A. Keller, *Makromol. Chem.*, **31**, 1, 1959.

40. F. R. Anderson, *J. Polym. Sci.*, **C3**, 123 (1963).

41. R. P. Palmer and A. J. Cobbold, *Makromol. Chem.*, **74**, 174 (1964).

42. A. Keller and S. Sawada, *Makromol. Chem.*, **74**, 190 (1964).

43. A. Miyagi and B. Wunderlich, *J. Polym. Sci.*, **A-2**, **10**, 2073 (1972).

44. H. D. Keith and F. J. Padden, Jr., *J. Polym. Sci.*, **41**, 525 (1959).

45. D. C. Bassett, A. Keller, and S. Mitsuhashi, *J. Polym. Sci.*, **A-1**, 763 (1963).

46. J. C. Hoffman, *SPE Trans.*, **4**, 315 (1964).

47. H. D. Keith, F. J. Padden, Jr., and R. G. Vadimsky, *J. Polym. Sci.*, **A-2**, **4**, 267 (1966).

48. P. J. Phillips and B. C. Edwards, *Polym. Lett.*, **14**, 449 (1976).

49. A. Peterlin and E. W. Fischer, *Z. Phys.*, **159**, 272 (1960).

50. A. Peterlin, *J. Appl. Phys.*, **31**, 1934 (1960).

51. J. I. Lauritzen, Jr. and J. D. Hoffman, *J. Chem. Phys.*, **31**, 1680 (1959).

52. F. P. Price, *J. Chem. Phys.*, **31**, 1680 (1959).

53. J. I. Lauritzen, Jr. and J. D. Hoffman, *J. Res. Natl. Bur. Stand.*, **64A**, 1 (1960).

54. F. P. Price, *SPE Trans.*, **151**, (July 1964).

55. F. P. Price, "Nucleation in Polymer Crystallization," in A. C. Zettlemayer, Ed., *Nucleation* Marcel Dekker, New York, 1969.

56. F. L. Binsbergen, *Kolloid Z. Z. Polym.*, **237**, 289 (1970).

57. F. L. Binsbergen, *Kolloid Z. Z. Polym.*, **238**, 389 (1970).

58. F. L. Binsbergen, *J. Cryst. Growth*, **16**, 249 (1972).

59. F. L. Binsbergen, *J. Polym. Sci.*, *Polym. Phys.*, **11**, 117 (1973).

60. B. Wunderlich, *Macromolecular Physics*, Vol. 2, Academic Press, New York, 1976.

61. D. Turnbull and J. C. Fischer, *J. Chem. Phys.*, **17**, 71 (1949).

62. T. Suzuki and A. Kovacs, *Polym. J.*, **1**(1), 82 (1970).

63. R. J. Roe and H. E. Baer, *Macromolecules*, **3**, 454 (1970).

64. F. P. Price, *J. Chem. Phys.*, **35**, 1884 (1961).

65. A. Keller and D. C. Bassett as quoted in *Proc. R. Soc.* (*London*), **A263**, 323 (1961).

66. T. Kawai, *Kolloid Z. Z. Polym.*, **229**, 116 (1969).

67. P. J. Makarewicz and G. L. Wilkes, *J. Polym. Sci.*, *Polym. Phys.*, **16**, 1559 (1978).

68. B. Wunderlich, *J. Polym. Sci.*, **A1**, 1245 (1963).

69. S. Matsuoka, *J. Appl. Polym. Sci.*, **4**, 115 (1960).

70. P. H. Geil, F. R. Anderson, B. Wunderlich, and T. Arakawa, *J. Polym. Sci.*, **A2**, 3707 (1964).

71. R. B. Prime and B. Wunderlich, *J. Polym. Sci.*, **A2**, 7, 2061 (1969).

72. B. Wunderlich and L. Melillo, *Kolloid Z. Z. Polym.*, **250**, 417 (1972).

73. T. Hatakeyama, T. Hashimato, and H. Kanetsuna, *Coll. Polym. Sci.*, **252**, 15 (1974).

74. D. C. Bassett and B. Turner, *Phil. Mag.* **29**(2), 285 (1974).

75. W. W. Doll and J. B. Lando, *J. Macromol. Sci.* (*Phys.*), **2**(2), 219 (1968).

76. R. Hasegawa, Y. Tanabe, M. Kobayashi, H. Tadokoro, A. Sawaoka, and N. Kawai, *J. Polym. Sci.*, **A-2**, **8**, 1073 (1970).

77. R. S. Porter, J. H. Southern, and N. Weeks, *Polym. Eng. Sci.*, **15**(3), 213 (1975).

78. F. J. Balta Calleja, D. C. Bassett, and A. Keller, *Polymer*, **4**, 269 (1963).

79. W. O. Statten and P. H. Geil, *J. Appl. Polym. Sci.*, **3**, 357 (1960).

80. W. O. Statten, *J. Appl. Phys.*, **32**, 2322 (1961).

81. E. W. Fischer and G. F. Schmidt, *Angew. Chem.*, **74**, 551 (1962).

82. V. F. Holland, *J. Appl. Phys.*, **35**, 59 (1964).

83. D. A. Blackadder and P. A. Lewell, *Polymer*, **11**(2), 148 (1970).

84. W. O. Statten, *J. Appl. Phys.*, **38**(11), 4149 (1967).

85. D. H. Reneker, *J. Polym. Sci.*, **59**, 539 (1962).

86. A. Peterlin, *Polymer*, **6**, 25 (1965).

87. A. Keller and D. J. Priest, *Polym. Lett.*, **8**, 13 (1970).

88. I. C. Sanchez, J. P. Colson, and R. K. Eby, *J. Appl. Phys.*, **44**(10), 4332 (1973).

89. H. E. Bair, R. Salovey, and T. W. Huseby, *Polymer*, **8**, 9 (1967).

90. E. Hellmuth and B. Wunderlich, *J. Appl. Phys.*, **36**, 3039 (1965).

91. E. W. Fischer and G. Hinrichsen, *Kolloid Z.*, **213**, 93 (1966).

92. S. Y. Hobbs, *Polymer*, **16**, 462 (1975).

93. M. Jaffe and B. Wunderlich, *Kolloid Z.*, **216**, 203 (1967).

94. P. L. Nealy, T. G. Davis, and C. J. Kibler, *J. Polym. Sci.*, **A-2, 8**, 2141 (1970).

95. D. V. Rees and D. C. Bassett, *J. Polym. Sci.*, **B7**, 273 (1969).

96. D. C. Bassett, *Polymer*, **17**, 460 (1976).

97. B. Wunderlich and L. Melillo, *Science*, **154**, 1329 (1966).

98. B. Wunderlich and L. Melillo, *Macromol. Chem.*, **118**, 250 (1968).

99. B. Wunderlich, *Polymer*, **5**, 611 (1964).

100. F. Tunistra and E. Baer, *Polym. Lett.*, **8**, 861 (1970).

101. J. A. Koutsky, A. G. Walton, and E. Baer, *Polym. Lett.*, **5**, 117 (1967).

102. S. Wellinghoff, F. Rybnikar, and E. Baer, *J. Macromol. Sci. (Phys.)*, **B10**(1), 1 (1974).

103. G. I. Distler and B. G. Obronov, *Nature*, **224**, 261 (1969).

104. T. Takehashi and N. Ogata, *Polym. Lett.*, **9**, 895 (1971).

105. T. Takehashi, M. Inamura, and I. Tsujimoto, *Polym. Lett.*, **8**, 651 (1970).

106. S. Y. Hobbs, *Nat. Phys. Sci.*, **234**(44), 12 (1971).

107. A. J. Pennings, J. M. A. A. van der Mark, and H. C. Booij, *Kolloid Z. Z. Polym.*, **236**, 99 (1970).

108. B. Wunderlich, L. Melillo, C. M. Cormier, T. Davidson, and G. Synder, *J. Macromol. Sci.*, **B3**, 485 (1967).

109. L. Melillo and B. Wunderlich, *Kolloid Z. Z. Polym.*, **250**, 417 (1972).

110. F. L. Binsbergen, *Polymer*, **11**, 253 (1970).

111. A. J. Lovinger, J. O. Chua, and C. C. Gryte, *J. Polym. Sci., Polym. Phys.*, **15**, 541 (1977).

112. A. J. Pemmings and A. M. Kiel, *Kolloid Z. Z. Polym.*, **205**, 160 (1965).

113. T. Nagasawa and Y. Shimomura, *J. Polym. Sci., Polym. Phys.*, **12**(11), 2291 (1974).

114. J. Peterman and H. Gleiter, *Polym. Lett.*, **15**, 649 (1977).

115. A. K. Fritzche and F. P. Price, *Polym. Eng. Sci.*, **14**(6), 401 (1974).

116. D. T. Grubb and A. Keller, *Polym. Lett.*, **12**, 419 (1974).

117. M. H. Theil, *J. Polym. Sci., Polym. Phys.*, **13**, 1097 (1975).

118. R. D. Ulrich and R. P. Price, *J. Appl. Polym. Sci.*, **20**, 1077 (1976).

119. R. D. Ulrich and F. P. Price, *J. Appl. Polym. Sci.*, **20**, 1095 (1976).

120. M. R. Mackley and A. Keller, *Polymer*, **14**, 16 (1973).

121, P. B. Bowden and R. J. Young, *J. Mat. Sci.*, **9**, 2034 (1974).

122. P. Allan, E. B. Crellen, and M. Bevis, *Phil. Mag.*, **27**, 127 (1973).

123. A. Peterlin, *J. Polym. Sci.*, **C18**, 123 (1967).

124. I. L. Hay and A. Keller, *J. Polym. Sci.*, **C30**, 289 (1970).

125. M. Bevis and E. B. Crellen, *Polymer*, **12**, 666 (1971).

126. H. Kiho, A. Peterlin, and P. H. Geil, *J. Polym. Sci.*, **B3**, 157 (1965).

127. W. Wu, A. S. Argon, and A. P. L. Turner, *J. Polym. Sci., Polym. Phys.*, **10**, 2397 (1972).

128. D. Hansen and J. A. Rusneck, *J. Appl. Phys.*, **36**, 332 (1965).

129. J. M. Peterson, *J. Appl. Phys.*, **39**(11), 4920 (1968).

130. P. H. Geil, *J. Polym. Sci.*, **A2**, 3813, 3835, 3857 (1964).

131. J. Petermann and H. Gleiter, *J. Polym. Sci.*, **10**, 2333 (1972).

132. A. Peterlin, *J. Polym. Sci.*, **C9**, 61 (1965).

133. K. Sakaoku and A. Peterlin, *J. Polym. Sci.*, **A-2, 9**, 895 (1971).

134. H. D. Keith and F. J. Padden, Jr., *J. Polym. Sci.*, **41** (138), 525 (1959).

135. H. D. Keith, F. J. Padden, Jr., and R. G. Vadimsky, *J. Polym. Sci.*, **A-2, 4**, 267 (1966).

136. R. G. Crystal and D. Hansen, *J. Appl. Phys.*, **38**(8), 3103 (1967).

137. K. O'Leary and P. H. Geil, *J. Macromol. Sci., Phys.*, **2**, 261 (1968).

138. R. J. Samuels, *Structured Polymer Properties*, Wiley-Interscience New York, 1974.

139. A. Peterlin, *J. Polym. Sci.*, **C18**, 123 (1967).

140. D. M. Bigg, *Polym. Eng. Sci.*, **16**(11), 725 (1976).

141. J. H. Southern and R. S. Porter, *J. Appl. Polym. Sci.*, **14**, 2306 (1970).

142. R. G. Crystal and J. H. Southern, *J. Polym. Sci.*, **A3, 9**, 1641 (1971).

143. P. Flory, *J. Chem. Phys.*, **17**, 303 (1949).

144. W. R. Krigbaum and R. W. Godwin, *J. Chem. Phys.*, **43**, 4523 (1965).

145. L. M. Alberino and W. W. Graessley, *J. Phys. Chem.*, **72**, 4229 (1968).

146. J. A. Semlyen and P. V. Wright, *Polymer*, **10**, 543 (1969).

147. H. Jacobson and W. H. Stockmayer, *J. Chem. Phys.*, **18**, 1600 (1950).

148, D. G. H. Ballard, G. D. Wignall, and J. Schelten, *Eur. Polym. J.*, **9**, 965 (1973).

149. H. Benoit, D. Decker, J. S. Higgins, C. Picot, J. P. Cotton, B. Farnous, G. Jannink, and R. Ober, *Nat. Phys. Sci.*, **245**, 13 (1973).

150. E. W. Fischer, J. H. Wendorff, M. Dettenmaier, G. Lieser, and I. Voight-Martin, *Polym. Prepr.*, **15**(2), 8 (1974).

151. R. G. Kirste, W. A. Kruse, and K. Ibel, *Polymer*, **16**, 120 (1975).

152. D. Y. Yoon and P. J. Flory, *Polymer*, **16**, 645 (1975).

153. D. Y. Yoon and P. J. Flory, *Macromolecules*, **9**(2), 299 (1975).

154. J. P. Cotton et al., *Macromolecules*, **7**, 863 (1974).

155. G. Lieser, E. W. Fischer, and K. Ibel, *Polym. Lett.*, **13**, 39 (1975).

156. J. Schelten, G. D. Wignall, and D. G. H. Ballard, *Polymer*, **15**, 682 (1974).

157. R. G. Kirste and B. R. Lehnen, *Macromol. Chem.*, **177**, 1137 (1976).

158. A. Maconnachi and R. W. Richards, *Polymer*, **19**, 740 (1978).

159. J. T. Bendler, "Conformational Fluctuations in Isolated Polyethylene Molecules," to be published.

160. M. Hoffman, *Makromol. Chim.*, **144**, 309 (1971).

161. M. Dettenmaier, E. W. Fischer, *Kolloid Z. Z. Polym.*, **251**, 922 (1973).

162. M. Dettenmaier, E. W. Fischer, *Macromol. Chem.*, **177**, 1185 (1976).

163. E. G. Ehrenburg, E. P. Diskareva, and I. Y. A. Poddubnyi, *J. Polym. Sci.*, **C42**, 1021 (1973).

164. G. D. Patterson and P. J. Flory, *J. Chem. Soc. Faraday Trans. II*, **68**, 1098 (1972).

165. T. G. F. Schoon and O. Teichmam, *Kolloid Z. Z. Polym.*, **197**, 35 (1964).

166. T. G. F. Schoon and R. Kretschmer, *Kolloid Z. Z. Polym.*, **197**, 45 (1964).

167. G. S. Y. Yeh and P. H. Geil, *J. Macromol. Sci.* (*Phys.*), **B1**(2), 235 (1967).

168. G. S. Y. Yeh and P. H. Geil, *J. Macromol. Sci.* (*Phys.*), **B1** (2), 251 (1967).

169. M. I. Kashmiri and R. P. Sheldon, *J. Polym. Sci.*, **B7**, 51 (1969).

170. W. Frank, H. Goddar, and H. A. Stuart, *J. Polym. Sci.*, **B5**, 711 (1967).

171. A. Siegman and P. H. Geil, "The Crystallization of Glassy Polycarbonate," paper presented at Meeting of APS, Philadelphia, Pennsylvania, March 1969.

172. G. S. Y. Yeh, *J. Macromol. Sci.* (*Phys.*), **B6**(3), 451 (1972).

173. S. H. Carr, P. H. Geil, and E. Baer, *J. Macromol. Sci.* (*Phys.*), **B2**(1), 13 (1968).

174. H. P. Zingsheim and L. Bachman, *Kolloid Z. Z. Polym.*, **246**, 561 (1971).

175. V. A. Kargin, *Pure and Appl. Chem.*, **12**, 35 (1966).

176. V. A. Kargin, *Pure and Appl. Chem.*, **16**, 303 (1968).

177. M. J. Richardson, *Proc. R. Soc.*, **279**, 50 (1964).

178. J. K. Ovchimikov and G. S. Markova, *Poly. Sci. USSR*, **11**, 329 (1969).

179. I. Voight-Martin and F. J. Mijlhoff, *J. Appl. Phys.*, **46**, 1165 (1975).

180. G. W. Longman, G. D. Wignall, and R. P. Sheldon, *Polymer*, **17**, 485 (1976).

181. A. Odajima, S. Yamane, O. Yodu, and I. Kuriyama, *Rep. Prog. Polym. Phys. Jap.*, **18**, 207 (1975).

182. E. W. Fischer, "The Physics of Non-Crystalline Solids" in G. H. Frischat, Ed., 4th International Conference, Clarsthal-Zellerfeld 1976, Trans. Tech. Publication, 34 (1977).

183. R. E. Robertson, *J. Phys. Chem.*, **69**, 1575 (1965).

184. R. E. Robertson, in R. A. Huggins, Ed., *Ann. Rev. Mat. Sci.* **5**, 73 (1975).

185. S. L. Aggarwal, Ed., *Block Polymers* Plenum Press, New York, 1970

186. J. A. Manson and L. H. Sperling, *Polymer Blends and Composites*, Plenum Press, New York, 1976.

187. W. H. Stockmayer, *J. Chem. Phys.*, **13**, 199 (1945).

188. T. Alfrey, Jr., J. J. Bohrer, and H. Mark, Copolymerization, Interscience, New York, 1952.

189. M. Gordon and J. S. Taylor, *J. Appl. Chem.*, **2**, 493 (1952).

190. T. G. Fox and S. Loshnek, *J. Polym. Sci.*, **15**, 371 (1955).

191. L. Mandelkern, G. M. Martin, and F. A. Quin, Jr., *J. Res. Natl. Bur. Stand.* **58**(3), 137 (1957).

192. G. E. Molau, *J. Polym. Sci., Polym. Lett.*, **3**, 1007 (1965).

193. B. Wunderlich and D. Poland, *J. Polym. Sci.*, **A1**, 357 (1963).

194. C. H. Baker and L. Mandelkern, *Polymer*, **7**, 71 (1966).

195. G. Natta, P. Corradini, D. Sianesi, and P. Morero, *J. Polym. Sci.*, **51**, 527 (1961).

196. P. J. Holdsworth and A. Keller, *Makromol. Chem.*, **125**, 82 (1969).

197. P. J. Holdsworth and A. Keller, *Makromol. Chem.*, **125**, 94 (1969).

198. P. J. Flory, Trans. *Faraday Soc.*, **51**, 848 (1955).

199. B. Wunderlich, *Polymer*, **5**, 611 (1964).

200. D. Bodily and B. Wunderlich, *J. Polym. Sci.*, **A-2**, **4**, 25 (1966).

201. P. J. Holdsworth and A. Keller, *J. Polym. Sci.*, **B5**, 605 (1967).

202. M. Matsuo, T. Ueno, H. Horino, S. Chujyo, and H. Asai, *Polymer*, **9**, 425 (1968).

203. K. Kato, *Polym. Lett.*, **4**, 35 (1966).

204. M. Matsuo, S. Sagae, and H. Asai, *Polymer*, **10**, 79 (1967).

205. T. Inoue, T. Soen, and H. Kawai, *Polym. Lett.*, **6**, 75 (1968).

206. T. Uchida, T. Soen, T. Inoue, and H. Kawai, *J. Polym. Sci.*, **A2**, 10, 101 (1972).

207. S. L. Aggarwall, *Polymer*, **17**, 938 (1976).

208. D. McIntyre and E. Campos-Lopez, *Macromolecules*, **3**, 322 (1970).

209. A. Keller, E. Pedemonte, and F. M. Willmouth, *Kolloid Z. Z. Polym.*, **238**, 385 (1970).

210. M. J. Folkes and A. Keller, *Polymer*, **12**, 222 (1971).

211. E. Pedemonte, A. Turturro, U. Bianchi, and P. Devetta, *Polymer*, **14**, 145 (1973).

212. L. Bi and L. J. Fetters, *Macromolecules*, **8**, 90 (1975).

213. P. R. Lewis and C. Price, *Polymer*, **13**, 20 (1972).

214. C. Price, A. G. Watson, and M. T. Chow, *Polymer*, **13**, 333 (1972).

215. B. Lotz, A. J. Kovacs, G. A. Basset, and A. Keller, *Kolloid Z. Z. Polym.*, **209**, 115 (1966).

216. B. Lotz and A. J. Kovacs, *Polym. Prepr.*, **10**, 820 (1969).

217. A. J. Kovacs, Genie Chim., **97**(3), 315 (1967).

218. A. E. Skoulios, G. Tsouladze, and E. Franta, *J. Polym. Sci.*, **C4**, 507 (1963).

219. R. G. Crystal, J. J. O'Malley, and P. E. Erhardt, *Polym. Prepr.*, **10**, 804 (1969).

220. R. L. Cormia, F. P. Price, and D. Turnbull, *J. Chem. Phys.*, **37**, 1333 (1962).

221. F. P. Price, paper presented at the I.U.P.A.C. Symposium on Macromolecules Wiesbaden, Germany, (1959).

222. M. Kojima and J. H. Magill, *J. Polym. Sci., Polym. Phys.*, **12**, 317 (1974).

223. D. G. LeGrand, *Polym. Lett.*, **9**, 145 (1971).

224. H. M. Li and J. H. Magill, *J. Polym. Sci.*, **16**, 1059 (1978).

225. A. J. Kovacs, J. A. Manson, and D. Levy, *Kolloid Z.*, **214**, 1 (1968).

226. R. N. Haward and J. Manni, *Proc. R. Soc.* (*Lond.*) **A282**, 120 (1964).

227. M. R. Grancio, *Polym. Prepr.*, **12**, 681 (1971).

228. H. Keskkula, A. F. Platt, and R. F. Boyer, *Encyclopedia of Chemical Technology*, 2nd ed., Vol. 13, Wiley-Interscience, New York, 1969, p.128.

229. E. M. Fettes and W. N. Maclay in H. Keskkula, Ed., "*Polymer Modification of Rubbers and Plastics*," *J. Appl. Polym. Sci.*, Symposia 7, Interscience (1967).

230. J. L. Amos, *Polym. Eng. Sci.*, **14**, 1 (1974).

231. E. H. Merz, G. C. Claver, and M. Baer, *J. Polym. Sci.*, **22**, 325 (1956).

232. J. A. Blanchette and L. E. Nielson, *J. Polym. Sci.*, **20**, 317 (1956).

233. G. E. Molau and H. Keskkula, *J. Polym. Sci.*, **A-14**, 1595 (1966).

234. H. Keskkula and P. A. Traylor, *J. Appl. Polym. Sci.*, **11**, 2361 (1967).

235. G. F. Freeguard, *Polymer*, **13**, 366 (1972).

236. M. R. Grancio and D. J. Williams, *J. Polym. Sci.*, **A-1**, 8, 2617 (1970).

237. C. B. Bucknall, *Toughened Plastics*, Applied Science Publishers, Barking, Essex, England, 1977.

238. H. Willerson, *Macromol. Chem.*, **101**, 296 (1967).

239. G. F. Freeguard and M. Karmarkar, *J. Appl. Polym. Sci.*, **16**, 69 (1972).

240. M. Baer, *J. Appl. Polym. Sci.*, **16**, 1109 (1972).

241. K. Kato, *Koll. Z. Z. Polym.*, **220**, 24 (1967).

242. K. Kato, *Polym. Eng. Sci.*, **7**, 38 (1967).

243. R. P. Kambour, *Polymer*, **5**, 143 (1964).

244. R. R. Durst, R. M. Griffith, A. J. Urbanic, and W. J. van Essen, *ACS Div. Org. Coat Plast. Prepr.*, **34**(2), 320 (1974).

245. S. G. Turley, *J. Polym. Sci.*, **C1**, 101 (1963).

246. C. B. Bucknall, *J. Mater.*, **4**, 214 (1969).

247. C. B. Bucknall and M. M. Hall, *J. Mater. Sci.*, **6**, 95 (1971).

248. S. Krause, *J. Macromol. Sci. - Rev. Macromol. Chem.*, **C7**(2), 251 (1972).

249. P. J. Flory, *Principles of Polymer Chemistry*, Cornell University Press, Ithaca, New York, 1953.

250. H. Van Oene, *J. Colloid Interface Sci.*, **40**, 448 (1972).

251. J. M. Starita, *Trans. Soc. Rheol.*, **16**(2) 339 (1967).

252. S. Danesi and R. S. Porter, *Polymer*, **19**, 448 (1978).

253. D. Heikens and W. M. Barentsen, *Polymer*, **18**, 69 (1977).

254. W. M. Barentsen, D. Heikens, and P. Piet, *Polymer*, **15**, 119 (1974).

255. W. M. Barentsen and B. Heikens, *Polymer*, **14**, 579 (1973).

256. S. L. Rosen, *Polym. Eng. Sci.*, **7**, 115 (1967).

257. Van Henten, *J. Plast.*, **25**, 4, 144 (1972).

258. C. W. Childers, U.S. Patent 3,429,951 (1969).

259. T. Inoue, T. Soen, T. Hashimoto, and H. Kawai, *Marcomolecules*, **3**(1), 87 (1970).

260. M. Niiomi, G. Akovali, and M. Shen, *J. Macromol. Sci.* (*Phys.*), **B13**(1), 133 (1977).

261. L. Toy, M. Niinomi, and M. Shen, *J. Macromol. Sci.* (*Phys.*), **B11**(3), 281 (1975).

262. A. R. Shultz and B. M. Gendron, *J. Appl. Polym. Sci.*, **16**, 461 (1972).

263. A. R. Shultz and C. R. McCullough, *J. Polym. Sci.*, **A-2**, **10**, 307 (1972).

264. E. P. Cizek (to General Electric Co.) U.S. Patent 3,383,435 (May 14 1968).

265. Seiji Koiwa, *J. Appl. Polym. Sci.*, **19**, 1625 (1975).

266. S. Y. Hobbs, unpublished data.

267. P. S. Pallai, D. I. Livingston, and J. D. Strang, *Rubber Chem. Tech.*, **45**, 241 (1972).

268. R. G. Bauer and P. S. Pillai, *ACS Div. Org. Coat. Plast.*, **34**(2), 332 (1974).

269. M. R. Kantz and R. D. Corneliussen, *Polym. Lett.*, **11**, 279 (1973).

270. D. G. Wahrmund, D. R. Paul, and J. W. Barlow, *J. Appl. Polym. Sci.*, **22**, 2155 (1978).

271. R. L. Imken, D. R. Paul, and J. W. Barlow, *Polym. Eng. Sci.*, **16**(9), 593 (1976).

272. T. Nishi and T. T. Wang, *Macromolecules*, **8**(6), 909 (1975).

273. E. Hirata, T. Ijitsu, T. Soen, T. Hashimoto, and H. Kawai, *Polymer*, **16**, 249 (1975).

274. S. Y. Hobbs, unpublished data.

275. J. S. Noland, N.N.-C. Hsu, R. Saxton, and J. M. Schmidt, *Adv. Chem. Ser.*, **99**, 15 (1971).

276. R. L. Imken, D. R. Paul, and J. W. Barlow, *Polym. Eng. Sci.*, **16**(9), 593 (1976).

277. T. Nishi and T. T. Wang, *Macromolecules*, **8**(6), 909 (1975).

278. R. L. Scott, *J. Chem. Phys.*, **17**, 279 (1949).

279. H. D. Keith and F. J. Padden, Jr., *J. Appl. Phys.*, **35**, 1286 (1964).

280. J. Boon and J. M. Azcue, *J. Polym. Sci.*, **A-2**, **6**, 885 (1968).

281. G. S. Y. Yeh and S. L. Lambert, *J. Polym. Sci.*, **A-2**, **10**, 1183 (1972).

282. L. M. Robeson, *J. Appl. Polym. Sci.*, **17**, 3607 (1973).

283. M. W. Williams, R. F. Landel, and J. D. Ferry, *J. Am. Chem. Soc.*, **77**, 3761 (1955).

284. F. Gornich and J. D. Hoffman in A. S. Michaels, Ed., *Nucleation Phenomenon*, American Chemical Society (ACS), Washington, D. C., 1966.

285. G. Natta, G. Allegra, I. Bassi, D. Sianesi, G. Caporiccio, and E. Torti, *J. Polym. Sci.*, **A-3**, 4283 (1965).

286. S. Ichihara, A. Komatsu, and T. Hata, *Polym. J.*, **2**(5), 640 (1971).

287. A. R. Shultz and G. I. Mankin, *J. Polym. Sci.*, *Symp.*, **54**, 341 (1976).

288. A. R. Shultz, private communication.

289. B. Maxwell, *J. Polym. Sci.*, **C9**, 43 (1950).

290. A. Weitz and B. Wunderlich, *J. Polym. Sci.*, **12**, 12 (1974).

291. J. H. Reinshagen and R. W. Dunlap, *J. Appl. Polym. Sci.*, **19**, 1037 (1975).

292. E. S. Clark, *Appl. Polym. Symp.*, **20**, 325 (1973).

293. E. S. Clark, *SPE J.*, **23**, 46 (July 1967).

294. Z. Mencik and A. C. Chompff, *J. Polym. Sci.*, *Polym. Phys.*, **12**, 977 (1974).

295. D. R. Fitchman and Z. Mencik, *J. Polym. Sci.*, *Polym. Phys.*, **11**, 951 (1973).

296. M. R. Kantz, H. O. Newman, and F. H. Stigale, *J. Appl. Polym. Sci.*, **16**, 1249 (1972).

297. S. Y. Hobbs and C. F. Pratt, *J. Appl. Polym. Sci.*, **19**, 1701 (1975).

298. H. D. Keith, F. J. Padden, Jr., N. M. Walter, and H. M. Wyckoff, *J. Appl. Phys.*, **30**, 1485 (1959).

299. C. J. Kuhre, M. Wales, and M. E. Doyle, *SPE J.*, **20**, 1113 (October 1964).

300. Z. Mencik and A. J. Chompff, *J. Polym. Sci.*, **12**, 977 (1974).

301. R. Giuffria, R. O. Carhart, and D. A. Davis, *J. Appl. Polym. Sci.*, **7**, 1731 (1963).

302. H. L. Goldsmith and S. G. Mason, in F. R. Eirch, Ed., *Rheology* Academic Press, New York, 1967, p. 87.

303. C. B. Bucknall, I. C. Drinkwater, and W. E. Keast, *Polymer*, **13**, 115 (1972).

304. E. M. Hagerman, *Plast. Eng.*, **29**, 67 (October 1973).

305. S. Y. Hobbs, *Polym. Eng. Sci.*, **14**(9), 621 (1974).

306. R. C. Thamm, *Rubber Chem. Tech.*, **50**(1), 24 (1977).

307. Z. Tadmor, *J. Appl. Polym. Sci.*, **18**, 1753 (1974).

308. J. Kruse, *Rubber Chem. Tech.*, **46**, 653 (1973).

309. D. A. Thomas, *J. Polym. Sci.*, *Polym. Symp.*, **60**, 189 (1977).

310. K. Hess, H. Mahl, and E. Gutter, *Kolloid Z.*, **155**, 1 (1957).

311. K. Hess, E. Gutter, and H. Mahl, *Naturwisschaften*, **46**, 70 (1959).

312. G. Kanig, *Koll. Z. Z. Polym.*, **251**, 782 (1973).

313. A. M. Hodge and D. C. Bassett, *J. Mat. Sci.*, **12**, 2065 (1977).

314. A. Peterlin, P. Ingram, and H. Kiho, *Macromol. Chem.*, **86**, 294 (1965).

315. K. Kato, *J. Electron Microsc.*, **14**, 220 (1965).

316. K. Kato, *Polym. Eng. Sci.*, **7**, 38 (1967).

317. E. H. Andrews, *Proc. R. Soc.*, **A227**, 562 (1964).

318. C. K. Rices and R. W. Smith, *J. Polym. Sci.*, **A1**, **9**, 2739 (1971).

319. M. Matsuo, C. Nozaki, and Y. Jyo, *Polym. Eng. Sci.*, **9**, 197 (1969).

320. W. M. Speri and G. R. Patrick, *Polym. Eng. Sci.*, **15**, 668 (1975).

321. A. C. Soldatos and A. S. Burhams, *ACS Adv. Chem. Ser.*, **99**, 531 (1971).

322. M. Niimoni, T. Katsuta, and T. Kotani, *J. Appl. Polym. Sci.*, **19**, 2919 (1975).

323. D. Stefan and H. Laverne Williams, *J. Appl. Polym. Sci.*, **18**, 1451 (1974).

324. N. C. Watkins and D. Hansen, *Text. Res. J.*, **38**, 388 (1968).

325. R. W. Smith and J. C. Andries, *Rubber Chem. Tech.*, **47**, 64 (1974).

326. W. M. Hess, C. E. Scott, and J. E. Callan, *Rubber Chem. Tech.*, **40**, 371 (1967).

327. V. A. Phillips, *Modern Metallographic Technics and Their Applications*, Wiley-Interscience, New York, 1971.

328. G. A. Bassett, *Phil. Mag.*, **3**, 1042 (1958).

329. G. A. Bassett, D. J. Blundell, and A. Keller, *J. Macromol. Sci. (Phys.)*, **B1**(1), 161 (1967).

330. B. J. Spit, *J. Macromol. Sci.(Phys.)*, **B2**(1), 45 (1968).

331. D. Krueger and G. S. Y. Yeh, *J. Macromol. Sci. (Phys.)*, **B6**(3), 431 (1972).

332. A. J. Kovacs and A. Gunthier, *Kolloid Z. Z. Polym.*, **250**, 530 (1972).

333. F. P. Reding and E. R. Walter, *J. Polym. Sci.*, **38**, 141 (1959).

334. J. S. Mackie and A. Rudin, *J. Polym. Sci.*, **49**, 407 (1961).

335. R. Giuffria, *J. Polym. Sci.*, **49**, 427 (1961).

336. K. Kubota, *Polym. Lett.*, **3**, 545 (1965).

337. L. Bartosiewicz and Z. Mencik, *J. Polym. Sci., Polym. Phys.*, **12**, 1163 (1974).

338. T. Davidson and B. Wunderlich, *J. Polym. Sci., A-2*, **7**, 2051 (1969).

339. V. J. Armond and J. R. Atkinson, *J. Mat. Sci.*, **4**, 509 (1969).

340. D. J. Priest, *J. Polym. Sci.*, **A-2, 9**, 1777 (1971).

341. F. R. Anderson and V. F. Holland, *J. Appl. Phys.*, **31**(9) 1960.

342. B. J. Spit, *Polymer*, **4**, 109 (1963).

343. E. L. Thomas and Y. Talmon, *Polymer*, **19**, 225 (1978).

344. R. J. Williams and R. W. A. Hudson, *Polymer*, **8**, 643 (1967).

345. C. B. Bucknall, I. C. Drinkwater, and W. E. Keast, *Polymer*, **13**, 115 (1972).

346. G. F. Freeguard, *Br. Polym. J.*, **6**, 205 (1974).

347. S. Y. Hobbs, *J. Polym. Sci., Polym. Phys.*, **16**, 1321 (1978).

348. H. Keskkula and P. A. Traylor, *Polymer*, **19**(4), 465 (1978).

349. S. Y. Hobbs and V. H. Watkins, unpublished data.

CHAPTER 7

Mechanical Behavior

C. G. SEEFRIED, JR.
J. V. KOLESKE

Union Carbide Corporation
Chemicals and Plastics
Research and Development Department
South Charleston, West Virginia

7.1   Introduction   297

7.2   Elastic Behavior   299

7.3   Rubber Elasticity   301

7.4   Modulus-Temperature Variation   306

7.5   Modulus-Time Behavior   309
      7.5.1   Linear Viscoelastic Behavior   309
      7.5.2   Nonlinear Viscoelastic Behavior   311
      7.5.3   Time and Temperature Correspondence   313

7.6   Phenomenological Characteristics   316
      7.6.1   Hysteresis and Stress-Softening   317
      7.6.2   Yield and Cold-Drawing Behavior   318
      7.6.3   Brittle and Ductile Failures   321
      7.6.4   Fracture Behavior   322
      7.6.5   Crazing   322
      7.6.6   Ultimate Properties   324
      7.6.7   Impact Strength   327
      7.6.8   Anisotropic Behavior   327
      7.6.9   Birefringence   329

7.7   Summary   331

7.8   Glossary of Symbols   332

7.9   References   333

7.1   INTRODUCTION

Engineering materials can be broadly classified as metals and nonmetals. The nonmetallics include ceramics, glasses, wood, and polymers. Polymers are of three types—thermoplastic, thermoset, and elastomeric. Each of these classes of materials has utility; and, at times, one type material can replace another type if the strength relationships that exist are understood. For example, rigidity can often be a limiting factor in the utility of plastics. Metals are about a hundred times more rigid than a plastic such as poly(methyl methacrylate). Even glass has a modulus about 10 times greater than rigid plastics. However, this rigidity limitation can be overcome in certain instances by part design—say, placing ribs in the part—and/or making the part somewhat larger when a plastic is used. If plastics are compared with metals or glass on a volume basis, as shown in Table 7.1, favorable comparisons result.

Chemists have various methods of altering molecular architecture and thus of designing new polymers. However, the practical utility of a polymer to a great extent depends on its mechanical attributes. New polymers that do not have a unique feature or features—mechanical, chemical, or other—are often termed "scientific curiosities." They may be useful in providing the research worker with a better understanding of the polymerization mechanism, polymer configuration, conformation,

Table 7.1 Comparison of Tensile Strength on Different Bases

Material	Tensile Strength (psi)	Tensile Strength Divided by Density
Polystyrene	7,000	6400
Nylon 6-6	11,000	9700
Polyethylene (high density)	4,300	4500
Acetals	9,500	6700
Steel	60,000	7500
Glass	10,000	4000

and so on, but the usefulness of a polymer, new or old, for commercialization is most often told by its properties. Its performance characteristics dictate whether the polymer has any, a particular, or a spectrum of end uses. This is apparent if certain characteristics of a few polymers are examined.

Polyethylene is a semicrystalline, translucent, flexible, tough, low glass transition temperature (T_g) polymer with a broad utility basis. However, it cannot replace, say polystyrene which is completely different in character. Polystyrene is an amorphous, transparent, brittle polymer that has a high glass transition temperature. Its relatively high-softening temperature and sparkling clarity make it useful for end uses in which one might have used glass. Molecular design comes into play when clear, brittle polystyrene is made into an opaque or translucent, tough polymer by introducing incompatible rubber into it during the polymerization process. Examination of the mechanical properties of this material shows that toughness is imparted by the rubber, which has a low T_g that is not altered because of the incompatibility factor. Its presence in polystyrene provides an energy sink that results in high-impact strength or toughness.

Polyurethane elastomers are tough materials that have a broad spectrum of properties that depend on selection of the chemical composition and/or molecular weight of a polyol, or soft segment, and the amount and/or nature of isocyanate and chain extender, or hard segment (1, 2). An understanding of the mechanical characteristics of

this class of materials allows end-use selection of a particular polymer from the myriad of possibilities. Once again, it is possible to design different properties into polyurethanes by making them into flexible or open-cell foams, rigid or closed-cell foams, or microcellular materials that have properties of both foams and elastomers. Mechanical characteristics of these materials again define which material is selected for a given end use.

The relatively new field of polymer blends allows a manufacturer to design a broad variety of products from known materials. Understanding the physical nature of polymer blends is important to their utility. Such blends can be crystalline in nature, truly compatible, incompatible, or mechanically compatible (3). Dynamic mechanical properties coupled with tensile properties provide the insight necessary to understand these materials so new, improved blends can be made for particular end uses.

Fillers, pigments, plasticizers, and other additives are utilized to alter the properties of polymers. How components such as these soften, harden, or reinforce polymers in sheet, film, fiber, or molded form is understood from the knowledge acquired by studying the mechanical characteristics of polymers.

Polymeric films are thin sections of polymers that can be prepared by a variety of techniques. Because of the great flexibility of films, thermoplastic polymers are ordinarily used in their manufacture. The main mechanical property difference of films compared to a polymer fabricated into other articles is flexibility, which varies inversely with film thickness. The physical properties of films are markedly improved when they are selectively oriented in one or two directions during the manufacturing process. The orientation process not only reduces thickness, but it also causes changes in crystallinity that enhance mechanical properties. For example, the tensile strength of extension cast polypropylene film ranges from 4,500 to 10,000 psi. When polypropylene film is biaxially oriented, its tensile strength ranges from 12,000 to 30,000 psi (4). Orientation also plays an important role in the properties of man-made fibers. The cold-drawing process orients the filaments, and strength enhancement results. Decorative and functional coatings or paints are another important use of polymers when they are in a thin-film form.

Certain physical properties of polymers are often poor when compared to nonpolymeric materials (5). Among these are relatively poor abrasion resistance, stress cracking, thermal expansion, flammability, permeability, and comfort when used in clothing and upholstery applications. These are areas in which a better understanding is necessary so that the performance characteristics can be improved. Molecular design (6) and measurement of the effect that the design changes impose on the mechanical properties of polymeric materials are the way to obtain the knowledge needed to meet the requirements of new applications.

7.2 ELASTIC BEHAVIOR

Theoretical considerations of the elastic behavior of materials are complex in nature and beyond the scope of this chapter. Rather than enter into the detailed discussion of stress analysis, a simplified explanation of deformation of isotropic materials is presented. The serious student of this subject will find a great deal of literature available (7–11).

Isotropic materials are those in which the properties are the same in all directions. This factor greatly simplifies the classical theory of elasticity which expresses Hooke's law in terms of six components of stress and six components of strain for a nonisotropic material. These relationships between stress and strain are markedly simplified and still very useful if only isotropic materials are considered. In this case, the elastic behavior can be described by four parameters, and only two of these are independent. The parameters are Young's modulus E, the shear modulus G, the bulk modulus B, and Poisson's ratio ν.

In a general sense, when a force (stress) is applied to a material, it causes a distortion (strain). A particular characteristic of polymers compared to solid, low molecular weight solids is their high elasticity. In comparison to many structural materials, such as metals, that can be elastically strained only small amounts ($< 1\%$), elastomeric polymers can be elastically strained 100% or more. According to Hooke's law, stress is proportional to strain, and when a material follows this law, it is behaving in an elastic manner. The force or stress imposed on a body can be simple tension, Figure 7.1a, in which the material is pulled; simple shear, Figure 7.1b, in which

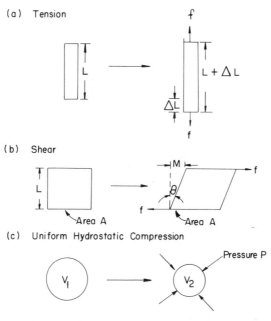

(a) Tension

(b) Shear

(c) Uniform Hydrostatic Compression

Figure 7.1 Simple deformation modes for materials.

opposing forces are applied to opposite faces of a material; or uniform hydrostatic compression, Figure 7.1c, in which the material is subjected to a crushing or compressive force (12). When these forces are applied to a material, there is a distortion that takes place as the material attempts to accommodate the force. The distortion which is dependent on the mode of deformation is termed the strain.

The stress σ in a sample is calculated by determining the force f (or F) per unit area A, and the strain ϵ is calculated by determining the change in length ΔL per unit length L. The modulus is defined as the ratio of the stress to the strain over the region in which these parameters are proportional, see Equation (7.1).

$$\text{Modulus} = \frac{\sigma}{\epsilon} = \frac{(F/A)}{(\Delta L/L)} \qquad (7.1)$$

Thus, in the case of tension, Figure 7.1a, the modulus of elasticity or Young's modulus E is defined as in Equation (7.2).

$$E = \frac{(F/A)}{(\Delta L/L)} \qquad (7.2)$$

For the simple shear case, Figure 7.1b, the shear modulus G is defined by Equation (7.3).

$$G = \frac{(F/A)}{(M/L)} = \frac{(F/A)}{\tan \theta} \qquad (7.3)$$

where M and θ are the shear elongation and angle, respectively. The bulk or compression modulus B results when a hydrostatic pressure P is imposed on a volume V_1 of material. That is, it is the ratio of hydrostatic pressure to the volume strain, as expressed in Equation (7.4).

$$B = P\left(\frac{V_1}{V_1 - V_2}\right) = \frac{PV_1}{\Delta V} \qquad (7.4)$$

The fourth parameter, Poisson ratio ν, is related to the fact that when a material is strained, its cross-sectional area as well as its length is changed. It is defined as the ratio of the relative lateral contraction (change in width per unit width) to the relative longitudinal extension (change in length per unit length) in a stressed, elastic body. A material that does not change in volume is termed an ideal rubber and has a Poisson ratio of 0.5. The interrelationship of these parameters for isotropic, elastic solids is given by the following relationship (13), see Equation (7.5).

$$E = 2G(1 + \nu) = 3B(1 - 2\nu) \qquad (7.5)$$

Polymers are intermediate in behavior to liquids and metals or elastic solids. That is, they are viscoelastic in character. Because of this behavior, it is difficult to measure the true Hookean moduli. A simple method used in industry to obtain the modulus in tension is to measure the ratio of the stress to the strain at a given low value of strain, which usually is 1%. This modulus is termed the secant modulus, and it is a very useful quantity when knowledge of the stiffness of a material is needed. When elastomers are studied in tension, the initial or low-strain modulus is of secondary importance. How the material behaves at 100% and 300% strain is of considerable interest. Thus one may hear of the 100% or 300% rubber modulus. However, this is not a modulus, but rather the tensile stress attained at the particular strain level. Many industrial tests have been standardized by the American Society for Testing Materials (ASTM), and their published procedures are useful for industrial laboratory testing.

There is a large variation in the stress-strain behavior of polymeric materials. They range from hard and brittle to elastomeric in character. The range of characteristics is shown in Figure 7.2. Brittle polymers undergo a small extension and then break. Ductile polymers may undergo uniform elongation (II) or they may exhibit a yield point (III), which is indicative of polymers that neck-down in cross section in a certain limited area and then undergo cold draw. The stress and elongation at the maximum are the yield stress σ_y and the yield elongation ϵ_y. Elastomers are characterized by great elongations and then fail (IV) or undergo strain-induced crystallization (V), which causes an increase in modulus at large elongations. The stress at the point of failure is termed the tensile strength at break σ_B and the

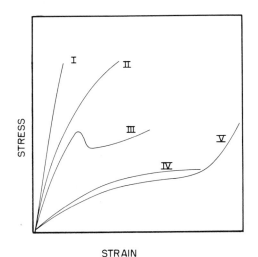

STRAIN

Figure 7.2 Generalized stress-strain behavior of polymeric materials: I, brittle; II, ductile; III, ductile with yield; IV, elastomeric without crystallization; V, elastomeric with crystallization.

strain at failure is the elongation at break ϵ_B. Usually these are merely termed tensile strength and elongation of a material.

The manner in which a polymer is processed can have a significant influence on its stress-strain behavior. This is especially important for crystalline polymers, which depend on the organizational arrangement between molecules for their bulk property responses. It has been shown that fibers based on isotactic polypropylene can be prepared so as to exhibit the extremes of mechanical property behavior available to a solid material, that is, brittle, ductile, tough, and elastic characteristics (14). Thus a single polymer can be made to possess a wide range of stress-strain responses without altering the basic compositional variables of molecular structure, molecular weight, or molecular weight distribution.

7.3 RUBBER ELASTICITY

Macromolecular chains of polymers are capable of existing in many different conformations owing to the ability of their segments to rotate with respect to each other. It is this conformational entropy that forms the basis for many of the molecular interpretations of the physical properties of polymers. The quantitative interpretation of rubber elasticity

is formulated on the use of Gaussian statistics to describe the behavior of a single polymer chain and then to treat the elastic network as an assembly of chains with particular restrictions (15, 16).

According to the first law of thermodynamics, as expressed in Equation (7.6)

$$dE = T\,dS - P\,dV + f\,dL \qquad (7.6)$$

where f represents stress, L is the length of the rubber sample, and the other symbols possess their usual familiar thermodynamic connotation. Hence, the stress or retractive force exerted by a rubber that has undergone deformation is given by Equation (7.7)

$$f = \left(\frac{\delta E}{\delta L}\right)_{T,V} - T\left(\frac{\delta S}{\delta L}\right)_{T,V} \qquad (7.7)$$

Since the internal energy change of the rubber is essentially negligible during deformation, the stress is associated principally with the change of conformation entropy, Equation (7.8).

$$f = -T\left(\frac{\delta S}{\delta L}\right)_{T,V} \qquad (7.8)$$

In applying statistical mechanics to the theory of rubber elasticity, four simplifying assumptions are made (17).

1. The statistical nature of an individual network chain obeys Gaussian statistics, that is, it is volumeless and freely jointed.

2. The total number of conformations for an isotropic network of these Gaussian chains is the product of the number of conformations of the individual chains.

3. The internal energy of the network is independent of the different conformation of the individual chains.

4. Cross-link junctions in the network are relatively fixed at their mean positions, and, upon deformation, these junctions transform affinely, that is, in the same ratio as the macroscopic deformation ratio of the rubber sample.

To calculate the entropy for the network chains, the Boltzmann relation is used, see Equation (7.9).

$$S = k \ln \Omega \qquad (7.9)$$

where Ω is the total number of conformations available to the network. For a volumeless and freely rotating chain, the number of conformations associated with a random chain of mean square end-to-end distance r_0^2 is given by the Gaussian distribution function, shown in Equation (7.10).

$$W(r) = c[\exp(-b^2 r_0^2)] \qquad (7.10)$$

where c is a normalization constant and $b^2 = 3/2 r_f^2$, r_f^2 being the mean square end-to-end distance of a free chain. For a network of N_0 chains, the total number conformations is the product of the individual chains as shown in Equation (7.11):

$$\Omega = \prod_{j=1}^{N_0} W(rj) \qquad (7.11)$$

and the conformational entropy of the network is given by Equation (7.12).

$$S = -k \sum_{j=1}^{N_0} b_j^2 r_j^2 \qquad (7.12)$$

Under affine transformation of the network chains during stretching, the entropy may be expressed as Equation (7.13)

$$S = -k \sum_{j=1}^{N_0} \frac{b_j^2 r_j^2 (\lambda^2 + 2/\lambda)}{3} \qquad (7.13)$$

where λ is the extension ratio (L/L_0) in the direction of uniaxial stretch. Substitution of this expression for the entropy in Equation (7.8) and performing the appropriate mathematical manipulations lead to the equation of state for rubber elasticity, expressed as Equation (7.14).

$$\sigma = N_0 RT \; \frac{\bar{r}_0^2}{\bar{r}_f^2} \left[\lambda - \left(\frac{1}{\lambda^2} \right) \right] \qquad (7.14)$$

The term $(\bar{r}_0^2 / \bar{r}_f^2)$, sometimes noted as the "front factor," can be regarded as the average deviation of the network chains from the dimensions they would assume if they were isolated and free from constraints. For an ideal rubber network, this front factor is unity.

The equation of state for rubber elasticity provides a molecular interpretation for the properties exhibited by rubbers. It recognizes the role of conformational entropy, as opposed to energy contributions, in polymers. It predicts that the elastic stress is directly proportional to the total number of chains in the network and to temperature. As contrasted to a single polymer chain, it predicts that the strain dependence of the elastic stress is non-Hookean. Figure 7.3 is a comparison of a

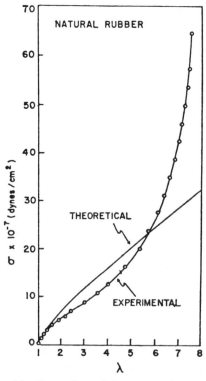

Figure 7.3 Comparison of the stress-strain curve for natural rubber with the theoretical curve calculated from the theory of rubber elasticity. Reprinted with permission from L. R. G. Treloar, *Trans. Faraday Soc.*, **40**, 59 (1944). Copyright by the Faraday Division of the Chemical Society, London, England.

theoretical and an experimental stress-strain curve (18). Owing to the assumed approximations, agreement between theory and the data occurs only in the region of low strain. It has been observed experimentally that the elastic stress varies directly with temperature and chain concentration as is predicted by theory.

Although the theory reveals many of the essential features of rubber elasticity, several important assumptions were made in its derivation (19). One of the inherent assumptions is that the polymer chains are volumeless, and the statistics employed do not consider the actual possibility that two or more segments of the polymer chain cannot occupy the space. The effect of considering finite polymer chains with excluded volume has been evaluated in regard to elasticity (20–23). The effect of excluded volume is to reduce the conformational entropy of the chain network and to yield a lower stress upon deformation than predicted by the Gaussian theory. Another assumption of the Gaussian theory is that the rubber sample remains isotropic under deformation. When a rubber is stretched, the real network becomes anisotropic owing to intermolecular obstructions. Analyses of the anisotropy effects show that the conformational energy is reduced and the elastic stress is lower than predicted by the simplified theory (24–26).

The statistical theory of rubber elasticity is based on the assumption that the elastic stress arises primarily from conformational entropy changes, and the energy associated with the polymer changes is relatively unaffected by deformations. The rotational isomeric model of chain molecules shows that there are rotational energy barriers that must be overcome to obtain conformational rearrangements of the chain segments (27). Hence new equations of state for rubber elasticity have been developed that incorporate the effects of intramolecular chain energies (28–32). Utilizing the thermodynamic identity, shown in Equation (7.15),

$$\left(\frac{\delta S}{\delta L}\right)_{T,V} = -\left(\frac{\delta f}{\delta T}\right)_{V,L} \qquad (7.15)$$

We may rewrite Equation (7.7) to provide the energy contribution of stress f_e and the entropy contribution of stress f_s, see Equations (7.16) and (7.17).

$$f = \left(\frac{\delta E}{\delta L}\right)_{T,V} + T\left(\frac{\delta f}{\delta L}\right)_{V,L} \qquad (7.16)$$

$$f = f_e + f_s \qquad (7.17)$$

By rearrangement, it is apparent that the relative energy contribution to the stress is given by Equation (7.18).

$$\frac{f_e}{f} = 1 - \left(\frac{\delta \ln f}{\delta \ln T}\right)_{V,L} \qquad (7.18)$$

It has been shown that this relative energy contribution is related to the temperature coefficient of the unperturbed chain dimensions (28), see Equation (7.19).

$$\frac{f_e}{Tf} = \frac{d[\ln (\bar{r}_f^2)]}{dT} \qquad (7.19)$$

This is interpreted to mean that temperature changes affect the chain dimensions through the rotational energy barriers, and the network chains expand or contract to produce variations in the elastic stress. Thus the energy effects in rubber elasticity arise mainly from the intrachain interaction energies of the network chains (33). It has been shown that the relative energy contribution may be determined according to Equation (7.20) (34–38) as follows:

$$\frac{f_e}{f} = 1 - \left(\frac{d \ln G}{d \ln T}\right) - \left(\frac{\beta T}{3}\right) \qquad (7.20)$$

where the temperature coefficient of the shear modulus G is obtained by plotting the material stress as a function of strain for a series of temperatures and β is the bulk thermal expansion coefficient of the material. Typical values obtained for the relative internal energy contribution to the elasticity of various polymers are given in Table 7.2

According to statistical theory, the relative internal energy contribution of the elastic

Table 7.2   Relative Internal Energy Contributions to Polymer Elasticity (39)

Polymer	f_e/f
Poly(ethylene)	− 0.47
Poly(styrene), atactic	0.16
Poly(isobutylene)	− 0.03
Natural rubber	0.18
Poly(isoprene), *cis*-1,4	0.12
Poly(isoprene), *trans*-1,4	− 0.09
Poly(dimethylsiloxane)	0.18
Poly(ethyl acrylate), atactic	− 0.17
Poly(n-butyl acrylate), atactic	− 0.68
Poly(vinyl alcohol), isotactic	− 0.68
Poly(vinyl alcohol), atactic	− 0.39

stress should be independent of deformation (40, 41). However, intermolecular interactions between the polymer chains affect the radial forces that maintain molecular separation. There are indications that the contribution of intermolecular forces to energy effects in rubber elasticity is not as significant as the intramolecular forces for most materials (42). However, the influence of specific intermolec-

ular forces has been observed, as, for example, in natural rubber. Figure 7.4 is a description of the relative internal energy contribution of elastic stress as a function of elongation ratios for natural rubber (43). The ratio f_e/f decreases with increasing λ (note: $\lambda = \alpha = L/L_0$ in Figure 7.4) at low strains, but becomes independent of λ at intermediate strains. The initial decrease in f_e/f is associated with a substantial contribution from intermolecular energy. The dramatic decrease of f_e/f at relatively large strains coincides with the occurrence of strain-induced crystallization. When the rubber is stretched to high strains, considerable anisotropy of the polymer chains occurs due to their orientation in the direction of stretch. This orientation leads to a favorable ordering of the molecules and may result in the formation of crystallites. The increased modulus associated with these crystallites results in an increase of the elastic stress, which becomes more pronounced at the higher extents of strain.

The retractive force in a rubber is propor-

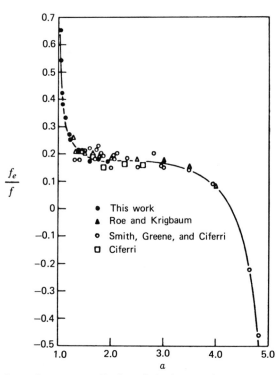

Figure 7.4   Relative internal energy contribution of elastic stress for natural rubber as a function of elongation ratios. (Note: $\alpha = \lambda = L/L_0$.) Reprinted with permission from M. C. Shen, D. A. McQuarrie, and J. L. Jackson, *J. Appl. Phys.*, **38**, 791 (1967). Copyright by American Institute of Physics.

tional to the concentration of network chains, $N_0 = \rho / \overline{M}_c$, where ρ is the material density and $\overline{M}_c$ is the average molecular weight of the chains between cross-links. Based on Equation (7.14), the shear modulus of the rubber may be expressed as Equation (7.21).

$$G_0 = \left(\frac{\rho RT}{\overline{M}_c} \right) \left(\frac{\overline{r}_0^2}{\overline{r}_f^2} \right) \qquad (7.21)$$

Since each linear polymer chain of number average molecular weight $\overline{M}_n$ contains free terminal ends that cannot support a stress, only the effective cross-linked chains that are formed should be considered. On this basis the shear modulus of the rubber is expressed as Equation (7.22):

$$G_0 = \left(\frac{\rho RT}{\overline{M}_c} \right) \left(\frac{\overline{r}_0^2}{\overline{r}_f^2} \right) \left[1 - \left(\frac{2\overline{M}_c}{\overline{M}_n} \right) \right] \qquad (7.22)$$

which reduces to Equation (7.21) as the degree of cross-linking becomes large (44). Hence the modulus of rubbers is directly proportional to their cross-link density. The effect of fillers on the elastic modulus of rubbers is usually described as follows in Equation (7.23) (45–47):

$$E_f = E_0 \left[1 + 2.5 \, \phi_f + 14.1 \, \phi_f^2 \right] \qquad (7.23)$$

where ϕ_f refers to the volume fraction occupied by the fillers. This representation of modulus enhancement by spherical filler particles is typically valid up to values of ϕ_f corresponding to about 0.3, as shown by the data in Table 7.3. Modulus enhancement by nonspherical fillers is best described by the inclusion of a shape factor that accounts for the asymmetric nature of the particles (48).

Since the stress-strain curves of rubbers are nonlinear and the equation of state for rubber elasticity is usually only valid at small strains, other representations are necessary to describe the behavior of rubbers over a larger region of strain. The phenomenological theory of continuum mechanics concerns itself with the response of isotropic solids without regard to molecular structures. Since a certain amount of work must be performed to attain a particular state of strain, the stored elastic energy may be expressed in terms of strain invariants (49, 50). For an incompressible solid, the strain invariants are interdependent since there is not volume change. Defining the strain-energy function as a power series of the strain invariants and retaining only the initial terms of the series expansion leads to the following expression, Equations (7.24) and (7.25), for uniaxial tensile deformations:

$$\sigma = 2C_1 \left[\lambda - \left(\frac{1}{\lambda^2} \right) \right] + 2C_2 \left[1 - \left(\frac{1}{\lambda^3} \right) \right] \qquad (7.24)$$

$$\sigma = 2 \left[C_1 + \left(\frac{C_2}{\lambda} \right) \right] \left[\lambda - \left(\frac{1}{\lambda^2} \right) \right] \qquad (7.25)$$

which is known as the Mooney-Rivlin equation with constants C_1 and C_2. It is seen that consideration of only the first term of Equation

Table 7.3   Modulus Enhancement of Vulcanized Natural Rubber by Carbon Black Filler (48)

Volume Concentration of Carbon Black ϕ_f	Young's Modulus E (psi)		Modulus Enhancement E_f / E_0
	Experimental	Calculated[a]	
0	192.0		1.000
0.015	209.9	199.8	1.093
0.0305	216.9	209.2	1.130
0.0535	236.8	225.4	1.233
0.0840	243.4	251.5	1.267
0.1382	288.2	310.1	1.501
0.2085	424.4	409.8	2.210

[a] Data calculated from Equation (7.23).

(7.24) corresponds with the Gaussian theory of rubber elasticity, that is, $C_1 = G_0/2$. The physical significance of the constants C_1 and C_2 have been evaluated for several types of polymers (51). However, plots of the Mooney-Rivlin equation $\sigma/[\lambda - (1 - \lambda^2)]$ versus $(1/\lambda)$ more closely represent the stress-strain curves of rubberlike materials than the single-term equation of the statistical theory. Several other attempts have been made to obtain a more realistic mathematical formulation of the elastic properties of rubbers from a purely phenomenological viewpoint (52, 53). These developments each exhibit varying degrees of success in describing the stress-strain curves of rubbers under different modes of deformation.

7.4 MODULUS-TEMPERATURE VARIATION

In a general sense, polymers can be classified as either amorphous or semicrystalline. Most analyses of polymers are based on amorphous materials and, strictly speaking, are not applicable to crystalline polymers. However, many polymers are partially crystalline, and often the techniques devised for amorphous polymers are utilized for them.

A simple, useful method of characterizing polymers is to measure the elastic modulus E as a function of temperature (10). When this is done at some particular, relatively narrow temperature range, a marked change in mod-

ulus occurs. This change is on the order of a few decades in magnitude, and it is related to the glass transition temperature T_g, of the polymer. The T_g is one of the most important and useful parameters for defining the behavior of a polymer. Below T_g, a polymer is hard and glasslike in character. Exceptions to this statement do exist, and they are discussed later. As the temperature of a polymer is increased above the point of glass behavior, it takes on a leatherlike character in the transition region as it passes through T_g. With further temperature increase the polymer next behaves as a rubber; and finally in the terminal, or flow region, it behaves as a liquid. The temperature of marked modulus change and the temperature range for each of these characteristics depends on both chemical composition and molecular weight. A plot of modulus as a function of temperature for a hypothetical, high molecular weight polymer is shown in Figure 7.5. If the molecular weight of the polymer is decreased, but maintained sufficiently high, the major change in behavior would be a decrease in the extent of the rubbery region and an onset of the flow region at a lower temperature as shown in this figure. T_g is obtained by determining the inflection point $dE/dT = 0$ in the modulus-temperature curve in the transition region. Simple, pictorial descriptions of the effect of crystallinity on modulus-temperature behavior are shown in Figure 7.6. The main effect of the crystalline

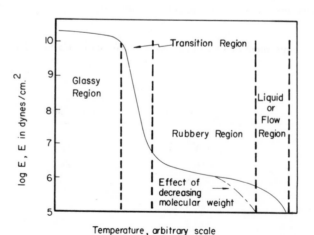

Temperature, arbitrary scale

Figure 7.5   Illustration of modulus-temperature behavior and regions of property variation.

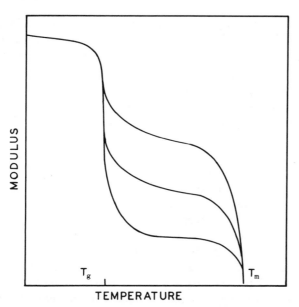

Figure 7.6   Typical effect of crystallinity on modulus-temperature behavior of polymers. The amount of crystallinity increases from bottom curve to the top curve. (T_g is the glass transition temperature, and T_m is the melting point.)

material is an enhancement of the modulus in the rubbery region.

By keeping the molecular weight "sufficiently high" in our analysis, the T_g was not affected. However, T_g is affected when low molecular polymers are considered. The relationship defining this variation is described by Equation (7.26) (54)

$$T_g(M) = T_g(\infty) - \left(\frac{K}{M_n}\right) \qquad (7.26)$$

where $T_g(\infty)$ is the limiting value of T_g and K is a positive constant with a value of about 10^5. Thus if T_g for various molecular weights of a given polymer is plotted as a function of $(1/M_n)$, the data extrapolate to the T_g of a high molecular weight polymer. Because the reciprocal of molecular weight is involved, $T_g(\infty)$ is rapidly approached as M_n increases. In fact, when M_n is in the range of about 5000–10,000, $T_g(M) = T_g(\infty)$ for most practical purposes (55, 56).

It should be readily apparent that homopolymers of different compositions will have different T_g's. Values for many polymers can be found in a variety of books and handbooks. However, the glass transition is a frequency or

time-dependent quantity (57). As the test frequency increases, T_g is shifted to higher temperature values. Therefore, when using literature reference values of T_g, it is important to note the test frequency and only to compare values for different polymers at comparable test frequencies.

Copolymers have glass transition temperatures and other properties that are intermediate to those of the respective homopolymers. The interrelationship of these parameters can be described in various ways. One of the more useful relationships was devised by Fox (58) and is shown as Equation (7.27).

$$\frac{1}{T_{g_{1,2}}} = \frac{w_1}{T_{g_1}} + \frac{w_2}{T_{g_2}} \qquad (7.27)$$

In Equation (7.27), T_{g_1} and T_{g_2} are the glass transition temperatures in degrees Kelvin (i.e., absolute temperature) of the respective homopolymers, and w_1 and w_2 represent the weight fraction of comonomer in the copolymer of $T_{g_{1,2}}$. The Gordon-Taylor relationship (59) is particularly useful in a rearranged form (60) when the T_g of a series of copolymers and of only one homopolymer is known, see Equation (7.28).

$$T_{g_{1,2}} = T_{g_1} + k \ \frac{(T_{g_2} - T_{g_{1,2}}) \, w_2}{w_1} \qquad (7.28)$$

In Equation (7.28) k is the ratio of the difference in thermal expansion coefficients between the rubbery and glassy states for homopolymer 1 and 2. However, when using Equation (7.28), k is treated as a constant and is obtained as the slope of a plot of $T_{g_{1,2}}$ versus $(T_{g_2} - T_{g_{1,2}}) \, w_2/w_1$. When obtained in this manner, k alone has little or no real known utility. When this plot is made, T_{g_1} or the unknown T_g is obtained from the intercept.

The same relationships described for copolymers can be used for combinations of a polymer and a plasticizer or other compatible diluents or for blends of two compatible polymers. Thus it should be apparent that Equation (7.27) is a powerful tool for ascertaining the utility of a variety of compatible polymeric systems including copolymers, plasticized polymers, and polymer blends. As one obtains an understanding of such mechanical and relaxation parameters, they can be utilized to intuitively make approximations or judgments of the time effects that occur in polymer behavior.

When polymers are cross-linked, the motion of the individual molecular chains is restricted. This restriction of motion shifts the glass transition to higher temperatures (54, 61–63). From the increase in T_g, it is possible to calculate the average molecular weight between cross-links M_c or the average number of carbon atoms in the polymer backbone between cross-links N_c, see Equations (7.29) and (7.30).

$$T_{gx} - T_g = \frac{3.9 \times 10^4}{M_c} \qquad (7.29)$$

$$T_{gx} - T_g = \frac{788}{N_c} \qquad (7.30)$$

In these empirical expressions, which Nielsen (64, 65) obtained by averaging literature data, T_{gx} and T_g refer respectively to the cross-linked and noncross-linked polymer. When a polymer is cross-linked, the flow region shown is eliminated and the magnitude of the modulus in the rubbery region is increased, as shown in Figure 7.7.

The actual "feel" one has for the importance of these properties is acquired as one works in the field. However, T_g is a property that can be quickly understood if one performs a simple experiment with a low molecular weight compound β-d-glucosepentaacetate (66). This inexpensive, crystalline compound can be easily made in glassy form, cooled to 10° or so

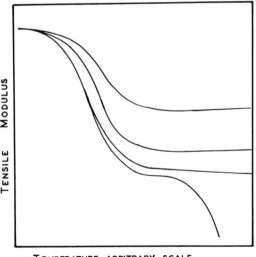

Figure 7.7   Typical effect of cross-linking on modulus-temperature behavior of amorphous polymers. The cross-link density increases from bottom curve to top curve.

below room temperature, and worked with one's hands to feel the actual glass to rubber transition since the material has a T_g at room temperature. It has an abbreviated rubbery modulus zone; but owing to slow crystallization above room temperature, the rubbery region is markedly increased. The compound also has utility in visualizing the frequency dependence of T_g and the difference in refractive index between amorphous and crystalline materials of the same composition.

Thus far the T_g has been related to modulus-temperature variation. However, there are a number of ways in which this quantity can be determined. The classical method is to determine the specific volume as a function of temperature (67, 68). The temperature at which this parameter has a change in slope is taken to be T_g. Other techniques include dynamic mechanical behavior (69–71), differential thermal analysis (72, 73), refractive index (74, 75), compressibility (76–78), β-radiation absorption (79), dielectric constant (80–82), heat capacity (83–85), penetration (86), ultrasonic and sonic vibrations (87, 88), nuclear magnetic resonance (89–91), and electrical resistivity (92). The technique to be used often depends on a particular investigator's preference. Many times more techniques than one, say dynamic mechanical and differential thermal analysis, are used to more fully understand and elucidate behavior.

It has been indicated that below T_g a polymer is in a glassy state. In effect, the motions of the main or backbone chain are "frozen-in," and they are essentially incapable of movement below T_g at the given test frequency. However, side chains do have freedom of rotation below T_g, and these give rise to smaller modulus decreases than those shown in Figure 7.5. Usually these rotations have little effect on the properties of the polymer. Yet, impact resistant polymers have a small transition markedly below T_g that can arise from molecular motion (polycarbonates) (93, 94); or from the presence of an incompatible, low T_g polymer (95) that has been incorporated into the polymer (as impact polystyrene).

In the next section, the effect of time on the mechanical properties is discussed. Because polymers are both viscous and elastic in character, that is, viscoelastic, time is an important factor that must be understood and kept in mind when one considers polymers for a given end use.

7.5 MODULUS-TIME BEHAVIOR

7.5.1 Linear Viscoelastic Behavior

Previously, the topic of isotropic elastic solids was considered along with the various ways a polymer can be deformed and characterized. For the linear, elastic body, the ratio of stress to strain was constant. That is, $d\sigma/d\epsilon = E = $ constant. Similarly, for a liquid that is Newtonian is character, that is a linear viscous liquid, the ratio of stress to strain rate, $\dot{\epsilon}$, is also constant; and it describes the viscosity of the material. A linear viscoelastic material is one in which the ratio of stress to strain is a function of time (96, 97). This ratio is a function of the deformation-time history and is independent of the magnitude of the strain. Such response is only observed at very small or infinitesimal strain levels, and the degree of deviation from linearity depends on the nature of the material being investigated.

Linear viscoelastic properties are specified by a variety of time functions. Dynamic mechanical properties in which a material is subjected to sinusoidally varying strains yield storage and loss moduli and mechanical loss (98–100). Stress relaxation (101) and creep (102) are also used to determine time-dependent quantities through the relaxation processes that occur during the test. Obviously, if relaxation phenomena are involved, factors such as time or frequency and temperature affect the results. In fact, the time-temperature superposition principle is the way in which these two variables are interrelated (103). In a simple sense, a decrease in temperature is equivalent to a decrease in time at a particular temperature. It is important to realize that a linear viscoelastic response does not mean that the initial portion of a stress-strain curve is linear.

Often arrays of linear or Hookean springs and linear or Newtonian dashpots, the Voight and Maxwell models, are used as mechanical analogs for understanding linear viscoelasticity (97, 100). The Voight element (Figure 7.8), which is a spring and a dashpot connected in parallel, is the basis of the Voight model. This model consists of a number of Voight elements connected in series. In addition, an isolated spring which represents the instantaneous elastic response and an isolated dashpot which accounts for irrecoverable deformation are added for completeness of the model.

The Maxwell element (Figure 7.8) is a spring and a dashpot connected in series. The Maxwell model consists of a number of these elements connected in parallel. A spring in parallel is added to represent the equilibrium response of the material being simulated. Each spring in the final model has a particular modulus or compliance and each dashpot a particular viscosity. When the models are made from a number of elements, a broad range of viscoelastic behavior can be described. However, to describe the real behavior of a material, an infinite number of elements from both models would be required; and, except for aiding one in visualizing viscoelastic response, the models have little practical utility.

The response or deformation of a material to stresses that vary with time, that is, periodically, can be measured by dynamic mechanical means (104). A test specimen is deformed by applied stresses that are generally varied in a sinusoidal manner at a characteristic frequency. From such measurements, an elastic or storage modulus E' or G', a viscous or loss modulus E'' or G'', and an energy dissipation factor tan δ or Q^{-1} can be determined. The inter-relationship between the modulus components is usually expressed in complex number notation, as shown in Equation (7.31).

$$G^* = G' + iG'' \qquad (7.31)$$

where G^* is the time or frequency-dependent complex modulus and i is the square root of minus one. The dissipation factor is the ratio of the energy dissipated to the energy stored per cycle of deformation or the ratio of the loss modulus component to the elastic modulus component, see Equation (7.32).

$$Q^{-1} = \frac{G''}{G'} \qquad (7.32)$$

The moduli determined will be either shear G^* or tensile E^* depending upon the particular type of device employed to make the measurements. By obtaining dynamic mechanical data as a function of temperature, it is possible to obtain useful, theoretical, and practical information about a polymer. Typical data are described in Figure 7.9. Data over a wide temperature and/or frequency range can be

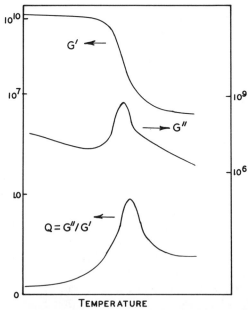

Figure 7.9 Typical curves obtained from dynamic mechanical measurements made as a function of temperature. G' and G'' values are in dynes/per square centimeter. Q^{-1} is a dimensionless quantity.

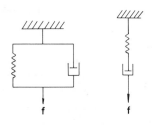

Voight Element   Maxwell Element

Figure 7.8 Mechanical elements used to introduce and simulate viscoelastic behavior.

obtained in much less time than is required for creep or stress relaxation experiments.

Practical information obtained from dynamic mechanical data is the modulus, the degree of energy dissipation, and the glass transition temperature. The maximum in the mechanical loss curve is usually taken to be the glass transition temperature. In addition, the maximum in the G'' curve is also the T_g, but it occurs at a slightly lower temperature (2–3°) when taken from this parameter (105). From a theoretical viewpoint, dynamic mechanical properties provide information about crystallinity, number of phases, compatibility of mixtures, aging effects, cross-linking, and other details of polymers and blends (57).

7.5.2   Nonlinear Viscoelastic Behavior

The theory of linear viscoelasticity represents the mechanical behavior of polymeric materials when certain conditions are met, which usually only occurs at sufficiently small stresses or strains. The limits of linear viscoelastic response are rather restrictive, however, with regard to an analysis of the performance of a polymer over a wide range of stresses and strains. In addition, polymers that exhibit linear viscoelastic behavior for small stresses or strains at short time periods may show significant nonlinear behavior for long time periods.

For linear viscoelastic materials subjected to creep tests (measurement of strain response to constant stress), the creep compliance is only a function of time and is defined as in Equations (7.33) and (7.34).

$$D(t) = \frac{\epsilon(t)}{\sigma} \qquad (7.33)$$

$$D(t) = D_0 + \Delta D(t) \qquad (7.34)$$

In these expressions D_0 represents the initial or immediate elastic contribution to the tensile compliance, and $\Delta D(t)$ corresponds to the transient component of the compliance. When $D(t)$ is known, the viscoelastic strain response to the arbitrary stress input can be calculated by utilizing the Boltzmann superposition integral, Equation (7.35).

$$\epsilon(t) = D_0 \sigma + \int_{-\infty}^{t} \Delta D\,(t - u) \left[\frac{d\sigma(u)}{du} \right] du$$

$$(7.35)$$

In Equation (7.35), $(t - u)$ is the difference between the time when the strain is measured and the time when a particular stress increment is applied. Based on this linear viscoelastic theory, three creep responses necessarily follow.

1.   Creep curves obtained for different levels of single-step stress loadings provide a unique creep-compliance curve.

2.   For a creep experiment in progress, the addition of an increased stress at a stated time results in additional creep, which is identical to the amount of creep that would have occurred had the stress been applied at the stated time without any previous loading.

3.   Creep and recovery responses are identical in magnitude for a given stress level, that is, the creep under stress is identical to the recovery from creep under the same stress.

Polymeric materials that exhibit nonlinear viscoelastic responses under creep conditions do not satisfy one or more of the consequences of the Boltzmann superposition principle. For example, plasticized polyvinyl chloride (106) and polypropylene (107) show higher initial rates of recovery from a given stress than initial rates of creep under the same stress. Experiments with polypropylene fibers indicate that the additional creep obtained in two-stage loadings is in excess of that obtained in a single-stage loading and that the initial or instantaneous creep observed with the second-stage loading is similar to that observed with recovery curves (108). To describe the behavior of nonlinear viscoelastic materials, it is necessary to evaluate the surface response of various stress-strain-time relationships. This characterization process usually involves the development of stress-strain or constitutive relations and the utilization of experimental data to determine the material property functions in these equations.

A formal extension of the Boltzmann superposition principle has been developed for nonlinear viscoelastic materials (108–115). These stress-strain relations are in the form of a sum of multiple integrals and allow approximations to be made with respect to any order of nonlinearity. Consider a creep loading history for which incremental stesses $\sigma_1(u_1)$, $\sigma_2(u_2)$, $\sigma_3(u_3)$ . . . are added at specified times $u_1, u_2, u_3, \ldots$. According to the Boltzmann superposition principle for a linear system, the deformation $\epsilon(t)$ at time t is given by Equation (7.36)

$$\epsilon(t) = \sigma_1 D_1(t - u_1)$$
$$+ \sigma_2 D_2(t - u_2) \cdots \text{ and so on} \quad (7.36)$$

where $D_1(t)$ is the creep compliance function. Additional terms are now introduced that correspond to the joint contributions of the stress loading steps to the final strain. These interaction terms are of the form as expressed as follows in Equation (7.37):

$$+\sigma_1\sigma_2 D_2(t - u_1, t - u_2)$$
$$+ \sigma_1\sigma_2\sigma_3 D_3(t - u_1, t - u_2, t - u_3)$$
$$+ \cdots \text{ and so on} \quad (7.37)$$

where D_2, D_3, and so on are the memory functions of the differences in time $t - u_1$, $t - u_2$, $t - u_3$, and so on, between the time when the strain is measured and the time when a particular stress increment is applied. For a continuous load history, the deformation $\epsilon(t)$ can be described as a sum of multiple integrals, see Equation (7.38).

$$\epsilon(t) = \int_{-\infty}^{t} D_1(t - u)\left[\frac{d\sigma(u)}{du}\right] du$$

$$+ \int_{-\infty}^{t} \int_{-\infty}^{t} D_2(t - u_1, t - u_2)\left[\frac{d\sigma(u_1)}{du_1}\right]$$

$$\left[\frac{d\sigma(u_2)}{du_2}\right] du_1\, du_2 + \cdots \text{ and so on} \quad (7.38)$$

The first term in this expression is the linear term of the Boltzmann superposition principle. This representation is not limited to a particular class of materials, but it becomes rather cumbersome analytically for materials with strong nonlinear responses. In addition, it does not take advantage of certain simplistic relationships observed in polymers. For example, the time or frequency dependence of a mechanical response in the nonlinear region can often be expressed in terms of linear viscoelastic properties.

Generalized single-integral constitutive equations have been developed from thermodynamic principles (116–119). These constitutive equations are similar in form to the Boltzmann superposition principle and require only time-dependent properties that exist in the linear viscoelastic range. For describing nonlinear responses, this thermodynamic theory yields Equation (7.39) as follows:

$$\epsilon(t) = g_0 D_0 \sigma + g_1 \int_0^t \Delta D\,(\Psi - \Psi')\left(\frac{dg_2\,\sigma}{du}\right) \quad (7.39)$$

where Ψ is the so-called reduced time given by the following Equations (7.40) and (7.41):

$$\Psi = \int_0^t \frac{dt'}{a_\sigma}\,[\sigma(t')] \quad (7.40)$$

$$\Psi' = \Psi(u) = \int_0^t \frac{dt'}{a_\sigma}\,[\sigma(t')] \quad (7.41)$$

For creep behavior at constant stress, these equations yield the following relationships for D_n, the nonlinear creep compliance, Equation (7.42).

$$D_n = g_0 D_0 + g_1 g_2\,\Delta D\left(\frac{t}{a_\sigma}\right) \quad (7.42)$$

The material properties g_0, g_1, g_2, and a_σ are functions of stress and have specific thermodynamic significance (119). When these material properties equal unity, Equation (7.39) reduces to the Boltzmann superposition principle. The creep compliance expression of

Equation (7.42) can be related to empirical power laws (102) and modified extensions of the Boltzmann superposition principle (103, 120, 121) which have been used to describe nonlinear viscoelastic creep behavior. The thermodynamic constitutive theory has been applied to the characterization of various nonlinear materials under uniaxial and multiaxial loadings in terms of linear viscoelastic creep and relaxation functions (119). The present theory may be extended to include the temperature dependence of nonlinear mechanical properties (117, 118).

A semiempirical extension of linear viscoelastic theory has been used to describe the nonlinear behavior of elastomers, especially at large strains. When an elastomer is stretched at a constant rate, the stress-strain response is the result of two competing processes. First, there is the progressive deformation of network chains which gives rise to an increase in stress. Second, at the same time, there is a continuous relaxation of the network chains which results in a reduction of stress (122). Deviations from equilibrium responses become more pronounced as the test temperature is decreased or the strain rate is increased. Analysis of stress relaxation properties for elastomers at large strains showed that the stress $\sigma(\epsilon, t)$ could be represented by an expression, Equation (7.43), which separated the strain and time dependence (123–128).

$$\sigma(\epsilon, t) = \Gamma(\epsilon) E(t) \qquad (7.43)$$

Linear viscoelastic theory shows that a stress-strain curve determined at constant strain rate $\dot{\epsilon}$ for an elastomer with a continuous distribution of relaxation times $H(\tau)$ is described as follows by Equation (7.44):

$$\sigma(\epsilon, t) = E_e \dot{\epsilon} t$$

$$+ \epsilon \int_{-\infty}^{\infty} \tau H(\tau) (1 - e^{-t/\tau}) d(\ln \tau) \qquad (7.44)$$

The slope of the stress-strain curve evaluated at $t = \epsilon / \dot{\epsilon}$ corresponds to the stress relaxation modulus $E(t)$, see Equation (7.45).

$$E(t) = \frac{d\sigma}{dE} = \frac{1}{\dot{\epsilon}} \frac{d\sigma}{dt}$$

$$= E_e + \int_{-\infty}^{\infty} H(\tau) e^{-t/\tau} d(\ln \tau) \qquad (7.45)$$

Substitution of the equivalent form of the strain rate ($\epsilon = \epsilon / t$) in Equation (7.44) yields the following, Equation (7.46):

$$\frac{\sigma(\epsilon, t)}{\epsilon}$$

$$= E_e + \frac{1}{t} \int_{-\infty}^{\infty} \tau H(\tau) (1 - e^{-t/\tau}) d(\ln \tau)$$

$$(7.46)$$

where $\sigma(\epsilon, t)/\epsilon$, which is a function only of time, is termed the constant strain rate modulus $F(t)$. This quantity is related to the stress relaxation modulus $E(t)$ by Equation (7.47).

$$E(t) = F(t) \left\{ 1 + \left[\frac{d \log F(t)}{d \log t} \right] \right\} \qquad (7.47)$$

Since Equation (7.43) shows that $E(t)$ can be represented by separate function of strain and time, a similar expression, Equation (7.48), was developed as follows for $F(t)$:

$$F(t) = \frac{g(\epsilon)\sigma(\epsilon, t)}{\epsilon} \qquad (7.48)$$

where $g(\epsilon)$ is a function of strain that approaches unity as the strain approaches zero. This approach has been used successfully to characterize the nonlinear stress-strain and stress relaxation behavior of elastomers (128, 129).

7.5.3 Time and Temperature Correspondence

Viscoelastic materials exhibit properties that are dependent on time and temperature. These functional relationships are equivalent to the extent that time-dependent data obtained at one temperature correspond to or can be superimposed upon time-dependent data obtained at another temperature (103). Hence it is possible to convert stress relaxation data over a range of temperatures to a single, composite curve that covers an extended time range at a particular reference temperature. Such behavior of amorphous polymers is decribed by the concepts of the time-temperature superposition principle.

As an example of the time-temperature relationship for polymers, consider the stress relaxation modulus data shown in Figure 7.10 as a function of temperature (130, 131). Once a suitable reference temperature has been chosen for superposition of the time-dependent stress relaxation data, the appropriate modulus values are reduced (132). This is accomplished by the utilization of a vertical shift factor that relates the temperature T and density ρ of the polymer to the values at the selected reference temperature (T_0 and ρ_0), see Equation (7.49):

$$E_r(t) = \left(\frac{T_0 \rho_0}{T \rho}\right) E(t) \qquad (7.49)$$

where $E_r(t)$ is the reduced modulus at various times of relaxation. This correction factor accounts for the dependence of the modulus on temperature, as prescribed by the theory of rubber elasticity, and on density or the concentration of molecular chains per unit volume. Once these vertical correction factors have been applied to the data, the stress relaxation curves are shifted horizontally on a logarithmic time scale to form a composite curve at the selected reference temperature. The master curve derived from the stress relaxation data of Figure 7.10 is shown in Figure 7.11 for the reference temperature of 25° C.

The extent to which each reduced modulus

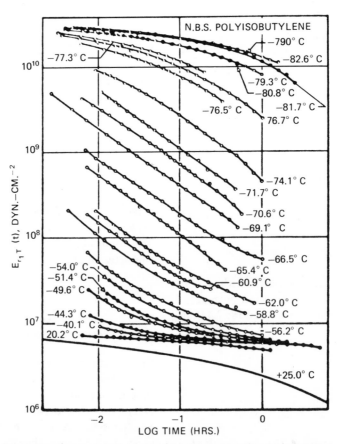

Figure 7.10 Stress relaxation data obtained at various temperatures for polyisobutylene. Reprinted with permission from E. Catsiff and A. V. Tobolsky, *J. Colloid. Sci.*, **10**, 375 (1955). Copyright by Academic Press, Inc., New York.

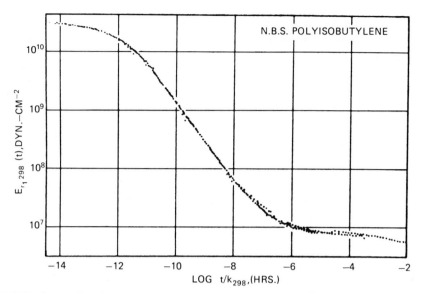

Figure 7.11 Stress relaxation master curve for polyisobutylene at the reference temperature of 25°C. Reprinted with permission from E. Catsiff and A. V. Tobolsky, *J. Colloid Sci.*, **10**, 375 (1955). Copyright by Academic Press, Inc., New York.

curve has to be shifted along the logarithmic time axis, the horizontal shift factor a_T, is a function of temperature. The relationship between the shift factor and temperature is approximately identical for all amorphous polymers and may be described by the WLF Equation (7.50) (133, 134):

$$\log a_T = -\frac{C_1(T - T_0)}{[C_2 + (T - T_0)]} \quad (7.50)$$

where C_1 and C_2 are constants and T_0 refers to the reference temperature. When T_g is selected as the reference temperature, the WLF equation is usually written as Equation (7.51).

$$\log a_T = -\frac{17.44(T - T_g)}{[51.6 + (T - T_g)]} \quad (7.51)$$

Although the values of C_1 and C_2 are sometimes considered as universal constants, they do vary slightly for different polymer systems (100). A somewhat better description of the shift factor is provided by Equation (7.52)

$$\log a_T = -\frac{8.86(T - T_s)}{[101.6 + (T - T_s)]} \quad (7.52)$$

where T_s is the reference temperature that generally corresponds to $T_g + 50°$ (133).

A fundamental interpretation of the WLF equation may be given in terms of the concepts of polymer free volume (17). The viscosities of monomeric liquids may be described by the following Doolittle Equation (7.53) (135):

$$\eta = A \exp\left\{\frac{B}{[(1/f) - 1]}\right\} \quad (7.53)$$

where η is the viscosity, A and B are constants, and $f = V_f/V$, the fractional free volume. It may be shown that the horizontal shift factor a_T is related to the ratio of relaxation times or viscosities of the polymer at two different temperatures (100), see Equation (7.54).

$$a_T = \frac{\tau(T)}{\tau(T_g)} = \frac{\eta(T)}{\eta(T_g)} \quad (7.54)$$

If it is assumed that the fractional free volume of the polymer increases linearly above the glass transition temperature, then one can write Equation (7.55)

$$f = f_g + \alpha_F(T - T_g) \quad (7.55)$$

where f_g is the fractional free volume at T_g and

α_f is the thermal coefficient of expansion of the fractional free volume above T_g. The combination of the preceding equations yields Equation (7.56).

$$\log a_T = \log \frac{\eta(T)}{\eta(T_g)}$$
$$= \frac{-B}{2.303 f_g} \left[\frac{T - T_g}{(f_g/\alpha_f) + (T - T_g)} \right]$$

$$(7.56)$$

Comparison shows this expression to be identical with the WLF equation where $C_1 = B/2.303 f_g$ and $C_2 = f_g/\alpha_f$.

The time-temperature superposition principle provides a description of the time dependence of properties at constant temperature that is similar to the description of the temperature dependence of properties at constant time or frequency, as usually determined by dynamic mechanical measurements. At temperatures below the polymer glass transition temperature, deviations from the predictions of the WLF equation are to be expected (136). Similar to the relationship between the temperature dependence of viscosity and the activation energy for flow, an apparent activation energy ΔH_a for viscoelastic relaxation may be derived from the WLF equation, see Equations (7.57) and (7.58).

$$\Delta H_a = R \left[\frac{d \ln a_T}{d(1/T)} \right] \qquad (7.57)$$

$$\Delta H_a = \frac{2.303\, R C_1 C_2 T^2}{[C_2 + (T - T_g)]^2} \qquad (7.58)$$

Contrary to reaction rate theory, this quantity is a function of temperature, increasing with decreasing temperature, and is independent of chemical structure except through T_g (100). The WLF equation has been extended to include an activation energy term influenced by chemical structure (137–139).

7.6 PHENOMENOLOGICAL CHARACTERISTICS

Qualitative descriptions of polymers are usually made in terms of the behavior of the materials during simple tensile loadings. A variety of stress-strain responses can be obtained for various types of polymers (140). Correspondingly, since polymers are viscoelastic in nature, their mechanical properties are dependent on the experimental conditions of testing, for example, rate of load application, temperature, amount of deformation, and previous strain history. The elastic theory of materials is based on a direct linear relationship between stress and strain in which all deformations are completely reversible. The theory of plasticity has been developed to describe materials that undergo a process that results in permanent or irreversible deformations. Polymers that exhibit rupture or failure in elongations within or not far removed from the elastic region of stress are termed brittle. Such materials possess a high modulus and a relatively low elongation at break. In contrast, ductile polymers exhibit permanent or plastic deformation at some critical stress level beyond the elastic region. A yield point is typically observed for these materials and the stress decreases immediately prior to failure. These polymers are qualitatively described as being hard and strong. Under certain conditions, considerable necking and cold-drawing of a polymer may occur at elongations beyond the yield point. The process that results in this response provides rather high extensions, and such materials are termed as being tough. If a polymer does not exhibit appreciable cold-drawing characteristics after yielding, the yield stress will approximate the maximum stress the material can support prior to rupture. Cold-drawing can result in an ultimate strength that is significantly higher than the yield stress. The rubberlike behavior of polymers occurs for relatively soft materials, especially at temperatures above the respective glass transition.

For a material exhibiting plastic deformation, a typical stress-strain curve obtained for a homogeneous specimen of uniform cross section subjected to uniaxial tension is shown in Figure 7.12. Initial elongation (AB) results in homogeneous deformation and the typical stress-strain response. At point B yielding occurs, and the specimen exhibits necking or thinning to a reduced cross section in a

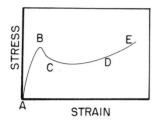

Figure 7.12 Typical stress-strain curve for a polymer that exhibits yielding (B), necking (BC), cold-drawing (CD), and strain-hardening (DE).

particular region. The yield point is associated with a reduction in the stress at subsequently higher strains (BC). Further elongation (CD) results in a generalized stress plateau region where the neck region is extended along the length of the specimen. At point D, a change in the morphology of the polymer chains occurs which causes an increase in the stress values at higher elongations. This region (DE) is associated with work-hardening or strain-hardening of the polymer.

7.6.1 Hysteresis and Stress-Softening

The recovery of polymers from elongations greater than the yield point is relatively poor as compared with recovery from elongations less than the yield point. Measurements of polymer hysteresis loss can be related to the recovery or resiliency properties of the ma-

terial. Figure 7.13 shows a typical hysteresis response obtained by stretching and relaxing a specimen in tensile deformation. The area under the loading portion of the stress-strain curve ($OXZO$ or area $A + B$) represents the energy per unit volume applied to the specimen during the deformation process. The area under the unloading portion of the stress-strain curve ($YXZY$ or area B) represents the stored or elastic portion of the energy that is recoverable from the deformed specimen. The hysteresis loss corresponds to the area between the loading and unloading curves ($OXYO$ or area A) and represents that portion of the applied energy lost during deformation. Point Y on the strain axis represents the amount of irrecoverable deformation or permanent set that remains in the specimen.

The stress-softening response or Mullins effect that occurs owing to repetitive tensile deformations of certain polymers is related to the hysteresial properties of the molecular chains. This phenomenon is particularly evident in filled homopolymers (141–146) and all heterophase elastomers, including blends, graft copolymers, and block copolymers (147–155).

Figure 7.14 is a description of the typical stress-softening effect for a urethane elastomer (156). Initial prestraining of the polymer results in reduced stress values for subsequent

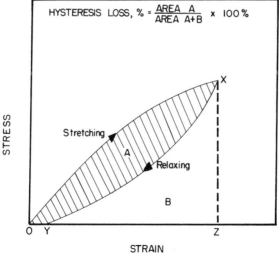

Figure 7.13 Typical hysteresis response for tensile deformation. Area $A + B$ is the energy per unit volume applied to the specimen during the deformation process; area A is the hysteresis loss per unit volume during the deformation process; and area B is the energy stored per unit volume during the deformation process.

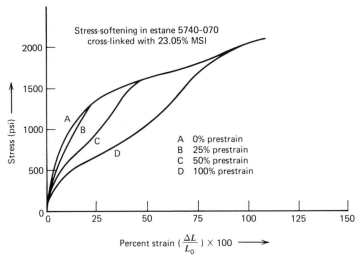

Figure 7.14   Effects of prestrain history on stress-softening characteristics of a urethane elastomer. Reprinted with permission from S. L. Cooper, D. S. Huh, and W. J. Morris, *Ind. Eng. Chem. Prod. Res. Dev.*, 7, 248 (1968). Copyright by the American Chemical Society.

stress-strain cycles. The lowering of the stress properties occurs only in the extension range of the prior prestrain history. At strain levels greater than the prestrain value, the stress values are coincident with those obtained for the unmodified stress-strain curve. The extent of stress-softening in various polymers is sensitive to composition, sample preparation, and previous history (157).

The physical interpretations ascribed to the stress-softening properties of filled polymers include softening of a hard region, slippage of the molecular chains over the filler particles, and chain breakage (158–162). The mechanism of this stress hysteresis for multiphase elastomers has been related to the morphological characteristics of these materials. Their stress-softening has been related to a breakdown of extended hard-phase associations (147–150) or a reordering of the paracrystalline arrangements for hard and soft domains (153–155, 163, 164). The effects of strain-softening may be removed by annealing the polymers at elevated temperatures. Stress-softening of polymers leads not only to a reduction of stress and modulus for repeated applications of strain, but also a substantial heat buildup can occur during repetitive stressings due to internal energy dissipation within the polymer. This can result in reduced wear resistance and performance for these products.

7.6.2   Yield and Cold-Drawing Behavior

The necking and cold-drawing characteristics of a polymer are associated with a difference in the stress-strain responses along the length of the test specimen. The initiation of a neck may occur owing to a particular segment of the polymer being subjected to a higher stress because of a reduction in its effective cross-sectional area. Correspondingly, localized variations in polymer properties may result in a reduced stress for yielding in a particular region. In either case, when one element of a polymer has reached its yield stress, continued deformation occurs more readily in this region, and this localized straining can result in the formation of a neck as shown in Figure 7.15 (165). This increased ease of localized deformation will occur until strain-hardening of the region develops. The strain-hardening increases the effective stiffness of the localized necking region and causes adjacent segments to attain their yield conditions. Hence the initial necking region of reduced stiffness is stabilized, contiguous regions yield, and the neck is propagated along the length of the specimen. As a greater proportion of the polymer is brought to a condition of strain-hardening, the test specimen exhibits a response of overall stress intensification.

During the necking process, the effective cross section of the specimen decreases due to

Figure 7.15   Necking of a cold-drawn polyethylene fiber that demonstrates neck formation resulting from localized straining. Reprinted with permission from R. Hill, Ed., *Fibers From Synthetic Polymers*, 1953, p. 267. Copyright by Elsevier Scientific Publishing Company, Amsterdam, The Netherlands.

the yield response. Hence the apparent stress (tensile force divided by original specimen cross section) of the specimen may be constant or even decreasing, while the true stress (tensile force divided by the actual effective cross section of the specimen at any time) of the specimen may be increasing. Substantial insight with regard to the necking and cold-drawing of polymers may be obtained by evaluating true stress conditions and applying the formalism of the "Consider construction" (166–168). The yield stress of ductile polymers is determined by the stress-strain response without any reference to the ultimate strength properties. This parameter defines the practical performance of plastics in service to a greater degree than the ultimate rupture stress value. The concepts of plasticity and yielding are applicable to forming, rolling, and drawing processes utilized in the processing of various polymers.

Many physical phenomena, such as yielding, that are dependent on the state of material stress are not functions of the stress components defined with respect to some arbitrary coordinate system. These phenomena must be analyzed in terms of functions of the stress components that are independent of the chosen coordinate system. Therefore, these functions must be invariant with respect to the coordinate system. There are three such functions that may be regarded as the basic forms of the invariants for the deviatoric stress system (11, 168–170). To a first approxima-

tion, a polymer that has yielded continues to deform at essentially constant volume. Various criteria have been proposed to define the critical stress value that results in yielding. The Tresca yield criterion states that yielding occurs when the maximum shear stress attains a critical value (171, 172). The von Mises criterion for yielding is based on the root mean square of the maximum shear stress reaching some critical value (173). The Coulomb yield criterion states that the critical shear stress for yielding to occur in any plane increases linearly with the pressure applied normal to this plane (174). The Coulomb criterion also specifies the direction in which yielding will occur, that is, the material yields in shear in the plane where the shear stress attains the critical value.

The yield results for a number of isotropic glassy polymers have been analyzed successfully with regard to the Coulomb criterion (175, 176). An oriented polymer subjected to extension in a direction not parallel to the initial draw direction sometimes exhibits a strain region concentrated into a narrow deformation band. The development of this deformation band is associated with a yielding condition and is analogous to the necking process in an unoriented polymer (177). The yield behavior of various oriented polyethylenes (178–184), nylons (185), and polyethylene terephthalate (177, 186–188) have been investigated. It was found that the direction of the deformation band was not the same as the initial drawing direction. For these oriented polymers, the yield stress varied significantly as a function of the angle between the initial draw direction and the tensile testing direction, as shown by the typical results plotted in Figure 7.16. When the tensile axis coincided with the initial draw direction of the polymer, the highest values of yield stress were obtained. For polyethylene terephthalate (187), the yield stress results were consistent with the Hill yield criterion for an anisotropic material with orthorhombic symmetry (169). The Hill yield criterion is a generalization of the von Mises yield criterion (173), and it reduces to this latter criterion for vanishingly small anisotropic conditions. In addition, it was found that the compressive yield stress parallel to the draw direction was significantly lower than the tensile yield stress in a similar direction. This result implies that it requires more stress to extend a previously oriented structure than to contract it.

The yielding and cold-drawing behavior of amorphous polymers is restricted by the fracture properties at low temperatures and by the glass transition at high temperatures. Hence temperature and strain rate have a substantial influence on these processes. Increasing strain rate results in increased yield and draw stresses, while increasing temperature produces lower yield and draw stress values. Plots of the yield and draw stress values for poly-(methylmethacrylate) as a function of temperature at various strain rates show a series of straight lines converging to zero stress in the temperature region of the glass transition temperature (168). At this temperature, the polymer stretches homogeneously without yielding.

At sufficiently high strain rates, it has been shown that the drawing process is influenced by the generation of adiabatic heat, while the yield process is not influenced (189). In polyethylene terephthalate, the results demonstrate that the yield stress is a linear function of strain rate, while the draw stress increases and then drops significantly at a particular strain rate. At low strain rates, any heat generated by the drawing process will be conducted away from the neck so that there is no increase in temperature. As the strain rate is increased sufficiently, the drawing process approaches adiabatic conditions, and the effective temperature at which drawing occurs is increased. The adiabatic process of heat transfer through the shoulders of the neck lowers the yield stress of the undeformed polymer and stabilizes the propagation of the neck at a reduced drawing stress. The effects of adiabatic heating become prominent at strain rates in excess of about 10^{-1}/min. and may cause localized temperature increases approaching 50° C.

Apparent activation energies derived for the yield and cold-drawing processes are quite high and have been related to the long-range, cooperative nature of the molecular chain responses (168). One molecular interpretation of the yield process is based on the internal viscosity theory of Eyring (190–197). According to this argument, the internal viscosity of the polymer decreases with increasing stress and results in molecular flow. The yield stress corresponds to the reduction in the internal viscosity, which allows the applied strain rate to be equivalent to the plastic strain rate $\dot{\epsilon}$, predicted by Equation (7.59):

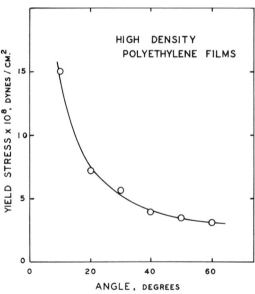

Figure 7.16 Dependence of yield stress on the angle between the tensile axis and the initial orientation draw direction. Data obtained from M. Kurokawa and T. Ban, *J. Appl. Polym. Sci.*, **8**, 971 (1964).

$$\epsilon = \frac{K \sinh V\tau}{2kT} \qquad (7.59)$$

where K is a constant, τ is the shear stress, and V is the "Eyring volume," which represents the volume of a polymer segment that has to move in a cooperative manner to produce plastic deformation. More sophisticated versions of this internal viscosity theory have been developed to describe the yield process of polymers (198, 199). The yield process also has been related to the increase in free volume under stress (200, 201). This proposal argues that the free volume increases until it attains its value at the glass transition temperature, and flow subsequently occurs with polymer deformation.

For cold-drawing to occur, strain-hardening of the polymer is a necessity. This may arise because of strain-induced crystallization at high degrees of extension or some general effect of inducing a degree of molecular alignment parallel to the draw direction. Strain-hardening increases rapidly as the polymer chains attain their limiting extensibility. The plastic deformation of crystalline polymers results in alterations of their morphological structure. For example, the molecular orientation of polyethylene changes from a spherulitic type to a fibrillar type as the degree of plastic deformation increases (202–204). However, the specific type of orientation response for crystalline polymers appears to be related to particular morphological structures (205, 206).

7.6.3 Brittle and Ductile Failures

The brittle behavior of polymers is associated with failure at the maximum stress and comparatively low levels of strain. Additional extension beyond the maximum stress at the yield point results in ductile failure. Differences in the dissipation of energy at rupture and the type of fracture surface are also evident for these two modes of failure.

Brittle fracture is proposed to occur when the yield stress of a material exceeds a critical level (207). As shown in Figure 7.17, it is assumed that brittle failure and plastic deformation are separate phenomena which possess individual curves for the temperature dependence of brittle fracture strength σ_B and tensile yield stress σ_y (208). At a particular

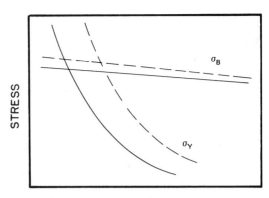

TEMPERATURE

Figure 7.17 Effect of temperature and strain rate on the brittle strength σ_B and yield stress σ_y of polymers. The solid lines represent a relatively low strain rate, and the dashed lines represent a relatively high strain rate.

temperature and strain rate, the process, either brittle fracture or yield, that exhibits the lower stress value will dominate. The intersection of the brittle-strength and yield-stress curves corresponds to the temperature for a change in the mode of failure, that is, the brittle-ductile transition. It is seen that the brittle strength is relatively insensitive to variations in strain rate and temperature. However, the yield stress is strongly dependent on these parameters, increasing with increasing strain rate and decreasing with increasing temperature. Hence the brittle-ductile transition will be shifted to a higher temperature as the strain rate is increased.

The application of a hydrostatic pressure influences the brittle-ductile transition for polymers. Consider a material that exhibits a brittle strength σ_B. When a hydrostatic pressure P is applied, the stress required for brittle failure becomes $\sigma_B + P$. If this stress is now greater than the yield stress σ_y, the polymer will fail in a ductile manner. It has been shown that although the yield stress increases for various polymers subjected to hydrostatic pressures, the occurrence of ductile failure is considerably enhanced (209–211).

An increase in the molecular weight of polymer chains reduces the brittle strength of the material, but it does not appear to directly influence the yield stress. It has been proposed that the tensile strength of a polymer is dependent on its number-average molecular weight $\overline{M}_n$, according to the relationship defined in Equation (7.60) (212).

$$\sigma_B = A - \frac{B}{\overline{M}_n} \qquad (7.60)$$

The brittle strengths for different polymers of various molecular weights have been evaluated (208, 213–215).

7.6.4 Fracture Behavior

The observed brittle strengths of polymers are typically orders of magnitude smaller than theoretical estimates of the stresses required for fracture of the molecular chains. The primary source of this discrepancy is related to the presence of flaws or cracks in the material that act as localized regions of stress concentration. The Griffith theory of rupture has been applied to the fracture mechanics of polymers (216). This energy release rate analysis considers the occurrence of rupture to produce a new surface area, and the increase in energy required to produce this new surface is balanced by a decrease in elastically stored energy. However, the elastically stored energy is not uniformly distributed throughout the material, but it is concentrated near small cracks. Hence rupture occurs due to the spreading of cracks that originate from pre-existing microscopic flaws.

If an infinite plate of uniform stress σ has a crack of length $2C$ introduced normal to the stress, then the rate of change of strain energy per unit thickness to propagate the crack is given as follows by Equation (7.61):

$$-\left(\frac{dW}{dC}\right) = \gamma\left(\frac{dA}{dC}\right) \qquad (7.61)$$

where $-dW$ is the change in elastically stored energy, γ is the surface energy per unit area of new crack face, and dA represents the increment of new crack surface area. Solving for the change in elastically stored energy results in the stress criterion for fracture expressed in Equation (7.62)

$$\sigma_B = \left(\frac{2\gamma E}{\pi C}\right)^{1/2} \qquad (7.62)$$

where σ_B is the stress at fracture and E is the Young's modulus of the material. The inverse relationship between fracture strength and crack length has been demonstrated for various polymers, and it has been shown that the derived surface energy is independent of the sample dimensions (217, 218). The experimental surface energies obtained for polymers are greater than the theoretical values calculated from molecular considerations. It has been suggested that the experimental surface energy represents the sum of an elastic and plastic component, that is, a contribution from the energy expended in generating new surfaces by cleavage and a contribution associated with localized plastic deformation (219–222). It has been shown that the fracture surface energy exhibits a reciprocal dependence on polymer molecular weight (223, 224). A theoretical relationship between fracture surface energy and molecular weight has been derived on the assumption that only polymer chains exceeding a critical molecular weight can contribute to the work of plastic deformation (225).

7.6.5 Crazing

Crazing is a phenomenon that occurs in polymers when cracklike discontinuities are formed containing fibrils connecting the two faces (226). Certain polymers subjected to a uniaxial tensile stress above a minimal value exhibit opaque striations perpendicular to the direction of applied force. However, the induced crazes are not cracks in the usual sense of open void cleavages. They contain polymer of reduced density (227) and reduced refractive index (228–231) as compared to the surrounding polymer. The formation of specially widened crazes on the surface of a polycarbonate specimen is shown in Figure 7.18.

Since the crazes always develop in a plane normal to the applied stress, the crazed matter represents oriented polymer aligned in the direction of the tensile stress. The occurrence of a craze is due to the lateral expansion of a

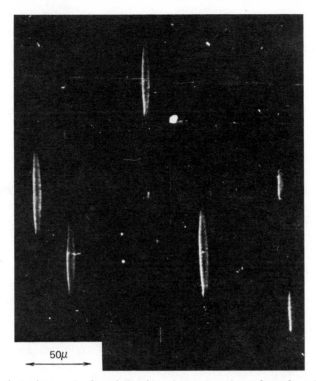

Figure 7.18 Optical micrograph of specially widened crazes on the surface of a polycarbonate film. Reprinted with permission from O. K. Spurr and W. D. Niegisch, *J. Appl. Polymer Sci.*, **6**, 585 (1962). Copyright by Interscience Publishers, New York.

layer of polymer of finite thickness. The presence of a large number of voids (232–234) in the crazed matter is a natural consequence of volume conservation during expansion, and they result in the reduced density and lower refractive index of the crazes. Crazing not only affects the visual characteristics of a plastic material, but also it is an incipient threat of brittle fracture (235–237).

Fracture mechanics criteria appear to govern the threshold conditions and propagation velocities for crazes (238–242). The craze matter is most probably created by a mechanism in which localized portions of the bulk polymer are converted into cavitated material as a result of a stress concentration. Craze initiation is often associated with inhomogeneities and flaws, for example, dust particles, gel, salts, catalysts, scratches, bubbles, voids, or other internal defects. Crazing is much enhanced by residual stresses or strains formed during the polymer or product preparation as a result of molding, processing, or handling operations.

The threshold conditions observed for craze formation are attributed to a cavitation criterion that includes both the work necessary to produce plastic yielding and the work required to create the void interfaces. A necessary requirement for craze formation is the ability of the polymer to undergo cavitation at stress levels lower than those required to enlarge the plastic zone symmetrically (243). The propagation behavior of a craze in interpreted according to the theory of fracture mechanics in a fashion similar to the formulation developed for crack propagation (244). However, the anisotropic craze matter confers load-bearing ability to the discontinuity that is distributed over the craze boundaries. A wide range of responses may be encountered if the craze is hysteresial and subject to load cycling.

Plastics generally are more susceptible to crazing when an organic liquid or vapor is present (245–252). The enhancement of crazing by organic fluids has been attributed to a reduction in the surface tension of the polymer by the liquid, which decreases the fracture energy, and to local plasticization of the polymer in the craze regions, which reduces the energy absorbed by the crazing deformation processes. The influence of organic liquids on craze formation has been related to the solubility parameter of the solvent (243), the size of the solvent molecules (251), and the ability of the liquid to reduce the glass transition temperature of the polymer (252).

Organic liquid also have the ability to heal crazes formed by tensile stress. Exposing stress-crazed specimens of poly(methylmethacrylate), polystyrene, or polycarbonate to a chloroform vapor atmosphere at room temperature results in the disappearance of crazes (247). Similar to craze formation, solvent healing is related to the effect of plasticization, which softens the polymer and results in retraction of the oriented craze matter. Heating of stress-crazed polymers to temperatures above their respective softening temperatures typically heals the materials (227). This healing mechanism may be related to the reversion of oriented craze matter to a nearly random state. It has been shown that the minimum surface work required to propagate a craze decreases as the temperature increases to a critical temperature and then remains constant at higher temperatures. This response is attributed to a combination of a yield stress effect and a polymer/solvent interfacial energy effect (243). It also has been related to the glass-rubber transition for the respective polymers (251).

Stress-whitening of polymers is a phenomenon similar to crazing and differs from it only in matter of degree (253). The opaque whitening of stressed areas in polymers, such as rubber-reinforced or high-impact polystyrene, is due to craze bands of relatively small size and large quantity. Similar to crazes, these stress-whitened regions are birefringent, of lower refractive index, capable of bearing load, and healed by annealing treatments. Thus the stress-whitening or high concentration of micro-crazes in rubber-reinforced polystyrene accounts for its toughened impact strength. This impact toughness is temperature dependent and is related to the occurrence of stress-whitening or the ability of the rubber particles to lower the craze initiation stress relative to the fracture stress.

7.6.6 Ultimate Properties

The ultimate tensile strength σ_b and ultimate elongation ϵ_b at rupture of elastomers vary significantly with the experimental conditions of testing. These mechanical properties depend markedly on both the temperature and experimental time scale of testing. At temperatures above their respective glass transition temperature, the viscoelastic properties of amorphous polymers may be interpreted in terms of various molecular theories (253–256). In a similar manner, the ultimate properties of materials may be treated with regard to their time and temperature dependences (257–259). The ultimate tensile strength and ultimate elongation for a given material are described as a universal function of a reduced time or a reduced strain rate, except at relatively short times or high-strain rates where the polymer approaches glassy behavior. To superpose ultimate property data obtained at different temperatures, a shift factor is required that arises from the temperature dependence of the frictional factor for the polymer segmental mobility. This is the same shift factor (WLF equation) used to superpose viscoelastic data determined at small deformation (260).

To obtain composite curves, the ultimate properties of the polymer are obtained over a wide range of strain rates at several temperatures (261–268). The σ_b and ϵ_b are superposed along the time axis using values of the shift factor $\log a_T$, given by the WLF equation. Typical composite curves obtained for an SBR gum vulcanizate are shown in Figures 7.19 and 7.20 for the respective variations of tensile strength and ultimate elongation with reduced strain rate (262). It is observed that increasing strain rate results in an increasing tensile strength and an ultimate strain or elongation that passes through a maximum value. Hence it is possible to predict the ultimate properties of this polymer at any strain rate and temperature. From these results it has been concluded that the ultimate properties vary with temperature in accordance

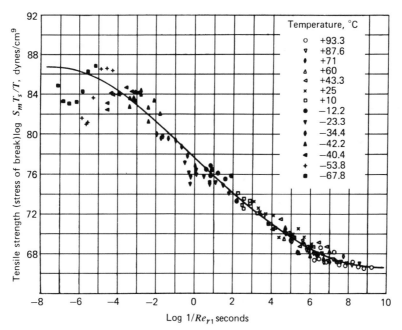

Figure 7.19 Variation of ultimate tensile strength with reduced strain rate. Reprinted with permission from T. L. Smith, *J. Polym. Sci.*, **32**, 99 (1958). Copyright by Interscience Publishers, New York.

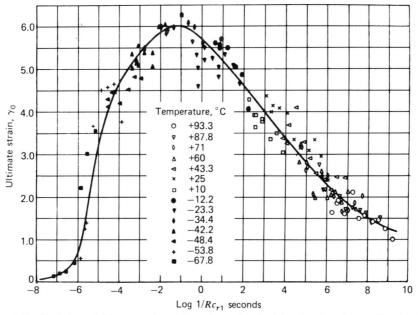

Figure 7.20 Variation of ultimate strain or elongation at break with reduced strain rate. Reprinted with permission from T. L. Smith, *J. Polym. Sci.*, **32**, 99 (1958). Copyright by Interscience Publishers, New York.

with the temperature dependence of the internal viscosity or molecular friction coefficient; the ultimate properties vary with strain rate due to the increase of the viscous resistance to network formation with increasing strain rate; and the temperature dependence of molecular parameters, such as rate of chemical bond rupture, do not significantly influence the temperature dependence of ultimate properties.

A basic method of characterizing the ultimate tensile properties of a polymer is in terms of a failure envelope, which is independent of time and temperature. The failure envelope prescribed by the data obtained from the study of the SBR elastomer is shown in Figure 7.21 (263). As either the strain rate is increased or the temperature is decreased, the respective rupture values move counterclockwise around the curve. The failure envelope illustrates the typical wide range of values that may be observed for the ultimate properties of amorphous elastomers.

Other studies have shown that the "energy input to break" obtained from tensile stress-strain curves is related exponentially to the reciprocal of absolute temperature (269–272). In addition, the "energy input to break" over a temperature range is related by a square law to the elongation at rupture. These results appear to be applicable to amorphous, crystalline, and cross-linked polymers, including filled and unfilled materials (273). Similar relationships have been found for the rupture energy per unit volume and the deformation energy dissipated in stretching an elastomer to its breaking elongation (274).

Tear propagation in elastomers represents a complex distribution of stresses and strains in the immediate vicinity of a tear tip. The mechanism of tear growth has been analyzed from an energy point of view (275). Using the Griffith criterion for crack growth as a starting point, it is found that growth of a cut in imperfectly elastic systems requires more energy than that obtained from the gain in surface free energy. This additional energy includes energy dissipated in plastic flow processes and energy dissipated irreversibly in viscoelastic processes. This concept assumes that the tearing energy is independent of the shape of the test specimen and is independent of the manner in which the forces are applied to the edges of the specimen, see Equation (7.63).

$$-\left(\frac{dW}{dc}\right) = \mathrm{T}t \tag{7.63}$$

Hence where $-dW$ represents the reduction

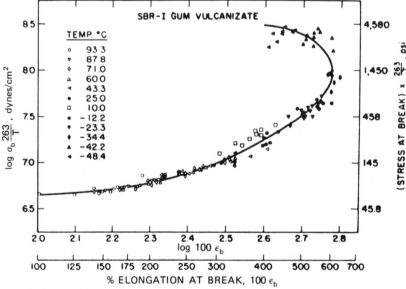

Figure 7.21   Failure envelope for ultimate properties determined at various strain rates and temperatures. Reprinted with permission from T. L. Smith, *J. Polym. Sci.*, Part A-1, 3597 (1963). Copyright by Interscience Publishers, New York.

in elastically stored energy, *dc* corresponds to an incremental increase in the length of the cut, *t* is the specimen thickness, and T denotes the characteristic tear energy. A number of different types of test pieces have been evaluated to verify that the tearing energy is independent of particular shapes or dimensions (276-279). A wide variety of fracture measurements have been correlated in terms of this tear energy concept.

The tear energy of a particular material is dependent on the temperature and rate of tear (280–282). Observed values of tear energy are considerably higher than the theoretical threshold fracture energy required to break all of the polymer molecules in a random plane. The reason for the enhanced strength lies in the process by which the energy is dissipated at the tear tip.

7.6.7 Impact Strength

Impact strength is the measure of a polymer's resistance to fracture under conditions of high-strain rates. Many impacts tests measure the energy required to break a standard specimen under certain specified conditions. The measurement of impact strength is usually made with the Izod test (a pendulum apparatus that employs a notched specimen struck on an unsupported end), the Charpy test (a pendulum apparatus with a notched specimen supported at the two ends and struck in the middle), the falling-weight test (a standard weight dropped from a prescribed height), and the high-speed stress-strain test.

The test specimens for impact strength evaluations are typically notched. Notching tends to improve the quality of the data by reducing the effects of random flaws. Impact strengths generally decrease as the sharpness of the notch radius is improved. Notching also influences the brittle-ductile behavior of materials by significantly raising the yield stress (283). Thus a material may exhibit ductile fracture for unnotched specimens, but fail in a brittle manner when a notch is introduced.

Reasonable correlation has been obtained between the falling-weight impact energy and the energy obtained from stress-strain curves at medium-strain rates (284). The impact strength of polymers is temperature dependent and generally increases as the glass transition temperature is approached. Polymers that possess a pronounced low-temperature, secondary relaxation in the glassy state tend to exhibit increased impact strengths (285-289). However, it should be noted that the presence of such a secondary relaxation maximum does not necessarily mean that a polymer will have high-impact resistance. Yet, the absence of a secondary maximum always appears to indicate that the polymer will have low-impact resistance. These dynamic mechanical loss processes are more effective when they are associated with molecular motions within the main chain of the polymer as opposed to side-chain motions. Crystalline polymers have high-impact strengths if their respective glass transition temperature is well below the test temperature. However, their impact strength decreases as the extent of crystallinity increases or the size of the spherulites increases. High-impact strength does not necessarily mean that a polymer is a tough material, and brittle fracture cannot be readily associated with low-impact strength. For example, composite and oriented materials can exhibit high-impact strengths while fracturing in a brittle mode. The impact properties or energy-absorbing characteristics of cellular polymeric materials have been related to the geometric structure and the physical properties of the matrix polymer (290–295).

7.6.8 Anisotropic Behavior

When an isotropic polymer is stressed, the molecular chains undergo a rearrangement that tends to orient the molecules in the direction of the applied stress. Under certain conditions, amorphous polymer chains may exhibit strain-induced crystallization. For polymers possessing varying degrees of crystalline structures, the response to orientation is much more complex. The crystalline regions may change their direction without a modification of internal structure or the orientation may involve a complete reorganization of the crystal structure (296). Oriented polymers have properties that vary as a function of direction or are anisotropic. Orientation is generally accomplished by deformation of a polymer at a temperature above its glass transition. Fixation of the orientation is achieved by cooling the polymer below its glass transition temperature, where the molecules do not have an opportunity to return to their random orientations.

Uniaxial forming processes involving drawn fibers or films can result in polymers that show isotropy in a plane perpendicular to the direction of drawing. Other commercial one-way draw processes, such as rolling or rolling and annealing, may yield polymer films that possess orthorhombic rather than transverse isotropic symmetry. Biaxial orientation occurs when a polymer sheet is deformed simultaneously in two directions perpendicular to each other. The polymer chain segments tend to orient parallel to the plane of the sheet, but in more or less random directions within the plane. The anisotropy of biaxially drawn films is markedly less than that of uniaxially drawn films (13). The number of independent elastic constants required to define the mechanical behavior of anisotropic materials will depend on the degree and type of orientation (297).

The independent elastic constants have been determined for monofilament fibers of oriented polyethylene terephthalate, nylon, low- and high-density polyethylene, and polypropylene (298–300). The development of mechanical anisotropy in these fibers is dependent on the details of their chemical composition and the exact nature of the drawing process. An increased draw ratio provides a higher degree of molecular orientation and results in an increased tensile modulus measured in the direction of the drawn axis of the fiber. The extensional modulus for the highly oriented fiber is greater than the transverse modulus. Similar results have been obtained for a series of oriented low-density polyethylene sheets prepared by uniaxial stretching of isotropic sheets (301). It was observed that the oriented sheets showed the lowest tensile modulus in a direction making an angle of about 45° to the initial draw direction. More complex behavior was obtained for the directional dependence of tensile modulus for rolled and annealed polyethylene sheets of orthorhombic symmetry (302, 303).

The anisotropic mechanical properties of oriented polymers will depend on the nature of the molecular arrangements, including the crystalline morphology and the molecular mobility. The aggregate model for mechanical anisotropy regards the unoriented polymer as an aggregate of anisotropic units with elastic properties comparable to those of the highly oriented polymer (304, 305). The average elastic constants for the aggregate can be derived by the assumption of either uniform stress or uniform strain throughout the polymer. The actual values for a random aggregate, which depends on molecular orientation for the development of mechanical anisotropy, lie between the calculated bounds predicted by these alternative assumptions (306).

While orientation increases the elastic modulus in the drawn direction, the transverse shear modulus is not greatly affected (307). Dynamic mechanical studies have considered the effects of orientation for several uniaxially drawn polymers, with and without subsequent annealing (307–311). The mechanical loss damping peak associated with the glass transition of a polymer tends to become broader, and the temperature of the damping maximum is shifted to higher values (312). Figure 7.22 shows the behavior observed for a sheet of drawn and annealed polypropylene. Both components of the complex elastic modulus exhibit higher values in the direction parallel to drawing at low temperatures and higher values in the direction perpendicular to drawing at high temperatures. This behavior can be interpreted on the basis of a lamination model for the disposition of the amorphous and crystalline regions (308). It has also been attributed to an interlamellar shear process for the crystalline structures (303).

Fabrication of polymers by mechanical means is accomplished with the macromolecules in various physical states that have been previously discussed. As described by Alfrey (313) during the fabrication process, alterations in the polymer structure may occur. When the fabricated article is in use, it is usually subjected to multiaxial stress, and the spatial arrangement of the molecules in the article will have a significant effect on the end-use durability.

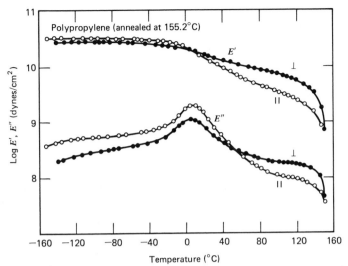

Figure 7.22 Temperature dependent of dynamic mechanical moduli parallel (○) and perpendicular (●) to the draw direction for oriented polypropylene. Reprinted with permission from M. Takayanagi, K. Imada, and T. Kajiyama, *J. Polym. Sci.*, **C15**, 263 (1963). Copyright by Interscience Publishers, New York.

Crystallizable polymers may enter the process in a metastable condition and undergo crystallization during deformation with the resultant product being crystalline and oriented in character. The crystallizable polymer may have been stretched at a temperature between the melting point and the glass transition, which would require major reorganization of the crystalline morphology. It could also have been oriented in the melt with the crystallization taking place in the oriented molten mass. The resultant properties would be different in each of these cases.

Amorphous polymers can be fabricated in the viscoelastic fluid state, the rubbery solid state, or in the glassy state. Since the molecular structure of the polymers has been established during the polymerization process, the alteration in properties that is obtainable during fabrication is due to the spatial arrangement of the molecules in the final product. For example, if polystyrene is uniaxially oriented, a highly anisotropic product results. It has a high tensile strength, extensibility, and resistance to environmental stress crazing and cracking in the direction of drawing or orientation. However, in a direction normal to the orientation direction, it is more susceptible to environmental stress crazing and is even weaker than the polymer in an unoriented state. Since tensile strength increases and

elongation at break first increases and then decreases with increased orientation, an optimum level of orientation exists for a balance between these characteristics. In contrast to this behavior, polystyrene that is uniformly, biaxially oriented at a temperature just above the glass transition temperature is tough, strong, and stress-crack resistant relative to unoriented polystyrene in all directions in the plane of orientation.

7.6.9 Birefringence

As light passes through a plastic material, it may be slowed down or refracted owing to the interaction between the electric field of the light wave and the electrons of the polymer chain. The index of refraction of an isotropic material measures the degree of velocity reduction for the passage of light and is dependent on the extent of polarizability or electron mobility within polymer segments. Optically anisotropic materials possess polarizabilities that vary with direction and exhibit birefringence or a difference in index of refraction with direction. Birefringence also is evidenced by the ability of a material to rotate the plane of polarized light. For low levels of anisotropy, the expression for birefringence Δ is given by Equation (7.64)

$$\Delta = \left(\frac{2}{9}\right) \pi \left[\frac{(n^2 + 2)^2}{n}\right] (P_1 - P_2) \quad (7.64)$$

where n is the index of refraction for the isotropic material and $\Delta = n_1 - n_2$ is the difference in index of refraction for two principal directions with polarizabilities P_1 and P_2 (314).

Birefringence in a stretched rubber may be determined by measuring the differences in refraction with crossed polaroids (16, 315). Transmission measurements may be made at various sample elongations (316, 317) and used to observe dynamic birefringence for an oscillational strain at a particular frequency (318, 319). Special techniques are necessary to determine birefringence in biaxially oriented samples (320–322). Polymer solutions exhibit birefringence if orientation is affected by external forces, such as flow (flow birefringence), rotation, electric fields (Kerr effect), or magnetic fields (Cotton-Moulton effect).

The theory for birefringence in a stretched rubber network is based on the statistical segment model (16, 323–325). If a rubber is stretched, λ ($\lambda = L/L_0$) the birefringence is given by Equation (7.65).

$$\Delta = \left(\frac{2}{45}\right) \pi \left[\frac{(n^2 + 2)^2}{n}\right] N (b_\parallel - b_\perp)$$

$$\left[\lambda^2 - \frac{1}{\lambda}\right] \quad (7.65)$$

In this equation n is the index of refraction for the rubber in the isotropic state or unstretched condition, N is the number of network chains per unit volume, and $b_\parallel$ and $b_\perp$ represent the principal polarizabilities of the statistical segment parallel and perpendicular to the contour of the polymer chain (326). The stress σ (based upon the actual cross-sectional area of the sample) of an ideal rubber network is given by Equation (7.66).

$$\sigma = NkT \left[\lambda^2 - \left(\frac{1}{\lambda_i}\right)\right] \quad (7.66)$$

An additional factor based on chain dimen-

sions in the stretched and unstretched states (15) recently has been included in the derivation of the equations for birefringence and stress of rubber networks (314). Dividing Equation (7.65) by Equation (7.66) yields Equation (7.67).

$$C = \left(\frac{\Delta}{\sigma}\right) = \left(\frac{2\pi}{45kT}\right)\left[\frac{(n^2 + 2)^2}{n}\right](b_\parallel - b_\perp)$$

$$(7.67)$$

which is termed the stress-optical coefficient C. Deviations from the ideal theory predictions that the stress-optical coefficient is independent of cross-link density and degree of strain have been evaluated (16, 314, 327–331). The temperature dependence of the stress-optical coefficient is proportional to the dependence of $(b_\parallel - b_\perp)$. This latter dependence for the statistical segment of the polymer is usually greater than that calculated for the monomer unit (16, 326) and reflects differences in polymer chain flexibility.

It has been shown that uncross-linked elastomers also exhibit birefringence (332-335). A general relationship based on birefringence measurements has been derived that allows an estimate of the draw ratio for fibers of various chemical compositions (336). Significant changes in the birefringence of polymers occur in the vicinity of the glass transition or in going from the rubbery to the glassy state (314). The birefringence changes sign at the glass transition temperature for poly(ethyl acrylate) (337), polyvinylchloride (338–341), and polystyrene (342).

The birefringence of a copolymer is dependent on an average value of the term $(b_\parallel - b_\perp)$ for the statistical segments of the polymer chain (331, 343, 344). For two-phase systems, the total birefringence is an additive property based on the volume fraction of the individual refracting components and includes a factor termed form birefringence, which arises from the discontinuity of the index of refraction at the interference of the components (314). This general additive expression for birefringence has been applied to filled rubber systems (345–347), phase segregated block copolymers (348–

350), and crystalline polymers (317, 336, 351–361).

7.7 SUMMARY

The detail and in-depth background necessary to an understanding of the mechanical properties of plastic materials cannot be completely summarized in the few pages of this chapter. The major goal was to familiarize the reader with various, important aspects of the topic and to provide a number of useful source references. Plastics are continually gaining acceptance as the materials of choice for numerous applications, with requirements ranging from decorative and esthetic to functional and high-performance serviceability. When compared to metals and wood, plastic products typically have advantages in ease of fabrication, energy savings, and reduction in weight. In addition, superior physical properties can be achieved with polymers when they are compared with those of other materials.

The fundamental properties of a polymer are determined by the chemical structure of its constituent molecules, while the bulk properties of a plastic material are controlled by the organizational arrangement of the molecules. The molecular structure is largely developed during the polymerization stage, but the spatial arrangement of the molecules is established by the fabrication process. Mechanical processing variables can influence the molecular orientation in amorphous polymers, the morphology and orientation in crystalline polymers, and the geometrical distribution of the phases in various multiphase materials such as fibrous and laminar composites and cellular foams. The supramolecular structure of polymeric articles has profound effects on the mechanical properties and end-use characteristics of one-dimensional fibers, two-dimensional films, and three-dimensional solid products.

New fabrication processes are being developed that provide plastic products with unique combinations of physical properties. Multi-layer thermoplastic film laminates containing several alternating layers of polymers can be prepared by coextrusion from the melt (362,

363). The mutual mechanical reinforcement between the "hard" and "soft" layers results in materials that undergo large ductile deformations. Films can also be combined to obtain bondable surfaces that accept adhesives. Examples of these combined films are polyolefins laminated to polyester films and fluorocarbons laminated to polyimide films.

Reaction injection molding is a process that involves the high-pressure impingement mixing of two or more reactive liquid components. After mixing, the liquid stream is injected into a closed mold at low pressure to produce finished parts. Molded urethane products ranging from hard, high-modulus plastics to soft, flexible elastomers can be prepared by the reaction injection molding process and are characterized by outstanding toughness, abrasion, and impact resistance over a wide temperature range (364, 365). Photocurable coatings and inks are applied as thin films of reactive monomers that contain a photoinitiator. Impinging ultraviolet light induces polymerization in a fraction of a second without heat or volatilization of solvents (366). Satisfactory physical properties of the coatings, including adhesion to a range of substrates and good heat and light stability, can be achieved by the proper selection of the photopolymerizable reactive monomers and oligomers.

An understanding of the mechanical properties of plastics is a basis for the development of products to meet specific application requirements. For example, the acetals are a class of polymers characterized by high stiffness, excellent fatigue properties, good resistance to creep, and impact strengths that are relatively unchanging with temperature variations. These properties, combined with low surface friction and low abrasion, have resulted in the use of acetals for moving parts like gears, bearings, cams, and ratchets (4). Many mechanical properties of plastics are affected significantly by part design. In the case of impact resistant, automobile bumpers, which are required to absorb a 5 miles per hour impact without acquiring a permanent set, impact resistance coupled with part design is of utmost importance (367). The material design engineer is interested in whether or not the plastic will act in a brittle or a ductile manner when subjected to the strain rate and temperature typical of its intended end use. For the automobile bumpers, the impact at the 5 miles per hour strain rate must be absorbed at subambient conditions. An understanding of the frequency dependence of the glass transition temperature and other relaxation phenomena of polymers is an important key to the successful design of plastics for such applications.

Engineering information to predict the serviceability performance of plastic materials can be obtained from a combination of several mechanical properties. A study of specific mechanical properties provides direction for the design of polymers with new chemical structures or of multicomponent systems based on existing polymers to satisfy particular end-use requirements. The quality control of materials and manufacturing processes for plastic products routinely includes the determination of specific mechanical properties during production operations. Knowledge of the mechanical properties of plastics is relevant only to the extent that one desires to understand the performance of products that are utilized in our everyday activities and will be important in the future. This relevancy is becoming increasingly more important as new applications are developed for plastics.

7.8 GLOSSARY OF SYMBOLS

A	Surface area
a_T	Horizontal shift factor of WLF equation for time-temperature superposition or ratio of relaxation times at two different temperatures
a_σ	Constant in nonlinear viscoelastic theory
B	Bulk modulus or characteristic constant
$b_\parallel, b_\perp$	Principal polarizabilities of statistical chain segment
C	Crack length or stress optical coefficient
C_1, C_2	Characteristic constants
c	Cut length in tear specimens
$D(t)$	Time dependent tensile creep compliance for linear viscoelastic theory
D_0	Elastic component of tensile creep compliance
$D_n(t)$	Time dependent tensile creep compliance for nonlinear viscoelastic theory
E	Elastic modulus, Young's modulus, or enthalpy
E^*	Complex tensile modulus
E'	Elastic or storage component of complex tensile modulus
E''	Viscous or loss component of complex tensile modulus
$E(t)$	Tensile stress relaxation modulus
$E_r(t)$	Reduced tensile stress relaxation modulus
E_f	Elastic or Young's modulus for filled polymers
$F(t)$	Constant strain rate modulus
f	Force or fractional free volume
f_g	Fractional free volume at the glass transition temperature
f_e	Energy contribution to retractive force for rubbers
f_s	Entropy contribution to retractive force for rubbers
G	Shear modulus
G^*	Complex shear modulus
G'	Elastic or storage component of complex shear modulus
G''	Viscous or loss component of complex shear modulus
G_0	Shear modulus of rubber
g_0, g_1, g_2	Constants in nonlinear viscoelastic theory
$g(\epsilon)$	Strain function in nonlinear viscoelastic theory
$H(\tau)$	Distribution function of relaxation times
K	Characteristic constant
k	Characteristic constant or Boltzmann's constant
L	Length
M_n	Number-average molecular weight
M_c	Average molecular weight of chains between cross-links
N_0	Moles of network chains
n	Index of refraction
P	Pressure
P_1, P_2	Molecular polarizabilities in principal directions of plane

Q^{-1}	Energy dissipation factor or mechanical loss	Ψ	Reduced time function for nonlinear viscoelastic theory
R	Gas constant	Ω	Total number of network conformations
$\bar{r}_0^2$	Mean square end-to-end distance of network chain		
$\bar{r}_f^2$	Mean square end-to-end distance of isolated chain	$\tan \delta$	Energy dissipation factor
S	Entropy		
T	Temperature		
T_g	Glass transition temperature		
T_{gx}	Glass transition temperature of crosslinked polymer		
T_0	Reference temperature for reduced variables		
T_s	Standard reference temperature for WLF equation		
t	Time or thickness		
u	Arbitrary time parameters		
V	Volume		
V_f	Polymer free volume		
W	Elastic stored energy		
w	Weight fraction of monomers in copolymer		
α_f	Coefficient of expansion of fractional free volume		
β	Bulk thermal expansion coefficient		
$\Gamma(\epsilon)$	Strain function in nonlinear viscoelastic theory		
γ	Surface energy per unit area of crack face		
Δ	Birefringence		
$\Delta D(t)$	Transient component of tensile creep compliance		
ΔH_a	Apparent activation energy		
ΔL	Strain length		
ϵ	Strain		
$\dot{\epsilon}$	Strain rate		
ϵ_b	Strain at break		
ϵ_y	Yield strain		
η	Viscosity		
λ	Extension ratio		
ν	Poisson's ratio		
ρ	Density		
σ	Stress		
σ_b	Stress at break		
T	Tear energy		
τ	Relaxation time or shear stress		
ϕ_f	Volume fraction of filler		

7.9 REFERENCES

1. C. G. Seefried and J. V. Koleske, *J. Appl. Polym. Sci.*, **19**, 2493 (1975).
2. C. G. Seefried and J. V. Koleske, *J. Appl. Polym. Sci.*, **19**, 2503 (1975).
3. D. R. Paul and S. Newman, *Polymer Blends*, Academic Press, New York, 1977.
4. C. A. Harper, *Handbook of Plastics and Elastomers*, McGraw-Hill, New York, 1975.
5. R. D. Deanin, *Plast. World.* **46**, December, 1976.
6. R. L. McCullough, *Polym. Eng. Sci.*, **16**, 294 (1976).
7. T. Alfrey, Jr., *Mechanical Properties of Polymers*, Interscience, New York, 1948.
8. J. E. Ashton, J. C. Halpin, and P. H. Petit, *Primer on Composite Materials*, Technomic Publishing Co., Inc., Connecticut, 1969.
9. L. R. Calcote, *The Analysis of Laminated Composite Structures*, von Nostrand Reinhold, New York, 1969.
10. A. V. Tobolsky and H. F. Mark, *Polymer Science and Materials*, Wiley-Interscience, New York, 1971.
11. J. G. Williams, *Stress Analysis of Polymers*, John Wiley, New York, 1973.
12. G. R. Smoluk, *SPE J.*, **27**, 19 (December, 1971).
13. L. E. Nielsen, *Mechanical Properties of Polymers*, Vol. 1, Mercel Dekker, New York, 1974.
14. R. J. Samuels, *Chemtech*, **169** (March, 1974).
15. P. J. Flory, *Principles of Polymer Chemistry*, Cornell University Press, Ithaca, New York, 1950.
16. L. R. G. Treloar, *Physics of Rubber Elasticity*, Oxford University Press, Oxford, 1958.
17. J. J. Aklonis, W. J. MacKnight, and M. Shen, *Introdution to Polymer Viscoelasticity*, Wiley-Interscience, New York, 1972.
18. L. R. G. Treloar, *Trans. Faraday Soc.*, **40**, 59 (1944).
19. G. Allen, *Proc. R. Soc. London*, A**351**, 381 (1976).
20. T. Alfrey, Jr. and W. G. Lloyd, *J. Polym. Sci.*, **62**, 159 (1962).
21. B. Mukherji and W. Prins, *J. Polym. Sci.: Part A*, **2**, 4367 (1964).
22. S. F. Edwards, *Proc. Phys. Soc. (London)*, **85**, 613 (1965).
23. H. Reiss, *J. Chem. Phys.*, **47**, 186 (1967).
24. H. M. James and E. Guth, *Proceedings, Third*

Rubber Technology Conference, Heffner, London, 1954.

25. E. A. DiMarzio, *J. Chem. Phys.,* **36**, 1563 (1962).

26. J. L. Jackson, M. Shen, and D. A. McQuarrie, *J. Chem. Phys.,* **44**, 2388 (1966).

27. P. J. Flory, *Statistical Mechanics of Chain Molecules,* Interscience, New York, 1969.

28. P. J. Flory, C. A. J. Hoeve, and A. Ciferri, *J. Polym. Sci.,* **34**, 337 (1959).

29. A. V. Tobolsky, D. W. Carlson, and N. Indictor, *J. Polym. Sci.,* **54**, 175 (1961).

30. W. R. Krigbaum and M. Kaueko, *J. Chem. Phys.,* **36**, 99 (1962).

31. R. J. Roe and W. R. Krigbaum, *J. Polym. Sci.,* **61**, 167 (1962).

32. K. J. Smith, Jr., A Greene, and A. Ciferri, *Kolloid-Z.,* **194**, 49 (1964).

33. C. Price, *Proc. R. Soc. London,* **A351**, 331 (1976).

34. M. Shen and P. J. Blatz, *J. Appl. Phys.,* **39**, 4937 (1968).

35. M. Shen, *Macromolecules,* **2**, 358 (1969).

36. E. H. Cirlin, H. M. Grebhard, and M. Shen, *J. Macromol. Sci.—Chem.,* **A5**, 981 (1971).

37. T. Y. Chen, P. Ricica, and M. Shen, *J. Macromol. Sci.—Chem.,* **A7**, 889 (1973).

38. Y. Kuwahara and M. Shen, *J. Macromol. Sci.—Chem.,* **A10**, 1205 (1976).

39. J. E. Mark, *Rubber Chem. Tech.,* **46**, 593 (1973).

40. A Ciferri, C. A. J. Hoeve, and P. J. Flory, *J. Amer. Chem. Soc.,* **83**, 1015 (1961).

41. J. E. Mark, *J. Polym. Sci.: Macromol. Rev.,* **11**, 135 (1976).

42. M. Shen, W. F. Hall, and R. E. DeWames, *J. Macromol. Sci., Rev. Macromol. Chem.,* **C2**, 183 (1968).

43. M. C. Shen, D. A. McQuarrie, and J. L. Jackson, *J. Appl. Phys.,* **38**, 791 (1967).

44. P. J. Flory, *Chem Rev.,* **35**, 51 (1944).

45. E. Guth, R. Simha, and O. Gold, *Kolloid-Z.,* **74**, 266 (1936).

46. H. M. Smallwood, *J. Appl. Phys.,* **15**, 758 (1944).

47. E. Guth, *J. Appl. Phys.,* **16**, 20 (1945).

48. L. Mullins and N. R. Tobin, *J. Appl. Polym. Sci.,* **9**, 2993 (1965).

49. M. Mooney, *J. Appl. Phys.,* **11**, 582 (1940).

50. R. S. Rivlin, *Philos, Trans.,* **A241**, 379 (1948).

51. L. R. G. Treloar, *Rep. Prog. Phys.,* **36**, 755 (1973).

52. J. E. Mark, *Rubber Chem. Tech.,* **48**, 495 (1975).

53. L. R. G. Treloar, *Proc. Soc. London,* **A351**, 301 (1976).

54. T. G. Fox and S. Loshaek, *J. Polym. Sci.,* **15**, 371 (1955).

55. G. Pezzin, F. Zilo-Grandi, and P. Sanmartin, *Eur. Polym. J.,* **6**, 1053 (1970).

56. F. E. Bailey, Jr. and J. V. Koleske, *Poly(ethylene oxide),* Academic Press, New York, 1976.

57. C. G. Seefried and J. V. Koleske, *J. Test. Eval.,* **4**, 220 (1976).

58. T. G. Fox, *Bull. Am. Phys. Soc.,* Series 2, **1** (3), 123 (1956).

59. M. Gordon and J. S. Taylor, *J. Appl. Chem.,* **2**, 493 (1952).

60. J. V. Koleske and R. D. Lundberg, *J. Polym. Sci.: Part A-2,* **7**, 795 (1969).

61. K. Ueberreiter and G. Kanig, *J. Chem. Phys.,* **18**, 399 (1950).

62. G. M. Martin and L. Mandelkern, *J. Res. Natl. Bur. Stand.,* **62**, 141 (1959).

63. H. D. Heinze, K. Schnieder, G. G. Schnell, and K. A. Wolf, *Rubber Chem. Tech.,* **35**, 776 (1962).

64. L. E. Nielsen, in G. B. Butler and K. F. O'Driscoll, Eds., *Reviews in Macromolecular Chemistry,* Vol. 4, Marcel Dekker, New York, 1970, p. 77.

65. L. E. Nielsen, *J. Macromol. Sci.—Rev. Macromol. Chem.,* **C3**, 69 (1969).

66. J. V. Koleske and J. A. Faucher, *J. Chem. Ed.,* **43**, 254 (1966).

67. N. Bekkedahl, *J. Res. Natl. Bur. Stand.,* **13**, 411 (1934); **43**, 145 (1949).

68. P. Manaresi and V. Giannella, *J. Appl. Polym. Sci.,* **4**, 251 (1960).

69. L. E. Nielsen, *Rev. Sci. Inst.,* **22**, 690 (1951); *SPE J.,* **16**, 525 (May, 1960).

70. J. R. McLoughlin and A. V. Tobolsky, *J. Colloid Sci.,* **7**, 555 (1952).

71. A. F. Lewis and J. K. Gillham, *J. Appl. Polym. Sci.,* **7**, 685 (1963).

72. M. L. Dannis, *J. Appl. Polym. Sci.,* **7**, 231 (1963).

73. S. Strella, *J. Appl. Polym. Sci.,* **7**, 569 (1963).

74. R. H. Wiley and G. M. Brauer, *J. Polym. Sci.,* **3**, 455 (1948); **11**, 221 (1953).

75. R. Kaneka, *Kobunshi Kagaku,* **24**, 272 (1967), C.A. 68:69481d.

76. A. H. Scott, *J. Res. Natl. Bur. Stand.,* **14**, 99 (1935).

77. J. M. O'Reilly, *J. Polym. Sci.,* **57**, 429 (1962).

78. U. Bianchi, A Turturro, and G. Basile, *J. Phys. Chem.,* **71**, 3555 (1967).

79. E. Eiermann and K. Hellwege, *J. Polym. Sci.,* **57**, 99 (1962).

80. E. W. Anderson and D. W. McCall, *J. Polym. Sci.,* **31**, 241 (1958).

81. Y. Ishida, O. Amano, and M. Takayanagi, *Kolloid Z.,* **172**, 126 (1960).

82. H. Kramer and K. E. Helf, *Kolloid Z.,* **180**, 114 (1962).

83. G. T. Furukawa, P. E. McCoskey, and M. L. Reilly, *J. Res. Natl. Bur. Stand.,* **55**, 127 (1955).

84. F. S. Dainton, *Polymer,* **3**, 286 (1962).

85. E. Passaglia and H. K. Kevorkian, *J. Appl. Phys.,* **34**, 90 (1963).

86. O. B. Edgar, *J. Chem. Soc.,* 2638 (1952).

87. B. Baccarelde and E. Butta, *Chem. Ind. (Milan),* **44**, 1228 (1962).

88. H. A. Waterman, *Kolloid Z.,* **192**, 1 (1963).

89. Y. Wada, *J. Phys. Chem. (Jap.),* **16**, (6), 1226 (1961).

90. J. A. E. Kail, J. A. Sauer, and A. E. Woodward, *J. Phys. Chem.,* **66**, 1292 (1962).

91. J. A. Faucher, J. V. Koleske, E. R. Santee, J. J. Stratta, and C. W. Wilson, III., *J. Appl. Phys.,* **37**, 3962 (1966).

92. R. E. Barker and C. R. Thomas, *J. Appl. Phys.,* **35**, 87 (1964).

'93. F. Krum and F. H. Müller, *Kolloid Z.,* **164**, 8 (1959).

94. F. P. Reding, J. A. Faucher, R. D. Whitman, *J. Polym. Sci.,* **54**, 556 (1961).

95. F. W. Billmeyer, Jr., *Textbook of Polymer Science,* Interscience, New York, 1962.

96. T. L. Smith, *Polym. Eng. Sci.,* **5**, 270 (1965).

97. T. L. Smith, *Polym. Eng. Sci.,* **13**, 161 (1973).

98. R. Buchdahl and L. E. Nielsen, *J. Appl. Phys.,* **21**, 482 (1950).

99. J. D. Ferry, *J. Am. Chem. Soc.,* **72**, 3746 (1950).

100. J. D. Ferry, *Viscoelastic Properties of Polymers,* John Wiley, New York, 1970.

101. W. N. Findley, *SPE J.,* **16**, 192 (February, 1960).

102. W. N. Findley, *SPE J.,* **16**, 57 (January, 1960).

103. H. Leaderman, *Elastic and Creep Properties of Filamentous Materials and Other High Polymers,* The Textile Foundation, Washington, D. C. (1943).

104. L. E. Nielsen, *SPE J.,* **16**, 525 (May, 1960).

105. P. R. Pinnock and I. M. Ward, *Proc. Phys. Soc.,* **81**, 260 (1963).

106. H. Leaderman, *Trans. Soc. Rheol.,* **6**, 361 (1962).

107. S. Turner, *Polym. Eng. Sci.,* **6**, 306 (1966).

108. I. M. Ward and E. T. Onat, *J. Mech. Phys. Solids,* **11**, 217 (1963).

109. A. E. Green and R. S. Rivlin, *Arch. Ration Mech. Anal.,* **1**, 1 (1957).

110. A. E. Green, R. S. Rivlin, and A. J. M. Spencer, *Arch. Ration. Mech. Anal.,* **3**, 82 (1959).

111. H. Leaderman, F. McCracken, and O. Nakada, *Trans. Soc. Rheol.,* **7**, 111 (1963).

112. V. V. Neis and J. L. Sackman, *Trans. Soc. Rheol.,* **11**, 307 (1967).

113. W. N. Findley and K. Onaran, *Trans. Soc. Rheol.,* **12**, 217 (1968).

114. D. W. Hadley and I. M. Ward, *J. Mech. Phys. Solids,* **13**, 397 (1965).

115. I. M. Ward and J. M. Wolfe, *J. Mech. Phys. Solids,* **14**, 131 (1966).

116. R. A. Schapery, Proceedings Fifth U.S. National Congress of Applied Mechanics, ASME, 511 (1966).

117. R. A. Schapery, *Proceedings IUTAM Symposium on Thermoinelasticity,* Springer-Verlag (1969).

118. R. A. Schapery, Purdue University, Rept. No. 69-2 (February, 1969).

119. R. A. Schapery, *Polym. Eng. Sci.,* **9**, 295 (1969).

120. W. N. Findley and J. S. Y. Lai, *Trans. Soc. Rheol.,* **11**, 361 (1967).

121. A. C. Pipken and T. G. Rogers, *J. Mech. Phys. Solids,* **16**, 59 (1968).

122. T. L. Smith, *Trans. Soc. Rheol.,* **6**, 61 (1962).

123. E. Guth, P. E. Wack, and R. L. Anthony, *J. Appl. Phys.,* **17**, 347 (1946).

124. A. V. Tobolsky and R. D. Andrews, *J. Chem. Phys.,* **13**, 3 (1945).

125. R. D. Andrews, N. Hofman-Bang, and A. V. Tobolsky, *J. Polym. Sci.,* **3**, 669 (1948).

126. R. F. Landel and P. J. Stedry, *J. Appl. Phys.,* **17**, 1885 (1960).

127. J. T. Bergen, D. C. Messersmith, and R. S. Rivlin, *J. Appl. Polym. Sci.,* **3**, 153 (1960).

128. T. L. Smith, *Trans. Soc. Rheol.,* **6**, 61 (1962).

129. T. L. Smith and R. A. Dickie, *J. Polym. Sci.: Part A-2,* **7**, 635 (1969).

130. E. Catsiff and A. V. Tobolsky, *J. Colloid Sci.,* **10**, 375 (1955).

131. E. Catsiff and A. V. Tobolsky, *J. Polym. Sci.,* **19**, 111 (1956).

132. A. V. Tobolsky and J. R. McLoughlin, *J. Polym. Sci.,* **8**, 543 (1952).

133. M. L. Williams, R. F. Landel, and J. D. Ferry, *J. Am. Chem. Soc.,* **77**, 3701 (1955).

134. M. L. Williams, *J. Phys. Chem.,* **59**, 95 (1955).

135. A. K. Doolittle and D. B. Doolittle, *J. Appl. Phys.,* **31**, 1164 (1959).

136. K. C. Rusch and R. H. Beck, *J. Macromol. Sci.,* **B3**, 365 (1969).

137. P. Macedo and T. A. Litovitz, *J. Chem. Phys.,* **42**, 245 (1965).

138. H. S. Chung, *J. Chem. Phys.,* **44**, 1362 (1966).

139. G. C. Berry and T. G. Fox, *Adv. Polym. Sci.,* **5**, 261 (1967).

140. C. C. Winding and G. D. Hiatt, *Polymeric Materials,* McGraw-Hill, New York, 1961.

141. W. L. Holt, *Rubber Chem. Tech.,* **5**, 79 (1932)

142. L. Mullins, *J. Rubber Res.,* **16**, 275 (1947).

143. L. Mullins, *J. Phys. Colloid Chem.,* **54**, 239 (1950).

144. L. Mullins and N. R. Tobin, *Trans. Inst. Rubber Ind.,* **33**, 2 (1956).

145. A. F. Blanchard, *J. Polym. Sci.,* **14**, 355 (1954).

146. A. F. Blanchard and D. Parkinson, *Ind. Eng. Chem.,* **44**, 799 (1952).

147. J. F. Beecher, L. Marker, R. D. Bradford, and S. L. Aggarwal, *Polym. Prepr.*, **8**, 1532 (1967).

148. F. M. Merrett, *J. Polym. Sci.*, **24**, 467 (1957).

149. C. W. Childers and G. Kraus, *Rubber Chem. Tech.*, **40**, 1183 (1967).

150. J. F. Henderson, K. H. Grundy, and E. Fischer, *J. Polym. Sci.: Part C*, **16**, 3121 (1968).

151. G. S. Trick, *J. Appl. Polym. Sci.*, **3**, 252 (1962).

152. E. M. Hicks, Jr., A. J. Ultee, and J. Drougas, *Science*, **147**, 373 (1965).

153. R. Bonart, *J. Macromol. Sci.*, **B2**, 115 (1968).

154. R. Bonart, L. Morbitzer, and G. Hentz, *J. Macromol. Sci.*, **B3**, 337 (1969).

155. D. Puett, *J. Polym. Sci.: Part A-2*, **5**, 839 (1967).

156. S. L. Cooper, D. S. Huh, W. J. Morris, *Ind. Eng. Chem., Prod. Res. Dev.*, **7**, 248 (1968).

157. G. M. Estes, S. L. Cooper, and A. V. Tobolsky, *J. Macromol. Sci.*, **C4**, 313 (1970).

158. F. Bueche, *J. Appl. Polym. Sci.*, **4**, 107 (1960).

159. F. Bueche, *J. Appl. Polym. Sci.*, **5**, 271 (1961).

160. G. Kraus, C. W. Childers, and K. W. Rollman, *J. Appl. Polym. Sci.*, **10**, 229 (1966).

161. B. B. Boonstra, in G. Kraus, Ed., *Reinforcement of Elastomers,* Chapter 16, Wiley-Interscience, New York, 1965.

162. E. M. Dannenberg and J. J. Brennan, *Rubber Chem. Tech.*, **39**, 597 (1966).

163. D. Puett, K. J. Smith, and A. Ciferri, *J. Phys. Chem.*, **69**, 141 (1965).

164. D. Puett, K. J. Smith, A. Ciferri, and E. G. Kontos, *J. Chem. Phys.*, **40**, 253 (1964).

165. C. W. Bunn, in R. Hill, Ed., *Fibres from Synthetic Fibers,* Elsevier, New York, 1953, Chapter 10.

166. A. Nadai, *Theory of Flow and Fracture of Solids*, McGraw-Hill, New York, 1950.

167. E. Orowan, *Rep. Prog. Phys.*, **12**, 185 (1949).

168. I. M. Ward, *Mechanical Properties of Solid Polymers,* Wiley-Interscience, New York 1971.

169. R. Hill, *Mathematical Theory of Plasticity,* Clarendon Press, Oxford, 1950.

170. W. Johnson and P. B. Mellar, *Plasticity for Mechanical Engineers,* D. Van Nostrand, London, 1962.

171. H. Tresca, *C. R. Acad. Sci. (Paris)*, **59**, 754 (1864).

172. H. Tresca, *C. R. Acad. Sci. (Paris)*, **64**, 809 (1867).

173. R. von Mises, *Göttinger Nachr., Math. -Phys. Klasse*, 582 (1913).

174. C. A. Coulomb, *Mem. Math. Phys.*, **7**, 343 (1973).

175. W. Whitney and R. D. Andrews, *J. Polym. Sci., Part C.* **16**, 2981 (1967).

176. P. B. Bowden and J. A. Jukes, *J. Mater. Sci.*, **3**, 183 (1968).

177. N. Brown and I. M. Ward, *J. Polym. Sci.: Part A-2*, **6**, 607 (1968).

178. M. Kurakawa and T. Ban, *J. Appl. Polym. Sci.*, **8**, 971 (1964).

179. A. Keller and J. G. Rider, *J. Mater. Sci.*, **1**, 389 (1966).

180. V. B. Gupta and I. M. Ward, *J. Macromol. Sci.-Phys.*, **B2**, 89 (1968).

181. Z. H. Stachurski and I. M. Ward, *J. Polym. Sci.: Part A-2*, **6**, 1083 (1968).

182. G. R. Davies, A. J. Owen, V. B. Gupta, and I. M. Ward, *J. Macromol. Sci.-Phys.*, **B6**, 215 (1972).

183. C. P. Buckley and N. G. McCrum, *J. Mater. Sci.*, **8**, 928 (1973).

184. M. Kapuscinski, I. M. Ward, and J. Scanlan, *J. Macromol. Sci.-Phys.*, **B11**, 475 (1975).

185. D. A. Zaukelies, *J. Appl. Phys.*, **33**, 2797 (1961).

186. N. Brown and I. M. Ward, *Phil. Mag.*, **17**, 961 (1968).

187. N. Brown, R. A. Duckett, and I. M. Ward, *Phil. Mag.*, **18**, 483 (1968).

188. C. Bridle, A. Buckley, and J. Scanlan, *J. Mater. Sci.*, **3**, 622 (1968).

189. S. W. Allison and I. M. Ward, *Br. J. Appl. Phys.*, **18**, 1151 (1967).

190. Y. S. Lazurkin, *J. Polym. Sci.*, **30**, 595 (1958).

191. R. E. Robertson, *J. Appl. Polym. Sci.*, **7**, 443 (1963).

192. C. Corwet and G. A. Homes, *Appl. Mat. Res.*, **3**, 1 (1964).

193. C. Bauwens-Crowet, J. A. Bauwens, and G. Homes, *J. Polym. Sci., Part A-2*, **7**, 735 (1969).

194. J. C. Bauwens, C. Bauwens-Crowet, and G. Homes, *J. Polym. Sci., Part A-2*, **7**, 1745 (1969).

195. J. A. Roetling, *Polymer*, **6**, 311 (1965).

196. R. N. Haward and G. Thackray, *Proc. R. Soc. London,* **A302**, 453 (1968).

197. D. L. Holt, *J. Appl. Polym. Sci.*, **12**, 1653 (1968).

198. R. E. Robertson, *J. Chem. Phys.*, **44**, 3950 (1966).

199. R. A. Duckett, S. Rabinowitz, and I. M. Ward, *J. Mater. Sci.*, **5**, 909 (1970).

200. G. M. Bryant, *Text. Res. J.*, **31**, 399 (1961).

201. M. H. Litt and P. Koch, *J. Polym. Sci.*, **B5**, 251 (1967).

202. I. L. Hay and A. Keller, *Kolloid-Z.*, **204**, 43 (1965).

203. A Peterlin, *J. Polym. Sci.*, **69**, 61 (1965).

204. P. H. Geil, *J. Polym. Sci.*, **A2**, 3835 (1964).

205. I. K. Richardson, R. A. Duckett, and I. M. Ward, *J. Phys. Div.: Appl. Phys.*, **3**, 649 (1970).

206. F. C. Frank, A. Keller, and A. O'Connor, *Phil. Mag.*, **3**, 64 (1958).

207. E. Orowan, *Rep. Prog. Phys.*, **12**, 185 (1949).

208. P. I. Vincent, *Polymer*, **1**, 425 (1960).

209. S. B. Ainbinder, M. G. Laka, and I. Y. Maiors, *Mekh. Polim.*, **1**, 65 (1965).

210. K. D. Pae, D. R. Mears, and J. A. Sauer, *J. Polym. Sci.,* **B6**, 773 (1968).

211. S. Rabinowitz, I. M. Ward, and J. S. C. Parry, *J. Mater. Sci.,* **5**, 29 (1970).

212. P. J. Flory, *J. Am. Chem. Soc.,* **67**, 2048 (1945).

213. J. P. Berry, *J. Polym. Sci.: Part A,* **2**, 4069 (1964).

214. R. P. Kusy and D. T. Turner, *Polymer,* **15**, 394 (1974).

215. R. P. Kusy and D. T. Turner, *Polymer,* **17**, 161 (1976).

216. A. A. Griffith, *Phil Trans. R. Soc., London* **221**, 163 (1921).

217. J. J. Benbow and F. C. Roesler, *Proc. Phys. Soc.,* **B70**, 201 (1957).

218. J. P. Berry, *J. Appl. Phys.,* **34**, 62 (1963).

219. D. K. Felbeck and E. Orowan, *Welding J.,* **34**, 570 (1955).

220. E. Orowan, *Welding J.,* **34**, 1575 (1955).

221. G. R. Irwin and J. A. Kies, *Welding J.,* **31**, 995 (1952).

222. G. R. Irwin and J. A. Kies, *Welding J.,* **33**, 1935 (1954).

223. J. P. Berry, *J. Polym. Sci.,* **A2**, 4069 (1964).

224. R. P. Kusy and M. J. Katz, *J. Mater. Sci.,* **11**, 1475 (1976).

225. R. P. Kusy and D. T. Turner, *Polymer,* **17**, 161 (1976).

226. W. D. Niegisch, *J. Appl. Polym. Sci.,* **5**, 59 (1961).

227. O. K. Spurr and W. D. Niegisch, *J. Appl. Polym. Sci.,* **6**, 585 (1962).

228. R. P. Kambour, *Nature,* **195**, 1299 (1962).

229. R. P. Kambour, *Polymer,* **4**, 143 (1963).

230. R. P. Kambour, *J. Polym. Sci.,* **A2**, 4159 (1964).

231. R. P. Kambour, *J. Polym. Sci.,* **A2**, 4165 (1964).

232. R. P. Kambour, *Polymer,* **5**, 143 (1964).

233. R. P. Kambour and A. S. Holik, *J. Polym. Sci., Part A-2,* **7**, 1393 (1969).

234. R. P. Kambour and R. R. Russell, *Polymer,* **12**, 237 (1971).

235. J. P. Berry, *J. Polym. Sci.,* **A3**, 2027 (1965).

236. R. P. Kambour, *J. Polym. Sci., Appl. Polym. Symp.,* **7**, 215 (1968).

237. J. Murray and D. Hull, *Polymer,* **10**, 451 (1969).

238. M. Braden and A. N. Gent, *J. Appl. Polym. Sci.,* **3**, 90 (1960).

239. M. Braden and A. N. Gent, *J. Appl. Polym. Sci.,* **3**, 100 (1960).

240. M. Braden and A. N. Gent, *J. Appl. Polym. Sci.,* **6**, 449 (1962).

241. A. N. Gent and H. Hirakawa, *J. Polym. Sci., Part A-2,* **6**, 1481 (1968).

242. G. P. Marshall, L. E. Culver, and J. G. Williams, *Proc. R. Soc. London (A),* **319**, 165 (1970).

243. E. H. Andrews and L. Bevan, *Polymer,* **13**, 337 (1972).

244. R. S. Rivlin and A. G. Thomas, *J. Polym. Sci.,* **10**, 291 (1953).

245. B. Maxwell and L. F. Rahn, *SPE J.,* **6**, 9 (1950).

246. R. L. Bergen, *SPE J.,* **18**, 667 (1962).

247. H. A. Stuart, G. Markowski, and D. Jeschke, *Kunstoffe,* **54**, 618 (1964).

248. R. L. Bergen, *SPE J.,* **24**, 77 (1968).

249. E. J. Andrews and L. Bevan, *Physical Basis of Yield and Fracture Conference,* Oxford University Press, London 1966.

250. G. A. Bernier and R. P. Kambour, *Macromolecules,* **1**, 393 (1968).

251. B. L. Earl, R. J. Loneragan, J. H. T. Jones, and M. Crook, *Polym. Eng. Sci.,* **13**, 390 (1973).

252. R. P. Kambour, C. L. Gruner, and E. E. Romagosa, *J. Polym. Sci.: Polym. Phys. Ed.,* **11**, 1879 (1973).

253. G. B. Bucknell and R. R. Smith, *Polymer,* **6**, 437 (1965).

254. P. E. Rouse, Jr., *J. Chem. Phys.,* **21**, 1272 (1953).

255. F. Bueche, *J. Chem. Phys.,* **22**, 603 (1954).

256. B. H. Zimm, *J. Chem. Phys.,* **24**, 269 (1956).

257. F. Bueche, *J. Chem. Phys.,* **26**, 1133 (1955).

258. F. Bueche and J. C. Halpin, *J. Appl. Phys.,* **35**, 36 (1964).

259. J. C. Halpin, *J. Appl. Phys.,* **35**, 3133 (1964).

260. M. L. Williams, R. F. Landel, and J. D. Ferry, *J. Am. Chem. Soc.,* **77**, 3701 (1955).

261. T. L. Smith, *J. Polym. Sci.,* **30**, 89 (1956).

262. T. L. Smith, *J. Polym. Sci.,* **32**, 99 (1958).

263. T. L. Smith, *J. Polym. Sci., Part A,* **1**, 3597 (1963).

264. T. L. Smith, *J. Appl. Phys.,* **35**, 27 (1964).

265. T. L. Smith and J. E. Frederick, *J. Appl. Phys.,* **36**, 2996 (1965).

266. T. L. Smith and R. A. Dickie, *J. Polym. Sci.: Part A-2,* **7**, 635 (1969).

267. T. L. Smith and W. H. Chu, *J. Polym. Sci.: Part A-2,* **10**, 133 (1972).

268. T. L. Smith, *Polym. Eng. Sci.,* **17**, 129 (1977).

269. J. A. C. Harwood and A. R. Payne, *J. Appl. Polym. Sci.,* **12**, 889 (1968).

270. A. R. Payne and R. E. Whittaker, *Composites,* **1**, 203 (1970).

271. A. R. Payne and R. E. Whittaker, *J. Appl. Polym. Sci.,* **15**, 1941 (1971).

272. J. A. C. Harwood, A. R. Payne, and R. E. Whittaker, *J. Appl. Polym. Sci.,* **14**, 2183 (1970).

273. R. E. Whittaker, *Polymer,* **13**, 168 (1972).

274. K. A. Grosch, J. A. C. Harwood, and A. R. Payne, *Nature,* **212**, 497 (1966).

275. R. S. Rivlin and A. G. Thomas, *J. Polym. Sci.,* **10**, 291 (1953).

276. H. W. Greensmith and A. G. Thomas, *J. Polym. Sci.*, **18**, 189 (1955).

277. H. W. Greensmith, *J. Polym. Sci.*, **21**, 175 (1956).

278. A. G. Thomas, *J. Appl. Polym. Sci.*, **3**, 168 (1960).

279. A. G. Veith, *Rubber Chem. Tech.*, **38**, 700 (1965).

280. L. Mullins, *Trans. Inst. Rubber Ind.*, **35**, 213 (1959).

281. H. K. Mueller and W. G. Knauss, *Trans. Soc. Rheol.*, **15**, 217 (1971).

282. A Ahagon and A. N. Gent, *J. Polym. Sci.: Polym. Phys. Ed.*, **13**, 1903 (1975).

283. E. Orowan, *Rep. Prog. Phys.*, **12**, 185 (1949).

284. R. M. Evans, H. R. Nara, and R. G. Bobalek, *Soc. Plast. Eng. J.*, **16**, 76 (1960).

285. J. Heyboer, *J. Polym. Sci.*, **C16**, 3755 (1968).

286. R. F. Boyer, *Polym. Eng. Sci.*, **8**, 161 (1964).

287. W. L. Jackson, Jr., and J. R. Caldwell, *J. Appl. Polym. Sci.*, **11**, 211, 227 (1967).

288. L. M. Robeson and J. A. Faucher, *J. Polym. Sci.*, **B7**, 35 (1969).

289. E. Sacher, *J. Macromol. Sci.-Phys.*, **B9**, 163 (1974).

290. K. C. Rusch, *J. Appl. Polym. Sci.*, **13**, 2297 (1969).

291. K. C. Rusch, *J. Appl. Polym. Sci.*, **14**, 1263 (1970).

292. K. C. Rusch, *J. Appl. Polym. Sci.*, **13**, 1433 (1970).

293. E. A. Meinecke and D. M. Schwaber, *J. Appl. Polym. Sci.*, **14**, 2239 (1970).

294. E. A. Meinecke, D. M. Schwaber, and R. R. Chiang, *J. Elastoplast.*, **3**, 19 (1971).

295. N. C. Hilyard and L. K. Djiauw, *J. Cell. Plast.*, **7**, 33 (1971).

296. M. Horio, *Proceedings, Fourth International Congress of Rheology*, Part 1, Interscience, New York, 1965.

297. J. F. Nye, *Physical Properties of Crystals*, Clarendon Press, Oxford, 1957.

298. D. W. Hadley, I. M. Ward, and J. Ward, *Proc. R. Soc.*, **A285**, 275 (1965).

299. P. R. Pinnock, I. M. Ward, and J. M. Wolfe, *Proc. R. Soc.*, **A291**, 267 (1966).

300. D. W. Hadley, P. R. Pinnock, and I. M. Ward, *J. Mater. Sci.*, **4**, 152 (1969).

301. G. Raumann and D. W. Saunders, *Proc. Phys. Soc.*, **77**, 1028 (1961).

302. I. L. Hay and A. Keller, *J. Mater. Sci.*, **1**, 41 (1966).

303. V. B. Gupta and I. M. Ward, *J. Macromol. Sci.-Phys.*, **B2**, 89 (1968).

304. I. M. Ward, *Proc. Phys. Soc.*, **80**, 1176 (1962).

305. H. H. Kausch-Blecken Von Schmeling, *Kolloid-Z.*, **237**, 251 (1970).

306. J. Bishop and R. Hill, *Phil. Mag.*, **42**, 414, 1298 (1951).

307. J. H. Wakelin, E. T. L. Voong, D. J. Montgomery, and J. H. Dusenbury, *J. Appl. Phys.*, **26**, 786 (1955).

308. M. Takayanagi, K. Imada, and T. Kajiyama, *J. Polym. Sci.*, **C15**, 263 (1966).

309. M. Takayanagi, *Pure Appl. Chem.*, **15**, 555 (1967).

310. Z. H. Stachurski and I. M. Ward, *J. Polym. Sci., Part A-2*, **6**, 1083, 1817 (1968).

311. Z. H. Stachurski and I. M. Ward, *J. Macromol. Sci.*, **B3**, 427, 445 (1969).

312. D. W. van Krevelen, *Properties of Polymers*, Elsevier, Amsterdam, 1972.

313. T. Alfrey, Jr., *Appl. Polym. Symp.*, No. 24, 3 (1974).

314. R. S. Stein, *Rubber Chem. Technol.*, **49**, 458 (1976).

315. S. Clough, M. B. Rhodes, and R. S. Stein, *J. Polym. Sci., Part C.* **18**, 1 (1967).

316. D. A. Keedy, R. J. Volungis, and H. Kawai, *Rev. Sci. Instr.*, **32**, 415 (1961).

317. R. S. Stein, R. S. Finkelstein, D. Y. Yoon, and C. Chang, *J. Polym. Sci., Polym. Symp.*, **46**, 15 (1974).

318. S. Onogi, K. Sasaguri, D. A. Keedy, and R. S. Stein, *J. Appl. Phys.*, **34**, 80 (1963).

319. R. Yamada and R. S. Stein, *J. Appl. Phys.*, **36**, 3005 (1965).

320. R. S. Stein, *J. Polym. Sci.*, **24**, 383 (1957).

321. H. De Vries, *Angew. Chem.*, **74**, 574 (1962).

322. H. De Vries, *Ned. Tijdschr. Natuurkd.*, **31**, 68 (1965).

323. W. Kuhn and H. Grun, *Kolloid Z.*, **101**, 248 (1942).

324. W. Kuhn, *J. Polym. Sci.*, **1**, 380 (1946).

325. L. R. G. Treloar, *Trans. Faraday Soc.*, **43**, 277, 289 (1947).

326. A. V. Tobolsky, *Properties and Structure of Polymers*, John Wiley, New York, 1960.

327. D. W. Saunders, *Trans. Faraday Soc.*, **52**, 1414 (1956).

328. A. N. Gent and V. V. Vickroy, *J. Polym. Sci., Part A-2*, **5**, 47 (1967).

329. T. Ishikawa and K. Nagai, *J. Polym. Sci., Part A-2*, **7**, 1123 (1969).

330. A. N. Gent, *Macromolecules*, **2**, 262 (1969).

331. M. Fukuda, G. L. Wilkes, and R. S. Stein, *J. Polym. Sci., Part A-2*, **9**, 1417 (1971).

332. R. S. Stein and A. V. Tobolsky, *Text. Res. J.*, **18**, 201, 302 (1948).

333. R. S. Stein, in H. A. Stuart, Ed., *Physik der Hochpolymeren*, Vol. 4, Springer, Berlin (1956).

334. B. E. Read, *Polymer*, **3**, 143 (1962).

335. T. K. Su and R. S. Stein, *Bull. Am. Phys. Soc., Ser. 2*, **20**, 340 (1975).

336. H. De Vries, *J. Polym. Sci.*, **34**, 761 (1959).

337. R. S. Stein, S. Krimm, and A. V. Tobolsky, *Text. Res. J.*, **19**, 8 (1949).

338. I. M. Daniel, *Trans. Soc. Rheol.*, **10**, 25 (1966).

339. A. Utsuo and R. S. Stein, *J. Polym. Sci., Part A-2,* **5**, 583 (1967).

340. R. D. Andrews and Y. Kazama, *J. Appl. Phys.*, **39**, 4891 (1968).

341. J. G. Rider and E. Hargreaves, *J. Phys. D., 3*, 993 (1970).

342. H. Kolsky, *Nature*, **166**, 235 (1950).

343. Y. Shindo and R. S. Stein, *J. Polym. Sci., Part A-2,* **7**, 2115 (1969).

344. R. J. Morgan and L. R. G. Treloar, *J. Polym. Sci.: Part A-2,* **10**, 51 (1972).

345. T. Kotani and S. S. Stearnstein, in S. Newman and F. Chompff, Eds., *Polymer Networks, Structure and Mechanical Properties,* Plenum Press, New York, 1971.

346. C. Ong and R. S. Stein, *J. Polym. Sci., Polym. Phys. Ed.,* **12**, 1899 (1974).

347. C. Ong, D. Y. Yoon, and R. S. Stein, *J. Polym. Sci., Polym. Phys. Ed.,* **12**, 1319 (1974).

348. J. F. Henderson, K. H. Grundy, and E. Fischer, *J. Polym. Sci.: Part C,* **16**, 3121 (1968).

349. G. L. Wilkes and R. S. Stein, *J. Polym. Sci.: Part A-2,* **7**, 1525 (1969).

350. M. J. Folkes and A. Keller, *Polymer*, **12**, 222 (1971).

351. R. S. Stein and F. H. Norris, *J. Polym. Sci.,* **21**, 381 (1956).

352. F. A. Bettelheim and R. S. Stein, *J. Polym. Sci.,* **27**, 567 (1958).

353. R. S. Stein, *J. Polym. Sci.,* **31**, 327, 335 (1958).

354. R. S. Stein, *J. Polym. Sci.: Part C,* **15**, 185 (1966).

355. Y. Fukui, T. Asada, and S. Onogi, *Polym. J.,* **3**, 100 (1972).

356. R. S. Stein, *Acc. Chem. Res.,* **5**, 121 (1972).

357. G. Kraus and J. T. Gruver, *J. Polym. Sci.: Part A-2,* **10**, 2009 (1972).

358. C. Chang, D. Peiffer, and R. S. Stein, *J. Polym. Sci., Polym. Phys. Ed.,* **12**, 1441 (1974).

359. R. J. Samuels, *Structured Polymer Properties,* Wiley-Interscience, New York, 1974.

360. M. Hashiyama, R. J. Gaylord, and R. S. Stein, *Makromol. Chem. Suppl.,* **1**, 579 (1975).

361. R. S. Stein, *Polym. Eng. Sci.,* **16**, 152 (1976).

362. W. J. Shrenk and T. Alfrey, Jr., *Polym. Eng. Sci.,* **9**, 393 (1969).

363. W. J. Shrenk, *Appl. Polym. Symp.,* **24**, 9 (1974).

364. W. R. Danien, *Plast. World,* **34**, 32 (1976).

365. R. M. Gerkin and F. E. Critchfield, Paper No. 741022, presented at the Automobile Engineering Meeting, Society of Automotive Engineers, Toronto, Canada, October 21-25, 1974.

366. J. H. Wilson, *Mod. Paint Coatings,* **67**, (3), 45 (1977).

367. J. M. Starita, *Plast. World,* **35**, 58 (1977).

CHAPTER **8**

Dielectric and Photoconductive Properties

J. M. POCHAN
D. M. PAI

Xerox Corporation
Webster, New York

8.1	Introduction	341
8.2	Dielectric Phenomena	342
	8.2.1 Polarization	342
	8.2.2 Dielectric Dispersions	343
	8.2.3 Linear Response Theory	346
	8.2.4 Distribution of Relaxation Times	348
	8.2.5 Temperature Dependence of Relaxation Spectra	350
	8.2.6 DC Conductivity	353
	8.2.7 Interfacial Polarization	354
	8.2.8 Instrumentation	356
	8.2.9 Applications	356
	8.2.9.1 Molecular Relaxations— Transition Maps	356
	8.2.9.2 Sorbed Water	358
	8.2.9.3 Sorbed Gases	358
	8.2.9.4 Plasticizers	359
	8.2.9.5 Mixing Phenomena—Block Copolymers, Polymer Blends, and Copolymers	360
	8.2.9.6 Process Control	361
	8.2.9.7 Aging	362
	8.2.9.8 Miscellaneous	364
8.3	Photoconductivity	364
	8.3.1 Background and Definitions	364
	8.3.2 Transient Photoconductivity	367
	8.3.3 Photogeneration	371
	8.3.3.1 Poly(*N*-vinylcarbazole) (PVK)	372
	8.3.3.2 Charge Transfer Complex of PVK with 2,4,7 Trinitro-9-Fluorenone (TNF)	376
	8.3.3.3 Solid Solutions of Triphenylamine in Bisphenol-*A*-Polycarbonate	378
	8.3.4 Charge Transport	379
	8.3.4.1 PVK and Charge Transfer Complexes of PVK and TNF	381
	8.3.4.2 TNF Dispersed in Polyester	383
	8.3.4.3 Triphenylamine (TPA) in Polycarbonate	384
	8.3.4.4 Mixed Systems of TPA and NICP in Polycarbonate	385
	8.3.4.5 Substituted Triarylmethanes in Polycarbonate	387
8.4	Summary	390
8.5	Glossary of Symbols	391
8.6	References	391

8.1 INTRODUCTION

The use of polymers for electrical applications in industry has increased dramatically in the last two decades. Prior to this time, their main function was that of an insulating material in capacitors and conductive cable. Now, polymers are being used in such diverse fields as conductors, photoconductors, electrical materials, and even as highly insulating, stable, cable coatings. Because of this increased use, topics such as AC and DC conduction, photogeneration and transport, dielectric polarization, electret formation, dielectric breakdown, weak and strong field effects, and tribo electric charging of polymers are being actively studied. It is beyond the scope of this chapter to provide information concerning all the aforementioned subjects, so this writing is limited to the polarization and photoconduction characteristics of polymers. The former topic encompasses dielectric polarization effects as studied by way of relaxation techniques. These approaches are presented in a simple, it is hoped understandable, form, which will enable the reader to interpret dielectric data as well as to design experiments. This is done by formulating the theory surrounding experimental techniques and by providing their applications. The latter topic covers recent developments of our understanding of photogeneration and transport in polymer systems. This chapter is not meant as an up-to-date review of the fields covered, but rather as an instructional guide to understanding certain aspects of the electrical properties of polymers.

8.2   DIELECTRIC PHENOMENA

The dielectric theory of polymer relaxations can be approached from a number of theoretical directions. The initial portion of this section defines the terms needed to describe dielectric materials and prepares the way for the application of linear response theory to the dielectric process. This is followed by a discussion of the deviation of polymeric systems from linear theory and the effect of highly conductive materials and dispersions on obtained results. Instrumentation and general applications then follow.

8.2.1   Polarization

Insulator or dielectric materials are usually classed as those exhibiting resistivities in the neighborhood of 10^8 ohm-cm or higher. Below this value, the materials are considered semiconducting. The 10^8 ohm-cm designation is somewhat arbitrary, and what is considered an insulator for some uses may be a conductor for others. The most useful property of di-

electric materials is their ability to store an electrical charge. This ability is directly related to their dielectric polarizability, which is a function of electronic, atomic, and orientational polarizabilities respectively, that is, $\alpha_T = \alpha_e + \alpha_a + \alpha_0$.

What is polarizability? It is the ability of a material to react to an applied electric field by either:

Deformation of its electron fields α_e.

Motion of its atoms (vibrational) α_a.

Gross motion of the dipolar segments α_0.

This reaction can be thought of in terms of applied AC voltages. When a DC field is applied to a material, all of the preceding components can react to the applied field. Alternatively, at an extremely high frequency, none can react. Gross overall dipolar motion or chain segment motion in polymers occurs at relatively low frequencies (10^{-4}–10^5 Hz). If the applied AC frequency is above these values, dipolar motion can no longer follow the field, and α_0 becomes very small. The same phenomenon applies for α_a and α_c as the measurement frequency is increased through the vibrational and x-ray ranges, respectively. Thus, as frequency is increased at a constant temperature, the ability of a dielectric to store charge will decrease in some stepwise fashion as each type of reaction to the applied field is eliminated.

For a simple capacitor (see Figure 8.1), the dielectric constant ϵ of the medium is the ratio of the charge stored in the capacitor containing the dielectric to that of the charge stored in vacuum with the identical geometric configuration. Thus the more polarizable the dielectric, the higher its dielectric constant will be. The polarization of polymeric materials can be affected by either gross changes in the internal (variations in semicrystalline phases) or external (particular fillers) morphology of the polymers. This subject is covered later.

In Figure 8.1, a term called the dielectric displacement D can also be defined. For a unit area of the capacitor, the difference in surface charge density with and without the dielectric

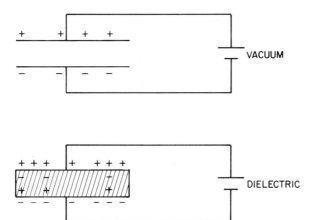

Figure 8.1 Schematic of a capacitor/voltage current setup.

material is the polarization P, as defined in Equation (8.1),

$$P = \sigma_D - \sigma_v \qquad (8.1)$$

where σ_D and σ_v are the charge density per unit area with and without the dielectric in the capacitor, that is $\sigma_v = q_v/A$ and $\sigma_D = q_0/A$ where $q_v = \epsilon' q_D$.

A is the capacitor area, q is the charge on the capacitor, and ϵ' is the dielectric constant of the material placed in the capacitor. By definition $\epsilon' = C_D/C_V$, where C is the capacitance of the capacitor. The electric field strength and dielectric displacement are then defined as Equation (8.2).

$$E = 4\pi\sigma_v$$
$$\qquad (8.2)$$
$$D = 4\pi\sigma_D$$

The value of 4π results from a units conversion and a spherical integration. In some cases (depending on the units used), it is ignored (1).

From Equation (8.1), one obtains the relationship expressed in Equation (8.3).

$$D = E + 4\pi P \qquad (8.3)$$

Equation (8.3) can then be manipulated to give $D = \epsilon'E$, which shows that the displacement is larger than E by the factor ϵ'.

The addition of the dielectric material to the capacitive system increases the displacement (total charge) of the capacitor. In other words, this displacement is nothing more than a true measure of the charge that a capacitor will accept.

Other quantities should be defined before proceeding with the discussion of dielectric dispersion. It can be readily shown (1) that the capacitance of a parallel plate capacitor can be written as Equation (8.4),

$$C = \frac{\epsilon_0\epsilon'A}{l} \qquad (8.4)$$

where A is the area of the capacitor, l is its thickness, ϵ' is its dielectric constant, and ϵ_0 is the permittivity constant of vacuum (8.85×10^{-12} F/m). Obviously, if A and l change, different capacitances can be obtained. The energy stored in the capacitor is $\frac{1}{2}CV^2 = \frac{1}{2}\epsilon'\epsilon_0E^2$ (1). If the dielectric material in a capacitive system has infinite resistance, all charge placed in the capacitor is stored. However, all dielectrics exhibit some propensity to conduct charge and, therefore, some of the electrical energy stored in the system is used to overcome the resistive nature of the material, and the material becomes "lossy." This effect is discussed later.

8.2.2 Dielectric Dispersions

Previous discussion has related to the various types of polarization that can occur in polymeric media and to the relationship between the natural frequencies (electronic, atomic, and polar) that occur in the molecules and the effective ability to store charge. This is illustrated in Figure 8.2. The characteristic frequencies for various motions result in decreases in ϵ' when the frequency is surpassed. Also notice that a dispersion (a dissipation of energy) occurs with each change in the dielectric constant. This dispersion is known as the dielectric loss ϵ'' and will be interpreted in terms of linear response theory. The electronic contribution to ϵ' and ϵ'' arises from an absorption dispersion (which could be considered a change in momentum of the electron). In the former case, $\epsilon' = \eta^2$ above the dispersion (η = the index of refraction of the material). In this region the dielectric properties are dominated by the electron contributions. The low-frequency relaxation is a

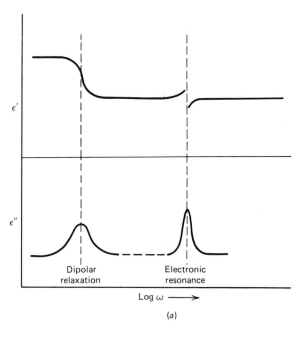

(a)

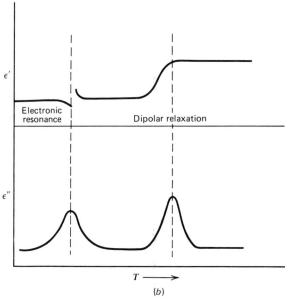

(b)

Figure 8.2 (a) ϵ' and ϵ'' versus log ω. (b) ϵ' and ϵ'' versus T.

result of dipolar motion within the polymer and a relaxation phenomena, that is, a phenomenon in which the change in orientation of the dipole is observed as a function of time after an applied field.

Clausius and Mosotti were able to derive a relationship between α_e, the density of the material ρ and ϵ', shown as follows in Equation (8.5):

$$\frac{\epsilon' - 1}{\epsilon' + 2} = \frac{4\pi}{3} \frac{\rho}{M} \alpha_e \qquad (8.5)$$

where M is the molecular weight. This equation is normally used at very high frequencies where dipolar reorientation is not a factor. For cases of dilute gases, Debye derived the relation given in Equation (8.6) as follows:

$$\frac{\epsilon' - 1}{\epsilon' + 2} = \frac{4}{3} \pi N \left(\alpha_e + \frac{\mu_0^2}{kT} \right) \qquad (8.6)$$

where k is Boltzmann's constant, μ_0 is the dipole moment of the molecule, and N is the Avogadro's number (2). This equation has been used to determine the dipole moment of polar materials in dilute solution (3, 4, 5), but it is not useful when specific interactions between dipolar molecules occur. Also, for Equation (8.6) to be applied to determine the dipole moment of a specific molecule, the dielectric frequency of measurement must be low enough to permit dipolar response. When condensed phases of dipolar matter are considered, Equation (8.6) cannot be used and, thus, the theoretical interpretation of the dielectric data becomes more complicated. This subject has been addressed by Onsager (6), and Kirkwood and Frohlich (7) and is described in detail elsewhere (8).

As mentioned before, motions considered in polymeric materials are usually low frequency in nature. In the case of polar polymers, the dipole contribution to the polarizability is important and is the major concern of this writing. In nonpolar polymers, the electronic contributions to the dielectric constant are important and can be estimated from atomic polarization and density data (9).

In considering Figure 8.2, the question must be asked: why does a dispersion take place in a particular frequency region for a given polymer? The answer lies in the dynamics of chain motion in the solid state. Consider a simple amorphous vinyl polymer with a dipole connected to the main chain. At any temperature, there is a frequency or characteristic band of frequencies associated with the motion of the dipole. If the polymer is in the glass transition T_g region, then the motions associated with this characteristic frequency are a micro-Brownian motion of the polymer main chain. Below T_g, localized motions of the main chain can be operative. If the sample is held at a constant temperature and the frequency is scanned, the dipoles of the polymer are able to maintain alignment with the applied electric field at the very low frequencies. This occurs because the relatively long time frame of the experiment allows the dipoles that are attempting to maintain their alignment with the electric field to find the path of least resistance to this motion. For these low frequencies, dipole polarization α_0 is in effect, the dielectric constant ϵ' of the media is high, and it is labeled the "relaxed" dielectric constant ϵ_r. As the frequency is increased, a point is reached where dipolar motion is impeded and electrical energy is needed to overcome the potential barriers that the dipole is experiencing. Part of the electric energy supplied by the electric field is therefore dissipated, and the ability to store charge is reduced somewhat. In this frequency regime, the amplitude of the dipolar motion is slightly reduced as compared to those at low frequency, and the dipole motion will lag the applied field by some phase angle δ. As the frequency of the measurement is increased, the dipole/matrix interaction is increased (this may be considered a viscous drag on the dipole motion) and the dielectric loss increases. When frequency is finally attained where dipole motion can no longer follow the applied field, dielectric loss, the dielectric constant and phase angle (the angle between the applied electric field and the molecular dipole moment) decrease. At extremely high frequencies, no dipolar motion exists, and the dielectric constant becomes the "unrelaxed" dielectric constant ϵ_μ, which is in reality only a function of atomic and electronic polarization.

A constant frequency experiment can also be done in which the temperature is varied. (See Figure 8.2.) When the temperatures are low, the dipoles are immobile owing to the low translational temperature of the system. The dielectric constant ϵ_μ, the dielectric loss ϵ'', and the phase angle δ do not change, and ϵ'' and δ are small. As the system is heated, molecular motion begins, but electrical energy must be dissipated to overcome the high internal viscosity of the system, and ϵ'' and δ increase. At higher temperatures, molecular mobility is facile, and the dipoles can respond to the applied field. At this point, ϵ' is maximized, and ϵ'' and δ diminish to near zero. Thus in either experiment (T constant/ν variable or ν constant/T variable) a dispersion as exemplified by a peak in ϵ'' or tan δ is observed, indicating that some form of dipolar molecular motion is taking place. As is shown later, these motions can be of a general variety (main chain) or of a localized motion along the chain.

8.2.3 Linear Response Theory

Linear response theory can be used to derive the variation of ϵ', ϵ'' and/or tan δ with frequency. As an illustration, the following is an application of the formalism described by McCrum et al. (8) to the response of a dielectric medium. Prior to any perturbation, $\epsilon(t) = \epsilon_r$ (the relaxed dielectric constant). If a perturbation is applied at $t = 0$, $\epsilon(t)$ becomes ϵ_μ. The equilibrium displacement D (the actual field created by the addition of the dielectric) is written as follows in Equation (8.7):

$$D(t = 0) = \epsilon(t = 0) E = \epsilon_r E \quad (8.7)$$

If a perturbation is applied (such as a change in field), D becomes time dependent, and any change in D can be equated to a displacement from equilibrium of the internal structure of the polymer. Thus

$$D(t) - D(t = 0) = \lambda P \quad (8.8)$$

where λ is a constant and P represents the degree to which the molecular rearrangement

is different from the equilibrium value. Therefore, Equation (8.8) may be written

$$\epsilon(t)E - \epsilon_\mu E = (\epsilon(t) - \epsilon_\mu) E = \lambda P \quad (8.9)$$

If it is assumed that any change in P, (dP/dt) is proportional to the difference between the instantaneous value of P and its average value $\bar{P}$, then for a function of time $[\psi(t)]$ that may describe P's return to equilibrium Equation (8.10) can be written

$$\frac{dP}{dt} = (P - \bar{P}) \psi(t) \quad (8.10)$$

which provides, upon integration, Equation (8.11),

$$P = P_0 [1 + \exp(\int \psi(t) \, dt)] \quad (8.11)$$

or Equation (8.12)

$$\lambda P = \lambda P_0 \phi(t) \quad (8.12)$$

Using Equation (8.9) and letting $\lambda \bar{P} = (\epsilon_R - \epsilon_\mu) E$ yields Equation (8.13),

$$[\epsilon(t) - \epsilon_\mu] E = (\epsilon_r - \epsilon_\mu) E \phi(t) \quad (8.13)$$

or Equation (8.14),

$$\epsilon(t) = \epsilon_\mu + (\epsilon_r - \epsilon_\mu) \phi(t) \quad (8.14)$$

In Equation (8.14), $\epsilon_r - \epsilon_\mu$ is usually termed the dielectric increment ($\Delta\epsilon$), and $\phi(t)$ may be considered a dielectric decay function.

Consider now this formalism when the electric field is varied. Equation 8.7 can now be written as Equation 8.15.

$$D(t) = \epsilon(t) E = [\epsilon_\mu + (\Delta\epsilon)\phi(t)]E \quad (8.15)$$

If E is changed by a value dE at time μ, then the response equation for Equation (8.15) becomes Equation (8.16).

$$dD(t) = [\epsilon_\mu + (\Delta\epsilon)\phi(t - \mu)]dE \quad (8.16)$$

Equation (8.16) can be integrated by parts employing the superposition principle. This principle, discovered by Hopkinson (11), states that for a system responding linearly, a small change in perturbation produces a small incremental change in the measured quantity. The total change of the system is the sum of the incremental changes. The integration results in Equation (8.17).

$$D(t) = \epsilon_\mu E(t) + \Delta\epsilon \int E(\mu)\phi(t - \mu) \, d\mu \tag{8.17}$$

Equation (8.17) is a general equation relating the dielectric displacement to time-dependent changes in the applied electric field. The equation contains two terms: an instantaneous term $[\epsilon_\mu E(t)]$ and a term that describes the time-dependent response of $D(t)$. In order to describe a system's response, only $\phi(t)$ is needed. If $\phi(t)$ is chosen as a single relaxation time exponential, $\phi(t) = e^{-t/\tau}$. Thus differentiating Equation (8.17) with respect to (t) gives Equation (8.18)

$$\tau \frac{d[D(t)]}{dt} + D(t) = \tau\epsilon_\mu \frac{dE(t)}{dt} + \epsilon_r E(t) \tag{8.18}$$

This is the general differential equation that must be solved for the dielectric response of the system. It may be solved for time-independent electric fields and displacement currents that yield characteristic capacitance charging equations. For instance, for a constant electric field $E = E_0$, $dE/dt = 0$ yields Equation (8.19).

$$D(t) = E_0[\epsilon_\mu + \Delta\epsilon (1 - e^{-t/\tau})] \tag{8.19}$$

or

$$\frac{D(t)}{R} = \frac{E_0}{R}[\epsilon_\mu + \Delta\epsilon (1 - e^{-t/\tau})] = i(t)$$

the displacement current for capacitive charging, R is the resistance of the capacitor. More important for this consideration is the application of a sinusoidal electric field ($E = E_0 e^{i\omega t}$) and an assumed response of the displacement $D = D_0 e^{[i(\omega t - \delta)]}$, which from Equation (8.18),

leads to a complex dielectric constant as shown in Equation (8.20)

$$\epsilon^* = \epsilon_\mu + \frac{\Delta\epsilon}{1 + i\omega\tau} \tag{8.20}$$

Equation (8.20) can be separated into real and imaginary parts, Equation (8.21),

$$\epsilon' = \epsilon_\mu + \frac{\Delta\epsilon}{1 + \omega^2\tau^2} \qquad \epsilon'' = \frac{\Delta\epsilon\,\omega\tau}{1 + \omega^2\tau^2} \tag{8.21}$$

with the tangent of the phase angle δ as given in Equation (8.22)

$$\tan\delta = \frac{\epsilon''}{\epsilon'} = \frac{\Delta\epsilon\,\omega\tau}{\epsilon_r + \epsilon_\mu\,\omega^2\tau^2} \tag{8.22}$$

These functions are plotted in Figure 8.3. This relaxation behavior leads to a peak in ϵ'' at $\omega\tau = 1(d\epsilon/d\omega = 0)$. It is also seen that $\tan\delta$ has a maximum value that is displaced to the high frequency side of ϵ''. Thus, as was decribed previously, the ability of a dielectric to store and dissipate charge can be phenomenologically related to relaxation phenomena. It is also obvious from Equations (8.21) and (8.22) that the response to the applied electric field vectorially lags the applied field, with the ability to store charge in-phase and the ability to dissipate charge lagging the applied field by 90°. Other terms associated with the phase angle delta are used in scientific literature. These are the dissipation factor, $1/Q$, and the power factor $\sin\delta$. $\omega\epsilon''E^2$ is the power dissipation per unit volume where $\omega\epsilon''$ is known as the AC conductivity. More important from the experimentor's point of view, the ratio of ϵ' to ϵ' is independent of sample dimensions [see Equation (8.4)], and relaxation mechanisms in polymers can be defined independently of geometry.

The complex dielectric constant can also be written in terms of the dielectric decay function as Equation (8.23),

$$\frac{\epsilon^* - \epsilon_\mu}{\epsilon_r - \epsilon_\mu} = \int_0^\infty -e^{-i\omega t}\,\dot{\phi}(t)\,dt \tag{8.23}$$

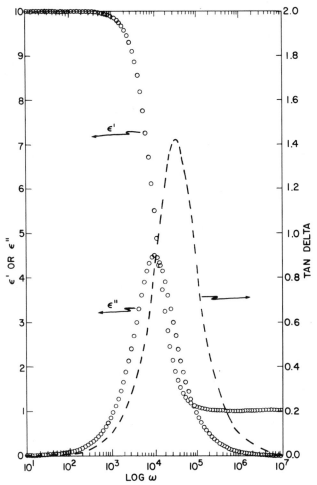

Figure 8.3 ϵ', ϵ'' and tan δ versus log ω for a single relaxation time process with $\epsilon_r = 10$, $\epsilon_\mu = 1$, and $\tau = 1.E\text{-}0.4$.

where ϕ is written in terms of time-averaged dipole interactions, Equation (8.24), (13).

$$\phi(t) = \frac{\langle\, \mu(0)\mu(t)\,\rangle}{\langle\, \mu(0)\mu(0)\,\rangle} \qquad (8.24)$$

In this way $\phi(t)$ is referred to as the dipole moment autocorrelation function and describes the projection of an average dipole moment upon a reference axis as a function of time after a perturbation is applied.

In the simplest case of a single relaxation time, $\phi(t) = e^{-t/\tau}$, and Equations (8.21) and (8.22) are derived. Within the formalism of Equation (8.23) one need only describe the system by its autocorrelation function and the dielectric relaxation spectra will be derived.

This is an extensive subject in itself, and the reader is referred to the writings of Williams (12), Kubo (17, 18), Zwanzig (18), Berne et al. (19), Glarum (20), and Boon and Rich (21) for a detailed analysis of the subject.

8.2.4 Distribution of Relaxation Times

If Equation (8.22) is considered, the half width at half height of the dielectric loss dispersion is a little over 1 decade on the frequency scale. Most polymer and monomer relaxations cannot be described by a single relaxation time curve (see Figure 8.4, for example). In Figure 8.4 are plotted reduced curves $\epsilon''/\epsilon''_{max}$ so that a comparison of line shapes can be made, regardless of the ϵ'' intensity. It is seen in the

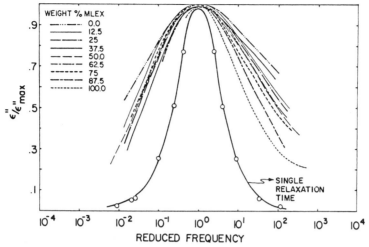

Figure 8.4 $\epsilon''/\epsilon''_{max}$ versus $\log \nu/\nu_{max}$ for the glass transition of bisphenol-A polycarbonate mixtures. A single relaxation time curve is included for reference. Reprinted with permission for *Macromolecues*, **11**, 165 (1978). Copyright by the American Chemical Society.

figure that monomeric and polymeric materials exhibit experimental line shapes much broader than the single relaxation time. The polymeric curve also demonstrates a high-frequency broadening rather than the smooth, sharp curve of a single relaxation peak. This type of broadening is characteristic of most polymeric systems and indicates that there are many high-frequency components to the motion in the polymer that are more facile than predicted by a single relaxation time theory.

An easy method of checking a system for a single relaxation time distribution is the Cole-Cole plot (22). Cole and Cole have shown that Equation (8.21) can be arranged to give Equation (8.25).

$$\left(\epsilon' - \frac{\Delta\epsilon}{2}\right)^2 + (\epsilon'')^2 = \left(\frac{\Delta\epsilon}{2}\right)^2 \quad (8.25)$$

This is the equation of a semicircle with a radius $\Delta\epsilon/2$ and a center at $(\epsilon_r + \epsilon_\mu)/2$. An

example is shown in Figure 8.5 along with experimental data for poly(vinyl acetate) (23). If a Cole-Cole plot is not a semicircle, a single relaxation time does not describe the system being studied.

Solutions to the linewidth and " skewedness" problem in explaining dielectric relaxations have been approached empirically by a number of researchers who modified Equation (8.20). Cole and Cole (22), Davidson and Cole (24), Fuoss and Kirkwood (25), Scaife (26) and Havriliak and Negami (27, 28) have proposed modifications of Equation (8.20) to curve-fit experimental data. The most generalized form is that of Scaife shown as Equation (8.26):

$$\frac{\epsilon^* - \epsilon_\mu}{\epsilon_r - \epsilon_\mu} = [1 + (\omega\tau)^{1-a}]^b \quad (8.26)$$

where a and b are constants. The Cole-Cole ($b = 0$) and Davidson-Cole ($a = 0$) are single parameter equations. Williams and Watts (29) proposed an empirical relaxation function

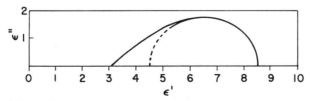

Figure 8.5 Cole-Cole plot for a single relaxation time and experimental data for polyvinylacetate (23).

$\phi(t)$ as expressed in Equation (8.27):

$$\phi(t) = e^{-(t/\tau)^{B}} \qquad (8.27)$$

to be used in Equation (8.23). All of these empirical approaches provide no information on the real molecular behavior of the system. This can be obtained from the time-dependent dipole moment correlation function for the polymer and is not discussed here. The reader is referred to References 12, 30, and 31 and to references therein for a discussion of this topic.

Although many empirical forms of Equation (8.20) can be used to fit experimental data, the experimental facts indicate that the data must be explained by a distribution of relaxation times rather than by a single relaxation. Polymer structure is sufficiently complicated so that one can assume that potential functions caused by the various orientations of the polymer chain can be slightly different. Thus an envelope of relaxation times is needed to describe observed data. In this case, a relaxation time distribution function is used instead of $\phi(t)$ to formulate the system's dielectric response (8). The relaxation time is used as a variable, that is, certain relaxation times would be weighted differently than others to correspond to populations of polymer chains in different conformations (32). In the case of a relaxing polymer, the maximum

in the loss measurement could indicate the most probable relaxation time.

8.2.5 Temperature Dependence of Relaxation Spectra

Thus far in this discussion it has been shown that linear response theory leads to dispersion equations that describe relaxation spectra. No mention has been made of the temperature dependence of individual polymer relaxations. In most polymers the thermal dependence of the relaxations is governed by two types of response. These are shown in Figure 8.6 in the form of a transition map, which is a plot of the frequency of the loss maximum observed in dielectric frequency scans as a function of inverse temperature.

The lables α, β, and so forth on the map refer to the various relaxations as they are observed as a function of increasing frequency at a constant temperature. In the case of a totally amorphous sytem, the map might appear as in Figure 8.6. In a semicrystalline polymer, a higher temperature relaxation due to the crystalline component of the polymer might also be observed. This is usually designated α_c.

The causes for the relaxation in Figure 8.6 are many. But the α with its curvature is indicative of the glass transition of the polymer and can be used in identifying molecular

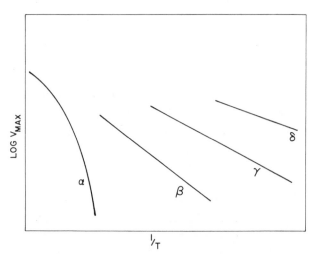

Figure 8.6 Schematic representation of a transition map.

motion due to T_g. The other relaxations that produce Arrhenius-type behavior cannot be identified with a specific polymer motion by inspection of the transition map but must be verified in other ways. These lower temperature relaxations (which are usually much broader than the T_g relaxation on a frequency scale) are usually associated with localized motion within the polymer, that is, side-group rotation, short-order motion in the main chain, and so on.

The Arrhenius-type behavior of Figure 8.6 is easily understood and implies the relationship shown in Equation (8.28)

$$\tau(T) = \tau_0 e^{E_a/kT} \qquad (8.28)$$

which indicates that the reorientation process is usually controlled by way of a barrier mechanism, with E_a being the activation energy for the transition. The activation energy is easily obtained from the transition map.

The transition map for T_g (the α in Figure 8.6) indicates a temperature-dependent relaxation time, which becomes infinitely long when one approaches the dilatometric glass transition. At higher temperatures, τ approaches a limiting value τ_0. This temperature dependence is usually described by a semi-empirical WLF (Williams, Landel, and Ferry) Equation (8.29) (33)

$$\ln\left(\frac{\tau}{\tau_0}\right) = \frac{-A(T - T_0)}{B + T - T_0} = \ln(a_\tau)$$

$$(8.29)$$

where a_τ is called the shift factor, A and B are constants relating to polymer void volume and coefficient of expansion, and T_0 is a reference temperature, many times defined as the dilatometric T_g. Equation (8.29) is a consequence of a principle discovered by Leadermann (34) who found that mechanical creep curves generated at various temperatures could be superimposed to produce a master creep curve. Examples of the superposition principle are given in Reference 8. The shift factor can be calculated by way of the use of Equation (8.29). Values of $A = 17$ and $B = 51$ have been found to predict accurate transition

maps for a variety of polymers. However, to assume these values for any given polymer is tenuous. Equation (8.29) is only valid for temperatures above T_g and indicates that the Leadermann superposition principle is not valid below T_g. This is probably due to the nonequilibrium state of the glass in this temperature region. Rusch (35) has circumvented this by introducing an effective temperature below T_g to be used in Equation (8.29).

The temperature dependence of the relaxation time has been demonstrated in Equations (8.28) and (8.29). Figure 8.7 shows an example of the glass transition as a function of temperature. It is readily seen that increasing the temperature shifts the relaxation to higher frequencies. The change in the intensity of the dielectric loss is due to a temperature de-

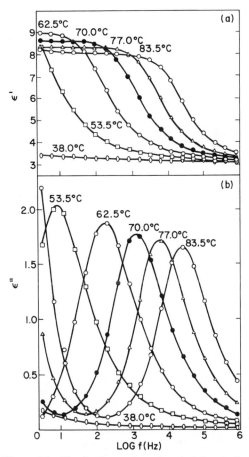

Figure 8.7 The frequency dependence of ϵ' and ϵ'' at various temperatures for the glass transition in polyvinylacetate. After Ishida et al. (23).

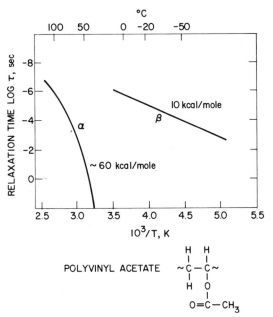

Figure 8.8   Log ν_{max} versus $1/T$ for T_g of polyvinyl acetate.

pendence of the dielectric increment that has been shown to be approximately inversely related to temperature (6,14). In fact, if one is to check the superposition principle with a system that exhibits a temperature-dependent $\Delta\epsilon$, one method is to plot a reduced ϵ'' (ϵ''/ϵ'' max) versus log ω. This type of plot will eliminate the $\Delta\epsilon$ temperature dependence and provide data for obtaining the shift factor. Shown in Figure 8.8 is a transition map for polyvinylacetate in which the WLF curve is

observed for T_g. There are many ramifications of the glass transition and its kinetics (volume relaxation, Ehrenfest relations) that can affect dielectric response to some degree. The reader is referred to References 8, 32, and 36 for details.

Thus far only frequency has been considered as a variable. However, many times in industrial applicaitons it is easier to use temperature as an experimental variable. An example of this is shown in Figure 8.9 for

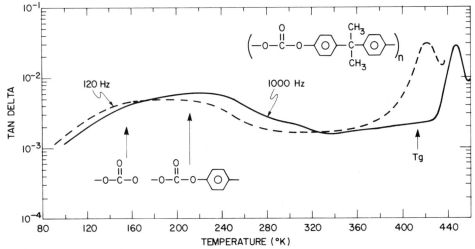

Figure 8.9   Tan δ versus temperature for bisphenol-A polycarbonate at two frequencies. Chemical structure and structural assignments are shown in the figure. Reprinted with permission from *Macromolecues*, **11**, 165 (1978). Copyright by the American Chemical Society.

bisphenol-A polycarbonate. This figure shows the relaxation structure of the polycarbonate. It is seen that various relaxations in the polymer are associated with certain structural groups. (See References 37 and 38 and the references therein.) Also note that higher frequency measurements shift relaxations to higher temperatures. A point should be made. Nmr (39) and dynamic mechanical measurements (40) indicate an additional relaxation associated with methyl group rotation in the region of $120°$ K. Since the methyl group rotation is about a symmetry axis on which the constitutent dipole resides, this particular motion does not provide dipole motion and will not be dielectrically active. Similar results have been observed for motion about the symmetry axis of polystyrene (41, 42) and indicate that sometimes complimentary relaxation experiments are needed to identify molecular motions in a polymeric system.

The temperature dependence of ϵ' or ϵ'', discussed earlier, can be taken into account theoretically by considering Equations (8.21), (8.28), and (8.29). For a single relaxation time, Equation (8.28) can be substituted into Equation (8.21), giving Equation (8.30).

$$\epsilon''(T) = \frac{\Delta\epsilon\,(T)\omega\tau_0 e^{-E_a/kT}}{1 + \omega^2 \tau_0^2 e^{-2E_a/kT}} \quad (8.30)$$

for the case where $\Delta\epsilon$ is not a function of temperature, Read and Williams (43) have shown that for a constant ω, Equation (8.30) can be integrated to yield Equation (8.31).

$$\int_0^\infty \epsilon'' d'\left(\frac{1}{T}\right) = \Delta\epsilon\,\frac{k}{E_a}\left[\frac{\pi}{2} - \tan^{-1}\omega\tau_0\right]$$

$$= \frac{\Delta\epsilon\,k}{E_a}\left(\frac{\pi}{2}\right) \quad (8.31)$$

for ω much less than one.

Thus the activation energy for a single relaxation time process can be determined from the temperature plots of the dielectric loss. In cases where $\Delta\epsilon$ is a function of temperature or a distribution of relaxation times is needed to describe the dipolar relaxation, equations almost identical to Equation (8.21) can be derived (8, 43). In the former

case, $\Delta\epsilon$ is now defined as $\epsilon_{rTmax} - \epsilon_{\mu Tmax}$, and is determined at the maximum in the loss curve. In the latter case, instead of E_a being obtained from the equation, an average value of $(1/E_a)^{-1}$ is derived. The form of the equation is otherwise identical.

Of the two methods for obtaining E_a (frequency scans versus temperature scans), the former is simpler. It does not require accuracy in sample geometry determination, since the peak position of the loss is desired and not its magnitude. However, in cases where frequency measurements can not be done, temperature scans can provide a method of obtaining a dipolar relaxation activation energy.

It has been mentioned previously that the temperature dependence of polymeric motions is important in determining relaxation spectra. Why is this so? If one is building electronic equipment or apparatus incorporating any polymeric materials, it is important to know if the dielectric constant and the loss values will be changing in the temperature domain in which the equipment will be operating. In certain cases, a minimization of these types of effects would be beneficial, and one usually would choose a nonpolar material. Then, even if relaxations occurred, the material would appear to be an almost perfect insulator. Examples of this behavior are polyethylene and polystyrene. If the desire is to control ϵ' and/or ϵ'' as a function of temperature, then a polar polymer exhibiting a rich room temperature dispersion might be desirable. An example of such might be polyvinylacetate.

So far only dipolar relaxations within the dielectric media have been considered to be a source of dielectric dispersions. However, there are other effects that will produce changes in the relaxation spectra, which might, without a complete analysis of the dielectric, be construed as a dipolar dispersion. Two of these effects are DC conductivity and interfacial polarization.

8.2.6 DC Conductivity

Since dielectric loss is a measure of the electrical dissipation in the dielectric, it should be expected that a material exhibiting a high conductivity would affect the dielectric loss.

This has been shown to be true (9, 14), and the total dielectric loss can be expressed as Equation (8.32).

$$\epsilon'' = \epsilon''_d + \epsilon''_c \qquad (8.32)$$

where ϵ''_d is the dipolar contribution and ϵ''_c is the conductive contribution. ϵ''_c is a function of the capacitance C and the resistance R of the sample (45, 46). [See Equation (8.33)].

$$\epsilon''_c = \frac{1}{\omega RC} \qquad (8.33)$$

RC is the effective time constant of an equivalent parallel R-C circuit. For a given RC value, ϵ'' will increase as ω is reduced. An example of this is shown in Figure 8.10 where low-frequency conduction in a polystyrene (PS)/polyethylenoxide (PEO) block copolymer is observed above the T_g of the PS. Since the resistance, or resistivity, of a polymer can usually be characterized by Equation (8.34)

$$R = R_0 e^{-E_a/kT} \qquad (8.34)$$

the temperature dependence of ϵ''_c can easily be obtained. In a temperature-scanning dielectric experiment, it would be expected that increasing the temperature would result in a catastrophic rise in ϵ''_c for a conductive material. (The rise is catastrophic in the sense that some dielectric equipment will not measure exceptionally high loss.)

8.2.7 Interfacial Polarization

In heterogeneous dielectrics, the possibility exists that polarization charge can build up at the interface between the phases of the medium. This is shown schematically in Figure 8.11 for spherical particles embedded in a continuous media. The interfacial buildup should occur only when the response of the embedded media is faster than the matrix, that is, if the embedded media conductivity is much higher than the matrix. The theory for predicting the effect of such dispersion has been studied for a variety of dispersed shapes (44–46) by Maxwell, Wagner, and Sillars. Interfacial effects are usually referred to as Maxwell-Wagner or Maxwell-Sillars-Wagner effects because of their initial work in the field. The simplest theoretical treatment of the effect is to treat the heterogeneity as a two-layer capacitor with different ϵ', conductances σ, and thicknesses d. The frequency dependence of the composite can be shown to be as expressed in Equations (8.35) and (8.36) as follows:

$$\epsilon'(\omega) = \bar{\epsilon}_\infty + \frac{\bar{\epsilon}_0 - \bar{\epsilon}_\infty}{1 + \omega^2 \tau_{mw}^2} \qquad (8.35)$$

$$\epsilon'(w) = \frac{(\bar{\epsilon}_0 - \bar{\epsilon}_\infty)(\omega \tau_{mw})}{1 + \omega^2 \tau_{mw}^2} + \frac{\sigma}{\omega} \qquad (8.36)$$

with terms defined as in Equations (8.37) through (8.40).

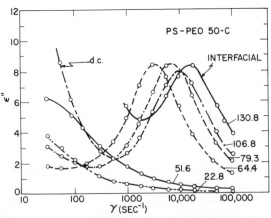

Figure 8.10 Dielectric loss versus frequency for a 50:50 polystyrene/polyethyleneoxide block copolymer. Reprinted from Reference 47 with the permission of Plenum Publishing Corporation.

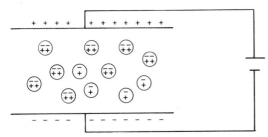

Figure 8.11 Schematic representation of interfacial polarization effects.

$$\bar{\epsilon}_0 = \frac{d(\epsilon_1' d_1 \sigma_2^2 + \epsilon_2' d_2 \sigma_1^2)}{(\sigma_1 d_2 + \sigma_2 d_1)} \qquad (8.37)$$

$$\bar{\epsilon}_\infty = \frac{d\epsilon_1' \epsilon_2'}{\sigma_1 d_2 + \sigma_2 d_1} \qquad (8.38)$$

$$\tau_{mw} = \frac{\epsilon_1' d_2 + \epsilon_2' d_1}{\sigma_1 d_2 + \sigma_2 d_1} \qquad (8.39)$$

$$\bar{\sigma} = \frac{d\sigma_1 \sigma_2}{\sigma_1 d_2 + \sigma_2 d_1} \qquad (8.40)$$

Although the dependence if ϵ' and ϵ'' of the composites are complicated functions of the individual components, the basic dispersion Equations (8.35 and 8.36) are identical to the single relaxation time equations derived previously, with the exception of a DC conductivity term in the ϵ'' equation. Thus, since most materials exhibit relaxation linewidths greater than that of the single relaxation time, one means of identifying the interfacial effect should be by its characteristic linewidth. It should be noted that τ_{mw} is a function of the conductivities, and so on, of the individual components and that the higher the conductivity of one of the components, the smaller the τ_{mw}. This would imply a higher frequency of observation in the dielectric experiment.

Semicrystalline polymers, as well as most filled systems, are more complicated than the simple series capacitor model. Sillars (44) has considered the problem of varying the shape of a conducting heterogeneity in a dielectric. One of his results is shown in Figure 8.12. Obviously, any degree of dielectric dispersion to the ultimate limit of DC conductivity can be attained. However, in each case, the linewidth remains constant at a single relaxation time.

Because most polymeric systems are not composed of well defined sample geometry, the equations derived from Sillars may not be acceptable in describing the observed effects. In these cases, it is many times easier to identify the effect by its characteristic linewidth and then to propose a mathematical model to describe the observed results.

The interfacial model is typically applied for studies of polymer systems such as carbon black-rubber (48), fused silicate-polymers (49), and other dispersions such as sols (30). There are even cases where microcracks or inclusions have caused interfacial polarization effects (51). What should be remembered in all these cases are the large differences in conductivity of the media and the interfacial boundary between the media of the dispersion.

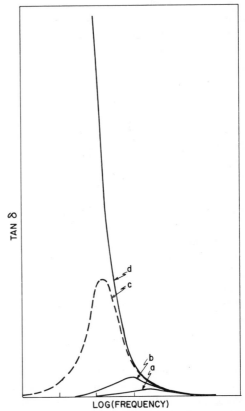

Figure 8.12 Relative tan δ for a conducting material C dispersed in a dielectric media D. a = parallel sheets of materials C and D; b = isolated spheres of C in D; c = cylinders of C dispersed parallel to the measuring electrode; d = cylinder of C dispersed perpendicular to the measuring electrode. *Note:* Same value fraction for all components.

The interfacial effect is also manifest when the permittivity of the samples is considered. Usually τ_{mw} is large and the interfacial dispersions occur at low frequencies, giving rise to large values of permittivity ϵ'. In these cases the effect can be identified if ϵ' is much larger than the values predicted on the basis of equations such as Equation (8.6). This will provide an estimate of the dipolar contribution to ϵ'. An extreme case of this may be the "hyperelectronic polarization" studied by Pohl (52) and co-workers for a variety of highly polarizable polymers that exhibit large effective dipole moments.

8.2.8 Instrumentation

Dielectric relaxation spectra can be taken over a variety of experimental frequency ranges from 10^{-4} to 10^{10} Hz. As discussed previously, with such a large frequency range, analysis of loss curves in terms of fundamental theories can be undertaken. In many cases, though, it is desirable to compare dielectric results with other experimental techniques such as mechanical relaxation techniques, differential scanning calorimetry, and thermally stimulated discharge (TSD) (53) experiments where temperature is considered as a variable. In such experiments effects of annealing volume (54), molecular relaxations (55), and crystal structure (56, 57), as well as scanning rate (53), can be studied and compared.

It becomes obvious, then, that a variety of techniques and instruments are needed to accomplish the desired results. In fact, there is no single dielectric apparatus capable of the full frequency range described. Thus instrumentation for dielectric studies can be the subject of entire chapters. Rather than undertake such a description of these instruments and techniques in a cursory fashion, the reader is referred to References 8 and 32 for discussions of classic bridge techniques and recording techniques for measuring dielectric relaxation spectra. In the former, the classic techniques developed through 1967 are described in some detail. In the latter, various "new" techniques such as time domain (Fourier transform) (58–62), direct current (63, 64), autobalancing

direct recording, computerized direct recording, and microwave techniques are discussed. In addition, related subjects such as TSD, thermomechanical, differential scanning calorimetry (DSC), and pulsed nmr are discussed and related. In the case of Fourier transform and the DC method of dielectric relaxation, a frequency range of 10^{-4}–10^{10} Hz is obtainable with proper experimental setup. Usually these techniques are unreliable in the high-frequency regime, and they are used below the 10^4 Hz range with other high-frequency techniques used for the 10^5–10^{10} range.

In the writers' opinion, there is little doubt that the advent of microprocessors will facilitate data acquisition even more. In fact, the use of microprocessing equipment to control autobalancing bridges and recorders is a relatively inexpensive method of obtaining dielectric data as a function of temperature (60).

8.2.9   Applications

The application of the dielectric technique requires only that the material be dielectrically active or that the capacitance resolution of the measuring instrument be such that volume effects can be taken into account. This latter effect is particularly true in cases of nonpolar systems where volume coefficients of expansion change as various molecular relaxations take place. Within these restraints, dielectric relaxation measurements have been used in a variety of scientific and industrial applications. These are reviewed in the following section along with some suggested ideas by the authors.

8.2.9.1   *Molecular Relaxations—Transition Maps*

Although this topic has been mentioned previously, another example is included here for completeness and an in-depth analysis of the motion of a common polymer. Shown in Figures 8.13 and 8.14 are the transition map and tan δ versus temperature for polymethylmethacrylate (PMMA). It is obvious from Figure 8.13 that four relaxation modes of the polymer have been assigned. The γ and δ have been assigned to methyl group rotations by way of nmr and dynamic mechanical tech-

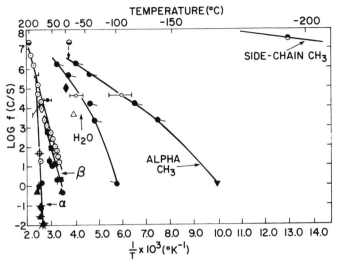

Figure 8.13   Log ν_{max} versus $1/T$ for polymethylmethacrylate (PMMA).

niques (67). The dielectric spectra would only exhibit the α and β relaxations (no dipolar motion due to methyl group rotation). In the higher methacrylate analogs the side-chain alkyl group motion is observed dielectrically (68). The transition map indicates that free rotation of the side-chain methyl group is very facile with a low activation energy. It also indicates that the methyl group of the polymer is somewhat constrained by its proximity to the polymer backbone and thus occurs at higher temperatures with a higher activation energy. The α relaxation has been assigned to T_g as the result of plasticization experiments (69) and DSC measurements.

Although the β relaxation has been assigned to side-group motion of the polymer, there is some evidence that the motion may be coupled to short-order main-chain (backbone) motion

with the polymer. This becomes evident when higher alkyl analogs are studied and the α and β relaxations merge. Even in the case of PMMA (see Figure 8.13), the α and β merge into one motional process at high temperatures. Such mixing of the α and β process to give a combined $\alpha\beta$ relaxation has been treated by Williams and Watts (70) by way of correlation theory and by Hoppleman (71) using a spring and dashpot mechanical analog. In the former case it was shown that increasing pressure on PMMA caused the $\alpha\beta$ process to separate into the α and β process. In the latter, Hoppleman was able to show that decreasing the temperature of measurement would cause the same effect (as one would expect from the transition map). The coalescence of the α and β processes is not unusual and has been observed in a variety of polymeric systems.

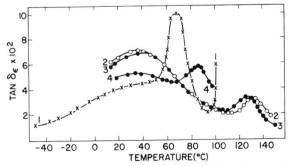

Figure 8.14   Tan δ versus temperature for PMMA at 28 Hz. Curve 1 isotactic, curve 2 syndiotactic, curve 3 processes PMMA (66).

This observation has led many researchers to conclude that, in polymeric systems, the β relaxation is many times a localized motion (i.e., resulting from motions of only a portion of the polymeric chain). It is certain that the motion of the carbonyl side chain takes place during the β relaxation rather than during the α.

The effect of tacticity on the relaxation spectra of PMMA is shown in Figure 8.14. It is seen that the tacticity can greatly affect T_g, as well as the position and intensity of the β relaxation. These results are similar to those for poly(t-butylmethacrylate) (72) and indicate that the steric constraints in the isotactic form of PMMA are such that the β motion becomes more hindered (a decrease in the dielectric loss intensity). In the case of all isotactic alkyl methacrylates studies this seems to be true. An increase occurs in the α relaxation intensity with a shift to lower temperatures, and a corresponding decrease in β relaxation intensity occurs with a shift to lower temperature. These studies also indicate that either fewer carbonyl moieties are taking part in the β relaxation, or that their motional amplitude is constrained while barriers associated with the α motion are lowered. These results indicate that tacticity must always be considered in preparing a polymeric system and that measurements of T_g may be a method of estimating it (73).

Thus the transition map will provide information on the effective modes of motion in a polymeric system and the molecular energetics governing them. The maps always indicate that, at extremely high temperatures, all motions within the polymer merge and become cooperative in nature. This information, coupled with ϵ' and ϵ'' data, will provide the experimentor with pertinent information that can be used in designing electrical devices, as well as in determining any impurities within the system.

8.2.9.2   Sorbed Water

As is seen in Figure 8.13, the presence of sorbed water in a polymer system can also produce dielectric relaxation. These water-induced relaxations have also been observed in polysulfones, polyamides, and polycarbonates. In these materials, the sorbed water usually creates at least one additional relaxation peak in the 170–220° K range at intermediate dielectric frequencies (10^2–10^5 Hz). In some cases more than one peak is observed (74). Freely rotating water has a relaxation time of approximately 10^{-11} sec. If the water molecule produces a new relaxation in a polymer system, the relaxation must be due either to a plasticized polymer relaxation or a bound state of water such that the water participates in the relaxation. The latter is suggested because of large increases in the dielectric loss of the water-induced relaxation (74). Thus it would appear that various bound states of water exist in the polymer matrix (probably due to hydrogen bonding in amide linkages or interactions with polymer hydroxyl end groups). As an illustration, Froix and co-workers (75) were able to distinguish many types of bound water by way of pulsed nmr experiments in cellulose. In cases where single water-induced relaxations occur, it should be feasible to monitor the water content of the samples by way of a single-frequency-single-temperature experiment.

8.2.9.3   Sorbed Gases

Unless there are specific interactions between sorbed gases such as CO_2, O_2, or N_2 and a polymer matrix, it might not be expected that an atmospheric environment of these gases would affect polymeric relaxation spectra. Two cases have been found (76–78) where sorbed oxygen produces a dielectrically active relaxation. There are for sub-T_g relaxations in polystyrene (PS) and poly(N-vinylcarbazole) (PVK). Both polymers exhibit γ relaxations near 170° K that are dielectrically active in the presence of sorbed oxygen but become dielectrically inactive when oxygen is removed by way of vacuum heat treatments (see Figure 8.15). Because the relaxation was observed by way of dynamic mechanical measurements and in the presence of sorbed oxygen, it was assigned to a rotational libration of the PS phenyl ring about the symmetry axis connecting it with the polymer backbone, and a similar result was obtained for PVK. Pulsed

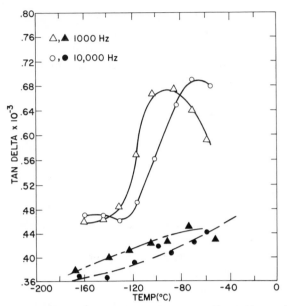

Figure 8.15   Tan δ versus temperature for PS. Open symbols: PS equilibrated in pure O_2 at 353° K for 48 hr after vacuum treatment. Closed symbols: PS at 10^{-6} mm Hg, 353° K for 216 hr, Δ = 1 KHz, O = 10 KHz. Reprinted from J. M. Pochanand, D. F. Hinman, *J. Polym. Sci. Polym. Phys. Ed.*, **14**, 1871 (1976) with permission of John Wiley and Sons.

nmr and dielectric experiments indicated that sorbed oxygen molecules diffused through the polymer lattice creating spheres of influence where an off-symmetry axis-induced dipole existed in the carbazole and phenyl groups. These induced dipoles made the rotational libration motion of the polymer dielectrically active.

8.2.9.4   Plasticizers
The effect of plasticizers on polymer properties is industrially important. Usually the plasti-

cizers are used to transform hard, brittle materials into tough, flexible materials. Many times a study of the relaxation mechanisms can provide insight into these mixtures on a molecular level. The plasticization effect can be studied by its effect on T_g, as well as on sub-T_g relaxations. The latter experiments can produce an understanding of the interactive mechanisms between polymer and plasticizer.

Two types of T_g relaxations are observed in polymer-plasticizer systems—correlated and uncorrelated. These are shown in Figure 8.16.

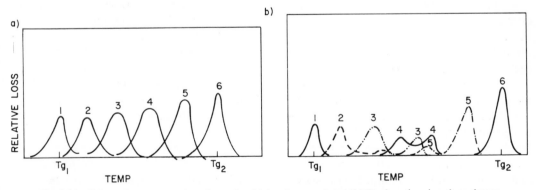

Fig. 8.16   Schematic representation of correlated (a) and uncorrelated (b) T_g relaxations in polar polymer-polar plasticizer systems.

In the correlated case the T_g of mixtures is described by a single relaxation peak that varies smoothly with composition between the two components. In the latter case each component affects the T_g of the other component but exhibits its own characteristic relaxation time. The latter system might be expected to exhibit phase-separated structures with enriched volumes of each component, whereas the former would appear intimately mixed on a dielectric time scale. An example of the former system would be polycarbonate/bisphenol-A-diphenylcarbonate (79), and an example of the latter is the polyvinylchloride/tricresylphosphate system (80). There are countless examples in the literature of systems that exhibit characteristics of both these examples depending upon the relative concentrations of the components. The dielectric spectra of such systems (especially temperature scans at constant frequency) provide information on the relative stability of the mixtures (time dependence of the spectra), as well as the effect of the plasticizer on T_g. Theories and empirical formulas relating the T_g's of the pure components are numerous and have been reviewed (32). A simple equation to estimate the T_g of mixtures is the Fox equation (8.41) (81) as follows:

$$\frac{1}{T_{gT}} = \frac{C_1}{T_{g1}} + \frac{C_2}{T_{g2}} \qquad (8.41)$$

where the C_i are either volume or the weight fraction of each component and T_{gi} are the T_g's of the individual components. Equation (8.41) has been shown to provide a reasonable estimate of the T_g of a mixed system. However, the reader is cautioned that it is only an estimate and that a direct measurement of each sample should be taken.

Little dielectric work on the sub-T_g's of mixed systems has been done. The authors feel that these studies provide insight into the molecular mechanisms taking place in plasticized systems. Dielectric evidence of direct molecular interactions between plasticizers and polymers has been obtained in the system N-butyl-4,5,6-trinitrofluorenone-2-carboxylate/Lexan (82). In that system it was shown

that low-temperature local modes of the polymer are affected by the monomer in such a way that can only be explained by direct electronic coupling of the compounds. Other systems [such as poly(methylacrylate)/tricresyl phosphate] (83) exhibit classic plasticization effects, that is, the β relaxations are shifted to lower temperatures, and the activation energy for the process is decreased. PMMA plasticized by toluene or di-n-butylphthalate exhibits similar effects (84). In the case of bisphenol-A-diphenylcarbonate/polycarbonate solid solutions, a combined phenyl ring-carbonyl motion was shown to be unaffected by the addition of the alter component, be it monomer or polymer, whereas another low temperature relaxation was plasticized.

Thus far not enough of these systems (polymer-diluent) have been studied that a general correlation between observed sub-T_g trends and other observables (T_g, mechanical properties, etc.) can be made. The authors feel that the key to understanding many of these systems in terms of other measurable properties lies in understanding the nature and interactive mechanism of the sub-T_g relaxations. For those with further interest in plasticization phenomena, Reference 32 and Reference 9, Chapter 8 are cited.

8.2.9.5 Mixing Phenomena—Block Copolymers, Polymer Blends, and Copolymers

In many polymer systems it is difficult to promote molecular mixing of two homopolymeric materials. For instance, homopolymers of widely different polarities, such as PS and PEO, may be soluble in a mutual solvent, but will phase separate upon drying. Properties such as molecular weight, solubility parameter and crystallinity may also affect the homopolymer's mutual solubility. Systems such as PEO/PS (85) and poly(2-vinylpyridine)/polyurethane (86) have been shown to be mutually soluble, but they are the exception rather than the rule. In cases where mutually soluble homopolymers are observed, the glass transition is usually at some intermediate value between the pure homopolymer components, and the T_g analysis is amenable to the Fox equation. In the case of block copolymer systems, the individual polymer components may be insoluble and phase separate. However, owing to the intimate coupling that exists in block copolymer systems, a "grey" region of mixing may occur between the pure homopolymer phases. This mixed phase usually will exhibit a T_g at some intermediate value between the T_g's and the homopolymers. The dielectric intensity of this intermediate peak is dependent upon the volume and the dipoles participating in the mixed phase (87). The dielectric method can thus be used to investigate the mixing character of compatible or noncompatible systems.

The ultimate "blend" of two incompatible materials will arise when copolymers are made of the components. In these cases, the glass transitions of the copolymers have been shown to vary from the simple additive functions of the homopolymer T_g's[the Fox (88) or Gordon-Taylor (89) approach] to the very complicated interactive functions of the component T_g's (90). In the former case the T_g varies between the T_g of the pure components. In the latter case specific interactions or changes in mixing volume are proposed to produce T_g's that are greater than or less than the component T_g's. The dielectric spectra will provide a simple method of obtaining the T_g data and may, by studying the sub-T_g relaxation in the copolymer system, provide insight into the molecular mechanisms of the system.

The reader is cautioned to realize that using the Fox and/or the Gordon-Taylor equations is appropriate only in total randomized copolymers. Copolymerization that leads to sequence distributions rather than randomization can produce totally unexpected T_g's.

8.2.9.6   Process Control
Process control is an important part of any industrial development of polymer materials. The following subjects deal with various aspects of processing and possible techniques to study these aspects.

Retained Solvent or Plasticizer.   Provided some caution is taken, the dielectric method can be used in monitoring the retained solvent in a polymeric system. This is shown in Figure 8.17. Curve (*a*) is considered to represent the T_g of an unplasticized polymer, and curve (*b*) represents a plasticized polymer in which some degree of conductivity is introduced. Two methods of monitoring the solvent may be made.

1.   By measuring tan δ at T_g, a possible direct correlation of the residual solvent with tan δ could be obtained. This might be made by correlating measured tan δ with residual solvent studies by way of gel permeation chromatography.

2.   A correlation of the DC component of tan

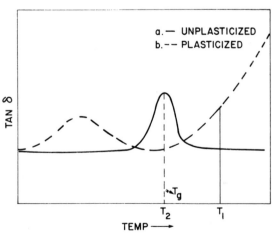

Figure 8.17   Schematic of tan δ versus temperature for plasticized and unplasticized polymer systems at constant frequency.

δ at T_1 might also be made in an identical manner.

With autoranging dielectric equipment such as the General Radio 1688 bridge, a process monitor might be installed. It should be noted that a method of obtaining proper electric contact for measurements is needed or that the use of the dielectric air gap technique (91) might be appropriate with known controls included. Retained plasticizer might also be monitored in the same way. Results of such studies would hinge on the molecular characteristics of the system being studied, that is, when uncorrelated motions involving T_g are observed, a direct correlation of dielectric data with retained solvent would be tenuous. In the case of a correlated T_g motion, a constant temperature test apparatus might be constructed to measure tan δ. A correlation of tan δ with residual plasticizer could be made by way of dielectric runs (tan δ versus temperature) on known samples. In order to gain the most sensitivity from such an experiment, measurements should be made in the temperature region of T_g. These same types of experiments (tan δ versus temperature) are also useful in studying phase separation in blended systems. A well mixed system will usually exhibit one dynamic glass transition. Any phase separation would lead to multiple components in the dielectric scan owing to the phase-separated components. Even in a well mixed system, if the components are not totally compatible, heating the blend at or near T_g will usually produce a phase-separated structure. It should be emphasized that the time dependence involved in phase separation will be a function of the microviscosity of the system, as well as the thermodynamic force for phase separation (87).

Process Monitoring of Reactions. Process monitoring by way of dielectric techniques of reaction and correlation of the observed data with desired mechanical or other attainable properties has become important in the area of adhesives (epoxies, etc.; 92–94). This is particularly true in the area of epoxies where curing time and so forth are important variables. In the cited studies tan δ was observed as a function of the time and the temperature of the cure and results were correlated with observed mechanical properties. In the dielectric studies, solvent removal, as well as chemical cure, was to be investigated for epoxy resins. In Allen's work (92), a direct correlation between mechanical properties and dielectric results was obtained for a glass-reinforced epoxy.

The studies of the curing or cross-linking of rubber (95, 96), epoxy (97–99), phenol-formaldehyde resins (100), polyester resins (101), and wood-plastic combinations (102) have been well documented in Chapter 6 of Reference 10. The reader is referred to that work for a detailed discussion of the findings. Each case provides different types of dielectric results with dielectric constant, tan δ and AC conductivity being used to determine the degree of curing. As expected, in many of these systems a detailed analysis of the dielectric response (ϵ', tan delta, or AC conductivity versus temperature as a function of time or vice versa) is required before the analysis can be used in any type of processing arrangement. Radiation cross-linking can also be studied similarly, with the radiation dosage replacing temperature as one of the variables.

8.2.9.7 Aging

Aging can be particularly important in plasticized systems, blends (phase separation), composites, and in reactive polymers where oxidation may change the polymer's mechanical properties. Blends and plasticizers are easily studied because of the appearance of separate dielectric relaxations owing to the pure components appearing in the blend relaxation spectra. Composites and reactivity require the studying of individual systems to ascertain the various types of changes that polymers can undergo when exposed to an oxidative environment. These oxidative environments become particularly important in polymeric systems exposed to ambient conditions, such as sun light, where ultraviolet can initiate reactions with the polymer.

Oxidation will usually lead to the introduction of oxidized sites (dipolar) and will provide for increased dielectric activity. This process can easily be studied in nonpolar systems such as PS (103) and PE (104, 105), where the addition of oxidized sites within the polymer produces measurable dielectric relaxation intensities. In the unaged polymer (particularly PE), these would not be observed. In the case of PE, relaxations that are mechanically active become dielectrically active with low oxidation levels. An example of oxidation is shown in Figure 8.18 (105). It is seen that the oxidation of PE by two different means produces different dielectric relaxations. The differences in the observed spectra are attributed to the different methods of oxidation. In the UV case the radiation is used to excite the system to oxidize. Reactions are limited by the availability of oxygen which is controlled by the oxygen diffusion within the PE. Oxygen diffusion is much faster in the amorphous segments of the PE. It would be expected that the relaxations observed were due to the amorphous PE. In the latter case oxidation took place in the amorphous state of the polymer and then, when the sample was crystallized, all components had equivalent numbers of oxidized groups. In that case the process assigned to relaxations within the PE crystal lattice becomes more intense.

Care must be taken when analyzing systems such as these in terms of observed relaxations. There is no reason to expect that the lattice dynamics of the polymer chain about some introduced dipolar group would be identical to systems lacking these dipolar groups. The relaxation assignments, as well as the activation energies calculated, may not be the same as in the nonpolar systems. The results in this polyethylene system are indicative of polymer oxidation. The reader should be aware that, in accelerated aging studies, many polymers are artificially aged by way of radiation bombardment (106) and that this bombardment can change the AC and the DC properties of the polymer (107).

In other cases of aging, UV radiation and mechanical stress can cause decomposition of the polymer. This can lead to main chain scission (a reduction in molecular weight) and elimination reactions. In the case of the main chain scission, the dielectric effects could be two-fold.

1. Dipolar groups could be formed at the chain ends, leading to an increased dielectric strength of the observed relaxations, or possibly new relaxations, if the chain-end density becomes sufficiently high.

2. The molecular weight could be sufficiently low so that the molecular weight dependence of T_g is manifest in the T_g relaxation.

In the case of an elimination reaction, new dipolar species may be formed due to the formation of HCl from the degradation of polyvinylchloride (PVC). In cases where ionic impurities such as HCl are formed, AC conduction could provide a convenient variable for monitoring the polymer. Hedvig and coworkers (109), as well as others (110), have been able to monitor the kinetics of the degradation by way of dielectric techniques. If the polar species formed are conductive and become occluded within the polymer matrix, a Maxwell-Wagner type relaxation might occur. If this happens in insulating materials, dielectric losses higher than those predicted may occur in the material and limit the technique's

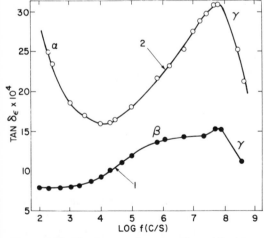

Figure 8.18 Tan δ versus frequency for oxidized low density polyethylene. Curve 1 = UV aging at 25° C; Curve 2 = milled at 160° C for 3 hr. Reprinted from C. A. F. Tuijman, *Polymer*, **4** (1963) by permission of the publishers, IPC Business Press, Ltd. ©.

usefulness for certain applications. The degradation products may also be water sensitive and further enhance interfacial polarization effects.

Applications for dielectric measurements are numerous. In conceiving these applications it is important that the experimentor design his research carefully so that results do not lead to a pitfall due to incomplete experimental data. Many times this will require a complete scientific study to insure that any correlation with other experimental data or a process determined by that correlation is correct.

8.2.9.8 Miscellaneous

The following is a brief summary of some special topics that we feel should be included as a part of dielectric analysis.

Orientation Effects. The dielectric technique can be used to study the effect of orientation in polymer films (111–113). Usually these studies will show shifts in observed relaxations, as well as changes in their relaxation strength. In many cases a change in the draw direction (perpendicular or parallel to the electric field) will also affect these results. If one is interested in controlling a process or finding the degree of orientation on a molecular scale, correlations of samples with known orientations could be made with those that are unknown. These types of experiments may be vital in certain industrial applications since orientation is known to affect electrical and mechanical properties of polymers.

Defect Structure in Amorphous Se Glasses. Recent studies by Abkowitz and Pai (114) have shown that amorphous Se glasses exhibit dielectric activity in the neighborhood of T_g. This is unusual in that these glasses are comprised of Se atoms and would not, a priori, be expected to exhibit dipolar relaxations. In the cited study, experimental evidence (ϵ' versus temperature) and AC conductivity versus temperature for the volume relaxation of the Se glass are also shown. The origin of the dipolar activity is questioned, and it is proposed that defect charge distributions and/or dynamic charging (displace-ment-induced charge redistribution) cause the effect. At this time the true nature of the dielectric activity is unknown. But, the study does indicate that the dielectric technique can provide valuable information concerning systems that many times would be considered dielectrically inactive.

Composite Analysis. In cases where, as has been pointed out in Section 8.2.7, a composite consists of a conducting media dispersed in a nonconductor, the Maxwell-Wagner approach to the relaxation spectra can produce results that indicate the degree of dispersion and its particle-shape distribution. Such approaches have been used by Wertheimer et al. (115) for polymer-mica composite and rubber-carbon black composites (116, 117). In these cases the Maxwell-Wagner-Sillars formalism was used to describe the observed low-frequency properties of the composites. Lukomskaya and Dogadklin (118) have reviewed the rubber-carbon black literature and have shown that the dielectric properties are functions of the conductivity, the Maxwell-Wagner effects, the chemical structure of carbon black, and the dipolar contributions. The latter study demonstrated the feasibility of investigating a complicated system and ascertaining to what degree each phenomenon contributes to observed dielectric properties.

Dielectric techniques have even been used to investigate the relationship between the dielectric constant, the density, and the free water content of snow (119). Bipolymer and natural products are being studied more often (120–122).

8.3 PHOTOCONDUCTIVITY

8.3.1 Background and Definitions

Photoconductivity is the enhancement of the conductivity of matter when it is exposed to a radiation source. This increase in conductivity occurs when absorbed radiation raises charge carriers from their nonconducting state to their conducting state with the subsequent freedom to migrate. In the dark the thermally generated carriers are in equilibrium among the available energy states and obey Fermi statistics. When the material is exposed to a radiation source, the free carrier concentration increases and reaches a steady-state value. The phenomenon of photoconduction therefore involves the absorption of radiation, photo-generation of carriers, and their ultimate recombination. In between the photogener-ation and the recombination processes, the excited carriers may undergo other transport processes such as trapping, trap release, drift, and diffusion. The study of photoconductive phenomena has therefore played a major role in the understanding of the physics of the solid state.

The task of describing photoconductivity in amorphous organic polymers and solid solu-tions of organic monomers in "inert" binders within the restrictions imposed by the length of this chapter is formidable. We would not attempt to catalog the list of organic materials in which photoconductivity has been reported (123). We happen to believe that, given ade-quate light sources and current measuring devices, it should be possible to detect photo-conductivity in most materials. It is sufficient here to describe the most recent developments in the understanding of photogeneration and charge transport processes in some typical, well characterized polymeric and monomeric materials. The section on photoconductivity begins by describing a few basic concepts that facilitate a clearer understanding of the exper-imental data to be presented. These concepts were developed in connection with the study of photoconductivity in crystalline covalent materials where the charge carrier transport occurs through allowed energy bands. It should be noted that the concepts of energy bands is of limited validity in an amorphous molecular solid owing to the weak interactions between the molecules. Although some of the tradi-tional concepts of photoconductivity are in-dependent of the nature of the solid, others are not. The subsection on the concepts of photo-conductivity also includes some of the exper-imental techniques that have been instrumental for a more meaningful evaluation of these materials.

Traditionally, the observation of photo-conductivity involves the detection of a current when a material with appropriate contacts is connected in a current-measuring circuit and is exposed to a radiation source. Though simple in concept, the results obtained with this measurement are the most difficult to interpret on account of the limitations imposed by the contacts. Consider, therefore, a photo-conductor without any contacts. If the photo-conductor is exposed to a monochromatic radiation source of intensity I, the rate of increase of carrier density n as a function of time is given by Equation (8.42) as follows (124):

$$\frac{dn}{dt} = g_0 - \frac{n}{\tau} \qquad (8.42)$$

When solved Equation (8.42) results in Equa-tion (8.43)

$$n(t) = g_0\tau(1 - e^{-t/\tau}) \qquad (8.43)$$

where g_0 is the rate of production of mobile carriers per cubic centimeter and τ is the recombination lifetime of the carriers. The rate of production g_0 is proportional to the absorbed radiation intensity I, with the quan-tum efficiency of photogeneration as the pro-portionality constant. The absorption of rad-iation produces positive (hole) and negative (electron) charge carriers. It has been assumed that only one of these charges gives rise to conductivity, while the other is instantly trapped. In general, however, both carriers could give rise to increased conductivity. According to Equation (8.43) the concentration of free carriers increases with time, with a time constant τ ultimately reaching a steady-state value of $n_0 = g_0\tau$. If the radiation source is now turned off, the charge carrier density reaches its value in the dark according to Equation (8.44).

$$n(t) = g_0\tau \exp(-t/\tau) \qquad (8.44)$$

Equations (8.42) and (8.43) represent the simplest case of a photoconductor with no trapping sites, whose number of recombination centers is large compared to the total number of carriers in the steady state.

If however, the photoconductor has trapping states, then the rate Equations (8.45) and (8.46) for free carrier density and the trapped carrier density n_t are as follows:

$$\frac{dn}{dt} = g_0 - \frac{n}{\tau_t} + \frac{n_t}{\tau_r} - \frac{n}{\tau} \qquad (8.45)$$

$$\frac{dn_t}{dt} = \frac{n}{\tau_t} - \frac{n_t}{\tau_r} \qquad (8.46)$$

where τ_t is the free carrier trapping time, and τ_r is the trapped carrier release time. For covalent solids with charge transport occurring in energy bands, the trapping time is given by $1/\tau_t = 1/N_t Sv$ where N_t is the density of trapping levels, S is the capture cross section of the trapping centers, and v is the thermal velocity of the carriers. This equation is easily derived when one equates the volume $Sv\tau_t$ to the volume $1/N_t$ containing one trapping center. $Sv\tau_t$ is the volume traversed in time τ_t by a carrier with a velocity v and a cross section S. This expression for the trapping time is not applicable to amorphous molecular solids where the charge transport mainly results from carriers hopping through a network of localized states. Even the definition of a trapping site may take on a new meaning. As we show later, the transport through localized states is an appropriate sum of the series of release times (hopping times) from these states. If the release time from one of these states is very large (long hop), then this state could be classified as a trap within the time frame of a particular experiment. In most amorphous organic materials, this release or trapping time is seen to be strongly electric-field dependent. Therefore, within a defined time frame, whether or not an "isolated" state is a trap would depend on the applied electric field (125). The more appropriate expression for the trapping time τ_t in systems with hopping transport is derived from diffusion considerations. Since in these systems the transport is diffusion controlled, the distance r traveled by a carrier within a trapping time τ_t is given by $(D\tau_t)^{1/2} = r$. The volume of the solid traversed

by a carrier within a trapping time τ_t is given by $(4\pi/3)r^3$ and is equal to $1/N_t$, the volume containing a trapping center. Substitution and rearrangement leads to the following expression:

$$\tau_t = \left(\frac{3}{4\pi}\frac{1}{N_t}\right)^{2/3}\frac{1}{D}$$

where D is the diffusion coefficient.

The release time from the traps is generally given by Equation (8.47) as follows:

$$\frac{1}{\tau_r} = v \exp\left(-\frac{\Delta_t}{kT}\right) \qquad (8.47)$$

where v is the attempt to escape frequency and Δ_t is the trap depth.

By solving the rate Equations (8.45) and (8.46) for the photoconductor with a set of trapping levels, it can easily be shown that the steady-state concentration of free carriers is still $g_0\tau$. At first glance, it would seem odd that the steady-state concentration is not changed from the case described earlier in which the photoconductor has no trapping states. This equivalence results from the steady-state invariance of the concentration of the carriers in the trapping states (126). This implies that the rate of carrier trapping into the trapping states equals the rate of carrier release from the trapping states. In other words, the trapping states are invisible as far as the steady-state concentration is concerned. However, the introduction of trapping states does increase the response time needed to reach steady state when the excitation is turned on or off. This increase in response time results from the extra time required to fill the traps when the excitation is applied to the photoconductor material or from the time required to empty the trapped carriers when the excitation source is turned off. If the trapping is due to a discrete trap level, then the trap depth can be determined by measuring the photoconductor decay time as a function of temperature. In the case of photoconductors with a distribution of trap levels, the trap distribution can be determined by measuring the initial rate of decay after the excitation is turned off. This procedure is outlined in Reference 129.

The conductivity increase due to illumination is given by the relation $\sigma = en\mu$ where e is the electronic charge and μ is the microscopic mobility. The measurement of DC conductivity requires that the material with at least two contacts be connected in a circuit containing a voltage source. The act of connecting the solid across the battery source may result in a change of the carrier density from its open circuit value. Therefore, for the measured current to have any meaning, the contacts should not disturb the free carrier density that existed before the solid was connected into the circuit. Contacts with this property are called ohmic contacts. They maintain a constant carrier density within the material by injecting carriers at a rate equal to the rate of exit of carriers at the other contact. Under conditions labeled space-charge-limited currents (discussed later), the carrier density within a solid with an ohmic contact will be higher than the value determined by generation, recombination, and trapping kinetics described by Equations (8.42) and (8.45).

In some instances the study of the dependence of the charge carrier density with the wavelength of the exciting radiation has been used to gather information concerning the nature of the transition involved. In experiments involving photoconductors overcoated with metal electrodes, it is possible, by the proper selection of the wavelength of the excitation, to emit carriers from an illuminated contact and to study the difference in energy levels between the metal and the photoconductor. This method of exciting carriers in the metal contact followed by their emission into the solid is termed internal photoemission. Although this type of experiment can be carried out with an illumination of fixed intensity, the sensitivity of these experiments can be improved by employing a source whose intensity is modulated at low frequencies. The study of conductivity, either by steady-state or intensity-modulated illumination sources, gives information on the product of the free carrier density and the drift mobility. Other experiments that independently determine

either the free carrier density or the charge carrier mobility are required. For example, the free carrier density can be determined with electron spin resonance experiments. The mobility values have been traditionally determined by Hall effect type measurements. These are not discussed in this chapter. We do, however, describe a transient photoconductivity experiment that has been successfully employed to study the efficiency of photogeneration and drift mobility in many low-mobility organic and inorganic materials (127, 128).

8.3.2 Transient Photoconductivity

Consider a film of thickness L which is either solvent coated or vapor deposited on an appropriate conductive substrate. A sandwich configuration is now formed by vapor depositing a semitransparent metallic electrode on the free surface of the film. This sandwich device is connected in a circuit shown in Figure 8.19a. Suppose the sample is exposed to a short flashlight flash which is absorbed in a region of the material close to the top electrode. If the substrate is biased negatively, a sheet of holes is injected into the bulk of the material, and these holes will eventually drift to the substrate. If the RC time constant of the circuit is made much larger than the transit time t_T of the sheet of the carriers, and if we assume that the sheet of carriers drifts with a constant velocity $\mu_d E$, a ramp type of signal appears across the resistance R. C is the capacitance of the sandwich device and is equal to $\epsilon\epsilon_0 A / L$ where $\epsilon\epsilon_0$ is the absolute dielectric constant and A is the surface area of the device. The amplitude of the voltage signal produced by the drifting carriers is given by Equations (8.48) and (8.49) as follows:

$$\Delta V(t) = \frac{\Delta Q(t)}{C} = \frac{\eta Fe}{C} \frac{t}{t_T} \qquad \text{for } t < t_T$$

$$(8.48)$$

$$\Delta V(t_T) = \frac{1}{C} \eta Fe \qquad \text{for } t > t_T$$

$$(8.49)$$

where η is the quantum efficiency (QE) of

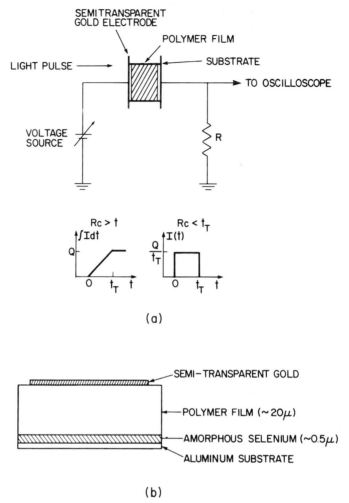

Figure 8.19 (*a*) Schematic of the experimental setup for the time of flight experiment. Figure shows the idealized signals in the voltage or open-circuit mode ($RC < t_T$). (*b*) Schematic of the multilayer structure employed to study charge transport in polymer with low QE of photogeneration.

photogeneration, F is the number of absorbed photons from the flash, and ΔQ is the number of charges in the drifting sheet.

Several assumptions are made in the derivation of Equations (8.48) and (8.49).

The number of carriers injected from the photogeneration region into the bulk is equal to the number of photogenerated carriers. In other words, after the photogeneration step, the carriers do not undergo loss processes such as free carrier recombination or loss at the surface. In the presence of these losses in the absorption region, the η in Equations (8.48) and (8.49) would be the QE of injection rather than the QE of photogeneration.

There is no loss of carriers due to bulk trapping as the injected sheet drifts across the film. This requires that the transit time $t_T = L/\mu_d E$ at the operating field E be much smaller than the trapping time τ_t.

The experiment is carried out under small signal operating conditions, that is, the charge in motion ΔQ is smaller than CV_0, the charge on the electrodes. This assures that all the carriers in the charge packet are drifting in a uniform field $E_0 = V_0/L$.

The contacts are blocking rather than ohmic. Blocking contact would assure that the photogenerated carriers are swept by the field only once across the sample and, therefore, only the "primary photocurrent" is measured. There is always a problem of excessive charge buildup and polarization in the samples having blocking contacts. These effects can be minimized by depolarizing the sample by shorting it and by exposing it to light, followed by dark "resting" the sample for extended periods between measurements.

The preceding experiments in which the charge traversing the sample is measured by integrating the signal are sometimes called the open circuit or voltage mode method of operation. The measured charge across the resistance would be less if the drifting sheet of carriers were to undergo loss by bulk trapping. This would be the case when the trapping time τ_t is smaller than the transit time t_T. The voltage signal under these conditions is given by Equations (8.50) and (8.51), (129).

$$\Delta V(t) = \frac{\eta e F}{C} \frac{\tau_t}{t_T} (1 - e^{-t/\tau_t}) \qquad \text{for } t < t_T$$

$$(8.50)$$

$$\Delta V = \frac{\eta F e}{C} \tau_t \frac{\mu E}{L} \qquad \text{for } t > t_T \quad (8.51)$$

The charge amplitude for $t > t_T$ is inversely proportional to the film thickness, as compared to the earlier case of no bulk trapping where the charge amplitude was independent of the film thickness. This test has been extensively employed to detect the presence of bulk trapping.

If, however, the same experiment is performed with a circuit employing a short time constant ($RC < t_T$), it is called the short circuit or current mode of operation, or the "time-of-flight method." The time-of-flight method is a powerful technique to measure drift mobilities in these materials. In the absence of bulk trapping, the current through the resistance R is given by Equation (8.52)

$$i = \frac{\Delta Q}{t_T} = e \eta F \frac{\mu_d E}{L} \qquad \text{for } t < t_T$$

$$(8.52)$$

$$= 0 \qquad \text{for } t > t_T$$

The current reaches its final amplitude in a short time (with a time constant RC) and stays constant at that value as the charge sheet is swept through the sample. At a time corresponding to the transit time t_T, the sheet reaches the counter electrode and the current drops to zero.

The drift mobility measured by this technique is to be distinguished from the microscopic mobility μ_0, which is the mobility of the carrier in a solid in the absence of any trapping events. This is related to the drift mobility by the expression shown as Equation (8.53):

$$\mu_d (n_0 + n_{t0}) = \mu_0 n_0 \qquad (8.53)$$

where n_0 and n_{t0} are the free and trapped carrier densities respectively. The significance of this expression is that, if the total carriers $n_0 + n_{t0}$ were to move with the drift mobility, then the conductivity would be identical to that observed when the free carriers move with the microscopic mobility. In the absence of trapping the drift and the microscopic mobilities are equal. In the presence of trapping the drift mobility is reduced. One can easily write Equation (8.54)

$$\mu_d = \mu_0 \left(\frac{N_c}{N_t}\right) \exp\left(-\frac{\Delta_t}{kT}\right) \qquad (8.54)$$

where N_c is the effective density of states in the conduction or valence band, N_t is the number of trapping levels, and Δ_t is the energy separation between the conducting and the trapping states. Clearly, these relations between the microscopic and drift mobilities are derived using concepts of charge transport through energy bands. Their validity in the case of amorphous molecular solids, where transport takes place through a network of localized states, is not straightforward. It would not be inconceivable to have trap-controlled hopping transport when there are two sets of hopping levels separated in the energy space. This has clearly been demonstrated by studying transport in solid solutions containing two transporting molecules with different ionization potentials.

It is difficult to perform time-of-flight measurements by direct photoexcitation of carriers in materials with a very small QE of photo-generation. Since most of the materials are UV absorbing, light flashes of high intensity in the UV region are required to photogenerate sufficient carriers to be able to detect the current. A simple device shown in Figure 8.19b has been very successfully employed in many instances to circumvent this problem. The material of thickness L is overcoated with a thin layer of a very sensitive photoconductor such as amorphous Se. This double-layer device is sandwiched between the substrate and the semitransparent electrode. In this case, the light flash is absorbed in the over-coating, and the photogenerated carriers are injected into the solid being studied. The drift of the charge sheet is time resolved.

A xerographic technique with underlying principles similar to those discussed in the previous section, is gaining acceptance as one of the most powerful ways of studying trans-port properties of disordered organic systems (129). In this method, the film on a conductive substrate is corona charged to a potential V_0 and is then exposed to an appropriate illum-ination source. In the absence of bulk trapping or loss due to recombination processes in the photoabsorption region, the rate of discharge in the emission-limited case is given by Equa-tion (8.55)

$$\frac{dQ}{dt} = C\frac{dV}{dt} = -\eta Ie\,\frac{t}{t_T} \qquad \text{for } t < t_T$$

$$C\frac{dV}{dt} = -\eta Ie \qquad \text{for } t > t_T$$

$$\text{(8.55)}$$

where η is the efficiency of photogeneration at the applied field E, and C is the capacitance of the device. As the name implies, the emission-limited discharge rate is governed by the rate of photogeneration. This condition requires that the total charge transiting the sample at any time be less than the surface charge. The efficiency of photogeneration can be deter-mined by measuring the rate of discharge for the film.

The preceding technique is also useful in determining charge carrier mobility and its field dependence. This experiment requires a light source capable of generating a free charge equal to the surface charge CV_0 in a time much shorter than the transit time. If the system under study is a very inefficient photo-conductor, the double-layer structure in which the organic film is overcoated with a thin layer of an efficient photocarrier generator is em-ployed. This discharge, when a CV_0 of free charge is injected into the bulk of the material, is called space-charge limited. It is the max-imum rate of discharge that can be obtained. Under these conditions the potential and the rate of decay of the potential are given by Equations (8.56) and (8.57) (130) as follows:

$$V(t) = V_0 - \frac{1}{(w+2)}\left(\frac{V_0}{L}\right)^{w+2}\mu_0 t$$

$$\text{(8.56)}$$

$$\dot{V}(t) = -\frac{1}{(w+2)}\,\mu_0\left(\frac{V_0}{L}\right)^{w+2} \qquad \text{for } t < t_T$$

$$V(t) = L\left(\frac{w+1}{w+2}\right)\left(\frac{L}{\mu_0 t}\right)^{1/w+1}$$

$$\text{(8.57)}$$

$$\dot{V}(t) = -\frac{L}{w+2}\left(\frac{L}{\mu_0}\right)^{1/w+1}\left(\frac{l}{t}\right)^{w+2/w+1}$$

$$\text{for } t > t_T$$

where the parameter w describes the electric field dependence of the mobility and is given by

$$\mu\,(E) = \mu_0 E^w$$

Under space-charge limited conditions, the rate of discharge plotted as a function of time is independent of time and is proportional to V_0^{w+2} for $t < t_T$. For $t > t_T$, the rate of discharge is independent of V_0 and is propor-tional to

$$t^{[-(w+2)/(w+1)]}$$

Expressions more general than those given by Equations (8.56) and (8.57) have been derived for discharge rates under space-charge perturbed conditions in which the charge injected at $t = 0$ is less than CV_0. The limitation of these expressions is that they assume that the electric field dependence of the mobility has a power law dependence rather than a very general functional dependence.

8.3.3 Photogeneration

Very little data are available concerning photogeneration for the majority of amorphous organic systems contained in the literature. The measurements that have usually been done have dealt with the detection of a photoconductivity value, which generally includes the photogeneration process, as well as the loss processes associated with transport, such as trapping and recombination. Measurements in which precautions have been taken to separate the photogeneration and the transport effects have been carried out on only a few systems. These include three diverse disordered organic systems: the polymer-poly(N-vinylcarbazole) (125, 131–133); charge transfer (CT) complex of poly(N-vinylcarbazole) and 2, 4, 7 trinitro-9-fluorenone (134); and a solid solution of a monomer-triphenyl amine in bisphenol-A-polycarbonate (135).

In these cases the thrust of the measurements has been to determine the absolute quantum efficiency of photogeneration and its electric field dependence. From the measured data in these systems, the photogeneration process can viewed as a two-step process. The first step involves the excitation of the electron from its ground state to its excited bound state, from which it can either lose energy by nonradiative recombination or by recombination processes producing fluorescence (or a combination of both), or thermalize into a "continuum state." The electron in this state is still under the coulombic influence of the hole, and the fraction of these hole-electron pairs is designated at ϕ_0. These pairs either recombine or dissociate under the combined influence of thermal excitation and applied electric field. The dissociation leads to photogeneration of free carriers. Assuming $g(r, \theta)$ as the initial distribution of electron hole pairs and $p(r, \theta, E)$ as the probability that an ion pair thermalized with an initial separation r and at angle θ with the electric field direction will escape initial or geminate recombination, the overall photogeneration efficiency is given as follows by Equation (8.58):

$$\eta(E) = \phi_0 \int p(r, \theta, E) g(r, \theta) \, d^3r \quad (8.58)$$

The measured electric field dependence of the generation efficiency in the three systems referred to earlier has been explained on the basis of this formulation with $p(r, \theta, E)$ as determined by Onsager (136).

In the Onsager formulation, the theory of geminate recombination reduces to the problem of Brownian motion in the presence of coulomb attraction and the applied electric field. With the appropriate boundary conditions, the Onsager solution to the probability of dissociation is given by Equation (8.59).

$$p(r, \theta, E) =$$

$$\exp(-A) \exp(-B) \sum_{n=0}^{\infty} \sum_{m=0}^{\infty} \frac{A^m}{m!} \frac{B^{m+n}}{(m+n)!} \quad (8.59)$$

where

$$A = \frac{e^2}{4\pi\epsilon\epsilon_0 kTr} \quad \text{and} \quad B = \frac{eEr}{2kT}(1 + \cos\theta)$$

with k as the Boltzman constant and T as the absolute temperature.

In explaining the photogeneration data in all these systems, it is assumed that the initial distribution of thermalized pairs is an isotropic δ function so that one can write Equation (8.60) as follows (137, 138):

$$g(r, \theta) = \frac{1}{4\pi r_0^2} \delta(r - r_0) \quad (8.60)$$

where r_0 is a characteristic thermalization distance. From Equations (8.58), (8.59) and (8.60) the expression for the escape (or generation) efficiency is given by Equation (8.61)

$$\eta(r_0, E) = \phi_0 \frac{kT}{eEr_0} \exp - \left(\frac{eEr_0}{kT}\right) \cdot$$

$$\sum_{m=0}^{\infty} \frac{A_0^m}{m!} \sum_{n=0}^{\infty} \sum_{l=m+n+1}^{\infty} \left(\frac{eEr_0}{kT}\right)^l \frac{1}{l!} \quad (8.61)$$

where

$$A_0 = \frac{e^2}{4\pi\epsilon\epsilon_0 kTr_0}$$

If one now defines a critical Onsager distance $r_c(T)$(139) as that distance at which coulomb energy is equal to kT, then one can write Equation (8.62)

$$r_c(T) = \frac{e^2}{4\pi\epsilon\epsilon_0 kT} \quad (8.62)$$

The first few terms of Equation (8.61) can be written as shown in Equation (8.63) (140)

$$\eta(r_0, E) = \phi_0 \exp(-r_c(T)/r_0) \left[1 + \frac{e}{kT}\frac{1}{2!} r_c E \right.$$

$$+ \left(\frac{e}{kT}\right)^2 \frac{1}{3!} r_c(r_c/2 - r_0) E^2$$

$$\left. + \left(\frac{e}{kT}\right)^3 \frac{1}{4!} r_c \left(r_0^2 - r_0 r_c + \frac{r_c^2}{6}\right) E^3 + \cdots \right]$$

$$(8.63)$$

The parameters that enter into the expression are r_0 (the initial distance between the oppositely charged carriers) and ϕ_0 (the fraction of the absorbed photons that result in thermalized pairs of oppositely charged carriers). The photogeneration data in the next few figures are explained in terms of the Onsager mechanism.

8.3.3.1 Poly(N-vinylcarbazole) (PVK)

The structure and absorption spectrum of PVK are shown in Figure 8.20 (131). PVK is a clear material, and the high absorption is mainly confined to the UV region of the spectrum. The easily identifiable bands in the regions 350–300, 300–270, and 270–250 nm are the first, second, and third singlet-excited states, respectively (141). Figure 8.21 shows the QE of the supply data for PVK films of four different film thicknesses (142). The crosses, circles, squares, and triangles are the measured supply efficiencies on a sandwich structure formed by an aluminum substrate overcoated with a PVK film and then vapor deposited with a semitransparent gold electrode. The number of holes transiting the sample is measured in the integrated mode of the time-of-flight technique when a microsecond flash of 2540 Å wavelength from a xenon flash is incident on the sample.

The independence of the measured points with film thickness indicates that bulk trapping is negligible (in the indicated applied electric field range) and that which is measured is, in fact, the supply efficiency. The measured supply efficiency is seen to be a strong function of the electric field and approaches a value of 0.025 at a field of 10^6 V/cm. The solid line is the Onsager dissociation efficiency computed from Equation (8.61) for $\phi_0 = 0.11$ and $r_0 = 26$ Å. The experimental results are in fair agreement with those predicted by the Onsager mechanism. The meaning of this fit is that only 11% of the absorbed light produces electron hole pairs, with an average separation of 26 Å at thermalization (other measurements have obtained values in the range of 26–30 Å). At finite fields, only a fraction of this 11% of these pairs dissociate, giving rise to free holes, and the rest recombine. The maximum photogeneration efficiency corresponding to the dissociation of all the electron hole pairs at an infinite electric field is 0.11.

Figure 8.22 shows the dependence of the measured photogeneration efficiency at 2×10^5 V/cm with the wavelength of the exciting radiation (132, 133). The measured efficiency increases in steps as the wavelength of the excitation is reduced. A correspondence of these discontinuities with the first, second, and third singlet excited states (compare Figure 8.20) is evident. Within each band the photo-

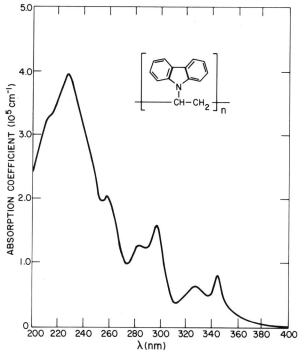

Figure 8.20 The structure and absorption spectrum of poly(N-vinylcarbazole) (PVK).

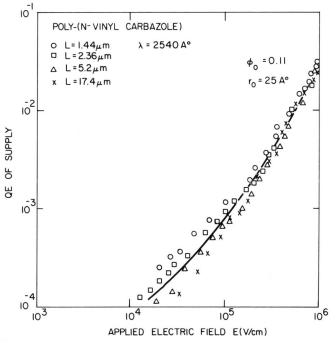

Figure 8.21 QE of supply versus electric field for PVK films of four thicknesses. The excitation wavelength is 2540 Å. The solid line is the theoretical computation from Equation (8.61) for $\phi_0 = 0.11$ and $r_0 = 26$ Å.

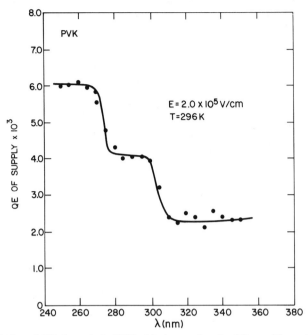

Figure 8.22 Variation of QE of supply in PVK with the wavelength of the exciting radiation. Reprinted from Reference 133 with permission of the American Institute of Physics.

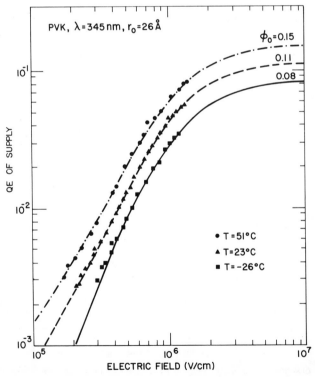

Figure 8.23 Temperature dependence of QE of supply in PVK at an excitation wavelength of 345 nm and for the fields indicated. Solid lines are the theoretical computation of Equation (8.61). Reprinted from Reference 133 with permission of the American Institute of Physics.

generation efficiency is independent of the exciting wavelength. By fitting the Onsager expression with the field dependence of photogeneration within each band, ϕ_0 is observed to be 0.11 and remains constant through the three regions. The stepwise increase in photogeneration efficiency through the three bands is caused by a stepwise increase in the value of the initial separation r_0. Figure 8.23 shows the temperature dependence of the photogeneration efficiency at an excitation wavelength of 345 nm and applied fields as indicated (133). This data, together with the electric field dependence of photogeneration at different temperatures, suggests that the thermalization distance r_0 remains invariant with temperature, whereas ϕ_0 is thermally activated with an activation energy of 0.05 eV. It should be pointed out that even in the absence of a temperature dependence of r_0 and ϕ_0, Equations (8.61) and (8.62) result in an activation energy for the Onsager quantum efficiency. The zero field activation energy under such conditions is given by $\Delta_0 = e^2/(4\pi\epsilon\epsilon_0 r_0)$. The activation energy predicted by Equations (8.61) and (8.62) decreases as the electric field is increased (140). The temperature independence of r_0 and the activation energy of 0.05 eV for ϕ_0 results in a zero field activation energy of $e^2/(4\pi\epsilon\epsilon_0 r_0) + 0.05$ eV for the photogeneration efficiency.

The fit of the Onsager mechanism with the measured photogeneration efficiencies deals only with the final step in the photogeneration process. The fit suggests that an electron hole pair thermalized with a separation distance of 26 to 30 Å (depending on the wavelength of the excitation) undergoes recombination as a result of coulombic attraction or dissociation into free carriers in the presence of the applied field. At room temperature only, 11% ($\phi_0 = 0.11$) of the absorbed light gives rise to electron hole pairs capable of producing free carriers. Eighty-nine percent of the absorbed light leads to excited bound states that decay back to the ground state by fluorescence or other nonradiative recombination processes. The application of the Onsager mechanism answers some questions, but raises several others. Which one of the many excited states—

excimer, exciplex, or charge transfer—participate in the dissociation process (143)? Since no electron transport has been observed in PVK, what happens to the electron? Is the electron trapped at an acceptor-type impurity such as residual solvent, dissolved oxygen, oxidized carbazole units, or other impurity? Are these impurities an essential part of the photogeneration process in that with their absence the photogeneration efficiency would be reduced well below 0.11? Assuming r_0 (obtained by fitting the Onsager expression to the measured data) to be a meaningful parameter, why is it independent of temperature?

The excited state that undergoes dissociation has been postulated by the following scheme for a system of PVK doped with small quantities of dimethyl terepthalate, an acceptor (144). A similar mechanism may be operative with PVK and other organic systems.

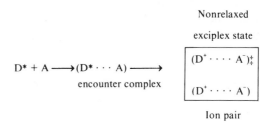

In this picture an encounter complex is produced when a singlet excited state of PVK migrating through the PVK chromophore interacts with an accepter A at comparatively large distances. The electron transfer from the excited donor D to the acceptor molecule takes place at this large distance, resulting in a nonrelaxed exciplex state. The excess kinetic energy of this state is then dissipated, and an electron hole pair at a separation distance r_0 is formed. The geminate recombination of this state gives rise to relaxed fluorescent exciplex states $D^+ A^-$ or else the dissociation of the ion pair gives rise to free carriers with the field dependence, as calculated by the Onsager expression. The recombination of the carriers in the relaxed exciplex state gives rise to exciplex fluorescence. The increase in dissociation of ion pairs with electric field leads to a corresponding quenching of the exciplex fluorescence.

In this scheme the independence of r_0 with temperature is explained by assuming that the thermalization involves emission of excess kinetic energy of phonon emission (145). The motion during the thermalization process is diffusive. The temperature dependence of r_0 is related to the temperature dependence of the diffusion constant. In crystalline materials the temperature dependence of the mean free path resulting from the fluctuations of the lattice density causes D to have an inverse temperature dependence. The thermalization distance which is proportional to $D^{1/2}$ is expected to have $T^{-0.5}$ dependence (146). However, in a disordered material, such as PVK, the mean free path is expected to be small. The effect of phonon scattering may also be small (140).

8.3.3.2 Charge Transfer Complex of PVK with 2, 4, 7 Trinitro-9-Fluorenone (TNF)

PVK photogenerates charge carriers when excited by radiation in the ultraviolet region of the spectrum. It is possible to make a photoconductor in the visible part of the spectrum by complexing the donor PVK with an acceptor such as TNF. The charge transfer (CT) complex reaction is given by the following scheme (147).

$$D + A \rightleftarrows D^+ - A^-$$

The bonding of the electronic ground state of these electron donor-acceptor complexes is controlled by dipole-dipole and dispersion forces with a partial transfer of charge from the donor to the acceptor. The small ionic contribution in the ground state is increased substantially when radiation is absorbed. A complete transfer of charge from the donor to the acceptor in the ground state results in the formation of ionic salts. The CT complexes are characterized by the appearance of wave-length bands longer than either of the starting materials. As shown in Figure 8.24, the absorption bands of PVK and TNF are in the UV region, and the formation of the complex results in a broadband absorption extending over the visible (148). By measuring the absorption coefficients of PVK-TNF films of various compositions, it has been shown that the absorption coefficient increases as the TNF concentration is increased and begins to saturate when the PVK-TNF concentration reaches 1:0.6. This property results from the variation in degree of complexing as the TNF concentration is increased. Its understanding is relevant for explaining the transport properties discussed in the next section. By performing absorption and electroabsorption mea-

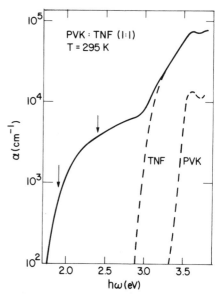

Figure 8.24 Solid curve shows the absorption spectrum of PVK-TNF. The dashed curves represent the absorption edges of a PVK film and TNF in a solution.

surements, is has been shown that a film of 1:1 PVK-TNF is actually a three-component film containing free TNF, uncomplexed carbazole, and CT complexes (149). The degree of complexing is the highest with the lowest concentration of TNF and varies from 72 to 40% when the PVK-TNF film composition is varied from 1:0.1 to 1:1.

As is the case in PVK, the efficiency of the photogeneration of free carriers in films of PVK-TNF is found to be a function of the electric field. The photogeneration process in the complex is viewed as a two-step process. In the first step, the absorbed radiation excites the electron from the ground state to some excited bound state. The electron may either thermalize and produce free carriers by dissociation from the coulombic influence of the opposite carrier, or the electron may decay back to the ground state. The process of dissociation and, hence, the electric field dependence of the photogeneration efficiency,

is explained on the basis of the Onsager mechanism.

Figure 8.25 shows the photogeneration efficiency in a film of 1:1 PVK-TNF when excited by a flash source of 5500 Å wavelength (150). Measurements were performed on a 15 μ thick film of the CT complex solvent cast on an anodized aluminum substrate. The experimental technique consisted of corona charging the film to a potential V_0 and exposing it to a strobe flash of 4 μ sec duration at 5500 Å wavelength. The voltage drop was measured by a capacitively coupled probe. The efficiency is given by $\eta(E) = C \, \Delta V / e F$ where F is the light flux in photons per cm^2. To assure small signal conditions, the light flux is adjusted so that the voltage drop V is a small fraction of V_0. This method of measurement differs slightly from the technique described in the previous section. In that procedure, the initial rate of discharge was measured when the corona-charged film was subjected to a

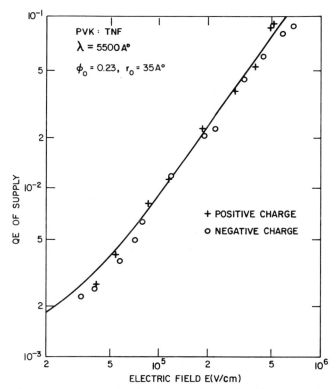

Figure 8.25   QE of supply versus electric field in a film containing 1:1 TNF to PVK molar ratio. The solid line represents the theoretical computation of Equation (8.61) for $\phi_0 = 0.23$ and $r_0 = 25$ Å. Reprinted from Reference 134 with permission of the American Institute of Physics.

light source of constant illumination. The measured efficiency is seen to be equal for both polarities of corona charging, indicating that an equal number of holes and electrons are photogenerated. The photogeneration efficiency is seen to vary sharply with the applied electric field. The solid line in the figure is the Onsager expression [Equation (8.61)] for $r_0 = 35$ Å and $\phi_0 = 0.23$. A factor of 5 reduction in QE occurs in going from 1:1 to 1.0:0.6 molar ratio PVK-TNF concentration. The Onsager computation of the photogeneration data for a 1:0.6 composition shows that ϕ_0 remains unchanged at 0.23 and that the reduction in efficiency is the result of a decrease in r_0 to 25 Å. These results are consistent with a scheme in which ϕ_0 is determined by local processes within a PVK-TNF complex. It is therefore not affected by the concentration of the complex, which varies according to the PVK–TNF molar ratio. However, r_0 is determined by the thermalization processes of the electron in the excited bound state and is, therefore, determined by the nature of the extended states. The increase of r_0 with the increase in the TNF concentration is not surprising.

8.3.3.3 Solid Solutions of Triphenylamine in Bisphenol-A-Polycarbonate

A third type of disordered organic system in which very accurate photogeneration data is available is a binary solid solution of triphenylamine (TPA) and bisphenol-A-polycarbonate (135). Films are prepared by dissolving well purified triphenylamine and polycarbonate in the appropriate ratios in a solvent, such as dichloromethane, and then coating a film on a conductive substrate. The residual solvent is removed by heating the samples at 100° C for approximately 48 hr under a nitrogen atmosphere.

The photogeneration measurements are carried out by corona charging the film and measuring the initial discharge rates when exposed to a UV source of 3000 Å wavelength that corresponds to the first singlet-excited state of TPA. The selection of this wavelength assures that the radiation is absorbed in a thin surface region. Therefore the photogenerated

holes discharge the film when positively corona charged.

Figure 8.26 shows the measured QE for films of four different thicknesses containing 40% TPA. The independence of the measured signal strength with film thickness indicates the absence of permanent bulk trapping of photogenerated holes in the course of their drift to the substrate. The measured efficiencies are very small and once again have a strong dependence with electric field. The solid line in the figure is the Onsager expression of Equation (8.61) for $r_0 = 26$ Å and $\phi_0 = 0.026$. The primary QE of 0.026 indicates that the maximum efficiency (at infinite electric field) obtainable with this system is 0.026. This low efficiency makes this a very poor photoconductor from the point of view of carrier photogeneration. However, these systems can be effectively employed to transport charge injected from an external source. This external source could be a thin film of an efficient organic or inorganic photogenerator of charge. The efficient photogenerator is overcoated onto the binary solid solution.

As the TPA concentration is varied from 15 to 45%, the primary QE, ϕ_0, varies from 0.014 to 0.03 and the thermalization distance r_0 varies from 22 Å to 27 Å. Both ϕ_0 and r_0 are independent of temperature in the range of 220 to 320° K. The agreement of measured photogeneration efficiency with the Onsager expression suggests that the photogeneration in this solid solution is also a two-step process described for the PVK-TNF CT complex. The increase of ϕ_0 with the concentration of TPA is to be contrasted with the independence of ϕ_0 observed for the CT complex when the concentration of TNF is increased. The existence of preferred sites for pair formation has been advanced as a reason to explain the increase of ϕ_0 with TPA concentration. These preferred sites could be surface locations or possibly those having a particular TPA-polycarbonate or TPA-TPA configuration. The efficiency of intermolecular energy transfer to these sites increases with the concentration of the TPA. The increase in r_0 with TPA concentration results from the thermalization process of the bound excited electron. If the thermalization

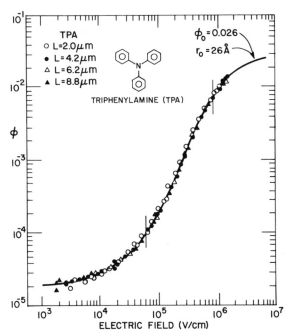

Figure 8.26 QE of supply versus electric field for films containing 40% TPA in polycarbonate. Excitation wavelength is 3000 Å. The solid line represents the theoretical computation of Equation (8.61) for $\phi_0 = 0.026$ and $r_0 = 26$ Å. Reprinted from Reference 135 with permission of the American Institute of Physics.

process is diffusive, the increase in r_0 results from the increase in the diffusion constant with the TPA concentration.

8.3.4 Charge Transport

The quantities of interest in the study of electronic transport are the mobility of the carriers and their lifetime against deep trapping and recombination. The insulating nature of the materials under consideration, coupled with their ability to easily form thin films, make them ideal candidates for studying charge transport by way of the application of electroded time-of-flight and xerographic space-charge-limited current (SCLC) techniques. Extensive time-of-flight and SCLC measurements have been conducted on both PVK (125, 131, 150) and CT complexes formed by doping PVK with various concentrations of TNF (151). Molecularly doped polymers, such as TPA (152, 153), *N*-isopropyl carbazole (NIPC) (154), and substituted triphenylmethanes (TPM) (155) in polycarbonate have also been studied. In all

these systems, the charge carrier sheet can either be photogenerated directly in the material or injected from an external source such as a metal electrode or an overcoated sensitive photoconductive layer. Unlike the inorganic crystalline state, where charge transport occurs through bands, the charge transport in an organic disordered state is characterized by charge hopping through localized states. In the case of the molecularly doped systems, these localized states are associated with the dopant molecule. This is easily established if one recognizes that essentially no charge transport is observed in undoped polycarbonate films. The charge transport is observed once the polycarbonate is doped with TPA, NIPC, or TPM, and so on. The charge carrier mobility is observed to increase sharply with the dopant concentration.

Charge transport polymers, such as PVK, are characterized by saturated polymer chains that ensure low dark conductivity. The photoexcitation is associated with large pendant planar structures with extended π-electron systems. Weak interactions of the π-electron systems along saturated chains make them analogous to the solid dispersions of similar monomer units in an inert binder. By comparing the charge transport in films of polycarbonate doped with NIPC to the charge transport in PVK, it has been established that the vinyl backbone is not essential for charge transport. The absorption spectrum of PVK in the solid state is also seen to be remarkably similar to that of the free molecule (141).

The hopping mechanism through localized states associated with the molecule or the pendant group can be visualized as an oxidation-reduction reaction between the radical cation and the neutral molecule in an applied field. Consider a film of a donor molecule D, such as TPA or NIPC, dispersed in an inert matrix, such as polycarbonate. Injection of a hole from a sensitizer S involves the transfer of an electron from the donor molecule creating a cation radical.

$$\overset{e}{\overbrace{S D}} D D \longrightarrow S^- \overset{e}{\overbrace{D^{\ddagger} D}} D \longrightarrow S^- D D^{\ddagger} D$$

Further migration of the hole towards the negative electrode in the applied field involves a series of one-electron oxidation-reduction reactions of the type that follows.

$$D \xrightarrow{-e} D^{\cdot +} \quad \text{(oxidation)}$$

$$D^{\cdot +} \xrightarrow{+e} D \quad \text{(reduction)}$$

A parallel set of reactions occurs for charge transport through acceptor-type molecules. These reactions suggest that only hole transport can be observed in films of molecularly doped polymers in which the dopant is an electron donor. However, electron transport is to be expected if the films are prepared with an acceptor-type dopant such as TNF.

A distinct feature of transport in the disordered organic solid state is the dispersion observed when a sheet of photogenerated charge drifts in an applied electric field. The fluctuations in the positional and orientational disorder lead to a wide distribution of hopping times owing to strong localization of the charge carrier. This wide distribution results in some features that are vastly different from the conventional Gaussian case observed in crystalline solids (156).

In the Gaussian case, the dispersion σ of the carrier sheet and the mean displacement l from the illuminated or injecting surface obey the relation $\sigma \propto t^{1/2}$ and $l \propto t$. For times shorter than the transition time t_T, this Gaussian spreading does not manifest itself in the current trace of the time-of-flight technique. Rather, the dispersion manifests itself as a tail of the same width as the dispersion when the mean of the charge distribution reaches the substrate. It is found experimentally that Gaussian statistics are not applicable to charge dispersion in a disordered organic state. The experimental transient current traces demonstrate a significant spreading of the hole packet as it drifts through the film. The dispersion and the mean displacement obey the same time dependence, that is $\sigma/l =$ constant as opposed to $\sigma/l = t^{-1/2}$ for Gaussian statistics.

The first theoretical framework to explain the non-Gaussian dispersive transport used the formulation of continuous time random walk (CTRW) (157). In the CTRW theory the distribution of events or hopping times is assumed to be a slowly varying algebraic function $\psi(t) \sim t^{-(1+\alpha)}$ where the disorder parameter α assumes a value between zero and unity. In comparision, Gaussian transport is characterized by a single event time τ and, therefore, $\psi(t) \sim e^{-t/\tau}$. The time dependence of the the transient current shape in the time-of-flight experiment under the assumption of algebraic event time distribution conditions is given by Equation (8.64) as follows:

$$i(t) \sim \begin{cases} t^{-(1-\alpha)} & \text{for } t < t_T \\ t^{-(1+\alpha)} & \text{for } t > t_T \end{cases} \tag{8.64}$$

with the transit time given by the following expression shown as Equation (8.65):

$$t_T = \left[\frac{L}{l(E)} \right]^{1/\alpha} \exp\left(\frac{\Delta_0}{kT} \right) \tag{8.65}$$

where L is the thickness, Δ_0 is an activation

energy, and $l(E)$ is the mean displacement per hop in the electric field direction. These expressions reveal some unusual features for the transient current shape and for the thickness and electric field dependence of t_T. On a log i-log t plot, the transient current gives rise to two straight lines. The sum of the power exponents describing the time dependence of the current for times shorter and longer than the transit time t_T is 2. The transient pulse under Gaussian spread conditions, however, is essentially a rectangle with a slight tail that forms when the charge carrier packet arrives at the counter electrode. This transit time results in a mobility that is independent of film thickness. For the non-Gaussian case, the shape of the transient current, given by Equation (8.64), and the thickness and the field dependence, predicted by Equation (8.65), are related by the disorder parameter α.

The use of an algebraic distribution function for hopping times may be very restrictive. More complicated functions may be required to explain the features over broad material and experimental ranges. Other formulations based on a generalized multiple trapping model (158, 159) and percolation theory (160) have also been proposed to explain non-Gaussian transport phenomena. Some of the time-of-flight data to be presented in the next few pages are obtained by the traditional method where the transient current is displayed on a scope as a linear i versus t scale. The transit time, in this case, is characterized by a kink or a fiduciary point in the transient current pulse. The data on solid solutions of TPA in polycarbonate (152) and mixtures of TPA and NIPC in polycarbonate (161) are obtained by a log i-log t display. The point at which the current shows a sharp change in slopes determines the transit time. The latter method is especially useful when the transit current pulse in a linear i-t display is very dispersive, and a fiduciary point is not easily identified.

8.3.4.1 PVK and Charge Transfer Complexes of PVK and TNF

Figure 8.27 shows the electric field dependence of the hole mobility in a series of PVK-

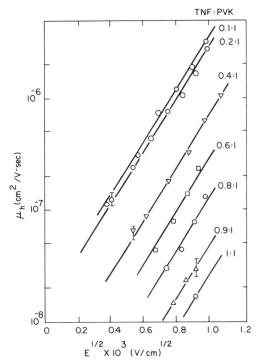

Figure 8.27 Logarithm of hole mobility as a function of the square root of the applied electric field for a range of TNF:PVK molar ratios. Reprinted from Ref. 151 with permission of the American Institute of Physics.

TNF films of varying compositions (151). The carrier drift mobilities are low in comparison to crystalline solids. The mobilities increase sharply with the electric field and obey the empirical relationship shown as follows as Equation (8.66):

$$\mu = \mu_0 \exp -\frac{(\Delta_0 - \beta E^{1/2})}{kT_{\text{eff}}} \quad (8.66)$$

where

$$\frac{1}{T_{\text{eff}}} = \frac{1}{T} - \frac{1}{T_0}$$

and the prefactor μ_0, the activation energy Δ_0, and a temperature T_0 are obtained by fitting the field and temperature dependence to the measured data. β is 2.7×10^{-5} eV(V/m)$^{-1}$ and is found to be the same for films of all compositions. As discussed in the photogeneration section, PVK-TNF films have been shown to be three component films containing

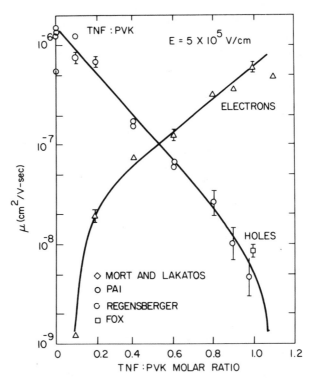

Figure 8.28   Hole and electron drift mobilities as a function of TNF-PVK molar ratio. Reprinted from Reference 151 with permission of the American Institute of Physics.

free TNF, uncomplexed carbazole, and CT complexes (149). It has been shown that transport in PVK is dominated by holes, whereas, in TNF it is dominated by electrons. This is in agreement with the explanation based on an oxidation-reduction reaction of charge transport that was described before. PVK is a donor and TNF is an acceptor. In PVK-TNF films both holes and electrons are observed to be mobile. Figure 8.28 shows the variation of hole and electron drift mobilities with the PVK-TNF molar ratio (151). The important conclusion to be drawn from the figure is that hole transport is associated with uncomplexed carbazole, while electron transport is associated with both free and uncomplexed TNF. The extremely low value of the mobility, coupled with the steep electric field dependence, indicates a high degree of localization. It also shows that the charge transport proceeds by way of an intermolecular hopping process. The reduction of the hole mobility with increasing TNF concentration results from the increase in the average

distance between the monomer units of uncomplexed PVK.

The empirical relationship of Equation (8.66) obtained by fitting the experimental electric field dependence to the measured data suggests a Poole-Frenkel (PF) mechanism (151). The PF effect deals with a carrier trapped in a coulombic barrier (162). The applied field causes a reduction of $\beta_{PF} E^{\frac{1}{2}}$ in the barrier height, and therefore the probability of escape is increased in the presence of the applied field. The slope β_{PF} predicted by the theory is $[e/(\pi\epsilon\epsilon_0)]^{\frac{1}{2}}$. The calculated value is 4.08×10^{-5} for an ϵ of 3.5 as compared to the experimental value of 2.72×10^{-5}. The PF theory can be improved by incorporating both the details of the shape of the coulomb potential and the microscopic diffusion mechanism of the charge escape (162). It is then possible to derive an expression whose slope is lower than β_{PF} and which is therefore in a better agreement with the measured β.

Several models have been proposed (151) to explain the origin and details of the PF mechanism operative in these materials. One is the traditional trap-controlled drift mobility mechanism where the trap depth is modulated by the applied field. This would require a large number of charged coulombic centers to maintain the neutrality of the charge. Since the slope of the field dependence remains invariant with the differing ratios of PVK and TNF, it is difficult to rationalize the presence of these charged centers in materials ranging from PVK through the CT complex of PVK-TNF at all molar ratios to pure TNF.

Another mechanism that has been proposed is that the carriers move as small polarons (163). A small polaron is a carrier that is "self-trapped" in a well created by the lattice distortion. This lattice distortion is formed when a carrier stays sufficiently long in a position to polarize the medium around it. The applied field can lower the polaron barrier in a PF fashion and increase the mobility.

The polaron transport model is attractive in that the mobilities in this mechanism are not critically dependent on the sample preparation.

8.3.4.2 *TNF Dispersed in Polyester*

Since the charge transport through the PVK-TNF films is determined by electron hopping from one localized TNF state to another (complexed or free), it is not surprising to find that substituting an inert binder (polyester) for PVK would not affect the electron mobilities. This assumes, of course, that molecular dispersion is not affected when PVK is replaced by polyester. Figure 8.29 shows the strong concentration dependence of electron mobility as a function of the average distance R between TNF molecules in TNF-polyester films (164). The average distance is calculated from molecular weight and Avagadro's number (assuming a uniform dispersion). The

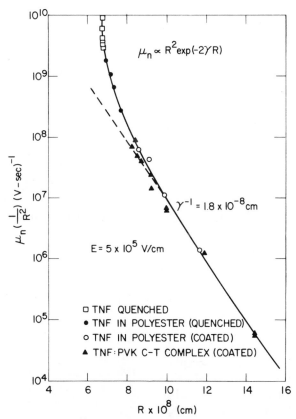

Figure 8.29 Electron mobility as a function of the average separation of TNF molecules in TNF-polyester and TNF-PVK films. Reprinted from Reference 151 with permission of the American Institute of Physics.

polyester used in this case is Goodyear Vitel PE200. Thin films of TNF dispersions in polyester were coated from tetrahydrofuran solutions onto aluminum substrates. Films containing higher concentrations of TNF are found to crystallize when fabricated in this fashion and were therefore prepared by quenching from the melt. The charge carrier sheet for the time-of-flight technique was generated by a high-intensity pulse of UV light. Figure 8.29 also shows the concentration dependence of the electron mobility of PVK-TNF films. There is excellent agreement between the two sets of measurements. The electron mobility in TNF-polyester films also has a strong electric field dependence, with a large zero field thermal activation energy ($\Delta_0 = 0.7$ eV) that decreases exponentially with field.

The strong dependence of μ_n/R^2 versus R indicates that the charge transport occurs by way of an intermolecular hopping transport through localized states. The extent of the overlap of the wavefunction between neighboring dopant molecules is given by the exponent γ in Equation (8.67).

$$\mu_n = R^2 \exp(-2\gamma R) \exp\left(-\frac{\Delta}{kT}\right) \quad (8.67)$$

A value of $\gamma^{-1} = 1.8 \times 10^{-8}$ cm is obtained for electron transport in films containing a low TNF concentration. A steep departure from the exponential is seen when the TNF concentration is increased. The average distance R in films containing large concentrations of TNF is seen to be comparable to the size of the TNF molecule. The simple expression derived on the basis of pointlike localized states is not expected to be accurate under these conditions.

8.3.4.3 Triphenylamine (TPA) in Polycarbonate

In its neutral state TPA is a donor molecule, as opposed to TNF which is an acceptor. Therefore, based on the oxidation-reduction scheme, hole transport is to be expected in films containing TPA dispersed in a binder. The absence of complex formation is proved by the study of the absorption spectra of films

by TPA, polycarbonate, and films of the solid solution. The field and temperature dependence of the hole mobility for a film of 0.5:1 (by weight) composition is shown by the circles in Figure 8.30 (165). The solid line is a phenomenological expresion that best describes the data and is given by Equation (8.68)

$$\mu_h(E) = L^{-1/\alpha} \sinh\left(\frac{eRE}{2kT}\right)^{1/\alpha} \exp\left(-\frac{\Delta_0}{kT}\right)$$

$$(8.68)$$

where R is the intersite distance between TPA molecules, Δ_0 is an activation energy, and α is the disorder parameter obtained by analyzing the shape of the transient current on the log i-log$t$ (CTRW) plot. This is given by Equation (8.69)

$$\alpha = \frac{\alpha_0}{1 - (T/T_0)} \quad (8.69)$$

T_0 therefore connects the temperature dependence of $i(t)$ with the field dependence of the mobility. The signals get more dispersive (lower α) as the temperature is lowered, and therefore the expression predicts a steeper field dependence of the mobility.

An extensive evaluation of the field, temperature, and concentration dependence showed the following characteristics:

The hole mobility is thermally activated with an activation energy Δ that increases with decreasing TPA concentration. At 70 V/μ, Δ increases from 0.32 eV to 0.49 when the TPA-PC concentration is varied from 0.5:1 to 0.1:1 by weight of TPA to PC.

The activation energy Δ decreases with the applied field and is given by $\Delta = \Delta_0 - \beta E$.

The hole mobility has a strong concentration dependence. The mobility decreases from 10^{-5} cm^2/V-sec to 3×10^{-9} cm^2/V-sec when the concentration is reduced from 50 to 10%. An μ_h/R^2 versus R plot results in a value of $\gamma^{-1} = 1 \times 10^{-8}$ cm for the localization radius γ^{-1} at $E = 50$ V/μ.

The localization parameter for hole trans-

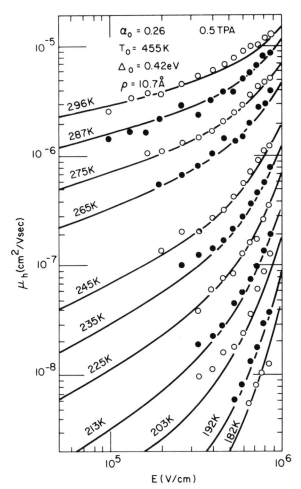

Figure 8.30 Electric field dependence of the hole drift mobility in a TPA-polycarbonate film at various temperatures. The solid line represents the theoretical computation of Equation (8.68) for the listed parameters. Reprinted from Reference 152 with permission of the Physical Review.

port increases from 1×10^{-8} cm at 296° K to 0.71×10^{-8} cm at 255° K.

The explanation for these phenomena is based on the sensitivity of the overlap term for the relative orientation of the molecules. The rotation of the molecule is temperature dependent, and therefore the overlap term would be expected to decrease as the temperature is decreased.

8.3.4.4 Mixed Systems of TPA and NIPC in Polycarbonate

There has been some discussion concerning trapping as a result of conformational considerations. It is conceivable that hopping from one molecule to its nearest neighbor is

difficult owing to the relative orientation of the molecule. It might be easier for the charge to hop to a molecule several molecular distances away but favorably situated as regards orientation (166). Since the molecular rotations are thermally activated, this type of "conformational" trapping is a temperature-dependent phenomenon. A second type of trapping that has been identified is impurity related. To be a trap in this case, the ionization potential of the impurity must be smaller than the ionization potential of the host molecule. This phenomenon is best illustrated in Figure 8.31 in which hole mobility in films containing TPA, NIPC, and polycarbonate is plotted (161, 165). NIPC is a donor with an ionization potential slightly higher than that of TPA.

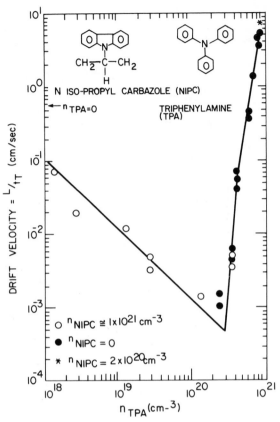

Figure 8.31 The hole drift mobilities (circles) as a function of TPA concentration in films containing TPA and NIPC dispersed in polycarbonate. The NIPC concentration is held constant at $10^{21}/cm^3$. Solid circles represent hole drift in films of TPA dispersed in polycarbonate ($n_{NIPC} = 0$). Reprinted from Reference 152 with permission of the Physical Review.

The solid circles represent the mobility values of films containing only TPA dispersed in polycarbonate.

A strong exponential concentration dependence of the type discussed earlier is seen for these films. The open circles represent mobility values at $E = 50$ V/μ for a three-component system containing TPA, NIPC, and polycarbonate. In this system, the concentration of NIPC was constant at 1×10^{21} molecules/cm^3. The drift velocity when $n_{TPA} = 0$ for this system is also indicated in the figure. The hole mobility in films of NIPC-PC (with NIPC $= 10^{21}/cm^3$) is seen to decrease when small quantities of TPA ($n_{TPA} \sim 10^{18}$–$10^{19}/$ cm^3) are present in the film. The value for hole mobility reaches a minimum when $n_{TPA} = 3 \times 10^{20}/cm^3$. For higher concentrations of TPA, the mobility values are identical to those obtained in TPA-PC films (solid circles). In

other words, in this region ($n_{TPA} > 3 \times 10^{20}/$ cm^3), NIPC is inactive and acts as part of the binder. The variation of the mobility with n_{TPA} at a fixed concentration of n_{NIPC} ($10^{21}/cm^3$) is explained as follows.

In films containing no TPA, the charge transport occurs by hopping between localized states associated with NIPC. Since the ionization potential of TPA is lower than that of NIPC, when small concentrations of TPA are introduced in the film, the holes hopping through NIPC states can get trapped at the TPA sites. As a result of the low concentrations of TPA, the overlap between TPA molecules is small. The charge residing at TPA sites has to be released to the NIPC sites, and this trap release has to overcome an energy barrier $\Delta \approx (I_{PNIPC} - I_{PTPA})$. I_{PNIPC} and I_{PTPA} are the ionization potentials of the NIPC and TPA molecules respectively. This barrier needed to release the charge from the TPA sites to the NIPC sites is in addition to the barriers (activation energy) associated with charge transport through the disordered materials. At intermediate concentrations of TPA, the charge trapped at TPA sites can either be released to the NIPC sites, overcoming the trap barrier, or can hop through the TPA sites. At high concentration of TPA, the overlap between TPA molecules is large. The TPA-TPA hopping dominates the transport, and, therefore, NIPC in these films is essentially "inert." These experiments demonstrate the power of the organic disordered state created by molecular doping in understanding the microscopic transport phenomena. The concept of molecular doping provides opportunities for independently controlling the generation, transport, and mechanical properties of the organic solid state.

8.3.4.5 Substituted Triarylmethanes in Polycarbonate

Structural dependence of charge transport through an organic disordered state has been studied in a system consisting of a solid solution of substituted triarylmethanes in polycarbonate (155). The structure of the triarylmethane (TPM) is shown in Figure 8.32. The time-of-flight measurements performed on TPM in polycarbonate have all the general features of transport in molecularly doped systems. The inverse transit time can be expressed by the relationship shown in Equation (8.70)

$$\frac{1}{t_T} = \text{constant} \left[R^2 \exp\left(-2\gamma R\right) \right]$$

$$\left[g\left(\frac{E}{kT}\right) \right] \exp\left(-\frac{\Delta}{kT}\right) \quad (8.70)$$

where the parenthetic term describes the concentration dependence, the second term describes the electric field dependence, and the second and third parenthetic terms describe the field dependence. The electric field dependence of the mobility varies between linear and square over most of the field regions in which the fiduciary point is easily recognized on the linear i-t plot for the time-of-flight signal. The electric field dependence of the

X = H

(a)

(b)

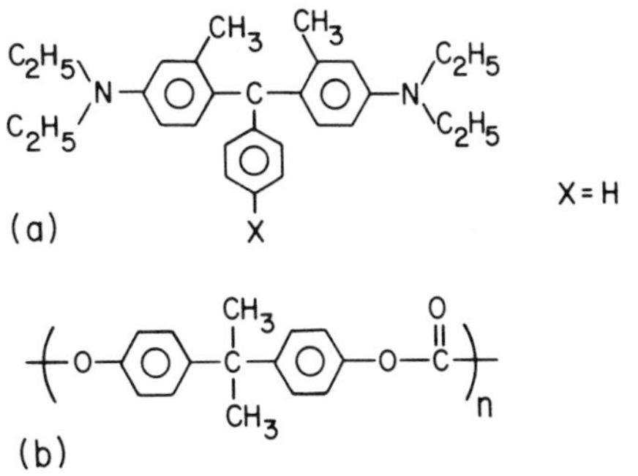

Figure 8.32 Molecular structures of substituted triarylmethanes and bisphenol-A-polycarbonate.

mobility is similar to that of PVK and for other molecularly doped systems. As with other molecularly doped systems, μ / R^2 is seen to vary exponentially with intersite distance R with a localization radius of $\gamma^{-1} = 1.72 \times 10^{-8}$ cm. This functional dependence of the mobility on concentration is consistent with a hopping process ensuing from an overlap of the wavefunctions of the neighboring sites. On the basis of this model, the localization radius of TPM is larger than the value reported for other donors, such as uncomplexed carbazole in the CT complex of PVK and TNF (164).

An attempt has been made to correlate the localization radius with the ionization potential I_p in the case of hole transport in other systems (167). Assuming a square well potential, the localization radius is related to I_p by the relationship shown as Equation (8.71),

$$\frac{\gamma^{-1}}{2} = \frac{2}{h} \sqrt{2mI_p} \quad (8.71)$$

where h is Planck's constant. From the concentration dependence of mobility for the uncomplexed carbazole (in the CT complex of PVK and TNF), the localization radius γ^{-1} is found to be 0.76 Å. This gives an ionization potential for carbazole of 6.6 eV and is a reasonable value. However, a localization radius of $\gamma^{-1} = 1.72$ Å for TPM results in an ionization potential of 1.3 eV, which is obviously too low.

However, the most interesting result with TPM is the dependence of charge carrier mobility on the structure of TPM (155). Figures 8.33 and 8.34 show the variation of

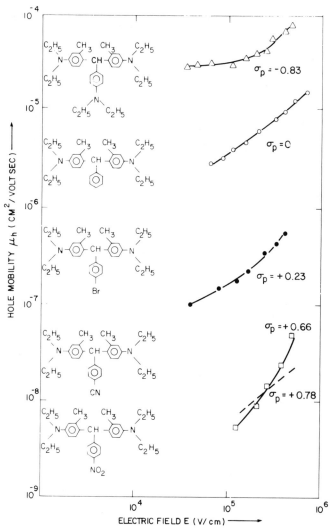

Figure 8.33 Effect of substituent groups $N(C_2H_5)_2$, Br, CN, and NO_2 on hole mobility of triarylmethanes dispersed in polycarbonate.

mobility of the films as the substituent x in the para position of the TPM is varied. The substituents $N(C_2H_5)_2$, OH, OCH_3 are electron donating and have Hammett's constant σ_p of -0.083, -0.37, and -0.27, respectively. The electron-withdrawing groups Br, CN, and NO_2 have positive Hammett's constant of 0.23, 0.66, and 0.78, respectively. The mobility results can be rationalized in terms of the molecular structure of the triarylmethanes, that is the substituent effect.

The molecular frameworks of these triarylmethanes differ only in the substituents in the p-position of one phenyl ring. If it is assumed that the transport mechanism in-

volves the exchange of a hole between an oxidized and a neutral triarylmethane molecule, then one can write the following reaction.

$$(TPM)^{+}_{a} + (TPM)_{b} \longrightarrow (TPM)_{a} + (TPM)^{+}_{b}$$

The ease of formation and the stability of the cation radical is enhanced by the presence of electron-releasing groups. It is hindered by the substitution of electron-withdrawing groups. The magnitude of these effects is related to the Hammett's constants. The mobility values of Figure 8.33 are consistent with the predictions based on the Hammett equation. However, the mobility values for the

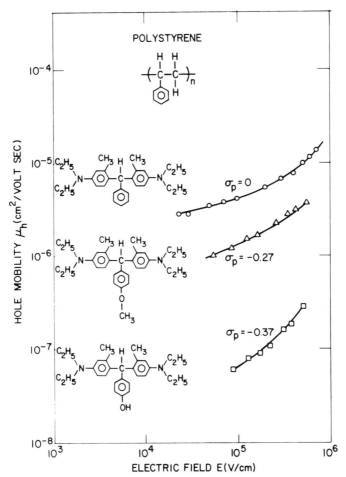

Figure 8.34 Effect of substituent groups OCH₃ and OH on hole mobility in triarylmethanes dispersed in polystyrene.

substituent OCH₃ and OH are not consistent with this theory. Since both OCH₃ and OH are electron releasing (based on Hammett's equation), the hole mobility should have been higher than that measured when the substituent is H. Instead, the measured mobilities are lower. These results with OH can be rationalized if one assumes hydrogen bonding between OH substituent and the carbonyl group of the polycarbonate (Figure 8.32b). However, measurements performed on films of these compounds dispersed in polystyrene (Figure 8.34) instead of polycarbonate give mobility values that are inconsistent with the Hammett prediction. The OH group is not expected to form any hydrogen bonding with polystyrene. Self-bonding between the OH groups of the substituted TPM may explain

this anomaly. These types of studies are at too rudimentary a stage to make definitive conslusions at this time. Factors such as molecular ionization potential, molecular size and shape, and the electron delocalization in the neutral and oxidized molecules are all important.

It has been tacitly assumed in these and other related studies on molecular dispersions in polymeric binders that the solutions are ideal. That is, the active molecules are distributed in a random array separated by an inert matrix with the average molecular separation depending only on concentration. This picture is obviously idealized. Even within the same molecular class, different substituent groups will affect the solubility characteristics with a given matrix polymer. In fact, Froix and co-workers (168) have shown that, in solid solutions of bisphenol-A diphenyl carbonate in polycarbonate, although the solutions appear homogeneous on a calorimetric, or an nmr T_1 time scale, they show phase separate structure on a spin-spin (T_2) nmr time scale. There is much to be learned about the solid state structure of these layers before any meaningful molecular and structural correlations with electronic transport can be realized.

8.4 SUMMARY

What is the future for the field of dielectric processing? With the advent of microprocessing, the capabilities of the experimentor using dielectric or polarization techniques are expanding. What is considered a difficult analysis can now be programmed and has become commonplace. Such analytical capability lends itself to the novice entering the field or desiring to use its equipment without an in-depth knowledge of how it functions. It would be hoped that programmable packages or equipment will be available in the near future for the very-high ($>10^6$ Hz) and very-low ($<10^0$ Hz) frequency ranges. With the advent of such equipment, research in the area of biopolymers will become more important, especially in the understanding of how molecular mechanisms of motion control the transport properties of the biological system. The use of sophisticated dielectric equipment to control chemical and physical processes will become more routine. It may also be in the realm of feasibility to approach our understanding of simple systems in such a way as to clarify the interpretation of data from other techniques. The realm of possibility is immense and will require only the inventive and scientific means to enrich the future.

Broad principles of charge generation and transport have been described for representative examples of well characterized, disordered, organic insulators. No attempt has been made to catalog the innumerable polymers that have been synthesized and studied in a cursory fashion. Rather, we have limited this chapter to the discussion of systems that

photogenerate and transport charge rather than systems that conduct in the dark. The category of conducting materials includes conjugated polymers in which the conductivity results from extensive delocalization of the charge along the polymer backbone. For charge transporting polymers that have insulating properties in the dark, the backbones are saturated with large planar pendant structures that have extended π-electron systems and are easily photoexcited. Although this book deals primarily with the properties of organic polymers, we have tried to generalize the chapter to include systems in which inert insulating polymers are doped with charge transporting organic molecules. They have the technological potential of providing a method for independently selecting desired photogeneration, charge transport, and mechanical properties.

Polymers have traditionally been known for their insulating properties. The application of organic photoconducting systems in electrophotography has provided the necessary motivating force for a sustained research effort in their photogenerating and charge transporting properties. It is now recognized that the range of electrical properties in the disordered organic state is large. In this chapter the photogeneration and charge transport have been discussed in three diverse types of disordered organic systems—a polymeric system, a charge transfer complex, and a solid solution of a monomer in an inert binder.

The photogeneration process has been explained as involving two steps. In the first step the absorbed photon excites the electron from its ground state to an excited bound state. From the excited bound state the electron may either undergo thermalization and dissociation or geminate recombination. The dissociation process has been shown to be adequately described by the Onsager formulation. The basis for the field dependence of QE in this formulation is the coulombic attraction due to the proximity of the oppositely charged carriers following the photoexcitation and thermalization steps. The closer the oppositely charged carriers find themselves, the higher is the probability of recombination and the lower the QE of photogeneration. Since the motion of the carriers before thermalization is governed by diffusion, the low initial separation r_0 (and hence low QE) between the carriers is the result of low values for the diffusion constants in these materials. The diffusion constant is related to the charge carrier mobility through the Einstein relation. Controlled and more sophisticated experiments on well purified samples must be carried out to determine the steps involved between the photoabsorption and the photodissociation and to shed further light on the nature of the photoexcited state that undergoes dissocation.

By employing molecularly doped polymers, it has been clearly demonstrated that charge transport involves hopping between localized states associated with the molecule. The charge transfer between molecules broadly involves the one electron oxidation-reduction phenomenon so well established in chemistry. However, the variation of the mobility with electric field and the origin of the large activation energy for charge transport are issues that are still speculative. Experimental and theoretical progress in these areas should come rapidly in the next decade, with a fundamental understanding of the processes involved and the chemical methods of controlling them.

8.5 GLOSSARY OF SYMBOLS

α	Polarizability
P	Polarization
σ	Charge density per unit area, conductivity
A	Area
E	Electric field strength
D	Dielectric displacement, diffusion constant
ϵ or ϵ'	Dielectric constant
ϵ''	Dielectric loss
l, L, r	Thickness
ϵ_0	Permittivity constant (8.25×10^{-12} F/m)
C	Capacitance
V	Voltage
ρ	Density
M	Molecular weight

μ	Dipole moment
N	Avogardo's number
t	Time
δ	Phase angle
T	Temperature
τ	Relaxation time, recombination time of carriers
R	Resistance
ω	Angular frequency
E_a	Activation energy
k	Boltzman constant
T_g	Glass transition temperature
W	Weight fraction
I	Intensity
n	Carrier density
g_0	Rate of production of mobile carriers
S	Capture cross section
v	Thermal velocity
D	Diffusion constant
r	Distance
Δ_t	Trap depth
ν	Frequency
e	Electronic charge
μ, μ_d, μ_h	Mobility microscopic, drift, hole
Q	Charge
F	Light flux
N_c	Density of states
η	Efficiency of photogeneration
I	Light intensity
ϕ_0	Parameter in Onsager equation
I_p	Ionization potential
t_T	Transit time
τ_r	Trapped carrier release time
τ_t	Free carrier trapping time
n_t	Trapped carrier density
g_0	Rate of production of mobile carriers

8.6 REFERENCES

1. D. Halliday and R. Resnick, *Physics*, John Wiley, New York, 1961.

2. P. Debye, "Polar Molecules,"*Chemistry Catalog, New York, 1929.*

3. J. M. Pochan, J. E. Kuder, J. Y. C. Chu, D. Wychick, and D. F. Hinman, *Can. J. Chem.*, **53**, 578 (1975).

4. E. A. Guggenheim, *Trans. Fararday Soc.*, **45**, 714 (1949).

5. J. E. Kuder, W. W. Limburg, J. M. Pochan, and D. Wychick, *J. Chem. Soc., Perkin Trans. II*, 1643 (1977).

6. L. Onsager, *J. Amer. Chem. Soc.*, **58**, 1486 (1936).

7. J. F. Kirkwood, *J. Chem. Phys.*, **7**, 911 (1939).

8. N. S. McCrum, B. E. Read, and G. Williams *Anelastic and Dielectric Effects in Polymeric Solids*, New York, 1967.

9. A. D. Jenkins, Ed., *Polymer Science*, American Elsevier, New York, 1972, p. 1285.

10. D. W. Van Krevelen, *Properties of Polymers, American Elsevier, New York, 1972*.

11. J. Hopkinson, *Phil. Trans. R. Soc.*, **167**, 465 (1977).

12. G. Williams, *Chem. Rev.*, **72**, 1, 55 (1972).

13. R. H. Cole, *J. Chem. Phys.*, **42**, 637 (1965).

14. H. Frolich, *Theory of Dielectrics*, 2nd ed., Oxford University Press, London, 1958.

15. M. F. Manning and M. E. Bell, *Rev. Mod. Phys.*, **12**, 215 (1940).

16. J. J. O'Dwyer and E. Harting, *Prog. Delect*, **5**, 143 (1963).

17. R. Kubo, *J. Phys. Soc. Jap.*, **12**, 570 (1957).

18. R. Zwanzig, *Ann. Rev. Phys. Chem.*, **16**, 67 (1965).

19. B. J. Berne and G. D. Harp, *Adv. Chem. Phys.*, **17**, 63 (1970).

20. S. A. Glarum, *J. Chem. Phys.*, **33**, 1371 (1960).

21. J. P. Boon and S. A. Rice, *J. Chem. Phys.*, **47**, 2480 (1967).

22. R. H. Cole and K. S. Cole, *J. Chem. Phys.*, **9**, 341 (1941).

23. Y. Ishida, M. Matsuo, and K. Yamafuji, *Kolloid Z.* **180**, 108 (1962).

24. D. W. Davidson and R. H. Cole, *J. Chem. Phys.*, **18**, 1417 (1950).

25. R. M. Fuoss and J. G. Kirkwood, *J. Amer. Chem. Soc.*, **63**, 385 (1941).

26. B. K. Scaife, *Proc. Phys. Soc. Lond.*, **81** 124 (1963).

27. S. Havriliak and S. Negami, *J. Polym. Sci.*, **14**, 99 (1966).

28. S. Havliak and S. Negami, *Br. J. Appl. Phys.*, **2**, 1301, (1969).

29. G. Williams and D. C. Watts, *Trans. Faraday Soc.*, **66**, 80 (1970).

30. G. Williams, *Trans. Faraday Soc.*, **62**, 2091 (1966).

31. M. Cook, D. C. Watts, and G. Williams, *Trans. Faraday Soc.*, **66**, 2503 (1970).

32. P. Hedvig, *Dielectric Spectroscopy of Polymers*, John Wiley, New York, 1977.

33. M. L. Williams, R. F. Landel, and J. D. Ferry, *J. Am., Chem. Soc.*, **77**, 3701 (1955).

34. I. H. Leaderman, *Elastic and Creep Properties of Filamentous Materials and Other High Polymers*, Textile Foundary, Washington, D.C., 1943.

35. K. C. Rusch, *J. Macromol. Sci., Phys.* **B2**, 2, (1968).

36. R. N. Hayward, *The Physics of Glassy Polymers*, John Wiley, New York, 1973.

37. J. M. Pochan, S. R. Turner, and D. F. Hinman, *J. Appl. Phys.*, **47** (10) 4245 (1976).

38. J. M. Pochan, H. W. Gibson, M. F. Froix, and D. F. Hinman, *Macromolecules*, **11**, 165 (1978).

39. S. Matsuoka and Y. Ishida, *J. Polym. Sci.*, **C14**, 247 (1966).

40. K. H. Illers and H. Breuer, *Kolloid Z.*, **176**, 110 (1961).

41. A. J. Curtis, *Soc. Plast. Eng. Trans.*, **18**, 82 (1962).

42. J. M. Pochan and D. F. Hinman, *J. Polym. Sci. Polym., Phys. Ed.*, **14**, 1871, (1976).

43. N. G. McCrum and G. Williams, *Trans. Faraday Soc.*, **57**, 1979 (1961).

44. R. W. Sillars, *J. Inst. Elect. Eng.*, **80**, 378 (1937).

45. K. W. Wagner, *Arch. Elektrotech.*, **2**, 371 (1914).

46. J. C. Maxwell, *Electricity and Magnetics*, Vol. 1, Clarendon Press, 452 (1892).

47. J. M. Pochan and R. G. Crystal, in Ed., F. Karas, *Dielectric Properties of Polymers*, Plenum Press, New York, 1972 p. 313.

48. A. I. Lukomskaya and B. A. Dogadkin, *Kolloid Z.*, **22**, 9576 (1960).

49. S. Yu Lipatov, and F. Y. Fabulyak, *J. Appl. Polym. Sci.*, **16**, 2131 (1972).

50. C. T. O'Konski, *J. Phys. Chem.*, **64**, 605 (1960).

51. J. Van Turnhout, *Polym. J.*, **2**, 173 (1971).

52. R. Rosen and A. H. Pohl, *J. Polym. Sci.*, **A-1**, 1135 (1966).

53. J. van Turnhout, *Thermally Stimulated Discharge of Polymer Electrets*, Elsevier, New York, 1975.

54. G. Rehage, *Ber Bunsenger*, **74**, 796 (1970).

55. W. Reddish, *Polym. Prepr.—ACS*, **6**, 571 (1965).

56. G. P. Makhailov and B. I. Sazhin. *Sov. Phys. Tech. Phys.*, **1**, 1670 (1957).

57. M. Takayanagi, *Mem. Fac. Eng. Kyushu Univ.*, **23**, 1,1 (1963).

58. A. M. Bottreau, Y. Dutuit, and J. Moreau, *J. Chem. Phys.*, **66**, 8, 3331 (1977).

59. R. H. Cole, *J. Phys. Chem.* **79**, 14, 1458 (1975).

60. M. J. C. van Gemert, *J. Chem. Phys.*, **62**, 7, 2720 (1975).

61. T. S. Clarkson and G. Williams, *Chem. Phys. Lett.*, **34**, 3, 461 (1975).

62. G. E. Johnson, E. W. Anderson, G. L. Link, and D. W. McCall, *ACS Div. Org. Coat. Plast. Chem. Pap.*, **35** 1, 404 (1975).

63. G. Williams, *Trans. Faraday Soc.*, **58**, 1041 (1962).

64. B. V. Hamon, *Proc. Inst. Elect. Eng.* **99**, Part 4, Monograph 27 (1952).

65. M. Ishler, R. Ladonna, and J. M. Pochan, to be published.

66. G. P. Mikhailov and T. I. Borisova, *Polym. Sci., USSR*, **2**, 387 (1961).

67. H. Hendus, G. Schnell, H. Thurn, and K. Wolf, *Ergeb. Exakt. Naturwiss.*, **31**, 5, 220 (1959).

68. E. A. W. Hoff, D. W. Robinson, and A. M. Willbourn, *J. Polym. Sci.*, **18**, 161 (1955).

69. G. P. Mikhailov, T. I. Borisova, and D. A. Dmitrochenko, *J. Tech. Phys. (USSR)*, **26**, 1924 (1956).

70. G. Williams and G. P. Watts, in P., Diehl, E. Fluch, and R. Kosfeld, Eds., *Natural and Synthetic High Polymers, NMR, Basic Principles and Progress* Vol. 4, Springer Verlag, Berlin, 1971, p. 271.

71. J. Koppelman, *Kolloid Z.*, **175**, 97 (1961).

72. G. P. Mikhailov, *Physics of Non Crystalline Solids*, North Holland, Amsterdam, 1965.

73. H. C. Elias, *Macromolecule*, Plenum Press, New York, p. 410.

74. T. Allen, J. McAinsh, and G. M. Jeffs, *Polymer*, **12**, 85 (1971).

75. M. F. Froix and A. O. Goedde, *Macromolecules*, **9**, 428 (1976).

76. J. M. Pochan and D. F. Hinman, *J. Polym. Sci., Polym. Phys. Ed.*, **14**, 1871 (1976).

77. J. M. Pochan, D. F. Hinman, and R. Nash, *J. Appl. Phys.*, **46**, 10, 4115 (1975).

78. M. F. Froix, D. J. Williams, J. M. Pochan, and A. O. Goedde, *Polym. Prepr.*, **16**, 2, 576 (1975).

79. J. M. Pochan, H. W. Gibson, M. F. Froix, and D. F. Hinman, *Macromolecules*, **11**, 165 (1968).

80. F. Wurstlin, *Kolloid Z.*, **113**, 18 (1949).

81. T. G. Fox, *Bull. Am. Phys. Soc.*, **11**, 123 (1956).

82. J. M. Pochan, D. F. Hinman, and S. R. Turner, *J. Appl. Phys.*, **47**, 10, 4245 (1976).

83. G. P. Mikhailov, L. L. Burshstein, S. I. Klemin, V. P. Malinovskaya, A. N. Cherkasov, and L. A. Chivaev, *Vysokomol. Soed.*, **A10**, 556 (1968).

84. A. M. Lobanov, D. M. Mirkamilov, and P. Platonov, *Vysokomol. Soed.* **A-10**, 1116 (1972).

85. W. J. MacKnight, J. Stoelting, and F. F. Karasz, *Adv. Chem. Soc.*, No. 99, 29 (1971).

86. J. L. Mincer and J. M. Pochan U.S. Patent No. 4, 007, 293.

87. J. M. Pochan, T. J. Pacansky, and D. F. Hinman, *Polymer*, **19**, 431 (1978).

88. E. F. Jordan, G. R. Reiser, W. E. Parker, and A. N. Wrigby, *J. Polym. Sci.*, **A-2**, 975 (1966).

89. M. Gordon and J. S. Taylor, *J. Appl. Chem.* **2**, 493 (1952).

90. S. Akijama, *Bull. Chem. Soc. Jap.*, **45**, 1381 (1972).

91. "Impedance Measurement Applications Note." EID II, General Radio, Concord, Massachusetts.

92. J. D. Allen, *Natl. SAMPE Symp. Exhib. Pap.*, **20**, 270 (1975).

93. R. A. Pike, F. C. Douglas, and G. R. Wisner, *Polym. Eng. Sci.*, **11**, 6, 1971.

94. D. Hudson, *Composites*, **5**, 6, 247 (1974).

95. R. H. Norman, *Proc. Inst. Elect. Eng.*, **100**, II, A, 341 (1953).

96. A. H. Scott, A. T. McPherson, and H. L. Curtis, *J. Res. Natl. Bur. Stand.*, **11**, 373 (1933).

97. E. M. Blyachman, T. I. Borisova, and Z. M. Levitzkaya, *Vysokomol. Soed.*, **A12**, 1544 (1970).

98. E. N. Haran, H. Gringas, and D. Katz, *J. Appl. Polym. Sci.*, **9**, 3505 (1965).

99. S. N. Antonov, I. M. Gourman, V. V. Kovriga, and G. A. Lushscheikin, *Plast. Massy*, No. 4, 38 (1966).

100. L. K. H. Van Beck, *J. Appl. Polym. Sci.*, **8**, 2843 (1964).

101. R. W. Warfield and M. C. Petree, *Makromol. Chem.*, **58**, 139 (1962).

102. T. Czvikovszky, *At. Energy Rev.*, **6**,(3), 3 (1968).

103. R. Greenwood and N. A. Weir, *Am. Chem. Soc. Symp. Ser.*, **25**, 220 (1976).

104. W. Reddish and J. T. Barrie, *I.U.P.A.C. Sym. Macromol.* Weisbaden, Kurzmetteiling, I.A. 3, 1959.

105. C. A. F. Tuijman, *Polymer*, **4**(a) 259, (b) 315, (1963).

106. H. Sasakura, *Jap. J. Appl. Phys.*, **2**, 66 (1963).

107. D. Gavrila, M. Diaconescu, and D. Iordache, *Rev. Rum. Chem.*, **21**, 1, 155 (1976).

108. D. Heinze, *Kolloid Z., Z. Polym.*, **210**, 45 (1966).

109. P. Hedvig and M. Krisbeny, *Angew. Makromol. Chem.*, **7**, 198 (1969).

110. R. Salovey, R. V. Albarino, and J. P. Luongo, *Macromolecules*, **3**, 314 (1970).

111. E. Ito, K. Sawamura, and S. Saito, *J. Colloid Polym. Sci.*, **253**, 480 (1975).

112. M. A. Bogirov, S. A. Abasov, V. P. Malin, and A. Ya. Jalilov, *J. Appl. Polym. Sci.*, **20**, 1060 (1976).

113. T. Yemni and R. H. Boyd, *J. Polym. Sci., Polym. Phys. Ed.*, **14**, 499 (1976).

114. M. Abkowitz and D. M. Pai, *Phys. Rev. Lett.*, **38**, 24, 1412 (1977).

115. M. R. Wertheimer, L. Paquin, H. P. Schreiber, and S. A. Boggs, *IEEE Trans. Electr. Insul.*, **EI-12**, 2, 137 (1977).

116. P. Thurion and R. Chasset, *Trans. Inst. Rubber Inc.*, **27**, 364 (1951).

117. W. C. Carter, M. Magat, W. G. Schmieder, and C. P. Smyth, *Trans. Faraday Soc.*, **42A**, 213 (1946).

118. A. S. Lukomskaya and B. A. Dogadkin, *Kolloid Z.*, **22**, 376 (1960).

119. W. Ambach and A. Denoth, Zeitschrift fur Gletscherhunde und Glazialgeologe, Bd. VIII, Heft, 1–2, 113 (1972).

120. J. Magoski and Y. Magoski, *J. Polym. Sci., Polym. Phys., Ed.*, **13**, 1347 (1975).

121. M. F. Froix, J. M. Pochan, A. O. Goedde, and D. F. Hinman, *Polymer*, **18**, 1213 (1977).

122. R. Michel, G. Saytre, and M. Maitrot, *J. Polym. Sci., Polym. Phys. Ed.*, **13**, 1333 (1975).

123. M. Stolka and D. M. Pai, *Polymers with Photoconductive Properties*, Advances in Polymer Science, Vol. 29 Springer Verlag, Heidelberg, 1978.

124. For details see, for instance, A. Rose *Concepts in Photoconductivity and Allied Problems*, Interscience Publishers, New York, 1967, and S. H. Ryvkin, *Photoelectric Effects in Semiconductors*, Consultants Bureau, New York, 1954.

125. D. M. Pai, *J. Chem. Phys.*, **52**, 2285 (1970).

126. D. M. Pai, "Photoconductivity," in A. V. Patsis and D. A. Seanor, Eds., *Polymers*, Technomic Press, Westport Conn. 1976, Chapter 2.

127. R. G. Kepler, *Phys. Rev.*, **119**, 1226, (1960).

128. W. E. Spear, *Proc. Phys. Soc. London Sec. B*, 669 (1975).

129. For details of experimental techniques described here, see for instance, F. K. Dolezalek, in J. Mort and D. M. Pai, Eds. *Photoconductivity and related phenomena*, Elsevier, New York, 1976 pp. 27–63.

130. I. P. Batra, K. K. Kanazawa, and H. Seki, *J. Appl. Phys.*, **41**, 3416 (1970).

131. P. J. Regensburger, *Photochem. and Photobiol.*, **8**, 429 (1968).

132. G. Pfister and D. Williams, *J. Chem. Phys.*, **61**, 2416 (1974).

133. P. M. Borsenberger and A. I. Ateya, *J. Appl. Phys.*, **49**, 4035, (1978).

134. P. J. Melz, *J. Chem. Phys.*, **57**, 1694 (1972).

135. P. M. Borsenberger, L. E. Contois, and D. C. Hoesterry, *J. Chem. Phys.* **68**, 637 (1978).

136. L. Onsager, *Phys. Rev.*, **54**, 554 (1938).

137. N. E. Geacintov and M. Pope, in E. M. Pell, Ed., *Proceedings of the Third International Conference on Photoconductivity*, Pergamon, Oxford, 1971, pp. 289.

138. R. H. Batt, C. L. Braun, and J. F. Hornig, *J. Chem. Phys.*, **49**, 1967 (1968).

139. A. Mozumdar, and J. M. Magee, *J. Chem. Phys.*, **47**, 939 (1967).

140. D. M. Pai and R. C. Enck, *Phys. Rev.*, **11**, 5163 (1975).

141. G. E. Johnson, *J. Chem. Phys.*, **62**, 4697 (1975).

142. D. M. Pai, Proc. Photogr. Soc. Meet. Cambridge (September 1976).

143. J. Pearson, *Pure Appl. Chem.*, **49**, 463 (1977).

144. M. Yokoyama, Y. Endo, and H. Mikawa, Chem. Phys. Letters, *34* 597 (1975); K. Okamoto, S. Kasabayashi and H. Mikawa, *Bull. Chem. Soc. J.* **46**, 2324 (1973).

145. J. E. Knights and E. A. Davis, *J. Phys. Chem. Solids*, **35**, 543 (1974).

146. M. Silver and R. C. Jarnagin, *Mol. Cryst. Liq. Cryst.* **3**, 461 (1948).

147. R. M. Schaffert, *IBM J. Res. Develop.*, **15**, 75 (1971).

148. S. J. Fox, *Photoconductivity in Polymer—An Interdisiplinary Approach*, A. V. Patsis and D. A. Seanor, Eds., Technomic Press, Westport, Conn. 1976, Chapter 10 (see Reference 10).

149. G. Weiser, *J. Appl. Phys.* **43**, 5028 (1972).

150. J. Mort and A. I. Lakatos, *J. Non-Cryst. Solids*, **4**, 97 (1970).

151. W. D. Gill, *J. Appl. Phys.*, **43** 5052 (1972).

152. G. Pfister, *Phys. Rev.*, **B16**, 3676 (1977).

153. P. M. Borsenberger, W. Mey and A. Chowdhary, *J. Appl. Phys.*, **49**, 273 (1978).

154. J. Mort, G. Pfister, and S. Grammatica, *Solid State Commun.*, **18**, 693 (1976).

155. W. Limburg, D. M. Pai, J. M. Pearson, D. Renfer, M. Stolka, S. R. Turner, and J. Yanus, *Org. Coatings Plas. Chem.*, **38**, 534 (1978).

156. H. Scher, *Photoconductivity and Related Phenomena*, J. Mort and D. M. Pai, Eds. Elsevier, 1976, Chapter 3 pp. 71–114.

157. H. Scher and E. W. Montroll, *Phys. Rev.*, **B12**, 2455 (1975).

158. F. W. Schmidlin, *Solid State Commun.*, **22**, 451 (1977).

159. J. Noolandi, *Solid State Commun.*, **24**, 477 (1977).

160. M. Pollak, *Phil. Mag.* **36**, 1157 (1977).

161. G. Pfister, J. Mort and S. Grammatica, *Phys. Rev. Lett.*, **37**, 1360 (1976).

162. D. M. Pai, *J. Appl. Phys.*, **46**, 5122 (1975).

163. S. Fox, in A. V. Patsis and D. A. Seanor, Eds., *Photoconductivity in Polymers*, Technomic Publishing Co., Westport, Conn., 1976, p. 253.

164. W. D. Gill, in J. Stuke and W. Brenig, Eds., *Proceedings of the Fifth International Conference on Amorphous and Liquid Semiconductors*, Garmisch-Partenkirchen, 1973, Taylor and Francis, London, 1974, p. 901.

165. J. Mort and G. Pfister, (to be published).

166. J. Slowik, *Bull. Am. Phys. Soc.*, **22**, 434 (1977).

167. H. Seki, in J. Stuke and W. Brenig, Eds., *Proceedings of the Fifth International Conference on Amorphous and Liquid Semiconductors*, Garmisch Partenkirchen, 1973, Taylor and Francis, London, 1974, p. 1901.

168. M. F. Froix and J. M. Pochan, *Polym. Prep.*, **19**, 553 (1978).

CHAPTER 9

Surface and Interfacial Characteristics

J. J. BIKERMAN

Department of Chemical Engineering
Case Western Reserve University
Cleveland, Ohio

C. C. DAVIS

Department of Science and Mathematics
General Motors Institute
Flint, Michigan

H. SCHONHORN

Bell Laboratories
Murray Hill, New Jersey

9.1	Introduction — Nature of Polymer Surfaces	396
	9.1.1 Characterization and Structure of Polymer Surfaces	397
	9.1.2 Surface Characterization Techniques	398
	9.1.3 Transcrystallinity and Polymer Surface Morphology	399
	9.1.4 Surface Region Morphology and Adhesion	401
9.2	Liquid Polymers	402
	9.2.1 Data on Surface Tension	404
	9.2.2 Interfacial Tension	406
	9.2.3 Unimolecular Layers	406
	9.2.4 Foams	408
	9.2.5 Theoretical Approaches to Polymer Surface and Interfacial Tension	408
9.3	Adsorption of Polymers	410
	9.3.1 Effect of Solvent	417
	9.3.2 Effect of Adsorbent	418
	9.3.3 Effect of Temperature	420
	9.3.4 Rate of Sorption	420
	9.3.5 Structure of the Adsorption Film	421
9.4	Adsorption by Polymers	423
9.5	Solid Polymers	424

	9.5.1 Surface Energy	425
	9.5.2 Wettability	426
	9.5.2.1 Formation of Interface	429
	9.5.3 Temperature Dependence of Wetting	430
	9.5.4 Surface Regions	431
	9.5.5 Fracture	433
	9.5.6 Friction	433
	9.5.7 Adhesion	434
	9.5.7.1 Adhesive Joint Strength	435
	9.5.7.2 Weak Boundary Layers	435
	9.5.7.3 Chromic Acid Etch of Polyolefins	436
	9.5.7.4 R. F. Glow-Discharge Modification of Polymers	436
	9.5.7.5 Fluorination of Polyethylene	437
	9.5.7.6 Other Surface Treatments	437
9.6	Environmental Aspects of Adhesion and Adhesive Joint Strength	438
	9.6.1 Calculation of the Thermodynamic Work of Adhesion	439
	9.6.2 Corrosion	440
	9.6.3 Wear	441
9.7	Adhesives and Coatings	441

J. J. Bikerman is deceased.

9.7.1   Surface Preparation   441
   9.7.1.1   Degreasing   441
   9.7.1.2   Substrate Removal   443
   9.7.1.3   Surface Conversion   444
9.7.2   Joint Design   446
9.7.3   Adhesive Selection   447
   9.7.3.1   Bonding Processes   448
9.7.4   Structure-Property Relationships   448
   9.7.4.1   Molecular Weight   448
   9.7.4.2   Molecular Weight Distribution   453
   9.7.4.3   Chain Geometry   453
   9.7.4.4   Intermolecular Attraction   453
   9.7.4.5   Crystallinity   454
9.7.5   Coatings   454

9.8   Summary   455

9.9   Glossary of Symbols   456

9.10   References   458

9.1   INTRODUCTION—NATURE OF POLYMER SURFACES

Polymer surfaces play a principal role in many areas of current technology. The nature of the polymeric surface is involved in the phenomena of adhesion, adhesive joint strength, wettability, and spreading of a polymer in contact with another phase. There is a large body of experimental information on the wetting of polymers, polymer melt surface tension, interfacial tension between different pairs of polymer melts, and the modification of polymer surfaces by additives and chemical reactions. This discussion endeavors to include the performance of bonded structures as reflected in their contact with potentially hostile environments. Ultimately, interfacial properties of the polymer-substrate interface becomes of utmost importance.

Special problems associated with polymers arise from their molecular weight heterogeneity and the existence of many classes of polymer chain defects due to compositional, optical, geometrical, and head-to-tail isomerism, as well as inadvertent branching. An immediate consequence of this architectural irregularity of the backbone chains of polymers is the possibility of several distinct modes of solidification from the melt which may result in different surface structures.

Bulk property of polymers exhibits time-dependent mechanical properties including a remarkable degree of viscoelasticity. Dynamic mechanical experiments yield both an apparent elastic modulus (in tension or shear) E and the mechanical damping $\tan \delta$, where δ is the loss angle. At a convenient frequency of, say 100 Hz, a typical linear amorphous polymer in the glassy region below T_g has a modulus of about 10^{10} dynes/cm^2. As the temperature is increased, the next region of mechanical response is the transition region including T_g, which can be taken to be, in first approximation, the point of inflection of the modulus curve or the maximum in the damping curve. In this region the modulus drops by a factor of 1000. The third region is called the rubbery plateau where the modulus remains roughly constant. As the temperature is further increased, this is followed by a region of elastic or rubbery flow and finally liquid flow with very little elastic recovery. In these regions the damping curve again begins to increase, and the modulus may drop below 10^5 dynes/cm^2. In a cross-linked (amorphous) polymer the last two regions are absent; since there is no flow, the cross-links prevent the chains from slipping past each other. The presence of crystallinity increases the modulus significantly above T_g, with a sharp drop in the modulus and increase in mechanical damping curve at the melting point T_m. Thus in crystalline polymers the demarcation point between the elastomeric (rubberlike) "solid" and rubbery "melt" can be taken to be given by T_m, while for amorphous solids it lies in the vicinity of the end of the rubbery plateau region.

The fact that the demarcation between the melt and the solid is far from sharp reinforces a need for a detailed study of the surface and interfacial tensions of polymer melts as related to adhesion phenomena. The complicated dynamic mechanical responses of polymers to applied stresses preclude the reliable estimation of surface tensions or surface free energies by such methods as mechanical necking experiments, which have been applied to other classes of solids. Ultimately it is the many internal degrees of freedom implied by their macromolecular nature that are responsible for this extensive and varied time dependence of mechanical and thermal properties.

Synthetic organic polymers are basically van der Waals or molecular solids in terms of their local surface properties. The fact that even the bulk polymer is nonuniform in composition, molecular weight, and structure leads to many new ways in which the interfacial region can differ from the bulk. One must deal with "contamination" by surface active substances from within as well as without. With some notable exceptions, most polymers that one considers are hydrophobic, and polar sites in such a hydrophobic matrix present special problems. Last but not least, polymer surfaces are usually not pretreated extensively unless they are being prepared for adhesive bonding.

The surface tension of liquids or interfacial tension of liquid pairs can be reliably measured experimentally. This is true also of the viscous polymer melts. On the other hand, *there exists currently no direct, reproach-free method of measuring the surface tension or specific free energy of a macroscopic surface of a polymeric solid.* The surface free energies that are discussed, and possibly even measured, of a crystalline nucleus of a polymer within an amorphous matrix have little relation to the surface tensions or specific surface free energies of a macroscopic polymer surface in contact with vapor or a simple nonpolar liquid.

Polymers introduce further problems since the definition of a polymeric substance allows for possibilities of isomerism and molecular weight heterogeneity. Thus the surface tension of an isotactic polymer melt should (and can) differ from that of the syndiotactic polymer melt. Even in the absence of isomerism, a heterodisperse polymer melt will not possess precisely the same molecular weight distribution in the immediate vicinity of a phase boundary as in the bulk owing to a difference of configurational entropy in the interface region of shorter chains as compared to longer chains. Furthermore, polymer solids can be quenched or otherwise prepared in metastable states that can exhibit hysteresis in surface as well as in bulk properties. In short, the large size, the extensive range of mechanical viscoelastic relaxation times, and the molecular inhomogeneity of polymer molecules aggravate characteristic difficulties in the application of continuum, phenomenological theory (hydrodynamics or thermodynamics) to the description of experimental surface phenomena. When a liquid spreads on a solid surface, the advancing edge may be quite irregular and not reproducible, and it may merge into a thin film whose profile shows many steps in height differing by many orders of magnitude in scale down to the molecular (1–5). What is the physical significance of the disjoining pressure when an apparent steady state in spreading has occured? Can one extend the concept of a contact angle to the edge of the film? Ordinarily, macroscopic observations using a low-power (10 ×) microscope are used in determining the contact angle at the apparent intersection of the three phase boundaries of vapor and a liquid and a solid. Even the temporal attainment of a situation in which the contact angle may possess a simple thermodynamic significance is fraught with ambiguity. Current studies of the hydrodynamic motion of fluids in the immediate neighborhood of a moving contact line show that the Stokes equations (with plausible boundary conditions) are insufficient to describe the motion (6, 7).

9.1.1   Characterization and Structure of Polymer Surfaces

To characterize an interface fully, one is interested in the values of various physical and chemical properties of the matter composing it, first, as a function of the depth into the interface and, second, as a function of two directions normal to the depth. This requires combinations of many analytical tools and techniques. Each experimental technique provides data on a specific property (or properties) that are an average over a characteristic depth (of sampling). In the case of polymer surfaces one is interested in the elemental and molecular compositions: the distribution in molecular weight; cross-link density, extent of branching and chain isomers; variation in density and percent of crystallinity; changes in morphology or extent of ordering; the nature and extent of preferential orientation of the molecular segments; the distribution of reions of varying surface free energy (surface energetics), patches of high polarity, and so on; the superficial surface area (particularly of finely divided polymer samples); and the microtopography of the surface.

9.1.2 Surface Characterization Techniques

Some techniques, for example, low energy electron diffraction (LEED) or Auger spectroscopy, have not yet been successfully applied to polymer surfaces for various reasons. The high-energy electron beams used to excite Auger emission may cause radiation damage to polymer surfaces, but, nevertheless, permit the composition of the polymer interface to be determined on an atomic scale (20 Å). This has also limited the application of electron and ion microprobe techniques. The nature of the nuclei composing organic polymers prevents the use of Mossbauer spectroscopy except in certain polymers such as the polysiloxanes. Newer techniques appear very promising but have not actually been developed for applications to polymer surfaces; examples are secondary ion mass spectroscopy or ion neutralization spectroscopy. Certain techniques, such as ellipsometry (8, 9), have been very successfully applied to the study of monolayers of adsorbed polymers but have not been applied directly to a bulk polymer surface. Finally, one must not forget that in studying, say, a polymer-metal interface, the presence of metal or metal oxide can be detected by a variety of techniques (x-ray fluorescence or electron microprobes, atomic absorption spectroscopy, and so forth). Barring effects due to voids or nonpolymeric impurities and so on such studies provide some insight into the distribution of the polymer in the interface, basing the argument that regions containing high concentrations of metal or metal oxide are largely devoid of polymer. Some useful techniques for the characterization of adherend surfaces (to polymers) have been recently reviewed by Hamilton (10).

A well known technique to determine surface composition, which also yields limited information on orientation and/or degree of crystallinity, is attenuated total reflectance (ATR) infrared spectroscopy (9). This technique is often augmented by comparison of its results with results of standard and modified infrared transmission studies on bulk films (approximately 1 mil or more in thickness). Following Harrick (11), in ATR the penetration of the incident beam into the surface region of a polymer film adhering to a KRS-5 crystal is approximately one-fifth of the wavelength, or about $2\,\mu$ for radiation of frequency $720\ \mathrm{cm}^{-1}$. It thus has a range of about $1\,\mu$ and measures a relatively thick surface layer. Good contact is required between the sample and the spectrometer prism, which may be a source of difficulty with rough or inflexible films.

An example of the application of this technique is Luongo and Schonhorn's (12) infrared study of the surface region of polyethylene (PE) nucleated on gold or aluminum (high-energy surfaces) and polytetrafluoroethylene (PTFE, a low-energy surface). The chemical composition of the surface region and the bulk appears spectroscopically to have the same chemical nature, but there are characteristic differences in the intensities of certain infrared bands that are known from studies of bulk samples (13) to be suitable for following relative changes in crystallinity.

The Henke-design x-ray tube (14) generates long wavelength or "soft" x-rays and thereby extends the usefulness of x-ray fluorescence analysis down to atomic number 5, by what is termed *soft x-ray spectroscopy*. The effective depth of analysis is about $1\,\mu$ into the surface (15, 16). It is a valuable technique in its own right as well as a useful complement of the infrared ATR technique.

X-ray photoelectron spectroscopy (XPS or ESCA), developed by Siegbahn and co-workers, allows chemical structure analysis in polymers in a range a thousand times less deep than either ATR or soft x-ray spectroscopy. Basically, this is because photoelectrons ejected from atoms deep in the sample are either self-absorbed by the sample or lose part of their kinetic energy and contribute only to the spectral background. The effective escape depth varies with the electron kinetic energy and surface structure but is of the order of a few tens of angstroms, (15–17). Besides providing a semiquantitative elemental analysis, ESCA spectra exhibit small shifts (chemical shifts) due to the environment of an atom in a molecule. Thus, for example, in the carbon spectrum of poly(vinyl fluoride) there is a doublet corresponding to the carbon atoms in CHF and CH_2 groups. The electron density is less around the fluorine-substituted carbon than around the carbon with only hydrogen substitutents, and the binding energy is correspondingly increased. Clark and co-workers (18) have exploited the ESCA technique with particular reference to the surface treatment of hydrocarbon and fluorocarbon polymers and the locus of failure to adhesive joints. This is explored in a later section.

Certain spectroscopic techniques can be applied only to the much higher area internal surfaces attained at the interface between polymer and (dispersed) small filler particles. A variety of nuclear magnetic resonance studies have been carried out on filled polymers; usually the filler employed was carbon black (19). Proton spin resonance studies support the view that a portion of polymer in the vicinity of the filler is immobilized.

Lipatov and Fabuliak (20) had earlier reported large shifts with temperature in the minima in the nmr signal of poly(methyl methacrylate) and polystyrene filled with silica. Somewhat smaller effects were observed with these polymers when filled with finely dispersed PTFE. These authors interpret their results in terms of decreased segmental motion, mainly determined by a decrease of the number of conformations in the interface rather than due to the significant changes of interaction energy at the surface.

One can now make the following general observations. Current surface characterization techniques appear to fall broadly into two categories—those whose effective range is measured in angstroms and those whose effective range is about a thousand times larger. Clearly the wetting behavior of a surface such as the polyfluorocarbons follows the ESCA characterization of a "surface layer" measured in angstroms. On the other hand, significant effects on the elastic properties of a thin crystalline film of PE can arise owing to a

modified degree of crystallinity in a surface region whose extent is measured in microns or tens of microns.

The lack of understanding of properties that appear to respond to an intermediate range, say of several hundred angstroms, is clearly in part a consequence of the lack of suitable probes with that range. [This intermediate region may possess unusual properties since the average diameter (radius of gyration) of a polymer coil has this dimension.]

Thin and thicker surfaces have to be separately investigated—the thin surface layer may have a different composition or structure from the thicker surface region, and both regions may have properties that differ from that of the bulk polymer (15, 21).

9.1.3   Transcrystallinity and Polymer Surface Morphology

The existence of an interface can change the morphology of polymer molecules lying in its vicinity. This is not only due to the anisotropy of the intermolecular forces, whose range cannot be excessive, but also because new longer-ranged cooperative phenomena can come into play, such as different heterogeneous modes of nucleation of molecular order. Such order can also be of various kinds — it can be an ordering in the center of mass of molecules within some space lattice, as in crystallization, or preferential ordering of the orientation of molecules that may not involve a first-order phase transition. In principle, an amorphous polymer that is on the verge of being able to crystallize could exhibit some degree of crystallinity when it is allowed to cool on a suitable substrate. This has not yet been reported, but the modification of the crystallinity of crystalline polymer on a variety of substrates has been investigated.

Although the bulk (or interior) of most semicrystalline polymer samples is spherulitic, there are several possibilities for the surface region, among them: spherulites smaller or equal in diameter to those in the interior; elongated spherulites originating at the surface and propagating normal to the surface, that is, transcrystallinity; and oriented lamellar crystals, originating at the surface and extending into the interior. The first two categories can sometimes be observed under certain conditions in injection molded samples where molecular orientation exists in the melt (21). Apparently, in transcrystalline specimens, the cooling of the surface of a polymer melt causes spherulites to nucleate very closely together on the surface. The lamellae that grow from these nuclei can propagate only in one principal direction normal to the surface since growth in the lateral directions is restricted by neighboring spherulites. These conditions lead to the formation of a surface region, as shown in Figure 9.1, in which only very narrowly divergent spherulitic sectors develop and give an overall rodlike appearance (22).

These transcrystalline regions, as observed in polyamides (23, 24), polyurethanes (23), and polyolefins (22, 25–27), have thicknesses estimated from photomicrographs often in excess of 10 μ. The exact value of the thickness of the transcrystalline region (if any) is dependent on factors such as thermal properties and history of the polymer film (e.g. its melt temperature, crystallization temperature, and to some extent cooling rate) and the nature of the surface and the polymer in contact with it

(22–28). It appears plausible that the occurrence of transcrystallinity can be attributed to massive heterogeneous nucleation induced at the mold surface (25). To be effective the mold surface should have a nucleating efficiency equal to or greater than that of adventitious nuclei present in the polymer. For polyolefins (21), as the crystallization temperature approaches the melting point, the activity of the mold surface is found to increase, leading invariably to transcrystalline formation. The degree of activity of various mold surfaces correlates with the known activity of specific dispersed nucleating agents having similar chemical structures.

Such transcrystalline layers of 10 μ or more in thickness are effectively macroscopic. One expects therefore, that in thin films (such as in adhesive layers) where the surface zone is a substantial portion of the total volume, the contribution of the transcrystalline region to the overall physical properties of the film may become significant. Experiments have shown that these transcrystalline regions differ from the bulk (or interior) of the polymer in mechanical properties, diffusion and solubility coefficients, and surface properties.

Figure 9.1 Transcrystalline regions.

These results have immediate possible implications for the cohesive failure of composites in which a semicrystalline polymer is used an adhesive. Criteria of failure derived from simplified versions of linear theory, such as Griffiths' for brittle fracture and the critical stress in rupture, relate directly the tensile stress in brittle failure to $(E')^{1/2}$ and the critical stress to E', respectively. Unfortunately these criteria are not applicable to real polymers, without extensive modification or reinterpretation, since ultimate properties of polymers involve highly nonlinear and history-(or memory-) dependent response. If indeed the transcrystalline regions prove on direct experimental test to be somewhat "stronger" and "tougher" that the interior regions, this should be used in designing adhesive combinations. Clearly, more direct determinations of the ultimate strength properties of the transcrystalline as opposed to the interior regions of polymer samples would be very desirable.

X-ray diffraction studies (26) support the view that the differences in diffusion behavior between the surface and interior regions are due to differences in orientation of the lamellar boundaries. In the transcrystalline region the lamellar boundaries, along which diffusion is strongly preferred, are oriented predominantly normal to the surface, as opposed to the interior region. Diffusion should be more rapid because the lamella boundary diffusion paths are relatively long and direct in the surface region (26). These differences would have some very specific implications for those adhesive composites involving a transcrystalline polymer film and a tacky liquid or another polymer solution to which the Voyutskii-Vasenin diffusion theory of adhesion has particular bearing. If the rate of interdiffusion of the adhesive and adherend, on initial wetting, is the key factor in producing an adhesive (or autoadhesive) joint, as is claimed in certain cases, then differences in diffusion (and solubility) coefficients become directly applicable, for example, the specific work of peeling is predicted to be directly proportional to the square root of the diffusion coefficient in the surface region. Also, environmental or aging effects in adhesive

failure may involve mass transport of moisture, oxygen, plasticizers, and so on. Factors such as structural change, accumulation producing a weak boundary layer, or chemical reaction deteriorating and enhancing the adhesive joint are immaterial as long as the kinetics of the process are to some extent diffusion limited.

9.1.4 Surface Region Morphology and Adhesion

In relating transcrystalline phenomena to adhesion and adhesive joint strength, one must distinguish carefully between the categories of possible adhesive joint failure. The joint failure may be understood in terms of the interface, mechanical properties, and environmental factors. A material adhesive must wet the substrate (adherend) sufficiently to result in intimate molecular contact at the interface. This condition is only necessary but not sufficient. Measured mechanical strengths of adhesive bonds may not have any relation to the surface energies of adhesive and adherend but may instead depend only on the "condition" of the joint and the mechanical properties of the components of the adhesive joint. Weak boundary layers, though, may be an important reason why adhesive joints do not achieve ideal strengths. One can further distinguish the nature of the failure. Failure may occur in bulk by more or less brittle fracture under constant load or under repeated (cyclic) stressing that mimics fatigue, to give just two examples. Environmental effects may be important, such as the presence or absence of stress, humidity, or oxygen. In semicrystalline polymers the correlation between morphology and physical properties has been the subject of intensive investigations. The fact, furthermore, that oriented structure can be produced near the surface of various semicrystalline polymers, for example, polyamides (23, 24), polyurethanes, and polyolefins (21, 22, 26, 27) has been recognized for some time. Since polymer physical properties (in turn) affect any or all the factors mentioned, it follows that morphology, and in particular surface morphology and orientation, affect adhesive joint formation with semicrystalline polymers.

While the bulk (or interior) of most semi-crystalline polymer samples is spherulitic, the surface region can consist of several different possibilities. Some of these possibilities previously noted relate to the following morphological arrangements:

Spherulites smaller or equal in diameter to those in the interior.

Elongated spherulites originating at the surface and propagating normal to the surface, for example, transcrystallinity.

Oriented lamellar crystals originating at the surface and extending into the interior.

Heterogeneous nucleation and crystallization of some polymer melts against certain high-energy surfaces (e.g. metals, metal oxides, and alkali halide crystals) have been found to result in marked changes in both the surface region morphology and increased wettability of these polymers, even though the chemical constitution of the polymer is unchanged (27). The critical surface tension γ_c of a variety of polymers nucleated against gold is considerably in excess of commonly accepted values. These conclusions are illustrated in Figure 9.2, which is a plot of the variation of contact angle θ, cos θ, and computed γ_c as a function of molding time. The metal substrates were removed by dissolution, the PTFE composite separated on cooling. Particularly interesting is the relatively large reduction of the contact angle on the PE surface with extensive transcrystalline regions obtained by nucleating the PE against gold. A similar influence on the surface layer of polymers has also been observed for some low-energy mold surfaces (29). This increase of wettability of transcrystalline surface regions to organic liquids, as observed in some instances, has immediate application to adhesion since it affects one of the necessary factors (molecular contact at the interface) for adhesion.

Finally, it may be noted that surface morphology of semicrystalline polymers certainly is expected to relate to their "adhesiveness." Much further experimentation is required before definitive conclusions concerning the magnitude of these effects on adhesive strengths can be made.

9.2   LIQUID POLYMERS

Surface tension γ is one of the most important properties of liquid surfaces, and liquid poly-

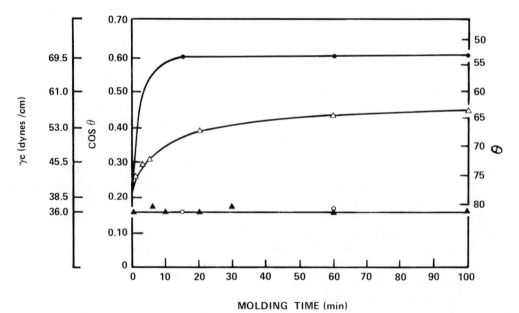

Figure 9.2   Variation of θ, cos θ, and γ_c.

mers would be expected to possess it in the same manner as liquids of a low molecular weight. Also, in principle, all the numerous methods of measuring γ are applicable to polymeric liquids. In reality, many of these methods are not suitable, mainly because of the high viscosity (or viscoelasticity) of polymers.

The procedures used for measuring γ may be classified as static methods involving no solid, static methods involving solids, dynamic methods involving no solid, and dynamic methods involving solids. For static methods the system must be in equilibrium, but the achievement of this is greatly retarded by high viscosity of the liquid because it is also difficult to be sure that the equilibrium in fact has been reached. In dynamic tests the effects of high viscosity usually so overshadow those of surface tension, that γ cannot be calculated from their results. Whenever a solid is an essential ingredient of the system, the equilibrium around the three-phase line (in which vapor, liquid, and solid meet) must be present in addition to that along the vapor-liquid interface, so that high viscosity becomes an even greater obstacle.

The viscosity of polymeric, as also of other liquids, generally decreases as temperature rises so that the measurement of γ at higher temperatures might be expected to be easy. Unfortunately, common polymers are not perfectly stable at high temperatures. Even when their oxidation is precluded by performing experiments in an inactive gas, scission of chains or continous polymerization may occur and affect many properties of the material.

The difficulties indicated in the preceding paragraphs should be kept in mind when quantitative data on the γ of viscous liquids are considered. Probably none of these data is as reliable as the γ values of ordinary liquids (water, benzene, etc.) Two recent reviews of the γ of polymers (30, 31) may be consulted for additional information.

A dynamic method based on complete wetting, that is, the method of the maximum bubble pressure (unsuitable according to the preceding classification), has, however, been used. In one of the earliest publications (32), this meant the greatest pressure "at which no bubble was formed within 3 minutes." The γ of nylon 66 was 35.1 at 285°C; nylon 6, 36.1 at 265°C; nylon 610, 37.0 at 265°C; nylon 18, 22.6 at 225°C; and polyethylene 22.8 g/sec^2 (or dyne/cm or mN/m) at 150°C. A simultaneous investigation (33) dealt with poly(caproamide), that is, nylon 6. The value of γ rapidly increased with the degree of polymerization. When this was raised from 20 to 120, the γ at 257°C rose from about 39 to 62 g/sec^2. The viscosity η naturally increased at the same time, and it may be surmised that at least a part of the change attributed to γ was caused by a change in η.

Apparently more reliable results have been obtained with poly(isobutylenes) (34) although the materials were used as received; the maximum pressure was deemed to be reached in 10–15 min. At 100°C, γ depended on the viscosity-average molecular weight M_v, as indicated in Table 9.1.

The rise of γ with M_v points out another difficulty of measuring the surface tension of polymers with confidence. Practically all polymers contain chains of different lengths. As shorter chains give rise to smaller γ values, and since components having a lower γ tend to concentrate in the surface layer and to lower the γ of the mixture, the experimental values probably correspond to a molecular weight smaller that indicated. For instance, a liquid consisting only of particles of $M = 77500$ is likely to have markedly higher γ than 28.1 listed in Table 9.1.

The dependence of γ on M_v was almost absent in a series of poly(tetrahydrofurans) (35); again the maximum bubble pressure was measured. The γ at 40°C was 39°C ± 0.6 g/sec^2 for three polymers of molecular weights 1100, 1300, and 2000. Also the γ of

Table 9.1 Surface Tension of Molton Poly(isobutylene) (34)

Molecular weight M_v	140	790	1560	2340	17100	38200	77500
Surface tension (g/sec^2)	20.5	23.1	24.1	24.5	26.3	27.7	28.1

poly(oxypropylene diols) at 40°C was little affected (32.9–35.1 g/sec^2) when the molecular weight varied between 500 and 3000. The maximum bubble pressure method was used also (36) to monitor the progress of polymerization of styrene containing small additions of benzoyl peroxide.

Another dynamic method is described in Reference 37. A gas bubble is introduced into a closed tube filled with molten polymer. When the tube is rapidly rotated, the axis being perpendicular to the tube length, the bubble is extended along the tube. The effect is caused by the centrifugal force, but it is counteracted by the capillary pressure at the gas-liquid boundary. Thus γ can be calculated from the shape of the bubble in the steady state. Unfortunately, this state often is not reached even after 3 hr of spinning. Some examples of the data obtained by this method are: polyethylene (*Epolene N*) 23.0 at 202°C, Zytel nylon resin 29.0 at 290°C, poly(ethylene terephthalate) 24 to 30 g/sec^2 at 290°C.

The classical method of capillary rise in narrow tubes is too time-consuming when the liquid is very viscous. It has been used, nevertheless, for poly(chorotrifluoroethylene) of number-average molecular weight M_n 1280 (38), several poly(isobutylenes) (34), and polystyrene (39). The ring method, whose theoretical justification is doubtful, was employed for polyethylene (40) and polypropylene (41).

The methods used most frequently are those of capillary pull (also known as the Wilhemy method), of sessile bubbles, and of pendant drops. The first-named procedure is predicated on the contact angle θ being zero, but this is not a serious drawback, because usually a zero contact angle can be achieved simply by slightly raising the vertical plate (such as microscope cover slide) more or less wetted by the liquid. In the other two procedures, the θ need not be known: the γ is calculated from the profile of the drop or bubble far from the solid. In all instances the experimenter must be able to test whether the equilibrium was in fact reached. This check is performed by applying the equations, which can be found in Reference 42, to more than two points on the profile.

The main results obtained with these three methods are arranged in Section 9.2.1 according to the chemical nature of the polymer (or its main constituent).

9.2.1 Data on Surface Tension

Polyethylene (PE) is probably the simplest of the polymers and is the first to be discussed here. A comparison of sample descriptions demonstrates an additional difficulty in studying surface properties of macromolecular liquids: some of the samples were used "as received," some were purified, and some others were deliberately contaminated with, for example, 0.4% of an antioxidant. Thus different specimens are not directly comparable with each other. Nevertheless, very similar γ values were obtained for different polyethylenes.

A supposedly linear polyethylene of the weight-average molecular weight M_w of 67,000 had $\gamma = 25.9$ at 186°C (43), 28.1 at 150°C (44), and 28.8 g/sec^2 at 140°C (45). The temperature coefficient calculated from these three values is $d\gamma/dT = -0.063$ g/sec$^2 \cdot$ deg, and temperature coefficients between -0.057 and -0.064 have been measured directly by these investigators. The agreement between the different test series is more remarkable because in Reference 43 capillary pull was used, and in the others pendant drops were used.

Polyethylenes of molecular weights 2000 and 7000 gave γ of 25.8 and 26.7, respectively at 150°C, and a highly polymerized polyethylene (M_w-340,000) had $\gamma = 26.6$ at 167°C (43). A branched polyethylene containing 5.2 methyl groups per 100 carbon atoms, gave $\gamma = 26.5$ at 150°C (44). Thus γ appears to increase with molecular weight and to be lowered by branching (46). In an earlier study, the γ of liquids of M_v near 69,000 and 80,000 was found (47) to be 27.5 at 148°C and 23.7 g/sec^2 at 200°C; the resulting temperature coefficient, -0.073, is significantly greater than those reported by the later authors. The method of sessile bubble was used. Also the ring method (40) afforded $d\gamma/dT = 0.076$ between 125 and 193°C. For additional data see Reference 48.

For two different polypropylenes, γ was

22.5 at 165°C (41) and 22.1 at 150°C (44). The value of $-d\gamma/dT$ was only 0.040 and 0.056 g/sec^2 · deg, respectively.

For poly(isobutylenes) the method of pendant drops afforded the following results: 25.1 at 150°C (44); 25.8 at 140°C ($M_n = 2700$)(45); $\gamma = 28.4, 29.6, 31.1, 32.1$, and 33.4 at 24°C for M_n of 410, 500, 740, 970, and 2840, respectively (49).

A polystyrene of $M_v = 44,000$ had $\gamma = 32.1$ at 140°C with $d\gamma/dT = -0.072$ (50); the pendant drop method was employed in this and the other series summarized in this paragraph. For polystyrenes of M_n equal to 1680, 2910, and 9290, the γ at 148°C was 29.3, 30.3, and 31.0 respectively, and $d\gamma/dT$ was about -0.064 up to near 220°C (51). A commercial polystyrene of an unknown molecular weight gave (11) 27.6 and 21.8 g/sec^2 at 200 and 300°C, respectively, so that $d\gamma/dT$ was near -0.058.

Five polystyrenes with reportedly very narrow distributions of molecular weights afforded the results condensed in Table 9.2 (52). It is seen that the temperature coefficients have, in this series, greater absolute values than in the earlier studies. A polychloroprene had $\gamma = 33.2$ g/sec^2 at 140°C(45). The γ of a poly(ethylene oxide) also designated as poly(ethylene glycol) of nominal molecular weight $M = 6000$ was 33.0 at 150°C, with $d\gamma/dT = -0.076$ (44). When M was varied between 400 and 6000, a minimum of γ was observed at $M = 3000$ (53, 54). At 25°C γ was 43.5 and 31.2 g/sec^2 for M equal to 300 and 2024, respectively (55). The original paper (56) was not available.

Poly(propylene oxides) at 40°C had a γ decreasing from 35.1 to 33.0 when M increased from 500 to 3000 (35). At 25°C, γ was 31.4 and 30.7 g/sec^2 for polymers of M near 400 and near 4100; $d\gamma/dT$ was approximately -0.08 (57).

Poly(tetrahydrofuran) had, at 40°C, a γ between 38.4 and 39.6 g/sec^2 for M varying from 1100 to 2000 (35), but a value of 27.7 at 25°C is given for $M = 2000$ (55). For M_n of 32,000, γ was 24.0 at 150°C (58), and for $M_n = 43,000$ it was 24.6 g/sec^2 at 140°C (46).

Dimethyl ethers of poly(ethylene oxide) have been studied at 24 (59) and 25°C (57). Data for various esters of poly(ethylene oxide) can be found in References 35, 60, and 61. Urethane formation raises the γ (62). Values ranging from 24 to 30 g/sec^2 are given for a poly(ethylene terephthalate) at 290°C (37, 56).

The γ of poly(vinyl acetate) was 25.9 at 180°C independently of the molecular weight (10^4–10^5) (50) and 28.0 g/sec^2 at 150°C for $M_n = 4400$, with $d\gamma/dT = -0.076$ (58, 63).

At 150°C (interpolated), the γ of poly(methyl methacrylate) of $M_v = 3000$ was 31.2 g/sec^2 (50). The rate of hydrolysis of this ester by sulfuric acid can be judged by monitoring the surface tension of the mixture (64). The γ at 140°C of poly(n-butyl methacrylate) was 24.1 at $M_v = 37,000$ (50), of poly(isobutyl methacrylate) 24.1 at $M_v = 35,000$, and of poly(t-butyl methacrylate) 23.3 at M_v of 6000 (46).

According to Reference 33, the γ of poly(Σ-caproamide) (or nylon 6) increased at 257°C with the degree of polymerization P to 62 g/sec^2 at $P = 120$. For other data see References 32 and 37.

None of the mentioned materials was, of course, a pure chemical individuum. Even less uniform were the copolymers studied, for instance References 55, 58, and 63.

Surface tension of liquids consisting of small molecules depends above all on the difference between the critical temperature T_c of the substance and the temperature of the measurement. For polymer liquids, T_c cannot be determined as decomposition takes place at

Table 9.2 Surface Tension of Molten Polystyrenes (52)

Molecular Weight	2100	4000	10,000	20,400	37,000
T(°C)	110	137	152	163	193
γ(g/sec^2)	34.9	33.3	32.9	32.3	30.8
T(°C)	152	189	230	206	254
γ(g/sec^2)	31.3	28.6	26.2	28.9	26.7

much lower temperatures. Still, in some instances, it should be possible to use the melting point T_m or the glass transition temperature T_g as a substitute for T_c; thus different polymers should be compared not at a constant T but as a constant ratio T/T_m or a constant ratio T/T_g. Such a comparison has apparently not been attempted yet.

Since many solutions of polymers have a much lower viscosity than the corresponding polymer melts, the measurement of the γ of the former systems is easier than that of the latter. Many determinations of γ have been performed on surfactant solutions, but the properties of polymeric surfactants are not reviewed in this chapter.

Surface tension was measured for polystyrene solutions in tetralin, decalin, and hexadecane at several temperatures (65), in toluene at several concentrations (66), in toluene and xylene (67), and in benzene at different ages of the solution and a range of molecular weights (68).

The γ of dilute aqueous solutions of five fractions of poly(ethylene oxide) in a range of temperatures was also determined in (69). Solutions of several esters of poly(ethylene oxide) in styrene (70) or dimethyl formamide (71) also have been studied. The critical micelle concentration of copolymers of ethylene oxide and propylene oxide could be detected from the γ of their solutions (72). For solutions of several poly(ethylene oxide) dipates in formamide see Reference 73.

Aqueous solutions of various formals and acetals of poly(vinyl alcohol) have been repeatedly studied (74–77). The γ of aqueous solutions of a copolymer of acrylic acid and butyl acrylate (78) and of copolymers of alkyl acrylates and methacrylic acid (79) also was reported. Dilute aqueous solutions of a poly (acrylamide) exhibited minima of γ at some temperatures (80).

9.2.2 Interfacial Tension

The foregoing data all refer to the boundaries of a polymer melt or a polymer solution against a gas phase. The interfacial tension γ_i between two polymer melts was not neglected either. Here the difficulties are even greater

than the melt-gas systems. First, both phases, not just one, have inconveniently high viscosities. Second, the experimental values of γ_i are likely to be affected by the mutual solubility of the two polymers. It is true that this solubility is small, but it should be remembered that γ_i depends on miscibility also when the two phases are as incompatible as alkanes and water (42).

Data for γ_i are available for about 40 binary systems (31). Only a few examples are recorded here. The tension along the boundary of two molten hydrocarbons usually is small. Thus in the system polyethylene-polystyrene at 180°C γ_i was 5.1 g/sec² (50). Unexpectedly, the γ_i of polyethylene-polychloroprene was even smaller: 3.4 at 180°C (46). At the boundary with polyethylene, poly(ethylene oxide) had $\gamma_i = 9.5$, and poly(tetramethylene oxide) 4.1 g/sec², and the pair poly(ethylene oxide)-poly(tetramethylene oxide) still had 3.8 g/sec², all at 150°C (58). Poly(vinyl acetate) had, at 180°C $\gamma_i = 10.2$ against polyethylene (45) and 3.5 against polystyrene (46); at 150°C γ_i was 4.5 against poly(tetramethylene oxide) (58). The γ_i at the interface of polystyrene and poly(methyl methacrylate) was 1.2 g/sec² at 180°C.

9.2.3 Unimolecular Layers

Because the solubility of so many polymers in water is extremely small, it is easy to find polymers suitable for spreading as unimolecular layers on the surfaces of aqueous solutions. If the surface tension of the uncontaminated liquid is γ_o, and that of the surface contaminated with a polymer film is γ_1, then the "surface pressure" Π is $\Pi = \gamma_0 - \gamma_1$. The concentration of the polymer on the surface usually is expressed in the terms of the area occupied by the unit weight of the film or by a segment of its chain. Thus, if x mg of a polymer is spread over 1 m² of the aqueous surface, then the area $A_1 = 1/x\,\text{m}^2/\text{mg}$. If the molecular weight of the segment (or monomer) is M g, then 1 mg of the polymer contains $0.001\ N/M$ segments, N being the Avogadro number. Hence the area per segment is $A_2 = 10^3\ M/xN$

Probably the most remarkable result of the

numerous studies of unimolecular polymer films is that the curves of Π as a function of A_2 are very similar to those obtained with monolayers of nonpolymerized materials such as potassium stearate or octadecyl acetate. It looks as if a segment, such as $-CH_2CH_2O-$, "does not know" whether it is alone or is a part of a long chain. It was believed that tension Π is analogous to osmotic pressure and thus could be used to determine the molecular weight of the material spread on the surface; but the molecular weight of polymers has only a small effect on the Π of their films. For instance, when the degree of polymerization of poly(vinyl acetate) varied from 10 to 330, the area per segment at $\Pi = 10$ g/sec^2 remained constant and equal to 18–19 Å^2 (81). Another series of poly(vinyl acetates) in which the molecular weight varied between 60,000 and 410,000 had at $A_1 = 7$ m^2/mg, tensions Π varying only between 0.142 and 0.157, and, at $A_1 = 2.7$ m^2/mg, Π was almost constant at 2.13–2.19 g/sec^2 (82).

As reproduction of the experimental Π-A_1 or Π-A_2 curves would demand too much space, only some values for the limiting area A_0 and the collapse pressure Π_m are listed here. The curve $\Pi = f(A_2)$ at small values of A_2 usually is an almost straight line; hence it can with some confidence be extrapolated to $\Pi = 0$; the area corresponding to this extrapolated value is A_0. The collapse pressure in many instances is a more definite quantity. When the "monolayer" is compressed its Π gradually increases, but it ceases to increase as soon as the value of Π_m is reached. Further compression causes thickness of the film without raising the tension of the latter.

The limiting area A_0 for a poly(vinyl butyral) in water was 30 Å^2 (83), for a poly(vinyl benzal) 32 Å^2 (83) and for a half-hydrolyzed poly(vinyl benzal) it was only 23 Å^2 (84) per segment. Also poly(vinyl benzals) of other degrees of hydrolysis and several mixtures of these polymers have been investigated (84). For data on other acetals of poly(vinyl alcohol) see Reference (81).

Poly(vinyl acetate) probably is the most popular substance for determining Π and A_2. However, at least some samples of this material showed marked hystersis (on 0.01 N HCl);

at a constant A_2, the difference between the Π values during compression and during expansion reached as much as 3 g/sec^2. The area A_0 was 24 Å^2 for these films (85). For another sample of this polymer (mol. wt. 4000) spread on water, the A_0 was 26 Å^2 (86). Additional data on poly(vinyl acetate) and poly(vinyl stearate) can be found in Reference 87.

Several determinations of Π-A_2 curves are available for poly(alkyl acrylates); the collapse pressure Π_m varied only between 20 and 25 g/sec^2 (81). The A_0 of a poly(methyl methacrylate) on water was 22 Å^2 (88); for a poly(ethyl acrylate) on 0.01 N HCl the A_0 was 30 Å^2, and the collapse pressure was 22 g/sec^2 (85). A poly(n-butyl acrylate) had $A_0 = 29$ Å^2 and $\Pi_m = 22$ g/sec^2 (89). The effect of partial hydroylsis on the Π-A_2 curves was described in References 86 and 87.

The values of A_0, on water, of poly(methyl acrylate) and poly(methyl methacrylate) were greatly influenced by the solvent used to spread the "monolayers." Thus, for the latter polymer, A_0 was near 5Å^2 when acetone or dioxane was the solvent, while chloroform afforded the greatest area A_0, namely over 20 Å^2 (90).

Poly(methyacrylic acid) films were studied on water acidified to pH 1.6 (87). The increase of A_0 with the length of the alkyl group in methacrylates was observed: A_0 was 15, 19, 24, 27, and 30 Å^2 for polymethacrylates of methyl, ethyl, propyl, n-butyl, and octadecyl (91). The A_0 of a poly(methyl methacrylate), molecular weight near 140,000, was 18 Å^2 (92); that is markedly greater than the value given by an earlier investigator (91). For other data see Reference 87. The effect of hydrolysis on the Π-A_2 curve of poly(methyl methacrylate) was studied in Reference 64.

The A_0 and Π_m values for poly(ethylene succinate), poly(pentamethylene succinate), and poly(neopentyl succinate) were, respectively, 60–70 Å^2, 90 Å^2, and 62 Å^2, as well as 4, 12, and 18 g/sec^2(93). Poly(ethylene adipate) had $A_0 = 75$ Å^2 and $\Pi_m = 11$ g/sec^2, whereas neither poly(isopropylene adipate), that is, $[-(CH_2)_4COOCH(CH_3)(CH_2OCO]_n$, nor poly-(trimethylene adipate) exhibited a definite collapse pressure (94). The Π-A_2 curves for poly(ethylene adipate) and the urethane formed by this compound with 2.4 toluene-diisocyanate were obtained in Reference 62.

Considerable differences have been observed between differed stereoisomers. Thus atactic, isotactic, syndiotactic, and stereo-block types of poly(methyl methacrylate) have been compared. At large areas above 1 m^2/mg, the isotactic isomer has the highest, and the syndiotactic the smallest value of Π (95, 96). Linear, branched, and cross-linked insertion polymers of methyl methacrylate had A_0 of 17, 17, and 8 Å^2 per segment, respectively, and Π_m of 25, 32, 56 g/sec^2 (97). In a later work (98), it was found that Π was much greater for a linear than for a branched poly(vinyl acetate) of an almost identical molecular weight; all at a constant area A_2.

In one of the few publications on the temperature dependence of Π-A_2 curves, it is claimed (99) that in the interval between 15 and 45°C the Π of poly(vinyl acetate) has a minimum near 25°C and that of poly(vinyl alcohol) has a maximum near 35°C.

The curve of Π versus A_2 on pure water can be shifted to higher pressures, or can be made steeper or shallower by additional solutes. The mechanism of these three effects is discussed in Reference 100.

Some polymers can be spread also in an interface between an aqueous solution and a liquid ("oil") immiscible with water. Comparison between the Π-A_2 curves in an air-water and a water-oil interface have been performed, for instance, with decane (101) and petroleum ether (86). For the behavior of poly(vinyl alcohol) and some copolymers in the interface between water and paraffin oil see References 102 and 103. Unimolecular films of polymer mixtures have also been studied. See References 84, 85, 88, 92, 104, 105.

Surface potentials, that is the changes in the potential difference between a liquid and its vapor, caused by the spreading of a unimolecular polymer film on the liquid surface, are noted in References 81, 86, 91, and 101.

9.2.4 Foams

Many polymer solutions are capable of foaming. This is true above all for two classes of polymers that are not further considered in this chapter, namely polymer surfactants and proteins. Substances such as $C_9H_{19} \cdot C_6H_4 \cdot O\,(CH_2CH_4O)_nH$ and $C_9H_{19} \cdot C_6H_4 \cdot O\,(CH_2CH_2O)_n \cdot SO_2 \cdot ONa$ are popular foaming agents, while the foaming ability of gelatin or egg albumin solutions is well known (106).

Observation that deserve to be repeated and expanded deal with the foam fractionation of polymers. When a solution of poly(vinyl alcohol), in which 0.5% of the hydroxyl groups were displaced by acetyl groups, was foamed, the first foam fraction contained 0.6% of acetyl radicals, and the last only 0.35% of these; thus separation of differently substituted polymers appeared possible (107). Also a separation according to the molecular weight was claimed: a partly hydrolyzed poly(vinyl acetate) gave an early foam fraction of molecular weight near 44,000 and a later fraction of $M \simeq 21,000$ (107).

Another fractionation of poly(vinyl alcohol) was achieved when a solution of this material was shaken with air and then allowed to stand for several hours. The surviving foam gave a more intense color reaction with iodine than the initial sample; this difference is believed to be caused by preferential accumulation of syndiotactic (as compared to isotactic) isomers in the foam. No separation according to the molecular weight was achieved in these experiments (108).

In a copolymer of methacrylic acid and methyl metharylate, different chains have different ratios of carboxyl to carboxymethyl. When such a polymer was foamed, the first foam fraction had a higher ratio than the initial polymer (109).

9.2.5 Theoretical Approaches to Polymer Surface and Interfacial Tension

Attempts have been made to apply to polymers most of the correlations and theories originally directed to understanding the surface tension of small-molecule liquids. While qualitative agreement with experimental values has been achieved, it appears that high-polymer liquids do not conform quantitatively to these theoretical predictions anymore than do most nonpolymeric liquids. More complete and precise data, for both surface and bulk properties of polymer liquid, are needed to assess the significance of the discrepancies between theory and experiment.

As predicted by Laplace, the temperature coefficient of γ is determined mainly by that of the density ρ of the liquid. For the polymer melts examined so far, γ is proportional to ρ^n, the exponent n having a value between 3.0 and 4.4 (31). In principle, of course, the difference $\rho - \rho_v$ rather than ρ alone should be used, but the vapor density ρ_v of the common polymers is too small to require consideration.

The effect of molecular weight or degree of polymerization on γ has been repeatedly mentioned. Apparently, data on additional systems are needed before the problem is clarified. The effect of the polymer composition on γ seems to be similar to that in monomers; in particular, the idea of parachor can be applied to polymer melts, as indicated, for instance, in References 30, 31, and 110.

If the atomic parachors for the elements in the monomer are known, it may be possible to calculate the γ of the polymer, assuming additivity of the parachors (111, 112).

Schonhorn (113) has used the scaled particle theory of rigid spheres to develop an equation for the surface tension of polymers in terms of the cohesive energy density; the result resembles Hildebrand's empirical relation between surface tension and solubility parameter (114). Schonhorn also noted a correlation between the temperature dependence of surface tension and viscosity (115). Roe (116) and Patterson and Rastogi (117) have discussed the application of the principle of corresponding states and cell theories of liquids to polymer surface tensions. The latter authors have suggested that a direct test of the corresponding states principle for surface tension is provided if the quantity $\gamma \beta^{2/3} \alpha^{1/3} / k^{1/3}$ is a universal function of αT (β = isothermal compressibility, α = thermal expansion coefficient, k = Boltzmann constant). They found that data for several polymers including polymethylenes, polyisobutylene, poly(dimethylsiloxanes), polyethylene oxide, and polypropylene oxide, as well as other polyatomic liquids, all fitted a single curve when plotted in this way. Data for polystyrene

also fit this correlation with about $\pm$ 10%, if the compressibility data of Matsuoka and Maxwell (118) are used. On the other hand, the reduction parameters used by Roe to obtain a corresponding states plot (based on the Prigogine-Saraga cell theory) overestimate the surface tension of polystyrene by 20–30% (51). Patterson and Rastogi also found that cell model calculations agreed reasonably (but not quantitatively) with their empirical corresponding states plot.

Slow and Patterson (119) have compared the solubility parameter, parachor, and corresponding states approaches in regard to their ability to predict $d\gamma/dT$ and the $M^{-2/3}$ molecular weight dependence. They concluded that these concepts are closely related, in each case yielding temperature and molecular weight dependency through changes in free volume. The solubility parameter relation does not predict either $d\gamma/dT$ or the molecular weight relationship well. The parachor and corresponding states correlations, however, do yield fairly good predictions for the slopes of γ versus T and $M^{-2/3}$, even though the absolute magnitudes of the calculated surface tensions are only approximately correct.

LeGrand and Gaines (120) also attempted to correlate the molecular weight dependence of surface tension with bulk properties through a crude free volume argument. Essentially, this assumed that the molecular weight variation of the surface tension and the glass transition temperature T_g both reflected free volume changes, and hence should be related. Agreement with experiment was poor reflecting (at least in part) difficulties with the assumed T_g-molecular weight relationships (120).

Steward and von Frankenberg (121) have applied the significant structure theory to the surface tension of polyethylene; again, agreement is only qualitative.

Mixtures of smaller oligomeric chain molecules seem to be capable of treatment by lattice-model equations (122, 123), but the significance of this observation is not yet clear. Rastogi and St. Pierre (124) applied several regular solution equations to their data on the surface tensions of mixtures of polyethylene oxide, polypropylene oxide, and poly(epichlorohydrin) without success. No attempt has yet been made to develop a theory for such mixtures.

Based on current understanding, it would appear that the surface tension of polymer fluids can be treated approximately within the context of existing theories of the liquid state, and qualitative estimates of surface tension values can be made, at least in some cases, from measurements of other physical properties. (Probably the discrepancies so far observed between theory and experiment reflect at least as much deficiencies in our understanding of bulk polymer liquids as in the extension of that understanding to surface phenomena.)

With regard to interfacial tensions between molten polymers, there are again some approximations originally developed by analogy to theories for low molecular weight fluids. Much analysis has so far attempted to apply the theories of Good and Girifalco (125) and Fowkes (126), or modifications of them (127), which state essentially that the interfacial tension is related to the surface tensions of the two separate liquids by expressions of the form $\gamma_{12} = \gamma_1 + \gamma_2 - D$. The deviation terms D have different significance in the several variants of these theories, but they are in practice evaluated by comparisons between different systems. Wu (128) has compared directly measured interfacial tension between polar polymer pairs with those that could be calculated on the basis of these theories from measurements of the interfacial tensions of the respective polymers against polyethylene. While qualitative correlation was achieved, the quantitative discrepancies were in some cases greater than 50%, and the theoretical values were often too low as too high.

Recently, Helfand and Tagami (129) formulated a statistical-mechanical theory of the interface between immiscible polymers A and B. The other theories discussed are outgrowths or adaptations of free volume, cell, hole, or significant structures theories of low molecular weight fluids. These theories are based on some lattice model of the liquid state augmented by some form of random mixing hypothesis of the connected statistical segments of the polymers. The scaled particle theory approach of Schonhorn hardly takes into account the connectedness of the polymer segments. The approach of Helfand and Tagami is more fundamental, being based on a self-consistent field that determines the configurational statistics of the polymer molecules in the interfacial region. At the interface, energetic forces (determined essentially by the polymer A-polymer B inter–action parameter) tend to drive the A and B molecules apart, but this separation must be achieved in such a way as to prevent a gap from opening between the polymer phases. The energetic force on an A molecule must be balanced by an entropic force describing the tendency of A molecules to penetrate into the B phase, because of the numerous configurations of the A molecule.

Currently one of the weaknesses of the theory is that it has been developed only for the case where A and B polymer molecules are so similar that they possess identical degrees of polymerization z, effective length of the monomer units b (thus the mean end-to-end distance zb^2), density p_0, and compressibility k. Helfand and Tagami recommend the use of the geometric mean when these properties are not actually identical. Under the circumstances, and with the considerable difficulty of assigning values to the parameters required, the practical utility of the theory at its present state of development is very limited.

9.3 ADSORPTION OF POLYMERS

Adsorption of polymers at solid and liquid surfaces is a very popular field of study. A book (130; see also Reference 131), written in 1971 contains 261 references, and probably over 100 additional papers have appeared since. One of the reasons for this popularity is the belief that adsorption of resins on pigments is of great importance for the stability of paints. As, however, the composition of customary paints is too complex for deriving rules of general validity, most papers on this subject are disregarded in the following discussion.

In discussions of the adsorption of compounds of a low molecular weight, usually the first question to be answered is: is the sorption reversible? A considerable irreversible uptake of chlorinated poly(vinyl chloride) and some other resins was observed on Prussian blue from xylene (132). Also the sorption of two polystyrenes by disperse silica (Aerosil) from carbon tetrachloride and from toluene was only partly reversible (133). Poly(ethylene glycols) of M 600 and 6000, adsorbed on Aerosil from CCl₄, were not released by diluting the supernatant liquid with pure carbon tetrachloride (134). Thus, sorption of polymers appears to be irreversible more often than that of organic compounds of a low molecular weight. Such a behavior might be expected. A long chain usually is attached to the adsorbent at many points, and it is improbable that all these numerous bonds would snap at once to liberate the immobilized molecule. Unfortunately, the problem of reversibility has not been considered in many experimental studies of polymer adsorption. Obviously, it is not safe to apply a theory based on reversibility, such as the familiar Langmuir theory, or the thermodynamic treatments of adsorption heat, to experimental results that correspond to no equilibrium.

As polymer adsorption has been investigated from solutions only (that is, not from vapor), the adsorbed amounts calculated in the customary manner (that is, from the decrease in the concentration after the introduction of the adsorbent in the solution) refer only to the "apparent" adsorption and require a correction for the simultaneous adsorption of the solvent.

Let the initial solution consist of P g of polymer and S g of solvent; thus the weight-to-weight concentration of the solute is $P/(P + S)$. If the adsorbed amount of the polymer is A_p and that of the solvent is A_s, then the concentration of the equilibrium liquid is $(P - A_p)/(P + S - A_p - A_s)$. The apparent adsorption A_a is then given by Equation (9.1).

$$A_a = P - \frac{(P - A_p)(P + S)}{P + S - A_p - A_s}$$

$$= \frac{SA_p - PA_s}{P + S - A_p - A_s} \quad (9.1)$$

which is not equal to A_p. As given by Equation (9.2),

$$A_p = A_a \frac{P + S}{A_a + S} + \frac{P - A_2}{A_a + S} A_s \quad (9.2)$$

shows that A_p is not significantly different from A_a as long as $P \ll S$ and $A_s \ll A_a$. Otherwise, the neglect of the difference $A_p - A_a$ is not justified, although it is not uncommon in the published literature. (See also Reference 42.)

The difference between the true and apparent sorptions is particularly clear whenever A_a is negative. In the instance of polystyrene (of M_v near 290,000) taken up by an Aerosil (pretreated with dilute hydrochloric acid) from a solution in toluene this was the case, that is, toluene was taken up preferentially to polystyrene (133). Also several poly(isobutylenes) of various M were negatively sorbed from benzene and carbon tetrachloride on an oxidized channel black (135), and poly(butyl methacrylate) was negatively sorbed from 2-butanone and butyl acetate by rutile, and from toluene and chloroform by a channel black (136).

Perhaps the difference between A_a and A_p can account also for at least a part of the unexpected, but not unusual, observation that, in many systems, A_a is too small for measurement. This was the case for poly(methacrylic acid) and several of its salts, all dissolved in water, in the presence of fine glass fibers (130) for poly(ethylene oxide) in dioxane and dimethyl formamide on Aerosil (137), and so on. It follows from Equation (9.1) that $A_a = 0$ whenever $SA_p = PA_s$ or when Equation (9.3) is applicable.

$$\frac{A_p}{P} = \frac{A_s}{S} \quad (9.3)$$

This means that, if equal percentages of the two ingredients are taken up by the sorbent, the apparent adsorption is zero. Equation (9.3) may be a more likely explanation of the preceeding experimental findings than the hypothesis of $A_p = 0$ advocated in some original publications. See the discussions in References 138 and 139.

The apparent adsorbed amount of an ingredient of a binary solution often shows a maximum and a minimum (in which this amount is negative) when the concentration varies from 0 to 100%. Maxima of A_a have been noticed many times for polymers as well, but usually is remained unknown whether the true adsorption also was maximal at this composition. For instance, TiO_2 took up more alkyd resin from its solution in xylene when the resin concentration was 3 wt% than at any other composition tested (140). In other systems, maxima on the adsorption isotherms have been observed (141) for polyesters (from ethylene glycol and dibasic aliphatic acids) on TiO_2 in chloroform or toluene, for a poly (methyl methacrylate) on alumina in a mixture of toluene and xylene (142), for polystyrene on glass fibers in toluene (143), and so on.

Unexpected minima on adsorption isotherms (in which the amount of polymer adsorbed is still positive) are claimed in Reference 144. Various polyacrylamides were sorbed by kaolinite and other clay minerals from water, and the sorption was completely irreversible.

A small step from the apparent to the true adsorption can be made by determining the adsorption of the solvent, in a separate experiment, from the vapor phase. Thus the apparent uptake of a poly(ethylene oxide) of M near 15,000 was more than twice as great from chloroform as from methanol, and, at a constant relative vapor pressure p/p_s of the solvent, the sorption of methanol was, for example, twice as great as the uptake of $CHCL_3$; thus the difference between the two A_a values may have been mainly due to the difference between the corresponding A_s values (137). For anologous data referring to poly (ethylene oxide) and activated charcoal see Reference 145.

Repeated attempts from 1955 (146) to 1974 (147) have been made to determine the thickness δ_f of the adsorbed film by hydrodynamic measurements. The volume rate of flow of a liquid through a cylindrical capillary, at constant external conditions, is proportional to r^4, r being the internal radius of the cylinder. If the capillary walls are coated with an adsorbed film, then the radius of the bore available to

the liquid is only $r - \delta_f$, and the rate of flow u_a is in the ratio $(r - \delta_f)^4/r^4$ less than that u in the absence of adsorption. Thus one may write Equation 9.4

$$\frac{u_a}{u} = \frac{(r - \delta_f)^4}{r^4} \qquad (9.4)$$

The values of δ_f obtained by this method usually are in the range between 200 and 800 angstroms (148). The calculation is based on the assumption that the film is rigid and disregards the surface roughness and the swelling of the walls; thus the results are not likely to be exact. Moreover, the meaning of a hydrodynamic thickness from the molecular viewpoint is not clear; this circumstance is mentioned again later.

An analogous method, subject to identical doubts, utilizes the rate of sedimentation of coated spheres. If coating raises the sphere radius from $r + r + \delta_f$ cm, the preceding rate is lowered approximately in the ratio $r/(r + \delta_f)$. The calculated δ_f on polystyrene balls was, for instance, near 50 Å for an oligomer (149) and near 140 Å for poly(vinyl alcohol) of $M = 28,000$ (150). For a comparison of δ_f values obtained by the methods of sedimentation, rate of diffusion, and electrophoresis see Reference 151.

A bold attempt of measuring A_p, see Equation 9.2, directly is described in Reference 152. It is a common method in gas adsorption studies to suspend a sample of solid adsorbent in the gas and to measure the weight of the former at appropriate intervals; the observed weight increase is caused, almost exclusively, by the condensation of gas on the sorbent. This method of direct weighing was applied to metal foils suspended in polymer solutions. The values of A_p obtained in this manner greatly exceeded the amounts of A_a observed in similar systems. Thus Γ_p, that is A_p referred to unit area of sorbent, was, for a poly(vinyl acetate) of $M = 700,000$ on a copper foil in benzene at 25° C as high as 45 mg/m^2, and at 45° C as high as 140 mg/m^2. As examples quoted in the later sections of this chapter demonstrate, the usual magnitude of Γ_a (i.e., A_a per unit area) is 1 mg/m^2. The measurable weight of a foil in a liquid is greatly influenced by buoyancy and by capillary pull. Apparently these obstacles were so great that the weighing method never became popular.

Not only the thickness of the adsorbed film but also the composition of the latter are supposed to be obtainable from ellipsometry. If a mirror is coated with a film, then the reflection of elliptically polarized light from the mirror is altered in two respects. First, the ratio of the two reflection coefficients, one for the electric vibrations in the plane of incidence and the other for perpendicular electric vibrations, is changed. Second, the phase difference between these two polarizations is shifted. These two changes, directly given by the experiment, can be predicted if the film thickness δ_f and the refractive index n_f of the film material are known. Unfortunately, the inverse operation is mathematically much more difficult. It is almost impossible to derive, without ambiguity, the values of δ_f and n_f from the experimental data.

Even more disturbing is the difference between the systems for which the theory (Drude, 1889) was derived and those of interest for polymer sorption. The theory is valid for films that are perfectly homogeneous and form sharp and plane surfaces of separation from the mirror on one side and the medium (liquid or gas) on the other. Adsorbed polymer films can be uniform in exceptional circumstances only, their boundary with the mirror is not plane because of surface roughness, and, above all, there usually is no real boundary with the solution from which adsorption has taken place; the average concentration gradually decreases from the wall toward the interior of the liquid. Consequently, the physical meaning of the values of δ_f and n_f, derived from the measurements, is not particularly clear (153).

As a rule, the experimentalist is more interested in the chemical composition of the film than in its refractive index. To obtain the former, it is simply assumed that the concentration c_f in the film is identical with that of the bulk solution whose refractive index is n_f. The product of this concentration and the thickness δ_f ought to be equal to the true adsorbed amount Γ_p. Unfortunately, the values of Γ_p needed for comparison are difficult to obtain, and a comparison of $c_f \delta_f$ with Γ_a is not easy because of the relatively small area (of the mirror) on which sorption takes place.

In spite of the uncertainty clouding the meaning of c_f and δ_f deduced from ellipsometry, several typical values are presented in this and the two following paragraphs. Polystyrene was sorbed on chrome ferrotype plates, stainless steel mirrors, and some other polished metals from solutions in cyclohexane near 34° C, which is the theta temperature for the system, as explained in Chapter 2 of this book. When the polystyrene concentration was raised, both c_f and δ_f increased to a plateau that was, for instance, for $M \approx 1,370,000$, near 0.16 g/cm^3 and 550 Å (154). It is seen that δ_f here is in the range derived earlier from filtration measurements. On the other hand, the value of c_f shows the major part (to be exact, 84%) of the film consisted of the solvent, so that it is impossible to visualize a sharp boundary between the film and the solution. In the range of molecular weights between 0.54 and 3.3 × 10^6, the δ_f along the foregoing plateau increased with the M.

Chrome mirrors, that is, chromium metal coated with a thin (<30 Å) film of a chromium oxide, have been used in Reference 154. Polystyrene from cyclohexane at 36° C gave δ_f between 100 and 200 Å as long as the concentration of the equilibrium solution was less than 0.002 g/cm^3. The c_f, in the equilibrium concentration range between 0.0001 and 0.002 g/cm^3, was roughly between 0.15 and 0.30 g/cm^3, depending on the M of the polymer, that is, the "adsorbed layer" in all instances contained considerably more of the solvent than of the solute. Even less dense layers were observed with a poly(ethylene glycol). At the equilibrium concentration of 0.015 g/cm^3, δ_f appeared to be between 100 and 230 Å, and c_f was only 0.04–0.10 g/cm^3. However, another, not clearly described, pretreatment of the mirrors resulted in a film thickness of about 30 Å. The ellipsometric method has been applied also to the polystyrene adsorption on liquid mercury (155). Again, cyclohexane at 35° C was the solvent used. The film thickness was near 300 Å independently of the molecular weight of the polymer. The c_f appeared to be about 0.1 g/cm^3; thus the film was mainly cyclohexane. No independent determination of Γ_p for comparison with $c_f\delta_f$ could be found in the paper.

A method that appears to afford the actual number of polymer segments in direct contact with the sorbent surface is based on the study of infrared spectra (156). Poly(tridecyl methacrylate) was adsorbed on a nonporous silica powder from dodecane. A thin layer of the powder was spread between two crystals of sodium chloride, and the extinction coefficient of the light passing through the system was determined for the wavelengths characteristic for the carbonyl group in methacrylate. The absorption peak for the unaffected methacrylate was near 1736 cm^{-1}, but the powder containing the polymer had also a peak near 1714 cm^{-1}, and this peak is attributed to the CO groups affected by the contact with the silica surface or with the hydroxyl groups on the latter. From the relation of the light absorptions at these two wavelengths, the percentage p of the affected carbonyls can be readily calculated. It proved to be between 37 and 53% when the Γ_a (determined in the customary manner) was roughly 0.001 g/m^2. As, from Equation (9.1), the true adsorption Γ_p is usually greater than Γ_a, the true value of p ought to be less than those found in the literature.

A similar silica powder adsorbed poly(vinyl acetate) from trichloroethylene, and the ratio p varied, almost irregularly, between 0.30 and 0.46 when the amount taken up by 1 g of SiO$_2$ varied between 0.01 and 0.12 g (157). When an identical powder was used for sorbing poly (isopropyl acrylate) from trichloroethylene, no clear difference could be observed between the A_a values for two atactic, two isotactic, and a syndiotactic isomer; in all instances, at equilibrium concentrations between 1 and 3 g/l, the A_a was near 0.13 g/g SiO$_2$. However, the isotactic isomers often showed higher p values (above 0.50) than the other three (usually below 0.40) (158).

Higher values of p, sometimes exceeding the theoretical limit of 1.0, have been observed (159) for the sorption of polyesters such as the poly(tetramethylene glycol of ester of adipic acid). In this study, not only the changes caused by the adsorption in the wave number and intensity of the characteristic carbonyl absorption were determined but also those in the band due to the OH groups on the adsorbent surface. The adsorbent again was a nonporous SiO$_2$ carrying one OH on each 40 Å^2 of the surface. Carbon tetrachloride was the solvent used. The technique adopted in Reference 159 was extended also to a fine dispersed alumina sorbent, with less definite results (160).

Also for the adsorption of poly(vinyl acetate) by Aerosil from carbon tetrachloride, not only the relative number of carbonyl groups in contact with the sorbent, but also the concentration of hydroxyl groups on the silica surface, modified by this contact, was determined; p seemed to be near 0.2 for the first ratio (161).

Nmr was used as well to detect the amounts of polymer immobilized by the sorbent (162). The proton signals clearly visible in a solution of isotactic poly(methyl methacrylate) in deutero-chloroform disappeared when four parts of silica powder (of surface area near 190 m^2/g) were added to one part of polymer. Thus it seems that the mobility of all adsorbed chains is greatly restricted. The shift in the CO frequency, caused by the proximity of SiO$_2$, was observed also, but no calculation of p from the infrared spectrum is presented.

In many reports on polymer adsorption only the apparently sorbed amount of A_a was determined. The dependence of A_a on concentration, on the nature of the polymer, solvent, and sorbent, and other parameters is reviewed in the following paragraphs.

The dependence of A_a on the concentration of the final (equilibrium) solution at a constant temperature usually is presented graphically as an adsorption isotherm. Because the polymers employed almost always are mixtures of chains of varying lengths, the coordinates of these curves are generally expressed as weight-to-weight and weight-to-volume ratios. Thus A_a is commonly shown as grams (or milligrams) of polymer taken up by 1 g of the sorbent, and the equilibrium concentration c_0 of the solution is given as grams of solute in one l of solution, or in related units.

Some adsorption isotherms of polymer solutions are indistinguishable from those of solutions of small molecules. In many systems, however, the rise of A_a with c_0 at very small values of c_0 is steeper than is common for ordinary solutes, and the dependence of A_a on c_0 at greater c_0 is less marked; in other words, the isotherms of polymers often look like a broken reed, whereas those of small molecules are like a gently bent rubber hose. An example of the first kind is shown in Figure 9.3. It refers to the uptake of a polystyrene ($M = 250,000$) by Graphon from butanone (upper curve) and toluene (lower curve) (163); it is seen that the first few experimental points of each curve practically fall on the ordinate axis.

As far as the nature of polymer is concerned, unfortunately, no clear picture emerges from the information available. For instance, in the sorption of esters by powdered glass from toluene solutions, A_a decreased in the series

poly(methyl acrylate) $>$ poly(ethyl acrylate) $>$ poly(propyl acrylate) $>$ poly(butyl acrylate), and the A_a of the last named polymer was practically zero (164). This was also the order of A_a values for the adsorption on an alumina sample, still from toluene solutions. However, it was inverted for poly(alkyl methacrylates) on Al_2O_3 so that the A_a of poly(butyl methacrylate) was much greater than that of poly(methyl methacrylate). At a constant c_0, for example, 1 wt %, the A_a of poly(methyl acrylate) was about 11 times that of poly(methyl methacrylate), but the difference was small for the two butyl esters (165). Substitution of a carbon black for alumina again altered the behavior (166). The A_a along the plateau of the isotherm, see Figure 9.1, when toluene was the solvent, increased in the order poly(butyl acrylate) $<$ poly(ethyl acrylate) $<$ poly(ethyl methacrylate) $<$ poly(methyl methacrylate) $<$ poly(butyl methacrylate) $<$ poly(methyl acrylate). The order of polymer adsorption from other solvents (e.g., acetone) was not identical with that from toluene.

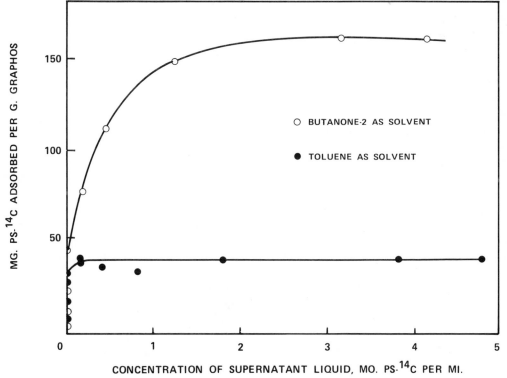

Figure 9.3   Absorption isotherms of polystyrene on Graphon from butanone and toluene.

Polyesters of adipic acid and the so-called neopentyl glycol were taken up by rutile from benzene solutions. The A_a was greater when the polymers contained an excess of the acid than when an excess of glycol was present (167). Analogously, the A_a, on an anatase sample, of some polymers of adipic acid and hexamethylene glycol was large and positive when the end groups of the chain were carboxyls, but small and negative when these were esterified (168). Perhaps the findings in Reference 169 afford a third example. When fatty polyamides of the general formula $HO[OCRCONH(CH_2)NH]_nH$ were sorbed on $\alpha - Fe_2O_3$ from cyclohexanone, the A_a at a given c_0 was greater—the greater the amine value of the polymer—determined by titration with HCl; unfortunately, the molecular weight of the polymer was smaller—the greater this value—so that the series observed experimentally may have been influenced by the trend in the M.

Also, when the polymer does not contain active groups such as—$COOH$ or —NH_2, the A_a strongly depends on its composition. Two polystyrenes of very different molecular weights (40,000 and 500,000) were adsorbed by silica from tetrachloroethylene in much greater amounts (nearly three times as much) as two polybutadienes of $M \approx$ 190,000 and 390,000 (170). Confirming this observation, polystyrene of $M \approx$ 78,000 was taken up by a porous activated carbon from toluene in a greater amount than a polybutadiene ($M \approx$ 93,000), which was followed by a copolymer of styrene (48%) and butadiene, and finally by poly(methyl methacrylate) of $M \approx$ 88,000 (171).

Sorption by fine dispersed silica was greater the more oxygen was present in the polymer (172). At comparable molecular weights, poly(oxyethylene) had A_a about five times as great as poly(oxypropylene); and copolymers of a tetrahydrofuran and poly(oxypropylene), in which the chains had hydroxyl groups at both ends, showed a greater A_a than the polymers with only one end hydroxyl.

The A_a appears to depend not only on the composition but also on the internal structure of the polymer. For instance, Graphon took up from a solution in benzene a markedly greater amount of a block copolymer of polystyrene and poly(methacrylate) than of a random copolymer of styrene and methyl methacrylate, at the constant ratio of the two monomers in the chain (173).

The influence of stereoisomerism on adsorption has been mentioned (158). In a similar comparison (173) it was found that isotactic and conventional poly(methyl methacrylates) were equally sorbed by silica from toluene or 1.2-dichloroethane, whereas, unexpectedly, the isotactic isomer was positively adsorbed (e.g., $A_a = 0.01$ mg/g) and the conventional isomer was negatively adsorbed from a solution in acetonitrile.

The molecular weights have been repeatedly mentioned because the difference in M may overshadow the effect of the chemical composition.

Two opposite effects of M can easily be predicted. A polymer chain is immobilized whenever even one of its segments is sorbed. The probability of this adsorption evidently is greater the longer the chain. Hence A_a should have the tendency to increase with M. Several investigators proposed a quantitative correlation, namely Equation (9.5) (174, 175),

$$A_a = KM^\alpha \qquad (9.5)$$

K and α being constant. It was expected that α would be near zero when every segment of the chain is attached to the solid, and near one when every chain is attached by one segment only.

This reasoning can apply to only those sorbents whose whole surface is available to chains of any length, that is, to essentially nonporous solids. When the adsorbent is porous, the polymers, which in many dilute solutions are present chiefly as nearly isometric coils or bundles (see Chapter 2), may be too voluminous to penetrate into a narrow pore. Thus on a fine, porous sorbent the A_a of short chains may be expected to exceed that of longer chains. In these systems the α of Equation (9.5) would be negative.

Experimentally, the value of α covers a wide range from about $+ 1$ to below $- 1$ and to some systems Equation (9.5) cannot be applied. Thus, when poly(methyl methacrylates) were sorbed by glass from toluene (164), the highest A_a (along the plateau seen in Figure 9.3) was observed for $M = 6 \times 10^6$; polymers having $M = 10^7$ and $M = 2 \times 10^4$ had lower values of A_a. Polyesters from adipic acid and 1.6-hexamethylene diol, adsorbed by anatase from ethyl acetate (168), gave $A_a = KM^{-1.15}$, that is, a negative exponent although the anatase used was supposed to be nonporous. In such instances it is particularly important to make sure that the equilibrium was reached: a small A_a for very long chains may be apparent also in the sense that the slow-moving large molecules did not have time to reach the sorbent surface when the measurement was performed. When copolymers of ethylene and vinyl acetate were sorbed by glass spheres from benzene, the α was near 0.3 as long as M was less than 180,000, but A_a was almost independent of M for longer chains (176).

An example in accord with the expectation is found in Reference 177. Two polystyrenes, of $M = 43,000$ and 290,000, were compared in toluene solutions. When the silica sorbent had pores of about 250 Å, the A_a of the first was nearly twice that of the second specimen, but the A_a values on the sorbent with pores of 1000 Å were almost identical for the two polymers. Also, the systems studied in Reference 140 behaved as expected. The A_a along the plateau of the isotherm was almost independent of the M of poly(ethylene oxide) when this was adsorbed from five different solvents by porous charcoal; presumably the amount sorbed by unit area increased with M, but the area available for sorption was smaller the greater M was. The behavior on a nonporous adsorbent (Graphon) was different: when the M of poly(ethylene oxide) was raised from 4700 to 18,000, the A_a from an aqueous solution rose from 48 to 62 mg/g, and a similar increase occurred also from a solution in methanol.

Three more recent examples of the increase of A_a with M are referred to in this paragraph. When poly(vinyl alcohol) samples of M_v ranging from 7000 to 67,000 were adsorbed on polystyrene from water, the A_a along the plateau was proportional to $M^{0.5}$ (150). The greatest amount (Γ_a mg/m^2) sorbed by unit area of graphitized carbon from a poly(ethylene glycol) solution in water was near 0.3, 0.6, and 0.8 when M was 300, 3000, and 15,000 respectively (178). A glass powder sorbed, from acetone or benzene, more of β-cyanoethylated poly(vinyl alcohol) i.e., [—CH$_2$ CH(O · CH$_2$CH$_2$CN)—]$_n$ and poly(β-cyanoethyl methacrylate) the higher was the M_v of the polymer (179).

9.3.1   Effect of Solvent

The amount apparently adsorbed greatly depends on the solvent employed. This effect has at least two obvious causes. First, the solvent competes with the solute for the area available on the surface of the sorbent. Second, the configuration of the polymer chain end and consequently its effective volume and mobility are strongly affected by the "goodness" of the solvent, as explained in Chapter 2. In special systems the sorbent may swell in the solvent or may react with some components in it. Thus it is often impossible to predict what effect (on the value of A_a) will be achieved by substituting one for another solvent.

Probable instances of the competition between the solute and the solvent are mentioned in the preceding paragraph. This competition was made responsible also for the observation that the A_a of poly(ethylene oxides) on charcoal increased from dimethylformamide to chloroform to dioxane to benzene to methanol to water (145). Water is known to have a very small "affinity" to charcoal whereas HCON(CH$_3$)$_2$ seems to have a particularly strong one; thus water did not, and dimethylformamide did, prevent the uptake of the polymer. The A_a of fatty polyamides on iron oxide from cyclohexanone was positive, as mentioned, but it was negative from isopropyl alcohol which, presumably, was sorbed preferentially to the amides (169). Displacement of poly(methyl methacrylate) from silica surface by methanol was recorded in Reference 162. When the solvent is as similar to the polymer as is toluene to polystyrene, the smaller molecules seem to occupy the sorbent surface before the large molecules can reach the surface; consequently, the adsorption of polystyrene on rutile, which is positive and reaches 0.7 mg/m^2 from carbon tetrachloride, is negative and very small from toluene (180). Aerosil sorbed more poly(vinyl acetate) from CCl$_4$ than from CHCl$_3$; this was attributed to a greater sorption of chloroform in comparison with carbon tetrachloride (161).

However, the second of the aforementioned effects, that is, that of the "goodness" of the solvent, seems to be more common than the competition stressed above. When the adsorbed molecules are small, the role played by the solvent usually is clear: the greater the solubility the less the adsorption. It is difficult or impossible to apply this rule to polymers because their thermodynamic solubility often cannot be measured. In the instance of polymers, a good solvent is not one in which more of the polymer can be dissolved, but rather one in which the intrinsic viscosity [η] is greater, as explained in Chapter 2. Thus, instead of a relation between A_a and solubility, there is a relation between A_a and intrinsic viscosity of the general form shown in Equation (9.6) as follows:

$$A_a = f([\eta]) \qquad (9.6)$$

A specific example of this relationship is shown in Equation (9.7), (176).

$$A_a = k([\eta])^{-0.67} \qquad (9.7)$$

which was approximately valid for the uptake of copolymers of ethylene and vinyl acetate on glass spheres from CS$_2$, CHCl$_3$, CCl$_4$, and C$_6$H$_6$. In the sorption of a poly(vinyl acetate) by iron powder (obtained by decomposition of iron carbonyl), the A_a appeared to be proportional to $[\eta]^{-1}$ when the solvents were chloroform, 1.2-dichloroethane, benzene, and carbon tetrachloride (181).

A linear dependence of A_2 (along the isotherm plateau) on [η] was observed (168) for a polymer of 1.6 dihydroxy-hexane and adipic acid during adsorption from nitromethane, p-xylene, trimethyl orothoformate, benzene,

toluene, ethyl benzoate, nitrobenzene, and 1.2-dichloroethane for which Equation (9.8) is valid,

$$\Gamma_a = 6.0 - 26([\eta]) \qquad (9.8)$$

when Γ_a is expressed in milligrams per square meter and [η] in deciliters per gram. Unfortunately, four common solvents did not satisfy Equation (9.8). A relation of this type seems, however, to be valid for the sorption of poly(vinyl acetate) by glass from CHCl$_3$, benzene, and toluene (182).

In many other systems the effect of the solvent on A_a is pronounced but cannot be correlated with [η] or another quantity characterizing the quality of the solvent.

The nature of the solvent affects not only the amount of A_a but also the thickness of δ_f and the concentration of the adsorbed film, as determined by ellipsometry, discussed previously. Thus this concentration in the system of poly(vinyl pyrrolidone) and chromium mirrors was greater at small, and smaller at large equilibrium concentrations of the solution when methanol was the solvent, as compared with water (183).

Adsorption from mixed solvents has been repeatedly measured, for example, References 157, 184, 185, but the results are not reviewed here because the addition of another ingredient complicates the system even more. At any rate, Equation (9.6) is confirmed for some mixtures.

9.3.2   Effect of Adsorbent

The difference between the behaviors of porous and nonporous sorbents has been mentioned in Section 9.3. A clear instance of this difference is described in Reference 186. Polystyrene of $M \sim 290,000$ was sorbed from solutions in CCl$_4$. At c_0 equal to, for instance, 3 mg/g, two silica gels whose most common pores were 120 to 140 Å wide, took up practically no polymer. An otherwise very similar silica gel, but with pores of 250 Å, sorbed about 0.4 mg/m^2 but needed several months to reach this state. When the pore diameter was near 750 Å, or when the SiO$_2$

sample was nonporous (Aerosil), the Γ_a was near 0.8 mg/m^2, and the equilibrium was attained in a few days. Two other nonporous adsorbents (rutile and a carbon black kept in hydrogen at 3000°) had adsorption isotherms indistinguishable from that of Aerosil, so that Γ_a was independent of the chemical nature of the sorbent. On the other hand, a polystyrene of $M \sim 48,000$ had a maximum Γ_a (-1.1 mg/m^2) when the pore diameter was near 410 Å and smaller Γ_a at the diameters of 550 and 800 Å (177).

A series of silica adsorbents with even finer pores was studied in Reference 187. When the M_n of a poly(oxyethylene glycol) was only 400, the greatest amount adsorbed (from water) by 1 g was 0.15–0.18g, independent of the pore diameter; this uptake corresponded to about 1 mg/m^2. When the M_n was 15,000, the greatest sorption was only 0.012 g for a pore diameter of 22 Å, but it was still 0.175 g for a diameter of 200 Å.

In a sample of activated alumina containing pores of very different diameters, some pores apparently were too small for polystyrene of $M \sim 60,000$. The Γ_a of this polymer, from cyclohexane at 35°C, was about 0.05 mg/m^2, whereas Γ_a on a nonporous sample of aluminum coated with alumina was 1.8 mg/m^2 (188). For a polypropylene of $M = 1,800,000$, the two Γ_a values were 0.013 and 7 mg/m^2; thus the obstruction caused by the small lumen of the pores was more severe in the instance of larger molecules.

When a solution of polydisperse polystyrene of an average M of about 300,000, in carbon tetrachloride, was filtered through a silica sorbent of an average pore diameter of 550 Å, the M in the filtrate for a time was above that of the unfractioned polymer, thus showing the fine-porous sorbent took up shorter chains preferentially to the long ones (189).

In principle, it is impossible to eliminate the effect of porosity by using either nonporous sorbents or polymers of low molecular weight. In an attempt of the second kind (169), a polyamide of $M \sim 3000$ was adsorbed from cyclohexanone on α-Fe$_2$O$_3$, anatase, and a carbon black. The Γ_a for the three solids was,

for instance, near 2.5, 1.0, and 0.15 mg/m^2. Thus the chemically active iron oxide was a more avid sorbent than a (supposedly) nonpolar carbon. Unfortunately, no proof is available for the assumption that all pores in the latter, accessible to nitrogen (used to determine the surface area of the sorbent), were accessible also to the polymer chains.

The ingratiating idea of the capacity of the adsorbent rising with the concentration of active sites on the adsorbent's surface was not confirmed in two other series of experiments. An Aerosil sample contained, on the average, one hydroxyl group for 12 Å^2 of its surface. When the sample was heated until the area for one OH increased to 100 Å^2, the Γ_a of a polystyrene, dissolved in carbon tetrachloride, practically did not change and was equal to 1.1–1.2 mg/m^2, corresponding to approximately 1 segment per 16 Å^2 (190). Also when a channel black was heated in a hydrogen stream to remove practically all hydroxyl, carbonyl, and carboxyl radicals from its surface, it took up nearly as much polystyrene (again from CCl$_4$) as carbon blacks containing, for instance, one OH per 50 Å^2, on CO per 7 Å^2, and one COOH per 170 Å^2, namely $\Gamma_a = 0.3$ mg/m^2 (191).

When the amount adsorbed is referred to unit area of the adsorbent, it is implied that adsorption occurs on the surface or in a thin surface layer (see Reference 42, p. 315). But some sorbents markedly swell in the polymer solution employed, and in these instances no line can be drawn between adsorption and absorption (or dissolution). A suitable example of this quandary is supplied by the uptake of poly(acrylic acid) from water or methanol by powdered nylon 66 (192).

Numerous studies have been reported on adsorption of polymers by modified sorbents. Thus a solid would be coated with substance B and then immersed in a solution of polymer P. As usual, it is not known whether P can displace B from the solid surface, and if it can, how much. The sorbent must be considered to be interacting simultaneously with three substances, namely B, P, and the solvent. The visible results of such an interaction are difficult to account for, consequently, they are not further discussed in this chapter. They are important for the formulation of paints because many pigments used in commerical paints are pretreated before being incorporated in the vehicle.

9.3.3 Effect of Temperature

The effect of temperature T on polymer adsorption is difficult to predict. An increase in T enhances the mobility of the adsorbed molecule and its tendency to escape; this is a phenomenon well known from the adsorption of gases. The rise of T also alters the solubility of the solute and the competition between the solute and the solvent for the surface available; these effects are familiar from the adsorption from binary solutions. In the instance of polymers, a third influence may be expected, namely that of the configuration of the polymer chain, which is also a function of temperature.

When the thickness δ_f and concentration c_f, see Section 9.3, were determined separately (183), it was found for poly(ethylene glycol) and poly(vinyl pyrrolidone) in water, adsorbed on a chromium mirror, that the δ_f increased and c_f decreased on a temperature rise from 25 to 45°C; as would be expected, the adsorption layer became more diffuse.

Several recent examples reported confirm that the temperature dependence of A_a is different for different systems and so far cannot be satisfactorily accounted for. The A_a of poly(methyl methacrylate) sorbed by glass fibers from chloroform more than doubled when T was raised from 19 to 40°C (193), but the A_a of this polymer taken up by an aluminum silicate from benzene was almost equal at 30° and 60° (194). When aqueous solutions of poly(methacrylic acid) were brought in contact with alumina (of surface area 35 m²/g), the A_a increased, for instance, from 50 to 70 to 95 mg/g on an increase in T from 20 to 40 to 80°C (195). On the other hand, the A_a of a polymer either of vinyl alcohol and β-cyanoethanol in contact with crushed glass or channel black in acetone was by about 30 or 50% smaller at 45 than at 25°C (196).

The Γ_p determined by direct weighing, as described in Section 9.3, exhibited a minimum for the system of poly(vinyl acetate), benzene, and copper foil (153). It was, for instance, 105, 45, and 140 mg/m² at 12, 25, and 45°C. It is remarkable that no minimum occurred in the same temperature range when toluene was substituted for benzene.

9.3.4 Rate of Sorption

As soon as a solid is immersed in a polymer solution, several processes start. Polymer molecules diffuse toward the solid surface; this is a process that can be accelerated by stirring and shaking. The next step is penetration of these molecules into the crevices on a nonporous sorbent or into pores of a porous one. Then they have to displace some molecules of the solvent already attached to the surface; only then is polymer sorption achieved. Moreover, the sorbent also may change during these processes. For instance, agitation and the heat of adsorption may cause breakdown of conglomerates of powder particles and ensuing increase in the surface area available. The adsorbent may also swell, as already mentioned. Thus the overall rate of (apparent) adsorption dA_a/dt, t being time, is a complex quantity, not easy to account for in any particular instance.

In some instances, an additional complication arises, caused by the poly dispersity of practically all polymer samples employed in adsorption measurements. Short chains diffuse more rapidly than the long ones; consequently they reach the solid surface first and become attached to it. Later on, big molecules invade the interfacial region and tend to displace the previous occupants because a long chain usually has more points of attachment than a short one; the probability of desorption rapidly decreases when the number of these points rises.

Such a displacement apparently took place during adsorption of poly(methyl acrylate) by glass powder from toluene (197). The A_a reached a practically constant value after 4 hr of shaking, but the viscosity of the filtrate continued to decrease for at least six additional days, and the total decrease during this time was equal to about one-third of the initial viscosity. Obviously, the lowering of the mean molecular weight was considerable. Apparently, an analogous displacement occurred in polystyrene adsorbed on silica from trichloro-ethylene (198), in polystyrene from toluene and poly(methyl methacrylate) from buta-none on activated carbon (171).

The importance of the foregoing diffusion step referred to has been stressed in many publications. The mass transfer can be accelerated by agitation, and this was confirmed for adsorption on carbon in Reference 165 and for other systems in independent articles. Diffusion into pores presumably determined the marked difference between the overall rates of sorption in porous and almost pore-free sorbents. The final A_a of polystyrene solutions in cyclohexane was achieved in a few hours when the sorbent was an aluminum powder with an area of 1 m^2/g (from nitrogen adsorption) but was not fully completed in two days when alumina of 310 m^2/g was employed (188). Analogous results were obtained when polystyrene dissolved in carbon tetrachloride was brought in contact with silica adsorbents. On the nonporous Aerosil, the gradual rise of Γ_a lasted for about 1 hr, and shorter chains ($M \approx 43,000$) reached equilibrium earlier than the longer molecules ($M \approx 290,000$), whereas on a silica sample containing many pores of 350 Å, the final state was attained after more than 20 days.

Quantitative correlation between the diffusion coefficients of polymers in solutions and their rates of adsorption from the same liquids was attempted in Reference 199. Poly(ethyleneimines) were sorbed, from solutions in very dilute sodium hydroxide, by regenerated cellulose fibers. The experimental curves of the function $A_a = f(t)$ have the expected shape, but the absolute rate dA_a/dt was far too slow. However, both cellulose and poly(ethyleneimine) are ionized in water, and electrostatic attractions and repulsions between the two should affect the rate of their mutual approach. Then, too, cellulose swells in water, and at least a part of the diffusion path presumably lies in the swollen layer.

Perhaps a less complicated system will be selected for analogous experiments in the future.

The common gradual rise of A_a in time has a few exceptions. Thus in the system of polyesters of 1.6-hexanediol and adipic acid, ethyl acetate, and anatase, A_a increased with t for several minutes but then slowly decreased during a day or a few days (168). Similar results were obtained with an unusual sorbent, namely ammonium chloride in dichloroethane (136). The A_a of a poly(aminostyrene) rose for 5 to 10 min, reached the high value of 0.25 g/g and then gradually decreased to 0.17 g/g. The equilibrium concentration used also was higher than in most other tests reported here, namely 120 g/l.

The rate of sorption of poly(vinyl alcohol) by silver iodide crystals in an aqueous solution affected the rate of coagulation of AgI sols by electrolytes (200).

9.3.5 Structure of the Adsorption Film

Many descriptions of the adsorption layer are empirical, that is, based on observations and measurements. For instance, some absolute values of Γ_a (Γ_p are likely to be even greater !) cannot be reconciled with the popular idea of unimolecular adsorption. A polystyrene chain, spread over a flat surface so that every —CH$_2$— and every —CH = group would be in direct contact with the solid, would result in much smaller adsorptions than those reported in the previous sections of this chapter. Is this an instance of multilayer sorption or are not all segments (of the chain) clinging to the surface?

Many theoretical discourses, only four referred to here (130, 201–203), have attempted to predict the preceding structure. The task is difficult because a multitude of parameters must be known or postulated.

As far as the sorbent is concerned, the theoretician, before he derives his first equation, must answer at least the following questions. Is the sorption a purely superficial effect or does dissolution in the surface region of the sorbent affect the experimental values of A_a? Is the sorbent altered (e.g., swollen) as a result of the sorption? Is the solid surface energetically uniform, or is it dotted with active sites? Are these sites all equally active? If not, what is the frequency of each kind? How rough is the surface? How porous is the adsorbent? Is the adsorbent finely dispersed so that one polymer chain can be simultaneously attached to two or several solid particles?

As far as the polymer is concerned, we need probably even more information. Is, for instance, the affinity of all segments of the chain to the solid identical or are active groups immobilized preferentially to the others? What percentage of the segments is in direct contact with the solid? What is the average distance between the points of attachment? Figure 9.4 indicates the imagined conformation of three chains in the vicinity of a solid plane represented as a shadowed space. The parts of the chain marked with the letters a, l, and t are referred to in the literature as, respectively, attached segments, loops, and tails. Thus the following questions can be formulated: How many regions are there and

how long are the two tails? How is the shape of the loop and the tail affected by the flexibility of the chain and by the attractions between segment and segment on one hand, and segment and solvent on the other?

A third group of problems deals with thermodynamics. What is the loss in free energy of the system when a chain becomes attached to the solid? How does this loss depend on the chemical composition of the various sections of the chain? What are the analogous energy changes associated with adsorption and desorption of a solvent molecule? What is the increase in entropy taking place when the chain that, in the bulk solution, possessed the most probable configuration is distorted to a less probable one because of the forces emanating from the solid and the geometrical restrictions imposed by the solid's presence? How does the proximity of an adsorbed chain effect the adsorbability of the nearest polymer coil? Are several coils so entangled with each other that clusters of coils are sorbed as such, rather than as individual chains?

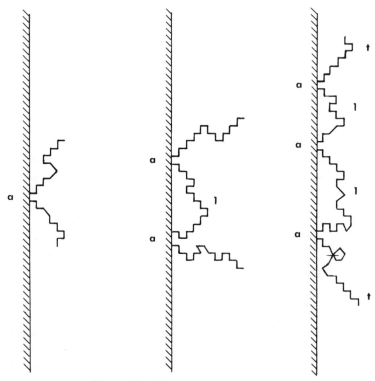

Figure 9.4 Adsorption of polymer chains.

As is clear from this incomplete list, the number of possibilities to be kept in mind is vast. Consequently it is not surprising that no theory of polymer adsorption has yet been universally accepted.

Some of the uncertainties mentioned are avoided when sorption of polymers at liquid-liquid interfaces is studied because these interfaces are more uniform and more mobile than those of solids. Also the determination of the area over which adsorption takes place is unambiguous. However, less information is available on these systems than on those with a solid sorbent. Because of space limitations only a few references are included here. The sorbent was mercury in References 204–206 and paraffin oil in References 207 and 208. The adsorption of poly(ethylene oxides) from their aqueous solutions at the interface with air was calculated in Reference 209.

9.4 ADSORPTION BY POLYMERS

An operation as common as the dying of textile fibers may be described as sorption of dyes by filamentous polymers. Numerous publications on this problem exist (210), but they are not reviewed here because, generally, it is unknown what importance surface phenomena have for the overall result. Another application of sorption on polymers involves polymeric ion exchange materials such as sulfonated polystyrenes, which immobilize many cations. These materials are deliberately prepared to have a large part of their mass, not just their surface, accessible to ions; thus the importance of surface phenomena usually is secondary only. A system in which ion-solid surface appeared to be more important than usual in ion-exchanges was studied in Reference 211. A macroreticular polymer of styrene and divinylbenzene was sulfonated, and the resulting material had a high specific surface area: $51 \, m^2/g$, as determined by adsorption of nitrogen. This polymer sorbed ethyl acetate from its vapor, and Γ_a was, for instance, 0.013 millimoles/m^2, so that the area occupied by one molecule of ester was approximately 13 $Å^2$. This area is too small, especially since the area accessible to nitrogen is likely to exceed

that accessible to the ester. Hence, some occlusion of ethyl acetate is likely here. When the polymer was titrated with aqueous sodium hydroxide, the number of moles consumed was two to three times as great as the number of ester moles sorbed, although NaOH would be expected to react with $—SO_3H$ groups only, whereas ethyl acetate might be immobilized by the entire surface of the polymer.

In the preceding example, one of the problems referred to in Section 9.3.5 emerges again: is the adsorbed material attached exclusively to the outer surface of the sorbent or does it penetrate the surface region of the latter? This, and some other questions mentioned in the preceding section, namely whether the adsorbent is or is not altered (e.g., by swelling) during the sorption process, are particularly important when the sorbents are polymeric solids. This is so, because these solids, generally, offer less resistance to diffusion of foreign molecules and have a less rigid structure than the customary sorbents such as metals, glasses, or various carbon blacks.

Real surface adsorption apparently was observed when films, freely suspended or desposited on glass beads, or poly(methyl methacrylate) were subjected to a vapor of n-decane mixed with air (212). The surface area of the films, calculated under the assumptions that adsorption occurred in a unimolecular layer and that a molecule of $C_{10}H_{22}$ occupied 70 $Å^2$, roughly agreed with the geometrical area of the adsorbent.

Quite different results have been obtained on a closely related system. Vapors of n-octane were admitted to a sample of poly-(octadecyl acrylate) suspended on a spring balance (213). The mass of the sample rose by several percent when the relative pressure p/p_s (p_s is the pressure of the saturated vapor, and p is the actual gas pressure) of octane was still only 0.5. Hence it is likely that some occlusion of the vapor took place. The intensity of the x-ray scattering by the crystalline part of the polymer was not altered by this sorption; consequently the dissolution must have been restricted to the amorphous part of the sample (about 25% of the whole). When poly(oc-tadecyl methacrylate) was substituted for the foregoing homolog, its x-ray scattering was weakened by sorbed amounts of octane as small as 2%, and the amount sorbed near $p/p_s = 1$ was about 60% of the mass of the sample. Obviously, dissolution of octane in the polymer, with the concurrent destruction of the crystalline structure, was taking place rather than superficial adsorption.

Only one example is described here to indicate the easy variability of polymer sorbents. When a mixture of methanol vapor and helium was kept in contact with a poly(vinyl chloride), the amount of methanol taken up in the steady state was twice as great for polymer samples before than after a mild heat treatment consisting of heating the sample to its T_g and cooling it to the temperature of the test (20 or 40°C) (214).

Complex sorbents made up of a solid of a high specific area and a thin coating of a polymer on it have repeatedly been investigated. Only one example is presented here to show that also in such systems true surface adsorption, as distinct from absorption, is not easy to measure. Carbon black and rutile were coated with poly(ethylene glycol) from aqueous solutions, and the degree of coverage Θ was determined. A complete unimolecular layer corresponds to $\Theta = 1$. These values of Θ were unfortunately doubtful because the area occupied by an adsorbed segment of the polymer had to be assumed, and the surface area determined by desorption of nitrogen presumably was different from that available to the polymer. On the other hand, this Θ was not an apparent coverage in the meaning of Section 9.3, since it was measured after the evaporation of the solvent (water) clinging to the substrate. Sorbents with Θ values ranging from approximately 0.2 to 5 (i.e. up to 5 unimolecular layers) were used to sorb vapors such as pentane and benzene (215). In all instances the volume of vapor taken up decreased when Θ rose from 0.2 to 1.0, and then increased on an increase of Θ from 1 to 5. The author explains this irregular curve by two effects: below $\Theta = 1$ adsorption decreases when polymer segments block a greater part of the actual adsorbent surface; above $\Theta = 1$ more hydrocarbon vapor is dissolved by the polymer, the thicker the film of the polymer.

Optically active polymers can sometimes be employed to separate the D- and the corresponding L-isomers. Thus a polymer of N-acrylyl-L-phenylalanine $C_6H_5CH_2CH$ $(NHCOCH_2:CH)CO_2H$ took up, from a buffered aqueous solution, more of L-phenylalanine than of the D-enanthiomer. The amount sorbed reached, for instance, 16 mg/g at a moderate concentration of the amino acid, was too great for the attachment to the surface only. The authors (216) consider this sorption to be a bulk effect.

Although so much of the reputed adsorption on polymers belongs to absorption and, consequently, does not belong to this chapter, a review of the phenomenon (217) may be referred to for further study.

9.5 SOLID POLYMERS

One of the most important properties of every solid surface is its roughness (42, p. 169). Generally speaking, the surface roughness of polymers is similar to that of harder materials (metals, glasses), but no systematic comparison between different types of solids could be found. In many instances the methods employed to determine roughness are identical for all these materials. For example, in Reference 218 slabs of poly(tetrafluoroethylene) were ground with silicon carbide papers of various degree of coarseness, and the mean height of elevations was measured with a profilometer. This height was near 0.01μ for a finely ground, and $0.5\ \mu$ for a coarsely ground sample. Films of poly(ethylene terephthalate) usually are covered with gentle hills, and the diameter of the hill base was found to be near $10\ \mu$ (219).

The absolute values listed in the preceding paragraph would be common also for a metal surface but the easier deformability of polymers (and plastics) causes surface roughness to be even more dependent on the previous history of the sample than is true for metals. The effect of stretching, for instance, was studied in Reference 220; the roughness was estimated from the external haze of extruded polyethylene rods and films. Data on the roughness of polyethylene before and after grafting can be found in Reference 221.

The difference between a ridge and a valley on a hilly surface, or between a crystalline and an amorphous region exposed, can be rendered visible by depositing gold, or another suitable metal, from the vapor on the polymer. For an example of this technique applied to polyethylene single crystals and to films of nylon 6 see Reference 222.

9.5.1 Surface Energy

Surface tension of liquid polymers was reviewed in Section 9.2. Unfortunately, no such review can be given for surface tension and surface energy γ_s of polymeric solids. At present no convincing method of measuring these two quantities is available (223; 42, p. 202). Of the many attempts to derive a value for the γ_s of polymers, only a few, of more recent vintage, are discussed here.

Perhaps the crudest hypothesis is that the rules applicable to liquids are equally valid for solids, and that consequently the γ_s can be calculated from the parachor values of the corresponding liquid (224). A similar assumption is that the surface energy of a solid is by about 20% greater than that of its melt just above the melting point.

Of the experimental methods popular for estimating the γ_s of metals, oxides, and so on, only one is fashionable for polymers at present. It is based on the comparison of thin and thick crystal lamellae. A thin plate has a greater surface-to-volume ratio than a thick crystal has; hence surface energy may be expected to influence the properties of the former to a greater extent than those of the latter. One of these properties is the melting point.

Four samples of poly(oxyethylene) having M_n between 1000 and 3300 have been prepared as lamellae of thickness ranging from 69 to 218 Å. The thicker the lamella, the higher was its melting temperature T_m (36–48°C). From a theory, which the present reviewer considers erroneous, and with the "incorrect" assumption that there was no other difference between different lamellae except in their thickness, it was concluded that γ_s was approximately 74 ergs/cm² (225).

Several data, obtained in an analogous manner, are available for polyethylenes. Polyethylene single crystals (still containing a high percentage of an amorphous phase) were prepared from various solvents at various temperatures so that the lamella thickness l varied for 100 to 290 Å (226). From the increase of T_m with l, the γ_s of the most extensive crystal faces appeared to be near 90 ergs/cm². In a very similar investigation (227), the free surface energy of folds was calculated to be 150 ergs/cm². This energy increased with the relative number of polypropylene segments in copolymers of ethylene and propylene so that, at the concentration of 30 side chains (methyl groups) per 1000 carbon atoms in the polymer, it reached 250 ergs/cm². For a modification of polyethylene stable at the pressure of 7000 bars, the γ_s of the long crystal faces appeared to be only 4 ergs/cm² (228). The weakness of the theory seems to be the most plausible explanation of these unexpected results.

Another doubtful result is that, for the interfacial energy of polyethylene folds in contact with various liquids, values of 60 to 120 ergs/cm² are deduced (229).

The preceding method was used also for poly(ethylene terephthalate) (230). When the lamella thickness was varied from 90 to 190 Å, the T_m rose from 257° to 266°, and the calculated γ_s was, on the average, 28 ergs/cm². It is difficult to think of a chemical reason for the surface energy of the terephthalate being so much smaller than that of polyethylenes.

The logarithm of the rate of growth of spherulites in polyethylene was, on the first stage of the process, nearly proportional to the undercooling of the melt, and, from the dependence thus obtained, the product of two surface energies, one for surfaces parallel and the other for surfaces perpendicular to the polymer chains, appeared to be approximately 300 (g/sec²)² (231). Thus for the geometric mean of the two energies the value of 17 ergs/cm² is obtained, smaller than anticipated.

When a multitude of small spheres is heated sintering occurs, and the heat taken up by the system must contain a term representing the decrease in the total surface energy, associated with the decrease in the combined surface area of the spheres. From this term, in principle, the free surface energy of the polymer ought to be obtainable. Measurements on the sintering of polystyrene spheres, 0.1–0.2 μ in diameter, at 100° resulted (232) in γ_s values from 17 to 27 ergs/cm^2.

9.5.2 Wettability

The surface energy of a solid polymer has been calculated also from the contact angles of various drops on this surface (see, for instance, References 233 and 234). However, this calculation involves arbitrary hypotheses and is not further considered in this chapter.

Numerical values of contact angles θ are available for several polymers and several liquids in air. Regrettably, the trustworthiness of some data is doubtful. The experimental θ depends on the chemical composition of the thin external layer (perhaps 10 or 20 Å thick) of the solid; hence the degree of cleanness or contamination of the sample is all-important. Unfortunately, insufficient attention was bestowed on this matter in some publications, and commerical samples (which may have been mixed with lubricants, antioxidants, and so on) have been employed in the tests. Moreover, the Θ on a given specimen usually depends on the procedure of forming the three-phase contact (between the vapor, the liquid, and the solid phase). Thus, if a little droplet is added to a large drop sitting on a horizontal solid plate, the visible Θ is greater than that obtained after withdrawing a small quantity of liquid from the large drop. The first contact angle belongs to the group of advancing contact angles θ_a, and the second, to receding angles θ_r; the whole phenomenon is known as hysteresis of wetting. The dif-

ference $\theta_a - \theta_r$ may be 45° for water on poly(tetrafluoroethylene)(235) or even higher; it is obvious that whoever disregards the hysteresis is likely to obtain incorrect results.

In view of the reservations formulated in the preceding paragraph, the agreement between different determinations of θ is surprisingly good. Table 9.3 lists some values published almost simultaneously for commercial polyethylene and polystyrene. The liquid employed is indicated above the data for θ (in degrees). In Reference 236 θ means the average of θ_a and θ_r; and in Reference 237 θ means an advancing contact angle said to be reproducible within ± 3°.

Polyethylene was a favorite material for studying the influence of various surface treatments on wettability. Extraction of low-molecular impurities from commercial polyethylene, which transformed the latter into a reasonably good adhesive, had almost no effect on θ (with water) (238), which remained in the range 102–105°. On the other hand a polyethylene film that, without cleaning or pretreatment, gave $\theta = 95°$ with water, showed $\theta = 48°$ after 20 min in a corona discharge in air (239). Oxidation of polyethylene films with HClO$_3$ depressed the θ for water from 96 to 68° (240). Activated argon plasma depressed the θ of polyethylene and water from 97 to 19° (241).

Halogenated polymers have wettabilities that depend on the nature of the halogen. Thus a poly(vinyl chloride) had $\theta_a = 83°$ with water and $\theta_a = 55°$ with formamide (237), whereas the angles on poly(tetrafluoroethylene) were even higher than on polyolefins. The θ_a for water and formamide on this polymer were 112 and 91° (237). A copolymer of tetrafluoroethylene and hexafluoropropy-

Table 9.3   Contact Angles between Two Polymers and Five Liquids in Air[a]

Polymer/Liquid:	Water	Glycerol	Formamide	$C_2H_6O_2$	$C_{10}H_7Br$	Reference
Polyethylene	95°	80°	75°	65°	35°	(236)
Polyethylene	95°	83°	81°	69°	33°	(237)
Polystyrene	92°	74°	69°	63°	?	(236)
Polystyrene	84°	71°	69°	60°	Swells	(237)

[a] $C_2H_6O_2$ is ethylene glycol.
$C_{10}H_7Br$ is 1-bromonaphthalene.

lene had θ of 108° for water and 91° for formamide at 20° (236). Additional information on wetting of highly fluorinated polymers can be found in Reference 242.

For a poly(vinyl alcohol) containing 3% acetyl groups and water, $\theta = 36°$ was found; when the polymer was reacted with formaldehyde up to 30% of aldehyde groups, the θ was 102° (243).

Drops of water on a polycarbonate and a poly(ethylene terephthalate) gave $\theta = 85°$; also θ of ethylene glycol drops on the two polymers were almost identical (approximately 58°) (236). The contact angles on poly(methyl methacrylate) seem to be a little greater than on poly(ethylene terephthalate), namely 74° and 71° for water, and 57° and 53° for formamide (237). Nylon 6,6 was more wettable than nylon 11. The θ_a of water drops on these amides were 65 and 75°, respectively, and the values for formamide were 48 and 60° (237).

Apparently the wettability of a polymer film depends on the solid surface on which film formation took place before testing (244, 245), but the cause of the phenomenon is still debated. The contact angles on a stretched polymer film may depend on the direction. When a film of a copolymer of butadiene and acrylonitrile was stretched 140%, the profiles of a sessile drop were different in the direction of extension and in the perpendicular direction. For instance, the θ for formamide, which was about 85° on an unstretched material, became near 55° in the stretch direction after extension; also, the perpendicular θ was smaller than before (246).

The importance of the specific surface free energies or surface tensions arises because many measurable equilibrium surface quantities of macroscopic surfaces can be expressed in terms of their differences, for example, differences in surface tensions. Simple instances are afforded by the measurable spreading pressures and contact angle θ. Thus the spreading pressure of a liquid on a solid Π_{SL} and a vapor on a solid Π_e can be expressed as Equations (9.9) and (9.10), where γ_S, γ_{SL}, and γ_{SV} are the solid, solid-liquid and solid-vapor interfacial tensions, respectively.

$$\Pi_{SL} = \gamma_S - \gamma_{SL} \qquad (9.9)$$

$$\Pi_e = \gamma_S - \gamma_{SV} \qquad (9.10)$$

The contact angle where a vapor, liquid, and a nondeformable solid meet can also be expressed in this fashion. Thus one obtains Equation (9.11).

$$\cos \theta = \frac{\Pi_{SL} - \Pi_e}{\gamma_{LV}} \qquad (9.11)$$

In terms of measurable entities, or using Equations (9.9) and (9.10), it can be expressed in terms of the three relevant surface tensions, by way of the so-called Young's Equation (9.12).

$$\cos \theta = \frac{\gamma_{SV} - \gamma_{SL}}{\gamma_{LV}} \qquad (9.12)$$

For stable equilibrium $\gamma_{LV} > 0$, $\Pi_e > 0$, but Π_{SL} may have either sign. If $\cos \theta \geqslant 1$, the solid is wetted completely by the liquid, which can spread freely over the surface; if $\cos \theta < -1$, the solid is not wetted. For intermediate values of $\cos \theta$ between ± 1 the cosine of the equilibrium contact angle is a direct measure of wettability. Equation (9.12) shows that the equilibrium contact angle and γ_{LV} determine the reversible work in replacing a unit area of solid-vapor interface by a solid-liquid interface, the adhesion tension $\gamma_{SV} - \gamma_{SL}$. This entity is involved in the reversible work of adhesion W_A between a solid and a liquid in the presence of adsorbed vapor.

$$W_A = \gamma_{SV} - \gamma_{SL} + \gamma_{LV} = \gamma_{LV}(1 + \cos \theta)$$
$$(9.13)$$

Equation (9.13) is referred to as the Harkins-modified Dupre relation.

Most measurements in the literature of the contact angle as a thermodynamic equilibrium property are actually measurements of the advanced (or advancing) contact angles within a minute or so of contact line displacement. The reported average deviations are commonly $\pm 2°$, though sometimes only $\pm 1°$. Often, on causing a rapid displacement of the contact line, one finds that the new static contact angle depends on the direction of the movement of the contact line (6). The difference between the advancing and the receding contact angle $\theta_a - \theta_r$ is called the contact angle hysteresis (247–249). Contact angle hysteresis can depend on the time interval between movement and measurement, on contamination, and on many aspects of the state of the substrate surface. In the case of solid surfaces these are known to include roughness, heterogenity, presence of viscoelastic stress, and the content of dissolved or dispersed liquid (248).

There are many regimes of contact angle hysteresis depending on the velocity of contact line displacement. Here one is concerned solely with the regime in which the contact line is caused to move sufficiently slowly (velocities of less than 1 mm/min) that the difference between advancing and receding contact angle is sensibly constant and reproducible. In this quasi-static regime the difference between advancing and receding contact angles of test liquids on a given solid surface may provide a rough insight into the distribution of surface heterogeneities as well as effects due to increasing surface rugosity. This technique is based on theoretical studies of a quasi-static advance of a liquid drop on model patchy surfaces (250–252). With heterogeneous or rough surfaces a conflict arises in the minimization of the total free energy of the system due to the differing quasi-static constraints at the vapor-liquid interface, where Laplace's equation of capillarity must be satisfied with Young's equation, Equation (9.12), at the three-phase boundary. This provides a mechanism of quasi-static hysteresis for patchlike surfaces (patches larger than $0.1\,\mu$ of different surface free energy), and surfaces with rugosities of the order of $0.5\,\mu$ or larger.

The work of Good (249), Johnson and Dettre (250–251), and Neumann and Good (252) is of considerable significance since it provides the basis of a novel method of surface characterization of polymers.

Zisman and co-workers have studied the wetting behavior of many solid surfaces by a variety of pure liquids, often members of chemically homologous series such as n-al-

kanes. If the cosines of the advancing contact angles θ_a for several liquids on a given solid surface (such as a solid polymer) are plotted against the liquid surface tensions γ_{LV}, it was observed that the points lie near a straight line or in a narrow rectilinear band. The intercept of the $\cos\theta$ versus γ_{LV} plot of this band or straight line at $\cos\theta = 1$ (zero contact angle) represents the liquid surface tension below which complete wetting is expected (at least for the class of liquids investigated) on the given solid surface. This limiting surface tension is called the critical surface tension for wetting γ_c. Mathematically, then, in the vicinity of $\gamma_{LV} = \gamma_c$ one has the empirical Equation (9.14),

$$\cos\theta = 1 + K(\gamma_c - \gamma_{LV}) \qquad (9.14)$$

with K as the limiting slope of the rectilinear band. As Fox and Zisman (253) have already emphasized ". . .the concept of γ_c must be used with caution since . . . . γ_c varies between liquid types (used to determine it). In particular, it is not valid to construe γ_c as a measure of γ_s." Clearly γ_c is only an approximate characteristic of the surface, unlike γ_s. [Rosoff (8) has summarized the arguments on these points in great detail.] Experimentally, Johnson and Dettre (254) in an elegant study of the wettability of low-energy liquid surfaces by a series of alkanes have shown that an analog of Equation (9.14) can be applied to liquid substrates, and the difference between γ_c and γ_s (here referring to the liquid substrate) ranged from 5 to 50 dynes/cm.

Even though γ_c does not have an absolute thermodynamic significance, it is an extremely useful figure of merit for classifying surfaces of organic polymers, among other materials, in regard to their wetting behavior. The observed values of γ_c (at 20–25° C) for organic polymers vary from about 50 to 60 dynes/cm for polar polymers as urea formaldehyde and polyamide-epichlorhydrin resins (255) to 10 dynes/cm for certain acrylic polymers bearing perfulorinated side chains (256). The limiting slope K also varies only over a narrow interval, roughly from 0.03 to 0.04 (257). The status of the contact angle as a thermodynamic entity is clearly dependent on the fact that it involves a difference in surface tensions, that is, the adhesion tension. Any inferences from contact angle data concerning the surface characteristic entity γ_s must be based on modelistic, extrathermodynamic assumptions (258).

Solid polymers surfaces are often classified into so-called low-energy and high-energy surfaces, where the term in principle refers to the specific surface free energy but is usually interpreted experimentally to mean low or high wettability. (The specific surface free energies of many solids are not accessible to direct measurement, but γ_c can be measured easily.) Most solid organic polymers are presumed to have specific surface free energies that are less than 100 ergs/cm (2) and so are classed as low-energy surfaces (8).

9.5.2.1 *Formation of Interface*

Two materials probably adhere, at least initially, because of van der Waal's attractive forces acting between the atoms in the two surfaces. Interfacial strengths, based on van der Waal's forces alone, far exceed the real strengths of one or other of the adhering materials. This means that interfacial separation probably never occurs to any sensible

extent when mechanical forces are used to separate a pair of materials that have achieved complete interfacial contact (probably a highly unlikely situation) or a number of separate regions of interfacial contact. It follows, then, that breaking the joint mechanically, in general, tells nothing directly about interfacial forces.

Van der Waal's forces are operative over very small distances. Hence, in order that materials adhere, the atoms in the two surfaces must be brought close enough together for these forces to become operative. If one had a piece of A (solid) and B (solid), and each had an absolutely clean, smooth (on an atomic scale), planar surface, and if these surfaces were brought together in a perfect vacuum, all attempts to separate them would result in failure in either A or B (Figure 9.5). But real surfaces differ from these ideal surfaces in that they are rough and contaminated, and both of these imperfections contribute to a greatly decreased real area of contact between the surfaces of A and B (Figure 9.5). In general, however, where they have achieved contact— that is, where they have adhered—and when they are separated mechanically a little of A remains on B, and B on A, depending on the

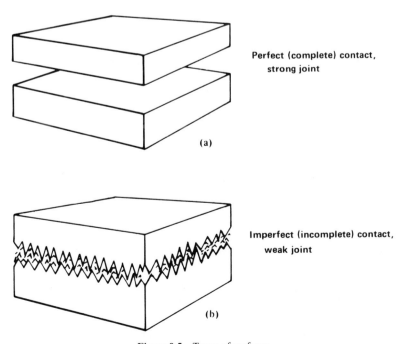

Perfect (complete) contact,
strong joint

(a)

Imperfect (incomplete) contact,
weak joint

(b)

Figure 9.5 Types of surfaces.

geometry in the neighborhood of each area of contact and the cohesive strength of each. The general assumption based on visual examination is that either the solids did not adhere, or the failure was in adhesion. The first statement, however, is incorrect, because surely some areas of A and B achieved interfacial contact, and the second is incorrect because where the surfaces were not in contact there was no adhesion, and it makes no sense to talk about the failure of something that did not exist.

This point of view leads to the conclusion that to get A and B to make a stronger joint, one needs to increase their real area of contact. This means that one or both of the materials must be made to conform better to the surface roughness of the other. This implies, in a practical sense, that one of the materials should be fluid when placed in contact with the other. However, that one of the materials be fluid is a necessary, but may not be a sufficient, condition, for if a high-viscosity fluid makes a sizeable contact angle with the solid, its tendency to create a large interfacial area of contact may be relatively poor. The result is that it may do a great deal of bridging, trap a great deal of air, achieve little penetration into the surface roughness of the solid, and form stress concentrations owing to the large contact angle involved. However, if the fluid member spontaneously spreads on the solid, the interfacial area of contact increases because the fluid can now flow more completely into the micro or submicro pores and crevices in the surface of the solid and can displace gas pockets and other contamination. In addition, the zero contact angle tends to minimize stress concentrations. The effect of creating a spontaneous spreading situation, then, is twofold: the real area of contact is increased and stress concentration is minimized (259).

Specifically, only van der Waal's forces have been mentioned in connection with the preceding treatment. This is not to be construed as meaning that other molecular forces may be excluded from participation in adhesion. In the initial process involving the establishment of interfacial contact, all the molecular forces involved in wetting phe-

nomena can be considered to be important. Chemisorption is not excluded, but if it is to occur, molecular contact must have already been established, that is, van der Waal's forces must already be operative. Therefore any such chemical reaction that does occur, occurs after adhesion has taken place. (Furthermore, since interfacial separation apparently does not occur under mechanical influences even when only van der Waal's forces are operative, it follows that chemisorption may not have any positive influence on the mechanical strength of an adhesive joint.) It may have a negative influence on strength if weak boundary layers (260) are formed. However, it is possible that chemisorption may increase the permanence of an adhesive joint by retarding or preventing destruction of the interfacial region as by moisture, low surface tension liquids, and so on. One of the authors and his associates have demonstrated (261–263) how strong adhesive joints could be obtained in the epoxy adhesive-polyethylene system and the epoxy adhesive-chlorotrifluoroethylene homopolymer system. Their results clearly demonstrate the importance of the surface tension of the adhesive, the T_g, and the surface roughness of the substrate.

9.5.3   Temperature Dependence of Wetting

It is also instructive to consider the temperature dependence of wetting at this point. As the temperature is increased, it is clear that both the liquid surface tension and the surface energy of the polymer should decrease. It is not clear that the value of θ will increase, decrease, or remain the same, since θ depends in a somewhat complicated way on the interplay between these phases when in contact. Barring increased compatibility, the meager data in the literature (264–272) would indicate that $d\theta/dT$ is essentially constant provided one is sufficiently far removed from the critical point of the wetting liquid. Near room temperature the variation is small in all cases so far studied, being typically ± 0.05–0.15 deg/°C. However, Johnson and Dettre found a sharp drop in the contact angles for both octane and hexadecane on fluorinated ethylene-propylene copolymer as the temper-

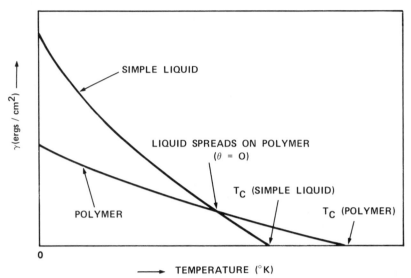

Figure 9.6 Critical temperature of wetting.

ature was raised to near the boiling point of the liquids. Petke and Ray (272) observed a similar sharp decrease for superheated water contact angles (120–160°C).

Schonhorn has described an approach based on the Katayama-Guggenheim equation (273), see Equation (9.15),

$$\gamma = \gamma_0 \left[1 - \frac{T}{T_c}\right]^{(11/9)} \tag{9.15}$$

(where γ_0 is a constant and T_c is the critical temperature) which has been applied to the data of Johnson and Dettre (248). The treatment agrees reasonably well with their rapidly changing values of θ for low molecular weight hydrocarbons in contact with a fluoropolymer. The sharp decrease in θ as the temperature is raised can be easily visualized by examining Figure 9.6. Since T_c of the liquid is lower than that of the polymer, there will be a point where spreading will occur, barring any degradation of the liquid or polymer. For n-octane on the fluoropolymer, the intersection temperature is approximately 100° C. Since $d\theta/dT$ depends strongly on the difference between γ_L and γ_P, it is not expected to change significantly except when $(\gamma_L - \gamma_P)$ approaches zero.

9.5.4 Surface Regions

Surface layers on solids generally differ from the bulk in their chemical composition (as oxide films on metals), their morphology (e.g., columnar structures in metal ingots), the frequency of dislocations, and so on. A review of these differences, which give rise to the "cuticular energy," can be found in Reference 42, p. 187. Analogous differences exist in polymers as well. They can be observed on polymer-gas interfaces, on boundaries between a polymer and another solid, and by comparing the properties of a thin film with those of the bulk polymer. The second kind of system is popular at present because of the commercial importance of reinforced plastics: it is important to know whether the fillers are embedded in an unchanged bulk polymer or are separated from the latter by a layer of modified material. In the third kind of study it is customary to assume the observed differences are caused only by thickness variation.

The columnar texture mentioned is found in polymers also. For instance, polypropylene solidified in contact with an aluminum alloy exhibited such a texture for a depth of 20–30 μ; and an even thicker oriented layer was seen in the surface that was in contact with another alloy during setting. The coefficient of friction of these layers (against steel) was different from that valid for propylene with the usual spherulitic structure (274).

Much smaller thicknesses δ of the surface layer were calculated from the rate of propagation of mechanical waves in a polymer block of an unknown solidification history (275). The modulus of elasticity of this layer appeared to be about four times as great as that of the bulk material, and δ was 290 Å for a sample of poly(methyl methacrylate). The δ values for a poly(tetrafluoroethylene), a polyurethane, and an isoprene rubber were, respectively, 160, 600, and 680 Å. The entropy of the surface layer, whose texture usually is more regular than that of the bulk, may be expected to be smaller than the entropy of the bulk. This expectation was confirmed by measuring the regularity of the propagation described (276); the difference between two entropies, per unit volume, was calculated to be near 14% of the mean value.

A difference in morphology exists also when no columns are visible in the superficial layer. Thus the average distance between the centers of two neighboring spherulites in the surface of a polypropylene bar was 120 μ, whereas it was 152 μ further from the surface (277).

A typical problem studied in the second kind of system, boundaries between a polymer and another solid, is that of glass transition temperature T_g. Does the vicinity of a filler alter the T_g of the adjacent polymer? Apparently, the answer depends on the system investigated. When generous amounts of carbon black were incorporated in a copolymer of styrene and butadiene, the T_g of the material did not change, whether it was measured dilatometrically or by differential scanning calorimetry. Similar or identical samples have been subjected also to an examination of two quantities, namely the temperature T' (related to T_g) and the absolute magnitude t' sec of the minimum spin-lattice relaxation time. If the polymer in the reinforced plastic consisted of two mechanically different materials (the unchanged bulk and the modified boundary layer), two values of each T' and t' would be expected. The experiment showed, however, only one value of each, and these values were identical with those of the unfilled copolymer (278).

On the other hand, the T_g of a poly(methyl methacrylate) was raised from 105 to 123° C by incorporating 10 g of Aerosil in 90 g of polymer; at the same time, the change in the

heat capacity at T_g decreased by nearly 20%. A similar addition of Aerosil to polystyrene did not alter T_g but markedly reduced the described change (130, p. 166). Perhaps the vicinity of the filler restricted the mobility of polystyrene chains so strongly that the polymer layer adjacent to the silica was in a glassy state also above the macroscopic T_g so that the heat capacity of the boundary layer did not change at this temperature.

Apparently, also, the thickness δ of the polymer film is important in this context. When glass beads coated with a polymer are used as the stationary phase in gas chromatography, the dependence of the retention volume V_g on the temperature T in many instances exhibits a singular point near T_g of the polymer; presumably, at lower temperatures, the main contribution to the uptake of gas stems from surface adsorption, whereas above T_g bulk adsorption predominates. The δ of polystyrene coatings was varied from 230 to 2830 Å, and a stream of nitrogen containing toluene vapor was forced through the column. The kink in the curves of $V_g = f(T)$ was observed near 120° C for the thinnest, and near 90° C for the thickest film; thus, T_g was a function of δ. At greater values of δ, thickness had no effect on T_g (279).

The coefficient of heat expansion of poly (vinyl acetate) filled with 10–60 vol% of finely dispersed quartz was, for instance, 25% lower than predicted by the additivity rule; consequently, it may be surmised that the filler restricted the volume increase of the adjacent polymer. The difference between the materials far from and adjacent to the filler was confirmed under microscope: the former was globular and the latter fibrillar (280). For an example of a surface region produced by irradiation consult Reference 281.

When a polymer sets in the presence of another solid, the chemical composition of the boundary layer may differ from that of the bulk; see, for instance, References 282 and 283. Recently, sensitive methods for detecting such differences have been worked out, and the chemical nature of surface layers thinner than 20 or 50 Å can now in many instances be ascertained. In all probability, more extensive application of ESCA (electron spectroscopy for chemical analysis) and related procedures will clarify the problem.

9.5.5 Fracture

A connection between fracture and surface properties of a polymer is established by a theory that is at least as old as Dupre's work (1867). When a bar is broken, two new (fracture) surfaces are created. If S is the area of each and γ_s is the specific surface energy, then $2\gamma_s S$ is the work required to produce these surfaces. Dupre believed, and many of his followers still believe, that $2\gamma_s S$ is essentially equal to the total work W of rupture, except the work spent on the extension of the bar to its total elongation (or maximum strain).

The newer view (42, pp. 209, 284) is that practically all work of rupture is spent on deformation leading to rupture. Both theoretically and experimentally it is possible to divide this work in two parts. One is needed for the elongation of the bar mentioned in the preceding paragraph. The other is spent on the analogous extension of the polymer in front of the growing crack. If this extension is elastoplastic, the second part will contain a term caused by plastic changes; for purely elastic extensions, as in many rubbers, only the elastic term remains.

At any rate, the work of fracture gives no information on the surface energy of the solids. This energy, if it exists, would be expected to be similar to the surface energy of liquids as far as the effect of the chemical composition on γ_s is concerned. As mentioned in Section 9.5.1, some scientists were bold enough to apply, unchanged, parachor calculations to γ_s. The work of fracture does not show any similarity with the γ of liquid in this respect. Changes in chemical composition are less important for it than are changes in the extensibility (maximum strain).

9.5.6 Friction

When a solid body of any composition is dragged over the surface of a polymer, or when a polymer block is dragged over the surface of another solid, polymeric or not, force is required to maintain the motion, and work is spent on it. These are the frictional force and the work of friction. For over 50 years many scientists and engineers believed and still believe that two solid surfaces, placed "in contact" with each other (slider and support), form adhesive bonds across the interface, and that frictional force is needed above all to break these bonds. If true, this hypothesis would be invaluable for determining atomic and molecular forces between identical or different polymers. Unfortunately, it is highly improbable.

Usually the main component of the frictional force is that needed to deform the polymer (support, slider, or both) to render the tangential motion possible (285). Figure 9.7 illustrates this concept. The slide (1 in Figure 9.7) is supposed, for the sake of simplicity, to be rigid so that its deformation

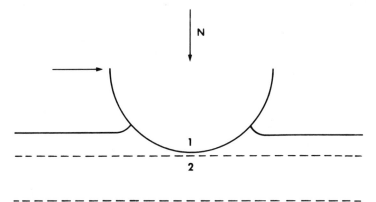

Figure 9.7 Deformation of a polymer plate by a rigid slider.

may be neglected. When a load N, acting vertically downward, is applied to it, the slider sinks a little in the polymer plate 2. All solid surfaces are rough, that is covered with hills and valleys, and the hemisphere shown in the figure may represent one of these hills. When the hemisphere is pushed to the right (without removing the load!), it must build a groove, the bottom of which is indicated by the upper dotted line. This drastic deformation of the surface layer causes, of course, deformation of neighboring volumes; thus the polymer is strained to the depth arbitrarily indicated by the lower broken line.

Of the several comparisons that demonstrate the superiority of the deformation theory of polymer friction as compared with the adhesion theory, only two can be presented here.

The deformation of one or both solids during their tangential displacement usually can be directly seen as long as at least one of the solids is a polymer. On the contrary, adhesion obviously is absent. If the hemisphere of Figure 9.7 is lifted vertically from the position indicated, the minimum force needed for this motion, as a rule, is practically equal to the weight of the slider so that no force attributable to adhesion can be discerned. See also Section 9.5.7.

For several polymers rubbing against other solids, it has been shown (286, 287) that the rate-temperature superposition explained in Chapters 5 and 7 is valid for frictional force and the work of friction as well. Molecular forces between slider and support are independent of the rate of sliding and little altered by moderate temperature changes, such as have been used in the experiments on polymer friction. For additional discussion consult Reference 288.

The morphology of surface layers can be altered by rubbing. For instance, the distance between spherulite centers mentioned in Section 9.3.2 was raised to that characteristic for the bulk of polypropylene by slow dragging of a polymer bar along a steel ring for 3 hr (277).

9.5.7 Adhesion

Organic polymers are widely used as adhesives. Hence the capacity to adhere often is con-

sidered to be a valuable property of polymeric substances. This capacity is a complex property, and some basic knowledge of the working of adhesive joints is necessary before it can be correctly understood.

It is convenient to classify all adhesive joints under three headings, namely, hooking, improper, and proper joints (260). In the first class the adherends (i.e. the solids to be glued together) are porous or fibrous, or at least have porous (fibrous) surface layers. When a liquid adhesive is spread over such a surface, it flows into the pores or around the fibers, solidifies, and thus forms "hooks." After solidification, the adhesive cannot be mechanically separated from the adherend without breaking either of the two or both. Thus the breaking stress of the joint is determined above all by the mechanical strength of the weaker of the two solids.

In improper joints the solidified adhesive has no, or very little, molecular contact with the adherend. The two are spatially separated by a weak boundary layer. When disruptive force is applied to the joint, separation proceeds in this layer; neither the adhesive nor the adherends is ruptured. Seven kinds of weak boundary layers may be distinguished, depending on their provenience and composition, but only two are discussed in this section.

The first is present whenever the adhesive does not displace air from the surface of the adherend in the course of application. This happens when the liquid adhesive does not wet the adherend or is too viscous to complete the wetting in the time allotted to the formation of the joint. In either case, an air-layer or a multitude of air bubbles remains between the adherend and the solidified adhesive. When rupture occurs, separation proceeds in the gas phase or from one air pocket to the next. An abundantly common example of weak boundary layers of the first kind is present between a slider and a support mentioned in Section 9.5.6; as both are solids, no displacement of the air separating them is possible.

In the second kind of weak boundary layer the material of the future weak layer is present in the adhesive. Practically all plastics in the laboratory and daily life contain low molecular ingredients such as oligomers or monomers corresponding to the main component [e.g., caprolactam in a poly(caprolactam)], lubricants employed in the drawing of films and fibers, and antioxidants. Some of these materials are readily soluble in the molten polymer but are squeezed out when the polymer solidifies and, especially, crystallizes. As a result of this process they concentrate in the interfacial layer (adhesive-adherend), which becomes the weakest part of the joint and breaks when disruptive forces are applied. The poor adhesiveness of commercial polyethylene is caused by these impurities.

In proper joints no weak boundary layer is present, and rupture takes place in the adherend or the adhesive. Fracture starts at the point in which the local stress first exceeds the local strength. Consequently, the stress pattern created by the external force is of fundamental importance for explaining and predicting the breaking stress of adhesive joints.

Rupture in proper joints does not proceed exactly along the interface, that is, true failure in adhesion is almost impossible, because every solid-solid interface is complicated, and the probability of a crack following all depressions and elevations on it is extremely small.

9.5.7.1 Adhesive Joint Strength

Kaelble (289) has enumerated the factors involved in the performance of adhesive-bonded structures. Table 9.4 lists two groups of prominent factors that would be of importance in evaluating adhesive joint strength. Generally joint strength decreases by way of mechanisms associated with inherent flaws (void or stress concentrations) and interfacial failure (weak boundary layer). Contrary, reversible (viscoelastic) and irreversible (plastic) deformation and fracture mechanism will ab-

sorb energy and tend to increase joint strength provided that the joint was properly prepared. By a proper joint one envisions the absence of a mechanically weak surface layer, either present or generated by reaction of the adhesive and adherend, and extensive interfacial contact (wetting). Systematic analysis of the interface by nondestructive means before and after joining should provide reasonable identification of the failure and the appropriate failure mechanism. Weak adhesive joints are susceptible to surface-sensitive analytical techniques. For strong joints, visual or standard microscopic techniques are sufficient to locate the locus of failure.

9.5.7.2 Weak Boundary Layers

The effect of wettability of epoxy adhesives on thermoplastics and the wetting of cured epoxy adhesives by molten thermoplastics have been demonstrated (259, 261–263). Although the wettability of both chlorotrifluoroethylene homopolymer and polyethylene are similar, strong joints can be made to the former but not to the latter at temperatures well below their respective melting temperatures. Polyethylene and probably many other melt-crystallized polymers seem to have associated with them a weak boundary layer, which precludes the formation of strong adhesive joints even though the extensive interfacial contact occurs between the polymer film and the epoxy adhesive (25, 291).

Apparently, at the solid-liquid interface (S-L, cured epoxy adhesive-molten polyethylene) on solidification of the polyethylene upon cooling, a region of high cohesive strength is generated by preferential adsorption and subsequent nucleation of the high-molecular-weight species. During cyrstallization from

Table 9.4 Factors Influencing Joint Strength (290)

Interfacial Factors	System Factors
Dispersion-polar interactions	Geometry of loading
Adherend wettability	Joint design
Adhesive-adherend cosolubility	Adhesive
Surface adsorption layers	Adherend rheology
Weak boundary layers	Residual stresses
Chemical bonding	Load induced stress concentration
Special electrostatic effects	

the melt the low-molecular-weight species are rejected from the interface into the bulk.

Contrary to the foregoing, a different process occurs at the liquid-air interface (polyethylene-air) or at the solid-liquid interface when the solid no longer acts as a nucleating site. Here low-molecular-weight species are rejected to the interface during the crystallization process, thereby resulting in a surface region of low mechanical strength, existing in an apparently amorphous configuration. Since many polymers are molded against nonnucleating surfaces (e.g. mold-release agents), thereby generating surface regions of low mechanical strength, one must be concerned with the removal of these weak boundary layers.

9.5.7.3 Chromic Acid Etch of Polyolefins

The chromic acid etching of low-density polyethylene and isotactic polypropylene films has been studied by XPS or ESCA. Oxygenated as well as sulfur containing ($-SO_3H$) species were detected. All the chromic acid specimens showed pronounced oxygen and sulfur peaks, while intensely treated specimens showed traces of chromium. Apparently only modest surface changes are required to affect significant increases in lap shear strengths. The presence of polar species (292) in the surface layer does not constitute proof for enhanced joint strength. Typically, many polar polymers such as nylon and polyesters require surface treatment while many polymers of low surface energy—polyvinylidene fluoride, and polychlorotrifluoroethylene—require no surface treatment prior to adhesive bonding.

When chromic acid treated polyethylene is exposed to temperature slightly higher than 80° C, there is a marked decrease in wetting with water, γ_{SL} increases to approach that of the untreated polyethylene (293). Presumably, the polar groups overturn to minimize the surface free energy of the polymers.

9.5.7.4 RF Glow-Discharge Modification of Polymers

The glow-discharge treatment of polymers is selectively controlled such that bulk properties of the polymer are unaffected. Conventional

spectroscopic techniques have proven ineffective in ascertaining the exact nature of changes in the surface region. For exposures that are sufficient to yield strong joints, wetting, ir, and ATR techniques have shown virtually no changes. Long exposure does induce changes, such as unsaturation, gel fraction, and oxidation, from trace accounts O_2 or H_2O present.

Clark and Dilks (294), based on their ESCA study, interpret their results in terms of a two-component direct and indirect radiative energy transfer. They suggest that the outermost monolayer cross-links by way of direct energy transfer from argon and metastables. A plausible mechanism for the CASING technique has been discussed by Clark (295). Production of radicals by initial argon bombardment can be shown as:

$$M^* + -CH_2-CH_2- \longrightarrow$$
$$-CH_2-CH- + \dot{H} + M$$

(where M* is an electronically excited species of Ar). The hydrogen atoms so produced because of the close proximity to neighboring hydrogens on $-CH_2-$ groups may abstract from a group next to the radical center leading to an unsaturated system, namely,

$$-CH_2-CH- + \dot{H}- > -CH = CH- + H_2$$

This would be particularly favorable because of the reduced C—H bond strength. It is conceivable that the M* impinging on the polymer could lead to direct elimination of hydrogen. Cross-linking, however, illustrates that although this could contribute to the overall reaction, radicals must be involved. Abstraction of hydrogen from a neighboring chain could lead to a radical combination and hence cross-linking:

$$-CH_2-\dot{C}H- \qquad -CH_2-CH-$$
$$-CH_2-\dot{C}H \qquad -CH_2-CH-$$

For a fully fluorinated chain (e.g PTFE), the corresponding process might be expected to be energetically less favorable because of the greater bond strength of the C—F as compared to the C—H bond and also because of the low-bond-dissociation energy of molecular fluorine. Indeed, it has been noted that PTFE films require extended treatment to improve their bondability (25).

9.5.7.5 Fluorination of Polyethylene

The differentiation of surface, subsurface, and bulk phenomena by electron spectroscopy for chemical analysis (ESCA) arises from the escape-depth dependence on kinetic energy for photoemission from core levels (corresponding to different binding energies). The surface fluorination of polyethylene is an ideal surface modification to explore since this constitutes available treatment for adhesive bonding (296). Recently Schonhorn and Hansen (297) showed by the fluorination of polyethylene that strong joints could be made to these surfaces even though the critical surface tension of wetting was reduced from 25 to 20 dynes/cm. While the predominance of work was performed in characterizing bulk properties, little information has been gathered concerning the earliest stages of fluorination, a situation favorable to ESCA analyses. Clark et al. (296) have performed a detailed ESCA examination of high-density polyethylene with particular emphasis on preparation, fluorination as a function of depth, stoichiometry of the surface, structure of the surface and subsurface, and the kinetics and mechanisms of the fluorination process. The spectra clearly indicate the formation of CF groups for an exposure of as little as 0.5 sec in a 10% fluorine in nitrogen mixture. The trace of O_{16} is considered due to the presence of adsorbed water rather than to oxidation of the polymer. Fluorination apparently proceeds by way of a chain reaction followed by appropriate termination reactions:

Initiation:

$$F_2 \underset{k_1}{\overset{k_{-1}}{\rightleftharpoons}} 2F\cdot$$

$$\Delta H = +37 \text{ kcal/mol}$$

$$F\cdot + -CH_2-CH_2- \longrightarrow -\dot{C}H-CH_2- +HF$$

$$\Delta H = -34 \text{ kcal/mol}$$

Propagation:

$$F_2 + -\dot{C}H-CH_2- \longrightarrow -CHF-CH_2 + F\cdot$$

$$\Delta H = -68 \text{ kcal/mole}$$

The low-bond-dissociation energy of fluorine (37 kcal/mol) coupled with the large negative enthalpy changes ensures an efficient chain reaction. Tedder (298) suggests that activation energy for hydrogen abstraction is low, inferring that fluorination is a relatively unselective process. Clark et al. (296) suggest from the ESCA data that the fluorination process is diffusion controlled.

9.5.7.6 Other Surface Treatments

Among the most extensively studied polymeric surfaces have been those of fluoropolymers, both control surfaces and those subjected to special surface treatments. These have been studied by combinations of infrared spectroscopy (ATR and transmission), ESCA, soft x-ray spectroscopy, contact angle hysteresis, and electron and optical microscopy (15, 17, 299–302). The polymers most studied are untreated commercial polytetrafluoroethylene (Teflon TFE) and tetrafluoroethylene-hexafluoropropylene copolymer (Teflon FEP) films. The same films have also been subjected to a variety of surface etchings to increase the adhesiveness of these to other polymers, for example, polyurethane, epoxy, or phenolic resin adhesives. These include treatment with sodium in naphthalene-tetrahydrofuran, sodium in liquid ammonia or molten potassium acetate at 325° C. The color of the fluorocarbon surface is changed to dark brown to black by these treatments. Microscopic measurements indicate the color change is localized within a surface region of a few tenths of a micron (301). The advancing contact angle of water (25° C) on PTFE changes from 108 for the unmodified film to 62–66 deg for the etched films (301). Similarly Teflon FEP exhibits a decrease of the advancing water contact angle from 109 to 52 deg (15). The infrared spectra of the original

PTFE (either of a composite layer built up from seven 5 μ films or the differential spectrum) is greatly changed in the region of 1600 cm^{-1} as a result of the three etching treatments. The presence of the intense absorption bond at 1600–1700 cm^{-1} has been ascribed to the presence of conjugated double bonds and/or the valence vibrations of C=O groups (301). Surface films subjected to Na in naphthalene-tetrahydrofuran or to potassium acetate treatment exhibit spectra that indicate the presence of the hydrophilic OH and CO groups as well as unsaturation. Films treated with Na in anhydrous ammonia exhibit indications of surface NH groups as well (301). These results are consistent with the work of Dwight and Riggs (15), Collins et al. (17), and Clark and co-workers (18), using among other techniques ESCA and soft x-ray spectroscopy. The presence of nitrogen in the sodium-ammonia etched films of both Teflon TFE and FEP is confirmed. The soft x-ray data indicate that the depth of etch after 60 sec of exposure is about 0.3 μ in PTFE and 0.07 μ in FEP.

9.6 ENVIRONMENTAL ASPECTS OF ADHESION AND ADHESIVE JOINT STRENGTH

Although a wealth of data exists for describing the deleterious effects of a variety of environments (e.g. humidity, temperature, stress) on the durability of composite structures (303) and polymers (304), few approaches appear to be of sufficient utility to predict mode of failure and, perhaps more important, to provide some insight into procedures to preclude premature failure.

Owens (305) and more recently Gent and Schultz (306) have considered the durability of adhesive joints consisting of two polymers A and B when the adhesive joint is exposed to a variety of environments. Owens has shown that in the presence of a particular liquid, if the thermodynamic work of adhesion of a particular composite A-B in contact with liquid L is near zero $W_{AB}^L \approx 0$, the adhesive joint will separate in a short period of time. In the absence of a liquid phase, the work of adhesion is large and positive $W_{AB} >> 0$, and the delamination may be retarded for an indefinite interval.

In the absence of chemisorption (307) and interdiffusion (308), it appears that at least two criteria are necessary to induce premature failure in the presence of stress and a fluid phase (305): the liquid must be immiscible with the polymer or other members of the composite structure, since miscibility of the liquid with one or both members of the composite would plasticize sufficiently that member and thus preclude failure; and the liquid or fluid phase must interact fully (i.e., in a surface-chemical sense) with either one or both of the members of the composite.

Andres and Kinloch (309) have described an adhesive failure energy θ that is based on the earlier efforts of Gent et al. (310). The adhesive failure energy is comprised of two major components, namely, the energy to propagate a crack through the unit area of interface in the absence of viscoelastic energy losses, that is, an "intrinsic" adhesive failure energy θ_0, and the energy dissipated viscoelastically within the adhesive in the propagating crack, again referenced to the unit area of interface. θ_0 should be rate and temperature independent since its value depends upon the nature of the bonding at the interface. The energy τ, dissipated viscoelastically, is by definition rate-temperature dependent.

Further, Andrews and Kinloch (309) relate to Equation (9.16)

$$\theta_0 = \frac{\theta \tau_0}{\tau} \qquad (9.16)$$

where τ is the cohesive failure energy. Methods are described for obtaining θ, τ, and τ_0 for a variety of rates and temperatures to permit a computation of θ_0. For cases where only secondary forces are operative it is shown (310) that $\theta_0 \approx W_A$, where W_A is the thermodynamic work of adhesion. When covalent boding occurs across the interface, $\theta_0 > W_A$ (311). Furthermore, θ_0 was found to be independent of test geometry.

A simple theory may be evolved that considers the features of the foregoing analysis. The work W per unit length required to separate an adhesive joint A-B may be represented as shown in Equation (9.17).

$$W = W_S + W_P + W_E + W_T \quad (9.17)$$

where W_S is the reversible work to form new surfaces, W_P the work expended in the plastic flow, W_E the loss in stored elastic energy, and W_T is associated with the work to desorb chemisorbed sites and to break tie molecules (312).

Schonhorn and Frisch have shown how the Griffith crack theory (313) in conjunction with Equation (9.17) yields Equation (9.18):

$$\left(\frac{\sigma_f^L}{\sigma_f}\right)^2 = \frac{W_{AB}^L}{W_{AB}} \quad (9.18)$$

where σ_f^L is the critical stress to failure in the presence of the liquid, σ_f is the critical stress to failure in the absence of liquid, and W_{AB}^L and W_{AB} are the thermodynamic work of adhesion in the presence and absence of liquid, respectively. Equation (9.18) is in accord with the qualitative results of Owens (305). From Equation (9.18) it is clear that if the liquid fully interacts with the composite, W_{AB}^L decreases

and $(\sigma_f^L/\sigma_f)^2$ is lowered. Clearly, when $\sigma_f^L \ll \sigma_f$ the composite will tend to delaminate spontaneously. On the contrary, minimal interaction of the liquid with either or both of the components in the composite results in $W_{AB} \approx W_{AB}^L$ and $\sigma_f^L \approx \sigma_f$, a condition that leads to composite stability in the environment, certainly from a stress cracking mode of failure. Recently, a large number of investigations have been undertaken to examine the effects of the environment on adhesive joints 290, 314–317.

9.6.1 Calculation of the Thermodynamic Work of Adhesion

Consider Figure 9.8, in which a composite A-B is in contact with either a vapor or a liquid. In the presence of a vapor, the thermodynamic work of adhesion is given by Equation (9.19),

$$W_{AB} = \gamma_{AV} + \gamma_{BV} - \gamma_{AB} \quad (9.19)$$

while in the presence of a liquid phase the work of adhesion of composite A-B is given by Equation (9.20).

$$W_{AB}^L = \gamma_{AL} + \gamma_{BL} - \gamma_{AB} \quad (9.20)$$

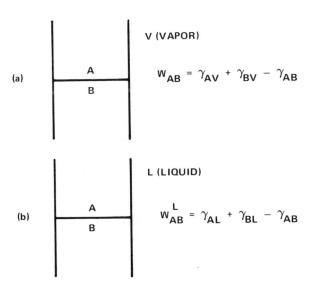

Figure 9.8 Thermodynamic work.

Fowkes (247) developed an expression for the interfacial tension between A-B phases, which was extended by Owens and Wendt (318), Dann (237), and Kaeble (289) into Equation (9.21).

$$\gamma_{AB} = \gamma_{AV} + \gamma_{BV} - 2[(\gamma_{AV}^d \gamma_{BV}^d)^{1/2} + (\gamma_{AV}^p \gamma_{BV}^p)^{1/2}] \quad (9.21)$$

By a judicious choice of liquids (319) values of γ_{AV}^d, γ_{AV}^p, γ_{BV}^d, and γ_{BV}^p can be determined and the interfacial tension γ_{AB} computed (290).

In the presence of a liquid phase, the interfacial tensions γ_{AL} and γ_{BL} can be calculated from the relations shown as Equations (9.22) and (9.23).

$$\gamma_{AL} = \gamma_{AV} + \gamma_{LV} - 2[(\gamma_{AV}^d \gamma_{LV}^d)^{1/2} + (\gamma_{AV}^p \gamma_{LV}^p)^{1/2}] \quad (9.22)$$

$$\gamma_{BL} = \gamma_{BV} + \gamma_{LV} - 2[(\gamma_{BV}^d \gamma_{LV}^d)^{1/2} + (\gamma_{BV}^p \gamma_{LV}^p)^{1/2}] \quad (9.23)$$

By combining Equations (9.21), (9.22), and (9.23), it is possible to calculate the work adhesion W_{AB} and W_{AB}^L.

It is evident from the data and the arguments in this section that the interfacial tensions at the metal oxide-polymer interface and the interaction with a humid environment may be responsible for premature delamination of the composite under low levels of stress. Modification of interfacial tensions of judicious use of particular silicone coupling agents, which not only chemisorb on to the metal oxide but may also modify the polymer surface by cross-linking, appears to be a valid explanation for the observed wet strength of the silicone-treated composite structure. Clearly, this approach may be extended to other combinations of adherends and guide one in a proper selection of surface treatments. Although the initial joint strength of surface-treated adherends ($\sigma_0 \gg W_A$) may exceed those where only interfacial (or secondary) forces are operative ($\sigma_0 \approx W_A$), in the presence of an adverse environment, as shown by Owens (305), these strong joints may deteriorate rapidly ($\sigma_0^L \ll W_A$). What appears to be a troublesome notion that high-energy

surfaces when surface treated by adhesion promoters to yield low-energy surfaces (i.e., typically with glass, coupling agents lower γ_c to low levels) often yield more permanent joints (higher wet strength), appears to be resolvable in the light of the arguments presented in this discussion. Although the same degree of wetting is achieved for both treated and nontreated substrates, it is clear that the interfacial structure, for the same extent of interfacial contact area is of prime importance.

9.6.2 Corrosion

Corrosion is closely related to adhesion; the appearance of corrosion is indicative of loss of adhesion between the paint and the underlying substrate. General corrosion mechanisms have been proposed to account for both anodic (oxidation) and cathodic (reduction) reactions (320). It has been shown that adhesion failure under anodic conditions is due to displacement of the primer by hydroxide ions electrochemically generated at the paint-metal interface (321). Resin composition markedly affects the reaction rate. Recently the use of photoelectron spectroscopy in corrosion science has been reviewed, including applications to passive films, solid state reactions, electrochemistry, and aqueous corrosion, concluding with examples of the concurrent use of ESCA and ion-beam etching (322).

Moisture accelerates the electrochemical reactions, and inorganic contaminants may draw moisture through a film as well as conduct current and react with a film or substrate. Recent ESCA work on the locus of a failure in the loss of paint adhesion specifically identified saponification of the polymer at the interface by hydroxide ions generated in anodic corrosion (323).

A thorough study of contamination and corrosion will require not only multiple energetic beam techniques but also in situ reactions to form identifiable derivatives. For example, ion profiling-Auger spectroscopy revealed the interactions between sulfur dioxide and modified 440C stainless steel at different temperatures (324). SO_2 adsorbed dissociatively with equal quantities of sulfur and oxygen on the surface, and the sulfur did not form a metallic surface phase. Instead, the oxide layer thickness increased the presence of SO_2, especially between 500 and 600° C where the oxide became thicker by a factor of 7.

The use of freshly cleaved in situ bulk samples as controls for contamination and surface studies is important. This was illustrated in the Auger analysis of E and S glass fibers (325). In S glass, magnesium and aluminum are concentrated on the fiber glass surface, whereas in E glass, chlorine, silica, and aluminum show surface segregation. In order to control progressive beam-induced depletion of cations in soda lime-silica glass analysis, profiles were obtained by making a linear series of analyses along the surface of a sloping ramp etched into the glass in a separate ion bombardment operation (326).

9.6.3 Wear

The interactions between polymers and metals in touch-and-sliding contact were elucidated by field-ion microscopy (FIM) and Auger spectroscopy. Strong adhesion between the organics and all metals (clean or oxidized) was observed (327). Using FIM with increasing voltage, electron-induced desorption in the vicinity of 20 keV indicated chemical bonding of polymer to the metal surface. ESCA studies in friction, lubrication, and wear are exemplified by a variety of experiments on wear conditions (328). Sulfide was formed at the expense of oxide under severe wear. In mild wear scars, however, there was not evidence of sulfide or mercaptide, but the oxide layer thickness doubled. Further, it was determined that surface chemistry was a function of wear rate rather than load.

9.7 ADHESIVES AND COATINGS

Film-forming materials that adhere to one or more surfaces are scientifically and technologically diverse. Their structures and properties are described in part by chemistry, polymer science, engineering mechanics, and physics. The technology is equally pluralistic as it relates to production, formulation, application, and processing.

This diversity is exemplified by the availability of over 2500 different adhesives and sealants (329). There were over 1400 patents issued on "adhesives" from 1976 to 1978, over 200 on acrylic coatings (330), and nearly 1150 on epoxy technologies (330).

Since it is impossible to document and describe the technology completely in a limited context, an overview of established practice and current trends are presented. The discussion is limited to synthetic materials used in industrial settings.

Article, patent, and presentation citations are confined essentially to those appearing after December 1969, which have been abstracted by *Chemical Abstracts* and *Engineering Index*. Reviews, monographs, and texts are time constrained only in that they document recent developments or established practice.

9.7.1 Surface Preparation

The single most important factor in a successful bond for both adhesives and coatings is surface preparation. Snogren (331) has explicity reviewed surface preparation, while Cagle (332) and Shields (333) have addressed the topic within an adhesive's framework. While "clean" and "cleaning" are nearly universal in this context, what is really implied is that the adherend's surface must be receptive and mechanically strong. To achieve this end, all surface materials with unsatisfactory adhesion-cohesion must be removed and/or replaced with an acceptable layer. The removal or cleaning process is subdivided into degreasing and substrate removal, while replacement is described as surface conversion. The particular sequence will depend on the adherends (metallic, ceramic, polymeric), the expected performance, the geometry, and the size.

9.7.1.1 Degreasing

It is necessary to assume that the surface of the material to be bonded or painted in its as-received condition is substantially different from the bulk material. In many cases the surface nature (mill scale, lubricants, processing aids, water) will be known, but the composition will, at best, be variable. In some cases both nature and composition will be unknown, namely, "dirt." In the broadest sense these surface contaminants are either foreign to or derived from the adherend. With metals the foreign substances would be typified by dirt, grease, and oils, while the derived components would be essentially oxides. Polymeric materials are prone to air-borne contamination due to their electrostatic nature, and they are often contaminated with mold releases. Derived contaminants in plastics and elastomers may be oxides, but this is complicated by the migration of nonpolymeric additives such as plasticizers and antioxidants. Ceramics and glasses will have foreign contaminants, but the derived layer is usually mechanically strong.

The foregoing distinction is significant in that it helps define cleaning strategy. Since foreign contaminants are, with the possible exception of silicones (334), typically less tenacious than derived ones, they may be removed by washing. In a physical sense washing involves wiping or flooding the surface with a liquid. The process is dynamic in that either or both parts and liquid are in motion. Chemically, the process is one of dissolving or suspending with minimal surface conversion. The latter distinction is important for aqueous, alkaline cleaners that saponify and/or neturalize contaminants such as die lubricants or processing aids.

Solvent cleaning or degreasing is typically accomplished with chlorinated hydrocarbons, although alcohols, ketones, toluene, and naptha are also used. The chlorinated hydrocarbons have good solvency and are nonflammable, but their use is constrained by environmental and toxicity considerations (335). Any solvent, whether it is used in cleaning or in the film-forming material, must be selected with due regard for environmental stress cracking (ESC) and dissolving. Solvent selection either for cleaning or formulating by means of solubility parameters has been re-

viewed by Barton (336) and by Gardon and Teas (337). Small (338) and Hoy (339) have developed estimative methods based on group additive constraints for polymers, while Fedors (340) has advanced an analogous method that within limitations obviates the necessity for the polymer density. While these strategies have mitigated the problems of solvent selection, ESC remains troublesome in that it involves the premature brittle failure of the substrate above some critical strain in the presence of a liquid that is generally a nonsolvent. Kambour (341), and Vincent and Raha (342) have delineated chemical parameters that affect the susceptibility of glassy plastics while Dunn and Sampson (343) have provided insight into the cracking mechanism in nylons. Vincent (344) has reviewed ESC in polyolefins; Singleton and co-workers (345) have investigated the mechanism.

In metallic substrates environmental stress cracking of titanium in the presence of chlorides or chlorine-containing solvents has been documented (346). Although the specific stress cracking of other adherends is less well documented, the premature failure of adhesive bonds by way of ESC has been reviewed (347) and may involve the near universal tendency of polymeric materials to undergo stress cracking. Much effort has gone into characterization of the environments and the correlation of critical strains with solubility parameters. Henry (348) has observed "three-dimensional" parameters to be a reasonable means of characterizing susceptibility. His approach was, however, two-dimensional in that the polar component as derived from Owens and Wendt (318) and Kaeble (301) was a combination of both polar and hydrogen bonding. While the studies of Wu (349) on nonpolar and polar interactions of molten polymers can also be applied, a separation of the polar and hydrogen-bonding components is necessary for a "rigorous" correlation. The phenomenon has also been studied as a mechanical event from a fracture mechanics viewpoint by Andrews (350), Kausch (351), and Marshall (352).

Although initial degreasing is *de rigueur* in surface preparation for adhesive bonding or coating, solvent clenaing is not unique. Degreasing with aqueous, alkaline solutions that contain detergents and inhibitors is common in the preparation of stamped steel for automobile exteriors (353). It is, however, pertinent to note that the contaminants in this case are amenable to saponification and/or neutralization. While alkaline degreasing is suitable for carbon steels, due care must be exercised with other adherends, particularly the amphoteric elements. The latter, which include aluminum, antimony, beryllium, lead, tin, and zinc and their oxides, are etched by strong bases.

9.7.1.2 Substrate Removal

As an intermediate step between degreasing and surface conversion, substrate removal or ablation serves to remove derived contaminants, increase surface bonding area, provide scarflike geometry, and increase the wetting tendency of the adhesive or coating (354). Care must be exercised so that all debris is removed and that gas entrapment is avoided. The removal of both contaminants and substrate, while it promotes adhesion, must be carefully controlled for thin or soft substrates, load-bearing or critical applications, and thin or high-gloss coatings.

Ablation without surface conversion of the surface or substrate is accomplished by physical or chemical means. The former as exemplified by cutting, chipping, or abrading is versatile but does not offer the control of chemical or etching methods. Of the physical methods, brushing, chipping, grinding, and scraping are the most rudimentary and are limited to gross contamination such as scale, rust, paint, and weld splatter. Even when power assisted, material removal is relatively slow and incomplete. Abrading with paper-supported or propelled abrasives is more common and usually more thorough, particularly with blast techniques. Paper-supported abrasives such as emery cloth of 100-grit and finer are quite common but are limited to open geometries and low-volume applications.

There are five basic types of blast abrasives (Table 9.5) that are propelled pneumatically, hydraulically, or centrifugally. Velocity is the

Table 9.5 Representative Blast Abrasives (331)

Type	Examples
Siliceous	Silica (sand or glass beads), garnet quartz
Carbides	Silicon, boron
Oxides other than silicon	Aluminina, cerium oxide, refactory slags
Metallic	Iron and steel shot or grit
Agricultural	Ground nut or fruit pit shells

most important process parameter, which is, in turn, dependent on abrasives' composition, weight, density, hardness, shape, size, and friability (355). Composition is important in that abrasives leave traces of themselves that must be removed or be compatible with the adhesive or coating. It thus follows that metallic shot should not be used with dissimilar metal substrates since any embedded remnants will become sites of galvanic corrosion. The shape of the abrasive determines the anchor pattern or profile produced. Shot or bead abrasives effect a peened profile, whereas angular ones produce a wedgelike pattern, which has been demonstrated to afford better bond strength in adhesive applications (354). The hardness, weight, specific gravity, shape, and size of the abrasive affect the profile height (Table 9.6), with coarser abrasives affording a deeper pattern.

The effect of roughness on bond strengths has been described in Section 9.7.1.2, and it is generally agreed that random roughening with cutting abrasives leads to increased bond strengths, while an optimum profile appears to be a function of both viscosity and the adherend (356). Since variable bond strengths may be associated with interfacial stress distribution (354), it is necessary to control work hardening, which can occur with physical cleaning.

Where optimum performance under extreme or adverse environments is necessary, chemical or electrochemical ablation is required. From a process standpoint, chemical cleaning is thorough, reproducible, amenable to complex geometries, and economical for high-volume production. Since the chemicals are toxic or hazardous, care in handling and

Table 9.6   Maximum Profile Height for Various Abrasives[a]

Abrasive	Maximum Particle Size	Maximum Height of Profile (mil)
Sand, very fine (Ottawa silica sand)	Through 50 mesh[b]	1.5
Sand, fine (Ottawa silica sand)	Through 40 mesh	1.9
Sand, medium (Ottawa silica sand)	Through 18 mesh	2.5
Sand, large (river sand)	Through 12 mesh	2.8
Iron Grit No. G-50[c]	Through 25 mesh	3.3
Iron Grit No. G-40	Through 18 mesh	3.6
Iron Grit No. G-25	Through 16 mesh	4
Iron Grit No. G-16	Through 12 mesh	8
Iron Shot No. S-230[d]	Through 18 mesh	30
Iron Shot No. S-330	Through 16 mesh	3.3
Iron Shot No. S-390	Through 14 mesh	3.6

Source.   Reprinted with permission from *Adhesives in Modern Manufacturing*, E. J. Bruno, Ed., Copyright 1970, Society of Manufacturing Engineers.

[a] Test criteria: Published in 1954 and expanded and reprinted in 1963. Pressure at machine: 80 psi., 0.516 in. orifice, nozzle length unknown.

[b] U.S. Sieve Series.

[c] Crushed iron grit.

[d] Operating mixture.

disposal is mandatory. As might also be expected, reproducibility is intimately associated with the control of concentrations, pH, time, temperature, and contaminants.

The technology of reactive chemical cleaning, that is pickling or etching, has been separately reviewed by Snogren (331), Cagel (332), and Shields (333). The most studied adherend is still aluminum, particularly for aircraft and aerospace applications. Etching of plastics as described in Sections 9.5.7.3 through 9.5.7.6 is considered from an operational standpoint in Section 9.7.1.3.

Acidic formulations are usually prescribed for iron, chromium, nickel, titanium, and their respective alloys. Caustic formulations are not nearly so common but are noted for aluminum, magnesium, and titanium (332). Beryllium is unique in that only alkaline cleaning is specified. Reactive chemical cleaning of ceramics and glasses is almost always accomplished with strong acids or acidic oxidant mixtures (331).

The material removal in the chemical cleaning of metals is usually ascribed to oxidation and/or hydration of oxides, hydroxides, and base metal. With hydrochloric or sulfuric acids the process is called pickling, and, as expected, hydrogen ion is the oxidant. As an oxidative etchant, hydroxide ion is generally limited to amphoteric metals and their oxides. In either case the spontaneity of the reaction is governed by the pH, temperature, and potential.

Although pickling is an effective and thorough cleaning method, the oxide-free surface is very reactive. For iron, chromium, nickel, titanium, and alloys containing major amounts of these metals, such reactivity is mitigated by passivation (357). This loss of reactivity is effected either *in situ*, or as a secondary operation with nitric, chromic, or hydrofluoric acid. Active metals such as aluminum and magnesium can also be effectively treated with acidic oxidizing mixtures like sulfuric-chromate, but the decreased reactivity is usually ascribed to explicit oxidation (333).

9.7.1.3   Surface Conversion

By convention, surface conversion is taken to mean anodizing, phosphating, coupling, or any of the wide variety of chemical modifications applied to low-surface-energy plastics. This distinction is essentially operational in that chemically resistant or mechanically strong surfaces formed "during" chemical cleaning are usually excluded.

The formation of a layer of zinc, iron, or manganese phosphate crystals on aluminum, steel, or zinc substrates for corrosion protection and improved adhesion is well established. A notable application is the zinc phosphating of mild steel used in automotive exteriors (335, 353). The process depends on the precipitation of metallic phosphates from strongly acidic solutions as the pH is raised with the latter directly related to hydronium ion consumption at the metal interface. In actual practice phosphating mixtures usually contain passivators and nucleating agents. Numerous preparations for both manual and automatic application are available as are detailed process descriptions (331, 332). The nature of the surface and its performance have also been studied (358).

Anodizing by chemical or electrochemical methods is best recognized in its application to aluminum, although magnesium and titanium can be so treated. Like phosphating, the process is well established, particularly for aircraft and aerospace applications. Although many variations exist, the net effect is usually the generation of $\beta\, Al_2O_2 \cdot 3H_2O$, as the faying surface (333). A detailed process and specifications for aerospace applications is available (329).

Coupling agents are analogous to amino acids in that they are difunctional. One of these functions is bonded or coupled to the adherend while the other is, at a minimum, compatible with the organic film. Reactive silanes as first developed in the 1940s and reviewed by Plueddemann (359) remain the *sine qua non* of the true coupling agents. Virtually all filamentous glass used as a reinforcing agent in polymer matrices is sized with a composition that contains silanes of the general structure $X_3Si(CH_2)Y$ where X is hydrolyzable and Y is either reactive or compatible with the organic film. While the labile or hydrolyzable group is usually ethereal, peroxidic groups may be used for heat-activated "adhesion promoters." The film compatible group is, with rare exception, tailored to the resin by way of amino, mercapto, ester, vinyl, and more recently azo (360). As expected the adhesion is most improved in cured resins with such groups as vinyl, epoxy, and mercapto. Silane coupling agents have also found application in surface conversion of metallics (361–363) and acetals (364). Although not immediately germaine, silination of fillers such as glass beads, clays, carbonates, talcs, and alumina trihydrate is also done (365, 366).

Adhesion promoters, according to definition, are equivalent to coupling agents in that they are chemisorbed to an adherend. A strict classification is, however, problematical in that physical and chemical sorbtion both affect adhesion strongly and are often difficult to distinguish. While chemisorption is apparent for reactive silanes (359, 365), any assumption that other adhesion promoters function by chemical reaction should be made only with careful consideration of the evidence. Coupling agents as adhesion promoters have been reviewed by Cassidy and Yager (367). As a class these also include carboxylic acids, sulfonic acids, phosphorous esters, chromium acid complexes, starch derivatives, organoborates (368), titinates (369), imides (370), and quinoxalines (371). The imide coupling agents are noteworthy in their application to graphite fibers (370), while titanates have achieved commercial significance with low-viscosity systems such as plastisols (366).

It is difficult to bond or paint satisfactorily plastics, particularly fluorocarbons, acetals, saturated elastomers, and olefins. In that these materials are characterized by low surface and cohesive energies, the usual strategy has been to improve wettability. To this end the surfaces have been treated with oxidizing agents, flames, plasmas, dissolved alkali metals, and irradiation. Since these treatments almost always afford a reduced water contact angle by way of oxidation, polar groups, and stronger surface interaction, the improvement is considered to be a surface phenomenon. The sufficiency of such rationale has, however, been questioned (see Section 9.5.7.2) in that polyethylene exposed to glow discharge in noble gases or persulfate oxidation and bonded with conventional epoxy adhesives exhibited

Table 9.7   Surface Preparation versus Lap-Shear Strength (372)

Group Treatment	Lap-Shear Strength (psi)
As received	444
Vapor degrease	837
Vapor degrease, 15% NaOH	1671
Vapor degrease, aluminum wool	1478
Solvent wipe, wet and dry sand, 240-grit only	1345
Solvent wipe, sand (not wet and dry), 120-grit	1329
Solvent wipe, wet and dry sand, wipe off with sandpaper (done rapidly)	1540
Vapor degrease, wet and dry sand, wipe off with sandpaper	1726
Vapor degrease, grip blast 90-mesh grit N_2-blown	1751
Unsealed anodized	1935
Vapor degrease, $Na_2Cr_2SO_4$, tap water	2756
Vapor degrease, alkaline clean, chromic-H_2SO_4, deionized water	2874
Vapor degrease, alkaline clean Na_2Cr_2O—H_2SO_4, tap water	2826
Vapor degrease, alkaline clean, Na_2Cr_2O—H_2SO_4, distilled water	2800
Vapor degrease, grit blast 90-mesh grit, alkaline clean, Na_2Cr_2O—H_2SO_4, tap water	2929
Vapor degrease, grit blast 90-mesh grit, alkaline clean, Na_2Cr_2O—H_2SO_4, distilled water	3091

bond strengths equivalent to sulfochromate treated polymer with virtually no change in wettability (296). Even with fluorine the common underlying effect was the production of a cross-linked skin. It is thus implied that dispersion interactions are of primary importance in adhesion with polar interactions being, if not minor, at least undefined.

The effect of surface preparation on bond strength is apparent from a study by Rodgers (372) on 2024T3 aluminum with a modified epoxy as noted in Table 9.7. The improvements, while dramatic, cannot answer the question of sufficiency. Only with careful evaluation of the environment and service requirements can this question be answered. Where the requirements are moderate and the environments mild, there may be no apparent advantage in surface preparation beyond effective degreasing.

9.7.2   Joint Design

Satisfactory adhesive bonding of properly prepared surfaces requires careful delineation of both physical and chemical environments. The physical or mechanical requirements as they determine joint design are briefly discussed in this section along with testing.

Environmental effects are noted in Section 9.6.

All joints, whether they are effected by mechanical fastening, fusion, or adhesives, must withstand both static and dynamic loading. In either case it is necessary to know or determine the stresses and their distribution. The important stresses in adhesive joint design are shear, tension, compression, peel, and cleavage. It is desirable to concentrate stresses in shear, with tensile, compressive, peel, and cleavage stresses being decreasingly desirable (373). Compressive loading is significant primarily in that any cracking or flow in the adhesive can lead to joint failure from other stress (372). A bond in pure compression would in all likelihood not need an adhesive. Offset loadings are undesirable as they effect stress concentration on the joint boundary line. The peel and cleavage stresses resulting from such offset loading are differentiated in that the peel involves a flexible adherend in an unrolling action.

If the joint is cyclically loaded or unloaded, a condition of stress known as fatigue occurs. Premature failure in dynamic environments has been related to the rate of crack propagation (374). Although a discussion of fracture mechanics is beyond the scope of this discussion, it is known that the crack propagation sensitivity of high-modulus, low-elongation adhesives is high (375).

While a uniform stress distribution is an often-cited advantage of adhesively bonded joints, it is known that the stresses are "never uniform" (329, 333). For a lap or surface joint in tension the stresses are concentrated at the ends of the adherends, with the center bearing little or no stress. Thus if one designs for an average stress, it will be too low. Some other "axioms" for joint design, as noted in *Adhesives Redbook,* 1978–1979 (329) are:

1. The failing load of lap joints increases as the width of the joint increases. A one-inch wide joint is twice the strength of a half-inch joint.

2. The failing load of lap joints does not increase proportionately with length of area. A two-inch long joint is not twice as strong as a one-inch joint. A four-inch long joint may not take an appreciably greater load than a two-inch long joint.

3. Although typically a stronger adhesive may produce a stronger joint, a higher elongation adhesive with lower strength can produce a stronger joint.

4. The stiffness of the adherends and adhesive influences the failure load of a joint. In general, the stiffer the adherend with respect to the adhesive, the less the failing load of the joint is influenced by joint geometry.

5. The adhesive layer thickness is not a strong influence on the strength of the joint. More important is a uniform and void-free adhesive layer.

6. The higher the E_t of the adherend, the stronger the joint. A 0.040 in. sheet steel lap joint is stronger than a 0.040 in. sheet aluminum joint. A 0.094 in. sheet aluminum lap joint is stronger than a 0.040 in. sheet aluminum lap joint.

Although there are only four basic types of joints, surface or lap, butt, angle, and tee, numerous variations exist (332, 333). If the stresses and their distributions are known, it may be possible to adopt one of these existing designs. The actual dimensions as a function of the stress-strain relationships of the adhesive and the adherend are often derived empirically, although Perry (376) has described a method for calculating maximum allowable stresses. Calculation methods, although de-

sirable, are limited by stress combinations and distributions. However, finite element analysis and photoelasticity studies have served to mitigate these problems (260, 377–379).

Most comparisons for both selection and quality control are made on the basis of a destructive test, with lap shear as described in American Society for Testing and Materials (ASTM) D1002 being the most common. Strict comparisons are not advisable unless the adherends, the surface treatment, the processing conditions, the test environment, and the duration are equivalent. Nondestructive testing is suitable for the detection of such bondline defects as chemical segration, delamination, gas entrapment, inadequate surface preparation, or cure variation (380, 381). As with most nondestructive techniques, success depends on a correlation with dependent variables. To this end, dielectric constants, dynamic and viscous modulii, thermal conductivity, and "reflection" have been used (333, 380, 381). Reflection techniques, particularly ultrasonic (382–384), have received the most attention. The ultrasonic techniques are used widely in aircraft manufacturing, and analytical models have been proposed (383). Strength classification is, however, difficult because of the extremely small signal variations attributable to "flaws" (383, 384). Although not specifically noted, fast Fourier transform signal analysis (385) may serve to improve reliability.

In vitro, testing of adhesives and coatings is directly related to the analysis and characterization of polymers and has been reviewed by Myers and Long (386). New developments include fast Fourier transform ir, pulsed Raman, pulsed ^{13}C nmr, neutron scattering, and electron scattering for chemical analysis (ESCA) (387). It is also pertinent to note the use of dynamic mechanical analysis, which has been reviewed by Murayama (388).

9.7.3 Adhesive Selection

While it is reasonable to assume that the specification of an adhesive for carefully prepared substrates in a joint that is designed for the expected stresses and environments should be straightforward, the existence of thousands of formulations and over 2000 suppliers (329) are serious complicating factors. This diversity has lead to development of many classification systems (333) with the most prevalent method according to chemical base (389). Other systems include: substrates, applications, mode of application, physical form, characteristic, and manufacturer/supplier. Classification as "structural" or "nonstructural" is generally arbitrary as there is no accepted specific definition of the former in terms of bond strength. Although such classification schemes are good and useful, final choices usually require supplier cooperation and in-house evaluation.

In reality, choosing an adhesive is a complex task that cannot be accomplished by any one material variable or classification. Thus the adhesive commentary noted in Table 9.8, as developed by Schneberger (390), should be used as a guide rather than a final arbiter.

9.7.3.1 Bonding Processes
The tripartate concerns of energy, environment, and economics have had a direct effect on bonding systems. In the drive to minimize cure energies and eliminate toxic, hazardous, or "polluting" solvents, while facilitating the bonding of impervious substrates, there has been much growth in pressure-sensitive, hot melt, powder, film, and "100% convertible" thermosetting adhesives. These advances in adhesive technology have been reviewed by Lee (391). Radiation curing as an alternative to forced-draft or autoclave heating has been reviewed by Kinstle (392) but, with the exception of pressure-sensitive adhesives (393), has been essentially limited to coatings and imaging compositions (394).

9.7.4 Structure-Property Relationships

The basic interfacial relationships in adhesion have been reviewed by Zisman (319) and are updated in previous sections of this chapter. While good adhesion is necessary for a satisfactory adhesive or coating, cohesion as the "sufficient" condition is considered in this section. The discussion traces the origins of cohesion as derived from molecular weight/

molecular weight distribution, composition/ arrangement, and crystallinity. The emphasis is on property trends as a function of structural variables for some of the more common polymer classes used in adhesives and coatings. The examples are essentially constrained to single correlations owing to formulation and generic diversity and incomplete theoretical correlations.

Within any substance cohesion is effected by attractive forces between atoms, ions, or molecules. These forces are electrostatic in that they are attributed to a nonuniform charge distribution within the particular specie. In the attraction continuum four basic modes of interaction are distinguishable: dispersion, dipole, hydrogen-bridge, and ionic. The last as characterized by "irreversible" electron transfer is of small importance in describing the cohesiveness of covalently bonded polymers. With other factors being equivalent, the cohesiveness or cohesive energy density is the sum of dispersion, polar, and hydrogen-bridge contributions.

9.7.4.1 Molecular Weight
Dispersion attractions as generated by so-called instantaneous dipoles are important in all polymer attractions, particularly the polyolefins. By virtue of their symmetry, these materials possess no time-average dipole and have very low dielectric constants, that is, they are nonpolar. In the absence of polar attraction, certain property changes are readily correlated, other things being equal, with an increase in dispersive interactions or, more simply, molecular weight. The most dramatic example is found in the normal alkanes in the transition from gas to liquid to solid to "engineering" material—methane to pentane to eicosane to polyethylene. As molecular weight is increased for any polymer, there is an initial exponential increase in softening point, modulus, tensile strength, and impact strength which plateau as the sum of the entanglements and the interactive forces becomes greater than the strengths of individual covalent bonds (395). The influence of molecular weight and molecular weight distribution on mechanical properties has been reviewed by Martin

Table 9.8 Adhesive Commentary (390)

Adhesive Type	Comments	Typical Cure Conditions
Polyesters and their variations	Used primarily for repairing fiberglass reinforced polyester resins, ABA, and concrete. Generally unsaturated esters are polymerized with a catalyst such as methyl ethyl ketone (MEK) peroxide and an accelerator such as cobalt naphthenate. A coreactant-solvent such as styrene may be present. Bonds are strong. Sometimes combined with polyisocyanates to control shrinkage stresses and reduce brittleness. Unreacted monomer, if present, viscosity low for application, provides good wetting, enhances crosslinking. Occasionally used on metals.	Minutes to hours at room temperature
Nitrile phenolic, neoprene phenolic	These adhesives are a blend of flexible nitrile or neoprene rubber with phenolic novolac resin. They combine the impact resistance of the rubber with the strength of the cross-linked phenolic. They are inexpensive, produce strong durable bonds that resist water, salt spray, and other corrosive media well. They are the workhorses of the adhesive tape industry although they do require high pressure, relatively long, high-temperature cures. They are used for metals and some plastics. Airframe components and automotive brakes are typical examples.	Up to 12 hrs at (120–150° C)
Alpha cyanoacrylate	These low-viscosity liquids polymerize or "cure" rapidly in the presence of moisture or many metal oxides. Thus most surfaces can be bonded. The bonds are fairly strong but somewhat brittle. Used widely for the assembly of jewelry and electronic components.	0.5–5 min at room temperature
Epoxy phenolic	A combination of epoxy resin with a resol phenolic. Noted for strength retention at 300–500° F (50–250° C), strong bonds, and good moisture resistance. Normally stored refrigerated. Used for some metals, glass, and phenolic resins.	1 hr (175° C)
Epoxy: amine, amide, and anhydride cured	As a class, epoxies are noted for high-tensile and low-peel strengths. They are cross-linked and in general have good-high temperature strength, resistance to moisture, and little tendency to react with acids, bases, salts, or solvents. There are important exceptions to these generalizations, however, that are often the result of the curing agent used. Primary amines give faster setting adhesives that are less flexible and less moisture resistant than is the case when polyamide curing agents are used. Anhydride cured epoxies generally have good high-temperature strength but are subject to hydrolysis, especially in the presence of acids or bases. Other important features of epoxies are their low shrinkage upon cure, their compatibility with a variety of	One part heat curable. Two part heat or ambient curable. Cure times and temperatures are variable as is pot life.

449

Table 9.8 Adhesive Commentary (390) (*Continued*)

Adhesive Type	Comments	Typical Cure Conditions
	fillers, their long life when properly applied, and their easy modification with other resins. Cross-link density is easily varied with epoxies; thus some control over brittleness, vapor permeation, and heat deflection is possible. These resins are widely used to bond metal, ceramics, and rigid plastics.	
Nylon epoxy	Tensile shear strengths above 6,000 psi (41.4 MPa) and peel strength above 100lb/in. (18 kg/cm) are possible when epoxy resins are modified with special low-melting nylons. These gains, however, are accompanied by loss of strength upon exposure to moist air, a tendency to creep under load and poor low temperature impact behavior. A phenolic primer may increase bond life and moisture resistance. Used primarily for aluminum, magnesium and steel.	1 hr (150–175° C)
Flexible adhesives, natural rubber, butadiene-acrylonitrile, neoprene, polyurethane, polyacylates silicones	These adhesives are flexible. Thus their load bearing ability is limited. They have excellent impact and moisture resistance. They are easily tackified and are used as pressure-sensitive tapes or as contact cements. Urethane and silicone adhesives are lightly cross-linked which gives them reasonable hot strength. They are also compatible with many surfaces but are somewhat costly and must be protected against moisture before use. They have good low-temperature tensile, shear and impact strength. The urethanes are two-part products that require mixing before use. Silicones cure in the presence of atmospheric moisture.	Pressure-sensitive tape or solvent cements. Low-temperature bake for urethane. Ambient cure for silicones.
Polyamides	These adhesives, which are chemically similar to nylon resins, have good strength at ambient temperatures and are fairly tough. They are available in a variety of molecular weights, softening ranges, and melt viscosities. Often applied as hot melts, they have good adhesion to a variety of surfaces. The higher molecular weight varieties often have the best tensile properties. Lower molecular weight polyamides may be applied in solution.	Hot melt—cure by cooling
Polyvinyl-phenolic	These resins, which combine a resol phenolic resin with polyvinyl formal or polyvinyl butyral, were the first important synthetic load bearing adhesives. A considerable range of compositions is available with hot strength and tensile properties increasing at the expense of impact and peel strength as the phenolic content rises. The durability of vinyl-phenolics is generally excellent. They are often selected for low-cost applications where heat and pressure curing can be used.	(150° C) 1 hr

Table 9.8 Adhesive Commentary (390) (*Continued*)

Adhesive Type	Comments	Typical Cure Conditions
Cellulose esters	Cellulose ester adhesives are usually high viscosity, inexpensive, rigid materials. They do not have high strength and are sensitive to heat and many solvents. Normally used for holding small parts or repairing wood, cardboard, or plastic items. Model airplane cement is a common example.	Air dry
Vinyl chloride-vinyl acetate	This is a combination of two resins that are sometimes used alone. They may be used as hot melts or as solution adhesives. Since thin films of vinyl chloride-vinyl acetate are somewhat flexible, they are often used for bonding metal foil, paper, and leather. A range of compositions is available with a corresponding variety of properties.	Cooling (hot melt) or solvent loss
Polyvinylbutyral	A tough transparent resin that is used as a hot melt or heat-cured solution adhesive. It has good adhesion to glass, wood, metal, and textiles. It is flexible and can be modified with other resins or additives to give a range of properties. Not generally used as a structural adhesive, although structural phenolics sometimes incorporate polyvinylbutyral to give better impact resistance.	Cooling (hot melt), heating under pressure.
Polyhydroxyether	These are resins based on hydroxylated polyethylene oxide polymers. Generally used as hot melts, they have only moderate strength, but are flexible and have fairly good adhesion.	Hot melt—cure by cooling
Polyvinyl acetate	This adhesive is generally supplied as a water emulsion (white glue) or used as a hot melt. It dries quickly and forms a strong bond. It is flexible and has low resistance to heat and moisture. Porous substrates are required when the resin is used as an emulsion.	Hot melt—cure by cooling Emulsion—air dry
Urea-formaldehyde, Melamine-formaldehyde, resorcinol-formaldehyde, phenol-formaldehyde,	These thermosetting resins are widely used for wood bonding. Ureaformaldehyde is inexpensive but has low moisture resistance. It can be cured at room temperature if a catalyst is used. Melamine-formaldehyde resins have better moisture resistance but must be heat cured. Phenol-formaldehyde adhesives form strong, waterproof wood-to-wood bonds. The resorcinol-formaldehyde resin will cure at room temperature while phenol-formaldehyde requires heating. These resins are often combined resulting in an adhesive with intermediate processing or performance characteristics.	Up to 149° C and 200 psi (1.38 MPa)
Polyacrylate esters	These resins are *n*-alkyl esters of acrylic acid. They have good flexibility and find frequent use for high-quality pressure-sensitive tapes and foams. They are not suitable for load bearing applications because of their poor heat	Pressure sensitive

Table 9.8   Adhesive Commentary (390) (*Continued*)

Adhesive Type	Comments	Typical Cure Conditions
	resistance and their cold flow behavior. Frequently used on flexible substrates.	
Polysulfides	These resins have good moisture resistance and can range from thermoplastic to thermosetting depending on the degree of cross-linking developed during cure. They are two- or three-part systems, the third part being a catalyst. Ventilation is generally required. They make excellent adhesive sealants for wood, metal, concrete, and glass. Polysulfide resins may be combined with epoxies to flexibilize the latter.	Low pressures, moderate temperature
Silicones See also flexible adhesives	These expensive adhesives have high peel strength and excellent property retention at high and low temperatures. They resist all except the most corrosive environments and will adhere to nearly everything. They are usually formulated to react with atmospheric moisture and form lightly cross-linked films.	Low pressure, room temperature
Ethylene-vinyl acetate	This copolymer is widely used as a hot melt adhesive because it is inexpensive, adheres to most surfaces, and is available in a range of melting points. It is widely used for bookbinding and packaging.	Hot melt—cures by cooling
Urethanes, rigid	Rigid urethanes are highly cross-linked. While somewhat expensive, they adhere well to most materials, especially plastics, and have good impact strength. "Structural" urethanes are two-part systems and have good low-temperature strength retention.	Low pressures up to 149° C
Acrylics	Acrylics are versatile load bearing adhesives that are becoming increasingly popular. They cure at room temperature and are applied as a conventional two-part system or by coating one substrate with the resin and the other with the catalyst. Impact resistance is controllable since acrylics may vary from rigid to flexible. Floor ventilation is often recommended because they may release heavier-than-air monomer vapors.	Room temperature or up to 54° C, 10–20 minutes with only fixturing pressure

(396). While there is general agreement that increased molecular weight affords increasing resistance to deformation and breakage (395), there is disagreement as to whether such ultimate properties correlate best with number, weight, or some other average molecular weight (397). The effect of increasing molecular weight on modulus is clear for polymers with a T_g near or below room temperature (398), while the practical correspondence is more explicable with creep and the higher temperature transition to rubbery flow (395).

Although cohesion can be improved by an increase in molecular weight, adhesion can actually decrease (399). Since the time-dependent spreading rate is controlled by the viscosity, which is in turn determined by the molecular weight (400), this decrease may be attributed to a spreading deficiency. Thus in order to "wet," an adhesive or coating must have sufficient fluidity, that is "low" molecular weight, which can compromise the attainment of optimum cohesive properties.

9.7.4.2 Molecular Weight Distribution (MWD)

Although the heterogeneity of polymers has been known for some time, experimental complexities have limited study of this variable. The advent of high-pressure size-exclusion chromatography (386) which allows "routine" determination of MWD has mitigated this difficulty and provided better fundamental understanding of the effect on properties (396, 401). To be sure, experimental difficulties still exist in that olefins must be analyzed at high temperatures in chlorinated aromatic solvents (402).

On a molecular basis an increase in MWD is directly associated with an increase in chain ends. This, in turn, is associated with greater free volume and mobility (395). The consequential lowering of the cohesive strength for polymers of equal average molecular weight has been demonstrated by Reynolds (399). External plasticization by monomeric liquids or moisture is analogous to an increase in MWD in that the increase in free volume and modulii, ultimate tensile strengths, compressive strengths, and tear strengths, while increasing ultimate elongation. As expected, plasticization has been studied extensively and has been independently reviewed (403–405). Such decreases in properties attributed to lowered cohesion by way of increased MWD or plasticization are not completely deleterious in that lowered viscosity affords better spreading while greater extensibility can actually lead to stronger bonds.

9.7.4.3 Chain Geometry

When polar interactions are absent or approximately constant, the chain flexibility of amorphous polymers will be governed by any and all restrictions to internal rotation. The torsional freedom of single covalent bonds is, in other words, limited by the proximity, position, and size of neighboring groups either on or within the polymer backbone. The basic relationships, as derived from conformal en-

thalpies of ethane-butane and expanded into the Gaussian model described by Flory (406), have been incorporated into polymer science textbooks (407–409). The stiffening effect of methyl substituents as reflected in increased glass and melt transition temperatures and modulii is well known for olefins (397), n-alkyl methacrylates and n-alkyl acrylates (410) and substituted polystyrenes (398). Chain stiffening by substitution of cyclohexyl or phenyl in the backbone of polyesters is equally well established (411–413). These latter studies also incorporated observations on the significance of regularity/close-packing ability and interchain attraction. The effect of molecular stiffness on thermal properties has been reviewed by Aharoni (414), who notes a direct relationship between polymer expansivity above T_g and the stiffness parameter of the chain. The effect of chain stiffness on solution thermodynamics predicted by Flory (415) has been verified by Miller (416).

9.7.4.4 Intermolecular Attraction

Isolating and quantifying intermolecular attractions arising from permanent dipoles and/or hydrogen bonding in the cohesion continuum that are not confounded with molecular inhomogeneity, chain geometry, or crystallinity is no small problem. A poignant example is the decrease in T_m and T_g in polyamide homologues as a function of increasing the methylene units in the diacid moiety (417). While the polar contribution is undoubtedly "diluted" the trend can also be rationalized in terms of decreasing chain stiffness (417). A somewhat better illustration is noted in Table 9.9. While the pendant groups (methyl, chlorine, and cyano) are approximately the same size, the crystallinity, molecular weight, and molecular weight distribution are unknown, and their contributions cannot be quantified. Wu (349) has separated polar and nonpolar interactions by way of molten surface tensions according to the "harmonic mean" equation and shown the correspondence with solubility parameters. Deanin (418) has chosen the solubility parameter as a measure of polarity and demonstrated correlation with T_g, which in turn is correlated with modulus,

Table 9.9   Effect of Intermolecular Forces

Polymer	δ^a	T_g (°C)^a	T_m (°C)^a	Δe_i (cal/mol)^b	ΔV_i (cm^3/mol)^c
Polypropylene	8	−20	165	1125	33.5
PVC	9.5	87	~190	2760	24
PAN	15.4	120	317	6100	24

^{a} See Reference 417.
^{b} Pendant group contribution to energy of vaporization (340).
^{c} Pendant group contribution to molar volume (340).

strength, heat deflection temperature, low-temperature embrittlement, dielectric constant, and impermeability.

9.7.4.5 Crystallinity

Order in macromolecules "has been under investigation virtually since the acceptance of polymers as molecular entities" (419). While single crystals had been obtained during polymerization prior to 1953, the obtaining of crystals from dilute solution by Schlesinger (420) resolved the question of molecular entanglement. From a macroscopic-microscopic viewpoint the lamellalike crystallite is a well substantiated morphological form. While light and electron microscopy have established such morphology and electron diffraction has indicated that the polymer chains are oriented normal to the face of the lamella, the nature of the molecular morphology is not yet fully understood (419). Although melting-crystallization is aptly described as a first-order phase transition (421), the extreme unlikelihood of equilibrium precludes the attainment of complete crystallinity. Thus crystallization kinetics assume high significance in the determination of the type and level of crystallinity (422), which in turn is paramount in the determination of physical and thermodynamic properties (423–425). The work and independent reviews by Mandelkern (421), Stein and Tobolsky (426), and Wunderlich (427) are particularly noteworthy in the evolution of understanding of order/disorder in macromolecules. Crystallinity-property relationships from an engineering viewpoint have been reviewed by Birley (428) and by Deanin (395). The latter study is noteworthy in that it addresses the correlative difficulties between crystallinity and engineering properties such as strength, stiffness, and toughness.

Since increasing crystallinity is generally associated with decreasing clarity, increasing brittleness, and anisotropic properties; glassy, amorphous polymers are generally preferred for coating applications. The desirability of isotropic mechanical properties in adhesive applications has also engendered a preference for amorphous polymers. However, the realization that polymers of moderate to high crystallinity have better cohesive strengths over a wider temperature range (429) has enhanced their utilization in hot-melt adhesives. The "use" of crystallinity as a property improver in adhesives is necessarily limited by the concomittant decrease in toughness as crystallinity is increased (430). To some extent this deleterious trend is mitigated by quenching (431) or by heterogeneous nucleation (432–434). The use of crystalline polyesters in hot-melt adhesives has engendered investigations into the rheological, thermal, chemical, and morphological variables affecting performance (435).

9.7.5   Coatings

The physiochemical dictums of adhesion, rheology, adsorption, and cohesion as noted in previous sections apply to any film-forming material, whether it is used as an adhesive or a coating. It is also logical, although not necessarily consequential, that paint binders are closely related to adhesives. Such technological cross-fertilization is readily apparent with epoxy resins as reviewed by May and Tanak (436) and with urethanes as reviewed by Barnat (437) and Backus (438). While the adhesion, flexibility, toughness, and durability of a polymer can, to some degree, be predicted and modified, the interactions with pigments, extenders, surfactants, driers, plasticizers, colorants, and solvent mixtures pose analytical problems at least as complex as those noted in the discussion of adsorption in Section 9.3. It is thus hardly surprising to find statistical methodologies used in formulation (439). Since Myers and Long (386) have reviewed coating technology through 1976, the remainder of this section is dedicated to a synopsis of current trends.

Electrophoretic deposition of paint as pioneered in the United States by Brewer and Burnside and described by Hays (440), Ranney (441), and Shanmugam (442) was first used on a U.S. production basis in 1961. The "traditional" processes involve the attraction, neutralization, and coagulation of a potassium or quaternary ammonium salt on a part or piece that is anodic. Since oxidation occurs at the part, there is some tendency for metal dissolution, which has been implicated in lowered corrosion resistance and staining (335, 443). In order to alleviate such problems, recent developments have been on reductive or cathodic systems (442). An allied development is that of self-depositing coatings that involve a chemical attack of the metal substrate and the use of reaction products to deposit a styrene-butadiene latex (444). It is a black-only formulation specific for steel that must be immersed (445). A Russian review (446) is available.

Under the areas of reducing emissions, pollutants, and/or energy input, extensive efforts are being devoted to waterborne and higher-solids coatings. Advances in these areas, which include aqueous and nonaqueous dispersions, aqueous emulsions, solutions and suspensions, conformal solvents, and powder coatings, are described in Symposia proceedings (447–449). High-solids coatings, in theory, are greater than 80% nonvolatiles by volume, while 60% is probably a practical lower limit (450). Schoff has reviewed (450) pigmented linseed oils, polyesters, alkyds, and acrylics with particular attention to rheology. Waterborne high-solids acrylics have been reviewed by Paul (451). Copolymers are described by Sherwin (452), and the effects of glycol variation on polyester coatings have been studied by Sheme (453). The synthesis and end use of waterborne polymers has been reviewed by

Gardon (447), as have advances in solvent selection for aqueous systems (454). The high viscosity of such systems pose real and potential problems in application, which have been reviewed by Devittorio and Scharfenberger (447).

Epoxy and polyester thermoset powder coatings are commercially significant in the finishing of pipe, metal shelving, furniture, electrical parts, wire goods, and appliances (455). Powder use for automotive exteriors has been limited owing to high-build films, poor acceptance of metal flake, low gloss, and color control problems (335). Thermoplastic powders such as polyethylene, cellulose acetate butyrate, PVC, polyamides, polyesters, and chlorinated polyethers are also available (455). As there is no need to dissolve the resin, a wider range of properties is available, and since higher molecular weights can be used, the finished coatings can be tougher and more solvent resistant (335).

9.8 SUMMARY

While quantitative studies of these surface-modifying treatments have been limited, it seems generally that relative layers of material are actually involved. In the preparation of surface-grafted materials, appreciable weight gains are observed. One important reason for chemical reaction modification is to improve receptivity to dyes; clearly, if easily detectable increases in color density are produced, the amount of dye involved (and hence presumably the depth of reaction) must correspond to more than a few molecular layers. Hence the treatments are "surface modifications" only in the sense that the bulk of the sample far from the surface is unaffected. It seems likely that in most, if not all, such processes, what is observed is a homogeneous chemical reaction occurring in a thin layer into which the reactants have diffused.

One has seen up to now that the local surface properties of organic polymers can be accounted for almost exclusively by treating them as a subclass of van der Waals or molecular solids of high molecular weight. In the zeroth order description of the surface phenomenology of organic polymers there are no qualitative differences that distinguish the polymers as such. Such effects presumably exist, but produce perturbations under ordinary conditions that are too small to observe readily with current experimental precision. Conceivably such effects might be most prominent in the vicinity of special points of the phase diagram associated with phase transitions. Here one might first distinguish effects on surface properties such as a small maximum in the contact angle of a liquid wetting a polymer surface owing to a phase transition in the bulk polymer. It is therefore significant that Neumann et al. (456) have reported such an effect for a polyfluorocarbon, occurring at the glass transition of the polymer. This, of course, has immediate consequences for very precise temperature extrapolations of liquid contact angles, but in a larger sense may herald the possibility that additional charac-

teristic transitions of bulk polymers may be reflected in unusual behavior of the local surface properties. The glass transition is (of course) not a special characteristic of polymer substances alone but is observed for many typical small-molecule van der Waals substances such as, for example, glycerine. Second, one might expect to distinguish phase transitions occurring solely within the interface region. For example, there does not yet appear to be to any extent data on the critical temperature below which a liquid will not spread on polymer surface perturbations of such a critical temperature; if such perturbations could be detected, they might provide a test of interfacial phase transitions. Even more speculative would be any discussion of a definitely characterized phase transition between different surface phases of a polymer, although such transitions may well be observed in the future.

9.9   GLOSSARY OF SYMBOLS

Symbol	Common Unit	Definition or Label	Section First Used
A_0	Å^2	Limiting area for unimolecular spreading	9.2.3
A_1	m^2/mg	Area for unimolecular spreading	9.2.3
A_2	Å^2	Area per segment	9.2.3
Å	10^{-10} m	Angstrom unit	
A_a	mg	Apparent absorption given by Equation (9.1)	9.3
A_p	mg	Amount of polymer absorbed	9.3
A_s	mg	Amount of solvent absorbed	9.3
b	cm	Effective monomer length	9.2.5
c_0	g/cm^3	Equilibrium concentration of absorbed solution	9.3
c_f	g/cm^3	Equilibrium concentration in film, of absorbed solute	9.3
D	dynes/cm; g/sec^2	Arithmetic deviation of $(\gamma_1 + \gamma_2)$ from γ_{12}	9.2.5
E	dynes/cm^2	Apparent elastic modulus	9.1
E'	dynes/cm^2	Dynamic modulus	9.1.4
E_t	dynes/cm^2	Young's modulus (in tension)	9.7.2
k	cm^2/dyne	Compressibility	9.2.5
k	$\text{erg}/{}^\circ\text{K}$	Boltzman's constant	9.2.5
k	—	Constant in Equation (9.7)	9.3.1
K	—	Constant in Equation (9.5)	9.3
K	cm/dyne	Constant in Equation (9.14)	9.5.2
l	angstroms	Lamella thickness	9.5.1
M	g/mole	Molecular weight	9.2
M_n	g/mole	Number average molecular weight	9.2
M_w	g/mole	Weight average molecular weight	9.2.1
M_v	g/mole	Viscosity average molecular weight	9.2
N	mole^{-1}	Avogadro's number	9.2.3

9.9 GLOSSARY OF SYMBOLS

Symbol	Common Unit	Definition or Label	Section First Used
n	—	Exponent in density-surface tension correlation	9.2.5
n_f	—	Refractive index of film	9.3
p	dynes/cm^2	Gas pressure	9.3
p_0	g cm^3	Density	9.2.5
p_s	dynes/cm^2	Pressure of saturated vapor	9.3
P	grams	Amount of polymer in Equations (9.1)–(9.3)	9.3
r	cm	Capillary radius	9.3
S	grams	Amount of polymer in Equations (9.1)–(9.3)	9.3
S	cm^2	Area	9.5.5
T	°C or °K	Temperature	9.1
T'	°C or °K	See Reference 278 for detailed discussion and definition	9.5.4
T_g	°C or °K	Glass transition temperature	9.1
T_m	°C or °K	Crystalline melting point	9.1
T_c	°K	Critical temperature	9.2.1
t	seconds	Time	9.3.4
t'	seconds	Minimum spin-lattice relaxation time	9.5.4
u	cm^3/sec	Flow rate without adsorption	9.3
u_a	cm^3/sec	Flow rate with adsorption	9.3
V_g	cm^3	Retention volume	9.5.4
W	ergs	Work of rupture	9.5.5
W_A	erg/cm^2	Reversible work of adhesion	9.5.2
W_E, W_P, W_S, W_T	ergs	Components of rupture work in Equation (9.17)	9.6
W_{AB}	erg/cm^2	Work of adhesion in composite $A–B$ in the absence of liquid	9.6
W_{AB}^{L}	erg/cm^2	Work of adhesion in composite $A–B$ in the presence of liquid	9.6
z	—	Degree of polymerization	9.2.5
α	°K^{-1}	Coefficient of thermal expansion	9.2.5
α	—	Constant in Equation (9.5)	9.3
β	cm^2/dyne	Isothermal compressibility	9.2.5
Γ_a	mg/m^2	A_a per unit area	9.3
Γ_p	mg/m^2	A_p per unit area	9.3
γ	dynes/cm; g/sec^2	Surface tension	9.3
γ_0	dynes/cm; g/sec^2	Surface tension of liquid uncontaminated by unimolecular polymer layer	9.2.3
γ_0	dynes/cm; g/cm^2	Constant in Equation (9.15)	9.5.3
γ_1	dynes/cm; g/sec^2	Surface tension of liquid contaminated by unimolecular polymer layer	9.2.3
γ_1, γ_2	dynes/cm; g/sec^2	Surface tensions of liquids 1 and 2	9.2.5
γ_{12}	dynes/cm; g/sec^2	Liquid-liquid interfacial tension	9.2.5
γ_c	dynes/cm; g/sec^2	Critical surface tension	9.1.5
γ_i	dynes/cm; g/sec^2	Interfacial tension between polymer melts	9.2.2
γ_L	dynes/cm; g/sec^2	Liquid surface tension	9.5.3

9.9   GLOSSARY OF SYMBOLS

Symbol	Common Unit	Definition or Label	Section First Used
γ_p	dynes/cm; g/sec^2	Solid polymer surface tension	9.5.3
γ_s	dynes/cm;	Solid polymer surface energy (surface tension)	9.5.1
γ_s	dynes/cm; g/cm^2	Solid polymer surface energy in Equations (9.9) and (9.10)	9.5.2
$\gamma_{SL}, \gamma_{SV}, \gamma_{LV}$	dynes/cm; g/cm^2	Solid-liquid, Solid-vapor, and Liquid-vapor interfacial tensions in Equations (9.9)–(9.14)	9.5.2
γ_{AL}, γ_{AV}	dynes/cm; g/cm^2	Solid A-liquid, Solid A-vapor interfacial tensions in Equations (9.19)–(9.23)	9.6.1
γ_{BL}, γ_{BV}	dynes/cm; g/cm^2	Solid B-liquid, Solid B-vapor interfacial tensions in Equations (9.19)–(9.23)	9.6.1
γ_{AB}	dynes/cm; g/cm^2	Solid A-Solid B interfacial tension in Equations (9.19)–(9.20)	9.6.1
$\gamma_{AV}^d, \gamma_{BV}^d, \gamma_{LV}^d$	dynes/cm; g/sec^2	Dispersion components of interfacial tensions in Equations (9.21)–(9.23)	9.6.1
$\gamma_{AV}^p, \gamma_{BV}^p, \gamma_{LV}^p$	dynes/cm; g/sec^2	Polar components of interfacial tensions in Equations (9.21)–(9.23)	9.6.1
δ	—	Loss angle	9.1
δ	Angstroms	Film thickness	9.5.4
δ_f	cm	Film thickness	9.3
η	poise	Viscosity	9.2
$[\eta]$	dl/g	Intrinsic viscosity	9.3.1
θ	degrees	Contact angle	9.1.5
θ	—	Degree of coverage	9.4
θ, θ_0	ergs/cm^2	Measured and "intrinsic" adhesives failure energies in Equation (9.16)	9.6
θ_a	degrees	Advancing contact angle	9.5.2
θ_r	degrees	Receding contact angle	9.5.2
$\theta_a - \theta_r$	degrees	Contact angle hysteresis	9.5.2
Π	dynes/cm; g/sec^2	Surface pressure	9.2.3
Π_e	dynes/cm; g/sec^2	Spreading pressure of vapor on a solid	9.5.2
Π_m	dynes/cm; g/sec^2	Collapse pressure	9.2.3
Π_{SL}	dynes/cm; g/sec^2	Spreading pressure of liquid on a solid	9.5.2
ρ	g/cm^3	Liquid density	9.2.5
ρ_v	g/cm^3	Vapor density	9.2.5
σ_f	dynes/cm^2	Critical stress to failure in the absence of a liquid	9.6
σ_0	dynes/cm^2	Fixed stress in absence of liquid	9.6.1
σ_0^L	dynes/cm^2	Fixed stress in presence of liquid	9.6.1
σ_f^L	dynes/cm^2	Critical stress to failure in the presence of a liquid	9.6
τ	erg/cm^2	Cohesive failure energy	9.6
τ_0	erg/cm^2	"Intrinsic" cohesive failure energy	9.6

9.10   REFERENCES

1. W. Hardy, *Proc. R. Soc.*, **A112** (1926).

2. D. H. Bangham, *J. Chem. Phys.*, **14**, 352 (1946).

3. B. V. Derjaguin and S. M. Zorin, *Proc. 2nd. Int. Congr. Surf. Act.* (*London*), **2**, 145 (1957).

4. A. Sheludko, *Adv. Colloid Interface Sci.*, **1**, 391 (1967).

5. J. F. Padday, *Spec. Discuss. Faraday Soc.*, **1**, 64 (1970).

6. C. Huh and L. E. Scriven, *J. Colloid Interface Sci.*, **35**, 85 (1971). See also G. E. P. Elliott and A. C.

Riddiford, *Recent Progr. Surface Sci.*, **2**, 111 (1965).

7. E. B. Dussan V., Ph.D. Thesis, Johns Hopkins University (1972).

8. M. Rosoff, in B. Carroll, Ed., *Physical Methods in Macromolecular Chemistry*, Marcel Dekker, New York, 1969, p. 1.

9. R. Peyser, D. J. Tutas, and R. Stromberg, *J. Polym. Sci.*, **(A-1)5**, 651 (1967).

10. W. C. Hamilton, *Appl. Polym. Symp.*, **19**, 105 (1972) and references cited therein.

11. N. J. Harrick, *J. Phys. Chem.*, **64**, 110 (1964).

12. J. P. Luongo and H. Schonhorn, *J. Polym. Sci.*, **(A-2)6**, 1649, (1968).

13. S. Krimm, C. Y. Liang, and G. B. B. M. Sutherland, *J. Chem. Phys.*, **25**, 549 (1956).

14. B. L. Henke, *Adv. X-ray Anal.*, **13**, 1 (1970).

15. D. W. Dwight and W. M. Riggs, *J. Colloid Interface Sci.*, (in press).

16. D. M. Hercules, *Anal. Chem.*, **44**, (5), 106 (1972); W. M. Riggs and R. P. Fedchenko, *Am. Lab.* **4** (11), 65 (1972).

17. C. G. S. Collins, A. C. Lowe, and D. Nicholas, *Eur. Polym. J.*, **9**, 1172, (1973).

18. D. T. Clark and W. J. Feast, *J. Macromol. Sci.— Rev. Macromol. Chem.*, **C12(2)**, 191–286 (1975).

19. G. Kraus, *Adv. Polym. Sci.*, **8**, 155 (1971).

20. Yu. S. Lipatov and F. G. Fabuliak, *Vysokomol. Soedin.*, **10**, 1605 (1968); *Acad. Nauk., SSR*, **10A**, 1605 (1968).

21. D. R. Fitchmun and S. Newman, *J. Polym. Sci.*, A-3, **8**, 1545 (1970).

22. T. K. Kwei, H. Schonhorn, and H. L. Frisch, *J. Appl. Phys.*, **38**, 2512 (1967).

23. D. Jenckel, E. Teege, and W. Hinrichs, *Kolloid Z.*, **129**, 19 (1952).

24. R. J. Barriault and L. F. Gronholts, *J. Polym. Sci.*, **18**, 3933 (1955).

25. H. Schonhorn and R. H. Hansen, *J. Appl. Polym. Sci.*, **11**, 1461 (1967).

26. R. K. Eby, *J. Appl. Phys.*, **35**, 2720 (1964).

27. H. Schonhorn, *J. Polym. Sci.*, **B2**, 465 (1964); H. Schonhorn, *Macromolecules*, **1**, 145 (1968).

28. H. M. Zupko, unpublished data.

29. R. E. Cuthrell, *J. Appl. Polym. Sci.*, **11**, 1495 (1967).

30. G. L. Gaines, *Polym. Eng. Sci.*, **12**, 1 (1972).

31. S. Wu, *J. Macromol. Sci.—Rev. Macromol. Chem.*, **10**, 1 (1974).

32. F. J. Hybart and T. R. White, *J. Appl. Polym. Sci.*, **3**, 118 (1960).

33. N. Ogata, *Bull. Chem. Soc. Jap.*, **33**, 212 (1960).

34. H. Edwards, *J. Appl. Polym. Sci.*, **12**, 2213 (1968).

35. G. P. Safonov and S. G. Entelis, *Vysokomol. Soed.*, **9A** 1909 (1967).

36. P. P. Pugachevich and A. G. Tokaev, *Makromol. Granitse Razdela Paz*, Naukova Dumka, Kiev, 1971, p. 24.

37. H. T. Patterson, K. H. Hu, and T. H. Grindstaff, *J. Polym. Sci.*, **C34**, 31 (1974).

38. H. Schonhorn, F. W. Ryan, and L. H. Sharpe, *J. Polym. Sci.*, **A2**, 538 (1966).

39. J. R. J. Hartford and E. F. T. White, *Plast. Polym.*, **37**, 53 (1969).

40. H. Schonhorn and L. H. Sharpe, *J. Polym. Sci.*, **A3**, 569 (1965).

41. H. Schonhorn and L. H. Sharpe, *J. Polym. Sci.*, **B3**, 235 (1965).

42. J. J. Bikerman, *Physical Surfaces*, Academic Press, New York, 1970, p. 13.

43. R. H. Dettre and R. E. Johnson, *J. Colloid Interface Sci.*, **21**, 367 (1966).

44. R. J. Roe, *J. Phys. Chem.*, **72**, 2013 (1968).

45. S. Wu, *J. Colloid Interface Sci.*, **31**, 153 (1969).

46. S. Wu, *J. Polym. Sci.*, **C34**, 19 (1971).

47. T. Sakai, *Polymer*, **6**, 659 (1965).

48. T. Hata, *Asahi Garasu Kogyo Gijutsu Shorei-kai Kenkyu Hokoku*, **13**, 603 (1967), through *Chem. Abstr.*, **69**, No. 52583 (1968).

49. D. G. Legrand and G. L. Gaines, *J. Colloid Interface Sci.*, **31**, 162 (1969).

50. S. Wu, *J. Phys. Chem.*, **74**, 632 (1970).

51. G. W. Bender and G. L. Gaines, *Macromolecules*, **3**, 128 (1970).

52. W. Yu. Lau and C. M. Burns, *Surface Sci.*, **30**, 478 (1972); *J. Colloid Interface Sci.*, **45**, 295 (1973).

53. Yu. S. Lipatov et al., *Vysokomol. Soed.*, Ser. B, **15**, 725 (1973), through *Chem. Abstr.*, **80**, No. 133946 (1974).

54. V. P. Privalko, Yu. S. Lipatov, and A. P. Lobodina, *J. Macromol. Sci.*, **11**, 441 (1975).

55. A. K. Rastogi and L. E. St. Pierre, *J. Colloid Interface Sci.*, **31**, 168 (1969).

56. A. Marcincin et al., *Chem. Vlakna*, **20**, 1 (1970), through *Chem. Abstr.*, **77**, No. 35084 (1972).

57. A. K. Rastogi and L. E. St. Pierre, *J. Colloid Interface Sci.*, **35**, 16 (1971).

58. R. J. Roe, *J. Colloid Interface Sci.*, **31**, 228 (1969).

59. G. W. Bender, D. G. Legrand, and G. L. Gaines, *Macromolecules*, **2**, 681 (1969).

60. A. E. Fainerman, Yu. S. Lipatov, and V. M. Kulik, *Kolloidn. Zh.*, **31**, 140 (1969).

61. G. Geipel, F. Wolf, and K. Loeffler, *Tenside*, **5**, 132 (1968), through *Chem. Abstr.*, **69**, No. 28768 (1969).

62. A. E. Fainerman, Yu. S. Lipatov, and V. K. Maistruk, *Dokl. Akad. Nauk SSSR*, **188**, 152 (1969).

63. T. Hata, *Kobunshi*, **17**, 594 (1968), through *Chem. Abstr.*, **69**, No. 78002 (1969).

64. J. Jaffe, G. Bricman, and C. Berliner, *J. Polym. Sci.*, **B5**, 153 (1967).

65. A. W. Neumann et al., *J. Macromol. Sci., Phys.*, 7, 525 (1973).

66. P. P. Pugachevich and I. I. Rabinovich, *Zh. Fiz. Khim.*, 45, 1313 (1971).

67. A. E. Fainerman et al., *Makromol. Granitse Razdela Paz*, Naukova Dumka, Kiev, 1971, p. 13.

68. K. Ueberreiter, S. Morimoto, and R. Steulmann, *Colloid Polym. Sci.*, 252, 273 (1974).

69. A. A. Kiriyanenko and L. P. Lyutin, *Izv. Sib. Otd. Akad. Nauk SSSR, Ser. Tekh. Nauk*, 1972, No. 3, 120, through *Chem. Abstr.*, 79, No. 19239 (1973).

70. P. P. Pugachevich, A. G. Tokaev, and R. M. Kamalyan, *Arm. Khim. Zh.*, 23, 376 (1970), through *Chem. Abstr.*, 73, No. 88303 (1970).

71. A. E. Fainerman, Yu. S. Lipatov, and V. M. Kulik, *Kolloidn. Zh.*, 32, 778 (1970).

72. R. A. Anderson, *Pharm. Acta Helv.*, 47, 304 (1972), through *Chem. Abstr.*, 77, No. 89020 (1972).

73. Yu. S. Lipatov and A. E. Fainerman, *J. Adhes.*, 3, 3 (1971).

74. R. A. Kul'man et al., *Kolloidn. Zh.*, 30, 860 (1968).

75. R. A. Kul'man, *Kolloidn. Zh.*, 33, 169 (1971).

76. R. A. Kul'man, *Makromol. Granitse Razdela Paz*. Naukova Dumka, Kiev, 1971, p. 31.

77. L. E. DeMillo et al., *Zh. Prikl. Khim.* (*Leningrad*), 43, 194 (1970).

78. N. B. Graham and H. W. Holden, *Polymer*, 11, 198 (1970).

79. R. E. Neiman, S. I. Taranovskaya, and T. P. Osipova, *Kolloidn. Zh.*, 37, 574 (1975).

80. A. A. Kiriyanenko, A. S. Basin, and L. P. Lyutin, *Izv. Sib. Otd. Akad. Nauk SSSR, Ser. Tekh. Nauk*, 1970, No. 3, 69, through *Chem. Abstr.*, 74, No. 88332 (1971).

81. D. J. Crisp, *J. Colloid Sci.*, 1, 49 (1946).

82. G. C. Benson and R. L. McIntosh, *J. Colloid Sci.*, 3, 323 (1948).

83. M. Ueno and K. Meguro, *J. Polym. Sci., Polym. Chem. Ed.*, 13, 815 (1975).

84. M. Koyama and K. Meguro, *Colloid Polym. Sci.*, 253, 485 (1975).

85. A. Labbauf, *J. Appl. Polym. Sci.*, 10, 865 (1966).

86. G. Gabrielli and M. Puggelli, *J. Colloid Interface Sci.*, 35, 460 (1971).

87. H. Hotta, *J. Colloid Sci.*, 9, 504 (1954).

88. S. Hironaka, T. Kubota, and K. Meguro, *Bull. Chem. Soc. Jap.*, 44, 2329 (1971).

89. S. Hironaka and K. Meguro, *J. Colloid Interface Sci.*, 35, 367 (1971).

90. I. I. Maleev, N. S. Tsvetkov, and I. E. Tvardon in Yu. S. Lipatov, Ed., *Thermodynamic and Structural Properties of Boundary Layers of Polymers*, Naukova Dumka, Kiev, 1976, p. 30.

91. D. J. Crisp, *J. Colloid Sci.*, 1, 161 (1946).

92. A. Labbauf and J. R. Zack, *J. Colloid Interface Sci.*, 35, 569 (1971).

93. W. M. Lee, J. L. Shereshefsky, and R. R. Stromberg, *J. Res. Natl. Bur. Stand.*, 65A, 51 (1961).

94. W. M. Lee, R. R. Stromberg, and J. L. Shereshefsky, *J. Res. Natl. Bur. Stand.*, 66 A, 439 (1962).

95. N. Beredijk and H. E. Ries, *J. Polymer Sci.*, 62, S64 (1962).

96. N. Beredijk in B. Ke, Ed., *New Methods in Polymer Characterization*, John Wiley, New York, 1964, p. 677.

97. A. Blumstein and H. E. Ries, *J. Polym. Sci.*, B3, 927 (1965).

98. J. Jaffe and J. M. Ruysschaert, *J. Polym. Sci.*, C23, No. 1, 281 (1968).

99. J. Llopis and D. V. Rebollo, *J. Colloid Sci.*, 11, 543 (1956).

100. C. H. Giles and N. Melver, *J. Colloid Interface Sci.*, 53, 155 (1975).

101. E. G. Cockbain, K. J. Day, and A. I. McMullen, *Proc. 2nd Int. Congr. Surf. Act., London*, 1, 56 (1957).

102. J. M. G. Lankveld and J. Lyklema, *J. Colloid Interface Sci.*, 41, 466 (1972).

103. J. Th. C. Boehm, "Meded. Landbouwhogesch. Wageningen 1974," thesis through *Chem. Abstr.*, 81, No. 126993 (1974).

104. H. E. Ries and D. C. Walker, *J. Colloid Sci.*, 16, 361 (1961).

105. G. Gabrielli, M. Puggelli, and F. Facciolli, *J. Colloid Interface Sci.*, 37, 213 (1971).

106. J. J. Bikerman, *Foams*, Springer-Verlag, New York, 1973.

107. C. Devin and M. Minfray, *C. R.*, 255, 116 (1962).

108. K. Imai and M. Matsumoto, *Bull. Chem. Soc. Jap.*, 36, 455 (1963).

109. K. Bolewski, T. Tomaszkiewicz, and M. Olbracht, *Polimery*, 15, 15 (1970), through *Chem. Abstr.*, 73, No. 45943 (1970).

110. M. M. Tarnorutskii, *Kolloidn. Zh.*, 36, 995 (1974).

111. R. J. Roe, *J. Phys. Chem.*, 69, 2809 (1965).

112. S. M. Yagnyatinskaya, S. S. Voyutskii, and L. Ya Kaplunova, *Kolloidn. Zh.*, 34, 132 (1972).

113. H. Schonhorn, *J. Chem. Phys.*, 43, 2041 (1965).

114. J. H. Hildebrand and R. L. Scott, *The Solubility of Non-Electrolytes*, Reinhold, New York, 1950, pp. 402, 431.

115. H. Schonhorn, *J. Chem. Eng. Data*, 12, 524 (1967).

116. R. J. Roe, *Proc. Natl. Acad. Sci.*, 56, 819 (1966).

117. D. Patterson and A. K. Rastogi, *J. Phys. Chem.*, 74, 1067 (1970).

118. S. Matsuoka and B. Maxwell, *J. Polym. Sci.*, 32, 131 (1958).

119. K. S. Slow and D. Patterson, *Macromolecules*, 4, 26 (1971).

120. D. G. LeGrand and G. L. Gaines, Jr., *J. Colloid Interface Sci.,* **31**, 162, 430 (1969).

121. C. W. Stewart and C. A. vonFrankenberg, *J. Polym. Sci.,* Part A-2, **6** 1686 (1968).

122. D. G. LeGrand and G. L. Gaines, Jr., *A. C. S. Polym. Prepr.,* **11**, 1306 (1970).

123. R. Aveyard, *Trans. Faraday Soc.,* **63**, 2778, (1967).

124. A. K. Rastogi and L. E. St. Pierre, *J. Colloid Interface Sci.,* **31**, 168 (1969).

125. L. A. Girifalco and R. J. Good, *J. Phys. Chem.,* **61**, 904 (1957); **64**, 561 (1960).

126. F. M. Fowkes, *J. Phys. Chem.,* **66**, 32 (1962); *Ind. Eng. Chem.,* **56**(12), 50 (December 1964); *Adv. Chem. Ser.* **43**, 99 (1964).

127. R. E. Johnson, Jr. and R. H. Dettre in E. Matijevic, Ed., *Surface and Colloid Science,* Vol. 2, 285 Wiley-Interscience, New York, 1969; M. Rosoff in B. Carroll, Ed., *Physical Methods in Macromolecular Chemistry,* Vol. 1, Marcel Dekker, New York, 1969 p. 69; R. J. Good and E. Elbing, *Ind. Eng. Chem.,* **62**(3), 54 (March 1970).

128. S. Wu, *Polym. Prepr.,* **11**, 1291 (1970).

129. E. Helfand and Y. Tagami, *Polym. Lett.,* **9**, 741 (1971); *J. Chem. Phys.,* **56**, 3592 (1972).

130. Yu. S. Lipatov and L. M. Sergeeva, *Adsorption of Polymers* (Russian), Naukova Dumka, Kiev, 1972. English translation: Halsted Press, New York, 1974.

131. Yu. S. Lipatov and L. M. Sergeeva, *Adv. Colloid Interface Sci.,* **6**, (1976).

132. S. N. Tolstaya, S. A. Shabanova, and A. B. Taubman, *Lakokras. Mater. Ikhtiol. Primen.,* **1966**, No. 6, 8, through *Chem. Abstr.,* **66**, No. 116755 (1967).

133. E. K. Bogacheva, A. V. Kiselev, and Yu. A. Eltekov, *Kolloidn. Zh.,* **31**, 176 (1969).

134. E. Killman and H. J. Strasser, *Angew. Makromol. Chem.,* **31**, 169 (1972).

135. E. Davidson and J. J. Kipling, *Proc. 4th Int. Conf. Surf. Act. Subst.,* Bruxelles 1964, **2**, 1029 (1965).

136. I. D. Kuleshova, S. N. Tolstaya, and A. B. Taubman, in *Makromol. Granitse Razdela Paz,* Naukova-Dumka, 1971, p. 86.

137. G. J. Howard and P. McConnell, *J. Phys. Chem.,* **71**, 2974 (1967).

138. A. Tarnawski, *Polimery,* **13**, 152 (1968), through *Chem. Abstr.,* **69**, No. 87700 (1968).

139. N. Shiraishi et al., *Mokuzai Gakkaishi,* **15**, 20, 24 (1969), through *Chem. Abstr.,* **71**, No. 129094, 129095 (1969).

140. K. Rehacek and H. Schuette, *Plast. Kautsch.,* **16**, 773 (1969), through *Chem. Abstr.,* **71**, No. 126071 (1969).

141. E. Porowska and Z. Hippe, *Rocz. Chem.,* **44**, 635 (1970), through *Chem. Abstr.,* **73**, No. 67147 (1970).

142. I. I. Maleev et al., *Makromol. Granitse Razdela Paz,* Naukova Dumka, Kiev, 1971, p. 69.

143. L. M. Sergeeva et al., *Makromol. Granitse Razdela Pax,* Naukova Dunka, Kiev, 1971, p. 73.

144. N. Schamp and J. Huylebroeck, *J. Polym. Sci., Polym. Symp.,* **42**, Part 2, 553 (1973).

145. G. J. Howard and P. McConnell, *J. Phys. Chem.,* **71**, 2981 (1967).

146. E. O. Uhrn, *J. Polym. Sci.,* **17**, 137 (1955).

147. Ph. Gramain, *C. R. Acad. Sci.,* Ser. C **278**, 1401 (1974).

148. F. W. Rowland and F. R. Eirich, *J. Polym. Sci.,* Part A-1, **4**, 2401 (1966).

149. R. H. Ottewill and T. Walker, *Kolloid-Z. Z. Polym.,* **227**, 108 (1968).

150. M. J. Garvey, Th. F. Tadros, and B. Vincent, *J. Colloid Interface Sci.,* **49**, 57 (1974).

151. M. J. Garvey, Th. F. Tadros, and B. Vincent, *J. Colloid Interface Sci.,* **55**, 440 (1976).

152. F. Patat and C. Schliebener, *Makromol. Chem.,* **44–46**, 643 (1961).

153. R. R. Stromberg, D. J. Tutas, and E. Passaglia, *J. Phys. Chem.,* **69**, 3955 (1965).

154. E. Killmann and M. V. Kuzenko, *Angew. Makromol. Chem.,* **35**, 37 (1974).

155. R. R. Stromberg and L. E. Smith, *J. Phys. Chem.,* **71**, 2470 (1967).

156. B. J. Fontana and J. R. Thomas, *J. Phys. Chem.,* **65**, 480 (1961).

157. C. Thies, *Macromolecules,* **1**, 335 (1968).

158. R. Botham and C. Thies, *J. Colloid. Interface Sci.,* **31**, 1 (1969).

159. G. R. Joppien, *Makromol. Chem.,* **175**, 1931 (1974).

160. G. R. Joppien, *Makromol. Chem.,* **176**, 1129 (1975).

161. A. V. Kiselev et al., *Kolloidn. Zh.,* **30**, 386 (1968).

162. T. Miyamoto and H. J. Cantow, *Makromol. Chem.,* **162**, 43 (1972).

163. M. J. Schick and E. N. Harvey, *J. Polym. Sci.,* Part B, **7**, 495 (1969).

164. T. M. Polonskii et al., *Vysokomol. Soed.,* **8**, 1901 (1966).

165. I. I. Maleev, T. M. Polonskii, and M. N. Solbys, *Vysokomol. Soed.,* **10**, 2122 (1968).

166. N. P. Sen'kov, V. P. Zakordonskii, and T. M. Polonskii, *Dopov. Akad. Nauk Ukr. SSR,* Ser B, 724, (1970).

167. V. T. Crowl and M. A. Malati, *Discuss. Faraday Soc.,* **42**, 301 (1966).

168. R. Worwag and K. Hamann, *Phys. Chem.,* **71**, 291 (1967).

169. T. Sato, *J. Appl. Polym. Sci.,* **15**, 1053 (1971).

170. R. A. Botham and C. Thies, *J. Colloid Interface Sci.,* **45**, 512 (1973).

171. G. S. Sadakne and J. L. White, *J. Apply. Polym. Sci.*, **17**, 453 (1973).

172. A. I. Kuzaev and S. G. Entelis, *Makromol. Granitse Razdela Paz*, Noukova Dumka, Kiev, 1971, p. 92.

173. E. Hamori, W. C. Forsman, and R. E. Hughes, *Macromolecules*, **4**, 193 (1971).

174. R. Perkel and R. Ullman, *J. Polym. Sci.*, **54**, 127 (1961).

175. S. Ellerstein and R. Ullman, *J. Polym. Sci.*, **55**, 123 (1961).

176. K. Hara and T. Imoto, *Kolloid-Z. Z. Polym.*, **237**, 297 (1970).

177. E. K. Bogacheva et al., *Vysokomol. Soed.*, **A10**, 574 (1968).

178. A. V. Kiselov et al., *Vysokomol. Soed.*, **A14**, 2343 (1972).

179. V. P. Zakordonskii and T. M. Polonskii in Yu. S. Lipatov, Ed., *Poverkh. Yavleniya v Polim.*, Kiev, 1970, p. 58.

180. E. K. Bogacheva, A. V. Kiselev, and Yu. A. Eltekov, *Kolloidn. Zh.*, **27**, 793 (1965).

181. J. Koral, R. Ullman, and F. R. Eirich, *J. Phys. Chem.*, **62**, 541 (1958).

182. K. Mizhuhara, K. Kara, and T. Imoto, *Kolloid-Z. Z. Polym.*, **229**, 17 (1969).

183. E. Killman and H. G. Wiegand, *Makromol. Chem.*, **132**, 239 (1970).

184. T. M. Polonskii and V. P. Zakordonskii, *Kolloidn. Zh.*, **33**, 721 (1971).

185. A. Ulinska and M. Soltys, *Rocz. Chem.*, **41**, 967 (1967), through *Chem. Abstr.*, **68**, No. 87668 (1968).

186. Yu. A. Eltekov, *J. Polym. Sci.*, Part C, **16**, 1931 (1967).

187. A. I. Kuzaev, in Yu. S. Lipatov, Ed., *Thermodynamic and Structural Properties of Boundary Layers of Polymers*, Naukova Dumka, Kiev, 1976, p. 41.

188. H. Burns and D. K. Carpenter, *Macromolecules*, **1**, 384 (1968).

189. E. K. Bogacheva et al., *Vysokomol. Soed.*, **A11**, 2180 (1969).

190. Yu. A. Eltekov in Yu. S. Lipatov, Ed., *Poverkh. Yavleniya v Polim.*, Kiev, 1970, p. 43.

191. E. K. Bogacheva and Yu. A. Eltekov in Yu. S. Lipatov, Ed., *Poverkh. Yavleniya v Polim.*, Kiev, 1970, p. 52.

192. D. Cole and G. J. Howard, *J. Polym. Sci.*, Part A-2, **10**, 993 (1972).

193. Yu. S. Lipatov and L. M. Sergeeva, *Kolloidn. Zh.*, **27**, 217 (1965).

194. V. M. Patel et al., *Angew. Makromol. Chem.*, **18**, 39 (1971).

195. M. N. Soltys et al., in Yu. S. Lipatov, Ed., *Poverkh. Yavleniya v Polim.*, Kiev, 1970, p. 65.

196. T. M. Polonskii, V. P. Zakordonskii, and M. N. Soltys, *Makromol. Granitse Razdela Paz*, Naukova Damka, Kiev, 1971, p. 62.

197. T. M. Polonskii et al., *Vysokomol. Soed.*, **8**, 1901 (1966).

198. C. Thies, *J. Phys. Chem.*, **70**, 3783 (1966).

199. W. A. Kindler and J. W. Swanson, *J. Polym. Sci.*, Part A-2, **9**, 853 (1971).

200. G. J. Fleer and J. Lykleman, *J. Colloid Interface Sci.*, **55**, 228 (1976).

201. A. Silberberg, *J. Chem. Phys.*, **46**, 1105 (1967); **48**, 2835 (1968).

202. F. Th. Hesselink, *J. Phys. Chem.*, **75**, 65 (1971).

203. D. Chan et. al., *J. Chem. Soc., Faraday Trans.*, (2)**71**, 235 (1975).

204. T. Yoshida, T. Ohsaka, and Y. Iida, *Bull. Chem. Soc. Jap.*, **45**, 3241 (1972).

205. T. Yoshida, T. Ohsaka, and M. Suzuki, *Bull. Chem. Soc. Jap.*, **45**, 3245 (1972).

206. L. Marszall, *Colloid Polym. Sci.*, **252**, 335 (1974).

207. J. M. G. Lankveld and J. Lykleman, *J. Colloid Interface Sci.*, **41**, 454 (1972).

208. J. Th. C. Boehm and J. Lykleman, *J. Colloid Interface Sci.*, **50**, 559 (1975).

209. J. E. Glass, *J. Phys. Chem.*, **72**, 4459 (1968).

210. I. D. Rattee and M. Breuer, *The Physical Chemistry of Dye Adsorption*, Academic Press, New York, 1974.

211. R. Komers and D. Tomanova, *Collect. Czech Chem. Commun.*, **37**, 774 (1972).

212. D. G. Gray and J. E. Guillet, *Macromolecules*, **5**, 316 (1972).

213. L. Ya. Kurdyukova et al., *Vysokomol. Soed.*, **A15**, 1948 (1973).

214. C. Jeaneret and H. F. Stoechli, *Helv. Chim. Acta*, **56**, 2509 (1973).

215. Yu. A. Eltekov et al., *J. Colloid Interface Sci.*, **47**, 795 (1974).

216. Y. Ihara et al., *J. Polym. Sci., Polym. Chem. Ed.*, **10**, 3569 (1972).

217. D. J. R. Laurence, *Physical Methods of Macromolecular Chemistry*, Vol. 2, Marcel Dekker, New York, p. 91 (1972).

218. A. P. Boyes and A. B. Ponter, *J. Appl. Polym. Sci.*, **16**, 2207 (1972).

219. E. Sacher, *J. Macromol. Sci., Phys.*, **9**, 521 (1974).

220. T. Fujiki, *J. Appl. Polym. Sci.*, **15**, 47 (1971).

221. V. Ya. Kabanov, N. M. Kazimirova, and A. E. Chalykh, *Vysokomol. Soed.* **A14**, 2042 (1972).

222. B. J. Spit, *J. Macromol. Sci. Phys.*, **2**, 45 (1968).

223. J. J. Bikerman, *Phys. Status Solidi*, **10**, 3; **12**, K 127 (1965).

224. S. M. Yagnyatinskaya, L. Ya. Kaplunova, and S. S. Voyutskii, *Zh. Fiz. Khim.*, **44**, 1445 (1970).

225. P. A. Spegt et al., *Makromol. Chem.*, **107**, 29 (1967).

226. R. J. Roe and H. E. Bair, *Macromolecules*, **3**, 454 (1970).

227. E. Martuscelli and M. Pracella, *Polymer*, **15**, 306 (1974).

228. Yu. A. Zubov, A. N. Ozerin, and N. F. Bakeev, *Dokl. Akad. Nauk SSSR*, **221**, 121 (1975).

229. A. Nakajima et al., *Kolloid-Z. Z. Polym.*, **222**, 10 (1968).

230. A. Wlochowicz and W. Przygocki, *J. Appl. Polym. Sci.*, **17**, 1197 (1973).

231. G. Nachtrab and H. G. Zachmann, *Ber. Bunsenges. Phys. Chem.*, **74**, 837 (1970).

232. T. G. Mahr, *J. Phys. Chem.*, **74**, 2160 (1970).

233. A. Schwarcz and R. S. Farinato, *J. Polym. Sci.*, Part A-2, **10**, 2025 (1972).

234. S. K. Rhee, *Mater. Sci. Eng.*, **16**, 45 (1974).

235. F. Vargara and B. Lespinasse, *Bull. Soc. Chim. Fr.*, **1970**, 3227.

236. F. D. Petke and B. R. Ray, *J. Colloid Interface Sci.*, **31**, 216 (1969).

237. J. R. Dann, *J. Colloid Interface Sci.*, **32**, 302 (1970).

238. J. J. Bikerman, *Adhes. Age*, **2**(2), 23 (1959).

239. K. Bright and B. A. W. Simmons, *Eur. Polym. J.*, **3**, 219 (1967).

240. A. Baszkin and L. Ter-Menassian-Saraga, *J. Colloid Interface Sci.*, **43**, 190 (1973).

241. R. R. Sowell et al., *J. Adhes.*, **4**, 15 (1972).

242. M. Bernett, *Macromolecules*, **7**, 917 (1974).

243. R. A. Kulman et al., *Kolloidn. Zh.*, **30**, 860 (1968).

244. K. Hara and H. Schonhorn, *J. Adhes.*, **2**, 100 (1970).

245. Y. Kitazaki and T. Hata, *Nippon Secchaku Kyokai Shi*, **7**, 289 (1971).

246. G. M. Bartenev and L. A. Akopyan, *Plaste Kaut.*, **16**, 655 (1969).

247. W. A. Zisman, "Contact Angle, Wettability and Adhesion," *Advan. Chem. Ser.* **43**, 1 (1964).

248. (*a*) R. E. Johnson, Jr. and R. H. Dettre, in E. Matijevic, Ed., *Surface and Colloid Science*, Vol. 2., Wiley-Interscience, New York, 1969, p. 85. (*b*) T. D. Blake and J. M. Haynes, *Progress in Surface and Membrane Science*, Vol. 6, Academic Press, New York, 1973.

249. R. J. Good, *J. Am. Chem. Soc.*, **74**, 504 (1952).

250. R. E. Johnson, Jr. and R. H. Dettre, *Adv. Chem. Ser.* **43**, 112 (1964).

251. R. E. Johnson, Jr. and R. H. Dettre, *J. Phys. Chem.*, **68**, 1744 (1964).

252. A. W. Neumann and R. J. Good, *J. Colloid Interface Sci.*, **38**, 341 (1972).

253. H. W. Fox and W. A. Zisman, *J. Colloid Sci.*, **7**, 109 (1952); and W. A. Zisman, *Advan. Chem. Series*, **43**, 1, (1964).

254. R. E. Johnson, Jr. and R. H. Dettre, *J. Coll. Interface Sci.*, **21**, 610 (1966).

255. H. D. Feltman and J. R. McPhee, *Text. Res. J.*, **34**, 654 (1964).

256. M. K. Bernett and W. A. Zisman, *J. Phys. Chem.*, **66**, 1207 (1962).

257. A. W. Adamson, *The Physical Chemistry of Surfaces*, Wiley-Interscience, New York, 1967.

258. A. W. Adamson and I. Long, *Adv. Chem. Ser.*, **43**, 57 (1964); A. W. Adamson, *J. Colloid. Interface Sci.*, **44**, 273 (1972) and references cited therein.

259. L. H. Sharpe and H. Schonhorn, *Adv. Chem. Ser.*, **43**, 189 (1964).

260. J. J. Bikerman, *The Sciences of Adhesive Joints*, Academic Press, New York, 1961.

261. H. Schonhorn and L. H. Sharpe, *J. Polym. Sci.*, **B2**, 719 (1964).

262. H. Schonhorn and L. H. Sharpe, *J. Polym. Sci.*, **A3**, 3087 (1965).

263. L. H. Sharpe, H. Schonhorn, and C. J. Lynch, *Int. Sci. Technol.*, 26, (April 1964).

264. F. M. Fowkes and W. D. Harkins, *J. Am. Chem. Soc.*, **62**, 3377 (1940).

265. N. K. Adam and G. E. P. Elliot, *J. Chem. Soc.*, 2206 (1926).

266. A. W. Neumann, *Proc. 4th Int. Congr. Surf. Act.* (*Brussels*), (1964).

267. R. E. Johnson, Jr. and R. H. Dettre, *J. Colloid Sci.*, **20**, 173 (1965).

268. H. Schonhorn, *Nature*, **210**, 896 (1966).

269. C. A. Sutula, R. Hautala, R. A. Dalla Betta, and L. A. Michel, *Abstr. 153rd Meet., Am. Chem. Soc.*, April 1967.

270. R. A. Dalla Betta, C. A. Sutula, R. Hautala, and L. A. Michel, *Abstr. 153rd Meet Am. Chem. Soc.*, April 1967.

271. J. F. Padday, *J. Colloid Interface Sci.*, **28**, 557 (1968).

272. F. D. Petke and B. R. Ray, *J. Colloid Interface Sci.*, **31**, 216 (1969).

273. H. Schonhorn, *J. Adhes.*, **1**, 38 (1969).

274. V. V. Nevzorov et al., *Mekh. Polim.*, 839 (1974).

275. P. K. Tsarev and Yu. S. Lipatov, *Mekh. Polim.*, 195 (1972).

276. P. K. Tsarev and Yu. S. Lipatov, *Mekh. Polim.*, **A17**, 717 (1975).

277. S. B. Ainbinder and K. I. Alksne, *Mekh. Polim.* 661, (1972).

278. M. A. Waldrop and G. Kraus, *Rubber Chem. Technol.*, **42**, 1155 (1969).

279. Yu. S. Lipatov and A. E. Nesterov, *Macromolecules*, **8**, 889 (1976).

280. G. I. Krus and A. T. Sanzharovskii, *Mekh. Polim.*, 62 (1972).

281. H. Schonhorn and F. W. Ryan, *J. Appl. Polym. Sci.*, **18**, 235 (1974).

282. R. G. Azrak, *J. Colloid Interface Sci.*, **47**, 779 (1974).

283. Yu. S. Lipatov, A. N. Kuskin, and L. M. Sergeeva, *J. Adhes.*, **6**, 259 (1974).

284. J. J. Bikerman in A. J. Chompff and S. Newman, Eds., *Polymer Networks*, Plenum Press, New York, 1971, p. 111.

285. J. J. Bikerman, *J. Macromol. Sci.—Rev. Macromol. Chem.*, **11**, 1 (1974).

286. K. A. Grosch, *Proc. R. Soc.*, **A 274**, 21 (1963).

287. S. Bahadur and K. C. Ludeman, *Wear*, **18**, 109 (1971).

288. J. J. Bikerman, *Wear*, **39**, 1, (1976).

289. D. H. Kaelble, *Physical Chemistry of Adhesion*, John Wiley, New York, 1971.

290. E. H. Andrews and A. J. Kinlock, *J. Polym. Sci.*, Sym. No. 46, 1 (1974).

291. R. H. Hansen and H. Schonhorn, *J. Polym. Sci.*, **B4**, 203 (1966).

292. D. Briggs, D. M. Brewis, and M. B. Konieczka, *J. Mater. Sci.*, **11**, 270 (1976).

293. A. Baszkin, N. Nishino, and L. Ter-Minassian-Sars, *J. Colloid Interface Sci.*, **54**, 516 (1977).

294. D. T. Clark and A. Dilks, *J. Polym. Sci.*, *Polym. Chem. Ed.*, **15**, 232 (1977).

295. D. T. Clark, in Lieng-Huang Lee, Ed., *Advances in Polymer Friction and Wear*, Plenum Press, New York, 1974, p. 241.

296. D. T. Clark, N. J. Feast, W. K. R. Musgrave, and I. Ritchie, *J. Polym. Sci.*, *Polymer Chem. Ed.*, **13**, 857 (1975).

297. H. Schonhorn and R. H. Hansen, *J. Appl. Polym. Sci.*, **12**, 1231 (1968).

298. J. M. Tedder, *Q. Rev.*, **14**, 336 (1960).

299. H. Schonhorn and F. W. Ryan, *J. Adhes.*, **1**, 43 (1969).

300. K. Hara and H. Schonhorn, *J. Adhes.*, **2**, 100 (1970).

301. F. K. Borisova, G. A. Galkin, A. V. Kiselev, A. Y. Korolev, and V. I. Lygin, *Kolloidn. Zh.*, **27**, 320 (1965).

302. D. H. Kaelble and E. H. Cirlin, *J. Polym. Sci.*, Part A-2, **9**, 363 (1971).

303. E. H. Andrews, *Fracture in Polymers*, American Elsevier, New York, 1968.

304. J. B. Howard, in R. A. V. Raff and K. W. Doak, Eds., *Crystalline Olefin Polymers*, Part II, Interscience, New York, 1964, p. 47.

305. D. K. Owens, *J. Appl. Polym. Sci.*, **14**, 1725 (1970).

306. A. N. Gent and J. Schultz, paper presented at 162nd Meeting, American Chemical Society, Organic Coatings and Plastics Chemistry, September 1971, *Preprints*, **31**, 113 (1971).

307. G. Saloman, in R. Houwink and G. Saloman, Eds., *Adhesion and Adhesives*, Vol. 1., Elsevier, Amsterdam, 1965.

308. S. S. Voyutuskii, *Autohesion and Adhesion of High Polymers*, Wiley-Interscience, New York, 1963.

309. E. H. Andrews and A. J. Kinlock, *Proc. R. Soc. (London)*, **A332**, 385 (1972); 401 (1973).

310. A. N. Gent and J. Schultz, *J. Adhes.*, **3**, 281 (1972).

311. G. E. Koldunovich, V. G. Epshtein, and A. A. Chekana, *Soc. Rubber Technol.*, **29**, 22 (1970); E. B. Trostyanskaya, G. S. Golovkin, and G. V. Komarov, *Sov. Rubber Technol.*, **25**, 13 (1966).

312. H. D. Keith, F. J. Paddin, and R. G. Vadimeky, *Science*, **150**, 1026 (1965); *J. Polymer Sci.*, **4**, 267 (1966).

313. H. Schonhorn and H. L. Frisch, *J. Polym. Sci.*, *Polym. Physics Ed.*, **11**, 1005 (1973).

314. R. A. Gledhill and A. J. Kinlock, *J. Adhes.*, **6**, 314 (1974).

315. M. Gettings, F. S. Baker, and A. J. Kinlock, *J. Appl. Polym. Sci.*, **12**, 375 (1977).

316. M. Gettings and A. J. Kinlock, *J. Mater. Sci.*, **12**, 2511 (1977).

317. E. H. Andrews and N. E. King, in D. T. Clark and W. J. Feast, Eds., *Polymer Surfaces*, John Wiley, New York, p. 47, 1978, and references therein.

318. D. K. Owens and R. C. Wendt, *J. Appl. Polym. Sci.*, **13**, 1741 (1969).

319. W. A. Zisman, in P. Weis, Ed. *Adhesion and Cohesion*, Elsevier, Amsterdam, 1962.

320. E. K. Koehler, *Corrosion*, **33**, 209 (1977).

321. A. G. Smith and R. A. Dickie, *Ind. Eng. Chem. Prod. Dev.*, **17**, 42 (1978).

322. J. E. Castle, *Surf. Sci.*, **68**, 583 (1977).

323. J. S. Hammond, J. W. Holubka, and R. A. Dickie, *Polym. Prepr.*, **39**, 506 (1978).

324. J. Ferrante, NASA Technical Note TN D-7933, Washington, D.C., (1975).

325. J. P. Rynd and S. K. Rastogi, *Ceram. Bull*, **53**, (9), 631 (1974).

326. R. A. Chappell and C. T. H. Stoddart, *J. Mater. Sci.*, **12**, 2001 (1977).

327. D. H. Buckley and W. A. Brainard, in L. H. Lee, Ed., *Advances in Polymer Friction and Wear*, Vol. 5A, Plenum Press, New York 1974, p. 315.

328. D. R. Wheeler, *Wear*, **47**, 243 (1978).

329. *Adhesives 1978/1979, The International Plastics Selector*, Cordura Publications, Inc., San Diego, California, 1977.

330. "Chemical Abstract Condensates," 86–90 (22), 1977–1979 through Lockheed Information Systems, Palo Alto, California.

331. R. C. Snogren, *Handbook of Surface Preparation,* Palmerton, New York, 1974.

332. C. V. Cagle, Ed., *Handbook of Adhesive Bonding,* McGraw-Hill, New York, 1973.

333. J. Shields, *Adhesives Handbook,* 2nd ed., Newnes-Butterworths, London, 1976.

334. G. L. Schneberger, private communication.

335. G. L. Schneberger, *Understanding Paint and Painting Processes,* Hitchcock Publishing Co., Wheaton, Illinois, 1975.

336. A. F. M. Barton, *Chem. Rev.,* **75,** 731 (1975).

337. J. L. Gardon and J. P. Teas, "Solubility Parameters" in R. R. Myers and J. S. Long, Eds., *Treatise on Coatings,* Vol. 2, Part 2, Marcel Dekker, New York, 1976, p. 413.

338. P. A. Small, *J. Appl. Chem.,* **3,** 71 (1953).

339. K. L. Hoy, *J. Paint Technol.,* **42,** 76 (1970).

340. R. F. Fedors, *Polym. Eng. Sci.,* **14,** 147 (1974).

341. R. P. Kambour, E. E. Romagosa, and C. L. Gruner, *Macromolecules,* **5,** 335 (1972).

342. P. I. Vincent and S. Raha, *Polymer,* **13,** 283 (1972).

343. P. Dunn and G. F. Sampson, *J. Appl. Polym. Sci.,* **14,** 1799 (1970).

344. P. I. Vincent, "Fracture" in N. M. Bikaies, Ed., *Encyclopedia of Polymer Science and Technology* Vol. 7, Interscience, New York, 1967, p. 261.

345. C. Singleton, E. J. Roche, and P. H. Geil, *SPE 32nd Annu. Tech. Conf.,* **20,** 217 San Francisco (1974).

346. F. W. H. Smyrl and M. J. Blackburn, *J. Mater. Sci.,* **9,** 777 (1974).

347. W. D. Bascomb, S. Gadomski, and R. L. Jones, *9th Natl. SAMPE Tech. Conf., Atlanta,* p. 121 (1977).

348. L. F. Henry, *Polym. Eng. Sci.,* **14,** 167 (1974).

349. S. Wu, *J. Adhes.,* **5,** 39 (1973).

350. E. H. Andrews "Fracture of Polymers" in *MTP Int. Rev. Sci., Phys. Chem.* Ser. 1, **8,** 227 (1972).

351. H. H. Kausch, *Kunstst. Ger. Plast.,* **66,** 583 (1976).

352. G. P. Marshall, N. H. Linkins, L. E. Culber, and J. G. Williams, *Soc. Plast. Eng., Tech. Paper,* **18** (Part 1), 40 (1972).

353. P. Weiss, "Automotive Applications" in N. M. Bikales, Ed., *Encyclopedia of Polymer Science and Technology,* Suppl. Vol. 1, Interscience, New York, 1976, p. 53.

354. C. W. Jennings, "Surface Roughness and Bond Strength of Adhesives" L. Lee, Ed., in *Recent Advances in Adhesion,* Gordon and Beach, London, 1971, p. 469.

355. E. J. Bruno, Ed., *Adhesives in Modern Manufacturing,* Society of Manufacturing Engineers, Dearborn, Michigan, 1970.

356. F. D. Petke, "Structure-Property-Performance Relationships in Synthetic Polymeric Adhesives," in L. Lee, Ed., *Adhesion Science and Technology,* Part A, Plenum Press, New York, 1975, p. 177.

357. M. G. Fontana and N. G. Greene, *Corrosion Engineering,* 2nd ed., McGraw-Hill, New York, 1978, p. 319.

358. A. Neuhaus and M. Gebhardt, "Epitaxy and Corrosion Resistance of Inorganic Protective Layers on Metals," in P. Weiss and G. D. Cheever, Eds., *Interface Conversion for Polymer Coatings,* American Elsevier, New York, 1968.

359. E. P. Plueddemann, "Interfaces in Polymer Matrix Composites," in L. J. Broutman and R. H. Krock, Eds., *Composite Materials,* Academic Press, New York, 1974.

360. K. Dawes and R. J. Rowley to Malaysian Rubber Producers Association, German Patent 2704506.

361. E. P. Plueddemann and G. L. Stark, *Proc. SPI 29th Conf. Reinf. Plast.,* 21-B (1973).

362. K. D. B. Faulkner and J. K. Harcourt, *Aust. Dent. J.,* **20,** 86 (1975).

363. I. Ferguson to ICI Ltd, German Patent, 2337641.

364. A. A. Shurman, T. I. Reznichenko, and V. A. Pinchuk, *Plast. Massy,* **5,** 73 (1977) through *Chem. Abstr.,* **87,** 24312a (1977).

365. J. G. Marsden and S. Sterman, "Organofunctional Silane Coupling Agents," in I. Skiest, Ed., *Handbook of Adhesives,* Van Nostrand Reinhold, New York, 1977, p. 640.

366. Anon., *Plast. Eng.,* **35** (1) 25 (1979).

367. P. E. Cassidy and B. J. Yager, *A Review of Coupling Agents as Adhesion Promoters,* Tracor, Inc., NASA Contract NASA-24073, 1969.

368. M. M. Fein, B. K. Patnaik, and F. K. Chu, to Dart Industries, U.S. Patent 4073766 (1976).

369. S. J. Monte and G. Sugerman, *Soc. Plast. Eng., Tech. Paper,* **22,** 27 (1976).

370. R. N. Griffin, *Further Development of High Temperature-Resistant Graphite Fiber Coupling Agents,* General Electric, NASA Contract, NASA-134987 (1976).

371. W. Chen, D. W. Dwight, and J. P. Wightman, *A Fundamental Approach to Adhesion: Synthesis, Surface Analysis, Thermodynamics and Mechanics,* Virginia Polytechnical Institute, NASA Contract, NASA-157228 (1978).

372. N. L. Rogers, "Surface Preparation of Metals for Adhesive Bonding," *Appl. Polym. Symp.,* No. 3, Wiley-Interscience, New York, 1966, p. 327.

373. G. L. Schneberger, *Understanding Adhesives,* 2nd ed. Hitchcock Publishing Co., Wheaton, Illinois, 1974.

374. W. D. Bascom and S. Mostovoy, *Am. Chem. Soc. Div. Org. Coatings Plast. Chem.,* **38,** 152 (177).

375. R. L. Patrick, "The Use of Scanning Electron Microscopy," in R. L. Patrick, Ed., *Treatise on Adhesion and Adhesives,* Vol. 3, Marcel Dekker, New York, 1973, p. 163.

376. H. A. Perry, "How to Calculate Stresses in Adhesive Joints," in *Product Engineering Design Manual,* McGraw-Hill, New York, (1959).

377. L. J. Hart-Smith, *19th SAMPE Natl. Symp. Exhib.*, **19**, 727, (1974).

378. S. S. Wang, J. F. Mandell, and F. J. McGarry, *Int. J. Fract.*, **14**, (1), 39 (1978).

379. R. D. Adams, J. Coppendale, and N. A. Peppiatt, *J. Strain Anal. Eng. Des.*, **13**, (1), 1 (1978).

380. J. L. Rose, *J. Appl. Polym. Sci.*, **21**, 389 (1977).

381. T. C. Clarke, *Mater. Eval.*, **29A**, 25 (1971).

382. R. J. Schiekemann, *Non-Destr. T.*, **5** (part 2) 144 (1972).

383. P. A. Meyer and J. L. Rose, *J. Appl. Phys.*, **48**, (9), 3705 (1977).

384. J. L. Rose and G. H. Thomas, U. S. NTIS AD-A053771, (1978).

385. F. J. Harris, *Proc. IEEE*, **66**, (1), 51 (1978).

386. R. R. Myers and J. S. Long, Ed., "Characterization of Coatings: Physical Techniques," in *Treatise on Coatings*, Vol. 2, Marcel Dekker, New York (1976).

387. S. Stinson, *Chem. Eng. News*, **57**, (2), 20 (1979).

388. T. Murayama, *Material Science Monographs, Dynamic Mechanical Analysis of Polymeric Materials*, Vol 1, Elsevier, Amsterdam (1978).

389. I. Skeist, Ed., *Handbook of Adhesives*, Van Nostrand Reinhold, New York, 1977.

390. G. L. Schneberber, *Choosing an Adhesive*, Tables to Accompany, GMI, Flint, Michigan, 1978.

391. L. Lee, Ed. "Significant Advances in Developments in Adhesion and Adhesives," in *Adhesion Science and Technology*, Plenum Press, New York, 1975, p. 1.

392. J. F. Kinstle, *J. Radiat. Curing*, **2**, (4), 15(1975).

393. K. C. Stueben, *Adhes. Age*, **20**, (6), 16 (1977).

394. M. W. Ranney, Ed., *Specialized Curing Methods for Coatings and Plastics*, Noyes, Park Ridge, New Jersey, 1977.

395. R. D. Deanin, *Polymer Structure, Properties and Applications*, Cahners, Boston, 1972, p. 53 ff.

396. J. R. Martin, J. F. Johnson, and A. R. Cooper, *J. Macromol. Chem.*, **8**, (1), 87 (1972).

397. M. L. Miller, *The Structure of Polymers*, Reinhold, New York, 1966.

398. L. E. Nielsen, *Mechanical Properties of Polymers*, Reinhold, New York, 1962.

399. G. E. J. Reynolds, *Aspects Adhes.*, **6**, 96 (1971).

400. A. N. Gent, *J. Polym. Sci.*, Part A-2, **9**, 283 (1971).

401. J. F. Fellers and T. F. Chapman, *Am. Chem. Soc. Div. Org. Coating Plast. Chem. Prep.*, **38**, 406 (1977).

402. L. Wild, R. Ranganath, and A. Barlow, *J. Appl. Polym. Sci.*, **21**, (12), 3331 (1977).

403. J. R. Darby and J. K. Sears, "Theory of (PVC) Solvation and Plasticization," in L. I. Nass, Ed., *Encyclopedia PVC*, Vol. 1, Marcel Dekker, New York, 1977, p. 385.

404. J. J. Bernard and H. Burrell, "Plasticization" in A.

D. Jenkins, Ed., *Polymer Science*, North-Holland, Amsterdam, 1972, p. 537.

405. F. W. Ball, "Plasticization of Coatings" in R. R. Myers and J. S. Long, Eds., *Treatise on Coatings* Vol. 2, Marcel Dekker, New York, 1972, p. 171.

406. P. J. Flory, *Principles of Polymer Chemistry*, Cornell University Press, Ithaca, New York, 1953.

407. F. W. Billmeyer, *Textbook of Polymer Science*, 2nd ed., Wiley-Interscience, New York, 1971.

408. L. Mandelkern, *An Introduction to Macromolecules*, Springer-Verlag, New York, 1972.

409. F. Rodriguez, *Principles of Polymer Systems*, McGraw-Hill New York, 1970.

410. J. K. Stille, *Introduction to Polymer Chemistry*, John Wiley, New York, 1962.

411. R. E. Wilfong, *J. Polym. Sci.*, **54**, 385 (1961).

412. D. G. Bannerman and E. E. Magat, in C. E. Schildknecht, Ed., *Polymer Processes* Interscience, New York, 1956, p. 254.

413. R. J. W. Reynolds, in R. Hill, Ed., *Fibers From Synethic Polymers*, Elsevier, Amsterdam, 1953, p. 146.

414. S. M. Aharoni, *J. Appl. Polym. Sci.*, **20**, 2863 (1976).

415. P. J. Flory, *Proc. R. Soc.*, (*London*), **234**, 60–73 (1956).

416. W. G. Miller, *Polym. Prep.*, **18**, (1) 173 (1977).

417. R. W. Lenz, *Organic Chemistry of Synthetic High Polymers*, Interscience, New York, 1967, p. 46.

418. R. D. Deanin, *ACS Div. Org. Coating Plast. Prep.*, **37**, (2), 378 (1977).

419. L. Mandelkern, *Polym. Prepr.*, **20**, (1), 267 (1979).

420. W. Schlesinger and H. M. Leeper, *J. Polym. Sci.*, **11**, 202 (1953).

421. L. Mandelkern, *Crystallization of Polymers*, McGraw-Hill New York, 1964.

422. C. C. Chu, *Polym. Prepr.*, **19**, (2) 2, 773 (1978).

423. L. Mandelkern, *Polym. Eng. Sci.*, **9**, 255, (1966).

424. L. Mandelkern, J. M. Price, M. Gopalan, and J. G. Fatou, *J. Polym. Sci.*, A-2, **5**, 239(1967).

425. F. C. Stehling and L. Mandelkern, *Macromolecules*, **3**, 242(1970).

426. R. S. Stein and A. V. Tobolsky, "Crystallinity in Polymers," in A. V. Tobolsky Ed., *Polymer Science and Materials*, Interscience, New York, 1971.

427. B. Wunderlich, *Macromolecular Physics*, Academic Press New York, 1976.

428. A. W. Birley, *J. Polym. Sci., Polym Symp.*, **62**, 343 (1978).

429. Y. Aoki, *J. Polym. Sci.*, Part C, **23**, 855 (1968).

430. W. J. Jackson, Jr., T. F. Gray, Jr., and J. R. Caldwell, *J. Appl. Polym.*, **14**, 685 (1970).

431. K. Nakao, *J. Adhes.*, **4**, 95 (1972).

432. H. Schonhorn and F. W. Ryan, *Adv. Chem. Ser.*, **87**, 140 (1968).

433. H. Schonhorn and F. W. Ryan, *J. Polym. Sci.,* **7**, 105 (1969).

434. H. L. Frisch, H. Schonhorn, and T. K. Kwei, *J. Elastoplastics,* **3**, 214 (1971).

435. S. M. Aharoni and D. C. Prevorsek, *Int. J. Polym. Mater.,* **6**, (1–2) 39 (1977).

436. C. A. May and Y. Tanaka, Eds., *Epoxy Resins Chemistry and Technology,* Marcel Dekker, New York, 1973.

437. A. Barnat, B. Bircher, and A. H. Schaffer, *Prog. Rubber Technol.,* **38**, 59 (1975).

438. J. K. Backus, *High Polym.,* **29**, 642 (1977).

439. T. J. Deerlove, G. A. Campbell, and R. P. Atkins, *J. Appl. Polym. Sci.,* **22**, (4), 927 (1978).

440. D. R. Hayes and C. S. White, *J. Paint Tech.,* **41**, 461 (1969).

441. M. W. Ranney, Ed., *Electrodeposition and Radiation Curing of Coatings,* Noyes, Park Ridge, New Jersey, 1970.

442. N. Shanmugam and S. Guruswamy, *Met. Finish,* **75**, (9), 62 (1977).

443. F. M. Loop, SME Technical Paper, Ser. FC77, 641 (1977).

444. Anon., *Automot. Eng.,* **86**, (7), 52 (1978).

445. H. M. Leister, SAE Technical Paper Ser. 780188 (1978).

446. Yu. M. Kovalisko, K. Sprysa, I. F. Moravskaya, L. Ya. Nitka, and L. V. Pritsker, *Lakokras. Mater. IKH Primen.,* **5**, 80 (1978), through *Chem. Abstr.,* **89** 199117J (1978).

447. Anon. (Ed.), *Water-Borne and High Solids Coatings Symposium,* New Orleans, University of Southern Mississippi, Hattiesburg, 1976.

448. G. C. Wildman and G. B. Bufkin, Eds. *Proceedings of the 4th Water-Borne and Higher Solids Coatings Symposium,* Vols. 1–2, New Orleans, University of Southern Mississippi, Hattiesburg, 1977.

449. G. C. Wildman and G. B. Bufkin, Eds., *Proceedings of the 5th Water-Borne and Higher-Solids Coatings Symposium,* 1978.

450. C. K. Schoff, *Prog. Org. Coatings.,* **4**, 189 (1976).

451. S. Paul, *Prog. Org. Coatings,* **5**, 79 (1977).

452. M. A. Sherwin, R. A. Taller, J. W. Hagan, and C. H. Carder, *Mod. Paint Coatings,* **66**, (137), 33 (1976).

453. F. Sheme, S. Belote, and L. Gott, *Polym. Paint Color J.,* **165**, 787 (1975).

454. C. M. Hansen, *FATIPEC Congr.,* **14**, 97 (1978).

455. M. W. Ranney, Ed., *Powder Coatings Technology,* Noyes, Park Ridge, New Jersey, 1975.

456. A. W. Neumann and W. Tanner, *J. Colloid and Interface Sci.,* **34**, 1 (1970).

PART 2

Product Technology

## CHAPTER 10

Synthesis and Reactor Design

E. B. NAUMAN

Xerox Corporation
Rochester, New York

10.1	Introduction	471
10.2	Factors in Reactor Design	473
	10.2.1 Mechanism and Kinetics	473
	10.2.1.1 Condensation Polymers	473
	10.2.1.2 Vinyl Addition Polymers	475
	10.2.1.3 Other Polymerization Mechanisms	477
	10.2.2 Reaction Engineering	477
	10.2.3 Purification and Recovery	478
	10.2.3.1 Washing	478
	10.2.3.2 Filtration	478
	10.2.3.3 Coagulation	478
	10.2.3.4 Devolatilization	478
	10.2.3.5 Drying	479
	10.2.4 Mechanical Design	479
10.3	Reactor Options	482
	10.3.1 Single-Phase Polymerizations	482
	10.3.2 Heterogeneous Polymerizations	483
	10.3.2.1 Gas-Liquid	483
	10.3.2.2 Gas-Solid	483
	10.3.2.3 Liquid-Liquid	484
	10.3.2.4 Liquid-Solid	485
	10.3.2.5 Solid-Solid	485
10.4	Fundamentals of Reactor Design	486
	10.4.1 Transport Equations	486
	10.4.1.1 Equations of Motion	486
	10.4.1.2 Equation of Energy	488
	10.4.1.3 Equation of Continuity	488
	10.4.2 Reactor Analysis Techniques	488
	10.4.2.1 Residence Time Distributions	489
	10.4.2.2 Micromixing	491
	10.4.2.3 Thermal Time Distributions	493
10.5	Batch Reactors	494
10.6	Tubular Flow Reactors	495
	10.6.1 Fixed-Wall Devices	496
	10.6.1.1 Undisturbed Flow	496
	10.6.1.2 Flow Redistribution Techniques	498
	10.6.1.3 Optimal Wall Temperatures	500
	10.6.2 Moving-Wall Devices	501
10.7	Stirred Tank Reactors	503
	10.7.1 Thermal Design	503
	10.7.2 Residence Time Distributions	504
	10.7.3 Micromixing	504
	10.7.3.1 Yields	504
	10.7.3.2 Molecular Weight Distributions	506
	10.7.3.3 Copolymer Composition Distributions	508
	10.7.4 Transient Operation	509
	10.7.4.1 Startup and Shutdown	510
	10.7.4.2 Periodic Operation	511
10.8	Overview and Extrapolations	511
10.9	Glossary of Symbols	514
10.10	References	514

10.1 INTRODUCTION

Polymer reaction technology is a subset of the presumably much larger field of chemical reaction technology. Yet, the span of specific technologies needed to cover all aspects of polymer manufacturing is so large that it

E. B. Nauman is currently at the Department of Chemical Engineering and Environmental Engineering, Rensselaer Polytechnic Institute, Troy, New York.

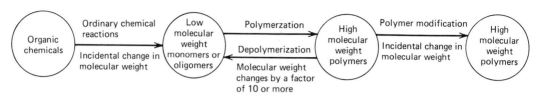

Figure 10.1 Classification of Polymerization Reactions.

closely approaches the technology base of the entire chemical industry. One reason for this complexity is that many definitions of the polymer industry include the manufacture of monomers and low molecular weight oligomers.

Figure 10.1 illustrates a useful division of polymer reactions into two classes, polymerizations (or depolymerizations) and polymer modifications and helps distinguish both of these from ordinary chemical reactions. We define polymerizations as processes that start with monomers or low molecular oligomers and end with high polymers. The average molecular weight typically changes by several orders of magnitude, and a factor of 10 change is an arbitrary but useful minimum that distinguishes polymerizations from ordinary chemical reactions. Polymer modification reactions start and end with high molecular weight polymers, the primary intent of the reaction being to modify structure rather than to change molecular weight. Cross-linking reactions that produce two- and three-dimensional structures are classed as polymerizations if they begin with low molecular weight materials and as polymer modifications if the starting materials are already polymers.

With these definitions, many common polymer processes can be relegated to the class of ordinary chemical reactions. For example, the industrial processes for phenolic resins (phenol + formaldehyde) and epoxy resins (bisphenol A + epichlorohydrin) both yield low molecular weight oligomers that become polymers only after a subsequent reaction step. This distinction in terms of molecular weight also makes sense from a reaction engineering viewpoint. The reactor design aspects for phenolic and epoxy resin processes are not notably different from countless other chemical reactions. The physical properties of the reactants and products are similar to those

of ordinary chemicals, and the reaction environment is well within the span of normal chemical processes.

Another reason for the apparent complexity of polymer reaction technology is the vast number of mechanisms by which polymerizations or polymer modifications can occur. In conventional chemical reactor engineering, reactions are first classified according to whether they are homogeneous or heterogeneous and then according to the phases involved: gas, liquid, or solid. With the major exception of homogeneous gas-phase reactions, polymerization examples can now be found for every class.

It is thus clear that the overall field of polymer synthesis and reactor technology is impossibly broad to be covered in a single chapter. The choice must be made between giving survey descriptions of important but specific polymerization schemes or of presenting fundamental principals of reactor technology that have broad applicability but are necessarily incomplete. The latter approach is taken here since descriptions of the major commercial processes are available elsewhere (1–5). Emphasis is placed on those physical aspects of polymer systems that make them unique in a reactor engineering sense. Specifically, these aspects are the high viscosities and moderate to low diffusivities associated with polymer melts or concentrated polymer solutions.

This tentative limitation of subject matter is too severe since it would exclude some of the most widely practiced polymerization techniques. Emulsion, suspension, and gas-phase polymerizations all use a low-viscosity phase as a carrier or suspending medium to avoid direct confrontation with the high viscosities of polymer melts. Because of their importance, these kinds of processes are discussed briefly. However, major emphasis remains on those situations where it is impossible to avoid treating the polymer in bulk, molten form. This is one aspect of polymer reaction engineering that lends itself to generalized and reasonably comprehensive treatment.

10.2 FACTORS IN REACTOR DESIGN

The industrial development of any polymer process is a highly interactive and often iterative procedure that begins with the bench chemist and, when successful, ends with the production engineer. From a technology viewpoint, there are four major factors that determine the final design: mechanisms and kinetics, reaction engineering, purification and recovery, and mechanical design. This chapter is primarily concerned with reaction engineering and especially those design aspects associated with polymer melts and concentrated solutions. However, other facets of the overall technology are discussed to give background and perspective.

10.2.1 Mechanisms and Kinetics

The number of specific polymerization mechanisms is limited only by the imagination of the polymer chemist and seems virtually inexhaustable. However, a simplified classification system that emphasizes the difference between condensation polymers and vinyl addition polymers is adequate for most purposes.

10.2.1.1 Condensation Polymers
Condensation polymers constitute the earliest and probably largest group of industrially important plastics. Examples of condensation polymers include nylons (diacids reacted with diamines), polycarbonates (bisphenols reacted with phosgene), polyesters (diacids reacted with diols), polysulfones (dichlorodiphenylsulfone reacted with bisphenols) and polyphenylenesulfide (dichlorobenzene reacted with sodium sulfide).

A typical condensation polymerization begins with two difunctional monomers that react in a sequence of steps to give a high polymer. The first reaction step can be represented as

$$X—A—X + Y—B—Y \longrightarrow$$
$$X—A—B—Y + XY$$

Subsequent steps have one of several forms:

$$X \sim X + Y \sim Y \longrightarrow X \sim Y + XY$$

$$X \sim X + Y \sim X \longrightarrow X \sim X + XY$$

$$X \sim Y + X \sim Y \longrightarrow X \sim Y + XY$$

$$Y \sim Y + X \sim Y \longrightarrow Y \sim Y + XY$$

Stoichiometry is of vital importance in most condensation polymerizations since a surplus of one monomer would soon give a nonreacting, low molecular weight system with all molecular species having the same end groups, either all $X \sim X$ or all $Y \sim Y$ depending on which reactant was in initial excess. The only situation where stoichiometry is unimportant is in certain self-condensations and ring-opening polymerizations where the single reactant has the form $X—A—Y$ so that perfect stoichiometry is assured. Otherwise, stoichiometry can be so critical that the polymerizations must be conducted in batch reactors to allow highly accurate initial weighings, possibly followed by subsequent titrations of the reactive mass to yield polymer of an adequate molecular weight. It should be noted that stoichiometry is rarely used to control molecular weight directly since the reaction rate becomes slow when one type of end group nears depletion. Instead, stoichiometry is adjusted to enable a very high molecular weight polymer, and the reaction is terminated before this ultimate molecular weight is achieved. Typically, remaining end groups are deactivated after the polymerization to avoid undesired further reactions during subsequent processing.

The disposition of the condensation product XY is also of critical importance in the manufacture of many polymers. Often the condensation reaction is equilibrium limited so that XY must be continuously removed to drive the forward reaction to completion. Most polyesters are in the equilibrium-controlled category, the condensation product being a small, soluble molecule such as water, and alcohol, or phenol depending on the specific chemistry. Removal of this relatively volatile component is easy at low molecular weights but becomes progressively more difficult as molecular weights and viscosities increase. Specialized extruder-like equipment is often used for the final stages of the reaction, the design of which is one unique and characteristic aspect of polymer reaction engineering.

Nonequilibrium-controlled condensation polymerizations typically result when XY is insoluble in the reaction medium. Examples of this are the precipitation of sodium chloride in the manufacture of polysulfones and polycarbonates. This instantaneous removal of condensation product eases the reactor design problem but leads to material of construction problems or to subsequent difficulties in removing salt from the viscous polymer solution.

Condensation polymerization kinetics are conceptually similar to ordinary chemical kinetics except that the number of molecular species is indefinitely large:

$$X—A—X \text{ and } Y — B —Y \quad \text{Monomers}$$
$$X— AB —Y \quad \text{Dimer}$$
$$X—AVA—X \text{ and } Y— BAB —Y \quad \text{Trimers}$$
$$X — ABAB —Y \quad \text{Tetramer}$$

and so on. We can denote the concentrations for these various species as follows:

A_i, $i = 1, 3, \cdots$,
        (molecules of the form $X \sim X$)

B_i, $i = 1, 3, \cdots$,
        (molecules of the form $Y \sim Y$)

M_i, $i = 2, 4, \cdots$,
        (molecules of the form $X \sim Y$)

and write the rate equations for each A_i, B_i, and M_i assuming all reactions are of the simple, second-order form. Thus at a very early stage in the polymerization, Equation (10.1)

$$\frac{dA_1}{dt} = -2K_{11}A_1B_1 \qquad (10.1)$$

will provide a good fit for experimental data since the only species present to any significant extent are the two monomers A_1 and B_1. However, other reactive species will soon be formed so that the complete rate expression for A_1 must include all the B_i and M_i species that can react with it as shown in Equation (10.2).

$$\frac{dA_1}{dt}$$
$$= 2K_{11}A_1B_1 - K_{12}A_1M_2 - 2K_{13}A_1B_3 - \cdots$$

$$= -2A_1 \sum_{i=1,3,\cdots} K_{1i}B_i - A_i \sum_{i=2,4,\cdots} K_{1i}M_i$$
$$(10.2)$$

Similar expressions must be written for every other A, B, and M. As an example consider Equation (10.3)

$$\frac{dM_2}{dt} = 2K_{11}A_1B_2 - K_{11}M_2A_1 - K_{21}M_2B_1$$
$$- K_{22}M_2^2 - K_{23}M_2A_3 - K_{23}M_2B_3 - \cdots$$
$$(10.3)$$

which shows that the concentration of dimer will increase during the early stages of the polymerization but will later decrease.

In the general problem, one obtains an unbounded set of ordinary differential equations, one equation for each reactive species. This set of equations can often be solved in closed form for the average molecular weight and molecular weight distribution as a function of time. However, the specifics of the solution will depend on the type of reactor used for the polymerization. This means that the reactor engineer must start with the unsolved set of differential equations whenever he wishes to employ a novel reactor type. For batch reactors and a few other idealized situations such as the perfectly mixed stirred tank, literature solutions are available. These are discussed elsewhere in this chapter and in Chapter 3. Also, see Peebles (6) for a review of existing results, and see Amundson and Luss (7) or Chappelear and Simon (8) for a summary of mathematical techniques used to derive such results.

For batch polymerizations that are not retarded by the condensation product, the set of differential rate equations can be integrated to give equations of the form

$$DP = DP\,(t, s, \mathbf{E}) \qquad (10.4)$$

where DP is the (number) average degree of polymerization, t is the time spent in the reactor, s is the stoichiometric ratio ($s = 1$ for perfect stoichiometry), and $\mathbf{E}$ is a vector representing environmental conditions such as temperature and pH. An explicit form for the function in Equation (10.4) is not generally possible since the environmental conditions will change as polymerization proceeds. For example, temperatures will usually rise during the polymerization, and this will affect the rate constants K.

Equation (10.4) gives the degree of polymerization as a function of time. It is also useful to know the total reaction rate, for example to estimate the reaction exotherm, and to have more information about the molecular weight distribution than just the first moment DP. With a given reactor type, this information can also be obtained from the set of differential equations. It is often adequate to characterize a molecular weight distribution by its polydispersity PD which is the ratio of weight-average to number-average molecular weight (see Chapter 3). For many condensation polymerizations the polydispersity is a function of the degree of polymerization alone and varies between the limits of $PD = 1$ for unreacted monomers to $PD = 2$ for high polymer. This situation corresponds to the well known Flory distribution (9) and arises whenever the rate constants K are independent of DP. The Flory distribution also arises in reversible polycondensations where the condensation product retards the forward reaction (10).

10.2.1.2 *Vinyl Addition Polymers*
Vinyl polymers are now the most important class of commercial plastics since they include all polyolefins, styrenics, polybutadienes, acrylics, and polyvinyl chlorides. There are two different polymerizations mechanisms that have found widespread use. Free radical polymerization gives a polymer of random molecular orientation and is the standard process for low-density polyethylene, polymethylmethacrylate, polyvinylchloride and copolymers, and polystyrene and copolymers including acrylonitrile-butadiene-styrene (ABS) and the styrene-butadiene rubbers. Coordination catalysis is the more recent development, dating from the late 1950's and gives polymer of highly controlled molecular structure. High-density polyethylene, polypropylene, polybutylene, and polybutadiene are all polymerized by this route.

For a simple vinyl homopolymerization, the prototypical chemical reaction can be expressed as

$$nM \longrightarrow M_n$$

where n is usually about 1000. This representation of the reaction completely ignores mechanism but is a reasonable engineering approximation of the situation. At any instant of time there are just two predominate molecular species, monomer M and polymer M_n. The lifetime of the growing polymer chain is so short that the concentration of active chains can be ignored. Concentrations of initiators, catalysts, poisons, and so on are lumped into the environment vector $\mathbf{E}$. Thus, as shown in Equation (10.5),

$$\text{polymerization rate} = R_M(M, \mathbf{E}) \qquad (10.5)$$

the reaction rate R_M is a function only of monomer concentration M and the reaction environment. This is also true for the average molecular weight and molecular weight distribution. From a reactor design viewpoint, most of the necessary mechanistic and kinetic considerations can be condensed into three expressions that give the rate, DP, and PD as functions of M and $\mathbf{E}$.

The relative simplicity of the kinetic equations for vinyl additions polymers stems from the fact that, to a first approximation, a polymer chain once grown does not enter into any subsequent reactions. When this assumption is violated, for example when chain transfer to polymer is an important process, it is necessary to use a formulation similar to that for condensation polymers. Free-radical polymerization involves three steps: initiation, propagation, and termination. The initiation step can be written as

$$I^{\cdot} + M \longrightarrow M_1$$

where $I^{\cdot}$ represents some "original" free-radical species which was formed spontaneously by thermal initiation or from a chemical initiator, typically a peroxide. The initiator radical reacts with monomer M to form living polymer of chain length one.

The propagation step then proceeds rapidly to give high polymer

$$M_n^{\cdot} + M \longrightarrow M_{n+1}^{\cdot}$$

The growing free-radical chain is terminated by any of several mechanisms as follows:

$$M_n^{\cdot} + M_m^{\cdot} \longrightarrow M_{n+m}$$
$$\text{(combination)}$$

$$M_n^{\cdot} + M_m^{\cdot} \longrightarrow M_n + M_m$$
$$\text{(disproportionation)}$$

$$M_n^{\cdot} + M \longrightarrow M_n + M_1^{\cdot}$$
$$\text{(chain transfer to monomer)}$$

$$M_n^{\cdot} + M_m \longrightarrow M_n + M_m^{\cdot}$$
$$\text{(chain transfer to polymer)}$$

$$M_n^{\cdot} + S \longrightarrow M_n + S^{\cdot}$$
$$\text{(chain transfer to solvent)}$$

All these termination mechanisms give inactive polymer chains, but only the first two decrease the total concentration of free radicals. Which of the termination mechanisms is important varies with the specific chemistry. Styrene polymerizations terminate by combination and, at the higher temperatures typical of thermal initiation, by chain transfer to monomer. Methylmethacrylate terminates almost entirely by disproportionation. Chain transfer to polymer is the branching mechanism in low-density polyethylene, and chain transfer to solvent is used to control molecular weight in ABS polymerizations. Theoretical

molecular weight distributions have been derived for a large number of specific reaction mechanisms (6, 11–13, see also Chapter 3). Theoretical calculations are also available for various idealized reactors and are discussed later in this chapter.

For simple vinyl copolymerizations, the prototypical chemical reaction can be expressed as

$$nA + mB \longrightarrow A_nB_m$$

where the notation A_nB_m is meant to imply a more or less random arrangement of the two constituents in a single molecular chain. For a 50/50 copolymer, a typical structure might be

$$ABBABAAABABB\cdots$$

The kinetic formulation for vinyl copolymerizations is similar to that for homopolymerizations except that there must now be two rate equations analogous in Equation (10.5), one for each monomer. Assuming simple bimolecular kinetics for the propagation steps, we can write Equation (10.6).

$$\frac{dA}{dt} = -K_{A^{\cdot}A}A^{\cdot}A - K_{B^{\cdot}A}B^{\cdot}A \quad (10.6)$$

and Equation (10.7)

$$\frac{dB}{dt} = -K_{A^{\cdot}B}A^{\cdot}B - K_{B^{\cdot}B}B^{\cdot}B \quad (10.7)$$

Equation (10.6) merely states that monomer A disappears through reactions either with polymer chains having A end groups (these active molecules being denoted $A^{\cdot}$) or with those chains having B end groups (denoted as $B^{\cdot}$). Any disappearance of A is due to initiation or termination steps is ignored, which is reasonable for long chain lengths. Equation (10.7) makes the similar statement about monomer B.

Turning to free-radical concentrations, it is usually appropriate to assume that the initiation and termination rates are equal so that the total population of free radicals is constant. Applying this pseudo-steady-state assumption independently to the A and B chains gives Equation (10.8)

$$\frac{dA^{\cdot}}{dt} = +K_{B^{\cdot}A}B^{\cdot}A - K_{A^{\cdot}B}A^{\cdot}B = 0 = \frac{dB^{\cdot}}{dt}$$

$$(10.8)$$

Equations (10.6), (10.7), and (10.8) can be combined to give the copolymer composition equation as follows:

$$\frac{dA}{dB} = \frac{A\,(\gamma_A A + B)}{B\,(A + \gamma_B B)} \qquad (10.9)$$

where γ_A and γ_B are constants known as copolymer reactivity ratios. These ratios determine both the overall concentrations of the two monomers in the polymer and also the probabilities for various sequences of A and B units within a single polymer chain. Provided the polymerization proceeds by a free-radical mechanism, it has been found that reactivity ratios are insensitive to the reaction environment (9). Ham (14) gives a comprehensive treatment of copolymerization theory and lists reactivity ratios for most monomers of industrial importance.

The reactivity ratios are related to the propagation rate constants. For example, γ_A is the ratio of $K_{A^{\cdot}A}$ to $K_{A^{\cdot}B}$ and thus measures the tendency for an $A^{\cdot}$ chain to react with monomer A compared to its tendency to react with monomer B. However, the relative monomer concentrations also affect reaction rates. The combined effect can be treated as a reaction probability, Equation (10.10)

probability of an $A^{\cdot}$ chain reacting with monomer $A = \dfrac{\gamma_A A}{\gamma_A A + B}$

$$(10.10)$$

Thus an $A^{\cdot}$ chain will react with its own monomer whenever γ_A is large or when the concentration of A is large. Conversely, it will react with the opposite monomer when γ_A is small or when the concentration of A is small (compared to the concentration of B).

If both $\gamma_A < 1$ and $\gamma_B < 1$, each monomer prefers to react with its opposite so that the copolymer will tend to have an alternating structure:

ABABABABABAB . . .

This situation is fairly common. For example, the system α-methyl styrene-acrylonitrile has $\gamma_A = 0.10$ and $\gamma_B = 0.06$ which will be strongly alternating if the monomer concentrations are approximately equal.

If $\gamma_A < 1$ and $\gamma_B > 1$, the natural tendency is for short segments of A to be followed by long blocks of B, but this natural tendency can be overcome by stoichiometry. Suppose, for example, that $\gamma_A = 0.5$ and $\gamma_B = 2.0$. Then to achieve a 50:50 copolymer, we set $dA/dB = 1$ in Equation (10.9) with the result that $2B = A$. Thus the concentration of monomer A must be twice that of B to achieve the 50:50 composition in the polymer. Under these conditions, however, the various reaction probabilities are all equal to 0.5, and the copolymer is completely random. Systems with $\gamma_A < 1$ and $\gamma_B > 1$ are again quite common. For example, α-methyl styrene-styrene has $\gamma_A = 0.10$ and $\gamma_B = 1.25$.

If both $\gamma_A > 1$ and $\gamma_B > 1$, both monomers tend to homopolymerize to give a block copolymer structure

AAA . . . AAABBB . . . BBB

It turns out, however, that such systems are extremely rare and that commercial block copolymers are made by more exotic and sequential reactions.

10.2.1.3   Other Polymerization Mechanisms

Other mechanisms are occasionally used for specialty polymers. A well known case is the oxidative coupling used in the polyphenylene oxide polymerization. This and most other novel polymerization schemes will give growing polymer chains with long lifetimes so that the kinetics and molecular weight distributions will usually be similar to those for condensation polymerizations.

10.2.2   Reaction Engineering

As used here, reaction engineering is a process that begins with specified mechanisms and kinetic equations and ends with the design for a reactor, this design providing an economically and mechanically attractive means for carrying out the reaction on an industrial scale. The design process has the general goal of providing a polymerization environment that is optimal in residence time, temperature, composition, and fluid mechanical shear. The design is optimal in the sense that it provides the best combination of product properties and product cost.

Before proceeding with this rather self-contained view of the design process, it is worth emphasizing that the reactor engineer should always challenge the constraints of his assignment. The first polymerization recipe furnished by a polymer chemist requires too much solvent, too many purification steps, too long a reaction time at too high a temperature, and uses a difficult, unscaleable recovery process. The commercial development of any new process or the evolution of an existing process requires a constant interaction between the chemist and the engineer. The chemist finds he can use lower-purity monomers, the engineer finds that exotic materials of construction are not really required, and so on until a finished process emerges. At its best, this interactive and iterative procedure can produce truly marvelous results. The original process for high-density polyethylene used an inefficient catalyst and large quantities of solvent to produce a polymer almost expensive enough to be considered an engineering plastic. The latest process for HDPE is surely the most elegant and potentially the cheapest polymerization yet devised, going from a low-pressure gas to a solid, particulate polymer in a single step.

10.2.3 Purification and Recovery

The common unit operations in the purification and recovery section of a polymer plant are washing, filtration, coagulation, devolatilization, and drying.

10.2.3.1 Washing
Washing refers to the normally aqueous extractions done on polymer solutions and coagulates to remove impurities, reaction by-products, and water-soluble solvents. Nonaqueous solvent extraction is considered a variant of washing. When done with polymer solutions, washing requires fairly low viscosities which can impose a serious limit on polymer concentrations. Most typically, washing is done on condensation polymers. It is also done on coagulated emulsion polymers to remove interfacial agents, but in this case the polymer is treated in semi-solid form.

10.2.3.2 Filtration
Filtration is sometimes used in condensation polymerization processes to remove solids such as salts. Coalescing filters may be used to remove a dispersed liquid phase from a polymer solution, typically as a cleaning step following washing. Both types of filtering are best accomplished at low polymer concentrations.

10.2.3.3 Coagulation
Coagulation refers to two distinct unit operations used to recover polymers. The first of these is the breaking of an emulsion, usually by adjusting the pH of the aqueous phase. The second definition is the one used throughout most of this chapter and refers to the precipitation of polymer from solution by adding a nonsolvent. This technique is commonly employed for laboratory preparations but presents a serious economic challenge when scaleup is attempted. For the process to be useful, the polymer must be obtained as a particulate solid, and this requires a low initial concentration of polymer in the solvent and a high ratio of nonsolvent to solvent.

10.2.3.4 Devolatilization
Devolatilization is the polymer industry term for the removal of volatile components from a polymer melt by subjecting it to elevated temperatures and reduced pressures. The volatile is typically residual monomer or a solvent used during the polymerization. When the solvent concentration is high, partial devolatilization can be accomplished just by boiling or flashing the polymer solution; but in the final stages, specialized devices such as vented extruders must be used. At low solvent

concentrations, devolatilization is diffusion controlled and is often the rate limiting process for the entire facility. The economics of devolatilization are usually important and drive the reactor design toward high polymer yields and lower solvent concentrations.

10.2.3.5   Drying

Drying refers to the removal of volatiles from particulate solids, which are typically coagulates or suspension beads. Although surface moisture is obviously removed in the process, the real aim of many drying operations is the removal of internally trapped components. The drying process is easy to understand and scaleup, but diffusivities are so low that very small or porous particles are necessary for economic feasibility. Fortunately, the coagulated fluff formed by precipitation with a nonsolvent is usually both small and porous.

Leaching is conceptually similar to drying, but the polymer particles are surrounded by a liquid rather than a gas. Leaching is occasionally used to diffuse reactants such as blowing agents into a polymer.

All the polymer purification and recovery processes have strong interactions with the reactor technology. Washing and filtration are two common purification steps that require low viscosities and consequently low polymer concentrations. Coagulation in the sense of nonsolvent precipitation is a combined purification and recovery step that also demands low polymer concentrations. Using drying as the final purification step, an entire polymerization process can be developed that deals only with relatively inviscid liquids and particulate solids. Unfortunately, this process is apt to be very expensive since internal solvent recirculation and recovery rates will be a factor of 10–100 higher than the polymer throughout.

Devolatilization is used for many bulk polymerizations. These typically require no clean-up steps other than removal of residual monomer or solvent. They are also capable of reaction to quite high polymer concentrations. Even when dilute solutions and clean-up are needed, devolatilization is still usually preferred over coagulation on economic grounds. However, most devolatilization techniques require high temperatures which may tax the stability of the resin.

10.2.4   Mechanical Design

In the early days of the polymer industry, the reactor engineer spent a major amount of time and energy designing his own specialized equipment for handling polymer melts. Gear and screw pumps, single- and twin-screw extruders, reactors, and special agitation schemes, all were designed and built on a one-of-a-kind basis. The patent literature is full of ingenious mechanical contrivances for accomplishing what is now done with more standardized components and mechanically simpler designs. This is clearly the trend of the polymer industry. If handling a polymer in its viscous melt form cannot be avoided, do it in simple, robust equipment with as few moving parts as possible. Specialized, custom equipment has too high a capital cost and too high a maintenance cost. It is hoped that some of the considerations voiced here will continue this trend toward simplicity rather than mechanical complexity.

Pilot plants are sometimes exceptions to the search for simplicity. The conventional pilot plant is dedicated to a single product and process and is thus a smaller-scale and somewhat more flexible version of the product facility that is to be built later. However, there is also the general-purpose pilot plant that is intended for the small-scale manufacture of many different polymers by many different processes. The typical aim in such pilot plants is not to develop the ultimate process for a known material but to make many new materials for property evaluations and market probes. Equipment for general-purpose pilot plants must obviously be very flexible and capable of operating under extreme conditions. Figure 10.2 shows an idealized reactor system suitable for the general-purpose pilot plant. Design features include the following:

Primary reaction vessel rated at 25 atm and 350°C.

Materials of construction suitable for strong acids, bases, and chlorides.

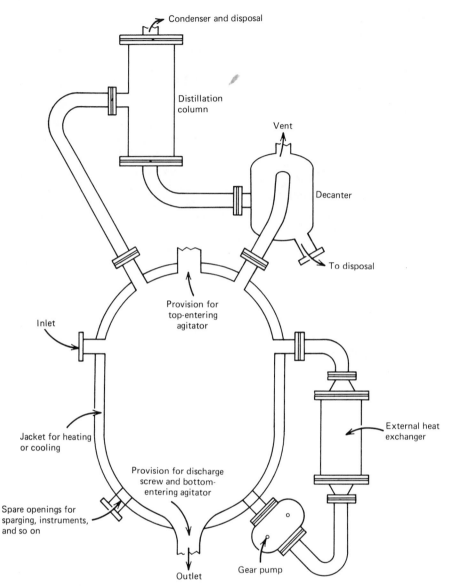

Figure 10.2 Polymerization reactor for general-use pilot plant.

Jacketing for heating or cooling.

Top-entering agitator (for turbines and other high-speed agitators).

Bottom-entering agitator with integral discharge screw (for helical ribbon and anchor agitators).

Overhead condenser with optional distillation column and decanter.

External circulation loop with heat exchanger.

Flexible flanging arrangements to allow in-strument and piping access throughout the vessel.

Such a vessel would rarely or never be operated with all components in place, but instead components would be selected as needed for a particular reaction. The reactor could not handle all possible polymerization schemes but could accomodate the more common ones. Batch or continuous operation and emulsion, suspension, solution, or bulk polymerizations would all be possible. The most notable limitations would be high vapor pres-

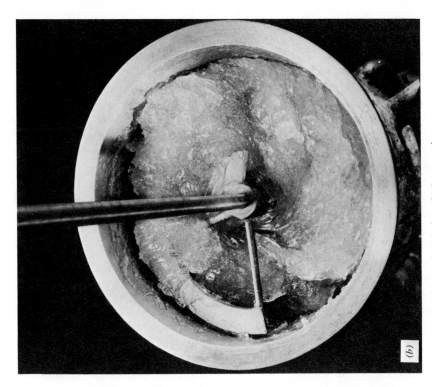

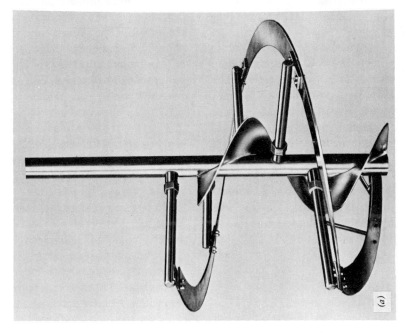

Figure 10.3 Helical ribbon agitator with reverse center screw. (*a*) Mechanical design. (*b*) In operation. Photographs Courtesy Mixing Equipment Company, Rochester, New York.

sures or very high polymer concentrations. Typically, general-purpose reactors are operated with polymer concentration of 75% or less so that downstream recovery equipment is required. A twin-screw extruder would be the best single choice, although the complete pilot plant would offer a variety of finishing devices.

The best all-purpose agitator even for polymerizations is the pitched blade turbine. Through the use of large turbine diameters (50–75% of the vessel diameter), quite viscous reaction masses can be accommodated. Above 100–1000 poise, however, the helical ribbon is preferred. Figure 10.3 illustrates the design and operation of a top-entering helical ribbon with reverse center screw. As suggested by Figure 10.2, bottom-entering agitators have some advantages for very viscous materials. When the agitator is rotated to give upflow at the walls, the corresponding downflow at the center helps to discharge the vessel. Indeed, the center screw flight can be continued beyond the bottom of the vessel so that the agitator shaft becomes a single-screw extruder.

With highly pseudoplastic or shear-thinning liquids, even the helical ribbon agitator may prove inadequate. It is entirely possible for the reaction mass to break loose at the walls and to rotate with the agitator, giving zero relative motion within the bulk of the fluid. The only solution to extreme problems of this type is a double-motion agitator such as the egg beater. Industrially, these are special and expensive designs, and usual practice is to alter operating conditions to avoid them. Usually, the reactor engineer has considerable freedom in picking a particular design option.

10.3   REACTOR OPTIONS

The simple monomers, ethylene, styrene, vinyl chloride, can be polymerized in myriad ways. Even complex polymerizations such as that for polycarbonate, although always a condensation process, can be carried out in several ways with several different starting materials. As a general rule, the vinyl addition polymers

have physical properties that are sensitive to the specific polymerization route, whereas those for condensation polymers tend to be insensitive. This section does not attempt a discussion of properties by way of process, but merely enumerates the various reaction schemes that are now available to the reactor engineer. The classification scheme is based on the number and kind of phases entering the reaction.

10.3.1   Single-Phase Polymerizations

The free radical polymerizations of styrene and methylmethacrylate are two industrially important examples of homogeneous, liquid-phase polymerizations. Monomers and polymers are mutually soluble over the entire composition range, and free radical initiation is either thermal or by a homogeneous catalyst so that the reaction remains truly single phase throughout the entire polymerization. Other bulk polymerizations are single phase over limited composition regions. For example, polyvinylchloride is about 1% soluble in vinyl chloride at 30° C while the reverse solubility is about 25% monomer in polymer. Thus the bulk polymerization of vinyl chloride is single phase from 0–1% conversion and again from 75–100% conversion. The high pressure process for low-density polyethylene can be single phase at high conversions, and several industrially important polycondensations are either truly homogeneous or else are homogeneous except for reaction byproducts which form a second, insoluble phase and are thus effectively removed from further participation in the reaction. Examples include the nylons, polyesters, and polysulfones.

The key point about all single-phase polymerizations is that the polymer exists as a concentrated solution in a solvent. The solution will have extremely high viscosities, typically in the range 10–1000 poise for the polymerization portion of the process and much higher in the devolatilization section. Such high viscosities lead to obvious difficulties in handling the fluid, in promoting good mixing, and in removing the heat of polymerization. Means of conducting reactions in highly viscous fluids are the primary topic of this chapter.

Coupled with the high solution viscosities are moderate-to-low diffusivities. A typical diffusivity of a monomer or solvent molecule remains moderately high at about 10^{-6} cm^2/sec, but the diffusivity of polmyer chains can easily be as low as 10^{-8} cm^2/sec. When the termination mechanism involves reactions between long-chain molecules, these low diffusivities can lead to a dramatic increase in polymerization rate, which is known as the gel effect (11, 15, 16). The initiation rate is unaffected by the gel effect, but the termination rate decreases, leading to a net increase in total radical concentration and to a consequent increase in the rate of polymerization. The usual rule in free radical polymerization is that any increase in reaction rate must be accompanied by a decrease in average molecular weight, but the gel effect is an exception. It represents the unique situation where viscosities and low diffusivities are an aid to reactor design rather than a hindrance.

Homogeneous gas and solid-phase polymerizations are not industrially significant. The gradual crystallization of glasses and amorphous semiconductors can be considered a type of solid-phase depolymerization, but such reactions are outside the traditional boundaries of polymer reaction engineering.

10.3.2 Heterogeneous Polymerizations

It is useful to divide heterogeneous polymerizations into two broad categories: those inherently heterogeneous by accident of chemistry and those intentionally devised to be heterogeneous to ease the difficulties of handling viscous polymer solutions. The most important example in the first group is the common situation where monomer and polymer have limited miscibility. The second category includes suspension and emulsion polymerizations. The following sections classify the industrially important polymerizations according to the types of phases involved. In most cases only two phases coexist. For the occasional three-or-more-phase system, classification will be based on the two phases that

are most important in a reaction engineering sense.

10.3.2.1 Gas-Liquid

The most important gas-liquid two-phase polymerization is the high pressure polyethylene process in which gaseous ethylene at about 2500 atmospheres is free-radical polymerized to give a liquid polymer phase. Industrial processes are continuous. They are usually conducted in a tubular reactor, but a stirred tank geometry is occasionally used. Copolymerizations with other vinyl monomers such as acrylic acid or ethylacrylate can be done in the same equipment. The liquid polymer phase contains substantial quantities of dissolved ethylene so the reaction may become single-phase at high conversions. The gaseous phase reappears, however, when the pressure is reduced. This pressure reduction is done step-wise to recover high pressure ethylene for recycling, and the economics of compression play a major part in the total conversion cost. With the last flash to near atmospheric pressure, liquid polyethylene is recovered that is substantially free of dissolved ethylene. Addition devolatilization may, however, be needed to remove less volatile comonomers.

Other gas-liquid polymerizations include condensation reactions that evolve gaseous by products such as water or HCl. The synthesis of polyphenylene oxide uses gaseous oxygen or air as the reactant for oxidative coupling.

In all the gas-liquid polymerizations, the liquid phase is a viscous polymer solution, and the characteristics of this solution dominate the reactor design problem.

10.3.2.2 Gas-Solid

The sole industrial example of a gas-solid polymerization is the fluidized bed process for high-density polyethylene. A heterogeneous coordination catalyst is suspended in gaseous ethylene. As polymer is formed, it creates a solid particle around the original catalyst nucleus. A rotary valve at the reactor exit gives free-flowing, solid polymer from gaseous ethylene in a single, elegant step. The reactor effluent is unstabilized and in fluff form but is otherwise finished polymer. One can ultimately expect that fabrication processes for high volume articles such as extruded pipe or blow-molded bottles will be coupled directly to the polymerization reactor.

The gas phase process for high-density polyethylene is a triumph of polymer reaction engineering and coordination metal chemistry. The catalyst must be sufficiently active so that residues can be left in the polymer without degrading properties. Also, it must yield polymer of the desired molecular structure. So far these combined characteristics have not been found with systems other than high-density polyethylene, although there is a recently announced process for making a low-density polyethylene structure using a low pressure, coordination catalyst route. If polymer physical properties are indeed similar to those obtained in high pressure polymerizations, this new process should eventually dominate the entire polyethylene business.

The analog of the gas-phase route to polyethylene has been commercially attempted for polypropylene but so far without success. The intrinsic polymerization rate for propylene is much slower than that for ethylene, and polypropylene is formed in two isometric types, syntactic and atatic, which must be separated after polymerization. An economically successful gas-phase process would require a catalyst combining improved yield with improved selectivity to avoid all separation and purification steps following the polymerization. This is a likely future development, and a fluidized polypropylene process can be forecast with some degree of certainty. However, owing to the relative ease of liquefying propylene, the ultimate process may be liquid-solid rather than gas-solid.

10.3.2.3 Liquid-Liquid
Liquid-liquid polymerizations are of three major types: monomer-polymer, polymer-polymer, and polymer-suspending fluid.

Monomer-Polymer. The monomer-polymer type results from the limited solubility of the polymer in its own monomer. The vinyl chloride polymerization is a typical example, having two liquid phases over most of the composition range. Polymerization typically occurs in both phases simultaneously, but, because the environments are quite different in terms of polymer concentration, the polymerization rate and molecular weight distribution will be different in the two phases.

Polymer-Polymer. The polymer-polymer class of liquid-liquid polymerizations is represented by various processes for making impact polystyrene. The reaction starts with a solution of polybutadiene rubber in styrene monomer. Free radical polymerization gives homopolystyrene, which is insoluble in the rubber-styrene solution and thus precipitates. However, styrene monomer is a good solvent for polystyrene so that the precipitated polymer does not remain a particulate solid but becomes a second polymer solution. The two-polymer solution phases coexist throughout the remainder of the polymerization. Initially, the rubber-styrene phase is continuous, and the polystyrene-styrene phase is distributed as discrete droplets. Continued polymerization increases the volume fractions of the polystyrene phase until phase inversion occurs. After this, the rubber phase is distributed and ultimately becomes the impact modifier in the finished polymer. At the point of phase inversion, which occurs when the polystyrene concentration is roughly equal to the rubber concentration, some polystyrene droplets may remain occluded in the rubber phase. Figure 10.4 illustrates this situation. To an extent, the existence of the occluded polystyrene is beneficial, and control of the number and size d of the occlusions is an important aspect of reactor design, as is control of the rubber particle size D. In bulk processes for impact polystyrene, reaction through phase inversion can be done as part of the overall, continuous process. In suspension processes, phase inversion is done in a bulk, batch polymerization prior to suspension.

Polymer-Suspending Fluid. The polymer-suspending fluid liquid-liquid polymerization is represented by suspension and emulsion

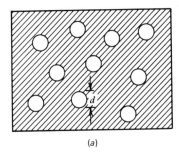

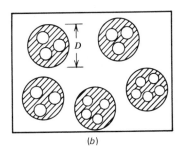

☐ Polystyrene-styrene solution

▨ Polybutadiene-styrene solution

Figure 10.4 Impact polystyrene process (*a*) before phase inversion, (*b*) after phase inversion.

processes. In suspension polymerizations, the suspending fluid plays little or no direct role in the chemistry. Salts and other surface active agents are used to stabilize the suspension and to achieve the desired bead size but have little effect on the polymerization kinetics. Each suspension bead acts as a small, batch reactor with negligible thermal gradients. The reaction exotherm is removed by way of the suspending fluid and, if desired, the entire system could be kept isothermal by external heat exchange. Usual practice, as typified by large-scale polystyrene processes, is to conduct a bulk, batch reaction up to about 20% total polymer concentration. This moderate-viscosity solution is then suspended and reacted to near completion using chemical initiators. A key engineering problem is to design the agitation system to keep the beads suspended (17, 18). At the end of the batch reaction cycle, the polymer content is so high that the polymer solution passes through the glass transition temperature and becomes a solid. The solid polymer is then separated, dried, and extruded.

In emulsion polymerization the suspending fluid does have a part in the chemistry. Free radicals are selectively formed in the aqueous phase where they react with dissolved monomer to give a growing polymer chain. This chain, which has very limited solubility in water, precipitates in the latex particles where it continues to grow until the free radical is terminated. The aqueous phase acts as a site for free radical generation, for some polymerization, and as a mass transfer medium for monomers, in addition to the heat transfer and particle suspension roles it shares with the aqueous phase in suspension polymerizations. Batch emulsion polymerization is used for many vinyl chloride, vinyl acetate, and acrylic coating resins and for the largest volume of ABS production. Continuous emulsion polymerizations are used for styrene-butadiene rubbers. In the rubber industry the polymerization train consists of 10 or more continuous stirred tank reactors in series so that the reaction kinetics closely approximate those for a batch reactor. The theory of emulsion has been treated in considerable detail, see for example Poehlein and DeGraff (19), Nomura et al. (20), and Thompson and Stevens (21).

10.3.2.4 Liquid-Solid

Liquid-solid polymerizations occur when the monomer solubility in polymer is so small that the polymer precipitates as a particulate solid rather than as a polymer solution. The polymer must be below its glass transition temperature or else be such a viscous, nontacky fluid as to be considered a solid. Solvent processes for polyolefins typically satisfy this criterion. Liquid-solid polymerizations also arise in the terminal stages of suspension and emulsion polymerizations. Some polycondensation processes involve solid reactants, for example, the sodium sulfide in the polyphenylenesulfide polymerization, or solid by-products, for example the sodium chloride formed in the polysulfone polymerization reaction.

10.3.2.5 Solid-Solid

Solid-solid polymerizations by way of grafting reactions occur in reactive coextrusion. An early example was the manufacture of impact polystyrene by extruding a blend of a styrene-butadiene rubber with homopolystyrene.

10.4 FUNDAMENTALS OF REACTOR DESIGN

This section discusses a variety of theoretical techniques that have proved useful in the design and analysis of polymer reactors. In principle, these same techniques can be applied to any reacting system whether or not it involves polymers. However, the high viscosities that complicate the physical design of polymer reactors actually lead to simplifications in the theoretical treatment. Turbulence is rare in polymeric systems. Steady, laminar flow is the rule, and this allows the equations of motion to be solved in a rigorous and comparatively simple manner.

10.4.1 Transport Equations

From an abstract viewpoint, a chemical reactor is completely defined when the physical properties, reaction kinetics, and thermochemistry are known and when the boundary conditions on the reactor are specified. Velocities, temperatures, and compositions at any point within the reactor can be found by solving the partial differential equations that govern momentum, heat, and mass transfer. Until recently, one made this theoretically precise statement and then immediately turned to correlations, approximations, and simplified models since the real system was hopelessly complex for mathematical rigor. Now, however, modern computers can generate numerical solutions for many practical problems in reactor design.

Fluid velocities are found by solving a set of three simultaneous, nonlinear, partial differential equations known as the equations of motion or the Navier-Stokes equations. Temperatures are found by solving a single partial differential equation known as the equation of energy, and concentrations are found by solving one partial differential equation governing diffusive mass transfer for each component. Detailed descriptions of these equations for a variety of coordinate systems may be found in specialized textbooks on transport phenomena (22).

Complete specification of an arbitrarily general reactor requires solution of $4 + (N - 1)$ coupled, simultaneous, partial differential equations where N is the number of different molecular species involved in the reaction. The factor $N - 1$ results from a stoichiometric constraint. The full set of $3 + N$ equations is still too complex for even the fastest computer, but no practical reactor requires the full set to be solved. Typically, only one or two velocity components need be considered, and it is often adequate to treat the fluid as if it consisted of only two chemical species, monomer and polymer. These simplifications give a set of three to five coupled equations that is quite possible to solve. Some literature results are discussed later; we now describe those general aspects of the transport equations that are of special interest to polymer reactor engineers.

10.4.1.1 Equations of Motion

In tubular reactors there is one dominant velocity component $V_z(r)$. If all other components can be ignored, the equations of motion reduce to a single ordinary differential equation.

$$\frac{-dP}{dz} = \frac{1}{r} \frac{d}{dr} (r\tau_{rz}) \qquad (10.11)$$

where $-dP/dz$ is the pressure drop per unit length of tube, r is the radial coordinate, and τ_{rz} is the only nonzero component of the fluid stress tensor. For Newtonian fluids, τ_{rz} is proportional to the radial velocity gradient.

$$\tau_{rz} = -\mu \frac{dV_z}{dr} \qquad (10.12)$$

which is known as Newton's law of viscosity. For constant μ, Equation (10.11) becomes

$$\frac{dP}{dz} = \mu \left[\frac{1}{r} \frac{dVz}{dr} + \frac{d^2 V_z}{dr^2} \right] \qquad (10.13)$$

The boundary conditions are $V_z(R) = 0$ and $dV_z/dr = 0$ at $r = 0$, and the solution is the well known parabolic velocity distribution for laminar, Newtonian flow in a circular tube

$$V = V_0 \left(1 - \frac{r^2}{R^2}\right) \qquad (10.14)$$

where $V_0 = (R^2/\mu)(-dP/dz)$ is the centerline velocity.

Most polymer melts and concentrated solutions are non-Newtonian, so Equation (10.12) should be replaced with a more complex description of the fluid rheology. A simple example is the power law fluid which gives velocity profiles of the form

$$V = V_0 \left[1 - \left(\frac{r}{R}\right)^{n+1/n}\right] \qquad (10.15)$$

where $n < 1$ for most polymer solutions. Power law behavior gives velocity profiles somewhat flatter than those for Newtonian flow, but the effect is usually rather small compared to other complications typical of polymer reactors. Indeed, polymerization reactors often have such large radial gradients in temperature and composition that the minor effect of non-Newtonian rheology is overwhelmed (23).

The equations of motion are coupled to the equations of energy and continuity through viscosity, which is usually a strong function of temperature and composition. The L/D ratio in tubular reactors is normally quite high so that the viscosity will vary rather slowly down the length of the reactor compared to its rate of change in the radial direction. If μ is regarded as a function of r but not of z, one-dimensional velocity profile results. For $\mu = \mu(r)$, Equation (10.12) is substituted into Equation (10.11) to give

$$\frac{dP}{dz} = \mu \left[\frac{1}{r}\frac{dVz}{dr} + \frac{d^2Vz}{dr^2}\right] + \frac{d\mu}{dr}\frac{dVz}{dr}$$

$$(10.16)$$

This contains a viscosity gradient term $d\mu/dr$ that did not appear in Equation (10.13). Whenever transport properties such as μ can vary from point to point, such gradients

appear, and it is necessary to derive the specific transport equations using the more general results presented in standard textbooks (22). It is not sufficient, for example, merely to substitute $\mu = \mu(r)$ into Equation (10.13).

It is perhaps worth noting that Equation (10.11) could have been integrated directly to give

$$\left(\frac{-dP}{dz}\right)\frac{r^2}{2} = r\tau_{rz} = -r\mu\frac{dVz}{dr} \quad (10.17)$$

where the constant of integration has been set to zero to satisfy the boundary condition that $dV_z/dr = 0$ at $r = 0$. However, the equivalent second-order Equation (10.13) better illustrates general properties of the Navier-Stokes equations.

Equations (10.16) and (10.17) are rigorously correct only if μ is a function of r but not of z. They also happen to be quite accurate approximations when μ varies slowly with z as will be true in most tubular reactors. The reason for this is that the entrance length in laminar flow is quite short

$$\left(\frac{L}{D}\right)_{\text{entrance}} - 0.035\, N_{Re} \qquad (10.18)$$

where N_{Re} is the Reynolds number. $N_{Re} < 1$ is typical for polymeric systems. Thus the velocity profile can change from a flat to a parabolic distribution in about 0.04 of a tube diameter. Whenever μ and thus V_z vary with radial position, there must be some nonzero radial flow V_r, but it will be small compared to V_z in most real situations. This means the two-dimensional equations of motion are not needed to obtain a good estimate of $V(r, z)$. Instead, the fluid continuity equation, Equation 10.19, can be used to find V_r as follows:

$$\frac{\partial V_r}{\partial r} + \frac{V_r}{r} + \frac{\partial Vz}{\partial z} = 0 \qquad (10.19)$$

Equations (10.16) and (10.19) together provide an estimate of the fluid dynamics that is sufficient for most reactor design purposes. If the two-dimensional equations of motion are really needed, numerical computation is still feasible but tends to be an order of magnitude more complex. Three-dimensional flow problems are at least two orders of magnitude more complex and are just now becoming tractable.

10.4.1.2 Equation of Energy

To determine viscosity as a function of position, it is necessary to find the temperature and composition as a function of position. The temperature is found from the equation of energy, Equation (10.20).

$$\rho C_v V_z \frac{\partial T}{\partial z} = \frac{\kappa}{r} \frac{\partial T}{\partial r} + \kappa \frac{\partial^2 T}{\partial r^2} + \frac{d\kappa}{dr} \frac{\partial T}{\partial r}$$

$$+ \mu \left(\frac{\partial V_z}{\partial r} \right)^2 \quad (10.20)$$

which is based on approximations similar to those used for Equation (10.11). One assumption is that $T(r, z)$ varies slowly with z so that conduction in the z direction can be ignored. The thermal conductivity κ tends to be a weak function of temperature and composition so that its gradient, $d\kappa/dr$, can usually be ignored. Also usually negligible is the viscous dissipation term $\mu \, (dV_z/dr)^2$, although it can be important in moving-wall devices such as extruders.

10.4.1.3 Equation of Continuity

The final partial differential equation is the equation of continuity which one here writes for component A as

$$V_z \frac{\partial C_A}{\partial z} = \frac{D_A}{r} \frac{\partial C_A}{\partial r} + D_A \frac{\partial^2 C_A}{\partial r^2}$$

$$+ \frac{\partial D_A}{\partial r} \frac{\partial C_A}{\partial r} + R_A \quad (10.21)$$

where C_A is the concentration of species A, D_A is the diffusion coefficient of A in the reacting mixture, and R_A is the reaction rate of A. In tubular reactors it is frequently possible to ignore diffusion so that the solution of Equation (10.21) is just the integral of the reaction rate down the fluid streamlines. For N components, $N - 1$ equations of the form of Equation (10.21) are required. For many cases, only two components, monomer and polymer, are treated. Typically, species A represents the monomer, and the polymer concentration is found by difference.

Equations (10.16), (10.19), (10.20), and (10.21), together with their boundary conditions, constitute an adequate description of most continuous flow polymer reactors that are reasonably tubular in nature. They have been written for the specifics of cyclindrical coordinates with θ symmetry and only one velocity component but they can be developed for other situations in a straightforward manner. The validity of the transport equations for describing heat transfer to non-Newtonian fluids in laminar flow has been confirmed experimentally (24, 25). Numerical results show excellent agreement with experimental heat transfer coefficients, while empirical correlations such as the Sieder-Tate equation show marked deviations, particularly with cooling at the tube wall.

In rectangular coordinates, correspondingly simple equations govern flow between flat plates, as in a parallel plate heat exchanger (26). Flow situations involving two and even three velocity components can be treated by using the full set of Navier-Stokes equations, and some results are cited later in this chapter. Batch reactors can also be treated in the same conceptual framework if the distance along the axis of the continuous flow tubular reactor is replaced by the time spent in the batch reactor.

10.4.2 Reactor Analysis Techniques

Given a detailed geometric design, the transport equations provide a rigorous method for estimating reactor performance. However, they offer little help in deciding what sort of reactor geometry is appropriate for such detailed study. Other approaches are needed to obtain a broad conceptual understanding of polymer reactor phenomena, to pick the reactor type best suited for a particular polymerization, to understand reactors that are still too complex for rigorous analysis, and to understand what went wrong when a reactor fails to operate as designed.

10.4.2.1 *Residence Time Distributions*

The yield of an isothermal homogeneous reaction can be closely approximated from knowledge of the batch kinetics and from knowledge of the residence time distribution in the reactor. A few polymer reactions do satisfy the isothermal and homogeneous conditions. Even when these conditions are violated, the residence time distribution provides a measure of reactor flow patterns without requiring solution of the detailed hydrodynamics. The theory of residence time distributions is also a necessary starting point to understand oher analytical techniques of broader applicability.

All continuous flow reactors have velocity profiles that lead to differences in the time material spends in the reactor. Danckwerts (27) characterized this distribution of residence times by a cumulative probability function,

$F(t)$ = fraction of exiting material that remained in the reactor for a duration less than t

$F(t)$ is a nondecreasing function with $F(0) = 0$, since no material stays in the reactor for a duration less than zero, and with $F(\infty) = 1$, since all the material will eventually leave. The derivative of $F(t)$ is known as the differential distribution or frequency function and is defined by

$f(t)\, dt = dF\,(t)$
       = fraction of exiting material that remained in the reactor for a duration between t and $t + dt$

The only restrictions on $f(t)$ are that it be nonnegative, and that

$$\int_0^\infty f(t)\, dt = 1 \qquad (10.22)$$

Residence time distributions have become a standard tool of chemical engineering analysis (28, 29, 30) and a recognized means of studying mixing in polymer reactors (31) so that only a few aspects are treated here.

Like other probability distributions, residence time distributions can be characterized by their moments μ_k' which is defined by

$$\mu_k' = \int_0^\infty t^k f(t)\, dt = k \int_0^\infty t^{k-1}\,[1 - F(t)]\, dt \qquad (10.23)$$

where $k = 1, 2, \ldots$. With $k = 1$ the mean residence time $\bar{t}$ is obtained as Equation (10.24).

$$\bar{t} = \mu_1' = \int_0^\infty t f(t)\, dt = \int_0^\infty [1 - F(t)]\, dt \qquad (10.24)$$

For incompressible fluids, Equation (10.25)

$$\bar{t} = \frac{V}{Q} \qquad (10.25)$$

has been shown to be a fundamental identity for real systems (27, 32–34). Moments higher than the first are usually measured about the mean, Equation (10.26)

$$\mu_k = \int_0^\infty (t - \bar{t})^k f(t)\, dt \qquad (10.26)$$

The variance of σ_t^2, of the distribution is obtained when $k = 2$, giving

$$\sigma_t^2 = \mu_2 = \mu_2' - (\bar{t})^2 \qquad (10.27)$$

Piston flow reactors represents a special and limiting case where all fluid elements have the same residence time so that the variance is zero. For piston flow, one may write Equation (10.28)

$$F(t) = 0 \qquad\qquad 0 \le t < \bar{t}$$

$$\quad\ = 1.0 \qquad\qquad t \ge \bar{t} \qquad (10.28)$$

$$f(t) = \delta(t - \bar{t})$$

$$\sigma_t^2 = 0$$

Another special case arises for continuous

flow stirred tank reactors (described in Section 10.7) which have an exponential distribution of residence times as shown in Equation (10.29)

$$F(t) = 1 - e^{-t/\bar{t}}$$

$$f(t) = \left(\frac{1}{\bar{t}}\right)e^{-t/\bar{t}} \qquad (10.29)$$

$$\sigma_t^2 = (\bar{t})^2$$

In conventional reactor analysis, the stirred tank reactor is sometimes considered the opposite extreme to a piston flow reactor since most flow systems have variances in the range 0 to $(\bar{t})^2$. However, polymeric systems often fall outside this range, as demonstrated later.

The mean of a distribution can be considered a measure of size whereas the variance is the lowest-order moment that gives a measure of shape. It is frequently useful to scale residence time distributions by their means to give a dimensionless distribution that is sensitive only to shape

$$\theta = \frac{t}{\bar{t}}$$

$$\Phi(\theta) = F(\theta\bar{t})$$

$$\phi(\theta) = \bar{t}f(\theta\bar{t}) \qquad (10.30)$$

$$\bar{\theta} = 1$$

$$\sigma^2 = \frac{\sigma_t^2}{(\bar{t})^2}$$

where all moments are now dimensionless numbers. The dimensionless variance σ^2 is frequently used to fit theoretical models to experimental residence time measurements (28), but this turns out to be a bad suggestion for polymer systems with laminar flow and low diffusivities. Variances tend to be very large in such systems; theoretically they would be infinite if there were no molecular diffusion (35). A better approach is to use the entire measured distribution and to calculate what sort of one-dimensional velocity profile would

be needed to give the measured results. In a cylindrical coordinate system the one-dimensional velocity profile is $V_z(r)$ and is related to the residence time frequency function by Equation (10.31)

$$\frac{V_z(2\pi r \, dr)}{\bar{V}(\pi R^2)} = \frac{\bar{t} \, 2r \, dr}{t \, R^2} = f(t) \, dt \quad (10.31)$$

with the boundary condition that $t = t_0 = L/V_0$ at $r = 0$. Equation (10.31) is solved to give t as a function of r. Then since $V_z = L/t$, V_z can be found as a function of r. As an example, suppose it is found that

$$f(t) = \frac{\bar{t}^2}{2t^3} \qquad t > \frac{\bar{t}}{2} \qquad (10.32)$$

Then using Equation (10.31) gives

$$\frac{r^2}{R^2} = 1 - \frac{\bar{t}}{2t} \qquad (10.33)$$

and substituting $t = L/V_z$ gives Equation (10.34)

$$V_z = \frac{2L}{t}\left(1 - \frac{r^2}{R^2}\right) = 2\bar{V}\left(1 - \frac{r^2}{R^2}\right) \quad (10.34)$$

which is the velocity profile for Newtonian flow in a circular tube. It is apparent that experimental residence time distributions can be used in the same way to provide a numerical velocity profile, and that obtaining this is a major step toward understanding flow phenomena in the reactor.

Laminar flow and low diffusivities give dimensionless variances greater than 1. Another way in which large variances can be obtained is through a phenomenon known as bypassing. Consider the dimensionless residence time distribution for two piston flow reactors in parallel:

$$\phi(\theta) = \alpha\delta(\theta - \theta_1) + (1 - \alpha)\delta(\theta - \theta_2)$$

$$(10.35)$$

$$\bar{\theta} = \alpha\theta_1 + (1 - \alpha)\theta_2 = 1 \qquad (10.36)$$

$$\sigma^2 = \alpha(\theta_1 - 1)^2 + (1 - \alpha)(\theta_2 - 1)^2 \tag{10.37}$$

$$= (1 - \theta_2)^2 \, \frac{1 - \alpha}{\alpha}$$

where α is the fraction of the total volumetric flow that is fed to reactor 1. The mean for reactor 2 is constrained by Equation (10.36) to the range $0 \leqslant \theta_2 \leqslant 1/(1 - \alpha)$. The variance has the range from zero $(\theta_2 = 1)$ to either $(1 - \alpha)/\alpha$ or $\alpha/(1 - \alpha)$, whichever is larger. Bypassing or short-circuiting is said to exist whenever some portion of the fluid has a residence time much shorter than the mean. This situation arises with the elongated velocity profiles sometimes found in tubular polymerizers.

10.4.2.2 Micromixing

Even for isothermal reactions, the residence time distribution is adequate to predict yields only if the reaction is first order. For other types of reactions, further information on mixing patterns is needed for a precise yield prediction although relatively close bounds on the yield can be calculated from knowledge of the residence time distribution alone. The theory behind these statements is called "micromixing theory." It springs from the observations of Danckwerts and Zwietering (33, 36, 37) that fixing the residence time distribution still leaves a degree of freedom on molecular level mixing possible within a reactor.

The residence time distribution merely de-fines what Levenspiel (28) has termed "macromixing" and does not define the molecular scale mixing that results from diffusion and that is called micromixing. Macromixing refers to those gross flow processes that cause different fluid elements to have different residence times. These same flow processes may or may not cause such intimate comingling of materials that diffusion becomes important so that micromixing occurs. Thus the beads in a continuous suspension process can show a distribution of residence times without any mixing on the molecular scale. Such a system is said to be "completely segregated," and complete segregation represents one limit on micromixing, that where there is no molecular mixing at all. There is also an upper limit on micromixing that corresponds to the maximum amount of molecular level mixing possible with a given residence time distribution. Zwietering (33) called this condition "maximum mixedness."

The state of complete segregation is easily imagined for any residence time distribution; simply consider the individual fluid elements to be encapsulated so that they experience the gross flow patterns of the vessel without any interchange of molecules. Alternatively, a completely segregated reactor can be modeled as a piston flow reactor with multiple side exits as shown in Figure 10.5a. These side exits can be so distributed as to match an arbitrary residence time distribution. From a yield viewpoint, this arrangement is equivalent to treating the system as a large number of piston

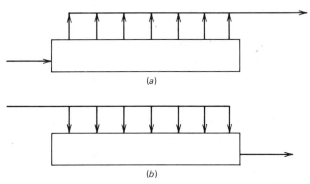

(a)

(b)

Figure 10.5 Piston flow reactor models. (a) Completely segregated reactor. (b) Maximum mixedness reactor.

flow elements in parallel and is thus a completely segregated reactor.

The other limit of micromixing, namely the condition of maximum mixedness, seems harder to visualize. Zwietering, however, showed that it could be modeled as a piston flow reactor with multiple side entrances. See Figure 10.5b.

These two piston flow models have the same residence time distribution but in general will predict different reaction yields. In the completely segregated reactor, mixing between fluid elements that have spent different times in the vessel occurs as late as possible, namely, in the exit stream. In the maximum mixedness reactor, this mixing occurs as early as possible, namely, at the various entrance points.

Zwietering showed that complete segregation and maximum mixedness represent bounds on the amount of mixing possible between molecules that have spent different times in the reactor. They also provide absolute limits on conversion under rather broad conditions. Assuming premixed feed, Chauhan, Bell, and Adler (38) have shown that complete segregation gives the highest possible conversion when the rate expression R_A is concave-up, $d^2 R_A / dC_A^2 \geqslant 0$; and maximum mixedness gives the highest conversion when R_A is concave-down, $d^2 R_A / dC_A^2 \leqslant 0$. If $d^2 R_A / dC_A^2 = 0$ the rate expression is linear, the reaction is first order, and yield is independent of micromixing.

Novad and Thyn (39) give generalized plots that show the complete segregation and maximum mixedness conversions for various reaction orders and residence time distributions. It turns out that the largest difference in conversion for a second-order reaction occurs in a stirred tank reactor and is only about 7% (40). This small difference explains the difficulties encountered by early workers (41, 42) who attempted to measure micromixing by performing second-order reactions in stirred tanks. The difficulty is aggravated by the fact that stirred tank reactors in the normal regime of turbulence and high diffusivities tend to be quite close to maximum mixedness. However, some degree of segregation can be found with unmixed feed, gentle agitation, or vary fast

reactions. Departures in yield from that of a maximum mixedness reactor provide a measure of micromixing that in turn can be used to evaluate parameters in a model.

A complete model of an isothermal continuous flow reactor must obviously predict the extent of both macromixing and micromixing. The usual approach to this problem is to use any conventional submodel for the residence time distribution and to devote one additional parameter for predicting levels of micromixing possible with this residence time distribution. The simplest way to create such a composite macro/micromixing model is just to treat the system as two different reactors in parallel. The two reactors have identical residence time distributions, but one is completely segregated and the other is in maximum mixedness. The micromixing parameter represents the division of entering material between the two reactors. This model, which was suggested by Methot and Roy (40), can be considered part of a general class of "two-environment" models first discussed by Ng and Rippin (41, 43) and Weinstein and Adler (44). However, in the usual formulation of the two-environment model, the division between the segregated and maximum mixedness regions is based on the age of a fluid element. Typically, fluid enters the reactor in the segregated environment and leaves from the maximum mixedness environment (45). Chen and Fan (46) have proposed a reversed two-environment model as more representative of polymer reactions. It is supposed that the reactants are initially well mixed but become segregated as viscosities increase.

There are other models of combined macromixing and micromixing that take a more mechanistic approach to mixing in turbulent flow. In the axial dispersion model (28), for example, the Peclet number simultaneously defines the residence time distribution and the degree of segregation. Within a limited range this model is also valid for laminar flow of non-Newtonian fluids (47, 48). The model of Manning, Wolf, and Keairns (49) is representative of those suitable for stirred tanks in the turbulent regime.

The theory of micromixing was really developed as a solution in search of a problem. Applications were relatively late in coming since the one-step reactions initially studied were insensitive to micromixing and since most systems tended to be near the limit of maximum mixedness except when operated outside the usual design range. For polymer reactors, however, performance criteria other than the conversion of monomer can be quite important, and for some of these criteria the level of micromixing is more significant than the level of macromixing. The theory of micromixing is particularly applicable to complex, multistep reactions such as polymerizations. It is also especially applicable to laminar flow systems with low diffusivities.

Micromixing is a phenomenon closely associated with the stirred tank reactor since the exponential residence time distribution gives the greatest freedom for variations in the extent of molecular mixing. The droplet diffusion model due to Nauman (31, 50) is well suited for studying segregation in stirred tank polymerizations. In common with the various two-environment models, a single parameter allows the system to range from complete segregation to maximum mixedness. However, Nauman's parameter, which is known as the segregation number, has a physical basis and can be approximated from *a priori* considerations. Also, this model allows not for just two environments but for a continuity of environments, although the nature of the polymerization process will usually lead to increasing degrees of segregation for those fluid elements that have been in the reactor for the longest times.

The segregation number is defined as Equation (10.38)

$$(N_{seg})_A = \frac{R^2}{\pi^2 D_A \bar{t}} \quad (10.38)$$

where R is a droplet diameter that characterizes the system and D_A is the diffusivity of species A. As suggested by the notation, the model is perfectly capable of treating situations where different molecular species have different diffusivities and hence different degrees of segregation. Thus a polymerization vessel may be well mixed with respect to monomers having typical diffusivities of 10^{-6} cm^2/sec but be quite segregated with respect to polymer chains having diffusivities of 10^{-8} cm^2/sec.

10.4.2.3 *Thermal Time Distributions*

In contrast to stirred tank reactors, which tend to be isothermal, tubular reactors usually have large temperature gradients both axially and radially. The residence time distribution remains useful to elucidate flow patterns but is not helpful in predicting yields. Indeed, improper use of residence time information can be misleading since the differences in temperature will usually have a larger effect on yield than differences in residence time. For reactions in nonisothermal systems, the thermal time distribution is the analog of the residence time distribution (51).

If the reaction can be characterized by a single activation energy, which is usually the case for polymerizations, then knowledge of the thermal time distribution can be used to predict the yield of a first-order reaction precisely and to closely bound the yield for reactions of order other than first.

The thermal time is defined as the integral of the Arrhenius temperature dependency along fluid streamlines,

$$t_T = \int_0^t \exp\left[\frac{-E}{R_g T(t')}\right] dt' \quad (10.39)$$

When $E = 0$ the thermal time is identical to the residence time of a fluid element following the same streamline. When $E > 0$ the thermal time weights temperatures and times spent along the streamline in exactly the manner that the reaction weights these factors.

The thermal time distribution for a reactor can be characterized by a cumulative probability function as follows:

$H(t_T) = $ fraction of exiting material that experienced a thermal time less than t_T

$H(t_T)$ is the analog of $F(t)$; it is nondecreasing with $H(0) = 0$ and $H(\infty) = 1$. The thermal time frequency function $h(t_T)$ is the derivative of $H(t_T)$ just as $f(t)$ is the derivative of $F(t)$. Similarily, moments of the thermal time distribution can be defined, and it can be shown (51) that the equations used for predicting or bounding the yield in a nonisothermal reactor are identical to those used for isothermal reactors except that $h(t_T)$ is substituted for $f(t)$. In fact, use of the thermal time distribution allows a system to be discussed and analyzed just as if it were isothermal. For example, the analog of bypassing exists and turns out to be quite common in fixed-wall heat exchangers.

A major difference between thermal time distributions and residence time distributions lies in their measurement. Residence times can be measured with tracers or can be calculated from theoretical flow models. No experimental technique is yet available for thermal times. They only can be calculated using combined thermal and flow models, that is, by solving the equations of motion and energy. This solution will tend to be specific for a particular system if the flow patterns or temperature profiles are strongly coupled to the extent of reaction; but even here the thermal time distribution is a useful conceptual tool for explanation and analysis of reactor phenomena. It becomes significantly more useful for that class of problems where the extent of reaction does not significantly affect the equations of motion or energy.

The field of polymer reaction engineering includes examples of both situations. Polymerizations, and particularly bulk polymerizations, usually show a dramatic influence of the extent of reaction on temperature and velocity profiles. Here, the thermal time distribution—like the residence time distribution—will be system specific and useful only in an interpretative mode. On the other hand, the processing of polymers in extruders, heat exchangers, and fabrication equipment may involve more subtle chemical reactions that have relatively little influence on the temperature and flow patterns. Industrially important examples include the modification of polymer end groups and side chains, thermal depolymerization, and the activation of chemical blowing agents.

10.5 BATCH REACTORS

Batch reactors are commonly employed for emulsion and suspension polymerizations, for the solution or bulk polymerization of most condensation polymers, and for a few vinyl polymers. The bulk and solution processes directly confront the issue of handling viscous polymer solutions. A typical batch cycle consists of charging the monomers at ambient conditions, heating to initiate reaction, polymerization accompanied by a strong exotherm, and either termination or transfer to other reaction equipment. Condensation polymers are often terminated in the primary reaction vessel by sparging in a liquid or gaseous reactant to cap active end groups. Sometimes it is useful to quench the reaction by dilution with cold solvent before end-capping. Vinyl addition polymers and equilibrium-limited condensation polymers are usually discharged from the primary vessel while still reacting. Early polystyrene processes used an agitated batch reactor followed by an unagitated plate and frame device where the polymerization went to near completion. Bulk batch reactors are still used for polymethylmethacrylate, for some varieties of polyvinylchloride, and for the initial polymerization in the suspension polystyrene process.

The reactants in batch polymerizations are usually well mixed before reaction begins. The entire contents of the vessel are heated, initially by external heating and later by the reaction exotherm; but until viscosities become very high, it is possible to maintain good mixing and a uniform temperature within the vessel. Thus the typical batch reactor is nonisothermal but internally uniform so that no concentration gradients develop. The mathematics of such reactors are identical to those for an ideal piston flow reactor that has axial temperature and composition gradients but no radial gradients. The reaction can be controlled by means of the initial stoichiometry, by adding reactants or dilutents, or by using a programmed temperature cycle. The last possibility introduces an opportunity for functional optimization, a topic discussed in the section on tubular reactors.

If a batch polymerization reaches the point where significant temperature gradients exist within the vessel, there is usually little choice but to let the reaction go on to completion. Typically, the contents are too viscous to terminate or discharge, and the final stages of polymerization are essentially uncontrolled. Although rupture discs may be used to vent reactors under runaway conditions, the safer approach is to design the vessel for containment. A runaway styrene polymerization will reach an equilibrium-limited end point at about 350° C and 25 atm, and thus styrene is about the most energic monomer for which adiabatic containment is a feasible strategy. Otherwise, surface to volume ratios should be made sufficiently high so that destructively violent runaways are not possible. It is worth noting that a continuous flow reactor usually becomes an unagitated batch reactor during a power failure.

Qualitatively, thermal and compositional inhomogeneities become important in a batch reactor when the batch blending time is commensurate with the reaction half-life. The batch blending time is a semiquantitative measurement based on visual observation of a tracer. Norwood and Metzner's correlations (52) include laminar flow and can thus be used for batch polymerizers. Studies on blending times, pumping capacities, and power requirements are available for non-Newtonian laminar flow in agitated vessels for a wide variety of agitator types and geometries. See Bates, Fondy, and Fenic (53), Gray (54), Johnson (55), and Sawinski, Havas, and Deak (56) for good summaries. These results also provide a starting point for specifying the fluid mechanical design of continuous flow reactors. Good agitation is most critical in a batch reactor when adding reactants to a partially polymerized mass, as when adjusting stoichoimetry or terminating the change. It can also be important in geometrically nonuniform initation schemes, for example, radiation-induced polymerizations (57). Generally, mixing requirements in the high-viscosity portion of the batch cycle will dictate the agitator design.

Isothermal batch reactors are the classic testing ground for theoretical polymerization kinetics and molecular weight distributions. Denbigh (58, 59) first showed that a batch reactor—or the mathematically equivalent isothermal piston flow reactor—will give a different molecular weight distribution than continuous flow reactors. For free-radial polymerizations where the lifetime of a growing polymer chain is short, the batch reactor gives a broader molecular weight distribution than (most) flow reactors. For condensation processes where the chain lifetime is long, batch reactors give the narrower distribution. The effects of macromixing and micromixing on molecular weight distributions are further discussed in Section 10.7.3.2.

10.6 TUBULAR FLOW REACTORS

The distinction between tubular reactors and stirred tank reactors is occasionally one of intent rather than realization. The usual design intent for a tubular reactor is to have progressive flow with little or no internal recycle, and to achieve a uniform distribution of residence times and thermal times. In exchange, the design engineer must accept large (axial) composition differences within the reactor that are often accompanied by large temperature differences. For stirred tank reactors, the usual design intent is to minimize composition and temperature differences by promoting mixing throughout the volume of the reactor. In exchange, the design engineer must accept high internal recycle rates and the exponential distribution of residence times, which is the nearly inevitable consequence of recirculation (60, 61).

In actual realization, it is all to easy to build a tubular reactor that has large variations in residence times or a stirred tank reactor that has large variations in composition. One theme of this chapter is to describe means by which the original design intent can be preserved. In the case of tubular reactors, this means to achieve a narrow distribution of residence times and thermal times.

The simplest tubular reactor is indeed just a straight, circular tube; or it can be the rectangular coordinate counterpart of a tube, which is the parallel plate heat exchanger. Such reactors with no moving parts are called fixed-wall devices and are certainly the easiest and cheapest to build. However, without turbulence to promote radial mixing, fixed-wall devices perform rather poorly as reactors.

There are a variety of techniques to promote radial mixing or flow redistribution in tubular reactors. Coiling the tube into a tight helix will introduce a secondary, centrifugally driven flow pattern that increases heat transfer and narrows the residence time distribution (62). There is a group of flow redistribution devices known as motionless mixers that achieve mixing by forcing the fluid through flow paths of cleverly chosen geometry. Surging caused by a periodic pressure release is used to redistribute flow in the high pressure polyethylene process (4), and wall temperatures can be used to change viscosities and reaction rates in order to narrow the residence time or thermal time distributions. Finally, of course, radial mixing can be achieved mechanically. This thought gave rise to the use of extruders as reactors and also generated a host of proprietary mechanical contrivances (63, 64). We class extruders and other agitated but more or less tubular vessels as moving-wall devices.

10.6.1   Fixed-Wall Devices

Fixed-wall devices are defined as tubes, parallel plates, and other flow geometries where every fluid boundary, meaning a point where $V = 0$, is stationary with respect to all other such boundaries. Simply put, there are no moving parts other than the fluid itself.

10.6.1.1   Undisturbed Flow
By undisturbed flow we mean reaction in a fixed-wall device where the reactor cross section remains constant in size and shape. This definition includes the ordinary circular tube and is the geometry that has been most widely studied. Literature references fall into two general classes.

1. Theoretical studies where the fluid mechanics are relatively simple and where the

real emphasis is on what might be called the fine structure of the polymerization, that is, molecular weight and copolymer composition distributions (65–67).

2. Papers where the main emphasis is on solution of the transport equations to give velocity, temperature, and composition profiles down the length of the tube. Composition can mean just gross yield of polymer (23, 68, 69, 70) but details of the molecular weight distribution are beginning to be considered (71, 72).

The first of these approaches is primarily useful for achieving conceptual understanding of polymerization phenomena, and no claim need be made that the assumed flow processes are especially realistic. Isothermal operation is also assumed so that the residence time and thermal time distributions are identical and can be calculated from the velocity profile using Equation (10.31). In the second approach, the transport equations must be solved simultaneously so that theoretical residence time and thermal time distributions are not available until the velocity, temperature, and composition profiles have already been calculated. However, they remain useful in an interpretative mode.

Polymerizations in tubular reactors are characterized by large radial gradients in composition and viscosity that lead to elongated velocity profiles. They are also characterized by large radial gradients in temperature and can exhibit thermal runaways with the reaction suddenly going to completion at very high temperatures (65, 68). The first systematic studies of these phenomena were done in industrial research organizations with results that were often discussed qualitatively but seldom published in detail (8, 31, 69). Lynn and Huff (23) were the first to give a detailed transport model for polymerization in a tube, and their results are typical for undisturbed flows in fixed-wall reactors. They more mechanistically oriented studies on polyethylene by Wallis, Ritter, and Andre (72) gave similar results. The overall conclusion is that incremental conversions above about 20% are infeasible in undisturbed tubular flow.

Even if tubular polymerizers could somehow be operated isothermally, differences in residence times are sufficient to give excessive composition differences between the tube wall and centerline. The composition differences in turn alter the velocity profile, giving larger differences in residence times and so on until the tube plugs. This type of flow instability exists in single tubes but is even more troublesome when scaleup is attempted by going to multiple tubes that are fed from a common header. The very real tendency is for unreacted monomer to flow through one tube while all other tubes are blocked with viscous polymer.

A numerical example may be useful to illustrate the kind of velocity profiles that can arise in tubular polymerizations. Suppose that viscosity varies exponentially with radial position so that

$$\mu = \mu_0 \exp\left(\frac{\alpha r}{R}\right) \qquad (10.40)$$

Equation (10.16) yields the velocity profile

$$V_z = V_0 \left\{ 1 - \frac{1 - [(\alpha r / R) + 1] \exp(-\alpha r / R)}{1 - (\alpha + 1) \exp(-\alpha)} \right\}$$

$$(10.41)$$

The range $10 < \alpha < 20$ is typical for a system with monomer at the centerline but with polymer at the walls; this corresponds to a viscosity ratio in the range 2×10^4 to 5×10^8. Figure 10.6 shows the velocity profiles calculated from Equation (10.34). Fluid at the centerline has velocities 17–67 times the average velocity and residence times of 1.5–5.9% of the mean. Such systems can certainly be said to exhibit bypassing.

Even in the absence of thermal effects, composition gradients can give rise to elongated velocity profiles and unfavorable residence times distributions. We now show the converse, that thermal effects in fixed-wall, laminar flow heat exchangers would give rise to very unfavorable thermal time distributions even if the velocity profile did not elongate but remained parabolic. Figure 10.7 shows the thermal time distribution for laminar flow

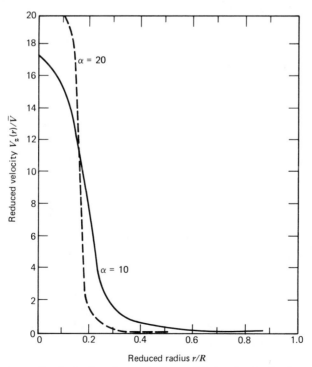

Figure 10.6   Velocity profiles for laminar flow and exponential viscosity distribution $\mu = \mu_0 \exp \alpha r / R$.

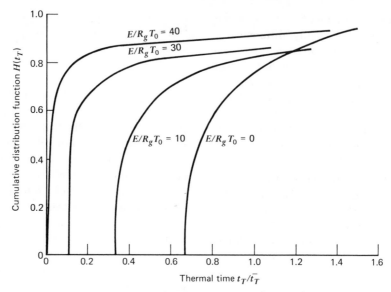

Figure 10.7 Thermal time distributions for parallel plate heat exchanger.

through a parallel plate heat exchanger. The parameters chosen for this literature example (51) correspond to only moderate heating: a 4.8% increase in average temperature (say 15° C) with a 20% wall temperature driving force (say 60° C). The various distribution curves in Figure 10.7 have all been scaled to have unit mean. The curve for $E/R_g T_0 = 0$ corresponds to the residence time distribution for isothermal flow between flat plates, whereas those with higher values of E/RT_0 are thermal time distributions. It is apparent that the nonisothermal environment is much worse from a reaction engineering viewpoint than would be expected based on just the velocity profile. This is due to an unfortunate coupling of temperatures and times that is characteristic of fixed-wall heat exchangers. Material near the wall experiences both the highest temperatures and the longest residence times while material near the centerline goes through so quickly and at such low temperatures that it effectively bypasses the reactor.

In gaseous reactors, molecular diffusivities are high enough to remove the deleterious effects of laminar flow. This is untrue for liquids and especially untrue for viscous polymer solutions. Lynn and Huff (23) found that diffusion has insignificant effects on yields and molecular weight distributions even when

unrealistically high diffusivities (10^{-5} cm^2/sec) were assumed. Merrill and Hamrin (73) proposed the general criterion that diffusion can be ignored in laminar flow reactors whenever

$$\frac{D_A \bar{t}}{R_{\text{tube}}^2} < 3 \times 10^{-3} \qquad (10.42)$$

With a typical monomer-in-polymer diffusivity of 10^{-6} cm^2/sec and with a 1 in. ID tube, this criterion indicates diffusion is negligible for mean residence times below 1.3 hr. With a 6 in. tube, the critical residence time becomes 2 days. For practical purposes, diffusion can be ignored in all but the smallest scale equipment.

10.6.1.2 Flow Redistribution Techniques

The foregoing considerations show that undisturbed flow in fixed-wall devices is a very poor way to conduct a polymerization. Some sort of flow redistribution is necessary, and the cheapest, most elegant way to achieve this is through the use of motionless mixers.

A motionless mixer is a static element installed in a tubular reactor that utilizes the energy of the fluid to induce mixing. There are now at least four types of commercial motionless mixers, and the operating principles and performance characteristics of some of these have been described in detail (74–81). The performance of motionless mixers is best understood by considering the two different kinds of mixing they promote: flow division and flow inversion. Figure 10.8a is a idealization of the flow division process as it occurs in a typical commercial unit. The fluid is divided, rotated, redivided, and so on in a process known as laminar simple mixing (82, 83). The striation thickness, which is the distance between layers of fluid that were initially adjacent, is halved after every stage of division and rotation. It varies according to the general formula, Equation (10.43).

$$\text{Striation thickness} = R_0(L)^{-m} \qquad (10.43)$$

Here R_0 is the initial striation thickness, typically the tube radius, m is the number of mixing elements, and L is a constant for a particular type of motionless mixer. Commercial examples have $L = 2$, 4, or 16, and even

with $L = 2$ it is apparent that the striation thickness can be reduced to molecular dimensions with a fairly small number of elements.

Flow Division.   The flow division type of motionless mixer provides an excellent reaction environment for single-phase polymerizations or polymer modifications. The major practical disadvantage is the high-pressure drop which can be a factor of 3–100 greater than for an open tube of the same diameter and with the same flow rate. As indicated in Figure 10.8a, commercial motionless mixers destroy radial and tangential symmetry so that rigorous theoretical treatment requires analysis of a three-dimensional flow. This lack of symmetry is not inherent to the flow division process as indicated by the conceptual possibility sketched in Figure 10.8b. However the practical difficulty remains.

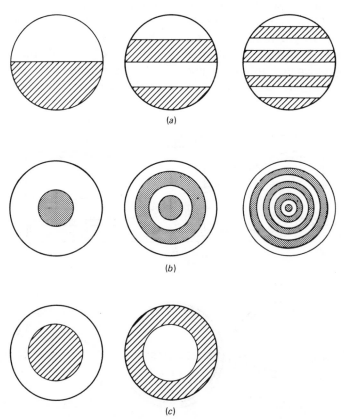

(a)

(b)

(c)

Figure 10.8   Motionless mixers. (a) Typical flow division process. (b) Flow division with radial and tangential symmetry. (c) Partial flow inversion.

Since motionless mixers cause high pressure drops, it is very tempting to use larger diameter pipes with only a few mixing elements or else to use short mixing sections followed by lengths of open pipe. Both these situations are difficult to analyze with commercial mixers. There is also a practical limitation with these approaches. If large composition or temperature gradients are ever allowed to exist, the commercial devices may not be able to rehomogenize the fluid. With large viscosity differences, even miscible fluids behave as if they were separate phases. The low-viscosity "phase" sometimes refuses to subdivide and instead channels down one side of the mixing element.

Flow Inversion. The other type of mixing possible with motionless mixers is flow inversion, the translation of fluid from regions near the tube wall to regions near the centerline and vice versa. In principal, a motionless mixer can be designed that completely inverts the flow, but the partial inversion represented by Figure 10.8c is the more practical case. Figure 10.9 shows one mechanical design for a partial flow inverter. A detailed description and performance calculations have been given else-

where (84). Flow inverters are used individually, interspersed between relatively long lengths of open tube. This means that pressure drops are moderate compared to the commercial motionless mixers. Reactor performance with flow inverters is also comparatively easy to analyze using the transport equations. For a single inverter the primary design parameter is R_I, the inversion radius illustrated in Figure 10.8c. Of course, the number and spacing of inverters must also be determined, and it is quite possible that the best R_I will vary with axial position.

10.6.1.3 Optimal Wall Temperatures

Accurate models of polymerization processes enable realistic optimization studies. The tubular reactor with flow inversion has a number of design parameters: the tube diameter and length, the wall temperature, and the number, spacing, and inversion radii of the flow inverters. The system will have constraints such as the maximum allowable pressure drop and the desired throughput and conversion, and there will be an objective function that is some suitable weighting of product properties (perhaps the polydispersity of the polymer) and product cost (perhaps the capital cost of the

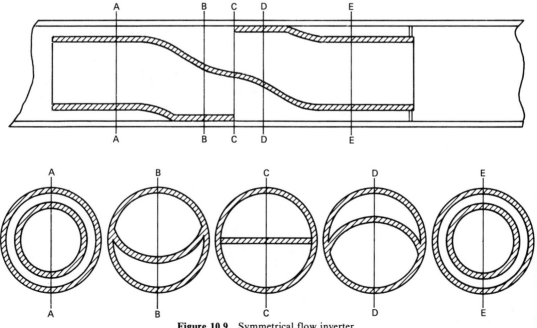

Figure 10.9 Symmetrical flow inverter.

reactor). Since the number of parameters is finite, standard multivariate optimization techniques can be applied to give the best possible design. Beveridge and Schechter (85) provide a comprehensive summary of such optimization techniques from a chemical engineering viewpoint.

If the wall temperature is allowed to vary continuously as a function of axial position, the problem becomes one of functional optimization rather than parameter optimization, and Denn's book (86) provides a good introduction, again with a chemical engineering flavor. In functional optimization, the aim is to specify a function $T_w(z)$ that will yield the best possible value for the objective function. In theory, specification of a function requires an infinite number of parameters since $T_w(z)$ can have a different value at every point in the range $0 < z < L$. However, it often turns out that the optimal function $T_w(z)$ has a very simple form, as given in Equation (10.44)

$$
\begin{aligned}
T_w(z) &= T_{max}, && 0 \leqslant z \leqslant z_s \\
T_w(z) &= T_{min}, && z_s \leqslant z \leqslant L
\end{aligned}
\tag{10.44}
$$

which is known as bang-bang control. The function represented by Equation (10.44) can be specified using only three parameters T_{max}, T_{min}, and z_s so that the functional optimization problem reduces to one of parameter optimization. A rigorous proof that bang-bang control is indeed optimal can sometimes be constructed using Pontryagin's maximum (or minimum) principle (86). However, the preferred numerical route is to assume a simple trajectory like that in Equation (10.44), to determine the parameters using standard multivariate search techniques, and then to perturb the assumed trajectory to see if any improvement in the objective function is possible. The general approach of solving a functional optimization problem by converting to an n-dimensional parameter search is known as the Rayleigh-Ritz method (85).

Axial variation in wall temperature represents a potential way of overcoming the flow distribution problems associated with tubular reactors. How this approach compares to the use of motion mixers remain unclear, probably the two methods will prove complementary.

Literature studies are beginning to appear on the application of functional optimization techniques to tubular or batch polymerizations. Kwon, Evens, and Noble (87), Nishimura and Yokoyama (88), Yoshimoto et al. (89), and Lee and Crowe (90), have studied minimum time-temperature trajectories for free-radial polymerizations, and Ray and Gall (91) have explored the use of temperature programs to control copolymer composition distributions in batch and tubular reactors. It can be anticipated that functional optimization will become increasingly important, particularly for tubular reactors and particularly for industrial design.

10.6.2   Moving-Wall Devices

Mechanical agitation was one of the earlier methods used to achieve radial mixing in tubular polymerizers, and extruders or extruder-like devises are still commonly employed for the final stages of equilibrium-limited polycondensations. Extruders are also used for many polymer modifications. They are the preferred device when any of the following is true:

It is necessary to melt or flux the polymer prior to reaction.

Volatiles must be removed concurrently with the reaction.

The reaction requires a more uniform thermal or composition environment than is practical in fixed-wall devices.

With the advent of motionless mixers, the third reason has lessened in importance, but extruders still provide an excellent reaction environment.

The flow patterns in single- and twin-screw extruders have been analyzed in detail. Pinto and Tadmor (92) derived the residence time distribution for the melt-fed single-screw extruder, and Nauman (51) extended the analysis to thermal time distributions. His treatment used a flow model without pressure generation which is correct for partially filled sections of

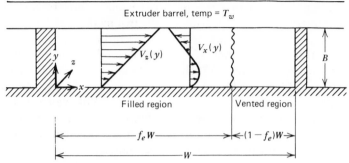

Figure 10.10   Vented section of single-screw extruder.

vented, single-screw extruders and is a reasonable approximation for the vented section of many multiscrew extruders. The polymer was heated through the barrel wall and also by viscous dissipation.

Figure 10.10 illustrates the flow model which was assumed to be independent of the thermal model. There are two velocity components: V_z (y) for flow down the screw channel and $V_x(y)$ for circulatory flow within the channel. In a partially filled screw, there is no pressure generated in the z-direction, and $V_z(y)$ is linear. The circulatory flow is a superposition of a linear profile due to the relative motion of screw and barrel and a parabolic flow due to pressure. The channel height B is assumed small compared to the filled width $F_e W$ so that the y-direction velocity components at each end of the filled

region can be neglected. Heat conduction to the polymer melt was approximated by a constant wall temperature at $y = B$ and a zero flux condition (neutral screw) at $y = 0$. Results with viscous dissipation used a physically plausible value for the Brinkman number.

Figure 10.11 illustrates Nauman's results for the thermal time distribution in an extruder with viscous dissipation. The curve for $E/R_g T_0 = 0$ is Pinto and Tamor's residence time distribution. This is narrower than the distribution for pressure driven flow between flat plates (compare Figure 10.7 with $E/R_g T_0 = 0$) but still shows a much broader distribution than the thermal time curves with $E/R_g T_0 = 5$ and 40. Short residence times in the extruder are coupled with higher temperatures and conversely. For $E/R_g T_0 = 5$ the overall result is a very close approximation to piston flow.

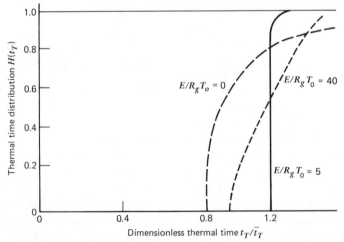

Figure 10.11   Thermal time in an extruder with viscous dissipation.

Thus single-screw extruders can provide a remarkable uniform reaction environment.

10.7   STIRRED TANK REACTORS

10.7.1   Thermal Design

Just as for tubular reactors, the first consideration in designing a stirred tank reactor is removal of the heat of polymerization, which is in the range of 10–26 Kcal/mol for vinyl monomers. Although it is occasionally possible to operate a reactor with a nearly neutral heat balance, provision must still be made to remove heat through the vessel walls, or through an external heat exchanger, or, if the contents can be allowed to boil, through a reflux condenser. It is also necessary to add heat during startups.

Like many other exothermic reactions in stirred tanks (93–95), three steady stages are possible for polymerizations beginning with monomer: a lower one where there is essentially zero conversion; an intermediate, metastable condition; and an upper, runaway condition where conversion to polymer is nearly complete. For polymerizers it is usually desired to operate at the intermediate, metastable condition. To illustrate this, assume that the polymerization kinetics are pseudo first order and write material and energy balances based on perfect mixing in the reactor. A material balance on monomer gives Equation (10.45)

$$M_0 = M - \bar{t}KM \qquad (10.45)$$

where M is the mass fraction monomer. The heat balance for the reactor is given by

$$T = T_0 + \frac{\Delta H_R}{Wc_p}\bar{t}KM - \frac{q}{pc_pQ} \qquad (10.46)$$

where W is the molecular weight and q is the rate of heat removal.

The following parameters are reasonable for the thermally initiated polymerization of styrene:

$$\frac{\Delta H_R}{Wc_p} = 400° \, C$$

$$K = 10^{10} \exp{\frac{-10000}{T}}, \, hr^{-1}$$

$$\bar{t} = 2 \, hr$$

$$M_0 = 1 \, (\text{monomer feed})$$

$$\frac{q}{pc_pQ} = 0 \, (\text{adiabatic operation})$$

$$T_0 = 27° \, C$$

With these values, Equations (10.45) and (10.46) can be solved simultaneously for M and T. The three solutions are:

$$T \approx 27° \, C, \quad M \approx 1 \qquad (0\% \, \text{polymer})$$

$$T \approx 131° \, C, \quad M \approx 0.74 \qquad (26\% \, \text{polymer})$$

$$T \approx 427° \, C, \quad M \approx 0 \qquad (100\% \, \text{polymer})$$

Operation at the middle steady state gives polymer of a commercial molecular weight and, broadly speaking, is representative of bulk polystyrene processes. The lower steady state gives no polymer, while the upper steady state is mechanically infeasible and also gives polymer of too low a molecular weight. In fact, thermal depolymerization of polystyrene becomes important above about 300° C, and runaway reactions become equilibrium limited at about 350° C.

Some form of thermal regulation is needed to maintain a reaction at a metastable operating point. In the preceding example, heat must be added (or the flow rate decreased) if the temperature falls below 131° C. Otherwise, the reactor will move to its lower steady state, and polymerization will essentially stop. Similarly, heat must be removed (or the flow rate increased) to avoid a runaway when the temperature exceeds 131° C. Since alternate heating and cooling is awkward, usual practice is to avoid operation near the adiabatic point. Typically, the reaction is designed for a positive exotherm, and the desired operating temperature is maintained by regulating the amount of cooling.

The results of Jaisinghani and Ray (96) confirm that three steady states are typical of cold monomer feed, although Bilous and Amundson (97) and Hoftyzer and Zwietering (98) have shown that up to five steady states are theoretically possible for chemically initiated polymerizations. With preheated feed, there is usually only one adiabatic steady state, which is the upper, runaway condition. Regulation by cooling is obviously required. Adding a substantial heat sink in the form of cooling introduces a new metastable state about which the process is controlled. If the heat sink is an external heat exchanger, the best design approach is to solve the transport equations (24). For heat transfer through the vessel walls, Suryanarayanan, Mujawar, and Rao (99) give experimental correlations of heat transfer coefficients plus a summary of the literature.

10.7.2 Residence Time Distributions

Most stirred tank reactors are designed to have uniform temperatures and compositions throughout the vessel. This is accompanied by high internal recirculations rates combined with thermal and molecular diffusion between portions of the recycled fluid. Thermal diffusivities are adequately high so that stirred tank reactors do tend to be fairly uniform in temperature. Molecular diffusivities are much lower, and large composition differences are possible. However, the residence time distribution in stirred tank reactors closely follows the exponential distribution of perfect mixers, Equation (10.29), even though mixing on the molecular scale may be far from perfect.

The exponential distribution of residence times is a direct consequence of recirculation (60, 61) and does not depend on molecular diffusion. Olson and Stout (100) found that the exponential distribution was a good approximation whenever the internal circulation rate was greater than about five times the net throughput. Szabo and Nauman (67) used a reaction criterion to determine that a recycle

ratio of 8 was adequate. Fu et al. (101) presented a quantitative measure for the convergence of $F(t)$ to its exponential limit. The conclusion from all these studies is that a reasonably designed agitation system can achieve an exponential distribution of residence times. If the reactor departs from the behavior expected of a perfect mixer, the reason is almost certain to be poor micromixing rather than poor macromixing. Although a number of studies (102–104) have noted departures from ideal flow when flat-bladed turbines were used in the laminar regime, such agitators are rarely used for polymer reactors. Anchor agitators, pitched turbines, and helical ribbons are more common, and at least the latter two tend to give much better pumping capabilities at the same power input.

Since stirred tank reactors tend to be isothermal, the thermal time distribution will closely approximate the residence time distribution.

10.7.3 Micromixing

Unlike most single-step reactions, polymerizations have kinetics complex enough to be significantly affected by micromixing. They also occur in laminar flow, low diffusivity environments where segregation is likely. Although a recent paper (105) discusses the possibility that intermediate degrees of segregation may affect the thermal design of a reactor by introducing unexpected, multiple steady states, the major influence of micromixing will be on yields, molecular weight distributions, and copolymer composition distributions.

10.7.3.1 Yields
The yield of a first-order reaction depends only on the residence time distribution, and since most addition polymerizations are at least pseudo first order, the effect of micromixing on yield is predictably small. Tadmor and Biesenberger (106) have shown that the conversion of monomer to polymer is insensitive to micromixing provided the polymer chains are long. However, there are several

possible exceptions to this generalization. Chemically initiated polymerizations are half order with respect to initiator concentration, and thermal initiation has been modeled as a second- or even third-order process (16) so that some effects of segregation may be detectable. A stronger dependence on the level of micromixing can be expected in systems showing a pronounced gel effect.

The gel effect arises when the termination of free radicals become diffusion controlled. Reaction rates and molecular weights become much higher than would be expected by extrapolating data at low conversions. One result is that segregated stirred tanks will behave differently from perfect mixers, giving higher yields when the mean residence time is low and lower yields when the mean residence time is high. To illustrate this phenomenon, consider the simplified kinetic scheme given by Equations (10.47) and (10.48).

$$\text{Rate} = K_0 M, \qquad M > M_g \quad (10.47)$$

$$\text{Rate} = K_g M, \qquad M \leqslant M_g \quad (10.48)$$

With these kinetics, the reaction yield can be found analytically for batch reactors, perfectly mixed stirred tanks, and completely segregated stirred tanks. Figure 10.12 shows the results when $K_g/K_0 = 2$ and $M_g = 0.5$. The batch reactor shows a change in slope at $K_0\bar{t} = 0.693$ and $M = M_g = 0.5$, which corresponds to the rate suddenly shifting from Equation (10.47) to Equation (10.48). For a perfect mixer, this change in rate gives a discontinuity at $K_0\bar{t} = 1$ and $M = M_g = 0.5$ when the gel effect suddenly begins. The yield from a segregated stirred tank is continuous since it behaves as a large number of batch reactors in parallel, some of which will always be at monomer concentrations higher than M_g while others will always be below M_g. The segregated stirred tank outperforms the perfect mixer for $K_0\bar{t} < 1$ since some fraction of the reacting mass is in the concentration region where the gel effect is important, whereas the perfect mixer operates completely outside this region. Conversely, the perfect mixer outperforms the segregated stirred tank for $K_0\bar{t} > 1$.

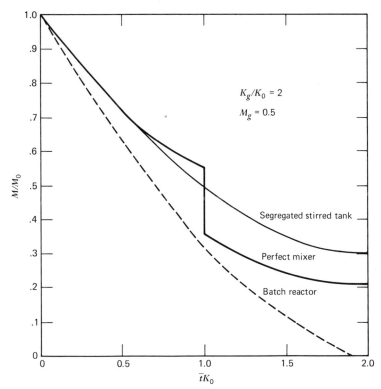

$K_g/K_0 = 2$

$M_g = 0.5$

Segregated stirred tank

Perfect mixer

Batch reactor

$\bar{t}K_0$

Figure 10.12 Gel effect in batch and stirred tank reactors.

Harada et el. (107) have found experimental evidence of segregation in a stirred tank polymerizer. They studied the azobisisobutyronitrile initiated polymerization of styrene at 80° C using 30% toluene as a diluent. At 75% conversion of styrene the viscosity of the reaction mixture was 35 poise, which is moderate by polymer engineering standards. The residence time distribution was measured and confirmed to be exponential, and the reactor was found to be reasonably isothermal. Batch reaction data were integrated to predict yield curves for perfect mixers and segregated stirred tanks. The results were qualitative similar to those in Figure 10.12 with a crossover point at about 40% conversion, below which a segregated reactor would give marginally higher yields than a perfect mixer. At about 40% conversion, the situation reverses and the curves diverge significantly. The residence time that gives a 90% yield in a perfect mixer would only give 70% in a completely segregated reactor. This 20% yield difference due to micromixing is the largest yet predicted using experimental batch kinetics. In their stirred tank experiments, the yield fell roughly midway between the two curves when a flat-bladed turbine was used for agitation and on the perfect mixing curve when a helical ribbon was used.

10.7.3.2 *Molecular Weight Distributions*

Figure 10.13 attempts a simplified graphical representation of the kinds of mixing possible in continuous flow reactors. Macromixing is one dimension of the two-dimensional mixing space. This dimension is characterized by the residence time distribution, and the dimensionless variance of the distribution σ^2 provides a possible—but not unique—quantitative measure of macromixing. Dimensionless variances range from zero to one in the normal region and correspond to residence time distributions ranging from the uniform distribution of piston flow to the exponential distribution of stirred tanks. Bypassing can give dimensionless variances greater than one; and there is no upper limit for macromixing when quantified in this fashion.

Micromixing is the other dimension depicted in Figure 10.13, and one possible quantification of this is $1 - J$ where J is the degree of segregation defined by Danckwerts (36) and Zwietering (33). The theoretical range for J is zero to one, but $J = 0$ is possible only with an exponential residence time distribution. All other distributions have some minimum possible J. Thus there is some maximum possible value for $1 - J$, and the mixing space shown in Figure 10.13 is bounded from above.

The triangular region corresponding to normal mixing has three idealized reactors at its vertices, and it has long been tacitly assumed that these three reactors provide limits on the molecular weight distributions possible in real systems (108). Reactors with bypassing may in fact give distributions that fall outside these limits; but the batch reactor, segregated stirred

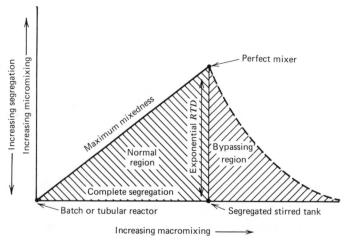

Figure 10.13   Mixing space for continuous flow reactors.

tank and perfect mixer do represent realistic limits for normal mixing situations. Consequently, the theoretical molecular weight distributions for these idealized reactors merit detailed study.

Batch reactors were the earliest studied (109, 110), and molecular weight distributions are now available for a wide variety of polymerization mechanisms (6, 11). Denbigh (58, 59) first analyzed polymerization in a perfect mixer and concluded that the molecular weight distribution will be narrower than for a batch reactor if the lifetime of a growing polymer chain is short compared to the mean residence time. Conversely, a perfect mixer gives a broader distribution when the life of a polymer chain is long. Other researchers, studying a variety of polymerization mechanisms and using several mathematical techniques, have confirmed Denbigh's generalization. Representative studies are those by Abraham (111), Amundson et al. (112–114), Beasley (115), Biesenberger et al. (116, 117), Duerksen and Hamielec (118, 119), Kilkson (120), Ray (121), Smith and Sather (122), and Graessley (123). Together, these studies define the types of molecular weight distributions that are possible in reactors operating at or near the condition of maximum mixedness, that is, reactors that lie on or near the line in Figure 10.13, which connects batch and perfectly mixed reactors. The studies illustrate the range of behavior possible by varying the residence time distribution at a fixed (maximum) level of micromixing. This range is quite large. For example, the polydispersity of a condensation polymerization in a perfect mixer goes to infinity as the molecular weight is increased (117).

However, an even broader range of molecular weight distributions is possible when the level of micromixing is varied at a fixed (exponential) residence time distribution.

Tadmor and Biesenberger (106, 108) were the first to explicitly consider the effect of micromixing on molecular weight distributions. They treated the segregated stirred tank for several polymerization mechanisms and compared their results to those for batch and perfectly mixed reactors. Figure 10.14 illustrates their results for a condensation polym-

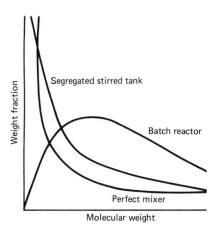

Figure 10.14   Molecular weight distributions for condensation polymers.

erization where the life of a growing polymer chain is long compared to the mean residence time. Figure 10.15 shows the case of a free-radical polymerization where the chain lifetime if short. Tadmor and Biesenberger generalized that, with short chain lifetimes, the molecular weight distribution is broader in a segregated stirred tank than in a batch reactor but narrower than in a perfect mixer.

Since Tadmor and Biesenberger's work, theoretical molecular weight distributions are now available, at least for the common polymerization mechanisms, for all three of the idealized reactors in Figure 10.13. Except for bypassing, which can give rise to bimodal distributions in condensation polymers, it is likely that these three reactor types do bound the performance to be expected in real systems. However, little work has been done to confirm this supposition by exploring the interior of

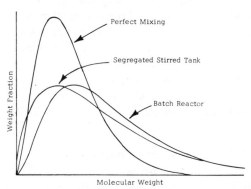

Figure 10.15   Molecular weight distribution for free-radical polymers.

the normal mixing region. Cintron-Cordero, Mostello, and Biesenberger (65) studied tubular reactors having power law velocity profiles, Equation (10.15). These systems lie along the complete segregation line in Figure 10.13 and have molecular weight distributions intermediate between batch reactors and segregated stirred tanks. A number of workers (116, 119, 124) have considered perfect mixers in series. These systems lie near (but not on, see Reference 33) the maximum mixedness line in Figure 10.13. Kilkson (120) studied condensation polymerization in a piston flow reactor with recycle, which is a maximum mixedness reactor (125). Again, intermediate molecular weight distributions were found.

Addition polymers with short chain lifetimes are particularly interesting from the viewpoint of micromixing theory. Molecular weight distributions are broadest in the segregated stirred tank and narrowest in the perfect mixer. Based on this fact, Dunn and Hsu (126) have proposed that the molecular weight distribution of free-radical polymers be used to measure micromixing in continuous flow reactors. The viscosity-average molecular weights of Harada et al. (107) indicated partial segregation, confirming their conclusions based on yield data. Other styrene polymerization studies have shown results in which segregation is confounded with the gel effect (118). Nagasubramanian and Graessley (127, 123) studied the vinyl acetate polymerization under conditions where branching was important. Their results, although subject to varied interpretations (128), showed segregation to be important. Branching caused by chain transfer to monomer is a special case of free-radical polymerization where a dead polymer chain can be reactivated and thus appears to be a long-lived species. The theory developed by Nagasubramanian and Graessley showed the highest polydispersity in perfect mixers, something one would normally expect for condensation polymers but not for free-radical polymers.

10.7.3.3 Copolymer Composition Distributions

In multicomponent polymerizations by a free-radical mechanism, polymer molecules will usually have a composition different than that of the monomer mixture from which they were formed. The polymer will be rich in those monomers with high reaction rates and deficient in monomers with low rates. For copolymerizations, the instantaneous compositions are related by Equation (10.9)

$$\left(\frac{A}{B}\right)_{polymer} = \frac{dA}{dB} = \frac{A\,(\gamma_A A + B)}{B\,(A + \gamma_B B)} \quad (10.49)$$

where A and B are the monomer concentrations.

In batch or piston flow reactors, Equation (10.49) can be integrated to give monomer and polymer compositions as a function of overall conversion to polymer. The composition of the polymer first formed is given by Equation (10.49) with initial values A_0 and B_0 substituted for A and B. The initial polymer will be rich in the more reactive component, but as polymerization proceeds, the monomer mixture will be depleted in this component, and compositions will drift toward the less reactive component. If a copolymer of uniform composition is desired, the more reactive monomer must be fed to the reactor as a function of time.

Azeotropes exist whenever γ_A and γ_B are both less than one or greater than one. The azeotropic composition is given by Equation (10.50)

$$\left(\frac{A}{B}\right)_{azeo} = \frac{1 - \gamma_B}{1 - \gamma_A} \quad (10.50)$$

For $A =$ styrene and $B =$ acrylonitrile, $\gamma_A = 0.41$ and $\gamma_B = 0.4$. The azeotrope lies at about 38 mol % acrylonitrile (24 wt %); at this monomer composition, the polymer composition is identical. A chemically uniform copolymer is easily manufactured at the azeotrope; indeed, large tonnage quantities of SAN copolymer and the matrix phase of ABS are made at or about 24 wt % acrylonitrile. However, certain product properties are favored at higher acrylonitrile contents, and commercial products are also made with average compositions of 28 to 36 wt % acrylonitrile. Close control of the distribution about the average composition is needed to ensure mechanical compatibility and good color of the polymer.

A perfect mixer gives chemically uniform copolymer. To obtain 28 wt % acrylonitrile in the polymer, the reacting mixture should contain about 36 wt % acrylonitrile according to Equation (10.49); feed composition to the reactor can be adjusted to achieve this. For other reactor types, some spread in copolymer composition distribution must be accepted.

Figure 10.16 shows the effects of both macromixing and micromixing on the styrene-acrylonitrile system. For these results, the feed composition was selected to give an initial polymer composition of 28 wt % acrylonitrile, and the mean residence time in each type of reactor was adjusted to give 60% overall conversion. The level of macromixing was adjusted by varying the residence time distribution from that of a perfect mixer to that of a piston flow reactor. Intermediate residence time distributions correspond to a stirred tank in series with a piston flow reactor and are characterized by the dimensionless variance σ^2. The magnitude of σ^2 represents the fraction of the system volume that is stirred. The perfect mixer gives a uniform composition distribution at the average value of 28% acrylonitrile while a batch reactor shows a spread with compositions ranging up to 31%. A segregated stirred tank gives the broadest copolymer composition distribution with the

tail of the distribution approaching pure poly-acrylonitrile. Intermediate results illustrate the effect of varying the level of macromixing at fixed levels of micromixing.

Other studies on copolymer composition distributions have also found perfect mixers and segregated stirred tanks to represent extremes within the normal mixing region of Figure 10.13. O'Driscol and Knorr (129) treated the three idealized reactors for copolymer reactivity pairs of $\gamma_A = 20, \gamma_B = 0.015$ and $\gamma_A = 0.5, \gamma_B = 1$. Mechlenburgh (66) also analyzes the three idealized reactors but extends the analysis to a segregated tubular reactor with a parabolic velocity distribution and to a piston flow reactor with backmixing by way of axial diffusion. Szabo and Nauman (67) analyzed a tubular reactor with power law flow, Equation (10.15), in a recycle loop. At high-recycle ratios this reactor approaches the performance of a perfect mixer and has served as a conceptual model for the design of agitation systems (31).

10.7.4   Transient Operation

The transient or unsteady-state operation of stirred tank reactors has received appreciable academic study, partly, one suspects, because the mathematics are interesting yet tractable and partly because stirred reactors are inherently more amenable to transient control strategies than tubular reactors.

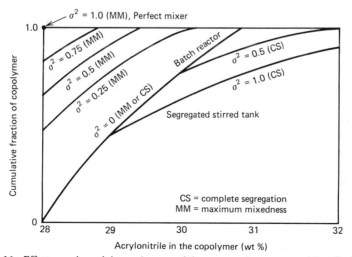

Figure 10.16   Effect on micromixing and macromixing on copolymer composition distributions.

When the stirred tank is a perfect mixer, the unsteady state material balance for a reacting component can be written as Equation (10.51)

$$Q_{in} C_{in} - Q_{out} C_{out} = KC^N + C\frac{dV}{d\tau} + V\frac{dc}{d\tau}$$

$$(10.51)$$

An equation similar to this can be written for each component; when summed for all components, the overall balance is

$$Q_{in} - Q_{out} = \frac{dV}{d\tau} \qquad (10.52)$$

In writing these equations, we have assumed constant fluid densities and simple Nth-order reaction kinetics, although these restrictions can be removed without overly complicating the mathematics. Equations (10.51) and (10.52) represent a set of first-order, ordinary differential equations that can be—and have been—solved for a host of specific control schemes.

When the reactor is completely segregated, the mathematics are more complex, but theoretical methods of solution are now available. One of these is the unsteady residence time distribution, introduced for stirred tank reactors by Nauman (130) and extended to general systems by Chen (131). For a stirred tank reactor in unsteady operation, the residence time distribution for material leaving the reactor is a function of real time τ as well as of residence time t,

$$F(\tau,t) = 1 - \exp\left[-\int_{\tau-t}^{\tau} \frac{Q_{in}}{V} d\tau'\right] \quad (10.53)$$

When Q_{in}/V is constant, the residence time distribution assumes its steady-state value, Equation (10.29). Constancy of Q_{in}/V does not necessarily imply steady-state operation for the reactor as a whole. One possibility is to vary the composition of the feed stream while holding the total flow constant. Another possibility, useful for maintaining product quality during reactor startups and shutdowns, is to hold Q_{in}/V constant while manipulating Q_{out}.

10.7.4.1 Startup and Shutdown

One way to start up a stirred tank reactor is to fill the system rapidly, then to batch react without discharge for a time $\bar{t}$ and then to begin steady-state flow with $Q_{in} = Q_{out}$. This simple strategy maintains a constant mean residence time for all material leaving the reactor, but other moments of the residence time distribution will change with real time, as will the exit concentration. More sophisticated startup techniques have been developed to minimize startup transients, but some departure from steady-state operation will always occur in kinetically complicated systems such as polymerizations. The details of the molecular weight distribution will depend on the higher moments of the residence time distribution; and steady-state values for these higher moments cannot be achieved instantaneously.

A similar—but far less studied—problem occurs during reactor shutdown. Suppose that the input flow is suddenly discontinued. At this instant the entire reactor is filled with material of the desired, steady-state composition, but, since it takes a finite length of time to discharge the vessel, some additional reaction will occur before the reactor is emptied.

The theory of unsteady residence time distributions offers a method for avoiding shutdown transients completely and for generating an arbitrary, small amount of off-specification material during startups (132, 133). Consider first the shutdown. Up to some time $\tau = 0$, the reactor has been operating at steady state so that exiting material has the fully developed, exponential distribution of residence times. At time $\tau = 0$ the discharge flow rate is marginally increased over its steady-state value, say $Q_{out} = 1.1\, Q_0$. The input flow rate is controlled so that Q_{in}/V remains at its old, steady-state value.

$$Q_{in}(\tau) = \frac{V(\tau)}{\bar{t}} \qquad (10.54)$$

The result of this is for both Q_{in} and V to gradually decrease while material in the reactor maintains an absolutely constant residence time distribution. At $\tau = 2.4\bar{t}$, the reactor goes empty.

A scheme similar to the preceding can be applied to several stirred tanks in series. Also, startups with minimum generation of off-specification material are accomplished by beginning with low initial values of V and Q, letting the reactor approach steady state at these values, and then by reaching full productivity while maintaining a constant $F(\tau, t)$. Details have been presented by Nauman (132), and an experimental verification of the technique for a styrene polymerization was reported by Nauman and Carter (133).

The method of constant Q_{in}/V supposes that the vessel remains well mixed in the sense that the steady-state residence time distribution would be exponential for all flow rates and active volumes encountered during transient operation. It is also necessary to assume that the level of micromixing is unaffected by the changes in flow rate and volume. For stirred tanks in series, intermediate feed streams can be allowed, provided they are ratio controlled to the main streams. The tanks need not be at the same temperature, but all temperatures must remain constant during the transients. In most practical situations, the method of constant Q_{in}/V ensures a uniform product over a broad range of operating conditions with completely arbitrary and even unknown reaction kinetics.

Other methods of reactor control during transients have been reported. Details tend to be specific for a given polymerization. The simplest startup is just to fill the tank quickly and then to set Q_{in} and Q_{out} to their steady-state values. The works of Hui and Hamielec (134) on free-radical polymers and of Smith and Sather (122) on polycondensations show typical results. Startup times of $4\bar{t}$ to $6\bar{t}$ are needed for the number- and weight-average molecular weights to line out. Hicks, Mohan, and Ray (135) have attempted to minimize these startup times through optimal controls on reactor temperature and flow rate.

10.7.4.2   Periodic Operation

By periodic operation we mean the situation where each system variable is periodic in time, for example,

$$V(\tau) = V(\tau + \tau_p) \qquad \text{for all } \tau \quad (10.55)$$

If different variables such as Q_{in} and Q_{out} have different periods, then τ_p will be the least common period for the system as a whole.

The purpose of periodic operation is to achieve a time-average value for a product property or productivity that is better than can be achieved in steady operation. Ray (136) and Laurence and Vasudevan (137) studied the effects of periodic feed concentrations on molecular weight distributions. For free-radical polymers, the distribution can be broadened in this manner; for condensation polymers it can be slightly narrowed. Neither paper demonstrated an industrially significant advantage for periodic control of feed composition.

Unlike feed composition controls where results tend to be specific to a given system, variable volume control allows some generalizations. Periodic systems have a time-average residence time distribution that governs yields in much the same manner as the ordinary residence time distribution does for steady-state reactors (138). As usually practiced, variable volume operation forces the time-average residence time distribution away from the exponential distribution of stirred tanks and toward the uniform distribution of batch reactors. This gives higher overall yields and narrows or broadens the molecular weight distribution depending on whether the lives of the growing polymer chains are long or short. The magnitude of this gain is suppressed because of physical constraints that give a reduction in mean residence time compared to that in the steady system (138). However, some gain in yield or product quality remains possible, and variable volume operation may be attractive for polymerizations that need a stirred tank for heat transfer or other reasons. Without such special reasons, one would simply build a tubular reactor to begin with.

10.8   OVERVIEW AND EXTRAPOLATIONS

The technical approach to developing a new polymer process often depends on the extent to which the polymer has been commercialized. Market projections for a new polymer, a new composition of matter, are notoriously inaccurate. Prudent management will insist that the market be thoroughly explored before substantial commitments are made to process development. Emphasis will be on rapid property and market evaluations while process elegance and ultimate manufacturing economics will be neglected or postponed. While this is a perfectly valid way to approach new business development, it places severe economic and time constraints on the reactor design engineer.

Figure 10.17 shows a generalized flow chart for process development activities. After the material and its probable markets have been identified, laboratory studies define the appropriate polymerization mechanisms and general chemistry. The polymerization reactor engineer then begins bench-scale studies and acquires enough engineering data to design a pilot plant. For new polymers this design should ordinarily utilize general purpose equipment such as that illustrated in Figure 10.2, and the polymerization scheme should usually be batch or semibatch. The primary purpose of the pilot plant is to prepare materials for evaluation and not to test process innovations. Thus the process is typically very similar to that originally devised in the laboratory, and even when the first pilot plant is scaled up for commercial sales and market development, it is probably best to minimize process changes.

Once commercial viability has been established, it is appropriate to repeat the development activity depicted in Figure 10.17.

Emphasis is now placed on product and process optimization and the upper path in Figure 10.17 is usually followed. Renewed laboratory and bench studies may suggest a process different from that used originally, and a pilot plant will be built especially to test this process.

The principles and techniques discussed in this chapter are primarily intended for process development activities that follow the upper path in Figure 10.17. Emphasis is on producing an economically optimum product by way of an economically optimum process. The achievement of this optimum requires a whole series of design choices, major among which are the following:

1. Polymer chemistry and structure.

2. Basic polymerization mechanism.

3. Batch or continuous mode of operation.

4. Polymerization reactor option.

5. Specific reactor design.

These are listed in approximate order of importance, but the first item is fixed for process development activities on existing polymers, and the second item is usually fixed as well. Start with a simple vinyl monomer and free radical polymerize it if you have a choice.

A continuous process is usually cheaper than a batch process at production rates typical of established polymers. Thus the specialized pilot plant shown on the upper path of Figure 10.17 is usually continuous,

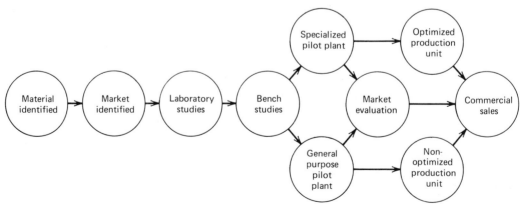

Figure 10.17 Polymer process development activity.

and this choice has a profound influence on the overall process design. In a batch process, individual steps are often combined in a single, versatile piece of equipment. Thus the individual operations are combined in space but are separate in time. In a continuous process, the individual operations coexist in time but are separated in space. These individual operations are accomplished in many separate and relatively simple devices that are each designed for a single purpose. Almost always this gives a cheaper process than the batch process, which uses fewer but more complex pieces of equipment.

Although it may require renewed laboratory and bench studies, the reactor design engineer can sometimes pick the number and kind of phases occurring in the reactor. The economic order of preference appears to be: solid polymer in a gas, solid polymer in a liquid, single-phase bulk, single-phase solution, and emulsion. This order is of course dictated more by the economics of purification and recovery than by those of the polymerization reactor itself. Just to polymerize, single-phase bulk reactors are usually the cheapest.

Finally, we come to the specifics of the reactor design. This subject occupies most of the chapter; it is the enabling technology for the entire process yet often represents a small fraction of the total system cost. When choices exist, the usual order of preference is tubular with fixed walls, then stirred tank, and last, tubular with moving walls.

Table 10.1 summarizes the various design choices and their usual order of preference. This table represents a type of logical sieve that can be applied to existing polymers. Polyethylene from a gas-fluidized reactor and polypropylene from a liquid-fluidized reactor emerge from the sieve as the ultimate low-cost polymers. For the polystyrenes, the sieve predicts that a continuous suspension process may yet prove the best. Until this process is developed, continuous bulk polymerizations should have the edge over semibatch suspension processes. The sieve suggests that vinyl suspension resins have superior economics compared to the other types and that the bulk ABS process will ultimately replace the emulsion process. All of these predictions seem consistent with known trends.

Table 10.1   Product and Process Design Choices

Design Choice	Usual Order of Preference
Polymer chemistry and structure	Simple monomer, homopolymer
	Simple monomers, single-phase copolymers
	Simple monomers, multiphase copolymers
	Complex monomers
Basic polymer mechanism	Vinyl addition
	Self-condensation
	Bimolecular condensation
	Other
Mode of operation	Continuous
	Batch
	Cyclic
Polymerization reactor options	Solid polymer in gas
	Solid polymer in liquid
	Homogeneous
	Emulsion
Specific reactor designs	Fixed-wall tubular
	Stirred tank
	Moving-wall tubular

10.9 GLOSSARY OF SYMBOLS

A	Concentration of monomer A or of polymer having $X \sim X$ form
B	Concentration of monomer B or of polymer having $Y \sim Y$ form
C_A	Concentration of species A
C_p, C_v	Heat capacities at constant pressure or volume
d	Rubber occlusion diameter
D	Diameter of tube or rubber particle diameter
D_A	Diffusivity of species A
DP	Number-average degree of polymerization
E	Activation energy
$\mathbf{E}$	Vector of environmental conditions
f	Residence time frequency function
f_c	Fraction of extruder channel filled
F	Cumulative residence time distribution function
G	Subscript denoting gel effect
h	Thermal time frequency function
H	Cumulative thermal time distribution function
i	Subscript denoting molecular length
k	Index for distribution moments
K	Reaction rate constants
L	Length of tube or parameter of motionless mixer
m	Number of mixing units in series
M	Concentration of monomer or of polymer having $X \sim Y$ form
M_0	Initial concentration of monomer
n	Power law viscosity index
N	Reaction order
N_{Re}	Reynolds number
N_{seg}	Segregation number
P	Pressure
PD	Polydispersity
q	Rate of heat removal
Q	Volumetric flow rate
r	Radial coordinate
R	Radius of tube or droplet
R_A	Reaction rate of species A
R_g	Gas law constant
R_0	Initial striation thickness
s	Stoichiometric ratio, A_0/B_0

t	Residence time or time in a batch reactor
t_T	Thermal time
$\bar{t}$	Mean residence time
T	Absolute temperature
T_0	Temperature of entering fluid
T_w	Wall temperature
V	Fluid velocity
V_0	Velocity at centerline
$\bar{V}$	Average velocity
W	Width of extruder channel or molecular weight
x, y, z	Coordinate directions
α	Constant
γ_A, γ_B	Copolymer reactivity ratios
ΔH_R	Heat of reaction
θ	Dimensionless residence time
κ	Thermal conductivity
μ	Viscosity
μ_k, μ_k'	Moments of residence time distribution
ρ	Density
σ^2	Dimensionless variance
σ_t^2	Variance of residence time distribution
τ	Real time in unsteady reactors
τ_p	Interval of periodicity
ϕ	Dimensionless frequency function
Φ	Dimensionless distribution function

10.10 REFERENCES

1. L. F. Albright, *Processes for Major Addition Type Plastics and Their Monomers*, McGraw-Hill, New York, 1974.

2. D. C. Miles and J. H. Briston, *Polymer Technology*, Chemical Publ. Co., New York, 1965.

3. M. V. Smith, *Manufacture of Plastics*, Reinhold, New York, 1964.

4. A. V. Raff and J. B. Allison, *Polyethylene*, Interscience, New York, 1956.

5. R. H. Boundy, R. F. Boyer, and S. M. Stoesser, *Styrene, Its Copolymers and Derivatives*, ACS Monograph Series, Havner, Darien, Connecticut, 1970.

6. L. H. Peebles, Jr., *Molecular Weight Distribution in Polymers*, Wiley-Interscience, New York, 1971.

7. N. R. Amundson and D. Luss, *J. Macromol. Sci.*, **C2**, 145 (1968).

8. D. C. Chappelear and R. H. Simon, *Adv. Chem.*, **91**, 1 (1969).

9. P. J. Flory, *Principles of Polymer Chemistry*, Cornell University Press, Ithaca, New York, 1953.

10. W. H. Abraham, *Chem. Eng. Sci.*, **25**, 331 (1970).

11. C. H. Bamford, W. G. Barb, A. D. Jenkins, and P. F. Onyon, *The Kinetics of Vinyl Polymerization by Free Radical Mechanisms*, Butterworths, London, 1958.

12. G. E. Ham, Ed., *Vinyl Polymerization*, Vol I: *Kinetics and Mechanisms of Polymerization*, Marcel Dekker, New York, 1967.

13. G. Odian, *Principles of Polymerization*, McGraw-Hill, New York, 1970.

14. G. E. Ham, Ed., *Copolymerization*, Interscience, New York, 1964.

15. K. Ito, *J. Polym. Sci.*, **A7**, 827, 2247, 2707, 2995 (1969).

16. A. E. Hamielec, *The Gel Effect in Vinyl Polymerization*, Proceedings of Symposium on Polymer Reaction Engineering, Lavel University, Quebec City, 1972.

17. J. M. Church and R. Shinnar, *Ind. Eng. Chem.*, **53**, 643 (1947).

18. R. Shinnar and S. Katz, *Adv. Chem.*, **109**, 56 (1972).

19. G. Poehlein and A. W. DeGraff, *J. Polym. Sci.*, **A9**, 1955 (1971).

20. N. Nomura et al., *J. Appl. Polym. Sci.*, **15**, 675 (1971).

21. R. W. Thompson and J. D. Stevens, *Chem. Eng. Sci.*, **32**, 311 (1977).

22. R. B. Bird, W. E. Stewart, and E. N. Lightfoot, *Transport Phenomena*, John Wiley, New York, 1960.

23. S. Lynn and J. E. Huff, *AIChEJ*, **17**, 475 (1971).

24. F. Popovska and W. L. Wilkinson, *Chem. Eng. Sci.*, **32**, 1155 (1977).

25. R. Mahalingam, L. O. Tilton, and J. M. Coulson, *Chem. Eng. Sci.*, **30**, 921 (1975).

26. W. H. Suckow, P. Hrycak, and R. C. Griskey, *Polym. Eng. Sci.*, **11**, 401 (1971).

27. P. V. Danckwerts, *Chem. Eng. Sci.*, **2**, 1 (1953).

28. O. Levenspiel, *Chemical Reactor Engineering*, 2nd ed., John Wiley, New York, 1972.

29. D. M. Himmelblau and K. B. Bischoff, *Process Analysis and Simulation: Deterministic Systems*, John Wiley, New York, 1968.

30. C. Y. Wen and L. T. Fan, *Models for Flow Systems and Chemical Reactors*, Marcel Dekker, New York, 1975.

31. E. B. Nauman, *J. Macromol. Sci.*, **C10**, 75 (1974).

32. D. B. Spalding, *Chem. Eng. Sci.*, **24**, 1461 (1958).

33. T. N. Zwietering, *Chem. Eng. Sci.*, **11**, 1 (1959).

34. B. A. Buffham and H. W. Kropholler, *Chem. Eng. Sci.*, **28**, 1081 (1973).

35. E. B. Nauman, *Chem. Eng. Sci.*, **32**, 287 (1977).

36. P. V. Danckwerts, *Chem. Eng. Sci.*, **8**, 93 (1958).

37. T. N. Zwietering, *Chem. Eng. Sci.*, **8**, 101 (1958).

38. S. P. Chauhan, J. P. Bell, and R. J. Adler, *Chem. Eng. Sci.*, **27**, 585 (1972).

39. Z. Novad and J. Thyn, *Collect. Czech. Chem. Commun.*, **31**, 3710 (1966).

40. J. C. Methot and P. H. Roy, *Chem. Eng. Sci.*, **26**, 569 (1971).

41. D. Y. C. Ng and D. W. T. Rippin, *Proc 3rd European Symposium on Chemical Reaction Engineering*, Pergamon Press, 1965.

42. G. R. Worrell and L. C. Eagleton, *Can. J. Chem. Eng.*, **42**, 257 (1964).

43. D. W. T. Rippin, *Chem. Eng. Sci.*, **22**, 247 (1967).

44. H. Weinstein and R. J. Adler, *Chem. Eng. Sci.*, **22**, 65 (1967).

45. R. W. Dunn and C. C. Hsu, *Adv. Chem.*, **109**, 56 (1972).

46. M. S. K. Chen and L. T. Fan, *Can. J. Chem. Eng.*, **49**, 704 (1971).

47. L. T. Fan and W. S. Hwant, *Proc. R. Soc.*, **A283**, 576 (1965).

48. K. D. P. Nigam and K. Vasudeva, *Chem. Eng. Sci.*, **32**, 673 (1977).

49. F. S. Manning, D. Wolf, and D. L. Keairns, *AIChEJ*, **11**, 723 (1965).

50. E. B. Nauman, *Chem. Eng. Sci.*, **30**, 1135 (1975).

51. E. B. Nauman, *Chem. Eng. Sci.*, **32**, 359 (1977).

52. K. W. Norwood and A. B. Metzner, A. B., *AIChEJ*, **6**, 432 (1960).

53. R. L. Bates, P. L. Fondy, and J. G. Fenic, in V. W. Uhl and J. B. Gray, Eds., *Mixing Theory and Practice*, Vol. 1, Academic Press, New York, 1966, Chapter 3.

54. J. B. Gray, in V. W. Uhl and J. B. Gray, Eds., *Mixing Theory and Practice*, Vol. 1, Academic Press, New York, 1966, Chapter 4.

55. R. T. Johnson, *I&EC Proc. Des. Dev.*, **6**, 340 (1967).

56. J. Sawinsky, G. Havas, and A. Deak, *Chem. Eng. Sci.*, **31**, 507 (1976).

57. H. T. Chen and J. Steenrod, *Polym. Eng. Sci.*, **15**, 357 (1975).

58. K. Denbigh, *Trans. Faraday Soc.*, **43**, 648 (1947).

59. K. Denbigh, *J. Appl. Chem.*, **1**, 227 (1951).

60. B. A. Buffham and E. B. Nauman, *Chem. Eng. Sci.*, **30**, 1519 (1975).

61. E. B. Nauman and B. A. Buffham, *Chem. Eng. Sci.*, **32**, 1233 (1977).

62. E. B. Nauman, *Chem. Eng. Sci.*, **32**, 287 (1977).

63. W. C. Brasie, U.S. Patent 3,280,899 (1966).

64. J. R. Crawford, U.S. Patent 3,206,287 (1965).

65. R. Cintron-Cordero, R. A. Mostello, and J. A. Biesenberger, *Can. J. Chem. Eng.*, **46**, 434 (1968).

66. J. C. Mecklenburgh, *Can. J. Chem. Eng.*, **48**, 279 (1970).

67. T. T. Szabo and E. B. Nauman, *AIChEJ*, **15**, 575 (1969).

68. J. A. Biesenberger, M. Sacks, and I. Duvdevani, "Polymerization in Tubular Reactors", Proceedings of Symposium on Polymer Reactor Engineering, Lavel University, Quebec City, 1972.

69. W. C. Brasie, "Some Design Considerations for Reactors for Mass and Solution Polymerization", Paper presented at 63rd National AIChE Meeting, St. Louis, February 1968.

70. W. C. Brasie, *Adv. Chem.*, **109**, 101 (1972).

71. S. Agrawal and C. D. Han, *AIChEJ*, **21**, 449 (1975).

72. J. P. A. Wallis, R. A. Ritter, and H. Andre, *AIChEJ*, **21**, 686 (1975).

73. L. S. Merrill Jr. and C. E. Hamrin Jr., *AIChEJ*, **16**, 194 (1970).

74. D. Pattison, *Chem. Eng.*, **76**(11), 94 (1969).

75. S. J. Chen and A. R. MacDonald, *Chem. Eng.*, **80**(7), 105 (1973).

76. T. Bor, *Brit. Chem. Eng.*, **16**, 610 (1971).

77. C. D. Grace, *Chem. Proc. Eng.*, **52**, 57 (1971).

78. S. J. Chen, L. T. Fan, and Watson, *I&EC Proc. Des. Dev.*, **12**, 42 (1973).

79. N. R. Schott, B. Weinstein, and B. LaBombard, *Chem. Eng. Prog.*, **71**(1), 54 (1975).

80. W. D. Morris and J. Benyon *I&EC Proc. Des. Dev.*, **15**, 338 (1976).

81. R. H. Wang and L. T. Fan, *I&EC Proc. Des. Dev.*, **15**, 381 (1976).

82. W. D. Mohr, R. L. Saxton, and C. H. Jepson, *Ind. Eng. Chem.*, **49**, 1855 (1957).

83. W. D. Mohr, E. C. Bernhardt, Ed., *Processing of Thermoplastic Materials*, Reinhold, New York, 1959, Chapter 3.

84. E. B. Nauman, *AIChEJ*, **25**, 246 (1979).

85. G. S. Beveridge and R. S. Shecter, *Optimization: Theory and Practice*, McGraw-Hill, New York, 1970.

86. M. M. Denn, *Optimization by Variational Methods*, McGraw-Hill, New York, 1969.

87. Y. D. Kwon, L. B. Evans, and J. J. Nobel, *Adv. Chem.*, **109**, 98 (1972).

88. H. Nishimura and F. Yokoyama, *Kagaku Kogaku*, **32**, 610 (1968).

89. Y. Yoshimoto, et al., *Int. Chem. Eng.*, **11**, 147 (1970).

90. S. I. Lee and C. M. Crowe, *Chem. Eng. Sci.*, **25**, 743 (1970).

91. W. H. Ray and C. E. Gall, *Macromolecules*, **2**, 424 (1969).

92. G. Pinto and Z. Tadmor, *Polym. Eng. Sci.*, **10**, 279 (1970).

93. R. P. Goldstein and N. R. Amundson, *Chem. Eng. Sci.*, **20**, 195, 449, 477, 501 (1965).

94. R. Aris, *Adv. Chem.*, **109**, 578 (1972).

95. D. C. Perlmutter, *Chemical Reactor Stability*, Prentice-Hall, Englewood Cliffs, New Jersey, 1972.

96. R. Jaisinghani and W. H. Ray, *Chem. Eng. Sci.*, **32**, 811 (1977).

97. O. J. Bilous and N. R. Amundson, *AIChEJ*, **1**, 513 (1955).

98. P. J. Hoftyzer and T. H. Zwietering, *Chem. Eng. Sci.*, **14**, 241 (1961).

99. B. A. Suryanarayanan, B. A. Mujawar, and M. R. Rao, *I&EC Proc. Des. Dev.*, **15**, 564 (1976).

100. J. H. Olson and L. E. Stout, in V. W. Uhl and J. B. Gray, Eds., *Mixing Theory and Practice*, Vol. 1, Academic Press, New York, 1960, Chapter 7.

101. B. Fu, H. Weinstein, B. Bernstein, and A. B. Schaffer, *I&ED Proc. Des. Dev.*, **10**, 501 (1971).

102. P. Zaloudk, *Brit. Chem. Eng.*, **14**, 657 (1969).

103. R. L. Stokes and E. B. Nauman, *Can. J. Chem. Eng.*, **48**, 723 (1970).

104. M. Moo-Young and K. W. Chan, *Can. J. Chem. Eng.*, **49**, 187 (1971).

105. M. P. Dudukovic, *Chem. Eng. Sci.*, **32**, 985 (1977).

106. Z. Tadmor and J. A. Biesenberger, *I&EC Fund*, **5**, 336 (1966).

107. M. Harada, K. Tanaka, W. Eguchi, and S. Nagata, *J. Chem. Eng. Jap.*, **1**, 148 (1968).

108. J. A. Biesenberger and Z. Tadmor, *Polym. Eng. Sci.*, **6**, 299 (1966).

109. P. J. Flory, *J. Am. Chem. Soc.*, **58**, 1877 (1936).

110. G. Gee and H. M. Melville, *Trans. Faraday Soc.*, **40**, 240 (1944).

111. W. H. Abraham, *Chem. Eng. Sci.*, **21**, 327 (1966).

112. S. I. Liu and N. R. Amundson, *J. Chem. Rubber Tech.*, **34**, 994 (1961).

113. R. J. Zeman and N. R. Amundson, *AIChEJ*, **9**, 297 (1963).

114. P. J. Zeman and N. R. Amundson, *Chem. Eng. Sci.*, **20**, 331 (1965).

115. J. K. Beasley, *J. Am. Chem. Soc.*, **75**, 6123 (1953).

116. J. A. Biesenberger and Z. Tadmor, *J. Appl. Polym. Sci.*, **9**, 3409 (1965).

117. J. A. Biesenberger, *AIChEJ*, **11**, 369 (1965).

118. J. H. Duerksen, A. E. Hamielec, and J. W. Hodgins, *AIChEJ*, **13**, 1081 (1967).

119. J. H. Duerksen and A. E. Hamielec, *J. Polym. Sci.*, **C25**, 155 (1968).

120. H. Kilkson. *I&EC Fund*, **3**, 281 (1964).

121. W. H. Ray, *Can. J. Chem. Eng.*, **47**, 503 (1969).

122. N. H. Smith and G. A. Sather, *Chem. Eng. Sci.*, **20**, 15 (1965).

123. W. W. Graessely, *AIChE—Inst. Chem. Eng. London Symp.*, **3**, 16 (1965).

124. E. G. Teaney and R. G. Anthony, *J. Appl. Polym. Sci.*, **14**, 147 (1970).

125. D. W. T. Rippon, *I&EC Fund*, **6**, 488 (1967).

126. R. W. Dunn and C. C. Hsu, *Adv. Chem.*, **109**, 85 (1972).

127. K. Nagasubramanian and W. W. Graessley, *Chem. Eng. Sci.*, **25**, 1549 (1970).

128. A. Chatterjee, W. S. Park, and W. W. Graessley, *Chem. Eng. Sci.*, **32**, 167 (1977).

129. K. F. O'Driscoll and R. Knorr, *Macromolecules*, **2**, 507 (1969).

130. E. B. Nauman, *Chem. Eng. Sci.*, **24**, 1461 (1969).

131. M. S. K. Chen, *Chem. Eng. Sci.*, **26**, 17 (1971).

132. E. B. Nauman, *Chem. Eng. Sci.*, **25**, 1595 (1970).

133. E. B. Nauman and K. Carter, *I&EC Proc. Des. Dev.*, **13**, 275 (1974).

134. A. W. Hui and A. E. Hamielec, *J. Polym. Sci.*, **C25**, 167 (1968).

135. J. Hicks, A. Mohan, and W. H. Ray, *Can. J. Chem. Eng.*, **47**, 490 (1969).

136. W. H. Ray, *I&EC Proc. Des. Dev.*, **7**, 422 (1968).

137. R. L. Laurence and G. Vanudevan, *I&EC Proc. Des. Dev.*, **7**, 427 (1968).

138. E. B. Nauman, *Chem. Eng. Sci.*, **28**, 313 (1973).

## CHAPTER 11

Rheology

G. H. PEARSON

Research Laboratories
Eastman Kodak Company
Rochester, New York

11.1   Introduction   519

11.2   General Concepts   520
   11.2.1   Stress-Deformation   520
   11.2.2   Flow Classification   521
   11.2.3   General Flow Behavior of Polymers   522
      11.2.3.1   Steady-Flow Behavior   522
      11.2.3.2   Transient-Flow Behavior   525
   11.2.4   Fluid Classification   527
   11.2.5   Constitutive Modeling   527
      11.2.5.1   Generalized Newtonian Fluid
          (Purely Viscous)   528
      11.2.5.2   Viscoelastic Fluids   529
   11.2.6   Temperature and Pressure
         Dependence of Viscoelastic Melts   537
      11.2.6.1   Concepts, Observations, and
          Theory   537
      11.2.6.2   Temperature Dependence in
          Viscoelastic Models   539

11.3   Rheological Characterization   541
   11.3.1   Viscometric Measurements   541
      11.3.1.1   Linear Viscoelastic Response   541
      11.3.1.2   Nonlinear Viscoelastic Response 543
   11.3.2   Elongational Flow   546
      11.3.2.1   High-Viscosity Devices   547
      11.3.2.2   Low-Viscosity Devices   548
   11.3.3   Characterization through
         Viscoelastic Models   549

11.4   Rheology and Composition   550
   11.4.1   Homogeneous Systems   551
      11.4.1.1.   Linear Homopolymers/
           Compatible Copolymers   551
      11.4.1.2   Branched Polymers   556
   11.4.2   Heterogeneous Systems   557
      11.4.2.1   Suspensions/Low Viscosity   557
      11.4.2.2   Suspensions/High Viscosity   558

11.5   Rheology and Processing   559
   11.5.1   Injection Molding   559
   11.5.2   Fiber Spinning   563

11.6   Summary   565

11.7   Glossary of Symbols   566

11.8   References   567

11.1   INTRODUCTION

Rheology, as a field of science, is concerned with the response of materials to deformation. The transformation of a polymeric material from pellet to finished article is accomplished only by considerable deformation of the polymer. So, the role of rheology in polymer processing is immediately established. Unfortunately, very few processes fit the realm of experimental rheology, which generally demands one-dimensional flow in controlled, isothermal experiments. Therefore the route from simple rheological measurements to predictions of processing performance is not clear. As an example, consider injection molding, a process that imposes first extrusion, then flow at high rates through narrow tubes to a cavity, and finally, under conditions of high pressure and cooling rate, cooling to the solid state. Each portion of the process could be viewed separately in terms of simpler

519

deformations; yet, the combination of effects determines the final product.

This chapter provides a pathway for understanding the role and influence of rheology on polymer processing. The chapter begins with a discussion of basic concepts, applicable particularly to polymeric materials, such as fluid and flow classification schemes, followed by an introduction to the behavior of viscoelastic materials and a section on fluid constitutive relations. Next, rheological characterization is described, covering the standard techniques such as the capillary rheometer but emphasizing the newer methods such as orthogonal-plate rheometry and extensional-flow experiments. Before rheology applied to processing is discussed, a section describing the relationship of rheological behavior to fluid constituents is given. A section on rheology and processing is next, reviewing much of the relevant work in the field, applied to some important processes listed as examples. Finally, a shopping list for the future is given to indicate where the field is and should be moving.

11.2 GENERAL CONCEPTS

As with any field of science or technology, rheology has its own specific set of symbols and terminology that must be comprehended before one is prepared to understand and to read the literature. This section is intended as a primer for the primitive concepts of rheology. An attempt has been made to use currently accepted symbols, a list of which is given at the end of the chapter. The units used for stress and viscosity are dynes per square centimeter $(dyne/cm^2)$ and poise, respectively, owing to the prevalence of these units in the literature. In the newer SI units stress is expressed as pascals where 10 $dyne/cm^2$ is 1 Pa, and viscosity is expressed as pascal-seconds (Pa-sec).

11.2.1 Stress-Deformation

Rheology is the study of the dynamics and kinematics of materials. One wishes to know the response of a given sample to an imposed stress or strain. The concept of stress is introduced quite well in most fluid-mechanics texts. The reader should consult Bird, Stewart, and Lightfoot (1), for instance, for further understanding. For the present chapter, stress is a force/unit area that must be expressed as a tensor quantity. Tensors are used as a formal language of rheology to ensure independence from a particular coordinate frame of reference. Because one deals in three-dimensional coordinate systems, there are nine components of the stress tensor $\boldsymbol{\sigma}$; see Equation (11.1)

$$\boldsymbol{\sigma} = \begin{pmatrix} \sigma_{11} & \sigma_{12} & \sigma_{13} \\ \sigma_{21} & \sigma_{22} & \sigma_{23} \\ \sigma_{31} & \sigma_{32} & \sigma_{33} \end{pmatrix} \quad (11.1)$$

The diagonal components are known as normal stresses, whereas the off-diagonal components are shear stresses. The stress tensor is symmetric $\sigma_{ij} = \sigma_{ji}$, and the normal stresses can be determined only within an isotropic term $p\mathbf{I}$, known as pressure. This reduces the number of unknown stress components to six and those of rheological significance to five, since only two normal stress differences need to be determined.

Deformation or strain is a more difficult quantity to determine because of the lack of a natural reference state. One must describe the change in shape of a quantity of material relative to some reference, either the present shape or a shape at a specified time in the past. Consider a rod as shown in Figure 11.1, which at time $t = 0$ had a length L_0 and at the present time t' has been strained to a length L. Obviously, different values for strain ϵ will be computed depending upon the reference state:

Reference: $t = 0$

$$\epsilon_1 = \frac{L - L_0}{L_0}$$

Reference: $t = t'$

$$\epsilon_2 = \frac{L - L_0}{L}$$

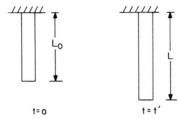

t=o   t=t'

Figure 11.1 Reference state for strain.

of-deformation tensor Δ with Cartesian components, as shown in Equation (11.4).

$$\Delta_{ij} = \left(\frac{\partial u_i}{\partial x_j} + \frac{\partial u_j}{\partial x_i}\right) \qquad (11.4)$$

The use of the strain and rate-of-strain tensors is demonstrated later as constitutive relations are considered.

For polymeric fluids, the accepted convention is to use the present configuration as a reference state. Unlike the situation illustrated in Figure 11.1, the geometric statement of strain is generally complex. The formalism used in continuum mechanics to describe strain is due principally to Noll (2). Picture a blob of fluid of arbitrary shape and label one point in the fluid X at the present time t. Then at time t, X has coordinates $x^i(t)$, and at some earlier time $t + s$, X had coordinates $\xi^i(t + s)$. Here s is a time running backward from the present $s = 0$ to the distant past $s \to -\infty$. The strain in the fluid, referenced to the present time, may be expressed by the Cauchy tensor (3) shown in Equation (11.2)

$$C_{ij} = g_{kl}(t + s)\frac{\partial \xi^k(t + s)}{\partial x^i(t)}\frac{\partial \xi^l(t + s)}{\partial x^j(t)}$$

$$(11.2)$$

or its inverse, the Finger tensor. For more information the reader should consult a reference textbook on tensors such as that by Aris (4) or works on continuum mechanics such as those by Astarita and Marrucci (5) or Lodge (6).

In addition to knowledge of the strain undergone by a fluid, the rate of strain is also important. The rate of strain is the rate of change of velocity between two points with position, $\partial u_i/\partial x_j$. It may be decomposed arbitrarily into a rigid body rotation and a deformation. The rigid rotation is described by the vorticity tensor ω with Cartesian components, Equation (11.3),

$$\omega_{ij} = \left(\frac{\partial u_i}{\partial x_j} - \frac{\partial u_j}{\partial x_i}\right) \qquad (11.3)$$

and the deformation is described by the rate-

11.2.2 Flow Classification

Deformations in polymer fabrication processes are often difficult to analyze. Generally, the flow is unsteady in both time and space, likely two- or three-dimensional, and commonly nonisothermal. Such behavior is in sharp contrast to the normal world of laboratory rheology, which relies principally on one-dimensional isothermal flow, steady in time and position. As a starting point in process analysis, one must force-fit the appropriate laboratory flows on the process flow and predict processing performance based on the known laboratory response. Two flow fields of profound importance in processing are shear flow (SF) and elongational flow (EF).

A simple shear flow is defined when the velocity v_1 is a function of, at most, time and x_2, a coordinate perpendicular to the direction of flow. Such is the case for flow in a cylindrical tube as shown in Figure 11.2. Flows of this type have a rate-of-deformation tensor with only two nonzero components,

a) Shear Flow

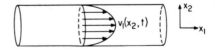

b) Elongational Flow

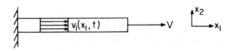

Figure 11.2 Flow classification illustrating (a) shear flow and (b) elongational flow.

Equation (11.5),

$$\Delta = \dot{\gamma} \begin{pmatrix} 0 & 1 & 0 \\ 1 & 0 & 0 \\ 0 & 0 & 0 \end{pmatrix} \qquad (11.5)$$

where $\dot{\gamma}$ is the shear rate. It can be shown that for a broad classification of fluids known as simple fluids, there are at most three material functions that characterize the response of the fluid in SF (3). If both the flow field and stress field are time-independent, then the three material functions are the shear viscosity, the primary normal stress coefficient, and the secondary normal stress coefficient as given by Equations (11.6), (11.7), and (11.8)

$$\eta = \frac{\sigma_{12}}{\dot{\gamma}} \qquad (11.6)$$

$$\Psi_{12} = \frac{N_1}{\dot{\gamma}^2} = \frac{\sigma_{11} - \sigma_{22}}{\dot{\gamma}^2} \qquad (11.7)$$

$$\Psi_{23} = \frac{N_2}{\dot{\gamma}^2} = \frac{\sigma_{22} - \sigma_{33}}{\dot{\gamma}^2} \qquad (11.8)$$

If these material functions are known as a function of shear rate, then their response in steady-shear flow is established. Much of our knowledge and predictive capability for fluid behavior is based upon the response in steady-shear flow.

Elongational flow is generated whenever the velocity v_1 varies in the direction of motion x_1. This flow is aptly illustrated in Figure 11.2 by the extension of a polymer rod. A pure uniaxial elongational flow involves changes in v_1 only in time and x_1 and results in a rate of deformation tensor with only diagonal non-zero components; see Equation (11.9):

$$\Delta = \dot{\epsilon} \begin{pmatrix} 2 & 0 & 0 \\ 0 & -1 & 0 \\ 0 & 0 & -1 \end{pmatrix} \qquad (11.9)$$

where $\dot{\epsilon}$ is the strain rate. If $\dot{\epsilon}$ is constant and the stresses are at steady state, the elongational viscosity can be defined by Equation (11.10)

$$\eta_e = \frac{\sigma_{11} - \sigma_{22}}{\dot{\epsilon}} \qquad (11.10)$$

which from simple fluid theory is the only characteristic material function in uniaxial elongational flow.

11.2.3   General Flow Behavior of Polymers

Before means of measuring the material functions are discussed, the normal flow characteristics of polymer systems are presented to provide both additional insight to responses when measurements are not feasible and guidelines for taking experimental data. Steady-flow properties are considered first, followed by a discussion of transient responses.

11.2.3.1   Steady-Flow Behavior
The steady-shear properties of polymers are undoubtedly the best known and studied. As examples data for polystyrene melts are used, but the same general behavior is found for most amorphous polymers above their glass transition temperature and for crystalline polymers above their melting temperature. The shear viscosity, shown in Figure 11.3 for polystyrene melts of various molecular weights (7), is constant at low shear rates and at higher $\dot{\gamma}$ decreases as some power function of shear rate. The low $\dot{\gamma}$ limit of η is the zero-shear-rate viscosity η_0, which is a function of the weight-average molecular weight and temperature within a given polymeric family (8), as shown in Equation (11.11).

$$\eta_0 = \lim_{\dot{\gamma} \to 0} \eta(\dot{\gamma}) = f(\overline{M}_w, T) \qquad (11.11)$$

The rate-dependent viscosity at high $\dot{\gamma}$ generally depends upon $\dot{\gamma}$ to a power between -0.5 and -0.85 for many commercial polymers (9), approaching higher powers as the polymer approaches a monodisperse distribution of molecular weights.

The normal stress coefficients Ψ_{12} and Ψ_{23} mimic the response of η although the power-law slope is higher. The primary normal stress coefficient is shown in Figure 11.4 for several polystyrene melts varying in molecular weight and molecular weight distribution (7). One

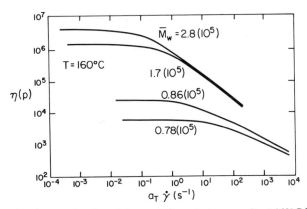

Figure 11.3 Shear viscosity as a function of shear rate for polystyrene melts at 160°C for weight-average molecular weights of 2.8(10^5), 1.7(10^5), 0.86(10^5), and 0.78(10^5).

can define a zero-shear-rate value for Ψ_{12}, as seen in Equation (11.12),

$$\Psi_{12}^0 = \lim_{\gamma \to 0} \Psi_{12} \qquad (11.12)$$

although for most materials, Ψ_{12} is very difficult to measure accurately. Few studies of Ψ_{23} have been made for lack of a good experimental technique. The most recent works of Tanner (10) and Miller and Christiansen (11) indicate that Ψ_{23} is negative and, to a first approximation, some fraction β of Ψ_{12}, where $0.1 \leqslant \beta \leqslant 0.3$. Figure 11.5 shows Ψ_{12} and Ψ_{23} for a polystyrene melt. Here β is roughly 0.15.

The significance of $\eta(\dot{\gamma})$ data for processing steps such as extrusion is obvious. The importance of Ψ_{12} and presumably Ψ_{23}, although little information is available owing to lack of data for Ψ_{23}, is that they indicate the elastic nature of the melt. As we develop a better feeling for the relationship between rheology and processing, this point will be explored more fully.

For elongational flow, only one property η_e is needed to characterize response. Unfortunately, η_e is available for most materials only at low- or zero-strain rate. The problem is one of designing a suitable experimental technique for measuring η_e, since for Equation (11.10) to hold, both stress and strain rate must be

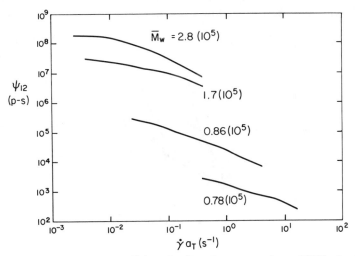

Figure 11.4 Primary normal stress coefficient Ψ_{12} for polystyrene melts at 160°C of weight-average molecular weights 2.8(10^5), 1.7(10^5), 0.86(10^5), and 0.78(10^5).

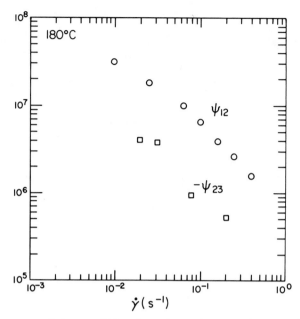

Figure 11.5 Primary normal stress coefficient Ψ_{12} and secondary normal stress coefficient $-\Psi_{23}$ for a polystyrene melt at 180°C.

independent of time. Generally, steady-state flow is not achieved during the experiment, leading only to transient responses. This problem receives more attention later in Section 11.3.3 where techniques for studying elongational flow for both solutions and melts are discussed.

The studies of Stevenson (12) on an isobutylene-isoprene copolymer, Ballman (13) on polystyrene, Vinogradov et al. (14) on polyisobutylene, and Meissner (15) on low-density polyethylene indicate that at low $\dot{\varepsilon}$,

Equation (11.13) applies

$$\eta_e = 3\eta_0 \qquad (11.13)$$

in uniaxial elongation. As the strain rate or tensile stress increases, it would appear that η_e may decrease, as reported by Münstedt for polystyrene (16), or be unmeasurable since no steady-state region of stress or strain rate can be obtained experimentally. Such is the case with the higher $\dot{\varepsilon}$ data of Meissner (15) shown in Figure 11.6. The extension of η_e data using

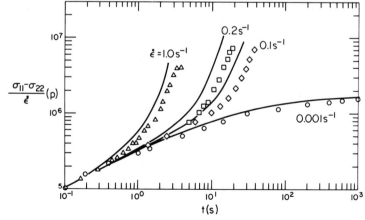

Figure 11.6 Tensile stress divided by strain rate for a low-density polyethylene melt from Meissner (15). The lines are predictions of the rubberlike liquid.

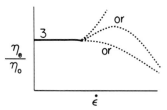

Figure 11.7 Possible elongational viscosity behavior as a function of strain rate.

constitutive relations is not advisable since, as Figure 11.7 demonstrates, the predicted η_e may increase without bound, initially increase but go through a maximum and decrease with increasing $\dot{\epsilon}$, or decrease as the shear viscosity. Specific models yielding each type of behavior are discussed in Section 11.2.5.2. The correct behavior is likely seen only when data over a wide range of rates are available. A complete description of $\eta_e(\dot{\epsilon})$ must await better techniques and more data.

11.2.3.2 Transient-Flow Behavior

Polymeric fluids possess a memory in the sense that they "remember" previous states of deformation. This memory is seen most clearly in the response to changing deformation patterns. Although fascinating, the transient response also is of great significance to processing since most processes subject the polymer to rapid changes in deformation state.

To illustrate the transient response, consider two deformation programs as illustrated in Figure 11.8. A typical response to a shear start-up experiment is illustrated for shear stress in Figure 11.9 using the data of Carreau

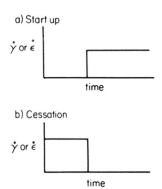

Figure 11.8 Sketch of two model transient programs, (a) start-up and (b) cessation.

(17). For convenience the data have been normalized to their steady-state values. Note particularly that as $\dot{\gamma}$ increases, the transient stress starts to overshoot and then returns toward the steady-state value. The time to reach the stress maximum for shear stress is a decreasing function of rate and occurs at a relatively constant shear strain $\dot{\gamma}t$, as shown in Figure 11.10. Here data for several polymer solutions are plotted together, normalized by a time constant for each fluid. The normal stress difference maxima generally occur at shear strains $\dot{\gamma}t$, roughly twice that at which the shear stress peaks, as given by Equation (11.14).

$$\dot{\gamma}t\,^{N_1}_{max} = 2\,\dot{\gamma}t^{\sigma_{12}}_{max} \qquad (11.14)$$

The response for start-up in elongational flow is similar to that in shear at low rates. From Figure 11.6, a monotonic approach to steady state is observed for low-density polyethylene at low $\dot{\epsilon}$, whereas at higher $\dot{\epsilon}$, steady state is not reached (15). Again more data are needed to make general statements concerning the general behavior. Constitutive relations do not help resolve the problem, for some predict no steady state (6), others predict monotonic increases to a large constant value of η_e with $\dot{\epsilon}$ (19), and still others predict stress overshoots at high $\dot{\epsilon}$ with a maximum in the $\eta_e - \dot{\epsilon}$ curve (20, 21).

When the flow is suddenly stopped, cessation, a gradual decay to zero stress, is observed for both shear flow and elongational flow. As illustrated in Figure 11.11, for shear flow there is a strong dependence on the shear rate prior to cessation. The higher the rate, the more rapid the decay. As with start-up experiments, N_1 lags σ_{12} in responding to cessation. Unfortunately the data are not available for cessation of elongational flow.

Transient effects can be generalized by considering three types of responses. Initially, the fluid behaves elastically; that is, a step change in strain is followed by a step change in stress. If a steady state is reached, the fluid is responding viscously; the stress depends only on the rate of deformation. In between, the fluid is schizophrenic, not knowing whether or when to be elastic or viscous. In elongational

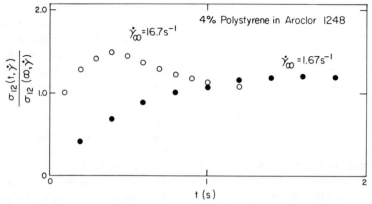

Figure 11.9 Examples of shear start-up response in a polymer solution, 4% polystyrene in Aroclor (17).

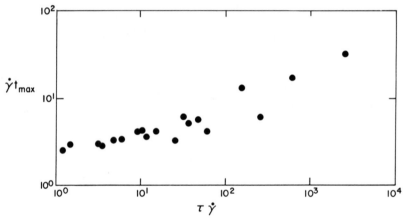

Figure 11.10 Correlation of shear strain at the stress overshoot versus Weissenberg number for polymer solutions.

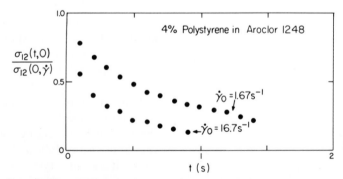

Figure 11.11 Example of stress relaxation in a cessation experiment for 4% polystyrene in Aroclor (17).

flow, the fluid quite often succumbs and breaks down before viscous behavior is realized.

11.2.4   Fluid Classification

It has been noted that polymeric fluids respond to a transient deformation according to a characteristic time scale for the fluid. The time constant can be viewed conceptually as indicating the time necessary to break and/or form temporary structural units within the fluid. For fluids such as water, the time constants are exceedingly small, $O(10^{-13}$ sec), beyond the normal range of measurement (22). For polymers of high molecular weight, the time constants may be quite large, from seconds to hours.

Polymeric fluids tend to exhibit various types of behavior in the extremes of transient response. As discussed in Section 11.2.3.2, the responses to step changes in shear rate are elastic at short times relative to some fluid time constant or relaxation time and viscous at long times. Pipkin (23) formalizes such behavior in the fluid classification scheme shown in Figure 11.12.

Along the abscissa the rate of deformation is plotted, normalized to some fluid relaxation time τ ranging from zero to infinity. On the ordinate the magnitude of the strain is plotted again from zero to infinity. For a given polymeric fluid there will be some region at small $\tau\Delta$ and γ where the response is Newtonian, $\eta = \eta_0$. As the $\tau\Delta$ product increases, the response becomes first linear viscoelastic and finally classically elastic as long as γ is small. In a start-up of steady-shear flow, suppose that the characteristic time is initially large at a

fixed shear rate, so the initial response at small strains is elastic $\sigma_{12} = \eta(\gamma)\dot{\gamma}$. One possible course for this imaginary deformation is plotted in Figure 11.12.

The relevance of fluid classification schemes such as Pipkin's becomes obvious at the extremes of fluid behavior, some of which are found in many polymer processing steps. Consider, for example, the behavior of a highly viscous sample of polyisobutylene at room temperature. Under most deformation conditions the material will behave as a rubber in that it is highly elastic and resists flow. It will flow in a viscous fashion if the deformation rate is small. Likewise, if a fluid of low viscosity is deformed quite rapidly, such that the time scale of deformation is much smaller than τ, it responds elastically much like a rubberlike fluid. It is wise, as well, to keep Figure 11.12 in mind when modeling flow behavior, since constitutive model assumptions will be determined by the position in the diagram for the rate and strain conditions of the problem at hand.

11.2.5   Constitutive Modeling

The concepts of the previous sections can be neatly unified by considering mathematical models of fluid behavior. A constitutive model attempts to describe once and for all the response of the fluid to any arbitrary deformation at any point within Pipkin's classification scheme. So far, man has not constructed the ultimate constitutive relation but instead has developed a number of approximate forms, some with strong theoretical lineage, while others are totally empirical. By far the largest class of empirical models may be cast as generalized Newtonian fluids (GNF) and these are discussed first. Following the models of viscous behavior, approximations to viscoelastic response are presented. The intent is to provide an overview of constitutive models, not an exhaustive review of the subject, so only representative examples are discussed. For more information the monographs of Astarita and Marrucci (5), Bird, Armstrong, and Hassager (22), and Lodge (24) should be consulted. The models are compared with

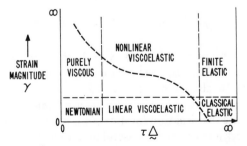

Figure 11.12   Fluid classification diagram.

data when possible, to indicate virtues and shortcomings of various approaches.

11.2.5.1 Generalized Newtonian Fluid (Purely Viscous)

The purely viscous fluid assumes that the stress is directly proportional to the rate of deformation as shown in Equation (11.15)

$$\sigma = \eta \Delta \qquad (11.15)$$

where η may be some arbitrary scalar function of Δ. A Newtonian fluid has a constant viscosity so that one has Equation (11.16).

$$\eta[f(\Delta)] = \mu = \text{constant} \qquad (11.16)$$

The more general case assumes only that η is a scalar function of Δ. There are three unique scalar functions of any tensor. These are the invariants: the first invariant is defined by Equation (11.17),

$$I_{\Delta} = tr(\Delta) \qquad (11.17)$$

the second invariant by Equation (11.18),

$$II_{\Delta} = \Delta_{ij}\Delta_{ji} \qquad (11.18)$$

where summation is indicated by a repeated index, and the third invariant by Equation (11.19).

$$III_{\Delta} = det(\Delta) \qquad (11.19)$$

If the fluid is incompressible, continuity of mass requires that I_{Δ} vanish, leaving the result shown in Equation (11.20).

$$\eta = \eta(II_{\Delta}, III_{\Delta}) \qquad (11.20)$$

The second and third invariants for the two simple flows, shear and elongation, are as follows:

	SF	EF
II_{Δ}	$2\dot{\gamma}^2$	$6\dot{\epsilon}^2$
III_{Δ}	0	$2\dot{\epsilon}^3$

All the models for the GNF developed so far

have been expressly designed to correlate shear viscosity data, so an assumption is made that η depends only on II_{Δ}; see Equation (11.21).

$$\eta = \eta(II_{\Delta}) \qquad (11.21)$$

One of the first GNF models was the power-law fluid of Ostwald (25) and deWaele (26). At high rates of shear, the viscosity becomes to a good approximation proportional to some power of $\dot{\gamma}$ as shown in Equation (11.22),

$$\eta = m\dot{\gamma}^{n-1} \qquad (11.22)$$

where m is the flow characteristic and n the power-law index. The fit of this model to $\eta(\dot{\gamma})$ data for a commercial polystyrene sample is shown in Figure 11.13. At high $\dot{\gamma}$, above $\dot{\gamma} = 1$ sec^{-1} the fit is adequate and demonstrates the utility of the model. Many processes are run in this deformation range, and good predictions can be made with a minimum of data fitting. One potential problem is also shown in Figure 11.13. Below 1 sec^{-1}, the power-law model begins to overpredict the viscosity and in fact as $\dot{\gamma} \rightarrow 0$, $\eta_{PL} \rightarrow \infty$, in contrast to the known zero-shear-rate response η_0. When using the model, one must be cautious and apply the approximation of power-law behavior only over a range of applicable shear rates.

Carreau (27) has corrected the power-law model's obvious deficiency by a simple adjustment as given in Equation (11.23).

$$\frac{\eta}{\eta_0} = [1 + (\tau\dot{\gamma})^2]^{(n-1)/2} \qquad (11.23)$$

At high $\dot{\gamma}$, the power-law form is retained as shown in Equation (11.24) when

$$(\tau\dot{\gamma})^2 \gg 1 \qquad \eta = \eta_0(\tau\dot{\gamma})^{n-1} \quad (11.24)$$

where $m = \eta_0\tau^{n-1}$. As $\dot{\gamma} \rightarrow 0$, the viscosity approaches η_0. The fit of this fluid to $\eta(\dot{\gamma})$ data is also shown in Figure 11.13.

The power-law and Carreau models illustrate the situation for $\eta = \eta(\dot{\gamma})$, but there are occasions where it is useful to have $\eta = \eta(\sigma_{12})$ instead. A popular model giving viscosity as a

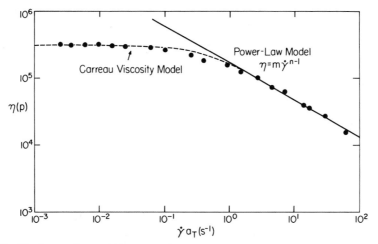

Figure 11.13   Fit of power-law and Carreau models to melt viscosity data for polystyrene, $m = 1.65(10^5)$ dyne/cm$^2 \cdot$ sec$^{.46}$, $n = 0.46$; $\eta_0 = 3.3(10^5)$ p, $\tau = 4.2$ sec.

function of shear stress is the Ellis fluid (28) as defined in Equation (11.25)

$$\frac{\eta}{\eta_0} = \left[1 + \left(\frac{\sigma_{12}}{\sigma_{12}*}\right)^{n-1}\right]^{-1} \quad (11.25)$$

where σ_{12} is the shear stress at which $\eta = 0.5\ \eta_0$. A fit of the same $\eta(\dot\gamma)$ data of Figure 11.13 replotted as $\eta(\sigma_{12})$ is shown in Figure 11.14 using the Ellis model. As Bird et al. (22) mention, the advantages of the Ellis model are the ease of use and the body of flow problems

that have been solved analytically using this model.

11.2.5.2   Viscoelastic Fluids

The models presented in Section 11.2.5.1 describe only the shear-viscosity behavior of polymeric fluids. To predict the elastic properties observed during the deformation of such fluids, more sophisticated models are needed. As for purely viscous fluids, many models have been proposed. Unfortunately, the viscoelastic constitutive equations do not

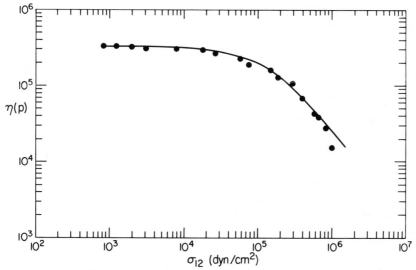

Figure 11.14   Fit of Ellis model to melt viscosity data for polystyrene, $\sigma*_{12} = 1.35(10^5)$ dyne/cm^2, $\eta_0 = 3.3(10^5)$ p, $(n - 1) = 1.24$.

have a uniformity of mathematical form; some exist as integral expressions, others are available as differential equations, some exist as both, and still others, approximations to behavior at extremes of Figure 11.12, are presented as algebraic equations. All are presented in tensor notation to satisfy the principle of material objectivity.

Although there is this diversity of form, initially very confusing to a beginner in the field, there is order as well. The models may be classified into two groups: those arising from simple fluid theory and those formed empirically by mechanical analogs of polymer dynamics. At present, both types are briefly reviewed giving predictions for many of the more popular forms and indicating directions for continued work in the field.

The theory leading to the concept of a simple fluid is the cornerstone of modern constitutive analysis for viscoelastic fluids. As stated by Noll (2), there are four major principles used to define a simple fluid: (1) determinism of stress, (2) local action, (3) nonexistence of a natural state, and (4) fading memory. Eliminating the mathematical formalism normally attached to these principles, they can be described in simple terms. A good discussion of simple fluid theory may be found in the work of Astarita and Marrucci (5). Determinism of stress states that the stress is specified by the entire history of deformation. Local action restricts the stress at any point to be determined only by the history of deformation within some small neighborhood of that point in the fluid. Nonexistence of a natural state reemphasizes the problem of determining strain in a fluid. There is no preferred natural state of a fluid. Finally, fading memory requires that present events are weighted most in the determinism of stress and that events occurring in the far-distant past are given least weight.

Although the general mathematical description of a simple fluid requires that stress is some functional over the past history of the deformation

$$\sigma = \underset{S=-\infty}{\overset{S=0}{H}} (strain) \qquad (11.26)$$

limiting cases of Equation (11.26) can be derived leading to meaningful models of real-fluid behavior. If the motion is slow in the sense that changes in the deformation are occurring much slower than some natural time scale for the fluid, then the simple fluid reduces to the fluid of grade n (29) shown in Equation (11.27)

$$\mathbf{T} = -p\mathbf{I} + \sigma = \sum_{n=0}^{N} S_n \qquad (11.27)$$

where $\mathbf{T}$ is the total stress and the first three summands are defined in Equations (11.28a), (11.28b), and (11.29c).

$$S_0 = -pI \qquad (11.28a)$$

$$S_1 = \alpha_1 A^{(1)} \qquad (11.28b)$$

$$S_2 = \alpha_2 (A^{(1)})^2 + \alpha_3 A^{(2)} \qquad (11.28c)$$

The tensor $\mathbf{A}$ is the Rivlin-Erickson deformation rate tensor (5). For the two basic flows, simple shear (shear flow, SF) and simple extension (SE) $\mathbf{A}^{(1)}$ and $\mathbf{A}^{(2)}$ are given in Equations (11.29a), (11.29b), (11.29c), and (11.29d).

$$\text{SF: } \mathbf{A}^{(1)} = \Delta = \dot{\gamma} \begin{pmatrix} 0 & 1 & 0 \\ 1 & 0 & 0 \\ 0 & 0 & 0 \end{pmatrix} \qquad (11.29a)$$

$$\mathbf{A}^{(2)} = \dot{\gamma}^2 \begin{pmatrix} 0 & 0 & 0 \\ 0 & 2 & 0 \\ 0 & 0 & 0 \end{pmatrix} \qquad (11.29b)$$

$$\text{SE: } \mathbf{A}^{(1)} = \Delta = \dot{\epsilon} \begin{pmatrix} 2 & 0 & 0 \\ 0 & -1 & 0 \\ 0 & 0 & -1 \end{pmatrix} \qquad (11.29c)$$

$$\mathbf{A}^{(2)} = \dot{\epsilon}^2 \begin{pmatrix} 4 & 0 & 0 \\ 0 & 1 & 0 \\ 0 & 0 & 1 \end{pmatrix} \qquad (11.29d)$$

The important models derived from this approximation for slow motion are the fluid of grade 1, Equation (11.30),

$$\sigma = \alpha_1 \mathbf{A}^{(1)} \qquad (11.30)$$

and the fluid of grade 2, Equation (11.31).

$$\boldsymbol{\sigma} = \alpha_1 \mathbf{A}^{(1)} + \alpha_2 (\mathbf{A}^{(1)})^2 + \alpha_3 \mathbf{A}^{(2)} \quad (11.31)$$

As long as α_1 is assumed constant, the fluid of grade 1 is equivalent to the Newtonian fluid with $\alpha_1 = \mu$. The fluid of grade 2 introduces the effects of fluid elasticity, predicting a constant shear viscosity, Equation (11.32),

$$\eta(\dot{\gamma}) = \alpha_1 = \mu \quad (11.32)$$

but nonzero normal stress differences in shear as shown in Equations (11.33a) and (11.33b)

$$N_1 = -2\alpha_3 \dot{\gamma}^2 \quad (11.33a)$$

$$N_2 = (2\alpha_3 + \alpha_2)\dot{\gamma}^2 \quad (11.33b)$$

and a strain-rate-dependent extensional viscosity given in Equation (11.34).

$$\eta_3 = 3\mu + 3(\alpha_2 + \alpha_3)\dot{\epsilon} \quad (11.34)$$

We know from shear flow experiments that N_1 is positive and $N_2 \approx -1.6\alpha_3$. These values for α_2 and α_3 imply that η_e is constant at low ϵ and increases as $\dot{\epsilon}$ increases.

While the slow-motion approximation leads to useful results as viscoelastic perturbations upon viscous flow models, transient behavior is not predicted. The limiting form for transient behavior is the integral expansion, first derived by Green and Tobolsky (30) and later formalized in single-integral form by Lodge as the rubberlike liquid (31), Equation (11.35),

$$\boldsymbol{\sigma} = \int_{-\infty}^{t} N(t - t') \, \mathbf{C}^{-1}(t, t') \, dt' \quad (11.35)$$

where $N(t - t')$ is the memory function for a fading memory expressed in discrete form as given in Equation (11.36),

$$N(t - t') = \sum_{n=1}^{N} \frac{G_n}{t_n} \exp\left[-\frac{(t - t')}{\tau}\right]$$

$$(11.36)$$

and $\mathbf{C}^{-1}$ is the Finger strain tensor. The memory function can also be described in continuous form as in Equation (11.37).

$$N(t - t') = \int_{-\infty}^{\infty} H(\tau) \exp\left[-\frac{(t - t')}{\tau}\right] d \ln \tau$$

$$(11.37)$$

Some important steps have been omitted in the presentation of Equation (11.35). Specifically, multiple integral expansions were not presented as a more general case. The practical fact remains that the rubberlike liquid with minor modifications provides the basis for several major constitutive relations in use today. To introduce nonlinear viscoelastic response, the memory function is modified such that it becomes a function of the state of deformation. Before discussing these other models, it is instructive to review the behavior of the rubberlike liquid.

In steady-shear flow, the rubberlike liquid predicts a constant viscosity as shown in Equation (11.38),

$$\eta(\dot{\gamma}) = \sum_{n=1}^{N} G_n \tau_n = \eta \quad (11.38)$$

a constant first normal stress coefficient given in Equation (11.39),

$$\psi_{12}(\dot{\gamma}) = 2 \sum_{n=1}^{N} G_n \tau_n^2 \quad (11.39)$$

and a zero second normal stress difference. The discrete spectrum has been used merely for convenience. If a combination of the Finger and Cauchy tensors had been used in Equation (11.35) $\left[(1 + \frac{\epsilon}{2})\mathbf{C}^{-1} + \frac{\epsilon}{2}\mathbf{C}\right]$ instead of simply $\mathbf{C}^{-1}$, a second normal stress coefficient is predicted. Owing to a lack of good measurement techniques for N_2, generally only $\mathbf{C}^{-1}$ is used. The elongational viscosity increases with $\dot{\epsilon}$ and becomes discontinuous at a finite extension rate: see Equation (11.40).

$$\eta_e(\epsilon) = \sum_{n=1}^{N} \frac{3G_n \tau_n}{(1 - 2\tau_n \dot{\epsilon})(1 + \tau_n \dot{\epsilon})} \quad (11.40)$$

For transient-shear flows, specifically start-up such as that shown in Figure 11.8, the rubberlike liquid predicts a monotonic rise to steady state at all shear rates, a behavior seen only for low $\dot\gamma$ with real fluids as illustrated in Figure 11.9. In cessation all relaxation occurs at the same rate, again unlike "real" behavior shown in Figure 11.11 for rate dependence of stress relaxation following sudden cessation. For start-up of simple extension, at low $\dot\epsilon$ below $1/(2\tau_1)$ where finite viscosities are obtained, a monotonic rise to steady state is predicted. Above $\dot\epsilon = 1/(2\tau_1)$, an unbounded monotonic rise in tensile stress is predicted.

The rubberlike liquid does provide a first approximation to transient-flow behavior. In shear flow, it is qualitatively correct at low rates. In extensional flow, Lodge and Chang (24) have successfully modeled the stress-growth data of Meissner (15) to reasonably high extension rates as shown in Figure 11.6.

The extension of Lodge's concept to predict nonlinear viscoelastic response more successfully has been achieved generally by changing the memory function as mentioned earlier. Many techniques for improving the constitutive assumption have been employed but these may be conveniently classified into models with either (1) rate-dependent memory, (2) strain-dependent memory, or (3) stress-dependent memory. Examples in each category are briefly discussed with an attempt to provide the spirit of the modification and not a detailed discussion of the reasoning that led to the model's form. The reader should consult the original references for further background.

The rate-dependent models introduce a scalar invarient of Δ, generally II_Δ, into $N(t - t')$, making either τ_n, G_n, or both decrease as II_Δ increases. The concept is that the fluid's memory is decreased by the rate of deformation. A prime example of a fluid with a rate-dependent memory is the Bogue-White model (32) expressed in Equations (11.41)–(11.44), retaining the general form of Equation (11.35) but replacing $N(t - t')$ by

$$N(t - t', \mathrm{II}_\Delta) = \sum_{n=1}^{N} \frac{G_n}{\tau_n^e} \exp \frac{(-t - t')}{\tau_n^e}$$

(11.41)

where τ_n^e is an effective relaxation time,

$$\tau_n^e = \frac{\tau_n}{1 + a_{BW}\tau_n(\langle \mathrm{II}_\Delta \rangle^{1/2}/2)}$$

(11.42)

and

$$\langle \frac{\mathrm{II}_\Delta}{2} \rangle^{1/2} = \frac{1}{s} \int_0^s \frac{\mathrm{II}_\Delta}{2}^{1/2} ds$$

(11.43)

where

$$s = t - t'$$

(11.44)

The steady-state properties of the Bogue-White model are given in Table 11.1, showing that unlike those of the rubberlike liquid, the shear viscosity is a function of shear rate, as is the primary normal stress coefficient. The elongational viscosity still becomes unbounded at some critical shear rate as long as the parameter $a_{BW} < 2/\sqrt{3}$, which is generally the case.

The Bogue-White model predicts transient response in shear reasonably well as demonstrated by Bogue and Chen (32) for start-up of steady-shear flow. Stress overshoots as well as a dependence of the approach to steady state on the shear rate are predicted, in qualitative agreement with data. The model seems adequate for extensional flow transients as seen in Figure 11.15 for tensile-stress transients of a polystyrene melt. It is worthwhile noting that the data do not indicate a region of steady-state stress, so the comparison of long-term response with the model's prediction cannot be made. For this specific material and strain-rate range the model predicts an infinite η_e.

The fluid's memory can be made strain dependent by including some function of the invariant of the strain tensor in the memory. A useful yet simple approximation is an exponential function shown in Equation (11.45)

$$N(t - t', \mathrm{I}_C) = \sum_{n=1}^{N} \frac{G_n}{\tau_n} \exp - \frac{(t - t')}{\tau_n}$$
$$\exp (- a_{BKZK} \sqrt{\mathrm{I}_C - 3})$$

(11.45)

This model was proposed by Kaye and Kennett (33) as a special form of the BKZ fluid (34).

Table 11.1   Steady-Shear Properties of Some Integral Constitutive Relations

Model	η	N_1
Rubberlike liquid	$\displaystyle\sum_{n=1}^{N} G_n \tau_n$	$\displaystyle 2\sum_{n=1}^{N} G_n \tau_n^2$
Bogue-White	$\displaystyle\sum_{n=1}^{N} \frac{G_n \tau_n'}{(1 + a_{BW}\tau_n\dot\gamma)}$	$\displaystyle 2\sum_{n=1}^{N} \frac{G_n \tau_n^2}{(1 + a_{BW}\tau_n\dot\gamma)^2}$
BKZK	$\displaystyle\sum_{n=1}^{N} \frac{G_n \tau_n}{(1 + a_{BKZK}\tau_n\dot\gamma)^2}$	$\displaystyle 2\sum_{n=1}^{N} \frac{G_n \tau_n^2}{(1 + a_{BKZK}\tau_n\dot\gamma)3}$
Structural	$\displaystyle\sum_{n=1}^{N} G_n^0 \tau_n^0 x_n^{2.4}$	$\displaystyle 2\sum_{n=1}^{N} G_n^0 \tau_n^0 x_n^{3.4}$

The steady-state flow properties for shear flow are listed in Table 11.1. They are not unlike those of the Bogue-White fluid, as might be expected since shear-flow measurements provide the first test of a constitutive assumption's worth. The correspondence in terms of predictive capability follows to transient-shear flow as well. Figure 11.16 illustrates the prediction of the BKZK model for start-up of a poly(methyl methacrylate) melt. Qualitatively the model responds properly, particularly at low shear rates. At higher shear rates, the model deviates from the data by not predicting a stress overshoot until $\dot\gamma$ approximates 10 sec^{-1}. The data at 0.1 sec^{-1}, on the other hand, exhibit an overshoot. This semi-quantitative fit of data is typical of the models presented here. Carreau (17) has developed highly sophisticated and specialized models that can predict transient response in specific shear tests more accurately. These models are cumbersome to manipulate for a general deformation, and as such it would seem unlikely that they would find use in process analysis.

Major differences arise when considering elongational flow. The elongational viscosity of the BKZK fluid is finite at all strain rates except when $a_{BKZK} = 0$ (rubberlike liquid behavior). This η_e prediction is shown in Figure 11.17 as a function of $\tau\dot\epsilon$ for a single relaxation time form with a_{BKZK} as a parameter. The transient elongational flow behavior is

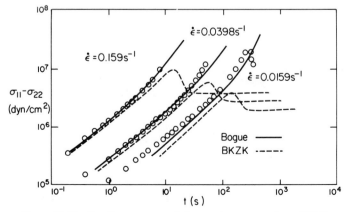

Figure 11.15   Tensile stress growth for a polystyrene melt at 130°C.

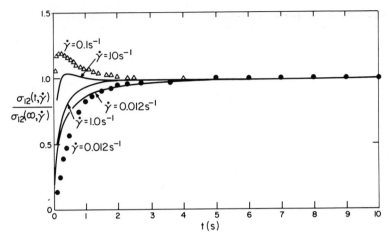

Figure 11.16 Stress growth for start-up of steady-shear flow for a poly(methyl methacrylate) melt at 210°C. The lines are predictions of the BKZK model.

shown in Figure 11.15 compared with data and the Bogue-White model. As in start-up of steady-shear flow, the stress is predicted to exhibit an overshoot and then approach steady state. Data are not available yet to verify this prediction. Note that both models, despite their differences in predictions of η_e, fit the short-time transient data well, indicating that either might be adequate for a highly transient extensional flow.

The Lodge rubberlike liquid, Bogue-White, and BKZK models can be cast in macromolecular terms in a qualitative sense. Considering a concentrated solution or melt as an entangled network with temporary junctions, the memory function is an indication of the probabilities of entanglement formation and destruction in the network, see Equation (11.46a)

$$N(t - t') = \sum_n L_n \exp\left(-\frac{(t - t')}{\tau_n}\right) \quad (11.46a)$$

where L_n is the probability of formation and τ_n^{-1} the probability of destruction. For the rubber-like liquid, L_n and τ_n^{-1} are constant, corresponding to G_n and τ_n^{-1}, respectively. The Bogue-White model reduces both L_n and τ_n^{-1} according to the rate of deformation. The BKZK model reduces L_n by a term proportional to strain; see Equation (11.46b)

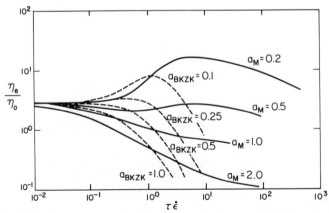

Figure 11.17 Elongational viscosity predictions for single relaxation time forms of the BKZK model and the structural model.

$$L_n = \frac{G_n}{\tau_n} \exp\left(-a_{BKZK}\sqrt{I_C - 3}\right) \qquad (11.46b)$$

Acierno et al. (35) have proposed a model, known as the structural model, with the network being formed by Brownian motion and destroyed by stress with a separate kinetic equation for entanglement concentration.

The structural model beings with a generalized statement of the Lodge rubberlike liquid shown in Equation (11.47)

$$\sigma = \int_{-\infty}^{t} \sum_{n=1}^{N} \frac{G_n(x_n)}{\tau_n(x_n)} \exp\left(-\int_{t'}^{t} \frac{dt''}{\tau_n(x_n)}\right)$$
$$\mathbf{C}^{-1}(t, t')\, dt' \quad (11.47)$$

where G_n and τ_n are assumed to be functions of the relative concentration of entanglement junctions x, $0 \leqq x \leqq 1$. The concentration of junctions is determined by a kinetic equation shown in Equations (11.48) and (11.49),

$$\frac{dx_n}{dt} = \frac{1 - x_n}{\tau_n} - \frac{a_m x_n \{\sqrt{[tr(\sigma)]/G_n}\}}{\tau_n} \quad (11.48)$$

with

$$G_n = G_n^0 x_n, \qquad \tau_n = \tau_n^0 x_n^{1.4}, \qquad \eta_n = G_n \tau_n$$
$$(11.49)$$

where the first term on the right-hand side indicates creation of junctions by Brownian motion and the second term, destruction of junctions by stress. The behavior of this model compared with experimental data seems worth the effort to solve the extra equations for x (35–37). The shear functions are listed in Table 11.1. For extensional flow, as with the BKZK model, η_e is predicted to exhibit a maximum as a function of strain rate, dependent upon the variable a_M as shown in Figure 11.17.

The fluid models discussed so far for viscoelastic fluids have been integral models. The first models of viscoelasticity, deriving their origin from mechanical analogs, were rate equations. If one considers a Newtonian response and adds to it an elastic term, then the Maxwell model results; see Equation (11.50)

$$\sigma + \tau\dot{\sigma} = \eta\Delta \qquad (11.50)$$

where $\dot{\sigma}$ is some properly formed tensor derivative of stress. The mechanical analogs have by and large fallen by the wayside, yet the rate models remain, primarily owing to their utility in solving flow problems. Most flow problems are posed in terms of differential equations, so with the constitutive form also a differential equation, solutions can be sought by classical methods. Using an integral model with a differential formulation of the physical problem creates difficulties.

A considerable body of literature is available concerning rate models; specifically, the reader should consult the work of Bird and his co-workers in this area (22,38, 39). As with the integral forms, several models are available using two of the accepted, but different, forms for $\dot{\sigma}$, the Jaumann or corotational derivative, Equation (11.51),

$$\frac{\mathcal{D}\,\sigma_{ij}}{\mathcal{D}\,t} = \frac{\partial\sigma_{ij}}{\partial t} + V_k \frac{\partial\sigma_{ij}}{\partial x_k} + \frac{1}{2}\left(\omega^{im}\sigma^{mj} - \omega^{mj}\sigma^{im}\right)$$
$$(11.51)$$

and the Oldroyd or codeformational derivative with contravariant components, Equation (11.52).

$$\frac{\mathcal{d}\sigma^{ij}}{\mathcal{d}t} = \frac{\partial\sigma^{ij}}{\partial t} V_k \frac{M\sigma^{ij}}{\partial x_k} \frac{1}{2}\Delta^{ik}\sigma^{kj} - \frac{1}{2}\Delta^{jk}\sigma^{ik}$$
$$(11.52)$$

In both Equations (11.51) and (11.52) a Cartesian coordinate system is assumed.

To illustrate the predictions of the two types of Maxwell models, corotational (CR) versus codeformational (CD), Table 11.2 was prepared listing the steady-shear and extensional-flow properties of each, assuming the model form of Equation (11.50). Beginning with the shear viscosity, the CD model predicts no shear rate dependence, whereas the CR model predicts too much dependence, $\eta \alpha \gamma^{-2}$. Similar comments apply to predictions for N_1. For the extensional viscosity, the η_e prediction of the CD model indicates an ever-increasing viscosity approaching infinity as $\dot{\epsilon} \to \frac{1}{2}\tau$, whereas the CR model predicts $\eta_e = 3\eta_0$ at all strain

Table 11.2 Properties of Codeformation and Corotational Maxwell Models (Discrete Spectrum)

Model	η	N_1	η_c
Codeformational	$\displaystyle\sum_{n=1}^{N} G_n \tau_n$	$\displaystyle 2\sum_{n=1}^{N} G_n \tau_n^{\,2}$	$\displaystyle 3\sum_{n=1}^{N} \frac{G_n \tau_n}{(1 - 2\tau_n\dot{\epsilon})(1 + \tau_n\dot{\epsilon})}$
Corotational	$\displaystyle\sum_{n=1}^{N} \frac{G_n \tau_n}{1 + (\tau_n\dot{\gamma})^2}$	$\displaystyle 2\sum_{n=1}^{N} \frac{G_n \tau_n^{\,2}}{1 + (\tau_n\dot{\gamma})^3}$	$\displaystyle 3\sum_{n=1}^{N} G_n \tau_n$

rates. For transient flow (illustrated in Figure 11.18 for both models) the CR model predicts damped oscillations about steady state for start-up of shear flow with overshoots at high $\dot{\gamma}$. These overshoots are considerably in excess of that observed (40). The CD model predicts a monotonic rise to steady state. In start-up of extensional flow, the CR model predicts a monotonic approach to steady state consistent with the CD model at low $\dot{\epsilon}$. At high $\dot{\epsilon}$, the CD model predicts extensional stresses which increase without bound.

The codeformational Maxwell model predicts results that are the same as those from the rubberlike liquid, as seen by comparison with Table 11.1. Indeed, Lodge (41) has shown that the two are equivalent, one simply the differential statement of the other. Although equivalent differential forms of most integral models should conceivably be tractable, few are known. The structural model discussed does have an equivalent using the codeformational framework; see Equation (11.53).

$$\frac{1}{G_n}\sigma_n + \tau_n \frac{\lozenge}{\lozenge t}\left(\frac{1}{G_n}\sigma_n\right) = \eta_n \Delta$$

$$(11.53)$$

The straightforward cure of the Maxwell model's deficiencies with either derivative form is to allow τ and η to be functions of II_Δ. The White-Metzner model (42), Equation (11.54),

$$\sigma + \tau \ (\mathrm{II}_\Delta)\frac{\partial\sigma}{\partial t} = \eta \ (\mathrm{II}_\Delta)\Delta \qquad (11.54)$$

has found great utility as a simple, yet in many cases accurate, constitutive representation. More recently the modified Jaumann form shown in Equation (11.55)

$$\sigma + \tau \ (\mathrm{II}_\Delta)\frac{\mathfrak{D}\,\sigma}{\mathfrak{D}\,t} = \eta(\mathrm{II}_\Delta)\Delta \qquad (11.55)$$

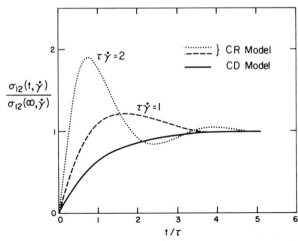

Figure 11.18 Start-up of steady-shear flow modeled by the CR and CD forms of the Maxwell fluid.

has been used to predict the stresses arising in collapse of bubbles within viscoelastic solutions (43).

11.2.6 Temperature and Pressure Dependence of Viscoelastic Melts

So far only the deformation dependence of various viscoelastic properties have been discussed. All the examples given have been for steady flow under isothermal conditions. In practice, rarely are materials processed isothermally, so a knowledge of the temperature dependence of the viscoelastic properties is essential to an analysis of most polymer fabrication operations.

11.2.6.1 Concepts, Observations, and Theory

Changes in temperature alter the time scale of the response of viscoelastic fluids by increasing or decreasing macromolecular mobility. Increases in temperature increase mobility and decrease both the viscosity and the relaxation time. Decreases in temperature reduce mobility and increase both the viscosity and the relaxation time. Empirically, it has been known for some time through the early work of Leaderman (44) that the full frequency or rate response of a material can be obtained by taking measurements at various temperatures over a fixed rate range. Then, as illustrated by Figure 11.19 for the loss modulus G'' (a linear viscoelastic property discussed in Section 11.3.1) in a poly(methyl methacrylate) sample (PMMA) as a function of frequency, a master curve can be constructed by shifting horizontally by the appropriate factor a_T to achieve superposition. The master curve of G'' versus ωa_T is shown in Figure 11.20 with the a_T shift factors given in Figure 11.21. As well as the horizontal shift, a small vertical shift is required as given by Equation (11.56)

$$G''_{T_{ref}} = G''_{T_{test}} \cdot \frac{T_{ref}\rho_{ref}}{T_{test}\rho_{test}} \quad (11.56)$$

where T_{ref} and T_{test} are given in an absolute temperature scale (45). The shift factor gives the change in zero-shear-rate viscosity or maximum relaxation time directly by Equation (11.57) or (11.58).

$$a_T = \frac{\eta_0 T_{test}}{\eta_0 T_{ref}} \cdot \frac{T_{ref}\rho_{ref}}{T_{test}\rho_{test}} \quad (11.57)$$

$$a_T = \frac{\tau_{T_{test}}}{\tau_{T_{ref}}} \quad (11.58)$$

The concept of using temperature to vary the time scale of response of the fluid and thereby

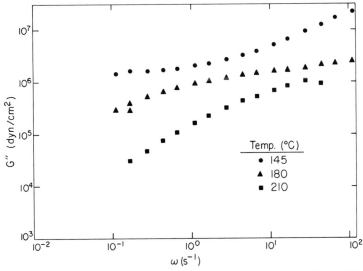

Figure 11.19 Loss modulus versus frequency at various temperatures for PMMA.

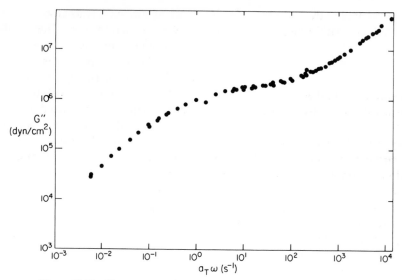

Figure 11.20 Master curve of G'' versus ωa_T for PMMA, $T_{ref} = 180°$C.

increase the frequency range of measurement is known as time-temperature superposition.

Fortunately, there is a theoretical framework for predicting the shape and quantitative magnitude of a_T versus T that is applicable to many polymer systems including homopolymers, copolymers, and compatible blends. If we assume that the viscosity of a system (all viscosities to follow are implied to be zero-shear-rate viscosities) is a function of the free volume available for macromolecular mobility, then the Doolittle Equation (11.46), first developed for low molecular weight liquids, yields Equation (11.59).

$$\ln \eta = \ln a + \frac{b(v - v_f)}{v_f} \qquad (11.59)$$

In Equation (11.59) the total volume v is assumed to be the sum of the occupied volume

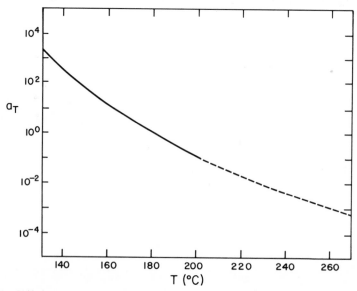

Figure 11.21 Shift factors for time-temperature superposition of PMMA data shown in Figure 11.20, (——) WLF equation, (---) Arrhenius equation.

v_0 and the free volume v_f. The temperature dependence of η from Equation (11.59) expressed as $\ln a_T$ is shown in Equation (11.60)

$$\ln a_T = b\left(\frac{1}{f_T} - \frac{1}{f_{T_{ref}}}\right) \quad (11.60)$$

where f is the free volume fraction v_f/v. If one assumes that f is linearly dependent upon temperature, then f depends on T as shown in Equation (11.61)

$$f_T = f_{T_{ref}} + \alpha_f(T - T_{ref}) \quad (11.61)$$

where α_f is the volume expansion coefficient, then the resulting form of Equation (11.60), known as the WLF equation (45), is given by Equation (11.62).

$$\ln a_T = -\frac{C_1 (T - T_{ref})}{C_2 + (T - T_{ref})} \quad (11.62)$$

For the special case of T_{ref} equal to T_g, the glass transition temperature, C_1 and C_2 have been determined empirically for several systems, a few of which are listed in Table 11.3. A more complete list is given by Ferry (45). As a first approximation for $T_{ref} = T_g$, C_1 is 17.44 and C_2 is 51.6. These numbers were obtained by fitting data from several polymer systems (45).

At temperatures roughly $100°$ C above T_g, the WLF equation does not predict observed a_T factors adequately. In this range an Arrhenius expression is more accurate, expressing η as an exponential function of absolute temperature; see Equation (11.63),

$$\eta = A \exp\left(\frac{\Delta E}{R_g T}\right) \quad (11.63)$$

where ΔE is the activation energy for flow.

The shift factor then can be expressed as shown in Equation (11.64).

$$a_T = \exp\left(\frac{\Delta E}{R_g}\left[\frac{1}{T} - \frac{1}{T_{ref}}\right]\right) \quad (11.64)$$

Both the WLF and the Arrhenius expressions have been used to fit the shift-factor data shown in Figure 11.21. The WLF form was used up to $200°$ C after which, for this PMMA sample, the Arrhenius expression was used.

Fewer data are available for the pressure dependence of viscoelastic properties, although, in general, the free-volume concept can be used to explain the observations. By increasing the pressure on a system, the free volume is decreased, thereby decreasing mobility and increasing viscosity and relaxation times. Analogous to a_T, a pressure shift factor a_p can be defined as seen in Equation (11.65).

$$a_p \equiv \frac{\eta_{0p}}{\eta_{0,P_{ref}}} \quad (11.65)$$

Generally, pressure effects on viscosity are ignored for practical considerations, until high pressures, 0 (10^4 psi), are reached where empirically it has been found that Equation (11.66) applies.

$$\eta_p = \eta_{P_{ref}} \exp b(p - p_{ref}) \quad (11.66)$$

A few data are available giving b for some systems. Representative numbers are listed in Table 11.4.

11.2.6.2   Temperature Dependence in Viscoelastic Models

Because of the nonisothermal nature of most polymer processing operations and the sensitivity of the viscoelastic properties of many

Table 11.3   WLF Constants for Some Representative Polymers $T_{ref} = T_g$ [a]

Polymer	T_g (°C)	C_1	C_2
Polystyrene	100	13.3	47.5
Polyisobutylene	−68	16.6	104.4
Poly (vinyl acetate)	32	15.6	46.8
Poly(methyl methacrylate)	115	32.2	80.0
Polyurethane	−35	15.6	32.6

[a] Reference 44.

Table 11.4 Pressure Dependence of Viscosity for Various Polymers

Polymer	b (bars^{-1} $\times$ 10^{-3})	T (°C)
Polystyrene[a]	7.3	130
Poly(methyl methacrylate)[a]	5.4	143
$Mw = 54000$	4.7	154
	4.0	166
Polypropylene[b]	1.2	232

[a] Reference 134.
[b] Reference 135.

polymers to temperature, consideration must be given to thermal effects in modeling. Temperature gradients serve to make constant-deformation-rate flows, such as steady flow through a tube, nonconstant rate from a material sense. This concept is illustrated in Figure 11.22, showing a hypothetical case of flow through a circular tube at a constant flow rate $\dot{Q}$ for the PMMA melt used in Figures 11.19–11.21. An inlet temperature of 150° C is assumed with linear temperature gradient along the tube exiting at 220° C. As can be seen from the material's view of the shear rate $\dot{\gamma} a_T$, the shear-rate profile is hardly constant, as would be assumed if thermal effects were not taken into account.

Nonisothermal flows can be included in viscoelastic models for a class of materials allowing time-temperature superposition, those known as "thermorheologically simple" fluids (47). To illustrate the modification of a viscoelastic model to predict nonisothermal effects, consider the rubberlike liquid model in

an isothermal form; see Equations (11.67) through (11.70).

$$\sigma = \int_{-\infty}^{t} \sum_{n=1}^{N} \frac{G_n}{\tau_n} \exp\left(-\frac{t-t'}{\tau_n}\right) \mathbf{C}^{-1}(t,t')\, dt'$$

(11.67)

We replace G_n by

$$G_n(T) = G_n(T_{\text{ref}}) \frac{T \rho_T}{T_{\text{ref}} \rho_{T_{\text{ref}}}}$$

(11.68)

and the nth relaxation time τ_n by

$$\tau_n(T) = \tau_n(T_{\text{ref}}) a_T$$

(11.69)

The nonisothermal model then is

$$\sigma(T)$$
$$= \int_{-\infty}^{t} \sum_{n=1}^{N} G_n(T_{\text{ref}}) \frac{T \rho_T}{T_{\text{ref}} \rho_{T_{\text{ref}}}} [\tau_n(T_{\text{ref}}) a_T]^{-1}$$
$$\exp \frac{-(t-t')}{[\tau_n(T_{\text{ref}}) a_T]} \mathbf{C}^{-1}(t,t')\, dt^3 \quad (11.70)$$

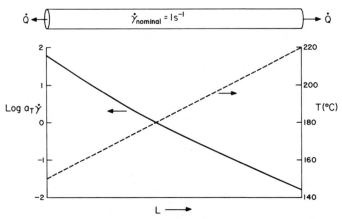

Figure 11.22 Illustration of the importance of thermal effects in flow modeling.

Equation (11.70) is a consequence of a straight-forward adaptation of the result of Section 11.2.6.1. Matsui and Bogue (48) have demonstrated the prediction of nonisothermal rheological parameters for step changes in temperature superposed onto melt extension experiments. They chose to use the Bogue-White constitutive relation modified similarly to Equation (11.70).

11.3 RHEOLOGICAL CHARACTERIZATION

Once the relevance of deformation to a polymer processing step has been established, rheological characterization of the fluid being processed becomes a primary objective. Before performing several experiments to obtain the response of the fluid in all the flow situations of Section 11.2, one should first determine the level of characterization needed. Would a single-point, single-shear-rate viscosity be enough, such as a melt-flow measurement? Many times this is sufficient for quality control or process analysis. Often more sophistication is necessary to obtain $\eta(\gamma)$ or elasticity behavior. When the goal is complete mathematical analysis with viscoelastic models or specific material analysis such as determining the extent of long-chain branching and its effect on processing, more tests are needed, including the linear viscoelastic response and behavior in elongational flow.

In this section characterization is dealt with as a tool, giving examples pertinent to the characterization of viscoelastic melts and solutions. Since the viscometric experiments are so well known, particularly those to measure shear viscosity, only a brief mention is made of them for completeness. More emphasis is placed on elongational flow techniques, as these are just now becoming routine and are of great importance both practically and theoretically. The section concludes with a brief illustration of the use of characterization data to fit material constants for viscoelastic fluids.

11.3.1 Viscometric Measurements

The viscometric experiments, such as small-amplitude sinusoidal shear and cone-and-plate steady shear, form the basis for rheological characterization. Although a processing flow may not mimic the experiment at all, the information gained through these classical experiments is correlated well with polymer composition as described in Section 11.4. For completeness and as examples, the small-amplitude sinusoidal shear experiment and the steady-shear experiment are described in Sections 11.3.1.1. and 11.3.1.2, respectively.

11.3.1.1 Linear Viscoelastic Response

A linear viscoelastic experiment is one in which the deformation is small so the material obeys the linear viscoelastic model, analogous to the case of gasses at moderate pressures and temperature obeying the ideal gas law. The rubberlike liquid, Equations (11.35)–(11.37), is a linear viscoelastic model.

Small-Amplitude Sinusoidal Shear (Cone and Plate). The experiment most often used for characterization of linear viscoelastic response is small-amplitude sinusoidal shearing. The material is placed between a plate and a cone as shown in Figure 11.23. The cone has a shallow angle, typically about 0.5–2° although larger angles up to as much as 6° may be used. The problem with larger angles is the likelihood that secondary flow will ruin the experiment. If the cone is forced to oscillate with a small amplitude, then the strain is given by Equation (11.71)

$$\gamma = A \sin \omega t \qquad (11.71)$$

where γ is the shear strain A/α, then the sample responds by a stress shown in Equations (11.72a) and (11.72b)

$$\sigma_{12} = B \sin (\omega t + \delta) \qquad (11.72a)$$

with

$$B = \frac{3T_p}{2 \pi R_p^2} \qquad (11.72b)$$

T_p in Equation (11.72b) is the torque developed in the sample resisting deformation. If the material were purely elastic, δ the phase

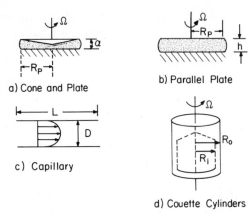

a) Cone and Plate

b) Parallel Plate

c) Capillary

d) Couette Cylinders

Figure 11.23 Sketches of four viscometric experiments: (a) cone and plate; (b) parallel plate; (c) capillary; and (d) couette cylinders.

angle would be zero, whereas a purely viscous material would be 90° out of phase $\delta = \pi/2$. The stress can be resolved readily into two components as shown in Equations (11.73)–(11.75)

$$\sigma_{12} = B \cos \delta \sin \omega t + B \sin \delta \cos \omega t \tag{11.73}$$

where the in-phase portion is the elastic or storage modulus G',

$$G' = \frac{B \cos \delta}{\gamma} \tag{11.74}$$

and the out-of-phase portion is the viscous or loss modulus G'',

$$G'' = \frac{B \sin \delta}{\gamma} \tag{11.75}$$

The viscous-to-elastic modulus ratio is known as the loss tangent as given in Equation (11.76).

$$\frac{G''}{G'} = \tan \delta \tag{11.76}$$

The experiment generally begins by a check of the linear viscoelastic behavior. One then takes measurements over a range of frequencies and temperatures allowable to the specific instrument used. A typical frequency range is 10^{-3}–10^{2} rad/sec. Walters (49) dis-

cusses the major sources of error including inertial effects, nonlinear effects, natural-frequency problems, and machine stiffness.

Small-Amplitude Sinusoidal Shear (Orthogonal Rheometer). An equivalent route to the storage and loss moduli is found with the orthogonal rheometer. This flow uses two parallel plates both free to rotate as shown in Figure 11.24. One member is driven at constant angular speed while the other follows. A small-amplitude sinusoidal deformation is created when the centers of rotation are made eccentric. A discussion of the flow field is given by Macosko and Davis (50).

The strain is calculated as the eccentricity a divided by the fluid gap h [see Equation (11.77)],

$$\gamma = \frac{a}{h} \tag{11.77}$$

and the moduli are separated into two independent stresses in the x- and y-directions as shown in Equations (11.78a) and (11.78b)

$$G'' = \frac{F_x h}{\pi R_p^2 a} \tag{11.78a}$$

$$G' = \frac{F_y h}{\pi R_p^2 a} \tag{11.78b}$$

for an eccentricity in the x-direction. The operation of the experiment is as with the cone-and-plate system except data analysis becomes much simpler since there are no phase-angle measurements to make. The viscous and elastic forces are resolved in the experiment. The orthogonal rheometer suffers from the same potential errors as the cone-and-plate, with the additional problem of instrument compliance playing a more important role.

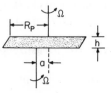

Figure 11.24 Sketch of the orthogonal rheometer.

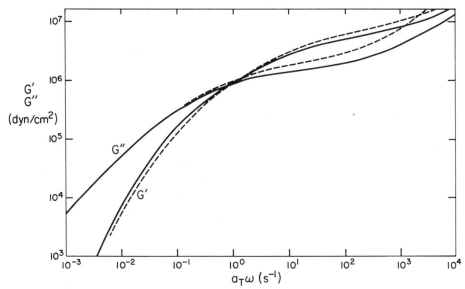

Figure 11.25 Effects of instrument compliance for PMMA studied by orthogonal rheometry, T_{ref} = 180°C (———) uncorrected, (•--•) corrected.

Macosko and Davis (50) have developed a theory to correct for machine compliance. Corrections are important at low freqencies for G' and high frequencies if G' and/or G'' are greater than 10^5 dynes/cm² and tan δ is much different from unity (51). Figure 11.25 shows data for G' and G'' before and after correction for a molding-grade PMMA. These data were taken on a Rheometrics mechanical spectrometer with a compliance of $3(10^{-9})$ cm/dyne.

11.3.1.2 Nonlinear Viscoelastic Response

When the polymer is strained from its rest configuration, nonlinear viscoelastic response is observed. In terms of the viscoelastic models of Section 11.2.5.2, the network is forced to disentangle by deformation. The basic experiment considered here is that of steady-shear flow leading to viscosity and normal stress differences as functions of shear rate.

Viscosity in Shear Flow.   The two foregoing experiments presented require special and often expensive instrumentation. They provide useful data, but often such sophistication is not available. More commonly, equipment is available or can be built to measure the shear viscosity of melts or solutions. Several geometries have been used, but of these, four remain

as primary experiments: (1) cone-and-plate, (2) parallel plate, (3) capillary, and (4) couette cylinders. A definitive sketch for each is given in Figure 11.23. The system equations and range of shear rates commonly obtainable are given in Table 11.5. The experiments differ in the range of usable shear rates as well as the uniformity of the shear field. Only the cone-and-plate flow results in uniform shear. The remainder induce a nonuniform shear rate and must be corrected by the suitable equation in Table 11.5.

Common experimental pitfalls include (1) attainment of steady-stress, (2) secondary flow in rotational devices, and (3) melt fracture in capillary flow. A specific concern for capillary flow is that sufficiently long tubes or slits are used to allow for neglect of entrance and exit effects. Rather than neglect these sources of pressure loss, one should use tubes of different lengths and correct the data according to the methods suggested by Middleman (3) or Bagley (52). The Bagley end correction can be recast as Equation (11.79) (53).

$$\Delta P_{total} = 4(\sigma_{12})_{wall} \frac{L}{D} + \Delta P_{entrance+exit}$$

$$(11.79)$$

A final note of caution concerns viscous

Table 11.5 System Equations and Shear-Rate Ranges for Steady-Shear Viscometry Using Various Geometries

Geometry	Shear Stress	Normal Stress Difference	Shear Rate	Range (per sec)
Cone/plate	$\dfrac{3T_p}{2\pi R_p^3}$	$\dfrac{2F}{\pi R_p^2} = N_1$	$\dfrac{\Omega}{\alpha}$	$10^{-3} < \dot\gamma < 10^2$
Parallel plate	$\dfrac{3T_p}{2\pi R_p^3}\left[1 + \dfrac{1}{3}\dfrac{d\ln T_p}{d\ln\dot\gamma}\right]$	$\dfrac{2F}{\pi R_p^2}\left[1 + \dfrac{1}{2}\dfrac{d\ln F}{d\ln\dot\gamma}\right] = N_1 - N_2$	$\dfrac{\Omega R_p}{h}$	$10^{-3} < \dot\gamma < 10^3$
Couette	$\dfrac{T}{2\pi R^2}$	—	$\dfrac{R_i\,\Omega}{(R_0 - R_i)}$ [a]	$10^{-1} < \dot\gamma < 10^3$
Capillary	$\dfrac{R}{2}\dfrac{\Delta P}{\Delta L}$	—	$\dot\gamma_{nom}\dfrac{3}{4} + \dfrac{1}{4}\dfrac{d\ln\dot\gamma_{nom}}{d\ln\sigma_{12}}$ $\dot\gamma_{nom} = \dfrac{4Q}{\pi R^3}$	$10^0 < \dot\gamma < 10^5$

[a] Approximation for narrow gaps; for more general forms consult Middleman (3).

heating as a source of error. In some geometries, viscous heating can be substantial, particularly at high viscosities and rates of shear. Although there are techniques for estimating the temperature rise (3, 49), an independent measure of temperature should be made, if possible, owing to the great sensitivity of some polymer melts to temperature.

Normal Stress Differences in Shear Flow. As we discussed in Section 11.2, viscoelastic fluids develop normal stresses in shear flow as well as shear stresses. In principle, any of the preceding devices discussed for measuring viscosity could be used to obtain the first normal stress difference N_1. In practice, the cone-and-plate and parallel-plate experiments provide the most direct route. In both of these flows the normal stress difference manifests itself as a force normal to the fixture attempting to move the upper and lower members apart. For the cone-and-plate geometry, N_1 is obtained directly from this normal force by Equation (11.80a).

$$N_1 = \frac{2F_z}{\pi R_p^2} \qquad (11.80a)$$

In the parallel-plate experiment, the difference $N_1 - N_2$ is obtained, correcting for the effect of nonuniform shear rate by Equation (11.80b)

$$N_1 - N_2 = \frac{2F_z}{\pi R_p^2}\left[1 + 0.5\,\frac{d(\ln F_z)}{d(\ln \gamma)}\right]$$

$$(11.80b)$$

Since $N_2 \approx \beta N_1$ where $0.1 < \beta < 0.3$, $N_1 - N_2$ is always larger than N_1. Performing both experiments, cone-and-plate and parallel plate, provides a method of obtaining both N_1 and N_2. The data for N_2 (Ψ_{23}) shown in Figure 11.5 were obtained in this manner.

Although the cone-and-plate experiment is the most common method of obtaining normal stress difference data, the shear rate attainable is limited at high rates by flow instabilities. Measurement of N_1 at high shear rates is difficult. All the available methods, jet-thrust (54), streaming-flow birefringence (55), exit-pressure drop (56), and extrudate swell (57), have potentially serious deficiencies such as fundamental material assumptions or assumptions made in the analysis of the experiment.

Extrudate swell can be used to estimate the magnitude of fluid elasticity and deserves further discussion since it is a relatively straightforward experiment. Elastic liquids upon exiting from a tube or die into an unconfined region will swell, sometimes several hundred percent of the inside tube diameter. The phenomenon is shown schematically in Figure 11.26 and illustrated in Figure 11.27 for a polystyrene melt. Tanner (58) has developed a model of die swell using a BKZ constitutive relation. He assumes that the elastic stresses developed during shear flow relax after exiting the die and the material recovers by changing dimensions. The final swell is related to the ratio of N_1 to σ_{12} evaluated at the wall according to Equation (11.81)

$$\frac{D_e}{D} = 0.1 + \left[1 + 0.5\left(\frac{N_1}{2\sigma_{12}}\right)_w^2\right]^{1\,6} \quad (11.81)$$

where 0.1 has been added to agree with results for Newtonian fluids. Figure 11.28 demonstrates the prediction for a polystyrene melt of die swell versus shear rate using Equation (11.81) and the data of Racin (7). Considering the assumptions made in obtaining the theoretical results, the model and data agree rather well. One can readily see that the die swell is directly related to N_1. Considerable care must be taken when swelling data are taken to ensure complete recovery of the polymer jet. White and Roman (53) discuss this problem in detail and develop a theory for extrudate swell with an external takeup force applied to the jet (a simulated fiber-spinning experiment).

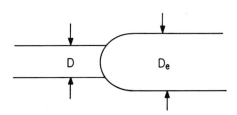

Figure 11.26 Schematic illustration of extrudate swell.

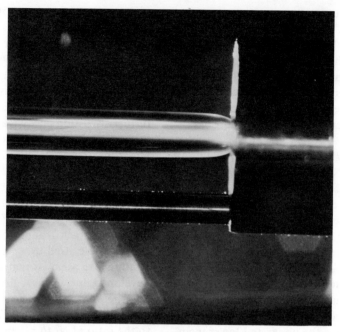

Figure 11.27 Extrudate swell for a polystyrene melt (7).

Finally, if quantitative estimates of N_1 are needed for characterization, but only means of obtaining $\eta(\dot{\gamma})$ are available, Abdel-Khalik, Hassager, and Bird (59) have suggested a semiempirical method of predicting N_1. They have found Equation (11.82)

$$N_1(\dot{\gamma}) = \dot{\gamma}^2 \, K \, \frac{4}{\pi} \int_0^\infty \frac{\eta(\gamma) - \eta(\gamma')}{\dot{\gamma}'^2 - \gamma^2} \, d\dot{\gamma}'$$

$$(11.82)$$

with $K = 2$ for solutions and 3 for melts is a reasonable relationship between $\eta(\dot{\gamma})$ and N_1. Equation (11.82) has been used to predict N_1 for a polystyrene melt with reasonable success as shown in Figure 11.29.

11.3.2 Elongational Flow

All the characterization tools discussed so far have been closely allied to shear flow. If the process contains a strong elongational-flow

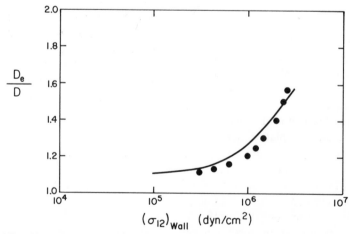

Figure 11.28 Extrudate swell versus shear stress for a polystyrene melt compared to the theory of Tanner.

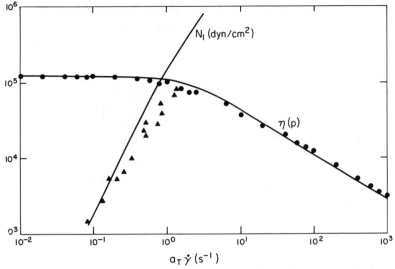

$$a_T \dot{\gamma} \ (s^{-1})$$

Figure 11.2e Prediction of the first normal stress difference for a polystyrene melt using the viscosity-normal stress difference relationship in Equation (11.82).

component such as is found in fiber spinning or blow molding, shear measurements may not suffice as predictions of performance. Elongational-flow experiments are coming of age; some are currently available in commercial instruments. The experiments break cleanly into two groups: high-viscosity methods and low-viscosity methods. Dealy (60) has written a review of extensional rheometers for high-viscosity systems that should be consulted for the wide range of schemes covered for uniaxial, biaxial, and general extensions. Here, we concentrate on only a few "successful" uniaxial experimental devices for melts and solutions.

11.3.2.1　High-Viscosity Devices

For melts with viscosities at zero-shear rate greater than 10^5 P, the primary experimental device involves, in some manner, extending a preformed, unoriented rod of the polymer. Several specific devices have been built, including the rotating clamps of Meissner (15), the post-winder experiment described by Macosko and Lornston (61) for the Rheometrics mechanical spectrometer, and the stationary clamp/movable clamp used in an Instron tensile tester, the Rheometrics extensional rheometer (62), and by Münstedt (16). These devices are sketched in Figure 11.30.

In these experiments the rate of extension is determined by the linear speed of the moving clamp(s), and the stress is calculated by dividing the measured force by the instantaneous cross-sectional area. The area is generally calculated from the extension-rate program rather than measured, a practice leading to large potential errors if sample nonuniformity develops (63). Aside from nonuniform draw-

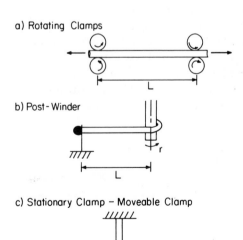

Figure 11.30 Extensional flow experiments for high-viscosity systems: (a) rotating clamps; (b) post-winder; and (c) stationary clamp—movable clamp.

ing, the primary experimental concern is rigid control of temperature, generally to within $\pm 0.2°$ C. This is particularly true for amorphous melts close to their T_g, where all rheological properties are extremely temperature sensitive. Nonuniform temperature during extension may lead to nonuniform extension.

Since steady-state flow regimes are not generally obtained in the rod-pulling devices except at low strain rates, extensional viscosity as a function of strain rate is not the final result of the experiment. Generally, one is faced with transient-stress data at a given rate, rate program, and temperature. The use of data such as these depends upon the characterization need. For distinguishing differences between batches of the same polymer, transient-stress data can be and have been used (64). For process analysis, the experiment should be designed to operate as close to the process in rate and rate program with time and temperature as possible, thus mimicking the process for a small sample of fluid in a controlled environment.

Before low-viscosity methods are discussed, one final technique should be mentioned. A converging flow is basically extensional. Cogswell (64) has proposed an extensional-flow rheometer based on converging flow, sketched in Figure 11.31. Everage and Ballman (65) have recently studied polystyrene with a device similar to Cogswell's. The major problems are (1) nonrest state at the start of the experiment and (2) approximations made in the analysis to separate viscous and elastic

effects. It remains to be seen if the converging-flow rheometer is useful as a fundamental tool; it has been shown useful as a practical device for exploring extensional flow of various melts, although not in the form shown in Figure 11.31. Shroff, Cancio, and Shida (66) have studied polystyrene, polypropylene, low-density polyethylene, and high-density polyethylene using an orifice die ($L/D = 0$) in a pressure-driven rheometer. By separating the shear deformation from the extensional deformation according to Cogswell's analaysis (67), the tensile stress is related to the pressure drop in the orifice die ΔP_0, by Equation (11.83)

$$T = \tfrac{3}{8}(n + 1)\,\Delta P_0 \qquad (11.83)$$

and the strain rate is given as Equation (11.84)

$$\dot{\epsilon} = \frac{4\sigma_{12}\dot{\gamma}}{3(n + 1)\,\Delta P_0} \qquad (11.84)$$

where n is the power-law slope of shear stress/shear rate in capillary flow, γ is the shear rate in a long tube of an equivalent diameter, and σ_{12} is the shear stress. Certainly differences in response among the melts were detected, and, as such, the method holds some promise as a simple tool for quality control. What is being measured is open to debate and must await data of $\eta_e - \epsilon$ from more conventional devices for resolution.

11.3.2.2 Low-Viscosity Devices

The lower limit of viscosity used in rod-pulling experiments is dictated by the ability of the material to maintain its dimensions before deformation without sagging. Even with a neutral-buoyancy oil bath, fluids must have an η_0 of 10^5 P or greater. For low-viscosity systems, the search for a suitable technique poses a challenging research problem. Two major classes of experiments have been studied. The first follows the deformation of a gas or insoluble liquid bubble in the flow. The second class extends a filament of fluid by various methods.

Pearson and Middleman (43) have shown that bubble collapse can be used to study extensional flow of solutions with zero-shear-

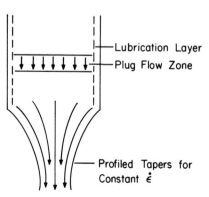

Figure 11.31   Converging-flow experiment.

rate viscosities as low as 10^2–10^3 P. A bubble is supported on the tip of a capillary tube. At $t = 0$ the pressure inside the bubble is lowered, causing collapse. The radius as a function of time is followed with high-speed photography. The stress at the bubble wall R, $\sigma^{11} - \sigma^{22}$, may be approximated as the pressure difference less the surface-tension stress [see Equation (11.95)]

$$\sigma^{11} - \sigma^{22}\big|_R = -\frac{3}{2}\left[P_g - P_{atm}\right] - \frac{2\gamma}{R}$$

(11.85)

and the strain rate is given by the collapse rate as Equation (11.86).

$$\dot{\epsilon} = \frac{-2\dot{R}}{R}$$

(11.86)

The major problem with the experiment as a method for measuring η_e is that of assuring oneself that steady state has been reached. The transients of the problem, perhaps more important practically anyway, are readily analyzable and have been modeled successfully with the BKZ-K constitutive equations (68).

Hsu and Flumerfelt (69, 70) have designed an experiment based on the rotating-drop method for interfacial tension developed by Princen (71) that has promise of applicability to low-viscosity systems. A drop of the fluid to be studied is placed along the axis of a long cylinder in another fluid that is immiscible in the first, of lower viscosity, higher density, and Newtonian. The cylinder is rotated to create a constant strain rate and can be modeled to obtain the tensile stress as well. Steady-state flow yielding near-constant $\eta_e(\dot{\epsilon})$ has been achieved for a polyisobutylene solution in decalin.

The fiber-extension experiments all mimic the fiber-spinning process. A viscoelastic fluid will sustain a flow between a stagnant reservoir and a vacuum line placed above the reservoir as sketched in Figure 11.32. This flow has been called Fano or inverse siphon flow. Several researchers (72–74) have studied Fano flow. The problem with the experiment is that neither stress nor rate of deformation is con-

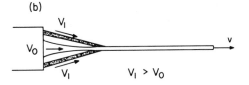

(b)

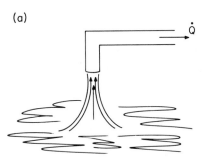

(a)

Figure 11.32 Low-viscosity extensional flow experiments: (*a*) inverse siphon or Fano flow; (*b*) triple jet.

stant along the flow length, so elongational viscosities are not readily attainable. A similar criticism holds for the triple-jet technique (75) where a strand of fluid, extruded from a circular die, is extended by two rapidly moving impinging jets of fluid, as pictured in Figure 11.32.

All the known techniques for studying extensional flow of solutions have serious deficiencies, most relating to the potential for steady-state flow. The problems should be recognized at the outset of characterization and a flow field chosen to mimic as closely as possible the process flow. In this manner, qualitative behavior pertinent to the process can be obtained.

11.3.3 Characterization through Viscoelastic Models

Before characterizing the rheology of a melt or solution, one should decide the amount and kind of information necessary. In many cases characterization is performed to determine the material constants of a given viscoelastic model. Although the parameter estimation is not difficult with the proper data at hand, a general procedure for obtaining material constants is not readily available in the literature.

As an example, consider the BKZ-K model with a discrete relaxation spectrum, Equation (11.87).

$$\sigma = \int_{-\infty}^{t} \sum_{n=1}^{N} \frac{G_n}{\tau_n} \exp\left(-a_{BKZK} \sqrt{I_c - 3}\right)$$

$$\exp\left[\frac{-(t - t')}{\tau_n}\right] \mathbf{C}^{-1}(t, t')\, dt' \quad (11.87)$$

The first step is to determine the relaxation spectrum $\{G_n, \tau_n\}$. One method is to use linear viscoelastic data $G'(\omega)$ and $G''(\omega)$ and convert this by the Ninomiya-Ferry (45) formulas to $H(\tau)$, the relaxation function, Equations (11.88) and (11.89),

$$H(\tau) = \left[\frac{G'(d\omega) - G'(\omega/d)}{2 \ln d}\right.$$

$$- \frac{d^2}{(d^2 - 1)^2} G'(d^2\omega) - \frac{G'(\omega/d^2)}{2 \ln d}$$

$$\left. - 2G'(d\omega) + 2G' \frac{\omega}{d}\right]_{1/\omega = \tau} \quad (11.88)$$

$$H(\tau) = \frac{2}{\pi}\left\{G''(\omega)\right.$$

$$- \frac{d}{(d - 1)^2}\left[G''(d\omega)\right.$$

$$\left. + G'' \frac{\omega}{d} - 2G''(\omega)\right]\right\}_{1/\omega = \tau} \quad (11.89)$$

where d is the spacing between points on the frequency axis with a spacing of $\log d = 0.2$ commonly used. Once $H(\tau)$ is determined, $\{G_N, \tau_n\}$ are obtained by integrating discretely Equation (11.90).

$$\eta_0 = \sum_{n=1}^{N} G_n\tau_n = \int_{-\infty}^{\infty} H(\tau) \cdot \tau\, d(\ln \tau)$$

$$(11.90)$$

This operation is illustrated in Figure 11.33. Normally, as $\tau \to \infty$, a truncation must be performed to satisfy Equation (11.90). Figure 11.34 illustrates the effect of changing the integration step size, which is equivalent to changing the number of relaxation times per decade, on the prediction of the model for G' and G''. As N decreases, the high-frequency prediction becomes progressively poorer.

With $\{G_n, \tau_n\}$ specified, the remaining constant is a_{BKZK}. This nonlinear parameter can be estimated from $\eta(\gamma)$ by first plotting the prediction for $\eta(a_{BKZK}\, \gamma)$. Then the data and prediction are shifted horizontally to obtain a reasonable fit. The horizontal shift fixes a_{BKZK}.

11.4 RHEOLOGY AND COMPOSITION

The rheology of a polymeric material must depend upon the overall composition of the system. Molecular weight and molecular weight distribution, concentration of solvent, extent of branching or cross-linking, and presence of a filler all must affect the deformation response of the material. Although this area remains an active one for research, some general guidelines are available, particularly for the terminal-zone properties η_0 and τ_{max}. This section contains a discussion of the composition dependence of these properties in terms of two broad system classifications: (i)

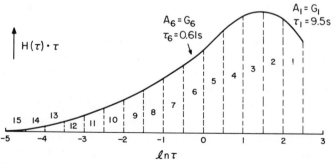

Figure 11.33 Integration procedure to determine the discrete relaxation spectrum $\{G_n, \tau_n\}$.

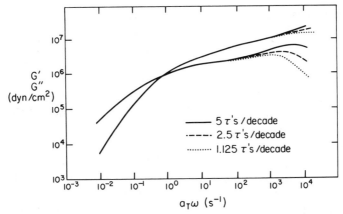

Figure 11.34 Predictions of G' and G'' using progressively fewer relaxation times per decade for a PMMA melt, $T_{ref} = 180°C$.

homogeneous systems, which include linear homopolymers, compatible copolymers, and branched polymers; (2) heterogeneous systems, which include incompatible blends and copolymers and filled materials.

11.4.1 Homogeneous Systems

The rheology of homogeneous systems has received much attention in the literature. In most cases the behavior in terms of system composition is well understood. This is true particularly for linear homopolymers. Slight modifications of the polymeric structure such as the introduction of small amounts of long-chain branching or the addition of ionic moieties to the polymer's backbone can lead to substantial alterations in properties. Generally even if a filler is to be added before use, the properties of the base system are of interest and as such are presented first as an introduction to the commercially important filled systems.

11.4.1.1 Linear Homopolymers/ Compatible Copolymers

For linear homopolymers, the material's macromolecular constituents are specified by its molecular weight distribution and some parameter proportional to the size of the macromolecule in solution, such as the intrinsic viscosity $[\eta]$ or the mean-square radius of gyration $\langle s^2 \rangle$. Molecular theories of macromolecules in solution attempt to predict the rheological properties from considerations of molecular size, intermolecular interactions, and intramolecular interactions. Several reviews of macromolecular theories are available that describe the progress in this field to date. Specifically, one should consult the monographs of Graessley (9) and Bird, Armstrong, Hassager, and Curtiss (76), as well as the review article by Williams (77). Rather than a review of this work, correlation techniques for η_0 and, when possible, J_e^0 or τ_{max} are presented, some methods based on theory, others rather empirical.

The zero-shear-rate viscosity can be correlated by several forms as a function of the solution concentration, molecular weight, and molecular size. At very dilute conditions, the polymer chains behave as discrete particles or coils with their size governed by the product of concentration and intrinsic viscosity. The product $C[\eta]$ is a measure of the degree of coil overlap and thus interaction in solution (78). At intrinsic viscosity concentrations η_0 should be directly proportional to $C[\eta]$; see Equation (11.91),

$$\eta_r = \frac{\eta_0}{\eta_s} = 1 + C[\eta] \qquad (11.91)$$

where η_r is the relative viscosity obtained by scaling the solution viscosity to the solvent viscosity η_s. Equation (11.91) follows directly from the definition of $[\eta]$. As the concentration is increased, higher-order terms are included, resulting in the series expansion shown in Equation (11.92).

$$\eta_r = 1 + C[\eta] + k_H C^2 [\eta]^2 + \cdots \quad (11.92)$$

Retaining the second-order term provides a useful model at low concentrations. The constant k_H, known as the Huggins constant, is relatively insensitive to molecular weight but is influenced by the thermodynamics of the polymer-solvent pair. In theta solvents, k_H ranges from 0.5 to 0.8, whereas for good solvents k_H assumes values from 0.3 to 0.4.

Two more equations are available that rely upon the coil-overlap interaction characterized by $C[\eta]$ to predict solution viscosity. The empirical Martin equation (79), given in Equation (11.93),

$$\eta_r = 1 + C[\eta] \exp [k_M C[\eta]] \quad (11.93)$$

can provide useful estimates of η_r once the constant k_M is determined. An improvement upon the Martin equation is the Lyons-Tobolsky equation (80) [see Equation (11.94)]

$$\eta_r = 1 + C[\eta] \exp \frac{k_{LT}[\eta]C}{1 - bC} \quad (11.94)$$

where b has been added to increase the concentration dependence of η_r. A recent study (81) has shown that data for several different polymer-solvent pairs could be adequately correlated by Equation (11.94) allowing k_{LT}, $[\eta]$, and b to be adjustable constants. The best-fit values of $[\eta]$ were slightly lower than the measured values, and the best-fit values of k_{LT} were slightly higher than the Huggins constant, k_H.

Chitrangad, Osmers, and Middleman (82) have attempted to correlate solution viscosity η_0 with Equation (11.92) to determine the limits of applicability of correlations based on the coil-overlap parameter. They found that, for polyisobutylene in various solvents, for $C[\eta] < 1$, all the data for η_r could be correlated using a plot of η_r versus $C[\eta]$. For $C[\eta] > 1$, the observed value of η_r became a function of the solvent and the polymer molecular weight. The effect of the solvent can be handled to some extent by Equations (11.92)–(11.94) through the k-parameters. Figure 11.35 demonstrates the predicted η_r from Equation (11.92) for a theta solvent versus a good solvent. The empirical Lyons-Tobolsky equation has been applied for values of $C[\eta]$ as high as 60 (81).

Eventually a second effect begins to dominate in determining the viscosity of solutions. As well as coil-overlap interactions there are segment-segment interactions, which can be characterized by the product of concentration

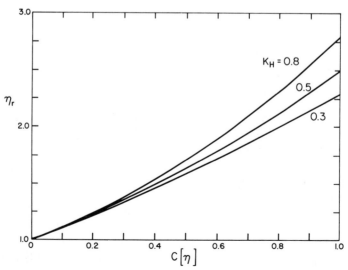

Figure 11.35 Effect of solvent type on solution viscosity predicted by Equation (11.92) in text.

and molecular weight (83). Williams (84) has suggested as a correlating parameter $CM^{5\,8}$ based on a molecular theory of polymer solutions. Chitrangad et al. (82) found the $CM^{5\,8}$ parameter useful for polyisobutylene solutions for $CM^{5/8} > 2(10^4)$.

At some combination of concentration and molecular weight an abrupt change in behavior takes place that has been associated with the formation of temporary entanglements among chains. Below the critical point, η_0 increases linearly as a function of $\bar{M}_w$. Above the $C\bar{M}_w$ product necessary for entanglements, a characteristic of the specific polymer, η_0 is proportional to $(C\bar{M}_w)^{3.4}$. This well-known behavior of concentrated solutions and melts is illustrated for a polystyrene melt above the critical molecular weight for entanglement M_c in Figure 11.36. Typical values of M_c for various undiluted polymers are given in Table 11.6.

Methods for correlating and predicting η_0 for polymers in solution or in bulk are readily available, as we have seen. Correlations proceed using either $C[\eta]$ or $C\bar{M}_w$ as the correlating parameter. Equations exist that can

Table 11.6 Critical Molecular Weights for Entanglement of Various Polymers[a]

Polymer	M_c
Polystyrene	31,200
1,4-Polybutadiene	5,900
Poly(vinyl acetate)	24,500
Polyethylene	3,800
cis-Polyisoprene	10,000
Poly(methyl methacrylate)	27,500
Polyisobutylene	15,200

[a] Reference 8.

predict η_0, given C and $[\eta]$ plus some additional parameters for the system such as b and k_{LT} in Equation (11.94).

For the elastic properties of solutions or melts, less help is available. For dilute solution, the models of Rouse (85) and Zimm (86) are helpful in providing predictions for the steady-state recoverable compliance J_e^0, defined as in Equation (11.95a).

$$J_e^0 = \frac{1}{2\eta_0^2} \lim_{\dot{\gamma}\to 0} \frac{N_1}{\dot{\gamma}^2} \qquad (11.95a)$$

J_e^0 is related to the terminal relaxation time τ_{max} by Equation (11.95b).

$$\tau_{max} = J_e^0 \eta_0 \qquad (11.95b)$$

Graessley (9) shows that τ_{max} defined by Equation (11.95b) is a weight-average relaxation time. The Rouse model assumes that the macromolecules may be replaced by a chain of $N + 1$ beads connected by N linear springs. All frictional resistances are localized at the beads. The model results in a prediction for J_e^0 for monodisperse polymers as shown in Equation (11.96)

$$J_e^0 = \frac{2}{5}\frac{M}{CRT}\left(\frac{\eta_0 - \eta_s}{\eta_s}\right)^2 \qquad (11.96)$$

The Zimm theory modifies Rouse's by allowing for hydrodynamic interaction between the beads, resulting in a somewhat different expression for J_e^0 when impenetrable coils are assumed; see Equation (11.97).

$$J_e^0 = \frac{0.206M}{CRT}\left(\frac{\eta_0 - \eta_s}{\eta_s}\right)^2 \qquad (11.97)$$

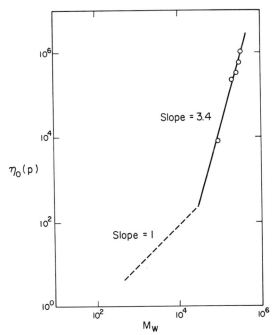

Slope = 3.4

Slope = 1

$\eta_0\,(\text{p})$

M_w

Figure 11.36 Illustration of η_0 versus $\bar{M}_w$ for polystyrene (90).

The Zimm model reverts to the Rouse model if there are no interactions, a situation known as the free-draining chain. Intermediate cases are obtained by changing the interaction parameter from 0 (Rouse) to values much greater than unity (impenetrable coils).

Data are available to test these theories. The Zimm theory seems to work fairly well for solutions of polystyrene (87) and poly(2-substituted methyl acrylates) in Aroclor 1254 polychlorinated polyphenyl (88). Comparisons between theory and data are readily made by defining a reduced steady-state recoverable compliance, J_{eR}^0, as in Equation (11.98).

$$J_{eR}^0 = \frac{CRT}{M}\left(\frac{\eta_0}{\eta_0 - \eta_s}\right)^2 J_e^0 \qquad (11.98)$$

For the Rouse model, J_{eR}^0 is given by Equation (11.99),

$$J_{eR}^0 = 0.4 \qquad (11.99)$$

and the Zimm model yields Equation (11.100).

$$J_{eR}^0 = 0.206 \qquad (11.100)$$

Both predictions are for monodisperse samples. Figure (11.37) demonstrates qualitative agreement with the Zimm theory for several polymer solutions at low values of $C\bar{M}_w$ and a gradual approach to Rouse-like behavior at higher values of $C\bar{M}_w$.

As the $C\bar{M}_w$ product increases, J_{eR}^0 approaches the Rouse model prediction of 0.4. The data for high concentration and/or high molecular weight samples support a form for J_e^0 independent of molecular weight for narrow-distribution samples (8) where b is commonly 2, Equation (11.101).

$$J_e^0 \propto \frac{1}{C^b T} \qquad (11.101)$$

Small amounts of high molecular weight material can drastically alter J_e^0, making measurements difficult. This fact also makes estimation of elasticity in commercially available polymers, which are commonly of broad distribution, impossible from the monodisperse theories. Theories and prediction schemes for J_e^0 of polydisperse samples are available, but none is completely satisfactory (9). The Rouse theory has been modified to include polymer polydispersity by Ferry, Williams, and Stern (89). Higher moments of the molecular weight distribution are used with the results for J_e^0 expressed in Equation (11.102).

$$J_e^0 = \frac{2}{5}\frac{\bar{M}_w}{CRT}\frac{\bar{M}_z\bar{M}_{z+1}}{\bar{M}_w^2} \qquad (11.102)$$

Equation (11.102) could be used as an estimate of J_e^0 in the absence of data, but experimental confirmation for the system at hand should be sought.

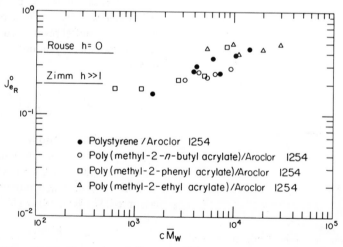

Figure 11.37 Illustration of J_{eR}^0 versus $C\bar{M}$ for various polymer solutions.

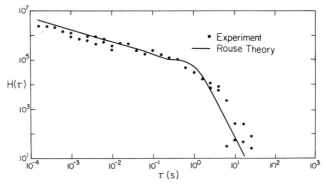

Figure 11.38 $H(\tau)$ versus τ for a polystyrene sample; dots are experimental points, the line is the prediction of the modified Rouse theory (90).

Use of the Rouse theory for polydisperse polymers was made by Pearson and Garfield (90) to predict the nonlinear rheology of polymer melts. Modifying the Rouse relaxation time τ_R by the appropriate polydisperse term, Equation (11.103),

$$\tau_M = \tau_R \frac{\pi^2}{15} \frac{\overline{M}_z \overline{M}_{z+1}}{\overline{M}_w^2} \qquad (11.103)$$

led to adequate predictions of linear viscoelastic behavior from molecular weight distribution data. Figure 11.38 demonstrates the agreement between the predicted and observed values for the relaxation spectrum $H(\tau)$ of a commercial polystyrene sample. Additional use of Graessley's model for shear viscosity in

polydisperse systems (91) provided sufficient information for the specification of the material constants of the BKZK model discussed in Section 11.2.5.2. Figure 11.39 shows the predicted effect of broadening the molecular weight distribution of a hypothetical sample of polystyrene. As the distribution is skewed to the high molecular weight side, departure from Newtonian behavior occurs sooner, indicative of the larger relaxation time. Graessley and Segal (92) have suggested using Equation (11.104)

$$\tau_{max} \approx \frac{1}{\dot{\gamma}_c} \qquad (11.104)$$

where $\dot{\gamma}_c$ is the shear rate at which $\eta(\gamma) = 0.8$

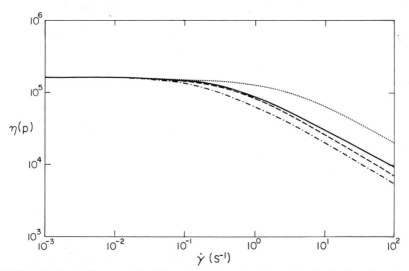

Figure 11.39 Predicted effect on $\eta(\dot{\gamma})$ of broadening the molecular weight distribution : ($\cdots$) narrow, (———) symmetrically broadened, (---) skewed-low, and (—·—) skewed-high distributions (90).

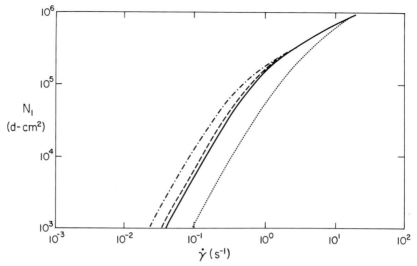

Figure 11.40 Predicted effect on N_1 of broadening the molecular weight distribution : ($\cdots$) narrow, (———) symmetrically broadened, (– –) skewed-low, and (—·—) skewed-high distributions (90).

η_0. Figure 11.40 demonstrates the effect of molecular weight distribution on N_1, showing a separation at low $\dot\gamma$ indicative of τ_{max} differences and a gradual approach to a single, distribution-independent response at high γ—foreboding problems in detecting differences in distribution by extrudate swell, which is a high-shear-rate measurement dependent on N_1.

11.4.1.2 Branched Polymers

For linear polymers the rheology is determined by frictional resistance along the chain, which may be conceptualized as local friction at bead centers, and by chain size or length. Branched polymers create additional problems since consideration of the degree of branching and the length of the branches must, of necessity, be made. The proper description of the terminal-zone properties η_0 and J_e^0 or τ_{max} of branched polymers from macromolecular models poses a difficult problem. Indeed, such a description is not entirely available for linear polymers. Use of branched polymers commercially continues to grow despite the lack of adequate means of molecular characterization because the branched systems offer in many applications reduced viscosity and enhanced elasticity.

If it is assumed that η_0 is proportional only to molecular size for both linear and branched

polymers, then a linear polymer with $\overline{M}_w > M_c$ behaves according to Equation (11.105)

$$\eta_0^{lin} = A\,\overline{M}_w^{3.4} \qquad (11.105)$$

and a branched polymer of identical molecular weight follows Equation (11.106)

$$\eta_0^{br} = A\,(g\,\overline{M}_w)^{3.4} \qquad (11.106)$$

where g indicates the reduction in size due to branching. The size of a branched polymer, measured by the mean-square radius of gyration $\langle s_b^2 \rangle$, is always smaller than that of a linear polymer of constant molecular weight; see Equation (11.107).

$$g = \frac{\langle s_{br}^2 \rangle}{\langle s_{lin}^2 \rangle} < 1 \qquad (11.107)$$

Equation (11.106) is a statement of Bueche's theory (93) for η_0 in concentrated polymer systems above the entanglement point. Graessley et al. (94) have shown agreement for star-branched polyisoprene with Bueche's theory at low molecular weights. Kraus and Gruver (95) also found reasonable agreement with Bueche's theory for star-branched polybutadienes of $\overline{M}_w$ below 50,000.

As the length between branches increases sufficiently to allow entanglements to be formed by the branches, the viscosity-molecular size relationship of Equation (11.106) fails to hold. Larger values of η_0^{br} are observed than would be predicted. Graessley and his co-workers (94) have characterized this increase in η_0 by an additional term as shown in Equation (11.108)

$$\eta_0^{br} = A\,(g\,\bar{M}_w)^{3.4}\,\Gamma \qquad (11.108)$$

where Γ is an enhancement factor. Enhancements of two orders of magnitude have been observed for four-arm star polyisoprenes (94). The attempt to predict enhancement based on an empirical expression of Berry and Fox (96) was not successful. More satisfying was an approach based on the molecular reptation model of DeGennes (97).

The elasticity of branched polymers does not seem to follow the orderly trend just presented for η_0. Most studies show enhanced elasticity while the viscosity may or may not be enhanced. Enhanced elasticity means higher normal stresses, larger values of J_e^0, greater die swell, and greater sensitivity of shear viscosity to shear rate. The study of star-branched polyisoprenes (94) shows high solution elasticity at branch lengths shorter than that necessary to entangle.

The molecular models cannot predict the rheology of branched polymers, particularly their elastic behavior. Qualitatively, the explanation of enhancement in η_0 due to entangling branches is satisfying. Why elasticity should be affected below a branch length sufficient for entanglement is unexplained. Better models as well as more data for well constructed and well characterized branched systems are needed.

11.4.2 Heterogeneous Systems

Much of the rheology discussed so far pertains specifically to homogeneous polymer systems. As used for fabrication of plastic articles, modification of the homogeneous system by particulate fillers or incompatible polymers is desirable to produce parts with unique properties in the solid state or specific coloration. The addition of fillers obviously influences the

melt rheology and therefore the processing performance of such systems. Despite great industrial importance, prediction of properties from constituents is not a realizable task. The discussion provided here is limited not for lack of importance but for lack of knowledge.

11.4.2.1 Suspensions/Low Viscosity

Einstein (98) considered the increase in viscosity caused by adding small spheres to a Newtonian fluid. He predicted, solely from volume effects, that the viscosity of the suspension should be expressed as shown in Equation (11.109)

$$\eta_s = \eta(1 + \frac{5}{2}\phi) \qquad (11.109)$$

where ϕ is the volume fraction of particles in the suspension. The result of Einstein holds for $\phi \ll 1$. As the concentration of particles is increased, the suspension viscosity can greatly exceed that predicted by Equation (11.109). Jeffrey and Acrivos (99) have reviewed the work on rheology of suspensions and discuss extensions of Equation (11.109) to higher concentrations. One of the more promising results is due to Frankel and Acrivos (100), who consider the hydrodynamic interactions between neighboring particles (spheres) and arrive at a relation useful at high concentrations, shown in Equation (11.110)

$$\frac{\eta_s}{\eta} = \frac{9}{8}\left[\frac{(\phi/\phi_m)^{1/3}}{1 - (\phi/\phi_m)^{1/3}}\right] \qquad \text{as}\;\frac{\phi}{\phi_m} \longrightarrow 1$$

$$(11.110)$$

where ϕ_m is the maximum concentration of spheres.

In addition to the increase in viscosity caused by the presence of particles, the shear sensitivity of the system may be altered. Commonly, the viscosity decreases with an increase in shear rate although the power-law slope may be close to -1.0, unlike polymer solutions, which typically have values of $n > -0.75$. If the particles can sustain order at high shear rates, shear thickening may result. Chaffey (101, 102) has investigated systems exhibiting either shear-thinning or shear-thickening behavior in an attempt to understand the mechanism for the latter response. Hoffman (103) has demonstrated very peculiar behavior in shear with monodisperse PVC spheres. He finds first a decrease in η, then a minimum, followed by an increase up to a critical shear rate where the viscosity becomes discontinuous. Hoffman attributes these results to the formation of planes of particles in the flow field.

At lower shear rates, the structure of the fluid is more evident. At sufficient particle concentrations, the structure is such that solidlike behavior is developed. Paints, pastes, and clay are classic examples of such fluids. Below a certain value of shear stress σ_y, they will not flow. Above the yield stress, they flow as normal fluids. The Bingham model (22) may be applied to such suspensions using Equation (11.111).

$$\eta = \infty \qquad \text{for } \sigma_{12} \leqslant \sigma_y$$

$$\eta = \mu + \frac{\sigma_y}{\dot{\gamma}} \qquad \sigma_{12} > \sigma_y \;\; (11.111)$$

The yield stress is commonly obtained by plotting the data as $\sigma_{12}^{1/2}$ versus $\dot{\gamma}^{1/2}$, in keeping with the model of Casson (104) proposed for printing inks, shown in Equation (11.112).

$$\sigma_{12}^{1/2} = \gamma_y'^{1/2} + k\dot{\gamma}^{1/2} \qquad (11.112)$$

Extrapolation to $\dot{\gamma} \to 0$ allows an estimation of σ_y. Onogi, Matsumoto, and Warashina (105) have extended Casson's equation to include non-Newtonian effects.

11.4.2.2 Suspensions/High Viscosity

Engineering plastics used for molding and extrusion quite often have substantial amounts of fillers added to them. The filler, now suspended in a very-high-viscosity continuous phase, still affects the rheology of the system in much the same way as experienced with lower-viscosity continuous phases.

Figure 11.41 demonstrates the effect upon $\eta(\dot{\gamma})$ of adding carbon black to a polystyrene melt (106). As the carbon-black loading is increased above 10%, the viscosity tends to increase without bound as $\dot{\gamma} \to 0$, indicative of a yield stress. In all cases, the general level of viscosity is increased by the filler. The other viscoelastic properties are also influenced by the filler in much the same way. When shear viscoelastic measurements are made, care must be taken to ensure that the strain is within the linear viscoelastic region. There is strong evidence (105) that the filler causes nonlinear behavior at small values of strain. Memory effects characteristic of viscoelastic fluids are still evident in filled systems. They appear to be enhanced in stress relaxation experiments (107). In die-swell experiments, melts with fillers do not swell as much as analogous systems without fillers (106). This is presumably due to a constraint to lateral expansion

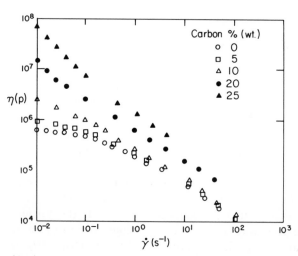

Figure 11.41   Shear viscosity versus shear rate for various loadings of carbon black in polystyrene at 170°C.

caused by the filler. Obviously, attempts to estimate N_1 from die swell would be futile for such systems.

Just as carbon black, glass beads, or TiO_2 can act as particulate fillers, an incompatible polymer phase existing in a polymer-polymer blend or in block copolymers can exist as a particulate domain. Studies of styrene-butadiene block copolymers by Gouinlock and Porter (108) and Vinogradov et al. (109) indicate typical dispersion behavior at low frequencies and shear rates at temperatures below the T_g of polystyrene. Above the T_g of polystyrene, normal homopolymerlike behavior develops as compatibility is improved between the phases. Such behavior is due to the existence of polystyrene domains acting as particulate fillers.

11.5   RHEOLOGY AND PROCESSING

After a thorough review of the fundamental concepts of rheology, the behavior of viscoelastic materials, and the relationship between the rheology and the macroconstituency of polymeric systems, the logical question of practical applicability must be broached. After all, does the shear viscosity at zero-shear rate matter in a molding process or is information at high rates of shear the only data needed? Then again, are shear data relevant or should extensional data be obtained?

The application of rheology and, in particular, of the properties and measurements reviewed in the previous sections can be made on different levels. Often a single-point property is useful to maintain process control or to identify a batch of polymer that might behave poorly. It is essential to remember that although this property is measured at some representative temperature and rate of deformation for the process, most of the properties discussed in Section 11.2 are highly dependent on temperature and rate. Identification of the proper rate and temperature for best correlation with the end-use behavior is imperative. Likewise, proper consideration of elastic versus viscous behavior must be made, since it is possible that two melts of the same polymer

may have identical shear-viscosity responses at high shear rates, yet different amounts of elastic response. Equally likely are melts with similar shear properties but different elongational properties.

The analysis of the process and its interaction with the polymer is essential. Before predictions can be made concerning the effect of rheology upon the final product, the controlling deformations must be analyzed either by mathematical modeling or by careful experimentation. Once the type and history of deformation are known as well as the temperature and pressure histories, decisions concerning which properties must be controlled to improve or control product uniformity will become clearer.

The route to empirical decisions, indicating which knob to turn for a difference in a given property or in prior performance, is difficult to describe and to discuss meaningfully. The modeling approach is more straightforward and visible in the open literature even though often the models simply verify what operators knew all along. The difficulty in describing modeling of polymer processes is the choice of representative processes, since many will soon be obsolete. Two examples have been chosen to illustrate the application of rheology, from viscous behavior to elongational-flow response. Injection molding has been chosen both because of its importance to industry and the extent to which the process pushes our ability to model its reality. The successful studies have relied primarily on our knowledge of shear-flow behavior. Fiber spinning is quite obviously an elongational-flow process and is equally important industrially. These two processes are described in terms of the interaction between the process and polymer.

11.5.1   Injection Molding

Injection molding is a primary method of forming plastic parts of all shapes and sizes, from automobile interior panels to photographic film cartridges. Present technology has made the production of highly intricate parts with thin sections possible to exacting tolerances as low as $\pm$ 0.0003 cm. The popu-

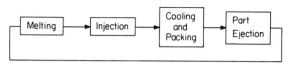

Figure 11.42   Flow chart for injection molding.

larity of the process evolves from its inherent efficiency once production parameters are set. Multiple-cavity molds allow several parts to be produced during one cycle.

The art of injection molding has come primarily from experimental development work. A clear understanding of the mechanics of the process, as the polymer is affected, is not yet available. The process, as depicted in Figure 11.42, begins by melting the pellets and extruding the molten polymer forward to a holding cavity. After a specified amount (shot) is extruded, the screw, which has a ram on its nose, is brought forward, generally at a constant speed, forcing melt into the mold cavity. Once the cavity is full, the ram maintains a high pressure on the part until the polymer solidifies at the cavity-filling point or gate. After solidification of the gate, further cooling is required before part ejection to solidify the runners and to maintain dimensional stability of the part. Once the part is ejected, the mold closes and the cycle begins again. Obviously, rheology of the polymer plays a dominant role in the process behavior. Flow under high temperature and pressure is required to fill the cavity. Higher pressures are used to "pack" more polymer in the cavity to allow for thermal contraction. At the same time, stresses inherent to flow are attempting to relax under these pressures and under high rates of cooling. Complicated as the process may seem in terms of the idealized experiments and responses discussed so far, considerable progress has been made to model various portions of injection molding and to predict performance.

Cavity filling is obviously a viscous-dominated phenomenon. The questions surrounding the filling process are: (1) Is there enough machine pressure available to completely fill the part? (2) How much clamping force is required to hold the mold halves together? (3) For a multicavity mold, is the flow evenly distributed? Reworded, the desired information is the relationship between pressure drop and flow rate for the system.

Several studies of mold filling and of flow through channels prior to entering the mold are available. Notable among these are the work of Williams and Lord (110, 111) and the Cornell Injection Molding Project (112–114). Williams and Lord considered steady-state viscous flow through circular or tapered channels with viscous heat generation and heat transfer to the mold walls. As a model for the viscosity, they chose an empirical form shown in Equation (11.113)

$$\ln \eta = A_0 + A_1 \ln \dot{\gamma} + A_{11} (\ln \dot{\gamma})^2 \\ + A_{12} (\ln \dot{\gamma}) T + A_2 T + A_{22} T^2 \quad (11.113)$$

found to be useful in the modeling of plastication during injection molding (115). Equation (11.113) does have the drawback of not properly predicting η_0 at low $\dot{\gamma}$. In general, the agreement between the model's predictions and experimental data for two melts, acrylonitrile-butadiene-styrene (ABS) and poly-(vinylchloride) (PVC) is quite good, as seen in Figure 11.43, although somewhat poorer for ABS. The model was also applied successfully by Williams and Lord to balance a multicavity mold.

The goal of the Cornell group is to provide a scientific basis for injection molding leading to computer-assisted design and manufacturing. They have developed finite-element programs, using a power-law viscosity model, capable of predicting front-position pressure drop and temperature rise during the filling of rather complicated molds involving two-dimensional flow around inserts with multiple gates. From the pressure drop, clamp force can be estimated, giving the required force to prevent mold flashing. Predictions obtained with the model are in reasonable agreement ($\pm$ 30%) with data from injection-molding tests.

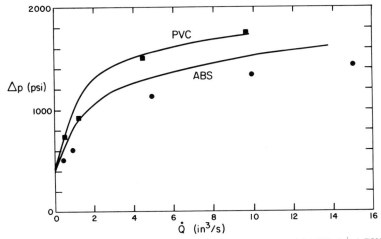

Figure 11.43 Prediction of pressure drop for flow in circular runners: (•) ABS; (▪) PVC (110).

To make the impact of their work more broadly applicable, the Cornell workers have created a system of design charts to predict injection pressure and clamp force for large plastic parts by a graphical procedure (114). The desired part is approximated by a series of circles or sectors of circles emanating from the gates(s). The area of the approximated surface must equal the area of one side of the part. The injection pressure and clamp force are then determined by calculating both for the power-law, isothermal case of filling a center-fed disk and then finding the appropriate correction on the design charts. The corrections were obtained using the computer model that included nonisothermal flow. Similar charts are being constructed for flow in the runners and sprues (112).

Although filling is predominantly a viscous phenomenon and can obviously be modeled well using purely viscous models, elasticity does play a role. The normal filling pattern for a small end gate is shown in Figure 11.44. The mold fills with a radially bounded advancing front. Sometimes the mold does not fill uniformly, but melt jets through the gate to the

opposite side and the mold fills in a swirling flow. Oda, White, and Clark (116) have associated jetting, an undesirable filling pattern, with die swell. They contend that jetting occurs when the melt swells insufficiently to contact the cold mold walls. The normal filling pattern occurs when contact is made as the polymer leaves the gate area. This work has obvious implications for gate and mold design, being especially critical for filled materials that tend to exhibit small amounts of die swell.

As the part is filled, stresses due to flow are present. After filling, these stresses and related orientation will relax. Since orientation greatly influences the mechanical properties of the part such as modulus, impact strength, and dimensional stability, a prediction of orientation or residual stress as a function of melt rheology and processing variables would prove useful. Dietz, White, and Clark (117) have modeled the residual stress problem for a rectangular part. They model filling with an isothermal power-law model to obtain the stress profile at the point at which the mold is full. They then use a linear viscoelastic model to account for stress relaxation while the melt is cooling.

The development of this type of model is interesting and provides valuable insight into the use of rheology and specifically fluid elasticity in solving a real problem. The analysis of Dietz, White, and Clark can be paral-

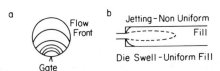

Figure 11.44 Normal filling pattern (a) and jetting (b) for a small end-gated disk.

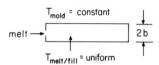

Figure 11.45 End-fed disk for residual stress model.

lelled by considering flow into an end-fed circular disk, shown in Figure 11.45. For a Newtonian fluid in isothermal flow, the velocity gradient some distance between the gate and the front is given by Middleman (118), as shown in Equation (11.114).

$$\frac{\partial u_r}{\partial z} = \dot{\gamma}(r, z) = -\frac{3}{4} \frac{Qz}{\pi b^3 r} \quad (11.114)$$

It is assumed that the slab half-thickness is much smaller than the radial part dimension. The initial conditions for stress relaxation after filling can be estimated by the structural model of Section 11.2.5.2. Assuming that shear flow predominates, the shear stress and first normal stress difference under steady-flow conditions are given by Equations (11.115)–(11.117) for the structural model with one relaxation time.

$$\sigma_{12} = \eta_0 \dot{\gamma} x^{2.4} \quad (11.115)$$

$$N_1 = 2\eta_0 \tau_0 x^{3.8} \dot{\gamma}^2 - \sigma_{11} - \sigma_{22} = \sigma_{11}$$

$$(11.116)$$

$$\frac{1 - x}{x^{2.4}} = a_M \dot{\gamma} \tau_0 \sqrt{2} \quad (11.117)$$

The subsequent relaxation of the stresses is governed by Equations (11.118)–(11.120)

$$\sigma_{12} \left(1 - \tau_0 x^{0.4} \frac{dx}{dt}\right) + \tau_0 x^{1.4} \frac{\partial \sigma_{12}}{\partial t} = 0 \quad (11.118)$$

$$\sigma_{11} \left(1 - \tau_0 x^{0.4} \frac{dx}{dt}\right) + \tau_0 x^{1.4} \frac{\partial \sigma_{11}}{\partial t} = 0 \quad (11.119)$$

$$\frac{dx}{dt} = \frac{1 - x}{\tau_0 x^{1.4}} \frac{a_M}{\tau_0 x^{0.4}} \left(\frac{\sigma_{11}}{\eta_0 \tau_0 x^{3.8}}\right)^{1/2} \quad (11.120)$$

To complete the model, one needs the temperature as a function of time during

cooling. For a parallel-sided disk with $b \ll R$, this is conveniently given by Bird, Stewart, and Lightfoot (1) as shown in Equation (11.121).

$$\frac{T_{mold} - T}{T_{mold} - T^\circ}_{melt}$$

$$= 2 \sum_{n=1}^{\infty} \frac{(-1)^n}{(n + 1/2\pi)} \exp \frac{(-n + 1/2)^2 \pi^2 at}{b^2}$$

$$\cos (n + 1/2) \frac{\pi z}{b} \quad (11.121)$$

The rheological constants must depend upon temperature and can be modeled as explained in Section 11.6.2

The preceding model described has been used to estimate effects of melt temperature, mold temperature, injection flow rate, part thickness, and relaxation characteristics on residual stress. The melt properties in isothermal tests are shown in Figures 11.20, 11.21, and 11.46. The predictions of the structural model for η and N_1 as functions of $\dot{\gamma}$ and for isothermal stress relaxation are shown in Figures 11.46 and 11.47 respectively. At a normalized radial position of 0.25, the residual shear stress as a function of position z from the center to a point near the wall is seen in Figure 11.48 to climb rapidly from zero to quite large values near the wall. This is because the melt has more time to relax at a high temperature in the center portion of the disk. Figure 11.48 illustrates the effect of changing melt temperature on the residual stress, and Figure 11.49

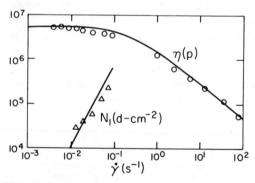

Figure 11.46 Viscosity and N_1 for PMMA, $T_{ref} = 180°C$. The lines are model predictions of the structural model $\eta_0 = 5(10^6)$ p, $\tau^\circ = 10$ sec, $a_M = 0.084$.

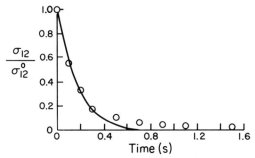

Figure 11.47 Results and structural model predictions for an isothermal stress-relaxation experiment, $\dot{\gamma}_0 = 2.5/\text{sec}$, $T = 220°\text{C}$.

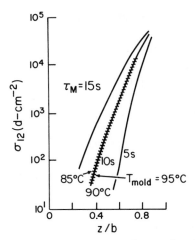

Figure 11.49 Residual stress predictions of model described in Section 11.5.1, $\dot{Q} = 0.16 \text{ cm}^3/\text{sec}$, $b = 0.08$ cm, $T_{melt} = 240°\text{C}$.

depicts the effect of changing the melt's relaxation time. Strong effects are indicated. Particularly interesting is the effect of relaxation time. Although the effect is fully anticipated, τ^0 is rarely monitored in a process situation. Generally only melt flow or viscosity is checked. Small changes in the molecular weight distribution may have a significant influence upon the relaxation characteristics without significantly affecting the viscosity (90).

An important characteristic of the models of injection molding that have been discussed is simplification. Various portions of the process are attacked as subprocesses. A coordinated program is required to link the subprocess model into a research and design tool eliminating some of the simplifications along the way, such as including the effects of high

pressure. The final link to part properties also must be developed to predict impact strength or distortion under load in the solid state.

11.5.2 Fiber Spinning

Although there are extensional-flow components in injection molding, notably in the gate region and the melt front, the deformation is largely a nonisothermal shear flow. A process that requires a nonisothermal extensional flow is fiber spinning. The melt is forced through a small die known as a spinneret, expands owing to die swell, and then is drawn to a small diameter while it cools. Although fiber spinning will be discussed specifically, many processes are analogous to it, and the models applied to spinning are readily extended to include such processes as film stretching or stretch coating.

The difficulties in modeling fiber spinning are twofold. First the aspect of nonisothermal flow must be handled. (This applies naturally to melt spinning; wet or solution spinning involves an equally difficult problem of accounting for solvent loss and its effect upon rheology.) The second problem concerns the unpredictable nature of viscoelastic melts and extensional flow. Which model and/or which $\eta_e(\dot{\varepsilon})$ relationship does one choose?

The basic questions requiring analysis for solution are: What properties are needed for

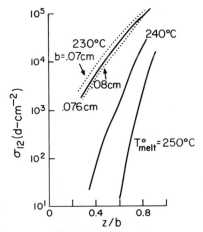

Figure 11.48 Residual stress predictions of model described in Section 11.5.1, $\dot{Q} = 0.16 \text{ cm}^3/\text{sec}$, $T_{mold} = 90°\text{C}$.

spinnability? What drawdowns (extensions) can be used for stable spinning? These not unrelated problems were first attacked for isothermal, Newtonian fluids. Later shear-thinning, but purely viscous, fluids in non-isothermal flow were studied. More recently, stability analyses and dynamic models have been developed for both isothermal and non-isothermal viscoelastic models.

The Newtonian analysis is a useful first exercise toward developing a mechanical model of spinning. It is possible to derive both the deformation rate $\dot{\epsilon}$ down the spin line, Equation (11.122),

$$\dot{\epsilon} = \frac{\partial u}{\partial x} = V_0 \frac{\ln D_R}{L} \exp \frac{(x \ln D_R)}{L} \quad (11.122)$$

and the force at take-up as shown in Equation (11.123)

$$F = \pi R^2 3\eta_0 \dot{\epsilon} \quad (11.123)$$

where R is the fiber radius, V_0 is the velocity at the point of maximum swell, and D_R is the draw ratio V_L/V_0 (118). A stability analysis (119) indicates that Newtonian fluids are stable to infinitesimal perturbations for $D_R <$ 20.5, a result verified experimentally by Donnelly and Weinberger (120).

Newtonian models are not particularly useful for predicting the behavior of viscoelastic fluids, since spinning, as it is practiced industrially, involves a highly transient deformation field from the fluid's view, so elasticity of the fluid must be properly considered. A considerable body of literature exists concerning spinning of viscoelastic fluids. Denn, Petrie, and Avenas (121) modeled the mechanics of steady spinning with a rate-type constitutive relation, Equation (11.124).

$$\boldsymbol{\sigma} + \tau \frac{d\boldsymbol{\sigma}}{dt} = \eta_0 \mathbf{A} + 2\tau\eta_0\nu\mathbf{A}^2 \quad (11.124)$$

This model can be applied successfully to mildly non-Newtonian melts in shear. The spinning analysis predicts that as the melt becomes highly elastic, the velocity profile down the spin line becomes linear, a result

observed experimentally by Spearot and Metzner (122) for three low-density polyethylenes. Attempts to predict the take-up force and compare the results of Spearot and Metzner with the model were not successful owing to the model's sensitivity to small changes in τ and the initial stress, two parameters not easily determined.

Matsui and Bogue (123) considered the case of nonisothermal spinning using an integral model shown in Equation (11.125)

$$\sigma_{ij} = p\delta_{ij}$$
$$+ \int_{-\infty}^{t} \sum_n G_n \frac{\exp\{-\int_{t'}^{t}[dt''/\tau_n^*(t'')]\}}{\tau_n^*(t')} C_{ij}^{-1} \, dt'$$
$$(11.125)$$

where

$$\tau_n^*(t) = \frac{\tau_n^\circ a_T}{1 + (a_{BM} II_\Delta^{1/2} \tau_n^\circ a_T)/2}$$

The model was compared with data of Dees and Spruiell (124) for various high-density polyethylenes using their measured threadline temperatures. As shown in Figure 11.50, the agreement obtained was quite reasonable. The

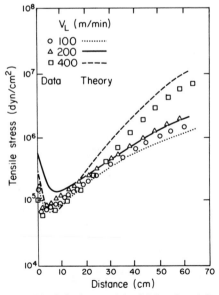

Figure 11.50 Spinning model of Matsui and Bogue (123) compared with data for high-density polyethylene of Dees and Spruiell (124).

initial dip in stress is due to the die swell upon leaving the spinneret.

Although the works of Denn, Petrie, and Avenas and of Matsui and Bogue are beginnings to a model of spinning and should allow predictions of stress and strain rate in the spin line, they do not yield predictions of spinnability. One of the largest problems faced when one wishes to achieve high throughput and small fibers is a phenomenon known as draw resonance. Draw resonance is an oscillation in the threadline diameter that may lead to fracture of the fiber, obviously deleterious to a spinning process. The theoretical and experimental limit for isothermal stability of Newtonian fluids is a draw ratio $D_R \leqslant 20.5$. Cooling generally aids stability, allowing for high draw ratios to be used (119). The effect of viscoelasticity is unclear. Petrie and Denn (125) reviewed the experimental work in the field and tabulated a wide variety of results, some showing increases in the critical D_R, some decreases for viscoelastic melts. They noted that for four sets of experiments, those by Donnelly and Weinberger (120), Cruz-Saenz et al. (126), Ishihara and Kase (127), and Zeichner (128), the critical draw ratio for instability in isothermal spinning was decreased by viscoelasticity except at very short fiber lengths for the poly(ethylene terephthalate) studies of Ishihara and Kase, where an upper plateau above $D_R = 50$ was found. This upper plateau was predicted by Fisher and Denn (129).

The resolution of the confusion surrounding draw resonance (spinnability) and viscoelasticity may be found in the extensional flow behavior of the melt. In other words, it all depends. Some very interesting work by White and Ide (130–133) on the spinnability of polymer melts and its dependence upon failure mechanisms in extensional flow indicates that spinnability may be influenced strongly by the reactions of the melt to extension through the memory function of the fluid. Using the nonlinear parameter in their model, a White-Metzner fluid with a deformation-dependent relaxation time, shown in Equation (11.126),

$$\tau = \frac{\tau_0}{1 + (a_{WM}\tau_0 II_\Delta^{1/2})/2} \quad (11.126)$$

they claim that spinnability increases as a_{WM} decreases. This is equivalent to saying that elastic fluids are easily spun whereas viscous fluids are not. They cite as evidence work with low-density polyethylene, a_{WM} approximately 0.1, and high-density polyethylene, a_{WM} approximately 1.0. The low-density polyethylene is readily spun whereas high-density polyethylene is not.

11.6 SUMMARY

It is hoped that the reader now has an appreciation for rheology and its foundations, virtues, and foibles. The intent at the outset was to provide a basis for further work and reading through an understanding of the fundamentals and a full list of references.

At this point it is fair to question the course of research in rheology for the future. Although to ask the question may be fair, to provide a long definitive discussion would be folly. There do seem to be some reasonable courses for future research. It is quite likely that considerable effort will be expended on property-composition relations, general elongational-flow phenomena, and direct application of rheology to polymer processes. Less work seems likely on the development of new viscoelastic models and their verification with viscometric flow.

In the field of property-composition relations, one would hope for a macromolecular model for high molecular weight polymer melts and solutions consistent with experiment and capable of property predictions as a constitutive relation. Equally important is work that would unravel our limited knowledge of composition modifiers such as long-chain branching, ionic groups copolymerized with the main chain, incompatible copolymers, and cross-linking agents. These studies will progress through a series of controlled, model polymers to a sensible approach to real systems.

Elongational-flow researchers must grasp the unattainability of steady-flow data at processing conditions and learn to deal with transient data. Commercial extensional rheometers are on the horizon, making the feasibility of such measurements very attractive. Before a mass of data is collected, the relevance of the data to process needs and the proper way to reduce the information to provide the needed input must be considered.

Perhaps the most exciting and difficult area for research is the application of rheology to polymer processes, qualitative as well as quantitative application. Numerical techniques, such as finite elements, are being developed for viscoelastic models, allowing more complete and versatile descriptions of process flows. The effects of elasticity are by necessity being given more consideration as new processes are developed or present processes are improved to meet tighter tolerances with specialty materials.

The understanding of the flow behavior of polymers has traversed a fair distance from the first recognition of elasticity and viscoelasticity. Much of the literature of rheology begins in the 1950s, making it a rather young science. It is young, but rapidly maturing. The developments mentioned will speed maturation and make rheology a highly useful branch of science.

11.7 GLOSSARY OF SYMBOLS

a	Eccentricity in the orthogonal rheometer
a_{BW}	Nonlinear constant in the Bogue-White model
a_{BKZK}	Nonlinear constant in the BKZK model
a_M	Nonlinear constant in the structural model
a_p	Pressure shift factor
a_T	Time-temperature superposition shift factor
a_{WM}	Nonlinear constant in the White-Metzner model
$\mathbf{A}^n$	nth Rivlin-Ericksen tensor
b	constant in the Lyons-Tobolsky equation
c	Concentration
$\mathbf{C}$	Cauchy strain tensor
$\mathbf{C}^{-1}$	Finger strain tensor
C_1, C_2	Constants in the WLF equation

d	Spacing for the Ninomiya-Ferry technique
D	Capillary diameter
D_e	Extrudate diameter
D_R	Draw ratio in stretching operations
f	Free volume
F_x, F_y	Forces in the orthogonal rheometer
F_z	Normal force
g	Size parameter for branched polymers
g_{kl}	Components of metric tensor
G'	Storage modulus
G''	Loss modulus
G_n	Discrete modulus
h	Plate separation in the orthogonal rheometer
$H(\tau)$	Relaxation function
$\mathbf{I}$	Unit tensor
J_e^0	Steady-state recoverable compliance
J_{eR}^0	Reduced recoverable compliance
k_H	Huggins constant
k_M	Constant in the Martin equation
k_{LT}	Constant in the Lyons-Tolobsky equation
L	Length
L_n	Probability of junction formation
m	Flow characteristic for power-law model
M	Molecular weight
$\overline{M}_n$	First moment of molecular weight distribution
$\overline{M}_w$	Second moment of molecular weight distribution
$\overline{M}_z$	Third moment of molecular weight distribution
$\overline{M}_{z+1}$	Fourth moment of molecular weight distribution
n	Power-law constant
N_1	First normal stress difference in shear flow
N_2	Second normal stress difference in shear flow
$N(t - t')$	Memory function
p	Pressure
$\dot{Q}$	Flow rate
R	Bubble radius
R_g	Gas-law constant
R_p	Plate radius

s	Time variable, backward running
$\langle s^2 \rangle$	Mean-square radius of gyration
t	Time
t_{max}	Time to stress overshoot in start-up of steady-shearing flow
T	Temperature
T_g	Glass transition temperature
T_p	Torque on plate in shear flow
u_i	Velocity components
x	Fraction of entanglement junctions in the structural model
x^i	Coordinate variable
α	Thermal diffusivity
α_f	Volume expansion coefficient
$\alpha_1, \alpha_2, \alpha_3$	constants in the fluid of grade 2
γ	Shear strain
$\dot{\gamma}$	Shear rate
Γ	Viscosity enhancement factor for branched polymers
δ	Phase angle
Δ_{ij}	Components of deformation rate tensor
ϵ	Elongational strain
$\dot{\epsilon}$	Strain rate
η	Viscosity
$[\eta]$	Intrinsic viscosity
η_e	Elongational viscosity
η_0	Zero-shear-rate viscosity
η_r	Relative viscosity
η_s	Solvent viscosity
μ	Newtonian viscosity
ν	Constant in Equation (11.124)
ζ^i	Components of hypothetical place in a blob of fluid
ρ	Density
σ	Stress tensor
τ	Relaxation time
τ_e	Effective relaxation time
τ_{max}	Maximum relaxation time
τ°	Undeformed relaxation time
Φ	Particle concentration
ϕ_m	Maximum particle concentration
Ψ_{12}	Primary normal stress coefficient
Ψ_{23}	Secondary normal stress coefficient
ω	Frequency
ω_{ij}	Components of vorticity tensor

11.8   REFERENCES

1. R. B. Bird, W. E. Stewart, and E. N. Lightfoot, *Transport Phenomena,* John Wiley, New York, 1960.

2. W. Noll, *Arch. Ration. Mech. Anal.,* **2**, 197 (1958).

3. S. Middleman, *The Flow of High Polymers,* Wiley-Interscience, New York, 1968.

4. R. Aris, *Vectors, Tensors and the Basic Equations of Fluid Mechanics,* Prentice-Hall, Englewood Cliffs, New Jersey, 1962.

5. G. Astarita and G. Marrucci, *Principles of Non-Newtonian Fluid Mechanics,* McGraw-Hill, Berkshire, England, 1974.

6. A. S. Lodge, *Elastic Liquids,* Academic Press, New York, 1964.

7. R. Racin, unpublished MS dissertation, University of Tennessee, Knoxville, 1978.

8. E. M. Friedman and R. S. Porter, *Trans. Soc. Rheol.,* **19**, 493 (1975),

9. W. W. Graessley, *Adv. Polym. Sci.,* **16**, 1 (1974).

10. R. I. Tanner, *Trans. Soc. Rheol.,* **14**, 483 (1970).

11. M. J. Miller and E. B. Christiansen, *AIChE J.,* **18**, 600 (1972).

12. J. F. Stevenson, *AIChE J.,* **18**, 540 (1972).

13. R. L. Ballman, *Rheol. Acta,* **4**, 137 (1965).

14. G. V. Vinogradov, B. V. Radushkevich, and V. D. Fikhman, *J. Polym. Sci.: Part A-2,* **8**, 1 (1970).

15. J. Meissner, *Rheol. Acta,* **10**, 230 (1971).

16. H. Münstedt, *Rheol. Acta,* **14**, 1077 (1975).

17. P. J. Carreau, *Trans. Soc. Rheol.,* **16**, 99 (1972).

18. G. Pearson and S. Middleman, *Trans. Soc. Rheol.,* **20**, 559 (1976).

19. N. Phan-Thien, *J. Rheol.* **22**, 259 (1978).

20. A. Kaye and A. J. Kennett, *Rheol. Acta,* **13**, 916 (1974).

21. D. Acierno, F. P. LaMantia, G. Marrucci, and G. Titomanlio, *J. Non-Newtonian Fluid Mech.,* **1**, 125 (1976).

22. R. B. Bird, R. C. Armstrong, and O. Hassager, *Dynamics of Polymeric Liquids,* Vol. 1, *Fluid Mechanics,* John Wiley, New York, 1977.

23. A. C. Pipkin, *Lectures on Viscoelasticity Theory,* Springer-Verlag, New York, 1972.

24. A. S. Lodge, *Body Tensor Fields in Continuum Mechanics,* Academic Press, New York, 1974.

25. W. Ostwald, *Kolloid-Z.,* **36**, 99 (1925).

26. A. deWaele, *Oil Color Chem. Assoc. J.,* **6**, 33 (1923).

27. P. J. Carreau, unpublished PhD dissertation, University of Wisconsin, Madison, 1968.

28. M. Reiner, *Deformation, Strain, and Flow,* Interscience, New York, 1960.

29. C. Truesdell in M. VanDyke, W. G. Vincenti, and

J. V. Wehausen, Eds., *Ann. Rev. Fluid Mech.,* Vol. 6, Annual Review Inc., Palo Alto, California, 1974.

30. M. S. Green and A. V. Tobolsky, *J. Chem. Phys.,* **14**, 80 (1946).

31. A. S. Lodge, *Rheol. Acta,* **7**, 379 (1968).

32. D. C. Bogue and I-J. Chen, *Trans. Soc. Rheol.,* **16**, 59 (1972).

33. A. Kaye and A. J. Kennett, *Rheol. Acta,* **13**, 916 (1974).

34. B. Bernstein, E. A. Kearsley, and L. J. Zapas, *Trans. Soc. Rheol.,* **7**, 391 (1963).

35. D. Acierno, F. P. LaMantia, G. Marrucci, and G. Titomanlio, *J. Non-Newtonian Fluid Mech.,* **1**, 125 (1976).

36. D. Acierno, F. P. LaMantia, G. Rizzo, and G. Titomanlio, *J. Non-Newtonian Fluid Mech.,* **1**, 147 (1976).

37. D. Acierno, F. P. LaMantia, and G. Marrucci, *J. Non-Newtonian Fluid Mech.,* **2**, 27 (1977).

38. R. B. Bird in M. VanDyke, W. G. Vincenti, and J. Wehausen, Eds., *Ann. Rev. Fluid Mech.,* Vol. 8, Annual Review Inc., Palo Alto, California, 1976.

39. R. B. Bird, O. Hassager, and S. I. Abdel-Khalik, *AIChE J.,* **20**, 1041 (1974).

40. G. Pearson and S. Middleman, *Trans. Soc. Rheol.,* **20**, 559 (1976).

41. A. S. Lodge in S. Onogi, Ed., *Proceedings of Fifth International Congress on Rheology,* University Park Press, Baltimore, Maryland, 1970.

42. J. White and A. B. Metzner, *J. Appl. Polym. Sci.,* **7**, 1867 (1963).

43. G. Pearson and S. Middleman, *AIChE J.,* **23**, 714 (1977).

44. H. Leaderman, *Text. Res.,* **11**, 171 (1941).

45. J. D. Ferry, *Viscoelastic Properties of Polymers,* 2nd ed., John Wiley, New York, 1970.

46. A. K. Doolittle and D. B. Doolittle, *J. Appl. Phys.,* **28**, 901 (1957).

47. F. Schwarzl and A. J. Staverman, *J. Appl. Phys.,* **23**, 838 (1952).

48. M. Matsui and D. C. Bogue, *Trans. Soc. Rheol.,* **21**, 133 (1977).

49. K. Walters, *Rheometry,* John Wiley, London, England, 1975.

50. C. W. Macosko and W. M. Davis, *Rheol. Acta,* **13**, 814 (1974).

51. W. E. Rochefort, unpublished MS Chem. Eng. thesis, Northwestern University, Evanston, Illinois, 1978.

52. E. B. Bagley, *J. Appl. Phys.,* **28**, 624 (1957).

53. J. L. White and J. F. Roman, *J. Appl. Polym. Sci.,* **20**, 1005 (1976).

54. A. B. Metzner, W. T. Houghton, R. A. Sailor, and J. L. White, *Trans. Soc. Rheol.,* **5**, 133 (1961).

55. J. L. S. Wales, *Rheol. Acta,* **8**, 38 (1969).

56. C. D. Han, M. Charles, and W. Phillippoff, *Trans. Soc. Rheol.,* **13**, 453 (1969).

57. C. D. Han, *Rheology in Polymer Processing,* Academic Press, New York, 1976.

58. R. I. Tanner, *J. Polym. Sci.: Part A-2,* **8**, 2067 (1970).

59. S. I. Abdel-Khalik, O. Hassager, and R. B. Bird, *Polym. Eng. Sci.,* **14**, 859 (1974).

60. J. M. Dealy, *J. Non-Newtonian Fluid Mech.,* **4**, 9 (1978).

61. C. W. Macosko and J. M. Lornston, *SPE Tech. Pap.,* **19**, 461 (1973).

62. J. M. Starita, personal communication, October 1978.

63. R. W. Connelly, L. J. Garfield, and G. H. Pearson, Society of Rheology Meeting, New York, February 1976.

64. F. N. Cogswell, *J. Non-Newtonian Fluid Mech.,* **4**, 23 (1978).

65. A. E. Everage, Jr. and R. L. Ballman, *Nature,* **273**, 213 (1978).

66. R. N. Shroff, L. V. Canico, and M. Shida, *Trans. Soc. Rheol.,* **21**, 429 (1977).

67. F. N. Cogswell, *Polym. Eng. Sci.,* **12**, 64 (1972).

68. G. H. Pearson and S. Middleman, *Rheol. Acta,* **17**, 500 (1978).

69. J. C. Hsu and R. W. Flumerfelt, *Trans. Soc. Rheol.,* **19**, 523 (1975).

70. J. C. Hsu and R. W. Flumerflet, unpublished.

71. H. M. Princen, J. Zia, and S. G. Mason, *J. Coll. Int. Sci.,* **23**, 99 (1967).

72. R. T. Balmer, *J. Non-Newtonian Fluid Mech.,* **2**, 307 (1977).

73. D. Acierno, G. Titomanlio, and L. Nicodemo, *Rheol. Acta,* **13**, 352 (1974).

74. S. T. J. Peng and R. F. Landel, *J. Appl. Phys.,* **47**, 4255 (1976).

75. R. Bragg and D. R. Oliver, *Nature,* **241**, 131 (1973).

76. R. B. Bird, R. C. Armstrong, O. Hassager, and C. F. Curtiss, *Dynamics of Polymeric Liquids,* Vol. 2., *Kinetic Theory,* John Wiley, New York, 1977.

77. M. C. Williams, *AIChE J.,* **21**, 1 (1975).

78. H. L. Frisch and R. Simha, "The Viscosity of Colloidal Suspensions and Macromolecular Solutions," in F. R. Eirich, Ed., *Rheology,* Vol. 1, Academic Press, New York, 1966, Chapter 14.

79. L. Utracki and R. Simha, *J. Polym. Sci.: Part A,* **1**, 1089 (1963).

80. P. F. Lyons and A. V. Tobolsky, *Polym. Eng. Sci.,* **10**, 1 (1970).

81. M. Rink, A. Pavan, and S. Roccasalvo, *Polym. Eng. Sci.,* **18**, 755 (1978).

82. B. Chitrangad, H. R. Osmers, and S. Middleman, *Polym. Eng. Sci.,* **17**, 806 (1977).

83. F. Bueche, *J. Chem. Phys.,* **20**, 1959 (1952).

84. M. C. Williams, *AIChE J.*, **12**, 1064 (1966); **13**, 534 (1967); **13**, 955 (1967).

85. P. E. Rouse, *J. Chem. Phys.*, **21**, 1272 (1953).

86. B. H. Zimm, *J. Chem. Phys.*, **24**, 269 (1956).

87. K. Osaki and J. L. Schrag, *Polym. J. Jap.*, **2**, 541, (1971).

88. J. W. M. Noordermeer, J. D. Ferry, and N. Nemoto, *Macromolecules*, **8**, 672 (1975).

89. J. D. Ferry, M. L. Williams, and D. M. Stern, *J. Phys. Chem.*, **58**, 987 (1954).

90. G. H. Pearson and L. J. Garfield, *Polym. Eng. Sci.*, **18**, 583 (1978).

91. W. W. Graessley, *J. Chem. Phys.*, **47**, 1942 (1967).

92. W. W. Graessley and L. Segal, *Macromolecules*, **2**, 49 (1969).

93. F. Bueche, *J. Chem. Phys.*, **40**, 484 (1964).

94. W. W. Graessley, T. Masuda, J. E. L. Roovers, and N. Hadjiichristidis, *Macromolecules*, **9**, 127 (1976).

95. G. Kraus and J. T. Gruver, *J. Polym. Sci.: Part A-2*, **8**, 305 (1970).

96. G. C. Berry and T. G. Fox, *Adv. Polym. Sci.*, **5**, 261 (1968).

97. P. G. DeGennes, *J. Phys. (Paris)*, **36**, 1199 (1975).

98. A. Einstein, *Investigations on the Theory of the Brownian Movement*, Dover, New York, 1956.

99. D. J. Jeffrey and A. Acrivos, *AIChE J.*, **22**, 417 (1976).

100. N. A. Frankel and A. Acrivos, *Chem. Eng. Sci.*, **22**, 847 (1967).

101. C. E. Chaffey, *J. Colloid Interface Sci.*, **56**, 495 (1976).

102. C. E. Chaffey, *Colloid Polym. Sci.*, **255**, 691 (1977).

103. R. L. Hoffman, *Trans. Soc. Rheol.*, **16**, 155 (1972).

104. N. Casson, "A Flow Equation for Pigment-Oil Suspensions of the Printing Ink Type," in C. C. Mill, Ed., *Rheology of Disperse Systems*, Pergamon Press, London, England, 1959, p. 84.

105. S. Onogi, T. Matsumoto, and Y. Warashina, *Trans. Soc. Rheol.*, **17**, 175 (1973).

106. V. M. Lobe and J. L. White, University of Tennessee Polymer Science Engineering Report No. 118, July 1978.

107. T. Matsumoto, C. Hitomi, and S. Onogi, *Trans. Soc. Rheol.*, **19**, 541 (1975).

108. E. V. Gouinlock and R. S. Porter, *Polym. Eng. Sci.*, **17**, 535 (1977).

109. G. V. Vinogradov, V. E. Dreval, A. Ta. Mallkin, Yu. G. Yanovsky, V. V. Barancheeva, E. K. Bonsenkova, M. P. Zabugina, E. P. Plotnikova, and O. Yu. Sabsai, *Rheol. Acta*, **17**, 258 (1978).

110. G. Williams and H. A. Lord, *Polym. Eng. Sci.*, **15**, 553 (1975).

111. H. A. Lord and G. Williams, *Polym. Eng. Sci.*, **15**, 569 (1975).

112. K. K. Wang, S. F. Shen, C. Cohen, C. A. Hieber, and S. Johanmer, Fifth Progress Report, "Computer-Aided Injection Molding System," Cornell University, New York, November 1978.

113. J. F. Stevenson, A Galskoy, K. K. Wang, I. Chen, and D. H. Reber, *Polym. Eng. Sci.*, **17**, 706 (1977).

114. J. F. Stevenson, R. A. Hauptfleisch, and C. A. Hieber, *Plast. Eng.*, **32**, No. 12, 34 (1976).

115. R. C. Donovan, *Polym. Eng. Sci.*, **14**, 101 (1974).

116. K. Oda, J. L. White, and E. S. Clark, *Polym. Eng. Sci.*, **16**, 585 (1976).

117. W. Dietz, J. L. White, and E. S. Clark, *Polym. Eng. Sci.*, **18**, 273 (1978).

118. S. Middleman, *Fundamentals of Polymer Processing*, McGraw-Hill, New York, 1977.

119. J. R. A. Pearson and Y. T. Shah, *Trans. Soc. Rheol.*, **16**, 519 (1972).

120. G. J. Donnelly and C. B. Weinberger, *Ind. Eng. Chem. Fund.*, **14**, 334 (1975).

121. M. M. Denn, C. J. S. Petrie, and P. Avenas, *AIChE J.*, **21**, 791 (1975).

122. J. A. Spearot and A. B. Metzner, *Trans. Soc. Rheol.*, **16**, 495 (1972).

123. M. Matsui and D. C. Bogue, *Polym. Eng. Sci.*, **16**, 735 (1976).

124. J. R. Dees and J. E. Spruiell, *J. Appl. Polym. Sci.*, **18**, 1053 (1974).

125. C. J. S. Petrie and M. M. Denn, *AIChE J.*, **22**, 209 (1976).

126. G. F. Cruz-Saenz, G. J. Donnelly, and C. B. Weinberger, *AIChE J.*, **22**, 441 (1976).

127. H. Ishihara and S. Kase, *J. Appl. Polym. Sci.*, **19**, 557 (1975).

128. G. Zeichner, "Spinnability of Viscoelastic Fluids," M. Chem. Eng. thesis, University of Delaware, Newark, 1973.

129. R. J. Fisher and M. M. Denn, *AIChE J.*, **22**, 236 (1976).

130. Y. Ide and J. L. White, *J. Appl. Polym. Sci.*, **20**, 2511 (1976).

131. Y. Ide and J. L. White, *J. Non-Newtonian Fluid Mech.*, **2**, 281 (1977).

132. Y. Ide and J. L. White, *J. Appl. Polym. Sci.*, **22**, 1061 (1978).

133. J. L. White and Y. Ide, *J. Appl. Polym. Sci.*, **22**, 3057 (1978).

CHAPTER 12

Processing

E. O. ALLEN

Eastman Kodak Company
Rochester, New York

12.1	Introduction	571
12.2	Extrusion	572
	12.2.1 Equipment	573
	12.2.1.1 Drive Mechanism	573
	12.2.1.2 Barrel	574
	12.2.1.3 Screw	574
	12.2.1.4 Screen Pack	576
	12.2.1.5 Pressure Control Valve	576
	12.2.1.6 Dies and Takeoff Systems	576
	12.2.2 Process	579
12.3	Spinning	582
12.4	Injection Molding of Thermoplastics	582
	12.4.1 Equipment	582
	12.4.1.1 Plunger Molding	582
	12.4.1.2 Reciprocating Screw Molding	583
	12.4.1.3 Molds	583
	12.4.1.4 Sizing Mold and Machine	586
	12.4.2 Process	587
12.5	Injection Molding of Thermosets	590
12.6	Reaction Injection Molding	591
12.7	Compression Molding	591
	12.7.1 Equipment	592
	12.7.1.1 Molds	592
	12.7.2 Process	592
12.8	Transfer Molding	593
12.9	Blow Molding	594
	12.9.1 Extrusion Blow Molding	594
	12.9.2 Injection Blow Molding	595
	12.9.3 Stretch Blow Molding	595
12.10	Rotational Molding	596
12.11	Thermoforming	596
	12.11.1 Equipment	596
	12.11.2 Process	597

12.12	Solid-Phase Forming	599
12.13	Calendering	599
12.14	Casting	600
	12.14.1 Solution Casting	600
	12.14.2 Monomer Casting	601
12.15	Foam Processing	601
	12.15.1 Expandable Polystyrene Molding	602
	12.15.2 Foam Extrusion	602
	12.15.3 Liquid Foam Processing	602
	12.15.4 Structural Foam Molding	603
12.16	Processing of Plastisols and Organosols	603
12.17	Radiation Processing	604
12.18	Machining	605
12.19	Bonding	605
	12.19.1 Mechanical Fastening	606
	12.19.2 Solvent Cementing	606
	12.19.3 Adhesive Bonding	606
	12.19.4 Melt-Phase Welding	606
12.20	Decorating	607
	12.20.1 Painting	607
	12.20.2 Printing	607
	12.20.3 Vacuum Metallizing	608
	12.20.4 Electroplating	608
12.21	Summary	608
12.22	References	610

12.1 INTRODUCTION

Many techniques are available for shaping polymeric materials into useful objects. These can be referred to as primary plastics fabrication processes. The selection of the proper processing method for a particular product is extremely important from both an economic and a quality point of view. Certain processes are better suited to very high unit-volume production because of short cycle times and ease of automation, but these techniques often require high initial costs for equipment and tooling. Other methods, requiring lower capital investments but sometimes resulting in lower productivity, can often be used when relatively few items are desired. Various processes affect the physical properties of the finished part differently. Processes that result in a high degree of molecular orientation, for example, will impart one set of mechanical properties to the molded article. Other methods will give drastically different characteristics to the same polymer. Certain processes are better suited to high-precision parts and the reproduction of fine details. Other techniques are preferred when dimensional considerations are less important and part geomtry is less complex.

Most primary plastic fabrication processes have three steps in common: make flowable, form, solidify. Plastic compositions can be made flowable by melting, dissolving, or sometimes simply by the application of sufficient force. The material is then formed by either rolling, squirting, pushing, pulling, spreading, packing, spraying, coating, or pouring into the appropriate shape. Solidification is accomplished by cooling, heating, evaporating, polymerizing, or cross-linking.

This chapter briefly describes some of these fundamental plastic fabrication techniques. The intent is to provide an understanding of the basic principles involved in each. Because of the commercial importance of extrusion and injection molding, these processes will be treated with significantly more detail. Many variations of each process have been developed, and many modern fabrication methods are actually hybrids of two or more of the fundamental techniques. In addition, a few of the secondary or finishing processes such as bonding and decorating are discussed briefly.

While safety considerations are not discussed repeatedly, it must be noted that many plastics processes deal with extremely hot surfaces and with polymer melts at several hundred degrees under pressure of several thousand psi. Some processes require moving mechanical parts with closing forces of many hundreds of tons. Other techniques include easily accessible rolling nips of powerful roll mills. Because of these characteristics, plastics processing can be hazardous. With the proper equipment, procedures, and attitudes, however, plastics processing can be a safe operation. Commercial processing equipment is usually provided with the appropriate guards and electrical, mechanical, and hydraulic interlocks. The practitioner should understand the operation of these safety devices and regularly verify their proper operation. Under no circumstances should these devices be bypassed or defeated. In addition, eye protection (sometimes full face shields), thermally insulated gloves where appropriate, and no jewelry or loose fitting clothing are advisable.

12.2 EXTRUSION

The principles of the plastic extruder form the basis for several important plastic fabrication processes. An extruder melts, compresses, mixes, and pumps plastic material to a forming operation. The forming operation is usually a die attached to the end of the extruder, which forms the melt into continuous products such as sheet, rod, pipe, and other shapes. Subsequent forming operations can also be performed on this initial shape. A cross-sectional diagram of a single-screw extruder is shown in Figure 12.1. A variable-speed motor turns a screw inside a heated barrel. Plastic material is fed by gravity from the hopper through an opening in the barrel wall. The plastic is conveyed down the screw and absorbs heat from the hot barrel as well as significant frictional heat. As the polymer melts, the air that surrounded the particles in the hopper is excluded and forced back out of the feed opening. Since the volume/unit length available for the plastic generally decreases as the material proceeds down the screw, the plastic is compressed, and a signif-

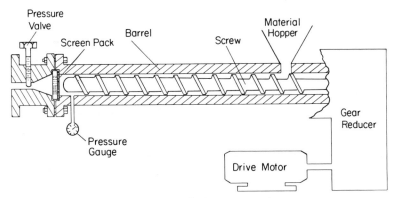

Figure 12.1 Single-screw extruder.

icant positive pressure is developed. It is often desirable to temporarily relieve that pressure at some location on the screw in order to vent the melt or to introduce certain additives into the melt stream. This is accomplished by a sudden increase in the volume available for the melt and by an opening in the barrel at that point. This technique is discussed in more detail later. At the discharge end of the barrel, the melt is generally filtered before being forced through a throttling mechanism and then the appropriate forming die.

Although this discussion is limited to the most common single-screw extruders, it should be noted that a variety of multiple-screw extruders is also available. The significance of these machines is growing, particularly in the area of polymer compounding. The twin-screw machine is by far the most common multiple-screw extruder, but four-screw devices are also available. The variety of designs includes both corotating and counterrotating intermeshing screws as well as nonintermeshing counterrotating types. Although the initial cost of multiple-screw equipment is substantially higher than that of an equivalent capacity single-screw device, this cost is sometimes justified by the major advantage of more positive and uniform feed resulting in more uniform extrusion pressure, regardless of the properties and condition of the starting materials.

12.2.1 Equipment

A typical extruder consists of a drive mechanism, barrel, screw, screen-pack, pressure control valve, and appropriate controls (temperature controllers for barrel heaters and drive-speed control). The addition of a material handling system, a die, cooling facilities, and takeoff equipment is necessary for a production extrusion line.

12.2.1.1 Drive Mechanism

The function of the drive mechanism is to provide sufficient power to the process at a speed within the proper range. Components of the drive include an electric motor, a speed control mechanism, and a gear reducer. Available drive motor sizes range from less that one to many hundred horsepower. While the required size depends on factors such as the type of material to be processed and the design of the screw to be used, the general guideline of 5–15 lb/hr of extruded product per each horsepower is sometimes used (1).

Speed control can be accomplished by either mechanical, magnetic, or electrical means. The mechanical type, usually a variable pulley, is low-cost, simple, and efficient. These units become excessively large, however, with the higher horsepower motors, and wear is sometimes a problem, particularly when the extruder is operated at the same speed for long periods of time. Eddy current couplings are frequently used with 50 hp and above alternating current (AC) motors. In this size range, the cost of these units is comparable to that of mechanical drives, with added advantages of compactness, a wider speed range, and the capability of relatively simple, closed-loop feedback control. The major disadvantages of these couplings include the need for cooling and the strong dependence of power transmission on speed. The direct current (DC) drives are usually considered to provide the best balance of speed control and efficiency at low speeds.

Typical extrusion speeds range from 5 to 200 rpm. In order to take full advantage of the capabilities of the drive motor, mechanical gear reduction is usually required. Belt-driven pulleys, chain-driven sprockets, and conventional gear boxes are used.

12.2.1.2 *Barrel*

Single-screw extruder barrels are simple, hollow steel cylinders with appropriate openings for feed and venting. Barrels must be designed to mechanically contain the pressure (several thousand pounds per square inch) and to resist wear and sometimes corrosive environments associated with the extrusion process. The inside surface of the barrel is sometimes nitrited or more commonly is a special alloy liner such as Xaloy. The barrel is heated, usually with electric resistance band heaters or inductive heating coils that are arranged in several zones along the barrel. Older extruders utilized circulation of steam or hot oil, which provided very uniform heating but had limited temperature ranges and are now much less common. It is usually necessary to pro-

vide cooling at the feed section of the barrel to prevent premature fusing of the material, which interferes with feeding. It is also frequently desirable to remove heat along the barrel, since too much frictional heat can be developed with some materials under certain extrusion conditions. This cooling is accomplished by circulation of water through a jacket and/or by directing air from blowers at the barrel.

Extruder barrels are described by two parameters: inside diameter and length to inside diameter ratio L/D. The throughput of an extruder increases with both diameter and L/D ratio. Standard single-screw extruder sizes range from 0.75 in. to 8 in. in diameter with L/D ratios from 16:1 to more than 36:1.

12.2.1.3 *Screw*

The use of the proper screw for a particular extrusion job is critical to the success of that operation. Much technology exists in the area of screw design. Computer-aided analysis utilizing mathematical models to predict screw performance is common. Only a very general discussion of screw design is possible here.

An extruder screw, with the usual nomenclature, is shown in Figure 12.2a. Clearance between the flights of the screw in the extruder

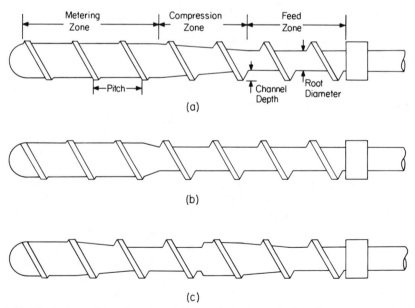

Figure 12.2   Three types of extruder screws: (*a*) general-purpose screw; (*b*) nylon screw; (*c*) two-stage (or vented) screw.

barrel is important. Excess clearance permits plastic melt to flow over the screw flight. This undesirable leakage reduces throughput, interferes with heat transfer to the bulk of the melt (plastic layer on barrel wall acts as thermal insulator), and can lead to overheating and degradation of that layer. Acceptable clearances are usually near 0.015 in./in. of barrel diameter. The most common screw has a constant pitch (or lead) equal to one screw diameter. A typical extruder screw contains three zones or sections: feed zone, compression zone, and metering zone. The geometric difference from one zone to another is simply a difference in root diameter of the screw. The root diameter is smallest (deepest screw channels) in the feed zone of the screw. The main function of this section of the screw is to convey the plastic along the barrel while the plastic is being heated. For amorphous or highly plasticized polymers, this zone can be very short or even nonexistent. With highly crystalline materials with sharp melting points, such as nylon, the feed zone must be sufficiently long to permit nearly complete melting. The root diameter must be small enough to provide a sufficient supply of material to completely fill the subsequent metering section of the screw, considering the density changes resulting from compaction and melting. Since the screw root diameter is the smallest in the feed zone, the mechanical ability of the screw to transmit torque is limited by that dimension and is an important consideration in screw design.

The root diameter of the screw gradually increases in the compression zone. In this zone, the softened polymer is compacted into a continuous melt. Pressure and frictional heat are developed in this section. Again, the length of the compression zone must be sized for the particular plastic being produced. For polymers such as nylon, for example, melt viscosity is relatively low, and compression can take place suddenly without excessive torque on the screw. A typical screw designed for nylon in which the compression zone is only one screw diameter in length is shown in Figure 12.2b. Other polymers such as vinyls, modified polystyrenes, and polyolefins, which soften more gradually, require longer compression zones.

The screw root diameter is again constant and is at the maximum in the metering zone. The main function of the metering zone is to provide a constant output from the extruder. Variations in that output result in severe problems with thickness and other characteristics of the subsequently formed product. Mixing and homogenization also occur in this zone, but excessive shearing in this region will cause overheating and degradation to some heat-sensitive polymers, such as vinyls.

Extruder screws are sometimes referred to in terms of a compression ratio, that is, the ratio of volume of the first feed zone channel to the volume of the last metering zone channel. While this is important information, it does not alone specify a screw. Channel depth of the feed and metering zones, as well as the length of each of the three zones, must be stated in order to better describe a particular screw.

A two-stage or vented screw is shown in Figure 12.2c. The two-stage screw is actually two complete extruder screws in line. The plastic is melted, compacted, and metered in the first stage as usual. When the melt reaches the second deep-channeled, or decompression zone, insufficient plastic is available to fill the channels, therefore, the pressure on the melt is relieved. The decompression zone is open to the atmosphere or even placed under vacuum. In this zone, air, water vapor, and other residual volatiles are removed from the melt. Frequently, a predrying step can be eliminated when vented extruders are used on polymers in which the only consequence of the presence of water is surface imperfection. For polymers that suffer degradation due to heating in the presence of water, however, significant damage to the material can occur in the first stage if predrying is not accomplished. Certain heat- or shear-sensitive additives can also be introduced in the decompression zone of a two-stage screw. Fiberglass, for example, suffers size reduction during the extrusion process. That size reduction can be minimized if the fibers are only required to traverse the second stage of the screw.

This discussion has concentrated on the more conventional full-flighted, constant-pitch screws. It should be noted, however, that many special-purpose screws are also available. Most of these special screws are aimed at providing improved mixing and melt homogeneity. Changing pitch, double flights, and small sections with a reverse spiral or fluted region are some of the techniques used for this purpose. One of the more sophisticated approaches utilizes a screw within a screw where the plastic is melted on one screw, and only after it is completely melted can it pass to the other screw for metering. In addition, output rate gains of as much as 30% over extrusion rates of conventional screw designs have been reported for specially designed screws (2).

12.2.1.4 Screen Pack
As the material exits the metering zone of the screw, it is common practice to force the melt through a screen pack or filter. The primary purpose of the screen pack is to filter foreign contaminants, screen possible slugs of unmelted polymer, and to reduce the spiral flow that was developed on the screw. The screen pack is generally composed of several layers of wire screen supported by a rigid breaker plate. One common concept for construction the screen pack is to start with a relatively coarse layer of screen to filter the larger particles, proceeding through several layers of increasing fineness, and then through one or more layers of increasing coarseness to provide support for the finest layer. The screens are supported by the breaker plate, which is usually a disc with many small holes. When film, sheet, or other flat products are being extruded, product quality can sometimes be improved by using a breaker plate with slits rather than holes. The breaker plate slits, of course, must be placed parallel to the width of the final product. These slits begin to establish the desired flow pattern before the material reaches the die.

As the filtering process continues, the screen pack begins to plug, pressure becomes excessive, and the screens must be changed. A variety of semi- and fully automatic screen changers is available for this purpose.

12.2.1.5 Pressure Control Valve
It is frequently desirable to provide some means of adjusting the operating pressure of the extruder. This adjustment should be rela-

tively independent of pressure caused by the screen pack and subsequent die. Common techniques include the use of a variable flow restriction either in the adapter, as with the plug shown in Figure 12.1, with an adjustable restricter bar in the case of sheet dies, or by some means of adjusting the clearances between a special screw tip and a seat at the end of the barrel. A streamlined design of the pressure control mechanism is a necessity in order to prevent flow stagnation and possible polymer degradation.

In general, increasing extruder pressure increases frictional heating and the degree of mixing, but also increases power requirements and decreases throughput.

12.2.1.6 Dies and Takeoff Systems
This brief discussion is not intended to be a primer for extruder die design but rather to illustrate some of the mechanical principles involved in producing a few of the limitless shapes and sizes of products that can be manufactured by the extrusion process. Common to all extruder dies is the need for streamlined flow. The correct size of the actual die opening depends on many factors. As material exits the opening, a phenomenon known as die swell occurs, and the cross section becomes larger. Die swell is a relaxation of stresses that were imposed on the polymer during the high-pressure extrusion process. At least two other factors, however, tend to make the product smaller: thermal shrinkage, as the plastic cools; and drawdown, which results from forces required to take the product away from the die. The exact manner in which these opposing effects interact depends on the polymer being used, extrusion conditions such as temperature, pressure and rate, die design parameters such as land length, cooling techniques, and takeoff system. In general, drawdown should be minimized in order to minimize the resulting uniaxial orientation.

Common to nearly all extruded products is the difficulty in handling the shape downstream of the die. As the polymer exists the die it is molten, sticky, droopy, and very difficult to handle without damaging the shape. The proper handling of the product usually requires an elaborate hauloff system. For flexible film and sheet, the takeoff equipment can be an arrangement of driven rolls, but for rigid products such as rod stock and pipe, pullers composed of two or more driven rubber caterpillar treads that squeeze the extrudate between them are popular. One of the most important characteristics of this system is the ability to maintain a constant speed once the desired speed is obtained. While film, wire, and flexible tubing can be wound on rolls or in coils, rigid products must be cut to some length. A traveling circular saw that moves along with the extrudate as it cuts and then returns to its starting position is most often used.

Extruded rod is produced with a die containing a simple hole. Maintaining the round cross section of the rod is the major problem since gravitational forces tend to produce an elipse during the cooling process. The rod is generally pulled through a water bath for cooling. If the finished product is to be plastic pellets, as in the case of an extrusion compounding line, the plastic is extruded through many small holes; the individual strands are either cut at the face of the die or are cooled in a water bath and continuously fed to a pelletizer. A die for such an operation is sometimes called a spaghetti die.

Pipe and tubing are produced by extruding through an annular opening. One arrangement of this type of die is shown in Figure 12.3. In this type, support of the core is a major concern and is usually accomplished with small ribs called spiders. The plastic must

separate and flow around the spiders and then flow back together before leaving the die. Improper extrusion conditions can result in weaknesses at these knit lines. Several different techniques are used for sizing extruded pipe. These methods include internal sizing mandrels, internal air pressure with external sizing rings, and vacuum sizers. The vacuum sizers are most widely used. In this case, the inside of the pipe is held at atmospheric pressure while a partial vacuum is drawn on the outside. The higher internal pressure forces the external pipe surface against sizing sleeves, which maintain the diameter during cooling.

Plastic sheet and some films are manufactured by extruding through a long narrow slit. Sheet dies are designed to minimize the tendency for more material to be extruded from the center of the die. The "coat hanger" design is a popular approach (Figure 12.4). The dies have lips that are adjustable by means of a row of adjusting bolts over its entire width. The molten sheet is either extruded onto the surface of a water-cooled chill roll or extruded into the nip of two rolls that assist with gauge control. The sheet is then either passed over additional cooling rolls, conveyed through a water bath, or simply air cooled. Thin flexible products are rolled, and thick sheets are cut to length.

Multilayered sheets and films can be produced through multimanifold dies designed to accept melt from two or more extruders and combine the layers prior to their exiting the die. Since flow is strictly laminar, little or no

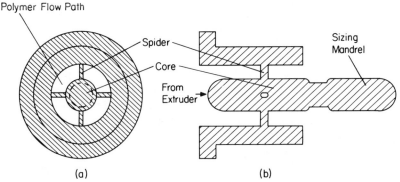

Figure 12.3 Pipe or tubing die. (*a*) End view showing spider support of the core. (*b*) Cross-sectional side view.

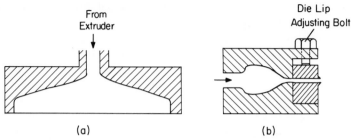

Figure 12.4 Sheet or flat film die. (*a*) Top view showing popular "coat-hanger" design. (*b*) Cross-sectional side view showing "tear drop" profile and adjustable die lips.

intermixing occurs at the interfaces. Such a multilayered structure is useful when certain properties are required in a film that cannot be obtained from a single polymer. A special packaging film, for example, might require one layer for toughness, one layer for gas or moisture barrier, and another layer for heat sealability.

When more than three layers are desired, multimanifold dies become very complex. A desirable alternative is the use of a feed-block assembly just upstream of the die. A relatively thick-layered, narrow vertical stack of the various polymer melts is introduced into the inlet of the feed-block. The geometry of the feed-block gradually changes from a vertical slit to a horizontal slit from inlet to outlet, which forces the polymer melts to flow in the width direction as they proceed through the feed-block. Again, laminar flow allows each layer to maintain its integrity during the process. Layered thicknesses are controlled by the relative volumetric extrusion rates of the individual components. The multilayered film or sheet then exists the device through conventional die lips. This technique has been used commercially to produce five-layered films from four different polymers. Experimentally, seven-layered sheet from five materials and films having over 100 alternating layers of two different polymers have been successfully produced (3). Careful consideration to the rheological compatability of the polymers used in this process is necessary for successful coextrusion by this technique.

In the extrusion coating process, a substrate such as paper or metal foil is passed under an extruder die that is very similar to the sheet or flat film die. A thin film of plastic is extruded

onto the substrate. The coated web in then usually passed into a nip formed by two rolls. The plastic side of the laminate contacts the larger chill roll which is usually polished or embossed metal; the substrate contacts the smaller roll which is usually coated with an elastomer. In this nip, the laminate is squeezed together as the resin begins to cool. The web continues around the water-cooled chill roll for additional cooling and is stripped from the roll. Familiar examples of products manufactured by this process include waterproof polyethylene coated paper. Multiple coating can be obtained by arranging several separate extruder lines in series. The extrusion coating process can also be used to laminate two layers of substrates together. In this case, the polymer acts as an adhesive. The first substrate is coated as described, but a second substrate is fed to the nip of the rolls on the plastic side of the web. This process is called extrusion lamination. Plastic temperature and extruder-die-to-roll-nip distance are the most important variables affecting the plastic-to-substrate adhesion in the extrusion coating and laminating process. Higher than normal extrusion temperatures are required for good adhesion. Some surface oxidation of the polymer is also necessary for adhesion, but an excessive die-to-nip gap will permit excessive cooling prior to the pressing step.

The blown film process utilizes a die similar to a pipe or tubing die (Figure 12.5). As the tubing is extruded upward, air is introduced through the die, which inflates a bubble of plastic. The formation of the bubble stretches and thins the melt. As the plastic cools and stiffens, it resists additional stretching. After sufficient cooling, the bubble is collapsed

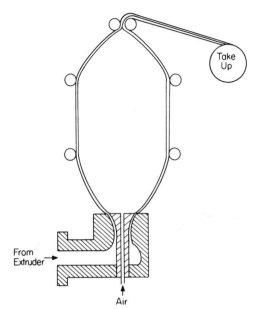

Figure 12.5   Blown film process.

The extrusion process is also used to place the insulation coating on wire and cable. The wire coating die (Figure 12.6) is very similar to a small tubing die except that an opening is provided from which the wire is supplied. It is usually desirable to preheat the wire to improve polymer-to-metal adhesion and to maintain optimum physical properties by minimizing the thermal shock to the plastic.

12.2.2   Process

For a particular material and screw-barrel combination, the three primary process, variables for the extrusion process are plastic temperature, plastic pressure, and output rate. The variables are controlled by machine parameters such as barrel temperature profile, screw speed, and drive motor power through a complicated relationship. A change in screw speed, for example, results in a change in all three of the primary variables as well as a change in drive motor power. Plastic temperature is usually measured by means of a thermocouple that protrudes into the melt and is located as close as possible to the die. Modern techniques using infrared thermometers are also available for measuring temperature of the extrudate. Plastic pressure is measured by means of a pressure gauge or electrical transducer usually located just upstream of the screen pack. Screw speed and power to the drive motor are monitored by means of an electrical or mechanical tachometer and an ammeter or wattmeter, respectively.

between rollers and wound up on a roll. This form of product is commonly called "lay flat tubing." Garment bags, for example, can be made by cutting off a length of the tubing and partially heat sealing one end. The tubing may also be slit to produce films much wider than would be possible by the slit die technique. These wide films are used primarily in agricultural and industrial applications. The combined effect of the drawdown due to overdriving the pinch rolls and the stretching due to bubble formation results in a biaxially oriented film, which offers unique properties. Techniques for producing multilayered films by this technique have also been developed.

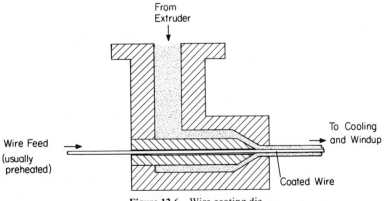

Figure 12.6   Wire coating die.

The three most important functions of the extruder are solids conveying, melting, and pumping. These functions may be treated separately in order to gain qualitative insights into the process. The solids-conveying zone is defined as the region between the hopper and the point where the surface of the solid plastic bed in contact with the barrel melts. Solids conveying depends on screw geometry, frictional coefficients of both the barrel and the screw surfaces, on the bulk density of the plastic, and on the screw speed (4). The frictional coefficients, of course, depend on the surface condition and temperature of the metal. In order to compact the solid bed of plastic during conveying, the pressure on the material must increase. Figure 12.7 shows the exponential rise in pressure with distance down the screw and the effect of different coefficients of friction of the barrel.

The melting process begins with the formation of a thin film of melted polymer on the barrel surface. This region on the screw (still in feed section) is sometimes called the delay zone. This zone extends to a point where the film thickness exceeds the clearance between the screw flights and the barrel. At that point, the melting zone begins. The advancing screw flight continuously scrapes the melted film from the barrel wall and forces the melt downward along the flight surface and into a circulating melt pool at the rear of the channel. Using a simplified Newtonian model, the basic relationships between the various process parameters can be described mathematically. The usual technique is to assume a stationary screw and a rotating barrel. For a constant screw channel depth, Tadmor and Klein (5) express the thickness of the melted film δ as follows:

$$\delta = \left\{ \frac{[2k_m(T_b - T_m) + \mu V_j^2]X}{V_{bx}\rho_m[C_s(T_m - T_s) + \lambda]} \right\}^{1/2} \quad (12.1)$$

Where k_m = thermal conductivity of melt film, μ = melt film viscosity, ρ_m = melt density, λ = heat of fusion, C_s = heat capacity of solid bed, T_b = temperature of inner barrel wall, T_m = melting temperature of the plastic, T_s = temperature of the solid bed, V_j = the vectorial difference between velocity of the barrel and of the solid bed, V_{bx} = the component of relative tangential velocity between barrel and screw in cross-channel direction, and X = width of solid bed in cross-channel direction. In addition, the rate of the melting per unit down channel distance ω can be shown as

$$\omega = \Phi X^{1/2} \quad (12.2)$$

where

$$\Phi = \left\{ \frac{V_{bx}\rho_m[k_m(T_b - T_m) + (\mu/2)V_j]^2}{2[C_s(T_m - T_s) + \lambda]} \right\}^{1/2}$$

$$(12.3)$$

As the value of Φ increases, the rate of melting also increases. While Equations (12.1) and (12.2) provide the general relationships, refinements to the model to allow for tapering channels and to correct for non-Newtonian behavior, channel curvature, flight-to-barrel clearances, and radial direction temperature

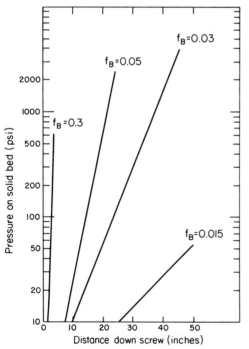

Figure 12.7 The effect of barrel coefficient of friction on pressure buildup in the solids-conveying zone of an extruder. Coefficient of friction for the screw equals 0.01 in all cases (4).

variations can be made in order to improve their accuracy (6).

After the polymer is melted, the remaining portion of the extruder can be treated as a melt pump. For isothermal, Newtonian melt behavior, output of the extruder is the difference between drag flow and pressure flow and can be expressed by Equation (12.4) (5).

$$Q = AN - B \frac{\Delta P}{\mu} \qquad (12.4)$$

where

$$A = \frac{n}{2} \pi D_b WH \cos \theta_b \qquad (12.5)$$

and

$$B = \frac{nWH^3}{12Z} \qquad (12.6)$$

and $Q =$ the net output of the extruder, $N =$ screw speed, $\Delta P =$ pressure rise, $n =$ number of parallel channels, $D_b =$ inside diameter of barrel, $W =$ width of screw channel, $H =$ channel depth, $\theta_b =$ helix angle at barrel surface, and $Z =$ helical length. Equation (12.4) is frequently referred to as the screw characteristic equation. Similarly, the die characteristic can be expressed as Equation (12.7) (5).

$$Q = k_d \frac{\Delta P}{\mu_d} \qquad (12.7)$$

where $Q =$ the output from the die, $k_d = a$ constant that depends on the die geometry, and μ_d is the average viscosity in the die.

From Equation (12.4), it can be seen that the output of an extruder is maximum when the pressure flow term is zero, therefore

$$Q_{max} = AN \qquad (12.8)$$

Since the output of the die must equal the output of the extruder, equating Equations (12.4) and (12.7) and solving for the pressure yields Equation (12.9).

$$\Delta P = \frac{\mu AN}{k_d + B} \qquad (12.9)$$

From Equation (12.9), it can be seen that the pressure is a maximum when the die constant is zero or

$$\Delta P_{max} = \frac{\mu AN}{B} \qquad (12.10)$$

Another important consideration is the power required to turn the screw. As derived by McKelvey (6), a simplified model of the power requirement is shown by Equation 12.11

$$\text{power} = E\mu N^2 Z + AN\Delta P \qquad (12.11)$$

where

$$E = \frac{\pi^3 D_b^3 \sin \theta_b}{H} (1 + 3 \sin^2 \theta_b) \qquad (12.12)$$

Surging is a condition of usually cyclic fluctuations in the extruder pressure that affects production rate and the product quality. Two of the causes of surging are slow melting and improper solids conveying (7). When slow melting occurs, the screw channels in the compression zone become plugged with unmelted plastic. This plug restricts the flow until the upstream pressure builds to a level sufficient to dislodge the plug, which then surges toward the extruder die; or in a worse case, the plug can cause melt to flow toward the hopper, coat the screw, interfere with solids conveying, and reduce the output to zero. The other solids conveying anomaly occurs when pressure buildup is inadequate to compact the solid bed. As the melt film forms, it penetrates the voids in the solid bed to varying degrees and, therefore, the volume of material in the channel changes with time. The channel continues to be semifilled until sufficient channel volume reduction occurs in the compression zone of the screw, and only then can pressure begin to build. Since the location of this point changes, the length of screw available for the melt pressure buildup also changes, as does the final pressure level.

The foregoing discussions assume a flood-feed condition in the extruder, that is, the plastic particles are free to flow from the hopper and enter the feed throat at whatever rate the extruder will accept them. Results from studies by Nichols and Kruder (8) and, more recently, by McKelvey and Steingiser (9) have shown advantages for metering the polymer to the extruder at a lower rate, which results in starve feeding. The reported advantages include: lower required energy per pound of material extruded; lower torque on the screw and drive mechanism; improved devolatilization; reduced surging; increased ability to handle high-viscosity polymers; and no sacrifice in throughput since the drop in output is offset by an increase in screw speed.

12.3   SPINNING

The technique used for the production of synthetic fibers intended for a subsequent weaving operation is very similar to the extrusion process and is known as fiber spinning. The object of the spinning operation is to produce a fiber with a very large length-to-diameter ratio and with very high tensile strength. The die for the spinning process is actually a plate containing many tiny holes. The die is known as the spinnerette. Either a polymer melt or a solution is forced through the spinnerette, and each hole in the spinnerette produces an individual fiber. The fibers are then solidified and stretched. The stretching results in high degrees of molecular orientation and crystalinity, which impart the desired tensile properties to the fibers. The individual fibers are then twisted together to form a thread.

For materials such as nylon the melt-spinning process is most often used. After the molten polymer extrudes from the spinnerette, it is cooled and solidified in air. In the solution-spinning process, solidification may be accomplished by either evaporating the solvent or by precipitating the polymer from the solution. These two variations are frequently referred to as dry spinning and wet spinning, respectively. Acrylic fibers such as Crestan and Orlon are usually produced by dry spinning. In this case, the fibers are bathed in warm air as they emerge from the spinnerette, and the solvent evaporates. For materials such as rayon, the fibers are passed through a nonsolvent liquid as they exit the spinnerette, and the polymer precipitates. In all cases the properties of the spun fibers are primarily determined by the starting solution or melt viscosity, the spinnerette geometry, the rate of solidification, and the degree of drying.

12.4   INJECTION MOLDING OF THERMOPLASTICS

Injection molding is the process of melting plastic, injecting the molten plastic into a closed mold under high pressure, cooling the plastic in the closed mold, and removing the rigid article from the mold. The process is useful for the production of both precision plastic products and for less critical parts.

12.4.1   Equipment

While there are many variations of the injection molding process, there are two basically different types: the plunger and the reciprocating screw. In both cases, injection molding equipment consists of a clamping unit, a plasticating and injection unit, a hydraulic pump driven by an electric motor, a mold, and the appropriate controls.

12.4.1.1   Plunger Molding

A schematic of a typical plunger injection molding machine is shown in Figure 12.8. The plastic material is introduced into the heated barrel. Heat is transferred to the plastic almost entirely by conduction. During the injection phase of the cycle, the hydraulic injection cylinder is pressurized, and the plunger is forced forward. The plunger pushes the plastic down the cylinder, around the torpedo, through the nozzle, and into the mold. When the plunger is returned, new plastic from the hopper falls into the cylinder. Owing to the relatively poor heat transfer characteristics of polymeric materials and the relatively simple flow in the cylinder, plunger molding machines tend to deliver nonuniform shots of material, particularly with respect to temperature. The incorporation of the torpedo in the barrel is an attempt to improve the uniformity, but is only partially successful. This nonuniformity is particularly trouble-

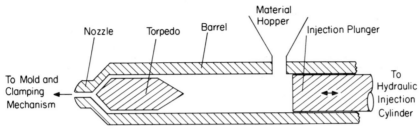

Figure 12.8 Barrel of plunger injection molding machine.

some when tight dimensional tolerances are required on the finished parts.

12.4.1.2 Reciprocating Screw Molding

The reciprocating screw injection molding machine incorporates a modification of the plastic extruder screw to accomplish the melting portion of the process. A simplified diagram of a typical reciprocating screw injection molding machine is shown in Figure 12.9. This "extruder" overcomes many of the disadvantages of the plunger injection molding machine. The screw can move back and forth as well as rotate. Rotation is accomplished with a drive motor. The motor is usually hydraulic, but some machine manufacturers supply electrical drives. A hydraulic injection cylinder is used to force the screw forward and inject the plastic.

Mold closing (and subsequent clamping) can be accomplished by a simple hydraulic cylinder. In this case, large volumes of hydraulic oil are required to move the movable platen at the desired speed. Hydraulic accumulators are often used to supply this volume of oil. With pure hydraulic clamp, clamping force can easily be determined by measuring the oil pressure and multiplying that pressure times the area of the hydraulic piston on which it acts. The hydraulically operated toggle is another common method of moving the mold and developing clamping force. Although many clever toggle design variations have been developed, the basic principles of the toggle clamp are shown in Figure 12.10. The toggle clamp makes use of a very large mechanical advantage as the toggle approaches the locked position. This principle can be easily verified by the application of simple trigonometry, which also shows that for a constant velocity of the hydraulic cylinder, the moveable platen automatically decelerates to zero as the toggle approaches the locked position. Toggle clamps generally tend to be faster than pure hydraulic clamps and require lower volumes of hydraulic oil. The setup of a toggle machine requires very precise adjustment to compensate for differences in mold thicknesses. Severe damage to machine and mold can occur if these adjustments are not correct. In addition, the toggle linkages become very massive and cumbersome in the larger machine sizes, and therefore many machine builders favor hydraulic units in machines above 400 tons. Measurement of clamp force with toggle clamps is accomplished by measuring tie rod elongation by way of a strain gauge or dial indicator and using strength-of-materials relationships to determine stress.

12.4.1.3 Molds

The proper design and construction of the injection mold is absolutely essential for the

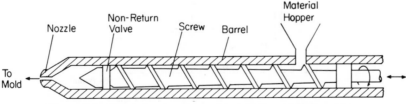

Figure 12.9 Barrel of reciprocating screw injection molding machine.

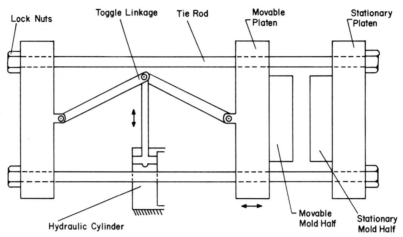

Figure 12.10   Toggle clamp mechanism.

commercial success of this process. A cross-sectional diagram of a typical cold runner, two-cavity mold is shown in Figure 12.11. The mold is clamped to the machine platens. The nozzle of the machine seats on the sprue bushing. The plastic flows from the nozzle, through the sprue bushing, runners, and gates and into cavity. While space does not permit even the most superficial treatment of the many critical elements of mold design here, a few items must be considered. The forces involved in the injection molding process are often underestimated, and consequently the

rigidity of the mold is inadequate. One can hardly expect to produce dimensionally accurate parts if the cavity walls deflect significantly during injection and packing. The size, shape, and positioning of the gate(s) are also very important aspects of mold design (10), (11).

Mold cooling also requires special attention since heat removal is one of the basic functional requirements of the tool. In fact, the heat removal step is almost always by far the largest contributor to the total cycle time and frequently dictates the overall economics of

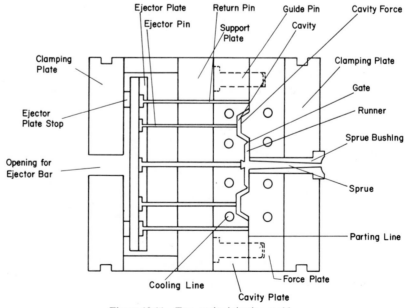

Figure 12.11   Two-cavity injection mold.

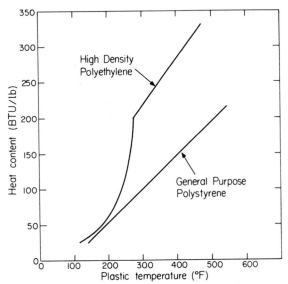

Figure 12.12 Heat content of high-density polyethylene and general-purpose polystyrene as a function of temperature (12).

the process. Mold cooling must also be uniform in order to minimize concentrations of molded-in stress. Prasad (12) provides some insights into the complex heat transfer problem of the injection mold with the following simplified treatment. The heat content of a plastic at various temperatures can be determined. For illustration, the heat contents of general-purpose polystyrene and of high-density polyethylene, as a function of temperature, are shown in Figure 12.12. From such a

curve the required change in heat content, as the part cools from injection temperature to the temperature at which it can be removed from the mold, can be determined. In addition, the rate of heat flow through a given plastic depends on the thermal characteristics of that material, which can be expressed as a curve relating minimum cooling time to part thickness (Figure 12.13). By combining the information in Figures 12.12 and 12.13 and the mass of the particular part, an approxi-

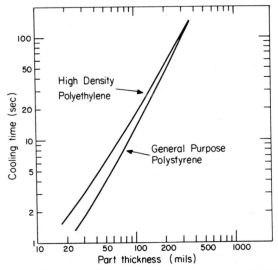

Figure 12.13 Minimum cooling time for high-density polyethylene and general-purpose polystyrene as a function of molded part thickness (12).

mation for the amount of heat per unit time that must be transferred to the cooling fluid can be calculated. The total heat transfer from the part to the coolant can be shown by Equation (12.13)

$$Q = GC_p(T - t)(1 - e^{-UL/GC_p}) \quad (12.13)$$

where Q is the total heat to be removed, G is the coolant flow rate, C_p is the specific heat of the coolant, T is the cavity surface temperature, t is the coolant inlet temperature, L is the total effective cooling length, and U is defined by Equation (12.14) (12).

$$\frac{1}{U} = \frac{1}{\pi \, dh} + \frac{1}{SK_s} \quad (12.14)$$

where U is the overall heat transfer coefficient, d is the cooling channel diameter, S is the channel shape factor, h is the heat transfer coefficient between the mold and the coolant, and K_s is the thermal conductivity of the mold material. Prasad also suggests that cooling channels should be positioned at a depth of one to two times their diameter from the molding surface and with a pitch of three to five times the diameter.

Care must be taken to provide an adequate path for air to exit the cavity as the melt flows into it. Inadequate venting severely restricts the flow of plastic and is frequently responsible for the inability to fill certain blind areas in the mold. In its worst case, inadequate venting can cause burn marks on the part by actually creating a small explosion of the hot, highly compressed gases traveling ahead of the melt front. The main venting of the cavity is usually accomplished on the parting line with channels that are shallow enough to prevent material from flashing into the vent, but deep enough that they do not collapse under clamping force. The channel must extend to the outside of the mold, and, as it continues away from the cavity, it is important that it be enlarged to minimize back pressure. It is sometimes necessary to vent the cavities all the way around. This is called continuous venting. In areas such as blind spots where parting line venting is not effective, either ejector pins or dummy pins can be used. The gases vent through the clearances around the pin, or slight flats can be ground on the pin to improve their effectiveness.

A very important property of a multicavity mold is the balance of the flow in the mold. It is highly desirable that all of the cavities fill simultaneously and that identical peak pressure is transmitted to each cavity. A properly balanced mold is necessary for producing identical parts from all cavities and for assuring the widest possible molding latitude. Factors that contribute to mold imbalance include unequal flow path length for the plastic, unequal venting, mold deflections resulting in unequal wall thicknesses, nonuniform mold surface temperatures, and unequal gate sizes.

Various methods have been developed to maintain the plastic in a molten state between shots everywhere except in the cavity. These methods such as hot runners, insulated runners, runnerless molds, and valve-gated molds are effective in eliminating the need to handle and regrind the sprue and runner. Since less heat must be removed during each cycle, these techniques usually result in shorter cycles. Precise temperature control of all portions of the material flow path is the key to effective utilization of these concepts.

12.4.1.4 Sizing Mold and Machine

Assuming that the surface area of the machine platens is sufficient to accept the mold and that the platens can be opened a sufficient distance to permit part ejection, three important considerations in sizing the mold and machine are clamp force, shot size, and screw plasticating rate. The clamp force required can be calculated from the well known relationship shown as follows in Equation (12.15):

$$F = PA \quad (12.15)$$

where F = clamp force, P = pressure in the cavity and A = the projected areas of the cavities and the cold runners. Unfortunately, the actual pressure in the cavity is usually unknown at the time of machine selection, and the industry has resorted to applying a rule of thumb. Two tons of clamp force per square inch of projected area is frequently used. This rule of thumb assumes a maximum average cavity pressure of 4000 psi and is usually satisfactory for thick-walled parts. Average cavity pressure depends on wall thickness and 2 tons/in.^{2} of clamp is grossly inadequate for thin-walled parts as shown in the following example. An actual 4 in. $\times$ 4 in. $\times$ 0.028 in. thick flat part requires cavity pressures of 10,000 psi near the gate and 6000 psi at the far end of the part. The part is produced in an eight-cavity mold. Estimating the average cavity pressure to be 8000 psi and multiplying by the projected area of 4 in. $\times$ 4 in. times eight cavities, or 128 in.^{2} yields a required clamping force of 1,024,000 psi or 512 tons. This is a minimum requirement of 4 tons/in.^{2} of projected area. Stevenson, Hauptfleisch, and Hieber (13) have proposed a simplified handbook method for estimating injection pressures and required clamp force.

The shot capacity of the molding machine barrel must be adequate for the total shot to be injected. The most repeatable performance of the molding machine is obtained, however, if the shot size being used is between 25 and 75% of the total barrel shot capacity.

Finally, the plasticating rate of the screw should be such that the screw can be made to return within the time allotted for part cooling.

12.4.2 Process

Although the description of the injection molding process is limited to the more common reciprocating screw machines, much of the process variable discussion also applies to plunger injection molding.

The machine is prepared for the first shot by rotating the screw. As the screw rotates, the plastic is conveyed down the screw and through the nonreturn valve. As the material meets the restriction in the nozzle, a slight pressure builds in the melt. This pressure acts on the screw in such a way as to cause it to move backward. The pressure necessary to force the screw backward is controllable by way of a back-pressure adjustment that imposes a varying restriction on the hydraulic oil flow from behind the ram. The screw rotates and translates backward until an adjustable back position is reached and sensed with some device such as a limit switch. At this point, the shot is prepared and is waiting at the front of the screw to be injected.

The sequential functions of the reciprocating screw injection molding machine are shown in Figure 12.14. The cycle begins with the closing of the mold. Since mold closing is a nonproductive phase of the cycle, it is desirable to close as rapidly as possible without excessive wear and tear on the machine and mold. With pure hydraulic clamp, it is also desirable to decelerate the system before the mold halves meet to prevent a destructive hammering of the mold. Toggle clamps decelerate automatically. Mold closing is usually accomplished with low oil pressure until the final stage. This low-pressure closing operates under the concept that a trapped part or other object will stop the forward motion of the mold, and since the clamp cylinder is being powered by low pressure only, no damage to the mold will result. The proper adjustment of the closing phase of the clamp motion is an important aspect of the mold protection. Because of the momentum associated with the

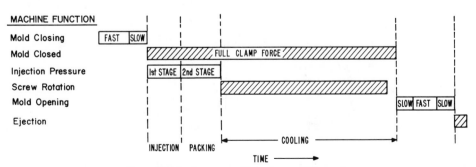

Figure 12.14 Sequence of injection molding cycle.

moving system, very high forces can be developed on the mold faces when closing on a trapped part, even though high-pressure clamp has not been activated.

As soon as full clamping force is developed, the injection phase of the cycle begins. First-stage injection pressure (sometimes called booster pressure) is activated, which forces the screw forward, closing the nonreturn valve on the end of the screw. The screw now acts as a plunger, forcing the plastic through the nozzle and into the mold. Although the rate at which the screw moves forward can be adjusted somewhat by means of a hydraulic flow control valve, it is desirable (particularly with thin-walled parts) to inject the plastic as rapidly as possible without adversely affecting the molded part. The hydraulic flow control valve is, therefore, frequently set wide open. At these conditions, the rate of flow into a cavity for thin-walled parts with small gates is a function of material viscosity rather than being limited by the ability of the hydraulic pump to deliver the required volume of oil. This situation can be easily verified by observing a change in fill rate with a change in

material temperature. The fill rate, therefore, is also strongly dependent on the injection pressure. The termination of first-stage pressure (booster cutoff) is normally determined by either a timer or by screw position as sensed by a limit switch or other position-sensing device. Neither technique is completely satisfactory. The timing method assumes a repeatable fill rate, and the screw position method assumes a repeatable shot volume. Both of these parameters are influenced by many other variables, however, and are not perfectly repeatable.

Details of the strategy for setting up the first-stage injection phase of the cycle depend on the actual part being molded, primarily the wall thickness and gate size. For thick-walled parts with large gates, peak cavity pressure is generally determined by the first-stage hydraulic pressure setting. In this case, first-stage pressure is maintained until the gate has sufficiently solidified to prevent a discharge of material from the cavity when the hydraulic pressure is reduced to some lower value (Figure 12.15). For thin-walled parts with small gates, the cavity is usually just filled on first-

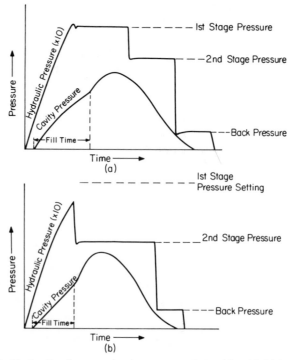

Figure 12.15 Typical hydraulic oil pressure/cavity pressure relationships; (a) thick-walled part; (b) thin-walled part.

stage pressure, and the shift to second-stage pressure occurs before full pressure is transmitted to the cavity (Figure 12.15b). Since the cavity pressure is increasing sharply at the time of booster cutoff, small changes in this cutoff time can have a gross effect on the resulting peak cavity pressure in parts being molded in this manner.

Second-stage, or holding, pressure is usually set lower than first-stage pressure. For thick-walled parts, this pressure simply maintains a supply of plastic to be filled into the cavity to take the place of the material that is shrinking. In the case of thin-walled parts, the holding pressure first determines the peak pressure and also supplies the extra material to compensate for shrinkage. If the second-stage pressure is too high, either the plastic pressure will overcome the clamping force and the mold will be forced open, or the gate end of the part will be overpacked leading to molded-in stresses. If holding pressure is too low, either the part will be incomplete or the sink marks will be excessive. Sink marks are depressions, usually opposite a change in cross-sectional thickness of the part, such as with a rib. Sink marks occur at the last point to cool. Plastic actually flows out of that area into the thinner sections which are cooling and shrinking faster. Holding pressure is maintained until material can no longer be forced through the gate to compensate for shrinkage, and when released, material does not back flow through the gate.

As the part continues to cool, the screw begins to rotate and prepares the plastic material for the next shot, as previously described. Screw speed should be adjusted such that the screw reaches its back position just before the time necessary to cool the part has elapsed. The setting of the back position of the screw is important. It is usually desirable to prepare a slightly larger volume of material that can be packed into the cavity during the next shot. This excess material is called a cushion. The existence of a cushion insures that sufficient material is available to fill the cavity and is also available for the subsequent packing step. Some back pressure is usually desirable in order to improve melt homogeneity. Too much back pressure results in material over-

heating and/or drooling at the nozzle during the next phase of the cycle.

It should be noted that a completely different injection strategy is sometimes employed when cycle time is the paramount concern, dimensional tolerances are not critical, and walls are thin and of relatively uniform cross section. An example of such a part is a thin plastic drinking glass such as those sometimes used on airlines. In this case the machine is operated without a cushion and is adjusted for maximum possible injection rate. Overpressurization of the cavity is prevented by the bottoming of the ram. This technique does not permit packing to compensate for shrinkage and is subject to many variables that affect the total mass of material injected into the cavity.

After the part has been sufficiently cooled, the mold is opened. It is usually desirable to begin the opening slowly to prevent damage to the part. Among other effects, parting the mold too rapidly can result in high forces on the molded part because of a partial vacuum that develops as the part releases from one mold half. The molds are usually designed so that the parts stay on the movable side of the mold. After the initial mold parting, high-speed opening is desirable. As the movable side reaches its full-open position, the ejector mechanism is actuated either mechanically by contacting a rigid (and stationary) knock-out bar or hydraulically by operating a hydraulic cylinder. It is usually necessary to maintain the mold in the open position for some small time while either the parts are removed by hand, by robot arm, or fall free of the mold. Once the parts have cleared the machine, the mold can be closed, which begins the next cycle. Total cycle time depends on many variables but, in general, can range from about 2 sec (e.g. plastic tumbler) to several minutes.

The primary variables of the injection molding process are: the plastic temperature as it flows into the cavity; peak plastic pressure in the cavity; rate of plastic flow into the cavity; and rate of cooling. All of the machine variables affect one or more of the primary variables. The conventional machine, however, gives very little information about the primary variables. The machine gives barrel temperatures, but barrel temperature is only one component of the total heat input. Mechanical heat input is an important factor and depends on back-pressure setting, screw revolutions per minute, material viscosity, cycle time, and more. Actual measurements of plastic temperature in the nozzle with a thermocouple that protrudes midway into the melt stream have shown that material temperature can range from 50°F below to 70°F above the highest temperature zone on the barrel as the preceding parameters are varied. Changes in plastic temperature result in changes in the level of molecular orientation in the molded part and, therefore, many of the physical properties of the part are affected (14).

The conventional machine gives the reading of hydraulic pressure behind the ram that pushes the screw that pushes the plastic. The hydraulic pressure gauge is usually difficult to read. Owing to the hydraulic multiplication, a 100 psi error in oil pressure is equivalent to a 1000 psi error in the plastic. Furthermore, the first-stage time often determines the amount of pressure that is transmitted to the cavity. Also, owing to the viscoelastic nature of the plastic material, significant pressure losses exist in the entire system. All of these factors suggest that actual cavity pressure cannot be accurately estimated from machine parameters and must be measured directly. Many instrument manufacturers offer a variety of equipment for this purpose. Typically, a load cell is placed behind an ejector pin in the mold; the plastic pressure exerts a force on the pin that is sensed by the load cell. For amorphous materials, peak cavity pressure almost solely determines the shrinkage of the plastic and, therefore, the final dimensions of the molded part. As peak cavity pressure increases, shrinkage decreases. Peak cavity pressure is only one of the variables that affect the shrinkage of crystalline type materials (14).

The rate of plastic flow in the cavity is important primarily because the plastic cools as it flows into the cavity; the slower it flows, the colder it becomes. As the amount of cooling increases, the difference in peak pressure for the gate end to the opposite end of the

part also increases. Since shrinkage is determined by the peak pressure, this difference results in a differential shrinkage across the cavity; shrinkage is highest at the greatest distance from the gate. When this shrinkage differential becomes excessive, high stresses are set up, and the part frequently warps. On conventional machines flow rate is generally calculated from fill time, which is manually measured with a stop watch by observing the ram travel. At best, this technique is inaccurate, particularly with small, thin-walled parts that are often filled in a few tenths of a second. Fill time can, however, be accurately determined from cavity-pressure curves with recording equipment with an appropriately expanded time scale.

Rate of cooling has little or no effect on thin-walled amorphous parts; lower cooling rates result in less orientation of thick-walled amorphous parts by providing additional time for relaxation; and rate of cooling has a strong effect on the nature and degree of crystallinity in crystalline type materials and, therefore, affects the shrinkages and physical properties of these parts. The cooling rate is typically controlled by controlling the temperature and flow rate of the cooling water being circulated in the mold. Recent developments in infrared temperature sensing of the plastic temperature in the cavity have shown some promise as a possible future method of measuring the cooling rate directly.

12.5   INJECTION MOLDING OF THERMOSETS

Although of far less commercial significance than that of thermoplastic molding, thermoset parts can also be produced by the injection molding process. With thermosets, the material is softened in the barrel and injected into a hot (about 350°F) mold where curing takes place. The major concern in the process is to prevent premature curing in the barrel. This is accomplished by controlling the barrel temperature by means of a water jacket rather than with electric heaters. The water jacket facilitates heat removal as well as input when necessary. Screws for thermosets usually have little or no compression ratio, to minimize mechanical heat input, and usually have no back-flow valves where material can hang up and eventually cure. The nozzle and screw tip are designed so that as much material as possible is cleared from the barrel when the screw is forced all the way forward. The process is normally run with no cushion. The barrels, screws, and nozzles are also designed for easy removal to facilitate cleaning if premature curing does occur. The molds are heated with electric cartridge heaters or by the circulation of hot oil. A subsequent flash removal step is frequently required.

Results of an investigation by Papoojian and Lynn (15) that studied the effect of process variables on the properties of injection molded phenolic parts showed that injection speed and mold temperature have the most significant effects. Specifically, impact strength increased as injection speed decreased; tensile and flexural strengths increased as mold temperature increased and injection speed decreased; and shrinkage decreased as mold temperature increased and injection speed decreased.

12.6 REACTION INJECTION MOLDING

Reaction injection molding (RIM), liquid reaction molding (LRM), or liquid injection molding (LIM) involve the thorough mixing of two reactive liquids and the injection of the mixture into a closed mold where a chemical reaction cures the part. Most RIM materials are currently based on urethane chemistry and provide a wide range of stiffness, hardness, and other physical characteristics. Typically, an isocyanate is stored in one tank and a polyol in another. The temperature of the materials in the tanks is precisely controlled. From the tanks, the components are pumped separately to the mixing head with constant-displacement pumps. The materials enter the mixing chamber through opposing orifices. Mixing is accomplished by impingement of the two streams, which creates turbulent flow. Once mixed, the material must be quickly and completely injected into the mold to prevent curing in the mixing head. Meanwhile, the

separate supplies of isocyante and polyol are recirculated to their respective storage tanks.

Cure time of RIM materials is usually less than 1 min. Cure times of about 20 sec are common. Unlike most plastic processes, cycle time is relatively independent of part-wall thickness. Since the viscosity of the material is low during mold filling, only relatively low (typically 100 psi) injection pressures are required. The cost of tooling and mold clamping mechanisms are, therefore, low. Important aspects of tool design include adequate mold venting and the proper placement of knock-out pins.

The reaction injection molding process appears to be particularly well suited for large automotive parts.

12.7 COMPRESSION MOLDING

Compression molding is the process of placing polymeric material between two open, heated mold halves and closing the mold. The charge of material is generally preweighed and can be in the form of pellets, granules, powder, or preformed tablets prepared by cold pressing. The material may be preheated, or all of the required heat can be supplied by the hot mold. As the mold closes, the material softens and is forced to fill all areas of the mold. In the case of thermosets, the most common class of materials used with this technique, the mold is held closed while additional heating occurs and curing takes place. During this curing or cross-linking reaction, the part develops rigidity and can be ejected at the mold temperature. If thermoplastics are used, the entire mold must be cooled after closing in order to sufficiently solidify the part for ejection. This requirement leads to very long cycles, which usually makes compression molding an uneconomical process for production molding of thermoplastics. The process is extremely valuable, however, for specimen preparation and prototype molding work.

Since flow distances are short and some relaxation time is inherently provided, the process produces relatively stress-free parts with very low levels of molecular orientation. In addition, the fibers in reinforced compositions are not subjected to the abuse that many other processes impose, and the fibers usually survive compression molding without excessive degradation. Compression molding produces little or no wasted material, which is particularly important when processing thermosets. Compression molded parts have no gate marks, therefore, secondary degating operations are not required.

12.7.1   Equipment

Compression molding equipment is relatively simple and low cost compared to that of most plastics processes. The compression molding press consists of two parallel platens with facilities for controlling the temperature of the platens and a mechanism for closing the platens with an appropriate force and rate of closing.

Heating is accomplished by circulating steam or hot oil, with electrical cartridge and/or strip heaters, and occasionally with a gas flame. The circulating liquids provide the most uniform heating, but their temperature range is usually limited. The electrical heaters provide quick temperature changes and a higher upper-temperature limit.

The clamping of compression molding presses is usually accomplished with a pure hydraulic cylinder, although some pneumatic presses are still in operation. Some presses utilize the principles of the toggle (discussed in Section 12.4.1.2) for clamping. In this case, the toggle can be powered either hydraulically or pneumatically. All modern presses are powered in both the opening and closing modes with speed controls in both directions.

12.7.1.1   Molds

Compared to injection molding tools, compression molds are generally simpler and less expensive. Since no allowances for sprues, runners, or gates are necessary, compression molding offers greater design flexibility. Modern compression molds contain ejector pins for semi- or fully automatic part removal. It is sometimes necessary to place ejector pins in both mold halves. The pins can be activated mechanically, pneumatically, or hydraulically. Although delicate mold parts are susceptible to damage from closing on cold material, compression mold maintenance costs are usually low. Owing to the short material flow path, mold wear is minimal even when abrasive materials are being processed.

Compression molds can be categorized as either flash, positive, or semipositive tools (Figure 12.16). In the flash mold, excess material is allowed to flow onto the parting line. With a positive mold, all of the material is trapped in the cavity. The semipositive mold permits some material to escape, but in a more controlled manner than with the flash tool.

12.7.2   Process

The temperature of the mold is one of the most critical variables in the compression molding process. Higher mold temperatures result in faster curing and shorter cycles, but if the mold temperature is too high, the material will cure prematurely before thin and intricate parts of the mold have been filled. Mold temperature uniformity is also important to encourage uniform curing. Nonuniform cure is sometimes a problem in parts with large changes in cross-sectional thickness.

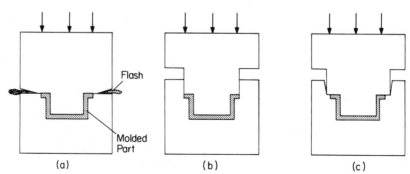

Figure 12.16   Three types of compression molds: (*a*) flash; (*b*) positive; (*c*) semipositive.

The primary variables controlling molded-part density are different depending on the type of mold being used. In the flash mold, material melt viscosity and the rate of closing determine the density. With the positive mold, the clamping force is critical. In the semipositive mold, the clearance between the male and female parts of the mold which form the flash channel, the material melt viscosity, and the rate of closing all affect the part density. In the positive mold, of course, final part thickness is dependent on the mass of the material charge. Preheating the material charge often offers several advantages. Cycle times can usually be reduced by preheating. Curing is more uniform, particularly in thick sections, and presoftening reduces the risk of damage to delicate mold parts during closing.

It is sometimes necessary to make special provisions for the escape of gases from the mold. Trapped gases result in voids and surface blemishes can be particularly troublesome when processing certain thermosets such as the melamines and the ureas. Mold breathing is a technique used to facilitate the escape of these gases and is accomplished by simply opening the mold slightly during the early phase of the curing cycle.

12.8 TRANSFER MOLDING

Transfer molding has many similarities to both injection and compression molding. In transfer molding, a measured, compacted, preheated charge of thermosetting material is loaded into a hot, open chamber in the mold. This chamber is called the pot. When the material is sufficiently softened, it is transferred, under pressure, through a sprue, runner, and gate into a closed cavity as in injection molding. The mold is held closed until the material cures. The mold is then opened, and the part is removed usually with the aid of ejector pins.

Although many variations of the transfer molding process have been developed, one of two distinct methods of applying the transfer pressure is most often used. The first method is called the integral pot type and uses a three plate mold in which a plunger is part of the upper plate (See Figure 12.17). This plunger precisely fits into the pot, which is located in the second plate. The mold cavity or cavities are located in the lower plate and are connected to the bottom of the pot with the sprue hole. A conventional compression molding press is used to open and close the mold. With the mold open, sufficient resin is loaded into the pot to provide a slight excess of material after the cavities are filled. When the material charge has been sufficiently heated, the press is closed. As the plunger enters the pot, the material is forced to flow through the sprue hole, runners and gates into the cavity. After the part is cured, the mold is opened, the parts and runners are removed from the lower parting line, and the sprue and excess material in the pot (cull) are removed from the upper parting line. The second method of applying transfer pressure involves the use of an auxiliary hydraulic ram that is completely inde-

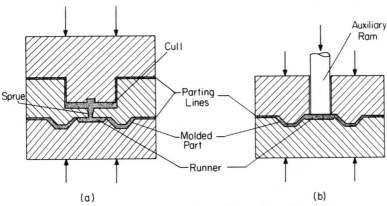

Figure 12.17 Two types of transfer molds: (*a*) integral pot; (*b*) auxiliary ram.

pendent of the closing action. In this case, a two-plate mold is used, and the auxiliary ram enters the pot in the upper plate during the transfer step. As before, the runners and cavities are contained in the bottom plate.

Preheating the material charge is usually desirable and results in shorter cycles and more uniform curing. Dielectric heating is the usual method of preheating, but a special variation of the process uses a screw plasticator (See Section 12.4.1.2). In this case, the screw is used to melt, homogenize, and meter the charge that is delivered to the pot.

Transfer molding is generally faster than compression molding, produces parts with dimensional accuracy and uniform density, and is particularly adaptable to insert molding. The process does, however, produce parts with molecular orientation in the direction of flow, parts with gate marks that often require a secondary operation, and unusable, cured, scrap in the form of the cull, sprue, and runners.

12.9   BLOW MOLDING

Blow molding is the process of inflating a molten, hollow thermoplastic preform called a parison inside a closed mold such that the outside surface of the parison contacts and conforms to the mold cavity (Figure 12.18). The part is cooled, the mold is opened, and the molded article is removed. The process is useful for producing an endless variety of hollow plastic parts such as bottles, drums, and tanks. Three basically different variations of the blow molding process are discussed

here. The first two differ primarily in the method of parison preparation.

12.9.1   Extrusion Blow Molding

In its simplest form, the extrusion blow molding process utilizes a conventional extruder (as described in Section 12.2) with a die very similar to a pipe or tubing die. The tube is extruded downward between two halves of the open mold. When the parison has reached a sufficient length, the mold closes and pinches off the parison at both ends. For products such as bottles air is injected at the open end, which forces the plastic out against the mold surface. For other products without a natural opening, the blowing air can be injected at any convenient location on the object by piercing the parison with a hollow, needlelike probe. After cooling, the mold is opened, and the part is ejected. Excess material at the pinch-off locations must be trimmed. In order to increase productivity and allow for the continuous operation of the extruder, it is usual practice to use more than one mold and clamping mechanism and to move the mold away from the extruder during the blowing, cooling, and unloading steps. Meanwhile, a different mold is moved under the extruder to accept the next parison, using one of many possible schemes. One popular technique is to mount four mold-clamping stations on a rotary table. The stations are spaced at 90° to each other. The table is indexed in 90° increments to successively move a mold from the extrusion position, to a blowing position, to a cooling position, and finally to an unloading (ejection) station. The use of multiple ex-

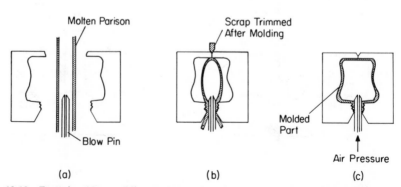

Molten Parison

Blow Pin

Scrap Trimmed / After Molding

Molded Part

Air Pressure

(a)   (b)   (c)

Figure 12.18   Extrusion blow molding process. (*a*) Parison in position. (*b*) Mold closed. (*c*) Parison inflated.

trusion heads in combination with multiple cavity molds further increases productivity.

In extrusion blow molding, parison formation is a relatively slow process. The force of gravity acting on the parison as it hangs from the die stretches the tube and causes it to sag. This effect generally limits the size of bottles that can be produced by the continous-extrusion blow molding process to one gallon or less. One approach to reducing the amount of parison sag is to reduce the amount of time that gravity acts on the molten tube. One technique for obtaining this effect is the use of a reciprocating screw plasticator (see Section 12.4.1.2). In this case, the forward motion of the screw is used to force the polymer through the tubing die rather than into an injection mold. With this technique, sufficient delivery rates are obtained to permit parison weights up to about 10 lb. If larger sizes are required, either the ram accumulator or the accumulator head designs are used. In both cases, an auxiliary hydraulic ram is used to force polymer, which has been accumulated from one or more conventional extruders, through the die at a high rate of speed. In the ram accumulator type, the polymer is stored in a separate reservoir, upstream of the die, between shots; in the accumulator head type, the polymer is stored in the specially designed die. Parison weights in excess of 50 lb can be obtained with these techniques. The accumulator head concept can also be used to obtain multiple-layer parisons by using coetrusion techniques.

If the diameter of the product being blow molded varies along its length, the larger areas will result in greater stretching of the parison in that area and, therefore, increased thinning. The resulting nonuniform wall thickness leads to a heavier product and increased resin usage in order to maintain some minimum wall thickness throughout the product. Nonuniform wall thickness can be minimized by using one of several techniques of parison programming. With these methods, the wall thickness of the parison is varied in the length direction to partially compensate for differences in drawdown during the blowing step. Parison programming can be accomplished by modulating the annular die opening either by moving a tapered mandrel in and out of a stationary die ring or by moving the die

ring over the stationary mandrel. Another method of parison programming involves varying the rate of parison delivery through a die of constant opening. This concept utilizes the principles of die swell to vary parison thickness and is most adapatable to reciprocating screw, ram accumulator, and accumulator head processes.

12.9.2   Injection Blow Molding

In the injection blow molding process, the parison, closed at one end, is actually injection molded around a core pin. The parison remains on the core pin and is transferred to the blow mold. The blowing, cooling, and stripping steps are essentially the same as in extrusion blow molding. Injection molding of the parison provides a no-scrap process, produces improved reproduction of details in the neck area, such as threads, and provides an easy and reproducible method of optimizing parison wall thickness. As with extrusion blow molding, multicavity molds and various schemes or multistation operation are used to improve the productivity of the injection blow molding process.

12.9.3   Stretch Blow Molding

Stretch blow molding is a process for producing hollow objects such as bottles with biaxial molecular orientation. This biaxial orientation provides enhanced physical properties, clarity, and gas barrier properties, all of which are particularly critical in products such as beverage bottles. Biaxial orientation is obtained by first mechanically stretching in the length direction a preformed parison that has been conditioned to a critical orientation temperature, and then blowing the parison, which provides stretching in the radial direction. Parisons can be formed by either extrusion or by injection molding. Parison formation can be either in-line or off-line (two-stage). When extrusion formation is used, it is necessary to mechanically grip each end of the parison in order to accomplish the lengthwise stretching; with injection molded parisons (closed at one end), a rod can be extended from the core pin after the parison has been positioned in the blow mold and is suspended by the neck area. Two-stage systems require rather large reheat ovens in order to bring the parison to the proper temperature. Polypropylene and thermoplastic polyesters such as polyethylene terephthalate are the most frequently used polymers in the stretch blow molding process.

12.10 ROTATIONAL MOLDING

If hollow plastic articles are to be produced, rotational molding is a process to be considered. In rotational molding (sometimes called roto-molding), a measured amount of polymer powder is placed inside a split mold, and the mold halves are closed and locked together. The mold is then heated while it is being rotated about two perpendicular axes simultaneously. The rotation distributes (by the action of gravity, not centrifugal forces) a uniform layer of plastic over the interior surface of the mold. This layer of polymer gradually melts and fuses. The mold is then cooled, and the part is removed.

The rotational molding process has found wide application in the toy industry, for example, the familiar hollow plastic rocking horse. The process is also used for the manufacture of some furniture, many shapes and sizes of storage tanks, and various automotive applications, such as certain duct work. Rotational molding is usually thought of as being most appropriate for relatively short runs of large parts.

Although results of recent work with some thermosetting materials have been encouraging, the rotational molding process continues to be most useful for thermoplastics. Although polyethylene is still the most frequently used polymer, other thermoplastics such as nylon and polycarbonate can also be used. The required particle size is a function of material melt viscosity and the complexity of the part being produced. Higher viscosity, thinner walls, and smaller part details require finer powders. Most products can be produced satisfactorily with about 35 mesh powder.

Since rotational molding does not involve high pressures or shear rates, the process produces relatively stress-free parts with very low levels of molecular orientation and utilizes low-cost molds and equipment. Molds, for example, are commonly of cast aluminum construction. In early rotational molding equipment, thermal cycling was obtained by first spraying molten salt or hot oil on the mold followed by a cold water spray. Modern equipment simply uses heated and cooled air or employs a jacketed mold for the circulation of a heat transfer liquid. The latter technique generally gives more precise temperature control and shorter cycles, but requires higher-cost molds and equipment that contain the appropriate rotary joints.

Typical rotational molding cycles range from 2 to 30 min. As with other plastic processes, several techniques have been developed to improve the productivity of rotational equipment. The most popular method is the use of a three-armed carousel with a mold on each arm. The carousel indexes each mold successively into a heating position, a cooling station, and finally into an unloading and loading station.

12.11 THERMOFORMING

Thermoforming is the process of softening a thermoplastic sheet with heat and forcing the sheet, by one or many techniques, to conform to the surface of a mold where it is cooled. The formed part is then trimmed by various possible techniques including blanking (Section 12.12).

12.11.1 Equipment

Thermoforming equipment may be broadly categorized as either discrete sheet fed or web fed. In addition, web-fed machines can use either coil stock, if the material is thin enough to be coiled, or operated in-line with a sheet extruder. Sheet-fed machines are usually either of the shuttle type or use a rotary carousel. Shuttle machines first move the sheet of material into the heating station and then into the forming station. For improved productivity, two or more forming stations can share a common oven. As with several previously discussed processes, the rotary carousel indexes a sheet from a loading station, to a heating station, to a forming station, and then to an unloading position. Several variations of this four-station scheme are frequently used to accommodate a particular manufacturing operation. Loading and unloading, for example, can be accomplished at the same station; or certain applications may require two heating stations in order to optimize the output of the machine. In web-fed devices, one of several techniques, such as two endless chains with spikes that pierce the sheet, is used to grip the edges of the sheet. The transport mechanism indexes the web through the temperature conditioning oven, the forming press, and on to the trimming operation. In modern equipment, the heating of the sheet to be thermoformed is usually accomplished with electrical heaters arranged to heat the material from both sides. Many types of electrical heaters are available. The proper selection depends on the material type and thickness to be used and the part to be formed. The success of any thermoforming operation usually depends on an ability to repeatedly heat the plastic to its critical forming temperature uniformly and without surface degradation.

12.11.2 Process

A number of different forming techniques have been developed primarily in order to maintain a uniform wall thickness in a variety of part configurations. The plastic thins out as it is being stretched during forming. In general, the last areas to touch the mold surface will thin the most. Corners, for example, often become very much thinner than the other areas of the part. Unless that situation is prevented with the proper forming technique, the overall thickness of the part would probably be increased to maintain acceptable corners, thus wasting material and probably increasing the required cycle time. In general, the heated sheet can be pushed into a cavity mechanically, by air pressure, or both. Figures 12.19–12.22 illustrate a few of the available thermoforming techniques. In mechanical forming (Figure 12.19) matched molds are simply closed on the softened sheet. In vacuum forming (Figure 12.20), a vacuum is drawn under the sheet, and atmospheric pressure forces the plastic into the cavity. More uniform wall thickness can be obtained by first stretching the sheet with a mechanical plug and then applying the vacuum (Figure 12.21). Additional improvements can be obtained by first blowing a bubble, which results in a uniform stretching, and then vacuum forming with or without a plug assist (Figure 12.22). Many other variations and combinations of these basic thermoforming methods are used depending on the particular part configuration to be produced.

Since high pressures are not involved in the thermoforming process, relatively inexpensive molds can be used. If sample parts or very short run production parts are required, wood molds can often be used. Epoxy molds, which can be quickly cast around a handmade model, are also popular in these cases. Some plastics have also been used. One of the major

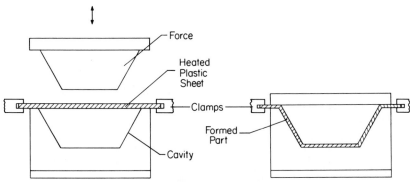

Figure 12.19 Mechanical forming.

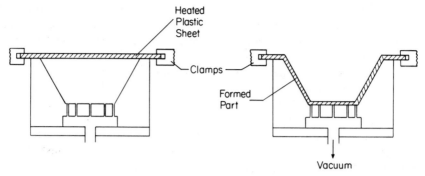

Figure 12.20   Vacuum forming.

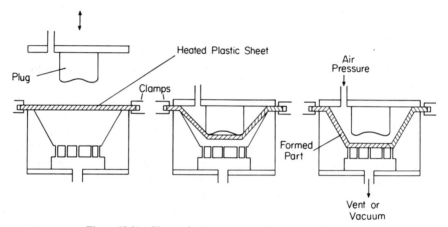

Figure 12.21   Plug-assisted pressure and/or vacuum forming.

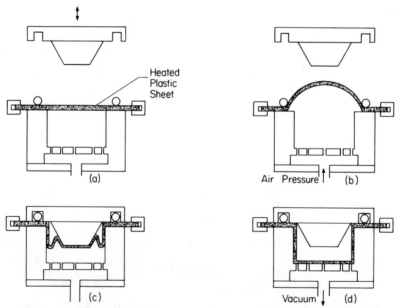

Figure 12.22   Pressure bubble, plug assisted, vacuum forming: (*a*) heated sheet ready for forming; (*b*) air pressure applied under sheet to give uniform stretching; (*c*) plug enters; (*d*) vacuum applied.

598

disadvantages of this type of tooling is the inability to control the mold temperature effectively. This condition often results in various part defects and excessively long cycles. Aluminum molds are usually the best choice with special attention to efficient cooling. This minimizes the risk of limiting the throughput of the thermoforming machine because of improper tooling. Where special cooling problems exist, beryllium copper mold parts can be used.

With respect to materials, thermoforming is a very versatile process. Essentially any thermoplastic material that can be extruded as sheet can be successfully thermoformed. Possible finished products are even more diversified and range from margarine tubs to recreational vehicle bodies.

12.12 SOLID—PHASE FORMING

Solid-phase forming covers a variety of plastic processing techniques that are characterized by the shaping of polymers at temperatures below the glass transition T_g of the material. Examples of solid-phase forming processes include forging, stamping, and blanking; all of which are in many ways analogous to the corresponding well known metal forming techniques. Since the process is not slowed by the need to remove heat from plastic, solid-phase forming techniques tend to be very fast.

In the plastics stamping process, extruded sheet, either in continuous web or cut sheets, is heated in a hot air oven with infrared heaters or by direct contact with heated plates or rolls. The heated sheet is placed between the open halves of a matched metal die. The die is closed by means of hydraulic or mechanical press. Forming pressures up to 5 ton/in.^{2} is not uncommon. Since solid-phase forming results from deformation rather than viscous flow, a high degree of molecular orientation is usually introduced into the product. Also, the polymer exhibits some degree of elastic recovery or "spring back" when the die is opened. Although this property makes the forming of precision parts very difficult by the solid-phase forming process, most of the recovery can be compensated for in the original die design. The majority of successful applications of the stamping process have been with polyethylene and polypropylene. To a smaller extent, some parts have also been produced from ABS, nylon, and polyvinyl chloride compositions. The major applications of the stamping process have been for parts with relatively uniform cross-sectional thickness. Protective shrouds, automotive fender liners, and certain simple containers are examples of successful applications.

The forging process is identical to stamping except that forging starts with a preformed billet rather than an extruded sheet. The billet permits the forming of much thicker parts, and forging can produce parts with substantial changes in cross-sectional thickness. One notable attribute of forging is the absence of sink marks at cross-sectional changes, which are characteristic of melt processes such as injection molding. In addition to the same materials that are useful for the sheet stamping process, difficult to process ultrahigh molecular weight polyethylene can be formed by the forging process. Flangeless pulleys, gears, and pipe flanges are examples of successful applications of the forging process.

Blanking is the process of cutting flat parts out of sheet stock or for separating preformed parts from precut sheets or continuous webs. The principles of plastic blanking are very similar to those of sheet metal blanking. The material is fed to a mechanical or hydraulic press where closing creates a shearing action in either a steel rule die or a more expensive matched metal die. In general, plastic blanking requires significantly tighter punch-to-die clearances than those commonly used for metal blanking, therefore, greater press accuracies are also required in the plastic operation. Other required characteristics in comparison to metal blanking include longer and faster cutting strokes.

12.13 CALENDERING

Calendering is a method of producing film and sheet products. The process is most frequently used for polyvinyl chloride (PVC)

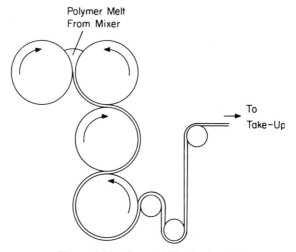

Figure 12.23 "Inverted L" calender stack.

but other materials such as acrylonitrile butadiene styrene (ABS) can also be used. A polymer melt is first prepared in an appropriate continuous or batch mixer (e.g. a Banbury mixer). In the mixer, the plasticizers or colorants and other components are carefully added. The melt is then delivered to the feed nip of a series or rotating, heated rolls (Figure 12.23). The melt is successively squeezed to the final sheet thickness in the subsequent nips. The figure shows the common "inverted L" calender stack, but other roll configurations are also available. Roll dimensions or typical calenders range from 2 ft in diameter by 5 ft long, to 3 ft in diameter by 8 ft long. Since pressures in the nips are extremely high (up to 6000 lb/lin in.), drive motors are very large, and calender frames are necessarily massive. Particular attention must be paid to the design, material, and lubrication of the roll bearings. Parallelism of the rolls is a major concern since deflections resulting from large and sometimes variable stress and thermal effects are often significant. A well built calendering line requires a large capital investment, but the process provides an extremely high throughput rate. Typical calenders run at speeds as high as 375 ft/min, processing up to 10,000 lb/hr or resin. The primary variables of the calendering process include the material viscosity and frictional characteristics, the temperature and temperature uniformity of

the material being discharged from the mixer, the roll temperatures, the nip openings, rotational speed of the rolls, and the speed of the take-up mechanism.

12.14 CASTING

Casting is the process of pouring, or otherwise placing, a liquid into a mold or onto a surface and then hardening the liquid into a rigid article. The liquid may be a polymeric solution or a catalyzed monomer system. For solutions, the hardening step involves evaporation of the solvent; for monomer systems, hardening is accomplished by polymerization.

12.14.1 Solution Casting

The primary importance of the solution casting process is in the manufacture of plastic film or sheet. Polymer is dissolved in an appropriate solvent, the viscous solution is spread onto a smooth surface, and the solvent is evaporated. Typically, the viscous solution is dispensed from a hopper and is spread by a doctor blade onto the highly polished surface of a slowly rotating casting wheel (Figure 12.24). The solvent is then evaporated from the cast sheet by warm air. When sufficient solvent has been removed to render the film self-supporting, the product is stripped from the wheel. Additional drying, if required, is

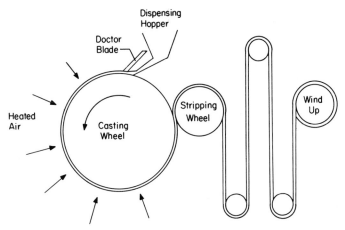

Figure 12.24   Solution casting process.

accomplished as the film is pulled through warm air ovens. The appropriate handling and recovery of the solvent vapors are required. Casting wheel diameters of 20 ft and widths of about 24 in. are typically used. Primary processing variables include solution viscosity, hopper feed rate, doctor blade to wheel clearance, wheel speed, and drying air temperature. High-quality cellulosic photographic film base is generally manufactured by the solution casting process.

12.14.2   Monomer Casting

In addition to the familiar poly(methyl methacrylate) slabs and sheets, this process is used for producing other acrylic shapes, such as rods or tubing, and for encapsulation. Although acrylic casting is used here for discussion purposes, it should be noted that the monomer casting process is also used for producing very large or thick nylon parts.

In its simplest form, a catalyzed acrylic ester is placed into a mold of the appropriate configuration and polymerized. Owing to the relatively low boiling point of acrylic monomers, extreme care must be taken to control the exothermic reaction and to prevent evaporation of the monomer. In an effort to shorten cycles and control shrinkage, partially polymerized casting syrups are frequently used. Syrups can be prepared by temporarily terminating the polymerization reaction or by actually dissolving polymer in

monomer. Syrups generally contain less than 50% polymer.

The batch process for the manufacture of high-optical-quality acrylic sheet consists of polymerizing a syrup between two glass plates that are separated by a flexible gasket. A slight clamping force is applied so that polymer-to-glass contact is maintained as the material shrinks. Cycles as long as 16 hr for $\frac{1}{4}$ in. thick sheet are not uncommon. For thicker sheets, best results are obtained by polymerizing in an autoclave. Sheets less than 1/2 in. thick can also be produced in a continuous process by polymerizing the syrup between two moving endless belts, usually stainless steel. Compared to the batch process, the continuous process generally produces sheet with improved thickness uniformity but with poorer surface and optical characteristics.

12.15   FOAM PROCESSING

Many processes are available for producing foamed plastic parts. Most of these techniques are based on, or are very similar to, previously described processes. All foam processes strive for the introduction of controlled voids or cells in the finished product, generally for the purpose of part weight reduction from both aspects of the end-use requirements and the materials savings. Uniform cell size is usually a concern since that condition leads to uniform and predictable properties. A special class of foams known as structural foams are characterized by a dense skin and an internal cell structure.

Foaming is achieved by introducing a blowing agent into the polymer melt, holding the melt under sufficient pressure to prevent premature foaming, and releasing the pressure at the proper point in the process to permit the blowing of tiny gas bubbles.

Blow agents are categorized as either chemical or physical. Chemical blowing agents degrade at a specific temperature and form gaseous products that expand to form the cells. Chemical agents are commonly preblended with the polymer pellets prior to processing. Physical blowing agents are simply gases at plastic processing temperatures and atmospheric pressures. These materials are generally injected into the melt stream after the polymer has been plasticated. Physical blowing agents are usually preferred when lower-density foams are desired. Regardless of the type of blowing agent, it is often desirable to nucleate the emerging gas bubbles in order to improve the control of the cell size and uniformity. Particulate fillers such as finely powdered talc are frequently used as nucleating agents.

12.15.1 Expandable Polystyrene Molding

Molding expandable polystyrene is a three-step process that consists of heating small solid polystyrene beads, which contain a blowing agent, in a pre-expander to obtain the desired final part density. The expanded beads are then cooled and conditioned. Finally, the conditioned beads are placed in a closed mold and heat is again applied to fuse the particles. The amount and uniformity of the blowing agent content and the temperature attained in the pre-expansion and fusing steps are of paramount importance. Heat in both the pre-expansion and molding phases is usually supplied by passing steam over the beads. The conditioning step, frequently requiring one or more days, is used to cool and harden the beads, to evaporate the moisture that condensed on the beads from the pre-expansion steam, and to relieve residual internal pressure. During heating in

the molding step, additional internal pressure is developed and is used to force the beads against each other for fusing. Familiar products such as insulation board, ice chests, ceiling tiles, floatation devices, packaging structures and dunnage, drinking cups, and thread spools are all manufactured by this process.

12.15.2 Foam Extrusion

When continuous lengths of plastic foam are desired, an extrusion process (see Section 12.2) can often be applied. Depending on the resin and blowing agent being used and the product being produced, the process can utilize one or more extruders. The extruder may be twin screw or single screw, and the extruders may or may not be vented. A common method utilizes two in-line extruders. The first extruder, usually a two-stage unit, melts and mixes a dry blend of polystyrene pellets and any additives, such as a nucleating agent, in the first stage. The blowing gas, either Freon or a hydrocarbon blowing agent, is injected, and mixing continues in the second stage of the first extruder. Pressure is maintained on the melt while it is pumped to the second extruder. This extruder is used to cool the melt and to pump it through the die. Although the use of a conventional flat sheet die is possible, a die and postextrusion handling very similar to those of the blown film process are most often used. As with the expandable polystyrene molding process, a conditioning period of one or more days is required before additional forming operations can be satisfactorily completed. A common forming operation for the extruded foam is the thermoforming process (Section 12.11). Foamed egg cartons and meat trays are frequently produced in this manner.

12.15.3 Liquid Foam Processing

A variety of thermosetting foams can be produced by combining two or more reactive chemicals in the presence of a blowing agent, injecting or spraying the mixture into the desired mold, and utilizing the heat of reaction during cure to decompose the blowing agent. Equipment for the liquid foam process can be identical to reaction injection molding equipment (RIM, Section 12.6) or can be equipped with a special spray gun for dispensing the foamable material. The spray gun can have an internal mixing chamber as with RIM equipment, or the separate chemical streams can be directed toward each other as they exit the gun.

Although several other material systems can be used and still others are under development, the most common commercially used liquid foam materials are based on urethane chemistry. In this case, the blowing agent is almost always delivered with the polyol component of the mixture. As with the RIM process, a very wide range of properties are available from very soft and flexible formulations to very hard and rigid structural products. Similarly, the many applications of the foamed compositions range from thermal insulation and packaging materials to furniture.

12.15.4   Structural Foam Molding

It should be noted that certain aspects of some of the structural foam molding processes to be discussed are currently protected by patents. Those aspects are not highlighted, and no implications as to freedom to use should be drawn. This discussion is intended only to provide a general description of the various techniques that are used to produce structural foam parts.

Structural foam molding can be broken down into two broad categories: high-pressure and low-pressure systems. In the low-pressure technique, a closed mold is partially filled with a foamable plastic melt where it is permitted to foam and thereby fill the cavity. In the high-pressure processes, the mold is completely filled and actually packed under pressure until the outer skin, which is in contact with the cold mold surface, freezes. The volume of the mold is then increased by either moving an internal component of the mold or by actually opening the clamp a specific distance. The latter volume-increasing technique makes use of molds with

telescoping parting lines to prevent flash. The increase in cavity volume permits the foaming of the core material.

Slightly modified conventional reciprocating screw injection molding equipment (Section 12.4.1.2) can be used to produce structural foam products. A shutoff nozzle, which prevents the discharge of material except at the time of injection, is necessary since the melt must be held under sufficient pressure in the barrel to prohibit premature foaming. Very-high injection rates are generally desirable, therefore, certain applications require special hydraulic modifications to attain the required velocities. Many structural foam parts are very large and require large shot sizes that can be satisfied by the use of melt accumulators. Another structural foam molding technique utilizes an injection molding machine with two separate barrels that feed a common nozzle. One barrel, which contains plastic material with no blowing agent, first injects a partial shot into the cavity. That shot is immediately followed by the injection of foamable material, which essentially inflates the solid skin.

Nearly all thermoplastic materials can be produced by these techniques. One of the most successful applications of the structural foam molding process has been in the lower-cost simulation of wood products.

12.16   PROCESSING PLASTISOLS AND ORGANOSOLS

Several techniques have been developed to process plastisols, which are dispersions of vinyl chloride homopolymer or copolymers or vinyl acetate in a plasticizer. The same techniques can also be used for organosols, which are plastisols that have been diluted with a volatile nonsolvent. The fundamentals of these techniques are to place the suspended polymer particles (frequently called pastes) into the desired shape, to fuse the particles with heat (also volatilizing the diluent in the case of organosols), and to cool the resulting product.

A variety of available mixer types can be used to prepare the dispersions. A successful mixing step requires sufficient shear to homogenize a paste. Extreme care is required to achieve the proper formulation, which frequently includes components such as resin, plasticizers, stabilizers, colorants, extenders, and viscosity modifier. The required formulation, of course, depends on the resin type, the subsequent forming process to be used, the configuration of the product, and the end-product requirements.

One very useful plastisol process involves applying a protective, decorative, or otherwise functional plastic coating to, for example, metal products. This process is accomplished by simply dipping the object, preferably preheated, into a tank of the plastisol. The tank is frequently agitated and also temperature controlled. After a specific time, the coated object is removed from the tank, and excess dispersion is allowed to drip off. The product is then placed in an oven, heated to the fusing temperature of the particular formulation, and then cooled. Coating thickness is determined by the plastisol viscosity, the preheat temperature of the object being coated, and the time in the tank. Familiar products that are coated by this technique include dishwasher racks and handles of pliers and other tools. Other applications of the dipping process involve coating and object and stripping the fused coating from the form. Surgical gloves, for example, can be produced by this technique.

Where very large objects are to be coated, the size of the tank and the quantity of paste required often make dipping impractical. In these cases, the plastisol or organosol can be sprayed onto the object prior to heating. Many other previously described processes can also be applied to plastisols and organosols. Hollow parts can be produced by rotational molding (Section 12.10), or by completely filling a mold with plastisol and then pouring out the excess after only a thin layer in contact with the mold surface has fused. The latter technique is referred to as slush molding. Solid parts can be produced by casting into a mold and heating, or coatings can be simply spread onto a substrate and fused.

12.17 RADIATION PROCESSING

The term radiation processing covers a diverse group of techniques used to cause desired changes in polymeric materials by exposure to certain forms of electromagnetic radiation. The desired changes are usually obtained by causing a chemical reaction such as polymerization, cross-linking, or vulcanization. One exception is the low-temperature radiation sterilization of medical and surgical plastic disposables which would be destroyed by the usual high-temperature autoclaving. Chemical reactions that would normally require many minutes by pure heat and chemical techniques frequently can be accomplished in less than 1 sec by processing with radiation. The mechanism of the radiation-induced reaction is the formation of free radicals, which are highly reactive sites for the reaction.

Useful radiation sources include ultraviolet lamps, high-energy electron beams from electron accelerators, and gamma radiation from certain radioactive materials such as cobalt-60.

Two important parameters in the radiation process are the depth of penetration and the exposure time required. Gamma radiation, for example, is highly penetrating, but relatively long exposure times are usually required. The penetrating ability of electron beams is determined by the accelerator voltage. Units with potentials up to several million volts are currently available. The processing rate attainable from a particular electron accelerator is determined by the beam current, usually expressed in terms of power. Although units with a wide range of power can be obtained, the largest typical commercial accelerators are rated near 200 kW.

In determining actual processing rate, one must consider the total energy that is absorbed by the product (dose), the total energy that is required for the desired reaction, and the efficiency of the process. The energy absorption of 100 erg/g of product is defined as 1 rad. A typical useful dose is in the megarad (Mrad) range.

The major applications of radiation

processing include: the cross-linking of plastic pipe, tubing, and the polymer coating on wire and cable to improve properties such as heat, stress-crack, and abrasion resistance: the rapid curing of organic coatings without solvent emission; the polymerization of residual monomer in food packaging applications: graft polymerizations to impart selected properties to products such as textiles; and the vulcanization of natural and certain synthetic rubbers.

12.18 MACHINING

It is frequently desirable to fabricate plastic parts by machining from stock prepared by extrusion, compression molding, and various other techniques. Situations that suggest the use of machining techniques include the need for very small numbers of parts, such as in model making, or the unavailability of the time necessary to build a mold. Machining of polymeric materials can be accomplished by proper application of conventional metal cutting techniques, by the use of high-velocity water jets, or with special lasers.

The successful application of metal machining procedures such as milling, turning, and drilling requires an understanding of some of the key property differences between metals and plastics. The relatively low softening temperatures and low thermal conductivity of plastics, for example, combine to require special considerations. The frictional heat generated during machining cannot be effectively dissipated into the workpiece and will be concentrated in the cutting tool and at the surface of the plastic. The polymer softening temperature and even its degradation temperature can easily be exceeded, which can result in poor surface finish and dimensional control. This situation is further complicated by the high coefficient of thermal expansion of plastics. Frictional heat generated by the machining operation results in expansion of the plastic, which increases tool drag and causes an increase in the frictional heat developed. Special attention to machining speeds, cooling fluids,

and to cutting-tool clearances will minimize the adverse effect of these characteristics. Additional considerations involve the relatively low modulus of elasticity of polymeric materials. This property permits the plastic to deflect as the tool exerts a force on it. The deflection and subsequent elastic recovery results in both dimensional inaccuracies and increased frictional drag on the tool. Proper support of the workpiece, minimization of the forces involved in the machining, and the use of tool geometries with sufficient clearances reduce these effects.

A large amount of data that provides starting condition recommendation for most plastics in the various machining operations is available (16).

The use of high-velocity water jets is primarily limited to cutting and drilling. Pressures of up to 50,000 psi are used to propel a fine stream of water at velocities greater than 2000 ft/sec (17). The impinging water jet literally wears the plastic away at a very rapid rate.

Similarly, polymer material can be removed by vaporization with the concentrated energy of a CO_2 laser. Although the depth of penetration can be easily controlled, the major practical applications for laser machining continue to be drilling and slitting. One limitation of laser machining is the conical shape of the focused beam, which tends to give tapered holes and cuts with tapered sides.

12.19 BONDING

It is frequently desirable to attach two or more plastic parts to form a finished assembly. Secondary fabrication techniques ranging from mechanical fastening to sophisticated ultrasonic welding methods are available for joining plastic components. The selection of the optimum joining technique depends on factors such as type of material, part configurations, desired production rate, and finished-part requirements.

Although not discussed here, the design of the joint is of great importance in virtually every bonding process. In general, joint designs that increase the surface area of the bond also increase the strength of the bond.

12.19.1   Mechanical Fastening

Both conventional and special screws, rivets, and other familiar fasteners are frequently used to join plastic parts. Special bosses are often molded into parts to accept self-threading screws, or the threads can be molded into the boss. Mechanical fastening can be used with all plastic materials and is frequently the best method of joining plastics to other materials such as metals. The major disadvantage of mechanical fastening is the high concentration of stresses around the fastener, which must be carefully considered.

12.19.2   Solvent Cementing

In the solvent cementing process, the surfaces of the two parts to be bonded are dissolved; the two surfaces are brought into contact; and a slight pressure is applied to the joint while the solvent evaporates. The pressure causes a slight flow and intermingling of the two softened surfaces which results in a cohesive bond. After complete evaporation of the solvent, the properties of the joint are very similar to those of a continous article.

The solvent cementing process is useful for highly soluble amorphous thermoplastics such as the styrenes, acrylics, and cellulosics. Dissimiliar plastics can be bonded by this technique, provided that they are both soluble and compatible.

A convenient method of selecting the proper solvent is to consult a solubility parameter table (18, 19) and match the parameter of the solvent with that of the polymer. Another consideration involves the boiling point of the solvent. Solvents with lower boiling points evaporate more quickly. While it is obviously desirable for rapid solvent evaporation to obtain short set times, extremely rapid evaporation often results in the formation of many tiny cracks (crazing) that weaken the bond. Blends of more than one solvent are, therefore, frequently used to obtain fast set times with minimum crazing.

A variation of the solvent cementing process involves the use of a dope or polymer solution in place of the pure solvent. This technique generally results in improved bond strength since crazing is reduced and the deposited solids fill the gaps of imperfectly mated parts. Solvents containing monomers that are polymerized in the joint can also be used.

12.19.3   Adhesive Bonding

This technique involves the application of a layer of adhesive material between the two surfaces to be bonded. The adhesive may be solvent based, water based, or a hot melt composition. With hot melts, the parts can be bonded immediately after the application of the adhesive or at some later time by reactivating the coating with heat. In all cases, the bond strength depends on the properties of the adhesive and frequently exceeds the strength of the substrates themselves.

Adhesive bonding can be used for all materials including thermosets, but materials such as polyethylene, polypropylene, and ethylene propylene diene rubbers usually require a special surface treatment prior to coating. Surface oxidation by flame treatment or exposure to corona discharge are common methods of preparing the surface of these materials for bonding.

The adhesive can be applied by many methods including brushes, squeeze bottles, trowels, printing techniques, rollers, spraying equipment, and by extrusion coating (see Section 12.2.1.6) in the case of hot melts.

Tables showing recommended adhesive types for various combinations of materials to be bonded are available (20) to aid in the selection of the proper adhesive for a particular application.

12.19.4   Melt-Phase Welding

All thermoplastics can be bonded by melting the surfaces of the two parts, bringing the surfaces into contact with each other, and applying a slight pressure while the joint solidifies by cooling. For thin film such as plastic bag material, the heat can be applied externally by pinching the film laminate with heated bars (heat sealing). For thicker parts, heat must be applied or generated at the actual surfaces to be bonded. The surfaces to be joined may simply be pressed against a hot plate until a molten polymer layer is formed and then pressed together until the joint solidifies. A process very similar to metal welding can also be used where additional material is supplied by means of a plastic "welding rod" which is melted into the joint by a hot gas jet or heated tool. Sufficient heat can also be generated by friction at the surface to be bonded. The most common friction welding technique is spin welding where one circular part is held stationary while the other is rotated. The rotating piece is pressed against the stationary part with sufficient force to produce the required frictional heat. Oscillatory or vibration joining techniques are also being used to friction weld noncircular parts. The use of high-frequency mechanical vibrations produced by an ultrasonic generator is a common method of obtaining very rapid welds. When ultrasonic welding is to be used, it is usually desirable to mold-in sharp (triangular cross-section) energy directors or concentrators in the joint area of one of the surfaces to facilitate melting. Certain materials that exhibit strong electric dipole moments, such as polyvinyl chloride and nylon, can also be welded by exposure to an alternating electric field (dielectric sealing). The polymer melts from the heat generated by the molecular agitation as the dipoles attempt to polarize with the alternating field.

12.20 DECORATING

It is frequently desirable to decorate the surface of a plastic part. The decoration may be for functional as well as purely aesthetic purposes. For example, external surface decoration can be used to improve abrasion resistance, electrical conductivity, and chemical and environmental resistance, as well as to mask molding imperfections such as splay marks. The most common of the many possible decorating techniques that can be used for plastic articles include painting, printing, vacuum metalizing, and electroplating.

12.20.1 Painting

Spray application of solvent-based lacquers that cure by evaporation continues to be the most popular method of painting plastic parts. Roller coating can also be used when painting relatively large flat surfaces. Special considerations when coating polymeric surfaces include surface preparation. All parts must be free of mold release agents, particularly silicone, and frequently must be cleaned with a detergent solution. Polyethylene and certain rubbers require a flame treatment or surface etch to obtain the proper adhesion. The solvent resistance of the plastic must be considered when selecting the solvent system for the paint. In the selection of the curing temperature, the distortion temperature of the plastic must also be considered.

Other painting techniques include the use of molded-in recesses such as letters or wood grain patterns. The entire surface is sprayed and then wiped to remove all paint except that which has filled the recesses.

12.20.2 Printing

Many methods are used to apply printed information and other designs to plastic products. Common to all of the processes is the need to obtain sufficient adhesion between the ink and the plastic. Some polymers, such as polyethylene, usually require a special surface treatment to obtain this adhesion. Again, a flame treatment or exposure to corona discharge can be used for this purpose. In many applications, such as blow molded bottles, the surfaces to be printed are not perfectly flat and are not necessarily rigid.

Screening, sometimes referred to as silk screen printing, is a common printing technique used for plastic parts. In this process the nonimage portion of a nylon, polyester, or wire mesh screen is made impervious to ink by the application of a stencil prepared by a photographic or mechanical process. A high-viscosity ink is then squeezed through the image portion of the screen and onto the plastic part. Multicolored decorating is accomplished by successive printing operations using a series of different screens.

In the gravure process, the pattern to be printed is engraved onto a cylinder, the cylinder is covered with ink, and the surface is wiped clean. The ink that remains in the etchings is then transferred to the product. This process is frequently used to transfer a wood grain pattern to a plastic panel.

Most other common printing techniques such as dry offset, flexography, and offset lithography are adaptable to plastic products provided that the transfer medium is sufficiently flexible to comply to the surface being printed.

12.20.3 Vacuum Metallizing

In the vacuum metallizing of plastic parts, a metallic coating is placed on the surface of the plastic for both functional and decorative purposes. Aluminum is the most frequently used metal in this process, but the use of many other metals and metallic salts is also possible. In the vacuum metallizing process, it is first necessary to prepare the plastic surface with a primer coating. This coating is usually a lacquer and is used to obtain adhesion, to smooth surface defects, and to reduce outgasing of the polymer during the subsequent vacuum step. After curing of the primer coating, the part is placed in a vacuum chamber, and the pressure is reduced to about 10^{-6} atmospheres. The metal is then vaporized on the tungsten filament and condenses on the plastic part. Depending on the application, a protective lacquer overcoat is sometimes used.

12.20.4 Electroplating

In general electroplating process, metal ions from a metallic salt solution are electrostatically attracted and subsequently deposited on the surface of an object by means of the electrolysis process (electrodeposition). This technique provides an alternate method of placing a metallic coating on certain polymeric articles. In comparison to vacuum metallizing, this process generally produces thicker coatings with improved adhesion and wear resistance. Although electroplating of many types of thermoplastic and thermoset materials is possible, acrylonitrile butadiene

sytrene materials are most frequently used. Many types of metals with both bright and satin finishes can be electroplated.

When applying the electroplating process to plastic materials, the ability to obtain satisfactory adhesion and to obtain electrical conductivity on the surface so that the object itself can act as an electrode in the electroylsis process are of primary concern. To those ends, a rather elaborate preplate operation is usually required. The plastic to be coated is first cleaned with a detergent solution and rinsed. The surface is then etched to remove the weakly bonded surface layer and to produce microscopic imperfections that improve adhesion. The part is again rinsed. The next step, neutralization, is used to reduce the troublesome hexavalent chromium contamination (present in etchant) to trivalent chromium. The part is again rinsed and then dipped in a catalyst that is usually an acid, tin-palladium solution, and then rinsed again. An accelerator is then applied to expose initiator sites for the subsequent chemical metal deposition (20).

The actual plating process begins with an electrolysis deposition of nickel, which renders the part conductive. The thickness and uniformity of this coating is then improved with an electrolytic strike. A relatively thick (0.5 mil) coating of ductile copper that levels and brightens the surface and acts as a cushion between the plastic and the subsequent coating is then generally electrodeposited. The copper plating is then followed by the deposition of the desired final surface.

12.21 SUMMARY

The basic principles of a few primary and secondary plastic fabrication techniques have been discussed.

Extrusion, which forms the basis of many methods, is used primarily for producing continuous lengths of thermoplastic sheet, film, rod, pipe, and other profiles. The extruder melts, compresses, mixes, and pumps the plastic material to the forming operation.

Synthetic fibers intended for a subsequent weaving operation are produced by the fiber spinning process. A polymeric melt or solution is forced through the tiny holes of the spinnerette, solidified, stretched, and formed into a thread.

Injection molding is the process of melting plastic, injecting the molten plastic under high pressure into a closed mold and solidifying the part in the mold. While major applications involve thermoplastics, the injection molding process can also be used for thermosets. The major attribute of injection molding is the ability to automatically produce precision parts.

Reaction injection molding (RIM) involves the thorough mixing of two reactive liquids and the injection of the mixture into a closed mold where a chemical reaction cures the part. This process most frequently utilizes urethane materials and is best suited for large, thick-walled parts.

Discrete plastic parts can also be produced by compression molding. Compression molding is the process of placing polymeric material between two open, heated matched-metal mold halves and closing the mold. In the case of thermosets, the most frequently used class of materials, the mold is held closed while additional heating occurs and curing takes place. If thermoplastics are used, the entire mold must be cooled to solidify the part for ejection. Compression molding produces relatively stress-free parts.

In transfer molding, a measured, compacted, preheated charge of thermosetting material is loaded into a hot, open, chamber (pot) in the mold. After additional heating, the material is transferred under pressure through a sprue, runner, and gate into a closed cavity. Although the process produces parts with molecular orientation in the direction of flow, transfer molding is generally faster than compression molding.

Blow molding and rotational molding are two techniques for producing hollow objects. In blow molding, a molten, hollow, thermoplastic preform (parison) is inflated inside a closed mold. Rotational molding is the process of distributing a layer of polymeric powder inside a closed mold and fusing the powder. Products from these processes range

from very small bottles to very large storage tanks.

Thermoforming is the process of softening thermoplastic sheet with heat and forcing the sheet, by one of many techniques, to conform to the surface of a mold where it is cooled. Virtually any thermoplastic material that can be extruded as sheet can be successfully thermoformed. Thermoformed products range from margarine tubs to recreational vehicle bodies.

Solid-phase forming covers a variety of plastic processing techniques that are charterized by the shaping of polymers at temperatures below the glass transition of the material. While the processes, such as forging and stamping, are not in widespread use today, perhaps these techniques offer substantial opportunities for the future, primarily because of the potential for very high-speed operations.

Certain film and sheet products can be produced by the calendering or solution casting processes. Calendering squeezes a polymer melt (usually PVC) in successive roll nips to final thickness. In solution casting, a polymer solution is spread onto a smooth surface, and the solvent is evaporated.

Slabs and other shapes of materials such as poly(methyl methacrylate) are produced by casting a catalyzed monomer into the appropriate shape and polymerizing.

Foamed plastic products can be produced in slightly modified extrusion, compression molding, injection molding, or reaction injection molding equipment.

Plastisol coatings can be obtained by dipping or spraying a dispersion of PVC particles in plasticizer and fusing with heat. Discrete parts such as surgical gloves can be produced by dipping an object in a plastisol and stripping the fused coating from the form. Solid PVC parts can also be produced by casting a plastisol into a mold and heating.

Radiation processing covers a diverse group of techniques used to cause desired changes in polymer materials by exposure to radiation. These changes are usually obtained by causing a chemical reaction such as polymerization, cross-linking or vulcanization. Applications of radiation processing include: the cross-linking of plastic pipe, tubing, and the polymer coating on wire and cable to improve properties such as heat, stress crack, and abrasion resistance; the rapid curing of organic coatings without solvent emissions; the polymerization of residual monomer in food packaging applications; graft polymerizations to impart selected properties to products such as textiles; and the vulcanization of natural and certain synthetic rubbers.

With special precautions, most of the common metal machining techniques can be applied to the fabrication of plastic parts.

Methods of joining two or more plastic parts include mechanical fastening, solvent cementing, adhesive bonding, and melt-phase welding. Both functional and decorative coatings can be applied to plastic parts by painting, printing, vacuum metalizing, or electroplating techniques.

This discussion considered only the more major plastic fabrication processes in use today. Many variations of each and combinations of two or more of these processes are also in use. Each process has advantages and disadvantages, and the selection of any one technique for a particular application is usually a compromise. A thorough understanding of each process, of the material to be used, and of the requirements of the finished product is essential for an optimum selection of a process.

One can expect the near future to bring continued refinements to these basic processes. The trend toward larger equipment will continue and will be spurred by the desire to produce lower-weight automobiles. Associated with ever-increasing labor costs and concerns for employee occupational health and safety, the plastics processing industry will continue to move toward complete automation. The application of robotics, for example, should become a common situation in even smaller plastics plants. We can expect to see equipment with increased versatility, but at the same time, the trend toward specialized equipment to optimize the process for certain applications will continue. A major concern in the coming years will be energy consumption. Machine manufacturers will undoubtedly

work toward equipment with increased energy efficiency. Perhaps the most significant near-term development will be the widespread use of improved control systems. Increasing demands on plastic products, coupled with recent developments in hydraulic valves, various transducers, microprocessor technology, and electronics in general have set the stage for a quantum leap forward in the plastics industry. The ability to measure actual process variables, rather than depending on machine parameters, and to eventually provide closed-loop feed-back control of the critical variables will offer the plastic processor new opportunities.

For the longer term, new processors featuring sharply higher production rates are required for continued improvement in productivity. Most of the current plastic processing techniques are limited by heat transfer constraints. In melt type thermoplastic processes, heat must first be added to melt the material and then it must be removed from the product; in most thermosetting processes, heat must be added to the material. Polymeric materials are inherently poor conductors of heat. Processes that would eliminate this dependence on heat flow would be highly desirable. Continued development of solid-phase forming techniques and of reactive materials, such as those used in reaction injection molding, but with an order of magnitude shorter cure times than those of current materials, are two of the possible approaches to this goal.

12.22   REFERENCES

1. A. L. Griff, *Plastics Extrusion Technology*, Reinhold New York, 1962.

2. C. I. Chung, *Plast. Eng.*, **33**(2), 34 (1977).

3. W. J. Schrenk, *Plast. Eng.*, **30**(3), 65 (1974).

4. I. Klein, *Plast. Des. Process.*, 20 October, (1973).

5. Z. Tadmor and I. Klein, *Engineering Principles of Plasticating Extrusion*, Van Nostrand Reinhold, New York, 1970.

6. J. M. McKelvey, *Polymer Processing*, John Wiley, New York, 1962.

7. I. Klein, *Plast. Des. Process.*, 22 September (1973).

8. R. J. Nichols and G. A. Kruder, Technical Papers,

20, SPE Annual Technical Conference, San Fransico, 1974.

9. J. M. McKelvey and S. Steingiser, *Plast. Eng.*, **34**(6) 45 (1978).

10. W. G. Frizelle, *Plast. Technol.*, 42, (June 1972).

11. E. G. Crosby and S. N. Kochis, *Practical Guide to Plastics Applications*, Cahners, Boston, 1972, p. 76.

12. A. Prasad, *Plast. Eng.*, **30**(6) 36 (1974).

13. J. F. Stevenson, R. A. Hauptfleisch, and C. A. Hieber, *Plast. Eng.*, **32**(12), 34 (1976).

14. E. O. Allen and D. A. Van Putte, *Plast. Eng.*, **30**(7), 37 (1974).

15. R. S. Papoojian and R. Lynn, *Mod. Plast.*, (November 1973).

16. *Machining Data Handbook*, 2nd ed., Machinability Data Center, Metcut Research Associates, Cincinnati, Ohio, 1972.

17. *Nontraditional Machning Guide, 26 Newcomers for Production*, MDC 76-101, Machinability Data Center, Cincinnati, Ohio.

18. J. H. Hildebrand and R. L. Scott, *The Solubility of Non-Electrolytes*, Reinhold, New York, 1950.

19. H. Burrell, "Solubility Parameters," *Interchem. Rev.*, **14**(3), 31 (1955).

20. *Modern Plastics Encyclopedia*, Vol. 54, No. 10A, McGraw-Hill, New York, (1977–1978).

General References

J.R.A. Pearson, *Mechanical Principles of Melt Processing*, Pergamon Press, Oxford, 1961.

F. Rodriguez, *Principles of Polymer Systems*, McGraw Hill, New York, 1970, Chapter 12.

I. I. Rubin, *Injection Molding Theory and Practice*, John Wiley, New York, 1972.

## CHAPTER 13

Additives

K. EISE

J. HARRIS

Werner and Pfleiderer Corporation
Ramsey, New Jersey

13.1   Introduction   613

13.2   Products   621
    13.2.1   Processing Agents (Catalysts and Initiators)   621
    13.2.2   Property Modification Agents   621
        13.2.2.1   Antistatic Agents   621
        13.2.2.2   Blowing Agents   621
        13.2.2.3   Colorants   622
        13.2.2.4   Flame Retardants   623
        13.2.2.5   Impact Modifiers   624
        13.2.2.6   Lubricants   624
        13.2.2.7   Plasticizers   625
        13.2.2.8   Other Additives   626
    13.2.3   Stabilization Agents   626
        13.2.3.1   Antioxidants   626
        13.2.3.2   Heat Stabilizers   627
        13.2.3.3   Ultraviolet Light Stabilizers   628
    13.2.4   Fillers and Reinforcements   628
        13.2.4.1   Fillers   628
        13.2.4.2   Coupling Agents   636
        13.2.4.3   Fiber Reinforcements   637

13.3   Compounding Methods (Equipment and Techniques)   645
    13.3.1   Direct Addition   646
    13.3.2   Multipurpose Concentrates   646
    13.3.3   How Much Concentrate?   647
    13.3.4   Conventional Versus Sophisticated Compounding Equipment   647
    13.3.5   Melt Conveying   647
    13.3.6   Commercial Twin-Screw Systems   647
    13.3.7   Case Histories   648
        13.3.7.1   Incorporation of Slip-Anti-block-antioxidant Concentrations   648
        13.3.7.2   Internal Lubricant Concentrates   649
        13.3.7.3   Compounding of Flame Retardants   649

13.4   Summary   650

13.5   References   651

13.1   INTRODUCTION

Additives are used for a variety of purposes in almost all polymers, ranging from aids in processing to modifying end products for improved properties, including impact strength, ultraviolet stability, mobility, and others.

The major additives and their uses are discussed here in some detail. Table 13.1 is a list of types with their use, concentrations, properties and characteristics. Included are additives to reduce or eliminate the various types of degradation (thermal, mechanical, chemical, environmental) that occur in polymer resins. These additives are normally used in very small amounts in the final product (usually in 100:1000 ppm ratio), and a uniform distribution in the polymer matrix is a prerequisite to achieve a well stabilized product.

The availability of improved processing equipment has had a major influence on the types and levels of additives used today. The improved dispersion obtained with multiscrew extruders has allowed a reduction in the

Table 13.1 Additives: Use, Concentrations, Properties, and Characteristics

POLYMERIZATION

CATALYSTS AND INITIATORS

TYPES	CHARACTERISTICS AND EFFECTS	ALKYDS	EPOXY RESINS	MELAMINES	PHENOLICS	POLYESTERS	POLYURETHANES	UREAS	ACRYLICS	CELLULOSICS	FLUOROPLASTICS	NORYL®	NYLONS	POLYACETALS	POLYBUTYLENE	POLYCARBONATE	POLYESTERS	POLYETHYLENES	POLYIMIDES	PPS	POLYPROPYLENE	POLYSTYRENE	VINYLS
Amines	Initiate action of peroxides; speed up reaction of isocyanate and polyol in urethanes						✓																
Organic peroxides	Principal source of free radicals in polyester curing; wide choice to suit requirements					✓																	
Peroxy carbonates	Generate free radicals at low temperatures for fast cycles					✓																	
Peroxyesters	Used in continuous processes					✓																	

ADDITIVES FOR PLASTICS

ANTISTATIC AGENTS

TYPES	CHARACTERISTICS AND EFFECTS	ALKYDS	EPOXY RESINS	MELAMINES	PHENOLICS	POLYESTERS	POLYURETHANES	UREAS	ACRYLICS	CELLULOSICS	FLUOROPLASTICS	NORYL®	NYLONS	POLYACETALS	POLYBUTYLENE	POLYCARBONATE	POLYESTERS	POLYETHYLENES	POLYIMIDES	PPS	POLYPROPYLENE	POLYSTYRENE	VINYLS
Amines	Effective in films and molded parts; nonvolatile; nonoxidizing			✓					✓				✓		✓			✓			✓	✓	✓
Anionics	Hydrophilic acid esters and sodium salts; sanctioned for food contact					✓	✓		✓				✓		✓	✓		✓			✓	✓	✓
Quaternary ammonium compounds	Effective but low in heat stability; not sanctioned for food-contact use					✓	✓		✓	✓			✓		✓	✓		✓			✓	✓	✓

BLOWING AGENTS

Azobisformamide	Nonstaining; nondiscoloring; nontoxic; flame-inhibiting; heat - stabilizer activated
Hydrazides	Moderate decomposition-temperature range in plastics, 130 to 140°C
Sulfonyl semicarbazide	Intermediate decomposition-temperature range, 240 to 250°C in plastics
Trihydrazide triazine	Has highest decomposition range in plastics— 265 to 290°C
Physical foaming agents	Compressed gases or soluble liquids, often used in crosslinking to control temperature

COLORANTS

Inorganic pigments	Dense colors; stable, low plasticizer absorption; insoluble; chemical resistant
Organic pigments	Bright colors; fair to good opacity; some have good stability; insoluble; some migrate
Dyes	Bright, transparent colors; limited stability; migrate; soluble; poor chemical resistance
Concentrates	Qualities relate to basic colorants; must be made with compatible resins
Special pigments	Fluorescents, metallics, pearlescents used for special effects; vary in stability

FLAME RETARDANTS

Organic additives: phosphates, halogens	Halogenated hydrocarbons, phosphates; brominated compounds; chlorparaffins
Inorganic additives: aluminum, antimony, tin, zinc	Oxides, bromides, borates; widely used antimony oxide a synergist with halogens
Reactive types: halogen compounds	Functional groups incorporated into polymer by chemical reaction with intermediate

Table 13.1 (Continued)

ADDITIVES FOR PLASTICS (continued)

IMPACT MODIFIERS

TYPES	CHARACTERISTICS AND EFFECTS	ALKYDS	EPOXY RESINS	MELAMINES	PHENOLICS	POLYESTERS	POLYURETHANES	UREAS	ACRYLICS	CELLULOSICS	FLUOROPLASTICS	NORYL®	NYLONS	POLYACETALS	POLYBUTYLENE	POLYCARBONATE	POLYESTERS	POLYETHYLENES	POLYIMIDES	PPS	POLYPROPYLENE	POLYSTYRENE	VINYLS
ABS/MBS	Impart impact strength to rigid PVC; also act as processing aids																						✓
Chlorinated polyethylene	Compatible with PVC; imparts flexibility; nonextractable																						✓
Nitrile rubber	Strengthens epoxy and phenolic molding and encapsulating compounds		✓																				

LUBRICANTS

TYPES	CHARACTERISTICS AND EFFECTS	ALKYDS	EPOXY RESINS	MELAMINES	PHENOLICS	POLYESTERS	POLYURETHANES	UREAS	ACRYLICS	CELLULOSICS	FLUOROPLASTICS	NORYL®	NYLONS	POLYACETALS	POLYBUTYLENE	POLYCARBONATE	POLYESTERS	POLYETHYLENES	POLYIMIDES	PPS	POLYPROPYLENE	POLYSTYRENE	VINYLS
Bisamide synthetic waxes	Internal with some external effect; light colored; some sanctioned for food contact	✓	✓	✓	✓									✓	✓								✓
Synthetic ester wax	Low melting; little effect on clarity; prevent stress whitening in rigid PVC			✓	✓	✓		✓		✓			✓	✓	✓		✓				✓	✓	✓
Paraffin-base waxes	Low-molecular-weight waxlike external processing aid for all PVC				✓													✓				✓	✓
Stearic acid	Calendering and extrusion aid																						✓
Metal stearates	Internal lubricants; heat stabilizers for some resins		✓	✓	✓	✓							✓	✓		✓		✓				✓	✓
Silicone fluids	Typically dimethylpolysiloxane; improves wear resistance; built-in mold release		✓				✓						✓	✓		✓		✓		✓	✓		
Fluoropolymers	Principally comminuted PTFE or FEP; reduce frictional coefficients of molded parts					✓	✓						✓	✓				✓	✓	✓	✓	✓	
Molybdenum disulfide	Blue-gray powder; low-friction laminar structure; solid-film lubricant	✓	✓			✓	✓						✓		✓	✓		✓	✓	✓	✓	✓	✓

616

Adipates	Primary plasticizers; good low-temperature properties; resist water extraction
Azelates	Low volatility and water extraction; high compatibility; good at low temperatures
Epoxidized oils	Linseed, soya bean, tall, and other oil derivatives; low volatility and migration
Epoxies	Nonmigrating, permanent types; used with monomeric plasticizers
Glycollates	Some compatible with nitrocellulose only; others across-the-board
Mellitates	Low volatility, nonfogging; add to processability; compatible; water resistant
Nitrile rubbers	Flexible, nonmigratory plasticizer for PVC; toughens thermosets
Phosphates	Flame-retardant plasticizers; trialkyls good at low-temperatures, have high volatility
Phthalates	Workhorses for flexible PVC with good overall balance of properties
Polyesters	Sometimes called polymeric, resinous, or permanent type; fair at low temperatures
Sebacates	Noted for good low-temperature properties; suitable for food-contact use
Terephthalates	Claim low soapy-water extraction; low marring; good low-temperature flexibility
Extenders	Aliphatic, chlorinated, other hydrocarbons; limited compatibility; may impair properties

Table 13.1 (*Continued*)

ADDITIVES FOR PLASTICS (continued)

OTHER ADDITIVES

TYPES	CHARACTERISTICS AND EFFECTS	ALKYDS	EPOXY RESINS	MELAMINES	PHENOLICS	POLYESTERS	POLYURETHANES	UREAS	ACRYLICS	CELLULOSICS	FLUOROPLASTICS	NORYL®	NYLONS	POLYACETALS	POLYBUTYLENE	POLYCARBONATE	POLYESTERS	POLYETHYLENES	POLYIMIDES	PPS	POLYPROPYLENE	POLYSTYRENE	VINYLS
Adhesion promoters	Improve bond between organic polymers and inorganic fillers and reinforcements	✓	✓	✓	✓	✓	✓	✓			✓	✓	✓	✓	✓	✓	✓	✓	✓	✓	✓	✓	
Antiblocking agents	Prevent adhesion between layers of film under pressure												✓		✓			✓			✓		✓
Bacteriostats and fungicides	Inhibit growth of microorganisms									✓								✓					✓
Odor control agents	Mask or absorb odors inherent in some polymers and additives																					✓	✓

STABILIZATION AGENTS

ANTIOXIDANTS

TYPES	CHARACTERISTICS AND EFFECTS	ALKYDS	EPOXY RESINS	MELAMINES	PHENOLICS	POLYESTERS	POLYURETHANES	UREAS	ACRYLICS	CELLULOSICS	FLUOROPLASTICS	NORYL®	NYLONS	POLYACETALS	POLYBUTYLENE	POLYCARBONATE	POLYESTERS	POLYETHYLENES	POLYIMIDES	PPS	POLYPROPYLENE	POLYSTYRENE	VINYLS
Alkylated phenols and polyphenols	Nondiscoloring and nonstaining, low odor and color; some sanctioned for food contact												✓	✓	✓			✓			✓	✓	✓
Thio-and dithio-poly-alkylated phenols	Effective at high temperatures; light-yellow or white crystalline powder				✓					✓			✓	✓	✓			✓			✓	✓	✓
Diphenyl amines	Prevent oxidation of nylon at high temperatures; off-white powders						✓			✓			✓	✓	✓			✓			✓	✓	✓
Dipropionates	Generally good clarity; used in clear compounds; nonstaining; synergistic								✓				✓	✓	✓	✓		✓					
Organic phosphates and phosphites	Clear and nonstaining		✓			✓			✓	✓						✓						✓	✓
Hydroquinones	White or light-green crystals; nonstaining; used principally with unsaturated polyesters				✓													✓		✓			

HEAT STABILIZERS

Barium/cadmium/zinc salts	Good clarity and color control; plasticizer compatible; may sulfide stain; toxic
Calcium/zinc salts	Moderately effective in heat and color stability; suitable for food contact
Epoxides	Essentially plasticizers with heat and light stability effects
Lead salts	Effective but limited in use by color, toxicity, and sulfur staining
Organotins	High clarity; some contain sulfur, others nonstaining; food contact sanctioned
Phenols	Heat and light stabilizers; principal use as antioxidants
Phosphites	Serve as chelators with barium/cadmium in PVC; also used as antioxidants

ULTRAVIOLET LIGHT STABILIZERS

Benzophenones	Effective in thin films; impart little color; low in toxicity
Benzotriazoles	Low toxicity; improve gloss and surface crazing
Nickel organics	Chelators; good color; low-volatile process stabilizers (antioxidants) in polyolefins
Pigments	Absorb or reflect UV radiation at the part surface; applies notably to carbon blacks
Salicylates	Excellent compatibility; have high energy absorption before color changes occur
Substituted acrylonitriles	Heat resistant; unaffected by heavy metals; resist oxidizing and reducing agents

Table 13.1 (Continued)

FILLERS, COUPLING AGENTS, AND FIBER REINFORCEMENTS

TYPES	CHARACTERISTICS AND EFFECTS	ALKYDS	EPOXY RESINS	MELAMINES	PHENOLICS	POLYESTERS	POLYURETHANES	UREAS	ACRYLICS	CELLULOSICS	FLUOROPLASTICS	NORYL ®	NYLONS	POLYACETALS	POLYBUTYLENE	POLYCARBONATE	POLYESTERS	POLYETHYLENES	POLYIMIDES	PPS	POLYPROPYLENE	POLYSTYRENE	VINYLS
Asbestos	Imparts strength, heat resistance, hardness; use being restricted by OSHA regulations	✓	✓	✓	✓	✓		✓					✓		✓		✓		✓		✓		✓
Boron	High-cost, lightweight fibers; used in high-performance aerospace parts		✓	✓	✓	✓											✓		✓				
Carbon and graphite	High-modulus fibers used in aerospace; lower cost may open other markets	✓	✓	✓													✓		✓	✓			
Glass	Used as rovings, chopped strands, milled fibers, mats; good chemical resistance	✓	✓	✓	✓	✓	✓	✓			✓	✓	✓	✓	✓	✓	✓	✓	✓		✓	✓	✓
Metal-oxide fibers	Hydrated alumina, magnesium and zirconium oxides make strong composites		✓	✓	✓										✓		✓				✓		
Natural fibers	Shredded cellulose and talc add impact strength to thermosets and thermoplastics	✓		✓	✓	✓		✓									✓				✓		
Synthetic fibers	Various man-made fibers strengthen laminates and molding compounds		✓	✓	✓	✓	✓										✓		✓				
Silicas and silicates	Control part shrinkage; improve dielectric properties, heat and moisture resistance		✓	✓	✓	✓	✓	✓					✓		✓		✓	✓	✓	✓			✓
Calcium carbonate	Improves physicals and extrusion of PVC; low marring; controls plastisol viscosity				✓	✓									✓		✓	✓			✓		✓
Metal oxides	Impart weather resistance, stiffness, hardness; some improve thermal conductivity		✓			✓	✓ ✓										✓	✓		✓			
Barium sulfate	Increases density and load-bearing strength; imparts drape and hand				✓		✓								✓								✓
Carbon black	Used in semi-conductive compounds; adds hardness, heat resistance; may embrittle				✓		✓								✓			✓	✓	✓	✓		✓
Cellulosics	Wood and shell flour reduce mold shrinkage, improve impact, flow, and gloss			✓		✓		✓								✓							
Wollastonite	Nonmetallic calcium metasilicate. Can increase flexural strength, tensile strength and modulus, improve dimensional stability	✓	✓	✓	✓	✓										✓	✓	✓			✓		✓

amount of additives required. This reduces additive costs and in many cases results in improved physical and processing properties.

Additives are usually referred to by function rather than by chemical composition. They are dispersed in a polymer matrix without changing the molecular structure, except in cross-linking formulations and as catalysts.

13.2 PRODUCTS

13.2.1 Processing Agents (Catalysts and Initiators)

Certain additives are used as catalysts or initiators both in the production of primary resins such as polyvinylchloride (PVC) and the polyolefins and in molding and in the processing of polyesters and polyurethanes. In urethane processing, amines are used to speed up the reaction of isocyanate and polyol and to initiate action of peroxides, especially in injection molding where fast cycles are required.

In polyesters, organic peroxides are widely used reaction initiators, and while they are sometimes termed catalysts, they do undergo change activated by amines. Their free radicals initiate the polymerization between the unsaturated polyester and a reactive monomer.

Certain factors must be taken into consideration in choosing a peroxide or perester for curing an unsaturated polyester. Among these are the permissible curing temperature, the processing system, final product properties desired, and the reaction of the various components of the resin system. Temperature control is crucial when using peroxides, as the most commonly available peroxides melt at relatively low temperatures. This can affect the molecular weight of the polyester. One way to overcome this difficulty is to use a combination of peroxides, which will also allow for lower molding temperatures and minimum thermal degradation of heat-sensitive additives and the polymer itself. Proprietary formulations are available that provide the same advantages of combinations of fast and slow initiators.

13.2.2 Property Modification Agents

13.2.2.1 Antistatic Agents

Unless they are protected, plastics can accumulate electrostatic charges at every step of their lives, in processing and during transit, storage, and at the consumer's. Dust attracted to the surface can create electrical hazards, even electrical shocks.

Use of either internal or external antistatic agents can dissipate the static charges by increasing the surface conductivity of products. Since the external agents are subject to wear and exposure, internal agents provide longer-term protection than do external agents. Because of their shorter-term durability, external agents require greater concentrations than internal agents. The external agents are usually applied by dipping or spraying and in this respect are not true additives. Internal agents, on the other hand, are compounded with the polymer. Although they must be compatible with the polymer, there must be a certain amount of incompatibility in order to migrate to the surface. On the surface, they attract moisture from the atmosphere to provide the surface conductivity. In contrast to the external agent, if any of the internal agent is wiped off, it is replaced from within the polymer mass.

Amines, anionics, and quaternary ammonium compounds are the principal antistatic agents for plastics, many of a proprietary nature. Selection of antistatic agents depends upon application, durability, effectiveness at low humidity, and the need for federal approval for food packaging materials. Polyolefins, nylon, polyesters, urethanes, cellulosics, acrylics, and acrylonitriles are particularly sensitive to static buildup.

13.2.2.2 Blowing Agents

Blowing agents are used to create a cellular structure in structural foam moldings and extrusions of acrylonitrile-butadiene-styrene (ABS), PVC, and impact polystyrene. They may also be used in making flexible sheeting and coated fabrics of PVC. While they may also be used in making polyurethane elastomers and foams, foaming in many urethane formulations is from carbon dioxide gas generated by the basic chemical reaction producing the polyurethane.

Cellular plastics may be produced by injecting air or nitrogen into the melt stream mechanically or by adding a solid or chemical blowing agent to generate the desired foaming gas when heated during melt processing. Blowing agents can be added by tumble mixing, by masterbatching, or by compounding. Direct metering into the machine hopper with a proportioning feed device may be used for the liquid type blowing agents. The amount varies with the application and the concentration is usually 2% or smaller. The physical properties of the product, the desired density, and whether it is rigid or flexible determine the concentration of blowing agent to be used.

Blowing agents are selected according to their gas yield, the melt viscosity and processing temperature of the resin, and the temperature at which the agent decomposes. It is necessary to correlate the temperatures of the agent and resin so that the agent will decompose when the melt is at the proper viscosity.

Azodicarbonamides, azobisformamide, and hydrazides and compounds that liberate nitrogen or carbon dioxide when heated form the basis of chemical blowing agents.

13.2.2.3 Colorants

Colorants are added to resins to impart a variety of end-product aesthetic properties. These include hue, opacity, transparency, fluorescence, iridescence, and pearlescence. These types of additives may be dyes, organic or inorganic pigments, and some specialty compounds.

Dyes are usually soluble, although not always when reacting with some resins. They are small-size particles and are used primarily to give brilliant transparent color to clear plastics. They many, however, be added to pigments to change their tonal value. The fine particle size of dyes enables them to be used in extrusions as they can reduce plugging of screen packs. Dyes have the advantages of migrating easily and readily bonding to plasticizers, but their disadvantages include poor resistance to chemicals and to processing

temperatures and a limited stability to light. They also have a tendency to bleed.

Pigments are large particles that can scatter light by diffraction and refraction. Organic pigments are made from dyes that have been made insoluble. They remain solid particles even after being applied and remain as suspensions in polymer matrices. Some of the organic pigments are azo pigments, with the disazos superior to monazos in heat stability, light fastness, and bleed resistance. Others are phthalocyanines, quinacridones, indolines, and perylenes. They differ widely in performance, and an evaluation of requirements should be made by the processor before making a choice. One of the most widely used is carbon black.

Inorganic pigments consist of salts and oxides of metal. The colors are not so brilliant as with dyes and organic pigments, but they have fair tinting strength and good hiding powers. Inorganic pigments tend to resist high processing temperatures, but they are readily dispersible. Widely used as an inorganic pigment is titanium, which has good whiteness and opacity. For yellow to red and tan, iron oxides provide good hiding properties and fair heat stability, except for the yellow. Brighter reds and yellows may be obtained with cadmiums, which, however, are more expensive.

Other metal oxides, chromates, molybdates, sulfates, sulfides, and salts of metals such as lead, zinc, mercury, strontium, and barium are also used as inorganic pigments. Metal flakes used in pigments include aluminum, copper, and bronze. The metal flakes are often used in the speciality colorants when mixed with other pigments such as pearlescent pigments based on bismuth, lead compounds, or fluorescent pigments. The combinations provide excellent hiding power and tinting strength. Iridescence is produced by contrasts between colorant and metallic hues. Pearlescent pigments are predispersed in polymers, whereas fluorescent pigments are generally dispersed in coatings and applied over a white base.

Pigments may be applied dry, as compounds, as color concentrates, or in paste form. The dry colors are particulate blends that contain no resin and may be used by molders and extruders in a dry-blending process wherein the resin pellets are tumbled with the pigment. While this method does not require so much labor, it is not so effective in color utilization because it depends on surface adhesion by the roughness of the pellets or by static charges generated by the process.

Compounds are pellets blended of pigment and resin by custom compounders. The total hot compounding of resins with colorants to produce precolored compounds is used where the ultimate in color uniformity is required. It is mainly used by small processors because it requires large inventories of different colored compounds, whether they are made in-house or purchased. The double hot processing, however, can be undesirable for the heat-sensitive materials and increases the amount of energy required.

Color concentrates contain higher loadings than the compounds and are let down or mixed with uncolored resin. The pellets are easy to weigh, and they have good dispersion capabilities because of the mixing that they have already undergone. Color concentrates are widely used by thermoplastic processors. The color concentrate is added usually by means of a metering device in proportion to the resin pellets. Several basic color concentrates or a combination of several colorants in a special concentrate may be used to achieve color matches in the final product. The amount of concentrate used can run as high as 5% of the total product, but is usually 1 or 2%.

Processors of vinyl or liquid thermosets are wide users of paste concentrates. These are dispersed in vinyl plasticizers or directly into the resins in the case of thermosets, such as polyesters, epoxies, and urethanes.

Proper loading of colorants is important for maximum effect. The viscosity loading depends upon the mixing equipment, and the viscosities, in turn, depend upon the oil absorption properties of the resin. Shear and dispersion during the process are determined by the range of loadings of pigment and resin. Temperature is an important processing condition in the use of colorants. Some are more heat sensitive than others, and the choice is

critical for resins that are processed at temperatures of 500° F—for example, the polycarbonates and nylons. Improvements in processing and final product can be effected by the use of other additives.

The use of liquid colorants for injection molding or extrusion can overcome the difficulties encountered with the double heating that is often required. Here, the pigments in a liquid can be pumped and metered into the natural resin pellets as they flow through the hopper into the feed channel of the extruder screw.

A saving can be obtained by the user of concentrates, either solid or liquid, by purchasing bulk quantities of natural resin and using bulk storage and handling instead of more costly manual operations.

13.2.2.4 Flame Retardants

The burning or decomposition of plastics in a fire can cause the generation of smoke, toxic fumes, and depletion of oxygen in an enclosure. This is, of course, true of most organic materials, but the characteristics of plastics in such situations has led to government regulations, particularly at the federal level with the passage of the Toxic Substances Control Act (TOSCA). This act, with particular emphasis on toxic substances, has led to a closer look at flame retardants and to the increased use of multiconcentrates, to make products of plastics more resistant to flames and to avoid long periods of waiting for government decisions.

The choice of a flame retardant for any given application depends upon the polymer, the effectiveness, the cost, and the ease of compounding. Another consideration is that flame-retardant additives cause deterioration of some polymer properties such as impact strength, tensile strength, and heat and light stability.

Flame retardants are of two types: additive and reactive. Phosphate ester, which is also used as a plasticizer for PVC, and chlorinated and brominated compounds are typical of the organic additives. Antimony oxide and hydrated alumina are among the inorganic additives. The combination of antimony compounds with chlorine-containing additives such as chlorinated paraffins produces the most effective synergistic flame retardance. The synergistic effect can also be increased by the use of halogens and brominated compounds combined with organic additives such as phosphate plasticizers. The use of halogens can lower the quantity of antimony oxide required when availability is limited and price is subject to fluctuation.

Reactive flame retardants such as chlorinated, brominated, or phosphated derivatives of the normally used polymer are substituted for all or part of one of the components of a polymer system. They are incorporated by chemical reaction and when carried into the polymer chain provide a high degree of flame retardancy.

The action of the flame-retardant additives is effective either by insulation in the case of gases; absorption of heat energy by reaction; coating by excluding oxygen, as in the case of phosphate esters; and interruption of the combustion process. There will be constant effort to improve the performance and reduce the hazards of flammability.

13.2.2.5 Impact Modifiers

Low-impact strength is an inhibiting factor in some polymers that might otherwise be used for many applications. PVC is one such material. Its use for many applications is complicated by the requirements for both impact strength and rigidity. Addition of a plasticizer will improve the impact strength but will reduce the rigidity. Such applications as pipe and fittings, electrical conduit, and bottles require both properties.

ABS, methyl methacrylate-butadiene-styrene (MBS), and nitrile rubbers are among the effective impact modifiers. ABS has proved to be suitable for use with PVC, but it must be of a low-modulus tensile strength to impart impact strength. ABS must also be such that it is not totally soluble in PVC, so that it will act as a reinforcement rather than a plasticizer. Its rubber particles act as shock absorbers against tendencies to fracture in the PVC. ABS polymers are also effective as processing aids. Clear, high-impact PVC compounds can also be obtained with the use of MBS polymers.

Reactive nitrile rubber can be used as a reinforcing agent for thermosetting resins, even when reinforced with glass, to overcome tendencies to crack under impact or because of thermal shock. Durability and structural strength are not improved, however. Crack resistance of glass-fiber reinforced polyesters as well as the bursting strength in pipes and tanks can also be improved by the use of nitrile rubber.

13.2.2.6 Lubricants

In many instances, engineering plastics are used in parts with very complex shapes. This sometimes results in high reject rates during the molding operation because of difficulties with mold ejection. For these applications, silicone fluid concentrates are available that improve material flow and act as mold release agents. The lubricity of the silicone eliminates part hangup and allows complete filling of cavities at lower pressure than would otherwise be possible (Figure 13.1).

Lubricants aid in the processing of many plastics and improve the appearance of end products. Such lubricants must be compatible with the resin being processed, and they must not adversely affect the end-product properties. The compatibility of the lubricant with the matrix depends upon melting point, molecular structure, and the degree of blending with the resin. The proper amount of lubrication must be taken into consideration—too much can cause excess slippage and lower output, while not enough lubrication can cause degradation and higher melt viscosities.

Lubricants are either internal or external, although some can be both. Most lubricants are used as processing aids, but some are used to reduce the coefficient of friction of finished articles, and these are commonly called slip agents. The internal lubricant must be compatible with the polymer to reduce the forces that make it difficult to process. External lubricants, on the other hand, must be able to migrate to the surface during heating in order to reduce the friction between the polymer and the heated surfaces of the processing machines, such as the barrel walls. Excess external lubrication can lead to slippage of the resin during processing.

EJECTOR PINS

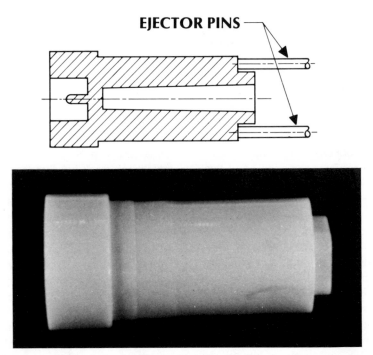

Figure 13.1 Freezer lock component molded with silicone/polycarbonate concentrate.

In addition to the silicone concentrates already mentioned, other lubricants are fatty acids, fatty acid amides, and esters, metallic stearates, hydrocarbon waxes, low molecular weight polyethylenes, and fluoropolymers.

Although some lubricants are used with many plastics, others have limited use. For ABS compounds, metal stearates with glycerol monostearate are required, whereas fatty acid amines are used for styrene acrylonitrile (SAN). Only small amounts of lubricants are needed for polyolefins, which are more or less self-lubricating. The heat sensitivity of rigid PVC requires adequate lubrication to prevent sticking. Most of the various lubricants can be used for PVC, and very often in combinations. Lubrication is also needed to prevent degradation of the resin.

The choice of lubricants for plastics that come in contact with food is subject to federal regulation, and this must also be taken into consideration in any process.

Slip agents, referred to briefly in preceding material, may be some of the same additives as lubricants. The surface slip agent allows like materials to slide against one another and reduces friction, wear, and operating temperatures of bearings and gears. It also increases the pressure-velocity limit of most thermoplastics. For these effects, silicone fluids are used for high-impact polystyrene, ABS, acetals, and nylons; molybdenum disulfide and fluoropolymers such as tetrafluoroethylene (TFE) and fluoroethylenepropylene (FEP) are used in many engineering thermoplastics and thermosets. They can improve existing good wear qualities in the basic resins.

13.2.2.7 Plasticizers

To impart such properties as flexibility, resilience, and softness to polymers, a range of materials that may be liquid or solid and that are chemically and thermally stable are added. These additives are called plasticizers, primarily because the molecules of the additive interspace between those of the resin to make the combination more flexible than the resin itself. The plasticizers must be compatible with the resin and must remain in the compound.

When the level of compatibility is high, the additive is a primary plasticizer; when only a small concentration can be used because of a low level of compatibility, the additive is a secondary plasticizer. Use of the secondary types with the primary plasticizers can impart specific properties and can reduce the overall compound cost.

PVC compounds are the largest users of plasticizers, accounting for about 80% of total consumption. A high proportion of the plasticizers are the phthalic acid derivatives (dioctyl and diisooctyl). Their qualities are good color, low toxicity and odor, good electrical properties, and fairly good low-temperature properties, as well as low volatility. Even better low-temperature properties are possible with aliphatic diesters such as adipates, azelates, and sebacates, when required. These materials replace only a part of the dioctyl, as their compatibility is not so high and their cost is greater.

Also used as plasticizers are epoxide materials derived from soybean and linseed oils. These are also heat and light stabilizers, which generally have low volatility and low oil and water extraction.

For long-term performance, nitrile rubber has been used as a plasticizer for PVC, although it is difficult to process the compounds.

Secondary plasticizers are used with primary plasticizers with cost reduction a major factor in their use. Compatibility is a primary concern, as is heat and light stability, which for most is somewhat limited.

13.2.2.8 Other Additives

A number of other materials can be combined with plastics to alter their processing or performance characteristics in various ways. Among these are adhesion promoters, which are used to improve the bond between organic polymers and inorganic fillers or reinforcements. Without the bond, failure of the composite can occur through moisture penetration. Monomeric silicon chemicals or silanes are of this type. Various specialty resins and synthetic and natural waxes are also used to promote adhesion, particularly in laminations.

To keep adjacent layers of thin films or nested molded parts from sticking together, waxes or finely divided powders added to resin compounds before processing can be used. These are antiblocking agents. They must be incompatible with the polymer at room temperature so that they migrate to the surface of the product. They permit easy separation of a single sheet of film or a bag from a stack, but unlike slip agents, they do not necessarily affect the coefficient of friction.

Plastics, especially plasticized PVC, are subject to attack by microorganisms and deterioration. In PVC, plasticizers are the subject of attack, along with other additives, while the resin itself is fairly immune. Shrinkage, staining, and embrittlement, as well as other physical property changes can occur. Shower curtains are particularly vulnerable because of moisture and soap residues. To combat this type of deterioration, bacteriostats and fungicides are used.

Another type of additive is used to control odors inherent in plastics compounds, from the resin or from the additives. Odors can also develop in the shipping container, and although they may sometimes dissipate after exposure to the atmosphere, odor-masking or absorbing modifiers may be added to the product to make the odor less offensive to the consumer.

13.2.3 Stabilization Agents

Additives are used to reduce or eliminate the various types of degradation of the basic polymer matrix, or resin, that occur either during processing or after completion, either environmentally or in action with other products. These additives are called stabilization agents. Some of the more common types are listed here.

13.2.3.1 Antioxidants

The antioxidants are substances that retard or inhibit oxidative degradation at normal or elevated temperatures. The type of antioxidant and the amount of concentration are dictated by the properties of the resin and end-product requirements.

The oxidation of polymers causes chain scission or cross-linking. Scission is destruction of the polymer chain and causes loss of molecular weight, increased melt flow, decreased toughness, and disintegration into powder. Cross-linking, on the other hand, increases the molecular weight, decreases the melt flow, and causes embrittlement.

To overcome this type of degradation, hindered phenols and amines or primary antioxidants and peroxide decomposers (or synergists such as thioesters and phosphates) are used. Synergists are used with primary antioxidants to aid stability of the polymer.

The polymers most sensitive to oxidation are high-density polyethylene, ABS, polypropylene, and high-impact polystyrene. Low-density polyethylene and crystalline polystyrene require little stabilization.

Because of high processing temperatures, higher molecular weight phenols are used for processing polypropylene. Again, the properties of the resin and the processing steps, as well as the end-product use, must be taken into consideration in the choice of an antioxidant additive. In all additive processing it is important to satisfy government toxicity requirements, especially for products in contact with food. Both the hindered phenols and thioesters satisfy these requirements.

13.2.3.2 Heat Stabilizers

PVC and vinyl chloride copolymers are particularly subject to degradation at the high temperatures of melt processing. ABS, SAN, polystyrene, chlorinated polypropylene, the polyolefins, and polyurethane are also vulnerable to heat, but not so much as PVC.

The thermal degradation of PVC is not prevented by heat stabilizers, but the action is delayed, and at the same time they retard the color development that is a measure of the polymer's degradation. Stabilizers prevent such changes as embrittlement and loss of electrical properties in vinyl compounds, and some heat stabilizers are also effective as light stabilizers.

Stabilizers are metal and acid compounds. Barium, cadmium, calcium, lead, magnesium, tin, and zinc metals are combined with various organic or inorganic acids such as lauric, maleic, naphthenic, octoic, ricinoleic, stearic, phosphorous, silicic, and sulfuric. Two or more metals are usually combined, as, for example, barium-cadmium-zinc or

calcium-zinc, to provide optimum performance. Often combined with other stabilizers, the phosphites are used principally as chelators in vinyl stabilization.

Most widely used are the barium-cadmium stabilizers, with or without zinc. Although the combination of the three produces better initial color stability, it often is at the expense of long-term stability and may lead to more rapid total blackening. The other additives in the compound and their possible interactions must be taken into consideration by the processor in selecting a barium-cadmium stabilizer for a particular compound. Important considerations are lubricity and effect on the processing equipment. Other factors to be considered are the cost and effectiveness of liquid compared with solid stabilizer systems.

The organotin compounds make up another group of stabilizers, among which are butyl-, methyl-, and octyl-tin mercaptides, sulfur-free tin maleates, dibutyl tin carboxylates, maleates, and dilaurates, with many variations.

Various factors must be weighed in the choice of tin mercaptides, which are among the most effective heat stabilizers. They are likely to have a strong sulfur odor and inferior light stability. These disadvantages are overcome with mercaptide-free tin stabilizers, but the heat stability is inferior to the mercaptides. The FDA has approved some octyl-tin stabilizers for use in food packaging, with the condition that less than 1 ppm of the stabilizer is extracted in tests for specific foods to be packaged.

Among the first stabilizers used by PVC processors were the lead stabilizers, and, where long-term heat stability is a principal concern, they are still used today, in the wire and cable industry. The newer barium-lead salts have joined the earlier dibasic lead stearate, basic lead carbonate and silicate, and others that have been used for some time, and they are all cost-effective. Extrusion of drain-waste-vent pipe is another application for these stabilizers.

For vinyl resins, many epoxy-type plasticizers also provide stability, especially with the barium-cadmium stabilizers with which they exhibit a synergistic effect. Of these, epoxidized linseed oil and soybean oil derivatives are the most widely used as stabilizers. These also have better heat stability, with a slight advantage for the linseed oil varieties.

Cost-effectiveness and choice of stabilizer for specific compound and process needs play an increasing role in their selection. Different stabilizers will be chosen, for example, for twin-screw or for single-screw extrusion, taking into consideration the greater thermal abuse that will be experienced in the single-screw extrusion.

Because of their availability in bulk and at lower cost, blends of epoxide plasticizer-stabilizer with barium-cadmium liquid stabilizers are finding favor with processors of flexible vinyl materials. However, there is a tendency to remove barium and cadmium from both liquid and solid stabilizers because of their high toxicity. Strontium-zinc combinations are finding favor because of their lower toxicity.

For high-volume production, PVC can be produced with the stabilizers added during the polymerization process, where it will do the most good. With the increased use of PVC, the numbers of stabilizer formulations will increase to satisfy the various markets—wire and cable, film, sheet, and pipe.

13.2.3.3 Ultraviolet Light Stabilizers

Another form of degradation to which polymers are subject is photochemical degradation, which is caused by exposure to ultraviolet radiation from the sun and fluorescent lighting. The internal bonds of the polymer are broken by absorption of the ultraviolet (UV) radiation, which forms free peroxides or causes energy transfer within the polymer, creating instability in the polymer molecules.

The results are color change, reduction in tensile and impact strength, increased stiffness and brittleness, shrinkage and cracking, and loss of electrical insulating properties. Most plastics, except for the fluoropolymers, are subject to this type of degradation but can be protected against breakdown by the addition of UV stabilizers.

To combat the two types of degradative activity, absorbers and energy-transfer agents are utilized to stabilize the product. The ab-

sorbers do just that—absorb the radiation. This absorbed energy reacts with the absorber and is converted into heat without damaging the polymer.

Energy-transfer agents take the excess energy into the stabilizer and convert it into thermal energy, also without damage to the polymer.

Most widely used as UV absorbers are substituted 2-hydroxy benzophenones, which because of their wide absorption band—from lower limit of solar radiation to the threshold of visible light, are particularly useful with polyolefins. Substituted 4-hydroxy benzophenones are more effective and can be used in lower concentrations, but because the radiation absorption extends into the visible, they impart some yellowness to the product. Substituted benzotriazoles are also effective with a lesser yellowing effect, and are particularly effective with rigid PVC.

With a narrow absorption band in the 290–320 nm range, the substituted acrylonitrile UV absorbers are effective with most plastics, and they impart little or no color. Where pigmentation is of no concern, carbon black can be used. It is one of the most effective UV absorbers. There are other pigments that can be used for this purpose, but their use is limited by the color they impart to the product.

The energy-transfer agents are nickel-organic or cobalt-organic compounds such as phosphonates, alkylphenones, or substituted phenols, whose use is limited to products like fibers and thin films in which color can be masked, since the metals impart a blue or green color to plastics. Here the energy level of carbonyl radicals is reduced by a process called "quenching" through resonant energy transfer, and the energy state decays harmlessly.

The choice of light absorbers and stabilizers depends upon overall performance, especially in staining and discoloration, compatibility, and toxicity.

13.2.4 Fillers and Reinforcements

13.2.4.1 Fillers

A wide variety of materials is used to enhance physical properties of plastics by acting as fillers or reinforcements, or sometimes both. Fillers may be active or inert, in which case they are primarily extenders. However, with the addition of an active material or by coating, the inert fillers can become reinforcements.

The most widely used extender filler for plastics is calcium carbonate, which is used in PVC, polypropylene, polyethylene, phenolic, epoxy, polyester, polyurethane, and ethylene propylene diamine monomer (EPDM) compounds. Experiments have been made with impact styrene and nylon compounds.

Calcium carbonate is available at low cost and in large quantities from abundant natural minerals, and energy requirements for preparing it for use are relatively low. Its properties include whiteness, natural purity, mild alkalinity, softness, resistance to heat, and controllable particle-size distribution. The qualities account for the high volume of use. Increasingly available new commercial grades include extra-fine grinds and surface modifications. Several new coatings have been developed for increasing the amount of filler loading in compounds without loss of properties.

Stearic acid, calcium stearate, and other coatings for calcium carbonate generally provide improved rheological properties. Purified calcite can be shipped in aqueous suspensions for aqueous polymer emulsions such as vinyl and acrylic emulsions. While this eliminates the use of energy to dry the filler, the water content adds to shipping costs.

Another approach to coating fillers is based on organic molecules such as salts of long-chain polyaminoamides or alkylolamines with unsaturated acids or high molecular weight acids. Both neutral and anionic agents are available and are promoted for lowering polyester viscosity during mixing with the filler, for improving properties of filled polypropylene, and for many other uses. Proprietary treatments that decrease absorption of plasticizer and stabilizer during the compounding and processing of calendered vinyl formulations have also been reported. These are said to modify surface energetics, reduce pore volume, and decrease dioctyl phthalate plasticizer absorption 50%.

For results of tests using titanate-coupling agents in polypropylene homopolymer and nylon 6/6 injection-molded compounds, see Tables 13.2 and 13.3.

Flexible PVC compounds, plastisols, and glass-reinforced polyester compounds make up the highest volume uses for calcium carbonate. Range of filler levels for flexible PVC is from 20 to 60 phr. For thinner calendered films, fine fillers provide better performance than the coarser grades with less of a decrease in physical properties. Loading is usually 20–100 phr for PVC plastisols and organisols, with a wide range of particle sizes used. Coarse particles produce plastisols of lower viscosity, but may form sediment during storage. They are most often used in carpet backing.

From 1 to 5 phr of 2–3 μm $CaCO_3$ is currently

Table 13.2   Titanate-Treated $CaCO_3$ in Polypropylene Homopolymer Injection-Molding Compound

Properties	Formulation, parts by weight	
	PP 40 CaCO₃ 60 titanate[a] 0.6 mineral oil 0.6	PP 40 CaCO₃ 60 titanate[a] 0 mineral oil 0.6
Unaged		
Tensile strength at yield, psi	3042	3659
Elongation, %	22	1
Unnotched Izod, ft-lb/in	4.3	1.6
Notched Izod, ft-lb/in	0.43	0.22
Gardner impact, ft-lb/in	75	3
Flexural modulus, psi	452,000	570,000
Aged in water 6 days at 93.3C (200F)		
Tensile strength at yield, psi	2952	2678
Elongation, %	37.5	15
Gardner impact, ft-lb/in	95.5[b]	80[c]

[a]KenReact TTS, Kenrich Petrochemicals.
[b]Drew apart, did not shatter.
[c]Shattered.

Table 13.3   Titanate-Treated CaCO$_3$ in Nylon 6/6
Injection-Molding Compound

Properties	Resin 100	Resin 60 CaCO$_3$[a] 38.8 titanate[b] 0.2	Resin 50 CaCO$_3$[a] 48.75 titanate[b] 0.25
Tensile strength, psi	9610	10,652	8517
Elongation, %	29	3.7	2.5
Flexural modulus, psi	426,000	704,000	943,000
Notched Izod, ft-lb/in	0.74	0.47	0.45
Unnotched Izod, ft-lb/in	no break	4.5	2.5

[a]Atomite, Thompson-Weinman & Co.
[b]KR 38S, Kenrich Petrochemicals

used in potable water pipe of rigid PVC, but it is possible that levels up to 15 phr of 1 μm CaCO$_3$ will eventually meet ASTM D 1874 requirements. Stearate-coated grades improve melt rheology and smoothness of extrusion, while uncoated CaCO$_3$ serves as an antiplateout agent. Other pipe and conduit use up to 40 phr, usually 1–3 μm CaCO$_3$. See Tables 13.4 to 13.7 for data on physical properties of rigid PVC.

For sheet-molding compound, median particle sizes range from 3 to 6 μm. Up to 200 phr loading can be used in low-viscosity one- and two-component polyester resins. Size and distribution are selected to meet requirements such as no shrinkage, microsmooth surface, and low-controlled viscosity. Moisture content must be less than 0.1% for controlled thickening during maturation.

Loadings for bulk-molding compounds are usually 250 phr or 100–150 phr for premix, preform, and mat. Over 400 phr is possible, however, in some cases. BMC uses 3–6 μm CaCO$_3$. Size, distribution, and purity are chosen to provide low-uniform-viscosity polyester.

For marine polyester, ground 3–5 μm CaCo$_3$ with loadings of about 175–200 phr meets ASTM D 1201 requirements of 0.15% maximum water absorption during 24-h immersion.

Low- and high-density polyethylene can be successfully loaded with ground calcite. Much effort has been devoted to increasing the physical properties of filled polyethylene by adding polar copolymers during polymerization; by adding coupling agent to filler, polymer, or formulation; and cross-linking the filled polyethylene.

Both polypropylene homopolymer and copolymers readily accept 1–3 μm CaCO$_3$, at loadings of 43–67 phr, or 30–50 wt % of the total formulation.

An increasing number of injection molded automotive parts are being made from mineral-filled polypropylene homopolymer. The flexural modulus of a typical unfilled resin is about 200,000 psi. Where stiffness is needed, the use of 40 percent calcium carbonate filler gives a flexural modulus of about 400,000 psi, about halfway between that of the virgin resin and that provided by addition of talc. (Table 13.8)

Calcium carbonate fillers can also be used with ABS, polyurethane foam, epoxy compounds, and alkaline-type phenolic resins.

Kaolin, or clay, is the second most commonly used extender pigment in plastics. It finds applications in a wide variety of polymers—PVC, nylon, polyethylene, polyester, and polyurethane, to name a few. The largest

Table 13.4   Rigid PVC Extrusion Compound

	Tensile strength at yield, psi	Impact strength, notched Izod, ft-lb/in	Gardner impact,[a] ft-lb/in
no filler	8355	0.44	297
5 phr Supercoat	7624	0.56	326
10 phr Supercoat	7906	0.52	307
15 phr Supercoat	7268	0.66	302
20 phr Supercoat	6874	0.88	267

Note: specimens were compression molded.
[a]8-lb. weight.

Table 13.5 Rigid PVC Extrusion Compound

	Tensile strength at yield, psi	Gardner impact % surviving specimens
no filler	8149	80
2.5 phr Atomite	7710	80
2.5 phr Kotamite	7683	78
2.5 phr Supermite	7788	82
2.5 phr Supercoat	7733	78
10 phr Atomite	7170	34
10 phr Kotamite	7120	62
10 phr Supermite	7486	80
10 phr Supercoat	7194	86
30 phr Kotamite	6083	0
30 phr Supercoat	6063	68

Note: specimens were compression molded.
[a] 8 lb-weight (144 in-lb).

Table 13.6 Rigid PVC Injection-Molding Compound

	Tensile strength at yield, psi	Gardner impact, ft-lb/in
no filler	7953	226
5 phr Kotamite	7948	167
7.5 phr Kotamite	7611	190
10.0 phr Kotamite	7514	120
5 phr Supercoat	8138	214
7.5 phr Supercoat	7633	315
10.0 Supercoat	7412	210

Note: specimens were injection molded.

Table 13.7 Rigid PVC Siding Compound

	Tensile strength at yield, psi	Gardner impact ft-lb/in
No filler	6542	220
5 phr Atomite	6430	126
5 phr Kotamite	6262	186
5 phr Supermite	6374	287
5 phr Supercoat	6354	230

Note: specimens were compression molded.

amounts are used in wire and cable, polyester sheet molding compound and bulk molding compound, and in vinyl-flooring compounds. This material is a hydrous aluminosilicate. Its properties, along with cost factors, determine its use in polymer systems. These properties are influenced by both the purity of the mineral deposit and subsequent processing operations. Kaolin deposits are cored and analyzed prior to mining to determine quality. Mined clays are processed by either air flotation (dry) or water fractionation (wet).

In the dry, or air-flotation, method the clay is hauled from the mine to the plant for drying, pulverizing, and air classification. The primary control on air-classified clays

is the 325-mesh grit for an upper limit in particle size; the lower limit is dependent on the type of crude.

Dry-processed kaolins consist of a wide range of particle sizes. Particle-size distribution depends somewhat upon the crude clay, but size generally ranges from 0.1 to 40 μ.

Air-classified clays contain a much wider range of impurities than do water-washed clays, and these impurities appear throughout the size range.

Impurity content changes with particle size: mica exists as larger particles, tending to concentrate in the coarse end; anatase forms small particles, tending to concentrate in the fine end. This variabilty of impurities results in less uniform physical properties.

Even though air-floated clays are widely used in the plastics industry, their primary function is to reduce cost with minimal loss in physical properties. Because of the dependence of properties on crude, a comprehensive discussion is not possible here.

In wet, or water-fractionation, processing the clay is amalgamated in water at the mine, using optimum amounts of dispersing agent, and degritted. The crude slurries, or "slip," are transported to the plant, where various crudes are blended to facilitate uniformity of feed materials.

Table 13.8 Polypropylene Homopolymer Compound

	Tensile strength at yield, psi	Elonga-tion, %	Unnotched Izod, ft-lb/in	Notched Izod, ft-lb/in	Gardner impact, ft-lb/in			Flexural modulus, 1000 psi	Melt flow,[c] L cm³/10 min
					23C[a]	0C[a]	0C[b]		
No filler	5741	24	18.32	0.30	6.7	1.5		200	4.20
40% Atomite	4479	24	7.03	0.43	36.8	5.3	6.1	391	2.04
50% Atomite	4144	22	3.69	0.42	41.2	4.7	6.1	445	1.39
40% Kotamite	4210	31	9.26	0.46	114.2	9.6	10.8	347	3.48
50% Kotamite	4090	26	7.17	0.46	116.2	7.4	9.6	444	2.70
60% Kotamite	3147	32	5.0	0.47	113			621	
40% Supercoat	4269	29	12.79	0.61	173.8	7.6	9.7	359	3.57
50% Supercoat	4089	29	8.35	0.47	138.4	5.9	11.2	450	3.21
60% Supercoat	3352	31	2.9	0.53	140			582	

Note: specimens were injection molded.
[a] 8-lb. hammer.
[b] 2-lb. hammer.
[c] Melt flow expressed in volume units, cubic centimeters (cm³) per 10 minutes, condition L.

The feed is fractionated with centrifuges to obtain specific particle sizes. The fractionated product is leached to reduce and solubilize iron-oxide impurities, which are removed by filtration. Following acid flocculaton and filtration, the kaolin can be dried and pulverized to produce acid-flocced grades. Predispersed or neutral-pH grades are spray dried and pulverized.

The water-fractionation process is used to produce four grades of kaolin: standard, delaminated, calcined, and surface modified. Standard water-washed kaolins range from 0.2 to 9.5 μ in median particle size. The variation in particle size is obtained by differing conditions of centrifugation.

Delaminated clays come in a variety of particle sizes. They are made by exposing the kaolin, prior to fractionation, to a shearing force that causes the kaolin stacks to cleave, forming thin platelets.

Calcined clays are also available in a variety of particle sizes. They are formed by dehydroxylating standard water-washed clays.

Surface-modified clays can be obtained by treating standard, delaminated, and calcined grades with surface-modification agents. The treatment can be performed by either the supplier or the end user.

These wet-processed clays tend to have more uniform physical properties than air-classified clays and are preferred by compounders looking for uniformity and specific properties, such as increased loadings and improved molding characteristics.

Selection of particle size for a given application depends on the physical or handling characteristic desired and, of course, cost-performance factors. Although little attention has been paid to the rheological influence of clays in polymer systems in the past, kaolin can be a very important rheological modifier.

Used in polyester primarily to reduce volumetric shrinkage and to impart a finishing characteristic, kaolin acts as a heat sink and reduces the peak exothermic temperature. This may also account for the reduction in surface crazing. The coarse-particle-size kaolins tend to give higher tensile strengths, whereas the finer-particle-size kaolins tend to give better compressive strength and improved surface finish.

The effect of clay particle size on the rheology of unsaturated polyesters is shown in Figure 13.2. These systems can be pseudoplastic, Newtonian, or dilatant. Rheology depends upon many factors, including resin characteristics and fillers, and the rheological characteristics required for processing depend upon the equipment to be used and the products to be produced. Figures 13.3, 13.4, and 13.5 show the effects of particle size on the rheology of mineral oil, dioctyl phthalate (DOP), and ditridecyl phthalate (DTDP).

Fine-particle-size clays tend to be higher in viscosity at low shear but are also more pseudoplastic, or thixotropic (see Table 13.9). The high-shear viscosity in a general-purpose medium-unsaturated polyester ranges from 7000 cps for a coarse kaolin such as Hydrite MP to 10,000 cps for a fine kaolin such as Hydrite UF; the low-shear viscosity varies from 4000 cps for Hydrite MP to 16,000 cps for Hydrite UF.

Proper selection of intermediate-particle-size kaolins, or blends with other fillers, provides an optimum balance between the extremes of vehicle separation and pigment decrease with fine particles and increase with coarse grades.

PVC plastisols and organosols provide an example of the influence of kaolin particle size. Organobentonites or fumed silica can control viscosity and thixotropy, but vehicle separation can be a problem. The proper selection of kaolin, for use in conjunction with other fillers, can help to optimize both thixotropy and vehicle or plasticizer separation. To obtain minimum separation and optimum thixotropy in both plastisols and polyesters, a fine-particle-size kaolin is the proper choice to use with calcium carbonate.

The effects of particle size on rheology or melt viscosities of thermoplastics are different from the effect on solvent-thinned plastics such as saturated polyester. The particle size and type of filler make little difference in the viscosity of polyethylene filled at a 30 phr level (Figure 13.6), neither does the size affect melt-flow properties or stability of plasticized PVC.

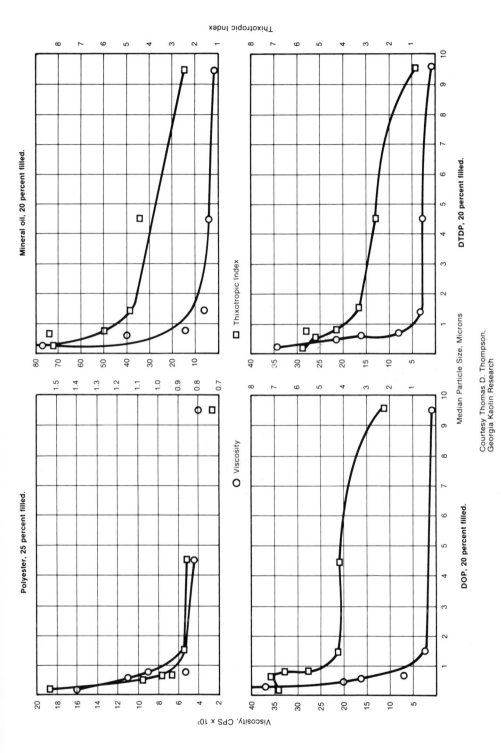

Figure 13.2–13.5 Effect of kalolinite particle size on viscosity and thixotropic index. Courtesy Thomas D. Thompson, Georgia Kaolin Research.

633

Table 13.9   Relationship of Viscosity to Properties

	Median particle size, microns	Oil adsorption; DOP	Surface area; methylene blue	Polyester viscosity	
				10 rpm	Thixotropic index
Standard kaolinites					
Hydrite UF	0.2	50	42	16,000	1.54
Hydrite PX	0.68	47	37	9,600	1.0
Hydrite R	0.77	44	31	5,600	0.93
Hydrite 121	1.5	41	22	5,600	0.87
Hydrite Flat D	4.5	39	16	4,800	0.85
Hydrite MP	9.5	39	10	4,000	0.71
Delaminated kaolinites					
Kaopaque 10	1.0	54	22	9,600	1.14
Kaopaque 20	2.0	52	19	8,000	1.05
Kaopaque 30	3.0	62	14	8,000	0.95

Each polymer system should be considered separately because polarity or wetting ability of the polymer as well as molecular weight influences the rheological character of the system. Water-washed clays, however, can be used as rheological modifiers in conjunction with other extenders to optimize flow characteristics and reduce costs.

Delaminated clays are supplied in a variety of particle sizes. They differ from the water-washed grades in that they are thin platelets of varying widths. Particle shape, like particle size, influences rheology of organic systems.

Oil adsorption, with delaminated clay, is related to neither particle size nor surface area, as is the case with water-washed grades. It has been shown that polyester viscosity is vaguely related to oil adsorption, but not to particle size or surface area (Table 13.10). There appears to be a relation between particle size (or surface area) and viscosity for mineral-oil or dioctyl phthalate plasticizer viscosities of delaminated kaolins. Each solvent-polymer system must be considered separately to obtain optimum rheology and handling properties.

At comparable particle sizes or surface areas, delaminated clays result in tensile and modulus strengths about 30% higher than standard clays, seen in unusual rheological behavior on polyester and reinforcement properties in rubber systems such as styrene butadiene rubber. This is believed to be due to increased surface reactivity rather than surface area. Delaminated clays tend to result in better color in polymer systems than standard kaolins, because of cleaner surfaces and different wetting characteristics.

Water-washed calcined clays vary in particle size and degree of calcination. Oil adsorption of calcined clays varies little with particle size and is approximately 20–50% higher than with standard or delaminated grades. Even with higher oil adsorptions, calcined clays are less reactive and give better stability and color in polymer systems.

The primary use of calcined clays is in

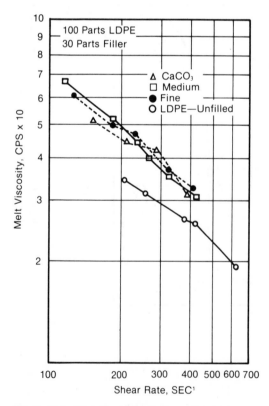

Figure 13.6 Viscosity of kaolinite- and calcium carbonate-filled polyethylene. Courtesy Thomas D. Thompson, Georgia Kaolin Research.

Table 13.10 Silane Coupling Agents: Structure and Recommended Polymer Systems

Silane[a]	Description	Structure	End-Use Polymers
A-151	Vinyltriethoxysilane	$CH_2 = CHSi(OC_2H_5)_3$	Unsaturated
A-172	Vinyl-tris (2-methoxy-ethoxy) silane	$CH_2 = CHSi(OCH_2CH_2OCH_3)_3$	polyester DAP, polyolefin
A-174	gamma-Methacryloxy-propyltrimethoxysilane	$CH_2 = \overset{\overset{\displaystyle CH_2 \; O}{}}{C}-\overset{}{C}-OCH_2CH_2CH_2Si(OCH_3)_3$	
A-1100	gamma-Aminopropyltri-ethoxysilane	$H_2NCH_2CH_2CH_2Si(OC_2H_5)_3$	Epoxy, phenolic, melamine,
A-1120	N-beta-(aminoethyl)-gamma-aminopropyl-trimethoxysilane	$H_2NCH_2CH_2NHCH_2CH_2CH_2Si$ $(OCH_3)_3$	polyamide, polyester, polycarbonate
A-1160	gamma-Ureidopropyl-triethoxysilane	$\overset{\displaystyle OH}{NH_2CNCH_2CH_2CH_2Si(OC_2H_5)_3}$	
A-186	beta-(3, 4-Epoxycyclo-hexyl) ethyltrimeth-oxysilane	$\overset{\displaystyle O}{O}$-$CH_2CH_2Si(OCH_3)_3$	Epoxy, phenolic, polyester
A-187	gamma-Glycidoxypro-pyltrimethoxysilane	$CH_2\text{-}CHCH_2OCH_2CH_2CH_2Si(OCH_3)_3$ O	
A-189	gamma-Mercaptopro-pyltrimethoxysilane	$HSCH_2CH_2CH_2Si(OCH_3)_3$	Epoxy, polysulfides

[a]Products of Union Carbide Corp.

wire and cable insulation. The use of small amounts of partially calcined clay in PVC can double the volume resistivity. Fully calcined clay results in slightly lower volume resistivities but gives substantially better color. Before aging, standard water-washed clays give higher volume resistivities than calcined clays; however, resistivities decrease with aging because of water adsorption due to hygroscopicity of the standard, hydrous kaolin.

Calcined clay can replace up to 20% of titanium dioxide when used in PVC and other polymers, without loss of optical properties. Brightness and opacity are important, and the degree of calcination and type of clay calcined are important in determining the color of the filled polymer. Fully calcined clay has high brightness and whiteness, whereas partially calcined clay is tan in color.

Oil adsorption with calcined clays is related to porous aggregate structure and organophilic surface rather than surface reactivity, as with other clays. Both partially and fully calcined clays give better thermal and color stability in PVC than do other water-washed kaolins, but higher amounts of plasticizer are required to maintain a low melt-flow index.

Treating clays with silane, titanate, polyester, and metal hydroxide for surface modification has been introduced with the objective of increasing filler loadings and

possibly improving physical properties such as melt viscosity, thermal stability, and reinforcement without loss of physical characteristics.

Most common surface-modifiers for clay are the silane and titanate coupling agents. Silanes are trialkoxy silicon compounds with an additional functional group; titanates are monoalkoxy titanium compounds with three additional functional groups. The choice of the proper functional group or type of modifying agent depends upon the polymer system and the desired change in the physical and/or rheological properties.

This modification can be performed by pretreatment done by the clay supplier or the end user and by mixing all ingredients together with the addition performed at the same time by the end user. Silane-treated calcined kaolins are used in electrical insulation, rubber products, and polyester, nylon, and other products. Electrical applications represent the largest use of surface-modified kaolin in plastics. Silane-treated calcined clay gives improved aging characteristics over untreated grades for electrical properties.

Polyester-treated clay and metallic-hydroxide-treated clay are primarily recommended for use in polymer systems when compatibility is of prime importance, regardless of additional reinforcement. In the case of thermoplastics, in which reinforcement is not obtained by cross-linking, clays that are compatible with the polymer give better physical properties.

Other silicates such as mica, nepheline syenite, talc, wollastonite, asbestos, and glass increase the hardness of the resins they fill and reduce swell. Mica has a Mohs hardness of 3.0. It may be made hydrophobic by heating with silanes. Composites containing mica exhibit excellent electrical and thermal properties and excellent dimensional stability. Mica is compatible with most resins. Nepheline syenite is an anhydrous sodium potassium aluminum silicate of exceptional uniformity. As a finely ground material, it is widely used in rigid and flexible PVC resins. It has virtually no effect on the transparency obtainable in the resins. It is extremely easy to wet and disperse in epoxy castings and polyesters. Loadings can be high.

Talc occurs naturally as magnesium silicate, whose fibrous structure produces stronger composites than other silicates. Polypropylene containing talc is creep resistant, retaining 90% of its rigidity at 170°F. Talc added to resins improves molding cycles with no added wear to machine parts.

Wollastonite, glass, and asbestos, as well as carbon graphite, are discussed more at length as reinforcements.

Among the silicas are sand quartz, tripoli, and diatomaceous earth, as well as novaculite, all of which have different average particle sizes and also differ in hardness and degree of crystallinity. Quartz has been used with phenolic and epoxy resins for high-performance aerospace applications. A quartz-epoxy composite with a 60% quartz filler has the same coefficient of expansion as aluminum or brass. The rate of crack formation, shrinkage, and moisture resistance of polyethylene is reduced by a 14–45% addition of quartz filler, while a 30–40% addition improves the dielectric, physical, and processing properties of polyacrylates.

Diatomaceous earth, which comes from skeletons of ocean organisms, is not nearly so hard as quartz. Its range on the Mohs' scale is from 1 to 1.5 as compared with 7 for quartz. It has application among other things as an antiblocking agent.

Novaculite has good moldability, extrudability, and low-compression set for use in elastomers and provides good flow, heavy loadings, low abrasive coefficient, and excellent dielectric properties in epoxies. These qualities are especially desirable for encapsulation of delicate electronic components. Its low specific heat improves the thermal conductivity of end products as well.

Wood flour, a cellulosic, can also serve as a filler at low cost. Obtained from milling of fibrous woods, wood flour results in easily wetted, uniform particles. It reduces mold shrinkage and gives good electrical properties and impact resistance to thermosetting resins. Shell flour, another cellulosic that comes from ground walnut, pecan, or peanut shells, imparts low shear strength and low impact resistance to products, as well as good flow properties and gloss and moisture resistance.

Also used as fillers are metallic oxides such as zinc oxide, aluminum oxide, and magnesium, titanium, and beryllium oxides. Zinc oxide imparts weatherability to polypropylene and provides superior polyester composites. It is also used in elastomers. Aluminum oxide increases the flame resistance of polyester. The magnesium and titanium oxides gives improved stiffness, hardness, and creep resistance to composites, while beryllium oxide, when added in very high loadings, imparts high thermal conduction to epoxy.

Molybdenum disulfide added to elastomers imparts antifrictional properties; barium ferrite makes polyvinyl chloride composites magnetic; and barium sulfate increases the specific gravity, functional resistance, and x-ray opacity of resins. Barium sulfate's particle sizes vary so they can be graded for specific applications, such as density, load-bearing properties, and processing requirements.

Metal powders of aluminum, bronze, or zinc also give improved processing or end-product qualities. A high loading of zinc powder gives a protective coating property to resins, whereas aluminum and bronze powders impart thermal and electrical conductivity to acetal or nylon moldings.

13.2.4.2 Coupling Agents

Use of coupling agents can control compound rheology, improve the resin wetting of inorganic components, and increase the strength of the final composite, both initially and after exposure to a hostile environment. They can overcome the problems that arise when components of differing chemical nature and physical form are combined; in the interfacial region where organic resin must wet and spread over the inorganic filler or reinforcement surface, the coupling agents, especially the silane agents, are able to react with both the organic and inorganic materials to modify the interface.

Silane coupling agents are used in a wide variety of composite products. The hydrolysis product provides a bond to the inorganic components and maintains good adhesion to most siliceous or metal oxide fillers and reinforcements even under environmental attack.

To take advantage of the strength that can be contributed by the inorganic filler-reinforcement in a composite, it is necessary to obtain maximum adhesion to the relatively low-strength resin. The silane mechanism results in chemical bonding from the resin through the silane coupling agent to the inorganic phase. Stress is thereby transferred from the low-strength resin to the higher-strength inorganic compound. The chemical bond also maintains good adhesion and mechanical strength when subjected to adverse environmental conditions, for example, exposure to water.

A further advantage of a well coupled composite is the provision of stable electrical properties, such as dielectric constant, dissipation factor, and volume resistivity, because the bonding minimizes ingress of water.

In addition to the ability to improve the mechanical and electrical properties of a filled and/or reinforced resin, perhaps even more important is the ability of the silane coupling agents to control the rheology of the compound. The capability of lowering viscosity provides the compounder with several options, which include increasing reinforcement level for higher strength, raising filler loading for improved economics, and improving flow for higher-quality moldings.

The variety of resin types used in commercial applications requires a range of silane coupling agents with matching func-

tionality. Table 13.10 shows some silane coupling agents that are commercially available and their suitability for use with various resins. In addition, there are proprietary dry silane concentrates that provide viscosity control and improved processability in filled thermosets.

The silane coupling agents have significant effects when used with a wide range of thermosetting and thermoplastic resins. When unsaturated polyester with various mineral fillers is treated with silane coupling agent, the viscosity of the entire system is reduced and both dry and wet flexural strength are improved. The viscosity reductions are particularly significant for sheet molding compounds when polyester resins are filled with alumina trihydrate treated with silane coupling concentrates (Table 13.11).

Epoxy- and amino-functional silane coupling agents provide a high level of reactivity with the matrix resin in filled epoxy resins. The effects may be seen in Table 13.12. Although this system had relatively little effect on dry strengths, wet properties are improved more than 100% with the use of epoxy-functional silane, and improvement is also noted in electrical properties.

Improvement is also obtained in thermosetting resins reinforced with glass. Mechanical properties are enhanced, particularly after exposure to water, when a silane coupling agent with functionality appropriate for the matrix resin is used. In this case the glass was fabric rather than the chopped glass fibers that are a more usual reinforcement (Table 13.13).

It has been shown also that the silanes react with relatively inert, high molecular weight thermoplastic resins. Property improvements similar to those in thermosetting-resin systems are obtained in many thermoplastics materials with fillers and glass fibers treated with functional silanes (Table 13.14).

13.2.4.3 Fiber Reinforcements

Reinforcing agents are added mainly to improve the tensile and flexural strength of a resin. Sometimes they are regarded as fillers, but fillers do not necessarily improve the tensile or flexural strength. Some reinforce-

Table 13.11   Effects of Silane-Treated ATH Filler in SMC Formulations (with Varying Bromine Levels) on Physical Properties of Polyester Composites

Formulation	Control A	Reduced bromine B DSC-20/ATH	Reduced bromine C Untreated ATH	No bromine D DSC-20/ATH
Resin component, parts by weight				
Isophthalate modified polyester resin	40	55	55	60
Brominated polyester resin (40% Br)	20	5	5	
BAKELITE LP 40A	30	30	30	30
Styrene	10	10	10	10
Bromine content in resin component, %	8	2	2	0
Other ingredients, parts by weight				
ATH	125	175	175	200
$CaCO_3$	25			
Zinc stearate	4	4	4	4
t-butyl perbenzoate	1.5	1.5	1.5	1.5
Black pigment dispersion				1.0
Sb_2O_3	3.0	3.0	3.0	
MgO	0.8	0.8	0.8	2.5[b]
Glass, 1-in fibers, soft finish, wt-%	25	25	25	25
Properties				
Shrinkage, mils/in	0.0	0.0	0.0	0.6
Tensile strength, psi	7,800	8,480	7,760	7,230
Flexural strength, psi	14,400	19,800	15,100	17,360
Notched Izod impact, ft-lb/in	11.5	13.0	12.0	8.3
Burning rating, UL-94[a] (55-mil thickness)	V-0	V-0	V-0	V-0

[a]UL-94 is a small-scale laboratory flame test. The flame spread rating is not intended to reflect hazards presented by this or any other material under actual fire conditions.
[b]MgO, 33% dispersion.

ments, such as carbon black, can also serve as pigments. Commonly used reinforcing materials include glass, graphite, asbestos, carbon, and ceramic, and synthetic fibers such as nylon, polyethylene, polyacrylonitrile, and polyaramid.

The predominant fibrous reinforcement used today is glass fiber in various forms such as ribbons, yarns, fabric, or chopped strands. Successful applications depend on a knowledge of the different types available and their optimum use in formulating and compounding for various processing methods.

For general-purpose usage, the type in widest use is electrical glass, which is a borosilicate type with good resistance to attack by moisture. The chemical glass com-

Table 13.12   Effects of Integrally Blended Silane Coupling Agents on Flexural Strength and Electrical Properties of Quartz-Filled Anhydride-Cured Epoxy Composites

	Control (no silane)	Silane and loading, wt-%[a] A-186 0.31	A-187 0.30	A-187 0.33	A-1100 0.28
Flexural strength, psi					
Dry, room temp	21,000			21,000	20,000
72-hour boil[b]	9,000			20,000	14,000
Dielectric constant[c]					
Dry	3.39	3.48	3.40		3.46
72-hour boil	14.60	3.52	3.44		3.47
Dissipation factor[c]					
Dry	0.017	0.016	0.016		0.013
72-hour boil	0.305	0.023	0.024		0.023
Volume resistivity,[d] ohm-cm					
Dry	8.4×10^{16}	8.0×10^{16}	$>8.2 \times 10^{16}$		$>8.1 \times 10^{16}$
72-hour boil	5.1×10^{11}	1.4×10^{15}	1.7×10^{15}		1.8×10^{15}
Dielectric strength,[e] volts/mil					
Dry	>381	>367	>357		>355
72-hour boil	103	>360	>391		>355

[a]Percent by weight loading based on weight of quartz filler.
[b]Tested at room temperature.
[c]Tested at 1 kHz, according to ASTM D 709-55T.
[d]Tested according to ASTM D 257-57T.
[e]Tested according to ASTM D 149-55T.

Table 13.13   Effects of Silane Coupling Agents on Flexural Strength of Thermosetting-Resin Glass-Fabric-Reinforced Composites

| | Flexural strength, psi | | | | |
| | Control | | | Silane finish | |
Resin system	Dry	Wet	Silane[a]	Dry	Wet
Unsaturated polyester	61,000	23,000[b]	A-174	87,000	79,000[b]
			A-186	101,000	66,000[c]
Epoxy (aromatic amine cure)	78,000	29,000[c]	A-187	87,000	56,000[c]
			A-1100	90,000	54,000[c]
			A-186	74,000	73,000[b]
Melamine	42,000	17,000[b]	A-187	91,000	86,000[b]
			A-1100	104,000	98,000[b]
Phenolic	69,000	14,000[d]	A-1100	80,000	55,000[d]

Note: All values based on 12-ply laminates.
[a] Silanes applied 0.5 wt-% based on fabric weight.
[b] 8-hour boil.
[c] 72-hour boil.
[d] Aged 100 hours at 500F and tested at 500F.

position that was developed for improved chemical durability shows better resistance to water and also to acid and alkali. Another type, which is closer to the standard window or bottle glass composition, exhibits fair resistance to attack by acid, but poor resistance to water and alkali. There are other types for various uses: some with improved electrical performance and lower density, and an alkali-resistant type more suitable for reinforcing concrete than for reinforced plastics.

General formulation and compounding are not affected by glass thermal constants since the highest temperatures used for curing plastics and resins are well below the minimum temperatures that affect the thermal stability of glass—about 1100°F.

Many combinations of number of filaments per strand, strand configurations, and fiber length-to-weight ratio are possible in glass fibers, with diameters varying from 0 to 18 μ. Most of the glass-fiber products contain sizing or a material for organic carrying

that is added either during fiber-forming or in a subsequent treatment. This sizing provides lubrication and film-forming capability in downstream fiber production processes. These sizings and finishes provide a film that bonds component filaments together, lubrication against filament breakage, and compatibility with the resin. Compatible glass fibers provide a strong inorgano-organic bond when the formulated glass resin mix is ultimately cured.

Some of the major glass-fiber products for reinforced plastics include yarns, fabrics, roving, chopped strands, mats, woven roving, milled fibers, or combinations of mat with woven roving.

Yarns are twisted and plied for further processing into tapes, fabrics, cord, and cabling. Glass-fiber strands in a variety of combinations are made into fabrics. In roving, continuous-filament strands are drawn together in a parallel, untwisted aspect into packages from 12 to over 450 lb. A variety of combinations is possible from one strand

Table 13.14   Effects of Silane Coupling Agents on Flexural Strength of Various Thermoplastic-Resin Glass-Fabric-Reinforced Composites

| | Flexural strength, psi | | | | |
| | Control | | | Silane finish | |
Resin system	Dry	Wet[b]	Silane[a]	Dry	Wet[b]
			A-186	44,800	39,800
Polycarbonate	38,600	24,100	A-187	45,200	37,400
			A-1100	51,100	37,900
Polystyrene	24,400	15,800	A-186	42,800	33,500
			A-174	48,400	31,000
Low-density polyethylene	10,200	6,400	A-151	15,700	16,600
			A-174	19,900	16,300
Nylon 6	26,200	18,400	A-1100	54,100	40,900

Note: All values based on 11-ply laminates.
[a] Silanes applied 0.5 wt-% based on fabric weight.
[b] Immersed in water at 50C for 16 hours.

Table 13.15 Glass Fiber Filament Diameters Most Widely Used in Reinforcing Products

Letter designation	Nominal filament diameter[a] Inches	Nominal filament diameter[a] Microns	Nominal yds/lb for a 204-filament strand[b]	Corresponding tex value for a 204-filament strand[c]
G	0.00038	9.46	13,615	36.3
H	0.00043	10.73	10,755	46.0
J	0.00048	12.07	8,625	57.3
K	0.00053	13.34	6,890	71.7
L	0.00058	14.61	5,740	86.1
M	0.00063	15.88	4,855	101.8
N	0.00068	17.15	4,163	118.8
P	0.00073	18.42	3,608	137.1

[a]In glass-fiber technology, filament diameters are determined by actual measurement on a calibrated screen of a projection microscope at 1000 X, or by filar micrometer eyepiece. It should be noted that filament diameters used in some specialty fabrics are smaller than G filaments.

[b]Yd/lb÷100 is also used to express strand count for each letter designation; yd/lb naturally varies inversely with both the filament diameter and the number of filaments per strand.

[c]The tex value, important in the establishment of a universal yarn-measuring system, is defined as the weight in grams/1000 meters; it may also be expressed as 494420÷yd/lb.

Table 13.16 Glass Fiber Types, Grades, and Sizings

Type of glass-fiber reinforcement	Description	Grades	Sizing[a]
Yarns	Single-end glass-fiber strands twisted on standard textile tube-drive machinery, can be plied with reverse twist on textile ply frames	B to K fiber, many fiber and yardage variations	Starch
Fabrics	Plied yarns woven into many cloth styles with various thicknesses and strength orientations	D to K fibers, 2.5 to 40 oz/yd² in wt	Starch size removed and compatible finish applied after weaving
Roving	Strands from forming cakes wound in parallel and untwisted manner into large packages using standard takeup machines; with more recent technology, unitized roving packages are wound directly in a single step	G to P fiber used. Strands 900 to 120 yd/lb and packages 5 to 450 lb wt (up to 24 x 24 in in size)	HSB, LSB and RTP
Chopped strands	Filament bundles bonded by sizing ingredients, subsequently cured and cut or chopped into short lengths for further processing	K to M fiber, 0.125 to 0.5 in or longer lengths, various yardages	LSB, RTP
Chopped strand mats	Fibers from forming cakes chopped and collected in a random pattern with additional binder applied and cured; some needled mat produced with no extra binding required	G to K fiber, weights 0.75 to 6 oz/ft²	HSB, LSB
Continuous strand mats	Fibers converted directly into mat form without cutting with additional binder applied and cured	Nominally K fiber, weights 0.75 to 4.5 oz/ft²	LSB
Woven roving	Coarse fabric, bidirectional reinforcement, mostly plain weave, but some unidirectional woven roving is produced	G to P fiber, fabric weights 15 to 48 oz/yd²	HSB
Mat/woven roving combinations	Chopped strand mat and woven roving combined into a usable reinforcement with additional binder applied	38 to 62 oz/yd²	HSB
Milled fibers	Fibers reduced by mechanical attrition to short lengths in powder or nodule form	Lengths range 0.001 to 0.125 in, several grades	None, HSB and RTP

[a]Designations here indicate reinforcements for room-temperature cure (HSB = high-solubility sizing), high-temperature cure (LSB = low-solubility sizing), or reinforcing thermoplastics (RTP).

with 2000 filaments to 6 or 12 strands with 200 filaments each, or 120 or more single strands. Weights may vary from 900 yd/lb to 120 yd/lb with variations in filament diameter and type and amount of sizing for varied effects in the compounds (Table 13.15).

Filaments may be cut into lengths of $\frac{1}{8}$, $\frac{3}{16}$, $\frac{1}{4}$ in. up to 1 in. or longer to make chopped strands to improve compounding efficiency in various processes and products.

Roving strands are woven in both warp and fill to provide a reinforcing material in weights from 15 to 48 oz/yd. Milled fibers are made of scrap fibers and have a 1:1–20:1 length-to-diameter ratio. They are widely used as reinforcing fillers in resin. The various types are shown in Table 13.16.

Wollastonite is a naturally occurring, nonmetallic calcium metasilicate. It is the only commercially available pure white mineral that is wholly acicular. Typical length-to-diameter ratios range from 3:1 to 20:1. Because of its chemical composition, particle structure, and utility as well as benefits derived from minerals in polymer systems, wollastonite use has increased over the last few years.

Wollastonite has broad application in most polymer systems such as engineering resins, thermoplastics, and thermosets. A newly introduced wollastonite has potential application in engineering resins, polyethylene and polypropylene film, and in rubber systems. It has a coefficient of expansion of 6.5×10^6 mm/deg C, a melting point of 1540° C and Mohs hardness of 4.5.

Wollastonite can reduce costs and increase flexural strength, tensile strength, and modulus; improve dimensional stability; while decreasing water absorption when used in nylon 6 and 6/6 as a reinforcing filler.

Physical strength is increased and wet-strength properties are maintained by adding an amino-functional silane coupling agent to the wollastonite (Table 13.17). Silane treatment enhances impact strength, as shown by tests where the improvement was from 5.7 in.-lb to 118 in.-lb at a 50% filler loading.

With a 50 wt % loading of wollastonite into polycarbonate at 50 mesh, strength properties, heat deflection (distortion) temperature equivalent to the base resin, and increased flexural modulus and tensile modulus were noted, while surface modification further enhanced physical properties. A treated wollastonite system shows a 15% increase in flexural strength over the base resin system (Table 13.18).

Rigidity modulus and dimensional stability of thermoplastic polyesters are improved by the addition of wollastonite.

One disadvantage in the reinforcing of

Table 13.17   Effect on Physical Properties of Untreated and Silane-Treated Wollastonite in Nylon 6 (Allied Chemical Plaskon 8201 Ground to 50 Mesh)

Property	Virgin resin	50% loading	
		Nyad 325 wollastonite untreated	325 Wollastokup 1100 0.5 treated
Flexural strength, psi			
Dry[a]	12,500	15,400	18,600
Wet[b]	6,700	8,900	12,000
Flexural modulus, psi x 10^5			
Dry	2.7	8.3	7.7
Wet	1.1	5.9	6.1
Tensile strength, psi			
Dry	9,000	8,900	10,500
Wet	6,700	5,500	7,500
Tensile modulus, psi x 10^5			
Dry	3.8	11.2	10.3
Wet	2.3	5.9	6.1
Heat distortion temperature, F[c]	133	302	284
Water absorption (wt %)[b]	3.14	1.36	1.23
Dart impact, in-lb	600	5.7	118

[a] Tested after conditioning at room temperature and 35% RH.
[b] Tested after immersing 16 hr in water at 50C.
[c] At 264 psi.

Table 13.18   Effect of 50 wt-% Untreated and Treated Wollastonite on Physical Properties of a Polycarbonate System

Property	Nyad 325 wollastonite	325 Wollastokup 1100 0.5	Base resin
Flexural strength, psi			
Dry	11,900	14,400	12,600
Wet	9,900	12,400	12,500
Flexural modulus, psi x 10^5			
Dry	9.4	9.2	3.0
Wet	8.7	9.2	3.0
Tensile strength, psi			
Dry	6,400	8,100	7,800
Wet	6,100	7,900	7,700
Tensile modulus, psi x 10$_5$			
Dry	11.5	11.9	2.8
Wet	10.4	12.3	2.6
Heat distortion temperature, F	266	271	268

thermoplastic polyester resin systems is the inherent incompatibility of resin and filler, which results in poor adhesion between the resin and reinforcement and causes reduced composite strength properties, increased sensitivity to moisture, and processing difficulties. These compounding difficulties can be reduced, however, by the addition of a compatible coupling agent.

The amount of untreated or modified wollastonite also has an effect on processing. With silane-treated wollastonite, there is no

Table 13.19   Comparison of Properties of Wollastonite, Kaolin Clay, and 325-Mesh Silica in Typical Epoxy Binder Composites

ASTM method	Property	Wollastonite	Clay	325-mesh silica
Compounds A and B, 30 vol-% filler, unmodified epoxy resin				
D 149-59	Dielectric strength, ⅛-in casting, volt/mil			
	Epoxy/amine binder (Compound B)	418	517	456
	Epoxy/anhydride binder (Compound A)	416	432	452
D 696-44	Coefficient of linear thermal expansion, -30 to +30C, in/in/deg C x 10^6			
	Epoxy/amine binder (Compound B)	35.8	27.9	33.3
	Epoxy/anhydride binder (Compound A)	40.7	50.9	57.4
Compound A, 28.9 vol-% filler, anhydride-cured flexibilized epoxy				
D 257-66	Volume resistivity at 25C, ohm-cm			
	Conditioned 240 hr at 70C and 100% RH	6.0 x 10^{13}	1.0 x 10^{14}	3.4 x 10^{13}
	Conditioned 48 hr at 25C and 50% RH	1.9 x 10^{15}	1.3 x 10^{15}	1.5 x 10^{15}
D 257-66	Volume resistivity at 200C, conditioned 48 hr at 25C and 50% RH, ohm-cm	1.2 x 10^{11}	9.1 x 10^{10}	4.8 x 10^{10}
D 570-59a	Water absorption, conditioned 240 hr at 70C and 100% RH, vol-%	0.91	1.13	0.81
D 484-66	Thermal shock resistance, -55 to +130C, cycles passed	10	2	0
D 1042-51	Linear shrinkage, in/in	0.0128	0.0138	0.0139
D 696-44	Gel time at 121C in shrinkage mold, min	55	80	70
D 150-59T	Dissipation factors at 100C			
	10^2 cps	0.0085	0.0315	0.033
	10^3 cps	0.0055	0.0165	0.012
	10^5 cps	0.027	0.0355	0.029
D 150-59T	Dielectric factors at 100C			
	10^2 cps	5.14	4.88	4.85
	10^3 cps	5.05	5.25	4.57
	10^5 cps	4.75	5.51	4.40
Compound C, 26 vol-% filler				
D 638-60T	Ultimate tensile strength, psi	8,590	6,615	8,920
D 638-60T	Elongation, %	1.5	0.7	1.2
D 638-60T	Tensile modulus, psi x 10^6	1.14	0.97	0.95
D 790	Ultimate flexural strength, psi	10,000	13,230	13,920
D 790-59T	Initial flexural modulus, psi x 10^6	1.06	0.98	0.98
D 695	Compressive yield strength, psi	16,380	14,520	14,580
D 695	Ultimate compressive strength, psi	22,750	20,510	25,090
D 695	Initial compressive modulus, psi x 10^6	0.18	0.22	0.18
D 570a	Water absorption, one month, vol-%	0.77	0.99	0.82

problem in injection molding with either a 50 or a 70% filling, but difficulty arises with untreated compound at the 70% level.

The acicular nature of wollastonite adds reinforcing properties to phenolic molding compounds. Tests have shown that physical properties of phenolics with wollastonite are superior to those of asbestos-filled phenolics. It also imparts improved physical and chemical properties to both anhydride- and amine-cured epoxy resin systems.

Outstanding features of these systems are excellent thermal shock resistance, low-dissipation factor, dielectric constant and loss factor through a range of temperatures up to 160°C, lower gel time in the mold, and good shelf life with no evidence of gellation.

Table 13.19 compares physical, thermal, and electrical properties of wollastonite with those of kaolin clay and 325-mesh silica in typical epoxy binder composites. It can be seen that wollastonite contributed the best balance of properties in the evaluated systems.

Silane-treated wollastonite products show significant strength and electrical property improvements over untreated wollastonite. The major improvement contributed by surface modification is maintenance of electrical and physical properties after prolonged moisture exposure.

In systems using unsaturated polyester as the base resin, wollastonite displays good reinforcement characteristics, good dimensional heat stability, and good electrical insulating properties when used as a functional filler. Current commercial uses include application as a functional filler in electronic components for electrical properties and dimensional stability and use as a flow aid for more uniform glass dispersion in injection-molded bar molding compound (BMC).

Surface modification with methacrylic silane increases initial strength properties and maintains these properties after exposure to moisture, enabling higher loading levels of filler. Silane-treated wollastonite has a 50% greater initial flexural strength than untreated, and the flexural strength after 8 hr boil is 25% greater than the initial flexural strength of the untreated wollastonite.

Wollastonite has been utilized in polyethylene compounds for its reinforcing value and for improving the electrical insulation properties of the system. At a 40% loading level in low-density polyethylene, Nyad 400 wollastonite exhibited a dissipation factor of 0.0017 compared to 0.1145 for talc and 0.0025 for asbestos. Dielectric strength was 1180 as compared to 1020 for talc and 920 for asbestos systems.

Wollastonite also exhibits some reinforcing characteristics when compounded into general-purpose polypropylene at 40 phr. The addition of 10 phr glass fiber to a 30 phr wollastonite polypropylene system shows increases in all tensile properties, flexural strength, impact strength, and heat-distortion temperature. This can be obtained at lower cost than with a glass-filled polypropylene system without wollastonite.

Wollastonite in polypropylene gives good heat-aging characteristics. Oven-aging tests of a stabilized polypropylene showed the wollastonite-filled system did not fail until 135 days or more, compared to talc-filled systems, which normally fail at 85 to 90 days.

Wollastonite can lower costs while maintaining physical properties when used in general-purpose polystyrene systems. Properties comparable to unfilled polystyrene were shown with a 30 wt % wollastonite-filled compound. The tensile strength was 8140 as against 8340, the flexural strength was 61.1 against 56.3, and Izod impact 0.28 as against 0.39.

Wollastonite has also been evaluated for use in thermoplastic urethanes, vinyl plastisols, and elastomers. In elastomers wollastonite can provide a good balance between low cost and property advantages such as low dielectric constant, low moisture absorption, and retention of mechanical properties at elevated temperatures. It is also being evaluated in high-impact polystyrene, ABS, and SAN. Surface-modified wollastonite is being evaluated in high-impact polystyrene for automotive and appliance applications.

Carbon-graphite is another reinforcement material in which use is increasing in plastics compounds. These materials have already had success in the aerospace industry. As compounders and molders become more aware of their capabilities and as their costs are reduced, they will be used for more applications. The filaments have been used with thermosetting resins, primarily epoxy resins, and they have great use in critical structural applications.

Although the cost of carbon-graphite filaments is higher than for glass filaments, for many end products their excellent properties make carbon graphite composites competitive in cost. They have found increasing application in general industrial parts, in sporting equipment, and marine and automotive parts.

Outstanding properties of these materials include high modulus and tensile strength, low density, high electrical conductivity, low thermal coefficient of expansion, low coefficient of friction. These also have excellent resistance to most environmental exposure conditions and chemicals. Carbon-graphite materials are available in fiber mats with low strength and modulus as well as in chopped fibers and continuous filaments that have high tensile strengths and moduli. Densities range from 1.6 to 1.9 g/cm^3.

These fibers have application with nylon, polysulfone, thermoplastic polyester, polyphenylene sulfide, polycarbonate, polypropylene, and the polyamide-imides. Compounds have carbon-graphite content ranging from 20 to 40 wt %. There are some applications with combinations of carbon and glass in nylon 6/6 (Table 13.20).

The tensile and flexural strengths are usually greater than those for glass-fiber-reinforced composites. Thermal expansion is generally lower. Melt shrinkage ranges from about one-half to one-fifth that of untreated resin, while thermal conductivity is about twice that of an equivalent glass-reinforced formulation.

One advantage for applications such as those requiring electrostatic painting or electromechanical components requiring dissipation of static charges is higher electrical conductivity. This, however, rules out its use as an insulator.

While the type of resin has an effect, generally, with more graphite, the coefficient of friction is lower. For example, it is 0.25 against steel compared with 0.8 for glass fibers. Parts wear is generally reduced compared with glass.

Carbon-graphite is bulky and fluffy and difficult to handle so that care must be taken and low-energy blending techniques are employed. Low-shear molding is recommended.

Asbestos, a generic term for naturally occurring fibrous, hydrated metal silicates, was one of the earliest reinforcements used for plastics. Various types include chrysolite—the most widely used, anthophyllite, and crocidolite. Chrysolite is available in pelletized form and is readily dispersible in

Table 13.20   Typical Properties of Nylon 6/6

Property	Test method ASTM or UL	Unreinforced	20% carbon	30% carbon	40% carbon	20% carbon 20% glass	40% glass
Specific gravity	D792	1.14	1.23	1.28	1.34	1.40	1.46
Water absorption (24 hr), %	D570	1.6	0.6	0.5	0.4	0.5	0.6
Equilibrium after continuous immersion		8.0	2.7	2.4	2.1		3.0
Mold shrinkage, (⅛-in. average section), 10^3 in/in	D955	15.0	2.0-3.0	1.5-2.5	1.5-2.5	2.5-3.5	3.5-4.0
Tensile strength, psi	D638	11,800	28,000	35,000	40,000	34,000	31,000
Tensile elongation, %	D638	10	3-4	3-4	3-4	3-4	2-3
Flexural strength, psi	D790	15,000	42,000	51,000	60,000	49,000	42,000
Flexural modulus, psi x 10^6	D790	0.4	2.4	2.9	3.4	2.8	1.6
Shear strength, psi	D732	9,600	12,000	13,000	14,000	13,000	12,000
Izod impact strength, ft-lb/in	D256						
notched (¼-in)		0.9	1.1	1.5	1.6	1.8	2.6
unnotched (¼-in)			8.0 .	12.0	13.0	16.0	19.0
Heat-distortion temp @ 264 psi, deg F	D648	150	495	495	500	500	500
Coefficient of thermal expansion, 10^5 in/in deg F	D696	4.5	1.40	1.05	0.80	1.15	1.4
Thermal conductivity, Btu-in/sq hr-ft/deg F	C177	1.7	5.5	7.0	8.5	6.4	3.6
Surface resistivity, ohms/sq		10^{15}	25-30	3-5	1-3		10^{15}
Flammability	UL 94						

thermoplastics using conventional processes. One of the major uses of asbestos is in vinyl flooring compounds to improve hardness, thermal properties, and impact resistance. It is also used with resins other than vinyl such as polypropylene, polyesters, nylons, epoxies, and silicones.

Asbestos has come under fire from the Occupational Safety and Health Admin-instration (OSHA) because breathing contaminated air has been shown to cause cancer in workers. This means that increased care must be taken in plants with regard to the handling of asbestos in order to comply with OSHA standards.

Boron filaments are used for epoxy resin structures primarily in the aerospace industry, but their cost is extremely high, which limits their use.

Ceramic fibers have been used as reinforcements for their high strength-to-weight ratios, high temperature resistance, and their ability to withstand environmental attack. They have about the same strength as glass fibers, but have four to six times the stiffness and twice the temperature resistance.

Some synthetic fibers such as nylon, polyethylene teraphthalate, and polyacrylonitrile are also used as reinforcements. Polyvinyl alcohol fibers provide excellent adhesion, superior impact resistance, and flexibility during processing among other qualities. Although inferior to glass fibers in tensile strength, when used with glass blends, they improve flow properties.

Other materials used as reinforcements are sapphire filaments, which have good tensile properties, and whiskers, which are single-crystal filaments with great strength. Their strength is provided by their small size. Whiskers are sapphire, alpha silicon carbide, beta silicon carbide, and silicon nitride. Although they have been used in both thermosets and thermoplastics, their use is limited because of their high cost.

13.3 COMPOUNDING METHODS (EQUIPMENT AND TECHNIQUES)

The availability of improved processing equipment has greatly influenced the types and levels of additives used today. The use of multiscrew extruders has improved dispersion so that a smaller amount of additive is needed for any given application, with a consequent reduction in additive costs. In many cases there is also improvement in physical and processing properties.

The equipment used to compound additives into a polymer system must be capable of performing some of the following process tasks:

1. Incorporate and homogenize the additives without exceeding degradation temperatures.

2. Generate high shear stresses for dispersion of nonreinforcing fillers or pigments.

3. Homogenize two or more materials of differing melt viscosities without creating a stratified or layered final mix.

4. Provide a uniform shear stress and heat history to each particle.

5. Allow precise control over the process ensuring narrow temperature distribution throughout the process and at discharge.

The control needed to ensure a narrow temperature distribution is particularly important; good control over residence time distribution is an essential feature of a continuous compounding system. To minimize heat history and to guarantee repeatability, residence time distribution must be short and uniform. Also, many additives are very sensitive to prolonged exposure to high temperatures or shear. The loss of additives due to unwanted devolatilizing effects may be critical.

In addition to these general requirements, certain types of additives present special compounding problems. Slip, antiblock, and wax-base lubricants are low-melting-point additives that present problems when compounded at master batch levels. They melt before the polymer has become molten and act as excellent lubricants. In a single-screw compounding system forward conveying is erratic or virtually impossible. Intermeshing twin-screw extruders are able to overcome this conveying problem because of their positive conveying characteristics. The ability to put work into the polymer for melting and homogenizing, however, is still greatly reduced by the lubricating effect of these additives. Optimum homogenizing of two components can be achieved when the components have nearly equal viscosities. In the case of liquid additives, this can be done effectively by feeding the additives after the polymer is fully molten. By melting the additive and injecting it downstream into the processing section after polymer melting takes place, high levels can be incorporated uniformly and homogenized in the twin-screw compounder. Kneading elements are also effectively used to introduce high amounts of energy at specific locations within the processing section. In addition, these kneading elements, by repeatedly changing the shear direction, overcome the striation problem inherent in less sophisticated mixing equipment.

A similar situation can exist with the stabilizer additives. These, being either all organic or part organic with some inorganic materials, can have significant viscosity differences between the organic component and the polymer matrix, especially at processing temperatures (zinc-stearate or plasticizers, for example).

13.3.1   Direct Addition

In most cases today, additives are added directly to the polymer matrix. The advantage of this method is that changes in formulation (additive package) can be made easily. However, there are also numerous drawbacks. For example, although this allows the processor to formulate the most effective proprietary additive package, it may not be possible to compound the mix on available processing equipment. Since the percentage of additives is usually very small, danger in inaccurate metering in continuous processing equipment can be a problem. To overcome this, a master mix is usually prepared, which requires an extra premixing step. Also in the direct method, the handling of powder-type additive ingredients may be difficult as well as unpleasant for plant personnel because of dust and odors.

13.3.2   Multipurpose Concentrates

One way to minimize the exposure of operating personnel to these undesirable odors and dust is to use recently developed multi-additive concentrates. When properly compounded, multipurpose concentrates can be used with a basic polymer to produce many different material characteristics and end-product properties. These multipurpose concentrates can be tailored to specific requirements; some of the common types available today include stabilizer and lubricant; stabilizer-processing aid; slip, antiblock, antioxidant, and others. Plastics processors (injection, extrusion, etc.) with limited technical resources have found that these multipurpose concentrates offer important production efficiencies and often substantial material cost savings. They also eliminate the need for the processor to handle and store a wide variety of additives.

Unless the processor has in-house compounding capabilities, the materials supplier or compounder prepares the additive package. Since the additive concentrates are normally composed of a standard polymer matrix, compatibility with the final end product must also be considered. It may be necessary to have more than one polymer additive system available, because of the broad spectrum of polymeric materials processed.

Future government restrictions through the Toxic Substances Control Act (TOSCA) will encourage most small-volume processors to turn to multipurpose concentrates to avoid lengthy and uncertain regulatory decisions. In addition, use of concentrates will remove from the final processor all the potential compounding problems that he might not be technically staffed to handle. All the processor has to be concerned about is that the concentrate level is within the range of his processing equipment's capabilities. It is, however, of no advantage to consider additive concentrates of extremely high levels and subsequently face problems during the let-down operation. For example, some slip-antiblock-antioxidant concentrates can be prepared in very high concentrations (50% or more) when using proper compounding methods. However, the final processor might have great difficulties in adequately dispersing this high-additive concentrate on conventional processing equipment since the final concentration required is usually below 1%. In this case, a more practical concentration is in the 20–30% range, which could be processed on conventional equipment.

13.3.3 How Much Concentrate?

The level of concentration is also an important factor in the efficiency and quality of the end product. The additives are most effective at their optimum concentrations. An excess amount of additive in a flame-retardant formulation, for instance, would result in reduced physical strength of the product. To minimize the loss of properties, other inorganic fillers such as glass fibers, talc, or $CaCO_3$ may be incorporated.

13.3.4 Conventional versus Sophisticated Compounding Equipment

Many additives are, of course, still compounded in discontinuous batch processes. However, to meet today's demands for higher product quality and increased volume requirements, continuous equipment is required. In continuous compounding systems, variation between batches is eliminated, and the possibilities of human error are minimized.

Conventional processing equipment is often represented by single-screw extruders. In the mixing aspect of the extrusion process, velocity distributions and, subsequently, stress distributions are the most important parameters. The stress distribution inside the screw channel determines the degree of mixing that can be achieved.

13.3.5 Melt Conveying

A considerable amount of information is available on melt conveying in single-screw equipment. By observing the velocity and stress profile of down channel and cross flow, it can be seen that a zero-shear stress point exists inside the channel. In this area very

little or no mixing will occur. This has been confirmed by actual experiments. Therefore considerable effort has been made to improve single-screw systems. This involved installing mixing devices that would shift the zero-shear stress point to various locations in the screw channel and, it was hoped, achieve an adequate mix provided that the equipment had a sufficient processing length. There are a number of single-screw systems available today with improved mixing devices developed to cope with the very demanding compounding of additives. Among these are the patented EVK, the Maillefer type, the fluted screw, and other designs.

13.3.6 Commercial Twin-Screw Systems

One way to overcome the limitation of equipment with poor stress distribution in the channel is to improve the processing equipment design concept. The available twin-screw equipment today is often considered the ultimate solution to many compounding problems. There is a variety of twin-screw designs commercially available, and these differ widely in their operating principles and functions (Figure 13.7).

The two basic twin-screw mechanisms, however, that are extensively used today commercially for compounding operations include intermeshing counterrotating and intermeshing corotating twin-screw equipment.

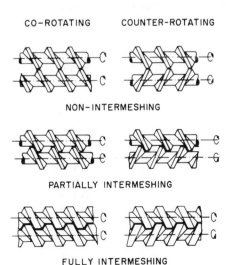

Figure 13.7 Types of twin-screw systems.

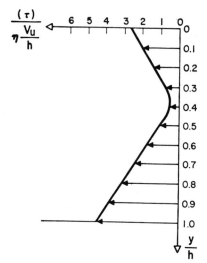

Figure 13.8 Counterrotating machines: stresses of cross- and down-channel flow components.

In corotating machines, the flow pattern of polymer melt is quite different. Depending on the number of screw flights, the degree of integrated shear rates is also different.

The material is actually conveyed in a figure-eight-shaped pattern. Zero-shear stress point can be influenced by changing the throughput and/or screw speed as well as by changing geometry. In the same analysis of shear-stress distribution for corotating machines, it is seen that zero-shear stress point can be influenced by operating variables. (Figure 13.9)

The most uneven stress distribution occurs at closed discharge, which is meaningless in actual operation. Since some additives are used in extremely small portions, it is imperative that uniform stress distribution is maintained even with increased throughput. Therefore, when more sophisticated twin-screw machines are considered for additive compounding, the degree of uniform dispersion is directly related to stress-strain distributions, particularly at very low concentrations.

In corotating machines, a proper selection and arrangement of individual screw sections and kneading elements further improves the uniformity of stress distribution.

13.3.7 Case Histories

13.3.7.1 Incorporation of Slip-Antiblock-Antioxidant Concentrations

A common formulation consists of the following:

In an analysis of the velocity and stress distribution in counterrotating twin-screw machines, it can be seen that there is actually no improvement regarding velocity profile and zero-shear stress-point location. The conveying efficiency achieved, however, is undoubtedly superior to that possible in single-screw machines. When the shear-stress distribution of down channel and cross flow (for a given coordinate system) are superpositioned, a characteristic minimum inside the screw channel in the area of $y/h = 0.38$ is observed (Figure 13.8).

Through analysis it is seen that the poor shear-stress region will always exist inside the channel regardless of operating conditions such as screw speed or throughput.

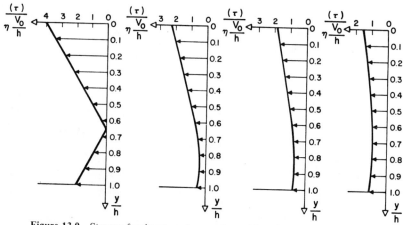

Figure 13.9 Stresses for down- and cross-channel flow in corotating screws.

1.   Erucamide slip agent.

2.   Butylated hydroxy toluene (BHT) antioxidant.

3.   Diatomaceous silica composition as antiblocking agent.

Materials 1 and 2 have melting points at 80°C and 70°C, respectively, while the third component does not melt at all. Feeding all the components into the feed port is limited to approximately 12–14% even when using more sophisticated processing equipment. At higher concentrations, the lubrication effect of additives is too high to achieve adequate mix.

To increase the level of additives, liquid injection of materials 1 and 2 was considered. Since the polymer is already melted before injection of the additives, viscosities are more compatible and 20–25% of additives can be incorporated before reaching the point of poor mix (Figure 13.10).

Since these kinds of additive packages are used primarily for large-volume commodity plastics (low-density polyethylene), direct addition into the melt stream of processing equipment is normally considered. These high-volume products are usually processed on conventional processing equipment. As a result, there is a limitation with respect to accepting high-concentrate levels while achieving good distribution in the final product. This kind of arrangement reaches its optimum at approximately 15% concentrate. Any higher concentration would result in deficient material quality or a reduction of throughput.

13.3.7.2   *Internal Lubricant Concentrates*

In many instances, engineering plastics are used in applications that involve very complex part shapes that result in high reject rates during the molding operation because of difficulties with mold ejection. Today, silicone fluid concentrates are available that improve material flow and also act as mold release agents. These concentrates can be used in a letdown ratio of 40:1 or higher. To achieve a high level of silicone in the concentrate, multiple injection is required, utilizing sophisticated processing equipment. On conventional equipment, it is difficult to achieve more than 10% concentration.

13.3.7.3   *Compounding of Flame Retardants*

Flame-retardant additives cause deterioration of some polymer properties (impact strength, tensile strength, etc.). These additives are primarily used to meet certain requirements set by government or industry. Since flame retardance is the key objective, it is imperative to have all the ingredients thoroughly mixed. Although these ingredients are supposed to retard flames, they can easily be degraded by heating during the compounding operation. This would lead to the reduction of the flammability rating or would require an increase of the additive level. Neither deficient material nor higher cost is desirable. In a controlled processing technique, however, the reduction of flame-retardant composition is possible without sacrificing the flammability rating. This reduction can be as high as 10–15%, which is

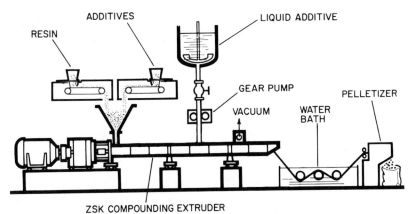

Figure 13.10   Compounding of solid and liquid additives.

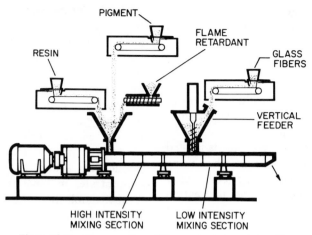

Figure 13.11   Compounding of flame retardant and glass fiber.

considerable savings in costs, as well as in physical properties. To offset the property loss caused by the flame-retardant additives, reinforcing fillers are added. In this case, two distinct compounding operations are accomplished in one step: high-intensity mixing of the flame retardant and low-intensity mixing of the reinforcing filler (Figure 13.11).

Although cross-linking does not come strictly within the scope of this chapter, some of the problems of compounding are similar to those that arise with additives for other purposes and the solutions for processing are similar. This is one of the specialty applications where it is necessary to employ a two-stage compounding system. The mixing and compounding operation is separated from the pumping, conveying, and pelletizing function. Compounding is accomplished in a twin-screw extruder followed by a single-screw discharge extruder. A typical 600 V cable formulation consists of low-density polyethylene with fillers, usually carbon black or calcium carbonate, plus one of several different peroxides with activation temperatures ranging from 120° to about 175°C.

Two approaches to compounding of cross-linkable polyethylene have been successfully demonstrated in commercial operations using corotating twin-screw equipment. One involves feeding a preblend of solid peroxide, fillers, and additives. The excellent materials conveying characteristics of this equipment

overcome any feeding problems due to the lubricity of the peroxide. The controlled high-shear input of the kneading blocks enables good mixing, homogenization, and dispersion of the fillers without generating a melt temperature that would initiate cross-linking. A variation of this requires preblending all of the components including polyethylene. A second approach involves metering the liquid peroxide downstream after polyethylene melting has occurred. In this manner the problem of handling solid peroxide is eliminated, as this has a great tendency to block. It may also be possible to eliminate the preblending operation by separately metering all of the solid components into the feed barrel.

13.4   SUMMARY

There are varying approaches possible to compounding high loadings or very small quantities of additives with widely varying viscosities. From the point of view of process engineering, each of these additive systems can be most efficiently incorporated into the polymer when they are at similar viscosities. To successfully carry out the process objectives, a compounding machine with excellent conveying characteristics is essential. Second, a compounding machine that allows maximum process flexibility such as multifeeding locations is also desirable. Without this flexibility, it may be necessary to sacrifice material quality to be able to compound at high levels. Finally, the ability to specify exactly where and how much shear input will be located within the processing section and the ability to control the degree of mixing intensity by using different screw arrangements are additional processing assets. With this versatility, it is possible to achieve optimum process conditions for even the most difficult mixing and compounding task and obtain the desired end-product properties.

As research continues and new consumer demands arise, technology will improve, and there will be not only new uses for old materials, but also new ways of combining additives to meet differing demands. Processing technology will be refined to overcome some of the problems now existing and to meet higher levels of production.

13.5 REFERENCES

1. A A. Schoengood, *Additives*, SPE Journal (April 1974).

2. Plastics Engineering Staff, "Additives: How and Why They Work," *Plast. Eng.* (March 1975).

3. Joseph R. Copeland and Olinda W. Rush, "Wollastonite: short-fiber filler/reinforcement," *Plast. Compd.* (November/December 1978).

4. Harry S. Katz, "Carbon graphite reinforcements are coming of age," *Plast. Compd.* (March/April 1979).

5. George Crowe and Paul E. Kummer, "Extending resins with calcium carbonate," *Plast. Compd.* (September/October 1978).

6. Thomas D. Thompson, "Applications for kaolin fillers," *Plast. Compd.* (May/June 1979).

7. James B. Marsden, "Functions, applications and advantages of silane coupling agents," *Plast. Compd.* (July/August 1978).

8. J. Gilbert Mohr, "Evaluating glass-fiber reinforcements," *Plast. Compd.* (July/August 1978).

9. Stan Jakopin, *Compounding of Additives*, SPE 37th ANTEC, New Orleans, Louisiana, 1979.

CHAPTER 14

Product Design

I. MARTIN SPIER

Beacon Plastic & Metal Products, Inc.
New York, New York

14.1	Introduction	653
14.2	Investigatory Design Criteria	654
	14.2.1 Ultimate Use and Cost	654
	14.2.2 Competitive Products and Performance Factors	655
	14.2.3 Production Requirements	662
14.3	Material Specifications	663
14.4	Manufacturing Considerations	668
	14.4.1 Processing and Design Technique Comparison	670
	14.4.2 Part Factors	673
	14.4.2.1 Size	673
	14.4.2.2 Shrinkage	678
	14.4.2.3 Draft or Taper	680
	14.4.2.4 Radii and Fillets	681
	14.4.2.5 Undercuts	682
	14.4.2.6 Movable Cores	683
	14.4.2.7 Molded Lettering	683
	14.4.2.8 Sink Marks	684
	14.4.2.9 Flow or Knit Lines	687
	14.4.2.10 Inserts	688
14.5	Service Performance	689
14.6	Finalization of Design	691
14.7	Design Case Studies	691
	14.7.1 Consumer Product Design	692
	14.7.2 Industrial Product Design	694
14.8	Summary	695
14.9	References	698

14.1 INTRODUCTION

Designing a part in plastic permits the designer to contemplate various similar or dissimilar materials and various methods of fabrication. Plastic can be made to look like wood, it can be sawed like wood, but it is not wood. Plastic suffers from creep—wood does not. Plastic may look like glass, but it is not glass—it has a lower melting temperature, it will yield, it will elongate, it is affected by acids, solvents, and so on. However, if dropped on the floor it will not break as readily as glass. Plastic can function as a sheet of metal, and it can be fabricated with metal working equipment. But again, it has memory and cannot be used completely interchangeably with sheet metal. Even though it has the advantages of built-in color, not requiring any secondary decorating operation, and the possibility of producing it without the sharp edges associated with metal stamping, it is not metal. Plastic can be cast as one would either a nonferrous or a ferrous alloy. However, it cannot be compared with cast metallics, as its temperature range is much lower. It is lighter in weight, which can be a distinct advantage or liability depending upon the application.

Plastics can be thought of as a film and woven. However, synthetic fabricated materials do not have the same qualities as natural fibers. Basically, they are nonhydroscopic and are more susceptible to heat and flammability, and not resistant to solvents. So although it is possible to use the plastic filament to create a fabric, it cannot be used completely interchangeably with natural fibers. It is far more suitable for use outdoors, such as a sail; a dacron sail will outlive a cotton sail. Frequently a nylon undershirt will leave a person chilled and clammy, whereas a cotton shirt will absorb moisture.

Plastics have generally replaced other materials because they are cheaper, lighter, more attractive, and because they have entered the market at a later date, replacing older products owing to superior and innovative design. The lower material cost of plastics in relation to natural materials, such as leather, made it possible for the product designer to do more creative thinking in the design of the product. To this, latitude was available by reason of the reduced initial cost of the raw material. It is the break in tradition from existing materials and the difference in processing that has permitted the uniqueness of a plastic product to be developed and exploited by the intelligent product designer. The ability to combine multiple functions in one piece or a simplified assembly has given rise to a number of unique new products that never would have existed had the introduction of plastics not permitted the complicated utilization of compound curvatures, and so on. The stimulus of plastic is almost a magical work, as exemplified by Dustin Hoffman's admonishment in *The Graduate*—"plastics."

14.2   INVESTIGATORY DESIGN CRITERIA

The design capabilities of plastics have opened new vistas to the design engineer. No longer is the design engineer hampered by tradition or with the problem of displacing an existing item or existing tooling. A new product can be presented, and fortunately, since there is no track record of any past performance of this product, one is able to do both imaginative and creative thinking and not be limited by any rigid manufacturing technology or discipline. An example of this is the development of plastic cameras which have only a fraction of the total component parts of the metal cameras that they have replaced. And whereas a slide made out of metal had multiple stampings and a difficult

assembly product, today slide carriers are molded of one single piece.

The standards of establishing the suitability or nonsuitability of a product to be produced in plastic are dependent upon several distinct factors. For instance, a product made out of a traditional material such as metal or wood could have enjoyed an excellent reputation and a good sales market. Then competition comes out with a plastic product that is lighter, more attractive, and cheaper than the product that is currently being manufactured. Suddenly, the part that has functioned properly for a long time now becomes obsolete because of the introduction of a competitive product at a lower price that has greater sales appeal. So the suitability of a product to be made in plastic can be dependent upon the part that it replaces and whether or not it will meet the service criteria that is required of the product.

14.2.1   Ultimate Use and Cost

The product designer should be able to view all the possible situations in which the product is going to be used, so that one can avoid the limitations and pitfalls inherent in designing with almost any material, be it plastics, metals, or other materials.

The ultimate use of a product is its most limiting factor. Obviously, if it is a small component that is going to be utilized in a computer, then the cost of materials and the cost of molding is completely irrelevant to its fail-safe capabilities, as one hour of downtime can run into thousands of dollars of loss compared to whether a product costs 10 or 15 cents to manufacture. Yet conversely, a disposable drinking cup would have a tenth of a cent as being the substantial difference between its economic saleability and its nonsaleability in this highly competitive market. Again, the ultimate cost of the product is a determining factor in the cost of the component. If one is designing an article of furniture and the component that one is involved in is a door knob or a handle, then again on a $200 (or $300) item of furniture whether the knob will cost 3 cents or 10 cents is not the significant contributing factor.

To the designer, selection of materials is only one of the considerations that is involved in creating a product.

There is a direct relationship between a product's size and its cost. Tooling and capital equipment expenses soar as the part gets larger, particularly for injection molding. Small does not always mean low cost, because of the high tooling cost for small and intricate instrument parts. But in general, designers of very large parts should be prepared to take note of the substantial increase in manufacturing costs. One of the factors accelerating the cost of large-part production is the time required. As the tool becomes larger, it requires a longer time to manufacture, to assemble, to test, and to modify than a smaller tool. In figuring costs, it has been common to take machine time as the controlling factor, not recognizing that many times a part can be molded in a relatively slow cycle while the operator uses the intervening time between cycles to perform secondary operations on the part. It is obvious that an operator's time cost is related to the productive capacity of the machine, whereas an operation that can be performed on an assembly line only carries the assembly line labor and not the overhead burden of the machine.

There are many hidden factors that go into determining the economics of a material that usually are not obvious when one looks at a plastic properties chart. For example, if a product had a breakage factor of 1 or 2%, it might be tempting to say it is cheaper to have a 2% breakage factor than to pay 10% more for the total cost of material. However, the added cost of the 2% breakage factor is not just on the cost of the product. The paperwork and the handling of a return from a store back to the manufacturer, the credit details, plus the dissatisfaction that occurs at the retail level can be very substantial.

As a common rule of thumb, in the instance of a unitary product that has no secondary or assembly operation, there is a relationship between the weight of a part and its value from the standpoint of both material and molding cost. Generally speaking, the cost of labor or molding is approximately the cost of the material.

Therefore, by reducing a part's wall thickness from $\frac{1}{4}$ to $\frac{1}{8}$ in., not only is the material cost cut in half but also the molding cycle would be substantially decreased, and this might amount to an even greater saving than the material cost itself. By redesigning a part, therefore, one can end up with a product that is superior because of the better quality of the material used (such as acrylonitrile butadiene styrene, ABS) and also costs less in production dollars than an older part produced with a cheaper material (such as general-purpose polystyrene).

With a higher molding temperature, the part must remain in the mold longer to cool, a lighter part may cost more to produce because it might require an injection molding machine of greater tonnage, which in turn carries a higher overhead burden. Although there may be a reduction in the cost of material or the amount of material used, the costs of the slower molding cycle and the larger machine necessary to take advantage of the thinner wall section are more than the cost of a heavier part, which usually will be superior anyway. Thinner does not equal cheaper in overall costs.

The interrelationship between the amount of material, molding cycle, hourly burden on the machine, and the variables discussed ultimately determines the economics of part production.

14.2.2   Competitive Product and Performance Factors

It is always rewarding to have had some prior experience directly from a similar part. Also, information can sometimes be made available from a competitive product if relevant facts and figures are published (which is not unusual). Although it is possible to predict with a fair degree of accuracy the manufactured costs of a product and its productive quantity capabilities, it is very difficult to predict as accurately the sales volume that will be generated. Again, the automobile industry, which has some of the most sophisticated and talented people associated with it, can reasonably project or extrapolate from sales trends whether 8, 9, or 10 million cars will be produced per year. But to determine whether or not a coupe or a station wagon or any one particular model will sell, and in what color, is a very difficult and an inexact science.

However, it is not a question of copying to relate contemplated design with past products or allied products. An example of this would be if one were involved in designing a new heat pump window unit in place of a conventional air conditioner. It would be a very good frame of reference to study the design criteria that had been used in successful air conditioning units, the material that had been employed, and all other conditions such as wall thickness, and so on. For although the heat pump would be an add-on feature not contained in the original air conditioner unit, so many factors are similar that once the requirement of the maximum thermal requirement was met in material selection, almost all the other requirements would be met by an air conditioner housing. As both products must meet Underwriters Laboratory Standards, they are both exposed to ultraviolet. There is mechanical vibration of the compressor, and condensation in the cooling cycle. All the air conditioning requirements that have been met satisfactorily apply to the heat pump, to which must be added the reverse cycle. It must be taken into consideration that now it will be used in the winter as well as during the summer, and so from a static position in low temperature, it is now going to be in an active position at low temperature. Therefore, its resistance to fracture and failure will be accentuated because of its low temperature environment, and at the very same time there will be an area of the part that will be subjected to an elevated temperature. However, these known variables can be anticipated and combined with the experience gained in producing an air conditioner.

If one were to think of designing an electric toothbrush and this was a new product for the designer, it would be possible to take an electric carving knife and study its material selection. A very close similarity would be established, since both of these instruments hold an electromechanical mechanism, both are held in one hand, both require switch control, and both basically are used indoors. So a material that sufficed and gave good satisfactory service as a carving knife would be an excellent base to relate from if one were designing an electric toothbrush.

It is not uncommon to see in the research and design laboratories every single competitor's product lined up to be examined carefully for their good and bad points. Then either an average or the sum total of the plus points is taken, and the obvious weak points are eliminated. This becomes the basis for the new product's design criteria. The usage of a material that met minimum requirements in the past need not necessarily be followed in the selection of the new design. The part could have been overdesigned, and it might be possible to use a cheaper material. Conversely, it could have been underdesigned, and then a superior material could justify itself if there was a substantial degree of part failure with the lower-cost material. It is always good to review the selection of a plastic material as opposed to alternate materials such as metal, wood, cardboard, or glass. Rarely is there only one material from which a product can be made. It is not only a question of economics, but there is also a traditional acceptance of a product. There is a competitive position, and consideration must be given to the possible choice of any one of the five or more different materials that could do the job. It is necessary to equate the alternate materials from a standpoint of availability, ease of fabrication, as well as total economic cost. Likewise, there are a good number of plastic materials in a similar economic range that could produce the part satisfactorily. Reasons for substitution could be material shortages, slight variation in price, or a familiarity in processing that one company prefers. Examples are high-density polyethylene and polypropylene, acrylics and cellulosics such as butyrate. Thus it can be seen that there is in a general purpose product more than one material that will meet the requirements.

The interior and exterior size is a good reference point in the first developmental thinking in plastic product design. Size is a limiting factor. If one were desirous of utilizing an injection molded housing, then one is limited by the reasonable available capacity of injection molding machines. Although injection molding machines do go up to 5000 tons, they are very few in number. It would be more advisable to think in terms of 3000 tons of clamping pressure, since there is a substantial number of such machines available in custom molding plants as well as from machine tool builders. With the rule of thumb of from 2 to 3 tons/in.^{2}, the 3000 ton machine would limit the size of a part to approximately 1000–1500 in.^{2}. Therefore, the maximum size part that could be readily contemplated would be 7 ft^2, with the possibility of 8 or 9 ft as the outer limit dependent on material, wall thickness, flow characteristics. Eight square feet or slightly more brings to mind a convenient rectangular size of a modular unit, such as 2 × 4 ft. The design of injection molding platens are basically square to rectangular and have a size ratio of less than 2:1. It would be impossible to mold a piece 6 in. × 16 ft (8 ft^2) long as there is no injection press with a platen that would accommodate such an extreme configuration. Thus, one has an economically limiting factor in the size of the part that can be produced by injection molding. Then again, a determination must be made on company policy as to whether parts to be produced must be able to be fabricated or manufactured in the plant with their existing equipment, or whether the company is willing to buy new equipment or have the part produced by a custom molder. It has been said that bankers are happy to lend money to people who can prove that they do not need it. This indicates a track record.

The same argument can be applied to the performance of a part in service. Just because parts have been made of a particular material or configuration does not mean that they have met with satisfactory service. So in determining the suitability of a product or material or design, if one can identify a similar product and get any information about its suitability and satisfactory performance, that is a plus factor in establishing one of the parameters necessary for intelligent product design. Many times a part will fail in service, owing not to any inherent physical limitation on the part of the product, but to the fact that it will be abused in service. It is, unfortunately,

a very difficult determination as to how far people will go in subjecting a product to use and abuse beyond normal contemplation. It is axiomatic that when something fails you try to do what you can to get it to work. When it doesn't work, you hit it. And if it still doesn't work, then you read the instructions.

An example of unanticipated abuse was the design of a "soap dispenser" used in industrial bathrooms. This was a crank-operated mechanism that had every single component molded. It had been engineered to the highest standards and performed excellently. It had a conical shaped filling cover that fitted snugly into the lower housing with no opportunity for manual removal of the cover. However, it was found that people use this nice round canister as a punching bag, and that when a sharp blow was applied to the container, the material, which was ABS flexed, caused the cap to fly off, thus exposing the soap to all the ambient dust, dirt, and so on. Although this might seem to be a far-fetched example, it is one that actually did take place. Its resolution was to design a bayonet-type of insertion that required a rotary action in addition to just a press fit. Why does a part fail in service? Usually not for reasons that are initially obvious. However, after an analysis of the cause, the solution does become fairly obvious. It is really the unanticipated problem that causes failure in service. Examples of this are numerous.

A polystyrene medicine cabinet frame that completely surrounds the mirror was designed with excellent appearance, good dimensional stability, high gloss, and met all the requirements so that it became a highly successful item in department stores. However, after a short time returns began to come into the stores, and the material looked crazed, warped, and so on. The cause: the designer did not realize that women use hair lacquer and that they spray their hair from the back facing the mirror and that there will be an overspray of the hair lacquer that will attack the plastic frame.

Another product was a "boarding ladder" used on boats, molded of structural polyethylene. It had all the desirable features. It was light, it was smooth. It would float—but it would also break. Although the part had been designed with apparent sufficient rigidity to support swimmers, the selection of conventional polyethylene without any ultraviolet inhibitor caused the failure. The structural foam without any ultraviolet inhibitor in the presence of salt water had limited life. And so it was the lack of recognition in material selection not to include the ultraviolet inhibitor that caused the failure of the product, and the bankruptcy of the company.

After the preliminary overview has been made, then one can establish a checklist of specific factors: number one would be the particular strength of the product; that is, what is required of the product in normal usage. Is it going to sit in a static or stationary position? Is the product going to be used as an article of manual handling? Is it a product that will be used as a container in shipment? These specific requirements of the ultimate strength of a product can be established. An example would be a shipping container, where possibly it would be subjected to an internal loading of 50 lb. The Interstate Commerce Commission imposes certain standards for shipping containers. There are drop tests, that is, the container must be loaded depending upon its cubic size and dropped 8 or 10 ft on a corner without breaking, or must withstand seven or six drops according to particular standards. This is relatively easy. When standards are set up the product designer is relieved of any personal determination. Although it may be difficult to achieve the standard, once it is met, the designer is off the hook.

Where there is a substantial visible flow-weld or knit mark, there is a reduction in the physical strength of the part. By gating such a product with multiple gates and allowing the hot material to enter directly around the core pin, it is reasonable to expect that the core pin area would have no visible flow marks and the part would be produced with more uniform physical strength. A properly designed foam part utilizes material more efficiently in terms of strength.

As are other materials, plastics are vulnerable to change and degradation by outdoor exposure. This may result in an unattractive appearance, loss of engineering properties, and even complete deterioration.

The moisture content of plastic directly influences certain physical properties of the plastic—electrical insulation resistance, dielectric losses, mechanical strength, appearance, and dimensions. When possible, to determine the effect of moisture on a plastic article, the whole article should be immersed, instead of specimens cut from it, especially when determining dimensional stability and surface effects. When determining moisture resistance on articles that are to be machined for use, the samples tested should also be machined, because machining removes the surface skin and exposes any filler. This often results in a machined piece absorbing more water than the unmachined molded piece.

The word stress as it affects plastics is used in two ways, both of which are important to the designer. "Molded-in stresses" are a function of product and mold design and processing variables, and are created during the molding operation. "Applied stress" refers to the mechanical or environmental strain or load inflicted on the product during use. Product stress failure often is the result of applied stress causing rupture or deformation in the areas of the part that contain molded-in stresses.

There are several things a designer can do to minimize molded-in stresses, which are caused by nonuniformity of flow pattern and wall thickness and improper mold venting. One may reduce the differences in wall thickness, eliminate sharp corners, and introduce ribs and bosses in strategic areas. In addition, adequate clearance should be provided in an assembly where a differential in thermal expansion is caused by the use of dissimilar materials. These steps can help eliminate many of the failures that result from molded-in stresses.

Applied stress, on the other hand, is not something that can be eliminated but must be considered when material selection and part design are initiated. Applied stresses are less likely to cause part failure if the material chosen to suit the application and molded-in stresses have been minimized.

It is important to know the environment to which the product is going to be subjected. It is necessary to know if it is going to be used in perfumery, a housing that will be in a manufacturing plant, a plating plant where there are acrid fumes, or in any other environment. A product designed for normal use will have a very short life if it is subjected to an inhospitable environment.

The various types of radiation may alter the physical or chemical properties of a plastic, which in turn will change its performance as a finished product. The constant testing and the peaceful uses of nuclear energy will create radiation that will continue to affect material. Until the exact level of exposure has been determined it is difficult to anticipate what effect in the degradation of the resin nuclear radiation will have. However, it is reasonable to assume that there will be some reduction in the physical or dielectrical characteristics of polymers by reason of exposure to nuclear radiation.

Exposure to sunlight and ultraviolet are two areas to which the product designer must give thought, even though at first blush the product might not be intended for outdoor exposure. Because fluorescent light simulates in many ways the ultraviolet characteristics of natural sunlight, the product may be exposed for a long time to fluorescent lighting with its range of ultraviolet.

An example of this situation was found in designing grills for fluorescent light fixtures; it was thought this was ideal for polystyrene. Polystyrene has good optical characteristics, could be molded in the egg-crate configuration as wanted, it would flow and fill the mold, and it was cheap. However, being exposed for a time to the x-rays existing in fluorescent lighting, it turned yellow and cracked. The result is that polystyrene has been replaced, and now polyvinyl chloride (PVC) as well as carbonate and acrylics with ultraviolet inhibitors are used in its place.

The dielectric strength of a material is directly related to the amount of electricity that the product is going to surround. If it is a cable for electrical wiring, then the standards have already been established by the Underwriters Laboratory for the type of material, the size of cable, and the like. However, there are housings that must be designed to accommodate both electronic and electrical equipment, and it is in this case, where the housing may function both as a housing and as an in-

sulator, that the question of dielectric resistance must be calculated. Fortunately, this is one area in which the material tables give excellent reliable information concerning the relative resistance and conductivity of various plastic materials and the specific dielectric resistance. Basically, thermoplastic materials are excellent insulators and as a rule can be considered to be nonconductive from an electrical standpoint. These materials find wide acceptance for electromechanical housings ranging from cords, appliances, and direct insulation for electrical-current-bearing wires. It is one of the limitations of plastic that its nonconductivity requires special modification to the part to incorporate conductive areas, as a housing made of plastic will be nonconductive. Plastic can be made conductive by structural alterations and compounding with metallics. Also, it is possible to apply a conductive surface to the plastic such as silver nitrate and then electroplate copper or nickel over the plastic either completely or in certain limited areas, as in printed circuit and circuit-board technology.

Creep is the one basic limitation inherent in a thermoplastic material not found in most other materials. Defined as the material's desire to return to its original state, given temperature, time, and pressure, creep is directly related to the elasticity of the resin. This relationship means that designers are confronted with a difficult choice. Shock absorbency, or ability to resist fracture, is directly related to the creep factor. The more resilient, the more pliable a material is, the more it is likely to exhibit creep. So the trade-off between creep on the one hand and impact absorption on the other must be weighed. There are certain known techniques for reducing creep's effects. For example, reinforcement can increase creep resistance in a thermoplastic molded part. Higher heat materials and improved processing technology also reduce the immediate and short-range effect of creep. But like death and taxes, creep can't be totally avoided.

While creep is a thorn in the designer's side, it can also be a valuable tool to indicate related property changes. Creep data can be utilized to monitor the effect of a change in a molding process; check anneal, or stress-relief time-temperature cycles; monitor the effect of different foaming agents on section modulus as pore size is altered in structural foam parts; determine deleterious effects on strength of plastics subjected to oxidation, chemical environments, or weathering conditions; and compare the strength of plastics before and after various surface treatments.

The effect of physical creep, in varying loading modes, should be of concern to the astute designer. Simple measurements and calculations will provide assurance of whether or not a speculated design will or will not work; whether a different, more creep-resistant plastic should be used; whether a metal reinforcement is needed—or, if the plastic must be substituted by metal in a specific application. A suggested engineering guide in this area is a paper by Dr. Herbert F. Rondeau (1) showing the means and manipulative techniques that will enable the implementor to predict plastic creep by a one-day test that is about 95% accurate when extrapolated to one or more years, when compared with actual tests conducted over the same period. As such, this is a powerful design tool that a plastics engineer should not overlook.

Solid-state viscosity (ν) of a polymer can be determined by dividing the fibril stress (σ) of the test specimen by the creep factor (β). In the cantilever testing technique, introduced by Rondeau, the required sensitivity could make this unique material property a reality. Equationally, $\nu = \sigma/\beta$, from which viscosity can be inferred dimensionally.

One of the attributes of a plastic product is toxicity. There are standards that must be met. The Food and Drug Administration (FDA) fortunately or unfortunately, as the case may be, has established certain standards for various food products. A film used to wrap meat or fresh vegetables could have a very limited shelf life, possibly limited to one week, whereas a container designed to hold spices might have a shelf life of a year or more. Since the relationship between the shelf life and the usage of the container is quite substantial, odor absorption and migration of plasticizer if one is used, can become a serious problem. Another example

would be the disposable drinking cup. Its use might be limited to only half an hour or an hour. However, from the time it was manufactured, distributed, and sold, to the time it was actually utilized, could be six months or a year.

Products that are associated with children must be guarded very carefully. To make plastic other than the olefins soft and pliable, it has become necessary to add plasticizers which are softening agents. And in the development of plasticizers, tin, and other metallics have been incorporated. From a designers point of view, today it is safe to say that there are formulas of materials available that will meet criteria of flexular strength and flexibility that do not appear to have a toxic stabilizer, or that could be considered toxic. Special care must be given to children's toys, as not only will they be chewed on, but will also be digested. As such, for a rather small stomach volume, a substantial amount of material can be eaten by a child. It is therefore absolutely essential in specifying a material that a child might get his hands on (and there are few materials that children don't get their hands on), that any tin-based plasticizer that has a degree of toxicity should be avoided at all costs. There is a threshold of toxicity—small amounts of a great many materials including poisons are not harmful at a low level, but it is the concentration that makes them dangerous: 0.1% of benzolative soda used as a preservative is not considered toxic, yet 10% of benzolative soda would be a toxic ingredient.

Most plastics are resistant to fungi and bacteria. Some of the thermosets are not. Additionally, plasticizers, stabilizers, and other additives may not be resistant. In the case of polyvinyl chloride used outdoors for pipe, building materials, transportation, furniture, and so on, microbiologic resistance is important. A biocide is added to the polymer in processing molded dishes and cosmetics. Again, the additives that are used in food products require FDA approval.

The heat range to which a product will be subjected must be considered in depth. In this definition, one recognizes or should recognize the singular subjection of a product to a positive ambient condition. If one is thinking in terms of a housing for television, where there is a known thermal rise, then it is possible to know the maximum temperature to which the part will be subjected. Venting, grills, and the like do reduce the entrapped heat. However, there are many products that in their initial design are not contemplated to be subjected to an ultra-high heat environment, nevertheless parts will be placed on a radiator or cabinetry in a kitchen near a stove where elevated temperatures will affect the service life. Fortunately, the material suppliers have constantly improved the heat capabilities of their materials, and again, this is one area in which the material specifications chart will give one an intelligent, reliable area in which to recognize the yield point and limitations of a specific resin.

Any saleable product is subjected to both shelf life and service life. There is no direct correlation between the two. Shelf life is a term applied to the interval of time from which a product leaves a factory and is used by the consumer or purchaser. Service life is the interval of time in which the product is used by the customer or user.

Examples of this would be as follows: a laundry bleach might sit in a container for one year. Starting with the time it was shipped, perhaps going to a central warehouse until it was sent to a store (and it may be put in the back of the warehouse), until it actually arrives on the dealer's shelf may be a period of one year. Yet, once it reaches the household, the container of bleach might possibly have only a service life of a couple of weeks. The importance of this consideration is made manifest by television commercials. Then we have a long period of time in a shelf life and a relatively short service life.

Other products in the appliance field might, because of styling and the like, have a shelf life of only several months, and yet would have a service life of 10 years or more. It is the understanding the difference between shelf life and service life that determines the material selection and other design criteria.

A further example would be the disposable cup: through channels of distribution and marketing and purchase and lying on a shelf in a household, the shelf life could be a substantial period of time. Yet once it is held in the hand and used for several drinks, it is disposed of within a matter of hours. It is very difficult, unless the product is understood and analyzed, to put any particular limitations on shelf life, where service life is a more predictable requirement.

Correlated to service life of a product is its usage, whether or not it is a reusable or disposable product. If the product is reusable, the service life can be anticipated for several years. If it is a disposable product then it means the interval of time from the first opening until the contents are completed. A disposable product could have a longer shelf life than a reusable product. For instance, a cotton swab or a tongue depressor might be in its package for several years before the entire contents were consumed. Yet during this time in its static state it must retain all its initial requirements, even though once used it might have a service life of minutes or less. Therefore it can be seen that just because a product is disposable in a limited period of time does not eliminate the need for a long shelf life.

The most generally accepted method for determining the level of odor of plastics involves comparing directly with a standard, using a group of judges who have keen senses of smell. Associated with odor is odor absorption and permeability. Absorption is the characteristic of a plastic product acquiring the odor of a surrounding environment.

Permeability is the amount of gas or water that is absorbed by either a film or a molded product. A common way of measuring permeability in film is through the volume of gas that will penetrate through a film of a given thickness during a 24 hr period at a given pressure differential. Water vapor permeability, on the other hand, is measured by putting the material to be tested over the mouth of a dish that contains either water or desiccant and putting the assembly into a controlled humidity and temperature cabinet. The change in weight is used to calculate the water vapor movement through the material. There are established ASTM standards for this particular type of testing.

One of the unusual features of plastic is that it is possible to add color in a full range, or the lack of color, and that it is possible to have certain resins to be completely transparent or translucent, with no additional cost in the processing operation. There is an insignificant increase in cost to have the resin colored that is completely disproportionate to the cost that would be required for a postmolding dipping, spraying, or fluidized bath operation. It is the recognition of color and its attention-getting qualities that has insured acceptance to plastics. It is a way of making one product look different by molding it in various colors.

There are a few materials that are transparent, more that are translucent, and the greatest number of materials are available as opaque. Even though they may start out life in a transparent condition, it is possible to pigment materials to make them opaque; unfortunately there is no reverse process. Materials and resins that are opaque to start with cannot be transmuted to translucency. Again, the thickness of a part can affect its transparency. Polyethylene in a very thin film is almost completely transparent, yet from $\frac{1}{16}$ in. up it becomes translucent, and then completely opaque in thicker sections.

14.2.3 Production Requirements

Annual, daily, or weekly peak requirements must be considered when a part is being designed: whether (or not) it can be produced in sufficient quantity with available equipment; wheter (or not) the economics of production will meet the minimum requirements.

Whereas an annual usage of 300,000 pieces might indicate a daily requirement of 1000 pieces, this need not be so, as seasonal requirements are not as predictable as annual requirements. The annual production of 300,000 pieces equates to 1000 pieces a day on a 300 day year. Seasonal variations and other forces work to upset this known or limited production quantity. For example,

suppose one were producing an ice cream container, and a heat spell developed for a long time with uncomfortably extreme temperatures. The consumption and sale of ice cream would zoom during this time and then would taper off to a more predictable sales volume. Whereas the overall annual volume might not materially change, during a short time the requirements could be substantially greater. Possibly 3000 or 4000 pieces per day would be required during the summer season and a very reduced amount in other periods. Accordingly, not only the annual requirements but also the peak requirements of a week or month should be taken into consideration. Variables such as annual period of shutdown for model change, the possible modification of design, shortages of other components, holidays, strikes, and others should be taken into account to arrive at a more realistic daily requirement with a safety factor of 2:1 or 2.5:1.

Where 250,000 units would be an annual requirement, 2500 pieces per day would average out to 12,500 pieces per week. Again, one has a safety factor to reckon with. There could be machine failure, power failure, an operator may not come in, mold breakage and breakdown, and so on. To produce 12,500 pieces per week, a 25% minimum safety factor must be built in, so that one would think in terms of 16,000 pieces per week and a new daily productive capacity of 3200 pieces per day. This assumes that 100% of production will meet inspection standards. From past experience it will be determined that some plants have a higher or lower rejection (or dropout) rate on production. So whereas one has now established a safe quantity requirement required for daily estimated production, to this must be added a factor or coefficient of rejection, and if this were taken as 10%, the approximate production requirement becomes 3500 pieces per day.

Injection molding plants, although operating on a 24 hr basis, should calculate a safety factor of operation by bringing it down to 20 hr per day, so that dividing 3500 pieces by 20, one gets an hourly production of 175 pieces per hour. Quickly converted, this means 20 sec per piece. If the part could be run on a 20 sec cycle, a single cavity mold would suffice as far as producing the minimum (or the practical) requirements is concerned. However, cycle time is not the only criterion in establishing the size of a mold to be utilized in producing a part. The economic advantages of multiple cavity molds must be recognized. Even though a single-cavity mold could meet the annual requirements with a reasonable safety factor, the cost consideration might justify the building of a four- or six-cavity mold, in spite of the fact that it might work for only 2 or 3 weeks per year as opposed to a single-cavity mold working 4 or 5 months per year. The determining factors of mold selection are that the mold must meet not only the annual requirements but also the peak requirements, and it must do so at the lowest possible cost consistent with the overall productive life of the part.

14.3   MATERIAL SPECIFICATIONS

The product designer has a wide range of plastic materials to select from. They fall in two broad categories, thermosets and thermoplastics. Thermoset materials, which are the older, are limited in color, heavier, and more fragile, but have the advantages of good dielectric strength, high heat resistance, and dimensional stability. They have wide acceptance in the electrical appliance field and other electronic applications, and account for approximately 15% of the total market. In the thermoset category the three most popular materials are phenolic, polyester, and urea. Table 14.1 lists the most frequently used compression molding materials (some thermosets can be injected molded) and the most important properties of the materials and give their relative value and characteristics. Thermoplastics are by far the largest source of material used by the product designer, both for consumer prod-

Table 14.1   A Material Selector for Certain Thermosets

Property	Thermoset			
	Phenolics	Ureas	Melamines	Polyesters
Specific gravity (density)	1.69–2.0	1.47–1.52	1.45–1.52	1.65–2.60
Tensile strength (psi)	5000–18,000	5500–13,000	5000–9000	8000–20,000
Elongation (%)	0.2	0.5–1.0	0.6	—
Compressive strength (psi)	16,000–70,000	25,000–45,000	33,500	15,000–30,000
Flexural strength (psi)	10,000–60,000	10,000–18,000	9000–11,500	10,000–36,000
Impact strength (ft-lb/in.)	0.3–18.0	0.25–0.40	0.27–0.36	7.0–22.0
Hardness, Rockwell	E54–E101	M110–M120	M115–M125	50–70 (Barcol)
Flexural modulus 10 (psi, 73°F)	20.0–33.0	13–16	—	—
Thermal conductivity (10 cal/sec · 1 °C · cm)	8.2–14.5	7.0–10.0	6.5–8.5	—
Deflection temperature (°F at 264 psi fiber-stress)	300–600	260–290	266	375–500
Dielectric strength short-time, ⅛ in. thickness (volts/mil)	140–400	300–400	350–400	380–450
Flammability burning rate (in./min)	0.13–0.5	—	0.34–0.80	—
Effect of sunlight	General darkening	Pastels grey	Slight color change	Slight
Effect of weak acids	None to slight depending on acid	Attacked	None to slight	None

Table 14.1 (*Continued*)

Property	Thermoset			
	Phenolics	Ureas	Melamines	Polyesters
Effect of strong acids	Decomposed by oxidizing acids; reducing and organic acids, none to slight effect	Decomposed or surface attacked	Decomposes	
Effect of weak alkalies	Slight to marked depending on alkalinity	Slight to marked	None	Slight
Effect of strong alkalies	Attacked	Decomposes	Decomposes	Attacked
Effect of organic solvents	Fairly resistant on bleed-proof materials	None to slight	None, proof colors	Slight
Cost ($/lb)	0.40	0.56	0.55	0.42
Cost ($/in³)	0.018	0.023	0.023	0.02
Qualities	Rigid, good dimensional stability, low moisture absorption, readily processed	Full range of opaque colors	Good range of colors, good dimensional stability, excellent surface hardness, resists fracture although can be broken	Good tensile modulus, greater resistance to breakage than other thersets, good range of colors, good dimensional stability, can be processed by injection molding
Limitations	Brittle, inserts are required for thread assemblies, limited to black and mahogany	Slightly lower heat resistance than phenolics	Opaque, although in very thin film grades transparency is possible	Expensive
Principal applications	Knobs, handles, bases, for kitchen cookware and electromechanical assemblies	Similar to phenolics except when lower heat and less impact are required but color is necessary	Dinnerware, handles for expensive cutlery, kitchen, and domestic items	Bathtubs, sinks, etc.
Why used	Cheapest, good heat resistance, oldest material	More attractive than phenolics owing to full color range, relatively inexpensive	Good surface hardness, does not scratch readily, good surface effects, particularly attractive for a consumer product	Decorative large areas, particularly adapted to sheet molding SMC process, ability be fabricated in large products

ucts and even industrial applications. They account for 85% of the total market. They are lighter in weight and can be generally processed much more rapidly. They have a full range of color and can even be molded translucent as well as transparent, have excellent mechanical properties, and good dielectric resistance. However, they have a lower heat distortion level than thermosets, and the one cardinal limitation of a thermoplastic material is its susceptibility to creep. The most popular material of the thermoplastics is low-density polyethylene, followed by polyvinyl chloride, high-density polyethylene, and polypropylene. Table 14.2 gives the characteristics and current prices. Other materials are pretty well distributed in small individual figures, with significant tonnage utilization. Specialized materials enjoy less than 1% of the total market. This, of course, reflects upon their commercial importance rather than their specific technical advantages. For special application in unusual environments, heat transparency, impact strength, resistance to chemicals, and so on; there is a justifiable use for these specialized materials. However, it is always advisable to employ a commercially accepted material as it has several distinct advantages. It is readily available, the molder is familiar with the processing of the material, it does not require any special order, and it can be purchased in large quantities at the lowest possible manufacturing cost.

The history of the plastics industry indicates that there is always a slight degree of imbalance between production and availability. Usually when a new material comes on the market, it is test marketed until a substantial market develops. At this stage it is available in limited quantities. When the production facilities jump from limited pilot production to full-scale production, a much greater supply of material is available. As competition comes into the market, adding still greater quantities, the material becomes so popular that its usage exceeds its productive capacity until shortages take place, followed by substantial increases in plant capacity, and the cycle starts all over again. Fortunately from the design engineer standpoint, since there are many companies producing similar materials, and they all have plant expansion capabilities that are not simultaneously executed, there is a continuous increase in availability of materials. Since most plastics are petroleum based, they suffer from the availability or lack of availability of some of the feedstocks and nomomers. Although it is possible to shift in midstream, from a high-density polyethylene to a polypropylene, it is important to recognize the differences in mold shrinkage, ultimate size, and so on to be assured that the part will be serviceable.

Engineering (or high performance) materials enjoy specific properties not found commonly in popular materials. An engineering material is tailor-made when it meets defined requirements of capacity, flexural strength impact strength, modulus of elasticity, dielectric strength, modified surface appearance, transparency or opacity, environmental resistance, and others. Usually they are more expensive, although some not necessarily substantially more expensive, but they do require a special order and possibly a waiting time for delivery. The determination of material selection is not just a question of chart position, inasmuch as all the available test data are considered absolute but only a relative rating of the performance of material when exposed to the identical testing standards. This is because rarely does one mold a spiral, produce a test bar with a notch in it to determine notch sensitivity, and so forth. Anyone of the properties of a part— its thickness, length, flow, surface appearance required, existence or nonexistence of cores and studs and ribs, location of gate— can materially change the ultimate strength of the part as much as 50% and more. Thus it might be possible by proper redesign of the part to eliminate the need for a high-strength engineering material. Instead a (more common) general purpose material could be utilized to increase the various parameters that are available in the design of the product. So, one can achieve the greater overall strength by means of design, rather than just by the added inherent strength in the particular resin.

New polymers are constantly being developed to meet the requirements of the industry. The product designer can feel confident that, if there is an available material that minimally meets the present requirements, from the time the design is finalized until the mold is built either additional or upgraded materials will be available that meet the requirements plus. From a practical standpoint it means that, if one can achieve at the time of initial design 100% utilization of the qualities of a given material with very little safety factor left, chances are that within the interval of approximately a year there will be a superior material available to give one an added safety factor. It is not recommended that one design a part where the material selected does not meet the cal-

Table 14.2 A Material Selector for Certain Thermoplastics

Property	Thermoplastic					
	Low-density polyethylene	High-density polyethylene	Polypropylene	Acrylonitrile butadiene styrene	Polystyrene	Polyvinyl chloride
Clarity	Transparent to opaque	Transparent opaque	Transparent, translucent, opaque	Translucent in thin sections	Transparent	Transparent to opaque
Burning rate (flammability) (in./min)	Very slow (1.04)	Very slow (1.00–1.04)	Slow	Slow	Slow < 1.5	Slow to self-extinguishing
Effect of weak acids	Resistant	Very resistant	None	None	None	None
Effect of strong acids	Attacked by oxidizing acids	Attacked slowly by oxidizing acids	Attacked slowly by oxidizing acids	Attacked by concentrated oxidizing acids	Attacked by oxidizing acids	None to slight
Effect of organic solvents	Resistant (below 60°C)	Resistant (below 60°C)	Resistant (below 80°C)	Soluble in ketones, esters, and some chlorinated hydrocarbons	Soluble in aromatic and chlorinated hydrocarbons	Resists alcohols, aliphatic hydrocarbons and oils. Soluble or swells in ketones and esters; swells in aromatic hydrocarbons.
Molding qualities	Excellent	Excellent	Excellent	Good to excellent	Excellent	Good
Mold (linear) shrinkage, (in./in.)	0.015–0.050	0.02–0.05	0.010–0.025	0.005–0.008	0.001–0.006 (C) 0.002–0.006 (I)	0.010–0.050 (varies with plasticizer)
Specific gravity (density)	0.910–0.925	0.941–0.965	0.902–0.910	1.01–1.04	1.04–1.09	1.16–1.35
Specific volume (in.^{3}/lb)	30.5–30.0	29.6–28.6	30.8–30.4	27.0–26.0	26.0–25.6	23.8–20.5
Tensile strength	600–2300	3100–5500	4300–5500	3500–6200	5000–12,000	1500–3500
Elongation	90.0–800.0	20.0–1000.0	200.0–700.0	5.0–60.1	1.0–2.5	200.0–450.0
Flexural yield	—	—	6000–8000	6000–10,000	8700–14,000	—
Impact strength (ft-lb/in.)	No break	0.5–20.0	0.5–2.0 (73°F)	6.0–8.0 (73°F), 2.0–4.0 (40°F) (⅛ × ½ in. bar)	0.25–0.40 (¼ in. bar) 1.6 (⅛ in. × ½ in.)	Varies depending on type and amount of plasticizer

	D41-D46 (Shore), R10	D60-70 (Shore)	R80-110	R75-105	M65-M80	50-100 (Shore A)
Hardness, Rockwell						
Resistance to heat (°F) (continuous)	180-212	250	225-300	140-210	150-170	150-175
Cost ($/lb)	0.50	0.48	0.52	1.21	0.52	0.55
Cost ($/in.³)	0.17	0.16	0.17	0.40	0.17	0.18
Qualities	Unbreakable; flexible; economical, readily processed	Inexpensive; semirigid; withstands boiling water	High heat resistance; light weight; inexpensive, good flow characteristics	High gloss, good dimensional stability, low molding shrinkage	Hard, glossy finish, good color, rigid, good dimensional stability	Chemical resistance, good surface finish, ranges from rigid to flexible, transparent to opaque
Limitations	Translucent, not transparent; poor paint retention; cannot withstand boiling water	Semiopaque, fair dimensional stability	Poor dimensional stability unless proper molding conditions maintained	Few; first in line of engineering materials	Poor solvent resistance, general purpose grade, fractures readily	Can be corrosive to injection molding barrels and screws, cannot be used with nonferrous molds, high specific gravity
Principal applications	Containers; injection blow molded products (squeeze bottles, jugs)	For boilable food containers, toys, containers including 5 gal. shippable	Coffeemakers, drinking cups, automotive, containers, domestic appliances	Pipe fittings, appliances, auto, telecommunications, luggage, sporting goods, boats	Air conditioner grills, radio-TV cabinets, business machine housings, toys, disposables, furniture. One of the most widely used resins	Conduit and pipe fittings, building components, small appliance components, phonograph records, toys
Why used	Inexpensive, unbreakable, consumer acceptance, fast molding cycle, chemically resistant	Does not rust, particularly advantageous in containers; chemically very resistant; break resistant; easily processed; consumer acceptance	Low cost, high heat resistance, easily processed, floats	Excellent abrasive resistance, flame retardant grades, competes favorably with metal	Can be foamed, low cost, easily processed, rigid, high gloss, can be molded in thin sections	Fire retardant, excellent clarity, good flexural strength, can be considered an engineering material, wide range of alloys

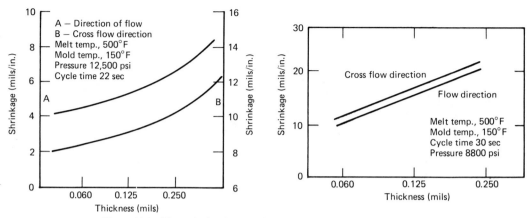

Figure 14.1 The variation between flow direction and cross-flow direction.

culated or contemplated strength required— and then hope that a new material will fit into place. But if one does have a material that does meet the satisfactory level even though there is a little margin of safety, the probability of an improved material being available is very substantial.

Material specifications, which are accumulated data under ideal test conditions, do not necessarily reflect the ultimate physical characteristics of a product. The thickness of a part will affect the strength of the part (See Figure 14.1). The flow direction of the material will affect the strength of the part as well as change the shrinkage (Figure 14.2).

In the design of a part the location of the gate is a critical consideration, and it is for this reason that material specification charts can only be used as a guide and not as an absolute; direction of flow, temperature,

pressure, cores, and configuration all effect the ultimate physical strength of a part.

From the experience gained in the development from the very inception of the plastics industry, the design guidance given in Table 14.3 is intended to provide a practical standard for economical processing of common commercial products. However, it is possible to modify these ground rules where special conditions prevail.

14.4 MANUFACTURING CONSIDERATIONS

Although thermoplastic and thermoset materials have different areas of disciplines for processing, there are areas in which they are interchangeable, and as such they have comparable processing technology.

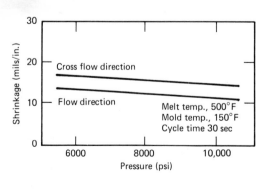

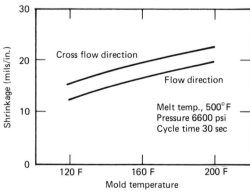

Figure 14.2 The effect of pressure and mold temperature on flow.

Table 14.3   Design guides

	Compression molding			Injection molding (thermo-plastics)	Cold press molding	Spray-up and hand lay-up
	Sheet molding compound	Bulk molding compound	Preform molding			
Minimum inside radius (in.)	$\frac{1}{16}$	$\frac{1}{16}$	$\frac{1}{8}$	$\frac{1}{16}$	$\frac{1}{4}$	$\frac{1}{4}$
Molded-in holes	yes[a]	yes[a]	yes[a]	yes[a]	no	large
Trimmed in mold	yes	yes	yes	no	yes	no
Core pull and slides	yes	yes	no	yes	no	no
Undercuts	yes	yes[b]	no	yes	no	yes[b]
Minimum recommended draft (in./°)		$\frac{1}{4}"$ to 6" depth: 1° to 3° 6" depth and over: 3° or as required			2° 3°	0°
Minimum practical thickness (in.)	.050"	.060"	.030"	.035"	.080"	.060"
Maximum practical thickness (in.)	1	1	0.250	0.500	0.500	no limit
Normal thickness variation (in.)	±0.005	±0.005	±0.008	±0.005	±0.010"	±0.020"
Maximum thickness build-up, heavy build-up, and increased cycle	as req'd.	as req'd.	2-to-1 max.	as req'd.	2-to-1 max.	as req'd.
Corrugated sections	yes	yes	yes	yes	yes	yes
Metal inserts	yes	yes	not recommended	yes	no	yes
Bosses	yes	yes	yes	yes	not recommended	yes
Ribs	as req'd.	yes	not recommended	yes	not recommended	yes
Molded-in labels	yes	yes	yes	no	yes	yes
Raised numbers	yes	yes	yes	yes	yes	yes
Finished surfaces	two	two	two	two	two	one

Source: Courtesy of Owens-Corning.
[a]Parallel or perpendicular to ram action only.
[b]With slides in tooling, or split mold.

All thermoplastic materials can be compression molded, and in this instance it would follow the technology of thermoset compression molding. Some thermosets can be injection molded, and in this instance the technology of injection molding is the criterion by which the thermoset material is considered. The most common and fastest growing technology for processing plastic products is injection molding. However, all products do not lend themselves to injection molding for economic, size, or technical reasons. The following are commercially available processing techniques:

Extrusion.   This is a very practical economical method of producing a profile or a product in linear form, such as piping, tub-

ing, or angles. In extrusion it is possible to design the external area to have any configuration that is linear, and it is possible to do the same with the interior. Through the medium of secondary operations of the extrusion line, it is possible to punch holes, cut off at uniform lengths, print or emboss, deform compress, and so forth operations or attachments to the most common extruder. It is possible to coat wire or any other product, to cover a substrate, and, by means of pultrusion, to have multiple colors and/or similar materials incorporated into a unitary product. The only limitation is a uniform cross section.

Thermoforming.   (Also known as vacuum forming).   This is a process that takes a sheet of material and by the application of heat and vacuum forms the sheet to the configuration of the mold which can be built inexpensively as it is only subjected to the pressure of one atmosphere (15 psi) and can be either male or female.

Blow molding. This employs a fixed mold. The external surface can have lettering and design, and it is possible by means of design to have recesses created in the part, such as the handle. The interior surface of a blow molded container does not lend itself either to specific wall thickness or to any secondary configuration, as the interior follows the outside with variations caused by uneven stretch and pressure.

Rotational molding. This is ideal for very large products, and again is limited to the exterior surface for both decoration and design effects. Since the process is controlled by heat transference through the medium of dissimilar mass of metal or the addition of a more conductive material, such as copper, in a specialized area, it is possible to have a controlled variation in the wall thickness.

RIM. Reaction injection molding employs a cavity and a core and, subject to the limitations of draft, it is possible to have design elements on the outside and the inside of the part.

LIM. Liquid injection molding similarly follows the design criteria for reaction injection molding.

Casting. It is possible to cast with a core and to have a product that will have design elements externally and internally. Potting is a variation of casting in which there are inserts or other products are inserted into the cavity and the material is poured around it. This is very common in the electronic industry. Casting of film does not permit any immediate design elements. However, on the takeup side of cast film it is possible to emboss the material.

Calendering. This does not in its first stage permit any design elements, but design is part of the entire process. Secondary calendering rolls can be employed that will emboss or create a linear configuration to the product such as corrugation.

Compression molding. This is employed for materials that cannot be injection molded, such as phenolics, ureas, melamines, and certain polyesters.

SMC. Sheet molding compound is capable of having external design elements incorporated as well as internal.

The preceding disciplines represent the most commonly utilized methods of processing plastic. There are specialized techniques which are derivatives or combinations of the preceding technologies that are adapted to specialized applications or high volume production.

14.4.1 Processing and Design Techniques Comparison

Of the several techniques available to process thermoplastics, practically all but injection molding require a secondary operation, as the product that emerges from the processing or manufacturing equipment is not complete in and of itself. In extrusion where there is a profile, the material is extruded in linear form and must be cut; secondary operations must be performed to make it into a saleable product. This, of course, even applies to pipe, and the like.

Blow molding is the technique to produce hollow vessels or containers such as bottles, jars, and barrels. An advanced technique known as injection blow molding utilizes an injection molded preform with a secondary blow molding function that is ideally suited for beverage bottles and containers. They usually require a secondary component such as a cap or closure or are fitted to a product such as a tank or pump. Film usually refers to thin gauge plastic material below 0.10 in. or 10 mil. It can be produced by blow molding or casting. Blow molding is the most common method. Above 0.10 in. material is calendered, a process where heated material is squeezed through large rollers and comes out in a continuous sheet. Both film and calendering material usually require secondary operations to change them into a useful end product. Rotational molding requires a secondary trim operation, and slush molding also requires further labor to be performed before the product is a saleable item.

It is with injection molding that parts can be collected as they fall out of the injection mold, moved to a conveyor, and then automatedly stacked and packaged as a finished product without any secondary operation (other than the physical counting and packaging of the material). For this reason injection molding enjoys a unique advantage; the product as it is produced from the mold need not be touched by the variables that result from individual hand operation. Injection molding assures uniformity of product, and modern onstream process technology and the computer can be employed to control both quality and quantity of product. Obviously, in a time of increasing labor costs, the injection molding process becomes the least labor-sensitive method of manufacture: it is possible for the industry and the product designer to achieve a product of which it can be said literally, "The product has not touched human hands." However, now the responsibility lies with the product designer and the mold engineer, the mold builder and the molder. Nothing is going to come out as an excellent or satisfactory product if certain conditions are not met. The requirements for the finished product must be formalized, established, or created in the original design; they must be transferred from the design to the mold; and standards of manufacturing technique must be established that will meet the requirements of the product. One cannot get out what one doesn't put in.

Because of the lack of any human contribution to the processing of an injection molded part, the critical responsibility is put on the design engineer and the mold builder. An analogy can be had from the automotive industry. When a modern automobile engine is put on a transfer machine, each and every part is interchangeable, with the tolerances established on the machine; this is because of the high standards of production control. In the case of some of the most expensive automobiles, like the Rolls Royce, where hand lapping is performed to achieve this close tolerance, one has a lack of interchangeability. This is caused by machining tolerances that are not as great as in the mass produced motors, the variance and tolerances being achieved by hand finishing.

Of all the process technologies available for thermoplastic materials, the injection molding process for large-volume parts is the most economical. However, there are certain products where other processes are preferred over injection molding. This is particularly true in thin-walled parts like disposable food containers and the like where thermoforming competes very favorably, and in some cases is more economical than injection molding. The reason for this is the reduction in the cost of material, since it is possible to form containers of 0.010 or 0.015 in., and it is not feasible to mold containers of this thickness. Not only would it be rather difficult if not impossible to mold, but the parts would be so stressed that they would break in normal commercial handling. This process is equally suited for very large objects such as refrigerators door liners and outdoor signs where the size is beyond the range of injection molding or where the quantities are minimal. Extrusion of long or linear parts is one of the most economical processes. However, extrusion is limited to the production of continuous cross section and should anything else have to be performed on the product (other than cutting off or punching holes), it becomes a secondary operation. Casting is a technique that lends itself to rather limited production but with very inexpensive tooling. It is excellent for prototypes and particularly advantageous in designing of displays, models, dioramas, and so on. However, for items of mass production, the technique of casting becomes prohibitively expensive in larger volumes, and conversely, it becomes most economical in very short runs. Most casting used today in production is in the electronics industry where the term "potting" of components is used. In this case a microelectric assembly is filled with a resin and left to cure. This eliminates the pressure normally associated with injection molding and permits the parts to maintain their own integrity and still be locked. It is really a sealing technique.

One of the other techniques that lends itself to design is rotational molding. In this case a female mold is made (which can be fabricated from thin steel or aluminum or any other material) with a cover, and a predetermined amount of material is introduced into the mold and then it is closed. It is then mounted in a rotational molding machine, and there it is biaxially rotated and subjected to the external application of heat. Heat may be supplied by infrared-ray or through the use of hot molten salts. The transfer of the heat from the outside of the mold to the resin permits the buildup and curing of the resin. In rotational molding the part that is in contact with the mold or the cavity has the surface finish, while the inside of the part varies in wall thickness and also does not have a smooth finished appearance.

Injection molding to casting offers a spectrum in terms of tool investment. Lower tooling cost relates to higher unit price and vice versa; therefore in part design, one must recognize the total quantity requirements before one selects the method of manufacture.

In addition, it is possible to process every thermoplastic material by compression molding. The advantages of compression molding over injection moldings are as follows: the part is strain free and can be molded of heavier sections without any visible shrinkage (which could not be achieved by injection molding). One can have sections of 2, 3, or more inches and have them bubble-free by compression molding. The mold cost is much less for compression molding than it is for injection molding. Particularly if the quantity is small, the cooling can be done on a more primitive form, and by increasing the molding cycle, thinner and lighter molds can be used. One is concerned with the relationship between pressure, heat, and time. The higher the pressure, the shorter the time and the lower the heat. The lower the pressure, the longer the molding cycle and the higher the heat required. In addition to the cost of the mold, there is the cost of equipment to be considered. Again, the expense of the mold is directly proportional to the cost of the equipment used to process the product.

Injection molding has the highest tooling cost and the highest machine cost, but over a large volume it ends with the lowest unit cost. At the present time the range of injection molding machines commercially available is in the 2500–3000 ton range. There are a small number of machines that weigh up to 5000 tons, but these usually are in captive plants, and it would be rather risky to design a part

without having a commitment from a molder or an agreement by management to purchase the necessary processing equipment. A 3000 ton press with (the rule of thumb of between) 2–3 tons/in.^{2} clamping pressure accommodates an area in excess of 1000 in.^{2}, which would be a common configuration of 24 in. × 48 in., which is 1152 in.^{2}. This is selected not as any binding dimension but represents a common module size, and the platen of a 3000 ton injection molding press would receive a mold of this size. A part 36 in. × 36 in., although having 1296 in.^{2} and having adequate wall thickness, could be filled since the flow length is uniform and shorter than the 24 in. × 48 in. base. However, a part 5 in. × 200 in., although containing only 1000 in.^{2} could not be molded owing to the length of flow and the unavailability of a press having a platen in excess of 20 ft. In addition, flow would create a very complicated molding problem requiring hot or insulated runners, and this configuration is not workable.

The thicknesses that can be processed in the various techniques are again related to the ultimate use and cost of the part. Structural foam parts need a minimum of 0.25 in. of wall thickness for the material to flow and fill the mold completely. Thermal forming can use thinner sheets, for example, some of the platforms used in candy boxes, cookies, and the like may be only 0.003 in. Containers used for food are in the 0.001 in. range. Disposable drinking cups can be injection molded from 0.012–0.015, and the economics of using the thinner wall is lost owing to the slower molding cycle. The minimum wall thickness recommended for a fairly small part where there will be not more than 4–5 in. of flow is 0.015 in. Oddly enough, in designing a part, contrary to what one would expect, the point or the gate should be at the thinner and not at the thicker section. It is easier to fill a thin mold going from thin to thick than otherwise. The filling process is similar to that of a funnel where material flows readily with a gradual increase in wall thickness. If one starts at a thicker point and reduces the wall thickness at the end to a substantially (proportionally) thinner section, the restrictions in flow will build up, and the part will freeze off before it can fill completely.

The product designer must recognize certain design limitations that will affect the problems of designing and building the mold. Actually what ultimately becomes a good design is in reality the most economical method of manufacture.

When there is a part with variations in its wall thickness, the molding cycle is determined by the thickness of the thickest section of the part. As an example, if a part had one-half of its area with a 0.100 in. wall and the other half with a 0.50 in. wall thickness, it might be possible to redesign the part to have a uniform 0.80 in. wall thickness. Although there would be a slight increase in the weight of the material, the improvement in the molding cycle would more than compensate for the added cost of the material. A part that is 10 in. × 10 in. that would have one section 10 in. × 5 in. with a 0.100 in. wall and another section 10 in. × 5 in. with a 0.050 in. wall would average out to a 0.75 in. wall. By making the part uniformly 0.80 in. thick we added approximately 6% to our material cost, and we could reduce our molding cycle by as much as 15%, or possibly more, ending with a superior product having uniform shrinkage and no distortion at a lower ultimate cost.

14.4.2   Part Factors

In the design of a part recognition must be given to the following specific considerations that affect the part and how they are resolved. If the design engineer is developing a new product he can incorporate all the basic good tenets of the state-of-the-art. However, if he has a redesign of an existing product where there are mating components and predetermined space limitations, it may be necessary to eliminate one or two desirable engineering features and substitute less desirable features to achieve the end results where not only is the part itself less functionable by reason of its deleterious part considerations but also usually these considerations entail a greater cost in mold building and create continuous problems with the part as it is being molded.

14.4.2.1   Size
The size of a part is one of the significant considerations as (stated elsewhere) the capability of an injection molding press to produce

the part is limited by the size of commercially available equipment. Although there are injection molding machines, few in number, that have been built up to 5000 tons, the Triulzi at 5000 plus tons is one of the largest injection molding machines ever built and other than of size, it is a conventional type horizontal injection molding press. It is fortunate that the man in front of the machine gives you a scale concept otherwise it would be difficult to comprehend the size of the machine.

The Windsor press is a special adaptation of two standard 1600 U.K. tons each presses (1700+ American tons) for a total of approximately 3500 tons. They were placed parallel and a single mold was hung on both platens and the presses functioned as a single unit. They were specially combined to produce the Topper Sailboat which is much longer than it is wide and required this unusual platen configuration. Most platens are 3-to-4 L/W or square but in this case they were approximately 4-to-1 L/W. The availability of a substantial amount of equipment is limited to the 2700–3000 ton category. And unless one had in-house capability beyond this tonnage, it would be wise to limit oneself in designing a part to approximately 1000 in.^{2} The basic rule of thumb is 2–3 tons/in.^{2} of developed area, so that a part 30 in. × 35 in. or 24 in × 48 in. would be a suitable configuration to be molded on a 3000 ton press. There are exceptions, and it is possible to design a part that would be within 1500 in.^{2} developed area, where design requirements would include special ribbing and flow canals to distribute the material to the extreme ends, at a reasonable molding cycle. Oddly enough, a structural housing that might appear as a unit 7 ft × 2 ft with a 0.100 in. wall thickness when expanded to 2 ft × 4 ft might only increase in wall thickness to 0.110 or 0.115 in. It is a problem of filling the part to have the plastic move a long distance from the sprue before it solidifies. This can be moderated, of course, by using multiple sprues, either by using a manifold or a three-plate mold, thereby reducing the runs from 24 to 12 in. However, the placement of the sprues must be considered, and any sprue mark will have a visible feature, which must be recognized as an ultimate consideration.

Figure 14.3   Examples of large size injection molding machines.

Although calculated shrinkage in a mold is linear, the ultimate shrinkage of the part (including post molding shrinkage) may not be linear. In molding a piece that is extremely long and narrow, as opposed to a circular part, one would find an (overall) greater shrinkage in the long thin piece relative to a rectangular or circular part. Dimensional tolerances are a factor of the particular resin used as well as the configuration of the part. Certain resins have a higher shrinkage factor than others. For example, polyethylene with a shrinkage factor of approximately 0.015 in. may be compared to Noryl (phenylene oxide based resin), which might have 0.002 in. shrinkage. Also, the variation that exists

within polyethylene is much greater than that which exists within Noryl. One could find a low-density polyethylene that would possibly shrink only 0.010 rather than 0.015 in., whereas in an engineered plastic material having 0.002 calculated shrinkage the variation might only be a few tenths of a thousand. A mold builder in calculating for mold shrinkage uses the top figure. This is the shrinkage factor that the material suppliers list on the specification sheet. It might be from 0.060 to 0.080 in., or 0.002 to 0.015 in. The mold builder takes the top figure because it is very easy to remove metal from the core if the part is too big, but it is very difficult to add material if the part is too small. How-

Figure 14.3 (*Continued*)

ever, many times a design engineer has specific experience with the material and knows from the use of this material, its configuration wall thickness, molding practice, and its precise shrinkage. In such an instance one can furnish this information to the mold builder who will then be covered by these specifics.

As stated previously, when the mold shrinkage is calculated it is basically in linear form. However, this applies to reasonable part thickness and reasonable dimensions. If one were to design a part 4 in. long with a $\frac{1}{8}$ in. wall in polyethylene, one could take 4×15 or 0.060 in. of mold shrinkage. However, if the part was now to have a $\frac{1}{2}$ in. wall thickness, the part would be so distorted in molding that it would be impossible to predict what the shrinkage would be.

Dimensional tolerances are usually speci-

fied when the part has a mating part or it is part of a subassembly. However, it should be recognized that as one decreases the tolerance limitations, one adds substantially to the cost of building the tool as well as the molding of the part. One should recognize, if the part is used as an assembly, what tolerances are available in the mating part. For example, if one were to employ a glass lens in a molded housing, the tolerance on glass is so much greater than that which is normal to even the most common standards of injection molding that it would be ridiculous to impose a rigid tolerance when the glass component falls all over the lot.

Maximum and minimum wall thicknesses are directly related to material, size of the part, function, service life, cost, and any other particular requirements of the finished prod-

(d)

(e)

Figure 14.3   (Continued)

(f)

GKN Windsor Ltd.,

78 Portsmouth Road, Cobham, Surrey KT11 1HY. Tel Cobham 7241

SPECIFICATION

SP 1600S Alpha and Gamma Injection Unit

INJECTION UNIT

UPPER UNIT

Plasticising capacity (Polyprop)	800 lb/h
Diameter of screws	5.75 in
Number of screws	2
R.P.M. of screws	0-60
Electric motor for screw drive (DC)	100 HP.

LOWER UNIT

Injection pressure	18000 lbf/in^2
Diameter of plunger ring	7.087 in
Stroke of plunger	28.5 in
Shot weight at 18000 lbf/in^2	30 lbs.polyprop.
Rate of injection at 18000 lbf/in^2	281 in^3/sec

ELECTRICS

Heating capacity (Upper & Lower Units)	133 kW

POWER UNIT – INJECTION END

HYDRAULIC
Vane & Piston pumps

Accumulators	344 gal/min
Accumulator pressure range	3200/2900 lbf/in^2

ELECTRICS

Electric motor	75 HP.
Electric motor (gearbox lubrication)	2 HP.

CLAMPING UNIT

Mould clamping force	1600 UKtonf.
Size of mould plates	70.86 in x 82.67 in
Distance between tiebars	49.21 in x 59.05 in
Mould space	70.86 in x 59.05 in
Minimum mould height	27.56 in
Stroke of platen	78.74 in
Maximum daylight	106.3 in
Tiebar diameter	8.66 in

POWER UNIT – CLAMPING END

HYDRAULIC
Vane & Piston Pumps

INTENSIFICATION:

Accumulators	190 gal/min
Accumulator pressure range	3500/3000 lbf/in^2

PLATEN MOVEMENT:

Accumulators	242 gal/min
Accumulator pressure range	850/500 lbf/in^2

ELECTRICS

Electric motor	75 HP.

(g)

Figure 14.3 (*Continued*)

uct. If one is designing a disposable or throw-away cocktail glass, then whether or not it is 4.0 ounces or 3.9 or 4.01 is not a critical factor, as people will never fill it exactly to the top, and as such a reasonable variation in thickness will not affect the function of the part. However, in producing a laboratory beaker or a graduated tube that would have delineation marks at 0.10 cc, obviously the tolerance is critical. Greater thickness does not necessarily increase the closeness of tolerance. Thin-walled parts rapidly molded suck up rather quickly in the mold and, with proper cooling, have a relatively uniform dimension, whereas thicker-walled parts with variations in the molding cycle entrap a greater amount of heat, which on ejection will cause postmolding shrinkage. It is impossible to set maximum or minimum wall thicknesses without knowing the ultimate requirement of a part. But it can be stated that a tumbler can be molded as thin as 0.012 or 0.015 in. in polystyrene, and it is possible to injection mold nonstructural foam parts of $\frac{1}{4} - \frac{3}{8}$ in. with a reasonable degree of efficiency. The limiting factor on the thickness of the part is the shrinkage and sink marks that will appear in the part as it becomes thicker. However, it is possible, with the addition of a slight amount of foaming agent, to produce a molded part that will have good dimensional stability in rather thick sections up to 3 or 4 in. square.

Both time and dimensional stability directly determine the ultimate thickness of a part. The minimum thickness of a part is determined by the length of travel that the plastic must flow from its injection to the extremeties. The longer flows limit the thinness of the part. For example, consider a molded housing connected to a die-cast base in an electromechanical assembly. The tolerances on the die-cast part usually will be greater than that required for injection molded piece. In addition, the thermal rise will have a greater effect on the die-cast material than on the injection molded material. If one does not allow for this differential in thermal expansion, it might be possible to have an assembly wherein the components fit together in their static performance, but once put to actual use the thermal rise and increase in dimension could cause the exterior housing to shatter.

Dimensional tolerances may be defined as follows. Allowances are the calculated dimensional differences to provide for proper fit. Tolerances are the uncalculated variances that occur during manufacture of molded plastics occasioned by dimensional disparity among multicavities by machining differences, variable shrinkage and warpage, and other factors of the molding cycle. Maximum and minimum dimensions define the tolerance. Basic dimensions limit the scope of allowable tolerance variation that can be calculated.

Thickness of a wall must be sufficient to provide for the essential mechanical requirements both during manufacture and service life. When calculating the wall thickness that falls in the scope of the allowable size, the following checklist will be handy: weight, strength, insulation, dimensional stability, structure, impact strength, and stress concentration. In addition to the use requirements, there are manufacturing limitations such as mold flow, ejection, assembly, and strength.

14.4.2.2   Shrinkage

All known substances with one exception shrink on cooling and expand on being heated. Only water disobeys this fundamental law of physics. In the case of polymers, this principle must be carefully taken into account during molding. When a polymer is heated in an injection molding machine, it expands as it enters the mold through its sprue and gate, and is also subject to a shearing action as it enters the mold cavity that again causes it to expand.

Because of the phenomenon of shrinking, parts are usually designed so that shrinking occurs on the core rather than on the cavity side; normally it is the cavity side that has the high polish or the decorative surface; ejection usually occurs at the core side. The shrinking of the part away from the surface of the mold permits reduction in scuffing, drag marks, and the like. Data on common thermoplastic polymers and their relative shrinkages are available from materials suppliers, although there is a variation in the calculable shrinkage of the polymer, and material supply companies will usually guarantee their materials only to within +10% of the specifications.

In addition, shrinkage (as stated previously) is calculated in a linear manner. Unfortunately, very few parts are just linear. The parts are characterized by varying con-figurations: shapes, masses, apertures, and have concentrated heavier sections such as bosses. It is these problems that the anticipated shrinkage does not take into consideration. Rather, the calculated shrinkage of the material by category is independent of the shrinkage that takes place by reason of the design of the part. (See Figure 14.4a and b.)

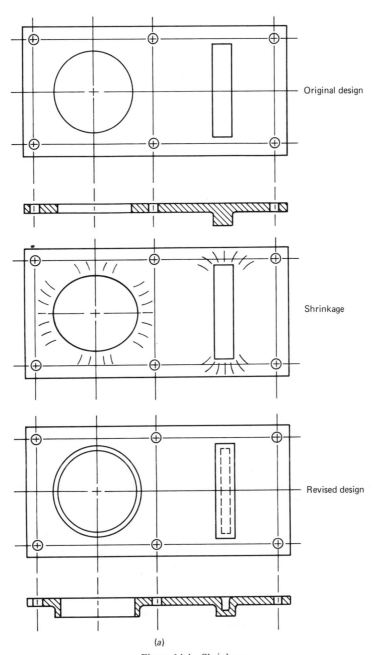

Original design

Shrinkage

Revised design

(a)

Figure 14.4   Shrinkage.

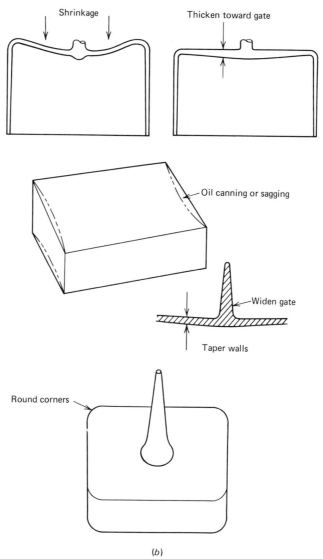

(b)

Figure 14.3 (*Continued*)

To overcome certain shrinkage problems a special shrink fixture may be used. As the warm part is taken out of the mold, it is forced over the shrink fixture; when it finally cools, it will approach the originally designed dimensions. During this process, part elongating or stretching occurs inducing additional strains. When the part reaches thermal equilibrium, there will be entrapped strains that at some future date may create problems of warpage. But good design and good molding can reduce the inherent problems caused by shrinkage.

14.4.2.3 Draft or Taper

Draft or taper are used interchangeably, even though there is a specific difference. Draft is the angle difference between one part of a part and the other. Taper usually refers to a cylindrical part, where the angle is the same completely around the part. Accordingly, it is possible to have a rectangular or square product with different degrees of draft on each side, whereas taper is uniform (see Figure 14.5). One degree of draft is approximately 0.017 in./in. Usually the deeper the part, the greater requirement there is for adequate

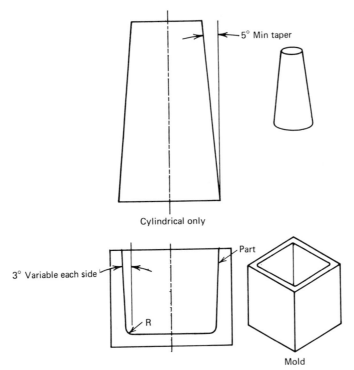

Figure 14.5 Draft or taper.

draft to avoid scratching and scuffing of a part as it is being removed from the mold. Even though molds for large parts are massive and one thinks of steel as being immovable, steel does flex and give, and it is not always constant. Therefore inadequate draft or taper put into the part could result in an unattractive finish. A part 24 in. in depth would require a minimum of 1 deg, preferably more, yet a part only $\frac{1}{4}$ in. in depth could be designed with only a 0.25 deg or even less taper without any problem of ejection. As a rule the maximum allowable taper will insure a better surface finish and ease of ejecting the part from the mold.

The normal method of injection molding is to have the cavity on the fixed platen and the core on the movable platen. Since plastic does shrink in the cooling cycle, it will shrink away from the cavity on to the core (where usually the ejection system is built). In most cases it is the outside of a product that will exert sales appeal and there it should look attractive. Therefore bosses, slots, studs, and so on should be placed on the inside or bottom of a product, so as not to mar the aesthetic appeal.

14.4.2.4 *Radii and Fillets*

Radii are essential in all molded parts as a sharp corner will create not only a stress concentration point but also in all probability will be nicked in usage and will create an undercut that will make it difficult to eject from the mold. (See Figure 14.6.) A ferret radius of less than one-fourth of the wall thickness of the part induces tremendous stress, and a radius greater than three-fourths of the wall thickness begins to lose any effective increase in material strength. It is always desirable to have the same internal as well as external radius for uniform flow. The molded part will be stronger and nearly free from stress. The elimination of sharp corners reduces cracking, which is the result of notch sensitivity, as well as its resistance to shock and sudden impact. Even a $\frac{1}{64}$ in. radius internally will greatly improve the strength of a part. It is possible to break the sharp corners of a core readily by using a hand file, as opposed to machining. However, in the cavity a $\frac{1}{64}$ in. radius can be produced only at a great expense; it might need the usage of electro erosion or an end plate, wherein it would be

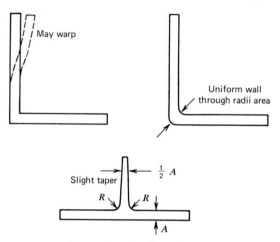

Figure 14.6   Radii and fillets.

possible to machine a 0.64 radius. But if one has a cavity that is 6 in. deep, trying to put in a solid piece of steel with a radius that small poses quite a substantial problem. It is necessary to use special mills, as no standard end mill 7 in. in length could possibly be produced with a radius of only $\frac{1}{64}$ in. As the end mill would be $\frac{1}{32}$ in. in diameter, the cavity would be set up on an angle plate and coming in with a fine ball-end mill, creating a radius and with it the problems of hand finishing, polishing, and so on. Unless there is some particular overriding requirement for a square external, then either create the exterior of the part by four inserted pieces of steel or else provide for a liberal radius. Plastic will flow much more readily around a rounded corner and induce less stress into the mold, thereby producing a part of more uniform density. The components of the mold itself are subject to increased internal stress and fatigue if there are restrictions to the smooth flow created by the elimination of radii.

14.4.2.5 Undercuts
Undercuts are an inherent problem in designing parts. An undercut is created, whether it be on the core or on the cavity, when the dimensions of the part at the bottom of the cavity exceed that at the top, making it impossible to withdraw the part. Also, undercuts are created when there are openings in the side walls of the part that are created by metal

interruptions contacting both the cavity and the core, thus excluding the material from where this metal contact is made. (See Figure 14.7.) When it is not possible to eliminate undercuts, one must provide for methods of permitting the part to be ejected from the mold. Many times the undercut or opening of the mold could be nothing more than a round hole or a vent hole. If it is a single-

Undercut — requires removable mold core

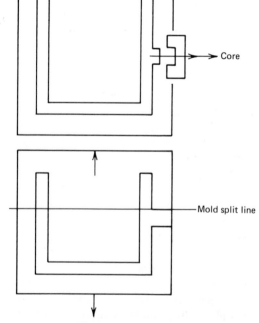

Figure 14.7   Undercuts.

cavity mold it is possible to use the time of the operator during the molding cycle to drill out the opening. In this case one has created the possibility of making an opening in the part, without any additionai cost to labor and at no extra cost in building the mold. In addition to drilling, it is possible to punch out configurations other than those that can be created by a drill press.

14.4.2.6   Movable Cores

It is necessary to have removable cores if movable cores are not utilized. For some particular reason the word "core" has been selected to indicate the shape of the opening, whether it is created by a movable part on the core of the mold or by a movable piece of steel from the cavity part of the mold. Initially, cores were operated by pins that fitted through a slot that were angled out from the core side of the mold, so that as the mold closed, the core's fingers engaged the slot in the movable core and pushed it into its final position. To avoid galling and to remove any excess stress on the core fingers a 30° angle has proven to be an effective design. This of course limits the extent of the core movement. For instance, if it was necessary to have the core travel at 10 in., and it was calculated how far this would be in relation to the length of the core fingers, or pickers, it can be seen that a very unsatisfactory mechanically designed part would result. To overcome the limitations of mechanical core movement, pneumatic or hydraulic cylinders are used. This way, the extent of the cylinder is only slightly more than the total stroke of the part, so that one has a rather compact mold with the possibility of having rather large core movement. Sometimes it is possible to recess the hydraulic or pneumatic cylinder into the mold frame itself, reducing the problem of damage when the mold is handled outside the press. However, by design it is possible to eliminate movable cores, and an example of this is the term "slotting off."

Where one has a tapered part, it is possible to have metal to metal contact that will shut off the flow of material around the interruptions and yet eliminate the necessity of movable cores. This is best illustrated by the typical laundry basket. In this example, air and light weight are achieved by just having a substantial taper. Also, interrupted areas of the steel are permitted to continue in contact with the plastic flow around the design, without producing an undercut upon withdrawal from the cavity. Wherever there is an opening, the flow of the resin will be around the opening, joining together behind the part. This causes what is referred to as weld marks, knit marks, or flow marks. It is not possible to eliminate completely any evidence of flow. It is only possible to minimize the flow. The part should be gated, so that the material is closest to the entrance point of the gate. Then, when the resin comes in contact with the core it is still rather warm and will flow readily together again. It is possible, by changing the configuration from a square or round opening to an airfoil section, to further reduce the flow of the material and achieve a more uniform mechanically welded structure. Not only are the flow marks aesthetically unattractive; but they also indicate that when the two materials came back together again they had a skinning effect, they were cooler and there is a substantiai reduction in mechanical strength in the area of the flow mark. In addition, not only do openings in a mold cause flow marks, but they also create problems in dimensional stability and in shrinkage.

14.4.2.7   Molded Lettering

Letters molded in the part, if designed in the bottom of the piece, will not create an undercut. If engraved on the side wall of the part it is necessary to have the part designed with substantial draft, so that the letters can be removed without creating an undercut.

Various kinds of lettering are used on plastic parts, usually for identification purposes. Often raised letters are less expensive than depressed letters because the raised letters are molded from letters recessed in the mold cavity, whereas depressed letters must have the mold steel cut all around them. (See Figure 14.8.) Wherever possible, hobbing of the mold often aids in reducing the total cost of making raised letters on the mold. In this case, the letters would be cut into the hob,

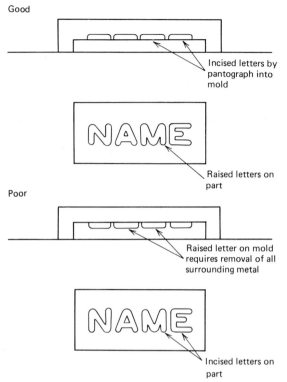

Good

Incised letters by pantograph into mold

Raised letters on part

Poor

Raised letter on mold requires removal of all surrounding metal

Incised letters on part

Figure 14.8 Molded lettering.

and the hob would be forced into a sifter steel that would become the mold after it hardens. Generally, raised decorations allow some savings in molding. Hobbing allows many types of letters and numerals and some types of decorative designs to be raised in the cavity. To improve legibility, depressed lettering filled in with paint sometimes is required.

14.4.2.8 Sink Marks

Sink marks occur in a product when there is an uneven thickness in the part. Many times it is necessary to install or design a rib to carry a substantial amount of load, reduce deflection, and in other instances to permit the rapid filling of the part by creating a flow channel. However, if the rib as it adjoins the wall is more than 50% of the thickness of its abutting wall, there will be a sink mark developed. Accordingly, it is necessary for the rib root to be less than one-half the thickness of the wall, this is to provide for the ability to create a fillet around the rib so that there will be no stress concentration area.

For ribs designed for either stiffening or for mechanical purposes, if the outside of the part's visible appearance is critical and sink marks would be objectionable, it is necessary to have the root thickness of the rib no more than one-half of the thickness of the wall to which it abuts. A slight radius sufficient to break any sharp corner and permit an easier flow is all that is required, and this will not affect the visual appearance of the rib. However, there is a limitation to the depth of a rib; for reasonable molding efficiency there should be an included taper of no less than 1 deg. This means 0.017 in./in. So that if the product is designed with 0.100 in. rib, it could have only a 0.050 in. root wall. And if the rib was over 2 in. in depth it would disappear into a razor-thin section, because 0.017 taken twice would be 0.034 which when subtracted from 0.050 would leave only an 0.016 in. area. It has also been demonstrated that the maximum amount of strength is achieved in a 4:1 or a maximum of 6:1 ratio (width of rib to depth of rib). Therefore, if one had a 0.050 in. base thickness, a rib of $\frac{3}{8}$ in. will give the maximum amount of strength, because, as one decreases the thickness of the rib, the antiflexing resistance decreases. It is necessary for this rib to have a radius in its base, so that one now finds nominally a $\frac{1}{32}$ in. thin section to the rib, a 0.050 in. root width, and a 1 deg. included taper. However, this formula need be followed only when the wall is going to be visible. If there is an invisible area, and if sink marks are not objectionable and strength is the paramount factor, then it is possible to have a root width of the rib equal to the wall or even more. One way to eliminate the sink mark effectively, economically, and aesthetically pleasingly is to raise a small rib on the opposite side of the wall. This means that one is removing plastic directly opposite the rib. (See Figure 14.9.)

Sinks and warpage are caused by differential volumetric shrinkage that takes place while the resin is cooling in the mold. Heavier areas have greater after-shrinkage, or post-molding shrinkage problems. Thinner wall sections have less postmolding shrinkage. In addition to just the volumetric difference, there is also a difference that takes place in the lack of uniform cooling when there is a greater mass in one area than another. A

very thin section cools and sets up very rapidly as the skin chills, takes on the high gloss, and is basically a very solid material. In a thick part there is not only skin but also an internal core structure. This core structure contains ambient heat or internally trapped heat that, after the part is removed from the mold, begins to radiate out to the surfaces. It is this desire to free itself from internal stress caused by the mass and differential of heat that causes both shrinkage and warpage. It is essential for the designer to recognize that it is readily possible after a mold has been built, if the part is determined to have too much flex, to add ribs or increase the wall thickness. This can all be done rather economically by removing metal from the core. However, if the part starts out too thick, it is very difficult to reduce the wall thickness of the part.

As noted before, it is possible to calculate mold shrinkage because all plastic materials have a known shrinkage, and this is determined by the material supplier; of course they have a tolerance between the minimum and maximum shrinkage, which is quite substantial. It is possible even where there are differentials in shrinkage to substitute one material for another, and still end up

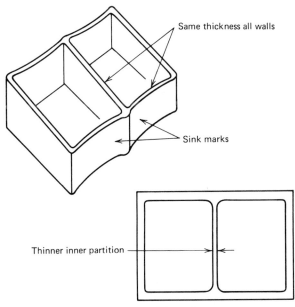

Figure 14.9   Sink marks.

with the same size part by molding variables such as temperature, injection speed, and packing. These are parameters that can overcome variations in dimensional configurations. Postmolding shrinkage is much more difficult to ascertain because it does not follow the same uniform laws as mold shrinkage. The wall thickness, openings, bosses, and so on all create problems that add to postmolding shrinkage.

Postmolding shrinkage can be reduced by the use of a shrink fixture (as stated previously). Basically a shrink fixture is a confining jib that holds the part outside of the mold to a given dimension, and permits it to cool at a slower pace. Many times a shrink fixture used in the part is dumped in a pan of recirculating water to allow for more rapid shrinkage, to keep the part dimensionally stable. However, if there are substantial differences in wall thickness, again the problem of stress can be generated. Shrink fixtures are specified and are not an indication of a badly designed part; on the contrary, it is the designer's recognition that stress and heat shrinkage exist, and this is a way of overcoming some of these inherent problems. In addition to maintaining dimensional stability, it is possible to use a shrink fixture to change the shape of the part, and thereby eliminate what would be a problem in molding. An example of this is a football helmet where the product is basically oval, the head part is smaller than the section across the ears, and then it again tapers in around the chin. If one were to mold this configuration, it would be necessary to develop a collapsable core. However, it is perfectly proper to mount the helmet on a shrink fixture and have the bottom lugs pressed in as a postmolding operation. This eliminates the complexity of a split core, with reliance put upon external cooling to form the part to its ultimately desired shape.

The sprue fills or permits the resin to enter the mold and fill the space between the cavity and the core, but to get the molded parts out of the mold, it is necessary to eject the part from the mold. This can be accomplished through several means. One consists of conventional ejector pins. The mold is designed with a movable ejector plate, to which ejector pins are physically attached. The pins are designed to come in contact with the molded part, either on the bottom of the piece or internally. As the mold opens and the ejector bar, which is built into the press, comes in contact with the movable ejector plate, it pushes the pins forward, at the same time pushing the molded part off the core. Where one has a part that has a relatively thin wall and is of a simple configuration, it is possible to use a second means, a stripper plate. This plate completely surrounds the core and is movable. As the mold opens it pushes the molded part forward off the core, completely surrounding the plastic material 360 deg. A stripper plate requires a deeper core and also requires a very close fit between the plate itself and the core, otherwise there will be seepage of material and flash will ensue. Gates need not be visible and need not be objectionable. When one realizes that many times a part will have differential configurations on the core it is possible to use this difference as a means of locating an ejector pin so that it will be almost unnoticeable.

There are many systems for actuating ejector plates; stripper plates can be motivated by cylinders, pull bars, bicycle chains, and so on. The designer must recognize that a part must be removed from the core to see where ejector pins can be placed so that even though there is evidence of their existence, it need not be objectionable. One means of ejecting a circular or closed part from the mold is to use air: a small valve, which need not be located in any particular position, but placed anywhere at the bottom of the core, actuated by a switch and moving as little as 0.300 in. to permit air to enter the core and rapidly eject the part from the mold. However, there must be no openings in the part and there must be a sufficient relationship of the depth of the part to the taper, so that the air pressure is not lost before the part is ejected. In addition to using the built-in ejection system provided by the movement of the press, it is possible to design special ejector systems operated independently of the stroke of the press, utilizing pneumatic hydraulic cylinders or other methods of movement. What comes out of a mold first had to come into a mold, and no part can be ejected if it hasn't been injected.

The injection process and location of gates, type of gates, and the like, are central to the aesthetic limitations of the design. It is impossible to eliminate completely a visible gate mark; it can in certain materials be reduced to a microscopic evidence. The gate mark can, in the instance of a tunnel or submarine-gated mold, appear on the side wall of the part, again very indistinguishable. The actual method of gating should best be left to the mold designer and the mold builder. The number of cavities, the size of the mold, the size of the part, the wall thickness, the material to be used, are parameters that must be recognized in designing the type of gating. But the design engineer must recognize that there will be an evidence of a gate, and if one can design the part aesthetically to incorporate the gate mark as an inherent feature of the product, one can reduce its otherwise objectionable feature. For example, a series of concentric rings could surround the initial small gate to give the appearance of an intentionally designed part.

14.4.2.9 *Flow or Knit Lines*

Whenever there is an opening in a molded part that is created either by means of an insert in the cavity or even by a movable core, the flow of the resin in the cavity when the mold is closed is cooled by the mass of steel that interrupts its flow. As it realigns itself, the surfaces that have been in contact with the steel are now chilled and are cooler than the rest of the material. Owing to this change in the pattern of flow a visible mark is created. If there is a short interruption area on the product, the interruption, although visible, does not create any substantial reduction in the physical strength of the part.

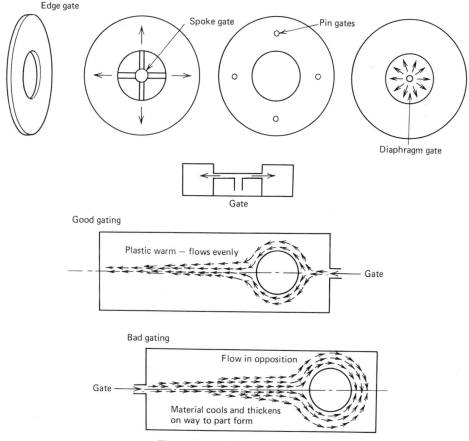

Figure 14.10 Flow or knit lines.

However, if a part has a large interruption in its configuration just before the end of the piece, and if by reason of gating this is at the end of the flow, then the knit line or weld line has a materially reduced physical strength. One method to eliminate the objections of the flow line both physically and structurally is to change the shape of the aperture so that the material flows in a more aerodynamic pattern. Another technique is to change the location of the gate (Figure 14.10) so that the hot material, as it enters the cavity, is closest to the aperature and not at the end of its flow where the material has had the opportunity to cool.

14.4.2.10 Inserts

Thermoset plastics were introduced approximately 50 years before thermoplastics. Thermoset materials, whether phenolic, urea, and even melamine, are basically brittle, and so if a screw had to be used to create an assembly, in the tightening process the strains would be transferred to the thin thread sections, and excess amount of torque would crack the part. To overcome this, metal inserts, aluminum, brass, steel, or stainless steel, could be inserted in the core of the mold. The resin itself can be strengthened by the use of fillers and reinforcing agents such as fiberglass, talc, microballoons, Kevlar, carbon and sapphire filaments, and so on. However, this increases the cost of the material for the total part, whereas an insert meets the mechanical requirements in a localized area.

Since thermosets molded by compression molding technique were done on vertical presses, it was very simple to drop the inserts into the bottom or core section of the hydraulic press. As the platens moved together the pressure was slowly built up around the inserts, and when the part was removed from the press one had a little section of metal, which could have nurls, concentric grooves, or anything to positively retain it in the part. Holding the insert internally was an internal thread. This way instead of bearing on one helix of the thread, or two, or three helices at the most, one was able to transfer the entire strain to the full binding capability of the insert. By using this technique fracture was

eliminated, also it was possible to remove the screws many times without any damage to the plastic.

With the introduction of injection molding and horizontal molding machines, it is much more difficult to keep the inserts on the horizontal, moving platen, and many attempts including the use of magnets and the like to hold inserts have been tried. Attempts have even been made to raise a conventional injection molding press from the horizontal to possibly 20 and 30 deg so that the inserts would be retained in the movable side. However, an injection mold that has been used with insert molding, suffers from dented mold faces due to inserts falling and crushing during the molding operation. It is much better today in the case of thermosets to provide for an undersized hole and then have the insert sonically bonded to the molded part. This permits the postmolding operation to proceed while the next part is being produced. There is no danger of inserts falling into the mold, also a much more rapid molding cycle is achieved, since no time is lost for the insertion of the inserts. Furthermore, by reason of improvements in materials it is now possible by designing bosses and the like to eliminate the necessity of inserts. In addition, it is possible by means of unscrewing devices to mold a thread directly into the part. Proper design can eliminate to a great extent the need for inserts. However, if it is necessary to have inserts in a thermoplastic part, it is better to do the same externally than as part of the molding cycle. There are certain specialized instances, such as in the manufacture of electrical appliance cords, where the metal tangs are dropped into the mold, the cable is inserted into the mold, and then the entire area is molded, encapsulating the tangs and the cable into one unitary structure.

To reduce the delays that would take place in a normal press, a sliding mold is built in which there are two cores, sets of cores, and one set of cavities. In this operation, while one part is being molded, the operator is able to load the next cores, set of cores and as the press opens the completely molded piece is ejected from the mold and the new mold already equipped with its inserts and cords now is ready to be processed by injection molding. While this part is being molded, the completed cord sets are removed. This is an expensive tooling and machine setup, and it has only become economically feasible when there are tremendous quantities to be produced of one particular item.

Since plastic is of a homogeneous nature, all of the part that is produced is of the same basic strength, even though there is a difference between the direct flow and cross-flow characteristics of the product. However, unless the part is distinctly linear, it is rather difficult to estimate and control the characteristics of the part by reason of gating, although this is sometimes used quite satisfactorily.

Most designers change the wall thickness of the part or add ribs to stiffen or strengthen a part. However, this affects both the molding cycle, the appearance of the part (since there might be sink marks), and warpage as differences in the thickness of the material affect shrinkage.

A way to overcome this problem is through the use of inserts. Inserts can be metallic or nonmetallic. Most often metallic inserts are used where threads are to be incorporated in a part. By the use of a threaded insert, one gets excellent bearing surface on the threaded portion, and it is not necessary to build an unscrewing mold to incorporate the thread.

Nonmetallic inserts are usually other thermoplastic materials that are inserted into the mold prior to the molding operation; then the entire product is molded together, and the different characteristics of the product are achieved at different areas of the part by reason of the location of the insert. An example of this technique has been used in a ski boot, diagrams of this technique are shown in Figure 14.11, and this concept is covered under a patented invention (4).

14.5 SERVICE PERFORMANCE

Service performance on a new product is very difficult to anticipate. Not only is a

Figure 14.11   Inserts.

product going to be subjected to the use that the designer has intended, but also it is going to be subjected to misuse and used in entirely other areas far from the realm of imagination. It is very akin to the problems in the space program if the personnel director had requested an "experienced female astronaut," when there had not been any female astronauts.

Service performance cannot be accurately simulated in the laboratory owing to the uniqueness of each and every product and the applications to which it will be put. If one is designing a part or a product that has a family relationship to an existing part or is just an upscaled or downscaled version of something that already has been put on the market, or if the product already exists on the market from another source, then it is possible to get a background of service performance and to correlate the differences between the experimental product and the one that actually is in service.

The question now is how does one anticipate and cope with service performance? One of the ways is to do a prototype manufacture of limited scale, put it in various test markets, and sit back and wait—not necessarily for the ceiling to fall down, for many times the response is more favorable than anticipated.

As mentioned in previous sections, the test data submitted either by material companies (or by the material specification charts) represents the performance of a single material under individual laboratory conditions for that distinct test. Each specific factor tested is unique and isolated from any other forces and is in a controlled environment, whereas the finished product is now subjected to many distinct factors that are now working on the product at the same time. For example, assume the case of an air conditioner. It is hung in a casement window, and there is vibration in the streets, particularly if it is in one of the larger cities—potholes create a tremendous problem in vibrations to buildings, not only to trucks and vehicles. There is a substantial thermal difference between the outside temperature and the inside temperature. The compressor is vibrating. There

may be a voltage drop that will again induce additional vibrations. In the room there could be smoke, which may be passing through the clogged filter. Someone might be spraying hair in front of the air conditioner with a lacquer spray that will certainly do havoc to the styrenic content in the molded housing. So instead of one static test, one now has the product subjected to 10 different forces all acting on it in unison relating to the sum total of 10 tests, each one interfering in a way with the statistical evaluation that has been achieved by the material being tested singly. It is not that the test data submitted are unreliable, it is the fact that the test data refer to test specimens dealing with spirals and test bars and not the finished product. In dealing with a molded product one should note that the character of the product, the design, the configuration of the part, the gating, and the molding technique can vary the physical specifications and characteristics of the materials as much as 50% or more.

A perfectly satisfactory material in a good working mold (in a good molding machine) could degrade as a result of malfunctioning of the heating element, thus completely changing the physical characteristics of the part even though there might not be much of a visual indication of the degradation of the material. There is an engineering maxim that roughly states that if it doesn't work, kick it. If it still doesn't work, read the instructions. This is not as far fetched as it seems. Many products fail in service owing to physical abuse. Refrigerator doors have failed because children have used them to swing on. A dishwasher door was never intended to be sat on. (In designing the door to be just opened and left in its open position without any load on it could cause failure in service since people will place heavy objects and even sit on the door of an open dishwasher.) However, from a limited reporting of failure or satisfactory performance of a sampling of the product in a test market, it is possible to deduce what might be expected of the product in actual performance.

The automotive industry in the United States represents the largest single manufacturing entity, that is, with its ancillary feed and subcontracting sources and having manufactured a product that has wheels and an engine for more than 75 years. Every time a new product comes on the market, even though is has several years of engineering development and has been tested on a track and in simulated use, until the product actually gets out into the market and is used by a large number is there any assurance that a recall will not be necessary? Therefore it is sometimes a rather unfair load imposed on the individual product designer, who develops a new product in a new field where there is no track record, to assume that he can, from scratch, come up with all the answers.

14.6 FINALIZATION OF DESIGN

After the product has been finalized it should be subjected to the following checklists to see that it conforms to the needs of both its function and its production. First note a checklist of background information independent of the product itself.

1. Ultimate use.
2. Ultimate cost of the product.
3. Annual, daily, or weekly peak requirements.
4. Prior and/or competitive production.
   a. Other materials.
   b. Other plastics.
   c. Similar or dissimilar form.
   d. Satisfactory or unsatisfactory performance.
   e. Evaluation of causes of previous failures.

Then see the following checklist of specific factors:

1. Particular strength of the product.
2. Environmental exposure.
   a. Weathering.
   b. Moisture.
   c. Vibrations, abrasion, physical stress.
   d. Creep.
3. Ambient acid and acrid fumes.
4. Exposure to sunlight and x-rays.
5. Specifications for dielectric strength.
6. Existence of insulation or electrical conductivity problems.
7. Toxicity.
8. Heat range subjection.
9. Service life of the product.
10. Shelf life of the product.
    a. Reusable.
    b. Disposable.
11. Odor absorption.
12. Permeability.
13. Color, lack of color, transparency, translucence or opaqueness.
14. Susceptibility to cracking and crazing.

Although these checklists are quite broad, there are specific areas where a part is used in an unusual application where other considerations must be given. These could be military or government specifications for specific material, as well as specific dimensions and tolerances, and this factor greatly simplifies the design engineer's problem since he has a positive standard to be able to meet. If the standard has been established, it is fair to assume that other products have been built that meet these standards. It does not require a Chinese copy to produce a part, but it certainly gives the design engineer a very positive direction to produce a good part.

14.7 DESIGN CASE STUDIES

Every product should have a check list. Product design covers the entire gamut of the performance of a part. There are products that require only restyling to keep pace with competition. There are parts that have very little styling, as they may be a replacement for a metal or other material. As such, most of the configurations have already been established, and it now becomes the work of the product engineer to incorporate the physical requirements of the use plus the principles of good engineering as applied to plastics such as radii, taper, uniform wall thickness, adequate coring, balanced flow, and all the textbook examples. However, these characteristics seldom fall like duckpins in a row in a product.

A product might have one or two desirable features, but usually there is a counterpoint. Whenever there are a lot of desirable features there is usually something that is hidden or is very difficult to achieve in the very same product.

The author, having been involved in approximately five hundred different products, can only recall fewer than a half-dozen that had uniform wall thickness, generous taper, no undercuts. Most of the problems or products that are designed usually require a violation (or an avoidance or overcoming) of some of the fundamental concepts of good design. It is not necessary to have a product with no limitations, but it is possible by intelligent design and compromise to overcome a lot of the problems that are created in the part by a recognition of their existence and designing around them.

In the development of society over the years, products have taken on common configurations: chairs are usually about 17 in. in height, a table is 30 in., a tumbler is slightly less than 3 in. in diameter and approximately 4 in. high. These dimensions have proven to be workable and anatomically acceptable and should be followed basically if there is no specific reason for deviation.

It is not that one should not attempt to improve or create new elements of design, but a freak design, in and of itself, although attention getting, will not last long in the marketplace. Change for the sake of change serves no economic purpose. Change for improvement should be the goal.

14.7.1 Consumer Product Design

This is a study of a good and bad design of a consumer product. We are looking at a wall-hanging flower pot that is a very popular and a much used product having aesthetic and practical applications and used for decorative purposes in homes, offices, stores, and so on.

Version A (see Figure 14.12), although aesthetically pleasing, is composed of two pieces requiring two separate molds. The assembly operation of gluing or bonding the two components together, with the possibility of mismatch in the bonding operation or some visible streaks of cement if they are mechanically assembled, is a problem requiring prompt assembly, as there will be substantial shrinkage between the two parts if they are not assembled immediately after molding.

The three openings in the top component to attach the hanging straps require movable cores. If the part is to be designed in one piece to eliminate any cementing operation as shown, movable collapsable cores would be required so that the part could be removed intact.

The difference between the thicker and thinner wall creates a problem in shrinkage, and the part being center gated through the core to eliminate any visible gate marks at the bottom of the piece would again create the necessity of having the core side on the injection half of the mold, which is the reverse or normal procedure and would require the designing of an ejection system into the mold rather than relying on the normal function created by the movable platen in an injection molding press.

The redesigned part (Figure 14.13) is one piece. There are no cores to create openings for tying the hanging supports. The part is dished downward to the center so that the ejection mark created by the sprue becomes a part of the design of the part, and the natural conical shape of the base resists shrinkage or canning and warpage. It requires only a single-cavity mold with normal ejection, and by means of either a hot sprue or a very fine capillary or pinpoint gate, this mold could be designed to run automatically. So we end up with a part that costs a fraction of the original cost to mold, no assembly costs, and a much less expensive mold cost.

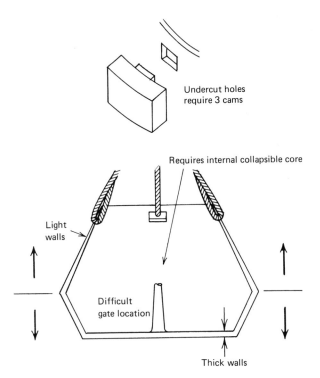

Undercut holes require 3 cams

Requires internal collapsible core

Light walls

Difficult gate location

Thick walls

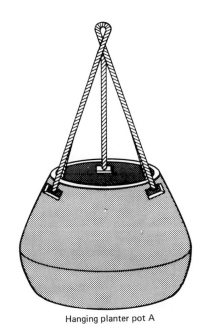

Hanging planter pot A

Figure 14.12 Consumer product design A.

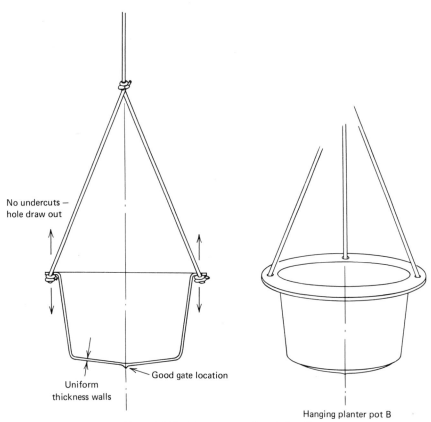

No undercuts —
hole draw out

Uniform
thickness walls

Good gate location

Hanging planter pot B

Figure 14.13   Consumer product design B.

14.7.2   Industrial Product Design

There are different parameters in industrial product design than exist in consumer products. In one area it can be said that "cost be damned." The cost of material of any component that would be utilized in a computer can be disregarded, as the most expensive material would be insignificant to the cost of failure of the part. Once the piece of equipment has been installed, its hourly cost is so great that a couple of hours downtime would exceed the total value of all the material put into the complete unit. So in any design consideration for a mechanical component for a computer terminal or any of the drive equipment and ancillary equipment associated with the computer, the cost consideration of material and molding can be eliminated, and the fail-safe product should be designed. Yet the ultimate value of the equipment, its earning capacity, and its

commercial sales price must be recognized before the material costs and molding considerations are thought to be insignificant.

In the pocket calculator instrument, which has now become a football, a 10-cent difference in the cost of material and molding would be a very substantial percentage of the total profit in an instrument of this type. So it is important for the design engineer to know all the ramifications of product before determining the ultimate design and specific material. In industrial design the product engineer usually has some basic concept of the total number of units that will be required on an annual basis, as there are positive sales projections and there is usually a track record of a similar product. There is the reasonable possibility of momentum, and the probability that the product, once designed, will continue for a reasonable market life. In a consumer product the demand may exceed the anticipated production capacity, in a short time production capacity may exceed demand, and in another interval of time there may be full production capacity but zero-limited demand.

One of the inherent problems in designing industrial parts is that such a part is a standard item used by many manufacturers and chosen from a catalog. In this case the part design has already been established, and there are several sources of supply. Now it is incumbent on the industrial designer to achieve the criteria that have been established, and to accomplish the very same functioning subassembly at a lower cost to meet the constant drive by purchasing agents and directors to meet management's insistence on reducing costs. Basically approved materials cost the same. Molding procedures are more or less the same. The way to reduce cost is the product designer's problem. In analyzing an existing subassembly, one might find that there are eight or nine components. Possibly, as originally designed, it might have been in a family mold so that all eight pieces were produced at one time, and they made up a complete assembly. The part worked. As the volume increased (and in keeping with the old saw, "don't make waves"), the economies were achieved by building a multifamily mold. Possibly two complete units or four complete units were used at a time, and in here the saving was made in molding cost rather than in material or in revised product design. However, if one were to analyze the subassembly, one would see that with all the material the same, each was not used to its best advantage. If the production requirements are such that it will be possible to build individual molds for individual components, it is then possible to design each component part to the best advantage of the most suitable material. This can be achieved in several different manners. Assuming that the selection of the housing was in ABS for visual as well as engineering requirements, the rigidity of ABS would preclude its being used in a spring application. However, it would be possible to use acetal resin for a plastic spring or latch mechanism that would deflect sufficiently and not create any failure. It would also be possible to produce the part more cheaply as it would elimi-

nate certain other components, including a metal spring and subassembly, to achieve the desired results.

A new technique (2) which has recently been patented, allows the use of two or more dissimilar materials in one mold at the same time. This involves a specialized machine having extra injection barrels located on the parting line. It could either be horizontal or vertical, and feeding is done by means of a runner on the parting line to multiple gates in the part. (See Figure 14.14.) The normal injection in a horizontal plane proceeds first, and by short shooting or by means of removeable cores, it is possible to position the first material; then the secondary and/or tertiary material enters the mold. One now has the unique situation of a multimaterial part, meeting the requirements of each component where it is necessary, still molded at one time without any secondary labor operations. This very same double-injection molding principle can be used in another patented technique of incorporating a torsion bar spring (3). This leads to a rigid assembly with an elastomeric torsion bar that would permit rotation of any device and still retain the rigidity of the outside housing, without the necessity of any multiple component assembly and the expense of the assembly. By the use of sophisticated design and by removeable cores in the mold, it is possible to create a physical undercut in the part so that the second material, although it is preferable to have chemical compatibility, could be employed where there was no chemical compatibility, but would rely completely on a mechanical lock. This would be the case of incorporating a thermoset polyester or even an injection molded thermoset material.

14.8 SUMMARY

It is difficult to project precisely the direction and length of time it will take to achieve the end results of an almost total plastic product. Ten years ago if one were to try to project the amount of plastic used in an automobile, it would have been a natural progression of part by part, where plastic had

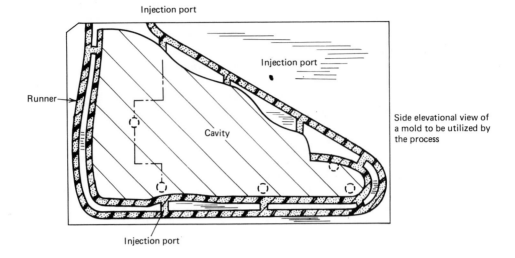

Injection port

Injection port

Runner

Cavity

Injection port

Side elevational view of
a mold to be utilized by
the process

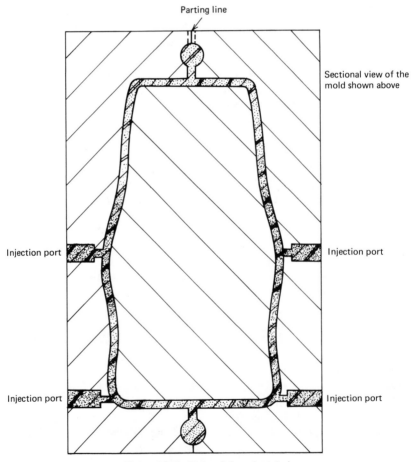

Parting line

Injection port

Injection port

Injection port

Injection port

Sectional view of the
mold shown above

Figure 14.14 Industrial product design.

to fight its way on the basis of a cost per cubic inch or total manufactured cost against steel, aluminum, and so forth. With the demand imposed by the energy crunch to achieve specific mileage, the emphasis has no longer been just on "is it cheaper to use plastic" but "can one save weight," and there are instances where plastic has been selected not because it was cheaper. In such cases, in all probability, it turned out to be cheaper anyway in the long run, owing to the creative designer's ability to combine multiple functions in a part and reduce the labor cost generally associated with injection molded parts.

The incredible rise in the demand for small boats has brought about utilization of plastic heretofore unanticipated in demand. It is not just the use of fiberglass for boat construction that has increased, but use has also increased of ABS, thermoformed sheets and injection molding of complete hulls. A recent example of this trend is the Topper boat molded in England where the dinghy is approximately 150 in. long, 59 in. wide and 31 in. deep. To achieve this size, it was necessary to combine two 1600 ton machines to accommodate the platen area. This is an example of design preceding equipment. Should additional items of this size be developed, a machine tool builder will build a single machine to achieve this requirement. This shows ability to improvise, meet design requirements, and develop a productive capacity that achieves the desired ends. There are new polymers constantly being developed which create design opportunities that heretofore were uncontemplated. TPX (methylpentene polymers) as a material permits clarity, boilability, autoclavability, high-impact strength, and distinctive other qualities such as vapor and oil transmission, which opens up a whole new area of plastic in elevated temperature situations for hospital, medical, food processing use.

The rise in leisure time has brought about a tremendous increase in demand for recreational vehicles. Here again, it is not necessarily its cost in relation to aluminum, but the fact that plastic is a better insulator than aluminum and will keep the camper body warmer in the winter and cooler in the summer that has accounted for its wide acceptance. It can be seen how particular requirements develop the application of the material. It is not only the fact that it is cheaper, but also the fact that plastics can do a better job than other materials in many applications. It is left to the imagination and the ability of the design engineer to convince management of the limitless possibilities available for the use of plastic in the foreseeable future.

There are very few areas in which the design engineer will be completely thwarted by the inability to design a product utilizing plastic, either for the main or component use. In oil rigs, at the present time, the structure and the drive mechanism to push a pipe down to a 3000–5000 ft hole still eliminates the use of plastics. However, storage tanks, splash tanks, splash zone guards, flexible couplings from the drilling platform to a supply ship, all these are areas where plastic can be utilized, and, again in this instance, it would not be the cost that is the problem, but the function. Because of the inherent hazard of spark on a drilling platform, particularly where gas, rather than crude oil is being pumped, the safety factor of the nonconductivity of sparking of plastic could overweigh the fact that it might have a limited-purpose life. It would have to be replaced more readily than a steel part, but its safety factor would permit it to be used without the balance of cost, since an oil rig is rather expensive, both in dollars and life.

As a guide, if from the resins that are available today, a designer can design a part where the minimum physical and/or chemical specifications are met, one can feel reasonably satisfied. At the time the design is made there may not be a built-in safety factor of any magnitude. However, the interval of time until the design is formulated, the processing of the building of the mold is done, to the actual production of the part is a minimum of one year. In this time, the probability of resin suppliers coming up with improved grades of materials is good, as material suppliers generally are constantly upgrading their materials. So if the project engineer submitted the requirements to material suppliers in this interval of time, they would develop a new resin that would have substantial improvement over the existing resin in at least some categories. For example, chemical inertness (or resistance) or dielectric resistance might be improved materially, but physical strength, shock, water absorption, and so on would certainly be areas in which improvement could be definitely anticipated.

The size of machines, particularly in injection molding, limit the physical size of the part. With the demand for lighter weight and complete redesign of parts, it will now be possible for the design engineer to think of substantially larger parts than heretofore used, as opposed to the necessity to adapt the components to an existing structure and method of manufacture. Should the design engineer now specify a part that would be 10 ft long, one can be reasonably assured that the machine tool builders could build a machine that would meet these requirements, as long as there is a demand for the machine.

As far as the physical size of presses are concerned, hydraulic presses with large platen areas, 15,000–25,000 tons clamping pressure, have been built primarily for the forging of nonferrous alloy parts for aircraft. With the technology of such press construction readily available, and with the development of extruder and screw barrels to sizes that are today uncommon, the combining of the processing techniques of plastication and hydraulic clamping forces should be able to produce presses of incredible size, once a demand for this size has been established. The example of joining the two Windsor presses to create the large platen area with a total clampage of slightly more than 3000 tons is an indication of design ingenuity.

14.9 REFERENCES

1. "You *Can* Predict Creep in Plastic Parts," Mach. Des., (March 11, 1976).

2. Injection Molding Process, U.S. Patent 3,950,483.

3. Torsion Bar Assembly and Method for Manufacturing the Same, U.S. Patent 3,490,756; Container Closure Assembly, U.S. Patent 3,235,145; Container Closure Assembly, U.S. Patent 3.140,020.

4. Ski Boot Construction and Method, U.S. Patent 3,718,994.

## CHAPTER 15

Computer Applications to Processing

IMRICH KLEIN

Scientific Process & Research, Inc.
Somerset, New Jersey

15.1 Introduction   699
    15.1.1   Terminology   699
    15.1.2   The State of the Art   701
    15.1.3   The Plastics Industry   702
    15.1.4   The Computer Programs   702

15.2 Description of Computer Programs   702
    15.2.1   Workup of Physical Properties   702
        15.2.1.1   Frictional Properties
              of Solids   702
        15.2.1.2   Flow Properties of Melts   704
        15.2.1.3   Thermal Properties   728
        15.2.1.4   Material Density   729
    15.2.2   Physical Models of Plastics
        Processing Operations   732
        15.2.2.1   Melt Extrusion   732
        15.2.2.2   Plasticating Extrusion   738
        15.2.2.3   Blow Molding   746
        15.2.2.4   Reciprocating Screw
              Injection Molding   746
        15.2.2.5   Melt Spinning   747
        15.2.2.6   Wire Coating and
              Cable Jacketing   750
        15.2.2.7   Crosshead Dies   751
        15.2.2.8   Sheet and Flat Film
              Formation   752
        15.2.2.9   Film Blowing   761
        15.2.2.10 Blow Molding Parison
              Dies   765
        15.2.2.11 Die Analysis   767
        15.2.2.12 Product Cooling   769
        15.2.2.13 Curing and
              Cross-linking   773
    15.2.3   Computer Aided Research (CAR)   775
        15.2.3.1   Design of Experiments   775

        15.2.3.2   Compounding and Formulation
              Development   776
        15.2.3.3   Modeling of Experiments   777

15.3 Summary   778

15.4 References   778

15.1   INTRODUCTION

15.1.1   Terminology

The following discussion of computer terminology is directed toward the reader whose exposure to computers in his or her everyday work has been limited. Before entering into a survey of calculation capabilities, it is necessary that the reader possess an understanding of the different devices that make up the modern computer. It is also helpful if the reader understands the various means by which access can be gained to computer systems.

The computer itself is a general purpose machine. It can be used to perform a wide variety of calculations or data manipulations, but to do so it requires a program or series of instructions that describe the precise action the computer is to perform. For this reason, computer systems are said to be comprised of two parts, HARDWARE and SOFTWARE.

Hardware, as the name implies, refers to the physical piece of machinery. In the case of computers, the machinery consists of a device with the ability to store and retrieve pieces of information and to perform some simple manipulations on the data as well. Typically, the hardware is capable of performing arithmetic operations, such as addition and subtraction, as well as logical operations such as AND and OR. A catalog of all the instructions that can be performed by the hardware is called the machine's INSTRUCTION SET. This instruction set represents the sum total of the computer's capabilities. Typically, these hardware instructions will not include complicated manipulations such as square roots or exponentiation. Such functions are usually left to software. This is true even with most of today's high-speed mainframe computers because these functions can be performed by combinations of other instructions fairly well.

Beyond the ability to perform any of the functions in the computer's instruction set, the hardware is also built with the ability to record and EXECUTE any of it's instructions in any order. In other words, the machine is PROGRAMABLE. When we PROGRAM the machine to execute its instructions in some particular order, we are creating software. It is the software that allows the machine to perform very complicated tasks, and since the software can be changed very easily, any computer can be used for a wide number of different applications.

Software is further broken down into two categories. One type of software is termed SYSTEMS SOFTWARE. Systems software includes operating systems, languages, editors, and other programs that help to make the computer itself more easily programable. The other kind of program is known as APPLICATIONS SOFTWARE. It is the applications program that most of us see every day in the supermarket, at the bank, or in computer games. Applications programs are generally written in such a way as to be easily used by a nonprogrammer.

When you use a computer you will inevitably be required to deal with both systems and applications programs. For instance, to play a computer game at the local pinball arcade, you are required to select options by setting various machine PARAMETERS, such as fast play or slow play, or in some cases to introduce variations in the game itself, and then you must start the program, often by simply inserting your quarter. These actions are part of machine operation, and the programs that you are dealing with are system programs. Once the game has begun, you are engaged with an applications program. By operating the controls you can vary the INPUT DATA to the applications program, thus effecting changes in program OUTPUT. Because you have the capability of varying your commands while the program is executing, we call this an INTERACTIVE operation.

With more complex applications, a wider range of capabilities must be supported and, therefore, a more detailed understanding of the nature of the computer and its software is required. For instance, the computer game is a DEDICATED APPLICATION, that is, only one game can be played on that particular machine at any given time. However when larger, more expensive, computers are used, this becomes a very inefficient mode of operation. Since the computer is very fast, and the user reads and types relatively slowly (a typical computer can read and write about 100,000 times as fast as you can, in fact) it is usually desirable for the machine to be programmed to interact with several users at the same time. Most mainframes use a system called TIME SHARING to accomplish this. In time sharing a certain portion of every second of operation is allocated to each user. The computer will execute your program for something like 1/30 sec before going on to someone else. The period during which the computer is working for you is called your TIME SLICE. Because the computer is set up to handle many different applications, the user must give rather detailed instructions to the system to specify which applications program he wishes to use. Also, because of the complexity of the applications programs, a great deal of input data is required.

In a time sharing system, the user communicates with the computer through a TERMINAL. The terminal is simply a typewriter keyboard and printer with the added ability to transmit what is typed through a telephone line. The terminal also receives the program output and prints it for the user. Each user also has a certain amount of permanent storage available to him on the computer system. ON-LINE STORAGE, that is information that is readily available for immediate use, is maintained on a device called a DISK. The disk on a computer system is similar to a phonograph record except that it is magnetic and turns at very high speed. Each disk is divided into a number of DI-RECTORIES or PARTITIONS allowing each user to maintain his own private FILES. Normally a set of passwords is required to gain access to a directory and to read the data stored there. The data are broken down into files because this allows the user to separate information for his own purposes.

Most of the information entered into an applications program is maintained in disk files. When the program is executed it asks for a file name and reads the data from the file. The amount of information that must be kept on file depends on the amount of information necessary for the particular applications program to make its calculations. Usually data files are quite long because most programs allow the user to set a large number of variables that describe the situation to be investigated.

The information often remains the same from one case to another except for one or two variations. It is preferable, therefore, if data files can be edited rather than being retyped. So using the computer usually involves becoming familiar with the system EDITOR. The editor is a systems program that operates on files and allows the user to modify them easily. The power of a particular system's editors is an important factor is selecting a time sharing system. If the editor is cumbersome or difficult to operate, it may be much more difficult to use an applications program on that system.

Applications programs, themselves, can be divided into business oriented and scientific programs. The vast majority of programs and program packages available today fall into the category of business software. These business packages can be either ALGORISTIC, such as accounting programs, payroll systems, and the like, or they may be INFORMATION MANAGEMENT packages. Information management involves the storage and retrieval of data according to complex specifications. Scientific packages of the straightforward algoristic type range from calculator programs to complex statistical packages. However, the major innovation that the advent of computers has provided in science and engineering has been the SIMULATION PROGRAM.

Computer simulation is a technique whereby the computer is programmed to act as if it were the process under study. This technique has proved invaluable in fields such as aeronautical engineering, geology, and nuclear physics. In the plastics processing field, simulation packages exist for most pieces of equipment, including extruders, injection molders, various types of dies, cooling and heating equipment, melt spinning takeup, and many other devices. The following sections discuss some of these programs in detail.

15.1.2 The State of the Art

Since the book *Computer Programs for Plastics Engineers* (1) was published in 1968 under the auspices of the Society of Plastics Engineers in their Polymer Science and Engineering Series, the use of computers has expanded considerably. This was primarily due to the great breakthroughs that have taken place in computer technology over the last decade. Computers became much bigger in capacity yet considerably smaller in size. They also became much faster and their operation virtually trouble free. Despite the general inflation that occured worldwide since 1968, the price of computers has dropped considerably. When considering the price of computers per task performed, a decrease of at least an order of magnitude, or ten to one, has already taken place. As a result, computers are no longer restricted to giant corporations, but are now available in most small companies as well.

Hand in hand with the advance of computer technology, communications technology has improved as well. Long-distance telephone connection to computers became better and cheaper with local dial-up service available even to very remote computers. At the present time such local dial-up facilities are available in the continental United States, Hawaii, and Puerto Rico, in 52 cities in Canada, and most of Western Europe. Similar services are offered in many places in Asia.

This development gave birth to a new industry: the specialized time sharing computer. Contrary to general purpose time sharing, which is on a decline. special purpose time sharing concentrates on a particular application. It is manned by people with specialized background in a limited field and concentrates on developing highly specialized computer programs. Such programs are under constant development, are supported by detailed users' manuals, and usually offer immediate telephone contact with the experts who developed the programs. The user can, therefore, get immediate expert advice on the judicious use of the computer programs.

15.1.3   The Plastics Industry

A similar development has taken place over the last 10 years in the plastics industry. The first computer software house was established in 1968 and grew into a complete time sharing computer network for the plastics industry (2). It operates its own time sharing computer with communication lines covering much of the western world. Most of this chapter concentrates on describing the computer programs offered to the plastics industry by this source.

15.1.4   The Computer Programs

The computer programs offered depending on their intended application, can be grouped as follows:

1.   Workup of physical properties.
     a.   Frictional properties of solids.
     b.   Flow properties of melts.
     c.   Thermal properties.
     d.   Density of melt and solid.
2.   Physical models of plastics processing operations.
     a.   Melt extrusion.
     b.   Plasticating extrusion.

c.   Blow molding.
d.   Injection molding.
e.   Melt spinning.
f.   Wire coating and cable jacketing.
g.   Crosshead dies.
h.   Sheet formation.
i.   Film blowing.
j.   Blow molding parison dies.
k.   Die analysis.
l.   Product cooling.
m.   Curing and cross-linking.

3.   Computer aided research (CAR).
     a.   Design of experiments.
     b.   Compounding and formulation development.
     c.   Modeling of experiments.

15.2   DESCRIPTION OF COMPUTER PROGRAMS

15.2.1   Workup of Physical Properties

15.2.1.1   Frictional Properties of Solids

It was found that to be able to predict the performance of an extruder or molding machine one must know the frictional properties of the solids to be processed. Since the solids are usually in the form of pellets, powder, beeds, flake, and the like, the technique must be specific and quite refined. The friction data can be conveniently measured on and SPR Friction Tester (See Figure 15.1). This instrument measures the frictional force produced when the normal force on the test surface is kept constant. The angular velocity of the specimen relative to the test surface can be controlled and measured on the tester and so can the surface temperature. In some models of this instrument the surface velocity and frictional force readings must be corrected from calibration curves. This is a tedious task particularly when it is remembered that 300 to 900 points are usually measured to fully characterize a solid feed material to an extruder or molding machine. The normal force applied must sometimes also be corrected. This is often caused by the weight of the specimen holder, which in newer models is

Figure 15.1 (*a*) Friction tester. (*b*) Specimen holder and test surface for friction tester.

normally suspended and does not therefore add to the normal force applied. In some cases, however, the weight of the specimen holder is directly applied to the normal force. The presence of this weight must therefore be reflected in the data and the data properly corrected.

The computer program called WORKUP was developed to handle many routine but tedious plastics processing-related applications. It has an option that reads the measured friction data and performs all corrections needed, column by column. It also generates a column giving the coefficient of friction for each measurement by dividing the corrected friction force by the normal force. A data file containing the corrected or "worked up" data is created by the program and placed on the disk of the time sharing computer.

The coefficient of friction was found to depend on surface velocity, pressure, and surface temperature. This relationship was also found to be very complex requiring a 19-term polynomial equation to express it. A regression analysis program was therefore assembled that automatically converts the three independent variables into the following form:

Instrument speed (rpm) into surface velocity V (in./min).

Normal force (lb) into pressure P (psi).

Surface temperature (°F) into (temperature − 70), T (°F).

The three variables are then processed by the regression program VISC, generating an equation of the following form:

$$f = a_0 + a_1 V + a_{11} V^2 + a_{12} VP + a_{13} VT + a_2 P$$
$$+ a_{22} P^2 + a_{23} PT + a_3 T + a_{33} T^2 + a_{111} V^3$$
$$+ a_{333} T^3 + a_{112} V^2 P + a_{113} V^2 T + a_{123} VPT$$
$$+ a_{223} P^2 T + a_{223} P^2 T + a_{221} P^2 V + a_{331} T^2 V$$
$$+ a_{332} T^2 P$$

Naturally, an equation of this complexity does not lend itself to easy visualization of the effect of the three independent variables on the friction factor. Both the VISC and the WORKUP programs therefore offer friction

evaluation. They give the coefficient of friction as a function of surface velocity in tabular form for each pair of pressure and temperature. Graphical output of these tables on a digital plotter is also available.

A sample case for the workup of frictional properties of solids in various forms, such as pellets or powder normally used as feed for an extruder or molding machine, is given in Table 15.1. The data are typed by the technician into the remote computer and stored on disk in the form of a data file. The data are stored in four columns representing the following information:

Data point number (#)

Speed of friction tester, rpm (RPM)

Normal force, lb (N)

Temperature, °F (DEG. F)

Frictional force, lb (F)

The whole procedure is run through automatically. It types out the input data; performs the regression analysis in a stepwise manner, that is, determines the constant term and the 18 coefficients defining the equation; and prints out a table comparing the coefficient of friction values computed for every data point with the values computed from the derived equation and the difference between the two. This is followed by an evaluation, that is, a tabular printout of coefficient of friction values at different surface velocities, pressures, and temperatures.

15.2.1.2 Flow Properties of Melts

The viscosity of plastic melt or other fluid is usually determined on a capillary rheometer. In such an instrument the plastic is charged into a heated cylinder where it melts. The melt is then pushed through a cylindrical die or capillary by a piston or ram. Each data point measured contains the following information:

Ram velocity (in./min).

Melt temperature (°F or °C).

Force used to make the melt flow at that velocity (lb).

Table 15.1

```
                    SAMPLE #1 LDPE
                    Thursday, August 31, 1978   2:53 PM
                    0270 POINTS IN FILE
                        1    5.   2.   89.6  0.1
                        2    9.   2.   89.6  0.2
                        3   30.   2.   89.6  0.4
                        4   50.   2.   89.6  0.5
                        5   90.   2.   89.6  0.8
                        6  150.   2.   89.6  0.9
                        7    5.   4.   89.6  0.1
                        8    9.   4.   89.6  0.3
                        9   30.   4.   89.6  0.8
                       10   50.   4.   89.6  1.1
                       11   90.   4.   89.6  1.5
                       12  150.   4.   89.6  1.7
                                  .
                                  .
                                  .
                      263   90.  30.  179.6 20.9
                      264  150.  30.  179.6 19.3
                      265    5.  36.  179.6 12.5
                      266    9.  36.  179.6 15.8
                      267   30.  36.  179.6 20.3
                      268   50.  36.  179.6 22.2
                      269   90.  36.  179.6 25.8
                      270  150.  36.  179.6 23.2


OK, VISC
GO
V I S C --   31-AUG-1978   SPR-RELEASE: JANUARY 16, 1978

  INPUT FILE FOR VISC = M-1
  OPTIMIZATION, FRICTION OR VISCOSITY FUNCTION?FRIC
  YOUR PROBLEM I.D.# 1
  NUMBER OF DATA POINTS=270
  WANT INTERMEDIATE STATISTICAL DATA? NO
  PREDICTED RESPONSES PRINTED? YES
  PRINT TRANSFORMED VARIABLES? NO

                    VARIABLE TRANSFORMATIONS
  ( Z(   1) +   0.000000)  *  (( 0) +   7.068583)  = W(  1)
  ( Z(   2) +   0.000000)  *  (( 0) +   1.086498)  = W(  2)
  ( Z(   3) + -70.000000)  +  (( 0) +   0.000000)  = W(  3)
  ( W(   1) +   0.000000)  = X(  1)
  ( W(   1) +   0.000000)  *  ( W(  1) +   0.000000)  = X(  2)
  ( W(   1) +   0.000000)  *  ( W(  2) +   0.000000)  = X(  3)
  ( W(   1) +   0.000000)  *  ( W(  3) +   0.000000)  = X(  4)
  ( W(   2) +   0.000000)  = X(  5)
  ( W(   2) +   0.000000)  *  ( W(  2) +   0.000000)  = X(  6)
  ( W(   2) +   0.000000)  *  ( W(  3) +   0.000000)  = X(  7)
  ( W(   3) +   0.000000)  = X(  8)
  ( W(   3) +   0.000000)  *  ( W(  3) +   0.000000)  = X(  9)
  ( W(   1) +   0.000000)  ** (( 0) +   3.000000)  = X( 10)
  ( W(   3) +   0.000000)  ** (( 0) +   3.000000)  = X( 11)
  ( X(   2) +   0.000000)  *  ( W(  2) +   0.000000)  = X( 12)
  ( X(   2) +   0.000000)  *  ( W(  3) +   0.000000)  = X( 13)
  ( X(   3) +   0.000000)  *  ( W(  3) +   0.000000)  = X( 14)
  ( X(   6) +   0.000000)  *  ( W(  3) +   0.000000)  = X( 15)
  ( X(   6) +   0.000000)  *  ( W(  1) +   0.000000)  = X( 16)
  ( X(   9) +   0.000000)  *  ( W(  1) +   0.000000)  = X( 17)
  ( X(   9) +   0.000000)  *  ( W(  2) +   0.000000)  = X( 18)
  ( Z(   4) +   0.000000)  /  ( Z(  2) +   0.000000)  = Y(  1)
```

Table 15.1 (*Continued*)

```
STEPWISE   REGRESSION

PROBLEM NO          1

NO OF DATA =        270

NO OF VARIABLES =        19

WEIGHTED DEGREES OF FREEDOM =        270.0000

F LEVEL TO ENTER VARIABLE =    0.00010

F LEVEL TO REMOVE VARIABLE =    0.00001

STANDARD ERROR OF Y =       0.170901
STEP NO.    17
   VARIABLE ENTERING        13
F LEVEL        0.00006
STANDARD ERROR OF Y =          0.03725
MULTIPLE CORRELATION COEFFICIENT          0.97749
  CONSTANT      0.6446294487E-01
```

VARIABLE		COEFFICIENT	STD ERROR OF COEF
X-	1	0.1021133037E-02	0.00008
X-	2	-0.1080557468E-05	0.00000
X-	3	0.4503457603E-05	0.00000
X-	4	-0.2624361514E-05	0.00000
X-	5	0.2662598155E-02	0.00140
X-	6	0.5675141438E-04	0.00003
X-	7	-0.1802521874E-03	0.00003
X-	8	-0.1568539534E-02	0.00083
X-	9	-0.1550198431E-04	0.00002
X-	10	0.4850383428E-09	0.00000
X-	11	0.2941587809E-06	0.00000
X-	12	-0.3060079745E-08	0.00000
X-	13	-0.3132592519E-09	0.00000
X-	14	-0.3891098999E-07	0.00000
X-	15	0.7437087106E-06	0.00000
X-	16	0.0	
X-	17	0.3244021940E-07	0.00000
X-	18	0.1493877789E-05	0.00000

```
                 PREDICTED VS ACTUAL RESULTS
        RUN NO.       ACTUAL       PREDICTED       DEVIATION
```

RUN NO.	ACTUAL	PREDICTED	DEVIATION
1.	0.05000	0.06333	-0.0133
2.	0.10000	0.08837	0.0116
3.	0.20000	0.19513	0.0049
4.	0.25000	0.26451	-0.0145
5.	0.40000	0.34205	0.0579
6.	0.45000	0.42867	0.0213
7.	0.02500	0.06397	-0.0390
8.	0.07500	0.08923	-0.0142
9.	0.20000	0.19692	0.0031
10.	0.27500	0.26692	0.0081
11.	0.37500	0.34490	0.0301
12.	0.42500	0.43018	-0.0052
.			
.			
.			
263.	0.69667	0.62105	0.0756
264.	0.64333	0.68006	-0.0367
265.	0.34722	0.37695	-0.0297
266.	0.43889	0.40562	0.0333
267.	0.56389	0.52771	0.0362
268.	0.61667	0.60591	0.0108
269.	0.71667	0.68415	0.0325
270.	0.64444	0.72947	-0.0850

```
WANT TO EVALUATE FRICTION FUNCTION? YES
```

Table 15.1 (*Continued*)

```
SPR FRICTION FUNCTION EVALUATOR   31-AUG-1977 RELEASE: FEB. 4,1977
PRESS= 1
TEMP= 70
```

V IN/MIN	P PSI	T DEG F	F
0.0	1.0	70.0	0.06718
50.0	1.0	70.0	0.11582
100.0	1.0	70.0	0.15939
150.0	1.0	70.0	0.19828
200.0	1.0	70.0	0.23285
250.0	1.0	70.0	0.26344
300.0	1.0	70.0	0.29044
350.0	1.0	70.0	0.31421
400.0	1.0	70.0	0.33510
450.0	1.0	70.0	0.35349
500.0	1.0	70.0	0.36973
550.0	1.0	70.0	0.38419
600.0	1.0	70.0	0.39723
650.0	1.0	70.0	0.40922
700.0	1.0	70.0	0.42052
750.0	1.0	70.0	0.43150
800.0	1.0	70.0	0.44252
850.0	1.0	70.0	0.45393
900.0	1.0	70.0	0.46612
950.0	1.0	70.0	0.47943
1000.0	1.0	70.0	0.49424

```
PRESS= 2
TEMP= 70
```

V IN/MIN	P PSI	T DEG F	F
0.0	2.0	70.0	0.07002
50.0	2.0	70.0	0.11887
100.0	2.0	70.0	0.16265
150.0	2.0	70.0	0.20172
200.0	2.0	70.0	0.23646
250.0	2.0	70.0	0.26721
300.0	2.0	70.0	0.29435
350.0	2.0	70.0	0.31824
400.0	2.0	70.0	0.33925
450.0	2.0	70.0	0.35772
500.0	2.0	70.0	0.37405
550.0	2.0	70.0	0.38857
600.0	2.0	70.0	0.40166
650.0	2.0	70.0	0.41369
700.0	2.0	70.0	0.42501
750.0	2.0	70.0	0.43599
800.0	2.0	70.0	0.44699
850.0	2.0	70.0	0.45838
900.0	2.0	70.0	0.47053
950.0	2.0	70.0	0.48378
1000.0	2.0	70.0	0.49852

```
PRESS= 4
TEMP= 70
```

V IN/MIN	P PSI	T DEG F	F
0.0	4.0	70.0	0.07602
50.0	4.0	70.0	0.12531
100.0	4.0	70.0	0.16949
150.0	4.0	70.0	0.20894
200.0	4.0	70.0	0.24402
250.0	4.0	70.0	0.27509
300.0	4.0	70.0	0.30251
350.0	4.0	70.0	0.32665
400.0	4.0	70.0	0.34787
450.0	4.0	70.0	0.36654
500.0	4.0	70.0	0.38303
550.0	4.0	70.0	0.39768

Table 15.1 (*Continued*)

PRESS= 4
TEMP= 70

V IN/MIN	P PSI	T DEG F	F
600.0	4.0	70.0	0.41087
650.0	4.0	70.0	0.42296
700.0	4.0	70.0	0.43432
750.0	4.0	70.0	0.44531
800.0	4.0	70.0	0.45629
850.0	4.0	70.0	0.46762
900.0	4.0	70.0	0.47968
950.0	4.0	70.0	0.49282
1000.0	4.0	70.0	0.50741

PRESS= 10
TEMP= 70

V IN/MIN	P PSI	T DEG F	F
0.0	10.0	70.0	0.09676
50.0	10.0	70.0	0.14736
100.0	10.0	70.0	0.19275
150.0	10.0	70.0	0.23333
200.0	10.0	70.0	0.26943
250.0	10.0	70.0	0.30144
300.0	10.0	70.0	0.32971
350.0	10.0	70.0	0.35460
400.0	10.0	70.0	0.37649
450.0	10.0	70.0	0.39573
500.0	10.0	70.0	0.41269
550.0	10.0	70.0	0.42773
600.0	10.0	70.0	0.44122
650.0	10.0	70.0	0.45351
700.0	10.0	70.0	0.46498
750.0	10.0	70.0	0.47599
800.0	10.0	70.0	0.48690
850.0	10.0	70.0	0.49807
900.0	10.0	70.0	0.50987
950.0	10.0	70.0	0.52266
1000.0	10.0	70.0	0.53681

PRESS= 30
TEMP= 70

V IN/MIN	P PSI	T DEG F	F
0.0	30.0	70.0	0.19542
50.0	30.0	70.0	0.25036
100.0	30.0	70.0	0.29980
150.0	30.0	70.0	0.34411
200.0	30.0	70.0	0.38365
250.0	30.0	70.0	0.41878
300.0	30.0	70.0	0.44987
350.0	30.0	70.0	0.47728
400.0	30.0	70.0	0.50138
450.0	30.0	70.0	0.52252
500.0	30.0	70.0	0.54108
550.0	30.0	70.0	0.55741
600.0	30.0	70.0	0.57188
650.0	30.0	70.0	0.58485
700.0	30.0	70.0	0.59669
750.0	30.0	70.0	0.60777
800.0	30.0	70.0	0.61844
850.0	30.0	70.0	0.62906
900.0	30.0	70.0	0.64001
950.0	30.0	70.0	0.65165
1000.0	30.0	70.0	0.66433

Table 15.1 (*Continued*)

PRESS= 1.
TEMP= 120

V IN/MIN	P PSI	T DEG F	F
0.0	1.0	120.0	0.00100
50.0	1.0	120.0	0.02752
100.0	1.0	120.0	0.06838
150.0	1.0	120.0	0.10447
200.0	1.0	120.0	0.13615
250.0	1.0	120.0	0.16380
300.0	1.0	120.0	0.18776
350.0	1.0	120.0	0.20841
400.0	1.0	120.0	0.22612
450.0	1.0	120.0	0.24123
500.0	1.0	120.0	0.25413
550.0	1.0	120.0	0.26516
600.0	1.0	120.0	0.27470
650.0	1.0	120.0	0.28311
700.0	1.0	120.0	0.29075
750.0	1.0	120.0	0.29799
800.0	1.0	120.0	0.30519
850.0	1.0	120.0	0.31271
900.0	1.0	120.0	0.32092
950.0	1.0	120.0	0.33018
1000.0	1.0	120.0	0.34086

PRESS= 10
TEMP= 120

V IN/MIN	P PSI	T DEG F	F
0.0	10.0	120.0	0.00100
50.0	10.0	120.0	0.01436
100.0	10.0	120.0	0.05617
150.0	10.0	120.0	0.09306
200.0	10.0	120.0	0.12542
250.0	10.0	120.0	0.15359
300.0	10.0	120.0	0.17795
350.0	10.0	120.0	0.19886
400.0	10.0	120.0	0.21668
450.0	10.0	120.0	0.23178
500.0	10.0	120.0	0.24451
550.0	10.0	120.0	0.25525
600.0	10.0	120.0	0.26436
650.0	10.0	120.0	0.27220
700.0	10.0	120.0	0.27913
750.0	10.0	120.0	0.28553
800.0	10.0	120.0	0.29174
850.0	10.0	120.0	0.29814
900.0	10.0	120.0	0.30509
950.0	10.0	120.0	0.31296
1000.0	10.0	120.0	0.32210

PRESS= 30
TEMP= 120

V IN/MIN	P PSI	T DEG F	F
0.0	30.0	120.0	0.00100
50.0	30.0	120.0	0.03961
100.0	30.0	120.0	0.08351
150.0	30.0	120.0	0.12220
200.0	30.0	120.0	0.15604
250.0	30.0	120.0	0.18540
300.0	30.0	120.0	0.21063
350.0	30.0	120.0	0.23211
400.0	30.0	120.0	0.25019

Table 15.1 (*Continued*)

PRESS= 30			
TEMP= 120			
V	P	T	F
IN/MIN	PSI	DEG F	
450.0	30.0	120.0	0.26525
500.0	30.0	120.0	0.27763
550.0	30.0	120.0	0.28772
600.0	30.0	120.0	0.29587
650.0	30.0	120.0	0.30244
700.0	30.0	120.0	0.30780
750.0	30.0	120.0	0.31231
800.0	30.0	120.0	0.31634
850.0	30.0	120.0	0.32025
900.0	30.0	120.0	0.32441
950.0	30.0	120.0	0.32917
1000.0	30.0	120.0	0.33490
PRESS= 1.			
TEMP= 170.			
V	P	T	F
IN/MIN	PSI	DEG F	
0.0	1.0	170.0	0.04646
50.0	1.0	170.0	0.09791
100.0	1.0	170.0	0.14416
150.0	1.0	170.0	0.18556
200.0	1.0	170.0	0.22248
250.0	1.0	170.0	0.25528
300.0	1.0	170.0	0.28432
350.0	1.0	170.0	0.30997
400.0	1.0	170.0	0.33259
450.0	1.0	170.0	0.35255
500.0	1.0	170.0	0.37020
550.0	1.0	170.0	0.38592
600.0	1.0	170.0	0.40007
650.0	1.0	170.0	0.41301
700.0	1.0	170.0	0.42510
750.0	1.0	170.0	0.43671
800.0	1.0	170.0	0.44820
850.0	1.0	170.0	0.45994
900.0	1.0	170.0	0.47228
950.0	1.0	170.0	0.48560
1000.0	1.0	170.0	0.50026
PRESS= 10			
TEMP= 170			
V	P	T	F
IN/MIN	PSI	DEG F	
0.0	10.0	170.0	0.05562
50.0	10.0	170.0	0.10729
100.0	10.0	170.0	0.15360
150.0	10.0	170.0	0.19494
200.0	10.0	170.0	0.23165
250.0	10.0	170.0	0.26410
300.0	10.0	170.0	0.29266
350.0	10.0	170.0	0.31769
400.0	10.0	170.0	0.33956
450.0	10.0	170.0	0.35862
500.0	10.0	170.0	0.37524
550.0	10.0	170.0	0.38979
600.0	10.0	170.0	0.40263
650.0	10.0	170.0	0.41412
700.0	10.0	170.0	0.42463
750.0	10.0	170.0	0.43452
800.0	10.0	170.0	0.44415
850.0	10.0	170.0	0.45389
900.0	10.0	170.0	0.46410
950.0	10.0	170.0	0.47515
1000.0	10.0	170.0	0.48740

Table 15.1 (*Continued*)

PRESS= 30
TEMP= 170

V IN/MIN	P PSI	T DEG F	F
0.0	30.0	170.0	0.15204
50.0	30.0	170.0	0.20417
100.0	30.0	170.0	0.25064
150.0	30.0	170.0	0.29182
200.0	30.0	170.0	0.32807
250.0	30.0	170.0	0.35976
300.0	30.0	170.0	0.38725
350.0	30.0	170.0	0.41090
400.0	30.0	170.0	0.43108
450.0	30.0	170.0	0.44816
500.0	30.0	170.0	0.46249
550.0	30.0	170.0	0.47443
600.0	30.0	170.0	0.48437
650.0	30.0	170.0	0.49264
700.0	30.0	170.0	0.49963
750.0	30.0	170.0	0.50570
800.0	30.0	170.0	0.51120
850.0	30.0	170.0	0.51650
900.0	30.0	170.0	0.52197
950.0	30.0	170.0	0.52797
1000.0	30.0	170.0	0.53486

PRESS= 1.
TEMP= 220

V IN/MIN	P PSI	T DEG F	F
0.0	1.0	220.0	0.48258
50.0	1.0	220.0	0.54762
100.0	1.0	220.0	0.60736
150.0	1.0	220.0	0.66218
200.0	1.0	220.0	0.71244
250.0	1.0	220.0	0.75851
300.0	1.0	220.0	0.80073
350.0	1.0	220.0	0.83949
400.0	1.0	220.0	0.87514
450.0	1.0	220.0	0.90805
500.0	1.0	220.0	0.93858
550.0	1.0	220.0	0.96710
600.0	1.0	220.0	0.99396
650.0	1.0	220.0	1.01953
700.0	1.0	220.0	1.04418
750.0	1.0	220.0	1.06828
800.0	1.0	220.0	1.09217
850.0	1.0	220.0	1.11623
900.0	1.0	220.0	1.14083
950.0	1.0	220.0	1.16631
1000.0	1.0	220.0	1.19306

PRESS= 10
TEMP= 220

V IN/MIN	P PSI	T DEG F	F
0.0	10.0	220.0	0.58237
50.0	10.0	220.0	0.64674
100.0	10.0	220.0	0.70568
150.0	10.0	220.0	0.75956
200.0	10.0	220.0	0.80874
250.0	10.0	220.0	0.85358
300.0	10.0	220.0	0.89445
350.0	10.0	220.0	0.93172
400.0	10.0	220.0	0.96573
450.0	10.0	220.0	0.99687
500.0	10.0	220.0	1.02549
550.0	10.0	220.0	1.05196

Table 15.1 (*Continued*)

PRESS= 10			
TEMP= 220			
V	P	T	
IN/MIN	PSI	DEG F	
600.0	10.0	220.0	1.07664
650.0	10.0	220.0	1.09989
700.0	10.0	220.0	1.12209
750.0	10.0	220.0	1.14358
800.0	10.0	220.0	1.16474
850.0	10.0	220.0	1.18593
900.0	10.0	220.0	1.20751
950.0	10.0	220.0	1.22986
1000.0	10.0	220.0	1.25332
PRESS= 30			
TEMP= 220			
V	P	T	F
IN/MIN	PSI	DEG F	
0.0	30.0	220.0	0.90176
50.0	30.0	220.0	0.96464
100.0	30.0	220.0	1.02179
150.0	30.0	220.0	1.07357
200.0	30.0	220.0	1.12034
250.0	30.0	220.0	1.16248
300.0	30.0	220.0	1.20033
350.0	30.0	220.0	1.23427
400.0	30.0	220.0	1.26466
450.0	30.0	220.0	1.29187
500.0	30.0	220.0	1.31625
550.0	30.0	220.0	1.33817
600.0	30.0	220.0	1.35800
650.0	30.0	220.0	1.37609
700.0	30.0	220.0	1.39282
750.0	30.0	220.0	1.40854
800.0	30.0	220.0	1.42363
850.0	30.0	220.0	1.43843
900.0	30.0	220.0	1.45333
950.0	30.0	220.0	1.46868
1000.0	30.0	220.0	1.48484

One only has to type in the following information:

Ram diameter (in.).

Capillary diameter (in.).

Capillary length (in.).

The FLOCAP program reads in a file containing all measured data from which it generates a new file and table containing:

Apparent shear rate at the capillary wall (1/sec).

Temperature (° F).

Apparent viscosity (lb · sec/in^2).

as can be seen in Table 15.2.

The data file given in Table 15.2 is processed through the WORKUP program which converts the raw capillary viscometer data into apparent shear rate, temperature, and apparent viscosity data. The sample case is shown in Table 15.3.

Sometimes the flow properties of plastic melt are determined by using more than a single die. In such a case each point in the data file must contain in addition to the ram or crosshead velocity, force and temperature and also the die diameter (in.) and die length (in.) used. Such more complex data offers the advantage that the effect of pressure on melt viscosity can also be determined. In addition to finding the effect of pressure, the data collected on two dies also permit one to determine the so-called end correction, that is,

Table 15.2

```
OK, TYPE DATAFL
GO
        RAM VEL  TEMP    FORCE
     1   0.05    320.     872.
     2   0.10    320.    1168.
     3   0.20    320.    1536.
     4   0.50    320.    2361.
     5   1.00    320.    3257.
     6   2.00    320.    4473.
     7   5.00    320.    8627.
     8  10.00    320.   14644.
     9   0.05    390.     136.
    10   0.10    390.     224.
    11   0.20    390.     344.
    12   0.50    390.     552.
    13   1.00    390.     760.
    14   2.00    390.    1024.
    15   5.00    390.    1464.
    16  10.00    390.    1865.
    17  20.00    390.    2345.
    18   0.05    460.      40.
    19   0.10    460.      72.
    20   0.20    460.     120.
    21   0.50    460.     200.
    22   1.00    460.     304.
    23   2.00    460.     456.
    24   5.00    460.     720.
    25  10.00    460.     960.
    26  20.00    460.    1320.
```

the shear stress corresponding to a die of zero length. This value must be subtracted from every measured shear stress value. (See Table 15.4.)

Preparing a Data File. The data file containing the raw data can be entered by the use of the SPR Column Editor (COLED). Table 15.5 illustrates the use of this editor.

The underlined numbers are typed in while all the others are typed by the computer. One must remember only a few simple commands. The S command starts and ends a sequence of numbers in a column. So the repetitious sequence of shear rates is entered only once. The C command in the first column defines the length of the column. The I2 command means that the sequence is to be inserted two more times. As can be seen, corrections can also be made easily. The replace command replaces a single member of the array or a whole subarray defined by the diagonal. For example R17,2:19,3 will replace in lines 17, 18, and 19 columns 2 and 3.

Next the whole data file is printed out by using the P command. Two more data points are missing at 20 in./min ram velocity, which are inserted by the insert (I) command, as shown in Table 15.6. When the data file is completed it is processed by the FLOCAP program generating both an output file and a table containing shear rate (1/sec), temperature ($°$ F), viscosity (lb $\cdot$ sec/in.^{2}), and pressure (psi).

Conversion of Viscosity Data into Functional Form. Measured viscosity data can not be

Table 15.3

```
OK, WORKUP
GO
 WORKUP 21-SEP-1978   SPR - RELEASE: SEPTEMBER 14, 1978
WANT VISPAK? NO
EXTRUD FILE VERIFICATION? NO
FRICTION EVALUATION? NO
WANT TO CORRECT RPM IN FRICTION DATA FILE? NO
TO COMPUTE HEAT OF FUSION? NO
FLOCAP ? YES
INSTRON CAPILLARY RHEOMETER DATA WORKUP    SPR RELEASE:AUG. 18, 1978
21-SEP-1977
# OF COLUMNS= 4
FILENAME= DATAFL
SINGLE DIE USED? YES
COLUMN NO.FOR RAM VELOCITY= 2
RAM DIAMETER-IN. =.3744
CAPILLARY DIAMETER-IN.= .05013
COLUMN NO.FOR FORCE=   4
CAPILLARY LENGTH-IN.=2.0058
COLUMN NO.FOR TEMPERATURE= 3
TEMP IN DEGREES "F OR C" ?F
OUTPUT FILE= DAOUT
```

Table 15.3 (*Continued*)

#	SHR RATE 1/SEC	TEMP F	VISCOSITY LB.SEC/IN**2
1.	0.74180E 01	0.32000E 03	0.73448E 00
2.	0.14836E 02	0.32000E 03	0.49190E 00
3.	0.29672E 02	0.32000E 03	0.32344E 00
4.	0.74180E 02	0.32000E 03	0.19887E 00
5.	0.14836E 03	0.32000E 03	0.13717E 00
6.	0.29672E 03	0.32000E 03	0.94189E-01
7.	0.74180E 03	0.32000E 03	0.72665E-01
8.	0.14836E 04	0.32000E 03	0.61673E-01
9.	0.74180E 01	0.39000E 03	0.11455E 00
10.	0.14836E 02	0.39000E 03	0.94337E-01
11.	0.29672E 02	0.39000E 03	0.72437E-01
12.	0.74180E 02	0.39000E 03	0.46495E-01
13.	0.14836E 03	0.39000E 03	0.32007E-01
14.	0.29672E 03	0.39000E 03	0.21563E-01
15.	0.74180E 03	0.39000E 03	0.12331E-01
16.	0.14836E 04	0.39000E 03	0.78544E-02
17.	0.29672E 04	0.39000E 03	0.49379E-02
18.	0.74180E 01	0.46000E 03	0.33692E-01
19.	0.14836E 02	0.46000E 03	0.30323E-01
20.	0.29672E 02	0.46000E 03	0.25269E-01
21.	0.74180E 02	0.46000E 03	0.16846E-01
22.	0.14836E 03	0.46000E 03	0.12803E-01
23.	0.29672E 03	0.46000E 03	0.96021E-02
24.	0.74180E 03	0.46000E 03	0.60645E-02
25.	0.14836E 04	0.46000E 03	0.40430E-02
26.	0.29672E 04	0.46000E 03	0.27796E-02

Table 15.4

```
OK, TYPE DAFL1
GO
```

	Ram Vel	Temp	Force	Cap.Diam	Cap.Length
1	0.05	320.	872.	0.05013	2.0058
2	0.10	320.	1168.	0.05013	2.0058
3	0.20	320.	1536.	0.05013	2.0058
4	0.50	320.	2361.	0.05013	2.0058
5	1.00	320.	3257.	0.05013	2.0058
6	2.00	320.	4473.	0.05013	2.0058
7	5.00	320.	8627.	0.05013	2.0058
8	10.00	320.	14644.	0.05013	2.0058
9	0.05	390.	136.	0.05013	2.0058
10	0.10	390.	224.	0.05013	2.0058
11	0.20	390.	344.	0.05013	2.0058
12	0.50	390.	552.	0.05013	2.0058
13	1.00	390.	760.	0.05013	2.0058
14	2.00	390.	1024.	0.05013	2.0058
15	5.00	390.	1464.	0.05013	2.0058
16	10.00	390.	1865.	0.05013	2.0058
17	20.00	390.	2345.	0.05013	2.0058
18	0.05	460.	40.	0.05013	2.0058
19	0.10	460.	72.	0.05013	2.0058
20	0.20	460.	120.	0.05013	2.0058
21	0.50	460.	200.	0.05013	2.0058
22	1.00	460.	304.	0.05013	2.0058
23	2.00	460.	456.	0.05013	2.0058
24	5.00	460.	720.	0.05013	2.0058
25	10.00	460.	960.	0.05013	2.0058

Table 15.4   (*Continued*)

```
OK, WORKUP
GO
 WORKUP 22-SEP-1978   SPR - RELEASE: SEPTEMBER 14, 1978
FLOCAP ? YES
CAPILLARY RHEOMETER DATA WORKUP   SPR RELEASE:AUG. 18, 1978
22-SEP-1978
# OF COLUMNS= 6
FILENAME= DAFL1
SINGLE DIE USED? NO
COLUMN # FOR CAPILLARY DIAMETER(IN) = 5
COLUMN # FOR DIE LENGTH(IN) = 6
COLUMN NO.FOR RAM VELOCITY= 2
RAM DIAMETER-IN. =.3744
COLUMN NO.FOR FORCE=   4
COLUMN NO.FOR TEMPERATURE= 3
TEMP IN DEGREES "F OR C" ?F
OUTPUT FILE= FLOU1
```

#	SHR RATE 1/SEC	TEMP F	VISCOSITY LB.SEC/IN**2
1.	0.74180E 01	0.32000E 03	0.73448E 00
2.	0.14836E 02	0.32000E 03	0.49190E 00
3.	0.29672E 02	0.32000E 03	0.32344E 00
4.	0.74180E 02	0.32000E 03	0.19887E 00
5.	0.14836E 03	0.32000E 03	0.13717E 00
6.	0.29672E 03	0.32000E 03	0.94189E-01
7.	0.74180E 03	0.32000E 03	0.72665E-01
8.	0.14836E 04	0.32000E 03	).61673E-01
9.	0.74180E 01	0.39000E 03	().11455E 00
10.	0.14836E 02	0.39000E 03	(.94337E-01
11.	0.29672E 02	0.39000E 03	0.72437E-01
12.	0.74180E 02	0.39000E 03	0.46495E-01
13.	0.14836E 03	0.39000E 03	0.32007E-01
14.	0.29672E 03	0.39000E 03	0.21563E-01
15.	0.74180E 03	0.39000E 03	0.12331E-01
16.	0.14836E 04	0.39000E 03	0.78544E-02
17.	0.29672E 04	0.39000E 03	0.49379E-02
18.	0.74180E 01	0.46000E 03	0.33692E-01
19.	0.14836E 02	0.46000E 03	0.30323E-01
20.	0.29672E 02	0.46000E 03	0.25269E-01
21.	0.74180E 02	0.46000E 03	0.16846E-01
22.	0.14836E 03	0.46000E 03	0.12803E-01
23.	0.29672E 03	0.46000E 03	0.96021E-02
24.	0.74180E 03	0.46000E 03	0.60645E-02
25.	0.14836E 04	0.46000E 03	0.40430E-02

used conveniently by a computer in the form of a file. This is so because a file contains many data points and occupies a lot of computer memory. Furthermore, viscosity values at shear rates and temperatures not actually measured can not be easily computed from it as a data file does not lend itself to interpolation and extrapolation. The data collected in a file must therefore be summarized as a function from which a viscosity value at conditions desired can easily be computed by the computer. A viscosity function must be general enough to contain data from a very low to a very high shear rate and must be capable of describing the effect of temperature and, if possible, also of pressure. A most commonly used viscosity function has the following form:

$$\ln VI = A + A_1 \ln G + A_{11} (\ln G)^2 + A_2 T + A_{22} T^2 + A_{12} T \cdot \ln G + A_{14} \cdot \ln G \cdot \ln P$$

where: VI = viscosity (lb · sec/in.^{2})
  G = shear rate (1/sec)
  P = pressure (lb/in.^{2})
  T = temperature (° F)

The last term of this equation represents the pressure dependence and can be derived only

Table 15.5

OK, <u>COLED DATAFL</u>			
GO			
FILE DATAFL NOT FOUND			
NUMERICAL DATA MODE			
INPUT			
<u>S</u>	1	.05	
	2	.1	
	3	.2	
	4	.5	
	5	1	
	6	2	
	7	5	
	8	10	
<u>S</u>	9	12	
	9	0.05	
	10	0.10	
	11	0.20	
	12	0.50	
	13	1.00	
	14	2.00	
	15	5.00	
	16	10.00	
	17	0.05	
	18	0.10	
	19	0.20	
	20	0.50	
	21	1.00	
	22	2.00	
	23	5.00	
	24	10.00	
	25	C	
<u>S</u>	1	0.05	320
<u>S</u>	2	0.10	17
	2	0.10	320.
	3	0.20	320.
	4	0.50	320.
	5	1.00	320.
	6	2.00	320.
	7	5.00	320.
	8	10.00	320.
<u>S</u>	9	0.05	390
<u>S</u>	10	0.10	17
	10	0.10	390.
	11	0.20	390.
	12	0.50	390.
	13	1.00	390.
	14	2.00	390.
	15	5.00	390.
	16	10.00	390.
<u>S</u>	17	0.05	460
<u>S</u>	18	0.10	17
	18	0.10	460.
	19	0.20	460.
	20	0.50	460.
	21	1.00	460.
	22	2.00	460.
	23	5.00	460.
	24	10.00	460.
	1	0.05	320.	<u>872</u>
	2	0.10	320.	<u>1168</u>
	3	0.20	320.	<u>1536</u>
	4	0.50	320.	<u>2361</u>
	5	1.00	320.	<u>3257</u>

Table 15.5   *(Continued)*

6	2.00	320.	<u>4473</u>
7	5.00	320.	<u>8627</u>
8	10.00	320.	<u>14644</u>
9	0.05	390.	<u>136</u>
10	0.10	390.	<u>224</u>
11	0.20	390.	<u>344</u>
12	0.50	390.	<u>552</u>
13	1.00	390.	760
14	2.00	390.	<u>1024</u>
15	5.00	390.	<u>1464</u>
16	10.00	390.	<u>1865</u>
17	0.05	460.	<u>2345</u>
18	0.10	460.	<u>72</u>
19	0.20	460.	<u>120</u>
20	0.50	460.	<u>200</u>
21	1.00	460.	<u>304</u>
22	2.00	460.	<u>456</u>
23	5.00	460.	<u>720</u>
24	10.00	460.	<u>960</u>
1	0.05	320.	<u>872</u>. (ESC)

EDIT
*R17,3/40
*P.
   17   0.05 460.   40.
*

from rheometer data collected on more than a single die.

An example of the conversion of the measured viscosity data into functional form utilizing the VISC program is given in Table 15.7 But before the viscosity data can be adequately represented by a viscosity function, the apparent shear rate that was computed by assuming that the material is Newtonain, that is shear-rate independent, must be converted into true shear rate by applying the Rabinowitsch correction. In order to do this the flow index N for each point must be determined. N is related to the shape of the $\ln (VI)$ versus $\ln (G)$ curve and can therefore be computed from a function. The VISC program fits a function to the raw data. From this function the program computes the flow index values for each measured point, performs the Rabinowitsch correction on each, and then fits a new function to the Rabinowitsch corrected data. The sample case given shows how the whole procedure is performed by the VISC program automatically. The coefficients generated from the Rabinowitsch corrected data (that is the

Table 15.6

```
*P
   1    0.05 320.      872.
   2    0.10 320.     1168.
   3    0.20 320.     1536.
   4    0.50 320.     2361.
   5    1.00 320.     3257.
   6    2.00 320.     4473.
   7    5.00 320.     8627.
   8   10.00 320.    14644.
   9    0.05 390.      136.
  10    0.10 390.      224.
  11    0.20 390.      344.
  12    0.50 390.      552.
  13    1.00 390.      760.
  14    2.00 390.     1024.
  15    5.00 390.     1464.
  16   10.00 390.     1865.
  17    0.05 460.       40.
  18    0.10 460.       72.
  19    0.20 460.      120.
  20    0.50 460.      200.
  21    1.00 460.      304.
  22    2.00 460.      456.
  23    5.00 460.      720.
  24   10.00 460.      960.
*I17
   17    20
   17   20.00    390
   17   20.00    390.    2345
*A
INPUT
   26    20
   27    C
   26    20.00    460
   26    20.00  460.    1320
EDIT
*P
   1    0.05 320.      872.
   2    0.10 320.     1168.
   3    0.20 320.     1536.
   4    0.50 320.     2361.
   5    1.00 320.     3257.
   6    2.00 320.     4473.
   7    5.00 320.     8627.
   8   10.00 320.    14644.
   9    0.05 390.      136.
  10    0.10 390.      224.
  11    0.20 390.      344.
  12    0.50 390.      552.
  13    1.00 390.      760.
  14    2.00 390.     1024.
  15    5.00 390.     1464.
  16   10.00 390.     1865.
  17   20.00 390.     2345.
  18    0.05 460.       40.
  19    0.10 460.       72.
  20    0.20 460.      120.
  21    0.50 460.      200.
  22    1.00 460.      304.
  23    2.00 460.      456.
  24    5.00 460.      720.
  25   10.00 460.      960.
  26   20.00 460.     1320.
*E

OK,
```

second regression) are automatically converted by the VISPAK program into the form customarily used. As can be seen, two options are available, one for temperature values divided by 10 and the other without it. The division is sometimes beneficial when intermediate output is requested. Without the division some of the output particularly relating to the square of temperature may not fit into the space allowed. A representative session is given in Table 15.8.

Since a viscosity function does not in itself show the viscosity behavior, or does it serve as a tool for comparing the flow properties of various fluids, the VISPAK program can generate the information in tabular form which can be easily interpreted. This was also illustrated in the sample case. Graphical output of the tables is also available on a digital plotter.

Expansion of Viscosity Data. It is seldom practical to measure viscosities at temperatures close to the melting point of the fluid or at very high shear rates. This is so because at low temperatures the shear stress is very high and the strain gauge of the instrument can get damaged. At very high temperatures, on the other hand, the plastic may degrade, and the measured viscosity may no longer represent the undamaged sample. It is therefore important to extrapolate or expand the measured data to cover the whole range of temperature and shear rate of interest. This is done by the VISPAK program which reads the data file that was either typed in by the user or generated automatically by the FLOCAP program from the raw flow property data measured on a capillary rheometer. A data file to be fed to VISPAK must contain the following information:

Shear rate (1/sec)

Temperature ($°F$).

Viscosity (lb · sec/in.^{2} or kilopoise).

The only additional information required is the melting point of the fluid or the lowest temperature at which the regularity of the viscosity data is to be insured. The program reads in the measured viscosity data and internally derives a relationship between the

measured viscosity, shear rate, and temperature. This function is extrapolated, and a new
file is generated containing the original viscosity data and the computer generated data.

Data is extrapolated to the melting point
(three shear rates) and to 150° F above the
highest temperature of the measured data file.
Data points are also generated at 100,000

Table 15.7

```
OK, VISC
GO
V I S C --    22-SEP-1978      SPR-RELEASE: JANUARY 16, 1978

   INPUT FILE FOR VISC = DAOUT
   OPTIMIZATION, FRICTION OR VISCOSITY FUNCTION?VISCOSITY
   WANT PRESSURE DEPENDENCE OF VISCOSITY? NO
   DATA TO BE EXPANDED IN VISPAK? NO
   USING RABINOWITSCH CORRECTED VISCOSITIES? YES
   IS IT IN ENGLISH UNITS OR KILOPOISE? ENGLISH
   YOUR PROBLEM I.D.# 1
   NUMBER OF DATA POINTS=26
   WANT INTERMEDIATE STATISTICAL DATA? NO
   PREDICTED RESPONSES PRINTED? YES
   PRINT TRANSFORMED VARIABLES? NO

                      VARIABLE TRANSFORMATIONS
      LN( Z(  1) +     0.000000)  = W(  1)
    ( Z(  2) +    0.000000)  = W(  2)
    ( Z(  3) +    0.000000)  = W(  3)
    ( W(  1) +    0.000000)  = X(  1)
    ( W(  1) +    0.000000)  *  ( W(  1) +    0.000000)  = X(  2)
    ( W(  1) +    0.000000)  *  ( W(  2) +    0.000000)  = X(  3)
    ( W(  2) +    0.000000)  = X(  4)
    ( W(  2) +    0.000000)  *  ( W(  2) +    0.000000)  = X(  5)
      LN( W(  3) +    0.000000)  = Y(  1)


   STEPWISE   REGRESSION

   PROBLEM NO           1

   NO OF DATA =         26

   NO OF VARIABLES =           6

   WEIGHTED DEGREES OF FREEDOM =      26.0000

   F LEVEL TO ENTER VARIABLE =    0.00010

   F LEVEL TO REMOVE VARIABLE =    0.00001


   STANDARD ERROR OF Y =      1.492984
   COMPLETED    5 STEPS OF REGRESSION.
   STEP NO.     5
      VARIABLE ENTERING        2
   F LEVEL          0.00000
   STANDARD ERROR OF Y =          0.13964
   MULTIPLE CORRELATION COEFFICIENT        0.99649
    CONSTANT      0.1829589462E 02
```

Table 15.7 (*Continued*)

VARIABLE		COEFFICIENT	STD ERROR OF COEF
X-	1	-0.5142824650E 00	0.12602
X-	2	-0.1895896345E-01	0.00877
X-	3	0.5527560133E-03	0.00026
X-	4	-0.8053931594E-01	0.00917
X-	5	0.7625402941E-04	0.00001

A1=X1
A2=X4
A11=X2
A22=X5
A12=X3
A14=X6

FOR TEMP/100---DIVIDE A2 & A12 BY 100, A22 BY 100**2

PREDICTED VS ACTUAL RESULTS

RUN	ACTUAL	PREDICTED	DEVIATION	E**ACTUAL	E**PRED	DEVIATION
1.	-0.30859	-0.42053	0.11194	0.73448	0.65670	0.07778
2.	-0.70948	-0.71618	0.00670	0.49190	0.48862	0.00328
3.	-1.12874	-1.03004	-0.09870	0.32344	0.35699	-0.03355
4.	-1.61510	-1.47290	-0.14220	0.19887	0.22926	-0.03039
5.	-1.98653	-1.82906	-0.15747	0.13717	0.16056	-0.02339
6.	-2.36245	-2.20345	-0.15901	0.09419	0.11042	-0.01623
7.	-2.62190	-2.72631	0.10441	0.07266	0.06546	0.00720
8.	-2.78591	-3.14299	0.35708	0.06167	0.04315	0.01852
9.	-2.16674	-2.19092	0.02418	0.11455	0.11181	0.00274
10.	-2.36088	-2.45975	0.09887	0.09434	0.08546	0.00888
11.	-2.62504	-2.74679	0.12175	0.07244	0.06413	0.00830
12.	-3.06841	-3.15420	0.08579	0.04649	0.04267	0.00382
13.	-3.44180	-3.48354	0.04174	0.03201	0.03070	0.00131
14.	-3.83678	-3.83110	-0.00567	0.02156	0.02169	-0.00012
15.	-4.39564	-4.31851	-0.07712	0.01233	0.01332	-0.00099
16.	-4.84668	-4.70838	-0.13831	0.00785	0.00902	-0.00117
17.	-5.31082	-5.11645	-0.19436	0.00494	0.00600	-0.00106
18.	-3.39050	-3.21402	-0.17647	0.03369	0.04019	-0.00650
19.	-3.49585	-3.45603	-0.03982	0.03032	0.03155	-0.00123
20.	-3.67818	-3.71625	0.03807	0.02527	0.02432	0.00094
21.	-4.08364	-4.08821	0.00457	0.01685	0.01677	0.00008
22.	-4.35808	-4.39073	0.03266	0.01280	0.01239	0.00041
23.	-4.64577	-4.71147	0.06570	0.00960	0.00899	0.00061
24.	-5.10530	-5.16343	0.05812	0.00606	0.00572	0.00034
25.	-5.51077	-5.52647	0.01570	0.00404	0.00398	0.00006
26.	-5.88545	-5.90773	0.02228	0.00278	0.00272	0.00006

0						
3 19.	-3.49585	-3.45603	-0.03982	0.03032	0.03155	-0.0012
4 20.	-3.67818	-3.71625	0.03807	0.02527	0.02432	0.0009
8 21.	-4.08364	-4.08821	0.00457	0.01685	0.01677	0.0000
1 22.	-4.35808	-4.39073	0.03266	0.01280	0.01239	0.0004
1 23.	-4.64577	-4.71147	0.06570	0.00960	0.00899	0.0006
4 24.	-5.10530	-5.16343	0.05812	0.00606	0.00572	0.0003
6 25.	-5.51077	-5.52647	0.01570	0.00404	0.00398	0.0000
6 26.	-5.88545	-5.90773	0.02228	0.00278	0.00272	0.0000

Table 15.8

```
OK, VISC
GO
V I S C --    22-SEP-1978    SPR-RELEASE: JANUARY 16, 1978

  INPUT FILE FOR VISC = DAOUT
  OPTIMIZATION, FRICTION OR VISCOSITY FUNCTION?VISCOSITY
  WANT PRESSURE DEPENDENCE OF VISCOSITY? NO
  DATA TO BE EXPANDED IN VISPAK? NO
  USING RABINOWITSCH CORRECTED VISCOSITIES? NO
  IS IT IN ENGLISH UNITS OR KILOPOISE? ENGLISH
  YOUR PROBLEM I.D.# 1
  NUMBER OF DATA POINTS=26
  WANT INTERMEDIATE STATISTICAL DATA? NO

  PRINT TRANSFORMED VARIABLES? NO


       VARIABLE TRANSFORMATIONS
     LN( Z(  1) )     = W(  1)
     Z(  2)           = W(  2)
     Z(  3)           = W(  3)
     W(  1)           = X(  1)
     W(  1) * W(  1)  = X(  2)
     W(  1) * W(  2)  = X(  3)
     W(  2)           = X(  4)
     W(  2) * W(  2)  = X(  5)
     LN( W(  3) )     = Y(  1)



  STEPWISE  REGRESSION

  PROBLEM NO          1

  NO OF DATA =        26

  NO OF VARIABLES =        6

  WEIGHTED DEGREES OF FREEDOM =      26.0000

  F LEVEL TO ENTER VARIABLE =   0.00010

  F LEVEL TO REMOVE VARIABLE =  0.00001


  STANDARD ERROR OF Y =     1.492984
  COMPLETED   5 STEPS OF REGRESSION.
  STEP NO.    5
     VARIABLE ENTERING        2
  F LEVEL          0.00000
  STANDARD ERROR OF Y =       0.13964
  MULTIPLE CORRELATION COEFFICIENT        0.99649
CONSTANT     0.1829589462E 02
         VARIABLE         COEFFICIENT        STD ERROR OF COEF

             X- 1   -0.5142824650E 00        0.12602
             X- 2   -0.1895896345E-01        0.00877
             X- 3    0.5527560133E-03        0.00026
             X- 4   -0.8053931594E-01        0.00917
             X- 5    0.7625402941E-04        0.00001
```

720

Table 15.8 (*Continued*)

```
      VARIABLE TRANSFORMATIONS
   Z ( 1 )                = X ( 1 )
   Z ( 1 ) * Z ( 1 )      = X ( 2 )
   Z ( 1 ) * Z ( 2 )      = X ( 3 )
   Z ( 2 )                = X ( 4 )
   Z ( 2 ) * Z ( 2 )      = X ( 5 )
   Z ( 3 )                = Y ( 1 )
```

STEPWISE   REGRESSION

PROBLEM NO   1

NO OF DATA =   26

NO OF VARIABLES =   6

WEIGHTED DEGREES OF FREEDOM =   26.0000

F LEVEL TO ENTER VARIABLE =   0.00010

F LEVEL TO REMOVE VARIABLE =   0.00001

STANDARD ERROR OF Y =   1.515134
COMPLETED   5 STEPS OF REGRESSION.
STEP NO.   5
   VARIABLE ENTERING   2
F LEVEL   0.00000
STANDARD ERROR OF Y =   0.13944
MULTIPLE CORRELATION COEFFICIENT   0.99661
  CONSTANT   0.1818773270E 02

```
           VARIABLE         COEFFICIENT        STD ERROR OF COEF

              X- 1      -0.5199369192E 00        0.12447
              X- 2      -0.1860041544E-01        0.00827
              X- 3       0.5416485947E-03        0.00026
              X- 4      -0.8047527075E-01        0.00917
              X- 5       0.7636137889E-04        0.00001
```
RUN VISPAK? YES
VISPAK 22-SEP-1978 RELEASE: NOVEMBER 30, 1977
```
A=       0.18187733E 02
A1=     -0.51993692E 00
A2=     -0.80475271E-01
A11=    -0.18600415E-01
A22=     0.76361379E-04
A12=     0.54164859E-03
A14=     0.00000000E 00
```

WANT TO PLOT VISCOSITY FUNCTION? YES
TYPE-IN TEMPERATURE: 230

| | T=   230.000 DEG. F | |
SHEAR RATE	VISCOSITY	N
0.001	260.189392	0.861616
0.010	171.426178	0.775958
0.100	92.727066	0.690300
1.000	41.179192	0.604642
10.000	15.013824	0.518984
100.000	4.494143	0.433326
1000.000	1.104445	0.347668
9999.994	0.222835	0.262010
99999.938	0.036912	0.176352

Table 15.8 (*Continued*)

```
TYPE-IN TEMPERATURE: 300
                                  T=      300.000 DEG. F
         SHEAR RATE              VISCOSITY         N
              0.001              12.173243        0.899532
              0.010               8.752037        0.813874
              0.100               5.165994        0.728216
              1.000               2.503460        0.642558
             10.000               0.996023        0.556900
            100.000               0.325342        0.471242
           1000.000               0.087247        0.385583
           9999.994               0.019209        0.299925
          99999.938               0.003472        0.214267
TYPE-IN TEMPERATURE: 400
                                  T=      400.000 DEG. F
         SHEAR RATE              VISCOSITY         N
              0.001               0.561497        0.953697
              0.010               0.457315        0.868039
              0.100               0.305791        0.782381
              1.000               0.167871        0.696723
             10.000               0.075661        0.611064
            100.000               0.027997        0.525406
           1000.000               0.008505        0.439748
           9999.994               0.002121        0.354090
          99999.938               0.000434        0.268432
TYPE-IN TEMPERATURE: 500
                                  T=      500.000 DEG. F
         SHEAR RATE              VISCOSITY         N
              0.001               0.119276        1.007861
              0.010               0.110049        0.922203
              0.100               0.083361        0.836545
              1.000               0.051841        0.750887
             10.000               0.026469        0.665229
            100.000               0.011095        0.579571
           1000.000               0.003818        0.493913
           9999.994               0.001079        0.408255
          99999.938               0.000250        0.322597
TYPE-IN TEMPERATURE:
```

1/sec shear rate for each of the temperatures at which viscosity data were collected. The expanded viscosity is given in the same units as the measured viscosity data. Table 15.9 is a sample run illustrating the conversion of the viscosity data file into a function and expansion of the data file to include extrapolated points.

The expanded data file containing the 26 measured data points and the six extrapolated points is given in Table 15.10. Such data should always be inspected, as proposed by the computer, to insure that the extrapolated points satisfy all requirements known to exist from independent knowledge of physical relationships. For example at constant temperature, increasing shear rate can not cause viscosity to rise. Furthermore, at constant shear rate, increasing temperature must al-

ways cause viscosity to drop unless a chemical reaction takes place and the plastic crosslinks. The data file given in Table 15.10 contains both the original 26 data points and the six extrapolated or expanded, that is computed, ponits. Three of these are at 260° F and three at 660° F. The points are given in order of increasing shear rates for each of the two temperatures. Whenever the function is not well behaved and does not compute proper viscosities at very high shear rates, points are also automatically added for each temperature at very high shear rates.

Splitting of Viscosity Functions. Even the seven-term viscosity equation does not always fully describe the flow properties of a melt. This deficiency is often caused by the lack of measured data at low shear rates. However,

Table 15.9

```
OK, TYPE DAOUT
GO
      Shear Rate     Temperature    Viscosity
  1.  0.74180E 01   0.32000E 03   0.73448E 00
  2.  0.14836E 02   0.32000E 03   0.49190E 00
  3.  0.29672E 02   0.32000E 03   0.32344E 00
  4.  0.74180E 02   0.32000E 03   0.19887E 00
  5.  0.14836E 03   0.32000E 03   0.13717E 00
  6.  0.29672E 03   0.32000E 03   0.94189E-01
  7.  0.74180E 03   0.32000E 03   0.72665E-01
  8.  0.14836E 04   0.32000E 03   0.61673E-01
  9.  0.74180E 01   0.39000E 03   0.11455E 00
 10.  0.14836E 02   0.39000E 03   0.94337E-01
 11.  0.29672E 02   0.39000E 03   0.72437E-01
 12.  0.74180E 02   0.39000E 03   0.46495E-01
 13.  0.14836E 03   0.39000E 03   0.32007E-01
 14.  0.29672E 03   0.39000E 03   0.21563E-01
 15.  0.74180E 03   0.39000E 03   0.12331E-01
 16.  0.14836E 04   0.39000E 03   0.78544E-02
 17.  0.29672E 04   0.39000E 03   0.49379E-02
 18.  0.74180E 01   0.46000E 03   0.33692E-01
 19.  0.14836E 02   0.46000E 03   0.30323E-01
 20.  0.29672E 02   0.46000E 03   0.25269E-01
 21.  0.74180E 02   0.46000E 03   0.16846E-01
 22.  0.14836E 03   0.46000E 03   0.12803E-01
 23.  0.29672E 03   0.46000E 03   0.96021E-02
 24.  0.74180E 03   0.46000E 03   0.60645E-02
 25.  0.14836E 04   0.46000E 03   0.40430E-02
 26.  0.29672E 04   0.46000E 03   0.27796E-02

OK, VISC
GO
V I S C --    22-SEP-1978    SPR-RELEASE: JANUARY 16, 1978

 INPUT FILE FOR VISC = DAOUT
 OPTIMIZATION, FRICTION OR VISCOSITY FUNCTION?VISCOS
 WANT PRESSURE DEPENDENCE OF VISCOSITY? NO
 DATA TO BE EXPANDED IN VISPAK? YES
 YOUR PROBLEM I.D.# 1
 NUMBER OF DATA POINTS=26
 WANT INTERMEDIATE STATISTICAL DATA? NO
 PREDICTED RESPONSES PRINTED? NO
 PRINT TRANSFORMED VARIABLES? NO


      VARIABLE TRANSFORMATIONS
      LN( Z(   1) )        = W(   1)
      Z(   2)              = W(   2)
      Z(   3)              = W(   3)
      W(   1)              = X(   1)
      W(   1) * W(   1)    = X(   2)
      W(   1) * W(   2)    = X(   3)
      W(   2)              = X(   4)
      W(   2) * W(   2)    = X(   5)
      LN( W(   3) )        = Y(   1)


  STEPWISE   REGRESSION

  PROBLEM NO          1

  NO OF DATA =        26

  NO OF VARIABLES =            6
```

Table 15.9 (*Continued*)

```
WEIGHTED DEGREES OF FREEDOM =        26.0000

F LEVEL TO ENTER VARIABLE =    0.00010

F LEVEL TO REMOVE VARIABLE =   0.00001


STANDARD ERROR OF Y =      1.492984
COMPLETED    5 STEPS OF REGRESSION.
STEP NO.    5
   VARIABLE ENTERING        2
F LEVEL           0.00000
STANDARD ERROR OF Y =           0.13964
MULTIPLE CORRELATION COEFFICIENT        0.99649
            EQUATION DERIVED FOR DATA EXPANSION

   CONSTANT      0.1829589462E 02
              VARIABLE         COEFFICIENT        STD ERROR OF COEF

              X-  1    -0.5142824650E 00          0.12602
              X-  2    -0.1895896345E-01          0.00877
              X-  3     0.5527560133E-03          0.00026
              X-  4    -0.8053931594E-01          0.00917
              X-  5     0.7625402941E-04          0.00001
RUN VISPAK? YES
VISPAK 22-SEP-1978  RELEASE: NOVEMBER 30, 1977
WAS TEMP. DIVIDED BY 100? NO
A=      0.18295895E 02
A1=    -0.51428246E 00
A2=    -0.80539316E-01
A11=   -0.18958963E-01
A22=    0.76254029E-04
A12=    0.55275601E-03
A14=    0.00000000E 00
WANT TO PLOT VISCOSITY FUNCTION? NO
WANT TO BREAK VISCOSITY FUNCTION? NO
WANT TO EXPAND MEASURED TEMP.& SHEAR RATE RANGE? YES
TYPE-IN MELTING POINT, DEG.F= 260
INPUT FILE FOR EXPAND= DAOUT
OUTPUT FILE= DAOUT1
TOTAL OF   32 POINTS IN NEW FILE
GOOD PRACTICE: TYPE OUT OUTPUT FILE NAMED :DAOUT1 & CHECK
CORRECTNESS OF VISCOSITY EXTRAPOLATIONS!
INPUT FILE FOR VISC = DAOUT1
OPTIMIZATION, FRICTION OR VISCOSITY FUNCTION?VISCOSITY
WANT PRESSURE DEPENDENCE OF VISCOSITY? NO
DATA TO BE EXPANDED IN VISPAK? NO
USING RABINOWITSCH CORRECTED VISCOSITIES? NO
IS IT IN ENGLISH UNITS OR KILOPOISE? ENGLISH
YOUR PROBLEM I.D.# 1
NUMBER OF DATA POINTS=32
WANT INTERMEDIATE STATISTICAL DATA? NO
PREDICTED RESPONSES PRINTED? NO
PRINT TRANSFORMED VARIABLES? NO


            VARIABLE TRANSFORMATIONS
            LN( Z(  1) )          = W(  1)
            Z(  2)                = W(  2)
            Z(  3)                = W(  3)
            W(  1)                = X(  1)
            W(  1) * W(  1)       = X(  2)
            W(  1) * W(  2)       = X(  3)
            W(  2)                = X(  4)
            W(  2) * W(  2)       = X(  5)
            LN( W(  3) )          = Y(  1)
```

Table 15.9 (*Continued*)

```
STEPWISE  REGRESSION

PROBLEM NO          1

NO OF DATA =        32

NO OF VARIABLES =         6

WEIGHTED DEGREES OF FREEDOM =        32.0000

F LEVEL TO ENTER VARIABLE =   0.00010

F LEVEL TO REMOVE VARIABLE =  0.00001


STANDARD ERROR OF Y =     1.995103
COMPLETED    5 STEPS OF REGRESSION.
STEP NO.    5
   VARIABLE ENTERING        2
F LEVEL          0.00000
STANDARD ERROR OF Y =         0.18200
MULTIPLE CORRELATION COEFFICIENT        0.99650
       EQUATION DERIVED FOR EXPANDED BUT UNCORRECTED DATA

   CONSTANT      0.1196611023E 02
            VARIABLE        COEFFICIENT        STD ERROR OF COEF

            X-  1   -0.5152546167E 00          0.10784
            X-  2   -0.1766673103E-01          0.00962
            X-  3    0.5122565199E-03          0.00013
            X-  4   -0.4615058750E-01          0.00220
            X-  5    0.3102615301E-04          0.00000


      VARIABLE TRANSFORMATIONS
    Z(  1)              = X(   1)
    Z(  1)  *  Z(  1)   = X(   2)
    Z(  1)  *  Z(  2)   = X(   3)
    Z(  2)              = X(   4)
    Z(  2)  *  Z(  2)   = X(   5)
    Z(  3)              = Y(   1)


STEPWISE  REGRESSION

PROBLEM NO          1

NO OF DATA =        32

NO OF VARIABLES =         6

WEIGHTED DEGREES OF FREEDOM =        32.0000

STANDARD ERROR OF Y =     1.994031
   VARIABLE ENTERING        2
F LEVEL          0.00000
STANDARD ERROR OF Y =         0.18196
MULTIPLE CORRELATION COEFFICIENT        0.99650
       EQUATION FOR RABINOWITSCH CORRECTED DATA

   CONSTANT      0.1185587692E 02
            VARIABLE        COEFFICIENT        STD ERROR OF COEF

            X-  1   -0.5221389532E 00          0.10873
            X-  2   -0.1732295379E-01          0.00909
            X-  3    0.5059291143E-03          0.00013
            X-  4   -0.4606224597E-01          0.00221
            X-  5    0.3106181975E-04          0.00000
```

Table 15.9 (*Continued*)

```
RUN VISPAK? YES
VISPAK 22-SEP-1978   RELEASE: NOVEMBER 30, 1977
A=       0.11855877E 02
A1=     -0.52213895E 00
A2=     -0.46062246E-01
A11=    -0.17322954E-01
A22=     0.31061820E-04
A12=     0.50592911E-03
A14=     0.00000000E 00

WANT TO PLOT VISCOSITY FUNCTION? YES
TYPE-IN TEMPERATURE: 280
                              T=      280.000 DEG. F
SHEAR RATE                  VISCOSITY        N
       0.001                24.418621        0.858847
       0.010                16.094608        0.779071
       0.100                 8.828041        0.699296
       1.000                 4.029704        0.619521
      10.000                 1.530760        0.539746
     100.000                 0.483911        0.459971
    1000.000                 0.127306        0.380196
    9999.994                 0.027871        0.300421
   99999.938                 0.005078        0.220646
TYPE-IN TEMPERATURE: 300
                              T=      300.000 DEG. F
SHEAR RATE                  VISCOSITY        N
       0.001                12.994459        0.868965
       0.010                 8.766695        0.789190
       0.100                 4.921965        0.709415
       1.000                 2.299672        0.629640
      10.000                 0.894166        0.549865
     100.000                 0.289331        0.470090
    1000.000                 0.077911        0.390314
    9999.994                 0.017459        0.310539
   99999.938                 0.003256        0.230764
TYPE-IN TEMPERATURE: 500
                              T=      500.000 DEG. F
SHEAR RATE                  VISCOSITY        N
       0.001                 0.092828        0.970151
       0.010                 0.079058        0.890376
       0.100                 0.056031        0.810601
       1.000                 0.033048        0.730826
      10.000                 0.016221        0.651050
     100.000                 0.006626        0.571275
    1000.000                 0.002252        0.491500
    9999.994                 0.000637        0.411725
   99999.938                 0.000150        0.331950
TYPE-IN TEMPERATURE: 600
                              T=      600.000 DEG. F
SHEAR RATE                  VISCOSITY        N
       0.001                 0.019922        1.020744
       0.010                 0.019063        0.940969
       0.100                 0.015180        0.861194
       1.000                 0.010060        0.781418
      10.000                 0.005548        0.701643
     100.000                 0.002546        0.621868
    1000.000                 0.000972        0.542093
    9999.994                 0.000309        0.462318
   99999.938                 0.000082        0.382543
```

Table 15.9 (*Continued*)

TYPE-IN TEMPERATURE: 650		
	T= 650.000 DEG. F	
SHEAR RATE	VISCOSITY	N
0.001	0.011651	1.046040
0.010	0.011817	0.966265
0.100	0.009974	0.886490
1.000	0.007006	0.806715
10.000	0.004095	0.726940
100.000	0.001992	0.647165
1000.000	0.000807	0.567389
9999.994	0.000272	0.487614
99999.938	0.000076	0.407839
TYPE-IN TEMPERATURE:		
ANY MORE TEMPERATURES ?!NO		

most melts are Newtonian at low shear rates. The derived viscosity function can therefore be split or broken along a line below which the equation describes a Newtonian fluid. The VISPAK program performs this utility and offers the following options:

Splitting takes place at a constant shear rate.

Splitting takes place along a line defined by two points.

Splitting takes place at the viscosity maxima for each temperature.

The example that follows describes the splitting of the derived viscosity function along the maximum viscosity values for each temperature. The new equation describes the New-

Table 15.10

OK, TYPE DAOUT1			
GO			
	Shear Rate	Temperature	Viscosity
1	0.741800E 01	320.000	0.734480E 00
2	0.148360E 02	320.000	0.491900E 00
3	0.296720E 02	320.000	0.323440E 00
4	0.741800E 02	320.000	0.198870E 00
5	0.148360E 03	320.000	0.137170E 00
6	0.296720E 03	320.000	0.941890E-01
7	0.741800E 03	320.000	0.726650E-01
8	0.148360E 04	320.000	0.616730E-01
9	0.741800E 01	390.000	0.114550E 00
10	0.148360E 02	390.000	0.943370E-01
11	0.296720E 02	390.000	0.724370E-01
12	0.741800E 02	390.000	0.464950E-01
13	0.148360E 03	390.000	0.320070E-01
14	0.296720E 03	390.000	0.215630E-01
15	0.741800E 03	390.000	0.123310E-01
16	0.148360E 04	390.000	0.785440E-02
17	0.296720E 04	390.000	0.493790E-02
18	0.741800E 01	460.000	0.336920E-01
19	0.148360E 02	460.000	0.303230E-01
20	0.296720E 02	460.000	0.252690E-01
21	0.741800E 02	460.000	0.168460E-01
22	0.148360E 03	460.000	0.128030E-01
23	0.296720E 03	460.000	0.960210E-02
24	0.741800E 03	460.000	0.606450E-02
25	0.148360E 04	460.000	0.404300E-02
26	0.296720E 04	460.000	0.277960E-02
27	0.741799E 01	260.000	0.393904E 01
28	0.545400E 02	260.000	0.148091E 01
29	0.296719E 04	260.000	0.132522E 00
30	0.741799E 01	660.000	0.437221E-02
31	0.545400E 02	660.000	0.245191E-02
32	0.296719E 04	660.000	0.488824E-03

Table 15.11

```
WANT TO BREAK VISCOSITY FUNCTION? YES
WANT CUTOFF AT CONSTANT SHEAR RATE? NO
WANT TO TYPE-IN NEW VISCOSITY FUNCTION? NO
WANT CUTOFF AT VISCOSITY MAXIMA FOR EACH TEMP.?YES
FUNCTIONS HAVE BEEN BROKEN ALONG THE LINE:
LN GAMMA= -15.070726 +    0.014603 T

A=   0.15790382E 02
A1= 0.0
A2= -0.53686969E-01
A11= 0.0
A22=  0.34755823E-04
A12= 0.0
THE ABOVE ARE THE COEFFICIENTS OF THE NEW VISCOSITY EQUATION.
```

tonian range at the lower shear rates, while the original equation directly obtained from the VISC program represents the viscosity data in the non-Newtonian higher-shear-rate range. An equation representing the location of the switch-over from one equation to the other is also given (See Table 15.11).

15.2.1.3 Thermal Properties

Chemical compounds normally exhibit a constant heat capacity below their melting points and another constant heat capacity above their melting points. At the melting point itself the heat capacity rises to infinity. The reason for this phenomenon is the mode by which heat capacity is determined. Heat capacity is defined as the amount of heat necessary to raise the temperature of one gram or one pound of the sample by one degree. For pure compounds precisely at the melting point the addition of heat does not result in any increase in temperature. The temperature difference is therefore zero. Thus an infinite amount of heat is needed to raise temperature by one degree. The amount of heat absorbed by a unit mass of the sample is its heat of fusion. Polymers do not exhibit such a simple relationship.

The heat capacity of a plastic is not a constant, it varies with temperature. A maximum value is exhibited at its melting point, if such can be accurately defined. The heat capacity values increase gradually below the melting point to that maximum value and similarly decrease gradually from this value above the melting point. The reason for this difference is that a polymer is really not a single compound. It represents a mixture of many different molecules, each with a different chain length. We call this a molecular weight distribution. The melting point of each kind of molecule is different from the other. A polymer behaves therefore in a manner similar to that of a mixture of various compounds. A differential scanning calorimeter (DSC) is normally used to determine the whole spectrum of heat capacity versus temperature values. Most computer simulations of plastics processing operations do not use point values for heat capacity, neither do they express heat capacity as a function of temperature. Instead they break up the heat capacity function into three terms as can be seen in Figure 15.2. The three terms represent one constant heat capacity for the solids and one for the melt while lumping the area difference between the curve and the three straight lines (two constant heat capacities and the vertical at the melting point) into a single term called heat of fusion.

This computation is normally performed graphically, which is not a very accurate procedure and is also very tedious. The WORKUP program performs this task much more accurately. It also enables the user to specify a melting point for the plastic and obtain a series of values as a function of melting point selected. The lowest temperature at which heat capacity is given represents the solids feed temperature and can also be altered. Heat of fusion values as a function of solids feed temperature can therefore be derived. This is an important consideration when the effect of storing raw material outdoors on the performance of extruders and

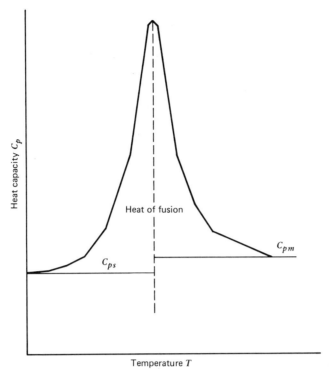

Figure 15.2   Heat capacity as a function of temperature. The diagram also shows the area representing the heat of friction when constant heat capacity values are used for both the solid C_{ps} and the melt C_{pm}.

molding machines is to be studied. A sample case for the workup of heat capacity data is given in Table 15.12.

15.2.1.4   Material Density

Both melt density and the bulk density of the solid are usually expressed as functions, the former as a function of temperature and pressure and the latter as a function of pressure alone. A general purpose regression analysis program is needed to derive the coefficients of these two equations.

Melt Density.   This material property is generally expressed in the following form:

$$D = R_0 + R_1 P + R_2 T + R_{12} PT$$

where: D = density (lb/ft^3)
  P = pressure (lb/in.^{2})
  T = temperature (°F)

The LABMAN system can be used to derive the four coefficients of the equation. The data are placed in a file with three columns where the first column ($Z1$) represents pressure (lb/in.^{2}), the second column ($Z2$) temperature (°F), and the third column ($Z3$) represents the measured density. The three terms of the equation, often referred to as the regression independent variables, are defined by transformation statements as follows:

$$Z1 = X1 \quad \text{defines the term } P$$
$$Z2 = X2 \quad \text{defines the term } T$$
$$Z1*Z2 = X3 \quad \text{defines the term } P \cdot T$$
$$Z3 = Y1 \quad \text{defines the dependent}$$
variable, that is, density

The data file is typed into the computer utilizing the column editor COLED. The transformation cards can also be included in the data by COLED. Table 15.13 illustrates the use of LABMAN in deriving an equation for melt density as a function of temperature and pressure. If the pressure was measured in atmospheres the program can be used to convert $Z1$ into pounds per square inch by multiplying it by 14.7. If temperature is expressed in degrees Celsius it can be converted

Table 15.12

```
OK, WORKUP
GO
 WORKUP 22-SEP-1978   SPR - RELEASE: SEPTEMBER 14, 1978
WANT VISPAK?
EXTRUD FILE VERIFICATION?
FRICTION EVALUATION?
WANT TO CORRECT RPM IN FRICTION DATA FILE?
TO COMPUTE HEAT OF FUSION? YES
HEAT OF FUSION COMPUTATION-22-SEP-1978   SPR  RELEASE:
                                      NOVEMBER 29, 1977
TEMP.IN DEG."F,C OR K"? C
WANT TO SET MELTING POINT?NO
TEMPERATURE & HEAT CAPACITY PT.BY PT
20    .45
30    .47
40    .49
50    .52
60    .55
70    .58
80    .62
90    .67
100   .77
110   .93
120  1.15
130  8.7
133   .59
140   .6
175   .62
210   .63

HEAT CAPACITY OF SOLID=     0.4522 BTU/LB.F
HEAT CAPACITY OF MELT=      0.6100 BTU/LB.F
MELTING POINT=        266.0
HEAT OF FUSION=    136.8547 BTU/LB
WANT TO CHANGE MELTING POINT?YES
MELTING POINT=125
HEAT CAPACITY OF SOLID=     0.4522 BTU/LB.F
HEAT CAPACITY OF MELT=      0.6100 BTU/LB.F
MELTING POINT=        257.0
HEAT OF FUSION=    135.4347 BTU/LB
WANT TO CHANGE MELTING POINT?YES
MELTING POINT=132
HEAT CAPACITY OF SOLID=     0.4522 BTU/LB.F
HEAT CAPACITY OF MELT=      0.6100 BTU/LB.F
MELTING POINT=        269.6
HEAT OF FUSION=    137.4227 BTU/LB
WANT TO CHANGE MELTING POINT?NO
```

into degrees Farenheit by multiplying $Z2$ by 1.8 and adding 32 to it. These conversions can be handled by the use of intermediate variables $W1$ and $W2$:

$$Z1*14.7 = W1 \quad \text{Intermediate variables}$$
$$Z2*1.8 = W2$$
$$W2 + 32 = W3$$

$$W1 = X1 \quad \text{Final transformations}$$
$$W3 = X2$$
$$W1*W3 = X3$$

$$Z3 = Y1 \quad \text{Response or Dependent Variable}$$

If density is to be converted from grams per cubic centimeter into pounds per cubic foot the last transformation statement can simply be altered to

$$Z3*62.43 = Y1$$

Bulk Density of Solids. This variable is mostly used in simulating the performance of

Table 15.13

```
OK, TYPE MLTDEN
GO
Z1=X1
Z2=X2
Z1*Z2=X3
Z3=Y1
END
    #   Z1    Z2    Z3
    1 1811. 250. 0.788
    2 4528. 250. 1.576
    3 9058. 250. 2.758
    4 1811. 300. 0.702
    5 4528. 300. 1.486
    6 9058. 300. 2.650

OK, MODEL
GO
MODEL     28-SEP-1978     SPR-RELEASE: APRIL 22, 1977

 INPUT FILE = MLTDEN
 TOLERANCE FOR MODEL=
 F LEVEL TO ENTER TERM=
 F LEVEL TO REMOVE TERM=
 YOUR PROBLEM I.D.# 1
 NUMBER OF OBSERVATIONS=
 EVERY POINT HAS THE SAME WEIGHT? YES
 EACH MODEL BUILDING STEP TO PRINT OUT? NO
 WANT INTERMEDIATE STATISTICAL DATA? NO
 PREDICTED RESPONSES PRINTED? YES
 INCLUDE A CONSTANT TERM? YES
 PRINT TRANSFORMATIONS? YES
 PRINT TRANSFORMED VARIABLES? NO


    VARIABLE TRANSFORMATIONS
    Z(  1)            = X(  1)
    Z(  2)            = X(  2)
    Z(  1) * Z(  2) = X(  3)
    Z(  3)            = Y(  1)


STEPWISE  REGRESSION

PROBLEM NO          1

NO OF DATA =        6

NO OF VARIABLES =        4

WEIGHTED DEGREES OF FREEDOM =        6.0000

F LEVEL TO ENTER VARIABLE =   0.00010

F LEVEL TO REMOVE VARIABLE =   0.00001


STANDARD ERROR OF Y =      0.883310
COMPLETED    3 STEPS OF REGRESSION
STEP NO.    3
    VARIABLE ENTERING        3
F LEVEL        0.00000
STANDARD ERROR OF Y =        0.04171
MULTIPLE CORRELATION COEFFICIENT        0.99955
 CONSTANT      0.7110115290E 00
```

Table 15.13 (*Continued*)

VARIABLE		COEFFICIENT	STD ERROR OF COEF
X-	1	0.2863566042E-03	0.00006
X-	2	-0.1572414767E-02	0.00135
X-	3	-0.6254519747E-07	0.00000

	PREDICTED VS ACTUAL MELT DENSITY		
RUN NO.	ACTUAL	PREDICTED	DEVIATION
1.	0.78800	0.80818	-0.0202
2.	1.57600	1.54373	0.0323
3.	2.75800	2.77009	-0.0121
4.	0.70200	0.72390	-0.0219
5.	1.48600	1.45095	0.0351
6.	2.65000	2.66314	-0.0131

WANT A PLOT? NO

extruders and molding machines. It is often expressed as depending exponentially on pressure as follows:

$$D = D_s - (D_s - D_{b0}) \cdot e^{-R_{b1} \cdot P}$$

where D = bulk density (lb/ft^3)

D_s = solid density (lb/ft^3)

D_{b0} = free flowing bulk density (lb/ft^3)

P = pressure (lb/in^2)

R_{b1} = Coefficient to be determined (in^2/lb)

The equation can be rearranged as:

$$\frac{(D_s - D)}{(D_s - D_{b0})} = e^{R_{b1} \cdot P}$$

or

$$\ln \frac{(D_s - D)}{(D_s - D_{b0})} = - R_{b1} \cdot P$$

Using the MODEL program of the LABMAN system the bulk density data can be typed into a data file containing pressure as the first variable and density as the second variable. Assuming that D_s = 64.5, D_{b0} = 30, and $D_s - D_{b0}$ = 34.5, the transformation statements would simply be:

$$Z2 - 64.5 = W1$$

$$\ln (W1/-34.5) = Y1$$

$$Z1 = X1$$

15.2.2 Physical Models of Plastics Processing Operations

15.2.2.1 Melt extrusion

Many large-diameter single-screw extruders are at present being used to convey plastic melt from a reactor usually to a pelletizing die. The function of the screw extruder is to deliver plastic melt to the die and to generate adequate pressure for the melt to flow through the die at the desired production rate. Simplified equations have been derived for melt extrusion, but they are most unreliable owing to the many simplifying assumptions used in deriving them. These simplifying assumptions are discussed as follows:

1. Extrusion is an isothermal process. This is never the case because heat is generated in the fluid processed, which must flow radially towards the barrel. Furthermore, not all heat generated is removed at the particualr axial location. As a result melt temperature usually rises, not only in the radial but also in the axial direction.

2. Plastic melts are Newtonian fluids. This is never the case, and the viscosity of plastics depends not only on temperature but also even more on shear rate and to certain extent even on pressure.

The EXTRUD® * computer program eliminates these simplifying assumptions and per-

*Registered Trademark of SPR, Inc.

forms the computations rigorously. It simulates on the computer the temperature rise that takes place along the extruder channel, the pressure rise, power consumption, heat transfer to and from the barrel, the effect of barrel and screw temperature profile, degree of mixing, and may other factors. One can therefore simulate and study the effect of screw speed and screw geometry and find the best operating conditions and screw geometry for the application. The simulation will also detect if the extruder is receiving adequate supply from the reactor or is being starve fed, a condition that usually causes undesirable side effects such as inadequate feeding, dead spots with associated long residence time, material degradation, and discoloration and even burning.

Input data to the EXTRUD® program contain the screw and barrel geometry, operating conditions, and the physical properties of the melt described in Section 15.2.1.

A sample computer simulation of the melt-fed extruder utilizing the EXTRUD® program is given in Table 15.14. As can be seen, the data required for the simulation are supplied by the user in response to questions by the EXTRUD® program. The physical properties of the polymer being extruded are specified by its I.D. number in the data file (-236). This is a

Table 15.14

```
OK, EXTRUD*
GO
EXTRUD*    29-SEP-1978           SPR RELEASE:   SEPTEMBER 26, 1978
 INPUT FROM DSK? NO
BATCH OR CONV?CONV
OUTDEV INDEV POISE IPLUG IPRES XTEST SCARD MIXREL IFRECS MLTCHL
1      1
USING NAMELIST?YES
 WANT A PLOT? YES
 WHEN EXTRUD IS DONE TYPE:GRAPH
PLT ID #FLT LEAD DBAR FLWD FLCL JJ DLY INJ COOL OPT MLFR MLTX SLF RNG
-5   1   1    6    6   .6   .006 3                   -236 1
 MELT EXTRUSION SPECIFIED
    L    H    L    H    L    H     L    H     L    H
18   .6   24  .2    24   .2
THCOND RHOS HETCAM HETCAS HETFUS TSOLID GAMAX
                             350
R0    TM0   R1    R2    R12    TM1   TM2

A          A1          A2          A11        A22       A12       A14

C0   C1  C2   P0  IPOE  ITOL  ISOL  NFRIC
350
F0   F1   F2   PARHOL HLAREA BREAK HEADPR ACDISV STOP
-1000
RPM OUTPUT RIGID? SOAK AXDTM RECLEN PRESIN
30  800
DESCRIPTION OF CASE:
  LDPE 12 M.I. PAPER COATING RESIN
DHOP  ROB  FRIB0  FRIS0  STFRIC  WBAR  TBW  DPEL
36                            1
TCONB TCONS AGREP HSH HOPT  DHOPX WBAR1 HOPST
25                   24  6         6
RB1   FRIB1  FRIB11  FRIB12  FRIB13  FRIB2 FRIB2

  FILE NAME=MLTEXT
```

For Glossary of Parameter Names used above see end of this section.

---

*Registered Trademark of Scientific Process & Research, Inc.

convenience feature giving the user access to a large data file of physical property data on several hundred plastic materials. Since all physical properties on material number 236 are automatically extracted from the central file by the program, the physical property data relating to the plastic, such as flow properties, melt densities, and heat capacities can be left blank in the input data. If one wants to perform a computer simulation on a resin grade that is not in the file, the data must be measured independently and entered in the spaces left blank in Table 15.14.

The actual computer simulation utilizing this data is given in Table 15.15.

Glossary of Parameter Names for EXTRUD®

A, A1, A2, A11, A22, A12, A14   Coefficients of viscosity equation

ACDISV   Axial distance to vent

AGREP   Angle of repose

AXDTM   Length of cooled barrel around hopper

BREAK   Location of solid bed breakup

C0, C1, C2   Barrel temperature versus axial location

COOL   Indicator for inefficient barrel cooling, adiabatic operation

DBAR   Barrel diameter

DHOP   Hopper diameter at top

DHOPX   Hopper diameter at bottom

Table 15.15

N= 30.0 RPM   OUTPUT= 800.00 LB/HR   SOAK TIME= 0.00 SECONDS

SCREW GEOMETRY

SECTION	LENGTH (IN)	CH.DEPTH (IN)	BAR.DIAM (IN)	NO OF HOLES	HOLE AREA (SQ. IN.)
1	18.0000	0.6000	6.0000		
2	24.0000	0.2000	6.0000		
3	24.0000	0.2000	6.0000		

L IN	TOUT DEG F	X/W INJ	X/W	GRATIO	PRESSURE PSI REAL	PRESSURE FULL	POWER HP	HOUT IN	TBAR DEG F	TVISC LB.SEC /SQ.IN	D.HEAT BTU/IN .MIN
1.00	350.3	.000	.000	.000	12.	12.	.135	0.600	350.0	0.105	3.4
6.00	351.6	.000	.000	.000	70.	70.	.805	0.600	350.0	0.103	3.6
11.00	352.8	.000	.000	.000	129.	129.	1.5	0.600	350.0	0.102	3.8
16.00	353.9	.000	.000	.000	186.	186.	2.1	0.600	350.0	0.100	4.0
18.00	354.3	.000	.000	.000	210.	210.	2.4	0.600	350.0	0.100	4.1
23.00	355.2	.000	.000	.000	268.	268.	3.1	0.517	350.0	0.097	4.3
28.00	356.1	.000	.000	.000	326.	326.	3.8	0.433	350.0	0.095	4.4
33.00	357.0	.000	.000	.000	370.	370.	4.5	0.350	350.0	0.093	4.4
38.00	358.2	.000	.000	.000	341.	341.	5.2	0.267	350.0	0.093	3.6
42.00	360.5	.000	.000	.000	141.	141.	5.8	0.200	350.0	0.090	1.0
47.00	364.7	.000	.000	.000	0.	0.	7.1	0.200	350.0	0.085	5.5
52.00	368.1	.000	.000	.000	0.	0.	8.5	0.200	350.0	0.081	6.1
57.00	371.0	.000	.000	.000	0.	0.	9.8	0.200	350.0	0.079	6.6
62.00	373.3	.000	.000	.000	0.	0.	11.0	0.200	350.0	0.076	7.0
66.00	375.0	.000	.000	.000	0.	0.	12.0	0.200	350.0	0.075	7.3

POWER BREAK-DOWN:
SOLIDS CONV. =   0.000 HP
MELTING   =   0.000 HP
MIXING   =   7.415 HP
FLIGHT CLEAR =   4.539 HP
COMPRESSION =   0.051 HP
MIXING DEVICE=   0.000 HP
PRESSURE FLUCTUATION=   163.63 PSI/CYCLE
TOTAL CONDUCTED HEAT THROUGH BARREL=   306.9 BTU/MIN

POWER DISSIPATED IN FLIGHT CLEARANCE MUST BE
REMOVED THROUGH BARREL.
OTHERWISE MELT & BARREL TEMPERATURE WILL BE HIGHER THAN SIMULATED.

Table 15.15 (*Continued*)

```
TYPE-IN INPUT
$INPUT HSEC=.6, 2*.35 $
LIKE A PLOT?YES
WHEN EXTRUD IS DONE TYPE: GRAPH

N= 30.0 RPM        OUTPUT=  800.00 LB/HR      SOAK TIME=  0.00 SECONDS


                              SCREW GEOMETRY
SECTION LENGTH  CH.DEPTH  BAR.DIAM NO OF HOLES  HOLE AREA
        (IN)    (IN)      (IN)                  (SQ. IN.)
   1    18.0000  0.6000    6.0000
   2    24.0000  0.3500    6.0000
   3    24.0000  0.3500    6.0000
```

L IN	TOUT DEG F	X/W INJ	X/W	GRATIO	PRESSURE PSI REAL	FULL	POWER HP	HOUT IN	TBAR DEG F	TVISC LB.SEC /SQ.IN	D.HEAT BTU/IN .MIN
1.00	350.3	.000	.000	.000	12.	12.	.135	0.600	350.0	0.105	3.4
6.00	351.6	.000	.000	.000	70.	70.	.805	0.600	350.0	0.103	3.6
11.00	352.8	.000	.000	.000	129.	129.	1.5	0.600	350.0	0.102	3.8
16.00	353.9	.000	.000	.000	186.	186.	2.1	0.600	350.0	0.100	4.0
18.00	354.3	.000	.000	.000	210.	210.	2.4	0.600	350.0	0.100	4.1
23.00	355.2	.000	.000	.000	268.	268.	3.1	0.548	350.0	0.098	4.3
28.00	356.0	.000	.000	.000	327.	327.	3.7	0.496	350.0	0.096	4.4
33.00	356.8	.000	.000	.000	385.	385.	4.4	0.444	350.0	0.094	4.6
38.00	357.6	.000	.000	.000	435.	435.	5.1	0.392	350.0	0.093	4.6
42.00	358.2	.000	.000	.000	465.	465.	5.7	0.350	350.0	0.092	4.6
47.00	359.0	.000	.000	.000	492.	492.	6.4	0.350	350.0	0.091	4.7
52.00	359.7	.000	.000	.000	519.	519.	7.1	0.350	350.0	0.090	4.8
57.00	360.3	.000	.000	.000	546.	546.	7.8	0.350	350.0	0.089	4.9
62.00	360.8	.000	.000	.000	574.	574.	8.4	0.350	350.0	0.089	5.0
66.00	361.1	.000	.000	.000	596.	596.	9.0	0.350	350.0	0.088	5.1

```
POWER BREAK-DOWN:
SOLIDS CONV. =    0.000 HP
MELTING      =    0.000 HP
MIXING       =    3.879 HP
FLIGHT CLEAR =    4.473 HP
COMPRESSION  =    0.645 HP
MIXING DEVICE=    0.000 HP
PRESSURE FLUCTUATION=       62.91 PSI/CYCLE
TOTAL CONDUCTED HEAT THROUGH BARREL=    291.2 BTU/MIN


POWER DISSIPATED IN FLIGHT CLEARANCE MUST BE
REMOVED THROUGH BARREL.
OTHERWISE MELT & BARREL TEMPERATURE WILL BE HIGHER THAN SIMULATED.


TYPE-IN INPUT
 $INPUT BEND=-1 $

OK, GRAPH
GO
 WHERE WOULD YOU LIKE THE GRAPH PRINTED?:
 (TTY,LPT,DSK)
TTY
```

Table 15.15 (*Continued*)

SCIENTIFIC PROCESS AND RESEARCH INC.EXPERIMENT NUMBER   1.00
OUTPUT= 800.00 LB/HR   SCREW SPEED= 30.00RPM

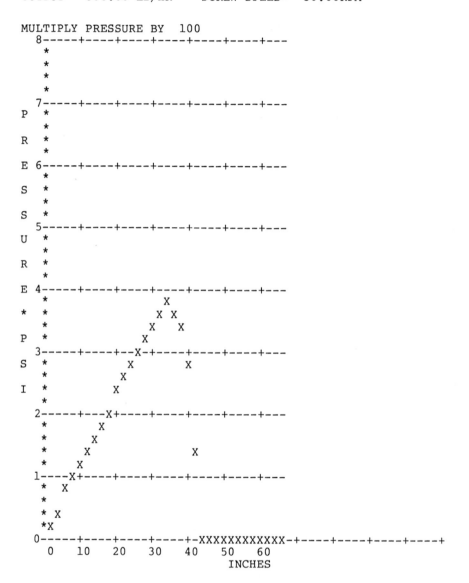

```
MULTIPLY PRESSURE BY  100
     8-----+----+----+----+----+----+---
       *
       *
       *
       *
     7-----+----+----+----+----+----+---
  P    *
       *
  R    *
       *
  E  6-----+----+----+----+----+----+---
       *
  S    *
       *
  S    *
     5-----+----+----+----+----+----+---
  U    *
       *
  R    *
       *
  E  4-----+----+----+----+----+----+---
       *                 X
  *    *               X X
       *               X   X
  P    *               X
     3-----+----+--X-+----+----+----+---
  S    *             X         X
       *             X
  I    *           X
       *
     2-----+---X+----+----+----+----+---
       *       X
       *      X
       *    X              X
       *   X
     1----X+----+----+----+----+----+---
       *  X
       *
       * X
       *X
     0-----+----+----+----+-XXXXXXXXXXXX-+----+----+----+----+
       0   10   20   30   40   50   60
                                INCHES
```

DLY   Indicator for delay in melting start computation

DPEL   Pellet diameter

#FLT   Number of parallel flights

F0, F1, F2   Screw temperature function

FLCL   Radial flight clearance

FLWD   Flight width

FRIB0   Coefficient of friction, barrel, constant term

FRIB1, FRIB11, ETC.   Coefficient of friction, barrel

FRIS0   Coefficient of friction, constant term, screw

GAMAX   Start of upper Newtonian viscosity range

H   Channel depth at end of each geometric section

HEADPR   Head pressure

HETCAM   Heat capacity melt

HETCAS   Heat capacity of solid

HETFUS   Heat of fusion

HLAREA   Hole area

HOPST   Height of untapered portion of hopper

Table 15.15 (*Continued*)

SCIENTIFIC PROCESS AND RESEARCH INC. EXPERIMENT NUMBER   1.01
OUTPUT=   800.00 LB/HR   SCREW SPEED=   30.00RPM

MULTIPLY PRESSURE BY   100

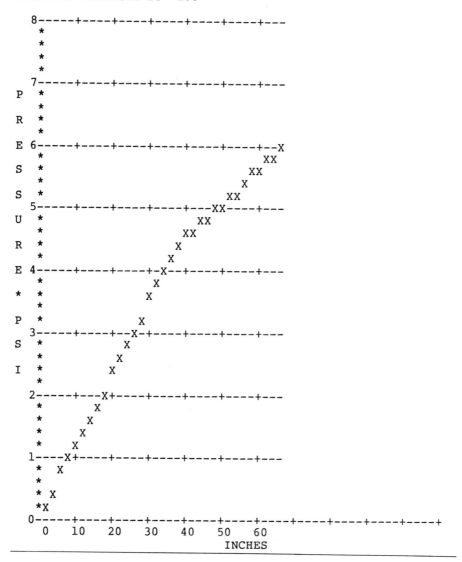

HOPT   Height of tapered portion of hopper

HSH   Height of solids in hopper

ID   I.D. number

IFRECS   Indicator for SPR's Solids Draining Screw—SDS

INDEV   Input device number

INJ   Indicator for injection molding

IPLUG   Instruction for computation in plugged channel

IPOE   Number of heater zones for barrel

IPRES   Instruction for printout when pressure is negative

ISOL   Indicator for solids conveying computation

ITOL   Tolerance level for computation

JJ   Number of geometric sections of screw

L   Length of each geometric section

LEAD   Screw lead or pitch

MIXREL   Indicator for mixing in a decompression section

MLFR   Indicator for multiple viscosity equations

or material number when data from SPR list is used

MLTCHL   Number of sections containing barrier

MLTX   Indicator for melt extrustion or semi-melted feed

NFRIC   Material number for frictional property data when SPR data base is used

OPT   Indicator for screw and barrel temperature optimization

OUTDEV   Output device number

OUTPUT   Production rate

P0   Inlet pressure

PARHOL   Number of holes in mixing device

PLT   Indicator for printout frequency

POISE   Indicator for viscosity not in English units

PRESIN   Injection pressure

R0, R1, R2, R12   Melt density equation

RB1   Bulk density of solids dependence on pressure

RECLEN   Reciprocating length for injection molding

RHOS   Density of solid bed

RIGID?   Indicator for rigid pellets

RNG   Indicator for mixing devices

ROB   Bulk density of solids

RPM   Screw speed

SCARD   Card number at which to stop reading input, diagnostic tool

SLF   Indicator for extruder-die combination and for automatic generation of simulations according to a statistical experimental design

SOAK   Soak time

STFRIC   Static coefficient of friction

STOP   Location of change for simulation

TBW   Temperature of cooling water on throat

TCONB   Thermal conductivity of barrel

TCONS   Thermal conductivity of solid plastic

THCOND   Thermal conductivity of melt

TM0, TM1, TM2   Melting point as a function of pressure

TSOLID   Feed temperature

WBAR1   Distance of thermocouple from inner wall of barrel

XTST   Fraction of channel to which solid bed is permitted to grow

15.2.2.2 Plasticating Extrusion

While the melt extruder processes molten plastic or other viscous fluids such as dough

and other food, the plasticating extruder processes solid plastic or other solid feed material. It is the primary task of the plasticating extruder to melt the solid feed for which it is particularly suitable. This is so because the high viscosity common to plastic melts is the cause of large internal viscous heat generation that replaces the slow process of heat conduction.

The plasticating extruder consists of several functional zones: the hopper, the solids conveying zone, the delay zone, the melting zone, and the melt pumping zone. A diagram explaining the location of the various geometrical sections of a plasticating extruder is given in Figure 15.3.

The Hopper.   The task of the hopper is dual: to serve as a reservoir for the feed material and to provide adequate inlet pressure for the solids conveying zone. Its performance depends on the hopper geometry, the height of solids column in the hopper, and the static coefficient of friction of the solid plastic on the hopper surface.

The Solids Conveying Zone.   This functional zone of the extruder is usually very short. It usually occupies the length of the extruder channel up to the axial location where the temperature of the barrel reaches the melting point of the plastic. This temperature rise is the result either of barrel heating in the area immediately downstream of the cooled hopper section of the barrel or of unusually efficient solids conveying resulting in a relatively high pressure rise along the solids conveying zone. High pressure also represents a high normal force and will therefore also result in a high frictional force on the barrel surface. This condition will result in excessive heat generation leading to melting of the plastic even on the cooled barrel surface. Screw geometry, operating conditions, length of the solids conveying zone, and the surface conditions greatly effect the ratio of the outlet to inlet pressure. Since the inlet pressure is really the hopper pressure, the hopper greatly affects the performance of the solids conveying zone.

The Delay Zone.   This zone starts at the axial location where the first traces of the melt

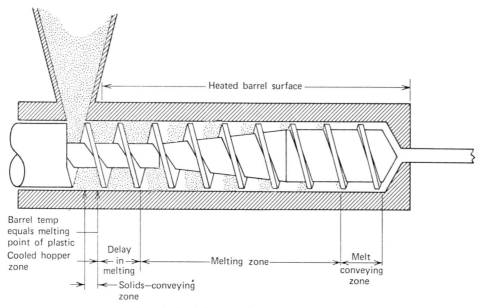

Figure 15.3   A diagram of a plasticating extruder.

appear at the barrel surface and ends where the melt film is fully developed and the melt pool appears in front of the pushing side of the flight. This zone is usually not very important except that it delays the start of the regular melting mechanism.

The Melting Zone.   This zone extends over most of the extruder channel from the formation of the melt pool until its disappearance because of complete melting. It is the most important functional zone of the extruder. Not only does this zone determine the extent of melting in the extruder and the presence of unmelted plastic in the extrudate, but also if it is not properly designed it can starve the rest of the extruder by forming a solid plug. Such a condition often also results in the output fluctuations commonly referred to as "surging."

The Melt Pumping Zone.   This zone usually occupies only a short portion at the discharge end of the extruder close to the die where all plastic has already melted.

The EXTRUD® computer program can simulate the performance of screws of any complexity and is not restricted to metering type screws having only three geometrical sections. Such screws can be single- or multiple-flighted, can have a variable screw

lead, can exhibit sudden changes in geometry at any axial location, and can even be worn in a complex manner. The screw can contain various mixing sections, including fluted sections of any complexity. It can also be of the barrier type and will show the effect of barrier flights on melting, temperature, and pressure rise, and on power consumption. A special feature of the program permits the simulation of the SPR Solids Draining Screw (SDS) consisting of an outer and an inner screw. In this screw type only fully melted plastic is permitted to reach the die. The unmelted plastic is returned or recycled through the inner screw to an upstream location in the outer screw. The unmelted plastic continues to melt in the inner screw. Because of the complete melting the main channel can usually be deeper than is the case in other screws, and the extrudate is therefore also considerably colder. Since the two screws interact, and computer simulation is also more complex than for regular screws.

A sample case for data entry and simulation of a plasticating extruder is given in Table 15.16.

Glossary of Terms used in Output Tables of EXTRUD® Program

D. HEAT   Heat conducted through barrel per unit length

Table 15.16

```
OK, EXTRUD
GO
EXTRUD    06-OCT-1978          SPR RELEASE:  SEPTEMBER 26, 1978
  INPUT FROM DSK? NO
BATCH OR CONV?CONVERSATIONAL
OUTDEV INDEV POISE IPLUG IPRES XTEST SCARD MIXREL IFRECS MLTCHL
1      1           1     1     .99
USING NAMELIST?YES
  WANT A PLOT? YES
  WHEN EXTRUD IS DONE TYPE:GRAPH
PLT ID #FLT LEAD DBAR FLWD FLCL JJ DLY INJ COOL OPT MLFR MLTX SLF RNG
-5  3   1   4.5  4.5  .45 .004 3  1              -239
    L   H   L    H    L   H    L   H    L    H
30    .6   35  .2    45    .2
THCOND RHOS HETCAM HETCAS HETFUS TSOLID GAMAX
                                 70
R0   TM0   R1   R2   R12   TM1   TM2
   A       A1        A2        A11       A22       A12

C0  C1  C2  P0  IPOE  ITOL  ISOL  NFRIC
450                   1     2
F0    F1   F2   PARHOL HLAREA BREAK HEADPR ACDISV STOP
-2000 100
RPM OUTPUT RIGID? SOAK AXDTM RECLEN PRESIN
100 600              5
DESCRIPTION OF CASE:
   PLASTICATING EXTRUSION SAMPLE CASE - LDPE .1 M.I.
DHOP ROB FRIB0 FRIS0 STFRIC WBAR TBW DPEL
36                          1.5  65
TCONB TCONS AGREP HSH HOPT  DHOPX WBAR1 HOPST
25          90    36  4.5         6
RB1   FRIB1 FRIB11 FRIB12 FRIB13 FRIB2 FRIB2

  FILE NAME=PLAST1
```

For glossary of Terms used above see section on Melt Extrusion.

The data typed in is now filed under the name PLAST1 and the execution started:

```
N=100.0 RPM      OUTPUT=  600.00 LB/HR      SOAK TIME=  0.00 SECONDS
```

SCREW GEOMETRY

SECTION	LENGTH (IN)	CH.DEPTH (IN)	BAR.DIAM (IN)	NO OF HOLES	HOLE AREA (SQ. IN.)
1	30.0000	0.6000	4.5000		
2	35.0000	0.2000	4.5000		
3	45.0000	0.2000	4.5000		

```
  ATTENTION - KEEP SOLIDS LEVEL IN HOPPER ABOVE   8.66 IN.
  BASE PRESSURE =       0.05 PSI
```

SOLIDS CONVEYING ZONE

L IN	PHI DEG	ROBP LB/CU.FT	FRIB	FRIS	TESOUT DEG F	P PSI	TSCREW DEG F
2.00	4.0	32.21	1.011	0.812	66.0	0.0	100.0
4.00	4.0	32.21	1.019	0.812	65.5	0.0	100.0
5.00	4.0	32.21	1.019	0.812	65.4	0.0	100.0

```
ATTENTION - SOLIDS CONVEYING CONTROLS OPERATION
REQUESTED OUTPUT IS UNATTAINABLE
```

Table 15.16 (*Continued*)

L IN	TOUT DEG F	X/W	X/W INJ	GRATIO	PRESSURE PSI REAL	FULL	POWER HP	HOUT IN	TBAR DEG F	TVISC LB.SEC /SQ.IN	D.HEAT BTU/IN .MIN
					DELAY ZONE						
7.37	388.1				94.	94.		0.600	450.0	0.035	
9.37	388.1				67.	67.		0.600	450.0	0.048	
11.37	388.1				73.	73.		0.600	450.0	0.058	
13.00	388.1				104.	104.		0.600	450.0	0.066	
15.00	388.1				106.	106.		0.600	450.0	0.073	
					MELTING ZONE						
16.50	382.8	.898	.898	.898	0.	94.	1.3	0.600	450.0	0.225	105.4
21.50	414.5	.815	.815	.815	0.	94.	7.7	0.600	450.0	0.148	24.1
24.50	424.1	.768	.768	.768	0.	969.	11.7	0.600	450.0	0.138	29.5
29.50	434.3	.694	.694	.694	0.	1438.	17.9	0.600	450.0	0.128	22.5
30.00	435.1	.687	.687	.687	0.	1484.	18.5	0.600	450.0	0.127	21.6
35.00	451.9	.672	.672	.608	0.	1967.	25.2	0.543	450.0	0.096	22.7
40.00	462.0	.658	.658	.533	0.	2484.	31.9	0.486	450.0	0.090	24.3
45.00	469.5	.644	.644	.460	0.	3045.	38.7	0.429	450.0	0.086	25.9
50.00	475.8	.631	.631	.391	0.	3650.	45.6	0.371	450.0	0.083	27.3
55.00	481.5	.619	.619	.324	0.	4259.	52.6	0.314	450.0	0.080	28.5
60.00	486.9	.607	.607	.260	0.	4865.	59.7	0.257	450.0	0.077	29.7
65.00	492.0	.597	.597	.199	0.	5401.	66.8	0.200	450.0	0.075	30.0
70.00	494.8	.433	.433	.144	0.	5878.	73.5	0.200	450.0	0.077	25.1
75.00	498.0	.302	.302	.101	0.	6334.	79.5	0.200	450.0	0.076	24.2
80.00	501.6	.201	.201	.067	0.	6769.	84.8	0.200	450.0	0.076	23.7
85.00	505.4	.127	.127	.042	0.	7188.	89.7	0.200	450.0	0.075	23.7
90.00	509.2	.074	.074	.025	0.	7594.	94.2	0.200	450.0	0.074	24.1
95.00	513.1	.040	.040	.013	0.	7989.	98.4	0.200	450.0	0.073	24.8
100.00	516.8	.020	.020	.007	0.	8377.	102.4	0.200	450.0	0.072	25.6
105.00	520.1	.006	.006	.002	0.	8759.	106.2	0.200	450.0	0.071	24.6
110.00	523.7	.000	.000	.000	0.	9135.	109.9	0.200	450.0	0.070	27.4

For Glossary of Terms used in the Output Tables see end of this section.

```
MINIMUM TEMPERATURE=    523.663 DEG F
TEMP FLUCTUATION=         0.000 DEG F
FLOW INDEX (N)=           0.373
GRATIO INJ.=              0.000
RESIDENCE TIME:
     MELT POOL=   0.018419005 HRS


POWER BREAK-DOWN:
SOLIDS CONV. =    0.007 HP
MELTING      =   53.026 HP
MIXING       =   32.486 HP
FLIGHT CLEAR =   18.878 HP
COMPRESSION  =    5.496 HP
MIXING DEVICE=    0.000 HP
PRESSURE FLUCTUATION=          622.85 PSI/CYCLE
TOTAL CONDUCTED HEAT THROUGH BARREL=    2480.2 BTU/MIN
```

```
VISCOSITY WAS CALCULATED IN THE FOLLOWING SHEAR RATE RANGE
SHEAR RATE        TEMP.        VISC.
 0.70674E 04    0.34050E 03   0.73192E-02
 0.39010E 02    0.45000E 03   0.21021E 00


VISCOSITY WAS CALCULATED IN THE FOLLOWING TEMPERATURE RANGE
SHEAR RATE        TEMP.        VISC.
 0.13998E 03    0.52364E 03   0.68559E-01
 0.88920E 03    0.20600E 03   0.42725E-01
```

Table 15.16 (*Continued*)

Since the solids conveying zone showed no pressure rise,
screw and barrel temperature optimization is saught in the
next simulation by typing in OPT=2:
TYPE-IN INPUT
$INPUT OPT=2 $
LIKE A PLOT?YES
WHEN EXTRUD IS DONE TYPE: GRAPH

N=100.0 RPM   OUTPUT=   600.00 LB/HR   SOAK TIME=   0.00 SECONDS

                              SCREW GEOMETRY

SECTION	LENGTH (IN)	CH.DEPTH (IN)	BAR.DIAM (IN)	NO OF HOLES	HOLE AREA (SQ. IN.)
1	30.0000	0.6000	4.5000		
2	35.0000	0.2000	4.5000		
3	45.0000	0.2000	4.5000		

        ATTENTION - KEEP SOLIDS LEVEL IN HOPPER ABOVE   8.66 IN.
        BASE PRESSURE =   0.05 PSI

FOR OPTIMUM SOLIDS CONVEYING KEEP: BARREL TEMP=   190.50
                                   SCREW TEMP=   60.00

                          SOLIDS CONVEYING ZONE

L IN	PHI DEG	ROBP LB/CU.FT	FRIB	FRIS	TESOUT DEG F	P PSI	TSCREW DEG F
2.00	4.0	32.21	1.126	0.799	169.5	0.0	60.0
4.00	4.0	32.21	1.365	0.799	185.3	0.1	60.0
5.00	4.0	32.24	1.402	0.790	191.2	0.5	60.0

L IN	TOUT DEG F	X/W	X/W INJ	GRATIO	PRESSURE PSI REAL	PRESSURE PSI FULL	POWER HP	HOUT IN	TBAR DEG F	TVISC LB.SEC /SQ.IN	D.HEAT BTU/IN .MIN
						DELAY ZONE					
7.37	388.1				98.	98.		0.600	450.0	0.035	
9.37	388.1				72.	72.		0.600	450.0	0.048	
11.37	388.1				75.	75.		0.600	450.0	0.058	
13.00	388.1				104.	104.		0.600	450.0	0.066	
15.00	388.1				106.	106.		0.600	450.0	0.073	
						MELTING ZONE					
16.50	382.8	.898	.898	.898	0.	94.	1.4	0.600	450.0	0.225	105.4
21.50	414.5	.815	.815	.815	0.	94.	7.8	0.600	450.0	0.148	24.1
24.50	424.1	.768	.768	.768	0.	969.	11.8	0.600	450.0	0.138	29.5
29.50	434.3	.694	.694	.694	0.	1438.	18.0	0.600	450.0	0.128	22.5
30.00	435.1	.687	.687	.687	0.	1484.	18.6	0.600	450.0	0.127	21.6
35.00	451.9	.672	.672	.608	0.	1967.	25.3	0.543	450.0	0.096	22.7
40.00	462.0	.658	.658	.533	0.	2484.	32.0	0.486	450.0	0.090	24.3
45.00	469.5	.644	.644	.460	0.	3045.	38.8	0.429	450.0	0.086	25.9
50.00	475.8	.631	.631	.391	0.	3650.	45.7	0.371	450.0	0.083	27.3
55.00	481.5	.619	.619	.324	120.	4259.	52.7	0.314	450.0	0.080	28.5

CHANNEL MAY BE ONLY PARTIALLY FULL DUE TO POOR SOLIDS CONVEYING

L IN	TOUT DEG F	X/W	X/W INJ	GRATIO	PRESSURE PSI REAL	PRESSURE PSI FULL	POWER HP	HOUT IN	TBAR DEG F	TVISC LB.SEC /SQ.IN	D.HEAT BTU/IN .MIN
60.00	486.9	.607	.607	.260	727.	4866.	59.8	0.257	450.0	0.077	29.7
65.00	492.0	.596	.596	.199	1267.	5406.	67.0	0.200	450.0	0.075	30.0
70.00	494.6	.432	.432	.144	1748.	5887.	73.6	0.200	450.0	0.076	25.1
75.00	497.9	.301	.301	.100	2214.	6353.	79.6	0.200	450.0	0.076	24.2
80.00	501.5	.200	.200	.067	2664.	6803.	85.0	0.200	450.0	0.075	23.7
85.00	505.2	.125	.125	.042	3100.	7239.	89.9	0.200	450.0	0.074	23.8
90.00	509.0	.073	.073	.024	3526.	7665.	94.4	0.200	450.0	0.073	24.2
95.00	512.9	.039	.039	.013	3945.	8084.	98.6	0.200	450.0	0.072	25.0
100.00	516.6	.019	.019	.006	4359.	8498.	102.5	0.200	450.0	0.071	25.9
105.00	519.9	.005	.005	.002	4770.	8909.	106.4	0.200	450.0	0.070	24.9
110.00	523.3	.000	.000	.000	5175.	9314.	110.1	0.200	450.0	0.068	27.7

Table 15.16 (*Continued*)

```
MINIMUM TEMPERATURE=      523.300 DEG F
TEMP FLUCTUATION=           0.000 DEG F
FLOW INDEX (N)=             0.372
GRATIO INJ.=               0.000
RESIDENCE TIME:
     MELT POOL=    0.018504538 HRS

POWER BREAK-DOWN:
SOLIDS CONV. =    0.109 HP
MELTING       =   52.989 HP
MIXING        =   32.521 HP
FLIGHT CLEAR  =   18.878 HP
COMPRESSION   =    5.602 HP
MIXING DEVICE=    0.000 HP
PRESSURE FLUCTUATION=           636.82 PSI/CYCLE
TOTAL CONDUCTED HEAT THROUGH BARREL=    2485.6 BTU/MIN


   VISCOSITY WAS CALCULATED IN THE FOLLOWING SHEAR RATE RANGE
   SHEAR RATE        TEMP.        VISC.
   0.71412E 04     0.34050E 03   0.72606E-02
   0.39010E 02     0.45000E 03   0.21021E 00


   VISCOSITY WAS CALCULATED IN THE FOLLOWING TEMPERATURE RANGE
   SHEAR RATE        TEMP.        VISC.
   0.14408E 03     0.52337E 03   0.67408E-01
   0.25534E 04     0.23100E 03   0.18126E-01
```

Since the barrel and screw temperature optimization with the frictional properties of material no. 2 (NFRIC=2) is still only capable of raising pressure to 0.5 psi which is inadequate to compact the solid bed, screw lead is shortened to 4 inches:

```
 TYPE-IN INPUT
 $INPUT SLEAD=4 $
 LIKE A PLOT?YES
 WHEN EXTRUD IS DONE TYPE: GRAPH

N=100.0 RPM       OUTPUT=  600.00 LB/HR      SOAK TIME=  0.00 SECONDS

                         SCREW GEOMETRY

SECTION LENGTH  CH.DEPTH   BAR.DIAM NO OF HOLES  HOLE AREA
        (IN)     (IN)       (IN)                 (SQ. IN.)
   1   30.0000   0.6000     4.5000
   2   35.0000   0.2000     4.5000
   3   45.0000   0.2000     4.5000

   ATTENTION - KEEP SOLIDS LEVEL IN HOPPER ABOVE      8.66 IN.
   BASE PRESSURE =       0.05 PSI

FOR OPTIMUM SOLIDS CONVEYING KEEP: BARREL TEMP=      190.50
                                   SCREW  TEMP=       63.50

                     SOLIDS CONVEYING ZONE

   L    PHI   ROBP    FRIB  FRIS  TESOUT      P      TSCREW
   IN   DEG  LB/CU.FT              DEG F      PSI     DEG F
  2.00   4.2  32.21   1.125 0.816 169.1      0.0      63.5
  4.00   4.2  32.22   1.360 0.813 185.1      0.2      63.5
  5.00   4.2  32.25   1.398 0.802 191.5      0.6      63.5
```

Table 15.16 (*Continued*)

L IN	TOUT DEG F	X/W	X/W INJ	GRATIO	PRESSURE PSI REAL FULL	POWER HP	HOUT IN	TBAR DEG F	TVISC LB.SEC /SQ.IN	D.HEAT BTU/IN .MIN
					DELAY ZONE					
7.38	388.1				65. 65.		0.600	450.0	0.035	
9.38	388.1				70. 70.		0.600	450.0	0.048	
11.38	388.1				74. 74.		0.600	450.0	0.058	
13.00	388.1				95. 95.		0.600	450.0	0.066	
					MELTING ZONE					
15.50	384.1	.907	.907	.907	0. 88.	1.4	0.600	450.0	0.277	98.0
20.50	414.1	.826	.826	.826	0. 88.	7.8	0.600	450.0	0.175	24.7
23.50	423.5	.779	.779	.779	0. 1013.	11.8	0.600	450.0	0.159	30.2
28.50	433.6	.705	.705	.705	0. 1510.	18.0	0.600	450.0	0.145	23.2
30.00	436.0	.684	.684	.684	0. 1655.	19.8	0.600	450.0	0.142	22.1
35.00	452.4	.670	.670	.606	0. 2175.	26.5	0.543	450.0	0.104	23.0
40.00	462.3	.657	.657	.532	0. 2726.	33.3	0.486	450.0	0.098	24.7
45.00	469.6	.645	.645	.461	0. 3315.	40.1	0.429	450.0	0.093	26.2
50.00	475.6	.634	.634	.392	0. 3925.	46.9	0.371	450.0	0.089	27.6
55.00	480.9	.623	.623	.326	120. 4520.	53.9	0.314	450.0	0.086	28.8
CHANNEL MAY BE ONLY PARTIALLY FULL										
60.00	485.7	.613	.613	.263	709. 5109.	60.9	0.257	450.0	0.084	29.8
65.00	489.7	.605	.605	.202	1193. 5593.	67.9	0.200	450.0	0.083	30.1
70.00	492.2	.442	.442	.147	1577. 5976.	74.5	0.200	450.0	0.083	25.0
75.00	495.2	.310	.310	.103	1932. 6332.	80.3	0.200	450.0	0.083	23.9
80.00	498.6	.208	.208	.069	2266. 6666.	85.6	0.200	450.0	0.082	23.2
85.00	502.1	.132	.132	.044	2583. 6983.	90.4	0.200	450.0	0.081	23.1
90.00	505.8	.078	.078	.026	2888. 7288.	94.8	0.200	450.0	0.080	23.5
95.00	509.5	.043	.043	.014	3185. 7585.	98.9	0.200	450.0	0.079	24.1
100.00	513.1	.022	.022	.007	3478. 7878.	102.8	0.200	450.0	0.078	25.0
105.00	516.5	.010	.010	.003	3768. 8168.	106.6	0.200	450.0	0.076	25.8
110.00	519.8	.000	.000	.000	4057. 8457.	110.2	0.200	450.0	0.075	26.8

```
MINIMUM TEMPERATURE=    519.774 DEG F
TEMP FLUCTUATION=         0.000 DEG F
FLOW INDEX (N)=           0.374
GRATIO INJ.=              0.000
RESIDENCE TIME:
    MELT POOL=    0.018499654 HRS


POWER BREAK-DOWN:
  SOLIDS CONV. =   0.120 HP
  MELTING      =  52.869 HP
  MIXING       =  31.325 HP
  FLIGHT CLEAR =  21.199 HP
  COMPRESSION  =   4.679 HP
  MIXING DEVICE=   0.000 HP
  PRESSURE FLUCTUATION=        475.97 PSI/CYCLE

TOTAL CONDUCTED HEAT THROUGH BARREL=   2490.7 BTU/MIN
VISCOSITY WAS CALCULATED IN THE FOLLOWING SHEAR RATE RANGE
  SHEAR RATE        TEMP.        VISC.
   0.70669E 04    0.34050E 03   0.73196E-02
   0.37605E 02    0.45000E 03   0.21514E 00


VISCOSITY WAS CALCULATED IN THE FOLLOWING TEMPERATURE RANGE
  SHEAR RATE        TEMP.        VISC.
   0.12758E 03    0.51963E 03   0.73944E-01
   0.25131E 04    0.23100E 03   0.18357E-01
```

Although the shortened screw lead did improve solids conveying by raising pressure at the end of the solids conveying zone from 0.5 to 0.6 psi,
this is still inadequate. Screw lead is therefore reset to the original
4.5 inches and the frictional properties are changed from those of material no. 2 to those of material number 8:

Table 15.16 (*Continued*)

```
TYPE-IN INPUT
$INPUT SLEAD=4.5, NFRIC=8 $
LIKE A PLOT?YES
WHEN EXTRUD IS DONE TYPE: GRAPH
```

N=100.0 RPM   OUTPUT= 600.00 LB/HR   SOAK TIME= 0.00 SECONDS

SCREW GEOMETRY

SECTION	LENGTH (IN)	CH.DEPTH (IN)	BAR.DIAM (IN)	NO OF HOLES	HOLE AREA (SQ. IN.)
1	30.0000	0.6000	4.5000		
2	35.0000	0.2000	4.5000		
3	45.0000	0.2000	4.5000		

ATTENTION - KEEP SOLIDS LEVEL IN HOPPER ABOVE   8.66 IN.
BASE PRESSURE =   0.05 PSI

FOR OPTIMUM SOLIDS CONVEYING KEEP: BARREL TEMP=   190.50
                                   SCREW TEMP=   128.50

SOLIDS CONVEYING ZONE

L IN	PHI DEG	ROBP LB/CU.FT	FRIB	FRIS	TESOUT DEG F	P PSI	TSCREW DEG F
2.00	4.0	32.51	0.413	0.136	173.3	3.3	128.5
2.90	3.7	34.24	1.025	0.188	232.0	23.2	128.5

L IN	TOUT DEG F	X/W	X/W INJ	GRATIO	PRESSURE REAL	PSI FULL	POWER HP	HOUT IN	TBAR DEG F	TVISC LB.SEC /SQ.IN	D.HEAT BTU/IN .MIN
							DELAY ZONE				
5.27	388.1				0.	0.		0.600	450.0	0.036	
7.27	388.1				0.	0.		0.600	450.0	0.050	
9.27	388.1				0.	0.		0.600	450.0	0.060	
11.27	388.1				0.	0.		0.600	450.0	0.069	
13.00	388.1				0.	0.		0.600	450.0	0.076	
							MELTING ZONE				
14.00	382.7	.898	.898	.898	0.	0.	3.0	0.600	450.0	0.225	105.4
19.00	414.5	.815	.815	.815	0.	0.	9.4	0.600	450.0	0.148	24.0
22.00	424.1	.768	.768	.768	875.	875.	13.3	0.600	450.0	0.138	29.5
27.00	434.3	.694	.694	.694	1346.	1346.	19.6	0.600	450.0	0.127	22.5
30.00	438.9	.651	.651	.651	1619.	1619.	23.2	0.600	450.0	0.123	21.8
35.00	453.4	.636	.636	.575	2097.	2097.	29.7	0.543	450.0	0.096	22.0
40.00	462.8	.620	.620	.502	2611.	2611.	36.3	0.486	450.0	0.090	23.6
45.00	469.9	.605	.605	.432	3171.	3171.	43.0	0.429	450.0	0.086	25.1
50.00	476.1	.590	.590	.365	3775.	3775.	49.7	0.371	450.0	0.083	26.5
55.00	481.7	.575	.575	.301	4388.	4388.	56.6	0.314	450.0	0.079	27.6
60.00	487.1	.560	.560	.240	5000.	5000.	63.5	0.257	450.0	0.077	28.8
65.00	492.2	.543	.543	.181	5552.	5552.	70.4	0.200	450.0	0.075	29.0
70.00	495.9	.389	.389	.130	5970.	5970.	76.9	0.200	450.0	0.074	24.3
75.00	498.9	.266	.266	.089	6444.	6444.	82.7	0.200	450.0	0.074	24.0
80.00	502.2	.173	.173	.058	6904.	6904.	87.9	0.200	450.0	0.073	23.7
85.00	505.8	.106	.106	.035	7354.	7354.	92.6	0.200	450.0	0.072	23.9
90.00	509.4	.060	.060	.020	7796.	7796.	97.0	0.200	450.0	0.071	24.5
95.00	513.1	.031	.031	.010	8231.	8231.	101.1	0.200	450.0	0.070	25.3
100.00	517.3	.015	.015	.005	8621.	8621.	105.1	0.200	450.0	0.069	25.7
105.00	521.5	.000	.000	.000	8982.	8982.	108.9	0.200	450.0	0.067	27.0
110.00	525.1	.000	.000	.000	9344.	9344.	112.7	0.200	450.0	0.066	27.8

MINIMUM TEMPERATURE=   525.146 DEG F   GRATIO INJ.=   0.000
TEMP FLUCTUATION=   0.000 DEG F   RESIDENCE TIME:
FLOW INDEX (N) =   0.372   MELT POOL= 0.019420866 HRS

Table 15.16 (*Continued*)

```
POWER BREAK-DOWN:
  SOLIDS CONV. =   1.698 HP
  MELTING      =  52.411 HP
  MIXING       =  33.636 HP
  FLIGHT CLEAR =  19.343 HP
  COMPRESSION  =   5.628 HP
  MIXING DEVICE=   0.000 HP
  PRESSURE FLUCTUATION=        596.58 PSI/CYCLE
  TOTAL CONDUCTED HEAT THROUGH BARREL=   2510.2 BTU/MIN

  VISCOSITY WAS CALCULATED IN THE FOLLOWING SHEAR RATE RANGE
  SHEAR RATE       TEMP.        VISC.
    0.71091E 04    0.34050E 03  0.72860E-02
    0.39010E 02    0.45000E 03  0.21021E 00

  VISCOSITY WAS CALCULATED IN THE FOLLOWING TEMPERATURE RANGE
  SHEAR RATE       TEMP.        VISC.
    0.14751E 03    0.52510E 03  0.65922E-01
    0.25533E 04    0.23100E 03  0.18126E-01
TYPE-IN INPUT
$INPUT BEND=-1 $

  Graphical output of the above simulations is obtained as follows:
OK, GRAPH
GO
  WHERE WOULD YOU LIKE THE GRAPH PRINTED?:
(TTY,LPT,DSK)
TTY                                                       •
```

FRIB   Friction coefficient on barrel surface

FRIS   Friction coefficient on screw surface

GRATIO   Fraction of unmelted plastic in flow rate

HOUT   Channel depth at location L

L   Location from downstream end of hopper

N   Screw speed

OUTPUT   Production rate

P   Pressure

PHI   Angle of advancement of solid bed

POWER   Power consumed

PRESSURE

ROBP   Bulk density of feed

SOAK TIME   For molding, time per cycle screw is stationary

TBAR   Barrel temperature

TESOUT   Temperature of inner barrel wall in contact with solid

TOUT   Melt pool temperature at location L

TSCREW   Screw temperature

TVISC   True viscosity of melt

X/W   Relative width of solid bed

X/W INJ   Relative width of solid bed at end of soak time

15.2.2.3 Blow Molding

A blow molding machine is really an extruder, and therefore all material relating to plasticating extrusion also applies to it. Its performance can be simulated on a computer the same way it is done for a plasticating extruder. If a blow molding machine operates intermittantly, its performance is similar to that of an injection molding machine and can be simulated by the EXTRUD® program utilizing that option.

15.2.2.4 Reciprocating Screw Injection Molding

Injection molding is also close to plasticating extrusion except that it is usually not a continuous operation and works intermittantly. The computer simulation of the plasticating extruder by the EXTRUD® program is therefore corrected for this intermittant operation, namely, for melting during the so called "soak time" when the screw is standing still. This is affected by the shot size, cycle time, and extrusion time. The data for the simulation is typed in the same way it is done for extrusion with a few minor exceptions:

Table 15.16 (*Continued*)

```
SCIENTIFIC PROCESS AND RESEARCH INC.EXPERIMENT NUMBER    3.00
OUTPUT=  600.00 LB/HR    SCREW SPEED= 100.00RPM

MULTIPLY PRESSURE BY 1000
      *                                                            X
    9-----+----+----+----+----+----+----+----+----+----+----+---X+
      *                                                           X
      *                                                          XX
      *                                                           X
      *                                                           X
    8-----+----+----+----+----+----+----+----+----+----+-XX-+----+
      *                                                        X
      *                                                        X
      *                                                        X
      *                                                       XX
    7-----+----+----+----+----+----+----+----+----+X---+----+----+
  P   *                                               X
      *                                               X
  R   *                                               X
      *                                               X
  E 6-----+----+----+----+----+----+----+----+X---+----+----+----+
      *                                            X
  S   *                                            X
      *                                           XX
  S   *
    5-----+----+----+----+----+----+----+X---+----+----+----+----+
  U   *                                        X
      *                                        X
  R   *                                        X
      *                                        X
  E 4-----+----+----+----+----+----+----+----+----+----+----+----+
      *                                       X
  *   *                                       X
      *                                       X
  P   *                                      X
    3-----+----+----+----+-X--+----+----+----+----+----+----+----+
  S   *                    X
      *
  I   *                   X
      *                   X
    2-----+----+----+--X-+----+----+----+----+----+----+----+----+
      *                 X
      *                 X
      *                X
      *               XX
    1-----+----+----+----+----+----+----+----+----+----+----+----+
      *               X
      *
      *
      *
    0-----+---XXX*****************************************----+-
      0    10   20   30   40   50   60   70   80   90  100  110
                              INCHES
```

1. The output is substituted by cycle time, extrusion time, and shot size.

2. The reciprocating length must be supplied if they screw is reciprocating.

From these input data the proper output is generated, giving the melting behavior and pressure behavior under both steady-state extrusion and nonsteady state that are molding conditions as seen in Table 15.17.

15.2.2.5 *Melt Spinning*

As the plastic melt leaves the die it is drawn by the takeup equipment and simultaneously

Table 15.16 (*Continued*)

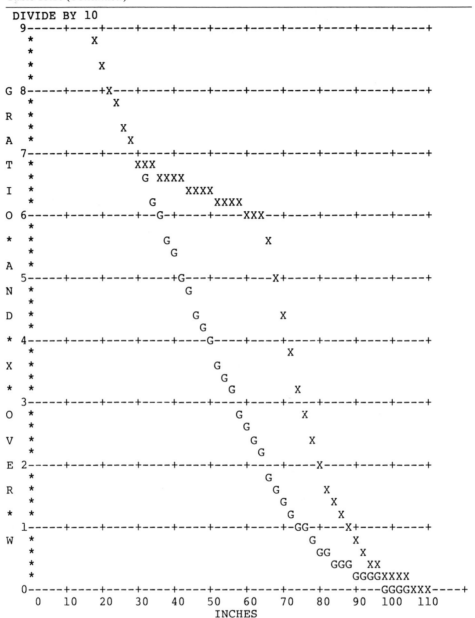

cooled. This process gives the monofilament its strength and its dimension. The fiber swells upon leaving the die and then gradually decreases the cross section with cooling and drawing time. During this process the temperature of the yarn decreases both in the radial and axial direction. This behavior is governed by heat capacity versus temperature property of the plastic of which the fiber is made, by air velocity, and by air temperature at various points in the equipment. Because of this complexity a complex computer program SPIN is needed to analyze fully the problem and accurately design the equipment. A sample session with the SPIN program is given in Table 15.18.

As can be seen in the example both a steady-state solution and a transient solution to the problem is obtained. The input data consist of operating conditions such as temperature and

Table 15.16 (*Continued*)

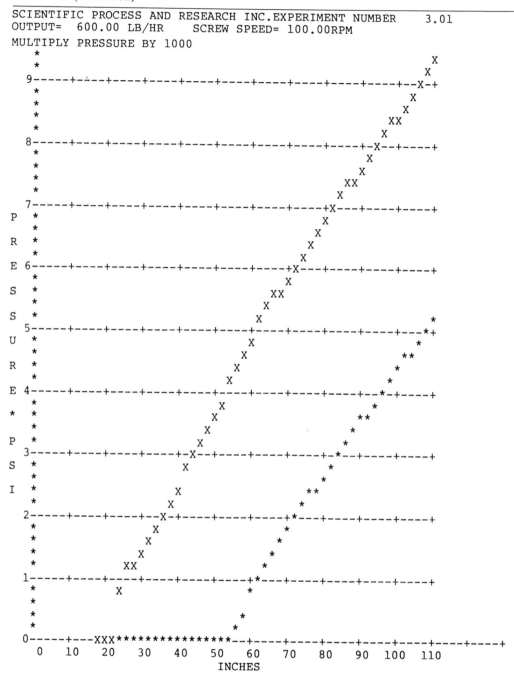

SCIENTIFIC PROCESS AND RESEARCH INC.EXPERIMENT NUMBER   3.01
OUTPUT=   600.00 LB/HR   SCREW SPEED= 100.00RPM
MULTIPLY PRESSURE BY 1000

air velocity profile, spinneret temperature, spinneret diameter, takeup speed and takeup denier desired. The physical property data of the polymer used for melt spinning are also required, such as density, heat capacity, coefficient of tensile viscosity, and its temperature dependence.

In the steady-state solution residence time is printed out together with thread temperature and thread cross-sectional area as a function of axial location from the spinneret.

In the transient solution the printout is the change in cross-sectional area (%), temperature, speed, and spinning tension in response

Table 15.16 (*Continued*)

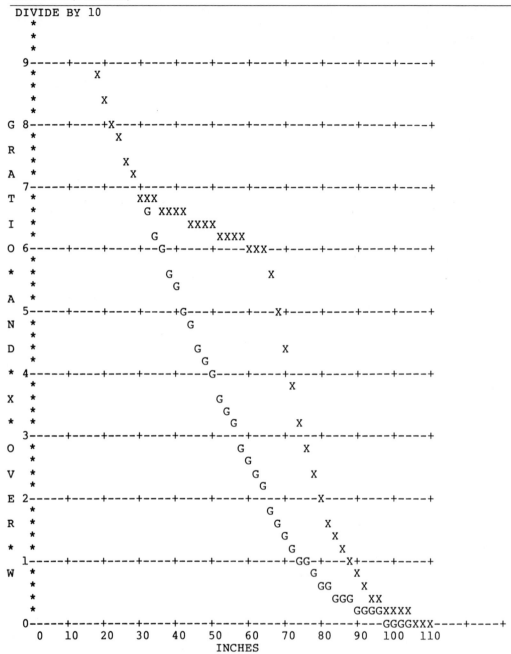

to a 1% change in the velocity of the cooling air from its steady-state condition. The print-out can be either at the takeup point or at any other location desired.

15.2.2.6   *Wire Coating and Cable Jacketing*

It is known that as plastic passes the tight clearances of a wire coating die, its temperature

rises appreciably. At high wire speeds this phenomenon is even more pronounced and is often responsible for the thermal degradation of the insulating material. This is particularly important when the plastic used is thermally unstable such as polyvinyl chloride. To reduce this danger the plastic is mixed with a so-called heat stabilizer, which is very expensive

Table 15.16 (*Continued*)

```
SCIENTIFIC PROCESS AND RESEARCH INC.EXPERIMENT NUMBER    3.02
OUTPUT=  600.00 LB/HR    SCREW SPEED= 100.00RPM
```

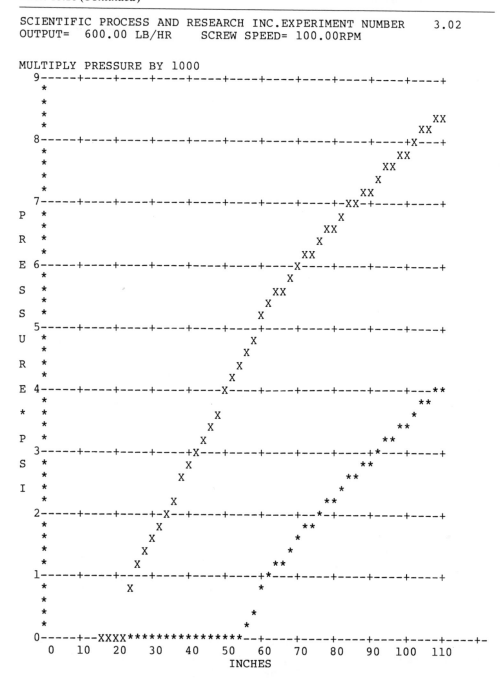

MULTIPLY PRESSURE BY 1000

and greatly effects the cost of the product. The use of the WIRE program permits the exact computation of the plastic temperature at every radial and axial position in the die. Therefore the die can be better designed to lower the melt temperature, and thereby also the need for thermal stabilizers, and to make the product cheaper. The program also applies to pipe, film blowing, and to other annular products.

15.2.2.7 *Crosshead Dies*

Many extruded plastic profiles are produced in so-called crosshead dies. The problem with

Table 15.16 (*Continued*)

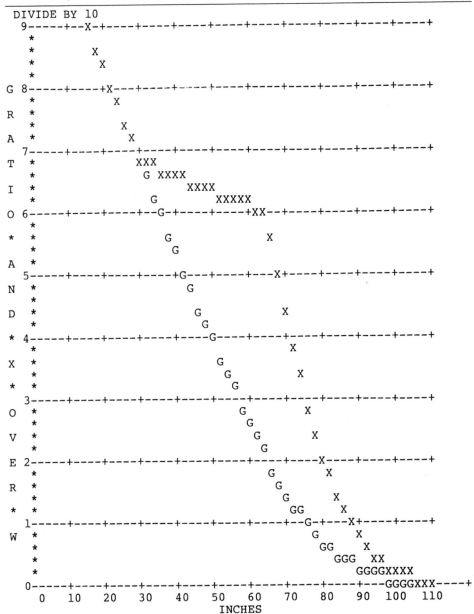

these products is often the lack of uniformity or of product thickness. The CROSSHEAD program designs the inner details of the die while taking into consideration the die temperature, the exact shear rate, and temperature dependence of the plastic being processed, the dimension of the product manufactured, and its desired production rate.

15.2.2.8 *Sheet and Flat Film Formation*

In the production of sheet the big problem is that the thickness of the sheet varies from the center to the sides. This problem is caused by the natural tendency of the melt to flow from the die directly downstream in the sheet and to resist flow sideways. As a result the flow rate is highest in the center of the sheet, and so is the sheet thickness. Since the minimum sheet thickness must be maintained, the product contains more plastic than would be the case with a fully uniform thickness. This difference in thickness usually increases as the sheet thickness decreases, making the loss in plastic exceptionally high for a thin film. The problem

Table 15.16 (*Continued*)

SCIENTIFIC PROCESS AND RESEARCH INC.EXPERIMENT NUMBER   3.03
OUTPUT=   600.00 LB/HR   SCREW SPEED= 100.00RPM

MULTIPLY PRESSURE BY 1000

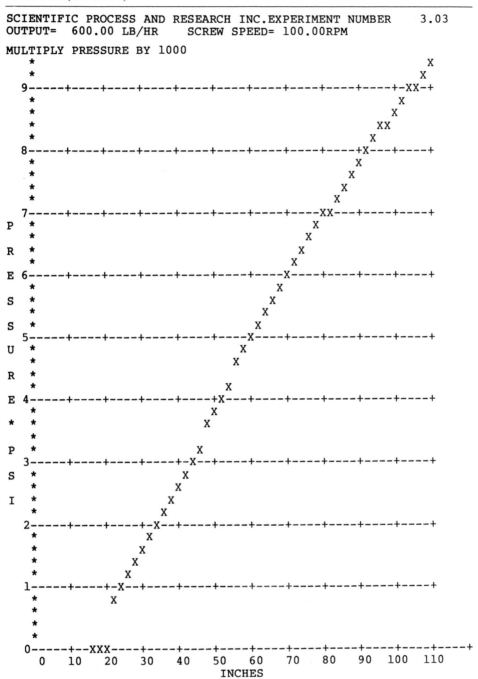

INCHES

---

is often corrected by installing a strainer bar in the sheet die. This device permits the application of a higher resistance to flow at locations where the flow rate is highest and tends to spread out the plastic more uniformly than would otherwise be the case. A similar device is the flexible lip which is only a strainer bar installed at the discharge end of the die.

However, corrective devices can seldom fully correct for the damage caused upstream of the flow. The problem must be eliminated at the point where it originates. This can be accomplished with the so-called Coat Hanger die. Two channels carry the plastic from the extruder toward the two sides of the sheet. Each channel forms an angle with the direction

Table 15.16 (*Continued*)

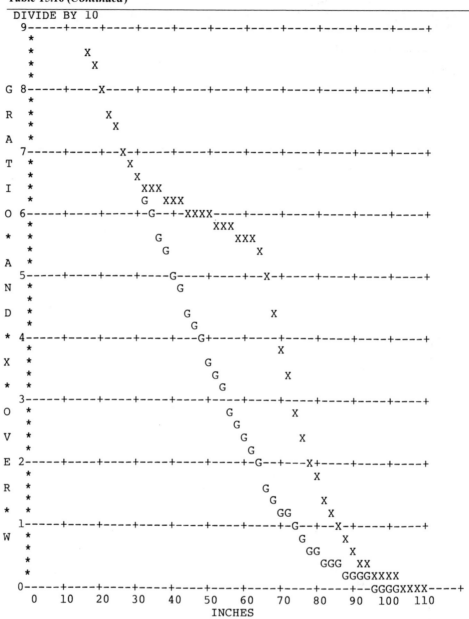

perpendicular to the direction to flow, thereby forming a shape resembling a coat hanger. The cross section of this channel or manifold varies from point to point to let out only the desired rate of plastic melt in the down-channel direction and causing the rest to flow further toward the sides of the sheet to lend it a high degree of uniformity.

The DIEDES program of SPR designs the inner details of coat hanger dies utilizing the flow properties of the plastic, its density, production rate requirement, and the desired sheet cross section. It yields the manifold cross section profile for a variety of manifold angles, for the distributor channel dimensions, and for the dimensions of the lip and strainer bar used. From this series of designs the user can select the one that can be most easily implemented in the actual die.

DIEDES also analyzes existing dies and shows

Table 15.17

```
OK, EXTRUD
GO
EXTRUD   22-DEC-1978      SPR RELEASE:  NOVEMBER 16, 1978
 INPUT FROM DSK? NO
BATCH OR CONV?CONV
OUTDEV INDEV POISE IPLUG IPRES XTEST SCARD MIXREL IFRECS MLTCHL
1      1
USING NAMELIST?YES
 WANT A PLOT? NO
PLT ID #FLT LEAD DBAR FLWD FLCL JJ DLY INJ COOL OPT MLFR MLTX SLF RNG
-5  2  1    2.5  2.5  .25  .003 3  1   1        2   -239
    L    H     L    H     L    H      L    H     L    H
20   .5     20   .12   10   .12
THCOND RHOS HETCAM HETCAS HETFUS TSOLID GAMAX
                                   65
R0   TM0   R1   R2   R12   TM1   TM2

    A       A1        A2        A11       A22       A12

C0  C1  C2  P0  IPOE  ITOL  ISOL  NFRIC
450                         1     2
F0   F1   F2  PARHOL HLAREA BREAK HEADPR ACDISV STOP
-2000
RPM OUTPUT RIGID? SOAK AXDTM RECLEN PRESIN
100 1             4
DESCRIPTION OF CASE:
 2.5 INCH INJECTION MOLDING Machine
DHOP ROB FRIB0 FRIS0 STFRIC WBAR TBW DPEL
36                          1.2
TCONB TCONS AGREP HSH HOPT  DHOPX WBAR1 HOPST
26               34  36     2.5         6
RB1  FRIB1  FRIB11  FRIB12  FRIB13  FRIB2 FRIB2

    SHOT       CYCLE     EXTIM
    3.2        15        7.5
 FILE NAME=INJ2

N=100.0 RPM     OUTPUT=  96.00 LB/HR     SOAK TIME= 7.50 SECONDS


                     SCREW GEOMETRY
SECTION LENGTH  CH.DEPTH  BAR.DIAM NO OF HOLES  HOLE AREA
        (IN)     (IN)      (IN)                 (SQ. IN.)
   1   20.0000  0.5000    2.5000
   2   20.0000  0.1200    2.5000
   3   10.0000  0.1200    2.5000

   ATTENTION - KEEP SOLIDS LEVEL IN HOPPER ABOVE   4.19 IN.
   BASE PRESSURE =        0.03 PSI


FOR OPTIMUM SOLIDS CONVEYING KEEP: BARREL TEMP=   190.50
                                   SCREW  TEMP=   190.50

                     SOLIDS CONVEYING ZONE
    L     PHI   ROBP    FRIB  FRIS  TESOUT     P     TSCREW
    IN    DEG LB/CU.FT              DEG F     PSI    DEG F
   1.80   2.5  32.94  1.235 0.556 217.1      8.0    190.5
   1.95   2.4  33.48  1.214 0.556 231.7     14.0    190.5
```

Table 15.17 (*Continued*)

L IN	TOUT DEG F	X/W	X/W INJ	GRATIO	PRESSURE PSI REAL	FULL	POWER HP	HOUT IN	TBAR DEG F	TVISC LB.SEC /SQ.IN	D.HEAT BTU/IN .MIN
					DELAY ZONE						
4.13	388.8				193.	193.		0.500	450.0	0.066	
					MELTING ZONE						
6.13	394.1	.848	.723	.788	228.	228.	.700	0.500	450.0	0.307	8.3
11.13	414.8	.611	.519	.570	228.	228.	2.4	0.500	450.0	0.180	-0.8
14.13	425.6	.490	.416	.459	1133.	1133.	3.4	0.500	450.0	0.162	2.0
19.13	438.1	.324	.274	.305	1620.	1620.	4.7	0.500	450.0	0.146	-0.2
20.00	439.8	.300	.253	.282	1703.	1703.	4.9	0.500	450.0	0.144	-0.1
25.00	451.0	.208	.169	.168	2248.	2248.	6.2	0.405	450.0	0.113	1.4
30.00	460.3	.123	.092	.080	2965.	2965.	7.4	0.310	450.0	0.099	2.8
35.00	470.8	.051	.032	.024	3931.	3931.	8.7	0.215	450.0	0.088	5.1
40.00	485.2	.000	.000	.000	5012.	5012.	9.9	0.120	450.0	0.078	8.0
45.00	495.5	.000	.000	.000	5887.	5887.	11.2	0.120	450.0	0.074	9.2
50.00	500.2	.000	.000	.000	6762.	6762.	12.5	0.120	450.0	0.072	9.8

```
MINIMUM TEMPERATURE=    500.198 DEG F
TEMP FLUCTUATION=         0.000 DEG F
FLOW INDEX (N)=           0.362
GRATIO INJ.=             0.000
RESIDENCE TIME:
     MELT POOL=    0.025114197 HRS


POWER BREAK-DOWN:
  SOLIDS CONV. =    0.342 HP
  MELTING      =    4.228 HP
  MIXING       =    4.693 HP
  FLIGHT CLEAR =    2.362 HP
  COMPRESSION  =    0.838 HP
  MIXING DEVICE=    0.000 HP
  PRESSURE FLUCTUATION=        691.16 PSI/CYCLE
  TOTAL CONDUCTED HEAT THROUGH BARREL=    165.7 BTU/MIN

  VISCOSITY WAS CALCULATED IN THE FOLLOWING SHEAR RATE RANGE
   SHEAR RATE       TEMP.         VISC.
   0.47354E 04    0.34050E 03   0.99653E-02
   0.26024E 02    0.45000E 03   0.27109E 00




  VISCOSITY WAS CALCULATED IN THE FOLLOWING TEMPERATURE RANGE
   SHEAR RATE        TEMP.         VISC.
   0.15182E 03    0.50079E 03   0.71750E-01
   0.24460E 04    0.23100E 03   0.18758E-01
  TYPE-IN INPUT
  $INPUT SHOT=4 $
```

Since pressure is so high and melting complete before the end of the screw, shot size can be increased to 4 oz.
  LIKE A PLOT?NO

N=100.0 RPM   OUTPUT= 120.00 LB/HR   SOAK TIME= 7.50 SECONDS

SCREW GEOMETRY

SECTION	LENGTH (IN)	CH.DEPTH (IN)	BAR.DIAM (IN)	NO OF HOLES	HOLE AREA (SQ. IN.)
1	20.0000	0.5000	2.5000		
2	20.0000	0.1200	2.5000		
3	10.0000	0.1200	2.5000		

Table 15.17 (*Continued*)

```
        ATTENTION - KEEP SOLIDS LEVEL IN HOPPER ABOVE    4.19 IN.
        BASE PRESSURE =        0.03 PSI


TO OPTIMIZE SOLIDS CONVEYING KEEP: BARREL TEMP=    190.50
                                   SCREW   TEMP=    190.50
```

```
                        SOLIDS CONVEYING ZONE
        L    PHI   ROBP    FRIB  FRIS  TESOUT      P     TSCREW
        IN   DEG LB/CU.FT                DEG F    PSI     DEG F
       2.00  3.2  32.52   1.240 0.612 204.4      3.6    190.5
       2.45  3.1  33.56   1.178 0.588 231.3     14.8    190.5
```

L IN	TOUT DEG F	X/W 	X/W INJ	GRATIO	PRESSURE PSI REAL FULL DELAY ZONE		POWER HP	HOUT IN	TBAR DEG F	TVISC LB.SEC /SQ.IN	D.HEAT BTU/IN .MIN
4.67	388.9				128.	128.		0.500	450.0	0.062	
					MELTING ZONE						
7.17	391.5	.858	.732	.797	142.	142.	.732	0.500	450.0	0.388	4.5
12.17	410.7	.664	.564	.619	142.	142.	2.4	0.500	450.0	0.227	-0.4
15.17	421.2	.561	.476	.525	1020.	1020.	3.5	0.500	450.0	0.200	2.4
20.00	433.1	.418	.354	.392	1479.	1479.	4.9	0.500	450.0	0.177	-0.2
25.00	445.8	.343	.280	.276	2018.	2018.	6.4	0.405	450.0	0.131	1.2
30.00	454.9	.268	.201	.173	2711.	2711.	7.9	0.310	450.0	0.115	2.2
35.00	463.6	.185	.117	.087	3568.	3568.	9.4	0.215	450.0	0.103	3.4
40.00	472.9	.090	.028	.025	4384.	4384.	10.8	0.120	450.0	0.096	4.6
45.00	484.0	.000	.000	.000	4904.	4904.	12.1	0.120	450.0	0.091	7.4
50.00	491.7	.000	.000	.000	5414.	5414.	13.3	0.120	450.0	0.088	8.4

```
MINIMUM TEMPERATURE=     491.690 DEG F
TEMP FLUCTUATION=          0.000 DEG F
FLOW INDEX (N)=            0.363
GRATIO INJ.=               0.000
RESIDENCE TIME:
     MELT POOL=    0.019399989 HRS

POWER BREAK-DOWN:
SOLIDS CONV. =    0.379 HP
MELTING      =    5.340 HP
MIXING       =    4.498 HP
FLIGHT CLEAR =    2.312 HP
COMPRESSION  =    0.745 HP
MIXING DEVICE=    0.000 HP
PRESSURE FLUCTUATION=       564.29 PSI/CYCLE
TOTAL CONDUCTED HEAT THROUGH BARREL=    124.1 BTU/MIN


VISCOSITY WAS CALCULATED IN THE FOLLOWING SHEAR RATE RANGE
SHEAR RATE       TEMP.        VISC.
  0.45136E 04    0.34050E 03  0.10338E-01
  0.21902E 02    0.45000E 03  0.30181E 00



VISCOSITY WAS CALCULATED IN THE FOLLOWING TEMPERATURE RANGE
SHEAR RATE       TEMP.        VISC.
  0.12034E 03    0.49151E 03  0.86451E-01
  0.23116E 04    0.23100E 03  0.19623E-01
TYPE-IN INPUT
$INPUT SHOT=4.5 $
```

Table 15.17 (*Continued*)

Since pressure is still high and all plastic fully melted before the end of the screw, an even larger shot size, 4.5 oz. is tried.

```
LIKE A PLOT?NO

N=100.0 RPM       OUTPUT=  135.00 LB/HR       SOAK TIME=  7.50 SECONDS


                           SCREW GEOMETRY
SECTION LENGTH  CH.DEPTH   BAR.DIAM NO OF HOLES   HOLE AREA
         (IN)    (IN)       (IN)                  (SQ. IN.)
   1    20.0000  0.5000     2.5000
   2    20.0000  0.1200     2.5000
   3    10.0000  0.1200     2.5000


    ATTENTION - KEEP SOLIDS LEVEL IN HOPPER ABOVE    4.19 IN.
    BASE PRESSURE =      0.03 PSI


FOR OPTIMUM SOLIDS CONVEYING KEEP: BARREL TEMP=    190.50
                                   SCREW  TEMP=    190.50


                      SOLIDS CONVEYING ZONE
     L    PHI   ROBP    FRIB  FRIS  TESOUT      P     TSCREW
     IN   DEG LB/CU.FT                DEG F    PSI     DEG F
    2.00  3.8  32.33   1.241 0.653 197.1      1.4    190.5
    2.90  3.6  33.70   1.152 0.608 232.0     16.1    190.5


  L     TOUT X/W  X/W GRATIO PRESSURE  POWER  HOUT  TBAR  TVISC D.HEAT
  IN    DEG F     INJ          PSI      HP    IN   DEG F LB.SEC BTU/IN
                            REAL  FULL              /SQ.IN  .MIN
                            DELAY ZONE
  5.15 388.9                122.  122.       0.500 450.0  0.060
  7.15 388.9                198.  198.       0.500 450.0  0.080
                           MELTING ZONE
  8.15 390.1 .873 .745 .811 198.  198. .820 0.500 450.0  0.458   2.3
 13.15 407.9 .697 .593 .650 198.  198.  2.5 0.500 450.0  0.261  -0.2
 16.15 418.1 .603 .512 .563 1054. 1054. 3.6 0.500 450.0  0.226   2.5
 20.00 427.9 .494 .419 .463 1413. 1413. 4.8 0.500 450.0  0.202  -0.3
 25.00 442.0 .433 .355 .347 1949. 1949. 6.4 0.405 450.0  0.142   1.3
 30.00 451.4 .372 .282 .240 2628. 2628. 8.0 0.310 450.0  0.124   2.2
 35.00 459.4 .304 .194 .142 3417. 3417. 9.7 0.215 450.0  0.112   3.0
 40.00 466.9 .216 .069 .058 4086. 4086. 11.3 0.120 450.0 0.106   3.1
 45.00 475.8 .041 .012 .011 4279. 4279. 12.7 0.120 450.0 0.102   4.7
 50.00 486.7 .000 .000 .000 4447. 4447. 13.8 0.120 450.0 0.097   7.3


MINIMUM TEMPERATURE=    486.695 DEG F
TEMP FLUCTUATION=         0.000 DEG F
FLOW INDEX (N)=           0.364
GRATIO INJ.=             0.000
RESIDENCE TIME:
     MELT POOL=   0.016739264 HRS


POWER BREAK-DOWN:
 SOLIDS CONV. =   0.467 HP
 MELTING      =   6.164 HP
 MIXING       =   4.368 HP
 FLIGHT CLEAR =   2.265 HP
 COMPRESSION  =   0.582 HP
 MIXING DEVICE=   0.000 HP
 PRESSURE FLUCTUATION=       426.59 PSI/CYCLE
 TOTAL CONDUCTED HEAT THROUGH BARREL=    98.0 BTU/MIN
```

Table 15.17 (*Continued*)

```
VISCOSITY WAS CALCULATED IN THE FOLLOWING SHEAR RATE RANGE
SHEAR RATE       TEMP.        VISC.
  0.43898E 04    0.34050E 03  0.10561E-01
  0.16924E 02    0.45000E 03  0.35402E 00

VISCOSITY WAS CALCULATED IN THE FOLLOWING TEMPERATURE RANGE
SHEAR RATE       TEMP.        VISC.
  0.10631E 03    0.48595E 03  0.95732E-01
  0.22423E 04    0.23100E 03  0.20104E-01
TYPE-IN INPUT
$INPUT SHOT=5 $
```

Increasing shot size further to 5 oz:

```
LIKE A PLOT?N?YES
WHEN EXTRUD IS DONE TYPE: GRAPH
```

N=100.0 RPM   OUTPUT= 150.00 LB/HR   SOAK TIME= 7.50 SECONDS

SCREW GEOMETRY

SECTION	LENGTH (IN)	CH.DEPTH (IN)	BAR.DIAM (IN)	NO OF HOLES	HOLE AREA (SQ. IN.)
1	20.0000	0.5000	2.5000		
2	20.0000	0.1200	2.5000		
3	10.0000	0.1200	2.5000		

```
ATTENTION - KEEP SOLIDS LEVEL IN HOPPER ABOVE    4.19 IN.
BASE PRESSURE =       0.03 PSI
```

FOR OPTIMUM SOLIDS CONVEYING KEEP: BARREL TEMP=   190.50
                                   SCREW TEMP=   190.50

SOLIDS CONVEYING ZONE

L IN	PHI DEG	ROBP LB/CU.FT	FRIB	FRIS	TESOUT DEG F	P PSI	TSCREW DEG F
2.00	4.3	32.25	1.229	0.686	193.2	0.5	190.5
3.45	4.1	33.64	1.144	0.623	232.0	15.3	190.5

L IN	TOUT DEG F	X/W	X/W INJ	GRATIO	PRESSURE PSI REAL	FULL	POWER HP	HOUT IN	TBAR DEG F	TVISC LB.SEC /SQ.IN	D.HEAT BTU/IN .MIN
					DELAY ZONE						
5.73	388.9				142.	142.		0.500	450.0	0.058	
7.73	388.9				95.	95.		0.500	450.0	0.077	
					MELTING ZONE						
8.73	388.9	.885	.756	.823	95.	95.	.784	0.500	450.0	0.537	0.0
13.73	405.3	.725	.617	.676	95.	95.	2.5	0.500	450.0	0.297	-0.1
16.73	415.1	.638	.542	.596	918.	918.	3.6	0.500	450.0	0.254	2.6
20.00	423.4	.551	.467	.515	1215.	1215.	4.6	0.500	450.0	0.227	-1.0
25.00	438.9	.505	.415	.404	1746.	1746.	6.3	0.405	450.0	0.152	1.4
30.00	448.4	.461	.352	.296	2403.	2403.	8.0	0.310	450.0	0.133	2.2
35.00	455.9	.417	.269	.194	3130.	3130.	9.8	0.215	450.0	0.121	2.9
40.00	462.4	.366	.119	.097	3642.	3642.	11.6	0.120	450.0	0.114	2.4
45.00	470.8	.112	.035	.031	3496.	3496.	13.1	0.120	450.0	0.111	2.6
50.00	480.8	.018	.005	.005	3296.	3296.	14.4	0.120	450.0	0.107	4.6

```
MINIMUM TEMPERATURE=    476.526 DEG F
TEMP FLUCTUATION=         4.242 DEG F
FLOW INDEX (N) =          0.363
GRATIO INJ.=              0.002
RESIDENCE TIME:
    MELT POOL=   0.014776366 HRS
```

Table 15.17 (*Continued*)

```
POWER BREAK-DOWN:
  SOLIDS CONV. =    0.432 HP
  MELTING      =    7.010 HP
  MIXING       =    4.296 HP
  FLIGHT CLEAR =    2.238 HP
  COMPRESSION  =    0.380 HP
  MIXING DEVICE=    0.000 HP
  PRESSURE FLUCTUATION=        272.49 PSI/CYCLE
  TOTAL CONDUCTED HEAT THROUGH BARREL=     74.5 BTU/MIN

VISCOSITY WAS CALCULATED IN THE FOLLOWING SHEAR RATE RANGE
SHEAR RATE       TEMP.          VISC.
  0.43633E 04    0.45000E 03    0.86283E-02
  0.13295E 02    0.45000E 03    0.41057E 00


VISCOSITY WAS CALCULATED IN THE FOLLOWING TEMPERATURE RANGE
SHEAR RATE       TEMP.          VISC.
  0.92685E 02    0.47974E 03    0.10719E 00
  0.21983E 04    0.23100E 03    0.20424E-01
TYPE-IN INPUT
$INPUT BEND=-1 $
```

Since solids content is only 0.5 % this simulation can be assumed to be a good design for the cycle. The pressure at the end of the screw is 3296 psi. With a 10 to 1 pressure ratio the hydraulic pressure must be maintained at 330 psi.

The fact that the screw also reciprocates was neglected in the above simulations. This can be corrected by simply typing in a value for RECLEN which will generate a simulation representing not only the screw at the start of the cycle, but also when the screw is retracted, that is at the end of the soak time.

```
OK, GRAPH
GO
 WHERE WOULD YOU LIKE THE GRAPH PRINTED?:
 (TTY,LPT,DSK)
TTY
          1
```

at which output or production rate an existing die will produce the most uniform sheet, how nonuniform the sheet will be at its highest uniformity or at any desired production rate, and how much of the nonuniformity an existing strainer bar or flexible lip can correct. It will also show how the corrective device should be set at every point across the sheet to fully correct the flow nonuniformity. This feature is particularly important when designing a die to be used for a variety of products and insuring that the die will be capable of producing the highest degree of uniformity for each. A sample session utilizing the DIEDES program is given in Table 15.19.

In the sample die design case it can be seen that certian key parameters are selected such as the production rate desired, the width and thickness of the sheet, the temperature of the melt entering the die from the extruder, and the manifold angle. The die design then gives the cross section profile of the manifold from the feed point or die center all the way to the side of the sheet. Naturally, if the pressure drop along the die is excessive or the manifold diameter impractical, some changes are made in the parameters used, and the design repeated until operating conditions and design become practical.

After the design is completed the computer program searches for the production rate at which flow uniformity across the sheet is most uniform. This is merely a check on the calculation procedure employed in the design.

Table 15.17 (*Continued*)

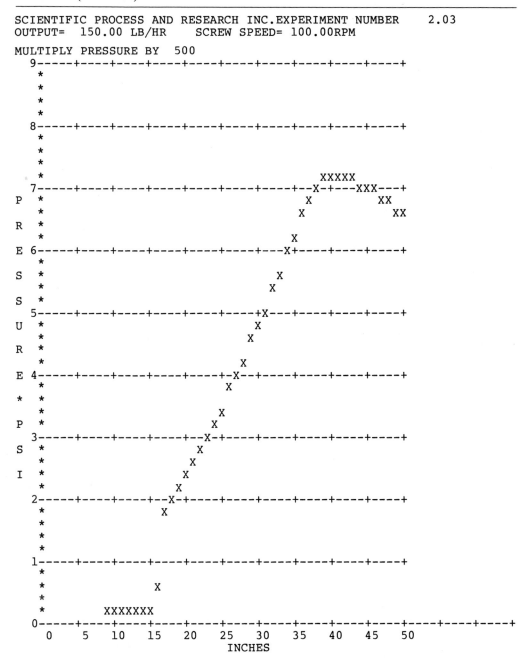

```
SCIENTIFIC PROCESS AND RESEARCH INC.EXPERIMENT NUMBER    2.03
OUTPUT=  150.00 LB/HR    SCREW SPEED= 100.00RPM

MULTIPLY PRESSURE BY   500
    9----+----+----+----+----+----+----+----+----+----+
    *
    *
    *
    *
    8----+----+----+----+----+----+----+----+----+----+
    *
    *
    *
    *                                      XXXXX
    7----+----+----+----+----+----+----+--X-+---XXX---+
P   *                                      X         XX
    *                                      X          XX
R   *
    *                                  X
E 6----+----+----+----+----+----+----+---X+----+----+----+
    *
S   *                            X
    *                            X
S   *
    5----+----+----+----+----+----+----+X---+----+----+----+
U   *                            X
    *                            X
R   *
    *                          X
E 4----+----+----+----+----+-X--+----+----+----+----+
    *                          X
*   *
    *                        X
P   *                        X
    3----+----+----+----+--X-+----+----+----+----+----+
S   *                      X
    *                      X
I   *                    X
    *                  X
    2----+----+----+--X-+----+----+----+----+----+----+
    *                X
    *
    *
    *
    1----+----+----+----+----+----+----+----+----+----+
    *
    *         X
    *
    *     XXXXXXX
    0----+----+----+----+----+----+----+----+----+----+----+----+
    0    5   10   15   20   25   30   35   40   45   50
                       INCHES
```

15.2.2.9 *Film Blowing*

In film blowing the product is extruded through a circular die and blown up internally by air to the desired diameter. It is then cooled, laid flat, and placed on rolls. The film is either used as is for large size bags or led through a heat sealing and cutting step where it is divided into a series of smaller bags. Often the double-walled film is cut on one or both sides to produce a flat film.

The problem of uniformity is the same as with sheet dies, with one exception, namely, that no strainer bar or flexible lip can be used. Control over uniformity is provided by moving the inner core of the die called a "mandrel" off center. This, however, is not a good enough

Table 15.17 (*Continued*)

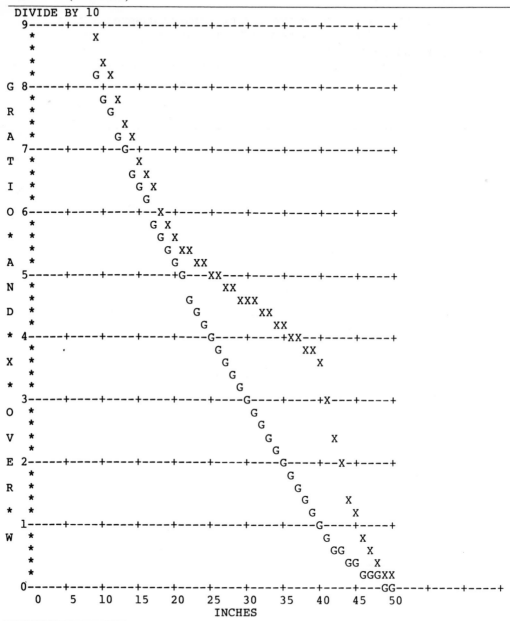

control tool as, contrary to a strainer bar that affects the gap size only at the location applied, moving the mandrel off center affects the gap size around the whole circumference of the die. The control over material distribution in a circular die is therefore considerably more complex and difficult than in a sheet or flat film die. However, the coat hanger principle can be effectively used for blown film dies as well. The plastic or other material used for the film is fed to the die from the side and is distributed over the periphery of the die through a channel in one or two directions. Here the channel or manifold geometry is accurately computed at every point along the die perimeter using SPR's DIEDES program. The manifold geometry will depend on the flow properties and density of the plastic, the production rate, film thickness, die diameter, and the manifold angle. It

Table 15.18

```
OK, SPIN
GO
   MELT SPINNING PROGRAM-MS3      RELEASE: OCTOBER 21,1977
     08-OCT-1978              SCIENTIFIC PROCESS & RESEARCH, INC.
                                EXPLANATION
   JW=100                    Take up to spinneret distance
   JN=400                    No.of time steps for computation
   NP=2                      Printing interval for transient solution
   NDIV=100                  No.of time increments to take up point
   DELTAX=2                  Length of axial increment
   SAVE DATA   FOR CABLE?YES
   PRINT OUT AT TAKE UP ? YES
   ROH=.83                   Polymer density
   CP=.7                     Heat capacity of polymer
   DEN=4.8                   Steady state take up denier
   TNOZ=270                  Spinneret temperature
   DNOZ=.6                   Spinneret diameter
   WDSP=500                  Take up speed
   BINF=.04                  Tensile viscosity coefficient
   E & R=3500   100          TENS.VISC.=BINF.exp(E/T)+R/T
   NO.OF AIR TEMP. POINTS=1
   X( 1),T( 1)               Axial location & temperature
      0      30

   NO.OF COOLING VELOCITIES=1
   X( 1),VY( 1)              Axial location & air velocity
      0      30
   ROH = 0.83 (GR/CUB.CM)      CP = 0.70 (CAL/GR/DEG)        DENIER = 4.80

   TNOZ = 270.(DEG.C)          WDSP=  500.(M/MIN)

   BINF= 0.039999999 (G.FORCE * SEC / SQ.CM)      E = 3500.(DEG.C)

   R= 100. (G.FORCE * SEC * DEG / SQ.CM)        F =0.17731

   NX=6    DNOZ =0.600


              STEADY STATE SOLUTION

   X(CM)    TIME(SEC)    T(DEG)    A(SQ.CM)

    0.00    0.00000    270.00    0.2827E-02
    2.00    0.60722    225.04    0.4241E-03
    4.00    0.71127    195.53    0.1331E-03
    6.00    0.74751    173.57    0.6100E-04
    8.00    0.76544    155.90    0.3502E-04
   10.00    0.77633    140.88    0.2329E-04
   12.00    0.78388    127.71    0.1714E-04
   14.00    0.78961    115.99    0.1356E-04
   16.00    0.79426    105.53    0.1132E-04
   18.00    0.79820     96.21    0.9817E-05
   20.00    0.80167     87.94    0.8765E-05
   22.00    0.80480     80.61    0.7995E-05
   24.00    0.80768     74.15    0.7412E-05
   26.00    0.81036     68.47    0.6955E-05
   28.00    0.81289     63.48    0.6589E-05
   30.00    0.81532     59.11    0.6426E-05

   FACTOR=-0.99773
```

In the transient solution below the printout columns mean the following:
Time (seconds)
Change in cross sectional area (%)

Table 15.18 (*Continued*)

Temperature (Deg.C)
Speed
Air spinning tension at the location specified after a 1 %
change in cooling air speed from steady state values.

TRANSIENT RESPONSE AFTER A STEPWISE INCREASE
IN NORMALIZED AIR SPEED VY-BAR BY UNITY.
THE UNITY STEP IS ASSUMED TO BE
UNIFORM ALONG THE WHOLE LENGTH OF THREAD LINE.

N=101   XS= 29.59 (CM)   DZ= 0.00815 (SEC)

J	TJ=J*DZ	A(N)	T(N)	V(N)	TF(TENSION)
0	0.0000	0.00000E 00	0.00000E 00	0.00000E 00	0.00000E 00
2	0.0163	-0.10264E-01	-0.32845E-01	-0.12988E-08	0.65663E-01
4	0.0326	-0.18617E-01	-0.78584E-01	-0.14712E-07	0.13281E 00
6	0.0489	0.72815E-02	-0.95339E-01	-0.57189E-07	0.18135E 00
8	0.0652	0.49879E-01	-0.77079E-01	0.10274E-07	0.20822E 00
10	0.0815	0.86171E-01	-0.49283E-01	0.10681E-07	0.22282E 00
12	0.0978	0.11181E 00	-0.25842E-01	0.89916E-07	0.23145E 00
14	0.1141	0.13340E 00	-0.33926E-02	-0.85420E-08	0.23731E 00
16	0.1304	0.14659E 00	0.10670E-01	-0.35070E-08	0.24259E 00
18	0.1467	0.15582E 00	0.20238E-01	0.31301E-07	0.24747E 00
20	0.1630	0.16349E 00	0.28112E-01	-0.43190E-07	0.25176E 00
22	0.1793	0.17068E 00	0.35666E-01	0.41779E-07	0.25524E 00
24	0.1956	0.17815E 00	0.43898E-01	0.27823E-07	0.25765E 00
26	0.2119	0.18674E 00	0.54125E-01	-0.25160E-07	0.25931E 00
28	0.2282	0.19325E 00	0.61841E-01	-0.72760E-08	0.26111E 00
30	0.2444	0.19766E 00	0.66742E-01	0.40978E-07	0.26305E 00
32	0.2607	0.20099E 00	0.70203E-01	0.17448E-07	0.26497E 00
34	0.2770	0.20378E 00	0.72990E-01	0.61133E-07	0.26681E 00
36	0.2933	0.20626E 00	0.75438E-01	-0.93380E-07	0.26854E 00
38	0.3096	0.20856E 00	0.77690E-01	0.59052E-07	0.27017E 00
40	0.3259	0.21070E 00	0.79783E-01	-0.10245E-07	0.27173E 00
42	0.3422	0.21271E 00	0.81742E-01	0.53071E-07	0.27322E 00
44	0.3585	0.21463E 00	0.83613E-01	0.52518E-07	0.27464E 00
46	0.3748	0.21650E 00	0.85446E-01	0.22919E-07	0.27600E 00
48	0.3911	0.21835E 00	0.87284E-01	-0.50699E-07	0.27729E 00
50	0.4074	0.22022E 00	0.89167E-01	0.36845E-07	0.27851E 00
52	0.4237	0.22210E 00	0.91101E-01	0.17331E-07	0.27968E 00
54	0.4400	0.22401E 00	0.93092E-01	-0.17419E-07	0.28080E 00
56	0.4563	0.22596E 00	0.95153E-01	0.44354E-07	0.28187E 00
58	0.4726	0.22797E 00	0.97302E-01	-0.39392E-07	0.28289E 00
60	0.4889	0.23004E 00	0.99546E-01	0.10565E-07	0.28387E 00
62	0.5052	0.23221E 00	0.10191E 00	-0.39290E-07	0.28480E 00
64	0.5215	0.23447E 00	0.10441E 00	-0.10084E-07	0.28570E 00
66	0.5378	0.23685E 00	0.10705E 00	0.79264E-07	0.28656E 00
68	0.5541	0.23936E 00	0.10984E 00	0.18612E-07	0.28738E 00
70	0.5704	0.24201E 00	0.11281E 00	-0.72760E-10	0.28817E 00
72	0.5867	0.24483E 00	0.11594E 00	-0.17608E-08	0.28893E 00
74	0.6030	0.24783E 00	0.11926E 00	0.20518E-08	0.28966E 00
76	0.6193	0.25102E 00	0.12275E 00	0.21537E-08	0.29037E 00
78	0.6356	0.25442E 00	0.12644E 00	-0.13126E-07	0.29106E 00
80	0.6519	0.25805E 00	0.13029E 00	0.24767E-07	0.29173E 00
82	0.6682	0.26193E 00	0.13430E 00	0.27707E-07	0.29240E 00
84	0.6845	0.26606E 00	0.13845E 00	-0.61584E-07	0.29307E 00
86	0.7008	0.27046E 00	0.14269E 00	0.56374E-07	0.29372E 00

Table 15.18 (*Continued*)

```
              TRANSIENT RESPONSE AFTER A STEPWISE INCREASE
              IN NORMALIZED AIR SPEED VY-BAR BY UNITY.
                 THE UNITY STEP IS ASSUMED TO BE
              UNIFORM ALONG THE WHOLE LENGTH OF THREAD LINE.
        N=101      XS= 29.59 (CM)      DZ=  0.00815 (SEC)
   J     TJ=J*DZ        A(N)              T(N)            V(N)        TF(TENSION)
  100    0.8148     0.34416E 00      0.21749E 00     0.86613E-07  0.29607E 00
  102    0.8311     0.27924E 00      0.92081E-01    -0.81200E-08  0.31378E 00
  104    0.8474     0.22700E 00      0.83475E-02     0.66328E-07  0.31619E 00
  106    0.8637     0.18903E 00     -0.42260E-01     0.70286E-07  0.31274E 00
  108    0.8800     0.16181E 00     -0.74986E-01     0.32218E-07  0.30892E 00
  110    0.8963     0.14186E 00     -0.98863E-01     0.27823E-07  0.30619E 00
  112    0.9126     0.12649E 00     -0.11798E 00     0.86729E-08  0.30458E 00
  114    0.9289     0.11371E 00     -0.13484E 00     0.83179E-07  0.30368E 00
  116    0.9452     0.10310E 00     -0.14933E 00     0.58208E-08  0.30292E 00
  118    0.9615     0.94373E-01     -0.16132E 00     0.59867E-07  0.30205E 00
  120    0.9778     0.87279E-01     -0.17098E 00     0.41618E-08  0.30106E 00
  122    0.9941     0.81512E-01     -0.17873E 00    -0.39872E-07  0.30014E 00
  124    1.0104     0.76483E-01     -0.18554E 00     0.17491E-07  0.29954E 00
  126    1.0267     0.71572E-01     -0.19254E 00    -0.23923E-07  0.29928E 00
  128    1.0430     0.66876E-01     -0.19945E 00     0.18044E-07  0.29913E 00
  130    1.0593     0.62575E-01     -0.20583E 00     0.74506E-08  0.29898E 00
  132    1.0756     0.58659E-01     -0.21165E 00     0.22526E-07  0.29883E 00
  134    1.0919     0.55066E-01     -0.21699E 00    -0.74506E-08  0.29867E 00
  136    1.1082     0.51734E-01     -0.22195E 00     0.32247E-07  0.29854E 00
  138    1.1245     0.48611E-01     -0.22661E 00     0.75495E-07  0.29844E 00
  140    1.1408     0.45678E-01     -0.23102E 00     0.27940E-08  0.29836E 00
  142    1.1571     0.42934E-01     -0.23515E 00     0.64058E-07  0.29829E 00
  144    1.1733     0.40363E-01     -0.23903E 00     0.23487E-07  0.29824E 00
  146    1.1896     0.37946E-01     -0.24269E 00    -0.10245E-07  0.29819E 00
  148    1.2059     0.35680E-01     -0.24613E 00    -0.21711E-07  0.29815E 00
  150    1.2222     0.33541E-01     -0.24938E 00    -0.13621E-07  0.29813E 00
  152    1.2385     0.31518E-01     -0.25247E 00     0.64465E-07  0.29812E 00
  154    1.2548     0.29605E-01     -0.25540E 00     0.55297E-07  0.29812E 00
  156    1.2711     0.27791E-01     -0.25818E 00     0.64902E-07  0.29812E 00
  158    1.2874     0.26078E-01     -0.26082E 00     0.10344E-06  0.29814E 00
  160    1.3037     0.24456E-01     -0.26332E 00     0.13737E-06  0.29816E 00
  162    1.3200     0.22914E-01     -0.26571E 00     0.27707E-07  0.29819E 00
  164    1.3363     0.21453E-01     -0.26798E 00     0.38621E-07  0.29822E 00
  166    1.3526     0.20067E-01     -0.27014E 00     0.27707E-07  0.29827E 00
  168    1.3689     0.18749E-01     -0.27220E 00     0.10105E-06  0.29831E 00
  170    1.3852     0.17492E-01     -0.27418E 00     0.70868E-07  0.29837E 00
  172    1.4015     0.16300E-01     -0.27606E 00     0.33237E-07  0.29843E 00
  174    1.4178     0.15154E-01     -0.27788E 00     0.76310E-07  0.29850E 00
  176    1.4341     0.14056E-01     -0.27964E 00    -0.16938E-07  0.29857E 00
  178    1.4504     0.13009E-01     -0.28131E 00     0.87195E-07  0.29865E 00
  180    1.4667     0.11996E-01     -0.28294E 00     0.92463E-07  0.29873E 00
  182    1.4830     0.11049E-01     -0.28447E 00     0.13883E-06  0.29879E 00
  184    1.4993     0.10193E-01     -0.28586E 00    -0.40163E-07  0.29881E 00
  186    1.5156     0.94992E-02     -0.28702E 00     0.25349E-07  0.29875E 00
```

will also depend on the gap size between the mandrel and outer wall of the die, which is selected by the use of the program. The length of the die depends on the manifold angle selected, which must be varied to insure that the designed manifold geometry is small enough to physically fit into the die. The manifold design is therefore repeated for several manifold angles.

Occasionally it happens that the desired production rate can not be obtained from a single manifold of practical size. In such a case several parallel manifolds must be used, resulting in a so-called "spiral mandrel" die.

15.2.2.10 Blow Molding Parison Dies

In a blow molding operation the extrudate leaves the extruder through a crosshead die

Table 15.18 (*Continued*)

```
          TRANSIENT RESPONSE AFTER A STEPWISE INCREASE
          IN NORMALIZED AIR SPEED VY-BAR BY UNITY.
             THE UNITY STEP IS ASSUMED TO BE
        UNIFORM ALONG THE WHOLE LENGTH OF THREAD LINE.
       N=101      XS= 29.59 (CM)      DZ=  0.00815 (SEC)
```

J	TJ=J*DZ	A(N)	T(N)	V(N)	TF(TENSION)
200	1.6297	-0.12032E-01	-0.32705E 00	0.62922E-07	0.30665E 00
202	1.6459	-0.15174E-01	-0.32834E 00	0.50990E-07	0.30372E 00
204	1.6622	-0.13904E-01	-0.32253E 00	0.81898E-07	0.30136E 00
206	1.6785	-0.11468E-01	-0.31689E 00	0.12858E-06	0.30031E 00
208	1.6948	-0.91605E-02	-0.31293E 00	0.66939E-08	0.30004E 00
210	1.7111	-0.73338E-02	-0.31031E 00	0.94122E-07	0.30003E 00
212	1.7274	-0.58624E-02	-0.30832E 00	0.80414E-07	0.29997E 00
214	1.7437	-0.44756E-02	-0.30636E 00	-0.61933E-07	0.29978E 00
216	1.7600	-0.31193E-02	-0.30431E 00	-0.59983E-07	0.29954E 00
218	1.7763	-0.18626E-02	-0.30239E 00	0.11511E-06	0.29936E 00
220	1.7926	-0.85885E-03	-0.30091E 00	0.17992E-06	0.29932E 00
222	1.8089	-0.20499E-03	-0.30004E 00	-0.70751E-07	0.29940E 00
224	1.8252	0.17736E-03	-0.29963E 00	0.55006E-07	0.29948E 00
226	1.8415	0.48699E-03	-0.29929E 00	0.28813E-08	0.29950E 00
228	1.8578	0.78457E-03	-0.29889E 00	0.85129E-07	0.29948E 00
230	1.8741	0.10696E-02	-0.29849E 00	0.71392E-07	0.29944E 00
232	1.8904	0.13141E-02	-0.29813E 00	0.15463E-06	0.29941E 00
234	1.9067	0.15107E-02	-0.29786E 00	-0.53085E-07	0.29939E 00
236	1.9230	0.16746E-02	-0.29763E 00	0.74186E-07	0.29937E 00
238	1.9393	0.18085E-02	-0.29745E 00	0.18947E-07	0.29936E 00
240	1.9556	0.19257E-02	-0.29730E 00	0.23458E-07	0.29934E 00
242	1.9719	0.20264E-02	-0.29716E 00	-0.34750E-07	0.29932E 00
244	1.9882	0.21017E-02	-0.29707E 00	-0.20285E-07	0.29931E 00
246	2.0045	0.21634E-02	-0.29699E 00	0.51252E-07	0.29930E 00
248	2.0208	0.22023E-02	-0.29696E 00	0.94529E-07	0.29929E 00
250	2.0371	0.22220E-02	-0.29695E 00	0.38504E-07	0.29928E 00
252	2.0534	0.22297E-02	-0.29697E 00	0.83586E-07	0.29928E 00
254	2.0697	0.22255E-02	-0.29700E 00	-0.64116E-07	0.29928E 00
256	2.0860	0.22137E-02	-0.29704E 00	0.13883E-07	0.29928E 00
258	2.1023	0.21878E-02	-0.29711E 00	-0.19791E-08	0.29929E 00
260	2.1185	0.21530E-02	-0.29719E 00	0.33673E-07	0.29929E 00
262	2.1348	0.21088E-02	-0.29728E 00	-0.11380E-06	0.29930E 00
264	2.1511	0.20551E-02	-0.29740E 00	0.46566E-09	0.29931E 00
266	2.1674	0.19987E-02	-0.29751E 00	0.73313E-07	0.29932E 00
268	2.1837	0.19311E-02	-0.29765E 00	0.75583E-07	0.29934E 00
270	2.2000	0.18549E-02	-0.29780E 00	0.52154E-07	0.29935E 00
272	2.2163	0.17756E-02	-0.29796E 00	-0.37428E-07	0.29937E 00
274	2.2326	0.16904E-02	-0.29813E 00	0.36962E-07	0.29939E 00
276	2.2489	0.16025E-02	-0.29831E 00	0.26834E-07	0.29940E 00
278	2.2652	0.15219E-02	-0.29848E 00	-0.57335E-08	0.29941E 00
280	2.2815	0.14653E-02	-0.29863E 00	-0.76892E-07	0.29941E 00
282	2.2978	0.14377E-02	-0.29874E 00	0.71304E-08	0.29939E 00
284	2.3141	0.14454E-02	-0.29882E 00	0.57626E-07	0.29939E 00
286	2.3304	0.14370E-02	-0.29896E 00	0.40600E-07	0.29944E 00

and generates a parison with an annular cross section. The free flowing hot parison is cut off by the closing mold and blown up to fully expand into the mold cavity. The wall of the bottle, container, or other blow molded part must have a uniform wall thickness in order to possess the required strength. On the other hand, by keeping the wall thickness at the minimum level but uniform all around the part, the amount of plastic required can usually be reduced to half its normal level—an enormous saving without sacrificing product quality.

The only way uniform wall thickness can be achieved is by starting out with a parison having a uniform wall thickness. The details of a blow molding parison die, or rather of the crosshead, are automatically designed by

Table 15.18 (*Continued*)

```
          TRANSIENT RESPONSE AFTER A STEPWISE INCREASE
          IN NORMALIZED AIR SPEED VY-BAR BY UNITY.
          THE UNITY STEP IS ASSUMED TO BE
        UNIFORM ALONG THE WHOLE LENGTH OF THREAD LINE.
       N=101     XS= 29.59 (CM)      DZ=  0.00815 (SEC)
```

J	TJ=J*DZ	A(N)	T(N)	V(N)	TF(TENSION)
300	2.4445	-0.46042E-02	-0.30682E 00	-0.19732E-07	0.29940E 00
302	2.4608	-0.28219E-02	-0.30363E 00	0.38126E-08	0.29919E 00
304	2.4771	-0.17198E-02	-0.30221E 00	0.15498E-06	0.29936E 00
306	2.4934	-0.11316E-02	-0.30172E 00	0.45489E-07	0.29954E 00
308	2.5097	-0.77799E-03	-0.30143E 00	-0.13068E-07	0.29960E 00
310	2.5260	-0.46912E-03	-0.30106E 00	0.14750E-06	0.29957E 00
312	2.5423	-0.16920E-03	-0.30062E 00	0.35769E-07	0.29952E 00
314	2.5586	0.83473E-04	-0.30024E 00	0.10573E-06	0.29950E 00
316	2.5749	0.22649E-03	-0.30006E 00	0.38795E-07	0.29954E 00
318	2.5911	0.25395E-03	-0.30008E 00	0.71042E-07	0.29961E 00
320	2.6074	0.21543E-03	-0.30020E 00	-0.53580E-07	0.29965E 00
322	2.6237	0.18583E-03	-0.30027E 00	0.62719E-07	0.29966E 00
324	2.6400	0.20104E-03	-0.30024E 00	0.46392E-07	0.29964E 00
326	2.6563	0.24766E-03	-0.30015E 00	0.11729E-07	0.29961E 00
328	2.6726	0.29310E-03	-0.30007E 00	0.71013E-08	0.29960E 00
330	2.6889	0.33291E-03	-0.30000E 00	0.29948E-07	0.29959E 00
332	2.7052	0.36217E-03	-0.29996E 00	-0.72177E-08	0.29959E 00
334	2.7215	0.39534E-03	-0.29991E 00	0.64058E-07	0.29958E 00
336	2.7378	0.42490E-03	-0.29987E 00	0.25029E-07	0.29957E 00
338	2.7541	0.45343E-03	-0.29983E 00	0.49651E-07	0.29957E 00
340	2.7704	0.47937E-03	-0.29979E 00	0.36205E-07	0.29956E 00
342	2.7867	0.50066E-03	-0.29976E 00	0.10798E-07	0.29956E 00
344	2.8030	0.51423E-03	-0.29975E 00	0.74739E-07	0.29956E 00
346	2.8193	0.52593E-03	-0.29974E 00	0.12864E-07	0.29956E 00
348	2.8356	0.53512E-03	-0.29973E 00	0.19907E-07	0.29956E 00
350	2.8519	0.53753E-03	-0.29973E 00	0.88854E-07	0.29955E 00
352	2.8682	0.54483E-03	-0.29973E 00	0.47730E-08	0.29955E 00
354	2.8845	0.54179E-03	-0.29974E 00	-0.31054E-07	0.29955E 00
356	2.9008	0.54185E-03	-0.29975E 00	0.44354E-07	0.29956E 00
358	2.9171	0.53712E-03	-0.29977E 00	0.12122E-06	0.29956E 00
360	2.9334	0.52784E-03	-0.29979E 00	0.61700E-07	0.29956E 00
362	2.9497	0.51818E-03	-0.29982E 00	0.18685E-07	0.29956E 00
364	2.9660	0.50659E-03	-0.29985E 00	0.43190E-07	0.29957E 00
366	2.9823	0.48980E-03	-0.29989E 00	-0.14028E-07	0.29957E 00
368	2.9986	0.47050E-03	-0.29993E 00	0.10649E-06	0.29958E 00
370	3.0149	0.44772E-03	-0.29998E 00	0.83819E-07	0.29959E 00
372	3.0312	0.42427E-03	-0.30004E 00	0.37166E-07	0.29959E 00
374	3.0475	0.40055E-03	-0.30010E 00	-0.40163E-08	0.29960E 00
376	3.0637	0.37049E-03	-0.30017E 00	0.41880E-07	0.29961E 00
378	3.0800	0.33337E-03	-0.30026E 00	0.13708E-07	0.29963E 00
380	3.0963	0.26986E-03	-0.30040E 00	0.48400E-07	0.29967E 00
382	3.1126	0.13488E-03	-0.30066E 00	0.93423E-08	0.29975E 00
384	3.1289	-0.12773E-03	-0.30113E 00	0.83557E-07	0.29990E 00
386	3.1452	-0.58201E-03	-0.30189E 00	0.20140E-07	0.30011E 00
.					
400	3.2593	0.14833E-03	-0.30005E 00	-0.98080E-08	0.29950E 00

SPR's DIEDES program. Here again the input data required consist of the flow properties and density function of the plastic, the parison thickness, and production rate. One can also specify some desired geometry or design parameters of the die that will be considered by the program in designing the details of the crosshead.

15.2.2.11 Die Analysis

While the die design programs concern the uniform distribution of the plastic melt or of other fluids across a sheet die or along the periphery of a circular or annular die, one is often interested in finding the pressure drop in the axial direction along various dies. SPR's DIEAN program performs this function. It

Table 15.19

```
OK, DIEDES
GO
      DIEDES  08-OCT-1978 SPR  RELEASE: JUNE 20,1978
 DATA FROM DSK? NO

 PHYSICAL PROPERTIES
 DATA FROM SPR DATA BANK? YES
 MATERIAL NUMBER= 315

 GEOMETRY
 WIDTH OF DIE(IN)= 100
 LAND LENGTH= .5
 LENGTH OF STRAINING BAR= 1
 LENGTH OF DISTRIBUTING CHANNEL= 0
 MANIFOLD ANGLE CONSTANT ?YES
 MANIFOLD ANGLE(DEG) = 15
 LAND DEPTH= .01
 DEPTH OF DISTRIBUTING SEC= .1
 STRAINING BAR DEPTH= .5
 VARYING HEIGHT SECTION LENGTH(IF ANY)=
 DISCHARGE END IS CIRCULAR SEGMENT ?NO
 SIDE FED CIRCULAR DIE ? NO
 OPERATING CONDITIONS
 TOTAL MASS FLOW RATE(LB/HR) = 1000
 MELT TEMP (DEG F) = 470
 ESTIMATED HEAD PRESSURE(PSI) = 8000
 EXIT PRESSURE(SEE MANUAL) = 15
 SENSITIVITY FROM 2-20; SENSITIVITY= 20
 NEW DIE? YES

 C O A T   H A N G E R   D I E   D E S I G N

 SPR MATERIAL #  315

WIDTH= 100.000IN   LAND DEPTH=  0.010 IN  DISTRIBUTOR DEPTH=  0.100 IN
LAND LENGTH=  0.500 IN      MANIFOLD ANGLE= 15.000 DEG

STRAINING BAR LENGTH=  1.000 IN   DISTRIBUTOR LENGTH=  0.000 IN

STRAINING BAR DEPTH=  0.500 IN  VARYING HEIGHT SECTION LENGTH=  0.000 IN

TEMP=  470.0 DEG F    PRESS=   8000.0 PSI  FLOW RATE=  1000.00 LB/HR

EXIT PRESSURE=   15.0 PSI      DENSITY=  67.25 +  -0.01722 * TEMP
551 POINTS HAVE BEEN CALCULATED
EVERY NTH POINT    WILL BE PRINTED
N= 15
```

neglects the distribution or uniformity problem in the direction perpendicular to the direction of flow, or considers this distribution to be uniform.

DIEAN can handle a variety of die cross sections. They can be rectangular or circular. In the direction of flow the cross section can remain constant or be decreasing or increasing. The program can also handle multiple perforated plates such as breaker plates, strainer bars, or flexible lips. The program yields pressure drop along each of the geometric sections as well as their cumulative values. It is therefore useful for computing the pressure drop at various processing temperatures and production rates, that is, in deriving the die characteristic equation for a particular die. It analyzes the die and determines which geometric section is responsible for most of the pressure drop. As soon as this is determined, the geometry can be modified to perform in a more reasonable way before the die is actually built, or even before a detailed design is performed using the DIEDES program. Because

Table 15.19 (*Continued*)

FLOW RATE PER WIDTH =   9.999998 LB/HR/IN

POINT	POSITION FROM CENTER (IN)	LENGTH OF DISTRIBUTOR (IN)	MANIFOLD DIAMETER (IN)
1	0.0000	13.3972	0.885939E+00
16	1.3636	13.0318	0.849180E+00
31	2.7273	12.6665	0.843665E+00
46	4.0909	12.3011	0.838026E+00
61	5.4545	11.9357	0.832258E+00
76	6.8182	11.5703	0.826357E+00
91	8.1818	11.2049	0.820306E+00
106	9.5455	10.8396	0.814105E+00
121	10.9091	10.4742	0.807733E+00
136	12.2727	10.1088	0.801187E+00
151	13.6364	9.7434	0.794461E+00
166	15.0000	9.3780	0.787530E+00
181	16.3636	9.0126	0.780394E+00
196	17.7273	8.6473	0.773021E+00
211	19.0909	8.2819	0.765409E+00
226	20.4545	7.9165	0.757535E+00
241	21.8182	7.5511	0.749370E+00
256	23.1818	7.1857	0.740894E+00
271	24.5454	6.8204	0.732083E+00
286	25.9091	6.4550	0.722894E+00
301	27.2727	6.0896	0.713300E+00
316	28.6364	5.7242	0.703246E+00
331	30.0000	5.3588	0.692683E+00
346	31.3636	4.9935	0.681550E+00
361	32.7273	4.6281	0.669768E+00
376	34.0909	4.2627	0.657244E+00
391	35.4545	3.8973	0.643866E+00
406	36.8182	3.5319	0.629481E+00
421	38.1818	3.1665	0.613898E+00
436	39.5454	2.8012	0.596864E+00
451	40.9091	2.4358	0.578022E+00
466	42.2727	2.0704	0.556863E+00
481	43.6364	1.7050	0.532607E+00
496	45.0000	1.3396	0.503967E+00
511	46.3636	0.9743	0.468546E+00
526	47.7273	0.6089	0.420933E+00
541	49.0909	0.2435	0.342540E+00
551	50.0000	-0.0001	0.166958E+00

of its versatility the DIEAN program is applicable to a large variety of dies. A sample run with the DIEAN program is given in Table 15.20.

15.2.2.12   *Product Cooling*

As a plastic product leaves the extrusion die it is shaped into the desired cross section, but it is still a melt and not a solid. Thanks to the high viscosity of many plastics it takes time for the melt to flow in the absence of a pressure drop or other forces. Some finite time is therefore available before the plastic product is deformed when it leaves the die. This time must therefore be used to effectively cool the product so that it retains its shape.

The cooling problem is particularly complex in cable jacketing and wire coating applications where several layers of conductors and insulators are placed concentrically. Several of these layers have different starting temperatures when cooling begins. Furthermore, the inner layers are usually heated up from the hot outer layer just deposited by the extruder. Precaution must be taken to prevent the melting and deformation of these inner layers, as this will alter their electrical properties. At the same time the outer layer is cooled externally, usually by pulling it through a water trough that must be long enough to lower the cable temperature below the melting point of

Table 15.19 (*Continued*)

```
PRESSURE DROP ACROSS LAND =   5337.0
HEAD PRESSURE =   5902.0

TEMP=   470.0 DEG F      PRESS=   5902.0 PSI     FLOW RATE= ?

EXIT PRESSURE=    15.0 PSI     DENSITY=   67.25 +  -0.01722 * TEMP
```

FLOW RATE & UNIFORMITY WITH A DIE
WHICH IS WORKING WITHOUT THE STRAINING BAR

POINT	MANIFOLD DIAMETER (IN)	LENGTH OF DISTRIBUTOR (IN)	FLOW RATE PER WIDTH (LB/HR/IN)	MANIFOLD PRESSURE (PSI)	FLOW RATE (LB/HR)
1	0.886	13.397	10.000	0.5902E+04	0.5000E+03
16	0.849	13.032	10.000	0.5887E+04	0.4864E+03
31	0.844	12.666	10.000	0.5872E+04	0.4727E+03
46	0.838	12.301	10.000	0.5857E+04	0.4591E+03
61	0.832	11.936	10.000	0.5842E+04	0.4455E+03
76	0.826	11.570	10.000	0.5827E+04	0.4318E+03
91	0.820	11.205	10.000	0.5812E+04	0.4182E+03
106	0.814	10.840	10.000	0.5797E+04	0.4045E+03
121	0.808	10.474	10.000	0.5782E+04	0.3909E+03
136	0.801	10.109	10.000	0.5767E+04	0.3773E+03
151	0.794	9.743	10.000	0.5752E+04	0.3636E+03
166	0.788	9.378	10.000	0.5737E+04	0.3500E+03
181	0.780	9.013	10.000	0.5722E+04	0.3364E+03
196	0.773	8.647	10.000	0.5707E+04	0.3227E+03
211	0.765	8.282	10.000	0.5692E+04	0.3091E+03
226	0.758	7.917	10.000	0.5677E+04	0.2955E+03
241	0.749	7.551	10.000	0.5662E+04	0.2818E+03
256	0.741	7.186	10.000	0.5647E+04	0.2682E+03
271	0.732	6.820	10.000	0.5632E+04	0.2545E+03
286	0.723	6.455	10.000	0.5617E+04	0.2409E+03
301	0.713	6.090	10.000	0.5602E+04	0.2273E+03
316	0.703	5.724	10.000	0.5587E+04	0.2136E+03
331	0.693	5.359	10.000	0.5572E+04	0.2000E+03
346	0.682	4.993	10.000	0.5557E+04	0.1864E+03
361	0.670	4.628	10.000	0.5542E+04	0.1727E+03
376	0.657	4.263	10.000	0.5527E+04	0.1591E+03
391	0.644	3.897	10.000	0.5512E+04	0.1454E+03
406	0.629	3.532	10.000	0.5497E+04	0.1318E+03
421	0.614	3.167	10.000	0.5482E+04	0.1182E+03
436	0.597	2.801	10.000	0.5467E+04	0.1045E+03
451	0.578	2.436	10.000	0.5452E+04	0.9090E+02
466	0.557	2.070	10.000	0.5437E+04	0.7727E+02
481	0.533	1.705	10.000	0.5422E+04	0.6363E+02
496	0.504	1.340	10.000	0.5407E+04	0.4999E+02
511	0.469	0.974	10.000	0.5392E+04	0.3636E+02
526	0.421	0.609	10.000	0.5377E+04	0.2272E+02
541	0.343	0.244	10.000	0.5362E+04	0.9086E+01
551	0.192	-0.000	10.000	0.5353E+04	0.0000E+00

the plastic. The problem is that of a nonsteady-state heat transfer, which must be solved numerically.

SPR's CABLE computer program was developed to solve cooling problems of the kind described above. It can handle not only a single cooling-water temperature but also a whole series of temperatures between the die and takeup equipment. One can therefore perform several computer simulations each with a different cooling arrangement. As a result one will obtain an optimum cooling profile or cooling strategy to minimize the total cooling length. This will naturally minimize the floor space required for the extrusion line. Alternately, one can use the technique to minimize the cooling-water requirements. Naturally, the most important objective is to

Table 15.20

```
OK, DIEAN
GO
   D I E A N    09-OCT-78   SPR - RELEASE:  OCTOBER 12, 1977
ARE DATA FROM A FILE?NO
FLOW RATE, LB/HR  =250
TEMP, DEGREES F   =500
LIP WIDTH, IN.    =20
LIP OPENING, IN.  =.03
NUMBER OF NON-PIPE SECTIONS=8
IS LIP OPENING FIXED?YES
CALCULATIONS EVERY 0.5 IN?YES
WILL STRAINING BAR SETTING BE CHANGED?NO
 NO.OF PIPE SECTIONS =4
 NO.OF HOLES IN BREAKER PLATE =30
ENTER GEOMETRIC DATA OF EACH SECTION IN INCHES
LENGTH, DEPTH  & WIDTH  1=.5   .03
LENGTH, DEPTH  & WIDTH  2=.1   .06
LENGTH, DEPTH  & WIDTH  3=.2   .08
LENGTH, DEPTH  & WIDTH  4=.2   .1
LENGTH, DEPTH  & WIDTH  5=.3   .14
LENGTH, DEPTH  & WIDTH  6=10   .25
LENGTH, DEPTH  & WIDTH  7=18   .32
LENGTH, DEPTH  & WIDTH  8=12   .5
 LENGTH & RADIUS  1 =1.5    .8
 LENGTH & RADIUS  2 =2.2   1.2
 LENGTH & RADIUS  3 =23    2.6
 LENGTH & RADIUS  4 =16    2.6
 LENGTH OF BREAKER PLATE & HOLE RADIUS =.5    .15
ENTER R0, R1, R2, R12 & SHEAR RATE TO START UPPER
 NEWTONIAN RANGE :
58  .3388E-4  -.0260126  .1E-5
ENTER A0,A1,A2,A11,A22 AND A12 BELOW FOR EQ.NO.  1
4.85337 -.7941432 -.01245915 -.01083274 .784E-6 .9E-3
ENTER A0,A1,A2,A11,A22 AND A12 BELOW FOR EQ.NO.  2
4.85337 -.7941432 -.01245915 -.01083274 .784E-6 .9E-3
ENTER A0,A1,A2,A11,A22 AND A12 BELOW FOR EQ.NO.  3
1
WANT TO SAVE YOUR DATA? NO
```

The physical properties used in this program are similar to those described earlier in this chapter and used for extrusion and die design. There are three viscosity equations used, one describing the low shear rate range which is often Newtonian, the other the general shear rate dependent range, while the third equation represents the transition between the two. In the example given the two ranges do not differ. The same equation is therefore used. The transition between the two can be at any point.

```
OUTPUT
------

OUTPUT,LB/HR  =    250.00
LIP OPENING   =      0.0300
```

	LENGTH INCHES	OPENING INCHES	WIDTH INCHES	P PSI	PRES.DROP PSI
1	0.50	0.030	20.000	614.44	614.4
2	0.10	0.060	20.000	674.73	60.3
3	0.20	0.080	20.000	718.17	43.4
4	0.20	0.100	20.000	743.51	25.3
5	0.30	0.140	20.000	764.43	20.9
6	10.00	0.250	20.000	1023.62	259.2
7	18.00	0.320	20.000	1210.38	186.8
8	12.00	0.500	20.000	1267.77	57.4
9	1.50	0.800		1311.20	43.4
10	2.20	1.200		1315.59	4.4
11	23.00	2.600		1326.95	11.4
12	16.00	2.600		1328.30	1.4
13	0.50	0.150		1348.55	20.2

Table 15.20 (*Continued*)

```
WANT TO SAVE YOUR DATA? NO
 TYPE-IN INPUT
 $INPUT OUTMAX=300, TEMP=400 $
OUTPUT
------

OUTPUT,LB/HR   =     300.00
LIP OPENING    =       0.0300
```

	LENGTH INCHES	OPENING INCHES	WIDTH INCHES	P PSI	PRES.DROP PSI
1	0.50	0.030	20.000	1159.97	1160.0
2	0.10	0.060	20.000	1280.50	120.5
3	0.20	0.080	20.000	1376.24	95.7
4	0.20	0.100	20.000	1434.85	58.6
5	0.30	0.140	20.000	1485.79	50.9
6	10.00	0.250	20.000	2170.13	684.3
7	18.00	0.320	20.000	2704.74	534.6
8	12.00	0.500	20.000	2879.38	174.6
9	1.50	0.800		2965.85	86.5
10	2.20	1.200		2978.82	13.0
11	23.00	2.600		3014.98	36.2
12	16.00	2.600		3020.61	5.6
13	0.50	0.150		3072.44	51.8

```
WANT TO SAVE YOUR DATA? NO
 TYPE-IN INPUT
 $INPUT BEND=1 $
```

Table 15.21

```
OK, CABLE
GO
 SPR CABLE ANALYSIS - RELEASE: AUGUST 3, 1978
 09-OCT-1978
 DATA ON DISK? NO
 NO OF LAYERS INCL.CORE :2
 PRINTOUT EACH NO OF SECONDS :5
 K=A*EXP(E/RT) - PRINT A & E :5.E17   35
 DESIRED DEGREE OF CROSSLINKING ON INNER SURFACE,
 FINAL COOLING TEMP. DEG F:250
 STEAM TEMP. DEG F : 212
 SINGLE COOLING WATER TEMPERATURE ?NO
# OF WATER TEMPS= 3
 TYPE-IN PAIRS OF COOLING WATER TEMP.(DEG.F) & TIME(SECS)
 AT THAT TEMP.
110 5   95 10   80 25
 HEAT TRANSFER COEF. STEAM/COATING-BTU/HR.SQFT.F: 73888
 HEAT TRANSFER COEF.COOLING MEDIUM/COATING-BTU/HR.SQFT.F
       (-1 IF VERY HIGH):  -1
```

	RADIUS IN	TH.COND BTU/ FT.HR.F	DENSITY LB/CUFT	HEAT CAP. BTU/LB.F	INIT.TEMP DEG F
OF LAYER NO. 1:	.5	2	.08	.24	450
OF LAYER NO. 2:	.6	.19	60	.55	450

```
 WANT TO SAVE DATA?NO
  C A B L E   P R O D U C T I O N
  NO OF LAYERS= 2
```

Table 15.21 (*Continued*)

```
A(1/SEC)=0.500E 18   E(KCAL/GMOLE)=35.000
MINIMUM CROSS-LINKING=0.000    FINAL TEMP(F)=250.00
STEAM TEMP=212.00   WATER TEMP=110.00   HEAT TR.COEF.= 0.739E 05

RADIUS  TH.COND  DENSITY  HEAT CAP.  INIT.TEMP
  IN      BTU/    LB/CUFT  BTU/LB.F    DEG F
        FT.HR.F
0.5000   2.000    0.080   0.240000    450.0
0.6000   0.190   60.000   0.550000    450.0
HEAT CAPACITIES FROM FILE?NO
   TIME   T1    T2    T3   CROSSLINK  CORT  TL2  TL3  TL4    TL5
    SEC    DEGREES  F    1    2    3        D E G R E E S    F
    0.00  450.  450.  450.   0    0    0  450.
    0.01  450.  450.  213.  100  100    3  450.
    5.00  419.  372.   95.  100  100    3  419.
   10.00  342.  296.   95.  100  100    3  342.
   14.99  274.  238.   80.  100  100    3  274.
   17.16  250.  218.   80.  100  100    3  250.
WANT TO REPEAT RUN WITH HIGHER ACCURACY? NO
ANY MORE DATA? YES OR NO:NO
```

insure that the product is totally solidified and no longer flows.

The computer program requires input data that describe the cross section of the product. If its cross section changes with distance or time, as is the case for example in melt spinning, information on this is required. The thermal diffusivity or, alternately, heat capacity, thermal conductivity, and density, of which thermal diffusivity is composed, must be supplied. If available, heat capacity should be given as a function of temperature. This is important in plastics processing operations where solidification and melting are involved, as the function also describes the heat of fusion, which is particularly significant in the case of crystalline plastics. A sample run utilizing the CABLE program is given in Table 15.21.

15.2.2.13   Curing and Cross-Linking

In many cable jacketing operations the cable is not led directly through a cooling trough as it exists in the extrusion die but is subjected first to high temperatures. This is done in a chemical reactor usually referred to as a cross-linking oven, or CV tube, where the cable is surrounded by steam under pressure to insure high temperature. However, a chemical reaction is not instantaneous, but its progress is dictated by various rate constants. These in turn are effected by temperature. Since the temperature inside the cable varies with axial and radial location, the residence time in the

reactor is the result of a complex computation.

The objective of the computation is to insure that the desired degree of cross-linking is achieved at every radial location of the outer layer. The CABLE program described under Product Cooling performs this computation. It enables the user to simulate the progress of cross-linking under various sets of operating conditions such as steam or other heating medium where the temperature is in contact with the outer cable surface.

In reality the situation is even more complex, as cross-linking does not stop when the cable leaves the reactor. The temperature of the inner surface continues to rise merely by heat flow into it from outer radial locations where temperature is higher. Furthermore, even when the outer cable surface is cooled, it takes time before the various radial locations inside the layer cool down, and as a result the reaction continues.

The CABLE program not only serves to determine the optimum cable velocity, residence time, and reaction temperature to reach a particular degree of cross-linking in the process, but also it gives information on the temperature profile of inner layers of insulators as well. One can therefore also see if the temperature at any radial location exceeds the melting point of the plastic. Should this become the case, the cable can be redesigned by adding a film or other layer having high heat-barrier properties.

Since the CABLE program computes both

Table 15.22

```
OK, CABLE
GO
 SPR CABLE ANALYSIS - RELEASE: AUGUST 3, 1978
 09-OCT-1978
 DATA ON DISK? NO
 NO OF LAYERS INCL.CORE :2
 PRINTOUT EACH NO OF SECONDS :5
 K=A*EXP(E/RT) - PRINT A & E :5.E17   35
 DESIRED DEGREE OF CROSSLINKING ON INNER SURFACE,
 FINAL COOLING TEMP. DEG F:200
 STEAM TEMP. DEG F : 430
 SINGLE COOLING WATER TEMPERATURE ?YES
 COOLING WATER TEMP. DEG F : 75
 HEAT TRANSFER COEF. STEAM/COATING-BTU/HR.SQFT.F: 73888
 HEAT TRANSFER COEF.COOLING MEDIUM/COATING-BTU/HR.SQFT.F
      (-1 IF VERY HIGH):  -1

 RADIUS TH.COND  DENSITY  HEAT CAP. INIT.TEMP
   IN     BTU/   LB/CUFT  BTU/LB.F   DEG F
        FT.HR.F
 OF LAYER NO.  1:
 .146    225      620       .094        275
 OF LAYER NO.  2:
 .224     .1       60       .55         275
WANT TO SAVE DATA?NO
   C A B L E   P R O D U C T I O N
   NO OF LAYERS= 2
   A(1/SEC)=0.500E 18  E(KCAL/GMOLE)=35.000
   MINIMUM CROSS-LINKING=0.970   FINAL TEMP(F)=200.00
   STEAM TEMP=430.00  WATER TEMP= 75.00  HEAT TR.COEF.= 0.739E 05

 RADIUS TH.COND  DENSITY  HEAT CAP. INIT.TEMP
   IN     BTU/   LB/CUFT  BTU/LB.F   DEG F
        FT.HR.F

 0.1460 225.000  620.000  0.094000     275.0
 0.2240   0.100   60.000  0.550000     275.0
HEAT CAPACITIES FROM FILE?NO
   TIME   T1    T2    T3   CROSSLINK  CORT  TL2  TL3  TL4    TL5
   SEC      DEGREES  F    1   2   3          D E G R E E S    F
    0.00 275.  275.  275.   0   0   0 275.
    4.98 278.  309.  430.  46  79 100 276.
    9.97 285.  336.  430.  77 100 100 281.
   14.95 293.  348.  430.  94 100 100 288.
   16.92 296.  351.  430.  97 100 100 291.
   21.90 298.  279.   75. 100 100 100 297.
   26.88 288.  223.   75. 100 100 100 293.
   31.87 277.  199.   75. 100 100 100 284.
   36.85 265.  186.   75. 100 100 100 273.
   41.84 254.  178.   75. 100 100 100 262.
   46.82 244.  171.   75. 100 100 100 251.
   51.80 234.  165.   75. 100 100 100 241.
   56.79 224.  160.   75. 100 100 100 231.
   61.77 216.  155.   75. 100 100 100 222.
   66.75 207.  150.   75. 100 100 100 213.
   71.46 200.  146.   75. 100 100 100 205.
 WANT TO REPEAT RUN WITH HIGHER ACCURACY? NO
```

cross-linking, that is heating and cooling, the input data must contain, in addition to the geometry and thermal properties of each layer, also the temperature dependence of the reaction constant.

A sample case with the CABLE program computing the progress of cross-linking followed by cooling is given in Table 15.22.

15.2.3 Computer Aided Research (CAR)

15.2.3.1 Design of Experiments

In any research and development project it is important that the key factors or variables in the problem be identified before the start of the experimentation. Once the variables are identified, the whole experimental program can be scientifically planned and the experiments laid out in such a fashion that they reveal a maximum amount of information on the problem to be investigated. We refer to such a plan as an experimental design.

There are various designs recommended by statisticians. The most common among these is the two-level factorial in which each variable is selected at a low and high level and all combinations of these are explored leading to 2^N experiments, where N is the number of variables. Denoting the low level of each variable by a minus ($-$) sign and the high level by a plus ($+$) sign a four-variable problem will require the 16 experiments summarized in Table 15.23.

However, a two-factorial experimental design is very limiting and permits only the simplest statistical tests. It is certainly inadequate for prediction and optimization purposes, which is the objective of process studies. Therefore each variable must be selected at more than two levels. However, this magnifies the whole experimental program by greatly increasing the number of experiments to be performed. For example in the four variable case when three levels of each are selected, the experimental program increases greatly from 16 experiments enumerated above to $(3)^4$ or 81 experiments. When each variable is selected at four levels the number of experiments increases to $(4)^4$ or 256, while with five levels the experimental program expands to $(5)^4$ or 625 experiments. Naturally the increase in the experimental program is even more pronounced when the number of variables is greater.

To compromise between the limited two-level factorial experimental design described with 16 experiments for a four-variable case and the five-level design with 625 experiments, the so-called rotatable central composite design (RCCD) was developed. It adds to the 16 experiments selected at -1 and $+1$ levels for each variable the center of design and the eight star points, that is the points at -2 and $+2$ levels of each of the four variables, while the other three variables are maintained at their zero level.

SPR's DESIGN computer program designs experiments according to the rotatable central composite design (RCCD). The user must supply only the minimum and maximum level desired for each variable, and the whole "experimental program" is laid out automatically by the computer. In addition to the design itself, the computer program also checks experiments already performed and compares these with the design itself. It determines which of the points performed are close enough to a designed experiment in the "experimental space" or "variable space" so that it can be considered to represent that experiment. The DESIGN program also determines which of the experiments previously performed are too close to each other so that they are merely replicates and do not represent

Table 15.23

| | VARIABLES | | | |
	1	2	3	4
1	$-$	$-$	$-$	$-$
2	$-$	$-$	$-$	$+$
3	$-$	$-$	$+$	$-$
4	$-$	$-$	$+$	$+$
5	$-$	$+$	$-$	$-$
6	$-$	$+$	$-$	$+$
7	$-$	$+$	$+$	$-$
8	$-$	$+$	$+$	$+$
9	$+$	$-$	$-$	$-$
10	$+$	$-$	$-$	$+$
11	$+$	$-$	$+$	$-$
12	$+$	$-$	$+$	$+$
13	$+$	$+$	$-$	$-$
14	$+$	$+$	$-$	$+$
15	$+$	$+$	$+$	$-$
16	$+$	$+$	$+$	$+$

Table 15.24

	VARIABLES				
	1	2	3	4	
1	-1	-1	-1	-1	
2	-1	-1	-1	+1	
3	-1	-1	+1	-1	
4	-1	-1	+1	+1	
5	-1	+1	-1	-1	
6	-1	+1	-1	+1	
7	-1	+1	+1	-1	
8	-1	+1	+1	+1	
9	+1	-1	-1	-1	
10	+1	-1	-1	+1	
11	+1	-1	+1	-1	
12	+1	-1	+1	+1	
13	+1	+1	-1	-1	
14	+1	+1	-1	+1	
15	+1	+1	+1	-1	
16	+1	+1	+1	+1	
17	0	0	0	0	Center of Design
18	0	0	0	-2	
19	0	0	0	+2	
20	0	0	-2	0	
21	0	0	+2	0	Star Points
22	0	-2	0	0	
23	0	+2	0	0	
24	-2	0	0	0	
25	+2	0	0	0	

additional information on the relationship between the variables. It also informs the user which of the designed experiments has not yet been performed.

15.2.3.2 Compounding and Formulation Development

In the development of new formulations or compounding of rubber, vinyls, and other materials such as ink and paints, or in color matching, an enormous number of experiments must be performed before the final compound with the right properties is derived. This development process is becoming more and more difficult as the number of ingredients and the number of responses or compound properties that must be maintained grows.

A typical vinyl formulation problem, for example, will have several independent variables representing levels at which experiments must be performed to learn anything about the "system." These variables can contain information on the base resin—for example,

its molecular weight, molecular weight distribution, flow property, and information on additives such as plasticizer content and stabilizer content. The responses will represent the results of the compounding such as the various physical properties of the compound.

In the case of a rubber formulation problem, for example, the independent variables will contain composition variables such as carbon black, oil, Hi-Sil, sulphur, and accelerator. They may, however, also contain processing variables such as cross-linking time, cross-linking temperature and pressure. Typical responses in a rubber formulation case are: 300% modulus, reversion start, composition set, cost, run temperature, scorch, ring tear, shore hardness.

SPR's OPTICOMP computer program was developed to handle these problems. As input it requires a very small number of experimental data. For example, when the number of variables is four, that is the number of ingredients is five, it requires 16 experiments;

when the number of experiments is six, 29 experiments; and with 10 variables a mere 67 experiments are needed. Also, the experiments do not have to be properly distributed according to a rigid experimental design. The layout of experimental points proposed by the DESIGN program is merely a recommendation. The experimenter can select points from this layout or from any other or even use experiments that were not the result of a systematic layout. Naturally, compositions or experimental points selected from a statistically sound experimental design carry more information than do others.

Nevertheless, as discussed, not all designed points are needed when OPTICOMP is used. One can cut the number of experiments to a bare minimum. OPTICOMP is in reality a system that uses a very small number of experimental points and asks the experimenter for the objectives of his research project. Based on these objectives OPTICOMP then selects or proposes new experiments to be performed. The new compositions or experiments are predicted by OPTICOMP to yield results closest to the researcher's objectives. The experimenter then performs one or more of the newly proposed experiments and, if the results are not fully satisfactory, adds them to his other experimental points previously collected. These are then fed to OPTICOMP for further evaluation. Reinforced by this expanded data base OPTICOMP predicts further experiments or compositions to be tried. Should the results of these new experiments still not be fully satisfactory with respect to all objectives, the lastest experiments are again added to the data base and the procedure repeated until the objectives are fully satisfied. OPTICOMP represents, therefore, an experimental technique or procedure that speeds up an experimental program and minimizes the number of experiments to solve complex problems.

15.2.3.3 Modeling of Experiments

In research and development activities it is often very important that the results of experimentation be represented in the form of mathematical equations. Such equations lend themselves to differentiation, integration, and other complex mathematical manipulations. A statistical tool named regression analysis in its computerized form is often used for this purpose. However, most presently available regression analysis "packages" suffer considerable shortcomings.

The most often experienced shortcoming of regression programs is their limitation to a specific form for the equation to be fitted to experimental data. These forms are usually much too simple to be able to reliably represent complex experimental data.

Another shortcoming often experienced with regression analysis is the complexity by which proposed mathematical equations or models can be described. This process is often referred to as variable transformation, which must usually be supplied in coded form. Such coding is quite difficult and requires a specialist in just that one program. It is obvious that very few organizations are big enough to maintain specialists to perform the transformations for each scientist or engineer wishing to use such a computer program. Yet without such assistance only minimal use can be made of such canned programs.

Another very important shortcoming of regression analysis is that the goodness of fit of the equation or model is simply measured by statistical means, that is how close the multiple correlation coefficient is from 1, what the standard error of the response is, and what the standard error of each coefficient is. While these are necessary measures of the degree of fit, they reflect only on the closeness of the function to the measured point and do not give any clue on the behavior of the function derived outside of the experimental range—or even inside it but not in the vicinity of measured points.

SPR's LABMAN computer program tries to solve all these shortcoming. It makes variable transformations very easy so that the technique can be rapidly acquired by any engineer. As seen in the section on material density (Section 15.2.1.4) the transformations can be written directly as mathematical equations, requiring only a few minutes of explanation. Because of this simplicity the method is also easily memorized and recalled whenever needed without undue refreshing courses required each time.

Because of this simplicity of defining variable transformations, the user can simply observe the fit of the model to the experimental data and decide if it is adequate or if the form of the equation should be modified. This model modification can therefore be performed on line while making the decision dynamically as soon as the results of the fitting are seen. Naturally the procedure can be repeated as many times as needed until a good enough equation is found. This procedure is facilitated by a program called COLED, which is a column editor of data. It permits one to enter his data column by column and insert repetitious sequences by a single command. It also permits copying columns from other data files, shifting columns, dropping them, and mostly easily editing the data files and eliminating errors introduced during typing. A special feature of this editor is the transformation mode, which takes existing numerical data files and appends the variable transformation instructions to it.

After the model is finally arrived at the user can view the shape of the equation derived in graphical form and see if it satisfies known or desired physical constraints. For example, one may have collected data at a medium range of a variable such as speed where the function will fit very well, but it will defy reasonable behavior in the low and high ranges. Since these mathematical equations are frequently used to study behavior under conditions outside the measurable range, its reasonableness must be assured. One can therefore alter the shape of the function to exhibit a particular behavior at any level of the many variables usually investigated. Natu-

rally, this procedure would not be possible without access to graphical representation of the functions.

15.3 SUMMARY

Although computers have penetrated the plastics processing industry to a very significant degree, this process will no doubt increase even more rapidly in the next few years. More and more people will take advantage of the tools already available, making processing better, more economical, and also more innovative. The total approach to experimentation will change as it will be done widely on a computer instead of with costly machinery and plastics. This will no doubt lead to much better processing conditions and to better screw, extruder, die, and molding machine design.

In the area of research the new computerized techniques will lead to better compounds and to the ability to produce more tailor-made compounds for specific applications, both by better compounding and by the selection of better reaction conditions. All this will be facilitated by computers becoming even more powerful than they presently are and as a result even less expensive. When we add to this the rapidly improving communications technology, we see that access to special purpose computerized technology will become universal.

15.4 REFERENCES

1. I. Klein and D. I. Marshall, Computer Programs for Plastics Engineers, Reinhold, New York, 1968.
2. Scientific Process & Research, Inc., Somerset, New Jersey.

## CHAPTER 16

Physical Testing for
End-Use Applications

R. E. EVANS

American Cyanamid Company
Stamford, Connecticut

16.1	Introduction	780
	16.1.1 Rationale for Physical Testing	781
	16.1.1.1 Types of Test Methods	783
	16.1.1.2 Assessment of Standard Test Methods	784
16.2	Selecting Meaningful Physical Test Procedures	789
	16.2.1 Material for a Known Application	790
	16.2.2 General Evaluation of a Material	790
16.3	Interpretation of Results	791
16.4	Influences of Specimen Preparation, Sampling, and Geometry	791
	16.4.1 Specimen Preparation	791
	16.4.2 Specimen Sampling	794
	16.4.3 Specimen Geometry	794
16.5	Bulk Mechanical Properties	794
	16.5.1 Mechanical Spectrometry Measurements	797
	16.5.1.1 Instrumentation	798
	16.5.1.2 Thermomechanical Profiles	800
	16.5.2 Quasi-Static Measurements	803
	16.5.2.1 Mechanical Testing Machines	803
	16.5.2.2 Stress-Strain Plots	804
	16.5.2.3 Tensile Tests	807
	16.5.2.4 Flexural Tests	809
	16.5.2.5 Compressive Tests	810
	16.5.2.6 Shear Tests	811
	16.5.2.7 Biaxial Stress Tests	811
	16.5.2.8 Fracture Toughness Tests	814
	16.5.3 Impact and High-Speed Measurements	815
	16.5.4 Long-Term Measurements	820
	16.5.4.1 Static Creep Tests	820
	16.5.4.2 Stress Relaxation	823
	16.5.4.3 Stress Rupture and Endurance Limit	826
	16.5.4.4 Prediction of Long-Term Strength Behavior	828
16.6	Surface Mechanical Properties	829
	16.6.1 Indentation Hardness	830
	16.6.2 Scratch Hardness	833
	16.6.3 Abrasion, Mar and Wear Resistance	834
	16.6.3.1 Rubbing Contact	834
	16.6.3.2 Scraping or Cutting	835
	16.6.3.3 Erosion	836
	16.6.4 Friction	837
16.7	Electrical Properties	838
	16.7.1 Permittivity and Loss Factor	838
	16.7.2 Insulation Resistance and Resistivity	841
	16.7.3 Dielectric Breakdown	844
	16.7.4 Tracking Resistance	849
	16.7.4.1 Wet Tracking	849
	16.7.4.2 Arc Resistance	850
	16.7.5 Electrostatic Charge	850
16.8	Thermal Properties	852
	16.8.1 Thermal Expansion	852
	16.8.2 Thermal Conductivity	853
	16.8.3 Specific Heat	856
	16.8.4 Flammability	857
	16.8.4.1 Ignition	859
	16.8.4.2 Burning	860
	16.8.4.3 Heat Release	861
	16.8.4.4 Smoke Generation	862

16.9	**Optical Properties**	862
	16.9.1 Index of Refraction	862
	16.9.2 Reflection and Transmission—	
	General Analysis	864
	16.9.2.1 Goniophotometer	864
	16.9.2.2 Integrating Sphere	864
	16.9.3 Reflection and Transmission—	
	Specific Methods	864
	16.9.3.1 Gloss and Diffuse	
	Reflectance	866
	16.9.3.2 Transparency and Haze	867
	16.9.4 Color	868
	16.9.5 Illuminants and Spectral	
	Receivers	869
	16.9.5.1 Illuminants	869
	16.9.5.2 Spectral Receivers	869
16.10	**Other Physical Properties**	870
	16.10.1 Permeation	870
	16.10.1.1 Water Vapor	
	Transmission	873
	16.10.1.2 Gas Transmission	874
	16.10.2 Density	876
	16.10.3 Permanence	876
16.11	**Summary**	876
16.12	**Glossary of Symbols**	877
16.13	**References**	878

16.1   INTRODUCTION

The use of plastics as principal end-use materials for products in many different industries is increasing rapidly. Various applications in building and construction, apparel, packaging, medicine, agriculture, adhesive bonding, recreational goods, transportation, coatings, and appliances and housewares, and many others, all involve the use of vast quantities of plastics. The reason for such wide use of plastics is because extensive chemical possibilities have provided the polymer chemist with opportunities to tailor-make products having various combinations of physical attributes.

The value of any particular plastic for a specific application, however, can be recognized only by an understanding of the physical properties that are required for its successful use. The evaluation of physical properties from the viewpoint of obtaining critical engineering data must be understood. It is therefore an objective of this chapter to review various testing methods and to provide some guidelines for the methodology necessary for meaningful physical testing for evaluating end-use performance. It is hoped that this will provide the reader with a deeper understanding of the types of physical tests necessary to obtain properties beyond those described in "typical value" tables. This is necessary because the concept of physical testing has been oversimplified in such tables. It must be emphasized at the beginning that physical properties of polymers cannot be presented as unchanging values that universally describe material behavior. This is a major theme of this chapter.

One reason for this oversimplification is that many of the tests (and testing equipment) used to characterize plastic materials were originally adopted with minor modifications from techniques used for the traditional materials of construction, such as steel, wood, concrete, and paper. However, materials of construction based on high polymers require special techniques for characterization. Devoted work by physicists for the last 40 years has continuously demonstrated an acute awareness of this point. Such devotion is still required because when plastics are stressed mechanically, excited electrically, optically, or magnetically, or activated thermally, the molecules rearrange themselves at a rate dependent on both time and temperature as well as past history. In addition, the degree of molecular rearrangement is vastly altered by such common plasticizers as moisture and solvents. In mechanical tests, such molecular rearrangements are described as viscoelastic phenomena. In electrical measurements such molecular rearrangements are described as polarization phenomena. Other general phenomena such as microstructural changes, photodegradation, cross-linking, and polarization, result from optical and magnetic excitations. Unfortunately, many test technologists choose to overlook such phenomena in favor of obtaining oversimplified single-point data. In addition, the significance of the test may not be considered.

Another reason for the general oversimplification of plastics test methods has been the failure of the testing community to actively seek and integrate the input of designers and product engineers during the development of standards.

In this chapter the basic rationale necessary to develop a sound philosophy of testing is first developed. This is followed by an assessment of how to test and what to test for so that prime indicators of performance are obtained. Included in this discussion are the selection of test procedures, influences of specimen sampling and preparation, and the interpretation of data. The remainder of the chapter surveys each major area of physical testing to indicate both tests essential for establishing the suitability for various end-use applications and the precautions necessary to obtain meaningful data from various types of tests. No attempt is made to include specific information on a wide range of plastics because such "property" data are not always available, and if available, would constitute a book in itself. However, appropriate illustrative examples are presented.

16.1.1 Rationale for Physical Testing

Physical testing, in phenomenological terms, involves measuring the output response of a body under given boundary conditions and in a given environment when it is subjected to external input physical excitations, as shown in Figure 16.1. These excitations may be such everyday occurrences as force, displacement, voltage, heat, or luminous radiation. These may be in many simple forms such as step functions, sinusoids, ramps, pulses, and gates, or they may be more complex combinations of such functions. The amplitude may be small so that molecules are simply displaced or may be large enough to rupture the bonds between the molecules. Boundary conditions, which also influence the response, may be considered as geometric constraints placed on the specimen or any conditions that must be satisfied at the specimen faces. The time-form of the excitation functions, the boundary conditions, and the environment have enormous influence on the nature of the response, and therein lies the source of the apparent complexity and confusions associated with many of the test methods (1). It is one of the main tasks of the physicist working with plastics to express this complex behavior in terms of a few identifiable and measurable parameters. When a laboratory test fails to predict performance, the physicist has failed to correctly match the input excitation, boundary conditions, and environment to those of end use.

There are two distinct approaches available for defining a physical property; one can select a particular type of excitation function and

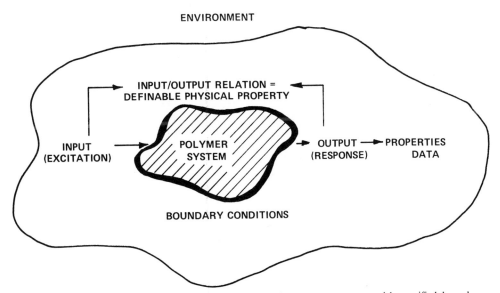

Figure 16.1 Physical testing can be represented as an input-output system with specified boundary conditions, after Turner (1). Either the output alone or the ratio of output to input may be considered as a definable physical property for given input excitation, boundary conditions, and environment.

merely display the response function as a "physical property" or one can measure the input-output relationship. For example, consider the measurement of tensile strength where the output response is displayed and modulus where the input-output relationship is measured. In any case, for a complete characterization of a physical attribute, the output response must be determined under various environmental conditions such as temperature, humidity, and radiation, and the observed responses interrelated. It is hoped that by using the integrated or systems approach all physical tests, regardless of how specific, can be described simply as input-output measurements under given boundary and environmental conditions and the vagueness surrounding many tests will be removed.

One reason for testing is to assess the behavior of a material or article in relation to the function it is to fulfill, that is to establish

the suitability for a given application or purpose. Most simply, the purpose may be performance related, endurance related, or sensory related as shown in Figure 16.2. More complex purposes may require a combination of any two categories or of all three categories. Other reasons for testing include quality control, obtaining standard reference or typical value data, and research to obtain a basic understanding of polymer structure. The difference between the first case and the other three is one of philosophy as well as one of technique.

Testing a plastic to determine suitability for a given purpose requires techniques that can be used to establish that a material or component fulfill its function by retaining all significant physical attributes throughout its expected period of service. These techniques may be standard methods (modified when necessary to conform to the service condition)

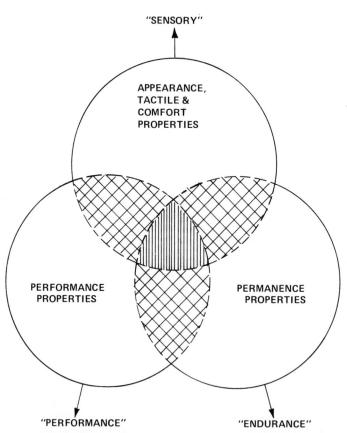

Figure 16.2 The suitability of a material or design for an end-use application is related to only its performance, permanence, or sensory properties or some combination thereof, after Von Krevelen (2).

or may require development of a unique method. The techniques may be designed to measure intrinsic polymer properties, processing properties, or product (article) properties.

Quality control tests and tests used to obtain standard reference data require techniques that can be used to assure process uniformity or to provide comparison data for materials under standard conditions. These tests are usually oversimplified and merely provide limited answers. Research of polymer structure requires techniques providing results that can be interpreted on a molecular scale. The relationship of these tests, however, may not be sufficiently close to the actual use requirements of plastics. While they have a definite place in the testing picture, their limitations must be carefully considered.

16.1.1.1 *Types of Test Methods*
Test methods employed in physical testing can be divided into three classes: fundamental tests, component or practical tests, and hybrid tests (3).

"Fundamental tests" yield data concerning basic physical properties. They permit reliable calculations under simple conditions that can be analyzed and expressed in absolute units. These tests provide a means for defining and evaluating and physical properties independent of the design and dimensions of either the test specimen or the test equipment that is, the boundary conditions. Fundamental test results may be dependent on the shape of the input excitation and on the environment. Uniaxial creep tests where a simple relationship of stress versus strain is obtained would be one example of a fundamental test.

"Component tests" yield data from a finished part to allow the prediction of material behavior in complex service conditions. Although this may be the most desirable way to test, component parts are usually not available or the component cannot be tested per se. Also, component tests are usually slow, expensive, and so inaccurate that many tests are required to give reliable average results. The main restriction, however, is that component properties may be related to more than one fundamental property, and when failure occurs it is not possible to determine which fundamental property must be improved. Component tests are vastly improved, however, when instrumentation is available for data acquisition so that combined effects may be isolated and specific properties calculated from the component data. Opening and closing a molded hinged lid at low temperatures until failure occurs is an example of a component test.

"Practical tests" are special forms of component tests. In practical tests a specimen is modeled so that the input excitation, boundary conditions, and environment of the test are identical with those to which the component would be exposed in service. They are valid for specific applications only. While not as expensive as component tests, they are still limited because failure may depend on more than one fundamental property. Bearing strength measurements, where load is applied to a panel by means of a pin having the same diameter and placement as the end-use application, is one example of a practical test. Practical tests may sometimes be used with caution for obtaining comparative data when the boundary conditions are the same as the component but the actual test specimen dimensions are not. Comparisons of the drop impact strengths of small clamped specimens in order to select a candidate material for a large cover plate is one example.

"Hybrid tests" yield data from special specimen geometries that are used for obtaining general information supposedly simulating particular service conditions. However, these tests have no physical validity because the boundary conditions imposed on the specimen are not the same as those imposed on the component during actual use. The measured physical property may be dependent on the design and dimensions of both the test specimen and test equipment. The hybrid test can be considered as a practical test for only one specific set of boundary conditions. Scratch-resistance tests using scribers cut from arbitrary materials and of arbitrary geometry are examples of hybrid tests.

The selection of the test method used is governed by the reason for testing. For example, in quality control testing the measurement of variability is more important than the relevance of the result. Thus, hybrid tests are acceptable. Methods to obtain typical values can also use hybrid tests, although fundamental tests are preferred. However, for research purposes and to establish suitability for a given application, fundamental or practical tests are recommended whenever possible, even if they are more time-consuming and more expensive than a hybrid test. A sound philosophy of engineering measurement is dependent on this.

Despite warnings, there still exists a great tendency among many to ascribe generality to hybrid tests for research and suitability testing. This practice is discouraged. The importance of avoiding the pitfalls associated with hybrid tests has long been recognized by other experimentalists (1, 4, 5). For example, the following is part of a 1937 commentary by Mooney (5):

A hybrid test is one that stands somewhere between the practical and the scientific. It does not represent faithfully any particular service condition; and it is not designed so as to permit the calculation of any basic physical property. Most such tests in common use have been designed with half an eye on service conditions, and an eye and a half on speed and accuracy of the laboratory test. If the measurements have any scientific value, it is largely a result of chance, or oversight.

Concerning the hybrid type of test, there is little, in the author's opinion, that can be said to justify it. If the test neither resembles closely any service conditions nor yields data concerning the basic physical properties, the results will neither predict with certainty the behavior of the material in complex service conditions, nor permit reliable calculations in simple conditions that can be analyzed.

An example illustrates how each of these four methods could be selected by an experimentalist trying to solve the following problem: A lavatory sink made of calcium carbonate–filled polyester may crack during its service life because of repeated thermal shock when first hot and then cold tap water impinges on the bottom of the bowl. The best resin-filler system must be selected to obtain the maximum cycles to failure. The following examples illustrate how these tests could be employed to investigate the problem:

Fundamental Test Small test coupons could be tested for fracture toughness, modulus, thermal expansion, and specific heat and combined to calculate thermal shock resistance.

Component Test The sink, or a model thereof, could be molded and subjected to thermal cycling by *partially filling* back and fourth with water at the two temperature extremes until failure occurs (6).

Practical Test The central area of a test coupon containing a hole could be thermally shocked with water between temperature extremes until a crack initiates in the part at the edge of the hole.

Hybrid Test A small test coupon could be cycled between hot and cold limits in an environmental chamber and the specimen examined periodically for cracks.

Obviously, the boundary conditions of the hybrid test do not represent a real situation since no mechanical constraint is offered against expansion or contraction and no temperature gradients exist. Thus any results obtained would have doubtful value. Component or practical tests represent tests with boundary conditions encountered in end use, but several hundred cycles may be required to cause rupture. The fundamental test, on the other hand, permits the calculation of thermal shock resistance using small laboratory specimens without the need to construct a large component.

This above example also serves to emphasize an important principle: Fundamental tests can be developed to measure in absolute units any clearly defined physical property possessed by the test material under normal conditions. The apparatus required for such tests may be more complicated than that needed for other types of tests but such tests also will give more complete and more reliable information.

16.1.1.2 Assessment of Standard Test Methods

The total number of organizations engaged in the development of standards in particular subject areas exceeds 400 (7) in the United States alone. In particular, the American

Society for Testing and Materials (ASTM) has issued more than 6000 standard specifications and test methods. These standards cover a diverse range and include fundamental, component, practical, and hybrid methods needed for the various reasons for testing.

The number of these standard test methods is so large, and their status within national and international commerce is so long established, that their influence sometimes extends inadvertently beyond their proper domain. An important case has been the use of single-point hybrid test results in general design calculations and predictions. Such results were intended originally as quality control or practical values and certainly have no more extended applicability. An example of how a single data point obtained in such a standard test can lead to an erroneous conclusion is shown in Figure 16.3. When polycarbonate is notched and the Izod impact strength measured, materials of both 3 mm and 6 mm thickness show a transition from brittle to ductile behavior dependent on notch radius. However, if only the standard ASTM test (8) was conducted on bars with a 0.25 mm (or 10 mil) radius notch, we would erroneously conclude that the material of 3 mm thickness is ductile while the material of 6 mm thickness is brittle.

Many other standard tests have been developed to evaluate plastics in terms of fundamental properties. While these methods are scientifically sound, they still can be mis-

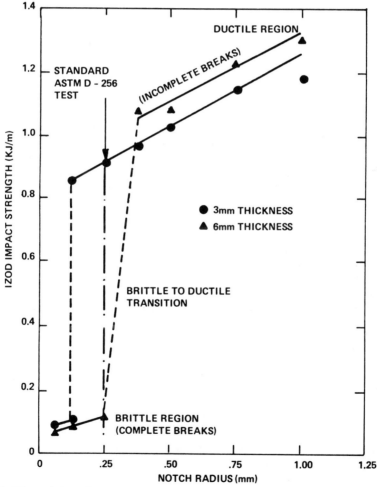

Figure 16.3 The variation of notched Izod impact strength with notch radius for two thicknesses of polycarbonate bars. The standard ASTM test would classify the 6 mm bar as brittle and the 3 mm material as ductile.

Table 16.1 ASTM Standard Test Methods Used of the Physical Testing of Plastics and Influencing Factors[a]

Property	ASTM Test Method	Influencing Factors	Fundamental Test	Component Test	Hybrid Test
		Bulk Mechanical Properties			
Bearing strength	D-953	A, B, C, D, E, H, J		X	X
	D-1602			X	X
Biaxial	D-2568	A, B, C, H, J		X	X
	D-2585				
Compressive	D-695	A, B, C, H, J	X		
	D-3410		X		
Creep (tensile, flexural, compressive)	D-2990	A, C, H, J, M	X		
Creep (compressive)	D-621	A, C, H, J, M	X		
Dynamic mechanical profile	D-2236	A, C, H, J, K	X		
	D-4065				
Fatigue (cyclic)	D-671	A, B, C, D, H, J, K, M	X		
	D-3479		X		
Fatigue crack propagation	E-647	A, B, C, D, H, J, K, M	X		
Flexural	D-790	A, B, C, H, J	X		
	D-747				
Fracture toughness	E-399		X		
High speed (instrumented tensile)	D-2289	A, B, C, H, J	X		
High speed (instrumented puncture)	D-3763	A, B, C, D, E, H, J, N		X	X
Impact	D-1709	A, B, C, D, E, H, J, N			X
(falling weight)	D-2444			X	X
	D-3029			X	X
	D-3420				X
Impact (izod)	D-256	A, C, D, E, H, J			X
Impact (tensile)	D-1822	A, C, H, J	X		
Shear strength	D-732	A, B, C, D, H, J, N			X
	D-2344				X
	D-2744				X
	D-3846				X
	D-3914				X
Static stress rupture (tensile, flexural, compressive)	D-2990	A, C, H, J, M	X		
Static stress rupture (hydrostatic pressure)	D-1598	A, C, H, J, M		X	
Stiffness-temperature profile (static loading)	D-1043	A, C, H, J, M	X		
Stress relaxation	D-2991	A, C, H, J, M	X		
Tensile	D-638	A, B, C, H, J	X		
	D-882		X		
	D-1180		X		
	D-3039		X		
	D-1708	A, B, C, E, H, J			X
		Surface Mechanical Properties			
Abrasion, Armstrong	D-1242	A, C, F, G, H, J, L, M			X
Abrasion, falling carborundum	D-673	A, C, F, G, H, I, J, L, M		X	X
Abrasion, taber	D-1044	A, C, F, G, H, I, J, L, M			X

Table 16.1 (*Continued*)

Property	ASTM Test Method	Influencing Factors	Fundamental Test	Component Test	Hybrid Test
Friction (against self or	D-1894	A, C, F, G, H, I, J, L, M, N			X
other surface)	D-3028			X	X
Indentation hardness,	D-2583	A, C, F, H, J, N			X
Barcol					
Indentation hardness,	D-785	A, C, F, H, J, M, N			X
Rockwell					
Scratch resistance	None	A, B, C, F, G, H, I, J, L, N		X	X
Wear (PV limit)	D-3702	A, B, C, F, G, H, J, L, M, N		X	X
		Electrical Properties			
AC dielectric constant	D-150	A, C, H, J, L	X		
and loss factor	D-669		X		
	D-1531		X		
	D-1673		X		
	D-2520		X		
	D-3380		X		
Corona resistance	D-1868	A, C, D, K, L, M, N		X	X
	D-2275			X	X
	D-3382			X	X
DC insulation resistance	D-257	A, C, H, J, L, M, N		X	X
DC resistivity (surface					
and volume)	D-257	A, C, H, J, L, M, N	X		
Dielectric breakdown	D-149	A, B, C, D, E, H, J, K, M, N		X	X
strength	D-3426			X	X
	D-3755			X	X
Electrostatic charge	D-3756			X	X
	D-2679	A, C, D, F, G, H, I, J, L, M, N		X	X
Tracking resistance, dry	D-3509			X	X
conditions	D-495	C, D, H, I, J, L, N			X
Tracking resistance, wet					
conditions	D-2132	C, D, H, I, J, L, N			X
	D-2302				X
	D-2303				X
	D-3638				X
		Thermal Properties			
Brittleness temperature					
	D-746	B, C, D, E, H, I, J, N			X
Burning, horizontal	D-1790				
	D-635	A, C, D, E, H, I, J			X
Burning, vertical	D-757				
Deflection temperature	D-3801	A, C, D, E, H, I, J			X
under flexural load	D-648	D, F, H, J			X
Heat release	E-162	A, C, D, J			X
Ignition, piloted	D-1929	A, E, D, J	X		
(convective heating and					
pilot flame)					
(Pilot flame)	D-3713	A, E, D, J		X	X
(Radiant heating and	D-229	A, D, E, J, N		X	X
electrical arc)					
(Radiant heating and	E-162	A, D, E, J, N		X	X
pilot flame)					
Ignition, unpiloted	D-1929	A, D, E, J	X		
(convective source)	E-136		X		
(Hot wire source)	D-3874	A, D, E, J	X		

Table 16.1 (*Continued*)

Property	ASTM Test Method	Influencing Factors	Fundamental Test	Component Test	Hybrid Test
Oxygen index	D-2863	A, C, D, E, J, N	X		
Smoke density	D-2843	A, C, D, E, M			X
Thermal conductivity	C-177	A, C, H	X		
Thermal diffusivity	D-2214	A, C, H			X
Thermal expansion	D-696	A, H, J	X		
	D-864		X		
	E-289		X		
	E-831		X		
Transition temperatures	D-3418	C, H, J	X		
	D-4065		X		
	E-289		X		
	E-831		X		
Specific heat	C-351	A, C, H	X		
	D-3418		X		
Vicat softening point	D-1525	D, F, H, J, N			X
Optical Properties					
Color	E-308	K	X		
Diffuse relfectance	D-589	E, H, J, L, N	X		
	E-97		X		
	E-179		X		
Gloss	D-523	E, H, J, L, N		X	X
	D-2457			X	X
	E-179		X		
Light redistribution	E-166	E, H, J, L	X		
(goniophotometer)	E-167		X		
	E-179		X		
Optical distortion	D-637	A, E, K, L		X	X
	D-881			X	X
Refracture index	D-542	A, K, N	X		
Transparency and haze	D-1003	E, F, H, J, L, N		X	X
	D-1494			X	X
	D-1746			X	X
	D-179		X		
Other Physical Properties					
Gas transmission rate	D-1434	A, C, D, F, H, J, M		X	X
	D-3985		X		
Specific gravity, density	D-792	A, C, G, H, J	X		
	D-1505		X		
Water absorption	D-570	A, D, E, M, J, M		X	X
Water vapor transmission	E-96	A, C, D, F, H, J	X		
rate	E-398		X		

[a] Key to factors that can influence test results:

A	Temperature	H	Molding conditions
B	Loading rate	I	Poor reproducibility
C	Moisture level	J	Environmental exposure
D	Sample size	K	Frequency
E	Sample shape	L	Quality of surface or mating material
F	Pressure	M	Test duration
G	Surface speed	N	Instrument/fixture geometry or boundary conditions

leading. This is because standard methods by nature are arbitrary. Thus the input excitation and environmental conditions of the test are controlled so that "materials properties" can be compared, or used as a basis for trade or acceptance. But these conditions are seldom relevant to the actual use conditions encountered. For example, while one may be able to use a standard test procedure to measure permittivity (dielectric constant), results will have little significance if the frequency, temperature, and environment are not similar to those encountered in use. However, it is not the test methods that are to be condemned, it is the failure by the experimentalist to extend these measurements over the correct time, temperature, or environment.

A summary of standard ASTM test methods and factors influencing results is given in Table 16.1. This table, an expansion of a similar table presented by Lamond (9), may be used to recognize the advantages and limitations of standard tests. One must be able to

recognize both the limitations and advantages of a test before one can draw logical conclusions concerning the results as they apply to the problem at hand. Regrettably, general training in this area is lacking.

Standard tests may not be available to cope with every situation. Appropriate specific tests may have to be designed for special applications. The references provided at the end of this chapter may be used to provide the reader with several sources of information concerning both standard and specific tests procedures. Schmitz and Brown (10) have also presented a list of periodicals, texts, handbooks, dictionaries, and encyclopedias that provide a summary of sources through 1967.

16.2 SELECTING MEANINGFUL PHYSICAL TEST PROCEDURES

Discretion must be used in selecting test procedures. Undoubtedly, too much physical

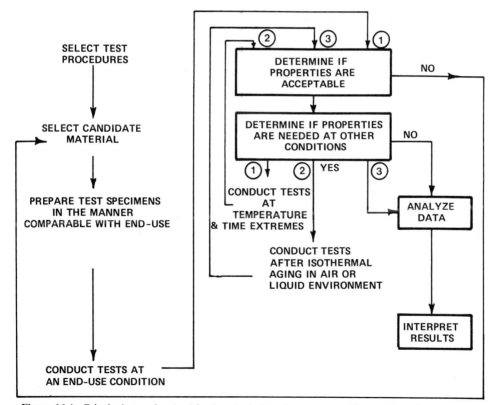

Figure 16.4 Principal steps involved in the physical testing sequence. The number of steps may be decreased when complete information is not needed. Reduced completely the testing technique becomes conducting a single test at standard condtions.

testing is performed. Beatty and Sitz (11) estimate that 70% of physical testing is useless or ignored! The chief reason for this is either a lack of analysis of what is required to solve the problem or a poor understanding of the limitations of the selected test methods—in other words, poor planning or understanding by the experimentalist of the problem at hand or the proposed solution. Such poor planning or understanding can also result in not running the critical test, a pitfall more serious than obtaining too much data.

The physical testing of a material may be thought of as a process or system with a number of steps. The actual number of steps may vary with the material and the nature of the testing desired. Because of the extremely wide use of plastics in a multitude of applications, we cannot define a priori a rigid test sequence to evaluate material performance. For any plastic, certain attributes should be emphasized more than others. However, a general flow chart showing the principal steps encountered is shown in Figure 16.4.

There are two evaluation classes where certain guidelines for selecting meaningful physical test procedures can be established. These are specific evaluation of a material for a known application and general evaluation of a material for various useful engineering attributes.

16.2.1 Material for a Known Application

This class is rather straightforward to outline although the actual test selection procedure may require much insight. It is encountered most often by the consumer of plastic materials. However, it is also increasing in importance with plastics manufacturers as they assume a larger role in design and engineering. An obvious rule for this class is that testing for properties having no relation whatsoever to the performance properties of the finished article must be avoided. A rational selection of a material would be based on the following points:

Analyze the application to determine the physical properties that will be important.

Estimate the shape and magnitude of the input functions, the boundary conditions and environment to which the material will be exposed, and the number of times the material must sustain the input, that is, the material's expected service life.

Fabricate the test specimens using the same processing techniques and the same thicknesses that will be used in the application.

Evaluate the physical properties of the material using fundamental or practical tests at these application-focused conditions and service times, or use appropriate predictive techniques.

Determine if these property values are acceptable for the desired performance.

If the material fails, determine if a change in composition (e.g., a change in the resin content of a polyester laminate) will provide a plastic with acceptable properties or if another would be more suitable.

Pang and Issacs (12) compare this above type of approach needed for the design of small electrical appliance products with the inadequate material property data furnished by the supplier.

16.2.2 General Evaluation of a Material

This class is more often encountered by the specialist concerned with supplying plastics and can best be evaluated by the following steps:

Develop a working hypothesis of the various desired physical attributes believed to be possessed by this material.

Select test procedures and specimens that can be employed over a wide range of input excitations; avoid tests that merely provide a single datum point or that are poorly defined in real physical terms. Select fundamental or practical tests whenever possible and avoid hybrid tests. When hybrid tests must be used, carefully compare the boundary conditions with those of various proposed uses or use test results for comparative purposes only.

Evaluate physical properties at "standard" conditions.

Determine physical changes over a range of environment (e.g., heat, moisture, radiation) and time. Establish bounds for acceptable and unacceptable performance.

Conduct all tests in such a manner that the sensitivity of a physical property can be determined as a function of material composition (e.g., plasticizer or filler content, molecular weight and distribution, cross-linking agent, etc.)

16.3 INTERPRETATION OF RESULTS

Too often the experimentalist, having conducted a series of tests, is unable to make any useful interpretation of the results. This is usually because the tests were not conducted in a manner that made a systematic presentation of the results possible. A systematic presentation of data is essential for a clear interpretation of test results. Careful consideration must thus be given to this presentation prior to actually conducting tests.

Graphical representations are usually the easiest way to present technical work. Tests should therefore be designed so that a measured property can be displayed graphically as a function of parameters of interest such as time, test temperature, composition, humidity, processing temperature, and magnitude of the input excitation. The more elaborate and comprehensive the experiment, the more parameters that should be considered. A typical representation, in this case for the mechanical loss factor of acrylic cast sheet (polymethylmethacrylate, PMMA) as a function of time and temperature, is shown in Figure 16.5. Such a detailed graph of the response function is essential if the objective is a fundamental understanding of the behavior. On the other hand, a much more abbreviated evaluation may suffice where the objectives are more practical or limited.

The main advantage of graphs is that they enable the user to have immediate access to data for any value of the independent variable and to examine the experimental results visually, thereby to see easily the effect of one variable on another, the presence of maxima, minima, critical inflection points, periodic features, and the like. Also, they allow one to formulate analytical relationships between the variables. The relationships that may be developed have the advantage of providing a means whereby property data taken at one set of conditions may be extended to a new set of conditions. Arrhenius plots are particularly useful.

The other important technique in the interpretation of results is the use of statistical procedures. Any measurement of a physical quantity is certain to contain errors in precision, and their fluctuations will lead to a distribution of numerical values if the same measurement is repeated several times. Statistical analysis of data should at least consist of calculating the mean and the 95% confidence limits. One can then determine statistically whether one set of values differs significantly from another, because, when there is significant scatter in results, intuitive judgements of differences between measured values are likely to be wrong. Notice, however, that if the data are badly scattered statistical procedures cannot improve their physical interpretation; they can only provide guidance as to the validity of any inferences.

More complex analysis, such as may be required when composition, processing and so on are to be studied requires the use of experimental design prior to starting evaluations. Texts on statistical analysis should be consulted for details concerning this technique.

16.4 INFLUENCES OF SPECIMEN PREPARATION, SAMPLING, AND GEOMETRY

16.4.1 Specimen Preparation

A specimen cannot be measured without also measuring the effects of specimen preparation. The various ways that a raw material can be processed and physical testing specimens obtained are shown in Figure 16.6. Generally, these effects of processing and/or machining history on test data are appreciable (13). Therefore we should try to fabricate the specimen in the same manner that the material will be fabricated in end use. Thus injection molded specimens, which are highly oriented

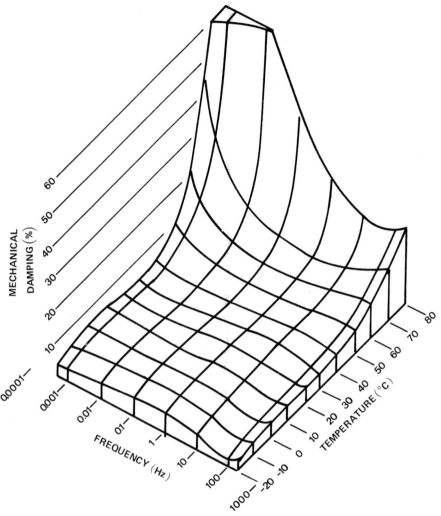

Figure 16.5 A general representation of a physical property, in this case the dynamic mechanical damping of PMMA, as a function of both frequency (time) and temperature. Practical data collection is usually abbreviated as necessary. Taken to the extreme, the curve degenerates into a single datum point.

in the length direction, should be avoided for mechanical tests unless the finished part is also to be injection molded. The effect of injection molding versus compression molding on the impact behavior of a typical toughened polystyrene is shown in Figure 16.7.

Extruded or injection molded plastics in general are not isotropic, that is, they do not have identical properties along the three principal axes. It is important, therefore, to cut specimens from two, or some cases, three orthogonal directions when determining uniaxial properties. The variation in tensile strength of an oriented polystyrene is shown in

Figure 16.8. Values between the transverse direction and orientation direction may differ by as much as 100%.

When specimens are cut or routed from sheet, moldings, or the actual component, it is necessary to avoid excess heating. This could change the degree of cure of a thermoset material and advance it sufficiently to change the physical properties. Other influences of machining are the possible flaws, stress concentrations, and so on caused by improper cutting tools or conditions.

Careful attention must also be given to the specimen surface to insure that it is comparable

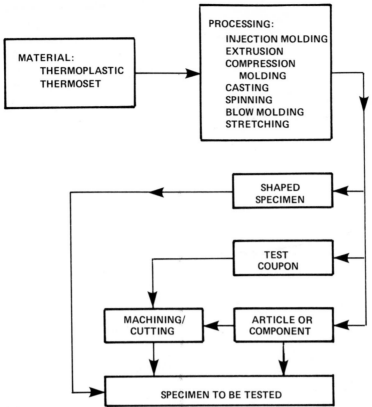

Figure 16.6 Various ways that a material may be processed and then formed into a specimen to be tested. The assessment of performance is concerned not only with the measurement of the isotropic properties of a particular grade of polymer but also with how its properties are modified by the method of fabrication and the processing conditions.

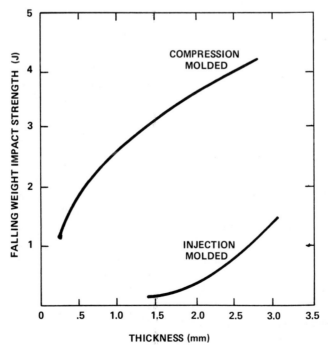

Figure 16.7 The effect of processing on the impact behavior of a typical toughened polystyrene, based on work done by Horsley (14).

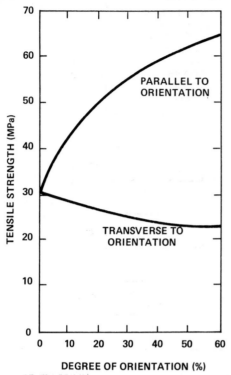

Figure 16.8 The effect of the test direction on the tensile strength of a polystyrene injection molded plaque, based on work done by Jackson and Ballman (15).

with the end-use form when friction and abrasion, electrical tracking, and optical properties are to be measured.

16.4.2 Specimen Sampling

Many times the test technologist is faced with obtaining test specimens from a finished article. Thus one has no control over the processing conditions. In addition to considering the machining or cutting operations to obtain test specimens, one must consider how the article is to be sampled.

Specimens used for most fundamental and practical tests are quite small, weighing in the order of 50 g or less, and usually a maximum of five specimens are used in any one physical measurement. It is obviously very important to insure that the essentially small quantity of material tested in the sample unit is representative of the whole. The concept of the sample unit must also take into account the intended use of the material. For instance, a single

bubble may seriously impair the usefulness of a large sheet of glazing material that is to be used in its entirety in military aricraft. If the sheet is to be cut into 300 mm diameter portholes, however, more defects presumably could be tolerated because they can be marked and then avoided in the cutting. Techniques used in its entirety in military aircraft. If the as elaborate as those for quality control; however, any technique should eliminate bias. For example, sheets, film, rod, tube, and pipe should be sampled randomly by division into a number of sample units, followed by selection of the areas to be tested with the aid of a table of random numbers as described in ASTM Method D-1899 (8). Specimens for component tests should likewise be randomly selected.

16.4.3 Specimen Geometry

Since most properties are reduced to units of length, area, or volume to yield the data for the material, it might be thought that the precise size of the test specimen is not important because all values can be normalized by calculation. However, a given material tested in one case in a thin section and in another case in a thick section may have different physical properties because of skin effects, orientation, stress distribution differences, moisture, and thermal transients, and the like. Also, some generally employed normalization techniques are invalid. A few illustrations will serve as a warning. Reverting to the example of impact strength, Figure 16.3 shows the thickness effect of polycarbonate. Considering electrical testing, dielectric breakdown strength is calculated in volts per millimeter by dividing the failure voltage of a flat specimen by its thickness. Figure 16.9, however, shows that there is not a linear relationship between breakdown voltage and thickness, and that the calculated value is no means independent of thickness. Such normalization is therefore erroneous.

16.5 BULK MECHANICAL PROPERTIES

Of all physical properties, mechanical properties often are the most important because

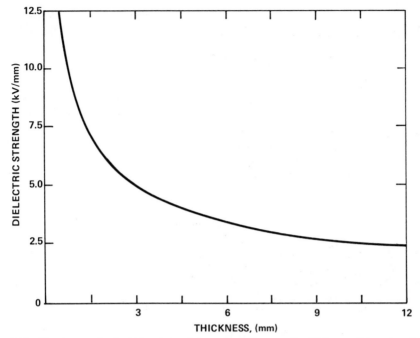

Figure 16.9   The decrease in the dielectric breakdown strength versus the thickness of the polycarbonate test specimen demonstrates the importance of considering specimen geometry in test design.

virtually all fabrication processes and service conditions involve some types of mechanical loading. Thus a large part of our picture of a material is based on the sum of its mechanical properties and the tests for measuring these properties.

The objective of mechanical testing is to supply information on a given plastic that will tell exactly how the material will react under a variety of loading conditions. The design of components in plastics, as in other materials, involves two major engineering problems arising from the loads they are expected to carry. One problem is deformation, which might have to be restricted for functional or aesthetic reasons. The other problem is the avoidance of failure by fracture, by irreversible loading, or by collapse through buckling.

The reasons for the use of various experimental techniques may be clearer if the types of deformations caused by various forces and the relationships between force and deformation are first reviewed. There are only three basic ways that applied forces can cause a material to deform. These are tension or compression, bending, and shear and are shown in Figure 16.10. Actual deformations are combinations of these three.

In the first case, parallel forces applied axially cause the material to stretch or compress. In such cases the displacement is uniform over the cross-sectional area. The strain ϵ is the change in length ΔL divided by the initial length L and is given by Equation (16.1).

$$\epsilon = \frac{\Delta L}{L} \qquad (16.1)$$

The stress σ is the force P divided by the cross-sectional area A perpendicular to the applied force and can be calculated using Equation (16.2).

$$\sigma = \frac{P}{A} \qquad (16.2)$$

In the second case, forces applied normal to the faces and acting at some distance apart impose a moment and cause the material to bend. The amount of deformation is a linear function of the distance from the neutral axis. Deformations on one side of the neutral axis are extensions, whereas deformations on the opposite side of the neutral axis are compressions. The neutral axis does not change in

TYPE OF FORCE	MODE OF DISTORTION	STRAIN EQUATION	STRESS EQUATION	MODULUS EQUATION
AXIAL (TENSILE OR COMPRESSIVE)		$\epsilon = \dfrac{\Delta L}{L}$	$\sigma = \dfrac{Fn}{A}$	$E = \dfrac{\sigma}{\epsilon}$
COUPLE PERPENDICULAR TO FACES (BENDING)		$\epsilon = \dfrac{c}{\rho}$	$\sigma = \dfrac{Mc}{I}$	$E = \dfrac{M\rho}{I}$
COUPLE PARALLEL TO FACES (SHEAR)		$\gamma = \dfrac{\Delta L}{Lo} = \text{TAN } \theta = \theta$	$\tau = \dfrac{Fs}{A'}$	$G = \dfrac{\tau}{\gamma}$

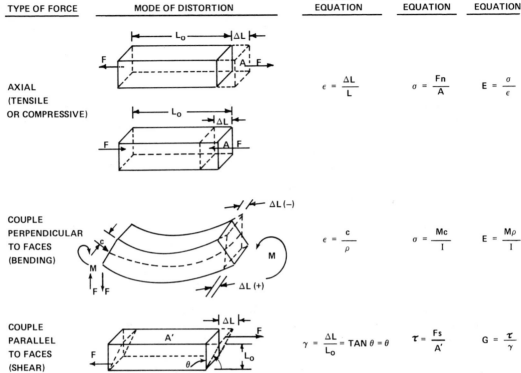

Figure 16.10 A summary of the various forces and the corresponding modes of distortion in mechanical testing. Although the stress and strain equations are valid for plastic materials, the modulus equation cannot be applied universally since plastics are viscoelastic materials.

length. The strain at any distance c from the neutral axis for a bar bent to a radius of curvature ρ is given by Equation (16.3).

$$\epsilon = \frac{c}{\rho} \qquad (16.3)$$

The stress is given by Equation (16.4).

$$\sigma = \frac{Mc}{I} \qquad (16.4)$$

where M = bending moment
I = moment of inertia of the cross section

Equation (16.4) is only valid for stresses below the yield stress and cannot be applied universally.

In the third case, parallel forces applied in the planes of the opposite faces a short distance apart cause the cross-section to rotate.

These shear forces are distinct from tensile, compression, or bending forces because shear forces cause a change in shape (rectangular to parallelepiped) whereas the former only cause a change in volume. In the case of shear strain γ, the distortion is expressed as the angle of deformation and is given by Equation (16.5).

$$\gamma = \frac{\Delta L}{L} = \tan \theta \qquad (16.5)$$

The shear stress τ is the force P_s divided by the cross-sectional area A parallel to the applied force and is calculated using Equation (16.6).

$$\tau = \frac{P_s}{A} \qquad (16.6)$$

Now that conventions have been established for expressing the forces and deformations (stresses and strains) that occur in materials, relationships between the two should be considered. For an ideal "elastic" material, the stress is proportional to the strain as shown in Equation (16.7)

$$\sigma = E\epsilon \qquad (16.7)$$

where the constant of proportionality E is the elastic modulus. This constant is independent of both time and temperature. Applying Equation (16.7) to the stress and strain equations developed above, the modulus for each mode of deformation can be expressed as Equations (16.8), (16.9), and (16.10).

Tensile of Compressive

$$E = \frac{PL}{A\,\Delta L} \qquad (16.8)$$

Bending

$$E = \frac{M\rho}{I} \qquad (16.9)$$

Shear

$$G = \frac{P_s}{A}\,cot\,\theta \qquad (16.10)$$

While no real material exactly obeys this law, the behavior of many metals, ceramics, glass, and other inorganic materials can be described by these equations with high accuracy.

Plastics, however, are viscoelastic; their behavior is thus dependent on both time and temperature. Most generally the relationship between the stresses and strains of a viscoelastic substance can be described by Equation (16.11).

$$\sigma = f(\epsilon, t, T) \qquad (16.11)$$

That is, stress is a function of strain ϵ, time t, and temperature T.

When the strains are limited to small values, usually 1% or less, then Equation (16.11) can be expressed as Equation (16.12)

$$\sigma = \epsilon f(t, T) \qquad (16.12)$$

in which stress is proportional to the product of strain and a function of time and tem-

perature. Expressing this function as a time and temperature dependent modulus one can write Equation (16.13).

$$\sigma = E(t, T)\,\epsilon \qquad (16.13)$$

Equations (16.11) and (16.13) are the controlling relationships that must be considered when an experimental procedure is needed to simulate an engineering application for a plastic. Modulus and strength will change with both temperature and time. Thus physical tests conducted over a few milliseconds, several minutes, or a thousand hours (e.g. impact, quasi-static and creep tests) should not be expected to give the same results or should they be expected to use the same test procedures.

16.5.1 Mechanical Spectrometry Measurements

The effect of temperature and strain rate on mechanical properties is usually first established to predict accurately polymer performance in any specific application. One way to determine the upper and lower limits of temperature and time is to identify the property or properties of significance in the intended use and then make appropriate measurements to follow the change of these values with a change in conditions.

However, it is much simpler to use a single test first that gives an overall picture of the mechanical behavior of a plastic as a function of temperature and time. Hybrid tests such as deflection temperature under flexural load (DTUL) (ASTM Method D-648) (8) and brittleness temperature (ASTM Method D-746) (8) have been commonly used to measure temperature effects but suffer from important defects that are further discussed later. More fundamental tests are also used to measure the modulus of a material at several temperatures, such as described in ASTM Method D-1043 (8).

However, a more useful but less commonly used tool is mechanical spectroscopy, also referred to as dynamic mechanical analysis. As Koo (16) notes, dynamic mechanical testing provides more insight and intrinsic information about a polymer than any other single measurement. In dynamic mechanical tests, the response of a viscoelastic material to externally applied varying excitations is measured. From such measurements it is possible to obtain both the modulus and the mechanical damping of the plastic. The modulus may be determined in tension, compression, flexure, or shear at different frequencies (strain rates) depending on the experimental equipment. The modulus is expressed by Equation (16.14)

$$E^*(\omega, T) = \frac{\sigma (\omega, T)}{\epsilon} \qquad (16.14)$$

where ω is the angular frequency (strain rate). This dynamic modulus E^* can be separated into two components as shown in Equations (16.15) and (16.16).

$$E' = E^* \cos \delta \qquad (16.15)$$

$$E'' = E^* \sin \delta \qquad (16.16)$$

One component E' represents the in-phase (elastic) stress versus strain; the other component E'' represents the out-of-phase component (viscous) stress versus strain where δ is the phase angle. The mechanical damping, the ratio of energy dissipated to the energy stored per cycle, gives a measure of the viscous component. The preceding equations lead to Equation (16.17).

$$\frac{E''}{E'} = \tan \delta \qquad (16.17)$$

The usefulness of these parameters is discussed later.

16.5.1.1 Instrumentation

Many types of instruments have been used to measure dynamic mechanical properties. They have the common feature that the imposed strains are small so that the material behaves in a linear viscoelastic manner, that is, Equation (16.13) is the controlling equation.

While Nielson (17) and others have discussed many of the instruments in use, all systems can be described as having one of

three types of input excitations: pulse; sinusoidal-resonance frequency; and sinusoidal-constant frequency. In pulse input excitation systems, such as the simply constructed torsional pendulum or torsional braid (TBA) apparatus shown in Figure 16.11a, displacement and the sinusoidal decay of the free end of the specimen (output) is followed after the fixed end is given a step input. Details of the specimen dimensions, test methods, and calculation of results are given in ASTM Method D-2236 (8). Gillham (18) has discussed other recent advances in torsional pendulum–torsional braid techniques. In sinusoidal-resonance frequency systems, the frequency of the input excitation necessary to keep the specimen in resonance changes as the modulus of the specimen changes. One such instrument, shown in Figure 16.11b, is the dynamic mechanical analyzer (DMA), an accessory to the DuPont differential thermal analyzer, where a flexural displacement is applied. This is described in detail by Blaine (19). In the sinusoidal-constant frequency instruments, one end of the specimen is simply driven with a force or displacement oscillating at a selected fixed frequency. In units such as the Rheovibron viscoelastomer, shown in Figure 16.11c, a small sinusoidal tensile strain is applied to the sample at one of several selectable fixed frequencies. This unit has been described in detail by Takayanagi (20) and Kenson (21). The Rheometrics mechanical spectrometer, described by Starita (22, 23), and the Dynastat mechanical spectrometer, described by Sternstein (24), are two other "sinusoidal/fixed frequency" instruments that are routinely being used to solve a wide spectrum of industry problems because of their extreme versatility. All equipment is or can be automated so that manual operation, once a factor limiting the routine use of dynamic mechanical analysis, is eliminated.

For the study of rate effects, constant frequency input systems which have selectable ranges from .1 to 200 Hz are recommended. The frequency range of resonant instrument can be changed somewhat, however, by changing the specimen size. For the study of temperature effects, the choice of systems is more a matter of convenience. When resonant

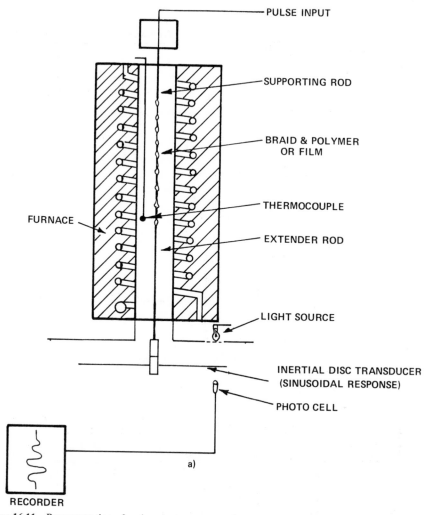

Figure 16.11 Representation of various instruments used to measure dynamic mechanical properties: (a) torsional braid apparatus (TBA); (b) dynamic mechanical analyzer (DMA); (c) Rheovibron. Each instrument subjects the polymer to a different input excitation.

frequency systems are used it is usually satisfactory to scan over the temperature range or to follow changes with time at isothermal conditions allowing the natural frequency to vary. This variation usually covers less than a decade for a given specimen size. Most dynamic measurements are made by following the change in physical properties as a function of temperature or time instead of the frequency because the change in properties as a function of temperature and time are often more important. For simulation of end-use conditions where mechanical vibrations will be imposed, the constant frequency instruments must be used.

A test involves either a series of measurements or a continuous measurement usually beginning at the lowest temperature or frequency that the equipment is capable of reaching. The specimen is heated at a constant rate until a temperature is reached where the specimen is too soft to support the load. ASTM Method D-4065 (8) discusses measurement techniques in more detail.

The behavior of a material starting from or changing to a viscous state can be measured, so cure or melting can be followed. This is accomplished by coating the material from a solution of uncured resin on a fiberglass braid having low torsional rigidity, a glass cloth

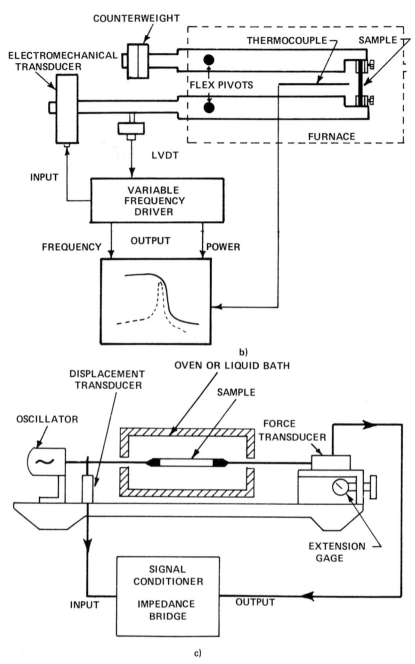

Figure 16.11 (*Continued*)

fabric having low flexural rigidity, or a quartz spring having low tensile rigidity. Accordingly, the behavior of the supported material can be determined as it undergoes a change of state.

16.5.1.2 Thermomechanical Profiles

Plotting the dynamic elastic and viscous components of a material versus temperature (or frequency) provides a graphic representation of rigidity and damping as a function of these parameters. Some typical shapes of thermomechanical profiles for polymers of different morphologies are shown in Figure 16.12. As shown in Figure 16.12 a, the modulus decreases with temperature in a nearly linear manner below the glass transition temperature T_g. A large abrupt decrease in modulus occurs over

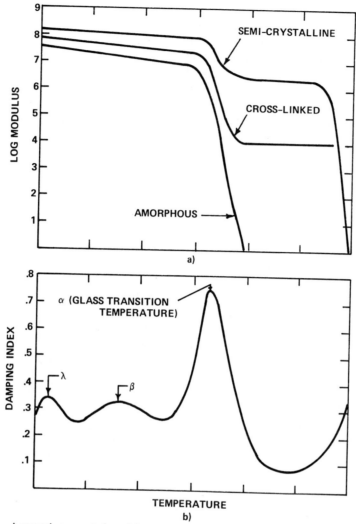

Figure 16.12 A general representation of thermomechanical profiles for amorphous, semicrystalline, and cross-linked plastics: (*a*) the modulus curves indicate that the materials have drastically different behavior above the glass transition temperature; (*b*) the damping index curve shows peaks at different temperatures which can be used to determine chemical structure and to predict physical behavior.

the T_g region since the polymer is undergoing a change of state, that is, from glassy to rubbery. The behavior of the polymer above the T_g will depend on whether the plastic is amorphous, semicrystalline or cross-linked.

As shown in Figure 16.12*a*, the modulus of the amorphous polymer decreases several orders of magnitude above the T_g. Semicrystalline polymers do not show such a percipitous drop in modulus. For such polymers, the modulus drops to a lower value that is dependent on the degree of crystallinity and then remains relatively constant up to the crystalline melting point. The plateau in modulus beyond the T_g is due to the retention of

crystalline domains that are unaffected by the transition in the amorphous domains.

Again as shown in Figure 16.12*a*, the cross-linked material softens at the T_g, and the modulus drops to a lower value depending on the degree of cross-linking. Because of the cross-linking, however, it cannot be remelted so that the modulus remains relatively high until the sample thermally degrades. If the material is sufficiently cross-linked, no glass transition will be evident.

A typical shape of the damping spectrum is shown in Figure 16.12*b*. The spectrum usually consists of a number of peaks of various heights and widths. The sharpest of these (α)

occurs at the T_g and represents a change in the flexibility of the polymer chain. This corresponds with the abrupt decrease in modulus in this region. Other broader peaks (β and λ) occur at lower temperatures and represent particular segmental motions and rotations of the polymer chain.

The modulus spectrum gives a rapid indication of the temperature range over which the stiffness is sufficient so that its use for a particular application can be further investigated. Such data are much more useful than standard ASTM tests that define brittleness temperatures and softening points of materials under arbitrary test conditions. For example, Figure 16.13 shows the thermomechanical profiles for two rubber modified acrylics. Both have the same heat deflection temperature determined by ASTM Method D-648 (8). From Figure 16.13 this might be expected since the standard ASTM hybrid test only measures the "relative" change in modulus from the room temperature value. Both curves have the same slope between room temperature and the DTUL temperature. However, Figure 16.13 shows that the T_g of the two materials differs by 18° C. Their moduli also differ. This would suggest that the material with the lower T_g would not be as dimensionally stable at higher temperatures.

Also, the standard ASTM Method D-648 (8) is very misleading if one assumes that a material with a higher DTUL has the higher heat resistance. Riddell (25) notes that an amorphous material, such as PVC, may have a higher DTUL temperature than a semicrystalline material, such as nylon. However, as can be seen from Figure 16.12, an amorphous material is unable to support a load above its T_g while the semicrystalline material will continue to be a stress-bearing solid until its crystalline melting point is reached.

The modulus spectrum also provides a means of assessing the effects of fillers, reinforcements, plasticizers, moisture, cure, and so on that may change performance. For example, Figure 16.14 shows how undercuring a melamine formaldehyde impregnated asbestos sheet would adversely affect its high temperature performance, as evidenced by the large drop in modulus.

Of equal importance with the modulus spectrum is the information provided by the damping spectrum since such properties as vibration dissipation, noise abatement, creep and stress relaxation, and heat buildup in fatigued parts can all be related to the magnitude and location of the mechanical damping peaks. Boyer (26) has also found a correlation between secondary (i.e. β and $|\lambda$)

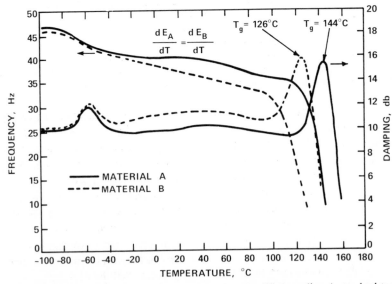

Figure 16.13 The thermomechanical profiles for two rubber modified acrylics. A standard test such as DTUL, ASTM Method D-648 (8) measures the "relative" change in modulus, ΔE. Since between RT and the DTUL temperature $dE_A/dT = dE_B/dT$, one could incorrectly assume that the heat resistances of the two materials were the same. In fact, Material A has a higher modulus and a T_g 18°C higher than Material B.

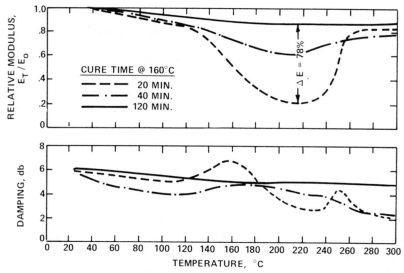

Figure 16.14   Thermomechanical profiles for melamine formaldehyde impregnated asbestos sheets cured for various lengths of time. While the moduli of the three sheets are comparable at room temperature, the undercured sheets show a large drop in modulus and a T_g not seen for the fully cured sheet.

transitions and the impact behavior of the polymer.

Depending on the end-use application, either high or low mechanical losses may be desirable. For example, a golf ball that absorbs large amounts of mechanical energy and transforms that energy into heat is not satisfactory. The same behavior is required, however, for a thermoplastic elastomer to perform effectively as an overload shock absorber on an automobile chassis.

16.5.2   Quasi-Static Measurements

Mechanical test methods can be classified into three other categories which have various shapes of input excitations: high speed (impulse excitations); quasi-static (ramp excitations); and long term (step or sinusoidal excitations). Specifically, a quasi-static or short-term test can be defined as a slow ramp input that leads to specimen failure in less than approximately 3–5 min but greater than a few seconds. There are numerous applications where the service condition imposes a similar input. Such short-term tests can therefore provide an adequate solution to many engineering problems when the dependence of temperature and history are considered. Also, these

measurements are the most straightforward tests that tell us something about the strength of a material or the conditions under which it will break or under which it can be loaded for short periods without permanent damage. However, data from such tests cannot be considered significant for applications differing widely from the load-time scale of the test employed. For example, under long-term loading a material may break at stresses only 20–30% of its quasi-static value.

Short-term tests can generally be made in a variety of uniaxial loading modes such as tension, flexure, compression, or shear. Other modes cause a crack to initiate and grow in a controlled manner. Tests can also be made in biaxial modes in which two of the uniaxial modes are combined. However, all quasi-static tests have two features in common: (*a*) some standard universal testing machine is normally used to extend the specimen at a constant rate; (*b*) the machine provides a plot of force versus extension that is, stress versus strain.

16.5.2.1   Mechanical Testing Machines
The testing machines used for conducting short-term tests are invariably used for several test modes—tension, flexure, compression,

shear, and so on—by using appropriate test fixtures. Such "universal" testing machines are commonplace today. While constant rate of loading and constant rate of extension machines are both available, the constant rate of extension machines are by far the more commonly used. Such commercial machines employ a motor-driven screw-gear mechanism that drives a moveable crosshead. This in turn holds the particular type of grip needed to rupture the specimen in the desired mode. A load is transmitted through the test specimen to a load cell. This transducer, either in the fixed or moving head, senses the force applied to the specimen. Crosshead speeds from 0.05 mm to 1.25 m are governed by a quartz crystal clock or by appropriate gearing of the driving motor. Full-scale loads from 10 mN to over 200 kN or more are governed by interchangeable load cells. Most machines are equipped to plot the load carried by the specimen versus the deformation in the specimen. Such a machine is represented schematically in Figure 16.15.

While there are many test machines available, accurate measurements are seldom possible with an improperly designed apparatus. This is mainly because in such machines the force-measuring device and crosshead may deform excessively, the speed of the drive

motor may change significantly according to the force applied to the specimen, and the range of the load and crosshead speed may not be adequate for the wide range of conditions encountered. Also the more elaborate machines permit measurements in various modes, such as both loading and unloading, important when properties such as resiliency are to be measured. They also permit direct integration of area under the load deformation curve, important when properties such as toughness are to be measured.

16.5.2.2 *Stress-Strain Plots*

As a sample is loaded, a force (stress) versus extension (strain) curve is recorded. According to Hooke's law, for an ideal elastic solid, stress is proportional to strain up to the point of rupture, and the stress-strain curve would be the same in tension, flexure, and compression. However, even ignoring the effects of temperature and time discussed at the beginning of this section, plastics deviate from this law even at small strains. When temperature and time are considered, the shape of curves vary even more. For example, the behavior of a modified acrylic as a function of time and temperature is shown in Figure 16.16. Plastics exhibit a whole spectrum of behavior from soft and weak to hard and tough. Carswell and Nason

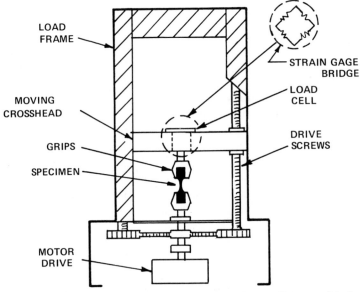

Figure 16.15   A universal test machine for conducting quasi-static tests. Courtesy of the Instron Corp.

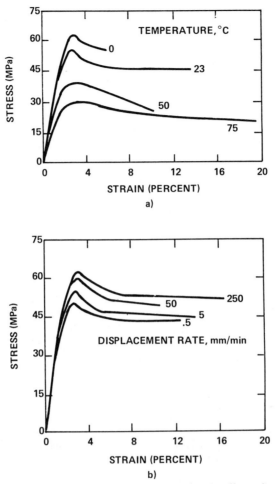

Figure 16.16 Stress-strain curves for a modified PMMA showing the effects of temperatures and strain rate, based on work of Roylance et al. (27).

(28) have presented the five forms usually encountered. A complete discussion of these curves is given by Nielson (17). They are represented in Figure 16.17 along with a description of the characteristics of the stress-strain curve.

The important features of a stress-strain curve are shown in Figure 16.18. Assuming Hookean behavior, the slope of the line OA, that is $\sigma_{PL}/\epsilon_{PL}$, is the (rate dependent) modulus of the plastic. Point A is called the proportional limit, since at this point the stress-strain curve starts to deviate from the straight line that characterizes the elastic (or linear viscoelastic) region. In practice this point is difficult to measure accurately. As the strain axis is magnified, the proportional limit decreases to lower values. More commonly, point B is read

at some stress where the stress-strain curve deviates from the straight line by a given amount, that is, 0.2% strain. This is the offset yield strength. The value is easily determined by constructing a line parallel to line OA that is separated from line OA by the desired amount. The intersection of this line with the stress-strain curve gives the offset yield stress, σ_{oy} and offset yield strain, ϵ_{oy}.

The elastic modulus can only be used to calculate deflections for a given load or vice versa if the strains are within the proportional limit. For design purposes it is often better to measure the secant modulus. The secant modulus is defined as the secant to the stress-strain curve from zero stress and strain to some defined stress or strain. This is represented as line OC in Figure 16.18. Secant moduli may be

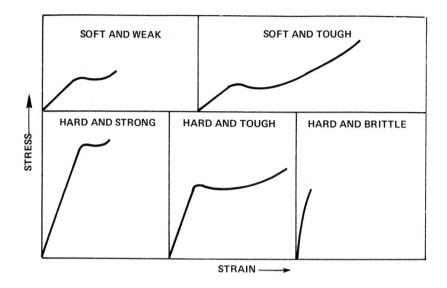

	CHARACTERISTICS OF STRESS-STRAIN CURVE			
DESCRIPTION OF POLYMER	MODULUS	YIELD STRESS	ULTIMATE STRENGTH	ELONGATION AT BREAK
SOFT, WEAK	LOW	LOW	LOW	LOW
SOFT, TOUGH	LOW	LOW	MODERATE	HIGH
HARD, STRONG	HIGH	HIGH	HIGH	MODERATE
HARD, TOUGH	HIGH	HIGH	HIGH	HIGH
HARD, BRITTLE	HIGH	NONE	MODERATE	LOW

Figure 16.17 Five general types of stress-strain curves that are typical of polymers in quasi-static tests and the characteristics of the stress-strain curve.

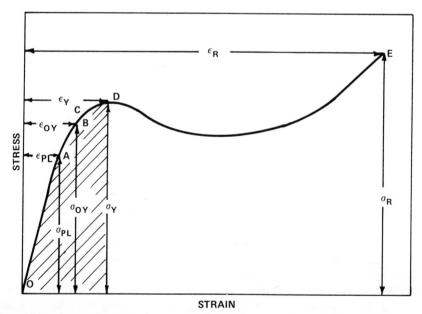

Figure 16.18 A general stress-strain curve showing the stresses and stains that may be calculated.

calculated at various points, and then the ratio of secant to initial modulus may be plotted versus either the stress or the strain at which the secant modulus was obtained. This gives a plot of the rate at which the stiffness changes above the proportional limit and gives an estimate of the stresses or strains to which a part can be subjected and still maintain sufficient stiffness. For example, assume that the stiffness of a part must not decrease by more than 15%. A PMMA part, with a proportional limit of 0.5% strain, could be strained to almost 2% before the modulus decreased by 15%. A 6-6 nylon (2.5% moisture content), with the same proportional limit, could only be strained to 0.75%. The use of the secant modulus concept allows estimation of the design error due only to the non-Hookean behavior of plastics. It does not include the error due to the time-dependence of the modulus. This is discussed later.

Point D on the stress-strain curve is defined as the yield point. In standard ASTM tests, the yield stress σ_y is defined as that stress at which no further increase in stress is necessary to increase the strain. According to this definition, "work hardening" materials (with stress-strain curves having only a knee) do not have a yield point. The yield point of such materials is usually determined by extrapolating the modulus curve and the slope of the work hardening region and reading their intercept as the yield point.

Above the yield point the material undergoes plastic flow until rupture occurs at point E in Figure 16.18. The area under the stress-strain curve up to rupture is directly proportional to the toughness of the material for the particular conditions of the test.

For structural applications, the features discussed for the region below the yield point represent the most important region of the stress-strain curve, shown as the shaded area in Figure 16.18. However, other than values for elastic modulus, values in this region are rarely reported in tables of physical properties. Elongation and strength at rupture values are also important for structural applications since catastrophic failure may be avoided if the material has a high elongation and high strength at rupture.

The suitability of plastics for other applications can also be assessed from the stress-strain curve. Resilient materials, such as those used in bumpers, are characterized by the large area under the stress-strain curve (high energy storage) up to the yield point. A highly ductile plastic, such as would be necessary to replace lead toothpaste tubes, conversely would have little or no resiliency but instead would have a very low yield point and high extension between the yield and rupture values. Marin (29) discusses other important interpretations of stress-strain curves.

16.5.2.3 Tensile Tests

Uniaxial tensile tests are the most widely used short-term tests. Such tests are recommended to assess the performance of the plastic when the application involves stretching. Tensile tests can also be used to monitor the progress of chemical or physical changes taking place in the polymer during service conditions. For example, polymers exposed to high heat or ultraviolet light can undergo depolymerization and oxidation reactions that will be reflected in changes in tensile properties. In a similar manner, when polymers are exposed to an aggressive chemical environment, they may become either more or less brittle, and the changes can be conveniently followed through measurement of tensile properties.

To carry out a tensile test, some form of specimen capable of being gripped at both ends is needed. In order to prevent failure in the grips, which would indicate a tensile strength lower than is characteristic for the material, a "dumbbell" specimen shown in Figure 16.19a is normally employed for bulk materials. The five types of specimens specified in ASTM Method D-638 (8) are of this geometry and combine negligible slippage at the clamps, a small size convenient for use with a limited quantity of experimental material, and a radius of curvature large enough for essentially no stress concentration in the fillets. The D-638 Type V specimen is especially useful for use with small amounts of material. However, the microtensile specimen described in ASTM D-1708 (8) has a sharp fillet radius that imposes a stress concentration 40% greater than in the reduced section,

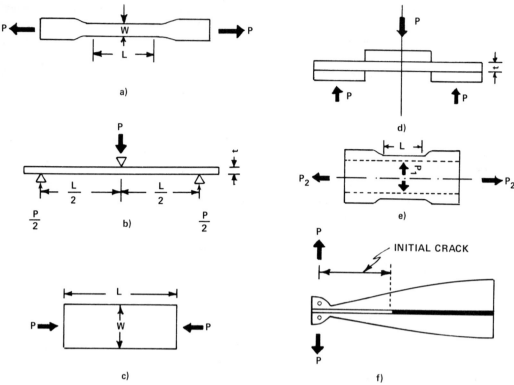

Figure 16.19 Methods of loading specimens for measuring quasi-static mechanical properties: (*a*) tension, (*b*) flexure, (*c*) compression, (*d*) shear, (*e*) biaxial, and (*f*) crack propagation.

and its use is not recommended. Thin, flexible films are usually tested using a straight-strip specimen, described in ASTM Method D-882 (8), since the effects of the grips can be minimized so a dumbbell shape is not needed. Other tests use specimens with geometries developed for very specific reasons, such as special specimens for composites and laminates described in ASTM Methods D-3039 (31) and D-3914 (31) or hoop specimens described in D-1180 (8) for bursting strength of rigid plastic tubing.

When the sample is tested, both load and elongation are measured. It might be thought that the easiest way to measure elongation would be to simply follow the relative movement of the jaws. However, the total deflection measured will depend on several factors: the extension of the material in the reduced-gauge section; the extension of the material in the fillets and slippage in the grips; slack in the jaws, the coupling, and/or the crosshead of the testing machine. Consider,

for example, the extension of a PMMA tensile bar having a modulus of 3.4 MPa and cross-sectional dimensions of 12.5×3.2 mm. A load of 690 N would be required to extend the sample to a 0.5% strain, which is the proportional limit of the material. In this elastic region, a 50 mm gauge length in the reduced section will extend only 0.37 μm/N. If a compliance of 0.16 μm/N is assumed for a rigid test machine and fixtures used for tensile testing, an error of 44% will be introduced even if slippage and extension in the fillets are ignored. Thus, for stiff materials no reliable figure for extension can be calculated by following the grip separation. There is no substitute for the use of an extensometer to define a gauge length and to follow the elongation accurately. For flexible materials, the force necessary to extend is relatively low and the last factor above can usually be ignored. Slippage from the grips can be limited in this case by using line-contact grips. Another technique is to measure the stress-strain properties at several grip separations and then extrapolate to infinite grip separation. Harris (30) discusses the various aspects of tensile testing in more detail.

A practical test using a tension load is the bearing-strength measurement. This measurement is designed to determine the maximum bearing load where rivets, bolts, or similar fastenings are used in joining members. Instead of gripping the specimen at both ends, as in a standard tensile test, a pin is placed through a hole near one end of the specimen and the load is carried by this pin. By measuring the load necessary to just deform the hole, a measurement of bearing strength is obtained. It is necessary that the pin dimensions and the distance to the edge of the specimen agree with the intended end-use application, otherwise the test is hybrid. ASTM Method D-953 (8) and Method D-1602 (31) describe bearing strength measurments in more detail.

16.5.2.4   Flexural Tests

Flexural tests are suggested to assess performance of the plastic when the application involves bending. A simple attachment to a standard tensile machine permits the specimen to be tested in flexure as a simple beam, as shown in Figure 16.19b. As the sample is bent, the face against the support posts is in tension; the face against the loading nose is in compression. For a bar with a rectangular cross section loaded at the midpoint (three-point loading), the equations presented in Figure 16.10 can be reduced to Equations (16.18) and (16.19)

$$\sigma = \frac{3}{2}\frac{PL}{bd^2} \qquad (16.18)$$

$$E = \frac{PL^3}{bd^3 y} \qquad (16.19)$$

where L = beam span
  P = force
  b = specimen width
  d = specimen thickness
  y = midspan deflection

The advantages of flexural tests are the elimination of an extensometer when bending forces are low enough to ignore machine deformation and the elimination of gripping requirements. Considering the example given in Section 16.5.2.3., a force of only 58 N would be required to deflect a bar on a 50 mm span to its proportional limit. Deflection will increase 11.4 μm/N. This represents magnification of greater than 30, compared to tensile tests. Other advantages for brittle materials are that slight misalignment of the grips does not have the exaggerated effect that it does in a tension test, and the zone of maximum stress is usually a molded face rather than a machined edge. Also, special high- or low-temperature extensometers are not necessary for measurements conducted at conditions different from ambient. Last, specimens that cannot be gripped in tension, such as unidirectionally fiber-reinforced rods, can be made to fail in the tensile face of a flexural specimen by proper machining to shift the neutral axis from the centroid of the cross section (32).

Specimen geometry must be carefully selected to insure that the flexural strength equation remains valid. The ratio of specimen test span to thickness must be great enough to insure that the specimen breaks in tension rather than shear. On the other hand, the span must be short enough to insure that failure occurs before large deflections are introduced. Otherwise values must be corrected to account for these deflections if accurate values are to be obtained (8). ASTM Method D-790 (8) recommends a 16:1 span:depth ratio for a homogeneous material and a 40:1 span:depth ratio for laminates. A limitation of the test is that materials that undergo strains greater than 5% before failure cannot be tested. Another limitation is when the specimen is tested in three-point (midspan) bending, the stress in the faces is at a maximum directly under the loading nose. This limitation can be overcome by the use of four-point loading; in such cases, the stress in the faces is uniform between the loading noses.

Even when all of the limitations are considered, however, the calculated flexural strength and modulus will probably not be the actual stress and modulus. This is because in the derivation of the flexural equation the neutral axis is placed at the center of the beam since it is assumed that the tensile and compressive moduli are equal. This is usually not true for plastics. For materials that have largely different tensile and compressive moduli, correction factors can be calculated (33) to account for this shift in the neutral axis. The maximum or breaking stress σ_B is then given by an equation of the form shown as Equation (16.20),

$$\sigma_B = \Omega \, \frac{3M}{bd^2} \qquad (16.20)$$

where for tensile stress one can write Equation (16.21),

$$\Omega = \frac{1}{2}\left[\left(\frac{E_T}{E_C}\right)^{1/2} + 1 \right] \qquad (16.21)$$

and for compressive stress one can write Equation (16.22).

$$\Omega = \frac{1}{2}\left[\left(\frac{E_C}{E_T}\right)^{1/2} + 1 \right] \qquad (16.22)$$

In these equations E_T = tensile modulus

E_C = compressive modulus

$M = PL/4$ for three-point bending

$M = PL/8$ for four-point bending

The flexural modulus E_F is given by the Equation (16.23).

$$E_F = \frac{4E_T E_C}{(E_T^{1/2} + E_C^{1/2})^2} \qquad (16.23)$$

Equations (16.18) and (16.20) imply an important but often overlooked fact stated as follows: Since the maximum tensile stress in flexure occurs at the specimen face, flexural tests may be more sensitive for detecting thermal and environmental changes and the contributions of skin reinforcement than tensile tests, which measure the average properties over the cross section. In addition, surface anisotropy, caused by a variability in cooling rate from the inside to outside during

sample preparation can lead to bending properties different from tensile or compressive properties for these same reasons. Loveless (34) has detailed several other features of the flexural test.

Discussion thus far has mainly concerned the flexural test as it applies to the determination of stress and material modulus. Beam stiffness, or the ability of a beam to resist flexure, is also of interest in many applications. For example, the importance of this property is well recognized in the design of reinforced beams and other structural members in which the amount of bending must be limited. It is also important in the opposite sense where, for a flexible material such as packaging film, stiffness must be kept to a minimum.

Bending stiffness B is the product of two parameters shown in Equation (16.24), namely, the flexural modulus of elasticity E of the beam and the moment of inertia I of the cross section about the beam's neutral axis.

$$B = EI \qquad (16.24)$$

It is thus not a material property. In simple terms, stiffness is dependent on both the material of construction and the geometry of the component and can be modified by adjusting either or both of these parameters. Techniques for measuring stiffness usually involve measuring the force-deflection relationship of a cantilever beam, as described in ASTM Method D-747 (8) or by Evans (35). Both techniques permit measurements in the linear region of the curve. Other measuring instruments that merely deflect a specimen of fixed amount and provide a single reading are questionable because the linear region may be unknowingly exceeded. That is, the secant modulus may be measured at an unknown strain.

16.5.2.5 Compressive Tests

Compressive tests are recommended to assess performance when the application involves deformation or pushing. Stress-strain tests should then be made in compression instead of tension or flexure since the stress-strain properties will be different. Materials under compression, for example, may be much less brittle than when they are under tension; many materials that are brittle under tension become ductile and show yield points when tested under compression.

To perform a compressive test, some form of specimen capable of being loaded uniformly at both ends is needed. Normally a rectangular specimen, such as shown in Figure 16.19c, is used. Some compromise must be made in selecting a test specimen, however. Long, slender specimens are often useless because they buckle rather than break. Short, wide specimens are under triaxial, rather than uniaxial stress, because of the frictional restraints at the end faces. The compromise made between these two extremes in ASTM Method D-695 (8) is a specification of a slenderness ratio of between 11:1 and 15:1. This is the ratio of the length of a column of uniform cross section to its least radius of gyration. Even with this compromise, it is extremely difficult to apply a truly axial force to the ends of a rigid specimen. When such data are required, the specimen ends must be carefully machined and special compression tools used. ASTM Method D-3410 (31) describes one such specimen holder needed to obtain correct information on composite materials. Sheet materials can be measured in some cases by stacking circular disks (poker chip fashion) between the platens of the compressive fixture.

As with tensile measurements, the total deformation measured by the machine will depend on many factors. Modulus measurements based on crosshead displacements will be in error by magnitudes similar to those in tensile measurements. Also, compressive specimens tend to broom at the ends. Therefore, a deflectometer is recommmended to measure the deformation of stiff materials.

16.5.2.6 Shear Tests
Shear tests are preferred to assess performance when the application involves parallel but opposite forces operating at a small distance apart. Such forces occur under various practical conditions such as lap joints, punching, stamping, or twisting. While tensile, flexural, and compressive forces cause a change in

extension, shear forces cause a change in shape. The measurement of shear properties that are not dependent on the boundary conditions is very difficult. While standard tests have developed, all measure an apparent value and are therefore hybrid tests. A fundamental measurement of shear can be obtained by placing a specimen in pure torsion by twisting it but no standard method is available for plastics. Shear strength of specimens machined in the shape of tubes or rods can be measured if the testing machine is equipped with a torsional fixture. Shear strength of sheets can be measured perpendicular to the plane of the sheet by using the punch shear test, Figure 16.19d, described in ASTM Method D-732 (8). In this test, a punch bears on a flat sheet of test material resting on a die. Special picture frame loading fixtures can also be used to apply in-plane shear stresses to sheets. Other means of measuring apparent internal shear forces, such as interlaminar shear strength of composites, involves loading of very short span:depth ratio specimens in flexure, as described in ASTM Method D-2344 (31); or pulling a lap joint in tension, as described in ASTM Method D-2744 (31), or compressing a notched bar or rod as described in ASTM Methods D-3846 (31) and D-3914 (31).

Shear modulus is more difficult to determine than tensile, flexural, or compressive modulus because high-precision extensometers are usually employed, especially in the measurement of lap joints (36). However, split ring specimens (37) can sometimes be used and the deflections measured directly from crosshead travel instead of from an extensometer since the deflections are large and the forces small.

16.5.2.7 Biaxial Stress Tests
The quasi-static tests already described are uniaxial tests in which a single force is applied to a specimen in one direction. Combined stress states are those in which stresses are applied along two or all three of the principal axes. The mechanical properties of polymeric materials under combined states of stress are important for the design of modern engineering structures and components since structures are rarely subjected to only uniaxial

stress. For instance, plastic pipes or tubes and plastic packaging bags and bottles are subjected to biaxial stresses when filled with liquid or powder. There are two problems to be considered in biaxial testing, as in uniaxial testing: resistance to deformation and prevention of failure.

In most cases the stresses and deformations in biaxial loading can be related to the deformations in uniaxial loading by using equations presented in theories of elasticity. The stress-strain relationships for a biaxial deformation of a material are given by Equations (16.25)–(16.27)

$$\sigma_1 = (\epsilon_1 + \mu\epsilon_2)\frac{E}{1 - \mu^2} \qquad (16.25)$$

$$\sigma_2 = (\epsilon_2 + \mu\epsilon_1)\frac{E}{1 - \mu^2} \qquad (16.26)$$

$$\tau_{12} = \gamma_{12}G \qquad (16.27)$$

where the subscripts refer to the directions of principal stresses and strains and μ is Poisson's ratio. An important implication of these equations can be seen by solving for the modulus, as shown in Equation (16.28).

$$E = \frac{\sigma_1(1 - \mu^2)}{(\epsilon_1 + \mu\epsilon_2)} \qquad (16.28)$$

Since the strain in one direction will be some fraction of the strain in the other direction, or $\epsilon_2 = k\epsilon_1$, Equation (16.28) can be reduced to Equation (16.29).

$$E = \frac{(1 - \mu^2)\sigma_1}{(1 + k\mu)\epsilon_1} \qquad (16.29)$$

This can be written in the form of an effective biaxial modulus $\bar{E}$, given in Equations (16.30) and (16.31):

$$\bar{E} = \frac{\sigma_1}{\epsilon_1} \qquad (16.30)$$

where

$$\bar{E} = \frac{E(1 + k\mu)}{(1 - \mu^2)} \qquad (16.31)$$

However, Poisson's ratio is always between 0.0 and 0.5. Thus Equation (16.31) will be positive for all numbers of $k\mu \geqslant -1$. The effective biaxial modulus will then always be greater than the uniaxial modulus in the tensile-tensile and compressive-compressive quadrants.

Since the various failure analysis theories discussed below do not predict a biaxial failure stress any greater than 115% of the uniaxial stress at failure, Equation (16.30) predicts that materials under biaxial tensile stress fail at lower elongations, that is, are more brittle, than materials under uniaxial stress. As an example, stress-strain curves of a tube of PMMA and the corresponding biaxial failure elongations are shown in Figure 16.20.

Several theories have been proposed for defining the conditions that cause failure in a material subjected to combined stresses. While such failure theories are not developed here, it is of interest to compare the failure loci of two of the most widely used, that is, the Tresca and the von Mises conditions. These are shown in Figure 16.21. These loci are based on Equations (16.32) and (16.33).

Tresca Condition

$$|\sigma_1 - \sigma_2|, |\sigma_2 - \sigma_3|, |\sigma_3 - \sigma_1| = \sigma_y \qquad (16.32)$$

Von Mises Condition

$$(\sigma_1 - \sigma_2)^2 + (\sigma_2 - \sigma_3)^2 + (\sigma_3 - \sigma_1)^2 = \sigma_y^2 \qquad (16.33)$$

Both of these equations assume that the tensile and compressive yield strengths σ_y are equal. A modified version of the von Mises yield condition has been proposed (39) to account for the differences in the tensile and compressive yield strengths. This may be expressed as Equation (16.34):

$$(\sigma_1 - \sigma_2)^2 + (\sigma_2 - \sigma_3)^2 + (\sigma_3 - \sigma_1)^2 + 2(\sigma_{yc} - \sigma_{yt})(\sigma_1 + \sigma_2 + \sigma_3) = 2\sigma_{yc}\sigma_{yt} \qquad (16.34)$$

where σ_{yc} and σ_{yt} represent the absolute values of the compressive and tensile yield strengths,

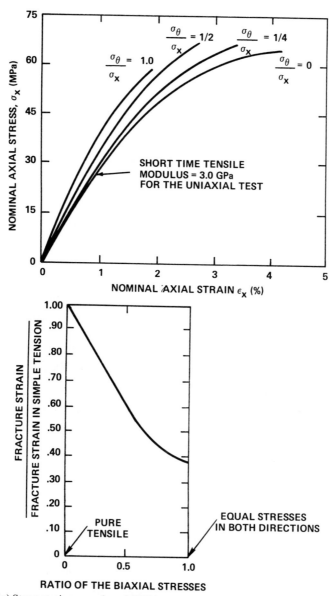

Figure 16.20 (*a*) Stress-strain curves for a PMMA tube when both axial stress, σ_x, and hoop stresses, σ_θ are present, (*b*) a comparison of the decrease in fracture strain due to increased stress biaxially for PMMA; based on work done by Thorkildsen (38).

respectively. This equation was verified for polyvinylchloride (PVC) and polycarbonate (PC) using the 0.3% offset yield stresses as a failure criteria.

Various test methods have been developed to experimentally apply biaxial stresses when more precise information is needed. For example, thin-walled cylinders with a reduced section at the test area can be subjected to either a constant internal or external pressure and simultaneous axial loading either in tension or compression (40), (41), as shown in Figure 16.19*e*. This type of specimen has the advantage that the biaxial stresses produced are quite uniform throughout the entire wall thickness, and these stresses can be calculated from simple formulas. Hydrostatic pressure can be applied with a universal testing machine by first filling the cylinder with a very high-viscosity hydraulic medium, such as a silicone

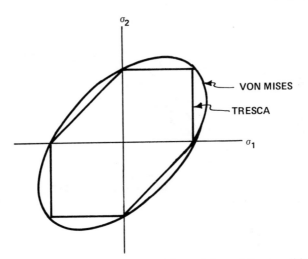

Figure 16.21   Comparison of the von Mises and Tresca failure conditions.

compound, or with a noncompressible elastomer, such as polyurethane, inserting close-fitting plugs into the ends, and then loading. Other useful techniques involve stretching films or plastic sheets simultaneously in two directions using a special mounting fixture consisting of a cable and pulley system on the test machine.

While various tests methods can be used to experimentally measure biaxial properties, one feature shown in Figure 16.21 is critical in test development: when combined tensile-tensile or compressive-compressive forces are applied, the failure stress is greater than or equal to the stress in uniaxial tension or compression; when combined tensile-compressive forces are applied, the failure stress is lower than in either uniaxial tension or compression. Thus any biaxial test design must match the biaxial stress condition in end use to be meaningful. ASTM Method D-2568 (31), for example, describes a hydrostatic compressive strength test of glass-reinforced plastic cylinders where compressive-compressive forces are imposed. ASTM Method D-2585 (31) describes a hydrostatic tensile strength of filament-wound pressure vessels where tensile-tensile forces are imposed.

16.5.2.8   Fracture Toughness Tests

Various physical abuses such as nicking, scratching, and cracking, can cause stress risers, or regions where the stress is much higher than in the rest of the body. Similarly, a part containing holes, sharp fillets, and so on or machining or molding imperfections will have built-in stress risers. Since the stress will be higher in these regions, failure will initiate there. A logical question concerns whether a material can contain a given size flaw without failing or can arrest to the progress of the growth of a crack. If this is not the case, the crack may grow at an uncontrolled rate and the material may fail catastrophically. The ability of materials to resist such catastrophic failure is the subject of fracture mechanics. Fracture toughness, which is a material property, must be considered in structural applications of plastics.

A complete discussion of fracture mechanics testing is not within the scope of this chapter. However, some consideration must be given to tests designed to determine fracture toughness. For most materials, such as acrylics, the best indicator of the relative resistance to catastrophic failure in service is furnished by a measure of the resistance to crack propagation which can be accomplished by measuring the load necessary to cause a crack to grow in a stable manner. The crack extension force or critical strain energy release rate G_c, which is twice the value of fracture surface work, is frequently referred to in fracture toughness investigations. This is defined as Equation (16.35):

$$G_c = \frac{PC^2}{2w} \frac{\partial C}{\partial l} \qquad (16.35)$$

where P_c = applied load at fracture
w = crack width
l = crack length
C = total specimen compliance at crack length

If the specimen is designed so that the compliance changes linearly with crack length, that is, $\partial C/\partial l$ = constant, the G_c depends only upon the failure load P_c provided that the crack width remains constant.

A commonly used specimen shape that meets these requirements and that is suitable for both brittle and ductile materials (42) is the cleavage specimen in the shape of a double-tapered cantilever beam as shown in Figure 16.19f. The specimen taper necessary so that $\partial C/\partial l$ = constant is given by Equation (16.36)

$$l = \{[Nh - (1 + \mu)]^{1/2} - 1.2\}\frac{h}{2} \quad (16.36)$$

where h = specimen height
N = constant
μ = Poisson's ratio

When the ends of the beams are deflected by pulling in the test machine at a constant crosshead speed, the crack grows at a constant rate proportional to that speed. The force to extend the crack will be nearly constant as shown in Figure 16.22 and can be directly related to fracture surface work by using Equation (16.35).

Various other specimen shapes used for fracture toughness, tests such as a notched beam and compact tensile specimen, are described in ASTM Method E-399 (43). These are suitable for plastics if the validity conditions discussed there can be satisfied. The validity conditions for testing high density polyethylene pipe are discussed by Flueler, et. al. (44). Hertzberg (45) discusses the deformation and fracture mechanics of engineering materials in detail and the reader should study this work before conducting fracture toughness experiments.

16.5.3 Impact and High-Speed Measurements

A fast buildup of forces (shock loading) may be encountered in plastic toys, refrigerator liners, and a host of consumer and industrial products. An increase in the speed of testing has an effect on mechanical properties equiva-

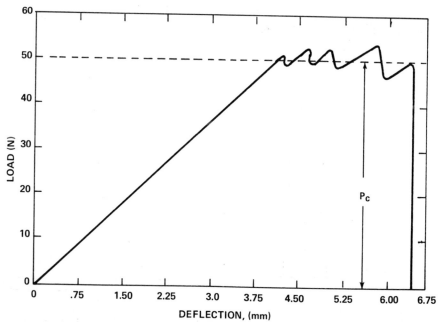

Figure 16.22 Typical load-deflection curve in a fracture toughness test using a double-tapered cantilever beam.

lent to a decrease in temperature. Thus, materials that were ductile in quasi-static tests may be brittle in high-speed tests.

Successful performance is achieved by polymers that remain "tough" over the entire service range, that is, they do not show a ductile-brittle transition. High-speed tests are therefore designed to measure the toughness of plastics. Such tests comprise both those previously described for quasi-static tests (particularly tensile, shear, puncture) and certain other biaxial impact tests such as impact tests employing a falling weight to simulate actual "blows" received in service. Typical loading modes are shown in Figure 16.23.

Standard impact testers obtain their energy through the acceleration by gravity of an impact head released from a fixed height. The head may be free-falling or may fall through an arc if attached to a pendulum. The speed of impact testers cannot be varied greatly; they all have speeds in the range of 250 m/min. In testers such as shown in Figure 16.24a, the amount of energy available is limited to the potential energy of the falling weight. When a free-falling weight test is used, failure can be defined in several ways: energy necessary to initiate a visible crack; energy necessary to produce a specific deformation (for ductile materials); or energy necessary to shatter or

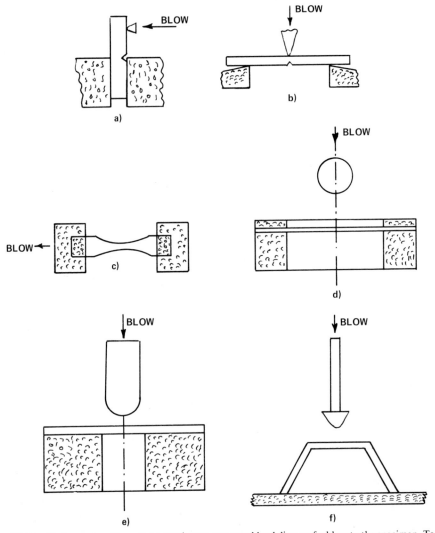

Figure 16.23 High-speed and impact properties are measured by delivery of a blow to the specimen. Test may be uniaxial flexure (*a*, *b*), uniaxial tensile (*c*), or biaxial (*d, e*). Test may also be on finished part (*f*).

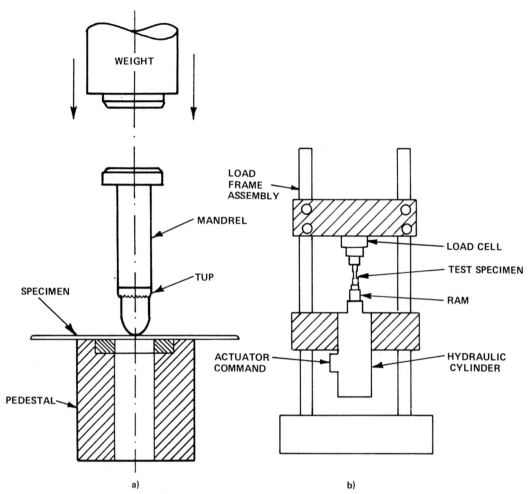

Figure 16.24 (*a*) Typical type of falling impact tester is relatively simple. More advanced types contain an instrumented tup. (*b*) Typical components of a servo-hydraulic high-speed test machine. The velocity of the high-speed machine can be varied over a wide range whereas the velocity of most impact machines is about 10,000 in./min.

puncture a particular specimen. When a pendulum tester is used failure is defined as the total energy necessary to initiate and propagate a crack.

A fundamental type of impact test using a pendulum machine is the tensile impact energy test described in ASTM Method D-1822 (8). By using two types of specimens, that is a type S or a type L specimen, two different strain rates are possible. One problem with other standard impact tests is that they are in general hybrid or at best practical and are more useful for comparison of materials than for absolute measurements. Special tests that simulate end-use conditions include the reverse impact of coated panels. Here failure is

defined as the energy necessary to cause cracking of the coating on the surface when impacted from the back side under standard condtions and gives, for example, an indication of the quality of a gel coating on a polyester laminate. Other impact tests that deliver blows similar to those received in service include ASTM Method D-3029 (8) for rigid materials, ASTM Method D-2444 (46) for pipe and fittings, and ASTM Methods D-3420 (31) and D-1709 (31) for plastic film. The magnitude of the values obtained in these impact tests depends upon the profile of the missile, the shape of the specimen support, and the method of clamping and should thus be considered as relative and not absolute.

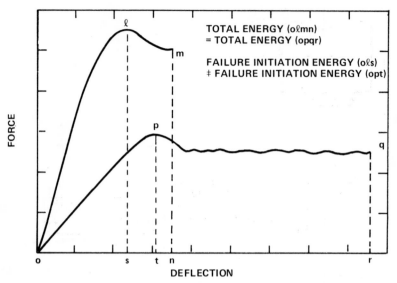

Figure 16.25 The impact energies (area under curve) of the two materials are measured as being equal on an noninstrumented impact tester. An instrumented impact tester or a servo-hydraulic high speed machine would show completely different stress-strain properties for the two materials.

The notch profile of the Izod or Charpy test described in ASTM Method D-256 (8) has little relation to most service applications. When such tests are used, notches of different radii should be used to change the severity of the test and to establish where or if the brittle/ductile transition occurs (Figure 16.3). Burnes (47) describes various impact test methods in more detail.

The biggest problem with impact testers is the manner in which the impact strength is defined. The impact strength is usually determined as the energy to rupture, either as a statistical 50% failure point in free-fall tests or as a scale reading in pendulum tests. However, two materials may have the same impact energy but have entirely different load-deformation curves, as shown in Figure 16.25. If one is interested in the amount of energy that a material can absorb without permanent damage, that is, without exceeding the proportional limit, there is a large difference between the materials. One means to overcome the limitation of standard impact tests is to use instrumented impact machines that plot data in the form shown in Figure 16.25. Such instrumented impact machines are commercially available. Ireland (48) and Rieke (49) discuss several applications of the instrumental impact test for evaluations of polymers and components.

The foregoing limitations of standard impact tests can also be overcome by using high-speed testing machines. In comparison with quasi-static testers operating at speeds up to slightly greater than 1 m/min, high-speed testers usually employ pneumatic or hydraulic drives to obtain speeds close to 1000 m/min as shown in Figure 16.24b. A slack adapter is most often used to allow the movable member to attain the desired test speed before impacting the sample. Since the speed can be varied, the rate sensitivity of a material can be measured by generating test data over a broad, usually logarithmic, time scale (51). For high-speed testers, there is an excess of energy available to insure rupture of the sample. Load versus deformation or load versus time may be recorded with a cathode ray or digital oscilloscope and recording camera, magnetic tape, computer, or other state-of-the-art techniques. In addition to toughness, most of the same type of information that is obtained from quasi-static stress-strain curves can be calculated. ASTM Method D-2289 (8) for high-speed tensile tests and ASTM Method D-3763 (8) for high-speed puncture tests should be consulted. Unfortunately, there has not been a wide-scale use of high-speed testing, mainly because testing equipment is expensive. However, to obtain fundamental properties over a wide range of

Table 16.2 Toughness of Plastics Determined by Various Methods

Material	Impact Properties		High-Speed Properties	
	Notched Izod (ft-lb/in.)[a]	Falling Dart (ft-lb)[a]	Tensile Energy (in.-lb/in.²)[a]	Puncture Resistance (lb/in.)[a]
Polycarbonate	17.9	34.8	4100	4150
ABS, high impact	9.5	35.0	3975	2500
ABS, general purpose	5.0	18.0	2600	1600
Polypropylene, high impact	1.8	1.0	2200	1800
Polystyrene, high impact	1.2	1.2	1400	950
Linear polyethylene	1.1	9.0	2000	1100
Nylon, general purpose	1.0	35.0	2850	800
Polyacetal	0.7	1.0	2100	950
Speed (in./min):	10,000	6,000–10,000	2000	2000

Source. F. J. Furno et al. (52)

[a]Conversion to SI units: 1 ft-lb/in. = 53 J/m; 1 ft-lb = 1.35 J; 1 in.-lb/in.² = 175 N-m/m; 1 lb/in. = 175 N/m.

conditions, an investment in equipment is recommended.

The problem of single-point impact testing can be seen by comparing the Izod impact, falling-dart impact, high-speed tensile strength, and high-speed puncture resistance of eight plastics as shown in Table 16.2. Different tests rank the materials differently. The reasons for these changes in relative ranking of test materials can be understood only by considering the fracture toughness and the profiles of rupture strength and energy versus impact speed. Such a comparison is shown in Figure 16.26. Here the puncture resistances of polycarbonate, nylon, and acetal are compared as a function of speed. From this figure one

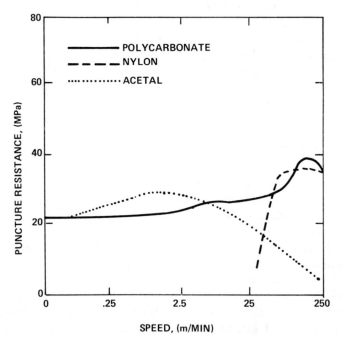

Figure 16.26 Comparison of the puncture resistance of three plastics as a function of impact speed, based on work done by F. J. Furno, et al. (52).

would expect the impact resistance of nylon and acetal to be comparable at an impact velocity of 50 m/min and to be much lower than the impact strength of polycarbonate. On the other hand, at 250 m/min the impact strengths of the nylon and polycarbonate should be equal, and the impact strength of the acetal should be much less. These observations are supported by the data in Table 16.2. The Izod impact strength of the acetal would be expected to be low because it has undergone a ductile-brittle transition before an impact speed of 250 m/min, that is, it is intrinsically brittle at this speed. However, nothing can be said about the relative Izod values of nylon and polycarbonate because the Izod test is a form of the fracture toughness measurement described in Section 16.5.2.8. Results will be dependent on the notch radius. Actually, both nylon and polycarbonate are quite notch sensitive; however, using the standard ASTM Method D-256 (8), notch radius causes nylon to fail in a brittle manner whereas it allows the 6 mm thickness polycarbonate to fail in a ductile manner.

If high-speed or instrumented impact testing capabilities are not available, the most realistic approach to impact testing of plastics at this time is to use several of the tests mentioned, bearing in mind the limitations inherent in each test, and then to qualitatively estimate the potential impact resistance of any material for a given application. If impact plays a critical role in the application, then prototype testing using a component tester such as described by Starita (53) is also recommended.

16.5.4 Long-Term Mechanical Properties

Plastic materials must be able to sustain long-term stresses or strains to be suitable for many uses, such as structural applications. Requirements may be endurance or sensory related. For example, if the walls of a plastic pressurized soda bottle creep during warehouse storage, the liquid level in the bottle will drop. Although this may not interfere with performance, the product may not sell because the customer may feel he is not getting a full measure.

In the most general terms, data are required to predict the behavior of a material throughout the life of the product. However, an important factor to recognize is that "long-term" is relative and depends on the application. Thus, in one case it may mean a time of less than one hour whereas in another case it may mean a time of more than one decade. While quantitative data can be obtained by applying the tests described in the previous sections, short-term properties do not determine if the polymer will deform to an unacceptable extent or fracture in long-term use at the stress levels encountered and in the environment proposed.

Assessment of material for many uses may be made from the results of long-term testing. Both changes in the dimensional stability caused by continued slow polymerization, crystallization, plasticizer loss, change in water content, and so on, and changes in structural integrity must be included in the measurement technique. Thus, when working at temperatures other than ambient, unstressed control specimens must be used to establish changes in stability.

In many situations materials or components will be subjected to stresses or strains that are steady (step input); steady for some period with intervening periods of negligible stress (gate input); or periodic (sine or square wave input).

16.5.4.1 Static Creep Tests
Creep tests involve stressing specimens of the same geometry used for short-term tests. In particular, ASTM Method D-2990 (8) specifies three loading modes: tension, compression, and flexure. The governing Equation (16.37) applies in all cases.

$$E(t) = \frac{\sigma_0}{\epsilon(t)} \qquad (16.37)$$

That is, all measurements of creep can be regarded as a simple matter of applying a static load input to a specimen and measuring the corresponding response deformation as a function of time. A typical set up of a creep testing machine used to apply such loads is shown in Figure 16.27. The essentials of such

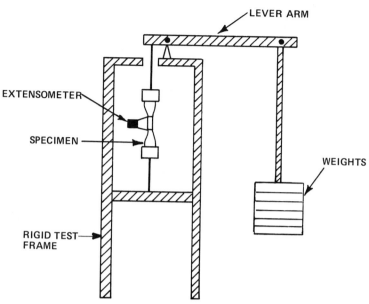

Figure 16.27 A long-term static creep and stress-rupture instrument.

machines are that the applied force should not cause extraneous bending and the strain should be measured in such a way that end effects are eliminated.

Like quasi-static tests conducted in tension or compression, large loads may be required to produce relatively small deformations. Usually a cathetometer of extensometer is needed to measure deformation. However, in flexure, even small loads can produce strains exceeding 5%, and dial gauges are usually suitable for following deformation.

The principles involved in creep measurements are the same as in short-term testing: defining stress-strain-time relationships over the whole range of expected service life. A family of creep curves constitutes the basis of a system of data on polymer deformation. For convenience these curves can be abstracted by the use of three families of orthogonal sections, as shown in Figure 16.28. A section at constant stress provides creep strain-time data, one at constant strain provides isometric stress-time data, and one at constant time provides isochronous stress-strain data.

The experimental work involved to obtain such curves is considerable, therefore the duration of the tests and the number of stress levels should be held to a minimum just to define the stress-strain-time relationship over

the whole range of expected service. Manipulations, such as by numerical extrapolation, can then be performed to evaluate data at the desired conditions.

Chambers (54) discusses the application of Findley's mathematical creep prediction models for use in structural design. The creep data provided by ASTM Method D-2990 (8) can be analyzed in a manner to obtain the necessary constants for this model.

Turner (55) has also outlined a straightforward testing regimen for creep evaluations. Using a similar approach, creep properties can be measured as described in the following sections.

Deformation. Determine the isochronous stress-strain curves at various times. A simple technique involves loading a specimen at stress σ_1 for time t, measuring the strain at time t, removing the stress and allowing the specimen to recover for time $4t$, and reloading the specimen at stress $\sigma_2(\sigma_2 > \sigma_1)$. An alternative is to load new specimens. The preceding stress and time sequence is repeated until a stress σ_n is reached. Usually three or four stresses are sufficient. In most materials testing, the isochronous stress versus strain data are usually arbitrarily obtained at times of 100 sec, 1hr, 100 hr, and 1000 hr. The applied

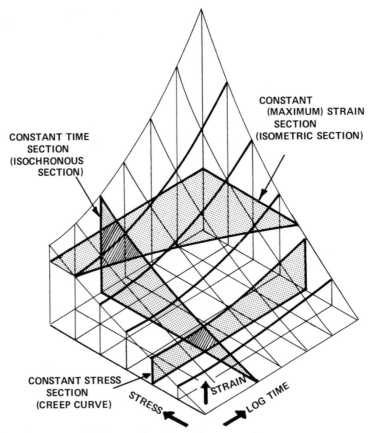

Figure 16.28   The stress-strain-time relationship obtained by static creep measurements. The stress is applied at zero time, based on work done by Turner (55).

stress plotted against the change in strain in time t represents the isochronous stress-strain relationship. The strain plotted against time t for each stress represents a creep curve. These curves can then be interpolated for other stresses in an exact correspondence to the isochronous data. Typical curves for a modified PMMA are shown in Figure 16.29. Calculations of this sort are readily performed by a computer. Also, from these curves isometric stress-time curves can be constructed and are especially useful when data obtained at several temperatures or after various aging times are presented on a single plot.

A traditional source of typical value data (56) has recently started to include creep charts that provide information on various trade name plastics.

Recovery. The discussion of isochronous curves has implied that these strains are

recoverable when the load is removed, assuming that imposed strains are below the yield point. Information on recovery after creep is particularly important for plastic components that are loaded only intermittently. Cumulative strain produced by an intermittently applied load is considerably less than the strain produced under the same load applied continuously for the same total time since the specimen has time to recover.

The most concise means of presenting recovery data is to plot fractional recovery of strain versus reduced time. These are defined in Equations (16.38) and (16.39)

$$F = \frac{\epsilon_R}{\epsilon_c} \qquad (16.38)$$

where F = fractional recovery of strain
ϵ_R = strain recovered at time t after removal of load
ϵ_c = total creep strain

and

$$R = \frac{t_R}{t_c} \qquad (16.39)$$

where R = reduced time

t_c = total time creep stress is applied

t_R = time that the stress has been removed.

By using these equations, data from different time periods and strains can be compared on the same plot. For example, Figure 16.30 compares the recovery of an unplasticized PVC after a short loading time at low stress with the same material after a long loading time at high stress.

16.5.4.2   Stress Relaxation

Stress-relaxation tests may be used as alternatives to creep tests as a fundamental means of studying time dependence. These tests are conducted at constant strain and the relaxation of stress follows. The governing Equation (16.40) is applicable in this case.

$$E(t) = \frac{\sigma(t)}{\epsilon_0} \qquad (16.40)$$

Thus these tests provide a more direct means of measuring many phenomena where the plastic is confined to a certain shape or must retain a certain shape during its use. For example, the determination of the time-dependent stress holding a metal insert in a

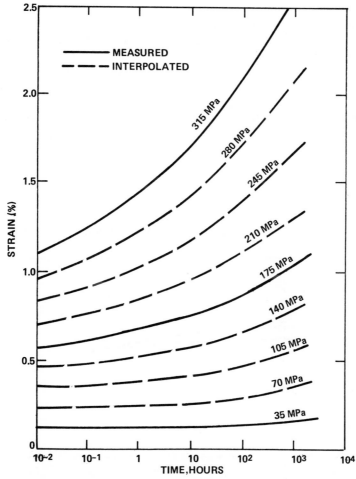

Figure 16.29   Complete family of creep curves for PMMA generated by first constructing three isochronous curves and then interpolating.

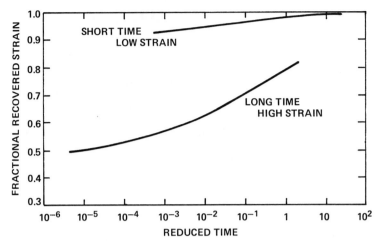

Figure 16.30 Recovery from creep for an unplasticized PVC iillustrate how data from different time periods and strains may be compared on the same plot by using the reduced time, based on work done by Ogorkiewitcz (57).

fabricated plastic part may be investigated by stress-relaxation tests. A system of data analogous to that discussed for creep testing can be generated for relaxation tests. The only difference in detail is that strain is prescribed instead of stress. However, creep modulus and stress-relaxation modulus are not equal except under special conditions owing to the strain-dependence of the modulus. Therefore one can write Equation (16.41).

$$\frac{\sigma_0}{\epsilon(t)} \neq \frac{\sigma(t)}{\epsilon_0} \qquad (16.41)$$

The creep experiment starts at a strain lower than the stress-relaxation experiment, ap-

proaches the strain of the stress-relaxation experiment, and may exceed it if the creep experiment is conducted for a long enough time. The time-dependent moduli obtained from the two experiments will appear as shown in Figure 16.31. The creep modulus E_c will start at a higher value than the stress-relaxation modulus E_{SR} owing to its lower strain, intersect the stress-relaxation modulus at approximately the strain of the stress-relaxation experiment, and then drop below E_{SR}.

ASTM Method D-2991 (8) describes a procedure for stress-relaxation measurements. Experimentally, however stress relaxation is more difficult to measure than creep. To

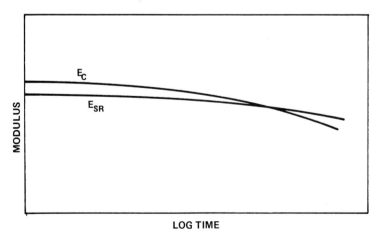

Figure 16.31 A comparison of the time-dependent creep and relaxation moduli, based on work done by Passaglia and Knox (58).

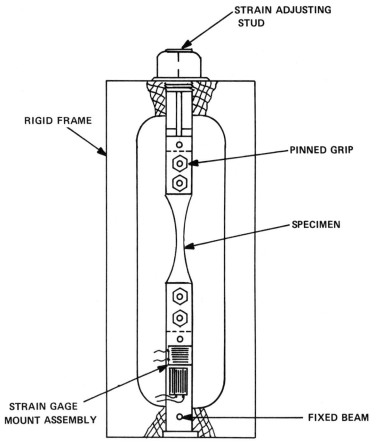

STRAIN ADJUSTING STUD

RIGID FRAME

PINNED GRIP

SPECIMEN

STRAIN GAGE MOUNT ASSEMBLY

FIXED BEAM

Figure 16.32 Stress-relaxation device with plastic specimen placed in the grips based on work done by Curran et al.

measure a decay in force as a function of time, if is necessary to have some type of load cell for each specimen. Also, the machine must be very stiff compared to the polymer stiffness if tensile of compressive tests are to be conducted. For example, Nielson (17) cites the case of a rigid specimen, initially 25 mm long and stretched 0.025 mm in a relaxation experiment. The load cell and the appartus must not deform by more than 250 μm if stress measurement accurate to 1% is to be achieved. A "square donut" apparatus suitable for such measurements, described by Curran, et al. (59) and shown in Figure 16.32, is suitable.

A dilemma that experimentalists must face, however, is the selection of a test specimen. If a dumbbell tensile specimen is used, then the initial strain must be set with an extensometer. It is obvious that the stress will be greater in this narrow portion of the length. This narrow section may creep, however, permitting the larger sectional areas of the specimen to retract and thus change the strain from its initially set value. If a uniform cross-sectional bar is used, then the grip slippage, machine deformation, an so on that were discussed in Section 16.5.2.3 can invalidate the results. One solution is to use a closed-loop or universal test machine with strain-holding capabilities. Here the output of the extensometer attached to the reduced section of the specimen is used to change the crosshead position to maintain the desired constant strain.

Another alternative is to conduct flexural measurements that have the same experimental advantages as discussed in Section 16.5.2.4. They also are subject to the same theoretical errors as discussed there.

16.5.4.3 Stress Rupture and Endurance Limit

Materials loaded for long times fail at lower stresses than the same materials loaded for short times. These stresses are often no greater than 20–30% of the quasi-static strength. The exact limiting stress will depend on both the time that the sample is under stress and the test environment. At high stresses, close to the short-term strength, failure takes place quickly, whereas at lower stresses, failure may not be likely for years. Increasing the temperature tends to cause the specimen to fail in shorter times. The endurance limit of a material at a specific test condition is the stress below which the material may be loaded for an infinite time in comparison with the end-use application without failure. Rupture and yield stress are not the only factors limiting the loads that can be imposed on plastics. This limit depends on the criterion of failure for the given component

such as crazing or elongation. Figure 16.33 compares three types of failure for a PMMA. Obviously, failure other than rupture is more difficult to observe. Regardless, tests have the common feature that the specimens are subjected to a stress, defined as some function of their short-term failing stress, and the time that elapses between application of load and failure is recorded.

The simplest case of long-term failure is subjecting a set of specimens to step excitations of different amplitudes and measuring the time to failure. This is often referred to as static fatigue. Plots of stress versus time to failure are called Wohler plots. ASTM Method D-2990 (8) discusses conducting stress-rupture tests on plastic sheets. The same test machines and test specimens used for creep testers are employed. ASTM Method D-1598 (46) describes a similar component test for plastic pipe under long-term hydrostatic pres-

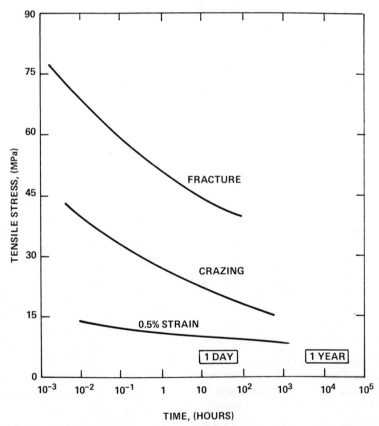

Figure 16.33 Stress creep fatigue in tension for PMMA versus various types of failure showing that fracture, crazing, or some limiting strain may be defined as failure.

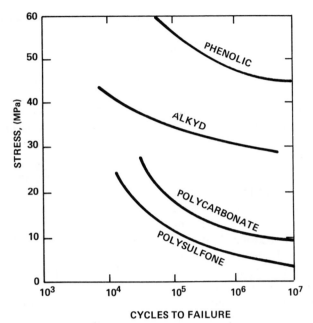

Figure 16.34   Fatigue life by some plastics measured using ASTM Method D-621, based on work done by Riddell (25).

sure. Other types of component or practical tests may be made where the importance or volume of the particular part warrants it.

The second case of long-term failure involves subjecting specimens to sinusoidal input excitations of different amplitudes or of different stresses. This is referred to as dynamic fatigue. In recent years there has been a general increase in the use of plastics in industrial and consumer equipment involving moving parts and mechanical components, such as nylon gears, where fatigue is one of the major modes of failure.

The principal object of most traditional fatigue testing of plastics has been the generation of S-N curves. These curves show the failure stress S against the number of cycles to failure N. The data are obtained by testing many specimens at different stress or strain levels. As with static fatigue, the life to failure increases as the stress is decreased.

While several types of commercial machines have been developed that operate in several modes, all use either a constant amplitude of force or a constant amplitude of deflection input excitation. In the former, rupture or a specified increase in strain is defined as failure. In the latter, a specified decay in specimen stiffness is defined as failure. The type of machine used should match the end-use condition. With constant amplitude of force machines, such as described in ASTM Method D-671 (8), a linear taper cantilever specimen is used to maintain a constant stress over the entire specimen face. Typical fatigue curves determined by ASTM Method D-671 (8) are shown in Figure 16.34. A limitation of most standard fatigue testers is that the sample can only be flexed at a constant frequency of 30 Hz. This is a speed frequently encountered in motor-driven mechanical parts. However, data obtained at this condition may not be applicable to other test frequencies.

This is because a major factor in fatigue of plastics is damping in which a significant portion of the applied energy is dissipated as heat. Many plastics have a high damping index (see Section 16.5.1) in specific regions of frequency and temperature. In addition, the heat generated also depends on how fast the external work is applied to the specimen, which depends on frequency. This can be summarized by an Equation (16.42) presented by Riddell et al. (60)

$$\frac{dT}{dt} = \frac{\pi \nu j''(f, T)\sigma^2}{\phi C_p} \qquad (16.42)$$

where dT/dt = temperature change per unit time

ν = frequency in cycles per second

j'' = loss compliance in $cm^2/dyne$ and is a function of the loss index

σ = amplitude of oscillating stress in dynes/cm^2

ϕ = density

C_p = specific heat

This factor, combined with the low thermal conductivity of plastics, may cause materials to rise in temperature under repeated stressing. The time available to transfer heat away between cycles is directly dependent on frequency. The above considerations favor the decrease of fatigue life with increasing test frequency and increasing damping index. Such a decrease is shown in Figure 16.35 for polytetrafluoroethylene (PTFE).

Since thermal effects are one of the controlling factors, fatigue life can also be increased by maximizing heat transfer to the surroundings. Since thinner specimens have higher rates of heat transfer than thicker specimens, thinner specimens will also have higher endurance limits (60).

Polymeric materials may also fail in fatigue because of crack initiation and propagation. Hertzberg and Manson (61) discuss the use of the relationship between crack growth per cycle and the fracture toughness of the material as a measure of fatigue resistance. The fatigue crack propagation of a particular polymer relative to another can be quickly determined

when data are presented in this manner instead of in S-N curves and then used to predict relative service life. A fatigue crack growth rate procedure described in ASTM Method E-647 (43) is suitable for plastics if the validity conditions described there are satisfied. In the development of relevant fatigue data, the application should be carefully analyzed and distinction between failure by thermal softening and by flaw propagation considered. Hertzberg and Manson's text (61) is recommended as a guide in conducting meaningful fatigue tests.

The test equipment and test specimens described in Section 16.5.2 can be used to conduct low-frequency cycling tests if the machine is equipped with load-cycle and extension-cycle controls. In such cases, the input is a sawtooth instead of a sinusoid. High-frequency cyclic tests, with various-shaped input excitations, can also be conducted using the servo-controlled high-speed testers described in Section 16.5.3. ASTM Method D-3479 (31) describes a tension/tension fatigue test for composites using a servo-hydraulic test machine.

16.5.4.4 Prediction of Long-Term Strength Behavior

The approaches used for creep and endurance limit determinations, for example, the construction of Wohler curves, are very slow, and some form of predictive techniques may be useful. Most predictive techniques involve extrapolation of results over a relatively short time, usually less than 1000 hr, to yield

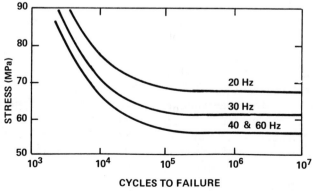

Figure 16.35 Effect of test frequency on the fatigue life of a PTFE type of plastic, based on work done by Riddell (60).

performance for 10,000 hr, 100,000 hr (11 years) or even longer. The extrapolation techniques described in ASTM Method D-2837 (46) or those discussed by Chambers (54) are recommended to determine strength regression over long periods of loading. One procedure for endurance limit that has been somewhat successfully used by experimentalists, however, is the Prot technique (62) where the maximum stress is increased at slow but constant rates until failure occurs.

The stress failure σ_R is given by Equation (16.43).

$$\sigma_R = \sigma_L + K_1 \left(\frac{d\sigma}{dt}\right)^{1/2} \qquad (16.43)$$

The equation defines a straight-line relationship between stress and the square root of the stress rate $d\sigma/dt$, which intersects the stress axis at the endurance limit σ_L. Applying this technique to a cyclic stress test, both Prot (using metals) and Lazar (63) (using plastics) reported estimated endurance limits that agreed well with Wohler values.

Loveless et al. (64) modified the Prot technique to obtain an equation for static fatigue:

$$\frac{(d\sigma/dt)t_B^2}{2} + (\sigma_0 - \sigma_L)t_B = K_2 \quad (16.44)$$

where t_B = time to break
σ_0 = preload stress

The endurance limit is derived from this equation by plotting $(d\sigma/dt)t_B^2$ versus t_B. This yields a straight line, the slope of which is -2 $(\sigma_0 - \sigma_L)$. Test data indicate that the endurance limit is preload (σ_0) dependent, which in turn is a function of the material under test. This technique has been used successfully to measure the static fatigue of polyesters (64).

Prot techniques, of course, do not permit measurement of the effects of environmental degradation. The influence of such factors should not be ignored when evaluating the load-bearing ability of materials over extended periods of time. However, such tests do provide some yardstick for quickly checking the anticipated long-time strength of plastics.

They at least provide an indication of the maximum endurance limit, and that in itself is very useful in many cases. For example, Loveless (65) used this technique to demonstrate that the proposed use of a glass reinforced polyester supporting cable for an electrical power transmission line would lead to failure.

16.6 SURFACE MECHANICAL PROPERTIES

Plastics are used in a number of applications where the mechanical properties of the surface are of interest. For example, plastic laminates are used extensively for kitchen table, bar, and counter-top applications. Plastics are being used in the construction of unlubricated bearings and gears. Their use in resilient floor coverings is growing. Other products in which plastics with outstanding surface characteristics are being used include windshield and glazing materials, photographic film, magnetic tape, cable covers, footwear, and tire chains.

However, mechanical testing of surfaces at best must be viewed with great skepticism. This is because all surface tests—indentation hardness, scratch hardness, abrasion and wear, and friction—have one common feature: they involve the measurement of the interaction of the plastic with an arbitrary external surface and are therefore practical or hybrid in nature. Interactions between the plastic and various external inputs are shown in Figure 16.36. In cases where it is possible to relate the particular disturbing mechanism to a fundamental property, then the prime test may be for that fundamental property. Otherwise the best procedure is to compare candidates in actual end use, that is to perform component tests. When this is not possible, however, we are forced to conduct practical or hybrid tests.

Because of their practical or hybrid nature, it should not be expected that results obtained from different kinds of surface tests will agree with each other. Sometimes this is the fault of the test, but more often it is caused by the differences in the particular mechanisms utilized. When properly conducted, the results of these tests are quite valid for the conditions

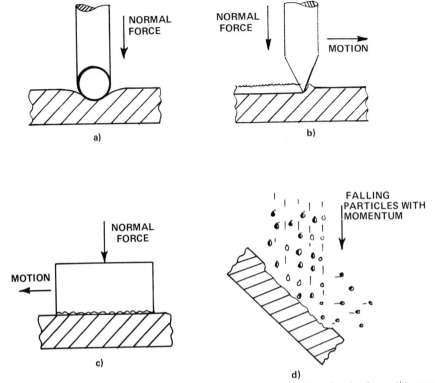

Figure 16.36 Various methods of measuring surface properties: (*a*) indentation hardness; (*b*) scratch hardness; (*c*) rubbing abrasion, wear, and friction; (*d*) erosion.

of the test, but the tests simply may not be relevant to certain end-use applications. No single test can simulate all the modes encountered in service. Consequently, in testing for a given surface property, it is important not to adhere unwaveringly to a single test mechanism. Rather, with each problem, it is essential to select a test that approaches the parameters of actual usage. Therefore, some of the various surface tests are summarized in the following sections.

16.6.1 Indentation Hardness

Indentation hardness, represented in Figure 16.36*a*, refers to the resistance of a surface to change by local mechanical action and carries with it the connotations of both a resistance to the disturbing deformation and the absence of residual effects when the disturbance is removed (66). Hardness is not a function of a single property but is related to several material parameters such as modulus, yield strength, recovery, and permanent deformation. The

nature and extent of these effects depend upon the amount and rate of deformation and the temperature. At present, let us consider only those cases where the deformation is normal to the surface, with no sliding along the face of the test material.

One simple method of measuring hardness uses a Barcol Impressor to determine indentation. It applies a step input function and follows the output response function. In ASTM Method D-2583 (8), the indentor is a hardened steel truncated cone. The force on the indentor is applied by a calibrated spring having an arbitrary spring constant. The indentor penetrates the plastic to the point where the spring force and reaction force of the material are in equilibrium. The measurement includes both elastic and plastic components, which cannot be separated.

Another procedure, the Rockwell test, described in ASTM Method D-785 (8), uses hardened steel balls ranging in diameter from 1.6 to 12.5 mm as indentors. The indentor is pressed into the plastic for 15 sec with a force ranging from 60 to 150 kg. This major load is then reduced to a minor load of 10 kg for another 15 sec. Rockwell numbers, determined from a difference in scale readings, are not measures of total indentation but rather of the recoverable indentation, or resiliency. Curves shown in Figure 16.37 illustrate the differences between the Rockwell hardness of a metal and a plastic. In Figure 16.37*a*, representing aluminum, the ball penetrates the specimen but after a few seconds remains constant regardless of whether or not the major load is removed because of the nonrecoverable yielding of the metal. In Figure 16.37*b*, representing PMMA, the ball continues to penetrate as long as the

major load is applied. When the major load is removed, considerable recovery occurs. To cover the entire range of plastics, several Rockwell scales are used, such as R, L, M, E, K. Each scale uses a different ball diameter-load combination.

An alternative to the preceding method is the modulus-related alpha (α) Rockwell method, also described in ASTM D-785 (8). This method uses a 12.5 mm diameter steel ball under a 60 kg load and measures the total indentation observed under load after 15 sec.

The differences between the R scale and α scale measurements can best be illustrated by considering hardness as a function of temper-

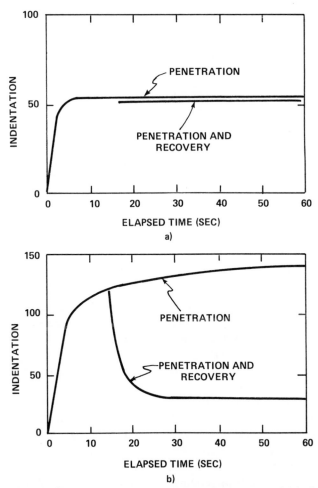

Figure 16.37 Curves illustrate the differences between Rockwell hardness of (*a*) aluminum and (*b*) PMMA. In (*a*) the ball penetrates in a few seconds and remains constant regardless of whether the major or minor load is applied. In (*b*) the ball continues to penetrate as long as the major load is applied. When the major load is removed, considerable recovery occurs.

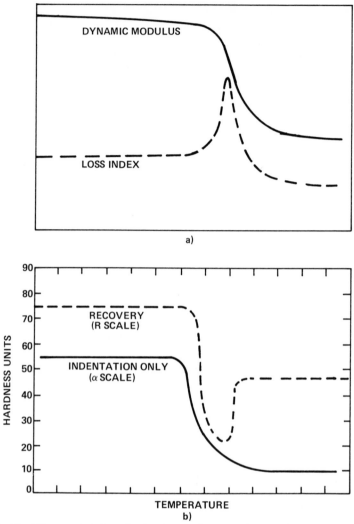

Figure 16.38   The differences in indentation hardness and recovery hardness in a Rockwell hardness test can be explained by considering the thermomechanical profile of the plastic.

ature. Consider the thermomechanical profile of a slightly cross-linked material that becomes rubbery above the T_g. Such a profile is shown in Figure 16.38a. The corresponding Rockwell R and α hardness curves are shown in Figure 16.38b. In the glassy region, the modulus is high so that total indentation, related to the α scale values, is low. Recovery is also high because the viscous component (loss index) is low. Both the R scale and the α scale Rockwell values are high. In the transition region, the modulus drops rapidly so that total indentation rapidly increases. The viscous component is at a maximum. In this region the R scale

Rockwell value reaches a minimum whereas the α scale value is intermediate. In the rubbery region, the modulus remains low and so do the Rockwell values. However, the viscous component decreases so that the R scale Rockwell value again increases.

Livingston (66) has suggested that the results of hardness tests that measure the resistance and resilience aspects be judiciously combined into one number. For example, the product of hardness values shown in Figure 16.38 would be high in the glassy region, low in the transition region, and somewhat higher than the α scale alone in the rubbery region. The

Table 16.3 Comparison of Rockwell Hardness Values[a]

Material	Rockwell Hardness Values		
	α Scale	R Scale	Combined Scale
Melamine–paper laminate	132	129	172
Phenolic, mineral filled	130	124	154
PMMA	102	127	129
Polyester	91	125	119
Cellulose acetate	68	115	100
Cellulose acetate–butyrate	5	82	66

Source: D. L. Livingstone, in J. V. Schmitz and W. E. Brown, Eds., *Testing of Polymers*, Vol. 3, Interscience, New York-London, 1967, p. 111.

wider separation of data obtained by a combined hardness value for some various materials is shown in Table 16.3.

16.6.2 Scratch Hardness

Scratch tests involve pressing a hard object or a precisely ground point into the polymer with a known normal force and then moving the object laterally, causing the object to deform or fracture the surface, as represented in Figure 16.36b. The simplest tests involve either increasing the load on the indenting tool or increasing the hardness of the tool until a scratch is produced; measuring the width of a scratch produced by a standard load or hardness indentor (67, 68); or measuring the haze produced from a standard tool (69). The problems associated with such tests, however, are the same as those with indentation hardness tests: results depend on the modulus, yield stress, and recovery time of the polymer, as well as surface lubrication and probe velocity.

Another feature that should be considered is the mechanism of failure that the scratch test is to simulate. For example, a scratch or score may be produced by a cylindrical probe (such as the end of a paper clip) that has a sufficient vertical load to cause the plastic to undergo irreversible deformation. The most resistant materials to this type of scratching are materials with a high modulus and those materials with low modulus but high recovery. Scratch-resistant coatings fall into one of these categories.

A scratch may also be produced by a cutting tool (such as a knife) if the shearing forces exceed the shearing stress of the polymer, and localized rupture occurs. The latter mechanism is somewhat analogous to machining operations. Here the toughness and tearing resistance of the material are most important. It can be easily seen how a scratch mechanism involved when using some "standard" tool could shift from tearing to reversible deformation depending on tool sharpness, geometry, angle, and so on.

It is therefore not surprising that results have been so scattered that the ASTM has no standard test for scratch hardness even though considerable work has been done to develop such a test. Wiinikainen (70) has described these efforts to develop a new standard.

At one time the Bierbaum (68) test appeared in the ASTM standards as Method D-1526, but it has since been discontinued owing to lack of reproducibility. The basic principle of the test is that a diamond is drawn, under a vertical load, steadily over a horizontal test surface. The Bierbaum scratch hardness (BSH) is calculated from Equation (16.45)

$$BSH = \frac{\text{Load (kg)}}{[\text{width of scratch (mm)}]^2} \qquad (16.45)$$

Because of recovery, Bierbaum scratch hardness numbers are time dependent.

Gouza (71) has reviewed some of the other scratch tests used for plastics. However, the selection of any scratch test must be based on the mechanism of failure. Carroll (72) has found that a spherical stylus scratch test produces results reliably related to trade experience with the thin coatings on photographic films. When a general assessment of scratch resistance is required, several different tests should be employed.

16.6.3 Abrasion, Mar and Wear Resistance

The abrasion resistance of a solid body, defined in ASTM Method D-1242 (8), is its ability to withstand the progressive removal of material from its surface as the result of mechanical action of a rubbing, scraping, or erosive nature. Wear mechanisms can be grouped in three cases: deformation wear, which includes abrasive and fatigue wear; interfacial wear, which includes adhesive wear and material transfer; and chemical wear, which includes material loss by melting or degradation. Abrasion is a weight or volume loss. Marring is a precursor to abrasion and is a change in optical properties such as loss of gloss or the development of haze. Abrasion and mar are represented in general by Figures 16.36c and 2.36d. Tests designed to measure abrasion, mar and wear resistance are judged by their ability to rank materials, with a satisfactory degree of reproducibility, according to the results obtained under specific service conditions.

There are literally hundreds of abrasion and mar machines in use that have different characteristics and that actually measure different qualities. It would be impossible to describe all of them. Instead the various types of abrasive action and an example of an instrument that measures the particular characteristics are discussed. Abrasive, mar and wear tests for polymers can be grouped in the following way: rubbing contact, scraping or cutting, and erosion. These groupings are based both on the mating surface and the input function. The type of abrasive mechanism selected must match the end-use condition. In any case, whether weight loss or a change in material appearance is to be used as a criterion for failure is completely governed by the end-use application. Eiss discusses a systems approach to wear testing and correlations with end use (73). The selection and use of wear tests are also discussed by Bayer (74).

16.6.3.1 Rubbing Contact

When two unlubricated surfaces are placed in contact, such as in plastic bearings or slide mechanisms, they do not touch over the whole of their geometric area, since on a molecular scale solid surfaces are relatively rough. Actual contact occurs only at the tips of the surface asperities. High pressures developed over these small areas may cause them to interact or weld. During sliding without lubrication, patches on one surface adhere to the other and are pulled off or plucked out.

Any polymer rubbing against another surface without lubrication at a specific ambient temperature and tested in a specific geometry has a limiting combination of pressure and velocity that will cause failure. This combination is called the pressure-velocity or PV limit. Since combinations of pressure and velocity falling below the PV limit for a given polymer represent useful operating conditions, one is interested in the wear of the polymer below this limit.

Both PV limit and wear factor can be evaluated using equipment that can provide a known pressure to a polymeric surface moving at a known velocity. A thrust-washer test apparatus, described by Lewis (75) and ASTM Method D-3702 (76) is especially well suited for such testing. The sample, in the form of a flat washer with a raised lip, is rotated at variable velocities and pressures.

A series of wear tests is used as one means to establish the PV limit of a polymer. To determine one point on the PV limit curve, one velocity and a series of pressures are selected. Then the wear test is conducted at each pressure, and a wear factor is calculated by Equation (16.46) as follows:

$$K = \frac{d}{PVt} \qquad (16.46)$$

where K = wear factor
t = time
d = thickness of material lost

A radical change in slope of a plot of wear factor versus pressure, as shown in Figure 16.39, is the limiting pressure. The wear factor should be determined at several combinations of pressure and velocity including high pressure–low velocity and low pressure–high velocity. The curve drawn through points of limiting pressure at the various velocities describes the PV limit.

Using the average wear factor within the PV limit, the thickness loss/unit time can then be calculated for various PV combinations. A typical presentation of wear rates and PV limit is shown in Figure 16.40. The preceding discussion emphasizes the proper selection of test conditions for wear evaluations. Should one arbitrarily select test conditions above the PV limit, data will have little meaning.

16.6.3.2 Scraping or Cutting

Multiple cuts, scratches, or scores occur when a rough hard surface or a soft surface containing hard particles slides over a softer surface and plows grooves in it. Since polymers are generally soft materials, they are especially subject to this type of abrasion and marring. An important case is that of measuring the resistance of clear plastics to a loss of transparency by exposure to mild abrasive conditions, such as cleaning and wiping.

One method for measuring this type of abrasion is the Taber abraser. This is a slip-angle machine in which two sliding and rotating abrasive wheels comprised of abrasive particles embedded in a rubber matrix wear a path in a rotating sample. Abrasive wheels of various coarseness are available, but they are not rigidly calibrated for their abrasive characteristics. ASTM Method D-1044 (8) describes a procedure for using the Taber Abraser to estimate the resistance of transparent plastics to scraping and cutting abrasion. This test provides a discontinuous abrasion, and the test piece is allowed to recover a few seconds between exposures to the abrasive,

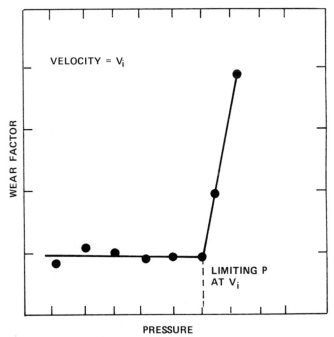

Figure 16.39 A plot of wear factor versus pressure will show an abrupt change at the PV limit. Meaningful wear factor data should be obtained below this PV limit.

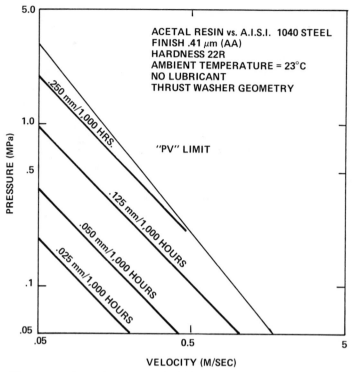

Figure 16.40   Wear rate results can be represented within the *PV* limit by a series of lines, based on work done by Lewis (75).

that is, the input is repeated gate. ASTM Method D-1242 (8) describes a similar test where a strip of abrasive rubs against a sample in a manner that fresh abrasive is always at the rubbing area. This latter test has the advantage that clogging of the abrasive paper with polymer is eliminated.

The Cyanamid Mar Tester (77) can also be used to simulate the action of wiping a dirty plastic window with a cloth. In this test, a spring-loaded cylinder presses a disc of 4/0 emery polish paper, which has abrasive particles about the same size as dirt particles collected on windows, against the plastic surface with a fixed pressure while the cylinder and disc are rotated manually for one revolution. The number of rings which can be seen with the unaided eye under approximately optimum observing conditions is a measure of sensitivity to marring. The abrasion is essentially continuous, that is, the input is a step function.

The same types of abrasion mechanisms that apply to scratch tests obviously apply to

scraping or cutting abrasion. Thus it is recommended that the selection of the abrasive be similar to the abrasive expected to be encountered in end-use including size, sharpness, and hardness.

16.6.3.3   *Erosion*

Another type of abrasion and marring to which polymers are frequently exposed in service is erosive wear where particles impact the surface at some velocity. For example, plastic windshields are marred by windblown sand. In such cases, the input is a repeated impulse. In standard tests, the particles are of varying shapes, and their size is fixed by screening. Momentum is gained either through free-fall or by air blasts. In all cases, the particles are randomly oriented when they hit the surface. Standard Ottawa sand of 20–30 mesh or silicon carbide are the two most commonly used abrasives. In one test, ASTM Method D-673 (8), No. 80 Carborundum is allowed to free-fall from a hopper down a tube of standard length and impinge on a plastic held at a 45 deg angle. In this method the change in gloss or haze is used as an end-point.

16.6.4   Friction

The frictional behavior of polymers governs their selection for a number of practical applications. For instance, it is desirable to have a high coefficient of friction between an airline serving tray and plate, but a low value of friction in a plastic bearing.

The coefficient of friction u is defined classically as the ratio of the tangential force f to the normal load N when the surface of one material is moved relative to another surface. Equation (16.47) defines the coefficient of friction.

$$u = \frac{f}{N} \qquad (16.47)$$

While Equation 16.47 defines the coefficient of friction as a constant, it is in fact dependent on speed, pressure, and the composition of the two materials in contact.

The important factor is to measure frictional forces at the conditions encountered in practice. The standard test described in ASTM Method D-1894 (8), where a plastic film is pulled over a second plastic film of the same material at a rate of 300 mm/min, tells little about the behavior of the film moving across metal processing rolls at several hundred feet per minute. In such cases, the variable speed frictionometer (78) described in ASTM Method D-3028 (8) is more suitable.

To measure the coefficient of friction of polymers at the service condition, the test apparatus must simulate the correct boundary conditions. For example, a modified friction and wear tester has been used to measure friction of polymer pellets under extruder conditions (79), necessary in extruder design and control. The curves shown in Figure 16.41 show the typical dependence of coefficient of friction on temperature and velocity for this case. This test apparatus is also suitable for

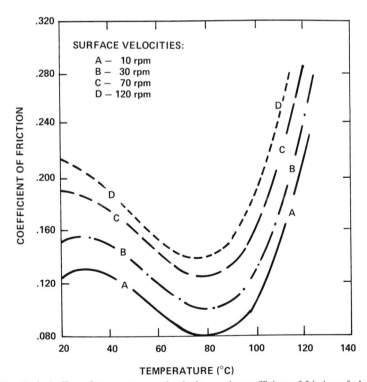

Figure 16.41   Typical effect of temperature and velocity on the coefficient of friction of plastic pellets. Courtesy of Scientific Process and Research, Inc.

studying the frictional properties of solid plastics as a function of pressure, velocity, and temperature.

Finally, it is possible for two materials to have the same average friction coefficient but have completely different stick-slip behavior that could signify differences in end-use behavior. For example, one material may chatter whereas a different material with the same coefficient of friction may run smoothly. To measure such a stick slip, it is necessary to use an autographic instrument.

16.7 ELECTRICAL PROPERTIES

While properly selected and performed mechanical tests may be the prime indicators of performance for a whole class of applications, relevant electrical tests are the prime indicators of performance for a whole class of other applications. Plastics are used in a wide variety of devices ranging from integrated circuit chips small enough to fit on the tip of a pencil to reinforced structures for antennae that measure many feet in diameter. The voltages involved may range from a fraction of a volt to perhaps 500,000 V. The primary electrical function of plastics is to act as insulators, that is, to separate metallic conductors at different potentials, often under severe ambient or design conditions of widely differing nature.

Important properties of electrical insulating materials may be obvious: low-voltage dielectric constant and loss factor, high-voltage breakdown resistance, high resistivity, and resistance to breakdown of the surface. What may not be obvious is the analogy to mechanical tests. First, electrical properties, especially in polar materials, are just as subject to time, temperature, thermal history, and moisture as are viscoelastic mechanical properties. Second, if electrical properties are compared to mechanical properties, dielectric breakdown voltage is equivalent to failure stress, AC dielectric constant and loss are equivalent to the storage modulus and damping index, electrical resistance is equivalent to viscous flow, and tracking and arc resistance are similar to mechanical abrasion and wear.

One may therefore expect that the most effective procedure for conducting electrical tests would be to follow the same guidelines used for mechanical tests, remembering the importance of stress, time, temperature, and moisture as well as the classification of each test as fundamental, component (or practical) or hybrid. Just as the grips or supports used to apply mechanical forces to the polymer impose boundary conditions, electrodes of various geometries also impose boundary conditions. Some of the various ways that electrical forces may be applied to a plastic through electrodes are shown in Figure 16.42.

16.7.1 Permittivity and Loss Factor

The ability of an insulator to resist the passage of alternating current or serve as a capacitor is determined by the permittivity (dielectric constant) and the dissipation factor. These properties are measured using low voltages so that bound charges are displaced but not ruptured. Such voltages are usually less than 500 V. Permittivity (dielectric constant) is a measure of the energy stored in a material subjected to electrical stresses. It is defined by Equation (16.48)

$$\epsilon' = \frac{C_p}{C_a} \qquad (16.48)$$

where ϵ' = permittivity (dielectric constant)
C_p = parallel capacitance of the material
C_a = capacitance of a dimensionally equivalent air (or vacuum) gap

The air capacitance C_a of a parallel capacitor having an area A and spacing t, is given by Equation (16.49)

$$C_a = 0.088542 \frac{A}{t} \text{ pF} \qquad (16.49)$$

if the dimensions are in millameters.

Loss index ϵ'' is related to the total energy dissipated by the system per cycle. The watts loss is given by Equation (16.50)

$$\text{watts loss} = 2\pi\nu C_a\epsilon'' E^2 \qquad (16.50)$$

where ν = frequency, Hz
E = voltage

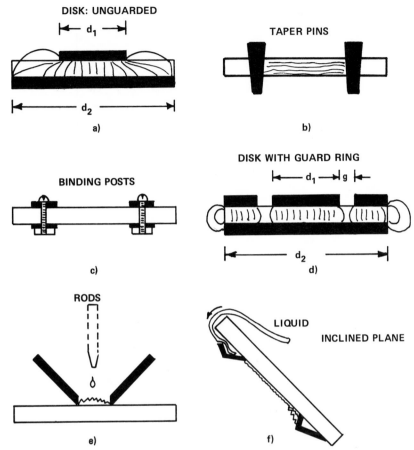

DISK: UNGUARDED

a)

TAPER PINS

b)

BINDING POSTS

c)

DISK WITH GUARD RING

d)

RODS

e)

LIQUID

INCLINED PLANE

f)

Figure 16.42 Various configuration used for electrical testing. Electrodes are used to cause current flow over the surface, through the volume, or a combination of both. Electrodes are also used to simulate end-use geometry. Liquids applied to the test surface simulate outdoor testing.

Dissipation factor is the ratio of the energy lost by the dielectric to the energy stored per cycle. This is defined by Equation (16.51)

$$\text{dissipation factor} = \tan \delta = \frac{\epsilon''}{\epsilon'} \quad (16.51)$$

and is analogous to the damping index defined for dynamic mechanical measurements.

Except for the field of capacitor design, where a high dielectric constant is desirable, the emphasis is toward low permitivity. Emphasis is almost always toward low loss index. This is especially true at high frequencies since the total energy loss is proportional to the product of loss index and frequency, as shown by Equation (16.50). One application where high dielectric losses are preferred is radio frequency heating where, for example, syn-

thetic resin adhesives used in furniture and cabinets may be cured by the application of strong electrical fields.

Permittivity and loss index are fundamental tests. However, for polar materials, the values of the permittivity and loss index are functions of time (frequency). Thus the common practice of providing one set of values at perhaps 10^3 Hz is inadequate. However, the change of the permittivity and the loss index as a function of temperature is usually more important than these changes with frequency, since a single, or at most, a narrow band of frequencies usually will be used in any particular application. Contour maps of permittivity and dissipation factor can give a three-dimensional plot versus frequency and temperature similar to that obtained from mechanical analysis. Figure 16.43 illustrates one means

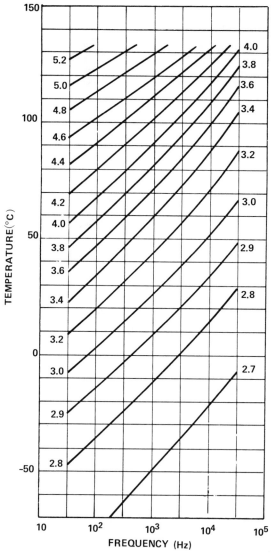

Figure 16.43 Permittivity versus frequency and temperature for PMMA. Results could also be presented using a three-dimensional plot such as shown in Figure 16.5, based on work done by Ogorkiewicz (57).

for presenting such data for PMMA. Such data may also be plotted as shown in Figure 16.5.

Both permittivity and dissipation factor are sensitive to the presence of moisture in plastic materials. Water, which has a high permittivity (about 80), readily encourages ionization that in some materials can change the dissipation factor as much as three orders of magnitude. The effects of moisture on PMMA, for example, significantly change its dielectric constant and dissipation factor. Moisture effects are not simple and must be considered in conjunction with other factors such as frequency and temperature.

The permittivity and the dissipation factor of a dielectric are so closely related that in practice many measuring circuits are designed to give both properties simultaneously. In any case, the determination of one property can be made with accuracy only by adjustment of a circuit component that compensates for the presence of the second property so that it is generally expedient to measure the two together.

To cover the dielectric measurement spectra,

permittivity and dissipation factors would have to be measured with sinusoidal inputs ranging from near DC to 10^{11} Hz. However, several types of bridges and networks are necessary to cover this range. For measurements from near DC to 10^8 Hz, ASTM Method D-150 (80) presents discussions of the necessary circuits. For measurements above 10^8 Hz, ASTM Method D-2520 (80) and ASTM Method D-3380 (80) describe necessary circuits.

For measurements up to 20 kHz, corresponding to the audio frequency range, either a Schering bridge or transformer ratio-arm bridge is ordinarily used. For measurements between 20 kHz and 100 M Hz, corresponding approximately to the radio transmission frequency range, resonant-circuit instruments, such as the Q-meter, are used. At frequencies above 10^8 Hz, corresponding to the microwave frequency region, measurements are conducted using either standing-wave methods or cavity-resonator techniques.

Design of electrodes is important. If the electrodes introduce an undetermined series capacitance, they will cause errors in both the measured capacitance and the dissipation factor. Electrodes for capacitance measurements are designed either to insure intimate contact between the specimen and measuring apparatus, or alternatively, to insure a precisely defined air or liquid gap between specimen and electrodes. These diverse methods may be referred to broadly as contacting and non-contacting systems respectively. The former include conducting paints, vacuum-deposited metallic films, and metal foils applied with grease. Of these, lead foils applied with petroleum jelly and rolled on very hard with a roller are convenient and simple and thus are the most widely used.

A difficulty with most contacting electrodes is that any preconditioning of the specimen must be carried out before their application, but this difficulty does not arise with non-contacting electrodes. With air-gap or liquid-displacement systems, the electrodes are spaced apart and the specimen introduced leaving a deliberate gap. Equations for calculating data when using noncontacting electrodes are given in ASTM Method D-150 (80).

In either case, the electrodes may be unguarded, as shown in Figure 16.42a, or guarded, as shown in Figure 16.42d. When unguarded electrodes are used, the values must be corrected for the fringe or edge capacitance, which is not the direct interelectrode capacitance as explained in ASTM Method D-150 (80) and by Endicott (81). ASTM Method D-150 and Method D-1531 (80) describe contacting and noncontacting electrodes in more detail. Tucker (82) describes other aspects of dielectric constant and loss measurements. Many excellent tables of dielectric constants and dissipation factors (56), (83), (84) have also been published for plastic materials.

16.7.2 Insulation Resistance and Resistivity

DC resistance is used to directly assess the general ability of a polymer to function as an insulator. It is also often used as an indirect measure of moisture content, degree of cure, mechanical continuity, and various types of deterioration.

To measure resistance R, a step voltage E_0 is applied to the specimen and the current output i is measured. Resistance is then calculated from Equation (16.52).

$$R = \frac{E_0}{i} \qquad (16.52)$$

Resistance can also be measured by using a Wheatstone bridge. Various circuits are described in ASTM Method D-257 (80). Scott (85) also describes these resistance measuring instruments in more detail. Because resistances of specimens may exceed 10^{12} ohms, very small currents are measured. Thus care must be taken to insure that stray currents do not influence the measurements. Electrically shielded test instruments and cables are mandatory if accurate measurements are to be made. Unshielded instruments are susceptible to interference from fluorscent lights, motors, and the like.

Selection of different boundary conditions permits integration or separation of leakage currents over the surface or through the volume of the material. Insulation resistance is the DC resistance measured between two electrodes without regard to geometry or the path of the current. Results can be used directly only when the test specimen has the same form as the end-use application and thus it is either a practical or hybrid measurement. Taper pins set into reamed holes for close contact, as shown in Figure 16.42b, can be used to simulate electrodes to measure insulation resistance and, for laminates, the interlaminar resistance. Washer electrodes, Figure 16.42c, placed under screws, simulate industrial circuit-mounting boards for resistance measuring purposes. Both are described in ASTM Method D-257 (80). Taper pin electrodes tend to emphasize volume resistance whereas bolt and washer electrodes tend to emphasize surface resistance.

Resistivity can be considered as resistance normalized to unit dimensions. The use of a third electrode, as shown in Figure 16.42d, called a guard ring, can be used to separate the volume resistance from the surface resistance and thus measure a fundamental property of the material. The circuitry necessary to do this is shown in Figure 16.44. Volume resistivity and surface resistivity can then be calculated from the specimen geometry and measured resistance, using Equations (16.53) and (16.54)

$$\rho_v = \frac{A}{d} R_v \qquad (16.53)$$

$$\rho_s = \frac{P}{g} R_s \qquad (16.54)$$

where R_v = volume resistance, ohm
R_s = surface resistance, ohm
ρ_v = volume resistivity, ohm-cm
ρ_s = surface resistivity, ohm
A = effective area of the measuring electrode
d = thickness of specimen
P = effective perimeter of the guarded electrode
g = guard gap width

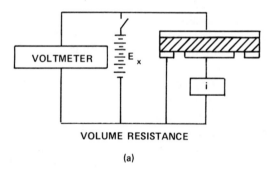

VOLUME RESISTANCE

(a)

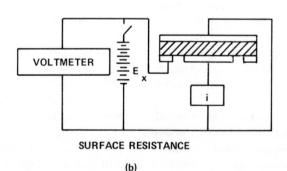

SURFACE RESISTANCE

(b)

Figure 16.44   Circuits used to separate volume resistance and surface resistance by the use of a specimen with a guard electrode.

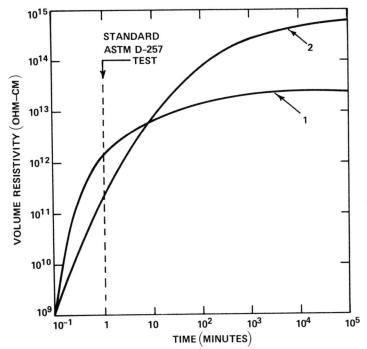

Figure 16.45   Dependence of the volume resistivity on time. If only the standard ASTM D-257 test is used, material 1 would be judged as having a higher resistance than material 2, based on work done by Blanck (86).

The volume resistivity of an insulator is above 10^8 ohm-cm.

When the insulation resistance or resistivity is measured as a function of temperature or relative humidity, electrodes must be able to withstand the environment without degradation and must not interfere with moisture conditioning of the specimen. For insulation resistance, stainless steel pins or washers are recommended. For resistivity measurements, either silver conducting paint can be applied to the specimen, or the specimen can be clamped between conductive rubber electrodes after conditioning. The latter has the advantage that the electrodes can be quickly applied and removed and thus do not interfere with moisture conditioning of the specimen.

Despite its inherent simplicity, the measurement of electrical resistance has been subject to more experimental error than perhaps any other electrical measurement. One reason is because when a DC electric potential is applied to an insulating material, polarization occurs that is time dependent. This time dependence is analogous to the flow of viscoelastic plastics under a constant mechanical stress. Volume resistivity of a material increases with time in an applied electric field, and it approaches an asymptotic value that is determined by the nature and conditions of the material and the environment. Because the change is slow, the ultimate resistivity is seldom, if ever, reported for a material. ASTM Method D-257 (80) stipulates the time of electrification as 1 min. During this interval the resistivity of one material may reach its ultimate value and another material, while having a much lower 1 min resistivity, may be more susceptible to dielectric-absorption effects and eventually reach a higher ultimate value, as shown in Figure 16.45. Thus it is not always possible to draw any conclusions about the relative long-term resistance of two materials in a specific component or system from a knowledge of their short-term resistivities. As a consequence values of apparent resistivity should be given as a function of time.

In addition to this time dependence, both volume and surface resistivity of a specimen may be voltage-sensitive, that is, these properties may not obey Ohm's law (87). In that case it is necessary to use the same voltage gradient if data obtained for different specimens are to be compared. Commonly specified test voltages are 100, 250, 500, 1000, 2500, 5000, 10,000 and 15,000 V (80). Of these, the most frequently used are voltages of 100 and 500 V.

Variations in temperature and humidity have considerable effect on resistivity. An increase in temperature generally causes a decrease in the volume resistivity of insulating materials, which can be represented (88) by Equation (16.55):

$$\rho_v = B \exp \frac{m}{T} \qquad (16.55)$$

where ρ_v = volume resistivity (or resistance)
        B, m = constants
        T = absolute temperature, $^\circ$K

This equation is valid only if materials do not undergo transitions in the temperature range over which measurements are made. The resistance-temperature behavior can be established experimentally by constructing an Arrhenius plot to determine m experimentally. Knowing the resistivity ρ_{v1} at any given temperature T_1, the resistivity ρ_{v2}, at another temperature T_2 can then be predicted from Equation (16.56).

$$\rho_{v2} = - \rho_{v1} \exp m \frac{\Delta T}{T_1 T_2} \qquad (16.56)$$

A typical plot of the resistivity of PMMA versus temperature is shown in Figure 16.46. Two linear regions are evident, one above and one below the glass transition temperature. Surface resistivity also varies with temperature, but moisture and contamination have such a marked effect that the effect of temperature may be obscured.

Moisture content in a material, resulting from immersion in water or exposure to a higher relative humidity, increases charge mobility and results in a decrease in resistivity. After materials reach equilibrium with respect to their moisture absorption, the resistivity may be generally expressed by an expression of the form shown as Equation (16.57)

$$\rho_{RH} = \alpha \rho_0 \exp \left(- k \cdot \mathrm{RH} \right) \qquad (16.57)$$

where ρ_{RH} = resistivity (or resistance) at a given RH
      ρ_0 = resistivity at 0% RH
      RH = relative humidity
      k = constant

It is important to remember that Equation (16.57) is only valid at steady-state conditions. Moisture equilibrium is reached very slowly. For example, it is not unusual for a material only 3 mm in thickness to require 60 days or more to come to moisture equilibrium. Not surprisingly, both the surface and volume resistivities will change with time if the specimen is not equilibrated. Thus, for a study of the effects of relative humidity, plots of resistivity versus time should be constructed and equilibrium times determined. Tests of resistivity under nonequilibrium conditions should be classified as hybrid or practical.

16.7.3   Dielectric Breakdown

Dielectric breakdown resistance governs the selection of materials in applications where a high-voltage stress may cause the component to fail. Measuring the dielectric breakdown strength of a solid material is analogous to measuring its mechanical strength. Like mechanical tests, dielectric breakdown strength tests can be classified into three categories that have various shapes of input excitations. These are very short time (impulse excitations); quasi-static (ramp or sinusoidal with increasing amplitude-cycle excitations); and long term (step or sinusoidal excitations). In the impulse test, failure occurs in seconds; in the quasi-static test, failure occurs in minutes; and in the long-term test, failure occurs somewhere between several minutes and several years. Such input shapes may correspond to stresses that develop, for example, when lightning strikes a standoff insulator or when a material is subjected to either a DC voltage stress or an AC voltage stress (at a particular frequency) for some time.

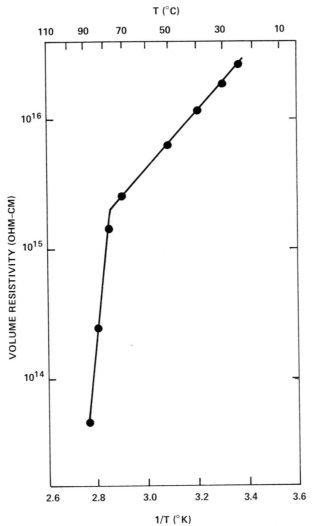

Figure 16.46   Resistivity versus the reciprocal of temperature $(1/T\,^{\circ}\mathrm{K})$ may usually be represented as a log-linear relationship if no transitions occur. When transitions do occur, there is an abrupt change in slope. The data presented here represent a PMMA.

Dielectric strength σ_v is usually expressed in terms of a gradient, volts per unit thickness, and is given by Equation (16.58):

$$\sigma_v = \frac{E}{d} \qquad (16.58)$$

where E = voltage at failure, volts
d = thickness

The test equipment used in dielectric testing should be capable of applying voltages up to 50–100 kV. The requirements of such equip-

ment and the conditions of test are defined in ASTM Methods D-149 (80), D-3426 (80), D-3755 (80) and D-3756 (80). Dakin (89) also discusses the necessary electrical circuitry.

As is discussed later, most dielectric strength tests are hybrid and are thus influenced by the boundary conditions. For example, the use of standard-shaped electrodes only permits comparison of materials in an arbitrary manner. Using the electrode configuration shown in Figure 16.47a, the voltage may cause flashover before breakdown occurs, that is, the current may flow across the surface instead of

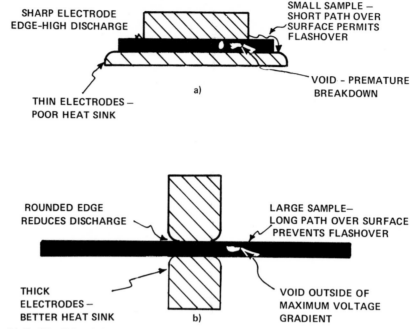

Figure 16.47 The dielectric-breakdown strength is dependent on the boundary conditions. In (a) the short interelectrode path, sharp electrode-edge radius, inclusion of voids, and so on could lead to premature failure. In (b) the electrode configuration would be expected to eliminate such problems, and the dielectric strength should be greater.

through the volume because the surface path between the electrodes is too short. If the sample is wide enough to prevent flashover and breakdown does occur, it may be at a lower voltage than the true value because the sharp electrode edge causes high voltage-stress concentrations leading to premature rupture. Also, the thin electrodes may not be effective heat sinks so that the temperature of the specimen may increase and thus the breakdown strength may decrease. Finally, by using the large-diameter electrode, it is more probable that a void will be in the test area. The situation is improved, but not completely remedied, by the electrode system shown in Figure 16.47b. Thus, for relevant tests, it is recommended that the size, the shape and the edge radius of the electrodes, and the ability of the electrodes (and plastic) to act as a heat sink, be the same as the application.

Another source of variation in dielectric strength testing is the surrounding medium. In standard tests, the electrodes and the specimen are generally immersed in a liquid to prevent flashover or excessive burning of the surface.

However, breakdown values obtained in a liquid, such as mineral oil, may not be comparable with those obtained in air. In oil, electrical discharges caused by the ionization of air (corona) that rapidly deteriorate the plastic cannot occur at the electrode edges. However, even values obtained in different types of oil are not equivalent. This is because the surrounding medium can affect the voltage gradient, heat transfer rate, external discharges, and field uniformity and thus influence the test results. The influence of the surrounding media on the dielectric strength of PMMA is shown in Figure 16.48.

In addition, dielectric strength values are also dependent on the thickness of the specimen. If electrical breakdown were a function of thickness alone, it would have a linear relationship, and dielectric strength (E/unit thickness) would be a constant. When measurements are made at very low temperature, especially when DC tests are conducted, such a relationship is found. However, in general, the dielectric strength decreases as an exponential function of thickness, as shown in

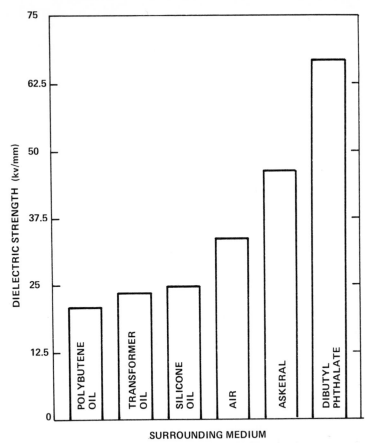

Figure 16.48 Influence of the surrounding media on the dielectric breakdown strength of 1/32 in thickness PMMA measured using ASTM Method D-149, based on work done by Blanck (86).

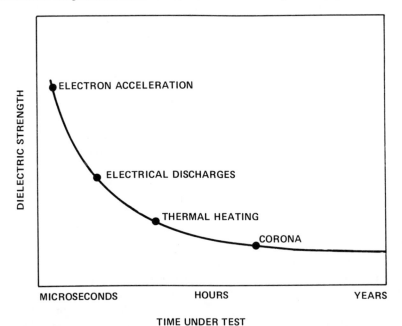

Figure 16.49 With a constant applied voltage, the dielectric breakdown strength decreases with time. Various failure mechanisms predominate, depending on electrification time. Duration of test depends on the end-use application.

Figure 16.9. Thus, the dielectric strength determined is meaningless unless the specimen thickness is given.

Finally, dielectric breakdown is dependent on the time that the voltage is applied. Such dependence is shown in Figure 16.49. This figure, which is analogous to mechanical fatigue plots, also shows the various failure mechanisms that predominate at different electrification times. The selection of a particular test duration is dependent on the intended application. For transient applications where only very short-time (microsecond) impulses are involved, establishing the intrinsic breakdown voltage may be sufficient. Intrinsic breakdown is a purely electrical stress effect and occurs when the field intensity becomes sufficient to accelerate electrons through the material. Intrinsic breakdown strength decreases as the temperature increases.

When voltages will be applied for a short time, that is, quasi-static test conditions are used, then failure should be measured using short-term tests. Quasi-static strength is lower than intrinsic strength. This is because of discharges at the edges of the electrodes or in internal voids. Also, internal heating of the specimen may occur according to Equations (16.59) and (16.60) as follows:

$$\text{watts loss} = \frac{E^2}{R} \quad \text{(DC voltage)} \quad (16.59)$$

or

$$\text{watts loss} = 2\pi\nu\epsilon'CE^2 \tan\delta \quad \text{(AC voltage)}$$
$$(16.60)$$

The former may cause various types of discharge-dependent breakdowns such as ionization-induced flashover and puncture through the thickness, rapid surface degradation induced by surface flashover, or arcing. The latter may cause the temperature to increase and thus lower the breakdown voltage if the heat cannot be conducted away at a rapid enough rate. Thermal runaway, that is, an uncontrollable temperature rise, may occur because of a cumulative effect of loss tangent (or resistance) and temperature. The loss

tangent causes the temperature to increase, which in turn causes the loss tangent to increase, and the rate of heating is accelerated. If the rate of dielectric heating ultimately exceeds the rate of cooling by thermal transfer, then thermal runaway will occur. Such dielectric heating is influenced by the specific heat and mass of the electrode, the specimen thickness and thermal conductivity, and the external cooling conditions.

The effects of both discharges and thermal heating are dependent on frequency and are less when a slowly applied DC voltage is used instead of a slowly increasing AC voltage. Thus DC dielectric strength tests will usually give values of breakdown voltage two to four times higher than equivalent AC tests.

When plastics will be exposed to an electric field for longer than a few minutes, long-term tests are recommended. Such tests are conducted using a step or a sinusoidal excitation that fatigues the plastic. This excitation voltage is low enough to prevent the catastrophic failure that occurs in the quasi-static test but not necessarily low enough to eliminate thermal heating and runaway. However, thermal equilibrium is usually established in several hours. Thus if thermal breakdown has not occurred in this time, it is safe to assume that the material is thermally stable at the selected test conditions. For many uses where lifetimes of several years are not expected, establishing a safe operating voltage where thermal runaway does not occur may be sufficient.

When the plastic is exposed to an electric field of low power that does not cause either catastrophic breakdown or thermal runaway, the field may still cause failure by chemical, mechanical, thermal, and electrical deterioration. Such low-power discharges from these fields that can cause ionization of the air are referred to as "corona discharges." In general failure caused by corona is not a material property but a property of an insulation system and includes electrode geometry.

For the effective design of insulating systems that will operate over several years, a knowledge of the maximum voltage that will not lead to corona attack, that is, a knowledge of the corona extinction voltage, is essential.

The failure produced by corona may be due to one of several possible factors: the corona may erode the insulation until the remaining insulation can no longer withstand the applied voltage; it may cause the insulation surface to become conducting; it may cause a "treeing" within the insulation, which may progress to failure; it may release gases within the insulation that change its physical dimensions; or it may change the physical properties of an insulating material, for instance, it may cause the material to embrittle or crack because of the presence of ozone and thus make it useless.

ASTM Method D-1868 (80) and ASTM Method D-2275 (80) describe techniques for measuring the corona extinction voltage and the voltage endurance limit. Mathes (90, 91), Bartnikas and McMahon (92), Dakin (89) also discuss factors important in dielectric breakdown and corona measurements.

As in mechanical strength predictions, dielectric strength and corona breakdown predictions are sometimes useful. Starr and Endicott (93) have used an accelerated test using steadily increasing voltage to measure dielectric breakdown. The procedure is similar to the Prot techniques discussed in Section 16.5.4.4 for mechanical fatigue tests. The expression used was Equation (16.61)

$$K = (E - E_2)^n \, t \qquad (16.61)$$

where E = failure voltage
E_2 = corona initiation voltage (voltage below which failure does not occur)
t = time
K and n = constants

Such a relationship will plot as a straight line on log-log paper.

If there is only negligible dielectric heating, the failure time of a corona test will vary inversely with the frequency. Thus tests can be accelerated in such cases by increasing the frequency of testing up to as high as several thousand Hertz. Since the amount of corona energy is essentially constant during each cycle, the failure time at 60 Hz, t_{60}, can be predicted from the failure time at 1200 Hz, t_{1200}. For example, $t_{60} = t_{1200} \, (1200/60) =$ $20 t_{1200}$. Where failure times exceed a few days, higher frequency tests are recommended to reduce the testing time if such an increase in frequency does not cause an increase in temperature.

16.7.4 Tracking Resistance

16.7.4.1 Wet Tracking

Tracking, that is, the formation of conducting paths on the surface of an insulator, may occur when an electric field is applied. Although wet tracking may occur in a variety of practical situations, common examples are the failure of line transformers and other materials that are exposed to dirty and/or moist environments. Tracking is considerably influenced by films of moisture and dust, since they alter surface leakage currents. Heat generated in a moisture film on the surface because of high electrical resistivity is sufficient to cause local evaporation. The resulting rapid break in the resistive film frequently results in electric sparking or scintillation that is repeated randomly over the surface of the plastic. Continued arcing leads ultimately to a conducting carbon path. When this occurs, the insulator has failed. Some materials, such as acrylics, erode rather than track. This results in a loss of volume in the electrical insulating material by the action of electrical discharges. Electrical tracking tests, like mechanical tests of surfaces, are classified practical or hybrid.

Various tests have been developed that use simulated service conditions. Standard test methods involve introducing conducting media such as salt solutions on the surface, as shown in Figure 16.42e and 16.42f, to produce an acceleration of the natural processes. The Dust and Fog Tracking Test, ASTM Method D-2132 (80), the Differential Wet Tracking Test, ASTM Method D-2302 (80), the Liquid-Contaminant Inclined Plane Test, ASTM Method D-2303 (80), and the Comparative Tracking Index Test, ASTM Method D-3638 (80), are a partial answer to the evaluation of the tracking capabilities of materials under outdoor exposure or a "dirty" environment. In ASTM Method D-2132 a synthetic dust consisting of SiO_2, clay, and salt is coated on the surface of the specimen. This specimen is then placed in a fog chamber and a voltage of 1500 V is applied. In ASTM Method D-2302 the specimen is soaked in a 1% ammonium chloride solution. A 1.5 mm gap is then subjected to various voltage levels. In ASTM Method D-2303 a liquid contaminant is allowed to flow over the inclined face of the specimen while a voltage stress between 0.5 and 6 KV is applied. Special test solutions can be used to meet specific service needs. In ASTM Method D-3638, an electrolyte, a 2% solution of ammonium chloride, is allowed to drop between two low-voltage electrodes. Failure is defined as the voltage that causes failure when 50 drops of electrolyte have fallen. All of these tests operate at a high current.

With the exception of Method D-3638, all of the preceeding tracking tests show good correlation with service when the boundary conditions are properly selected. With Method D-3638, agreement is fair. Method D-2302 requires the shortest test time whereas Method D-2132 requires the longest test time. All tracking tests, however, are notoriously variable with standard deviations as high as 40% not uncommon. Thus tests require careful monitoring to obtain reproducible test results. Mitchell (94) discusses the status of various standard methods.

No single-tracking test method is recommended to cover the complete voltage range of application. For the low-voltage application range (to 600 V), the Method D-3638 is proposed. For the medium-voltage range (600 to > 35,000 V), the inclined plane and differential wet tracking methods are used. The dust and fog test is used only at 1500 V.

Obviously, the exact tracking test used should match the end-use condition as closely as possible. When general information is required, the use of several tracking tests is recommended to obtain the relative ranking of materials under various conditions. It is frequently found that with many different contaminant solutions the relative order of a material's resistance to tracking is unchanged. Thus certain plastics are predisposed to tracking regardless of the test employed.

16.7.4.2 Arc Resistance

When an arc forms in air between electrodes on the surface of a plastic, it causes both thermal degradation and oxidations that may cause the surface to become conducting. In electrical applications such as automotive distributor caps and switchgear in general, an electrical arc of this description forms in air between electrodes that may be in contact with the surface of the plastic. Such arcs, although different than the scintillation involved in wet tracking, can cause failure. The standard arc test is ASTM Method (495) (80) where an interrupted high test voltage (15 kV) is applied between two chisel-shaped tungsten or stainless steel strip electrodes spaced 6 mm apart on the test surface. Currents between 10 and 40 mA are applied. Usually stainless steel strip electrodes will cause failure in a shorter time than tungsten electrodes because the arc is closer to the plastic surface. In both cases, arcing occurs closer to the surface when the electrodes are inverted, but the use of inverted electrodes is not included in the standard ASTM method.

The high-voltage, limited current input excitations of this ASTM test do not represent conditions of end-use and therefore the results of this hybrid test do not correlate with service performance. In addition, this test is not applicable to materials such as silicones that do not produce carbonous products (conductive paths) under the action of an electric arc or to materials such as PMMA that melt or form fluid residues that float conductive residues out of the active test area thereby preventing formation of a conductive path. More sophisticated evaluation techniques for arc resistance have been described by Mandelcorn (95) and Hoff and Springling (96). When there is a need for arc resistance tests, relevant tests should be developed based on these more sophisticated techniques.

16.7.5 Electrostatic Charge

Electrostatic charge is transferred between two dissimilar materials upon separation after they have been pressed together. When the materials separate each has an equal but opposite charge. Charges may be left on plastics for an appreciable time by such contact and separation if they do not dissipate. These charges may simply be a nuisance, such as dust collection during the shelf storage of bottles, or may be particularly dangerous in an explosive atmosphere, such as in a hospital operating room where oxygen is being used.

The electrostatic properties of a plastic material can be assessed for most practical purposes by making two measurements: the maximum charge that will be generated when the plastic is contacted with another material and then separated; and the rate of which the charge leaks away. In practice, the charge generated on a particular plastic varies considerably with the composition of the contact material, the pressure or area of the contact, and the relative humidity.

The charge generated at a given relative humidity may be represented by Equation (16.62) as follows:

$$Q(\text{RH}) = Q(0) - \alpha \cdot \text{RH} \quad (16.62)$$

where $Q(\text{RH})$ = charge generated at a given relative humidity

$Q(0)$ = charge generated at 0% relative humidity

RH = relative humidity

α = constant

The rate at which this charge dissipates is much more dependent on relative humidity. Charge dissipation can be expressed as a function of relative humidity by Equation (16.63):

$$\tau(\text{RH}) = \tau(0) \exp[\beta \cdot \text{RH}] \quad (16.63)$$

where $\tau(\text{RH})$ = half-life of the charge at a given RH

$\tau(0)$ = half-life of the charge at a zero RH

β = constant

There are usually several orders of magnitude difference between charge dissipation at 10 and 90% relative humidity. Practical considerations favor conditioning and testing at 20% RH and 50% RH, however, since measurements and lower and higher relative humidities would be exceedingly long or rapid respectively. The actual moisture in some materials at a general relative humidity may vary with the direction of approach, that is, whether the testing condition is approached from a wetter or dryer condition. Thus specimens may have to be measured after they have been brought to the test relative humidity from both higher and lower humidities. All of these humidity factors must be considered and controlled for any meaningful measurement of electrostatic charge.

Normally, the generation of electrostatic charge is performed using a precise mechanical device that permits either a solid plastic to be pressed against the second material in a reproducible manner or a powdered plastic to be slid down an inclined plane. Such tests should be as analogous to the particular application as possible. This charge is then measured by enclosing the specimen in a Faraday cage, as described in ASTM D-2679 (80), which is connected to an electrometer. If the plastic sample is connected to a metal chain or held in a metal frame connected to ground and suspended in the cage, the dissipation of charge can also be followed. Since this decay will be exponential, usually only the time necessary for the charge to decay to one-half of its original value, the half-life, is calculated. However, charge dissipation measurements conducted in such a manner should be used for comparisons only since the magnitude of the half-life will be dependent on the sample size. On larger parts that cannot be charged and lowered into the Faraday cage, the field strength due to surface charges can be measured with a probe as described in ASTM Method D-3509 (80).

Other techniques (97) discard the measurement of charge generation based on the argument that if the charge can be dissipated rapidly, the magnitude of the original charge is of little interest. In one technique, ASTM Method F-365 (98), which has been withdrawn because of lack of use, the charge mobility is measured when an uncharged plastic specimen is inserted into an electric field at 2 kV potential and a charge is first induced and then dissipated. The time necessary to induce and discharge the plastic provides a comparison of the relative-charge mobility. For such tests to be reliable indicators of antistatic performance, it must be established that the half-life at low relative humidity is very rapid and that in the application there will be an electrical path to ground. If these criteria cannot be satisfied then low-charge generation is needed to assure reliable performance.

Many types of antistatic agents may be added to plastics, and extensive lists of such materials are available (56). The basic mechanism for conventional antistatic activity is the attraction of moisture toward the polymer surface, facilitating the rate of static discharge.

16.8   THERMAL PROPERTIES

Thermal properties include thermal expansion, thermal conductivity, specific heat, and flammability. The importance of some of these thermal properties has been illustrated in part during the discussion of mechanical and electrical tests. Other information concerning the thermal behavior of a plastic permits one to determine the change in its physical dimensions caused by a change in temperature, its effectiveness as a heat insulator or conductor, the energy required to alter its temperature, or the conditions under which it will burn. These properties are important in any design that must function in a thermal environment. For example, the plastic lid of a Thermos® bottle must be an insulator to contain the heat or cold, it must not store heat, and it must not undergo large dimensional changes when cooled or heated that will cause the bottle to leak or be difficult to open. Conversely, the plastic coating on the windings of an electric motor should have a high thermal conductivity to allow heat to escape but must not expand when heated if there are small clearances. The performance of a plastic in a fire situation is always a serious safety consideration. The change in thermal properties with temperature can also

be used to determine the material morphology and transition temperatures similar to the dynamic mechanical analysis discussed in Section 16.5.1., as well as other basic thermodynamic properties.

Thermal expansion, thermal conductivity, and specific heat tests may all be classified as fundamental tests. Flammability tests are classified as hybrid or practical tests.

16.8.1   Thermal Expansion

Thermal expansion is the ratio of the change in the dimensions of a body to its original dimensions for a given temperature change. The coefficient of a linear thermal expansion is expressed as Equation (16.64):

$$\alpha = \left| \frac{\Delta L}{L_0 \, \Delta T} \right| \qquad (16.64)$$

where α = coefficient of linear thermal expansion
ΔL = change in specimen length
L_0 = original specimen length at temperature T_0
ΔT = temperature change

This coefficient is useful whenever the temperature of the structure or part is likely to vary. For example, because plastics are molded at elevated temperatures, the finished article will be somewhat smaller than the mold, and allowance must be made for this in the mold design.

Thermal expansion is particularly important where two dissimilar materials are used in the same structure or when a material may be constrained. Then the thermal expansion of components should be matched as closely as possible to prevent cracking or warping. For example, if a plastic bar is constrained at both ends and subjected to a temperature change, the residual stress σ_R is given by Equation (16.65)

$$\sigma_R = \int_{T_0}^{T} E(T) \cdot \alpha(T) \cdot dT \quad (16.65)$$

where $E(T)$ = temperature dependence modulus

$\alpha(T)$ = coefficient of thermal expansion

The development of such residual stresses should be considered both in component design and test specimen selection if premature failure and the like are to be avoided.

Most plastics have coefficients of expansion several times greater than metals or glass. Unfilled plastic materials have linear coefficients generally in the range of $5\text{--}15 \times 10^{-5}/°C$ compared to 1×10^{-5} for iron, 2×10^{-5} for aluminum, 1×10^{-5} for soda glass, and 0.045×10^{-5} for fused silica. Filled plastics have lower coefficients than unfilled plastics. The reduction is dependent on the percentage and type of filter. Unlike inorganic materials, the expansion of plastics is usually not linear with temperature over a large range. The expansion coefficient increases, for example, above the glass transition temperature. In fact, the measurement of expansion versus temperature is one means of determining the glass transition temperature (99). Expansion may simply be determined using a quartz dilatometer by observing the change in length of the specimen between two arbitrary limits within the service range, as in ASTM Method

D-696 (8), where measurements are made between $-30°C$ and $30°C$. Care must be taken, however, to insure that the thermal expansion over this region is linear and is the same as the region of interest. However, a better presentation of data is obtained by directly plotting the expansion or contraction curve against temperature. A suitable apparatus for this is a thermal mechanical analyzer (TMA) such as shown in Figure 16.50. Expansion or contraction causes a displacement of the core of a linear variable differenal transformer (LVDT), which along with the thermocouple output is displayed on and XY recorder (99). The use of this thermodilotometry technique is described in ASTM Method E-831 (123). Interferometry can also be used to measure thermal expansion using the technique described in ASTM Method E-289 (43).

16.8.2 Thermal Conductivity

Thermal conductivity may be defined as the ability of a material to transmit heat energy. The low thermal conductivity of plastics is generally advantageous, especially in the building industry. For example, expanded plastics are widely used to insulate older homes. However, low thermal conductivity

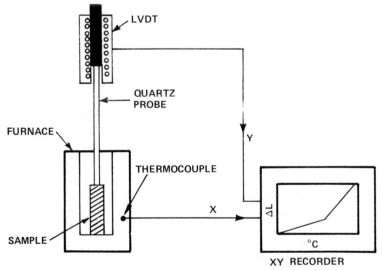

Figure 16.50 The thermal mechanical analyzer (TMA) apparatus as used to measure the change in length of a plastic sample as a function of temperature.

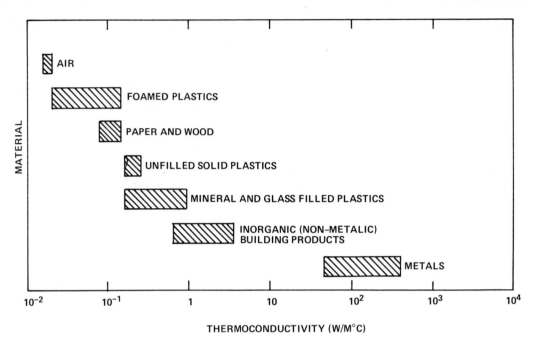

Figure 16.51   The thermal conductivity of plastics compared to other construction materials and to air.

may not be desirable in many applications. As previously discussed, because of their low thermal conductivity, plastics dissipate heat very slowly during fatigue. Thus, heat builds up locally which can lead to breakdown of both mechanical and electrical components. Also low thermal conductivity causes differential cooling in molded or extruded parts and can cause a thickness anisotrophy. Figure 16.51 compares the range of the thermal conductivities of plastics with those of other materials.

The very low conductivities of plastics mean that their measurement is difficult because of the small quantities of heat flow involved. Relatively small heat losses can introduce significant errors in the measured values. A sophisticated apparatus that permits precise control of the boundary conditions is therefore necessary for accurate thermal conductivity measurements. Design of such a thermal conductivity apparatus for solid plastics is usually based on specimens of simple geometry, such as a disc or square, whose areas are large relative to their thicknesses so that the heat flow through their planes is large compared with possible lateral heat losses from the edges.

There are almost as many methods for measuring thermal conductivity as there are experimentalists to do the measurements. However, all methods can be divided into two classes: those in which measurements are made at equilibrium or steady-state conditions; and those in which measurements are made at nonequilibrium, or transient conditions.

In steady-state measurements, one face is usually heated to a constant higher temperature and the other maintained at a constant lower temperature, and the problem resolves into a measurement of the heat and temperature difference. Under steady-state conditions, when temperatures are constant, the heat flow can be calculated. The thermal conductivity is given by Fourier Equation (16.66)

$$\frac{dq}{dt} = KA \frac{dT}{dx} \qquad (16.66)$$

where K = thermal conductivity
          A = area
     dT/dx = temperature gradient
     dq/dt = rate of heat flow

Under steady-state conditions, when temperatures are constant, the temperature gradient becomes T/x, and the heat flow becomes q/t.

A unique instrument for steady-state measurements is the Colora thermoconductometer described in detail by Schröeder (100), and shown schematically in Figure 16.52. In this instrument, the two flat ends of a cylindrical sample, about 18 mm in diameter and 1–30 mm in height, are placed in contact with two silver plates that are kept at constant temperatures by means of two boiling liquids; the boiling points of these two liquids differing by about 10–20°C. The time is measured in which a quantity of heat flows, in the steady state, through the sample. This quantity is determined very simply by the quantity of liquid condensed in a calibrated cylinder. Boiling liquids also maintain the sides of the disc at constant temperature of prevent lateral heat loss. Various liquid pairs can be used for different mean temperatures.

Low thermal conductivity materials such as cellular plastics can be measured using the guarded hot-plate technique described in ASTM Method C-177 (8). With this apparatus, heat losses from the sides of the specimen are eliminated by the use of a guard ring. Otherwise, the heat losses from the sides

of cellular specimens and the heat could approach the heat flow normal to the specimens and cause considerable errors.

The main problem with steady-state tests is the long time necessary to reach equilibrium. During this time the sample may change. For example, water may be driven off or the cure may advance. This is not a major problem with transient tests, however.

In transient measurements, one face of the test specimen is suddenly placed in contact with a constant temperature source, and the temperature rise of the opposite insulated face is measured. For transient conditions, the amount of heat the material can store as internal energy must also be considered. The thermal conductivity expression is then represented as Equation (16.67)

$$\frac{\partial T}{\partial t} = \frac{K}{\theta C_p} \frac{\partial^2 T}{\partial x^2} \qquad (16.67)$$

where θ = density, C_p is the specific heat, and K is the thermal conductivity.

One of the most commonly used transient techniques is the Cenco Fitch apparatus described in ASTM Method D-2214 (98). Here the rate of heat rise of a copper block is proportional to the amount of heat trans-

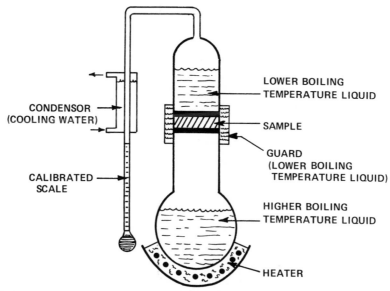

Figure 16.52   A unique means of measuring the thermal conductivity of plastics by using two liquids with slightly different boiling points.

mitted by the sample. This method is not recommended for measurements where high accuracy is required because there are many heat losses that are not considered, that is, the boundary conditions are not controlled. More accurate methods are described by Anderson and Acton (101).

With both steady-state and transient techniques, care must be taken to insure intimate specimen contact with the plates of the apparatus otherwise an air gap will be in series with the specimen. The use of a conducting graphite is usually helpful.

Typical thermal conductivity data for a wide range of polymers are available (56). Regrettably, there is often an enormous spread of experimental values for the thermal conductivity of even common materials. Majuery (102) has discussed the lack of consistent information of published data.

16.8.3   Specific Heat

Specific heat is the capacity of a material to store heat energy. That is, it is the heat content

per unit mass per unit temperature rise. More formally, it is defined as the amount of heat required to raise the temperature of a unit mass by one degree.

A knowledge of specific heat is important in the processing of plastics, calculating heat buildup in mechanically or electrically fatigued parts, calculating thermal diffusivity from thermal conductivity measurements, and so on. Also, a plot of specific heat versus temperature is valuable because various transitions as well as basic thermodynamic properties can be calculated. Changes in such properties can be related to such physical processes as quenching and cooling, crystallization, melting, and the glass transition and other secondary transitions found in plastics (99).

One means of measuring specific heat is by use of differential scanning calorimeter (DSC) (103). In this unit, shown in Figure 16.53, two identical calorimeters are instrumented to operate at the same temperature and are programmed to increase the temperature at the same rate. A sample is placed in one

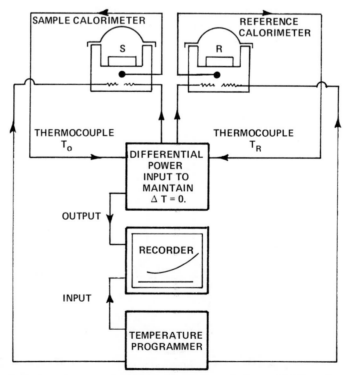

Figure 16.53   General arrangement of a differential scanning calorimeter (DSC) used to measure the specific heat of a plastic as a function of temperature.

calorimeter and the other is left empty. Instrumentation is provided to measure electrical power necessary to keep the two sample pans at the same temperature. To measure specific heat in this calorimeter, the sample pan and reference pan are heated at the same constant rate. Since the reference pan is empty, it will require a smaller amount of electrical power to achieve this rate. By comparing the electrical power input required to raise the sample temperature to the amount of electrical power required to raise the temperature of a known standard, the specific heat can be calculated. The method is a comparative rather than an absolute method so that standards of known specific heat must be available. When basic thermodynamic measurements are made, it is the change in specific heat that is of interest, so standards are not necessary. ASTM Method D-3418 (8) describes a procedure for measuring transition temperatures using DSC techniques.

Figure 16.54 shows the change in specific heat as a function of temperature for a typical thermoplastic. Since specific heat is a func-

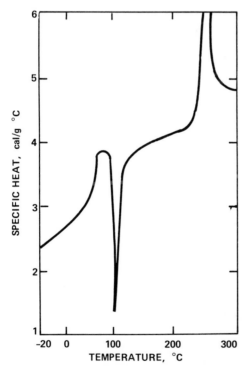

Figure 16.54 The variation in specific heat of a typical plastic as a function of temperature.

tion of unit mass of the material, the specific heat of cellular plastics is the same as for the solid material.

16.8.4 Flammability

The flammability of plastic materials is a frequent concern in design. Processes have been developed for molding and forming large parts for use in building construction, appliance housings, transportation equipment, and many other applications where safety and performance under fire conditions are of necessity a serious consideration. Also, the development of new materials with improved mechanical and thermal properties has led to their increased application as electrical insulation, which has in turn caused their exposure to electrical ignition hazards.

Flammability is a highly complex phenomena as it relates to a specific application. The general mechanism of flammability is shown in Figure 16.55. A specimen of arbitrary shape and mass is subjected to heat sources sufficient to cause pyrolysis. Thermal decomposition products are thus formed. When these products are combustible, either autoignition may occur or these products may be ignited by an ignition source. If sufficient oxygen is available, these products burn. Such burning generates more heat. Some fraction of the total heat generated is fed back to the material and causes further pyrolysis. The amount of feedback, in addition to being dependent on the total heat release, is dependent on the direction of burning, for example horizontal burning or burning from the bottom of a vertical specimen and convection (natural or forced air), conduction, and radiation. To be flame resistant, a material should have a high-decomposition temperature; form combustion inhibitors on decomposition; require a high concentration of oxygen to burn; and/or have a low heat release during combustion. In addition, any combustion products should produce minimum levels of smoke and should also be nontoxic.

The ultimate tests of acceptability of an article would be flammability tests conducted using the component. However, component

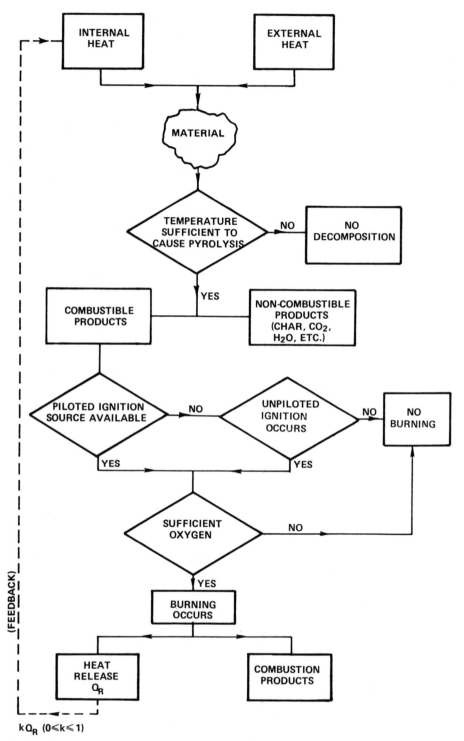

Figure 16.55 The general mechanism of flammability.

tests that successfully certify a material for one application must be repeated for every new application because factors influencing performance in different applications are rarely duplicated. Full-scale testing may be conducted and mathematical models developed for predicting and comparing the fire growth of various combustibles, such as those found in a room (104).

However, some laboratory screening processes are needed to provide guidelines for the best selection of materials for finished articles. Thus flammability evaluations are initially performed on small test specimens. However, because of the complex mechanism discussed, flammability cannot be described by any one single test. Rather it requires a number of different tests. Such flammabilty tests should answer three questions: How well will a material resist ignition? If a material does ignite, how easily does it burn and release heat? How much smoke and toxic gas is produced?

It is not appropriate in this chapter to provide a detailed discussion of flammability. Rather, a brief survey is presented of existing test methods that may be used to separate and study the various components of flammability. It should again be noted that actual performance in a fire situation depends on a complex combination of these factors. Hilado (105) provides an excellent discussion of such factors.

16.8.4.1 *Ignition*

Unpiloted ignition may occur by convective heating where air is passing around a specimen; conductive heating where the specimen is in contact with a heat source; or radiant heating where a specimen is irradiated by a heat source. Piloted ignition may occur if a flame or an electrical arc is present.

Unpiloted ignition using convective, conductive, and radiant heat sources may not occur at the same temperature because of differences in heat transfer for these different types of source. Piloted ignition temperatures will depend upon the time that the ignition source is in contact with the specimen, since the heat rise will be governed by Equation (16.67). In an actual fire situation, both un-

piloted and piloted heat sources may be available so that the ignition temperature will be lower than if either source alone was available.

Tests are available to measure each of the categories described. In ASTM Method D-1929 (8) originally developed by Setchkin (106), heated air is passed around the specimen placed in a furnace. Either unpiloted or piloted ignition can occur. Specimens are heated at the rate of 600°C/hr until unpiloted ignition occurs or, if lit, a small flame at the top of the furnance ignites the decomposition products. In a modification of this method, ASTM Method E-162 (107), specimens are placed in the furnace at 750°C and classified as noncombustible if there is no flaming or if the specimen reaches an equilibrium temperature between 750°C and 780°C. The response of a plastic to the direct application of a small flame is described in ASTM Method D-3713 (8). The maximum time that the flame can be applied to the bottom of a vertical specimen without it burning for more than 30 seconds after the flame is removed is defined as the "Ignition Response Index".

Other tests have been developed for unpiloted ignition resistance of a specimen in contact with a hot wire (108). A hot-wire ignition index, described in ASTM Method D-3874 (80), is expressed as the number of seconds needed to ignite a 12 mm wide bar wrapped with resistance wire that dissipates 26W/mm of heat. The heating is continued for a maximum period of 2 min. The length of time (in seconds) to cause ignition is recorded as the index by this method. Since the method is intended to simulate overloaded electrical circuits, only unpiloted ignition is measured.

Ignition of a panel exposed to radiant heat is described in ASTM Method E-162 (107). This test method employs a radiant heat source consisting of a 300 × 450 mm panel in front of which an inclined 150 × 450 mm specimen is placed. A pilot flame impinges on the top center of the specimen. An ignition factor is derived from the rate of progress of the flame front. The flame front is followed for 15 min.

Arc ignition tests used for electrical equipment have also been developed (108). In such tests either a high-voltage electrical arc or a high-current arc is energized on the specimen surface, and the time to failure is noted. An arc ignition test, described in ASTM Method D-229 (80) uses both a heat source (860° C) and an electrical arc to cause ignition. Specimens are mounted in the center of the coil, and a spark gap is positioned above the specimen. The test method is not applicable to thermoplastics because they melt below the temperature of the coil. It is recommended that the selection of an ignition test simulate the end-use application as closely as possible.

16.8.4.2 *Burning*

The various ignition tests discussed are designed to assess the ease with which a material will ignite. Burning tests, on the other hand, are designed to measure the relative ease with which the specimen will continue to burn after the external ignition source has been removed.

Unfortunately, there is no one area of physical testing where the considerations of boundary conditions and the failure mechanism have been ignored more than in the area of small-scale burning tests. Such methods have been proliferated in ASTM standards as well as elsewhere. One such test is ASTM Method D-635 (8). A specimen, 125 mm long by 12 mm wide, with a thickness normally used in the given application, is tested at an ambient temperature of 23° C. The specimen is clamped with its longitudinal axis in a horizontal position and its transverse axis at an angle of 45° to the horizontal. The specimen is ignited (for 30 sec) with a Bunsen burner. The average time of burning or the average extent of burning after the burner is removed is recorded. Until recently it was possible to classify materials as nonburning or self-extinguishing by this test. However, as Allen and Chellis (109) discuss, very few end-use applications call for long narrow parts to be used only in a horizontal position. Numerous materials that do not completely burn in this test configuration do continue to burn under less flame exposure if the specimens are mounted vertically. The test does not give any indication of end-use performance. Similar criticisms apply to the other Bunsen burner tests using a horizontal specimen.

Other open-flame tests may also be used, such as those used by Underwriter's Laboratory (108). Those tests are somewhat similar to ASTM Method D-635 (8) but are more stringent in that the tests are conducted vertically (with bottom ignition) as well as horizontally. An ASTM procedure, Method D-3801 (8), has recently been adopted that is very similar to the UL vertical test method. Various ratings are given dependent on the time to extinguish and/or whether the specimen drips flaming particles.

A further drawback of a number of the flammability tests is that they are run under nonequilibrium conditions, that is, the end point of the test is time, which has been arbitrarily set as satisfactory, but which may be too short for equilibrium to be achieved between the heat generated by the combustion and the heat fed back to the specimen and that used to heat the environment. Nonequilibrium conditions exist if the oxygen concentration is higher than that necessary to just support combustion.

The oxygen concentration that is needed to support combustion in candlelike burning can be determined in the oxygen index test, originally developed by Fenimore and Martin (110), and described in ASTM Method D-2863 (8). The test apparatus, which is shown in Figure 16.56, consists of a chimney where a regulated concentration of oxygen (mixed with nitrogen) flows over the specimen. An oxygen-nitrogen mixture is arbitrarily selected, and the specimen is well ignited by a burner. If it burns for a selected time or length, the oxygen concentration is reduced and another sample tested. If it does not burn, the oxygen concentration is raised. The oxygen index (OI) of the material is the concentration of oxygen that will just support combustion under steady-state candle-like burning. This is calculated using Equation (16.68).

$$OI = \frac{O_2}{N_2 + O_2} \qquad (16.68)$$

A well ignited specimen needs only 30 to 60 sec after ignition is complete to come to equilibrium. If the specimen is allowed this time to come to equilibrium, the required

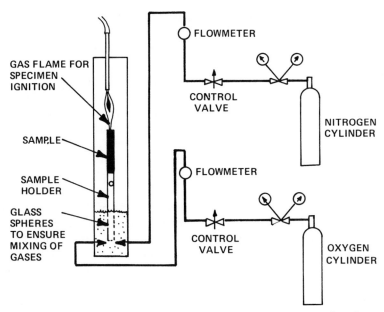

Figure 16.56 The oxygen index apparatus used to measure the minimum concentration of oxygen that will support combustion under candlelike burning.

oxygen concentration will be independent of the exact definition of end point. Thus, whether the end point is measured when the specimen is burning steadily for 3, 4, 5, or 6 min, or whether a limit is taken such that the specimen burns 75 mm, the final results are close to being the same (111).

The total heat available to the specimen to cause pryolysis is dependent both on internal and external sources. This, for the most part, has been overlooked in standard tests. Since the flammability of materials increases with ambient temperatures, the oxygen index may be expected to decrease with increasing temperature. Thus a material must be considered flammable at any temperature where the oxygen index is below 0.21. Johnson (112) has presented a plot showing the oxygen index retention as a function of temperature. A quantitative experimental expression for these data is expressed in Equation (16.69)

$$R = 100 - 0.00455 \, (T^{3/2} - 5144) \quad (16.69)$$

where R = percent of 25°C oxygen index retained

T = temperature, °K

The temperature can be derived at which any given oxygen index measured at room temperature will be reduced to 0.21. Flammability tables containing the oxygen index of many commercial plastics are available (56). In addition to temperature, the oxygen index value, like other flammability results, is still dependent on the direction of burning (which partially controls the degree of heat feedback) and on the size and construction of the specimen holder (which acts as a heat sink).

A good example of a heat sink adversely affecting results was ASTM Method D-1692 for cellular plastics. The test specimen is supported on a horizontal metal screen, which conducts large amounts of heat away, causing the material to extinguish prematurely. Again, no correlation with end-use performance should be expected because the boundary conditions were not matched. This method has been deleted as an ASTM Standard.

16.8.4.3 Heat Release

Tests for heat release provide a measure of the amount of heat produced by a material in the process of burning. Such tests generally provide information on the relative amount of heat produced or the equivalent amount of fuel required to produce similar results, when compared to other materials.

The relative amount of heat released during

combustion can be measured by such means as placing thermocouples in the stack used to remove gaseous products of ignition. This is done in ASTM Method D-162 (107). Smith (113) has also described a heat-release apparatus involving measurement of the temperature increase of stack gases. In addition to such approaches, heat of combustion of fuel values can be readily determined by means of a calorimeter similar to those used for specific heat measurements.

16.8.4.4 *Smoke Generation*

The smoke generated from burning (or smoldering) materials may obscure vision and hinder an occupant escaping from a room. The density of smoke produced by burning plastics is generally determined by burning the specimen in a closed box using a controlled ignition source and photometrically measuring the fraction of light absorbed or obstructed. The two most commonly used smoke chambers are the XP-2 chamber, developed by Rarig and Bartosic (114), and described in ASTM Method D-2843 (8), and the NBS chamber developed by Gross, et al. (115).

One critical difference in these chambers is the ignition source: the XP-2 chamber uses a Bunsen burner to insure that the specimen is well ignited; the NBS chamber uses either a radiant heat source only or a radiant heat source and a small pilot flame to obtain incomplete combustion. Other differences include the positioning of the specimen and the direction of the light beam through the smoke.

Smoke density tests can provide information in addition to the maximum smoke density. This additional information includes smoke accumulation rate, time to reach maximum smoke density, and time to reach a critical smoke density. This last property is called the obscuration time and is a measure of the time available before a typical occupant in a typical room would find his vision sufficiently obscured by smoke to hinder escape.

Another important measurement of the fire gases is toxicity. Many infrared, gas chromatographic, mass spectrometric, and calorimetric analyses techniques have been proposed, but no standard techniques have been accepted.

16.9 OPTICAL PROPERTIES

Plastics are chosen for their geometric optical attributes in areas such as lenses, building illumination, lighting, signs, and packaging and for their general spectral optical attributes. The reasons may be purely aesthetic appeal or may be to achieve a more utilitarian goal. In any case, plastics are used because of their ability to transmit, deflect, or absorb light rays. Each application requires the plastic to have a different combination of optical properties. Some of the geometric redistributions of light from a plastic, shown schematically in Figure 16.57, are distortion, gloss, haze, diffuse reflectance, and transparency. These geometric attributes of appearance form one category of appearance evaluation and are the main subject of this section. The other category of appearance evaluation is spectral attributes of appearance and involves color measurements.

Geometric measurements involve several distinct types of tests to characterize a material: refraction characterizes the change in direction of a light ray passing through the plastic; reflection describes the amount of light returned from the surface after striking it; transparency describes the quality of an image of an object viewed through the plastic in terms of contrast and resolution; and light transmission describes the characteristics of a beam after traveling through and exiting a plastic.

16.9.1 Index of Refraction

The index of refraction is used to calculate how light will be distorted, deviated, or magnified when passing through a plastic, such as shown in Figure 16.57a, and to determine how much light will be reflected specularly, assuming that light is mirror reflected from a perfectly flat surface (116). Geometrically, the index of refraction n is defined by the ratio of the sine of the angle of incidence i and the sine of the angle of refraction r. This is given in Equation (16.70).

$$n = \frac{\sin i}{\sin r} \tag{16.70}$$

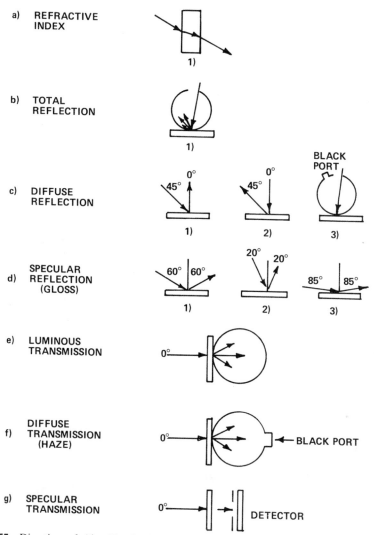

Figure 16.57 Directions of object illumination and view most frequently used for measurement of object-appearance attributes.

The index of refraction is a fundamental property but is dependent on the wavelength of the incident light. This is why white light is separated into its various components upon passing through a prism.

Various techniques suitable for measuring the refractive index are available with several types of refractometers (117). Particularly, one based on the standard Abbe' refractometer, is described in ASTM Method D-542 (8). In this method, a small test specimen replaces the auxiliary prism normally used to measure liquids. The Abbe' refractometer is capable of much greater precision than is ordinarily required except for optical design. A microscopical technique, described in ASTM Method D-542 (8), may also be used that is accurate to three significant figures. Here a specimen is placed on the measuring microscope table and the microscope focused first on the bottom surface and then on the top surface. The longitudinal displacement of the lens tube is the apparent thickness of the specimen. The index of refraction is found by dividing the actual thickness of the specimen by the apparent thickness.

Other techniques may be used to measure the optical distortion and the deviation of the line of sight through formed transparent plastic articles directly. In one technique, ASTM Method D-637 (8), a sharp image is projected onto a screen. The test specimen is then held perpendicular to the direction of projection in front of the projector. Movement of the projected image is noted. In a second technique, ASTM Method D-881 (8), a line of sight is established by focusing a telescope on a target. By placing the specimen to be tested in the line of sight, the apparent position of the cross hairs on the target is shifted. From the magnitude of the shift and the distance between the target and the specimen, the angular deviation of the line of sight is calculated.

16.9.2 Reflection and Transmission— General Analysis

Surfaces of materials vary from glassy smooth to granular and wavy with a variety of different textures. Surface departures from planarity only one-fourth the wavelength of light are large enough to cause scattering of specularly reflected light. Both these surface properties and the internal structure of the polymer can cause scattering of specularly transmitted light.

There are two methods to measure the amount of light reflected or transmitted by the plastic: one is to differentiate the light reflected or transmitted as a function of angle; the other is to integrate all or part of the light reflected or transmitted. These measurements are accomplished with either a goniophotometer or an integrating sphere hazemeter. Both can be considered fundamental tests.

16.9.2.1 Goniophotometer

The most complete way to measure reflectance or transmittance from a surface is by use of a goniophotometer, as shown in Figure 16.58a. A goniophotometer curve is produced by fixing the light source at a specific angle and measuring the light transmitted or reflected at different viewing angles, or vice versa.

The shape of the goniophotometer curve may change both with the direction of incident light and with the spectral quality of this light, that is, with the wavelength of the input excitation, as well as with the angle of viewing.

Because of this, a whole family of curves is necessary to describe the light-redistribution from a small specimen area. Suitable procedures are described in ASTM Method E-166 (118) and ASTM Method E-167 (118). However, complete goniophotometer analysis of light distributions by objects are too cumbersome and complex to be attemped in most normal circumstances. Goniophotometric data are generally used as guides for designing simpler measurement methods for a limited range of specific material characteristics. For example, Figure 16.57 (1 and 2), d, and g show some possible specific measurements derived from goniophotometric data. For such measurements some abridged form of the equipment is used.

16.9.2.2 Integrating Sphere

The method for measuring total (luminous) reflectance or transmittance employs an integrating sphere of the form shown in Figure 16.58b. The interior surface of the sphere is coated with magnesium oxide, which reflects approximately 98% of the light striking it. Any light that is transmitted through the instrument alone (specimen removed) or instrument plus specimen (specimen in) is collected by the photocell after multiple reflections from the highly reflective walls of the sphere. This gives a measure of the total reflectance or transmittance. Figures 16.57b and e illustrate the technique for measuring total reflectance or transmittance. The method for measuring diffuse reflectance or transmittance is the same as that used for total reflectance or transmittance except that a light trap is added to remove all specular light, as shown in Figures 16.57c (3) and f.

16.9.3 Reflection and Transmission— Specific Methods

By using the principles described in the preceding sections, abridged instruments have been developed and special relationships for reflectance and transmission established. Selection of geometric conditions for the measurement of reflectance and transmittance are described in ASTM Method E-179 (116) for

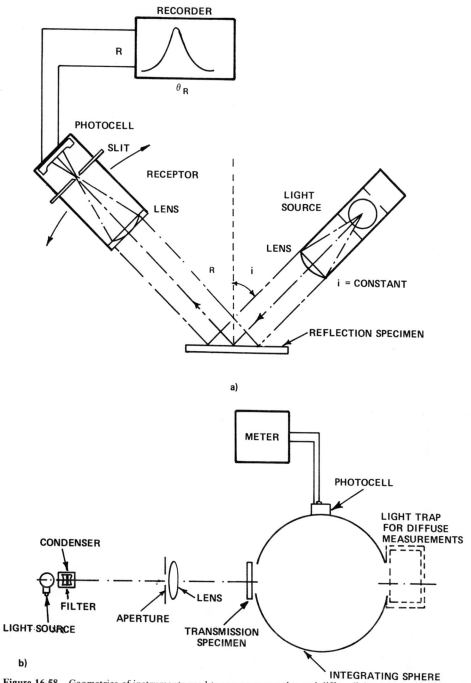

Figure 16.58 Geometries of instruments used to measure specular and diffuse light: (*a*) a goniophotometer; (*b*) an integrating sphere. Both geometries may be used to measure transmission or reflection by
changing either the location of the detector of the location of the specimen.

evaluating such appearance characteristics as glossiness, opacity, lightness, transparency, and haziness in terms of reflected or transmitted light. Some of the most important tests are discussed in the following sections.

16.9.3.1 Gloss and Diffuse Reflectance

Subjectively perceptible attributes of reflectance are dependent on the angle of viewing. Instrument evaluations must measure the reflected light at this same angle. Standardized methods for gloss measurements specify the character of the reflectance measurement and the exact geometric conditions used. Hunter (119) states that, "Gloss is the degree of which a surface simulates a perfect mirror in its capacity to reflect incident light." He concludes that there are at least six different visual criteria by which glossiness rankings are made. Correspondingly, there are at least six different procedures for making useful measurements of gloss that consider measurement of specular or diffuse reflectance or some ratio between the two.

These procedures are named and diagrammed in Figure 16.59. Each of the six criteria commonly apply to surfaces in a specific gloss range. In general, high angles from the vertical, that is, near grazing incidence, are best for low-gloss levels because specular reflectance is highest at the high angles. By contrast, high gloss is usually measured at low angles, that is, near normal incidence. The reason for this can be seen by considering the surface reflection of a perfectly flat specimen as originally described by Fresnel and now discussed in books on physical optics. For unpolarized incident light one can write Equation (16.71)

$$R_s = \frac{1}{2} \frac{\sin^2 (i - r)}{\sin^2 (i + r)} + \frac{\tan^2 (i - r)}{\tan^2 (i + r)} \qquad (16.71)$$

where R_s = specular reflectance
    i = angle of incidence
    r = angle of refraction
       = $\sin^{-1}(\sin i / n)$
    n = refractive index

R_s is plotted as a function of the angle of

GLOSS TYPE AND RANGE	DIAGRAM
SPECULAR, INTERMEDIATE	I $\quad i \mid r$ S $G_s \sim$ S/I
SHEEN, LOW	I $\quad i \mid r$ Sh $G_{sh} \sim$ Sh/I
CONTRAST OR LUSTER, LOW	I $\quad$ D $\quad$ S $\quad i \uparrow r$ $G_c \sim$ S/D
BLOOM OR HAZE, HIGH	I $\quad$ D $\quad$ B S $\quad i \uparrow r$ $G_b \sim$ (B–D)/I
DISTINCTNESS-OF-IMAGE, HIGH	$\quad$ dR/dθ $i \mid r$ $G_{di} \sim$ dR/dθ

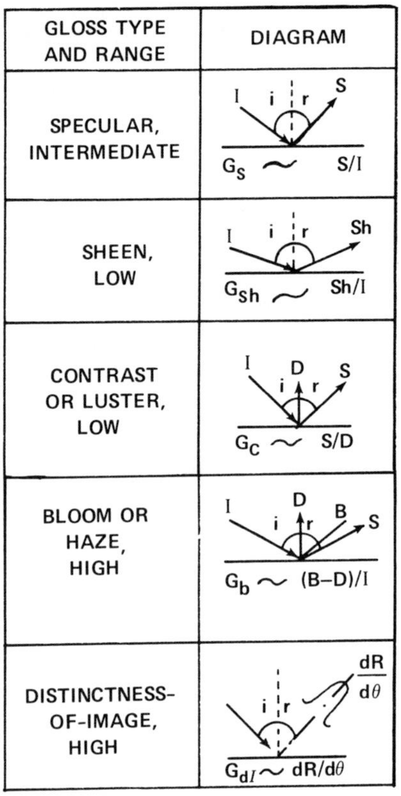

Figure 16.59 Physically perceptable attributes are dependent on the angle of viewing. Instrument evaluations must measure the reflected light at this same angle, based on work done by Hunter (119).

incidence for a nonmetallic mirror (such as black glass), for which $n = 1.530$, in Figure 16.60.

In practice, the precise manner in which a specimen surface will reflect an incident beam of light is determined by a complex set of factors not subject to measurement by a single number. However, the preceding equation and corresponding plot clearly indicate why gloss values vary as a function of the angle of incidence. Standard techniques for gloss measurements are ASTM Method D-523 (8) and ASTM Method D-2457 (8). In these methods, specular gloss is measured at 20, 60, or 85 deg.

Clarity of a reflected image is the capacity of the sample for allowing details in the object to be resolved in the image that it forms. An assessment of the clarity can be made by indicating the loss in angular separation of two points that can just be resolved in the image in comparison with the object. This

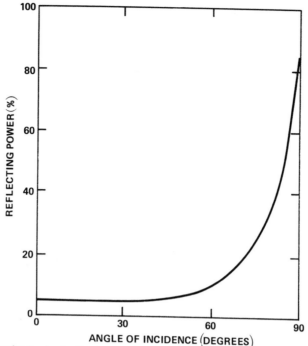

Figure 16.60 Specular (surface) reflection as a function of the angle of incidence. High angles (from the vertical) are recommended for materials with low gloss and low angles are recommended for materials with high gloss.

distinctness of image test measures loss of any light spreading 0.1 deg or more from the 30 deg specular direction. Commercial instruments can also be used to measure narrow-angle haze, 2 deg off specular; wide-angle haze, 5 deg off specular; and diffuseness, 15 deg off specular. Diffuse reflection (lightness) can be measured by illuminating the specimen at an angle of 45 deg and viewing it perpendicularly, as described in ASTM Method E-97 (118), or using an integrating sphere with a light trap to remove specular reflectance. Another technique, ASTM Method D-589 (120), is especially useful for opacity measurements. Here the total reflection is measured with an integrating sphere. Readings are taken with the specimen first backed with a white reflector and then backed with a black light trap.

16.9.3.2 Transparency and Haze

As with reflected light, tests using transmitted light consider either specular or diffuse components of the transmitted light or the total light transmitted. The specular transmission, that is, transparency, is usually defined as the percentage of incident light that is transmitted without deviation, and haze is usually defined as the percentage of incident light that is transmitted with more than certain angular deviation by forward scattering. Glazing material used for windows and diffusing material used for lighting fixtures must both have a high light transmittance, but the former must also be free from haze and very transparent, whereas the latter must have maximum diffusion and minimum transparency so that a bright light source cannot be seen through it. Film for many packaging purposes must also be transparent to resolve objects clearly through it. The most common technique used to measure diffuse light transmission and haze involves the use of an integrating sphere with a dark annulus of 1.3 deg as described in ASTM Method D-1003 (8). Here light that is greater than 1.3 deg off specular is measured. However, diffuse measurements by this test are hybrid when the specimen scatters the majority of light at low angles, since the 1.3 deg annulus is arbitrary and has no relation to the distinctness of image. For example, light is scattered from the surface of an abraded plastic at low angles, and most of this light may fall within the dark annulus, that is, the light trap, of the instrument.

Specular transmittance is measured by projecting a collimated beam of light on a receptor that measures only transmitted light that is within 0.1 deg of the specular source. Measurements are made with and without the specimen placed between the light source and receptor. Such a technique is described in ASTM Method D-1746 (8).

A simple component test for light transmission of reinforced plastic panels is described in ASTM Method D-1494 (31). The test uses a 0.6 m square panel. A photocell measures the transmitted light, and the transmittance is given as the ratio of photocell output with the specimen in and the specimen out.

16.9.4   Color

The second category of appearance evaluation is the measurement of the spectral attributes, or color. Light is defined as visually evaluated radiant energy having wavelengths from about 380 to about 770 nm. Different wavelengths have different colors. Measurements of the fractions of incident white light either reflected or transmitted at different wavelengths provide readings that relate to the color of the object. For example, the relative diffuse reflectances for white, blue, red, and black objects as a function of the illuminant wavelength are shown in Figure 16.61. The white object reflects light equally at all wavelengths. The blue object reflects light at the blue end of the spectrum but absorbs light at the red end, whereas the opposite is true for the red specimen. The black object absorbs all light

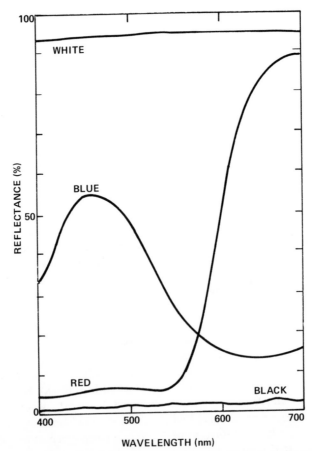

Figure 16.61   Color may be defined as the ability to absorb light at different wavelengths. White materials reflect all light equally whereas black materials absorb all light equally.

regardless of wavelength. The process of absorbing only certain wavelengths of light is called selective absorption and is the primary source of most of the color we see in everyday life. The specific instruments and color scales used instrumentally to describe color are not discussed here. However, the basis of color measurements involves a standard means for converting any spectral curve into three numbers called tristimulus values X, Y, and Z, that properly identify the color of the object in terms of a mixture of three primary lights that match it visually. The interested reader is referred to the excellent work by Hunter (121). ASTM Method E-308 (118) describes a spectrophotometric technique for measuring color.

16.9.5 Illuminants and Spectral Receivers

16.9.5.1 Illuminants

A factor of importance in reflectance and transmittance measurements is the wavelength distribution of the illuminant, that is, the input excitation. Standard illuminants have been established that have the characteristics close to the spectral distribution of natural light sources. For example, CIE (Commission International de l'Eclairage) Illuminant A represents light typical of that from an incandescent lamp, and Illuminant C represents average daylight from the north sky. The relative spectral distributions of these two sources are shown in Figure 16.62. Illuminant C is most widely used in color measuring instruments. Illuminant A is most widely used in instruments used to measure geometric attributes.

The importance of the wavelength distribution for color measurements is obvious. For example, a blue object will look darker using Illuminant A than using Illuminant C. The importance of the wavelength distribution for reflection and transmission is not as obvious, however. In fact, specular reflection occurs at the surface of the plastic and is not affected by the illuminant, except for differences that may occur because of changes in index of refraction. However, diffuse reflection (resulting from light scattering in the granular structure just beneath the skin) and transmission measurements involve penetration of light into the plastic. Thus selective absorption and scattering can occur. For example, a rubber-modified transparent PMMA may appear to scatter more light, that is, appear more hazy, when viewed by eye in daylight than when viewed in incandescent light. This is because the rubber particles scatter a disproportionate amount of light in the blue end of the spectrum. Thus, in this case, a blue filter in the haze-measuring instrument is recommended if correlations with end use are expected.

16.9.5.2 Spectral Receivers

Within the range of visible light, some wavelengths can be seen more easily than others. Luminosity is the property of light by which we define how easily we can see it. The

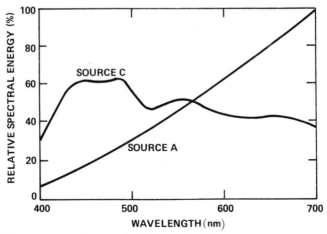

Figure 16.62 The appearance of a material will be dependent on the wavelength distribution of the illuminant. Illuminant A represents an incandescent lamp; illuminant C represents the average north sky.

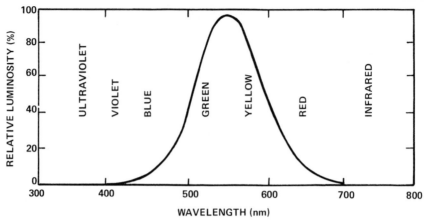

Figure 16.63 The luminosity function of the human eye. The light in the yellow-green portion of the spectrum around 550 nm is seen much more easily than elsewhere in the spectrum, after Hunter (121).

luminosity of various wavelengths differs; the eye is most sensitive to light at about 550 nm (green yellow light) and is relatively insensitive to radiation at the ends of the visible spectrum. Thus the eye is more sensitive to a given amount of energy at 550 nm than it is to the same amount of energy at 650 nm. The sensitivity of the eye to all wavelengths has been determined and can be simply summarized in the form of a graph called the luminosity function for the human eye. This graph is shown in Figure 16.63. One of the components Y of the tristimulus values corresponds exactly to the luminosity function.

For the spectral receiver (photocell) to "see" in a manner similar to the human eye, the response of the photocell should be similar to the curve shown in Figure 16.63. Of course, in many cases the properties of the material in a certain region either inside or outside the visual region is of interest. Measurements of the effectiveness of ultraviolet absorbers are examples. In general, the response characteristics of the spectral receiver should be considered in relation to the property of interest.

16.10   OTHER PHYSICAL PROPERTIES

Most of the important methodology for physical testing has been categorized in the preceding sections. These sections have attempted to describe how to measure the properties of thermoplastics and thermosets using funda-

mental techniques when applicable and practical or component techniques if not. The advantages and limitations of these various tests have been emphasized. It is hoped that this information will provide guidelines for product design and selection of materials.

There remain a number of important tests, however, that have not fallen logically into previous sections. These are discussed here.

16.10.1   Permeation

The resistance of polymeric materials to penetration and transmission of gases, vapors, and liquids is a primary factor affecting their use in many applications. For example, low permeability to moisture, oxygen, and carbon dioxide is essential when polymers are used as packaging films. Also, permeability of water through polymers is of extreme interest, since most or all of our clothes are made of polymeric materials, and water vapor transport is one of the principal factors of physiological comfort. Typical applications requiring water-vapor-permeation resistant plastics in thicknesses of 1.5–3.0 mm are found in encapsulation of electronic components, hermetic seals, and in protective liners in valves and pipes exposed to corrosive chemicals.

Under steady-state conditions, Fick's law has been found to hold, and the transmission follows the relationship shown as Equation (16.72)

$$q = D \frac{dc}{dx} \qquad (16.72)$$

where q = amount of gas diffusing through a unit area of the film in unit time

D = diffusion constant

dc/dx = concentration gradient across a thickness dx

Gas concentrations are normally measured in terms of the pressure p of the gas that is at equilibrium with the film. The vapor concentration c can be represented by means of Henry's law, defined as Equation (16.73),

$$c = Sp \qquad (16.73)$$

where S is the solubility coefficient of the gas in the film. Hence

$$q = D \cdot S \frac{dp}{dx} \qquad (16.74)$$

or upon integrating across the total thickness

of the film l one obtains Equation (16.75)

$$q = D \cdot S \frac{(p_1 - p_2)}{l} \qquad (16.75)$$

Since the product $D \cdot S$ is referred to as the permeability constant P, Equation (16.75) upon rearrangement leads to Equation (16.76).

$$P = \frac{q \cdot l}{(p_1 - p_2)} \qquad (16.76)$$

This equation assumes that gas is permeating under steady-state conditions. In general, there will be an interval from the time that the penetrant first enters the membrane until the steady-state flow is established. A typical plot of the total amount of penetrant that has passed through the membrane as a function of time is shown in Figure 16.64. The intercept on the time axis of the extrapolated linear steady-state portion of the curve is called the time lag τ. Under ordinary conditions the steady-state of flow is reached after a period of approximately three times the time lag.

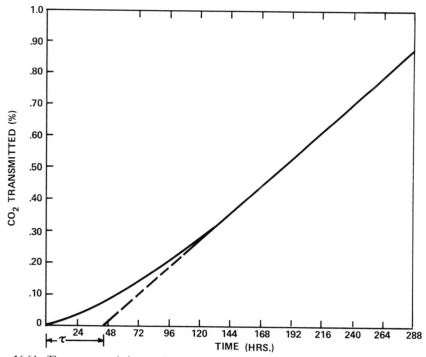

Figure 16.64   The gas transmission reaches a steady-state condition after approximately 3τ where τ is defined as the lag time. Values measured in times less than 3τ have little meaning.

In the most common experimental case, the membrane is initially free of penetrant and the diffusion flux occurs through the membrane into a reservoir of essentially zero concentration so that τ can be defined as in Equation (16.77)

$$\tau = \frac{l^2}{6D} \qquad (16.77)$$

The vapor transmission is a function of temperature. The permeability of vapor and gases follow an Arrhenius relationship shown as Equation (16.78)

$$P = P_0 \exp \frac{-E}{RT} \qquad (16.78)$$

For example, the oxygen transmission rate for two films as a function of temperature is shown in Figure 16.65.

Transmission rates are also influenced by the relative humidity. With hydrophilic films, the sorbed water acts as a plasticizer facilitating the diffusion of gas through the membrane. With hydrophobic films, the presence of water vapor has little effect on the gas transmission rate. Comparisons of hydrophilic and hydrophobic films are shown in Figure 16.66.

Permeability can be determined by several procedures. However, it is important to keep in mind the foregoing discussion to insure that the laboratory data will agree with those of end use.

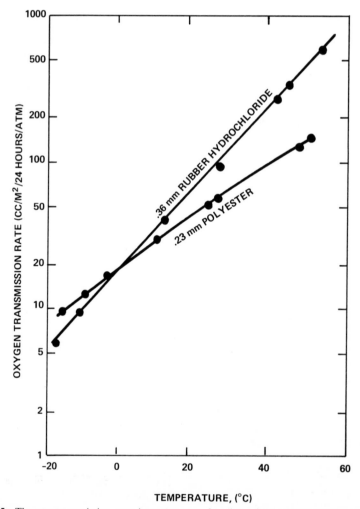

Figure 16.65 The gas transmission rate increases as a function of temperature. (Courtesy Modern Controls, Inc.)

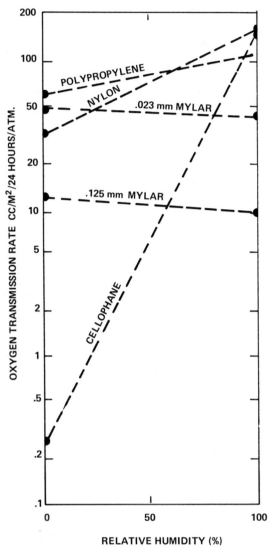

Figure 16.66   The influence of relative humidity on the gas transmission rate. With hydrophilic films, the sorbed water acts as a plasticizer facilitating the diffusion of gas through the membrane. With hydrophobic films, the presence of water vapor has little effect on the gas transmission rate.

16.10.1.1   Water Vapor Transmission

The simplest procedure for measuring water vapor transmission rate is the gravimetric procedure. The material to be tested is fastened over the mouth of a dish, which contains either a desiccant or water, as shown in Figure 16.67a. The assembly is placed in an atmosphere at constant temperature and humidity, and the weight gain or loss of the assembly is used to calculate the rate of water vapor movement through the sheet material. For meaningful results, the test condition must match the climatic condition (both temperature and humidity) in use. Several conditions are described in ASTM Method E-96 (8).

For high-barrier films or thick materials this technique is not practical because long testing times, in terms of weeks or even months, are necessary to reach equilibrium. In such cases, dynamic measurement of water vapor transfer is useful. One apparatus (122) consists of a two-compartment cell, as shown in Figure 16.67b, which permits a humidity gradient to be established across the speci-

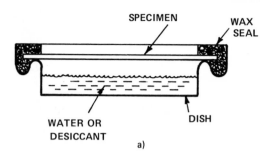

a)

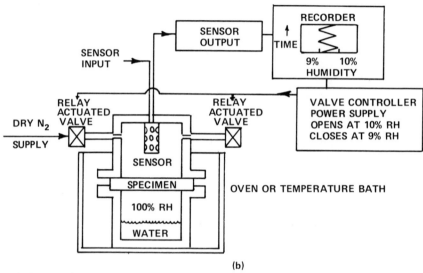

(b)

Figure 16.67 Techniques used to measure the water vapor transmission rate: (*a*) gravimetric; (*b*) dynamic.

men. The lower test cell compartment is maintained at 100% humidity. The sensor compartment is maintained at an average humidity of 9.5% by automatically cycling dry nitrogen purges between humidities of approximately 9 and 10%. The rate of accumulation of water vapor in the upper compartment is measured with a lithium chloride humidity sensor whose resistance is a function of cell humidity. Even for thick samples, equilibrium can be reached in 1–4 days. A similar apparatus is also described in ASTM Method E-398 (123).

Finished packages can be tested using the gravimetric technique by following weight change or by the dynamic technique. In the latter, the test cell consists of a single chamber with a humidity sensor into which the package is inserted

16.10.1.2 Gas Transmission

Many techniques have been used to measure the gas transmission rate through a polymer membrane that forms a partition between two chambers. Most of the methods for determining the amount of gas that has passed through the membrane in unit time involve measuring the increase in concentration, pressure, or volume of the gas as a function of time.

In simple tests, such as the standard ASTM test, Method D-1434 (8), a measurement is obtained of the gas transmission of a single pure *dry* gas through sheet materials. In these simple tests, the sample is mounted in a gas transmission cell forming a sealed semibarrier between two chambers, as shown in Figure 16.68*a*. One chamber containing the test gas is at a specific high pressure and the other chamber, at a lower pressure, receives the permeating gas. In the manometric cell, the lower pressure chamber is evacuated and the transmission through the specimen is indicated by increase in pressure. However, gas transmission is affected by conditions not specifically considered in these tests, such as moisture, temperature, plasticizer content, and nonhomogeneities. Interactions, which affect performance in applications where several gases are present, cannot be measured. Also, it is difficult to insure that steady-state conditions have been obtained, and lag time cannot be easily measured (124). Correlation of test values with any given use, such as packaging protection, must be determined by experience.

A more suitable technique (125, 126) permits the measurement of gas permeability as a function of temperature, pressure, gas content, and moisture. In the instrument shown in Figure 16.68b, test films are clamped between stainless steel shells to form a pair of outer chambers and a single inner one. Chambers

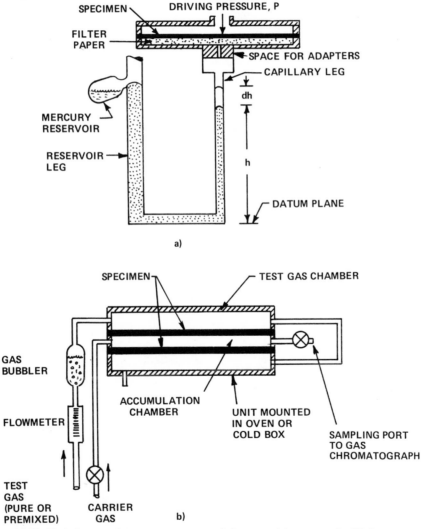

Figure 16.68 Techniques used to measure gas transmission rates: (*a*) manometric; (*b*) chromatographic. Manometric techniques should be used only to measure pure dry gas. Chromatographic techniques are useful for mixed or humidifed gases.

formed on either side of the film sample have valved outlets and inlets for gas supply to the outside and inside of the two films. By opening the valves, a carrier gas (e.g., nitrogen) at any desired humidity level flushes the center cell. While this gas is flushing the center chamber, test gas such as oxygen or a combination of gases at any desired relative humidity is flushing the outer cells. Both the test gas and carrier gas are at atmospheric pressure; there is no pressure differential across the sample. After purging, the valves to the center cavity are closed while the test gas continues to flow in the outer chambers. Temperature can be controlled above or below ambient. At intervals, a sample of inner-chamber gas is drawn for analysis by gas chromatography. The gas transmission for the various components can then be calculated by plotting the change in areas of the chromatograph peaks as a function of time. A similar unit, described in ASTM Method D-3985 (8), uses an oxygen-specific coulometric detector so that gas samples do not have to be withdrawn. The change of current in the coulometric detector is plotted as a function of time on a strip-chart recorder.

Instead of using a sample in the form of a membrane, the latter two techniques can be used to follow permeation through a finished container. The container is equipped with a seal or lid that permits introduction of the carrier gas to the interior. This whole assembly is placed is an outer chamber that is swept with the test gas.

Additional details of permeation measurements are discussed by Stannett et al. (127, 128) and Rodgers (129).

16.10.2 Density

Density may be an important criterion for specific applications where weight, thermal transfer, or buoyancy are factors. Density is also useful for calculating strength-weight and cost-weight ratios.

The simplest way to measure density on a large regular sample is to measure weight and volume directly. Other techniques are suitable for small or irregular-shaped specimens. The

sample can be weighed in air in a suitable liquid of known density and the density calculated from the change in weight, as described in ASTM Method D-792 (8).

In another technique, described in ASTM Method D-1505 (8), the sample can be placed in a liquid column with a density gradient, and the level to which it sinks can be compared with the level of standards of known density.

In a similar technique, the sample and standards of known density can be placed in a beaker containing a lower-density liquid and a second-higher-density miscible liquid titrated until, after stirring and allowing to come to rest, the sample is suspended. Densities can be calculated by observing the amount of higher-density liquid needed to cause the sample to become suspended compared to the amount needed to suspend standards.

16.10.3 Permanence

The permanence of most polymers is an extremely important property, but is difficult to assess. By definition, this category includes any change in the measurable properties of a material with time and environment as the prime controlling variables. This category embraces changes due to such commonplace environmental variables including humidity, temperature, various weather conditions (direct sunlight, etc.), and corrosive atmospheres.

A knowledge of permanence properties is necessary for the intelligent utilization of a polymeric material in any end-use application. The various physical tests described in this chapter can be used to ascertain many of these important permanence properties. Various procedures for exposing plastics to various environments before testing for the specific properties of interest are discussed in various ASTM methods on permanence (8).

16.11 SUMMARY

The advancement of relevant physical testing and evaluation during the next decade will depend on several things: the wider recognition of the criteria important for such testing;

the education of physical testing technologists to abondon (or at least supplement) "typical value data" test techniques and instead broaden their testing techniques; the development of standard methods of testing that play a bigger role in the science and engineering of polymers, accompanied by the removal of inadequate standards; and the automation and computerization of physical testing equipment to make acquisition of more meaningful test data both easier and less costly. To accomplish the development of methods for engineering performance the input of designers and product engineers must be secured and integrated with those of the physical evaluation physicist (130). Additionally methods intended for engineering performance must be clearly separated from those intended for purchase specifications and so on.

The principles of meaningful physical testing and the technology for equipment automation have been fully developed. All that is left is for the physical testing technologist to "utilize" the methods and the tools available to him or to develop new ones based on sound guidelines. The physical testing technologist who continues to provide or rely on only quality control or typical value data in place of engineering data will find in the next decade that such effort will no longer be acceptable. It is obvious that the use of plastics in highly specialized areas as well as traditional areas is rapidly continuing. It is recommended that the development and use of meaningful physical testing techniques keep pace. Finally, new product development can be greatly enhanced by using an interdisciplinary approach where material engineering and physical evaluation are integrated with chemical efforts at the initial stages.

16.12   GLOSSARY OF SYMBOLS*

Mechanical Properties

ϵ	Normal strain
L	Length

*The same symbol is used in different sections with different meanings. This has been done to maintain the nomenclature normally employed in the literature.

ΔL	Change in length
σ	Normal stress
P	Force
A	Area
E	Modulus of elasticity
c	Distance from neutral axis
ρ	Radius of curvature
M	Bending moment
I	Moment of inertia
τ	Shear stress
γ	Shear strain
θ	Angular measurement
G	Shear modulus
t	Time
T	Temperature
E^*	Complex dynamic modulus
ω	Angular frequency
E'	Dynamic storage modulus
E''	Dynamic loss modulus
δ	Phase angle
T_g	Glass transition temperature
α	Peak in damping curve associated with the Tg
β, λ	Peaks in damping curve not associated with the Tg
b	Specimen width
d	Specimen thickness
y	Beam deflection
Ω	Constant
B	Bending stiffness
μ	Poisson's ratio
k	Fractional value between -1 and 1
w	Crack width
l	Crack width
C	Specimen compliance
G_c	Strain energy release rate
F	Fractional recovery of strain
R	Reduced time
S	Stress at failure in a fatigue test
N	Number of cycles to failure in a fatigue test
ν	Frequency
C_p	Specific heat
ϕ	Density
j''	Loss compliance
K	Constant

BSH	Bierbaum scratch hardness
v	Velocity
p	Pressure
f	Frictional force
N	Normal force
u	Coefficient of friction
K	Constant

Electrical Properties

ϵ'	Dielectric constant
C_p	Parallel capacitance of material
C_a	capacitance of a dimensionally equivalent air gap
ϵ''	Loss index
E	Voltage
ν	Frequency
δ	Phase angle
A	Area
d	Specimen thickness
R	Resistance
i	Current
ρ	Resistivity
g	Guard gap width
P	Effective perimeter of the guarded electrode
RH	Relative humidity
α, K, k, n	Constants
σ	Dielectric strength
Q	Electrostatic charge
τ	Half-life of electrostatic charge

Thermal Properties

α	Coefficient of linear thermal expansion
ΔL	Change in specimen length
L	Length
T	Temperature
K	Thermal conductivity
A	Area
q	Heat flow
C_p	Specific heat
x	Distance through specimen
OI	Oxygen index
R	Percent of 25°C oxygen index retained

Optical Properties

n	Index of refraction
i	Incidence angle
r	Refraction angle
θ	Reflection angle
R	Specular reflectance

Other Physical Properties

q	Amount of gas diffusing through a unit area
D	Diffusion constant
c	Vapor concentration
x	Increment of thickness
S	Solubility
p	Pressure
P	Permeability constant
l	Specimen thickness
τ	lag time
E	Constant
R	Gas constant
I	Absolute temperature

16.13   REFERENCES

1.   S. Turner, *Mechanical Testing of Plastics*, CRC Press, Cleveland, 1973.

2.   D. W. Van Krevelen, *Properties of Polymers, Their Estimation and Correlation with Chemical Structure*, Elsevier, Amsterdam, Oxford, New York, 1976.

3.   R. E. Evans in R. E. Evans, Ed., *Physical Testing of Plastics—Correlation with End-Use Performance*, STP 736, ASTM, Philadelphia, 1981, p. 3.

4.   T. Alfrey, Jr., *Mechanical Behavior of High Polymers*, Vol. 6, Appendix IV, Interscience, New York 1948, p. 565.

5.   M. Mooney, *Symposium on Consistency*, 9–12, American Society for Testing and Materials, Philadelphia (June 29, 1927).

6.   J. M. Sherlock, *Plast. Eng.*, **35**, (March 1976).

7.   W. E. Brown, F. C. Frost, P. E. Willard, "Standards and Sources of Tests for Polymers," in J. V. Schmitz, Ed., *Testing of Polymers*, Vol. 1, Interscience, New York, London, 1965, p. 1.

8.   *ASTM Annual Book of Standards*, Part 35, American Society for Testing and Materials, Philadelphia, (1981).

9.   L. Lamond, *Plast. Des. Proc.*, 6 (March, 1976).

10. J. V. Schmitz and W. E. Brown, "Selected References on Sources of Standards and Tests for Polymers", in Schmitz and Brown, Eds., *Testing of Polymers*, Vol. 3, Wiley-Interscience, New York, London, Sydney, 1967.

11. J. R. Beatty and C. E. Sitz, "Physical Testing of Elastomers and Plastics," in J. Craner and R. W. Tess, Eds., *Applied Polymer Science*, American Chemical Society, Washington, D. C., 1976, p. 100.

12. F. S. Pang and J. L. Isaacs in R. E. Evans, Ed., *Physical Testing of Plastics—Correlation with End-Use Performance*, STP-736, ASTM, Philadelphia, 1981, p. 101.

13. L. A. Cohen, "Influence on Properties of Specimen Shape and Preparation", in J. V. Schmitz, Ed., *Testing of Polymers*, Vol 3, Interscience, New York, London, 1967, p. 15.

14. R. A. Horsley, "Assessment of Performance," in R. M. Ogorkiewicz, Ed., *Thermoplastics*: *Effects of Processing*, London Iliffe Books, Ltd., London (1969).

15. G. B. Jackson and R. L. Ballman, *SPE J.*, **16**, 1150 (1960).

16. G. P. Koo, *Plast. Eng.*, 33 (May 1974).

17. L. E. Nielsen, *Mechanical Properties of Polymers*, Reinhold, London, 1962.

18. J. K. Gillham and M. B. Roller, *Polym. Eng. Sci.* **11**(4) (1971).

19. R. B. Blaine and L. Woo, ACS Polym. Prep., **17**(2) 1 (August 1976).

20. M. Takayanagi and M. Yoshino, *ZAIRYOSHI— IKEN (J. Jap. Soc. Test Mater.)*, **8**, 308 (1959).

21. A. S. Kenson et al., *ACS Polym. Prepr.*, **17**(2), 7 (August 1976).

22. J. M. Starita and C. W. Macosko, *SPE J.*, **27**, 38 (November 1971).

23. J. M. Starita, Automated Dynamic Mechanical Techniques Seminar, SPE, September 29–30, 1977, Princeton, N.J.

24. S. S. Sternstein, in Reference 19b.

25. M. N. Riddell, *Plast. Eng.*, 71 (April 1974).

26. R. F. Boyer, *Polym. Eng. Sci.*, **8**(3) (July 1968).

27. M. E. Roylance, D. K. Roylance, and J. N. Sultan, "The Role of Rubber Modification in Improving High Rate Impact Resistance," in R. D. Deanin and A. M. Crugnola, Eds., *Toughness and Brittleness of Plastics*, American Chemical Society, Washington, D.C., 1976, p. 192.

28. T. S. Carswell and H. K. Nason, *Mod. Plast.*, **21**, 121 (June 1944).

29. J. Marin, "Mechanical Relationships in Testing for Mechanical Properties of Polymers," in J. V. Schmitz, Ed., *Testing of Polymers*, Vol. 1, Interscience, New York, London, 1965, p. 87.

30. W. D. Harris, "Measurement of Tensile Properties of Polymers," in W. E. Brown, Ed., *Testing of Polymers*, Vol. 4, Interscience, New York, London, 1969, p. 399.

31. *ASTM Annual Book of Standards*, Part 36, American Society of Testing and Materials, Philadelphia (1981).

32. H. S. Loveless, *J. Test. Eval.*, **1**(6), 532 (1973).

33. R. M. Jones, *J. Compos. Mater.*, **10**, (October 1976).

34. H. S. Loveless, "Flexural Tests," in J. V. Schmitz, Ed., *Testing of Polymers*, Vol. 2, Interscience, New York, London, 1966, p. 321.

35. R. E. Evans, *TAPPI*, **53**(12), 2255 (1970).

36. W. J. Renton, *Exp. Mach.*, p. 225, **16**(11) 409 (1976).

37. L. B. Greszczuk, "Shear Modulus Determination of Isotropic and Composite Materials," in *Composite Materials: Testing and Design*, STP 460, American Society for Testing and Materials, Phildelphia, 1969, p. 140.

38. R. L. Thorkildsen, "Mechanical Behavior", in E. Baer, Ed., *Engineering Design for Plastics*, Reinhold, New York, London, 1964, p. 277.

39. R. M. Raghava, R. M. Coddell, and G. S. Y. Yeh, *J. Mater. Sci.*, **8**, (1973).

40. L. J. Broutman, S. M. Kirshnakumar, and P. K. Mallick, *J. Appl. Polym. Sci.*, **14**, 1477 (1970).

41. M. G. Sharma and C. K. Lim, *Polym. Eng. Sci.*, 254 (October 1965).

42. T. Kobayashi and L. J. Broutman, *J. Appl. Polym. Sci.*, **17**, 1909 (1973).

43. *ASTM Annual Book of Standards*, Part 10, American Society of Testing and Materials, Philadelphia, 1981.

44. P. Flueler, D. R. Roberts, J. F. Mandell, and F. J. McGarry in R. E. Evans, Ed., *Physical Testing of Plastics—Correlation With End-Use Performance*, STP-736, ASTM, Philadelphia, 1981, p. 15.

45. R. W. Hertzberg, *Deformation and Fracture Mechanics of Engineering Materials*, John Wiley, New York, 1976.

46. *ASTM Annual Book of Standards*, Part 34, American Society of Testing and Materials, Philadelphia, 1981.

47. H. Burns, "Impact Resistance," in N. M. Bikales, *Mechanical Properties of Polymers*, Interscience, New York, London, 1971, p. 175.

48. D. R. Ireland in R. E. Evans, Ed., *Physical Testing of Plastics—Correlation With End-Use Performance*, STP-736, ASTM, Philadelphia, 1981, p. 45.

49. J. K. Rieke, in Reference 39a, p. 59.

50. H. Gonzalez, *J. Appl. Polym. Sci.*, **19**, 2717 (1975).

51. M. Silberberg and R. Supnik, *Encyclopedia of Polymer Science*, Vol. 13, John Wiley, New York, 1970, p. 601.

52. F. J. Furno, R. S. Webb, and N. P. Cook, *Prod. Eng.*, (August 17, 1964).

53. J. M. Starita, *Plast. Worls*, 58 (April 1977).

54. R. E. Chambers, "Behavior of Structural Plastics," *Structural Plastics Design Manual*, U.S. Government Printing Office, Washington, D.C., 1978.

55. S. Turner, "Deformation Data for Engineering Design," in W. E. Brown, Ed., *Testing of Polymers*, Vol. 4, Interscience, New York, London, 1969, p. 1.

56. *Modern Plastics Encyclopedia*, McGraw-Hill, New York, 1977.

57. R. M. Ogorkiewicz, Ed., *Engineering Properties of Thermoplastics*, Interscience, New York, London, 1970.

58. E. Passaglia and J. R. Knox, "Viscoelastic Behavior and Time-Temperature Relationships," in E. Baer, Ed., *Engineering Design For Plastics*, Reinhold, New York, 1964.

59. R. J. Curran, R. D. Andrews, F. J. McGarry, *Mod. Plast.*, **38**(3), 142 (November 1960).

60. M. N. Riddell, *Polym. Eng. Sci.* (October 1966).

61. R. W. Hertzberg and J. A. Manson, *Fatigue of Engineering Plastics*, Academic, New York, 1980.

62. E. M. Prot, *Rev. Matturgie*, **45**(12) (1948). English Translation by E. J. Ward, WADS Technical Report 52-148, Wright Air Development Center (September 1952).

63. L. S. Lazar, "Use of the Prot Accelerated Fatigue Test to Predict Endurance Limits of Plastics," Presented at the 11th Annual Meeting of the Society of Plastics Industry, Inc., Atlantic City, New Jersey, February 1956.

64. H. S. Loveless, C. W. Deeley, D. L. Swanson, *SPE Trans.*, **2**(2) (April 1962).

65. H. S. Loveless in R. E. Evans, Ed., *Physical Testing of Plastics—Correlation With End-Use Performance*, STP-736, ASTM, Philadelphia, 1981, p. 32.

66. D. L. Livingstone, "Indentation Hardness Testing," in J. V. Schmitz and W. E. Brown, Ed., *Testing of Polymers*, Vol. 3, Interscience, New York, London, 1967 p. 111.

67. E. C. Bernhardt, *ASTM Bull.*, **157**, 123 (March 1949).

68. C. H. Bierbaum, *Trans Am. Soc. Steel Treat.*, **18**, 1009 (1930).

69. J. F. Carroll and J. O. Paul, *Photogr. Sci. Eng.*, **5**(5), p. 288 (1961).

70. R. A. Wiinikainen, Mater. Res. Stand., (17 December 1969).

71. J. J. Gouza, "Methods of Test for Hardness and Wear of Plastics," in J. V. Schmitz, *Testing of Polymers*, Vol. 2, Interscience, New York, London, 1966, p. 225.

72. J. F. Carroll in R. E. Evans, Ed., *Physical Testing of Plastics—Correlation With End-Use Performance*, STP-736, ASTM, Philadelphia, 1981, p. 77.

73. N. S. Eiss, in R. E. Evans, Ed., *Physical Testing of Plastics—Correlation With End-Use Performance*, STP 736, ASTM, Philadelphia, 1981, p. 91.

74. R. G. Bayer, Ed., *Wear Tests for Plastics: Selection and Use*, STP-701, ASTM, Philadelphia, 1979.

75. R. B. Lewis, "Rubbing-Contact Evaluation of Polymers," in J. V. Schmitz and W. E. Brown, Ed., *Testing of Polymers*, Vol. 3, Interscience, New York, London, 1967, p. 203.

76. *ASTM Annual Book of Standards*, Part 25, American Society of Testing and Materials, Philadelphia, 1981.

77. A. E. Sherr, W. G. Deichert, and R. L. Webb, *Plast. Des. Proc.*, 24 (March 1970).

78. R. F. Westover and W. I. Vroom, *SPE J.*, **19**(10), 1093 (October 1963).

79. R. E. Evans, *J. Test. Eval.*, **3**(2), 133 (1975).

80. *ASTM Annual Book of Standards*, Part 39, American Society of Testing and Materials, Philadelphia, 1981.

81. H. S. Endicott, *J. Test. Eval.*, **4**(3), 188 (May 1976).

82. R. W. Tucker, "Dielectric Constant and Loss Measurement," in J. V. Schmitz, Ed., *Testing of Polymers*, Vol. 1, Interscience, New York, London, 1965, p. 237.

83. A. Von Hippel, *Dielectric Materials and Applications*, John Wiley, New York, 1954.

84. *Technical Data on Plastics*, Manufacturing Chemists Association, Washington, D.C. February, 1957.

85. A. H. Scott, "DC Dielectric Conductance and Conductivity Measurements," in J. V. Schmitz, Ed., *Testing of Polymers*, Vol. 1 Interscience, New York, London, 1965, p. 213.

86. A. Blanck, Electro-Technology, Engineering, and Science, Article No. 102, June 1967.

87. R. Hertou and R. LaCoste, "Sur La Me'sure Des Resistivities et L'Etude de Conditionnement des Isolantes en Feuilles," Report IEC 15-GT$_2$ (France), April 4, 1963.

88. E. Oachini and G. Maschio, *IEEE Trans. Power Appar. Syst.*, Vol. PAS-86, No. 3, March 1967.

89. T. W. Dakin, "High Voltage Electrical Testing of Polymers," in J. V. Schmitz, Ed., *Testing of Polymers*, Vol. 1, Interscience New York, London, 1965, p. 267.

90. K. N. Mathes, *Encyclopedia of Polymer Science*

and Technology, Vol. 5, Interscience, New York, London, 1966, p. 582.

91. K. N. Mathes, "Electrical Properties," in E. Baer, Ed., *Engineering Design for Plastics*, Reinhold, New York, 1964, p. 437.

92. R. Bartnikas and E. J. McMahon, Ed., Engineering Dielectrics, Volume I, Corona Measurement and Interpretation, STP-669, ASTM, Philadelphia, 1979.

93. W. Starr and H. Endicott, *A.I.E.E. Trans. Pure Appl. Syst.*, **80**, Pt. 3, 515 (August 1961).

94. G. R. Mitchell, *J. Test. Eval.*, **2**(1), 23 (1974).

95. L. Mandelcorn, *Trans. Am. Inst. Electr. Eng.*, Part 3, **80**, 481 (1961).

96. R. Hoff and G. Springling, *Trans. Am. Inst. Electr. Eng.*, Part 3, **80**, 486 (1961).

97. V. E. Shashoua, *J. Polym. Sci.*, **33**, 65, (1958).

98. *ASTM Annual Book of Standards*, Part 21, American Society for Testing and Materials, Philadelphia, 1975.

99. W. E. Wendlandt, *Thermal Methods of Analysis*, Interscience, New York, London, 1974, p. 428.

100. J. Schröder, *Rev. Sci. Instrum.*, **34**(6), 615 (June 1963).

101. D. R. Anderson and R. U. Acton, *Encyclopedia of Polymer Science and Technology*, Vol. 13, Interscience, New York, London, 1970, p. 764.

102. M. J. Majurey, *Plast. Rubber Int.*, **2**(3), 111 (1977).

103. DuPont Thermal Analysis Application Brief, No. 11 (1968).

104. B. Miller, Mod. Plast., 42, (November 1977).

105. C. J. Hilado, *Flammability Handbook For Plastics*, Technical Publication, Stamford, Connecticut, 1969.

106. N. P. Setchkin, *J. Res. Natl. Bur. Stand.*, **43**, 591 (1949).

107. *ASTM Annual Book of Standards*, Part 18, American Society for Testing and Materials, Philadelphia, 1981.

108. H. Reymers, *Mod. Plast.*, 92, (October 1970).

109. L. B. Allen and L. N. Chellis, "Flammability Tests," in J. V. Schmitz, Ed., *Testing of Polymers*, Vol. 2, Interscience, New York, London, Sydney, 1966, p. 349.

110. C. P. Fenimore and F. J. Martin, Mod. Plast., 141 (November 1966).

111. K. G. Goldblum, *SPE J.*, 50 (February 1969).

112. P. R. Johnson, J. Appl. Polym. Sci., **18**, 491 (1974).

113. E. E. Smith, Fire Technol. **8**(3) 237 (1972).

114. F. J. Rarig and A. J. Bartosic, *ASTM Spec. Tech. Publ.*, No. 422, 1967, p. 106.

115. D. Gross, J. J. Loftus, and A. F. Robertson, *ASTM Spec. Tech. Publ.*, No. 422, 1967, p. 166.

116. W. F. Bartoe, "Optical Properties," in E. Baer, Ed. *Engineering Design for Plastics*, Reinhold, New York 1964, p. 593.

117. A. C. Hardy and F. H. Perrin, *The Principles of Optics*, McGraw-Hill, New York, London, 1932, p. 362.

118. *ASTM Annual Book of Standards*, Part 46, American Society for Testing and Materials, Philadelphia, 1981.

119. R. S. Hunter and L. Boor, "Tests for Surface Appearance of Plastics," in J. V. Schmitz, Ed., *Testing of Polymers*, Vol. 2 1966, p. 279.

120. *ASTM Annual Book of Standards*, Part 21, American Society For Testing Materials, Philadelphia, 1981.

121. R. S. Hunter, *The Measurement of Appearance*, Wiley-Interscience, New York, London, Sydney, Toronto, 1975.

122. R. G. Hadge, M. N. Riddell, and J. L. O'Toole, ANTEC 1972.

123. *ASTM Annual Book of Standards*, Part 41, American Society of Testing and Materials, Philadelphia, 1981.

124. R. E. Evans, *J. Test. Eval.*, **2**(6) (1974).

125. S. G. Gilbert and D. Pegaz, *Package Eng.*, 66 (January 1969).

126. D. G. Pye, H. H. Hoehn, and M. Panar, *J. Appl. Polym. Sci.*, **20**, 287 (1976).

127. V. Stannett and Yasuda, "The Measurement of Gas and Vapor Permeation and Diffusion in Polymers," in J. V. Schmitz, Ed., *Testing of Polymers*, Volume 1, Interscience, New York, London, Sydney, 1965, p. 393.

128. V. Stannett, M. Szwarc, R. L. Bhargana, J. A. Meyer, A. W. Myers, C. E. Rodgers, *Permeability of Plastic Films and Coated Papers to Gases and Vapors*, TAPPI, New York, 1962.

129. C.E. Rodgers, "Permeability and Chemical Resistance," in E. Baer, Ed., *Engineering Design For Plastics*, Reinhold, New York, London, 1964, p. 609.

130. J. L. O'Toole in R. E. Evans, Ed., *Physical Testing of Plastics-Correlation* With End-Use Performance, STP-736, ASTM, Philadelphia, 1981, p. 109.

General References

E. Baer, Ed., *Engineering Design for Plastics*, Reinhold, New York, 1964.

N. M. Bikales, Ed. *Encyclopedia of Polymer Science and* science, New York, London, 1971.

N. M. Bikales, Ed. *Encyclopedia of Polymer Science and Technology*, Interscience, New York, London, Vols. I–XVI, 1964–1972.

Composite Materials: Testing and Design, STP 460, American Society of Testing and Materials, Philadelphia, 1969.

R. E. Evans, Ed. *Physical Testing of Plastics—Correlation With End-Use Performance*, STP-736, ASTM, Philadelphia, 1981.

R. S. Hunter, *The Measurement of Appearance*, Interscience, New York, London, 1975.

G. C. Ives, J. A. Mead, and M. M. Riley, *Handbook of Plastics Test Methods*, CRC Press, Cleveland, 1971.

I. M. Kolthoff, P. J. Elving, F. H. Stross, Ed., *Treatise on Analytical Chemistry*, Part 3, Vol. 3, Interscience, New York, London, 1976.

A. E. Lever and J. A. Rhys, *The Properties and Testing of Plastics Materials*, CRC Press, Cleveland, 1968.

L. E. Nielsen, *Mechanical Properties of Polymers*, Reinhold, New York, London, 1962.

L. E. Nielsen, *Mechanical Properties of Polymers and Composites*, Volume I and II, Marcel Dekker, New York, 1974.

R. M. Ogorkiewicz, Ed., *Engineering Properties of Thermoplastics*, Interscience, New York, London, 1970.

D. V. Posato and R. T. Schwartz, Ed., *Environmental*

Effects on Polymeric Materials, Vol. 1, *Environments*, Interscience, New York, London, 1968.

P. D. Ritchie, Ed., *Physics of Plastics*, D. Van Nostrand, Princeton, New Jersey, 1965.

W. J. Roff, *Fibers, Plastics and Rubbers*, Butterworths Scientific Publications, London, 1956.

J. V. Schmitz, Ed., *Testing of Polymers*, Vol. 1, Interscience, New York, London, 1965.

J. V. Schmitz, Ed., *Testing of Polymers*, Vol. 2, Interscience New York, London, 1966.

J. V. Schmitz and W. E. Brown, Ed., *Testing of Polymers*, Vol. 3, Interscience, New York, London, 1967.

W. E. Brown, Ed., *Testing of Polymers*, Vol. 4, Interscience, New York, London, 1969.

Structural Plastics Design Manual, U. S. Government Printing Office, Washington D. C., 1978.

S. Turner, *Mechanical Testing of Plastics*, CRC Press, Cleveland, 1973.

D. W. Van Krevelen, *Properties of Polymers, Their Estimation and Correlation with Chemical Structure*, Elsevier, Amsterdam, Oxford, New York, 1976.

E. B. Wilson, Jr., *An Introduction to Scientific Research*, McGraw-Hill, New York, Toronto, London, 1952.

CHAPTER 17

Environmental Resistance

B. BAUM

C. H. PARKER

Springborn Laboratories
Enfield, Connecticut

R. D. DEANIN

Lowell Technological Institute
Lowell, Massachusetts

17.1   Introduction   884

17.2   Physical Resistance   884
  17.2.1   Mechanical Stress   884
    17.2.1.1   Modes of Stress   884
    17.2.1.2   Stress-Strain Curves   885
    17.2.1.3   Elasticity   887
  17.2.2   Creep   887
    17.2.2.1   Major Variables Affecting Creep   887
    17.2.2.2   Graphical Analysis of Creep   887
    17.2.2.3   Effect of Plastic Composition   888
  17.2.3   Cyclic Mechanical Stress   888
    17.2.3.1   Molecular Mechanisms   888
    17.2.3.2   Graphical Representation: The S-N Plot   890
    17.2.3.3   Factors in Cyclical Mechanical Stress Fatigue   890
  17.2.4   Cryogenic Temperature   891
    17.2.4.1   Applications   892
    17.2.4.2   Materials   893
    17.2.4.3   Properties   893
  17.2.5   Thermal-Mechanical Behavior   894
    17.2.5.1   Modulus-Temperature Curve   894
    17.2.5.2   Practical Significance of the Modulus-Temperature Curve   896
    17.2.5.3   Effects of Polymer Structure   896
    17.2.5.4   Ideal Versus Practical Modulus-Temperature Curves   897
    17.2.5.5   Reversible Effects of Temperature on Other Properties   897
  17.2.6   Thermal-Chemical Degradation   898

17.2.6.1   Anaerobic and Oxidative Processes   898
17.2.6.2   Effects on Polymer Structure   898
17.2.6.3   Temperature and Useful Life   899
17.2.6.4   Efects of Structure and Composition   899
17.2.6.5   Ultra High-Temperature Plastics   901
17.2.6.6   Fire   901
  17.2.7   Electrical Degradation   901
    17.2.7.1   Mechanisms of Dielectric Breakdown   902
    17.2.7.2   Factors Affecting Dielectric Strength   902
    17.2.7.3   Other Effects of Dielectric Breakdown   906

17.3   Chemical Resistance   906
  17.3.1   Moisture   906
    17.3.1.1   Effects of Moisture on Materials   907
    17.3.1.2   Sources of Moisture in Polymers   907
    17.3.1.3   Structure, Composition and Moisture Resistance   907
    17.3.1.4   Effects of Water Absorption   908
    17.3.1.5   Water-Sensitive and Water-Soluble Plastics   911
  17.3.2   Chemical Environment   911
    17.3.2.1   Conventional Structural Materials   911
    17.3.2.2   Solvents   911
    17.3.2.3   Staining   913
    17.3.2.4   Oxidizing Agents   913
    17.3.2.5   Acids and Bases   913

883

17.3.2.6   Specific Salt Solutions   914
17.3.2.7   Applications   914
17.3.3   Environmental Stress Cracking   914
17.3.3.1   Mechanism   914
17.3.3.2   Factors Affecting
           Environmental Stress Cracking   915
17.3.3.3   Prevention of Environmental
           Stress Cracking   919

17.4   Atmospheric Resistance   920
17.4.1   Weathering   920
17.4.1.1   Conventional Materials   920
17.4.1.2   Components of Weather   920
17.4.1.3   Variables in Weathering   922
17.4.1.4   Accelerated Weathering   922
17.4.1.5   Criteria of Failure   923
17.4.1.6   Polymers and Compounding
           Ingredients   923
17.4.2   Effects of Plastics on Environment 924
17.4.2.1   Plastics Manufacture   924
17.4.2.2   Disposal of Plastics Products at
           the End of Their Useful Life   925
17.4.3   The Space Environment   926
17.4.3.1   Kinetics in Space   926
17.4.3.2   Radiation Degradation   926
17.4.3.3   Temperature Degradation   926
17.4.3.4   Vacuum Degradation   927
17.4.3.5   Physical Impact Degradation   928
17.4.3.6   Friction Degradation   928
17.4.3.7   Sealants in Space Environment   928
17.4.3.8   Multiple Aspects in Space
           Environment   929
17.4.3.9   Additional Reading
           Suggestions   929

17.5   Biomedical Resistance   929
17.5.1   Biodegradation   929
17.5.1.1   Biological Studies   929
17.5.1.2   Optimum Conditions for
           Biodegradation   930
17.5.1.3   Effects of Biodegradation and
           Properties   930
17.5.1.4   Biodegradation and Structure   930
17.5.1.5   Biodegradation and Additives   930
17.5.1.6   Prodegredents   931
17.5.2   Toxicology   931
17.5.2.1   Materials Involved   931
17.5.2.2   Route of Entry into the Human
           Body   932
17.5.2.3   Analytical Technique for
           Research, Development, and
           Quality Control   933

17.6   Summary   934

17.7   References   935

17.1   INTRODUCTION

Whenever structural materials are used in practical applications, they are subjected to a wide variety of environmental stresses that test them to the limits of their endurance. Their ability to perform under these conditions, and their ultimate useful lifetimes in these applications, depend both on the types of environmental stress and on the composition and structure of the material itself. This is true not only of synthetic plastics but also of older conventional materials such as metals, ceramics, wood, leather, textiles, and paper as well. The present discussion is concerned primarily with the major types of environmental stresses in different applications, the way in which plastics perform under such stresses, and the best ways of using plastics in such applications.

17.2   PHYSICAL RESISTANCE

The physical environments in which plastic products are used include simple mechanical stress, long-term stress resulting in creep, repetitive cyclic mechanical stress, low and high temperatures, higher temperatures, higher temperatures leading to chemical degradation, and high voltage and/or frequency resulting in electrical failure.

17.2.1   Mechanical Stress

Most plastic products are subjected to mechanical stress during use, and excessive mechanical stress is one of the leading causes of failure. In the simplest case, a single application of excessive mechanical stress causes immediate failure; this is discussed first. Two more complex cases, long-term stress (creep) and cyclic stress, are discussed later (Sections 17.2.2 and 17.2.3); and the combination of mechanical and chemical stress is discussed much later (Section 17.3.3).

17.2.1.1   Modes of Stress
In simple laboratory testing mechanical stress can be applied to the plastic in one of five ways (Figure 17.1) (1). Tension is important primarily in synthetic fibers and in plastic films and tapes. Compression is important under resting loads, as in flooring

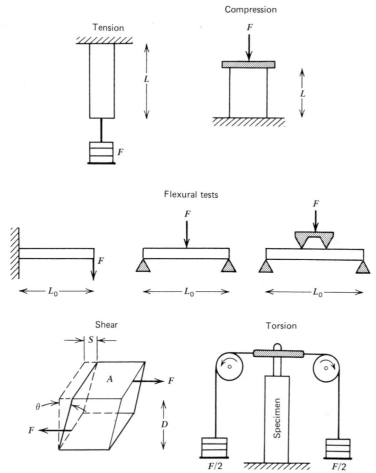

Figure 17.1 Types of mechanical tests (1). L = Length, L_0 = Initial length, F = force, S = linear shear deformation, 0 = angular shear deformation, A = area of shear plane, D = vertical displacement of shear force. Reprinted from L. E. Nielson, *Mechanical Properties of Polymers and Composites* by courtesy of Marcel Dekker, New York, N.Y.

and furniture, and in low-density foams. Flexure is the most universally important, both in the rigidity of hard plastic products and in the flexibility of soft ones. Shear is important in laminates and adhesives. Torsion is important in practical use of many flexible products and in failure of rigid engineering structures. In actual practice, however, most plastic products are subjected to combinations of these five simple modes, which are much more difficult for theoretical analysis or even laboratory testing.

17.2.1.2 *Stress-Strain Curves*
The effect of mechanical stress on plastics is conveniently analyzed by plotting stress (force) versus strain (deformation) throughout the test (Figure 17.2) (2). The figure plots a generalized stress-strain curve for plastics. The initial linear portion of the plot is a measure of the force required to deform the plastic and thus characterizes its rigidity or flexibility; this is known as the initial modulus of elasticity and is generally reversible and recoverable when the load is removed. At higher loading, the material may deform permanently and fail to recover when the load is removed; this critical point is known as the yield point—most often reported as the yield stress, and sometimes also as the yield deformation. If the material has a marked yield point, it may deform consid-

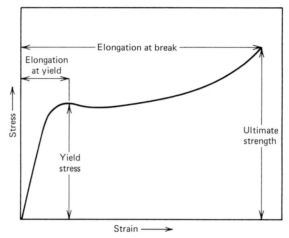

Figure 17.2   Generalized tensile stress-strain curve for plastics (2). Reprinting permission granted by John Wiley & Sons, Inc., New York, N.Y.

erably with no increase in stress, or even at a slightly lower stress; this is seen particularly in orientation of semicrystalline materials. Eventually, with or without a yield point, the material ruptures completely, breaking or tearing into two or more pieces.

Considering the wide variety of plastic materials and properties, stress-strain curves offer a convenient way of comparing their behavior under mechanical stress (Figure 17.3) (3). This figure shows stress-strain curves for several polymeric materials. Soft materials have a low initial slope and continue out to high deformation; if they are weak, the curve ends at rather low stress, whereas if they are tough, it rises gradually to moderate stress before breaking. Hard materials have a high initial slope; if they are brittle they break at very low deformation, whereas if they are tough, they show considerable deformation beyond the yield point. Thus these curves provide a simple way of visualizing the comparative behavior of different plastics under mechanical stress.

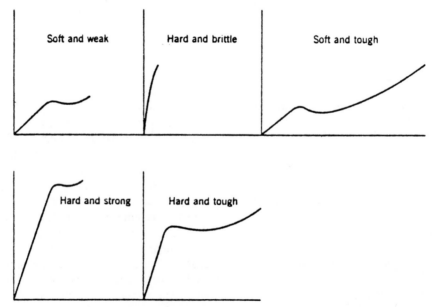

Figure 17.3   Tensile stress-strain curves for several types of polymeric materials (2). Reprinting permission granted by John Wiley & Sons, Inc., New York, N.Y.

17.2.1.3   Elasticity

It is generally assumed that the initial linear portion of the stress-strain curve is completely elastic and that when the load is removed the sample will recover its initial shape immediately with no permanent deformation. In rigid plastics this assumption is essentially useful. In flexible plastics and elastomers, high deformations require large molecular motions, which produce intermolecular friction—converting some of the mechanical energy into heat. When the load is released, the sample does not recover completely (2). This conversion of mechanical energy into frictional heat is called hysteresis.

In rubber tires hysteresis is slight, but the continual vibration gradually causes considerable heat buildup. When a rubber ball is dropped or a rubber pendulum oscillates, it is noted that the loss of mechanical energy in each cycle is more immediately obvious (3). In "deader" elastoplastics such as vinyls and polyurethanes, most of the mechanical energy is converted into intermolecular frictional heat, even in a single cycle, making them very useful for energy-absorbing applications such as crash padding.

17.2.2   Creep

When a solid material is subjected to a sustained mechanical stress for a long time, it tends to deform gradually in the direction of the stress. When this occurs in metals, it is believed to be due to sliding dislocations in the crystalline structure. When it occurs in plastics, it is due to molecules disentangling and sliding completely past one another. Either way, the process is generally referred to as creep.

More precisely, this can be analyzed as two distinct complementary processes as follows:

1.   When a material is held under constant "stress," it deforms gradually and continually, as in pipes and large architectural structures, and this is properly called "creep."

2.   When a material is held under constant "strain," this produces a high initial stress, but the stress decreases gradually and continually over a long time, as in bolts and flanges, and this is properly called "stress relaxation."

The two phenomena are precisely complementary in theory, and approximately inverse in practice.

17.2.2.1   Major Variables Affecting Creep

Obviously there are three main variables that affect the rate of creep: molecular mobility, stress level, and time span. Experimentally, increasing temperature generally increases molecular mobility—thus becoming a major variable itself.

17.2.2.2   Graphical Analysis of Creep

When a fixed stress is applied to a plastic product, and deformation (strain) is plotted

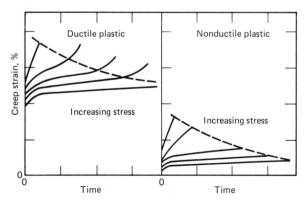

Figure 17.4   Typical creep and creep rupture behavior at constant temperature (5). Reprinted with permission from the *1969–1970 Modern Plastics Encyclopedia*. Copyright 1969 by McGraw-Hill, Inc. All rights reserved.

against time, the overall process is observed as an S-shaped curve that proceds in four successive stages: instantaneous elastic deformation; mixed elastic plus plastic deformation; "constant" rate plastic deformation; and accelerating plastic flow up to the point of rupture (4).

The first two stages may be too fast to be observed by the test instrument, and the fourth (rupture) stage may not occur or at least may be beyond the range of reasonable experimental testing; thus in many studies the major observation may be a fairly constant rate deformation, or at most a simple parabolic curve. When tests are run at a series of stress levels, these generate a family of creep curves, each one ending at rupture (Figure 17.4); and the curve through these rupture points forms a creep rupture envelope that defines the working range for the plastic material (5).

17.2.2.3 Effect of Plastic Composition

Creep and stress relaxation are measured by a great variety of nonstandard tests, so that data on different materials from different laboratories are very difficult to compare. Detailed tabulations are available annually through the *Modern Plastics Encyclopedia.* Typical data can be abstracted for specific comparisons—see Table 17.1 (6). Here, the creep properties of eleven polymer types are shown by measurement of apparent modulus after 1000 hr at 500, 1000, 2000, 3000, and 5000 psi.

In general, creep and stress relaxation depend directly on the ability of polymer molecules to slide completely past each other. Conversely, creep is minimized by preventing or at least minimizing such mobility. Thus high cross-linking in thermosets is very effective.

On a more gross mechanical level, reinforcement with glass fibers, which cannot creep, supports the organic plastic and greatly reduces creep—especially in filament-wound structures from continuous glass roving such as are used in pressurized gas cylinders. At the other extreme, even light cross-linking in vulcanized elastomers reduces creep, whereas thermoplastic elastom-

ers suffer considerably from it. In between these extremes, high crystallinity and molecular stiffness both contribute to immobilizing the polymer molecule, and distinguish engineering plastics from common amorphous thermoplastics by providing greater creep resistance. Further study of such relationships should provide considerable guidance in selection of plastic materials for engineering applications and in design of new plastic materials for higher performance under long-term load.

17.2.3 Cyclic Mechanical Stress

Ultimate strength properties are conventionally determined by a single stress-strain cycle. In most applications, however, failure occurs at lower stress-strain values. In many cases this is due to cyclic mechanical loading and unloading, or even alternating loading in reversing directions. According to one engineering estimate, 90% of all service failures are caused by such conditions (7). This was observed first in metals and wood, and more recently in plastics as well.

Such failure occurs most often in gears, bearings, and machinery, in general, where the frequency is often 1800 cps. Theories on the mechanism of abrasive wear often rely on cyclic mechanical stress to explain the process. Thermal cycling causes repeated expansion and contraction, which in turn produce cyclic mechanical stresses, and thus produce failure by a similar mechanism. The term fatigue failure is often applied to all of these. It is also applied to environmental stress cracking, involving the chemical environment as well (Section 17.3.3). It may also be applied to heat aging (Section 17.3.5 and Section 17.2.6) and to weathering (Section 17.4.1). This discussion is limited to cyclic mechanical stress.

17.2.3.1 Molecular Mechanisms

Mechanical deformation of a rigid plastic below the glass transition temperature requires more material transport than is possible through segmental motion of polymer molecules. It breaks the polymer molecule into free radicals. If these radicals cannot

Table 17.1 Creep Properties of Typical Plastics (7)[a]

Materials	Apparent Modulus After 1000 Hours At									
	500 psi		1000 psi		2000 psi		3000 psi		5000 psi	
	RT	140°F	RT	140°F	RT	140°F	RT	140°F	RT	140°F
ABS - High impact	–	–	210	–	250	–	–	–	–	–
- Heat resistant	–	–	245	–	210	–	–	–	–	–
Acetals -										
Homopolymer	250	120	250	120	210	100	–	–	–	–
GR homopolymer	650	300	550	275	500	190	–	–	775	–
Copolymer	270	70	270	45	–	–	220	–	–	–
GR copolymer	–	–	–	–	–	–	1150	–	750	–
Acrylics - general-purpose	–	175	245	–	–	–	–	–	–	–
Polyphenylene oxides -										
General purpose	–	–	245	90	230	89	230	–	–	–
GR self-extinguishing	–	–	–	–	1050	950	980	860	–	–
Nylons - Type 6/6	90	60	85	55	65	25	–	–	–	–
- Type 6/6 GR	–	–	–	–	–	–	585	480	625	410
Polycarbonates -										
General purpose	–	190	–	178	–	225	310	–	–	–
GR	–	–	–	–	–	–	760	950	1050	830
Polyethylenes -										
High density	55	–	40	–	–	–	–	–	–	–
GR	–	–	420	–	–	–	–	–	–	–
Polypropylenes -										
General-purpose	65	30	60	–	–	–	–	–	–	–
GR	–	–	–	–	825	–	–	–	–	–
Polysulfones -										
General-purpose	–	–	–	–	–	–	325	310	–	–
GR	–	–	–	–	–	–	–	–	1075	795
Polyphenylene sulfides										
General-purpose	–	–	–	–	–	–	550	–	–	–
GR	–	–	–	–	–	–	2100	–	1475	795
Vinyls - Type I	320	45	325	25	–	–	–	–	–	–
- Type II	160	–	250	–	215	–	–	–	–	–

[a] GR = glass reinforced.

recombine immediately, they stabilize into smaller molecules of lower strength, producing embrittlement of the polymer—particularly after repeated cycles. Atmospheric oxygen generally contributes to such degradation reactions. These are generally observed as brittle failure.

Mechanical deformation above the glass transition temperature is facilitated by larger segmental motion, but internal intermolecular friction still produces some hysteresis conversion of mechanical energy into heat, raising the temperature of the plastic and thus softening it still further. Repeated cycles then produce sufficient heat-softening to cause ductile failure.

In crystallizable polymers, mechanical deformation may help to align polymer molecules into a more regular conformation, which promotes further crystallization; and internal frictional heating may give the molecules enough mobility to move into the conformation required by the crystal lattice. Thus mechanical cycling may produce a gradual increase in crystallinity, making the material more rigid and thus leading to brittle failure. By analogy with similar phenomena in metals, this is sometimes referred to as "work-hardening."

17.2.3.2 Graphical Representation: The S-N Plot

The behavior of a material under cyclic mechanical stress is generally represented by plotting failure stress against number of cycles to failure, most often as log of cycles to failure, producing a semilog plot (Figure 17.5) (8). The figure relates fatigue stress to number of cycles to failure. Generally, high stress levels produce rapid ductile failure by heat-softening. Lower stress levels produce slower brittle failure. Below a certain characteristic stress level there may be no failure at all; this is called the endurance limit or the fatigue limit of the material. Some materials—such as ferrous metals and thermoset plastics—show such an asymptote. In other materials it is not always clear whether there is a true endurance limit or whether the failure point is simply beyond the limits of the test. Some studies have indicated that a log-log plot could give straight lines for more facile analysis and prediction (Figure 17.6). The figure shows fatigue curves for bonding and compression stresses.

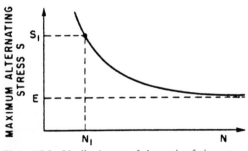

Figure 17.5 Idealized curve of alternating fatigue stress versus number of cycles to fracture (8). From *Engineering Design for Plastics* by E. Baer, © 1964 by Litton Educational Publishing, Inc. Reprinted by permission of Van Nostrand Reinhold Company.

17.2.3.3 Factors in Cyclic Mechanical Stress Fatigue

The endurance of a plastic product under conditions of cyclic mechanical stress will depend upon the quantitative characteristics of the stress itself and of the environment in which it occurs, upon the design and quality of the product, and upon the nature of the plastic material itself.

Environment. Mechanical stress level is the factor most often quoted in the literature, assuming that every cycle will apply the same load to the sample. As noted, increasing mechanical stress first passes the endurance limit of the material, producing fatigue failure; further increase produces brittle failure at shorter times; and still further increase produces thermal softening at still shorter times.

Strain level may be the controlling factor in applications where the amount of deformation is the same in every cycle, whereas the resulting stress will vary as the fatigue process progresses. Thus interpretations based on constant strain may produce correlations that are inverse to those based on constant stress. This has caused some confusion in the literature.

Frequency of cycling determines the rate at which stress and strain are applied. Since rate of test is very important in single-cycle tests, it is not surprising that frequency has important effects in mechanical fatigue studies. One common observation is the fact that higher frequency produces faster thermal softening and earlier ductile failure.

Mode of deformation may be by simple tension, compression, flexure, torsion, or shear; or in practical applications, by combinations of these simple types. Fatigue life varies somewhat from one mode of deformation to another.

Temperature is important because it favors thermal softening and ductile failure. Tests at higher temperature require less hysteresis heating to reach the glass transition temperature.

Chemical environment is a major complicating factor. In mechanical degradation studies, presence of atmospheric oxygen

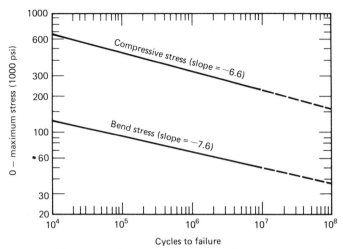

Figure 17.6 Spur gear fatigue curves for bending and compression stresses (7). Reprinted with special permission of the author, Dr. J. H. Faupel.

favors faster breakdown to lower molecular weight. In the presence of weak solvents and surfactants environmental stress cracking becomes important.

Time or number of cycles to failure is of course the standard dependent variable measured in most of these tests. On the other hand, the data are most often plotted with lifetime as the independent variable, which in turn determines the permissible stress or strain that can be applied in each cycle. Rate of loading is also mentioned in some of the literature, but not always clearly defined.

Sample: "Surface" quality is often described as critically important. All of the tress or strain tends to concentrate at any notch, scratch, or other surface imperfection, producing a very high stress that tears it open and produces continuing crack growth until complete failure occurs. Conversely, surfaces that are smooth, polished, and flawless tend to give much longer fatigue life (7).

Thickness is occasionally mentioned for two reasons. Hysteresis heating occurs because internal intermolecular friction produces heat faster than the sample can lose it to the surroundings. Since plastics are poor conductors of heat, thicker samples will soften and show ductile failure much sooner than thin samples. Stress-strain level depends on thickness in flexure and torsion

tests, where the stress and strain at the surface increase with distance from the midplane, producing tension at one surface and compression at the other. These effects deserve more detailed consideration in actual product design.

Materials. Cyclic mechanical fatigue life varies greatly from one plastic material to another. The great variety of laboratory tests in use seriously limits comparison between different materials in different laboratories, but occasionally precise comparisons are possible (Table 17.2). Data in Table 17.2 show fatigue properties in 11 polymer types by measuring stress for failure at 1800 cpm for cycles ranging from 10^4 to 10^7.

Glass transition temperature and hysteresis heat rise would obviously be important in comparing the fatigue life of different materials. High molecular weight has occasionally been observed to improve fatigue life, but theoretical opinions and predictions differ. Reinforcement with glass fibers generally lengthens fatigue life at constant stress, but lowers fatigue life at constant strain. Further study of such structural factors in fatigue life should be a great help in precise engineering design.

17.2.4 Cryogenic Temperatures

In the exploration and utilization of outer space, very low temperatures are a critical

Table 17.2 Fatigue Properties of Typical Plastics (7)[a]

Materials	Stress (psi) for Failure at 1800 cpm			
	10^4 Cycles	10^5 Cycles	10^6 Cycles	10^7 Cycles
ABS –				
High impact	–	2600	1650	1300
Heat resistant	4400	2700	2240	2100
Acetals –				
Homopolymer	6000	5400	5100	4900
Copolymer	6200	4100	3500	3300
GR copolymer	9000	7000	7000	7000
Acrylics –				
General purpose	–	–	–	1000
High impact	–	–	–	1500
Fluoroplastics –				
PTFE	2100	1800	1700	–
GR PTFE	4200	3700	3500	
Polyphenylene oxides –				
General purpose	4500	2600	2000	1800
GR	7200	5800	4800	4400
Nylons – Type 6 GR	7000	6000	5750	5750
– Type 6/6	4000	3500	3400	3300
– Type 6/6 GR	8000	6500	6000	5900
– Type 6/6 MR	18,000	13,000	10,000	8500
Polycarbonates –				
General purpose	5800	2500	1300	1000
GR	16,000		5500	5400
Polyesters –				
General purpose	5500	4500	3500	2850
GR	6500	6000	5400	5000
Polypropylene, GR	9000	8000	7000	6000
Polysulfones –				
General purpose	3700	2200	1400	1000
GR	14,000	6500	5000	4500
Polystyrenes –				
General purpose, GR	8000	7000	6000	5000
Acrylonitriles, GR	8500	7500	6500	5500

[a] GR = glass reinforced; MR = mineral reinforced.

problem for the materials engineer (Table 17.3) (9). Data in Table 17.3 show the temperatures obtainable in several cryogenic media. In the temperature range from −150°C down to absolute zero, generally defined as cryogenic, the properties and utility of structural materials often change drastically, and sometimes unpredictably, from the values generally measured in more conventional temperature ranges.

17.2.4.1 Applications

Major applications in space technology are primarily for thermal insulation, and rely largely on foamed plastics (9). Other important uses are for electrical insulation such as wire coating, potting, and encapsulation—which generally utilize plastic materials. Other important uses are primarily structural in nature; here weight and space are critical, so high-strength reinforced plastics

Table 17.3 Environmental Temperatures Frequently Encountered in
Cryogenic Engineering (9)[a]

Environment or Condition	Temperature [a]			
	°C	°F	°K	°R
Room temperature	25-27	77-81	298-300	537-541
Liquid oxygen (O_2)	-181.1	-297.4	90.1	162.3
Liquid fluorine (F_2)	-187.0	-304.6	86.0	154.8
Liquid nitrogen (N_2)	-195.8	-320.8	77.3	138.9
Liquid hydrogen (H_2)	-252.9	-423.4	20.4	36.3
Liquid helium (He)	-268.9	-452.0	4.2	7.7
Absolute zero	-273.16	-459.69	0	0

Reprinting permission granted by John Wiley & Sons, Inc., New York.

[a] For the cryogenic liquids in this list the temperatures represent the boiling points at 1 atm.

and sandwich construction are often desirable. In all of these the effects of low temperatures upon mechanical properties are critical.

17.2.4.2 Materials

Thermal insulation is most often provided by polystyrene and polyurethane foams (9). Solid polymers that have most often shown promise in these extreme conditions include the fluoropolymers, reinforced epoxies and unsaturated polyesters, polyethylene terephthalate, silicone, and organo-phosphonitrilic elastomers. For example: fluoropolymers offer excellent thermal, electrical, and chemical resistance and are frequently successful in resisting cryogenic embrittlement; polyethylene terephthalate film has shown promise as a pump bladder for feeding liquid hydrogen fuel; and silicone and organo-phosphonitrilic elastomers can retain their flexibility to lower temperatures than most other materials.

Reinforced epoxies and polyesters generally offer the maximum strength and strength-weight ratios for structural applications. Unfortunately, the coefficients of thermal expansion and contraction of the polymers are much greater than those of the inorganic reinforcing fibers used to strengthen them; during cryogenic cooling, this causes severe thermal shock, which lowers the strength of these materials considerably.

17.2.4.3 Properties

Since temperature is simply a measure of atomic and molecular motion, cryogenic cooling immediately implies a severe reduction in such motion (10). This in turn allows the molecules to contract toward each other, reducing the free volume between them and thus reducing their mobility. The reduction in free volume produces shrinkage, causing strains and stress concentrations in the solid material. The reduction in mobility produces a general hardening of mechanical properties, and eventually embrittlement. For example, when these structural changes were studied by infrared spectroscopy, polystyrene showed distinct transitions at 235°K and 50°K, whereas polyethylene terephthalate showed them at 220, 180, 120, and 50°K (11).

Mechanical Properties. Since cooling reduces molecular mobility, this generally produces a marked increase in modulus (10). In high molecular weight polymers, the reduction in molecular mobility also demands more force to pull molecules past each other and thus increases the strength of the materials. The strength of many polymers is approximately doubled at these low temperatures. Ductile materials, particularly, show higher yield strength at cryogenic temperatures. In a study of the yield strength of crystalline fluoropolymers, tensile yield strength increased three- to 12-fold (9, p. 773). Whereas crystallinity generally reduces molecular mobility and increases yield strength at ordinary temperatures, in this study it actually reduces cryogenic yield strength, suggesting that stress concentration in the amorphous interfaces is aggravated at high crystallinity and may thus become the controlling factor at these low temperatures.

Conversely, as low temperature reduces molecular mobility, this in turn reduces ductility and ultimate elongation (9). In studies of abrasive wear, the mechanism at room temperature involved transfer, surface cutting, and plastic deformation; but at cryogenic temperature of 77°K, the mechanism was primarily localized deformation and scoring, with evidence of surface fatigue (12).

In reinforced plastics, instead of improving strength at low temperatures, the difference in contraction of polymer matrix versus inorganic reinforcing fiber produces weakening of the structure, and the result can actually be a loss of strength in cryogenic behavior (9, p. 782).

In polystyrene and polyurethane foam, going down to cryogenic temperatures generally increases the modulus, yield strength, and compressive strength (9, p. 780), and lowers the ultimate elongation, as expected from theory (13).

Thermal Properties Thermal conductivity of foamed plastics has been found to correlate directly with temperature (9, p. 779), so that they insulate best under cyrogenic conditions. Specific heat and coefficient of thermal expansion both correlate directly with temperature in most polymer systems. In many applications, the same polymeric product must be used both at low cyrogenic temperature and at high temperatures, where softening, weakening, and thermal degradation can become serious problems; this makes the design and selection of such polymer systems particularly critical.

17.2.5 Thermal-Mechanical Behavior

The effects of heat on polymers are conveniently divided into two types: immediate reversible effects on physical properties; and gradual permanent chemical degradation of structure and properties. The reversible effects are discussed here; permanent chemical degradation is considered in Section 17.2.6.

17.2.5.1 Modulus-Temperature Curve
In the normal operating temperature range of a plastic material, the effect of temperature on mechanical behavior is conveniently described by the modulus-temperature curve (Figure 17.7) (10). In this figure, the effect

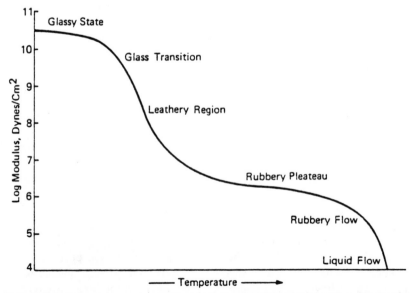

Figure 17.7 Effect of temperature upon modulus of a linear amorphous polymer (10, p. 89). Reprinted by permission of C.B.I. Publishing Co., Inc.; 221 Columbus Avenue, Boston, Massachusetts.

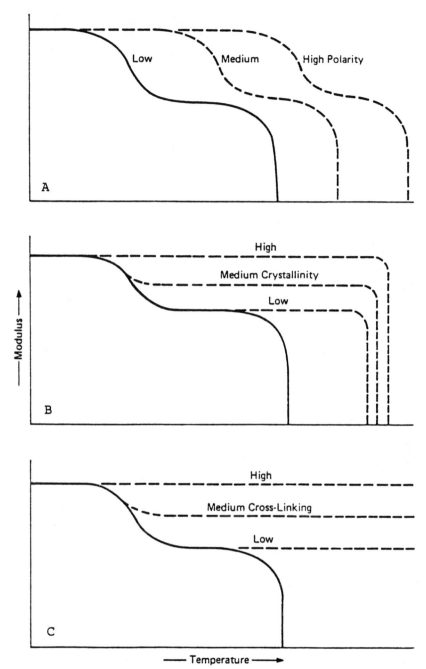

Figure 17.8 Effect of intermolecular bonding upon modulus versus temperature curves (*a*) (*b*) (*c*) (10, p. 342). Reprinted by permission of C.B.I. Publishing Co., Inc., 221 Columbus Avenue, Boston, Massachusetts.

of temperature on a linear amorphous polymer is shown. This curve applies generally to all amorphous, linear, homogeneous (single-phase) polymers; the specific temperature scale of the abscissa depends on the structure of the polymer molecule. Thus increasing polarity shifts the entire curve horizontally to higher temperatures (Figure 17.8), whereas addition of liquid plasticizers shifts the entire curve horizontally to lower temperatures. In Figure 17.8, the effect of intermolecular bonding upon modulus versus temperature is shown. More extreme effects are noted in Section 17.2.5.3.

17.2.5.2 Practical Significance of the Modulus-Temperature Curve

Regardless of polymer structure and the temperature abcissa, the horizontal plateaus and the transition regions between them serve to characterize the molecular flexibility and mobility of the polymer system and their effects on the fundamental reversible mechanical behavior of rigidity/flexibility/flow.

Rigid Plastics. Rigid plastics are normally on the glassy state plateau at room and use temperatures. If cooling to low temperature produces embrittlement, this may be seen as a further rise in the glassy plateau. On warming, increasing motion and mobility of molecular segments produces sufficient free volume so that the rigid plastic can deform under mechanical load; this is seen as the glass transition temperature and indicates the maximum useful temperature of a product made from such a rigid plastic.

On further warming, the material may show more or less of a rubbery plateau region, depending on its molecular structure and molecular weight. When it first reaches the rubbery flow region it is easily deformed, as in blow molding and thermoforming, and can be chilled to "freeze in" this structure in the finished product; on the other hand, in normal melt processing such rubbery flow produces serious distortions in the final product and must be avoided by using higher temperature, lower molecular weight, lower shear rate, or additives.

Ultimately the polymer can be heated to a temperature at which it is a fluid melt, and is optimally processed by liquid flow in extrusion or injection molding. Conversely, if a fluid melt process is carried out in the rubbery flow region, the frozen-in strains produce plastic memory, and when the product is warmed above the glass transition temperature it will distort to relieve these strains. Such distortion is essentially irreversible failure. When a thermoplastic is warmed to the fluid melt temperature, the finished product will simply revert to a molten mass or puddle.

Flexible Plastics. Flexible plastics and elastoplastics are normally on the rubbery plateau at room and use temperatures. On cooling to lower temperatures, they go through a leathery transition region in which they stiffen more and more down to the glass transition, below which they become rigid and brittle in the glassy state. Such low-temperature stiffening and embrittlement are characteristics of plasticized vinyls and other elastoplastics and elastomers, in general, and are used to characterize their minimum-use temperatures; still lower cryogenic temperatures have just been considered separately (Section 17.2.4). Turning to higher temperatures, the rubbery plateau continues up to the point at which the material begins to creep or stress-relax and deform permanently under mechanical stress, which marks the rubbery flow region and the maximum-use temperature of the material.

All efforts to extend the useful temperature range of such materials concentrate on lowering the stiffening-embrittlement end of the rubbery plateau and/or raising the rubbery flow end of the plateau. Above this point the range of true liquid flow is used in normal melt processing by extrusion or molding of thermoplastics and is used in the earlier stages of vulcanizable thermosets as well.

17.2.5.3 Effects of Polymer Structure

The shape of the preceding modulus-temperature curve applies generally to amorphous, linear, homogeneous polymers, as described in Section 17.2.5.1. Typical polymers include polystyrene, plasticized polyvinyl chloride, polyvinyl acetate, acrylics, cellulose esters and ethers, and thermoplastic polyurethanes. Further complications in polymer structure introduce strikingly different shapes for the modulus-temperature curve as follows.

Crystallinity. Crystallinity in the polymer molecules immobilizes them up to much higher temperatures, until the thermodynamic point at which the crystals melt suddenly (Figure 17.8).

Low crystallinity in sterorubbers and plasticized vinyls simply extends the rubbery plateau up to the melting point of the crystallites, giving a much broader useful temperature range. Medium crystallinity, as in low-density polyethylene, almost eliminates a continuous amorphous phase, making the material stiffly flexible to leathery. High crystallinity—as in high-density polyethylene, polypropylene, acetal, polyester, and nylons-fixes nearly all of the polymer firmly in place, providing the high-temperature performance of the engineering thermoplastics.

Rings. Rings in the polymer main chain—particularly aromatic rings and five-membered rings, which are flat and rigid—provide a great stiffening effect, retaining the rigid glassy state plateau up to much higher temperatures. These are typical of many of the new engineering thermoplastics and ultra-high-temperature resins such as polyphenylene oxides, polysulfones, polyoxybenzoates, aromatic nylons, polyphenylene sulfides, and especially the ultrahigh-temperature resins containing continuous resonating ring systems.

Cross-Linking. In thermoset plastics, cross-linking can be carried to very high degrees—as in polyester and epoxy, and especially in urea, melamine, and phenolic resins—immobilizing the segments of the polymer molecules and preventing any appreciable softening up to the thermal decompostion point of the resin (Figure 17.8).

Multiphase Systems. Multiphase polymer systems are generally either rubber-modified rigid thermoplastics in which the continuous rigid phase controls most properties but the dispersed rubber particles provide impact strength; or thermoplastic elastomers in which the continuous phase of rubber molecules is held in place by the separation of tiny discrete hard particles acting as "thermoplastic cross-links." In such systems each phase shows its own transition temperature; between them, the level of the most useful region—the intermediate plateau—is determined by the ratio of the "hard" and "soft" polymer structures in the system.

17.2.5.4 Ideal versus Practical Modulus-Temperature Curves

Although the foregoing graphs and concepts are clear and precise in theory, experimental curves for practical systems often blur the plateaus into gradual slopes of varying intensity. The basic concepts remain valid and useful in understanding practical systems but must be applied with caution in critical cases. A typical case is the heat-deflection temperature of nylon, which is surprisingly low but can be raised tremendously by reinforcement with glass fibers; examination of the modulus-temperature curves shows that a slight change in the level of the plateau produces a tremendous change in the critical approval limit.

17.2.5.5 Reversible Effects of Temperature on Other Properties

The primary effect of temperature on molecular mobility, in addition to its effects on modulus, also produces a number of secondary effects that are quite reversible. These include thermal expansion, electrical polarization, optical refraction, and absorption and diffusion of small molecules such as solvents.

Coefficient of Expansion. Since temperature is simply the vibration of atoms and molecules, and the free volume this requires, it is obvious that volume will change directly with temperature. This means that changes in temperature will create problems of dimensional instability. This is particularly serious when plastic parts must have precise dimensions—especially when they must match with mating metal or ceramic parts of lower coefficient.

It is also serious when cyclic thermal expansion and contraction create mechanical strains in the material, eventually resulting in weakening and cracking. It is best remedied by use of inorganic fillers and especially fibrous reinforcements. In some cases, dispersed rubbery microparticles can also add ductility and extend useful life.

Electrical Polarization. Also sensitive to temperature is electrical polarization. At very low temperatures the polar groups in polymers are immobile and cannot orient in an alternating electrical field, so the polymer has low dielectric constant. With increasing temperature, mobility increases, polar groups orient more easily, and dielectric constant rises. At still higher temperatures—particu-

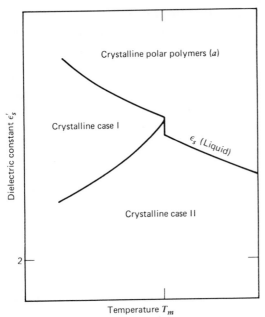

Figure 17.9 Effect of temperature upon dielectric constant of polar crystalline polymers (14, p. 540). From *Engineering Design for Plastics* by E. Baer, © 1964 by Litton Educational Publishing, Inc. Reprinted by permission of Van Nostrand Reinhold Company.

larly in the melt—Brownian mobility permits polar groups to disorient randomly faster than the electrical field can orient them, so dielectric constant drops again (Figure 17.9) (14). This figure indicates the effect of temperature upon dielectric constant of polar crystalline polymers. In the region where the dielectric constant is changing from low to high, partial orientation produces maximum hysteresis loss—conversion of electrical energy into internal intermolecular friction and thus to heat.

Optical Effects. Changing temperature in special polymer systems can produce unusual effects. For example, if a two-phase polymer system is made with matched refractive indexes in both phases to produce optical clarity, changing the temperature generally produces change at a different rate for each phase so that the refractive indices of the polymer phases become unmatched. As this process progresses, the material first becomes bluish, then cloudy, then translucent-to-opaque. When the system is allowed to return to room temperature, the matching and clarity return again.

Chemical Effects. Since increasing temperature increases free volume and segment mobility, it changes most chemical properties accordingly. For example, solvent molecules swell, permeate diffuse, and evaporate more readily. Polymers dissolve more rapidly. Chemical reactions, in general, proceed more rapidly and easily. These features are considered in Section 17.2.6.

17.2.6 Thermal-Chemical Degradation

Organic polymers are sufficiently stable for most long-term applications. Nevertheless, the strength of C—C, C—H, and other common C—X bonds is well within reach of common temperatures that can activate them to reactivity—especially to free-radical chain reactions. Such reactions may occur to a limited extend during processing, continue gradually during long-term use, and accelerate greatly in high-temperature use and crisis conditions.

17.2.6.1 *Anaerobic and Oxidative Processes*

In most of these degradative processes, atmospheric oxygen has been implicated, lowering the threshold temperature for the free-radical chain reactions to begin. Some processes may appear superficially anaerobic—such as extrusion or injection molding—yet prove to carry enough trapped oxygen or preformed peroxides to explain the reaction mechanism and products. Occasionally, careful studies under maximum exclusion of oxygen do demonstrate different mechanisms but generally give rise to considerable theroetical controversy. These subtle distinctions are not pursued further in this survey of current technology.

During thermal oxidative degradation the initial products formed include peroxide and hydroperoxide groups, which then decompose to hydroxyl, ether, carbonyl, carboxylic acid, and ester groups in the degraded polymer. Most of these processes are free-radical chain reactions.

17.2.6.2 *Effects on Polymer Structure*

Aside from producing more polar groups in the polymer molecule, these reactions can produce four gross changes in molecular structure:

Cross-linking between linear polymer molecules increases melt viscosity during processing sometimes forming an infusible gel, and requiring shutdown and cleanout of the process equipment. When cross-linking occurs gradually in an end product during use, it produces stiffening of flexible materials and embrittlement of rigids.

Depolymerization is an unzipping of the polymer molecule—the reverse of a chain polymerization reaction. It produces liquid or gaseous monomer, often without serious change in the molecular weight of the remaining solid polymer.

Chain scission is random cleavage of the polymer chain, producing lower and lower molecular weight in the degraded polymer. This may make the resulting polymer soft, tacky, and weak, or in some rigid polymers, it simply makes them more brittle.

Liberation of volatiles may occur without change in the size of the polymer molecule. This is characteristic in the way polyvinyl chloride loses HCl gas and the way polyethylene loses hydrogen gas when irradiated. The resulting unsaturation can produce other special effects.

The balance between these different processes varies greatly from one polymer to another (Table 17.4) (15). Data in Table 17.4 show degradation processes for each of seven polymers.

17.2.6.3 Temperature and Useful Life

Thermal degradation obviously accelerates with increasing temperature. It can be measured in many ways; but most important, it is the measurement of practical use properties against time. For polyvinyl chloride this is most often darkening of color. For some thermoplastics it is often small index versus time at process temperature. For most plastics it may be modulus, strength, impact strength, or loss of elongation versus time at use temperature.

Obviously the choice of test will define the "useful life" that is observed. This prevents easy comparison of different data from the literature and requires precision and critical judgment in reading manufacturers' literature. Figure 17.10 (16) illustrates a common average starting point for such comparisons.

17.2.6.4 Effects of Structure and Composition

Sensitive points in polymer structure may be $C=C$ double bonds: $C-H$ bonds that are alpha to activating groups—double bonds, oxygen, carbonyl, or halogen; and tertiary $C-H$ or $C-Cl$ groups. Thermal stability generally improves with increasing molecular weight for three reasons:

1. Most end groups are more reactive than the rest of the polymer chain;

2. Small molecules are more mobile,

3. Cleavage of high to medium molecular weight produces less loss of properties than does cleavage of medium to low molecular weight.

Table 17.4   Major Degradation Processes in Different Polymers (15)

Polymer	T_{50} * (°C)	Major Degradation Process
Polyoxymethylene	—	Depolymerization
Poly(methyl methacrylate)	330	Depolymerization
Polytetrafluoroethylene	510	Depolymerization
Polyethylene	400	Chain scission
Polystyrene	360	Chain scission
Poly(vinyl chloride)	260	Liberation of volatiles
Poly(butyl methacrylate)	—	Liberation of volatiles

*Fifty percent loss in weight after 30 min in vacuum.

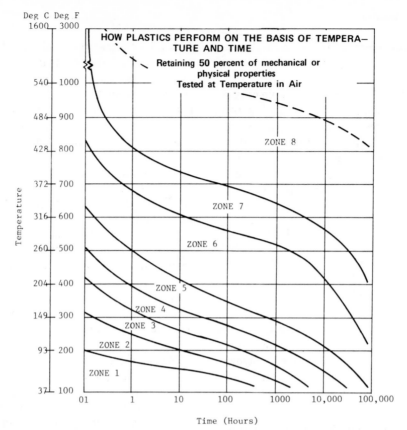

Figure 17.10 How plastics perform on the basis of temperature and time (tested at temperature in air). Note: Plastics retain 50% of their mechanical or physical properties. (16, p. 85). From *Markets for Plastics* by Rosato, Fallon, and Rosato, © 1969 by Litton Educational Publishing, Inc., reprinted by permission of Van Nostrand Reinhold Company.

ZONE 1
Acrylic
Cellulose acetate (CA)
Cellulose acetate-butyrate (CAB)
Cellulose acetate propionate (CAP)
Cellulose nitrate (CN)
Cellulose propionate
Polyallomer
Polyethylene, low-density (LDPE)
Polystyrene (PS)
Polyvinyl acetate (PVAC)
Polyvinyl alcohol (PVAL)
Polyvinyl butyral (PVB)
Polyvinyl chloride (PVC)
Styrene-acrylonitrile (SAN)
Styrene-butadiene (SBR)
Urea-formaldehyde

ZONE 2
Acetal
Acrylonitrile-butadiene-styrene (ABS)
Chlorinated polyether
Ethyl cellulose (EC)

Ethylene vinyl acetate copolymer (EVA)
Furan
Ionomer
Phenoxy
Polyamides
Polycarbonate (PC)
Polyethylene, high-density (HDPE)
Polyethylene, cross-linked
Polyethylene terephthalate (PETP)
Polypropylene (PP)
Polyvinylidene chloride
Urethane

ZONE 3
Polymonochlorotrifluoro-ethylene (CTFE)
Vinylidene fluoride

ZONE 4
Alkyd
Fluorinated ethylene propylene (FEP)
Melamine-formaldehyde
Phenol-furfural

Polyphenylene oxide (PPO)
Polysulfone

ZONE 5
Acrylic (thermoset)
Diallyl phthalate (DAP)
Epoxy
Phenol-formaldehyde
Polyester
Polytetrafluoroethylene (TFE)

ZONE 6
Parylene
Polybenzimidazole (PBI)
Polyphenylene
Silicone

ZONE 7
Polyamide-imide
Polyimide
S.A.P.

ZONE 8
Plastics now being developed using intrinsically rigid linear macromolecules' principle rather than usual crystallization and cross-linking principles.

"Volatile" ingredients such as plasticizer and stabilizer may be lost early in the life of a plastic, leaving it to stiffen and degrade faster afterwards. Impurities such as residual polymerization catalysts and the presence of transition metals such as iron and copper greatly speed oxidative and other free-radical chain degradation processes.

17.2.6.5   Ultrahigh-Temperature Plastics

Space-age demands for ultrahigh-temperature materials in aerospace vehicles, miniaturization of instrumentation and computers, and high performance in improved consumer applications all have prompted vigorous research to develop organic polymer molecules that will have longer life at higher temperatures. Such developmental effort can be summarized very briefly in ascending order, as follows:

Fluoropolymers contain C—F bonds that are much stronger than C—H bonds. In perfluorinated polymers containing no residual hydrogen at all, these can attain high thermal stability, as witnessed in their common use as nonstick coatings in cooking and baking.

Minimizing the hydrogen content of polymers often eliminates the C—H bonds at which thermal oxidation starts and improves the stability of many polymer classes. This is often the last word in improving a fairly high-temperature polymer.

Resonance stabilization is important in most ultrahigh-temperature polymers. The more benzene rings, or resonating heterocyclic rings, that can be built into the polymer molecule, the higher its thermal stability.

Heterocyclic rings containing oxygen, nitrogen, and sulfur often have stronger bonds than carbocyclic rings alone—particularly when stabilized by aromatic resonance.

Ladder structures are fused-ring molecules in which the breakage of any single bond cannot rupture the polymer molecule. Many such ruptures are required statistically before there can be any degradation in molecular weight and therefore in practical properties.

Semi-inorganic and inorganic polymers offer the hope that the extremely high thermal stability of inorganic chemicals can finally be built into useful polymers. This has been realized most fully in organo-siloxanes, but thier pendant organic groups still provide the weak point for thermal-oxidative failure.

Other semicommercial systems have included carboranes, phosphonitrilics, and similar polymers, with organic side groups modifying the performance of inorganic main chains. Obviously the extreme thermal stability of simple inorganic chemicals still remains far above any of these developments but continues to provide a challenging goal for research and development.

17.2.2.6   Fire

When organic polymers are heated far above their decomposition temperatures they release organic gasses that are highly combustible, and provide fuel for fire which then becomes autocatalytic. Although the polymer itself does not "start" the fire, it becomes the victim and culprit in its own demise. The increasing usefulness and utilization of organic polymers in appliances, building and construction, clothing, home furnishing, and transportation have accentuated the problem of flammability in such applications.

Polymer structures can be designed for increased resistance to burning. Thus halogens generally inhibit the free-radical oxidation reactions in the burning process, and heavy hydrogen halide gas blankets the fire and prevents access of oxygen for burning. Organophosphorus compounds inhibit the free-radical burning process and burn to phosphorus oxides, which char the plastic or coat it with an inorganic glass to prevent further burning. Addition of antimony oxide generally synergizes halogen inhibition of burning, possibly by forming $SbOCl$ or $SbCl_3$, which is the effective flame retardant. Much more work must be done in this area, particularly with respect to inhibiting liberation of smoke and toxic gases and to proper design and utilization of organic polymers in the total structure.

17.2.7   Electrical Degradation

Since organic polymers conduct little or no electrical current, they are very useful as electrical insulators. When a polymer is used at higher and higher voltages, however, it will reach a point at which its insulating ability breaks down, showing a sudden drop in resistance. The maximum voltage before this occurs is called the dielectric stength and is taken as a measure of the maximum useful voltage for that polymer under those conditions.

17.2.7.1 Mechanisms of Dielectric Breakdown

Dielectric breakdown may occur very rapidly due to an avalanche of electrons, fairly rapidly due to dielectric heating leading to chemical degradation, or more slowly due to electrical discharge leading to chemical degradation.

Intrinsic Breakdown. The intrinsic breakdown of the polymer may be approached by using thin samples, low temperatures, and direct current to eliminate most of the other mechanisms. It represents an avalanche of electrons flashing through the polymer in the time span of 10^{-8}–10 sec. In general, it is often higher for polar polymers (Table 17.5) (14). The table shows the intrinsic electric strength of eight polymers at room temperature and low temperatures (-90--$200°C$).

Thermal Breakdown. In many or most failures of plastic insulation, thermal breakdown is important. Hysteresis conversion of electrical energy into heat through flow of current and dielectric loss raises the temperature of the polymer. Since polymers are poor conductors of heat, this can result in rapid severe temperature rises, increasing normal conductivity, or even producing chemical breakdown to more conductive species. The time span is generally 0.1–10^4 sec.

Discharge-Dependent Breakdown. Discharge-dependent breakdown is caused by electrical discharge on or near the surface of the insulator, degrading the surface of the polymer and thus making it more conductive. Such discharge may be flash ionization of the gaseous atmosphere just above the surface of the polymer, rapid surface flashover or arcing, corona discharge causing slower local degradation of the polymer surface, or gradual surface conduction by contaminants resulting in tracking degradation of the polymer surface itself. These slower processes may proceed over a time span of minutes or hours, up to much longer lifetimes.

17.2.7.2 Factors Affecting Dielectric Strength

Dielectric strength depends upon composition, product design and quality, and environmental conditions—both physical and chemical. Their interaction can be distinguished

Table 17.5 Intrinsic Electric Strength of Various Plastics (14) (Peak Values)

Plastic	Room Temp. (Kv/cm)	Low Temperature Kv/cm	Low Temperature °C
Polyethylene	6.5×10^3	7×10^3	-200
Chlorinated polyethylene	6.5×10^3	11×10^3	-200
Polystyrene	6×10^3	7.2×10^3	-200
Polyisobutylene	1×10^3	6×10^3	$- 90$
Shellac	3.5×10^3	7.5×10^3	-180
Polyvinyl alcohol	3×10^3	15×10^3	-200
Plasticized polyvinyl chloride	6.5×10^3	12×10^3	-200
Polymethylmethacrylate	10×10^3	14×10^3	-200

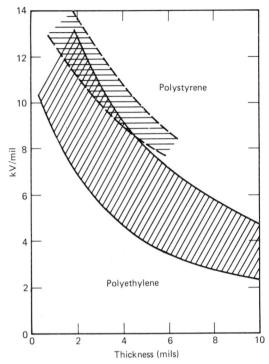

Figure 17.11 Short-time breakdown strength at room temperature (14). From *Engineering Design for Plastics* by E. Baer, © 1964 by Litton Educational Publishing, Inc. Reprinted by permission of Van Nostrand Reinhold Company.

from effects on both thermal and discharge-dependent breakdown.

Thermal Breakdown. The following pertinent factors relate to thermal breakdown.

Thickness of the insulator retards its ability to lose hysteresis heat to the atmosphere and thus aggravates temperature rise and thermal breakdown (Figure 17.11) (14). The figure shows breakdown strength at room temperature. This is a major factor in insulator design.

Time is obviously crucial in such heat buildup. The longer the voltage is applied, the greater the excess in heat generated versus low conduction to the surroundings, until thermal failure finally occurs (Figure 17.12) (14). In the figure thermal stability is shown as a function of applied voltage and time. Thus time of test must be specified, or lifetime at standard voltage can be taken as the test result. Either way, this limit is important in product design.

Temperature correlates directly with ease of breakdown because it increases conduction, reduces loss of heat from the polymer to its surroundings, and raises the base from which hysteresis heating continues. All three aggravate the thermal degradation of the polymer to form more conductive species. Even in intrinsic dielectric strength measurements the effect of temperature is very clearly seen.

Voltage is conveniently raised throughout the test until breakdown occurs, and this critical voltage is taken as the measure of dielectric strength. If, conversely, voltage is held constant and time or temperature are measured to failure (Figure 17.12), it is obvious that increasing voltage shortens lifetime to failure. This effect is also seen in studies of hysteresis by way of the dissipation factor (Figure 17.13), which in turn produces the heating to failure.

Frequency of alternating current determines how serious the hysteresis heating will be; generally high frequency limits the insulator to much lower voltage (Table 17.6). Data in Table 17.6 show the average breakdown stress in volts per mil of 10 filled and unfilled polymers at 60 cps, 2 MHz and 10 MHz.

Water absorption plasticizes polymer systems and provides a fluid polar medium for electrical conduction, so it can seriously aggravate dielectric breakdown (14). This is a serious consideration for insulators operating at high humidity, or especially in actual immersion.

Discharge-Dependent Breakdown. The factors that follow contribute to discharge-dependent breakdown.

Structural flaws such as cracks, fissures, voids, and lamelar defects—especially at the surface of the insulator—are generally the loci for surface flashover, arcing, and corona discharge, which leads to breakdown. These flaws may be present in the original insulator, or they may form during use due to factors such as mechanical stress and vibration. A particularly serious case is the use of foamed plastics, where the gas bubbles give higher volume resistance but lower the breakdown resistance.

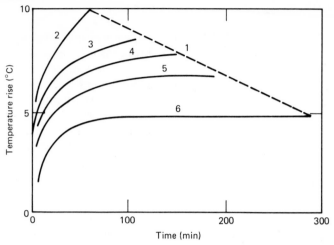

Figure 17.12 Thermal instability in PVC cable as a function of applied voltage and time. Inside diameter, 0.63 in.; outside diameter, 1.06 in.; ambient temperature, 20°C; outer electrode thermally insulated (14). From *Engineering Design for Plastics* by E. Baer, © 1964 by Litton Educational Publishing, Inc. Reprinted by permission of Van Nostrand Reinhold Company.

Curve	Total Voltage (kV)	Average Volts/Mil
1	(Boundary curve for thermal instability)	
2	26	121
3	23	108
4	21.9	102
5	20.7	96.5
6	19.5	91

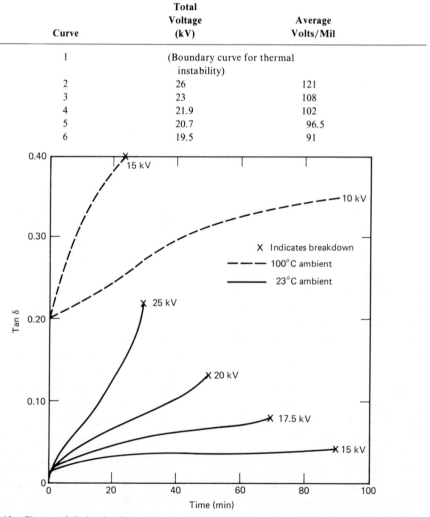

Figure 17.13 Change of dissipation factor (tan δ) of a ¼ inch thick phenolic-paper laminate prior to failure at different voltages (14). From *Engineering Design for Plastics* by E. Baer, © 1964 by Litton Educational Publishing, Inc. Reprinted by permission of Van Nostrand Reinhold Company.

Table 17.6 Average Breakdown Stress of Selected Insulating Material (1/8 In. Specimens) (14)

Material	Volts Per Mil		
	60 cps	2 mc	100 mc
Mica-filled phenolic molding	376	96	12
Asbestos-fabric phenolic molding	72	15	4.9
Polytetrafluoroethylene	566	348	130
Polyethylene	1338	336	98
Polystyrene	2275	352	100
Glass-silicone laminate	468	209	60
Glass-melamine laminate*	422-492	34-47	14-19
Paper-phenolic laminate*	642-729	22-30	8.2-10
Linen-phenolic laminate*	284-301	18-22	4.7-6.1
Canvas-phenolic laminate*	222-319	13-17	5.0-5.9

*Minimum and maximum of the average values are reported.

Contamination of the plastic by foreign inclusions or by collection of dirt, salt, and moisture during use can provide the physical inhomogeniety where flashover, arcing, and corona occur. They can also provide the initial conductive paths by which surface conduction, heating, and breakdown can occur, turning the polymer into more conductive species such as polar groups, ionic groups, or even pure carbon paths; this is called tracking.

Water is a particularly important contributor to surface tracking because it provides the mobile polar medium for ionic conduction which then leads to heating and chemical degradation of the polymer. The water can come from humid air, condensation, fog, rainfall, or actual immersion. The effect can be very serious, even on water-resistant polymers (Table 17.7). Date in Table 17.7 show flashover voltage for polystyrene, conditioned dry and under high humidity, at 60 cycles, 1 kc and 100 MHz.

Gas phase at the surface of the plastic insulator is important in ionization and corona discharge, which in turn lead to degradation of the polymer surface until it becomes conductive itself. Thus oxygen and air contribute to oxidation of polymers to more polar species, whereas nitrogen and especially vacuum prevent this, and vacuum additionally removes most of the ionizable gas itself. Practical data are complex, requiring interpretation of all of these factors.

Geometry. The design of the electrical part is important in determining the relative geometry of the conducting metal and the insulating polymer. Designs that can produce a uniform voltage stress gradient all over the

Table 17.7 Polystyrene—Flashover Voltage, kV (R.M.S) (14)

Conditioning	Voltage		
	60 cps	1 Kc	100 mc
As received (dry)	14.3	13.9	12.0
1 hr - 25°C, 100% RH	10.7	9.6	6.5
180 days - 25°C, 100% RH	5.5	4.5	3.0

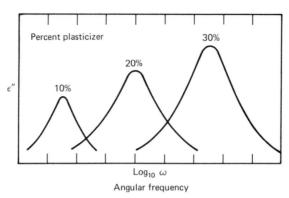

Figure 17.14   Effect of plasticizer upon dielectric loss of polar polymer. (10) ϵ'' = dielectric loss index; ω = angular frequency. Reprinted by permission of C.B.I. Publishing Co., Inc., 221 Columbus Avenue, Boston, Massachusetts.

polymer are more successful than designs that produce sharp voltage stress concentrations at specific points in the insulator structure. This requires careful geometric-electrical analysis.

Composition. Too little is known about the fundamental relationships between polymer structure and additives and their effects on dielectric strength. In general, high polarity seems to favor intrinsic dielectric strength but produces the hysteresis heating that contributes to the more important thermal breakdown. Addition of plasticizer generally increases the mobility of polar groups and their ability to orient in an alternating electric field; this shifts the dielectric loss peak to higher frequencies and greater losses (Figure 17.14), producing a greater hysteresis heating and thus contributing to thermal breakdown.

There is also a general belief that polymers with all-carbon backbones will degrade to conductive carbon tracks, whereas polymers with other (hetero) atoms in the main chain—such as oxygen or nitrogen—cannot do this and will therefore give higher dielectric strength. Also, as mentioned earlier, foamed plastics inherently introduce a high concentration of "structural flaws" and polymer-gas interfaces which aggravate breakdown seriously.

17.2.7.3   Other Effects of Dielectric Breakdown

Aside from the loss of electrical resistence, dielectric breakdown can result in serious overheating of the plastic insulator, causing softening and even—in extreme cases—ignition and fire hazards. Electrical breakdown may also produce concomitant physical deterioration and loss of physical properties, particularly weakening and embrittlement. These, in turn, can contribute further to electrical failure if the part has not already failed.

17.3   CHEMICAL RESISTANCE

The effects of chemical environment on plastic products may be observed in several distinct areas. Moisture is a major factor in performance of many products. Acids, bases, and oxidizing agents may lead to chemical attack and structural changes, whereas organic solvents may lead to swelling and extraction. Environmental stress cracking is a specialized phenomenon resulting from the combination of mechanical stress with mild solvent or surfactant action.

17.3.1   Moisture

The effects of moisture are best understood by considering its effects on structural materials in general, the sources of moisture in polymers, the effect of polymer structure on moisture resistance, the effects of water absorption on polymer properties, and the special used of highly hydrophilic polymers.

17.3.1.1 Effects of Moisture on Materials

Water has severe effects on conventional structural materials, and man has accepted this fact for thousands of years. Iron rusts, copper discolors, aluminum pits and chalks. Concrete spall, while limestone dissolves and erodes. Wood swells and shrinks as humidity cycles, then rots if wet or cracks if dry. Leather softens when wet, then hardens into irregular shapes when redried. Textiles are permeable to water, which makes them comfortable to wear; but the more water they absorb, the less they hold their shape and appearance. Paper deteriorates completely when wet.

By comparison, organic plastics are relatively water resistant, and many of their uses depend upon this. Their low water absorption is the basis of this superiority. They do not corrode like metals or swell and shrink like wood. Being less permeable than textiles and paper, they are useful for rainwear and packaging; in fact, resin treatment improves the shape retention of textiles and the wet strength of paper.

On the other hand, plastics still do absorb some moisture; this causes some losses of properties, and occasionally a few improvements. Even more serious, water can leach out plasticizer and stabilizer and even hydrolyze the polymer to lower molecular weight. For these reasons, the effects of moisture on plastics are important.

17.3.1.2 Sources of Moisture in Polymers

Water content of polymers is the result of aqueous processing, atmospheric humidity, weathering, and direct immersion of plastic products in water.

Aqueous Processing. Many plastics processes expose the polymer to water, and then fail to remove it completely. Polymerization in emulsion and suspension requires careful postdrying. Washing to remove emulsifier, suspension stabilizer, buffer, and catalyst all use copious amounts of water, which must be thoroughly removed in turn. Some polymers are solution-processed and then precipitated by water or aqueous solutions—acrylic fiber, cellophane, vulcanized fiber—requiring thorough postdrying. And many plastics are pelletized by extrusion into a cold-water bath before chopping, thus demanding redrying afterwards.

Relative Humidity. Most of the serious moisture problems arise from the humidity of the air in which polymers are stored and used. Moisture absorption is directly related to relative humidity. Since the plastic equiliberates only slowly with the atmosphere, the resulting changes occur gradually over a long time (17).

Weather. Plastics exposed to outdoor weathering absorb moisture from both humidity and rainfall, and lose it because of low humidity, wind, and heating from the sun. While plastics weather better than most structural materials, these changes eventually produce deterioration and failure.

Immersion. Because of their water resistance, plastics are often used in water-immersion applications, which test them to the limit of their endurance. In marine applications such as boat hulls, plastics resist salt water much better than metals. In tanks, pipe, and plumbing fixtures plastics must withstand not only water itself but also pressure and heat, and frequently detergents and solvents, which can cause stress cracking (Section 17.3.3). Housewares similarly may be exposed to hot water, detergents, and household chemicals. In sports equipment, the water resistence of plastics is generally highly desirable.

17.3.1.3 Structure, Composition and Moisture Resistance

Generally, water resistance depends primarily on low polarity and especially on low hydrogen bonding. Nonpolar polymers show the least water sensitivity, whereas polymers containing sites for hydrogen bonding are most affected by moisture and humidity. Although specific figures vary with different tests, grades, and suppliers, Table 17.8 is typical (18). The data show moisture resistance values of unfilled material for 36 polymeric products. In addition, the same factors apply to low molecular weight additives such as plasticizers

Table 17.8   Moisture Resistance Values of Unfilled Resins (18)

MATERIALS			
PTFE	0.0	Polysulfone	0.3
FEP	0.0	Nylon 11	0.3
Polyvinylidene fluoride	0.0	Acrylic	0.1-0.4
CTFE	0.0	SAN	0.2-0.4
Polymethylpentene	0.0	Acetal	0.2-0.4
Polypropylene	0.0-0.03	Polyimide	0.3-0.4
Polyethylene	0.0-0.03	Nylon 6/12	0.4
Polyphenylene sulfide	0.02-0.05	Nylon 6/10	0.4
Polyvinyl chloride	0.02-0.10	Epoxy	0.1-0.5
Polyphenylene	0.06-0.10	Polyester	0.1-0.6
PBTP	0.1	Polyurethane	0.5-1.6
Silicone	0.1	Cellulose acetate butyrate (CAB)	1.5-1.6
Polystyrene	0.0-0.2	Cellulose butyrate	1.0-2.0
Allylic esters	0.2	Cellulose propionate	1.3-2.4
Nylon 12	0.2-0.25	Cellulose acetate	0.1-4.6
Polycarbonate	0.1-0.3	Nylon 6/6	1.2-9.0
ABS	0.2-0.3	Nylon 6	1.6-11.0
PMMA	0.3		

and stabilizers. Secondary factors such as cytstallinity can reduce water absorption moderately but do not change the inherent effects of polarity and hydrogen bonding.

17.3.1.4 Effects of Water Absorption
Several factors are involved in the effects on plastics of water absorption.

Processing.   In melt processing of plastics, absorbed moisture can cause visual defects referred to as orange peel, splash marks, mica specks, or silver streaking. In more serious cases, physical defects and even visible bubbles can detract seriously from strength. Still worse, the combination of water and heat during processing can cause hydrolysis, resulting in serious property losses. Consequently, manufacturers often specify maximum moisture contents permissible for melt processing (Table 17.9) (17). Permissible ranges, for injection and extrusion molding, of 11 polymer materials are shown in Table 17.9.

Coatings.   In solution coatings such as lacquers, rapid evaporation of low-boiling solvents in humid air can produce cooling and moisture condensation, resulting in opaque whitening—which is called blushing. On the other hand, latex coatings dry by evaporation of water, and the remaining water must plasticize the polymer particles enough to make them flow and fuse into a continuous coating before the cycle is complete.

Dimensional Stability.   Plastics that have high hydrogen bonding and can absorb considerable moisture at high relative humidity will suffer from dimensional instability, creating problems in precise engineering design. While this is an acknowledged problem, the success of nylon as an engineering plastic suggests that it is not an insuperable one.

Modulus.   Plastics that absorb considerable moisture will be plasticized by it and suffer considerable loss of rigidity (Figure 17.15), as measured by indentation hardness or the

**Table 17.9   Permissible Moisture Content (%) for
Various Plastic Resins[a] (18)**

Plastic Resin	Moisture Content, %	
	Injection	Extrusion
ABS	0.10–0.20	0.03–0.05
Acrylic	0.02–0.10	0.02–0.04
Cellulosics	Max. 0.40	Max. 0.30
Ethyl cellulose	0.10	0.04
Nylon	0.04–0.08	0.02–0.06
Polycarbonate	Max. 0.02	0.02
Polyethylene, LD	0.05–0.10	0.03–0.05
Polyethylene, HD	0.05–0.10	0.03–0.05
Polypropylene	0.05	0.03–0.10
Polystyrene	0.10	0.04
Vinyls	0.08	0.08

[a] Where two figures on permissible moisture content are shown, the lower figure is for critical jobs and the higher is the maximum allowable.

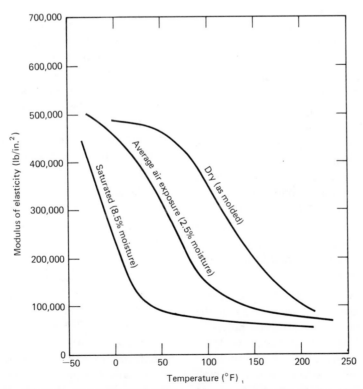

Figure 17.15   Effect of moisture absorption upon modulus of nylon 66 (10). Reprinted by permission of C.B.I. Publishing Co., Inc., 221 Columbus Avenue, Boston, Massachusetts.

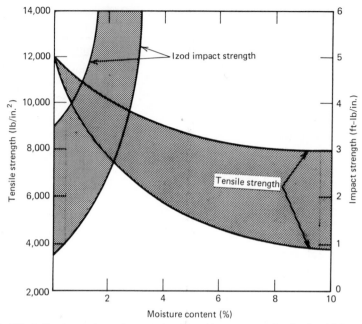

Figure 17.16   Effect of moisture absorption upon tensile and impact strengths of nylon 6 (10). Reprinted by permission of C.B.I Publishing Co., Inc.; 221 Columbus Avenue; Boston, Massachusetts.

modulus-temperature curve (10). Figure 17.15 shows this on nylon 66.

Strength.   Moisture absorption similarly lowers strength of hydrogen-bonding plastics such as nylon (Figure 17.16) (10). The figure shows the effect of moisture absorption upon tensile and impact strengths.

Impact Strength.   On the other hand, the plasticizing effect of moisture absorption can make a plastic much tougher at high humidity than at complete dryness (Figure 17.16). When absorbed water plus heat of processing go so far as to cause hydrolysis, however, this can result in severe embrittlement (19).

Friction/Lubricity.   In nylon gears and bearings, it has been reported that moisture absorption produces improved lubricity and abrasion resistance.

Abrasion by Rainfall.   Rainfall causes gradual abrasive wear, most often seen in the erosion of fiberglass-reinforced polyester panels in building and construction. Most dramatic is the severe effects of raindrops on plastic windows in aircraft, where the high velocity of impact makes the water act as abrasive particles.

Modulus versus Temperature.   As noted earlier, (Figure 17.15), moisture absorption produces plasticization, which increases molecular mobility, shifting the modulus-temperature curve to lower temperatures.

Flame Retardance.   The high heat of vaporization of absorbed water undoubtedly contributes to the superior flame retardance that natural fibers like cotton and wool show as compared with the more moisture-resistant synthetic fabrics. It also explains the flame retardance of vulcanized fiber.

Volume Resistivity.   Absorbed moisture increases mobility of ionic impurities in plastic insulators, lowering their insulation efficiency.

Polarization.   Absorbed moisture also increases the mobility of polar groups in the polymer molecule, permitting them to orient more readily in an electric field and thus increasing dielectric constant and loss.

Static Electricity.   Since plastics are nonconductors of electricity, they tend to build up a static charge when rubbed, or simply upon evaporation of water. This is a hazard in some manufacturing operations and in industrial products such as truck tires and conveyor belts; it is also a discomfort in consumer applications such as clothing and carpeting. In many cases the discomfort and hazard are minimized or even eliminated by higher humidity. The mechanism is probably solution of ionic impurities on the surface of the polymer, producing a means for electrolytic conduction to bleed off the static charge to ground or even in air. Some antistatic additives function simply by absorbing moisture from the air onto the surface of the plastic, whereas others also provide the ionic species needed to make the moisture conductive.

Optical Quality. Plastics and coatings in outdoor use gradually chalk and erode from continual weathering. A large part of this is due to rainfall and fluctuations in humidity.

Physical Deterioration. More serious is the complete decomposition of the plastic product due to the continual effects of moisture. Urea resins swell and shrink repeatedly until they crack and thus cannot be used in severe moisture conditions. Polyurethane foam in mattresses suffers from the combined hydrolytic effects of body moisture plus body heat until it fails under the repeated compressive load.

Discomfort. When plastics replace leather and natural textiles in wearing apparel and upholstery, their low moisture absorption and low water-vapor permeability sometimes make them relatively uncomfortable. Although there have been some studies to develop more permeable synthetics, most present practical solutions appear to be based on mechanical approaches such as inducing porosity, felting on the surface, blending synthetic with natural fibers, and so on.

17.3.1.5 Water-Sensitive and Water-Soluble Plastics

In specialized cases water sensitivity or actual solubility have been used to good advantage. For example, plastics based on hydroxyethyl methacrylate absorb enough water to be highly plasticized—making them useful in contact lenses, body-joint repair, and so on. Starch-based plastics may deteriorate fast

enough outdoors to avoid problems of litter and solid waste disposal. Carboxymethyl cellulose and other soluble plastics are useful thickeners in foods, latex paints, and other applications. Polyvinyl pyrrolidone has been useful as a blood substitute. These applications, and some earlier mentioned property improvements, indicate that the effects of moisture on plastics can be either favorable or unfavorable—depending on the application.

17.3.2 Chemical Environment

In comparison with conventional structural materials, plastics are most sensitive to organic solvents and to organic stains that can dissolve in them. Many plastics are also attacked by oxidizing agents, and some are also sensitive to acids, bases, or even specific salt solutions. For the most part, plastics replace conventional materials when they can offer improved resistance to inorganic corrosion.

17.3.2.1 Conventional Structural Materials

The conventional structural materials, in use for many centuries, all suffer seriously from many types of chemical attack (Table 17.10). Data on attack on seven structural materials (not including plastics) by four chemical agents are shown.

As stated before, by comparison, organic plastics offer much greater resistance to most types of inorganic corrosion, but are more likely to be attacked by organic solvents; some of them are also sensitive to acids, bases, and/or oxidizing agents as well. Since there are more than 80 families of organic polymers in use, it is possible to choose among them to obtain optimum chemical resistance for each desired type of application.

17.3.2.2 Solvents

The attack of organic solvents on plastics may be characterized by the relation between polymer and solvent structure, and the type and extent of attack that may occur.

Polymer and Solvent Structure. A number of structural factors determine the attack of

Table 17.10   Chemical Attack on Conventional Structural Materials[a]

Materials	Acids	Bases	Oxidizing Agents	Solvents
Metals	X	X	X	-
Concrete	X	-	-	-
Glass	-	X	-	-
Wood	X	X	X	-
Leather	X	X	X	-
Natural Fibers	X	X	X	-
Paper	X	X	X	-

[a] Code: X = yes; − = no.

organic solvents on polymers. These may be considered from least to most significant in their effects.

Molecular weight of the polymer is directly related to its resistance to solvent attack. This is due to the thermodynamic relation between low molecular weight and high entropy gain when the polymer dissolves in the solvent and kinetic relation between low molecular weight and low solution viscosity, producing faster mixing of solvent with polymer during the solution process. Thus, where solvent resistance of a plastic is borderline for a certain application, switching to a higher molecular weight grade of the same polymer may meet the desired performance specifications.

Polarity of polymer and solvent determines the degree of attraction or repulsion between them, and therefore the plastic's sensitivity or resistance to the solvent, respectively. This is most often judged by comparing the solubility parameters of polymer and solvent. When the two are fairly close together, solvent attack on polymer is very likely; whereas when the two are far apart, the plastic is much more likely to resist attack (20).

Hydrogen bonding or other highly specific attraction between functional groups in polymer and solvent can contribute powerfully to solubility. These effects are much stronger than simple polarity and can actually overshadow the effects of solubility parameters alone (20).

Crystallinity of a polymer produces such a high concentration of intermolecular attraction that is almost impossible for solvents to penetrate between polymer chains. Thus crystalline polymers are generally very solvent resistant at room temperature. When they are heated toward or past their melting points, however, enough energy is supplied that potential solvents can then attack and dissolve them. In some cases these crystalline polymers, once dissolved, may even stay in solution on recooling—particularly if there is strong hydrogen bonding between polymer and solvent (21).

Cross-linking is the ultimate structural factor in preventing a polymer from dissolving in a solvent. Although this cannot completely eliminate the effects of polarity and hydrogen bonding, it raises molecular weight to the size of an infinite network, preventing "individual" polymer chains from dissolving in the solvent. The higher the degree of cross-linking—particularly in thermosetting cure and vulcanization—the less free-volume and segmental mobility remain available in the polymer, so that solvent molecules can hardly penetrate the cross-linked network at all. In many practical applications thermoset plastics therefore often offer maximum solvent resistence.

Type and Extent of Solvent Attack.   Maximum attack of a solvent on a polymer produces complete dissolution. This is often desirable in processing but generally undesirable in end-use applications. Between complete dissolution and complete resistance, however, there are several intermediate levels of partial attack or partial resistance that are of major importance in most practical cases.

Swelling is the absorption of solvent into a polymer, without producing complete disintegration and dissolution of the plastic product. This occurs when the difference between polymer and solvent, in solubility parameter and/or hydrogen bonding, is too great to permit complete solution but too close to permit complete resistance. It is also particularly prominent in lightly cross-linked polymers—particularly vulcanized elastomers—where there is still considerable free-volume and segment mobility but the polymer molecules are all tied together loosely, preventing them from separating and dissolving completely. In such systems a moderate amount of solvent disperses in the solid polymeric product, swelling and softening it, changing dimensions, and usually making it too weak to support the design stress in practical applications.

Extraction occurs whenever a solvent cannot dissolve the basic polymer in the plastic product but can dissolve and remove low molecular weight components of the plastic composition. Such components may be simply the low end of the polymer molecular weight distribution curve; they also include many common monomeric additives such as plasticizers, stabilizers, and lubricants. In some products the loss of these components may not be an immediate cause of failure; but in most cases it reverberates into unsatisfactory performance in a variety of ways such as stiffening, discoloration, or loss of other essential properties.

Crazing and stress-cracking can result even when the solvent has very little solubilizing effect on the plastic product. These processes produce microvoid opacification or brittle fracture and are discussed separately in a later section.

17.3.2.3 Staining
When plastic products are exposed to organic colors, the plastic often dissolves the color,

producing disfiguring discoloration. When light-colored vinyl floor tile develops dark stains, it is generally believed that organic colors—such as asphaltic tar on shoe soles—are dissolving in the vinyl plasticizer to produce the dark stains. Spillage of food stains on vinyl clothing and upholstery often produces permanent staining as the organic color dissolves in the vinyl plasticizer. Such problems are sometimes remedied by changing to a plasticizer that has less solubilizing effect on the color, or even by changing to a copolymer that is less able to dissolve the color itself.

17.3.2.4 Oxidizing Agents
Strong inorganic oxidizing agents, oxidizing acids, and particularly organic peroxides and hydroperoxides tend to attack organic polymers, generally producing free-radical reactions that result in abstraction, cleavage, or cross-linking and lead to discoloration, embrittlement, or occasionally tackiness along with attendant loss of mechanical, electrical, and other useful properties. This attack varies not only with the oxidizing agent and the use conditions but also most markedly with the structure of the polymer and the use of stabilizers or of unstable additives in compounding. In general, vinyl groups—and hydrogen which is activated by being located alpha to a vinyl group or an oxygen atom—are the positions in the polymer most sensitive to oxidative attack. In some cases, use of antioxidant stabilizer systems can provide considerable protection. In others the materials engineer must simply choose among plastics to obtain the structure that give the oxidative stability needed in the product.

17.3.2.5 Acids and Bases
Whereas most conventional structural materials are subject to severe attack by many aqueous acids and bases, most organic polymers are relatively resistant to these corrosive environments. Only specific functional groups in the polymer may cause sensitization to such reagents. For example, the acidic phenolic hydroxyl group in phenolic resins remains sensitive to alkali even after final

cure. The acetal group in cellulosics, poly-formaldehyde, and polyvinyl acetals is very sensitive to hydrolysis by aqueous acid.

Polyesters, polyamides, and polyurethanes may be hydrolyzed by acid or alkaline catalysis; specific sensitivity varies from one structure to another and must be checked carefully before use. Although these specific cases are certainly troublesome, the fact remains that the majority of commercial polymers do not contain such hydrolyzable groups and are much more corrosion resistant than most of the older conventional structural materials.

17.3.2.6 Specific Salt Solutions

A few highly polar polymers are insoluble in most organic solvents but can be dissolved in very specific aqueous salt solutions for processing. For example: polyacrylonitrile is soluble in aqueous LiCl, $ZnCl_2$, NH_4ClO_4, NaSCN, $Ca(SCN)_2$, and quaternary ammonium salts (21); nylon 66 is soluble in aqueous $CaCl_2$ and $MgCl_2$; and so on.

17.3.2.7 Applications

Since organic plastics offer chemical environment resistance very different from and in many ways superior to older conventional structural materials, plastics have achieved much of their commercial growth in areas where this superior chemical environment resistance is of critical importance. A few major examples are agricultural equipment that will carry irrigation water and handle fertilizers without rusting; home appliances that are rustproof and easy to maintain; automotive parts that resist road salt in winter; aviation components that are lightweight and corrosion resistant; window frames that are not corroded by weathering; plumbing fixtures that do not corrode; swimming pool equipment; oil-resistant shoe soles; nonrusting fasteners; nonrusting gears and bearings; packaging equipment, pipe and fittings; road signs; tanks for chemical storage; and chemical plant construction, in general.

In addition to being used as solid plastics, many organic polymers are applied as protective coatings over corrodible or chemically sensitive conventional structural materials. The major examples are, of course, rust-proofing and corrosion-resistant coatings on metals. Protective coatings on concrete also provide prevention of salt and acid attack as well as freeze-thaw spalling and cracking. Polymeric coatings have always been used to protect wood and are being improved constantly. Coatings on textiles are becoming more common. Coatings on paper give it the chemical resistance required in such demanding applications as battery separator plates.

In all of these applications the chemical resistance of the organic polymer protects the chemically sensitive conventional structural material beneath.

17.3.3 Environmental Stress Cracking

Plastic materials are conventionally laboratory-tested for mechanical behavior by stress-strain analysis and for chemical resistance by exposure to various environments. Occasionally, when they are then used in end products that are subjected to a combination of mild stress plus mild chemical environment, these materials suffer unexpected brittle failure. This is generally referred to as environmental stress cracking. Similar effects have been observed in conventional structural materials (7), but the interaction of organic polymers with organic chemical environments is so wide spread in our everyday life that the subject is of particular importance in plastics technology. Major applications are in the packaging and handling of food and household products, and in piping and chemical equipment.

17.3.3.1 Mechanism

Theoretical understanding of the mechanism of environmental stress cracking begins with submicroscopic weak points in the plastic structure. In amorphous polymers these may be molded-in strains, density variations due to random irregular packing of polymer chains, microvoids, or invisible surface scratches. In crystalline polymers they are more prominently the weak amorphous areas that remain after spherulite growth has left these weak, strained boundaries between the crystalline regions. When the product is used under sustained mechanical stress—particularly tensile stress—the stress is concentrated and greatly magnified in these weak areas. The resulting localized strain makes these areas still lower in density, and aggravates any existing microvoids and microcracks.

Now introduction of a mild chemical environment produces attack on these strained areas by a number of mechanisms. Solvents that are too weak to dissolve the normal polymer may still be able to extract the low molecular weight fraction of the polymer, and particularly the monomeric compounding additives, leaving a solid polymer matrix that is still lower in density and richer in microvoids. Solvents that are too weak to dissolve the polymer may still be able to swell it, thus acting as plasticizers, lowering the glass transition temperature of the polymer and permitting it to deform further under the applied stress. Stress, itself, produces lower density, which permits greater attack by such weak solvents, so the two effects are mutually progressive and complement each other.

Whereas the creation of microvoids requires surface energy, if the chemical environment is a fluid of low surface energy, it will tend to spread over the new surface, lowering surface tension and thus stabilizing the system. This formation of microvoids often progresses to such an extent that the resultant low refractive index produces cloudiness or even opacity that may be the first gross indication of attack. This appearance is generally referred to as crazing—particularly when the microvoids are visible in a low-power microscope.

As these processes progress, the stress continues to concentrate at the tip of the growing crack, weakening it further and promoting the growth of the crack. At some point the remaining uncracked cross-section of the sample becomes too small to support the total stress, and catastrophic brittle fracture occurs.

17.3.3.2   Factors Affecting Environmental Stress Cracking

Environmental stress cracking of plastics depends upon the polymer, the chemical environment, and the stress, strain, temperature, and time in use under these conditions.

Polymer Structures. A number of features (described as follows) in the polymer structure have major effects on environmental stress cracking and resistance to it.

Molecular weight of the polymer is directly related to environmental stress crack resistance (Figure 17.17) (22, 23, 24). The figure shows the effect of melt index on stress cracking of polyethylene. This is because high molecular weight polymer has greater resistance to both solvents and mechanical stress.

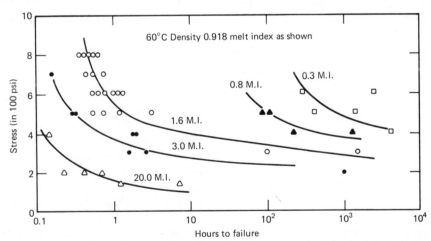

Figure 17.17   Effect of melt index (M.I.) on environmental stress-cracking of polyethylenes in constant load test (22). From *Engineering Design for Plastics* by E. Baer, © 1964 by Litton Educational Publishing, Inc., and reprinted by permission of Van Nostrand Reinhold Company.

Molecular weight distribution is important because it is the low molecular weight fraction that is most readily leached by weak solvents, leaving low-density and microvoid structures that are more subject to both swelling and stress deformation. Thus narrow molecular weight distribution and high number-average molecular weight are both useful in improving environmental stress crack resistance.

Crystallinity is an important factor in environmental stress-crack resistance, but its effects are complex. In most tests and practical applications the product is under long-term constant deformation; increasing crystallinity concentrates all of the stress in the residual interfacial amorphous areas, aggravating the environmental stress cracking behavior. On the other hand, in short-term tests under constant stress and in piping applications, for example, high crystallinity produces higher yield strength and thus higher stress crack resistance. Thus sound critical engineering judgment is required to pick the proper laboratory test method and optimum crystallinity for each specific application.

Texture of crystallinity is probably another significant factor in environmental stress-crack resistance. The effects of spherulite growth and interfacial morphology on stress cracking are probably important and deserve intensive study. This, in turn, will require more sophisticated control of nucleation and process conditions in the manufacture of the end products.

Multiphase polymer systems may be as helpful in preventing environmental stress cracking as they are in preventing conventional brittle failure. The dispersion of submicroscopic rubbery domains in a rigid glassy matrix gives it the power to stop the growth of cracks and thus prevent fracture. While this technique has been used primarily to improve the impact strength of rigid plastics, it has often been observed to improve environmental stress-crack resistance as well, particularly in polyolefin and polystyrene systems.

Microstructure of the polymeric product can be critical in explaining and controlling environmental stress cracking. Density variations and molded-in strains result from viscoelastic flow during melt processing and from shrinkage during cooling and crystallization. Surface scratches, often submicroscopic, can form the nuclei from which stress cracking starts. Microvoids may result from shrinkage during cooling or crystallization, from stress-strain deformation, or from leaching by weak solvents. All of these microstructures form the points at which stress concentrates and becomes grossly magnified, far beyond the failure point of the polymer. This is where crack growth starts and then continues until it results in total failure. Such microstructures can all be considered in analytical studies, and many of them can be controlled in practical processing and use.

Macrostructure in product design may provide powerful control over environmental stress cracking and its prevention. Sharp corners and angles generally act as stress concentrators, and produce brittle failure both under high-speed impact and in long-term environmental stress cracking. In most cases the design engineer can eliminate this problem by careful redesign. On a more sophisticated level, the radius of curvature of any corner or angle is inverse to stress concentration and thus directly related to environmental stress-crack resistance.

Environment. The chemical environment is critically important in determining whether the plastic will resist it or will fail under stress. Most failures occur as a result of immersion in specific organic liquids (Tables 17.11 and 17.12) (25). Data in Table 17.11 show comparison of liquid media, including solvents, that cause stress corrosion in 15 polymer compositions. Table 17.12 shows optimum exposure conditions in nine media for assessment of internal tension in moldings from nine polymer types to produce stress corrosion. In many cases, simple exposure to solvent vapors is sufficient to cause failure—particularly if the liquid is a strong enough solvent to dissolve the polymer.

The following factors in solvent structure and properties affect the ability of such solvent to cause stress cracking:

Table 17.11 Liquid Media (Vapor in the Case of Solvents) Causing Stress Corrosion of Plastics (25)

Polymer	Medium
Polyolefins (rigid and flexible PE, and to some extent PP)	Aqueous solutions of soaps and surfactants (alkyl sulphonates, alkylarylsulphonates, polyalkylphenol glycol ethers etc.); aliphatic acids and vegetable oils (oleic acid, linseed oil, cotton oil); transformer oil, ethereal oils, silicone oil, glycols (methyl ethyl glycol), technical lubricants, lower alcohols, e.g. C_1–C_4, benzene, toluene, acetone, n-heptane, esters.
Polystyrene	n-Heptane, hexane and other higher aliphatic hydrocarbons, aliphatic acids (oleic) and their derivatives, vegetable oils (linseed, cotton), isopropanol, n-butanol, hexanol, cyclohexanol, esters, ketones and aromatic hydrocarbons (acetone, benzene), lacquer benzine, petroleum ether, kerosene, ethylene glycol.
Styrene copolymers (ABS, SAN, SER)	n-Heptane, hexane, aliphatic acids and vegetable oils (linseed, cotton), methyl alcohol, isopropanol, hexanol, cyclohexanol, chlorinated hydrocarbons (CCl_4), ethylene glycol, glycerol, Freon, lacquer benzine, kerosene, aqueous solutions of soaps and surfactants.
Polycarbonates	n-Heptane, hexane, aliphatic acids (oleic) and their derivatives, vegetable oils (olive, linseed, cotton), methanol, isopropanol, butyl alcohol, chlorinated hydrocarbons (CCl_4), lacquer benzine, glycerol, ethylene glycol, turpentine, ketones and aromatic hydrocarbons (acetone, benzene, toluene), hot water (60° C)
Polyamides	(for the dried polymer) Benzine, benzene, mineral oils, methanol, ethanol, amyl alcohol, glycols, cyclohexane, trichloroethylene, CCl_4, acetone, tetrahydrofuran, ethyl acetate, saturated solution of $ZnCl_2$ (50° C), distilled water (for the wet polymer) Trichloroethylene, CCl_4, saturated solution of $ZnCl_2$ (50° C)
Polyurethanes (linear)	Ethyl alcohol, acetone, diethyl ether, chloroform, benzene.
High-impact P/S	n-Heptane, hexane, methyl alcohol, isopropanol
PVC	Aqueous solutions of surfactants (soaps, wetting agents, emulsifiers etc.), methyl alcohol, isopropanol, n-hexane, ethyl acetate, organic lubricants, naphthalene vapour.
PMMA	CCl_4, chloroform, aliphatic acids and their derivatives, vegetable oils, paraffin oil, lower alcohols (C_1–C_4), glycerol, n-heptane, cyclohexane, acetone, toluene, benzene, Freon, MMA monomer, turpentine, water.
Phenolic resins	Acetone
Poltesters, epoxides	Water at elevated temperatures
Poly(ethylene terephthalate)	Dibutyl phthalate, acetone, methyl ethyl ketone, aliphatic alcohols.
Poly(vinyl acetate)	Methyl alcohol
Poly(phenylene oxide)	Aliphatic and chlorinated hydrocarbons, ketones, esters, aliphatic acids and their derivatives, vegetable oils, methanol.
Polychlorotrifluoroethylene (Teflex)	Methyl alcohol, acetone, kerosene.

Reprinted by permission of International Polymer Science and Technology, a RAPRA produced periodical.

Table 17.12   Optimum Exposure Conditions for Assessment of The Internal Tension in Mouldings from Certain Plastics Due to Stress Corrosion (25)

Polymer	Medium	Temperature of medium, °C	Exposure time
PMMA	Carbon tetrachloride	20	5–10 min
	MMA monomer	20	max. 5-, 15 min
ABS copolymer	Linseed oil, cotton oil	55	48 h
	Conc. acetic acid	20	max. 1–5 min
High-impact PS	n-Heptane	20	24 h
SAN	Methyl alcohol	20	60–120 min
Polycarbonates	Carbon tetrachloride	20	max. 5 min
	Toluene + isopropanol (1:3)	20	max. 3 min
Poly(phenylene oxide) (Noryl)	Methyl alcohol + trichloroethylene (2:1)	20	20 h
Standard PS	Lacquer benzine	20	1–3 min
Polyamides (6 and 6.6)	Saturated aqueous solution of $ZnCl_2$	50–60	80 h
Polyolefins (rigid and flexible PE)	A 10–15% solution of a powerful surfactant in water, e.g. Hostapal HL, Igepal Co 630, and Marlophene 89	50–55	72–48 h

Reprinted by permission of International Polymer Science and Technology, a RAPRA produced periodical.

Polarity determines its ability to interact with the polymer.

Hydrogen bonding can be even more important in this interaction (26).

Low surface tension makes the liquid a good wetting agent and permits it to lower the surface energy of the craze surface, thus promoting the crazing process.

Low viscosity of the liquid promotes spreading and wetting of the surface and thus makes stress cracking progress faster.

Even with the recognition of all these factors, the great variety of liquids that cause stress cracking is still sufficiently bewildering to challenge much further research in this area.

Strain. Although the phenomenon is always referred to as stress cracking, it is really the resulting deformation or strain that produces the micromechanism by which crazing, crack growth, and failure actually result (27). In many studies it is thus the strain that is taken as the independent variable and studies as the casual effect. Although tensile strain is generally given prime consideration, other modes and complex combinations are actually important—particularly in practical end products.

Considerable literature centers on the distinction between monoaxial and multiaxial strain; this requires further study to define the basic differences between their effects. In practice, most products are subjected to a fixed mode and extent of deformation, and this—together with the chemical environment—determines their useful lifetime until catastrophic failure.

Stress. In most practical applications the sample product is deformed by an external stress. The stress may be applied and held constant or it may be applied initially and allowed to relax as plastic creep allows the polymer to comply gradually with the enforced deformation. These variables have critical effects on the test results, as already noted.

One particularly critical factor is the relation between applied stress and yield stress in the material: if the applied stress is well below the yield stress, the product will have a much longer useful life. In some cases the stress that causes stress cracking is not externally applied but is actually built into the sample by the manufacturing process. Molded-in stresses and strains are very common, particularly at high production rates that cause nonuniform viscoelastic flow and rapid cooling of the polymer. These internal stresses can be the actual cause of ultimate failure.

Temperature.   The mechanism of environmental stress cracking, as described, involves localized plastic deformation. Such viscoelastic creep will obviously increase with temperature. Thus it is generally observed that environmental stress cracking accelerates with increase in temperature. On the other hand, it should be noted that preliminary thermal annealing—before application of external stress or chemical environment—should relieve built-in stresses and thus make the resulting product more resistant to environmental stress cracking in the end-use application. Although such annealing is often recommended by the material manufacturer, processing economics generally pevent its widespread use. In applications where environmental stress cracking is a serious problem, however, annealing should be seriously considered as a practical solution.

Time.   In most laboratory studies and practical products, time to failure is the acutal measure of environmental stress-crack resistance. On the other hand, if useful life of the product or time of the laboratory test are specified as a fixed value, then one of the other variables becomes the criterion. Thus in some tests stress-crack resistance may be specified in terms of maximum stress, maximum strain, or maximum temperature level that can be tolerated before failure will occur.

17.3.3.3   Prevention of Environmental Stress Cracking

The plastics engineer is often faced with the problem of designing a product to resist environmental stress cracking or of redesigning a product to solve the problem after it has been observed. He has a number of variables among which to choose, depending upon the specific case, as follows:

Choice of polymer is often the first consideration. If sufficient data are available, he can choose the polymer that is most likely to resist the operating conditions.

Choice of test method is very crucial to predicting practical performance. From all that has been said before, the enigneer requires experienced wisdom to choose both test methods and performance standards that will properly predict product performance.

Molecular weight and distribution can often provide the most straightforward solution to the problem. High molecular weight and/or narrow molecular weight distribution—or simply elimination of a low molecular weight fraction—may be all the engineer has to do to obtain high performance.

Copolymerization can be used in several ways to modify the polymer and improve stress-crack resistance. In crystalline polymers, a little random copolymerization can reduce total crystallinity or the size of crystallites sufficiently to improve stress-crack resistance significantly. In polymers of low solvent resistance, copolymerization with a more solvent-resistant comonomer may be sufficient to improve stress resistance considerably—as in styrene-acrylonitrile, for example (24).

Polyblending can disperse soft rubbery microparticles throughout a rigid glassy matrix and leave such microparticles to stop growth of microcracks before failure can occur. Such effects have often been noted in polyolefins and impact styrene, among others.

Coating with a more solvent-resistant surface can sometimes prevent solvent from attacking the substrate and thus protect it against stress cracking. Perhaps a more extreme version would be metallizing, or laminating to a film that has high stress-crack resistance.

Redesign of the product can often lower the stress or strain that is applied to it below the level that causes stress cracking. Elimination of sharp corners and angles and use of larger radii of curvature are conventional examples. In specific products the design engineer may find other ways of reducing the critical stress and strain.

The environment itself may sometimes be the critical variable that the design engineer must change to solve the problem. This is not generally possible, but in some products it is perfectly feasible to redesign so that the polymer is not exposed to as severe an environment, and this may be the most practical solution to the problem in these cases.

Finally, environmental stress cracking is a frequent phenomenon when organic polymers are subjected to mild stress and strain plus exposure to mild chemical environment. It should be obvious from the preceding discussion that such problems are not final failures but simply greater challenges to the ingenuity of the polymer chemist and the design engineer.

17.4   ATMOSPHERIC RESISTANCE

Weather is a major consideration when plastic products are used outdoors. Plastics may affect the environment, both during manufacture and during disposal at the end of their useful life. In aerospace applications plastics are called upon to perform in very unusual environments.

17.4.1   Weathering

In comparison with conventional structural materials, plastics may perform quite differently toward the different factors involved in weathering. While "normal" outdoor weathering may vary greatly from place to place and even from time to time, accelerated weathering is based upon accentuating individual factors in weathering conditions. There are many criteria for evaluating weather resistance, depending upon the type of product and performance required. These in turn can be used to choose the optimum polymers and stabilizers for each application.

17.4.1.1   Conventional Materials

Many of the products that man makes are intended for long-term use out of doors. Thus the effects of weather on his materials of construction have a critical effect on the maintenance and ultimate life of such products.

Most conventional materials suffer severely from outdoor weathering. Metals corrode—requiring frequent painting and repainting and periodic replacement. Concrete spalls and cracks—requiring painting, patching, and eventually replacement. Limestone erodes and gradually dissolves. Wood swells and shrinks, discolors, cracks, and rots—requiring frequent painting and replacement. Lifetimes until replacement may be 5–10 years for painted sheet metal, 10–20 years for concrete highways, 10–20 years for unpainted wood, and 2–5 years for most paint jobs. After thousands of years of experience with such conventional materials, man simply accepts these lifetimes and these maintenance schedules without complaint.

Synthetic organic polymers are generally quite stable to outdoor weathering, but we tend to view their performance much more critically. Thus the effects of weather on them, and the ability to control them, are of great importance in choice of materials for outdoor applications.

17.4.1.2   Components of Weather

Outdoor weathering results from a combination of many environmental factors. These can be considered separately to analyze their combined effects.

Ultraviolet (UV) light is only 5% of the total solar energy reaching the earth, but these short wavelengths (290–400 nm) have the high-energy levels that can activate organic polymers so that they dissociate into free radicals, which initiate degradative chain reactions. These lead to conjugated unsaturation, causing discoloration; and to crosslinking or cleavage, which causes embrittlement. Over all, such UV degradation is the most important factor in weathering of plastics.

Oxygen is the most common chemical agent in polymer aging, including weathering. Where UV light alone would have little or no effect, entry of atmospheric oxygen into the chain reaction produces peroxides and hydroperoxides that accelerate the degradation processes at lower energy levels. In specialized systems—particularly unsaturated elastomers—atmospheric ozone is a powerful degradant, causing early cracking and embrittlement of rubber products. Incidentally, oxygen is also a major contributor to the corrosion of metals, as well.

Heat generally accelerates polymer aging processes (Sections 17.2.5 and 17.2.6). About 53% of the solar energy reaching the earth is infrared heat, and its absorption by polymers raises their temperature sufficiently to contribute significantly to the weathering process.

Water appears in the weathering process as humidity, rain, hail, and snow. Polar hydrogen-bonding polymers can absorb considerable moisture from humid air, causing swelling and shrinking with variations in relative humidity and leading to crazing and cracking (Section 17.3.1). Absorbed moisture can also lead to hydrolysis and thus permanent degradation. Rain and hail add the problem of abrasive wear, and snow loads add the problem of mechanical strength. Most polymers, however, are relatively resistant to water compared with the severe corrosive effects on metals and the severe swelling, shrinking, and cracking of wood.

Biodegradation of plastics often occurs during outdoor weathering, and involves many of the components in weather. This is discussed separately in Section 17.5.1.

Wind can subject plastics to considerable mechanical load—particularly in signs and building panels. Often it is wind blown dust—mostly inorganic particles of high hardness—that causes abrasive wear of the plastic surface. Most often it is dust settling out of the air, soiling plastic surfaces and making them unsightly, that leads to ultimate aesthetic failure and consumer rejection; this is most common in the repainting of houses,

but it is important in many other products as well.

Background of the test samples or plastic products is much more important than is commonly realized. Depending upon whether the background is vegetation, rock, asphalt, steel, wood, or water, it may absorb solar heat or reflect radiation. Distance from the background is important in determining how serious these effects may be. If the sample is mounted on it, failure of the substrate may cause failure of the plastic too.

Freezing of adsorbed moisture is most serious in concrete, wood, and asphalt, where it causes cracking and crumbling. In some cases, similar effects can be seen in plastics as well.

Cycling of weather components such as light, heat, water, and freezing often accentuates their effects. Free-radical processes, expansion and contraction, swelling and shrinkage, time for evaporation are all processes that become more evident in the cycle of day to night, and sometimes in the cycle of season upon season.

Atmospheric pollution has recently gained increasing attention as a contributor to weathering processes. Acid gases such as CO_2, SO_2, HCl, and so on, attack metals, concrete, stone, and wood, as well as hydrolyzable polymers. Ozone and oxides of nitrogen participate in free-radical processes that can cause degradation. Hydrogen sulfide causes severe staining of pigments and stabilizers containing lead, cadmium, and other heavy metals.

Salts from seawater and other sources are primarily harmful to metals but may sometimes have milder effects on plastics as well.

Other processes that occur in weathering include the continued cure of thermosets for several years, producing increase in strength and brittleness; the gradual crystallization of thermoplastics over a long time; evaporation of volatiles such as solvents, water, and even plasticizers producing gradual hardening; leaching of these small molecules by water producing similar effects; and chemical degradation of additives producing effects as important as weathering of the polymer itself—sometimes more so.

Time of course is implicit in all of these weathering processes. Experimentally it is possible to study the effect of time on any measured property, which is the most scientific approach; or time can be fixed and properties compared at that standard time. In practice it is the duration of useful life that is important, and this may be defined either aesthetically or technically, depending on the product.

17.4.1.3 Variables in Weathering

When plastics are compared with each other or with nonplastic materials in outdoor weathering, many of the preceding components of weather may cause confusion. Solar radiation varies with the angle at which the sample is mounted, with sunspots and cloudiness, and with seasonal cycles. Angle of mounting also determines whether moisture and dirt will lie on the sample or run off easily. Altitude above sea level affects temperature differently, depending on wind. Latitude affects radiation. Industrialization and population density affect air pollution. Thus different weathering studies may fail to control all of these variables and may come to opposing conclusions in their comparison of materials for outdoor products.

The thickness of sample is a factor that is too often neglected and is rarely given its full consideration. Most weathering effects occur at the surface and may never penetrate into the interior at all. In some cases deterioration of the surface causes failure of the entire sample; examples are surface defects that initiate mechanical failure and color change of consumer products. In other cases surface changes form a protective coating that stabilizes the interior; examples are the oxidation of aluminum and the opacification of plastics to UV light.

17.4.1.4 Accelerated Weathering

When materials and products are being designed for long-term outdoor use, normal weathering is much too slow and variable for research and development or even for quality control. Thus there is great interest in accelerating the normal weathering process. This is done by picking one (or more) normal-weather component(s) and accentuating it by technological manipulation. Following are some of the most common examples.

UV light may be accelerated by taking samples to a lower latitude such as Florida or Arizona. It is increased further by mounting samples more perpendicular to the sun's rays—particularly in a follow-the-sun machine. It can be increased still further by using mirrors or lenses to concentrate the sunlight falling on the samples. All of these methods leave the spectral distribution of sunlight the same and simply increase the amount reaching the samples. They do, however, accentuate UV light at the expense of other components of weather, thus unbalancing the normal effect. In general, this would age plastics faster than metals or concrete.

UV light can be further emphasized by use of laboratory sources such as the General Electric S-1, and more recently RS-4, lamps and more commonly by use of carbon-arc and the more recent Xenon-arc Fade-Ometers and Weather-Ometers. These methods can easily produce fast-aging results for feedback in research and development; on the other hand, their spectral distribution is quite different from normal sunlight, and they differ greatly among themselves. These differences can be modulated by use of filters, but this rarely succeeds in duplicating outdoor sunlight. In any case, such accelerated methods severely emphasize UV light at the expense of other weather components.

Heat can be supplied to accelerate oxidative aging. Many UV sources simultaneously heat the sample, with or without the knowledge of the experimenter, giving more accelerated results; but the value of this acceleration is seriously open to question, varying from one material to another. Some tests control the amount of heating, letting it work for them but always skewing the balance of normal weather components.

Other accelerated factors such as water immersion, freezing, salt spray, and even mechanical deformation are sometimes used to produce fast results in weathering. Warm humid climates such as Florida or New Orleans are used to include natural biodegradation, whereas industrial atmospheres sometimes show atmospheric pollution more important than southern sunlight in producing accelerated aging.

17.4.1.5 Criteria of Failure

The effects of weather on plastics can be judged by a wide variety of criteria, depending upon which give the earliest and/or sharpest indication of failure, or which are most important in the material or application. The following are the most common for plastics:

Appearance may represent the practical end of useful life in many consumer products. Darkening, occasionally bleaching, loss of gloss, loss of transparency, chalking, and erosion of a smooth plastic surface to reveal the internal fibrous reinforcement are common observations during weathering.

Mechanical properties represent the most common types of failure in engineering design. Stiffening, embrittlement, loss of strength, and especially loss of elongation are commonly observed, and a certain percent change must be set to represent the end of useful life of the product.

Physical failure such as cracking or fracture may be the practical end point, but it generally requires physical abuse to make it meaningful. When an undisturbed sample goes this far, it is far beyond the end of its useful life.

Electrical properties such as dielectric strength or resistivity may be the most important criterion in outdoor insulation such as wire and cable.

Chemical measurements: such as molecular weight change, cross-linking, and carbonyl formation are helpful tools in understanding the weathering process, but their application to predict practical weathering should be approached with greatest caution and limited to narrowly controlled systems in which the correlations are well proved.

17.4.1.6 Polymers and Compounding Ingredients

Polymers most often used out of doors are reinforced polyester, polyethylene, and vinyls. Polyesters retain mechanical properties quite well provided the glass fiber reinforcement is surface treated to bond it well to the resin; practical failure is more often due to unsightly discoloration, erosion, and opacification. Polyethylene wire insulation and pipe are sensitive to photooxidation. Plasticized vinyls suffer gradual leaching and evaporation of plasticizer; rigid vinyls suffer oxidative embrittlement of their rubbery impact modifiers; and all vinyls require careful stabilization to retard UV and thermal dehydrochlorination leading to discoloration and stiffening.

For transparent applications such as glazing, signs, and outdoor lighting, polymethyl methacrylate has long been the leading candidate because it retains its clarity and color despite gradual degradation of mechanical properties. Cellulose acetate butyrate similarly retains good appearance long enough to use in signs, even though mechanical properties deteriorate continuously. More recently polycarbonate has found favor because of its high impact resistance and flame retardance but requires careful stabilization to retard UV discoloration.

For engineering applications, thermoset plastics with fibrous reinforcement are frequently preferred for their greater retention of mechanical properties. Phenolics have the longest history, whereas melamine provides the most durable coatings on auto bodies, and reinforced polyester—as mentioned—is most common in building panels.

Additives have important effects in weathering. Plasticizers tend to leach and evaporate gradually and are themselves chemically attacked during weathering, leading to gradual stiffening of plasticized polymers. Reinforcements must be well bonded to the polymeric matrix to retard the attack of water, and organic fibers—particularly cellulosics—are water sensitive, as well as UV degradable and biodegradable. Some pigments are light sensitive, resulting in color changes; or H₂S sensitive, tending to darken in regions where high-sulfur fuels are used.

17.4.2 Effects of Plastics on Environment

In the past decade the public has become increasingly concerned with the effects of modern technology on the environment in which we live and work. This has led to many studies by scientists, engineers, and medical personnel, and to government activity such as the Environmental Protection Agency for the general public and the Occupational Safety and Health Administration for the industrial worker specifically.

Since plastics form a significant portion of modern technology, it is not surprising that they are involved in this subject in many ways. Ways in which plastics contribute to improving our environment include piping and equipment for water supply, purification, and sewage disposal; synthetic fabrics and ductwork for air purification; trash bags for garbage disposal; and many thousands of similar more specialized applications. Of more critical concern, however, are ways in which the manufacture and use of plastics may harm the environment and the ways in which polymer scientists and plastics engineers can eliminate or at least minimize these problems.

Plastics manufacture can cause pollution of air, water, and solid waste in a number of ways, and all of these can be controlled by the process engineer. Use of plastics rarely causes adverse effects on the environment, except in the area of toxicology, as discussed in Section 17.5.2. Of greater concern is the ultimate disposal of plastic products after their useful life is over, contributing significantly to the more general problem of solid waste disposal. These are considered here in more detail.

17.4.2.1 Plastics Manufacture

During original manufacture of plastic products, the manufacturing operation may cause problems of air, water, or solid waste pollution, which can generally be controlled by the manufacturer himself.

Air Pollution. Plastics manufacture can contribute to air pollution in a number of ways. Monomer vapors such as vinyl chloride, acrylonitrile, and styrene may occur in the manufacturing atmosphere owing to leakage or even open handling, and residual monomer may appear in stack gases exiting into the community; process engineers have shown that they can control these very effectively when required.

Plasticizers are considered nonvolatile, but they do vaporize to some extent during processing; this requires careful ventilation in manufacturing operations. Solvent vapors, especially in coating operations, are a significant contributor to metropolitan smog conditions, leading to growing government regulation and increasing use of low-solvent or nonsolvent coating systems. Solid fillers that may enter the atmosphere and cause pollution include channel black, whose manufacture had to be stopped entirely in this country, and asbestos, whose carcinogenic activity is requiring strenuous efforts to improve its handling safety at the present time.

Noise pollution should also be mentioned here, particularly such operations as grinding of plastic scrap, which require vigorous efforts to protect workers in or adjacent to the operations.

All of these are problems that can be solved and that are receiving the effort required to do so.

Water Pollution. The chemical process industries, including plastics, have contributed significantly to water pollution, and as these effects become apparent they require considerable expenditure to eliminate them and prevent them in the future. Most common is simple thermal pollution, where cooling water is returned to streams and rivers at higher temperature, inhibiting normal aquatic life. More serious, however, is the disposal of wash water and chemical wastes, in general, often causing serious poisoning problems.

Typical problems relating to plastics manufacture include residual monomer washed from finished polymer, heavy metals from catalysts and stabilizers, polychlorinated biphenyls and other halogenated organics used as plasticizers and flame retardants, and so on.

Solid Waste.   Solid waste in plastics manufacturing is primarily off-specification production of resins and plastics products. Most thermoplastic scrap is recycled, whenever possible in-plant—preferably by blending into virgin material in the same operation. Lower-grade thermoplastic scrap is either used directly for lower quality products at lower price or sold to independent compounders who convert it into useful plastic materials for less demanding sectors of the industry.

Thermoset scrap and some of the worst tailings of thermoplastic scrap prove completely unusable and are either dumped as solid waste in municipal landfills or are incinerated to dispose of them completely. In some cases incineration can actually be used to produce energy for plant operation, as discussed later.

17.4.2.2   Disposal of Plastic Products at the End of Their Useful Life

Total solid waste in the United States was 400 million tons in 1970 and may reach 700 million tons in 1980. Since plastics are a growing factor in modern technology, it is not surprising that they are a growing factor in solid waste. In 1970 plastics contributed about 3% of total solid waste; in 1980 it is predicted that they will rise to about 5%. If the United States should make a vigorous effort to reclaim metals, glass, and paper from solid waste, this would leave plastics as a much larger proportion of the remainder. Thus the role of plastics in this growing problem is important and deserves serious consideration.

Landfill.   Probably 90% of solid waste disposal is still by simple dumping in specified municipal or industrial dump sites; a small but growing proportion of this is going into sanitary landfill, where the solid waste and earth are mixed and turned over periodically. While plastics are still only a small proportion of this solid waste, their weather resistance, low density, and bright colors make them particularly noticeable to the casual observer, creat-

ing the impression that they are a major offender. Roadside and street litter—what has been labeled the "slob index" of our population—is a particularly outstanding example. Many suggestions, and some research and development, have considered the possibility of making and using plastics that will degrade and return to the ecocycle as readily as paper does. These include water-soluble and biodegradable polymers and fillers, as well as UV-degradable polymers produced by introducing UV-sensitive groups and/or UV-oxidation catalysts. Up to this point, however, commercial interest has been relatively mild.

Incineration.   Some medium-sized and larger cities, lacking convenient landfill sites, have resorted to incineration of solid waste. In the United States this disposes of about 10% of the solid waste; in Europe and Japan the proportion is necessarily higher. When plastics are added to this solid waste, there have been complaints or fears of molten plastic jamming the equipment; gases such as HCl corroding the equipment; and stack gases and smoke polluting the neighborhood. Careful study has proved that well designed, well run incinerators can solve all of these problems, and that such good incinerator practice is much more common in Europe and Japan than in the United States.

Aside from simple disposal of solid waste, incineration can actually be used to produce considerable useful energy from the heat of combustion; here plastics are particularly useful, since their heat of combustion is higher than most other solid waste—often as high as the raw petroleum from which they were made originally. Thus incineration could offer a major contribution to the industrial portion of the ecocycle in the future.

Recovery as Material.   Aside from simple dumping or incineration, it is quite possible that specific portions of solid waste could be recovered for use as materials. Past effort has concentrated mainly on metals, paper, and glass, but similar efforts could be extended to plastics as well. Recent analyses of the plastics in solid waste indicate that they contain about 55% polyolefins, 20% polystyrene and copolymers, 11% polyvinyl chloride, and 14% of all the other plastics. Several possibilities for use of these materials can be considered.

Separation of plastics from the remaining solid waste can permit their reuse as plastics. These can be further separated by density to yield polypropylene, polyethylenes, styrenics, and vinyls in fairly separate fractions. They might also be used in total blends by crosslinking them together or adding compatibilizers such as blending polymers.

Total solid waste has been treated for use as a building material in several different studies.

Pyrolysis of total organic solid waste can be routed to produce fuels, fertilizers, or commodity organic chemicals, some of which would then return into the plastics raw material stream again.

This short survey illustrates the major ways in which plastics interact with our environment, and some of the positive trends that could be followed in the years to come to keep plastics as an increasingly valuable part of the total ecocycle.

17.4.3   The Space Environment

With the advent of the space age, materials engineers found that their concept of material suitability needed to be sharply revised. Rosato (9) points out that, for example, a material such as an asbestos laminate is considered to be an excellent heat insulator under earth conditions; in the hard vacuum of space, it can become a heat conductor. A hard vacuum can make electric insulators become conductors; rubber becomes hard and brittle; and oils and plasticizers become volatile and vaporize. A material is asked to perform its intended function under the orbiting environment in addition to all those preorbital conditions and testing that designers have to consider.

17.4.3.1   Kinetics in Space
The kinetics of deterioration of materials change drastically under the extremes of the

heat, cold, high vacuum, electromagnetic radiation, dissociated and ionized gases, plasma, meteoric dust, and auroras and coronas found in the space environment. Deterioration that requires months or years on earth is accomplished in days under the conditions of space. Environments encountered by earth-orbiting satellites also include: acceleration, vibration and shocks, weightlessness, gas pressure, and micrometeorite fluxes.

17.4.3.2   Radiation Degradation
Penetrating radiation may come from a variety of sources, of which the most important are cosmic radiation, trapped radiation, auroral radiation, and solar-flare radiation (28). Damage to organic compounds—characterized primarily by breakage of covalent carbon-to-carbon and carbon-to-hydrogen bonds—is a function of the energy absorbed from the radiation source. The primary reactions are: the electron is raised to a high-energy level but remains bound to the parent nucleus; or the electron is removed from its paent nucleus, giving rise to a free electron and a positively charged (ionized) atom or molecule. Secondary processes following these primary ones result in cross-linking, chain scission, chain polymerization, unsaturation, and chain transfer. For example, radiation reduces the size of large crystallites in polyethylene while at the same time increasing the number of smaller ones, resulting in transparency. Most of the damage is produced by secondary atoms and electrons that have been displaced from their equilibrium positions by the primary radiation (28).

The approximate order of stability of commercial polymers to nuclear radiation is shown in Table 17.13. Nineteen polymers and/or elastomers are compared in Table 17.13.

17.4.3.3   Temperature Degradation
The temperature of a space vehicle is achieved through proper balance between absorbed and radiated heat—the spacecraft's thermal control system is designed to keep the spacecraft components at or near room temperature. The system must be stable in three en-

Table 17.13 Approximate Order of Stability of Commercial Polymers (9)

Commercial polymer	Nuclear energy for appreciable damage * (rad)	Comments
Polystyrene	10^9	Crosslinks but distorts under load at 80°C
Silicone (aromatic)		Crosslinks
Polyethylene		Crosslinks
Epoxy	10^8	
Melamine-formaldehyde		
Urea–formaldehyde		
Mylar		
Natural rubber (polyisoprene)		Crosslinks
Silicone elastomers (aliphatic)	10^7	Crosslinks
Polypropylene		Crosslinks
Polycarbonates (Lexan)		
Polyvinyl chlorides		Degrades via scission
Nylons		Crosslinks
Synthetic rubbers		
Kel-F	10^6	Degrades via scission
Polyurethanes		
Polymethacrylates		Degrades via scission
Polyacrylates		Crosslinks
Teflon	10^5	Degrades via scission

Reprinting permission granted by John Wiley & Sons, Inc., New York.

*> 25% change in a significant mechanical property.

vironments: prelaunch, launch or re-entry, and space, and must function unattended in the space environment for varying time periods, some of which may exceed one year.

The four basic thermal control surfaces are: solar reflector, solar absorber, flat reflector, and flat absorber. Internal spacecraft temperatures are predominantly controlled by the radiation characteristics of the external satellite surfaces. The parameter of greatest importance is the ratio of solar absorptance to infrared emittance. Thus the effect of heat on the temperature control in a space vehicle is measured as a balance between absorbed and radiated heat, and an adequate thermal control system becomes of great importance.

In the presence of air, thermal aging of polymers occurs through oxidation. Oxidation is a free-radical mechanism and the chemical reactions occurring during oxidation are accompanied by chain scission and cross-linking. During oxidation hydroperoxides and peroxides form, then decompose to aldehydes, ketones, acids, esters, and so on. Thermal aging with oxygen present leads to

changes in physical properties, for example, tensile, elongation, modulus, and compression set, the measurement of which provides an estimate of aging damage.

17.4.3.4 Vacuum Degradation

Frankel (29) discusses the degrading effects of vacuum on metals, alloys, lubricants, ceramics, coatings, thin films, and polymeric systems. Polymers will not volatilize as a result of vacuum alone, and the vacuum is no real problem in itself when the molecular weight of the polymer is reasonably high and the polymer is free of lightweight components. Obviously molecular weight distribution becomes important here for loss of volatile materials.

Polymers lose weight in vacuum by decomposition, that is, a degradation caused by the breaking down of long-chain polymers into smaller, more volatile fragments. Chain length, extent of branching, and cross-linking have direct bearing on the rate of decomposition, but unless a considerable loss of polymer mass occurs, there is very little effect of vacuum on changes in the engineering

Table 17.14   Approximate Decomposition Temperature of Polymers in Vacuum (10^{-6} Torr) (9)

Polymer	Estimated temperature for 10% weight loss per year (°F)
Polysulfide	100
Polyurethane	150–300
Nylon	80–340
Butyl rubber (isobutylene–isoprene)	250
Polystyrene (not crosslinked)	270–420
Natural rubber	380
Silicone elastomer (methyl)	400
Mylar	400
Polystyrene (crosslinked)	440–490
Polyethylene (low density)	460–540
Polypropylene	470
Polyethylene (high density)	560
Teflon	730
Phenyl methyl silicone resin	>740

Reprinting permission granted by John Wiley & Sons, Inc., New York.

properties. Small amounts of impurities and addition agents such as polymerization catalysts serve to accelerate decomposition. In general, polymers that are suitable for high-temperature service are best for service under high-vacuum environment.

Table 17.14 uses the unit torr, which is equivalent to a pressure of 1 mm of mercury. Under the pressure condition (e.g., vacuum) of 10^{-6} torr, the temperature required to cause a 10% weight loss in one year's exposure for the tested polymers is shown. It will be noted that the plastic materials all show higher "decomposition" temperatures under vacuum than any of the elastomer materials. The data show estimated temperature at 10^{-6} torr to yield 10% weight loss per year on 13 polymers and/or elastomers.

17.4.3.5   Physical Impact Degradation
In a consideration of physical impact phenomena, Neiger (30) states that damages to surfaces in the space environment as a result of particle collisions can occur in two basic forms that lead to erosion, penetration, and puncture. Surface erosion can be caused by the impact of meteoric particles, cosmic rays, solar corpuscular radiation, Van Allen Belt radiation, and atmospheric molecules. Erosion on a molecular or atomic scale is known as "sputtering"; atoms or molecules

are removed by impingement of high-energy particles. Penetration and puncture is a factor of importance in space vehicle design.

17.4.3.6   Friction Degradation
One of the major problem areas for advanced spacecraft systems is that of friction and wear of rubbing or sliding surfaces as encountered in the operation of gears, bearings, cams, rollers, and electrical contacts in which plastics are used. Clauss and Young (31) state that lubrication of spacecraft systems is complicated by many special problems such as the very high vacuum that exists in space, temperature extremes, and nuclear radiation.

17.4.3.7   Sealants in Space Environment
In examining sealants for aerospace applications, Mauri (32) discusses the effect of the space environment on plastics and elastomers. He states that variations in plastics properties arising from processing and cure variables are less than in the case of elastomers.

Plastics as a class of materials are regarded as superior to elastomers in mechanical properties and strength retention under temperature variations and high-radiation environments, but they do not possess the resilience and conformability required of optimum sealing materials. While plastic types are currently available that can operate continually in the temperature range of −423°F (−253°C) to +600°F (315.6°C) and at radiation exposures up to 5×10^9 rads, no single plastic material has the desired perfect combination of properties. For example, according to Mauri, fluorinated polymers, while having the best high- and low-temperature resistance among common polymers and having the highest degree of compatibility with rocket fuels and oxidizers, do not have acceptable radiation resistance. Hence compromises must still be made in the choice of materials for a given design.

17.4.3.8 Multiple Aspects in Space Environment

Rittenhouse, et al. (33) update data compiled by various contributors covering multiple aspects of the space environment. Covered in this *Space Materials Handbook* are: ascent aerodynamic environment; ascent vibration environment; structure of the upper atmosphere; effects of solar radiation, albedo, and earth radiation; penetrating radiation, including general effects of radiation on materials; physical impact phenomena; effects of magnetism, zero gravity, ionosphere, and radio noise as minor environments; and other subjects dealing with the behavior of materials—including plastics—in the general area of the space environment.

17.4.3.9 Additional Reading Suggestions

Rosato and Schwartz (34) in their book, *Environmental Effects of Polymeric Materials*, cover radiation, other properties and characteristics, service life versus environment, and the many test methods involved.

A few of the other authors whose work offers contributions in the matter of materials in the space environment include: Campbell (35) on the service life of insulating materials to be used in nuclear radiation environment; and Buckley and Johnson (36) on the degradation of polymeric lubrication compositions under high vacuum and high (593°C, 1100°F) ambient temperature.

For further reading, proceedings of two SAMPE meetings (37, 38) offer many subjects for study. Among these are: materials

and processes for environmental extremes; electronic applications; deterioration prevention; processing of materials in space; testing techniques involving ultrasonics, radioactive gas penetrants, and other nondestructive tests; ablative protection; composites; nuclear environments; and high-temperature control. Of the more than 100 papers presented, most are concerned with some aspect of plastics technology in space applications.

17.5 BIOMEDICAL RESISTANCE

The interaction of plastics with their biological environment may be considered in two very different categories: resistance of plastics to biodegradation and toxicological effects of plastics upon human life.

17.5.1 Biodegradation

A variety of microorganisms may attack plastic products causing degradation of properties and product failure, most often in warm humid climates. Such attack may affect physical, electrical, or optical properties, particularly in polymers of natural origin, or begin by attack on low molecular weight additives such as plasticizers, cellulosic, and fatty additives. For minimum impact in solid waste disposal, on the other hand, it is possible to design polymer structure and additives to promote such biodegradation and thus complete the ecocycle.

17.5.1.1 Biological Species

Plastic products sometimes fail because of attack by various biological organisms. Of the microorganisms, the most important are the fungi—especially species known as mold or mildew; these are simple plants that exist as spores in air, or grow as colonies in topsoil or on any material that provides nutrition. The other important class is bacteria—much simpler, more primitive structures found in air, water, soil, and as parasites in larger living species. Both of these classes of microorganisms exist in many thousands of different types; such microorganisms attack plastics by liberating enzymes that break down potential foods into assimilable chemical fragments.

Several types of macroorganisms attack plastics by simple physical means such as boring, gnawing, and abrasion. Insects such as termites bore into cellulosic materials for food and sometimes abrade other plastics in their search. Marine borers such as mollusks and crustaceans burrow into a great variety of materials in search for shelter. Rodents do not eat plastics for nutrition but may gnaw through plastic materials in their search for food or escape, or just out of simple curiosity.

17.5.1.2 Optimum Conditions for Biodegradation

Degradation by microorganisms is most often seen in the tropics because humidity and warmth are most conducive to their growth. Fungus requires at least 70% relative humidity, and prospers best at 95–100%. Bacteria require liquid aqueous environment, so their effect is seen only in more specialized applications. Although optimum temperature may be 20–30°C on the average, some species thrive in temperatures down to subzero and others up to 65°C. Atmospheric oxygen is essential to all fungi and to most bacteria, but some types of bacteria (anaerobic) grow only in the absence of air. Light is not particularly important, and UV light can actually destroy many microorganisms.

Nutrition is the critical consideration. As mentioned, microorganisms send out enzymes to break down potential food into assimilable fragments; each species has certain enzymes that permit it to grow on certain types of food sources. They thrive best on carbohydrates and proteins, requiring specific amounts of nitrogen and trace elements for healthy growth. They also use natural fats and oils readily, and can even digest small straight-chain aliphatic hydrocarbons.

17.5.1.3 Effects of Biodegradation and Properties

When microorganisms attack plastics the effects most often noticed are decrease in tensile strength and elongation, embrittlement, loss of electrical resistance, exudation, and discoloration. One of the most common problems—pink staining of vinyls—is caused by attack on either the vinyl or on an adjacent material such as fabric, producing pink by-products that dissolve in the vinyl and thus stain it, often without affecting physical properties.

17.5.1.4 Biodegradation and Structure

Natural polymers are most easily attacked—especially carbohydrates and proteins—also those polymers based on fats and oils. These include cellulose and to a lesser extent its partially substituted derivatives, casein plastics, alkyd paints, and natural rubber.

Among synthetic polymers, aliphatic polyesters are most readily attacked—particularly polycaprolactone and certain low molecular weight species; this sensitivity even extends to polyester polyurethanes. Some other synthetics that have suffered from biodegradation include low molecular weight phenolic resins, melamine resin, polyvinyl acetate, and other latex paint binders.

Polyethylene and other polyolefins showed some attack on the lowest molecular weight species. Copolymerization with increasing amounts of hydrolyzable comonomers such as vinyl acetate and acrylics—and especially with acrylamide—promoted biodegradability. Hydrophilic groups that help ionize enzymes also promote attack. Oxidizable groups also help. On the other hand, high molecular weight, branching, and crystallinity inhibit biological attack.

Attempts to modify polystyrene for greater biodegradability have been reported, but most of them met with no success. One commercial product, however, does claim to have achieved biodegradability in polystyrene foam.

17.5.1.5 Biodegradation and Additives

In most plastics the high molecular weight polymer itself does not serve as nutrient for microorganisms, but the low molecular weight additives used in compounding are much more susceptible. This is particularly true of the plasticizers in polyvinyl chloride. The resin itself is unaffected, but linear ester plasticizers are good nutrients for microorganisms. Long aliphatic monobasic acid esters are most reactive; dibasic esters are less so. Epoxidized oils, polyesters, and polyglycol esters are also quite sensitive. Sebacic esters with butyl or higher alcohols, adipic with decyl or higher, and phthalic with dodecyl or higher are also degradable; while lower phthalates, branched and cyclic structures, and most phosphates are quite resistant.

Cellulosic fillers in phenolic and melamine resins account for most of the degradation observed in tropical uses. Starch filler has actually been recommended as a biodegradable filler for disposable plastics.

Other additives based on fatty acids, their esters, amides, and soaps—stabilizers and lubricants—likewise show biodegradibility when used in stable polymers. Among antioxidants, dilauryl and stearyl thiodiproprionates are readily degradable, and some microorgnisms can even tolerate phenolic hydroxyls but have trouble with the tertiary alkyl groups commonly used to block the phenol. There is a difference of opinion on UV absorbers—probably based on the specific microorganisms used. Flame retardants mostly inhibit biodegradation because of their halogen, phosphate, and antimony contents.

17.5.1.6 Prodegradants
Recent concern with the accumulation of plastics in solid waste has led to some studies to accelerate the biodegradation of plastics. As outlined, it is possible to choose polymers and additives for maximum sensitivity to attack by microorganisms. In addition, it is recommended that a broad spectrum of microorganisms be used to maximize the possibility that one of them will produce an enzyme capable of breaking down the plastic and to permit symbiotic action between species. It is also recommended that, unless the polymer contains nitrogen, a nitrogen source such as NH_4NO_3 be added to provide this essential element for growth.

17.5.2 Toxicology

Although high polymers are generally nontoxic to human and animal life, many of their compounding additives, impurities, and degradation products produce a variety of toxicity problems that are a growing concern of polymer scientists and the medical profession. These are best summarized in terms of materials involved, route of entry into the human body, and analytical techniques for predicting and preventing such problems.

17.5.2.1 Materials Involved
Materials involved may be summarized in order of the quantities used in the plastic industry, noting the type and severity of the problems they cause.

Fillers and Reinforcements. Although fillers have for the most part been in use for thousands of years with little or no health problems, glass fibers have caused dermatitis problems in manufacturing workers, requiring protective covering of the skin, for many years. More serious is the recent concern over the long-term carcinogenic effect of inhaling very fine short fibers—partricularly asbestos—primarily in manufacturing but perhaps also in long-term use such as buffing of flooring and wear of brake linings. These fibers are the subject of intensive current study.

Plasticizers. Most of the plasticizers used in vinyls and cellulosics are extractable by organic solvents and even aqueous detergents. Thirty years ago some plasticizers, and some manufacturing impurities in others, caused serious poisoning problems in food handling and packaging; but these have long since been eliminated to provide approved lists of plasticizers for food, drug, and toy applications. Of more recent concern is the observation that the plasticizer in blood transfusion equipment—although insoluble in water—is extracted by lipids in blood, accumulating in the body and causing long-term mutational possibilities; this considerably restricts the list of acceptable plasticizers, and lengthens the time required for their approval.

Colorants. Most colorants were originally inorganic pigments of long-standing preindustrial acceptability. Of these, titanium dioxide white and iron oxide colors are generally harmless, but many of our brightest colors based on cadmium, chromium, and molybdenum are coming under increasingly critical scrutiny and are gradually being replaced by organic pigments with less potential for toxicity, both in food and toy applications.

Stabilizers. Of the halogen stabilizers added to polyvinyl chloride, calcium and zinc soaps, epoxidized fatty esters, and alkyl aryl, phosphites are generally considered harmless but relatively inefficient as stabilizers; more powerful systems based on lead compounds, barium and cadmium soaps, and organo-tins are coming under increasing scrutiny, and their use in the future may be more strictly controlled. Antioxidants used in most plastics are primarily hindered phenols plus thioorganic synergists; some of these are harmless, whereas other phenols, and particularly aryl amines, must be used with caution because of their toxicity. Use of biological preservatives is occasionally practiced in plastics compounding, but the balance between toxicity to microorganisms versus nontoxicity to humans is a difficult one to achieve and especially to optimize; the loss of organo-mercurials to the coatings industry is a particularly trying problem.

Processing Aids. Of the variety of known and proprietary chemicals added to plastics to improve their processing, the most common health problems come from heavy metal soaps used as lubricants. Since there are many harmless lubricants, this has not generally been a major area of concern.

Flame Retardants. Many flame retardants are potential health hazards, particularly the halogenated organics and antimony synergists in most common use. We are only beginning to sort these out and eliminate those that may cause long-term toxicity problems.

Catalysts. Polymerization catalysts and curing agents include many that are potential health problems. Residual peroxides and their decomposition products, heavy metal compounds, organic amines, and silicone catalysts are typical of additives that have caused toxicity problems. Thus these must be chosen and controlled with careful attention to the end use of the polymeric products.

Residual Monomer. No polymerization runs to 100% completion, and the residual monomer may be held quite tenaciously in the polymer. In the last few years several of these residual monomers have drawn the sharp attention of the medical profession, then required intensive effort by process engineers to eliminate their hazards—both during manufacture and during use of the end products. Although the long-term carcinogenicity of vinyl chloride monomer is the most newsworthy, some other problems include acrylonitrile in soft-drink packaging, vinyl carbazole in high-temperature electrical insulation, and methyl chloracrylate in flame-retardant acrylic sheeting. In the older plastics industries, long-term exposure to formaldehyde, amines, and solvents has caused skin sensitization and dermatitis as well as more general bodily allergenic reaction problems.

Degradation Products. Although high polymers themselves may be nontoxic, their degradation to low molecular weight species often does produce serious toxicity problems. Biodegradation of body implants can be the source of such problems and is becoming of greater concern as more such plastic implants are used in modern surgery. The fire hazard of plastics is largely due to smoky flames that obscure vision and thus prevent escape from the fire and is also often due to the production of toxic gases such as carbon monoxide and hydrogen chloride. Although some combustion products are not peculiar to synthetic plastics, and result from wood and paper as well, the critical attention they draw from the general public makes them a serious concern for plastics scientists and engineers.

17.5.2.2 Route of Entry into the Human Body

These potentially toxic ingredients of plastics can enter the human body through a variety of routes. Three major types of exposure are manufacturing hazards, handling and packaging of foods and drugs, and medical implants. The actual routes of entry into the body must be considered in more detail:

Breathing. Inhaling of toxic ingredients from plastics is a problem primarily in manufacturing and in fires. There have, however, been suggestions of such problems in the use of plastic products as well. These are generally difficult to prove, and require further study.

Swallowing. Most of the toxic effects of plastic are due to oral ingestion of foods and drugs that have extracted toxic impurities from the plastic materials used for handling and packaging them. This is a primary concern of the Food and Drug Administration in this country, and of similar government agencies in other industrialized countries of the world.

Injection and Other Pharmaceutical Dosage Forms. In the pharmaceutical industry the toxicity of plastic ingredients is most critical in parenteral infusions and injections directly into the blood stream or the body cavity because these can traverse the entire body in a minute or two, producing immediate and sometimes traumatic effects. Somewhat less critical, but still serious, are ophthalmic preparations administered directly to the eye. A step less critical are topical local applications to tender mucous membranes such as the ear, nose, throat, and rectum. Considerably less critical are solid oral dosages such as capsules and tablets. And least critical are applications directly to the skin. For each of these levels the pharmaceutical industry has a separate degree of testing and approval required for safe use of plastics.

Skin. Contact of chemicals on the skin can result in sufficient absorption to produce toxic effects. Many of these are local dermatitis, but some may have more widespread effects on the body functions such as allergenic reactions. In many cases initial contact is harmless, but repeated exposure produces increasing sensitivity until the material cannot be tolerated any more. In the plastics industry this has been noted most often with halogenated organics, amines, and formaldehyde. In preplastics times, even natural materials such as wool caused such effects in some individuals.

Surgical Implants. There is growing interest in the use of plastics to rebuild or even replace worn parts in the human body such as blood vessels, joints, bones, valves, and even parts of vital organs. In such applications, extractible monomeric additives, impurities, and degradation products may cause toxic effects, either immediately or after a long time, and these must be checked very carefully. Another recurrent problem is "compatibility," here meaning that the body may have an adverse reaction to the simple presence of a foreign substance; in some cases this can be resolved by changing from solid to porous or fibrous structure to prevent carcinogenesis, or by chemical treatment of the polymer surface to eliminate surface phenomena such as blood clotting. Such applications are growing rapidly in importance and are so critical that they justify great effort in devising solutions to any problems that may arise.

17.5.2.3 Analytical Techniques for Research, Development, and Quality Control

Considering the variety of complex problems that can be involved, the development of new plastic materials for products where toxicity may be involved and their quality control in routine manufacture pose many interesting challenges for the plastics engineer. These can be considered stepwise as follows:

Analysis for Known Toxic Ingredients. Where the plastic may contain ingredients of known toxicity, the polymer chemist can generally develop analytical techniques to identify and control these directly.

Animal Tests. New materials of unknown toxicity are almost always tested first on animals. Initial low-cost test on small animals, such as mice and rats, are later scaled up to fewer more expensive tests on larger animals that are closer to man in their responses to such problems. Initial observations of immediate effects are later followed up by time-consuming study of long-term effects, first on short-lived animals such as mice and later on longer-lived species to a more limited extent.

Tests on Humans. If all of the animals tests prove harmless, cautious testing on human subjects can begin, later followed by clinical tests with continued medical observation to pick up any problems as soon as they appear. Most recent studies on long-term toxic effects of plastics ingredients have come from study of the history of long-time employees in the manufacturing industries. This remains the most difficult and questionable part of the entire process.

Identification of Toxic Ingredients. Whenever toxicity problems appear, the next challenge is to identify the actual ingredient causing the problem. This can generally be done, but a more difficult question is the concentration and exposure that cause the problem versus safe lower levels that can be permitted without harm to the individual. At present these levels are often based on accelerated test on mice—using a large safety factor—to insure that no hazardous mistakes can be made in the process. In some cases, safe levels have been extremely out of line with practical data but must be accepted until firmer, more precise evidence is available.

Pharmaceutical Criteria. As pointed out, the packaging and handling of pharmaceutical preparations is more critical and the criteria are more detailed and severe. Again, the initial question is the safety of a new plastic material for use in packaging a specific pharmaceutical; this requires very extensive evaluation. A second stage is the change to a new dosage form and/or package design, which may require retesting or many of the same criteria because even such a seemingly minor change can be significant. A third stage is the expansion of existing products into new plants at other locations, and particularly in other countries, where the conditions and techniques may differ sufficiently to reintroduce some of the problems already solved earlier.

Quality Control. All of these foregoing considerations apply primarily to research and development of new plastic products. Once development is completed and human safety is assured, the remaining problem is one of routine quality control to insure that successive production batches meet the requirements for safety—and, in the case of pharmaceuticals, for production of the packaged material as well. Such quality control tests are generally the simplest laboratory tests that can be drawn from the earlier research and development studies and provide the most critical assurance to detect and prevent harmful variations from batch to batch.

Time and Cost. A major problem in plastics research and development is the time and cost involved in the subject studies. The prospective manufacturer must decide on the size of the market and its profitability, and the probability of success in the research project, before undertaking the time and expense of assuring complete safety in the use of the product wherever toxicity may be involved. In many cases the manufacturer of a new plastic has simply declined to go to this trouble. On the other hand, farsighted, enterprising companies that develop new products for such fields generally earn excellent rewards for their efforts and the respect of society as a whole for their contribution to human well-being.

17.6 SUMMARY

In designing a product for practical performance, it is important to start by defining the environmental conditions under which it will operate and the useful lifetime expected of it.

More specifically, it is of critical importance to identify the precise modes of stress and type of test conditions and data that come closest to predicting performance in the actual operating environment. At this point it is then possible to compare different plastic and nonplastic materials as candidates for use in a specific product.

With respect to plastics, in particular, one can choose from among 80 families of commercial polymers. Most of these are available in a range of molecular weights, copolymers, and compounds with various additives—all of these variables can have major effects on environmental resistance. Careful consideration of these variables, followed by experimental verification of reasoning and choice, is the surest route to selection of the best material for the application. Beyond this point, since no material is ideal or perfect, the role of the product designer is largely concerned with redesign to use the optimum material to best advantage by moderating the effects of the environment and the ways in which it impacts on the base material.

Since research and development, product development, and design are primarily geared toward future manufacturing and marketing, it is appropriate to consider how current trends in plastics science and engineering point to environmental resistance in the future. Over the years, one has seen continual improvement in the range of the polymers and plastic compounds available to withstand a wide variety of environmental conditions.

It is possible to identify the major current trends that appear to offer the most promise for continued development. Among these are the increasing control of crystalline morphology and orientation, the increasing variety of engineering thermoplastics based on cyclic main-chain structures, the increasing use of polyblending and reinforcing fillers to enhance base polymer properties, the increasing sophistication in the use of stabilizers, and, it is hoped, a new wave of progress in thermosetting plastics to combine their superior environmental resistance with the superior convenience and economics of the commodity thermoplastics.

These are the trends that the alert materials engineer should follow most closely to anticipate continuing improvement in environmental resistance of plastics in the years to come.

17.7. REFERENCES

1. L. E. Nielsen, *Mechanical Properties of Polymers and Composites*, Marcel Dekker, New York, 1974, pp. 7, 69, 71.

2. F. W. Billmeyer, Jr., *Textbook of Polymer Science*, 2nd ed., John Wiley, New York, 1971, pp. 126–130.

3. M. Morton, *Rubber Technology*, 2nd ed., Van Nostrand Reinhold, New York, 1973, p. 137.

4. E. Baer, Ed., *Engineering Design for Plastics*, Van Nostrand Reinhold Company, New York, 1964, p. 284.

5. J. L. O'Toole, *Mod. Plast. Encycl.*, **46**, (10A), 939 (October 1969).

6. *Mater. Eng.*, **78**, (5), 29 (October 1973).

7. J. H. Faupel, *Engineering Design*, John Wiley, New York, 1964, p. 642.

8. R. L. Thorkildsen, "Mechanical Behavior," in E. Baer, Ed., *Engineering Design for Plastics*, Van Nostrand Reinhold, New York, 1964.

9. D. V. Rosato, "Other Properties and Characteristics," in D. V. Rosato and R. T. Schwartz, Eds., *Environmental Effects on Polymeric Materials*, Vol. 1, Interscience, 1968.

10. R. D. Deanin, *Polymer Structure, Properties, and Applications*, Cahners (C·B·I) Publishing Company, Boston, Massachusetts, 1972.

11. Y. S. Huang and J. L. Koenig, *J. Appl. Polym. Sci.*, **15**, 1237 (1971).

12. W. A. Glaeser, J. W. Kissel, and D. K. Snedicker, *Am. Chem. Soc. Org. Coatings Plast. Prepr.*, **34**, (1) 388 (April 1974).

13. J. M. Arvidson, R. L. Durcholz, and R. P. Reed, *Adv. Cryog. Eng.*, **18**, 174 (1973).

14. K. N. Mathes, "Electrical Properties," in *Engineering Design for Plastics*, Van Nostrand Reinhold, New York, 1964, Chapter 7.

15. F. H. Winslow, L. D. Loan, and W. Matreyek, *Am. Chem. Soc. Org. Coatings Plast. Prepr.*, **31**, (1), 124 (1971).

16. D. V. Rosato and W. K. Fallon, *Markets for Plastics*, Van Nostrand Reinhold, New York, 1969, p. 85.

17. G. N. Ong, "Drying of Plastic Resins," thesis, Lowell Technological Institute, 1974.

18. M. J. Howard, *International Plastics Selector*, Cordura Publications, LaJolla, California, 1977.

19. J. M. Newcome, "Effects of Processing Conditions on Mechanical Properties of Injection Molded Polycarbonate," Mobay Chemical Company, Plastics and Coatings.

20. H. Burrell, "Solubility Parameter Values," in J. Brandrup and E. H. Immergut, Eds., Interscience, New York, 1966, p. IV–341.

21. K. Meyersen, "Solvents and Nonsolvents for Polymers," in J. Brandrup and E. H. Immergut, Eds., *Polymer Handbook*, Interscience, New York, 1966, p. IV–185.

22. J. B. Howard, "Stress-Cracking," in E. Baer, Ed., *Engineering Design for Plastics*, Van Nostrand Reinhold, New York, 1964, Chapter 11.

23. J. F. Fellers and B. F. Kee, *J. Appl. Polym. Sci.*, **18**, (8), 2355 (1974).

24. T. J. Stolki, *SPE J.*, **23**, (10), 48 (October 1967).

25. J. Hell, *Int. Polym. Sci. Tech.*, **2**, T/80 (Feburary 1975).

26. P. I. Vincent and S. Raha, *Polymers*, **13**, (6), 283 (June 1972).

27. G. Menges and H. Schmidt, *Plast. Polym.*, **38**, (133), 13 (February 1970).

28. E. E. Gaines and W. L. Imhoff, "Penetrating Radiations," in *Space Materials Handbook*, 3rd ed. Rittenhouse and Singletary, Lockheed Missiles and Space Company, 1968.

29. H. E. Frankel, Symposium, "Environments and Their Role in Spacecraft Technology," Sixty ERO Summer School, Noordwijk, The Netherlands. Volume 1, 1968.

30. J. B. Neiger, "Physical Impact Phenomena," in *Space Materials Handbook*, 2nd ed., Lockheed Missiles and Space Company, 1965, Chapter 8, Part 1.

31. F. J. Clauss and W. C. Young, "Materials in Lubricating Systems," in *Space Materials Handbook*, 2nd ed., Lockheed Missiles and Space Company, 1965, Chapter 12, Part 2.

32. R. E. Mauri, "Organic Materials," in *Space Materials Handbook*, 3rd ed., Lockheed Missiles and Space Company, 1968, Chapter 12, Part 2.

33. J. B. Rittenhouse and J. B. Singletary, *Space Materials Handbook*, 3rd ed., Lockheed Missiles and Space Company, 1968.

34. D. V. Rosato and R. T. Schwartz, Eds., *Environmental Effects on Polymeric Materials*, Vol. 1, *Environments*, Interscience, New York, 1968.

35. F. J. Campbell, "Combined Environments Vs. Consecutive Exposures for Insulation Life Studies," IEEE Transactions in Nuclear Science 123–128, November 1964.

36. D. H. Buckley and R. L. Johnson, "Degradation of Polymeric Compositions in Vacuum to 10^{-9} mm Hg in Evaporation and Sliding Friction Experiments," *SPE Trans.*, **4**, (4), 1–9 (October 1969).

37. Society of Aerospace Materials and Process Engineers (SAMPE), "Materials and Processes in the '70's," Vol. 15, Proceedings of the Fifteenth National SAMPE Symposium, April 1969.

38. Society of Aerospace Materials and Process Engineers (SAMPE), "Materials '71," Vol. 16, Proceedings of the Sixteenth National SAMPE Symposium, May 1971.

General References

Mechanical Stress

American Society for Testing and Materials, "Annual Standards: Plastics."

E. R. Booser, "Bearings and Gears," in E. Baer, Ed., *Engineering Design for Plastics*, Van Nostrand Reinhold, New York, 1964, Chapter 18.

R. D. Deanin and A. M. Crugnola, "Toughness and Brittleness of Plastics," *Am. Chem. Soc., Adv. Chem.*, **154** (1976).

R. D. Deanin, G. F. Sullivan, D. J. Doyle, R. A. Granoff, and A. J. Karszes, *Am. Chem. Soc., Org.; Coatings Plast. Prepr.*, **29**, (2), 533 (1969).

L. H. Lee, *Advances in Polymer Friction and Wear*, Plenum Press, New York, 1974.

Creep

J. H. Faupel, *Engineering Design*, John Wiley, New York, 1964, Chapter 12.

R. R. Kambour and R. E. Robertson in A. D. Jenkins, Ed., *Polymer Science*, Elsevier, New York, 1972, pp. 758–775.

Cyclic Mechanical Stress

C. B. Bucknall, K. V. Gotham, and B. I. Vincent, in A. D. Jenkins, Ed., *Polymer Science*, Elsevier, New York, 1972, Chapter 10.

N. Grassie, in A. D. Jenkins, Ed., *Polymer Science*, Elsevier, New York, 1972, Chapter 22.

J. K. Lancaster, in A. D. Jenkins, Ed., *Polymer Science*, Elsevier, New York, 1972, Chapter 14.

Cryogenic Temperatures

C. L. Tien and G. R. Cummington, "Cryogenic Insulation Heat Transfer," *Adv. Heat Transfer*, **9**, 349 (1973).

Thermal-Chemical Degradation

A. H. Frazer, *High Temperature Resistant Polymers*, John Wiley, New York, 1968.

W. E. Gibbs and T. E. Helminiak, in A. D. Jenkins, Ed., *Polymer Science*, Elsevier, New York, 1972, Chapter 25.

W. Lincoln Hawkins, *Polymer Stabilization*, John Wiley, New York, 1972.

L. I. Nass, *Encyclopedia of PVC*, Marcel Dekker, New York, 1976, Chapters 8 and 9.

J. W. Lyons, *The Chemistry and Uses of Fire Retardants*, John Wiley, New York, 1970.

Electrical Degradation

Modern Plastics Encyclopedia, McGraw-Hill, annually.

A. E. Molzon, "Electrical Properties of Plastic Materials," *Plastec*, **23**, Picatinny Arsenal AD-624922 (July 1965).

J. J. O'Dwyer, *The Theory of Electrical Conduction and Breakdown in Solid Dielectrics*, Oxford University Press, 1973.

Chemical Resistance

N. E. Hamner, *Corrosion Data Survey, Nonmetals Section*, 5th ed., National Association of Corrosion Engineers, 1975.

I. Mellan, *Corrosion Resistant Materials Handbook*, 2nd ed., Noyes Data Corporation, 1971.

J. H. Perry, *Chemical Engineers Handbook*, 4th ed., McGraw-Hill, New York, 1963, Section 23.

Environmental Stress Cracking

R. Fraser, *Plast. Poly.*, **43**, 165 (June 1975).

Weathering

Appl. Polym. Symp., **4**, (1967).

B. Baum and R. D. Deanin, *Polym. Plast. Technol. Eng.*, **2**, (1) (1973).

B. Baum, *SPE EPS DIVTEC*, 180 (October 1973).

R. B. Fox, *Encycl. Polym. Sci. Tech.*, **11**, 760 (1969).

W. L. Hawkins, *Polymer Stabilization*, John Wiley, New York, 1972.

S. H. Pinner, *Weathering and Degradation of Plastics*, Columbine Press, Manchester, England, 1966.

G. R. Rugger, in D. V. Rosato and R. T. Schwartz, Ed., *Environmental Effects on Polymeric Materials*, Vol. 1, Interscience, New York, 1968, Chapter 4.

J. B. Titus, "Environmentally Degradable Plastics," Plastec Note N24, Picatinny Arsenal, February 1973.

Biodegradation

W. Coscarelli, in W. L. Hawkins, Ed., *Polymer Stabilization*, John Wiley, New York, 1972, Chapter 9.

G. J. L. Griffin, in R. D. Deanin and N. R. Schott, Eds., *Fillers and Reinforcements for Plastics*, Chapter 16; *Am. Chem. Soc.*, *Adv. Chem.*, **134** (1974).

J. E. Potts, R. A. Clendinning, W. B. Ackart, and W. D. Niegisch, *Am. Chem. Soc. Polym. Prepr.*, **13** (2), 629 (August 1972).

N. M. Rei, Ventron Corporation, private communication.

E. L. Weinberg, *Mod. Plast. Encycl.*, **53** (10A) 217 (1976).

H. E. Worne, Enzymes, Inc., private communications.

Toxicology

Jack Cooper, "Plastic Containers for Pharmaceuticals: Testing and Control," World Health Organization, 1974.

R. D. Deanin, *Environ. Health Perspect.*, **11**, 35 (1975).

R. D. Deanin and J. Cooper, "Plastic Containers for Pharmaceuticals," *SPE NATEC*, (November 1977).

Org. Coatings Plast. Prepr., *Am. Chem. Soc.*, **30** (1) Symposium on Medical Applications of Plastics (May 1970).

World Health Organization, WHO/PHARM/76.487 (1976).

Effects of Plastics on Environment

B. Baum and C. H. Parker, *Solid Waste Disposal*, Vol. 2, Ann Arbor Science Publishers, Ann Arbor, Michigan 1973.

C. H. Parker, *SPE J.*, **23**, 12, 26 (December 1967).

Council on Environmental Quality, annual reports.

J. R. Lawrence, "Status Report on Plastics Recycling," *ASTM Special Publication 533*, 50 (1973).

H. F. Lund, *Industrial Pollution Control Handbook*, McGraw-Hill, New York, 1971.

INDEX

Abrasion, 11, 605, 607, 691, 779, 786, 794, 829, 830, 834, 836, 838, 910
Abrasive, 414, 835, 888, 921
Acetal (polyacetal), 8, 9, 14, 21, 281, 298, 331, 614, 616, 618, 620, 625, 636, 819, 836, 888, 892, 897, 900, 908
Acrylic (polyacrylate), 8, 9, 14, 20, 21, 304, 475, 582, 606, 614, 616, 618, 620, 621, 656, 892, 896, 900, 907–909
Acrylonitrile (ABS, AN, PAN), 1, 4, 6–9, 19, 23, 35, 41, 84, 271, 274, 454, 475, 476, 485, 508, 513, 560, 561, 599, 600, 608, 616, 618, 621, 624, 625, 627, 630, 638, 643, 655, 657, 666, 695, 697, 819, 889, 892, 900, 908, 909, 918
Activity coefficient, 39, 50, 51
Addition polymer, 2, 7, 82, 85, 251, 475, 481, 508
Additive, 2, 7, 11, 19, 298, 613, 614, 616, 618, 626, 645, 646, 649, 660, 922, 923, 930, 931
Adhesion (adhesive), 8, 14, 18, 21, 23, 30, 362, 395, 401, 427, 434, 439–441, 445, 447, 571, 605, 610, 618, 839
Aging, 341, 362
Alfrey-Price, 135, 328
Alkyd, 8, 19, 614, 616, 618, 620, 827, 900
Alloy, 2, 3, 7, 11, 12, 23, 24, 574, 653, 698
Allylic, 8, 908
Aluminum, 3, 9–11, 599, 636, 687, 697, 831, 853
American Society for Testing of Materials (ASTM), 300, 447, 629, 630, 642, 644, 662, 785, 786, 789, 794, 797–799, 802, 807–809, 811, 814–816, 818, 820, 821, 824, 826–831, 833–837, 841–843, 845, 847, 849–851, 855, 857, 859–865, 867–869, 873, 874, 876

Amino, 2, 7, 9, 14, 20, 23
Amorphous, 1–3, 7, 11, 306, 308, 320, 324, 326, 329, 363, 522, 575, 801, 894, 895, 897, 914
Anisotrophic, 327–329
Annealing, 250, 258
Antiblocking agent, 613, 618, 626, 636, 646, 648
Antioxidant, 613, 618, 626, 646, 648
Antistatic agent, 613, 614, 621, 911
Arc resistance, 779, 838, 850, 905
Arrhenius, 316, 351, 353, 383, 391, 493, 538, 539, 844, 872
Asbestos, 23, 620, 638, 643, 644, 803, 905
Atactic, 278, 304

Bacteriostat, 618, 626, 660
Banbury mixer, 8, 600
Bar molding compound (BMC), 630, 643, 669
Barrel, 571, 573, 574, 580, 698, 734, 756, 757
Batch polymerization, 475, 495, 501
Bending force, 795, 796, 810, 877
Biaxial orientation, 328, 595
Biaxial stress, 779, 786, 807, 811–813, 815
Biodegradation, 921, 924, 928, 930
Biomedical resistance, 884, 928
Birefrigerence, 329, 330
Blend (polyblend), 2, 7, 11, 12, 21, 23, 251, 277, 278, 298, 300, 311, 327, 331, 341, 360–362, 551
Blowing agent, 494, 602, 603, 613, 615, 621, 622
Blow molding, 13, 15, 20, 547, 571, 594, 595, 609, 670, 699, 702, 746, 765, 793
Boltzmann, 302, 311, 312, 345, 371
Bonding (covalent, hydrogen, intermolecular), 3, 4, 11, 15, 20, 23, 304, 395, 429, 430, 443, 448, 453, 551, 571, 605, 606, 692, 895, 907, 912, 913, 918, 921

Bosses, 669, 679, 681
Branching (long and short), 3, 63, 83, 122–125, 127, 159–161, 175, 186, 243, 407, 519, 556, 557, 927
Brass, 3, 5, 687
Breakdown, *see* Dielectric properties
Brittle, 1, 11, 17, 22, 23, 300, 301, 321, 322, 628, 688, 785, 787, 806, 815, 816, 820, 888, 890, 921
Brownian motion, 345, 371
Bulk polymerization, 13, 30, 479–481, 494, 513
Burning, 638, 667, 779, 787, 860

Cable, 699, 702, 750, 773, 774
Calendering, 7, 13, 571, 599, 600, 609, 670
Capacitance, *see* Electrical properties
Carbon, 11, 23, 411, 420, 558, 559, 620, 638, 644, 688
Casein, 8, 14
Casting, 7, 15, 571, 600, 601, 670, 672, 793
Catalyst, 13, 323, 475, 478, 483, 484, 608, 613, 614, 621, 901, 924, 928, 932
Cavity, 560, 584, 586–590, 593, 595, 597, 609, 663, 670, 682, 683, 687, 696, 766
Cellulosic, 7, 8, 14, 19, 606, 614, 616, 618, 620, 621, 636, 656, 833, 896, 900, 908, 909
Ceramic, 3, 11, 12, 638, 884, 897
Chain, 1, 3, 7, 11–13, 82, 241, 247, 249, 260, 283, 284, 303, 305, 314, 328, 330, 350, 395, 416, 453, 475, 476, 485, 495, 508, 511, 519, 551, 626, 728, 802, 894, 895, 899
Charge transfer complex, 341, 371, 375–377, 379, 381–385, 387
Charge transport, 341, 379
Charpy test, 327, 818
Chemical resistance, 15, 22, 698, 883, 884, 906, 911, 916, 922

Coating, 7, 8, 16, 18, 21, 23, 30, 299, 341, 395, 441, 453, 563, 572, 578, 605, 608, 609, 699, 750, 817, 892, 908, 914, 919, 923, 924

Colligative, 80, 83, 84

Color, 12, 14, 20, 21, 603, 613, 622, 623, 691, 780, 788, 868, 899, 913, 931

Compatibility, 274, 519

Composite, 7, 8, 18, 23, 24, 636, 637–639, 642, 644

Compounding, 7, 613, 699, 776

Compression molding, 8, 13, 571, 592–594, 609, 663, 669, 670, 792

Compressive strength, 13, 17, 642, 663, 779, 786, 795, 796, 803, 808, 810, 884, 885, 891, 894

Computer, 14, 654, 699–702, 704, 712, 713, 715–718, 720, 724, 726, 728–730, 733, 734, 748, 775, 777, 778

Concrete, 11, 912, 920

Condensation polymer (condensation polymerization), 2, 7, 17, 81, 83, 471, 473, 476–478, 494, 507

Configuration and conformation, 3, 38, 136, 140, 141, 147, 165, 257, 302, 889

Constant, see Dielectric properties

Contact angle, 402, 404, 426, 430

Copolymer (copolymerization), 1, 7, 134, 135, 159, 180, 230, 241, 261–264, 266, 269–271, 274, 307, 341, 354, 360, 361, 471, 476, 477, 496, 508, 509, 513, 524, 551, 889, 918, 919, 926

Cores, 653, 665, 670, 674, 678, 681–683, 685–687

Corrosion, 12, 395, 440, 918, 921

Corona, 787, 848, 905

Coupling agent, 23, 613, 620, 635–639

Covalent bonding, see Bonding

Crack, 12, 15, 20, 21, 23, 322, 329, 691, 792, 808, 814, 817, 826, 828, 884, 903, 913, 915, 916, 918–920

Crazing, 272, 322–324, 691, 826, 913

Creep, 13, 20, 23, 311, 636, 653, 659, 660, 691, 779, 786, 797, 802, 820–824, 826, 883, 884, 887–889, 896, 918, 919

Crosslink, 1, 3, 12, 13, 18, 83, 301, 305, 308, 326, 362, 572, 591, 603, 605, 609, 616, 626, 627, 650, 699, 702, 801, 888, 895,

897, 899, 912, 913, 920, 923, 927

Cryogenic, 883, 891, 893, 894, 896

Crystallinity (crystallization), 1–3, 7, 11, 57, 163, 207, 208, 210–214, 216, 241–243, 245, 247–250, 252, 254–256, 259–261, 264, 269, 281, 284, 287, 298–300, 304, 306, 311, 318, 319, 323, 326–329, 363, 395, 399, 400, 424, 453, 582, 590, 636, 801, 802, 887–889, 893–898, 908, 912, 914, 916, 921, 935

Cure, 13, 18, 224, 362, 591, 593, 594, 603, 610, 699, 702, 802, 803

Cycle time, 587, 589, 594, 596, 601, 621, 647, 655, 663, 668, 672, 673, 678, 688, 689

Damping, 798, 800–803, 877

Decorating (electroplating), 571, 607, 608

Deflection temperature underload (DTUL), 663, 797, 802

Deformation, 305, 321, 519–521, 525, 549, 796, 821, 885–888, 890, 894, 896, 918

Density (specific gravity), 3, 14, 16, 23, 35, 314, 322, 364, 510, 549, 593, 663, 666, 682, 702, 729, 730, 732, 749, 773, 780, 828, 855, 876, 877, 914, 916

Depolymerization, 483, 494, 807, 899

Diallyl phthalate (DAP), 19, 635, 900

Die, 571, 575–579, 581, 582, 594, 595, 599, 699, 701, 702, 704, 713, 717, 747, 750, 754, 760–762, 765–768, 778

Dielectric properties (breakdown, constant, loss, modulus, resistance, strength), 15, 16, 19, 219, 274, 276, 309, 341–343, 345, 346, 351, 356, 363, 636, 638, 642, 643, 658, 659, 663, 665, 698, 779, 787, 795, 838–840, 844–848, 878, 883, 902, 904, 905

Differential thermal analysis (DMA, DSC, DTA, TMA), 59, 356, 728, 786, 797–799, 803, 856, 857

Diffusion (molecular, thermal), 3, 14, 98, 733, 788

Dimensional stability, 16, 17, 20, 658, 663, 674, 675, 678, 685

Dioctyl phthalate (DOP), 632–634

Dipole, 342, 345, 363

Dissipation, see Electrical properties

Draft, 653, 669, 670, 680, 681

Drawing, 2, 316–320, 749

Ductility, 2, 11, 310, 319, 321, 785, 807, 815, 816, 820, 889, 894

Ejection, 584, 587, 678, 685, 692

Elastomer (elasticity), 1, 2, 5, 7, 8, 11, 264, 271, 272, 297–301, 303, 308, 311, 313, 314, 316, 324, 325, 330, 331, 519, 525, 527, 530–534, 537, 539–543, 549, 550, 557, 563, 565, 605, 643, 797, 798, 812, 878, 883, 887, 888, 893, 894, 896, 916, 918, 928

Electrical conductivity (resistivity), 3, 11, 341, 342, 353, 355, 362, 364, 365, 367, 370, 371, 379, 481, 488, 580, 586, 607, 636, 637, 642, 644, 659, 663, 773, 779, 787, 788, 828, 841–845, 852–855, 878, 898, 901, 910

Electrical properties (capacitance, dissipation, loss), 3, 13, 19, 20, 309, 341–343, 347, 637, 638, 658, 779, 787, 838, 839, 877, 901

Electroplating, see Decorating

Elongation, 1, 2, 13, 300, 316, 324, 521, 522, 525, 533, 546–548, 563, 565, 629–631, 642, 663, 666, 806, 807, 826, 886, 894, 923, 926

Embrittlement, 16, 22, 626, 627, 889, 896, 913, 920, 923

Emulsion polymerization, 480, 485, 494

Endurance, 779, 782, 826, 828, 829, 890

Entanglement, 1, 553, 556, 557

Entropy, 31, 33, 36, 39, 46, 72, 302

Environmental resistance, 3, 17, 658, 659, 665, 883, 884, 913–915

Epoxy, 1, 8, 9, 19, 24, 25, 472, 614, 616–618, 620, 629, 630, 635–639, 642, 643, 893

Erosion, 779, 830, 836

Ethylene, 1, 4, 8, 9, 15, 20, 274, 629, 900

Excluded volume, 40, 86, 303

Extender, 604, 617

Extrusion, 7, 8, 13–17, 19, 331, 474, 478, 479, 485, 486, 494, 496, 501, 502, 523, 543, 545, 546, 560, 571–581, 586, 588, 590, 595, 599, 601, 602, 604, 608, 609, 617, 621, 628, 631, 645–648, 650, 669–671, 698, 699, 702, 704, 712, 713,

732–735, 737–740, 742, 745, 747, 750–753, 755, 759–761, 765, 778, 793, 796, 817, 908, 915

Failure, 438, 807, 812, 814, 817, 818, 827, 849, 877, 889, 896, 915, 916, 918
Fatigue, 15, 446, 786, 828, 854, 877, 891
Feedstock, 3, 4, 9, 11, 14
Fiber, 1, 4, 6, 8, 11, 16, 17, 21, 23, 298, 328, 519, 545, 547, 563, 564, 575, 582, 591, 608, 613, 620, 637, 638, 640, 644, 647, 650, 654, 884, 893, 894, 907, 912, 913
Filament, 1, 11, 299, 641, 644, 645, 653, 688
Filler, 2, 17, 18, 20, 298, 318, 557–559, 613, 620, 628, 629, 637, 647, 650, 688, 784, 802, 931, 935
Fillet, 653, 681, 682, 684
Film, 8, 9, 15, 16, 20, 23, 298, 331, 407, 577, 579, 601, 609, 653, 670, 699, 702, 752, 761, 799, 810, 884
Flame retardant, 14, 613, 615, 623, 647, 649, 650, 910, 925, 932
Flammability (ignition, self-extinguishing, smoke generation), 3, 13, 16–23, 653, 663, 664, 779, 852, 857–859, 861, 862, 901
Flexibility, 1, 15, 16, 298, 802, 803, 896, 897
Flexural properties (modulus, strength, test, yield), 581, 591, 629–631, 637–639, 641, 642, 644, 663, 666, 699, 702, 704–707, 779, 786, 808–810, 814, 825, 885, 890
Flow (fluid), 1, 488, 490, 497, 520–522, 525, 530, 533, 535, 536, 543, 545, 546, 548, 563, 565, 578, 581, 597, 631, 704, 896
Flow index, 717, 743–745
Flow line (or mark), 653, 687
Fluoropolymers, 1, 8, 10, 13, 14, 19, 20, 249, 278, 279, 284, 331, 614, 616, 618, 620, 625, 808, 828, 892, 893, 900, 901, 905, 908, 917
Foam, 7, 17, 298, 395, 408, 571, 601–603, 621, 672, 854, 885, 892, 894
Forging, 599, 609
Fractionation, 104, 125

Fracture, 249, 322, 323, 327, 395, 433, 775, 779, 786, 813–815, 828, 916, 923
Free radical, 475–477, 481, 484, 501, 507, 888, 901, 921, 927
Free volume, 315, 316, 893
Frequency, 342, 343, 352, 792, 798, 800, 803, 838–840, 877, 878
Friction, 62, 326, 395, 433, 644, 702–704, 730, 779, 787, 794, 829, 830, 837, 884, 887

Gardner impact, 629–631
Gas transmission, 780, 874, 875
Gate, 584, 609, 672, 680, 687, 690, 693, 694
Gate mark, 683, 694
Gel, 270, 483, 505, 508, 642, 643
Gel permeation chromatography (GPC), 63, 78, 79, 82, 104–115, 117–119, 122, 125
Glass, 1, 3, 11, 12, 298, 308, 415, 559, 620, 637–640, 644, 647, 650, 653, 675, 779, 853, 854, 891, 905, 912, 923
Glass transition, 1–3, 59, 61, 202, 262, 279, 298, 303, 306, 307, 309, 311, 315, 324, 328, 330, 345, 352, 359, 396, 432, 485, 539, 548, 559, 599, 800–803, 832, 844, 877, 888, 891, 894
Gloss, 17, 685, 780, 788, 837, 862, 866
Graphite, 24, 638, 644
Graphon, 415, 416

Hardness (indentation, Rockwell, scratch), 17, 663, 667, 779, 786, 787, 829–833, 908
Haze, 780, 837, 862, 867
Heat capacity, 309, 580, 585, 728–730, 749, 773, 774
Heat distortion temperature, 641, 642, 644
Heat of fusion, 580, 729, 730
Heat resistance, 17, 23, 663, 779, 861
Heat stabilizer, 613, 619, 627, 750
Heat transfer, 488, 585, 586, 733, 774, 828
Homopolymer, 4, 308, 477, 483, 513, 519, 551, 629, 631, 889
Hookean, 302, 310, 805
Hydrogen bonding, see Bonding
Hysteresis, 426, 428

Ideal solution, 31–33
Ignition, 779, 859
Impact modifier, 276, 613, 616, 624

Impact strength (IZOD), 11, 13, 16, 17, 20, 327, 628–631, 636, 638, 641, 644, 663, 665, 666, 678, 779, 785, 786, 797, 803, 815, 816, 818–820, 910, 916, 928
Initiator, 613, 614, 621
Injection molding, 7–9, 13–17, 280, 519, 559, 560, 561, 563, 571, 572, 584, 585, 587, 589–593, 596, 603, 609, 630, 631, 646, 655, 657, 662, 669–673, 677, 681, 688, 695–699, 701, 702, 704, 712, 713, 746, 792, 793, 817, 896, 908
Insert, 653, 669, 687–689
Insulator, 3, 11, 341, 659, 697, 779, 903, 905
Interaction parameter, 46, 48, 52, 54
Interfacial properties, 3, 14, 324, 353–355, 395, 434, 637, 916
Intermolecular bonding, see Bonding
Ionomer (polyelectrolyte), 8, 11, 20, 100, 900
Iron, 3, 8, 10, 853
Isotactic, 143, 147, 278, 304, 415
Isotropic, 299, 391

Joint strength, 395, 436, 438, 446

Kinetics, 8, 222, 232, 247, 473, 474, 476, 477, 483, 485, 489, 504, 505, 510, 926
Knit lines (or marks), 653, 658, 687

Labman system, 729, 733, 777
Laminating, 18, 20, 23, 274, 331, 578, 790, 809, 833, 842, 905
Leather, 883, 894, 912
Light scattering, 63, 78, 84, 85, 115, 127, 258
Linear, 3, 7, 13, 82, 519, 551, 894, 895, 899
Loss, see Dielectric properties; Electrical properties
Loss factor, 779, 787, 838, 839, 877
Lubricant, 2, 613, 616, 624, 625, 645, 646, 649, 836, 910

Machining, 571, 605, 681, 792, 793
Macromolecule, 2, 8, 33, 134, 534, 551
Magnesium, 3, 9, 10
Mar resistance, 779, 834, 836
Mass polymerization, 13, 16, 30
Mechanical fastening, 571, 606, 610, 811
Melamine, 1, 8, 20, 594, 614, 616, 618, 620, 635, 639, 663, 664, 670, 688, 802, 833, 900, 905
Melting transition, 1, 251, 264

Metals, 3, 5, 8–12, 14, 297, 298,
    331, 599, 620, 636, 644, 653,
    656, 669, 674, 681–683, 687,
    691, 697, 831, 836, 853, 854,
    884, 890, 912, 920, 921, 924
Mixing, 8, 471, 481, 491–493, 496,
    498–500, 503, 504, 506, 509,
    575, 600, 733
Modulus (bulk, elastic, tensile,
    young), 2, 11, 17, 217, 256,
    299, 305–307, 310, 328, 396,
    537, 538, 542, 543, 641, 642,
    665, 782, 796, 798, 800, 802,
    803, 806, 810, 812, 813, 824,
    832, 877, 883, 885, 893–895,
    897, 908, 909, 927
    see also Dielectric properties;
    Flexural properties
Molded-in-stress and strain, 589,
    658, 914, 916, 919
Molded lettering, 653, 683, 685
Molding (blow, compression,
    injection, LIM, RIM, ROTO,
    slush, transfer), 7–9, 13, 14,
    16–18, 21, 259, 303, 323, 328,
    561, 571, 582–586, 591–596,
    602, 603, 605, 607, 609, 621,
    630, 631, 641, 643, 644, 654,
    663, 667–673, 677, 678, 681,
    682, 685, 686, 688, 689, 692,
    695, 696, 699, 732, 746, 747,
    765, 766, 778, 782, 792, 793,
    797, 805, 807, 811
Molecular weight (average,
    distribution), 1, 3, 7, 31, 33,
    41, 55, 63, 68, 78–89, 92–97,
    101, 109, 113, 249, 307, 322,
    395, 404, 409, 411, 412, 415,
    416, 453, 471–473, 475, 496,
    498, 506–508, 511, 522, 523,
    550, 552, 553, 566, 599,
    625–627, 728, 893, 896, 912,
    915, 916, 919, 923, 927
Monomer, 2, 7, 30, 36, 134, 331,
    472–477, 479, 481, 488, 495,
    497, 503, 504, 605, 609, 610,
    924, 933
Morphology, 1–3, 7, 11, 14, 57, 72,
    135, 163, 207, 208, 210–214,
    216, 240–245, 247–252,
    254–262, 264, 269, 278–282,
    284, 287, 298–300, 304, 306,
    308, 311, 318–320, 323, 324,
    326–329, 342, 363, 395, 399,
    400, 402, 424, 453, 522, 575,
    582, 590, 636, 801, 802,
    887–889, 893–898, 908, 912,
    914–916, 921, 935
Multiphase systems
    (multicomponent), 11, 251, 331,
    327

Newtonian, 310, 487, 488, 490, 492,
    519, 527, 528, 531, 535, 547,
    549, 555, 557, 562, 564, 580,
    581, 717, 727, 728, 732
Nitrile, 1, 8, 20, 616, 617
Noryl, 8, 21, 277, 614, 616, 618,
    620, 674, 889, 897, 900, 918
Notch, 665, 785
Nuclear magnetic resonance (NMR),
    134, 138, 150, 151, 155, 309,
    353, 356
Nucleation, 244, 247, 281
Nylon (polyamide), 1, 2, 5, 8, 9, 14,
    19, 20, 241, 281, 283, 284,
    328, 403, 405, 481, 574, 582,
    596, 599, 614, 616, 618, 620,
    621, 623, 625, 630, 635, 636,
    638, 639, 641, 644, 889, 892,
    897, 900, 908, 909, 914, 917,
    918

Oligomer, 331, 412, 472
Opacity, 12, 635, 665, 667, 691,
    898, 915
Organosol, 571, 603, 604, 632
Orientation, 1, 7, 12, 244, 263, 319,
    321, 324, 327–329, 364, 475,
    590, 591, 780, 792, 794, 898
Osmometry, 39, 40, 52, 56, 78,
    84–87, 89, 90
Oxygen index, 788, 860, 878

Packaging, 8, 14, 15, 19, 20–23,
    578, 604, 609, 613, 617,
    623–625, 634, 660, 802, 895,
    896, 901, 906, 907, 913, 915,
    923–925, 931
Paint (pigment), 8, 298, 571, 607,
    622
Paper, 3, 10, 14, 833, 854, 884,
    905, 912
Parison, 594, 595, 609, 699, 702,
    765
Particle size, 632–634
Permanence, 780, 876
Permeation, 12, 14, 19, 661, 691,
    780, 870, 871, 878
Phenolic, 1, 2, 9, 14, 17, 19, 362,
    472, 614, 616, 618, 620, 629,
    630, 635, 639, 663, 664, 670,
    688, 900, 904, 905, 917
Phenoxy, 2, 8
Photogeneration, 341, 342, 390
Plastic, 1–5, 8–16, 18–21, 23–25,
    323, 329, 331, 362, 475, 571,
    576, 579, 580, 586, 588, 590,
    594, 596, 597, 599, 601, 604,
    605, 614, 616, 618, 620, 621,
    622, 624, 637, 653, 657, 660,
    665, 673, 681, 683, 685, 686,
    695, 699, 704, 734, 751, 758,

    762, 780, 786, 790, 798, 804,
    807, 819, 825, 826, 828, 831,
    835, 837, 846, 854, 855, 868,
    885–888, 890, 892, 893, 896,
    902, 907, 911, 914, 924, 926,
    928, 929, 933
Plasticizer, 16, 298, 308, 341, 357,
    359, 361, 362, 575
Plastosol, 571, 603, 604, 609, 632
Platen, 673, 677, 681
Poissons ratio, 299, 300, 812
Polamide, 14, 19, 20, 635, 644, 900,
    914, 917, 918
    see also Nylon
Polarity, 895, 898, 905, 907, 918
Polarization, 13, 16, 30, 244, 329,
    330, 341–343, 369, 472, 475,
    476, 501, 504, 512, 513, 774,
    870, 897, 910, 928
Polyacetal, see Acetal
Polyacrylate, see Acrylic
Polybutadiene, 1, 6, 8, 19, 35, 41,
    64, 84, 158, 475, 485, 553
Polybutylene (polyisobutylene), 9,
    21, 41, 52, 64, 65, 136, 278,
    304, 403, 405, 475, 539, 549,
    553, 614, 616, 618, 620, 902
Polycarbonate (PCO), 1, 8, 9, 21,
    84, 260, 278, 324, 341, 349,
    352, 353, 360, 378, 384, 387,
    390, 473, 474, 481, 596, 614,
    616–618, 620, 623, 635, 639,
    642, 644, 795, 813, 819, 820,
    827, 889, 892, 900, 908, 909,
    917, 918, 923
Polydispersity, 83, 123, 475, 500,
    507
Polyelectrolyte, see Ionomer
Polyester, 2, 6–9, 14, 18, 21, 24,
    25, 331, 341, 362, 383, 414,
    473, 481, 596, 614, 616–618,
    620, 621, 629, 630, 632–635,
    638, 639, 641, 644, 663, 670,
    784, 790, 833, 892, 894, 897,
    900, 908, 914, 917, 923, 930
Polyether, 2, 6, 18, 19, 22, 84,
    269–271, 354, 360, 404,
    411–413
Polyethylene (HDPE, LDPE), 1, 2,
    4, 7–9, 11, 15, 16, 20, 24, 25,
    64, 65, 68, 84, 136, 158, 160,
    161, 175, 260, 264, 283, 284,
    297, 301, 321, 328, 353, 363,
    384, 395, 398, 402, 404, 424,
    430, 437, 475, 476, 478, 481,
    483, 484, 524, 525, 548, 553,
    564, 565, 578, 585, 596, 599,
    606, 607, 614, 616, 618, 620,
    625, 627, 629, 630, 632, 638,
    641, 650, 656, 658, 664, 665,
    667, 674, 675, 678, 815, 819,

873, 889, 897, 899, 900, 902, 903, 905, 908, 909, 915, 923, 930

Polyethylene terphthalate (PET), 1, 20, 245, 258–260, 278, 283, 319, 320, 328, 330, 398, 402, 404, 431, 596, 893, 899, 900, 917

Polyimide, 8, 20, 246, 331, 400, 614, 616, 618, 620, 644, 900, 908

Polyisoprene, 2, 84, 304, 331, 556, 557

Polymer, 1, 2, 4, 5, 8, 10, 12–14, 16, 19, 20, 33, 59, 81–83, 125, 134, 137, 138, 180, 219, 225, 256, 308, 309, 318, 321, 324, 327, 328, 331, 345, 390, 395–398, 402, 410, 423, 428, 436, 471–473, 475–479, 481, 488, 494, 497, 499, 502–504, 511, 513, 519, 526, 534, 551, 553, 565, 566, 571, 576, 578, 581, 595, 599, 600, 607, 608, 613, 614, 616, 618, 620, 625, 627, 628, 632, 635, 645, 659, 660, 679, 701, 780, 798, 799, 803, 806–808, 818, 834, 836, 856, 883, 888, 892–895, 901, 913, 920, 921, 927, 928

Polymer chain, see Chain

Polymerization, 2–4, 7, 13, 15–17, 30, 81, 134, 135, 224, 329, 471–473, 475–478, 480, 481, 483–485, 490, 493–496, 498, 499, 503–505, 512, 519, 604, 605, 610, 614, 901, 928

Polymethyl methacrylate (PMMA), 1, 20, 41, 65, 68, 136, 138, 142, 143, 144, 152, 158, 258, 278, 279, 320, 324, 356–358, 360, 404, 407, 411, 415, 420, 431, 475, 533, 534, 537–540, 543, 551, 553, 562, 601, 609, 790, 792, 805, 807, 808, 812, 813, 822, 823, 826, 831, 833, 840, 844–847, 869, 899, 902, 917, 918

Polymethyl pentene (TPX), 211, 697

Polyolefin, 16, 20–22, 56, 331, 400, 436, 475, 575, 621, 916–918, 930

Polyoxymethylene (POM), 7, 21, 256, 899

Polyphenylene oxide (PPO), see Noryl

Polyphenylene sulfide, 22, 473, 614, 616, 618, 620, 635, 644, 889, 897, 908

Polypropylene (PP), 1, 4, 7–9, 16, 23, 25, 84, 252, 328, 404, 454,

475, 540, 548, 596, 599, 606, 614, 616, 618, 620, 627, 629, 631, 636, 641, 643, 644, 656, 665, 667, 873, 889, 892, 900, 908, 909

Polysiloxane, 1, 84, 257, 269, 270, 304

Polystyrene (PS), 1, 4, 7–10, 17, 22, 25, 30, 35, 41, 54, 61, 64, 65, 68, 84, 257, 259, 263, 269–271, 278, 297, 324, 329, 330, 353, 354, 358, 360, 363, 404–406, 409, 411, 413, 415, 430, 475, 484–486, 494, 503, 522–524, 526, 529, 533, 539, 540, 547, 548, 553, 558, 559, 575, 585, 602, 614, 616, 618, 620, 625, 639, 643, 655, 657, 667, 678, 792–794, 819, 892–894, 896, 899, 900, 902, 903, 905, 908, 909, 916–918

Polysulfones, 2, 7, 8, 22, 474, 481, 489, 644, 827, 892, 900, 907, 908

Polyurethane (PU), 1, 6, 8, 9, 14, 23, 24, 84, 298, 318, 331, 400, 539, 591, 614, 616, 618, 620, 621, 629, 630, 643, 893, 894, 896, 900, 908, 914, 917, 930

Polyvinyl acetate (PVA$_c$), 8, 22, 84, 349, 351, 352, 404, 407, 412, 414, 539, 553, 896, 900, 902, 917

Polyvinyl alcohol (PVA), 22, 84, 283, 304, 412

Polyvinyl carbazol (PVK), 341, 358, 371–376, 379–381, 383–385

Polyvinyl chloride (PVC), 1, 7–9, 16, 22–25, 30, 35, 64, 65, 68, 72, 84, 138, 143, 151, 152, 158, 279, 283, 330, 363, 454, 475, 481, 560, 561, 599, 609, 616, 621, 623–626, 628, 630–632, 634–636, 660, 664, 667, 750, 813, 823, 896, 899, 900, 902, 904, 908, 917, 926, 932

Polyvinyl pyrrolidone, 420, 911

Power law, 527, 529

Printing, 15, 20, 571, 607

Processing (secondary finishing, solid state, uniaxial forming, vacuum metalizing), 3, 7, 12, 14, 16, 24, 328, 571, 599, 607–609, 669

Propylene, 3, 8, 16, 283

Pseudoplastic, 316, 632

Pyrolysis, 134, 180, 190, 191, 926

Quantum efficiency, 367, 368, 373, 376, 378, 391

Q-E scheme, 135, 328

Radiation processing, 571, 604, 610, 926

Radii, 653, 681, 682, 692

Radius of gyration (molecular dimensions), 67, 68, 85, 106, 124, 302, 556

Ram, 704, 712, 713, 817

Reactivity (principle, ratio), 81, 477

Reactor (design, engineering, type), 471, 473, 477–479, 486–496, 498, 504–509, 511–513

Reflection, 780, 864–866, 878

Refraction, 323, 780, 862, 878

Reinforcement, 7, 8, 11, 14, 271, 331, 613, 628, 635, 802, 868, 891, 931

Relaxation (mechanical, time), 344, 345, 347, 348, 350, 355, 356, 363, 526, 527, 532, 534, 551, 553, 558, 562, 563, 565, 576, 579, 590, 591, 786, 802, 824, 825, 888

Resin (commodity, engineering), 2, 3, 8, 9, 15–18, 24, 297, 604, 626, 637, 638, 641, 649, 661, 662, 665, 685, 698, 734, 784, 790, 836, 897

Resistance, see Dielectric properties; Electrical conductivity

Ribs, 665, 669, 684, 685, 689

Rigidity, 9, 11, 13, 16, 17, 298, 641, 695, 799, 800, 811, 888, 896

Rubber, 1, 5, 7, 8, 11, 39, 46, 264, 271, 272, 300, 301, 303, 304, 308, 314, 330, 531–534, 605, 894

Runners, 561, 584, 593, 609, 696

Rupture, 779, 786, 807, 821, 826, 888

Rutile, 411, 416

Scratch resistance, 21, 23, 323, 787

Screw, 7, 571, 573–575, 580, 586, 588, 590, 645, 647, 648, 650, 698, 699, 733, 737, 739, 742, 751–753, 756, 757, 761

Service (life, performance, temperature), 11, 18, 23, 653, 661, 689–691

Shear (thickening, thinning), 557, 558

Shear rate, 522, 523, 537, 541, 544, 545, 547, 551, 555, 558, 559, 632, 648, 712, 714, 715, 717, 718, 722, 723, 726–728, 741, 743, 744, 746, 756, 757, 759, 760

Shear strain, 299, 304, 305, 317, 520, 525, 526, 534, 542, 565, 779, 795–797, 804–807, 810,

813, 820, 822, 824, 825, 827, 877, 882, 885–887, 890, 891, 918

Shear strength, 644, 786

Shear stress, 1, 255, 299, 305, 317, 329, 442, 519, 520, 525, 529, 534, 544, 549, 563, 565, 637, 645, 648, 658, 691, 713, 779, 786, 795–797, 804–807, 810, 813, 820, 822, 827, 877, 883–887, 890, 891, 905, 914, 916, 918

Sheet, 8, 16, 577, 630, 638, 669, 670, 752

Sheet molding compound (SMC), see Sheet

Shelf life, 661, 691

Shrinkage, 576, 589, 590, 626, 636, 638, 642, 644, 651, 653, 668, 673, 675, 678–681, 685, 686, 689, 692, 893, 916

Shrink fixture, 680, 686

Silica (silane), 3, 112, 411, 414, 635, 637, 638, 641, 642

Silicone, 7, 8, 14, 23, 269, 270, 607, 893, 900

Sink mark, 653, 684, 685, 689

Slip agent, 254, 613, 625, 646, 648

Slots, 681, 683

Solid state processing, 571, 599, 609, 669

Solubility parameter, 34–37

Solvent (cementing, etching, resistance), 17, 18, 248, 278, 417, 442, 501, 606, 610

Sorption, 418, 420

Space environment, 884, 891, 892, 926

Specific heat, 219, 636, 644, 779, 788, 828, 852, 855, 856, 878

Specific volume, 309, 666

Spinning, 1, 519, 545, 547, 563–565, 582, 690, 701, 702, 747–749, 764, 793

Sprues, 561, 584, 609, 673, 686

Stabilizer (heat, UV), 2, 20, 363, 604, 613, 619, 620, 626, 658–660, 901, 907, 908, 921, 922, 924, 925

Stamping, 599, 609, 654, 811

Static charge, 644

Steel, 8–12, 14, 298, 644, 681, 683, 687, 688, 697, 836

Stereoregularity, 3, 8, 143, 150, 358, 416

Stiffness, 591, 628, 786, 810, 888, 896, 923

Stoichiometry, 473, 475, 486

Strength, see Dielectric properties; Flexural properties; Tensile properties

Stress, see Tensile properties

Stress cracking, environmental, 12, 15, 20, 21, 23, 792, 884, 913–915, 918–920

Stretch, 304, 563, 595, 793

Structure (cis, micro, trans), 3, 7, 8, 14, 41, 133, 304, 321, 477, 916

Studs, 665, 681

Styrene, 4, 14, 17, 18, 23, 36, 503, 606

Styrene acrolonitrile, 262, 508, 643, 900, 905, 918

Styrene butadiene rubber (SBR), 1–3, 6, 9, 23, 262, 266, 324, 475, 486, 900

Surface properties, 395, 396, 402–409, 411, 416, 424, 425, 429, 431, 437, 441, 443, 444

Suspension polymerization, 13, 16, 23, 472, 480, 485, 490, 494, 557, 558

Syndiotactic, 143, 150

Taper, 653, 680, 681, 692

Tensile properties (modulus, strength, stress), 1, 2, 11, 13, 15, 20, 24, 305, 310, 316, 319, 324, 325, 327, 553, 582, 591, 629–631, 637, 638, 641, 642, 644, 666, 779, 782, 786, 807, 910, 915

Test, see Flexural properties

Textiles, 8, 884

Thermal conductivity, 488, 580, 586, 636, 644, 663, 773, 779, 788, 828, 852–855, 878, 898

Thermal diffusivity, 733, 788

Thermal expansion, 11, 15, 20, 316, 644, 852, 853, 878

Thermal properties, 11, 16, 20, 24, 59, 216, 227, 229, 231–233, 699, 728, 787, 788, 798, 799, 853, 897

Thermodynamics, 3, 8, 12, 14, 31–34, 36, 37, 40, 41, 43, 69, 85

Thermoforming, 571, 596–598, 669, 697

Thermoplastic, 2, 7–9, 12–17, 25, 264, 273, 297, 331, 571, 582, 591, 596, 603, 635, 637, 641, 643, 659, 663, 664, 671, 793, 857, 870, 888, 896, 897, 921, 925, 935

Thermoset, 2, 7, 9, 13, 14, 16, 18, 25, 297, 571, 582, 590, 593, 623, 637, 641, 660, 663, 684, 793, 870, 888, 890, 896, 897, 899, 921, 925

Theta temperature, 39, 64, 66, 68, 101

Thixotropy, 632, 633

Titanium dioxide, 412, 559

Toughness, 15, 16, 298, 627, 815

Toxicity, 660, 691, 884, 931

Tracking resistance, 779, 838, 849, 905

Transparency, 12, 17, 665, 691, 780, 835, 864, 865, 867

Trinitro fluorenone (TNF), 341, 371, 376, 377, 381–383

Triphenyl amine (TPA), 341, 371, 378, 384, 385, 387

Triphenyl methane (TPM), 379, 387

Ultimate strength, 2, 806. 886

Undercut, 653, 669, 682, 683, 693, 694

Urea, 1, 8, 593, 663, 664, 670, 688, 900

Vacuum forming, 598, 669

Vacuum metallizing, 571, 607, 608

Vander Waal, 429, 430

Vinyl chloride, 4, 7, 9, 16

Vinyls, 2, 5, 6, 14, 16, 138, 475, 481, 494, 503, 575, 614, 616, 618, 620, 627, 776, 889, 896, 923

Viril coefficient, 4, 24, 41, 62, 88, 89, 95

Viscoelastic, 217, 309, 311, 313, 316, 324, 519, 527, 529, 530, 532, 537, 539, 540, 542, 543, 549, 550, 563, 565, 798, 916, 918

Viscometer (rheometer), 62, 103, 520, 541–545, 547, 712, 715, 717

Viscosity, 2, 14, 24, 30, 31, 41, 56, 62, 63, 65, 66, 68, 78, 84, 98, 100–103, 125, 310, 321, 326, 472, 481, 487, 494, 495, 497, 500, 519–522, 525, 527–529, 531, 532, 534, 535, 537–545, 547–552, 556, 558, 559, 562, 565, 580, 601, 604, 623, 633, 634, 636, 637, 704, 705, 712–715, 717, 718, 720, 722, 723, 726–728, 741, 743, 744, 746, 757, 759, 760, 769, 813, 899, 918

Voltage, 342, 343, 844, 848

Vulcanization, 7, 604, 605, 609, 896, 912

Wall thickness, 655, 673, 675, 689, 692, 694, 766

Warpage, 685, 689, 692

Water absorption (vapor transmission, wetting), 426, 430, 641, 642, 780, 873